ENCYCLOPEDIA OF PHYSICS

CHIEF EDITOR

S. FLÜGGE

VOLUME XLIX/4

GEOPHYSICS III

PART IV

EDITOR

K. RAWER

WITH 188 FIGURES

SPRINGER-VERLAG
BERLIN · HEIDELBERG · NEW YORK
1972

HANDBUCH DER PHYSIK

HERAUSGEGEBEN VON

S. FLÜGGE

BAND XLIX/4

GEOPHYSIK III

TEIL IV

BANDHERAUSGEBER

K. RAWER

MIT 188 FIGUREN

SPRINGER-VERLAG
BERLIN · HEIDELBERG · NEW YORK
1972

ISBN 3-540-05583-5 Springer-Verlag Berlin Heidelberg New York
ISBN 0-387-05583-5 Springer-Verlag New York Heidelberg Berlin

Contents.

Introductory Remarks.

Volume 49/4 deals with the magnetosphere and magnetospheric phenomena. The first two contributions consider the magnetosphere from different aspects. Quite naturally, there is some overlap and this may be helpful for better understanding of the phenomena. Some of these topics are also discussed in the contribution by PARKER and FERRARO in Vol. 49/3 from the viewpoint of magnetic disturbances. Cross-references have been introduced in several places.

The third contribution in this volume discusses a technique of observation, which might at first view be thought more appropriate to Vol. 49/3, where the classical geomagnetic variations are discussed. However, it has been shown that the rapid variations are mostly provoked by magnetospheric phenomena of a particular nature in which wave propagation in the magnetospheric plasma plays an important role. This is one of the subjects considered in the final contribution, dealing with plasma theory. It should be noted that the theory, as presented by K. RAWER and K. SUCHY in Vol. 49/2, refers almost entirely to cold plasma, while in this volume warm plasma is involved in the majority of cases.

Since it is intended that the different contributions to this volume should be usable as a unit, there was the old, inevitable dilemma of which *units to use* in writing equations. A clear trend towards the "Système International" (SI) is nowadays apparent; essentially the same system is referred to in the literature as m.k.s.A. or GIORGI units. The International Union of Pure and Applied Physics (IUPAP) has recommended the general use of SI units. In an increasing number of countries this system is being introduced as the only approved one and is exclusively tought in schools. Thus, being forward looking, we would prefer all equations to be written in SI units. The existing literature in the field of geomagnetism, on the other hand, makes use of one or the other of the c.g.s. systems. We finally decided to admit both types of units. The equations have been written in a form that is valid in all commonly used systems of units. The necessary transcription has been made by the editor and is thus primarily his responsibility.

This generalized way of writing equations precludes the use of the simplifications typical of c.g.s. systems. In these systems the permittivity of free space, ε_0, and the permeability of free space, μ_0, are chosen to be dimensionless (and made unity). Now we have to introduce ε_0 and μ_0 as physical quantities, which may in fact correspond to their nature. The three most generally used c.g.s. systems are obtained by specializing the numerical values of quantities as indicated in the following table where c_0 is the velocity of light in free space.

	Electrostatic	Electromagnetic	GAUSS
$\varepsilon_0 =$	1	$1/c_0^2$	1
$\mu_0 =$	$1/c_0^2$	1	1
So that the product:			
$c_0^2\,\varepsilon_0\,\mu_0 =$	1	1	c_0^2

In SI units, too, which are basically electromagnetic, we have

$$c_0^2 \varepsilon_0 \mu_0 = 1 .$$

Unfortunately, there is yet another difference between the systems of units. The c.g.s. systems most frequently used in the literature are non-rationalized while SI is rationalized (as is the special c.g.s. system introduced by H. A. LORENTZ). In a rationalized system of units the factor 4π appears only in spherical problems, for example, in COULOMB's law. In non-rationalized systems the natural factor 4π has been artifically eliminated from COULOMB's law, although it appears in planar problems. There are thus two alternatives, regardless of the choice of the constants ε_0, μ_0 and c_0. The two alternatives may be allowed for by a dimensionless numerical constant u, which assumes the values:

$$\mathrm{u} = 1 \text{ in rationalized systems,}$$
$$\mathrm{u} = 4\pi \text{ in non-rationalized systems.}$$

These rules for writing equations in a generalized way were used by RAWER and SUCHY in Vol. 49/2 of this Encyclopedia in the contribution entitled "Radio Observations of the Ionosphere". Detailed explanations are given in an appendix (pp. 535 and 536 of Vol. 49/2), in which the 'transformations' between two systems of units and the relevant 'invariants' (e.g. energy quantities) are also discussed.

Numerical values for the different constants are given in the following summary table:

System		u	ε_0	μ_0	$c_0^2 \varepsilon_0 \mu_0$
SI	= m.k.s.A.	1	$8.854 \cdot 10^{-12}$ $\mathrm{AV^{-1}\,sm^{-1}}$	$1.257 \cdot 10^{-6}$ $\mathrm{A^{-1}\,V\,sm^{-1}}$	1
GAUSS	= "symmetric" c.g.s.	4π	1	1	$c_0^2 = 9 \cdot 10^{20}$ $\mathrm{cm^2\,s^{-2}}$
	el.magn. c.g.s.	4π	$\frac{1}{c_0^2} = 1.113 \cdot 10^{-21}$ $\mathrm{s^2\,cm^{-2}}$	1	1
	el.stat. c.g.s.	4π	1	$\frac{1}{c_0^2} = 1.113 \cdot 10^{-21}$ $\mathrm{s^2\,cm^{-2}}$	1
LORENTZ	= rat. GAUSS	1	1	1	$c_0^2 = 9 \cdot 10^{20}$ $\mathrm{cm^2\,s^{-2}}$

It may be helpful to repeat the most important equations of electromagnetic theory. With the definitions

electric field intensity (field strength)	electric flux density (displacement)	magnetic field intensity (field strength)	magnetic flux density (induction)	current density
$\boldsymbol{E}$	$\boldsymbol{D}$	$\boldsymbol{H}$	$\boldsymbol{B}$	$\boldsymbol{J}$

there is

$$\boldsymbol{D} = \varepsilon \boldsymbol{E}; \quad \boldsymbol{B} = \mu \boldsymbol{H}; \quad \boldsymbol{J} = \sigma \boldsymbol{E},$$

and in vacuum

$$\varepsilon = \varepsilon_0; \quad \mu = \mu_0, \quad \sigma = 0.$$

MAXWELL's equations connecting the different field quantities are now written:

$$\frac{\partial}{\partial \boldsymbol{r}} \times \boldsymbol{H} \equiv \nabla \times \boldsymbol{H} = \frac{1}{c_0 \sqrt{\varepsilon_0 \mu_0}} \frac{\partial}{\partial t} \boldsymbol{D} + \frac{\mathrm{u}}{c_0 \sqrt{\varepsilon_0 \mu_0}} \boldsymbol{J}$$

$$\frac{\partial}{\partial \boldsymbol{r}} \cdot \boldsymbol{D} \equiv \nabla \cdot \boldsymbol{D} = \mathrm{u}\, \varrho$$

$$\frac{\partial}{\partial \boldsymbol{r}} \times \boldsymbol{E} \equiv \nabla \times \boldsymbol{E} = -\frac{1}{c_0 \sqrt{\varepsilon_0 \mu_0}} \frac{\partial}{\partial t} \boldsymbol{B}$$

$$\frac{\partial}{\partial \boldsymbol{r}} \cdot \boldsymbol{B} \equiv \nabla \cdot \boldsymbol{B} = 0,$$

The two systems of units most used in geomagnetism are that of GAUSS and SI. With regard to these two systems, we may say that in SI units $c_0^2 \varepsilon_0 \mu_0 = 1$ and $\mathrm{u} = 1$, so that the factors in MAXWELL's equations can be disregarded. In the GAUSS system the constants ε_0 and μ_0 can be omitted, but c_0 remains and we have the additional constant $\mathrm{u} = 4\pi$.

As this volume deals with magnetic phenomena, it appears necessary to discuss the definition of the magnetic moment $\mathbf{M}$ in particular. Two different definitions are unfortunately in general use. The potential energy of a dipole magnet in a magnetic field is given by the scalar product of a vector $\mathbf{M}$, characterizing orientation and strength of the dipole, and a field vector. The question now is whether $\boldsymbol{H}$ or $\boldsymbol{B}$ should be chosen as the field vector. The distinction between $\boldsymbol{H}$ and $\boldsymbol{B}$ has not been made clear in much of the older literature, in which GAUSSian units were commoly used; $\boldsymbol{H}$ and $\boldsymbol{B}$ have the same dimension in this particular system in which $\mu_0 = 1$. If, however, $\boldsymbol{H}$ and $\boldsymbol{B}$ are distinguished, the dimension of $\mathbf{M}$ depends on the choice of the field vector. IUPAP has decided in favour of $\boldsymbol{B}$, so that the energy must be written

$$\mathscr{E} = -\mathbf{M} \cdot \boldsymbol{B}$$

and $\mathbf{M}$ has the dimension Joule/Tesla $= \mathrm{Am}^2$ for SI units (not Vms as would occur with the other choice). The IUPAP definition is used throughout this volume.[1]

Note that by definition the product must be 'invariant' in the sense used by RAWER and SUCHY (Vol. 49/2, p. 535). If the procedure of these authors is adopted, the general transformation for $\mathbf{M}$ can be written as

$$\sqrt{\mathrm{u}\mu_0}\, \mathbf{M} = \sqrt{\mathrm{u}'\mu_0'}\, \mathbf{M}'.$$

The free space field of a dipole of negligible extension is

$$B_{\text{dipole}} = \mu_0 \frac{\mathrm{u\,M}}{4\pi\, \mathrm{r}^3} (1 + 3 \sin^2 \phi)^{\frac{1}{2}}$$

where M is the magnetic moment of the dipole, r the distance and ϕ a 'latitude' angle.

All equations are, of course, usable in any system of units because they are written in physical quantities. The accepted definition of a physical quantity is

(numerical value) · (dimension)

such that by dividing each term through the dimension a purely numerical equation can be obtained. We tend to write such equations, if at all, so that each

[1] The same definition is used in the well-known book "Electromagnetic Theory" by STRATTON, who puts forward arguments against the second choice. Unfortunately, H. EBERT's "Physikalisches Taschenbuch" uses the second choice in disagreement with the IUPAP definition.

physical quantity individually is divided by its own dimension. Where other units are to be used, the numerical change follows from an *algebraic substitution*, e.g.

$$c_0 = 3 \cdot 10^8\,\mathrm{ms^{-1}}; \quad 1\,(\mathrm{nt.mile}) = 1.853\,\mathrm{km}; \quad 1\,\mathrm{m} = \frac{10^{-3}}{1.853}\,(\mathrm{nt.mile}),$$

$$c_0 = 3 \cdot 10^8 \frac{10^{-3}}{1.853}\,(\mathrm{nt.mile})\,\mathrm{s^{-1}} = 1.619 \cdot 10^5\,(\mathrm{nt.mile})\,\mathrm{s^{-1}}.$$

A few remarks on *mathematical signs* may be in order.

Bar or stroke may be used to express division. According to IUPAP rules, the stroke / has priority over multiplication such that $a \cdot b/c \cdot d = \frac{a\,b}{c\,d}$. It is worth noting that this is not so in computer languages like ALGOL.

It is a special convention in this Encyclopedia that the natural logarithm is denoted by log (not by ln).

Differential operations in vector fields are normally expressed by means of the symbolic vector (or 'vector operator')

$$\frac{\partial}{\partial \boldsymbol{r}} \equiv \nabla$$

which in cartesian coordinates x_1, x_2, x_3 reads

$$\left(\frac{\partial}{\partial x_1}, \quad \frac{\partial}{\partial x_2}, \quad \frac{\partial}{\partial x_3}\right).$$

In the contribution GINZBURG and RUHADZE (p. 395) FOURIER transforms are currently used. In Chap. A of this contribution, a special designation[2] is used for the transformed field functions, namely:

$$\tilde{A}(\boldsymbol{k}, t) \equiv \int \mathrm{d}^3 \boldsymbol{r}\, A(\boldsymbol{r}, t) \exp(-i\, \boldsymbol{k} \cdot \boldsymbol{r})$$

and

$$\tilde{\tilde{A}}(\boldsymbol{k}, \omega) \equiv \int \mathrm{d}^3 \boldsymbol{r} \int \mathrm{d}t\, A(\boldsymbol{r}, t) \exp(i\,\omega\, t - i\, \boldsymbol{k} \cdot \boldsymbol{r}).$$

Consequently the dimension of $\tilde{A}$ is obtained from A by multiplying by m³ (in SI units) that of $\tilde{\tilde{A}}$ by multiplying by m³ sec. In the following chapters, however, the special signs are omitted. The distinction between original function $A(\boldsymbol{r}, t)$ and FOURIER transform $A(\boldsymbol{k}, t) = \tilde{A}$ and $A(\boldsymbol{k}, \omega) = \tilde{\tilde{A}}$ is only seen from the independent variables. Therefore the dimension of $A(\boldsymbol{k}, t)$ differes from that of the original $A(\boldsymbol{r}, t)$ by a factor m³ and that of $A(\boldsymbol{k}, \omega)$ by a factor m³ s.

IUPAP proposes two different ways of denoting tensors: use of sanserif letters as symbols (e.g. T), or analytical expression with reference to cartesian coordinates (e.g. T_{ik}).

In the latter case the summation rule is to be applied. Though symbolic writing is preferred, we give both presentations in most cases. The unit tensor is written as U or δ_{ik}. The tensorial product of two vectors $\boldsymbol{a}\,\boldsymbol{b}$ must be distinguished from the vector product $\boldsymbol{a} \times \boldsymbol{b}$ and the scalar product $\boldsymbol{a} \cdot \boldsymbol{b}$.

---

[2] See footnote on p. 398.

SI units (V, A, m, kg, sec etc.) are normally used for measured data and other numerical values, but magnetic fields (magnetic induction) are quite generally given in GAUSS (Gs or Γ) and gamma (γ).

The relation to the SI unit Tesla (Vsm^{-2}) is

$$1\ \mathrm{Gs} \equiv 1\,\Gamma = 10^{-4}\,\mathrm{T}; \qquad 1\,\gamma \equiv 10^{-5}\,\Gamma = 10^{-9}\,\mathrm{T}.$$

Special letters, e.g. D, $\mathcal{H}$, Z, are used for the measurable components of the geomagnetic field.

A few widely used designations should, in the opinion of the editors, be replaced by more appropriate terms. One of these is "magneto-hydro-dynamic" (or "hydro-magnetic"). It is misleading because it appears to refer to a fluid instead of a gas.

Incompressibility was, in fact, assumed in ALFVÉN'S original theory, and this justified the expression at that time. In the wide field of waves in plasmas, however, the plasma has the character of a compressible, gaseous medium, though strongly affected by the presence of a magnetic field. Therefore we prefer the terms "magneto-gas-dynamic", or "gas-magnetic" or simply "magneto-dynamic".

The word *collision frequency* is often used in plasma treatises. It gives the impression of being involved in a counting problem. For simple interpretation one uses to adopt the model of "rigid spheres", i.e. each particle is symbolized by something similar to a billiard ball. Considering one red ball and many white ones, the number (per unit time) of touches between the red ball and anyone of the white ones seems to be well defined. This would be the uncritical model interpretation of "collision frequency" which is basically used in the simple "mean free path theories". However, this term when applied to kinetic or transport problems, does not just mean the counting result, i.e. the simple number of touches.

What is needed in computations of *collision frequencies* is more a kind of efficiency function describing the effect of collisions in the process to be computed. This effect depends strongly upon the angle of deflection, which is related to the collision parameter. (This is the minimum distance between the straight line orbits of both particles before the interaction.) Take, for example, the problem of diffusion of one kind of gas through another kind. For problems of this kind backward collisions (with a deflection angle $>90°$) have greater effect than forward collisions (deflection angle $<90°$). So we should use a *weight-function* depending upon the deflection angle: Large angles should have more weight than small ones. The final result is then obtained by integration over all collision parameters. Unfortunately, there is no unique solution to the weighing problem: the weight-function depends largely on the physical process, which is considered. Inside certain limits we could even say that this is an arbitrary choice.

Furtheron there can be little doubt that the model of "rigid spheres" is an oversimplification in view of the real forces between colliding particles.

More realistic models use some continuous force/distance function usually with *a power-law* $U \sim r^{-p}$ for the potential. Except for Coulomb-forces the corresponding range of distances is well limited. If, however, two charged particles are colliding, the Coulomb-force is unlimited in distance so that the problem of "screening" appears. Anyway, by adopting an appropriate weight-function depending on the "collision parameter" one obtains a suitable approximation for any such problem.

Thus, the weight-function depends on both, the kind of problem to which the "collision-frequency" is to be applied and the physical model for particle interaction, thus on the kind of colliding particles.

Until now our considerations did not take account of the *relative velocity* between the colliding particles. To express the results of quantum mechanical interactions it is convenient to use different force/distance functions for different ranges of the relative velocity. Thus, in order to have a result usable in a real plasma, we must still admit one more averaging process viz. over different velocities, taking due account of the statistical distribution functions of the two kinds of particles. The corresponding integration has to be extended over both velocity (or momentum) spaces. Since the distribution functions may be different from maxwellian distributions, f^{M}, we have to admit correction terms. A practical way is to use generalized LAGUERRE polynomials of increasing order m.

Using the notations introduced by K. SUCHY (this Encyclopedia, Vol. 49/2, pp. 11—15) and considering colliding particles of kind h and k, we come to the following definitions:

$\nu_{\mathrm{hk}}^{(l)}(g)$ = *Transport collision frequency:* describes the specific collisional effect for a given relative velocity, g, by applying a LEGENDRE polynominal of order l as weight function.

$\hat{\nu}^{(l)}$ = *Most probable collision frequency:* the transport collision frequency (with a weight function of order l, see above) for the most probable thermal velocity:

$$\hat{\nu}^{(l)} = \nu^{(l)}(\hat{g}) \quad \text{where} \quad \hat{g}_{\mathrm{hk}} = \sqrt{\frac{2kT_{\mathrm{h}}}{m_{\mathrm{h}}} + \frac{2kT_{\mathrm{k}}}{m_{\mathrm{k}}}}\,.$$

(This term is sometimes called ν_{m} in the literature.)

$\nu_{\mathrm{hk}}^{(l|0)}$ = *Mean transport collision frequency:* is obtained by averaging transport collision effects (see above) over the different velocities assuming maxwellian velocity-distribution functions: $f_{\mathrm{h}}^{\mathrm{M}}$, $f_{\mathrm{k}}^{\mathrm{M}}$.

$\nu_{\mathrm{hk}}^{(l|m)}$ = *Averaged transport collision frequency:* (order m) is obtained by averaging transport collision frequencies (see above) over the different velocities assuming velocity distribution functions with corrections (against f^{M}) up to order m.

While these $\nu^{(l|m)}$ are defined with power terms g^{2m} in the corrections K. SUCHY and K. RAWER (J. Atmosph. Terr. Phys. **33**, 1853—1868 (1971), Eq. (3.5)) recently used generalized LAGUERRE polynomials, $L_m^{(l+\frac{1}{2})}(g^2/\hat{g}^2)$, instead. The averaged transport collision frequency obtained in this particular way is designated by $\nu^{(l\,m)}$. The two definitions are related as follows:

$$\nu^{(l|m)} = \sum_{\mu=0}^{m-l} \binom{m-l}{\mu} \nu^{(l\,\mu)}; \qquad \nu^{(l\,m)} = \sum_{\mu=0}^{m} (-1)^{m+\mu} \binom{m}{\mu} \nu^{(l|l+\mu)}.$$

Averaged transport collision frequencies are usually needed in plasma theory. The orders l and m must be adapted to the problem under discussion. In cases where it is not intended to discuss the conditions in detail, we may just write $\bar{\nu}$ for the *appropriate averaged transport collision frequency*. The term ν is reserved for the "mean collision frequency" as used in simple "mean free path theories".

Expressions for numerical computation under ionospheric conditions can be found in K. RAWER and K. SUCHY: this Encyclopedia, Vol. 49/2, Sect. 3; see also ibidem Sects. 4 and 9, in particular Fig. 38.

Freiburg, 22 July 1972 KARL RAWER

The Earth's Magnetosphere.

By

H. POEVERLEIN* **.

With 41 Figures.

1. Introduction. In this article it is attempted to give a fairly concise description of the Earth's magnetosphere, of essential processes occurring in it, and of the fundamental facts and theorems involved. More detailed reviews of special topics pertaining to the magnetosphere are presented by various authors in this and the preceding volume of the Encyclopedia of Physics.

Some ideas on motions imposed on the magnetosphere by the rotation of the Earth and the lower atmosphere (presented in Sects. 25 and 26) are largely those of the author. It may still seem desirable to have them confirmed or refuted by observations or by more elaborate theories. Two sections on the relationship of geomagnetic oscillations with the position of the plasmapause and on the current system of polar magnetic substorms are due to FUKUSHIMA. They have been added in revised form as a separate chapter (Chap. H, Sects. 33 and 34). A discussion of recent findings was added in proof-reading (Chap. I).

Much of the knowledge on the magnetosphere stems from satellite observations. One should be aware that the measured data frequently refer to individual satellite paths and it is not always possible to separate spatial from temporal variations or to establish two or three-dimensional spatial configurations without any conjecture. Derived models may generally appear too smooth. Some unexplained facts could be due to overlooked fine structures or irregularities. It may, however, be expected that most of the present statements, being based on numerous observations and often supported by theory, will persist, but additions and corrections may come up and surprises are possible in cases of poorly understood phenomena.

A. Basic views.

2. The outermost atmosphere.

α) In studying the *neutral gas* of the atmosphere one comes to the conclusion that in the exosphere, starting somewhere between 300 and 800 km, no processes of significance might take place. The bottom of the exosphere is defined by identity of the mean free path length of atoms or molecules with the scale height of the atmosphere. In the exosphere nearly no collisions take place; ascending atoms are turned back by gravity or escape. Thus no collective phenomena of gasdynamic nature can be expected in the exospheric neutral gas, but outflow and inflow of atoms through the bottom of the exosphere and exchange of atoms with outer space take place.

---

* Manuscript received March 1971.

** *Acknowledgment.* The author acknowledges support of this work during activity at the US Air Force Cambridge Research Laboratories and expresses his thanks to Dr. K. TOMAN and his former colleagues for many stimulating discussions.

β) The situation is quite different for the *ionized constituent* of the atmosphere. The ratio between ion density and neutral density is roughly 10^{-3} at F2 layer height (300 km), around 10^{-1} at 1000 km, and slightly higher far out. The ionized part thus seems to be a very tenuous gas; but it represents a highly conductive plasma whose behavior is decisively affected by the geomagnetic field. The collisions between charged particles, being due to Coulomb interaction, are much more frequent than those between neutral particles (cf. Sect. 7β). A plasma exosphere consequently would have its bottom at a much higher altitude.

With regard to processes in the plasma the outer areas (several Earth radii from the Earth's center) and the lower altitudes appear linked by the geomagnetic field; transport phenomena of colliding particles contribute very little to the linkage. The plasma with its high conductivity suppresses (or nearly suppresses) electric field components along the magnetic field lines. The short-circuiting magnetic field lines, on the other hand, cause the electric fields at various altitudes to be closely connected. The drift motions accompanying the electric fields also have to continue in a certain way through some altitude range. This is demonstrated by the theorem of frozen-in field lines (see Sect. 8).

γ) In comparing the *energy densities* of the thermal plasma, of the neutral gas, and of the geomagnetic field one finds the magnetic energy density predominant at all heights beyond the lower ionosphere; thermal and kinetic energy densities of the total medium (including plasma and neutral gas) are lower (cf. Sects. 8γ and 28). The geomagnetic field consequently controls the plasma motion and keeps the plasma enclosed. The *thermal* plasma can distort the magnetic field only very little. This is the condition of a "low" or "moderate" temperature plasma. In geomagnetic storms, however, the trapped low-energy particles (with energies between 200 eV and 50 keV) carry energy densities comparable to the magnetic energy densities of their locations (Sects. 3δ, 8γ, and 21α).

The situation in the atmosphere seems to be contrary to that in the solar corona, where the plasma is free to depart as "solar wind" because of the relatively low magnetic energy density (Sect. 5).

δ) The control of the processes by the magnetic field justifies the name "*magnetosphere*" for the outer part of the atmosphere. The name was coined by T. GOLD[1] (1959). Typical magnetospheric processes are of *magnetodynamic*[2] *nature* or, more generally, involve the *presence of plasma and a magnetic field*. With reference to the outer *neutral gas* the terms "*thermosphere*" and "*exosphere*" are more common. The thermosphere reaches from the temperature minimum near 80 km up to the exosphere.

3. The confined magnetosphere.

α) There is a continuous stream of plasma, the "solar wind", departing from the solar corona and traveling outward in interplanetary space (Sects. 4 and 5). This plasma stream prevents the geomagnetic field from extending out to infinity (or to extreme distances). The magnetodynamic theory postulates a limitation of the field to a definite space, which is the magnetosphere. The boundary of the magnetosphere or "*magnetopause*" separates the terrestrial magnetic field from the solar wind.

[1] T. GOLD: J. Geophys. Res. **64**, 1219 (1959).

[2] Unfortunately there is no consensus about the name for dynamic processes in a conductive medium with a magnetic field and the corresponding macroscopic theory (see Sect. 7). The words "magnetodynamic" and "magnetogasdynamic", used in this volume, clearly refer to dynamic processes controled or affected by a magnetic field, see "Introductory Remarks", p. 1. In the literature the adjectives "magnetohydrodynamic" or "hydromagnetic" are more common.

The idea of a confined magnetosphere appears in a paper of F. S. JOHNSON (1960)[1]. The confinement at a boundary is a consequence of a basic theorem of magnetodynamics stating that a highly conductive plasma (the solar wind) can not penetrate into an existing magnetic field or, vice versa, the magnetic field can not enter the plasma (Sect. 8). Strictly speaking, plasma and magnetic field can penetrate each other, but this needs very long time. In the magnetospheric situation the time constant for penetration is found to be near 10^6 years (Sect. 10). At the boundary there must be balance between the pressure of the incident solar wind, acting from outside, and the magnetic pressure of the geomagnetic field, perhaps enhanced by the pressure of some terrestrial plasma, from inside (Sect. 9). The incidence of the solar wind from one side causes a strong asymmetry of the magnetosphere. (JOHNSON suggested a tear-drop model.)

β) Many satellite observations made since 1960 led to a good picture of the magnetopause and gave evidence of various interesting *features of the magnetosphere*. The observational model of the magnetosphere is discussed in Sect. 14. The simple theoretical model requires some corrections; in particular, the separation between solar wind and geomagnetic field may not be so perfect (see Sect. 17). The position of the magnetopause is normally at 10 R_E geocentric height (i.e., 10 Earth radii from the Earth's center) in the subsolar point and farther out in other directions. It varies with the solar-wind pressure.

The temporary existence of a boundary corresponding to the magnetopause was already 1931 postulated by CHAPMAN and FERRARO in their theory of geomagnetic storms[2]. A solar particle stream, supposed to arrive at times of perturbations, was considered to be the cause of a compression of the geomagnetic field with a consequent geomagnetic storm. It is now known that the particle stream is permanently present, but enhanced in intensity during perturbations.

γ) The magnetosphere has no obvious boundary at *the bottom*. It remains a matter of *definition* whether part of the ionosphere is included in the magnetosphere or not. Toward lower altitudes one finds more and more phenomena of nonmagnetodynamic nature to play a role and the neutral gas to gain in importance. Near 120 km height collisions of ions with neutrals are so frequent as to prevent the ions from circling in the magnetic field. This puts an end to magnetodynamic theory. The plasma at this height can not be considered highly conductive. HINES [*16*] assumes the bottom of the magnetosphere at 150 km.

The predominance of the magnetic energy density is at an end around 130 km. Below 130 km the density of internal energy in the neutral gas exceeds the magnetic energy density (see Sect. 28).

The F layer according to this definition is a part of the magnetosphere; but in general one refers with "magnetosphere" to the space beyond the ionosphere, including the upper ionosphere only to the extent it takes part in the true magnetospheric processes.

δ) The *radiation belt* with its trapped fast particles, discovered 1958 by VAN ALLEN and cooperators, is an essential part of the magnetosphere. It is extensively treated by HESS in this volume of the Encyclopedia [*15*] and will only be briefly discussed in Chap. E of the present article. PARKER and FERRARO show in this Encyclopedia[3] that protons of a few keV in the radiation belt can be numerous enough to cause a significant depression of the geomagnetic field — the main

[1] F. S. JOHNSON: J. Geophys. Res. **65**, 3049 (1960).

[2] S. CHAPMAN: Solar plasma, geomagnetism and aurora, in [*1*], p. 373—502. — S.-I. AKASOFU and S. CHAPMAN [*24*]. — Contribution by E. N. PARKER and V. C. A. FERRARO: Vol. 49/3 of this Encyclopedia, [*28*] p. 131—205, in particular p. 203.

[3] [*28*], Vol. 49/3 of this Encyclopedia, p. 164—169.

phase of a geomagnetic storm (Sect. 29 of *this* article). This is in contrast to the statement on thermal particles in Sect. 2. Their low energy density prevents a notable effect on the magnetic field.

ε) The emphasis of the present article is on the *magnetodynamic description* of the magnetosphere, which is determined by the presence of the geomagnetic field and the thermal plasma. As a matter of fact, thermal and fast particles are not always clearly separable and transitions between the two categories play a role (see Sects. 21 and 22).

B. The solar wind.

4. The solar wind.

α) The idea that a *temporary stream* of charged particles from the Sun causes geomagnetic storms and auroras was mainly substantiated by BIRKELAND, CHAPMAN, and STOERMER [*7*]. The presence of a *continuous solar particle stream* was more or less implied by BARTELS (1949) when he considered the index of magnetic activity, K_p, as a measure of the varying intensity of this particle radiation, but BIERMANN (1951 and following years) was first in clearly proposing a permanent corpuscular stream [*7*]. He explained ionized comet tails ("type I tails") as the result of acceleration of cometary ions by a stream of charged particles that comes from the Sun [*5*], [*8*].

The explanation of geomagnetic storms by an intensification of the particle stream [*24*], [*27*][1] does not require a great modification of the old hypothesis in which a sudden onset of the stream was assumed. The emission of the continuous stream from the solar corona was explained by PARKER in a hydrodynamic theory [*7*], [*10*]. An outline of this theory will be given in Sect. 5. PARKER called the particle stream the "solar wind".

β) The *solar wind* [*5*], [*8*], [*28*] consists of protons, a small percentage of α-particles (4 to 5% normally), and electrons; it has, of course, to be neutral. A few heavier ions have been detected in it. The first direct measurements of the solar wind by means of a satellite were carried out by GRINGAUZ (on Lunik II, 1959) [*5*], [*8*]. Fig. 1 shows part of a continuous recording of the solar wind that was obtained on the Venus sonde Mariner 2 in 1962 by NEUGEBAUER and SNYDER [*8*], [*9*][2]. In the figure it is seen that all parameters are highly variable.

The plasma velocity in Fig. 1 varies between 300 and 800 km/sec; lower and higher values are occasionally observed[3]. A typical velocity in "quiet" periods is 320 km/sec (corresponding to protons of 530 eV). The average velocity is somewhat higher (perhaps 380 km/sec). At 320 km/sec a perturbation would need 5.4 days to travel from the Sun to the Earth. The travel time of particles causing magnetic disturbances has long been known to lie between 1 and 4 days [*25*].

In cases of heavier perturbations the delay between solar events and terrestrial effects amounts to 20 to 40 hours, indicating velocities from 1000 to 2000 km/sec. These velocities, however, are not necessarily particle velocities, they may relate to shock fronts traveling in the solar wind (cf. Sects. 12 and 29, [*5*], [*6*]).

The proton density at most times is between 0.1 and 100 cm^{-3}, in the average around 5 cm^{-3} [*5*], [*8*]. The temperature of the protons near the Earth ranges

[1] S. MATSUSHITA: Geomagnetic storms and related phenomena, in [*26*], Vol. 1, p. 455—483.
[2] M. NEUGEBAUER and C. W. SNYDER: J. Geophys. Res. **72** (7), 1823—1828 (1967).
[3] See also [*14*], 77—90 and [*5*].

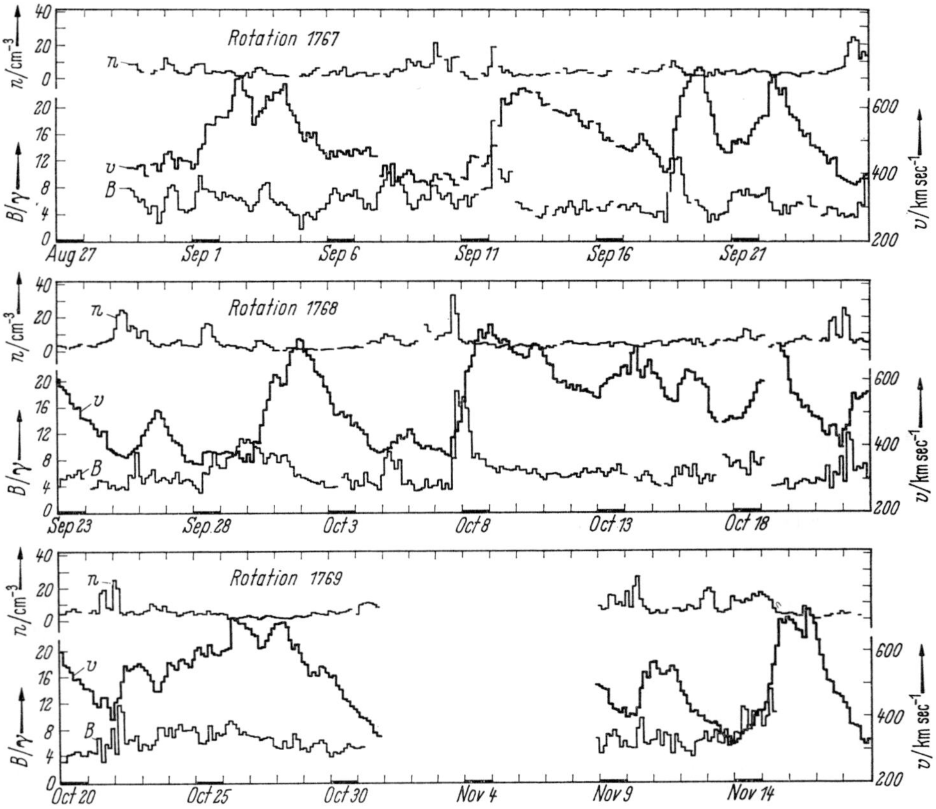

Fig. 1. Solar wind[4]. Proton density n, velocity v, and embedded magnetic flux density B are shown for three solar rotation periods in 1962.

from below 10^4 °K to almost 10^6 °K with an average near $1.5 \cdot 10^5$ °K. The proton temperature in the *quiet* solar wind was found[5] to be $4 \cdot 10^4$ °K; the electron temperature is somewhat higher[5], near 10^5 °K.

The solar wind contains a magnetic field [*8*], [*11*] whose magnitude usually lies between 2 and 8 γ, centered around 5 γ ($1\,\gamma = 10^{-5}$ Gauss $= 10^{-9}$ V sec m^{-2}, measuring the magnetic flux density B). The field direction (Fig. 2) is preferably near 45° from the Sun-Earth line in (or near) the ecliptic plane and can be away from, or toward the Sun.

The temperature of the protons is below that of the solar corona, which is 1 to $2 \cdot 10^6$ °K. This might be expected from a stream originating in the corona. The directed velocity, however, is higher than the thermal velocity in the corona (root mean square velocity at 10^6 °K: 160 km/sec). The solar wind thus is *supersonic* and does not seem to be the result of a simple evaporation of the corona. PARKER's hydrodynamic theory of the generation of the solar wind in the corona is similar to the theory of supersonic flow through nozzles.

---

[4] M. NEUGEBAUER and C. W. SNYDER: l.c., Fig. 1.

[5] A. J. HUNDHAUSEN: Space Sci. Rev. **8** (5/6), 690—749 (1968).

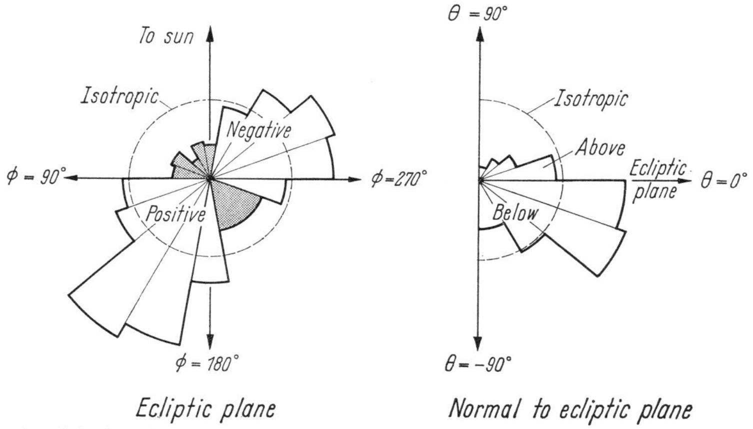

Fig. 2. Direction of the interplanetary magnetic field averaged over 3-hour intervals[6]. The distributions of the observed angles ϕ (in the ecliptic plane) and θ (normal to the ecliptic plane) are shown. The dashed circles represent isotropic distributions with the same number of cases. ϕ is located in the interval marked "positive" in 44%, in the interval marked "negative" in 34% of the cases; θ is above the ecliptic plane in 27%, below the ecliptic plane in 73% of the cases.

5. Theory of the solar wind.

α) PARKER's *hydrodynamic theory* [*7*], [*10*][1] will be outlined now. A static equilibrium of the solar corona would be possible only after a certain gas pressure in interplanetary space had been built up. The equilibrium pressure is derived as some 10^{-13} N m^{-2} ($=10^{-12}$ dyn cm^{-2}). This is obviously a much too high value to be postulated. An alternative would be a situation similar to that of the terrestrial exosphere: The mean free path length might exceed the scale height at some level in the corona with the result of a continuous escape or evaporation of plasma. The fact that the solar wind is supersonic indicates, however, that the departure of plasma from the Sun is of a different nature; it is according to PARKER a *hydrodynamic phenomenon*, requiring a *sufficiently short free path length*. The short free path length must be due to a randomizing effect of the magnetic field. Coulomb interaction would yield a too low collision frequency. Random motions of the particles in combination with a random structure of the magnetic field are expected to lead to momentum changes of particles like those commonly observed in collisions.

β) PARKER in consideration of an outward stream of plasma in the corona obtains a *choking process* and an *expansion process* in superposition so that at the beginning of the flow the choking is predominant and later the expansion prevails. This is basically the same as in a De Laval nozzle [*7*]. In both cases the succession of choking and expansion causes a supersonic flow under appropriate conditions. The processes that occur in departure of the plasma stream from the solar corona may be recognized from the basic equations of PARKER's theory [*7*].

The equation of motion is

$$\varrho\, v \frac{\mathrm{d}v}{\mathrm{d}r} = -\frac{\mathrm{d}p}{\mathrm{d}r} - \varrho \frac{G M_\odot}{r^2} \tag{5.1a}$$

or, after multiplication with dr and rearrangement of terms,

$$\mathrm{d}p = V_\mathrm{s}^2\, \mathrm{d}\varrho = -\varrho v\, \mathrm{d}v - \varrho \frac{G M_\odot}{r^2} \mathrm{d}r \tag{5.1b}$$

[6] J. M. WILCOX and N. F. NESS: J. Geophys. Res. **70** (23), 5793—5805 (1965); Fig. 3.
[1] E. N. PARKER: The dynamical theory of gases and fields in interplanetary space, in [*3*], p. 45—55.

and the hydrodynamic continuity condition is

$$\mathrm{d}\,(r^2 \varrho v) = 0 \tag{5.2}$$

with the following denotations: p pressure, ϱ density, V_s sound velocity, v upward stream velocity, $M_\odot$ mass of the Sun, r distance from the Sun's center, $GM_\odot/r^2$ gravitational acceleration at distance r.

Spherical expansion is assumed. Some heat supply is necessary to reach velocities sufficiently above sound velocity (see the following Sect. 5γ). In a simplification one may therefore assume isothermal change of state with V_s being the isothermal sound velocity.

Eqs. (5.1b) and (5.2) in combination yield

$$2\frac{\mathrm{d}r}{r} - \frac{GM_\odot}{V_s^2 r}\,\frac{\mathrm{d}r}{r} = \left(\frac{v^2}{V_s^2} - 1\right)\frac{\mathrm{d}v}{v}. \tag{5.3}$$

Two counteracting terms appear on the left-hand side of this equation. $2\,\mathrm{d}r/r$ is due to radial *expansion*, while the gravity term has the same effect as a *reduction of the cross section* (e.g., in the converging section of the De Laval nozzle). The reduction means a "choking" with an increase of the velocity (positive $\mathrm{d}v$ for $v < V_s$). At low altitude the gravity term on the left of Eq. (5.3) predominates. Ignoring the factor r^2 in Eq. (5.2) one sees that at low altitude the velocity v increases on the way up because of the density decrease.

The transition to predominance of the expansion effect takes place at a heliocentric height r_c at which

$$\frac{GM_\odot}{V_s^2 r_c} = 2. \tag{5.4}$$

This critical heliocentric height r_c is between 1.7 and 4 solar radii, depending on the temperature of the corona [*7*].

In the De Laval nozzle a supersonic stream is obtained when the sound velocity is passed just at the place of the closest constriction. In the present situation the *sound velocity* must be reached *at the critical height* r_c. Below r_c the velocity increases on the way up for $v < V_s$ and above r_c it continues to increase when now $v > V_s$. Far out the velocity asymptotically approaches a limiting value so that according to Eq. (5.2) $r^2\varrho$ becomes constant.

γ) The continued increase of the velocity beyond the critical radius despite the decelerating gravity force requires *supply of energy in form of heat.* A similar heat supply is necessary in the De Laval nozzle. Some heat supply definitely takes place in a layer at the base of the corona, thus generating the high temperature of the corona. A high thermal conductivity through the corona might transport the required heat to outer areas or the heat production mechanism of the corona has to continue to areas in the outer corona [*7*], [*10*]. This last alternative is thought of as the more likely.

There may be waves such as shock waves, solar atmospheric gravity waves, and magnetodynamic waves, coming from below and losing their energy in dissipation processes at various coronal heights. This type of extended coronal heating was postulated by PARKER. About earlier objections to the amount of supplied heat and other ideas on the solar wind see the literature [*7*], [*10*][2] and the following paragraph.

δ) The earlier *evaporative theory,* explaining the solar wind as the result of departure of solar plasma from the corona in near absence of collisions, was dis-

[2] L. M. NOBLE and F. L. SCARF: J. Geophys. Res. **67** (12), 4577—4584 (1962).

carded by many authors because it did not yield a supersonic wind velocity. In recent years, however, it became clear that the electric field bringing proton and electron fluxes in agreement may accelerate the protons sufficiently[3, 4] (work by JOCKERS, see DONAHUE[4]). This electric field is not the "Pannekoek-Rosseland-field", which corresponds to static equilibrium. The evaporative theory with consideration of the proper electric field does not seem to be so much in opposition to the hydrodynamic theory. The problems are closely related to those of the "polar wind" in the terrestrial atmosphere, which were reviewed by DONAHUE[4] (see also Sect. 24α).

ε) The *magnetic field* in the solar wind consists of solar magnetic field lines that are carried out by the departing plasma in accordance with the theorem of frozen-in magnetic fields (see Sect. 8). The direction of the field will be discussed in the next section. Validity of the hydrodynamic theory with the additional assumption of a magnetic field requires an energy density of the plasma larger (or much larger) than the magnetic field energy density [7][5], i.e.

$$\beta \equiv \frac{\mathscr{E}_{\text{kin}} + \mathscr{E}_{\text{therm}}}{\mathscr{E}_{\text{mag}}} = \frac{\frac{1}{2}\varrho\, v^2 + 3NkT}{\mathrm{B}^2/2\mathrm{u}\,\mu_0} > 1\,, \tag{5.5}$$

at the place of departure; a too strong magnetic field would hold the plasma back. In Eq. (5.5) the thermal energy of a particle appears multiplied by the ion (or electron) density N and by a factor 6, accounting for the combined degrees of freedom of ions and electrons.

With a field of 1 Gauss ($=10^{-4}$ V sec m^{-2}) one finds $v=100$ km sec^{-1} as the lower limit of the wind velocity (for $\beta=1$). The field thus cannot exceed 1 Gauss very much. On the other hand the interplanetary field near the Earth permits to compute the field at departure if one assumes a radial component $B_r \propto r^{-2}$. With $B_r = 4\,\gamma$ near the Earth one finds approximately 2 Gauss near the Sun. Apparently there is $\beta \approx 1$ in departure. The solar magnetic field thus may play a role in the release of the wind from the corona and a corresponding correction of the hydrodynamic theory seems desirable (see p. 80–88 of [*37*]).

6. Sector structure.

α) *Plasma* emitted from a particular location on the Sun in a radial direction will occupy a *spiral path* in interplanetary space because of the Sun's rotation. Each parcel of plasma moves radially outward, but the spiral — an *Archimedes spiral* — makes the impression of a rotation, its foot being anchored in the emitting source that rotates with the Sun. The *magnetic field lines* carried along with the plasma become *also Archimedes spirals*; they are directed outward or inward and take part in the rotation [*7*], [*8*], [*11*]. In fact, the field directions observed in interplanetary space show some distribution with preference of the expected spiral direction (Fig. 2). The angle of the spiral toward the radial direction should in the vicinity of the Earth be 45° for a wind velocity of approximately 400 km sec^{-1}.

β) An observation of the interplanetary magnetic field near the Earth during three successive rotation periods of the Sun resulted in the picture of Fig. 3 [*8*], [*11*]. There appear *alternating sectors* with opposite magnetic field directions. Only the part near the Earth was, of course, observed on a satellite. The signs in Fig. 3 denote outward or inward direction in the average over 3 hours. Paren-

[3] H. K. SEN: The electric field in the solar coronal exosphere and the solar wind, AFCRL-68-0506, Environmental Research Papers, No. 29, Air Force Cambridge Research Laboratories, Bedford, Mass., U.S.A., 1968.

[4] T. M. DONAHUE: Rev. Geophys. and Space Phys. **9** (1), 1–9 (1971).

[5] u = 1 in rationalized systems of units (e.g., V,A,m,kg,sec); u = 4π in non-rationalized systems of units (e.g., Gaussian system). See Introductory Remarks, p 1.

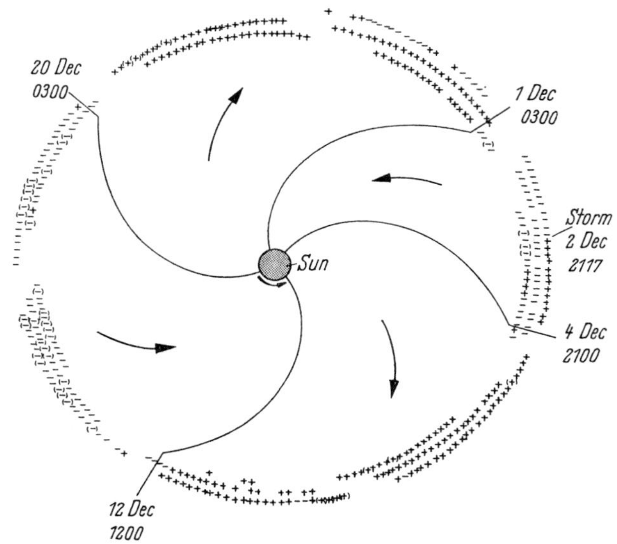

Fig. 3. Sector structure of the solar wind[1]. Four sectors with their predominant magnetic-field directions are shown. Symbols + and − indicate observed directions predominantly away from and toward the Sun during 3-hour intervals. Three solar rotations are included, beginning with 27 November 1963 at the top of the figure. Only the dates of the first rotation are noted. Symbols + or − without parentheses refer to angles ϕ (Fig. 2) between 90° and 230° or between 270° and 50°. Transgression of these limits for a few hours in a smooth and continuous way is indicated by parentheses.

theses around a sign indicate temporary stronger deviation from the specified direction. The observations were made near sunspot minimum. At this time four sectors were seen. Their widths were 2/7 or 1/7 of a full circle. The sectors stayed essentially unchanged over many solar rotations. It seems that the same sector structure persisted from late 1963 until early 1965, although the observations were not continued during the whole period. At times of higher solar activity only two sectors existed, which appeared to be not quite so invariable[2, 3].

On the Sun between 40° S and 40° N heliographic latitude there are extensive regions in which the photospheric magnetic field according to measurements of the Zeeman effect is directed outward or inward[4]. The polarity of these regions apparently corresponds to the field direction in the adjoining sector of the solar wind [*11*][4,5].

γ) Within the individual sectors a certain regular *structure of the plasma stream and the magnetic field* is found (Figs. 4 and 5) [*11*]: Both the plasma velocity and the magnetic field strength increase to a maximum in the early part of the sector and are lower in the later part; also plasma density and flux vary in a certain manner. This regularity and the correlation with the magnetic regions on the Sun indicate that the structure is determined by phenomena on the Sun. The similarity in the behavior of the plasma velocity and the magnetic field again suggests that there might be a connection between magnetic field and plasma release on the Sun, with the condition $\beta \approx 1$ possibly being of

[1] J. M. Wilcox and N. F. Ness: J. Geophys. Res. **70** (23), 5793—5805 (1965); Figs. 1 and 7.

[2] K. H. Schatten, N. F. Ness, and J. M. Wilcox: Solar Phys. **5**, 240—256 (1968).

[3] J. M. Wilcox and D. S. Colburn: J. Geophys. Res. **74** (9), 2388—2392 (1969).

[4] N. F. Ness and J. M. Wilcox: Astrophys. J. **143** (1), 23—31 (1966).

[5] A. Severny, J. M. Wilcox, P. H. Scherrer, and D. S. Colburn: Solar Phys. **15** (1), 3—14 (1970).

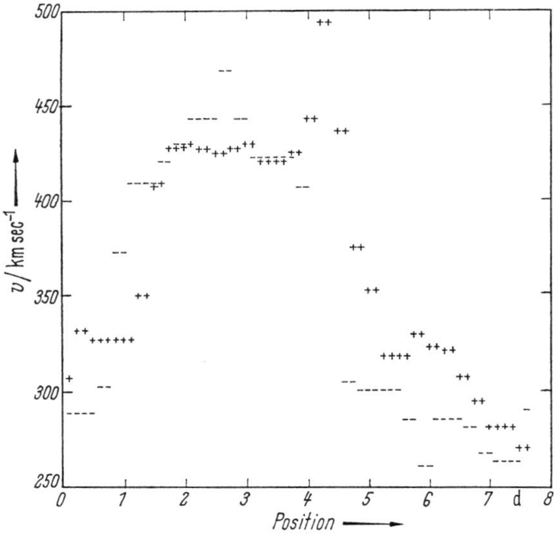

Fig. 4. Solar wind velocity as a function of position within $\frac{2}{7}$-sectors[6]. The position within a sector is denoted by the number of days from the beginning of the sector. The values are averages over all sectors with field away from the Sun (+) and with field toward the Sun (—); they are smoothed by superposed epoch analysis with 24-h epochs.

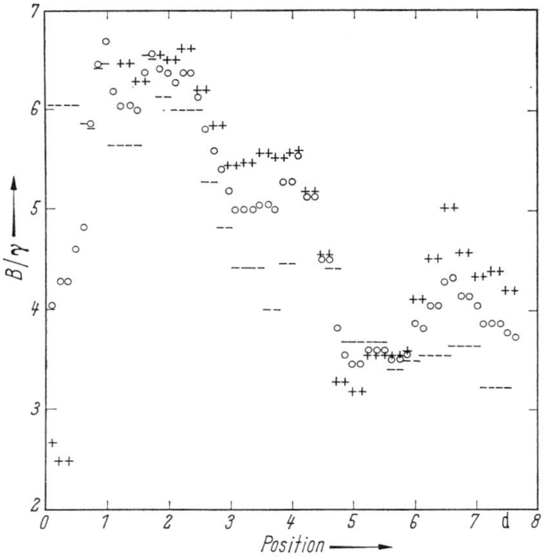

Fig. 5. Interplanetary magnetic field as a function of position within $\frac{2}{7}$-sectors[1]. Averaging and smoothing were the same as in Fig. 4. The sign ○ combines away-sectors and toward-sectors.

importance (cf. Sect. 5 ε). Attention may be called to the fact that the magnetic regions on the Sun contain large numbers of centers with increased field strength [11][4].

[6] [11], Fig. 11.

The boundary between sectors is definitely less than 150,000 km thick as satellite observations show [*11*]. This limit was set by the intermission between successive observations. A much lower limit was found in some more closely spaced observations. The parameters of the solar wind are in a number of ways related with variations of the geomagnetic field (Sects. 27, 29, 30).

C. Fundamentals of magnetodynamics.

7. Basic formulations. Three different theoretical approaches are in use in treating dynamic processes in a plasma with presence of a magnetic field[1]. The *microscopic approach* deals with individual particles (ions, electrons, eventually also neutrals), for which equations of motion are set up. In this approach one is mainly concerned with some average behavior of particles and tends to ignore random deviations of individual particles and random fields.

The second approach is the *"magnetodynamic"* (or *"hydromagnetic"*[2]) *theory*, in which the medium appears as a kind of fluid or gas. Macroscopic equations may be set up either for the total medium or separately for various constituents (plasma and neutral gas or ion gas, electron gas, and neutral gas).

The third approach is *statistical* and makes use of the distribution function that describes the particle distribution in phase space (coordinate space plus velocity space). The starting equation is of the type of the BOLTZMANN equation or FOKKER-PLANCK equation. This third approach is the most rigorous; it should yield all the results obtained by the other methods. The macroscopic approach of the magnetodynamic theory, however, lends itself to a very brief and elusive description of phenomena; it is therefore frequently preferred. The stability problem, i.e., the question whether a wave or perturbation under certain conditions decays or grows, is not solved in this macroscopic treatment, it is a matter of the third, the statistical approach[3]. In many practical cases questions of stability have not yet been uniquely answered.

A treatment of theoretical plasma physics, serving as an introduction to magnetodynamics and fast-particle phenomena in magnetosphere and outer space, is found in a book by ROSSI and OLBERT [*31a*].

α) *Microscopic formulation.* The equations of motion for ions (i) and electrons (e) may be written[4]

$$\left.\begin{aligned} m_{\mathrm{i}}\frac{\mathrm{d}\boldsymbol{v}_{\mathrm{i}}}{\mathrm{d}t}+\frac{1}{2}\nu_{\mathrm{i}}m_{\mathrm{i}}\boldsymbol{v}_{\mathrm{i}} &= q\left(\boldsymbol{E}+\boldsymbol{v}_{\mathrm{i}}\times\frac{\boldsymbol{B}}{c_0\sqrt{\varepsilon_0\mu_0}}\right)\\ m_{\mathrm{e}}\frac{\mathrm{d}\boldsymbol{v}_{\mathrm{e}}}{\mathrm{d}t}+\nu_{\mathrm{e}}m_{\mathrm{e}}\boldsymbol{v}_{\mathrm{e}} &= -q\left(\boldsymbol{E}+\boldsymbol{v}_{\mathrm{e}}\times\frac{\boldsymbol{B}}{c_0\sqrt{\varepsilon_0\mu_0}}\right)\end{aligned}\right\}\qquad(7.1)$$

with the following notations: m particle mass, $\boldsymbol{v}$ particle velocity, q elementary charge, $\boldsymbol{E}$ electric field strength, and $\nu_{\mathrm{i}}=\nu_{\mathrm{in}}$, $\nu_{\mathrm{e}}=\nu_{\mathrm{en}}$ frequencies of collisions with neutral particles that are at rest. The factor $\frac{1}{2}$ in the first equation expresses that only half the momentum is lost in a collision between particles of equal masses[5]. The magnetic field may always be described by the magnetic flux

[1] See the contribution to this volume by V. L. GINZBURG and A. A. RUHADZE [*35*], p. 395.

[2] With regard to nomenclature see footnote 2 of Sect. 2.

[3] This problem is discussed in the contribution by V. L. GINZBURG and A. A. RUHADZE, in this volume, Chap. D, p. 490, [*35*].

[4] The constants c_0, ε_0, μ_0 are the light velocity, permittivity, and permeability of vacuum in any conventional system of units. $c_0\sqrt{\varepsilon_0\mu_0}=1$ in SI-units (Système International d'Unités, using V, A, m, kg, sec), $=c_0$ in Gaussian units. See Introductory Remarks, p. 1.

[5] H. POEVERLEIN: J. Geophys. Res. **72** (1), 251–256 (1967).

density $\boldsymbol{B}$. Neglecting the collision terms and introducing angular gyrofrequencies as vectors one obtains

$$m_{\mathrm{i,e}}\left(\frac{\mathrm{d}\boldsymbol{v}_{\mathrm{i,e}}}{\mathrm{d}t}+\boldsymbol{\omega}_{B\,\mathrm{i,e}}\times\boldsymbol{v}_{\mathrm{i,e}}\right)=\pm q\boldsymbol{E} \tag{7.2}$$

with the two vectors[4]

$$\left.\begin{aligned}\boldsymbol{\omega}_{B\,\mathrm{i}}&=\frac{q}{m_{\mathrm{i}}}\,\frac{\boldsymbol{B}}{c_0\sqrt{\varepsilon_0\,\mu_0}}\\ \boldsymbol{\omega}_{B\,\mathrm{e}}&=-\frac{q}{m_{\mathrm{e}}}\,\frac{\boldsymbol{B}}{c_0\sqrt{\varepsilon_0\,\mu_0}}\,.\end{aligned}\right\} \tag{7.3}$$

In many cases it is recommendable to decompose the particle velocities into various parts: a velocity $\boldsymbol{u}$ representing circling in the magnetic field, a drift velocity $\boldsymbol{w}$ normal to $\boldsymbol{B}$, and a velocity $\boldsymbol{v}_{\|}$ parallel to $\boldsymbol{B}$ [*32*]. In *electric and magnetic fields* that are *constant in space and time* the following equations hold for the three types of velocities:

$$\frac{\mathrm{d}\boldsymbol{u}_{\mathrm{i,e}}}{\mathrm{d}t}+\boldsymbol{\omega}_{B\,\mathrm{i,e}}\times\boldsymbol{u}_{\mathrm{i,e}}=0, \tag{7.4a}$$

$$\boldsymbol{w}=\frac{c_0\sqrt{\varepsilon_0\,\mu_0}}{B^2}\,\boldsymbol{E}\times\boldsymbol{B}, \tag{7.4b}$$

$$m_{\mathrm{i,e}}\frac{\mathrm{d}\boldsymbol{v}_{\|\,\mathrm{i,e}}}{\mathrm{d}t}=\pm q\,\boldsymbol{E}_{\|}, \tag{7.4c}$$

$$\boldsymbol{v}_{\mathrm{i,e}}=\boldsymbol{u}_{\mathrm{i,e}}+\boldsymbol{w}+\boldsymbol{v}_{\|\,\mathrm{i,e}}. \tag{7.4d}$$

The equations are verified by inserting $\boldsymbol{v}_{\mathrm{i,e}}$ of Eqs. (7.4) in Eq. (7.2).

The drift velocity $\boldsymbol{w}$, which is the same for ions and electrons, is determined by the electric field and the magnetic field. The circling velocity $\boldsymbol{u}$ is frequently disregarded; the remaining velocity, $\boldsymbol{w}+\boldsymbol{v}_{\|\,\mathrm{i,e}}$, is the velocity of the centers of the circles or the "guiding center velocity" (according to ALFVÉN [*29*]).

The decomposition of the velocities shown in Eqs. (7.4) can still be used as an *approximation* in *nonhomogeneous fields*. If the inhomogeneity of the magnetic field is strong, however, the velocities $\boldsymbol{u}_{\mathrm{i,e}}$ according to Eq. (7.4a) include circling and a drift due to a nonconstant $\boldsymbol{\omega}_{B\,\mathrm{i,e}}$. A parallel velocity $\boldsymbol{v}_{\|}$ on curved field lines leads to an acceleration term $m\,\mathrm{d}\boldsymbol{v}_{\|}/\mathrm{d}t$ corresponding to a centrifugal force, which causes another drift motion. These drifts in an inhomogeneous magnetic field, which are essential for fast particles, will be discussed in Sect. 19β. They do not appear in the macroscopic description that is to follow now.

β) *Macroscopic formulation.* In the equations of particle motion, Eqs. (7.1) and (7.2), the acceleration terms are small and may be ignored in a first approximation for frequencies well below ion gyrofrequency. Neglect of the acceleration and collision terms yields

$$\boldsymbol{E}+\boldsymbol{v}_{\mathrm{i,e}}\times\frac{\boldsymbol{B}}{c_0\sqrt{\varepsilon_0\,\mu_0}}\approx 0 \tag{7.5}$$

or (approximately)

$$\boldsymbol{E}+\boldsymbol{v}\times\frac{\boldsymbol{B}}{c_0\sqrt{\varepsilon_0\,\mu_0}}=0 \tag{7.6}$$

with introduction of the plasma velocity

$$\boldsymbol{v}=\frac{m_{\mathrm{i}}\,\boldsymbol{v}_{\mathrm{i}}+m_{\mathrm{e}}\,\boldsymbol{v}_{\mathrm{e}}}{m_{\mathrm{i}}+m_{\mathrm{e}}}\,. \tag{7.7}$$

$\boldsymbol{v}$ is almost equal to $\boldsymbol{v}_{\mathrm{i}}$ as $m_{\mathrm{i}}\gg m_{\mathrm{e}}$. Eq. (7.5) suggests that $\boldsymbol{v}_{\mathrm{i}}$ and $\boldsymbol{v}_{\mathrm{e}}$ or at least the velocity components normal to the magnetic field are approximately equal.

In the presence of an electric field this cannot be exactly correct; there must be a current density

$$\boldsymbol{J} = N q (\boldsymbol{v}_{\mathrm{i}} - \boldsymbol{v}_{\mathrm{e}}) \tag{7.8}$$

with N representing the ion density and electron density, both being identical.

The two equations of particle motion, Eqs. (7.2), may now be multiplied by N and added. This leads to the macroscopic equation of motion of the plasma

$$\varrho \frac{\mathrm{d}\boldsymbol{v}}{\mathrm{d}t} = \boldsymbol{J} \times \frac{\boldsymbol{B}}{c_0 \sqrt{\varepsilon_0 \mu_0}} \tag{7.9}$$

with the plasma density

$$\varrho = N (m_{\mathrm{i}} + m_{\mathrm{e}}) \,. \tag{7.10}$$

Eqs. (7.6) and (7.9) are the macroscopic equations in the simplest form. Eq. (7.6) is nothing but Ohm's law with infinite conductivity. Ohm's law with a finite conductivity σ is

$$\boldsymbol{J} = \sigma \boldsymbol{E}^* \tag{7.11}$$

when $\boldsymbol{E}^*$ represents the electric field strength seen by an observer moving with the plasma. The conductivity σ may be a scalar or a tensor. The transformation formula of the electric field with disregard of the relativistic factor $\sqrt{1 - v^2/c_0^2}$ is

$$\boldsymbol{E}^* = \boldsymbol{E} + \boldsymbol{v} \times \frac{\boldsymbol{B}}{c_0 \sqrt{\varepsilon_0 \mu_0}} \,. \tag{7.12}$$

Infinite conductivity requires disappearance of $\boldsymbol{E}^*$. This disappearance is postulated by Eq. (7.6). It may already be noted that disappearance of the curl $\left(\equiv \frac{\partial}{\partial \boldsymbol{r}} \times\right)$ of Eq. (7.6) is the basis of the theorem of frozen-in magnetic field lines (Sect. 8).

The inference of disappearance of $\boldsymbol{E}^*$ from the severely clipped equations of motion leaves one somewhat unsatisfied. A better justification of the disappearance of curl $\boldsymbol{E}^*$ is given by SPITZER [6].

The equation of motion of the plasma in a more complete version has to include the pressure force, which is present in the plasma just as in a neutral gas, and the gravity force. The equation thus completed is [7]

$$\varrho \left(\frac{\partial \boldsymbol{v}}{\partial t} + \boldsymbol{v} \cdot \frac{\partial}{\partial \boldsymbol{r}} \boldsymbol{v} \right) = \boldsymbol{J} \times \frac{\boldsymbol{B}}{c_0 \sqrt{\varepsilon_0 \mu_0}} - \frac{\partial}{\partial \boldsymbol{r}} p + \varrho \boldsymbol{g} \tag{7.13}$$

[*32*], [*35*]. The pressure p results from the thermal motion in the same way as in a neutral gas [cf. Eq. (9.6) with remarks]. In more rigorous studies a symmetric tensor [8] has to be used in place of the scalar p. Linearized theory replaces ϱ and $\boldsymbol{B}$ by the stationary density ϱ_0 and the stationary magnetic flux density $\boldsymbol{B}_0$ and omits the term $\boldsymbol{v} \cdot \frac{\partial}{\partial \boldsymbol{r}} \boldsymbol{v}$.

In Eq. (7.13) no collision term is included. The equation refers to oscillatory processes with angular frequencies much greater than the collision frequency of *ions against neutrals*, which is the main cause of momentum loss. The pressure p, appearing in the equation, is, on the other hand, definable only with a high frequency of collisions between *charged particles*. In the F2 layer of the ionosphere in daytime some characteristic values for the decisive collision frequencies are [5] $\nu_{\mathrm{in}} = 1\ \mathrm{sec}^{-1}$, $\nu_{\mathrm{ii}} = 4\ \mathrm{sec}^{-1}$. The collision frequency between ions, ν_{ii}, decreases

---

[6] L. SPITZER: [*32*], p. 41–42.

[7] In most articles of this volume the symbol $\partial/\partial \boldsymbol{r}$ is used in the meaning of the Nabla operator ∇.

[8] See contribution by V. L. GINZBURG and A. A. RUHADZE [*35*], p. 395, in particular Sect. 15, p. 437.

much more slowly with height than the collision frequency of ions with neutrals, ν_{in}, because of the slow decrease of the plasma density.

Another interesting limiting case is that of an (angular) oscillation frequency well below all collision frequencies, in particular below ν_{ni}, which is ν_{in} multiplied by the ratio of the densities of plasma and neutral gas[5]. The maximum ν_{ni}, of the order of $10^{-3}\,\text{sec}^{-1}$, is found for the F2 layer peak in daytime, where the ion density is highest. A diurnal oscillation has $\omega \approx 10^{-4}\,\text{sec}^{-1}$. In the limiting case $\omega \ll \nu_{\text{ni}}$ neutral gas and plasma should move jointly and Eq. (7.13) should apply to the total medium with ϱ representing the density of neutral gas and plasma in combination.

Equations needed in addition to the equation of motion, Eq. (7.13), are: the hydrodynamic continuity condition

$$\frac{\partial \varrho}{\partial t} + \frac{\partial}{\partial \boldsymbol{r}} \cdot (\varrho \boldsymbol{v}) = 0 \tag{7.14}$$

or

$$\frac{\mathrm{d}\varrho}{\mathrm{d}t} + \varrho \frac{\partial}{\partial \boldsymbol{r}} \cdot \boldsymbol{v} = 0, \tag{7.15}$$

furthermore the relationship between the varying parts of p and ϱ, which may express adiabatic change of state in a (moving) plasma parcel, and Maxwell's equations. The displacement current in Maxwell's equation can in general be neglected.

γ) *Remarks on magnetospheric processes.* A plasma may be characterized as "hot" or "cold", depending on which of the two forces, $\frac{\partial}{\partial \boldsymbol{r}} p$ or $\boldsymbol{J} \times \boldsymbol{B}/c_0 \sqrt{\varepsilon_0 \mu_0}$, is predominant. The thermal plasma of the magnetosphere is in this sense "cold"; it has a small ratio $c_0 \sqrt{\varepsilon_0 \mu_0} \left| \frac{\partial}{\partial \boldsymbol{r}} p \right| / |\boldsymbol{J} \times \boldsymbol{B}|$ under plausible assumptions [see Eq. (7.18) with the subsequent remarks and the notes on energies in connection with Eq. (8.12) and in Sect. 28]. The negligible collision frequency between charged particles in the upper magnetosphere, however, would suggest to call the plasma "warm", because the collision frequency between charged particles is nearly proportional to $T^{-\frac{3}{2}}$, thus being low for high temperature T. In the literature there is no general agreement about the definition of "cold" and "hot" plasmas.

Denoting variable quantities by subscript 1, introducing the sound velocity $V_{\mathrm{s}} = (p_1/\varrho_1)^{\frac{1}{2}}$, and setting $\frac{\mathrm{u}\,\mu_0}{c_0 \sqrt{\varepsilon_0 \mu_0}} \boldsymbol{J} = \frac{\partial}{\partial \boldsymbol{r}} \times \boldsymbol{B}$, corresponding to a Maxwell equation without displacement current density[9], one finds for the ratio of force densities [see Eq. (7.13)]

$$c_0 \sqrt{\varepsilon_0 \mu_0} \frac{\left| \frac{\partial}{\partial \boldsymbol{r}} p \right|}{|\boldsymbol{J} \times \boldsymbol{B}|} \approx \mathrm{u}\,\mu_0 \frac{V_{\mathrm{s}}^2 \varrho_1}{B_0 B_1}. \tag{7.16}$$

Vacuum permeability μ_0 is supposed to be used in all formulas. It is now assumed that $\partial/\partial \boldsymbol{r}$ in both force terms refers to nearly the same length. With motions normal to the magnetic field, to which the principle of frozen-in field lines (Sect. 8) applies, there is

$$\frac{\varrho_1}{\varrho_0} \approx \frac{B_1}{B_0}. \tag{7.17}$$

From Eqs. (7.16) and (7.17) the ratio of force densities is obtained in the form

$$c_0 \sqrt{\varepsilon_0 \mu_0} \frac{\left| \frac{\partial}{\partial \boldsymbol{r}} p \right|}{|\boldsymbol{J} \times \boldsymbol{B}|} \approx \mathrm{u}\,\mu_0 \frac{V_{\mathrm{s}}^2 \varrho_0}{B_0^2} = \frac{V_{\mathrm{s}}^2}{V_{\mathrm{A}}^2} \tag{7.18}$$

[9] The factor u is 1 or 4π, depending on the system of units used (see Introductory Remarks, p. 1). In SI-units (V, A, m, kg, sec) there is $\mathrm{u} = 1$ and $c_0 \sqrt{\varepsilon_0 \mu_0} = 1$.

when V_A denotes the Alfvén velocity according to

$$V_A^2 = \frac{B_0^2}{u\,\mu_0\,\varrho_0}. \tag{7.19}$$

The role of V_A for waves will be discussed in Sect. 11. In the magnetosphere the Alfvén velocity V_A is in the range 150 to 6000 km sec^{-1} (approximately)[10]; the sound velocity determined from the plasma data is between 1 and 30 km sec^{-1}. The last value corresponds to hydrogen plasma at 30,000 °K. The ratio of Eq. (7.18) apparently is small. [Also see Ineq. (8.12).]

In the cold plasma the pressure gradient term in the equation of motion can be neglected in reference to motions normal to the field lines. It thus seems advisable to deal with motions parallel and normal to the field lines separately. The *motions normal to the field lines* represent an electrodynamic drift, which according to Eq. (7.6) requires an electric field. The force density for these motions (force per unit volume) is $\boldsymbol{J} \times \boldsymbol{B}/c_0 \sqrt{\varepsilon_0 \mu_0}$.

Motions along the field lines, in turn, require pressure forces quite generally. These motions may follow curved field lines, because $\boldsymbol{J} \times \boldsymbol{B}$-forces in general are strong enough to act as guiding force. The required current density results from minor deviations of positive and negative particles from the curved field lines. In many circumstances the necessary $\boldsymbol{J}$ will be small compared to the current density relating to motions normal to the field lines. The motion along the field lines will at higher altitudes, where collisions become too rare for a hydrodynamic pressure, turn into diffusion, to which the present equation of motion does not apply.

The different behavior of motions along and across the field lines suggests the use of the theory of magnetodynamic waves for oscillatory states of motions or even for nonperiodic variations[11]. In magnetodynamic waves the different motions are ascribed to different wave modes (see Sect. 11).

8. Frozen-in magnetic field lines.

α) *The theorem.* A well known theorem of fluid dynamics says that vortex lines in an ideal fluid move with the fluid. The theorem of frozen-in magnetic field lines is an analogue to this theorem. Its corresponding formulation is:

Magnetic field lines in a perfectly conducting fluid move with the fluid or, in other words, are "frozen-in" in the fluid (ALFVÉN, 1942, see [29]).

The highly conductive fluid may be a plasma permeated by a magnetic field. The magnetic field lines are supposed to take part only in the velocity components of the conductive fluid normal to the field direction One may imagine plasma parcels threaded on the field lines like beads on a string that are freely movable along the string, but can only move with the string in the normal direction.

The limitation of the theorem by the imperfect conductivity of the plasma will be discussed later (in Sect. 10). To ascribe an identity to magnetic field lines as it is done in the theorem means some arbitrariness. Nobody can tell which field line in a certain instant is some other field line in a different instant. The validity of the theorem consequently can not be checked from an observation of field lines, but the magnetic field variations, which are observable, have to conform with the idea of field lines that move in accordance with the theorem.

The theorem of fluid dynamics follows from the equation of motion of a homogeneous ideal fluid

$$\frac{\partial \boldsymbol{v}}{\partial t} + \left(\boldsymbol{v} \cdot \frac{\partial}{\partial \boldsymbol{r}}\right)\boldsymbol{v} = -\frac{\partial}{\partial \boldsymbol{r}}\,\Phi. \tag{8.1}$$

[10] A. DESSLER: Geomagnetism, in [2], p. 153—182.

[11] H. POEVERLEIN: J. Atmos. Terr. Phys. 28, 1111—1123 (1966).

The gradient term on the right-hand side is supposed to include all forces acting on the mass unit of the fluid. Transforming the second term on the left and taking the curl one obtains

$$\frac{\partial \boldsymbol{\Omega}}{\partial t} = \frac{\partial}{\partial \boldsymbol{r}} \times (\boldsymbol{v} \times \boldsymbol{\Omega}) \tag{8.2}$$

with the vorticity

$$\boldsymbol{\Omega} = \frac{\partial}{\partial \boldsymbol{r}} \times \boldsymbol{v}. \tag{8.3}$$

Eq. (8.2) is the equation for $\boldsymbol{\Omega}$ expressing that the vortex lines move with the fluid or are "frozen-in".

An equation analogous to Eq. (8.2) is found to hold for the magnetic flux density $\boldsymbol{B}$ in a highly conducting fluid [*29*], [*32*], [*33*] [1]. Taking the curl of Ohm's law for a perfectly conducting fluid, Eq. (7.6), one obtains

$$\frac{\partial}{\partial \boldsymbol{r}} \times \left(\boldsymbol{E} + \boldsymbol{v} \times \frac{\boldsymbol{B}}{c_0 \sqrt{\varepsilon_0 \mu_0}}\right) = 0. \tag{8.4}$$

With use of one of the Maxwell equations this becomes

$$\frac{\partial \boldsymbol{B}}{\partial t} = \frac{\partial}{\partial \boldsymbol{r}} \times (\boldsymbol{v} \times \boldsymbol{B}). \tag{8.5}$$

The equation obviously is the analogue of Eq. (8.2). This suggests the statement that the magnetic field lines "move with the plasma" or are "frozen-in". In the following subsection it will be shown that Eq. (8.5), in fact, expresses the field variation resulting from motion of field lines.

β) *Variation of the magnetic field.* The right-hand term of Eq. (8.5) will be transformed and decomposed into four parts comprehending all the effects of motion of the field lines on the magnetic field. In Eq. (8.5) the velocity vector $\boldsymbol{v}$ may be replaced by its projection into the plane normal to the field line, $\boldsymbol{u}$. The equation thus is written

$$\frac{\partial \boldsymbol{B}}{\partial t} = \frac{\partial}{\partial \boldsymbol{r}} \times (\boldsymbol{u} \times \boldsymbol{B}) \tag{8.6}$$

or, with use of $\partial/\partial \boldsymbol{r} \cdot \boldsymbol{B} = 0$,

$$\frac{\partial \boldsymbol{B}}{\partial t} = \left(\boldsymbol{B} \cdot \frac{\partial}{\partial \boldsymbol{r}}\right) \boldsymbol{u} - \boldsymbol{B}\left(\frac{\partial}{\partial \boldsymbol{r}} \cdot \boldsymbol{u}\right) - \left(\boldsymbol{u} \cdot \frac{\partial}{\partial \boldsymbol{r}}\right) \boldsymbol{B}. \tag{8.7}$$

A coordinate system is now chosen whose z-axis has the direction of $\boldsymbol{B}$ at the point under consideration. At a neighboring point there may well be B_x and B_y components and a u_z component. In the spatial differentiation it is of advantage to separate differentiation in the x, y-plane and in the z-direction. The two component vectors of which $\partial/\partial \boldsymbol{r} \equiv \nabla$ may be composed are denoted by $\nabla_{x,y}$ and ∇_z and are defined as $\nabla_{x,y} \equiv \boldsymbol{x}^\circ \frac{\partial}{\partial x} + \boldsymbol{y}^\circ \frac{\partial}{\partial y}$ and $\nabla_z \equiv \boldsymbol{z}^\circ \frac{\partial}{\partial z}$ with the unit vectors $\boldsymbol{x}^\circ, \boldsymbol{y}^\circ, \boldsymbol{z}^\circ$. Eq. (8.7) becomes for the location at which $\boldsymbol{B}$ has the z-direction

$$\frac{\partial \boldsymbol{B}}{\partial t} = (\boldsymbol{B} \cdot \nabla_z) \boldsymbol{u} - \boldsymbol{B}(\nabla_z \cdot \boldsymbol{u}) - \boldsymbol{B}(\nabla_{x,y} \cdot \boldsymbol{u}) - (\boldsymbol{u} \cdot \nabla_{x,y}) \boldsymbol{B}. \tag{8.8}$$

The substitutions

$$(\boldsymbol{B} \cdot \nabla_z) \boldsymbol{u} - \boldsymbol{B}(\nabla_z \cdot \boldsymbol{u}) = (\boldsymbol{B} \cdot \nabla_z) \boldsymbol{u}_{x,y} \tag{8.9}$$

and

$$(\boldsymbol{u} \cdot \nabla_{x,y}) \boldsymbol{B} = (\boldsymbol{u} \cdot \nabla_{x,y}) \boldsymbol{B}_z + (\boldsymbol{u} \cdot \nabla_{x,y}) \boldsymbol{B}_{x,y} \tag{8.10}$$

---

[1] W. A. Newcomb: Ann. Physics **3**, 347—385 (1958).

with rearrangement of terms in Eq. (8.8) yield

$$\frac{\partial \boldsymbol{B}}{\partial t} = -(\boldsymbol{u}\cdot\nabla)\boldsymbol{B}_z - \boldsymbol{B}(\nabla_{x,y}\cdot\boldsymbol{u}) - (\boldsymbol{u}\cdot\nabla)\boldsymbol{B}_{x,y} + (\boldsymbol{B}\cdot\nabla)\boldsymbol{u}_{x,y}. \tag{8.11}$$

Some unnecessary subscripts were omitted in this equation.

The four terms on the right-hand side of Eq. (8.11) represent various types of field changes due to motion of field lines with the normal velocity $\boldsymbol{u}$. They are illustrated in Fig. 6a–d. The magnetic field lines move only with the normal velocity of the plasma, $\boldsymbol{u}$, and are not affected by a velocity component of the plasma parallel to $\boldsymbol{B}$.

The first two terms on the right-hand side of Eq. (8.11) are in the z-direction, thus expressing a change of the magnitude of $\boldsymbol{B}$. The other two terms refer to a change of the field direction. The individual terms express the following: transport of the magnetic field with varying magnitude (first term, Fig. 6a), change of field magnitude by convergence or divergence of the motion in the normal plane or normal surface (second term, Fig. 6b), transport of field lines of varying direction (third term, Fig. 6c), and rotation of field lines due to inhomogeneity of the velocity field (fourth term, Fig. 6d). These are all the field changes to be expected from motion of the field lines with the plasma or, more strictly, with its normal velocity $\boldsymbol{u}$. Along the field lines the plasma may slide without effect on the magnetic field.

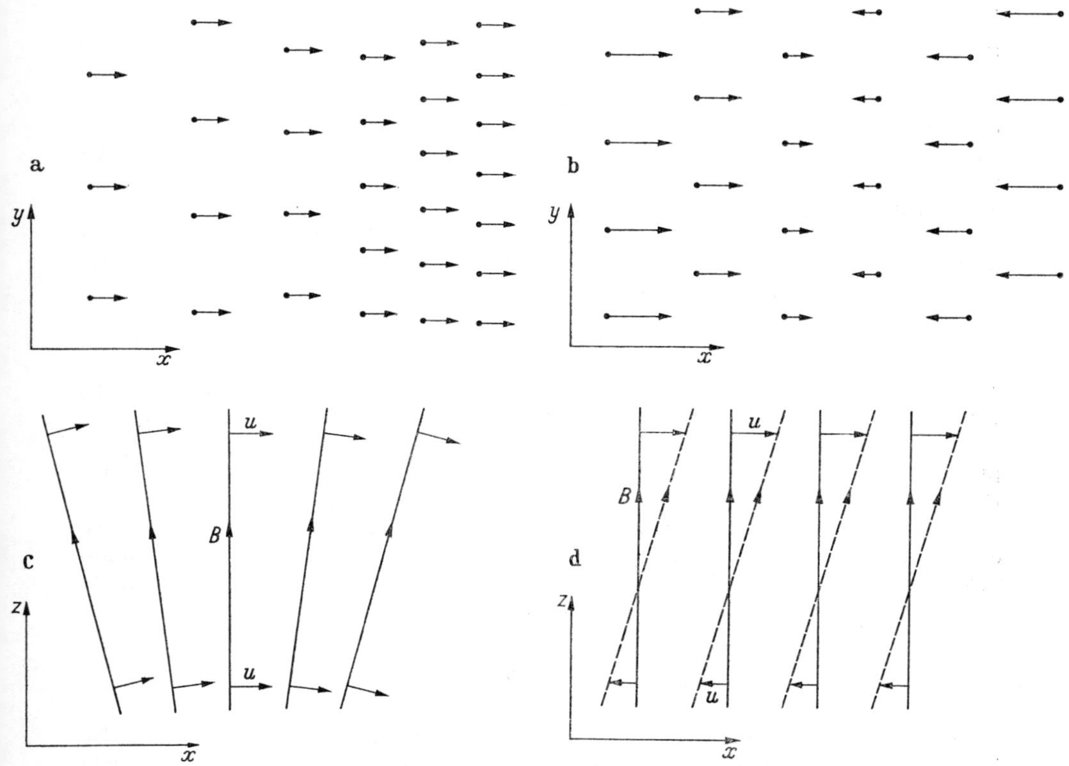

Fig. 6a–d. Changes of the magnetic field resulting from motion of field lines. (a), (b) Change in magnitude, (c), (d) change in direction, (a) transport in transverse plane, (b) divergence or convergence of motion in transverse plane, (c) transport of field lines of different directions, (d) rotation of field lines.

It may be repeated that the statement that the field lines move with the plasma (i.e., with the velocity $\boldsymbol{u}$) is based on an arbitrary identification of the field lines. Maxwell's equations do not assign an identity to magnetic field lines. The correct interpretation of Eq. (8.11) is: The temporal variation of the magnetic field is such as if the field lines do move with the plasma or are "frozen-in" in the plasma. It should also be recalled that Eq. (8.11) is only valid for a perfectly conducting fluid, to which Eq. (8.4) applies (see Sect. 10). In a fluid of low conductivity it is in general not possible to define a motion of field lines.

γ) *Application.* The theorem of frozen-in or moving field lines is useful in a large part of the magnetosphere (about limitations see Sect. 10). It may also be applied in the conductive Earth, but one has to be aware of the insulating space between ground and ionosphere, in which the field lines lose their identity [*16*][2]. Two systems of moving field lines, one in the magnetosphere, the other in the Earth, have consequently to be distinguished. In general no motion of field lines can be defined in the insulating interspace. The field lines, although appearing continuous in the interspace in each instant, are not connected through it in an invariable way; they are continually separated and reconnected differently.

Various possibilities of using the theorem of moving field lines may now be outlined. If the *plasma velocities* throughout some space are *known*, the theorem permits *computation of the variation of the magnetic field with time.*

A *knowledge of the magnetic field at all times* does *not* lead to a *unique velocity field,* but it imposes some *restrictions on the velocities.* The entire magnetic field may, for example, be known at two instants, t_1 and t_2. If some plasma parcel travels during the time interval $t_1 \ldots t_2$ from a point P_1 to a point P_2, then all plasma on the field line through P_1 at the time t_1 reaches the field line through P_2 at the time t_2.

An important *special case* is that of an *invariable magnetic field.* There may be motions in the invariable field, but they have to be such that always one field line replaces some other field line and the whole field line configuration is not changed. In talking about convection in the magnetosphere one deals preferably with this type of motion (Sect. 23).

In the *terrestrial magnetosphere* the *energy resides mainly in the magnetic field* (see Sect. 28). In contrast to the conditions on the Sun [Ineq. (5.5)] the ratio of energy densities β is in general small:

$$\beta = \frac{\mathscr{E}_{\mathrm{kin}} + \mathscr{E}_{\mathrm{therm}}}{\mathscr{E}_{\mathrm{mag}}} \ll 1 . \tag{8.12}$$

The only exception appears to be during the main phase of geomagnetic storms, when an abundance of trapped keV-particles is present (Sect. 21α). They may in the region of the ring current (several Earth radii high, near the equatorial plane) attain an energy density higher than that of the geomagnetic field[3] (for the theory of the main phase see [*28*] and Sect. 29 below).

In the case of no kinetic and small thermal energy density the ratio β is, except of a constant factor, identical with the square of the ratio of sound velocity to Alfvén velocity $V_{\mathrm{S}}^2/V_{\mathrm{A}}^2$, which was also seen to be the approximate ratio of force densities [see Eqs. (7.18) and (28.1)]

$$c_0 \sqrt{\varepsilon_0\, \mu_0}\, \frac{\left|\dfrac{\partial}{\partial \boldsymbol{r}}\, p\right|}{|\boldsymbol{J} \times \boldsymbol{B}|} \,.$$

As a consequence of Ineq. (8.12) the magnetic field is very little affected by thermal or macroscopic motion of the plasma. In computing the magnetic field

[2] T. GOLD: J. Geophys. Res. **64**, 1219 (1959).

[3] N. SCKOPKE: Verhandlungen der Deutschen Physikalischen Ges. (VI) **6** (8), 493—494 (1971).

one may ignore the effect of motions, though still assuming a high conductivity of the medium. Motions of the plasma, which may exist, have then to proceed so that the predetermined magnetic field is not modified by the co-motion of the magnetic field lines. In a nonvarying magnetic field the motion should lead to the just described exchange of field lines that leaves the magnetic field unchanged.

A closed contour in the plasma is now considered. The contour is identified by the plasma particles on it and is supposed to move with them. The magnetic flux through this plasma loop is invariable. This is evident from Figs. 6a and b, which demonstrate that the intersection points of the field lines with a normal surface (x, y-plane) move the same way as plasma particles in this surface. An additional component of plasma motion parallel to the field lines has no effect on the magnetic field. The fact that the magnetic flux through some moving plasma body does not vary is expressed by saying that the flow of plasma is "*flux-preserving*". The reversal of this is also correct: Plasma present in a field line tube remains enclosed in this field line tube.

That the flow of plasma is "*line-preserving*" is demonstrated by Fig. 6c, d. It is possible to specify a wider condition than Eq. (8.4) that entails line-preservation, but no flux-preservation [*33*][1]; the condition is

$$\boldsymbol{B}\times\left[\frac{\partial}{\partial \boldsymbol{r}}\times\left(\boldsymbol{E}+\boldsymbol{v}\times\frac{\boldsymbol{B}}{c_0\sqrt{\varepsilon_0\mu_0}}\right)\right]=0. \tag{8.13}$$

A flux-preserving flow, on the other hand, is always line-preserving.

The *magnetospheric boundary* (magnetopause) separates two plasmas, the atmospheric plasma with the geomagnetic field and the solar-wind plasma with the interplanetary field. If the two magnetic fields are not connected, i.e., if no merging of field lines occurs at the boundary, then the theorem of moving field lines postulates that no plasma moves across the boundary. The two plasmas, thus, will remain separated.

δ) *Neutral points.* There may be neutral points of the magnetic field, at which the field strength disappears. X-type neutral points, which are now considered, are seen on the magnetospheric boundary in the MEAD-BEARD model for example (Fig. 9). In an unperturbed situation, i.e. in an invariable field, the neutral points are stationary. If the plasma moves toward a stationary neutral point the theorem of moving field lines demands that each field line replace some neighboring field line during the motion. Because the spacing of the field lines becomes infinitely wide in approaching the neutral point, the velocity there should tend toward infinity; it should jump from plus to minus infinity [*33*]. The only other alternative is zero velocity at the neutral point[4] with a slow rise of the velocity in departure from the neutral point. This is consequently required by the theorem of moving field lines, if infinite or extremely high velocities are to be excluded.

On the other hand, the theorem is inapplicable at zero magnetic field. Merging of field lines, which might be expected at neutral points, will be described in Sect. 17. STERN [*33*] gave a more extensive account of neutral points of the magnetic field and the motions expected in their neighborhood, though theory and experimental evidence are still incomplete. The field geometry around neutral points may be decisively influenced by plasma motion, even if the plasma has little effect on the field in other areas. The coefficient β [Eq. (8.12)] tends toward infinity at a neutral point.

---

[4] H. POEVERLEIN: J. Atmos. Terr. Phys. **28**, 1111—1123 (1966).

9. Equilibrium at a boundary. The magnetopause is the boundary between the geomagnetic field and the solar wind — a plasma stream also carrying some magnetic field with it. The boundary is stationary as long as the solar wind is invariable. Boundaries separating different magnetic fields and plasmas of different properties are of general interest. Moving boundaries (discontinuity surfaces) will be discussed in Sects. 12 and 29.

Stationary boundaries of the type of the magnetopause are a special case of *stationary equilibria*, which may be studied with the help of the equation of motion, Eq. (7.13) (see [*31*], [*32*]). The general relationship for stationary equilibria [the following Eq. (9.3) or (9.5)] follows from the equation of motion with introduction of the momentum flux density $\varrho \boldsymbol{v}\boldsymbol{v}$, which is the dyadic product of the mass flux density $\varrho\,\boldsymbol{v}$ with the velocity $\boldsymbol{v}$.

The acceleration term on the left of Eq. (7.13) is transformed by adding $\boldsymbol{v}\frac{\partial \varrho}{\partial t}+\boldsymbol{v}\frac{\partial}{\partial \boldsymbol{r}}\cdot(\varrho\,\boldsymbol{v})$, an expression that vanishes according to the hydrodynamic continuity condition, Eq. (7.14). The acceleration term thus becomes

$$\varrho\frac{\mathrm{d}\boldsymbol{v}}{\mathrm{d}t}=\varrho\frac{\partial \boldsymbol{v}}{\partial t}+\boldsymbol{v}\frac{\partial \varrho}{\partial t}+\varrho\left(\boldsymbol{v}\cdot\frac{\partial}{\partial \boldsymbol{r}}\right)\boldsymbol{v}+\boldsymbol{v}\frac{\partial}{\partial \boldsymbol{r}}\cdot(\varrho\boldsymbol{v}) \tag{9.1}$$

or

$$\varrho\frac{\mathrm{d}\boldsymbol{v}}{\mathrm{d}t}=\frac{\partial}{\partial t}(\varrho\boldsymbol{v})+\frac{\partial}{\partial \boldsymbol{r}}\cdot(\varrho\boldsymbol{v}\boldsymbol{v}). \tag{9.2}$$

The last term represents the divergence of the momentum flux density.

In a *stationary* situation there is $\partial/\partial t=0$. The divergence term of Eq. (9.2) may consequently be used as the acceleration term in the equation of motion, Eq. (7.13). Finally replacing the current density $\boldsymbol{J}$ by $c_0\sqrt{\varepsilon_0\,\mu_0}\;\partial/\partial\boldsymbol{r}\times\boldsymbol{B}/\mathrm{u}\mu_0$ and neglecting the gravity force one obtains

$$\frac{\partial}{\partial \boldsymbol{r}}\cdot(\varrho\boldsymbol{v}\boldsymbol{v})+\frac{\partial}{\partial \boldsymbol{r}}p+\frac{1}{2\mathrm{u}\,\mu_0}\frac{\partial}{\partial \boldsymbol{r}}(B^2)-\frac{1}{\mathrm{u}\,\mu_0}\left(\boldsymbol{B}\cdot\frac{\partial}{\partial \boldsymbol{r}}\right)\boldsymbol{B}=0. \tag{9.3}$$

For a *plane stratification* Eq. (9.3) can be simplified under the assumption that the plasma passing through the stratification does not lose or gain any momentum of the direction parallel to the stratification. This assumption is frequently justifiable by reason of symmetry. A case in which it is not will, however, appear in Sect. 12γ as "rotational discontinuity".

The x-axis may now be the direction of the gradient of the plane stratification. It is assumed that the y and z components of the momentum of a particle do not change in progression in the x-direction. This means that $\frac{\partial}{\partial \boldsymbol{r}}\cdot(\varrho\,\boldsymbol{v}\boldsymbol{v})$ has an x-component only, which is $\frac{\partial}{\partial \boldsymbol{r}}(\varrho v_x^2)$. The first three terms of Eq. (9.3) thus have the x-direction. The last term can not have a component in this direction because of the divergence condition $\frac{\partial}{\partial \boldsymbol{r}}\cdot\boldsymbol{B}=0$; it must consequently disappear. This requires $B_x=0$ (unless $\boldsymbol{B}=$ const). The equation of the equilibrium obtained from Eq. (9.3) finally becomes

$$\frac{\partial}{\partial \boldsymbol{r}}\left(\varrho v_x^2+p+\frac{B^2}{2\mathrm{u}\,\mu_0}\right)=0 \tag{9.4}$$

or

$$\varrho v_x^2+p+\frac{B^2}{2\mathrm{u}\,\mu_0}=\text{const.} \tag{9.5}$$

Eq. (9.5) can be interpreted as saying that in a plane stratification the sum of kinetic pressure, thermal pressure, and magnetic pressure remains constant. The kinetic and thermal pressures are, in fact, of the same nature. Denoting the root mean square of the thermal velocity by $\boldsymbol{u}$ one obtains the thermal pressure in the same form as the kinetic pressure:

$$p = \varrho \mathrm{u}_x^2. \tag{9.6}$$

The most common case of plane stratification to which Eqs. (9.4) and (9.5) are applied is the transition between two homogeneous spaces. The transition may be arbitrarily rapid, it may be a *boundary*. (Some remarks about the minimum thickness of the transition or boundary follow later in this section.)

If the *idealized Ohm's law*, Eq. (7.6), is valid it yields an additional relationship between v_x and $\boldsymbol{B}$, because the tangential $\boldsymbol{E}$ component must be constant. The idealized Ohm's law and the continuity condition for the energy flux are, however, not considered now, but they are essential in the study of *traveling discontinuity surfaces* (see Sect. 12). The continuity of the energy flux in general involves a warming-up of plasma passing through the boundary.

Eqs. (9.4) and (9.5) are, of course, in agreement with *particle theory*. Summing the equations of motion of all particles up leads to Eq. (9.4). At a boundary it is easily seen that the gyromotions of the particles transgressing the boundary somewhat lead to opposite deflection of ions and electrons, in other words a current. The resulting force is the magnetic force at the boundary.

At the *transition between a field-free plasma stream and a magnetic field without plasma* one finds from Eq. (9.5) that the plasma pressure from one side and the magnetic-field pressure from the other side must be in balance. For plasma incident under an angle φ toward the normal of the "boundary" the pressure is

$$N\, m_i v^2 \cos^2 \varphi + p$$

(with the particle density N and ion mass m_i). This would be the pressure if the plasma just were to arrive and stop. The plasma stream, however, has to be returned if no persistent density increase takes place. In this case the pressure is doubled. The equilibrium condition thus is formulated

$$\mathrm{a}(N\, m_i v^2 \cos^2 \varphi + p) = B^2/2\mathrm{u}\, \mu_0 \tag{9.7}$$

with $\mathrm{a}=1$ (for stopping) or $\mathrm{a}=2$ (for reflection) and p denoting the thermal pressure of the arriving plasma stream alone.

The *magnetopause* as the *boundary between solar wind and geomagnetic field* has been studied under the simplifying assumptions that the solar wind exerts a kinetic pressure only and that this pressure is balanced by the pressure of the geomagnetic field (see Sect. 13). The present description, however, does not immediately give the position of the boundary. The position depends on the total magnetic field, which itself is affected by the boundary currents that are not known in advance.

The *thickness of the boundary or transition area* may be determined by the gyroradius of the ions (which is variable in the increasing field). The unmodified gyroradius of ions, however, can become apparent only when the charge separation due to the different ion and electron orbits is compensated by some thermal plasma. Otherwise the electrostatic force shortens the ion orbits by $\sqrt{m_e/m_i}$ (with the mass ratio m_e/m_i) and stretches the electron orbits at the inverse rate[1].

[1] See [*21*] and [*31*], p. 90—100.

A question not yet finally answered is that for the *stability* of a boundary [*14*], [*21*]. Various instabilities could be thought of as causing a dented structure of the boundary. The consequence might be a violation of basic magnetodynamic principles. A possible violation would be penetration of plasma from a field-free space into a magnetic field. The magnetopause on the solar side of the magnetosphere resembles theoretical models in which such complications are ignored (see Sect. 14). Some insufficiencies of the simplified models will, however, be recognized in Sects. 16, 17, and 24.

10. Limitations of magnetodynamic theory.

α) *Finite conductivity.* The most commonly discussed deviation from the simple magnetodynamic theory results from the *finite conductivity* of the plasma [*29*], [*32*], [*33*]. If the plasma has the finite (scalar) conductivity σ one has to use Ohm's law in the form of Eq. (7.11), which is now written

$$\boldsymbol{J} = \sigma\left(\boldsymbol{E} + \boldsymbol{v} \times \frac{\boldsymbol{B}}{c_0\sqrt{\varepsilon_0\mu_0}}\right). \tag{10.1}$$

The procedure of Sect. 8α with this version of Ohm's law yields

$$\frac{\partial \boldsymbol{B}}{\partial t} = \frac{\partial}{\partial \boldsymbol{r}} \times (\boldsymbol{v} \times \boldsymbol{B}) + \frac{c_0^2\,\varepsilon_0\,\mu_0}{\mathrm{u}\,\mu_0}\,\frac{1}{\sigma}\left(\frac{\partial}{\partial \boldsymbol{r}}\right)^2 \boldsymbol{B}. \tag{10.2}$$

Eq. (10.2) again has a hydrodynamic analogue relating to the vorticity $\boldsymbol{\Omega}$. Substitution of $\boldsymbol{\Omega}$ for $\boldsymbol{B}$ and of the kinematic viscosity η/ϱ (with the coefficient of viscosity η) for $\mathrm{u}\,\mu_0\sigma/c_0^2\,\varepsilon_0\,\mu_0$ leads to the corresponding equation of hydrodynamics, an expanded version of Eq. (8.2). The following interpretation of Eq. (10.2) is in accordance with this analogy.

Eq. (10.2) suggests to decompose $\partial\boldsymbol{B}/\partial t$ into two parts:

$$\left(\frac{\partial \boldsymbol{B}}{\partial t}\right)_1 = \frac{\partial}{\partial \boldsymbol{r}} \times (\boldsymbol{v} \times \boldsymbol{B}), \tag{10.3}$$

$$\left(\frac{\partial \boldsymbol{B}}{\partial t}\right)_2 = \frac{c_0^2\,\varepsilon_0\,\mu_0}{\mathrm{u}\,\mu_0}\,\frac{1}{\sigma}\left(\frac{\partial}{\partial \boldsymbol{r}}\right)^2 \boldsymbol{B}. \tag{10.4}$$

The first part represents the motion of the field lines with the plasma (of velocity $\boldsymbol{v}$) in the conventional sense. The second part may be ascribed to an additional motion of the field lines with a velocity $\boldsymbol{w}$ normal to the magnetic field, a motion in which the plasma does not take part. This means, the second part, $(\partial\boldsymbol{B}/\partial t)_2$, is set identical to $\partial/\partial\boldsymbol{r} \times (\boldsymbol{w} \times \boldsymbol{B})$. Introduction of this expression in Eq. (10.4) yields

$$\frac{\partial}{\partial \boldsymbol{r}} \times (\boldsymbol{w} \times \boldsymbol{B}) = \frac{c_0^2\,\varepsilon_0\,\mu_0}{\mathrm{u}\,\mu_0}\,\frac{1}{\sigma}\left(\frac{\partial}{\partial \boldsymbol{r}}\right)^2 \boldsymbol{B} \tag{10.5}$$

or

$$\frac{\partial}{\partial \boldsymbol{r}} \times (\boldsymbol{w} \times \boldsymbol{B}) = -\frac{c_0^2\,\varepsilon_0\,\mu_0}{\mathrm{u}\,\mu_0}\,\frac{1}{\sigma}\,\frac{\partial}{\partial \boldsymbol{r}} \times \left(\frac{\partial}{\partial \boldsymbol{r}} \times \boldsymbol{B}\right). \tag{10.6}$$

The velocity $\boldsymbol{w}$ has to be derived from this equation. It may be tried to omit the outer curl, thus reducing the equation to

$$\boldsymbol{w} \times \boldsymbol{B} = -\frac{c_0^2\,\varepsilon_0\,\mu_0}{\mathrm{u}\,\mu_0}\,\frac{1}{\sigma}\,\frac{\partial}{\partial \boldsymbol{r}} \times \boldsymbol{B}. \tag{10.7}$$

A velocity $\boldsymbol{w}$ is defined by Eq. (10.7) if and only if $\frac{\partial}{\partial \boldsymbol{r}} \times \boldsymbol{B}$ is normal to $\boldsymbol{B}$ or

$$\boldsymbol{J} \cdot \boldsymbol{B} = 0. \tag{10.8}$$

Assumption of no currents parallel to the field lines guarantees validity of this condition.

Introduction of a scale length l of the magnetic field according to

$$B/l \approx \left| \frac{\partial}{\partial \boldsymbol{r}} \times \boldsymbol{B} \right| \tag{10.9}$$

makes[1]

$$w \approx \frac{c_0^2\, \varepsilon_0\, \mu_0}{\mathrm{u}\, \mu_0} \frac{1}{\sigma l}. \tag{10.10}$$

Two limiting cases may be distinguished, the first being

$$v \gg w, \tag{10.11}$$

the second

$$v \ll w. \tag{10.12}$$

In the first case the field lines "move with the plasma" according to simple magnetodynamic theory (Sect. 8). In the second case the first term on the right-hand side of Eq. (10.2) is negligible; the equation thus assumes the form of a diffusion equation. The velocity $\boldsymbol{w}$ apparently describes a diffusion of field lines against the plasma (or of plasma against the field lines), a process connected with dissipation of energy. In general the two types of motion, corresponding to $\boldsymbol{v}$ and $\boldsymbol{w}$, are superimposed.

The magnitude of the ratio v/w with w taken from Eq. (10.10) is usable as a criterion which motion predominates. This ratio, which is[1]

$$R_{\mathrm{m}} = \frac{\mathrm{u}\, \mu_0}{c_0^2\, \varepsilon_0\, \mu_0} \sigma l v, \tag{10.13}$$

is called the "*magnetic Reynolds number*" (after ELSASSER) [*28*], [*29*], [*33*]. It plays the same role for $\boldsymbol{B}$ as does the hydrodynamic Reynolds number for $\boldsymbol{\Omega}$. The length l is the *characteristic length* of a problem: the length over which the magnetic field changes remarkably in magnitude or direction. If a characteristic time t is given, one may set

$$R_{\mathrm{m}} = \frac{\mathrm{u}\, \mu_0}{c_0^2\, \varepsilon_0\, \mu_0} \sigma l^2/t. \tag{10.14}$$

A different dimensionless quantity is obtained for a magnetic field that consists of a large stationary part ($\boldsymbol{B}_0$) and a small perturbation field ($\boldsymbol{B}_1$), corresponding to magnetodynamic processes of small amplitude [*29*]. In Eq. (10.2) the left-hand term and the last term are supposed to contain the variable field $\boldsymbol{B}_1$ only. The critical number deciding which of the two right-hand terms now predominates is [*29*]

$$L_{\mathrm{m}} = \frac{(\mathrm{u}\, \mu_0)^{\frac{1}{2}}}{c_0^2\, \varepsilon_0\, \mu_0} \frac{\sigma l B}{\varrho^{\frac{1}{2}}} \tag{10.15}$$

(introduced by LUNDQUIST). This expression is the magnetic Reynolds number, Eq. (10.13), with replacement of v by the Alfvén velocity V_{A}, Eq. (7.19)[2]. The replacement results from the fact that a weak perturbation of the magnetic field in a plasma is propagated as a wave phenomenon, traveling with a velocity near Alfvén velocity V_{A} (cf. next section). The ratio l/t now is this wave velocity.

[1] In the units V, A, m, kg, sec there is $w \approx 1/\mu_0 \sigma l$ and $R_{\mathrm{m}} = \mu_0 \sigma l v$.

[2] Some authors define L_{m} as *the* Reynolds number, for instance PARKER and FERRARO in [*28*], Eq. (5.2).

At heights above the vertex of the F 2-layer, where collisions between charged particles are predominant, the conductivity is nearly that of a pure plasma. The condition $R_m = 1$ with the conductivity of a proton-electron plasma becomes approximately[3]

$$\frac{t}{\text{sec}} = 10^{-9} \left(\frac{T}{°\text{K}}\right)^{\frac{3}{2}} \left(\frac{l}{\text{m}}\right)^2. \tag{10.16}$$

Values corresponding to the magnetopause are $T = 10^5$ °K and $l = 3 \cdot 10^7$ m, leading to $t = 3 \cdot 10^{13}$ sec $= 10^6$ a. [The coefficient in Eq. (10.16) may be approximately $2 \cdot 10^{-9}$ at lower heights.] The long time constant indicates the validity of magnetodynamic theory, if not small-scale irregularities (eventually caused by an instability) introduce a much smaller scale length.

Near the F 2 layer, Eq. (10.16) would set a more severe limitation to magnetodynamic theory, but here the value of L_m (i.e. R_m with $v = V_A$) is a more appropriate criterion, because of the reduced variability of the magnetic field. According to this criterion the magnetodynamic character of processes is preserved at all reasonable scale lengths.

The starting equation of the present considerations, Eq. (10.1), appears oversimplified in that it disregards the *anisotropy* of the medium with respect to *current flow*. A justification for the neglect of additional terms in Eq. (10.1) can be found in [4].

β) *Various influences*. There are many more factors that affect the validity of magnetodynamic theory. With *scale lengths comparable to a gyroradius or smaller* the particles undergo drift motions caused by the inhomogeneous magnetic field, thus violating the theorem of field lines and plasma moving jointly. Sect. 19β deals with these drift motions, which are of importance for fast, nonthermal particles.

The *Debye length*, inside which the plasma can not be considered to be neutral [*29*], [*32*], is in the magnetosphere quite small, below 1 m or perhaps a few m. It is computed at approximately 20 m for 1 cm^{-3} ions and electrons and a temperature of 10^5 °K, corresponding to the region beyond the plasmapause (Sect. 18). In the study of magnetospheric phenomena one is concerned with much longer distances.

Collisions between charged particles and neutrals quite generally cause dissipation of energy. If the collision frequency of ions exceeds their angular gyrofrequency, i.e., below 120 km height approximately, the ions are largely inhibited from a drift motion in an electric field such as postulated by Eq. (7.5). Below 120 km the drift of electrons without much participation of ions represents a Hall current [*16*][5]. In processes of long time constants, e.g. diurnal phenomena, the neutral gas at ionospheric heights may be enforced by collisions to move to some extent with the plasma (Sect. 7β). This enhances the apparent density of the magnetodynamic medium[5].

A severe violation of the magnetodynamic theory has to be expected from *electric fields parallel to the magnetic field* if they are produced in some way. This has been emphasized by ALFVÉN and his school[6]. The idea of moving field lines is usable only when the electric field component parallel to the magnetic field, $\boldsymbol{E}_{\parallel}$, disappears or, at least, fulfills the condition

$$\frac{\partial}{\partial \boldsymbol{r}} \times \boldsymbol{E}_{\parallel} = 0. \tag{10.17}$$

[3] [*32*], Eq. (2-38) with omission of factor 2.

[4] L. SPITZER: [*32*], p. 42.

[5] H. POEVERLEIN: J. Geophys. Res. **72** (1), 251—256 (1967).

[6] H. ALFVÉN: Ann. Géophys. **24** (1), 341—346 (1968).

In this case Eq. (8.4), on which the theorem of moving field lines is based, is satisfied by the velocity

$$\boldsymbol{v} = \frac{c_0 \sqrt{\varepsilon_0 \mu_0}}{B^2} \boldsymbol{E} \times \boldsymbol{B}.$$

This is, on the other hand, the drift velocity of the plasma, see Eq. (7.4b).

At heights at which collisions (Coulomb interaction) confine the individual particles within a narrow space the parallel electric field should, in fact, be suppressed by the high conductivity. In the outer magnetosphere, however, thermal particles may, like trapped fast particles, describe long helical paths along field lines. If positive and negative particles had different reflection points on a field line, an electric field along the field line would arise that brings the particles of the two signs together[6].

11. Magnetodynamic waves. Waves in a plasma with presence of a stationary magnetic field have a peculiar character at frequencies so low that ions oscillate as well as electrons; the waves at these frequencies are described by the magnetodynamic theory and are consequently named "magnetodynamic"[1]. In Sect. 7β it was stated that ions and electrons perform nearly the same drift motion in an oscillatory electric field at frequencies well below the gyrofrequency of the ions

$$\frac{\omega_{B\mathrm{i}}}{2\pi} = \frac{1}{2\pi} \frac{q}{m_{\mathrm{i}}} \frac{B}{c_0 \sqrt{\varepsilon_0 \mu_0}}. \tag{11.1}$$

The gyrofrequency of oxygen atom ions at 0.5 Gs (Gauss $= 10^{-4}$ V sec m^{-2}) is 48 Hz, that of protons at 0.3 Gs amounts to 460 Hz. The frequency of the waves is in the following assumed to be much lower.

The term "magnetodynamic" or "hydromagnetic waves" is commonly used at frequencies up to the ion gyrofrequency, but in approaching the gyrofrequency there are remarkable deviations from the idealized theory of magnetodynamic waves. Well beyond the gyrofrequency the ion motion is greatly reduced and the waves are no longer of the "magnetodynamic" nature.

Natural radio noises at extremely low frequencies and magnetic micropulsations with periods up to 100 sec or more are necessarily propagated as magnetodynamic waves if they pass through the magnetosphere and ionosphere[2]. Also other processes, e.g. traveling fronts in the magnetosphere or in the solar wind, may after a Fourier decomposition be described by the theory of magnetodynamic waves (see Sect. 12α), provided linearity is conserved.

α) *Waves of small amplitudes* are now considered and all formulas are linearized. Stationary quantities are denoted by subscript 0, e.g. $\boldsymbol{B}_0, \varrho_0, p_0$ (pressure). The oscillatory quantities, denoted by subscript 1, are $\boldsymbol{E}_1, \boldsymbol{B}_1, \boldsymbol{J}_1$ (current density), $\boldsymbol{v}_1, \varrho_1, p_1$. They are supposed to be proportional to $\exp(i\omega t - i\boldsymbol{k}\cdot\boldsymbol{r})$. The angle between the propagation vector $\boldsymbol{k}$ and the magnetic-field vector $\boldsymbol{B}_0$ may be φ. With a given direction of $\boldsymbol{k}$ the equations of the magnetodynamic theory yield three solutions for the propagation constant k, corresponding to three different modes of magnetodynamic waves.

The three propagation constants are obtained as the solutions of the two equations

$$k^2 V_{\mathrm{A}}^2 \cos^2\varphi - \omega^2 = 0, \tag{11.2}$$

$$k^4 V_{\mathrm{A}}^2 V_{\mathrm{s}}^2 \cos^2\varphi - k^2\omega^2 (V_{\mathrm{A}}^2 + V_{\mathrm{s}}^2) + \omega^4 = 0 \tag{11.3}$$

[1] or "hydromagnetic", "magnetohydrodynamic"; see Introductory Remarks, p. 1 and footnote 2 of Sect. 2.

[2] See contributions to this Encyclopedia by R. Gendrin: Vol. 49/3, p. 461–525, and by E. Selzer in this volume, p. 231–394.

[29], [32], [36][3]. These equations are the *dispersion equations* for plane magnetodynamic waves in a homogeneous medium under the assumption

$$\omega/\omega_{B\,\mathrm{i}} \ll 1 .$$

The velocity quantities in the two dispersion equations are the earlier [in Eq. (7.19)] defined Alfvén velocity V_{A} and the plasma sound velocity

$$V_{\mathrm{s}} = (p_1/\varrho_1)^{\frac{1}{2}},$$

which generally is taken for an adiabatic variation of state. Various possible mechanisms of energy dissipation, in particular, collisions between charged and neutral particles, are ignored in the present simplified formulation. (For more complete formulations see [35], [36].)

The phase velocities of the wave modes are ω/k; their refractive indices are

$$n = \frac{c_0\, k}{\omega} .$$

The magnetospheric plasma is a cold plasma in the sense of Sect. 7γ, characterized by $\beta \ll 1$ or $V_{\mathrm{s}} \ll V_{\mathrm{A}}$ [cf. Eq. (8.12)]. The *refractive indices* of the three modes under this condition become

$$n_{\mathrm{A}} = \frac{c_0}{V_{\mathrm{A}} \cos\varphi} , \tag{11.4}$$

$$n_{\mathrm{m}} = \frac{c_0}{V_{\mathrm{A}}} , \tag{11.5}$$

$$n_{\mathrm{s}} = \frac{c_0}{V_{\mathrm{s}} \cos\varphi} . \tag{11.6}$$

β) Of these *three modes* the first two modes are fast, traveling with a velocity equal or comparable to Alfvén velocity V_{A}. The third mode is slow. The *properties of the three modes* with respect to electromagnetic polarization and other parameters, derived from the magnetodynamic theory, are shown in Fig. 7.

The first mode, corresponding to Eq. (11.4), is known as *Alfvén wave*. The second mode has received various names: *fast magnetodynamic wave*, fast hydromagnetic wave, fast magnetosonic wave, modified Alfvén wave etc. Both modes involve plasma motion normal to the magnetic field, i.e., electrodynamic drift according to Eq. (7.4b). The two modes can be interpreted as of essentially electromagnetic character.

The third mode, the "*ion-acoustic wave*", shows acoustic character in its slow velocity and in the role of pressure forces, though its directional behavior is clearly anisotropic. The plasma velocity in it is along the magnetic field lines. The presently described ion-acoustic wave requires a frequency sufficiently below the collision frequency of particles (ions) just as any acoustic wave[4]. Under peculiar conditions collisions are not important, but then an oscillatory electric field parallel to $\boldsymbol{B}_0$ has to coordinate ion and electron motions [36].

"Longitudinal" plasma velocities, i.e. velocity components parallel to $\boldsymbol{k}$, appear in the fast magnetodynamic and ion-acoustic waves. The longitudinal velocity entails compression and expansion of the plasma, in particular in the slow ion-acoustic mode. The plasma motion in the Alfvén wave is purely transverse. The Alfvén wave consequently is not connected with a compression and

[3] R. LÜST: Introduction to plasma physics, in [34], p. 1—23.

[4] H. POEVERLEIN: J. Geophys. Res. 72 (1), 251—256 (1967).

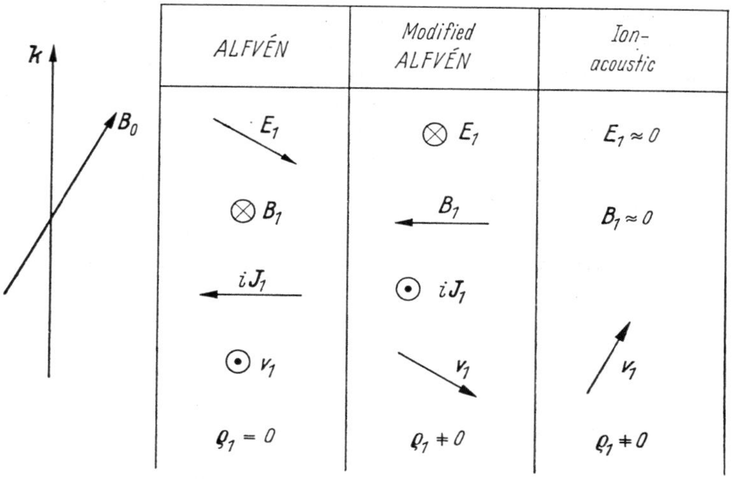

Fig. 7. Types of magnetodynamic waves and their properties for frequencies well below ion gyrofrequency. As for the terminology of the waves see text.

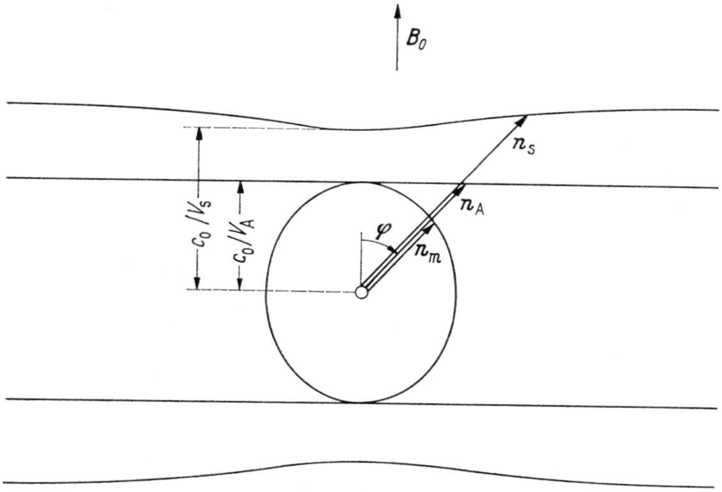

Fig. 8. Refractive indices of magnetodynamic waves vs. direction of the wave normal. A frequency well below ion gyrofrequency and an Alfvén velocity V_A larger than the sound velocity V_s of the plasma are assumed. The three wave modes are the Alfvén wave (A), the fast magnetodynamic or modified Alfvén wave (m) and the ion-acoustic wave (s), which essentially has sound wave character.

is independent of the compressibility of the medium. Values of the Alfvén velocity were noted in Sect. 7γ. Perturbations need between 1 and 2 min to travel with Alfvén velocity down through the entire magnetosphere (Sect. 29).

Refractive index surfaces representing the refractive indices as functions of the wave normal direction are shown in Fig. 8 for a V_s only slightly smaller than V_A. For $V_s \ll V_A$ the surfaces become planes and a sphere. Since the *ray direction*, i.e. the direction of propagation for a wave parcel, is generally normal to the refractive index surface, the Alfvén wave and the ion-acoustic wave have ray directions parallel to $\boldsymbol{B}_0$ (in both senses), whereas the fast magnetodynamic wave has isotropic character.

γ) The propagation of magnetic *micropulsations* seems fairly well described by magnetodynamic-wave theory; but their generation mechanisms are in most cases hypothetical. Even the location of their origin may be unknown. A variety of sorts of micropulsations with a wide range of periods (from 0.2 to 1000 sec) is distinguished. The generation may occur in turbulent regions (e.g., the magnetosheath), at some (perhaps unstable) surfaces or in particle streams at some height. For more information about micropulsations and other magnetodynamic-wave phenomena see the literature [*23*], [*36*][5–7] (also see Sect. 33).

The theory of magnetodynamic waves may also be applied to *magnetospheric processes with extremely large time constants or periods* (minutes, hours or one day), though with some modifications and precautions [*36*][4,8]. The gravity force (gravity acceleration $\boldsymbol{g}$) changes the character of ion-acoustic waves for $\omega V_s \gtrsim g$ decisively; they become evanescent and show some resemblance to atmospheric gravity waves[9]. Collisional interaction with the neutral gas leads not only to attenuation, but at the lowest frequencies also to participation of the neutral gas in the (oscillatory) motion (cf. Sect. 7β).

Plane-wave theory, of course, is only of very limited use when the wavelength exceeds the available dimensions. Severe *coupling* between wave modes is found at *wavelengths comparable to or larger than a scale height of the stratification of the medium*. Excitation of fast waves leads under this condition to secondary ion-acoustic waves, which are accompanied by motion along the field lines and significant density variations[8].

In a transfer of considerations made for plane waves to nonplane wave fields one has to be aware that a current normally filling a space of the size of a wavelength λ (or $\lambda/2\pi$) may be urged into a much smaller space. If large force densities can not be resumed the current becomes parallel to $\boldsymbol{B}_0$. The possibility of such currents should not be overlooked.

12. Discontinuity surfaces.

α) *General.* Discontinuity surfaces play an important role in the physics of the magnetosphere and solar wind. The quantities that may be discontinuous at some surfaces are the magnetic field (its intensity and direction), plasma density and pressure, and the plasma velocity. The magnetodynamic theory permits various types of discontinuity surfaces. Some of them are at rest with respect to the plasma; they are stationary if the plasma is at rest or they move with the plasma. Other discontinuity surfaces are propagated within the plasma. The magnetopause without traversal of plasma is an example of a discontinuity surface at rest with respect to the plasma. A surface propagated in the plasma may appear stationary if the plasma moves in the appropriate way.

Small-amplitude magnetodynamic waves were discussed in the last section. The refractive indices turned out to be independent of the frequency at the low frequencies assumed. This indicates the possibility of weak discontinuities that correspond to the various wave modes and are propagated like those [*31*][1]. Stronger discontinuities do not necessarily obey the small-amplitude theory.

Strong discontinuities may be the result of certain *gradual* transitions which are *steepened* during propagation. These discontinuities are the *shock fronts* in the

---

[5] V. A. TROITSKAYA: Micropulsations and the state of the magnetosphere, in [*3*], p. 213—274.

[6] W. H. CAMPBELL: Geomagnetic pulsations, in [*26*], Vol. 2, p. 821—909.

[7] E. SELZER, p. 231 in this volume.

[8] H. POEVERLEIN: Ann. Géophys. **24** (1), 325—332 (1968).

[9] About atmospheric gravity waves see the contribution by W. L. JONES in this Encyclopedia, Vol. 49/5.

[1] J. R. SPREITER, A. L. SUMMERS, and A. Y. ALKSNE: Planetary Space Sci. **14** (3), 223—253 (1966).

strict sense; because of their gradual development they are called "*evolutionary*" (Sect. 12δ).

A systematic survey of the various possible discontinuities is obtained from magnetodynamic theory [*6*], [*30*][1]. In the following the different types are briefly discussed without presentation of deductions. The deductions have to start out from the continuity conditions for the mass flux, the momentum flux density together with the magnetic tension tensor, the energy flux, and the electromagnetic quantities B_{n} and $\boldsymbol{E}_{\mathrm{t}}$ (with n and t referring to normal and tangential components). $\boldsymbol{E}_{\mathrm{t}}$ and the *normal components* of the above-mentioned flux quantities have to be *continuous in a coordinate system in which the front is at rest.* The theorem of moving field lines remains valid at the discontinuity surface.

β) *Discontinuities not propagated in the plasma.* With *contact discontinuity* one denotes a jump of density and temperature with continuity of pressure, magnetic field, and plasma velocity. The contact discontinuity remains at rest with respect to the plasma. That means that the plasma can have no velocity component, v_{n}, normal to the discontinuity surface. A tangential velocity $\boldsymbol{v}_{\mathrm{t}}$ is not excluded, but it has to be continuous. This is expressed by

$$[\boldsymbol{v}_{\mathrm{t}}] = 0, \tag{12.1}$$

using brackets to denote the variation of a quantity at the discontinuity surface. A discontinuous $\boldsymbol{v}_{\mathrm{t}}$ with some B_{n} would require a discontinuous $\boldsymbol{E}_{\mathrm{t}}$.

A *tangential discontinuity* is characterized by a tangential plasma flow and a tangential magnetic field (magnetic flux density $\boldsymbol{B}_{\mathrm{t}}$) with a jump of the magnetic field in direction and magnitude:

$$v_{\mathrm{n}} = 0, \qquad B_{\mathrm{n}} = 0, \qquad [\boldsymbol{B}_{\mathrm{t}}] \neq 0, \qquad [\boldsymbol{v}_{\mathrm{t}}] \neq 0. \tag{12.2}$$

There is no restriction to the jump of $\boldsymbol{v}_{\mathrm{t}}$. The density has to vary so that

$$\left[p + \frac{B_{\mathrm{t}}^2}{2\,\mathrm{u}\,\mu_0}\right] = 0. \tag{12.3}$$

This is the earlier discussed equilibrium condition [Eq. (9.4)] with absence of a kinetic pressure ϱv_{n}^2.

The magnetopause is a tangential discontinuity surface, to which Eq. (12.3) applies, if field lines do not penetrate through it and if the solar plasma arrives at it without a normal velocity as it should because of the slowing-down in the magnetosheath (see Sects. 14 and 15). Only in simplified theory the solar wind arrives at the magnetopause with a normal velocity v_{n} and a kinetic pressure ϱv_{n}^2 (see Sects. 9 and 13). About a magnetopause penetrated by field lines see Sects. 12ε and 17.

γ) *Rotational discontinuity.* The discontinuity surfaces to be discussed in Sects. 12γ and δ are *propagated in the plasma.* The rotational discontinuity is also called "transverse shock" or "Alfvén shock" although it is not a shock front in the sense of Sects. 12α and δ. It has the *character of an Alfvén wave.* In an Alfvén wave the magnetic perturbation field is normal to the stationary magnetic field and, of course, parallel to the wave planes (see Fig. 7). A *weak* rotational discontinuity shows the same behavior of the magnetic field: The increment of the field at the discontinuity is a (small) vector parallel to the discontinuity surface and normal to the field vector.

The rotational discontinuity, however, may be *strong.* Its propagation is not affected by the size of the jump. In the case of high intensity the variation of the field can be composed of many small increment vectors, each parallel to the discontinuity surface and normal to the instantaneous $\boldsymbol{B}$-vector. This means, *the $\boldsymbol{B}$-vector appears to rotate* on a cone around the normal to the discontinuity surface.

The rotation of the field lines means a change of the direction of the tangential magnetic field component $\boldsymbol{B}_t$ while its magnitude and B_n stay unchanged. The conditions of the rotational discontinuity are

$$[B_t]=0, \quad [\boldsymbol{B}_t]\neq 0, \quad [B_n]=0, \quad [\varrho]=0, \quad [v_n]=0. \tag{12.4}$$

$$[\boldsymbol{v}_t]=\pm\frac{1}{(\mathrm{u}\,\mu_0\,\varrho)^{\frac{1}{2}}}[\boldsymbol{B}_t]. \tag{12.5}$$

The discontinuity moves with respect to the plasma. Its velocity is the Alfvén velocity corresponding to the magnetic field $\boldsymbol{B}_n$,

$$\boldsymbol{V}=\mp\frac{1}{(\mathrm{u}\,\mu_0\,\varrho)^{\frac{1}{2}}}\boldsymbol{B}_n. \tag{12.6}$$

Eqs. (12.4) indicate that there is no compressional effect, just as in small-amplitude Alfvén waves.

The motion of the front can be interpreted as a motion of the field lines (including their kink) with the velocity $\mp V$ in the direction of $\boldsymbol{B}_n$. (This velocity is, of course, not normal to the field lines.) In order to stay attached to the field lines the plasma has to move with a velocity

$$\boldsymbol{v}_t=\pm\frac{\boldsymbol{B}_t}{(\mathrm{u}\,\mu_0\,\varrho)^{\frac{1}{2}}}$$

in the tangential direction. The velocity $\boldsymbol{v}_t$ changes its direction at the front in the same way as $\boldsymbol{B}_t$. It is permissible to superimpose on this tangential velocity an additional invariable tangential velocity.

δ) *Shock fronts.* There are two types of traveling discontinuities that are *shock fronts in the strict sense,* connected with a *compression of the plasma.* Such shock fronts are *evolutionary*; they develop out of a gradual transition between two different compressional states by steepening during propagation.

In neutral gases there exist evolutionary shock fronts of acoustic character. An example of a nonevolutionary front in the plasma is the rotational discontinuity; a corresponding gradual transition would not steepen, it would retain its shape.

The *two evolutionary shock fronts* possible in a plasma have the *character of the two compressional magnetodynamic wave modes*: fast magnetodynamic waves and ion-acoustic waves. *Weak* shock fronts thus are propagated in the plasma with the velocities of these waves. Eqs. (11.5) and (11.6), based on the assumption $V_A \gg V_s$, give the velocities as

V_A for a "fast" shock, and
$V_s \cos\varphi$ for a "slow" shock.

Both types of shock fronts involve an *increase of the plasma density.* The steepening of the front, i.e., its evolution, is a consequence of the fact that the velocity of a wave or front is higher in the denser plasma behind the front than in the preceding less dense plasma [*31*]. The *ion-acoustic* velocity increases with temperature and the temperature increases with the density. The velocity of *fast magnetodynamic waves* depends on the density and the magnetic field strength. Both quantities vary so that the wave velocity is again higher at the higher density.

The normal velocity of the plasma with respect to the front is discontinuous because the normal plasma flux ϱv_n must be continuous in a coordinate system moving with the front. The fast shock front, corresponding to the fast magnetodynamic wave, is accompanied by an increase of B_t without change of B_n and of the direction of $\boldsymbol{B}_t$. The entropy of the plasma passing through a shock front of finite step size increases.

A sudden density decrease does not occur at a front; it would flatten out in reversal of the steepening process. Moving with such a front — if it existed — one would find an entropy decrease in the plasma passing through the front.

Because the wave velocity is higher behind the front than ahead of it the velocity of the front is higher than the corresponding wave velocity in the medium before the front. The *front velocity* thus becomes a *function of the step size*.

The magnetodynamic theory does not indicate what processes take place in the shock front. In *hydrodynamic shocks* the increase of entropy is due to *viscous interaction and heat conduction*, processes that depend on collisions between particles. In the solar wind or in the outer magnetosphere the mean free path length exceeds all involved dimensions by far; in the solar wind it amounts to nearly one astronomical unit. The *magnetodynamic shock fronts* consequently have to be "*collisionless*" as in a high-temperature plasma. Dissipation arises from plasma instabilities and the structure or thickness of the shock front depends on gyroradii rather than free path lengths.

The theory of collisionless shock fronts does not yet appear completed; see the references [*31*] [2–4] and the macroscopic descriptions in the literature [*6*], [*30*], [*31*].

Among the oversimplifications of the present description is the assumption of an isotropic gaskinetic pressure, which is incorrect for the nearly collisionless plasma. Consequences of an anisotropic pressure, e.g. the possibility of an expansion front, were discussed by NEUBAUER[4].

ε) *Discontinuities in the solar wind and at the magnetosphere. Shock fronts in the solar wind* are expected when the plasma density increases by a remarkable amount and they have been observed (see Sect. 29). The steepening process causes a pronounced front, which is propagated within the wind, thus traveling faster than the plasma.

The *magnetosphere is surrounded by a stationary shock front* through which the solar wind streams before arrival at the magnetopause (Sects. 14 and 15). A hydrodynamic theory (outlined in Sect. 15β) gives a fairly good description of this front and the adjacent transition region (the "magnetosheath"), although a magnetodynamic theory would be more pertinent. The *hydrodynamic* formulations of Sect. 15β may serve to illustrate some of the above remarks on *magnetodynamic* shock fronts.

The *magnetopause* (magnetospheric boundary) and the *neutral sheet* in the tail of the magnetosphere are other examples of discontinuity surfaces (see Sects. 12β, 13, 14, 16, 17). All "surfaces", in fact, are *boundary layers* of a finite thickness, which is determined by a gyroradius or some other length parameter (cf. remarks in Sect. 12δ above and in Sects. 9 and 15γ). Magnetopause and neutral sheet depart according to some hypotheses from the surfaces discussed in a quite significant point: A *tangential plasma flow* might carry plasma out within the boundary layer. With this mass flux in the boundary layer (an areal flux of mass) the common condition $[\varrho v_n]=0$ at the boundary can be violated. At the magnetopause this behavior of the plasma is supposed to result from a

[2] A. KANTROWITZ and H. E. PETSCHEK in W. B. KUNKEL (ed.): Plasma Physics in Theory and Application, p. 147—206. New York: McGraw-Hill Book Co. 1966.

[3] C. F. KENNEL and H. E. PETSCHEK: Magnetic turbulence in shocks, in [*13*], p. 485—513.

[4] F. M. NEUBAUER: Z. Physik **237**, 205—233 (1970).

field line connection between the arriving solar wind and the magnetosphere. A similar field line connection takes place across the neutral sheet of the tail. (For hypotheses on the neutral sheet and on the "open" magnetosphere look up end of Sect. 16β and Sect. 17.)

D. Model of the magnetosphere.

13. Model derived from equilibrium at boundary. The idea of a confinement of the magnetosphere by the solar wind, which was outlined in Sect. 3, leads to a simple model of the magnetosphere. In this model the solar wind is supposed to remain undisturbed until it arrives at the magnetospheric boundary or "magnetopause", the surface separating the geomagnetic field from the solar wind. The boundary is considered to be a magnetodynamic boundary in equilibrium, corresponding to the description in Sect. 9 [cf. Eq. (9.7)]. The kinetic pressure of the solar wind, $\mathrm{a} N m_i v^2 \cos^2\varphi$, acting from outside, and the pressure of the geomagnetic field, $\mathrm{B}^2/2\mathrm{u}\,\mu_0$, acting from inside, have to be in balance if other pressures can be ignored. (See also Sect. 34 of the article by HESS [*15*].)

Neglect of the plasma pressure in the magnetosphere is almost perfectly permissible, because the magnetic pressure is there predominant by far. In Sect. 7γ it was noticed that the magnetospheric plasma is a cold plasma, corresponding to $V_A \gg V_s$ [cf. Ineq. (8.12)]. The solar wind, according to Ineq. (5.5), exerts mainly a kinetic pressure. Its thermal pressure is small because the wind is supersonic. Magnetic and thermal pressures are now neglected in the solar wind.

Magnetospheric models based on these simplifications were derived by various authors. Fig. 9 represents the model of MEAD and BEARD [*14*], [*15*][1,2]. The authors chose a coefficient $\mathrm{a}=2$ in the kinetic pressure, corresponding to reflection of the solar wind at the magnetopause [see Eq. (9.7)]. The scale of the figure is under the present assumptions correct for a solar wind of $N=5\ \mathrm{cm}^{-3}$ protons and $v=320\ \mathrm{km\ sec}^{-1}$. The conventional length unit of one earth radius, R_E, is used. The coordinate x_{SM} runs from the Earth's center toward the Sun. The dipole axis of the Earth was assumed normal to the solar wind and in the plane of drawing. In reality it may be tilted as much as 35 degrees, depending on the season and the instantaneous location of the geomagnetic axis with respect to the geographic axis. Fig. 9 is a composition of separate pictures of the magnetopause and the field line configuration that MEAD and BEARD gave. The composition required minor conjectural changes of the field line picture, in which inevitable computational inaccuracies seem to show up.

At the magnetopause there is a current system set up by the arriving charged particles, which penetrate somewhat into the geomagnetic field and are deflected in opposite directions. The resulting force density,

$$\frac{1}{c_0\sqrt{\varepsilon_0\mu_0}}\,\boldsymbol{J}\times\boldsymbol{B} \quad \text{or} \quad -\frac{\partial}{\partial \boldsymbol{r}}\,\frac{B^2}{2\,\mathrm{u}\,\mu_0},$$

provides the force balancing the solar-wind pressure. The equilibrium condition at the boundary or "magnetopause", Eq. (9.7), is quite trivial, but the computation of the shape of the boundary and of the current system is complicated by the fact that the field $\boldsymbol{B}$ just inside the boundary is not known a priori, but depends on the position and shape of the boundary.

---

[1] G. D. MEAD and D. B. BEARD: J. Geophys. Res. **69** (7), 1169—1179 (1964).

[2] G. D. MEAD: J. Geophys. Res. **69** (7), 1181—1195 (1964).

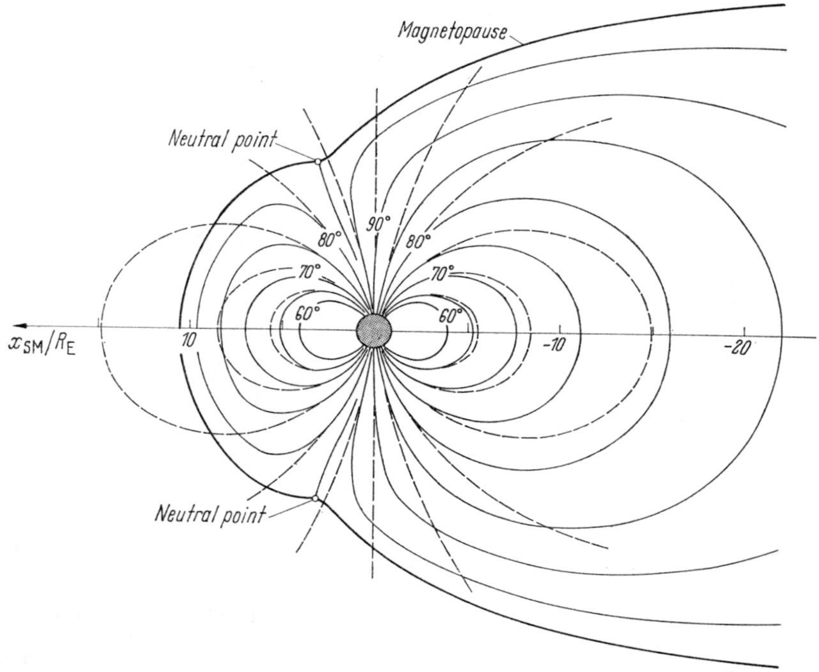

Fig. 9. Theoretical model of the magnetosphere (after MEAD and BEARD) [3]. Solid lines are computed field lines in the noon-midnight meridian. Dashed lines are dipole field lines.

MEAD and BEARD [*14*], [*15*][1] developed an iterative procedure for the computation. In the first step the conditions for a plane boundary are used in each point. The field right inside the boundary is assumed to be twice the tangential component of the dipole field and the surface current density is chosen corresponding to a jump of the field to zero. This first step yields a shape of the boundary and a current system on the boundary. The current system on the curved boundary permits an improved computation of the magnetic field. Now a second step follows, in which the corrected magnetic field is used to compute a corrected boundary shape and current system and another corrected field. Four steps of this type lead to a very good approximation as is seen from checking the disappearance of the outside field.

In the present model the solar-wind pressure disappears when the boundary becomes parallel to the Sun-Earth line. The boundary consequently can nowhere face the rear halfspace. A thermal pressure of the solar wind, however, would make this possible. The impossibility for a field line to penetrate through the boundary leads to two X-*type neutral points of the magnetic field* on the boundary. Zero field strength in the neutral points requires a tangent of the boundary parallel to the Sun-Earth line. The field lines reaching the neutral points start on the Earth at 83° magnetic latitude in the (geomagnetic) noon meridian.

To adjust Fig. 9 to *different solar-wind intensities* or to correct it for a non-reflected solar wind, one has merely to *change the scale*. The height of the boundary is, however, quite insensitive to such variations because the magnetic pressure varies with r^{-6}.

Fig. 10 shows the intersection of the magnetopause with the equatorial and N-S planes (according to SPREITER and BRIGGS, but with the scale made to match Fig. 9).

[3] G. D. MEAD and D. B. BEARD[1], Fig. 3. — G. D. MEAD[2], Fig. 4.

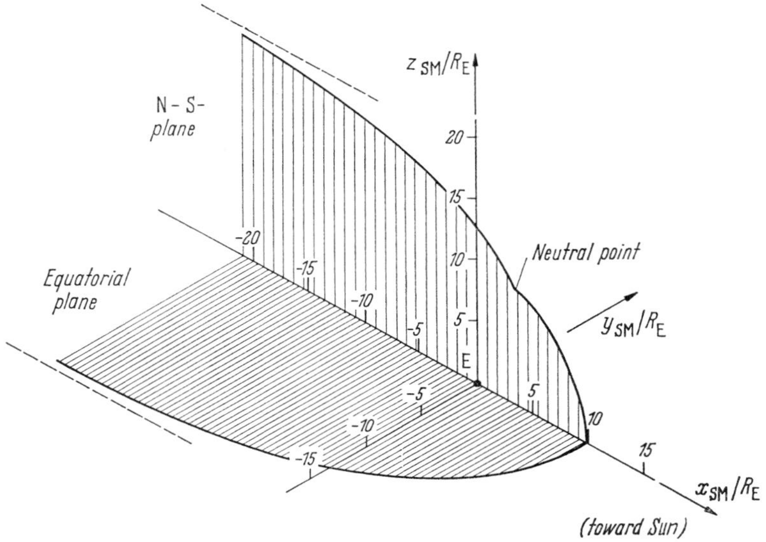

Fig. 10. Intersection of the magnetopause with the equatorial plane and N-S-plane of the magnetosphere (x_{SM},y_{SM} and x_{SM}, z_{SM} planes)[4,5]. The x_{SM}-direction points to the Sun; the magnetic dipole is assumed at E, parallel to the z_{SM}-axis.

The dipole is again supposed to be normal to the equatorial plane. The coordinates in the figure, x_{SM}, y_{SM}, z_{SM}, are those commonly used in the magnetosphere, called "solar-magnetospheric coordinates". The abscissa x_{SM} points toward the Sun; the solar wind is thought of as arriving in the $-x_{SM}$ direction. The dipole is assumed parallel to the z_{SM} axis. With an arbitrary direction of the dipole, the x_{SM}, z_{SM} plane would be chosen so as to contain the dipole.

The described model of the magnetosphere is highly idealized. The most significant *deviations* that will be discussed in the following sections are the presence of the *magnetosheath*, a space around the magnetosphere in which the solar wind is greatly perturbed (thermalized), and a *tail* of the magnetosphere. In order to add a tail in the Mead model one can assume a current sheet in the x_{SM}, y_{SM} plane on the nightside. The field configuration that Williams and Mead derived this way is shown in Fig. 41 of Hess's contribution to this volume [*15*].

Motions of the magnetospheric plasma were ignored in the Mead model and in similar models of the magnetosphere. Very little effect of the plasma motion on the magnetic field has to be expected because of the small ratio of energy density in the plasma to magnetic energy density at all altitudes beyond the ionospheric E layer. Currents at E layer altitudes may, however, affect the magnetic field in the magnetosphere. In addition, trapped particles of energies higher than thermal may at times lead to currents large enough for a magnetic-field variation. This is essential in the explanation of the main phase of a geomagnetic storm by Dessler and Parker [*28*] (see Sect. 29).

14. Model based on observations. Fig. 11 represents a model of the magnetosphere, largely based on satellite observations, but drawn to fit theoretical implications [*14*], [*17*]. The magnetic dipole of the Earth is assumed tilted. The plane shown is the magnetic noon-midnight meridian plane. The figure is somewhat schematic, with few details.

Fig. 11 makes clear that Mead's model (Fig. 9) is greatly simplified. The solar wind apparently is modified before arrival at the magnetopause; but the

[4] J. R. Spreiter and B. R. Briggs: J. Geophys. Res. **62** (1), 37 (1962); Fig. 7 (modified).
[5] G. D. Mead: Space Sci. Rev. **7** (2/3), 158—165 (1967); Fig. 2 (slightly modified).

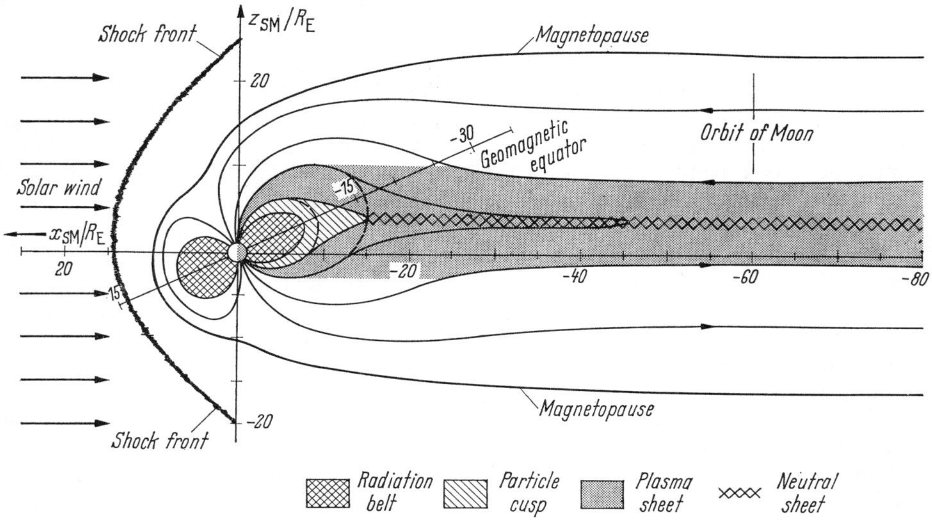

Fig. 11. The magnetosphere with tail and magnetosheath[1].

magnetopause remains the separation between solar wind and geomagnetic field, although the perfect separation might also be questioned (see Sect. 17). A *shock front* ("standing shock wave", "bow shock") is formed in the solar wind at a certain stand-off distance from the magnetopause (4 to 5 R_E in the subsolar point). The *transition region* between magnetopause and shock front, in which the solar wind appears modified, is called "*magnetosheath*". The next section will deal with the shock front and the magnetosheath in more detail.

The magnetopause and the shock front have been *observed* by many satellites and space probes; the first observation of the magnetopause was made on Explorer 10 in March 1961. IMP-1 in 1963 and 1964 determined the locations of the magnetopause and shock front in many crossings (Fig. 44 in [*15*] or Fig. III.8 in [*17*]).

The *typical behavior of the magnetic field* is shown in Figs. 42 and 43 of Hess's article [*15*] (also see [*14*], [*17*]). Inside the magnetopause the magnetic field is nearly the dipole field, but it is increased by approximately a factor two in approaching the magnetopause. In *traversal of the magnetopause* the magnetic field drops (frequently by a factor from $\frac{1}{2}$ to $\frac{1}{4}$) and changes direction. This jump occurs simultaneously with the onset of the plasma flux that represents the solar wind in the modified form. The thickness of the magnetopause ("jump" region) is 100 km or somewhat more; 100 km is the gyroradius of protons with 500 km sec^{-1} in a field of 52 γ.

On the passage through the *magnetosheath* the magnetic field fluctuates strongly in magnitude and direction (see the figures in [*15*]). The shock front is another discontinuity surface, at which the field falls somwhat. The field outside, i.e., the field in the solar wind, is weak and smoother again. It is particularly stable in its magnitude.

The *plasma flux* is also quite different outside and inside the shock front. Outside there is the unchanged solar wind, a smooth flow, coming from the Sun,

[1] [*17*], Fig. III.21.

with a narrow energy spectrum. Inside, the solar wind appears partly thermalized (see next section).

It is not unusual that a space probe or satellite on one outward or inward pass *crosses the same boundary several times.* These multiple crossings, already discovered when Explorer 10 as the first satellite observed the magnetopause in 1961, are explained by an alternating outward and inward motion of magnetopause or shock front. Velocities of 150 km/sec are observed in the motion of the magnetopause. The motions may have the character of waves traveling along the boundaries.

It is striking that the *Mead model* represents the *front part of the magnetosphere* (on the day-side) fairly well despite the neglect of the transition area, the magnetosheath. If the magnetosheath were very thin, one might expect that the unperturbed momentum flux through the shock front is nearly equal to the pressure acting from the magnetosheath on the magnetopause. This would justify the equilibrium condition of MEAD's theory, Eq. (9.7). Apparently the equilibrium condition with some value of a is approximately correct even with the realistic magnetosheath that has a significant thickness[2]. The computation on the basis of the hydrodynamic theory, outlined in the next section, indicates a pressure slightly below that of a nonreflected wind. For the subsolar point the coefficient in Eq. (9.7) has been derived[3] as $a = 0.84$ or $a = 0.88$. The first value relates to a ratio of specific heats $\gamma = 2$, the second to $\gamma = \frac{5}{3}$ (see Sect. 15β).

In Fig. 11 it is seen that the magnetosphere has an extensive *tail,* in which the field lines are stretched out in the antisolar direction. Between the field lines of opposite directions in the tail there is necessarily a region of field reversal, the so-called "*neutral sheet*". A *plasma sheet* is found to surround the neutral sheet. A more detailed description of the tail will be given in Sect. 16.

Fig. 11 also shows the radiation-belt area, in which energetic particles are trapped (see Sect. 20). The particle cusp is a transition between radiation belt and plasma sheet in the tail that is fairly well discernable from these regions (see Sect. 21β).

15. Shock front and magnetosheath.

α) *General.* In the preceding section some observational facts about the shock front and magnetosheath were mentioned. The magnetosphere corresponds to a blunt, nonrigid body in a supersonic stream, the solar wind. It is known from supersonic aerodynamics that in this situation a shock front forms, which encloses the body. The stream is heavily perturbed from the shock front on. The solar wind is such a stream, unaffected outside the shock front, but considerably modified in the magnetosheath, i.e., in the area between the shock front and the magnetosphere [*14*], [*17*], [*20*][1].

The plasma flow in the magnetosheath has a wide energy spectrum and a wide angular distribution; it seems to be nearly isotropic although with preference of the direction coming from the sun. The plasma flux in this region fluctuates strongly in both magnitude and direction of incidence. The wide energy spectrum may indicate thermalization. The electrons have attained an energy of the order of 100 eV (cf. Fig. 17), whereas in the solar wind they had the same *velocity* as

---

[2] For further notes on the validity of the $\cos^2 \varphi$-law see Sect. 15α of the contribution by PARKER and FERRARO in Vol. 49/3 of this Encyclopedia, p. 131–205, [*28*].

[3] J. R. SPREITER, A. L. SUMMERS, and A. Y. ALKSNE: Planet. Space Sci. **14** (3), 223–253 (1966).

[1] J. R. SPREITER, A. Y. ALKSNE, and A. L. SUMMERS: External aerodynamics of the magnetosphere, in [*13*], p. 301–375.

the protons. The irregularity of the magnetic field observed in this region seems to be another sign of the turbulent motion.

A shock front is formed for the same reason as in hydrodynamics or aerodynamics. The solar wind is both supersonic and super-Alfvénic (see below).

A weak perturbation can not travel upstream in the unperturbed solar wind. At the magnetopause, however, a perturbation is created. The perturbation has to travel a few earth radii upstream to the shock front. This requires that the perturbation is so heavy as to reduce the Mach number (or Alfvén Mach-number) in the perturbed area (magnetosheath) to a value below unity. The heavily perturbed area, separated from the unperturbed wind by the shock front, is the magnetosheath. The wave velocity is supposed to exceed the wind velocity in some (middle) part of the magnetosheath, but not in all of it. Far behind the obstacle (the magnetosphere) the shock front should become the Mach cone.

With a temperature of $5 \cdot 10^4$ °K in an electron-proton mixture (cf. Sect. 4) one finds a sound velocity of 46 km sec^{-1} and an Alfvén velocity of 49 km sec^{-1} in the solar wind. Mach numbers and Alfvén Mach-numbers between 6 and 10 thus seem likely.

A shock front ahead of the magnetosphere in the solar wind was suggested by Žigulev (1959), Axford (1962), Gold (1962), and Kellogg (1962) [*14*], [*15*] [2,3]. Various authors contributed to the more complete hydrodynamic theory. Figs. 12 and 13 and some of the following notes are based on the treatment by Spreiter et al. [*14*][1,3]. The magnetic field is neglected in an essential part of this theory, although the Alfvén Mach-number seems to be usable in place of the acoustic Mach number.

β) *Hydrodynamic theory.* A number of *assumptions and simplifications* that have to be introduced in the hydrodynamic theory should be noted first. The formulations are macroscopic. There are obviously randomizing effects causing transformation of ordered forms of energy into statistically distributed energy, i.e., thermal energy. Their nature is not definitely known. Collisions are much too rare; gyromotions in the magnetic field, various instabilities with resulting irregularities or wave phenomena must provide the randomization. The shock front is "collisionless" (cf. Sect. 12*δ* on *magnetodynamic* shock fronts).

Effects of the magnetic field besides the randomization are ignored. The gas pressure in the plasma is considered isotropic. The shape of the magnetopause is supposed to be given, corresponding to the equilibrium condition with the unperturbed solar wind (cf. Sect. 13). The computations were facilitated by the assumption of axial symmetry around the Sun-Earth line. The equatorial profile was used as the profile of the magnetosphere in all planes containing the Sun-Earth line. For the ratio of specific heats the values 2 (corresponding to two degrees of freedom) and 5/3 (for three degrees of freedom) were tried. A control of the motions by the magnetic field with limitation to the directions normal to the magnetic field suggests $\gamma = 2$, the value assumed in the present figures. A definite decision about the value of γ has not yet been made[1].

Some justification for the various simplifications can be given on theoretical grounds[3], but the best justification seems to be the fairly good agreement between theory and observations (see the figures and the later part of this section). Shape and position of the derived boundaries, shown in Figs. 12 and 13, correspond to those observed (cf. the schematic Fig. 11, which is supposed to fit the observations). The structure of the shock front, however, does not so well agree with theoretical predictions (see remarks in Sect. 15*γ*).

The *hydrodynamic theory* of a shock front is based on the continuity postulates for a front at rest (cf. Sect. 12*α*): the "Rankine-Hugoniot equations" [*14*], [*15*].

[2] W. I. Axford: J. Geophys. Res. **67**, 3791—3796 (1962).

[3] J. R. Spreiter, A. L. Summers, and A. Y. Alksne: Planet. Space Sci. **14** (3), 223—253 (1966).

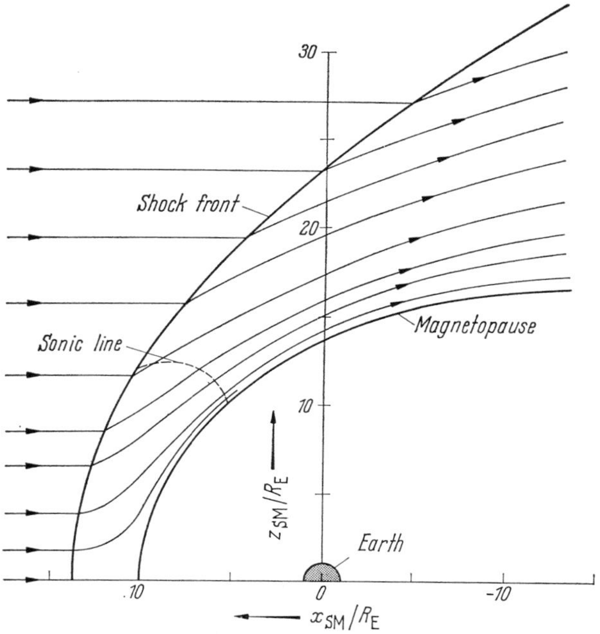

Fig. 12. Streamlines in the magnetosheath[4] according to hydrodynamic theory with $M=8$ and $\gamma=2$.

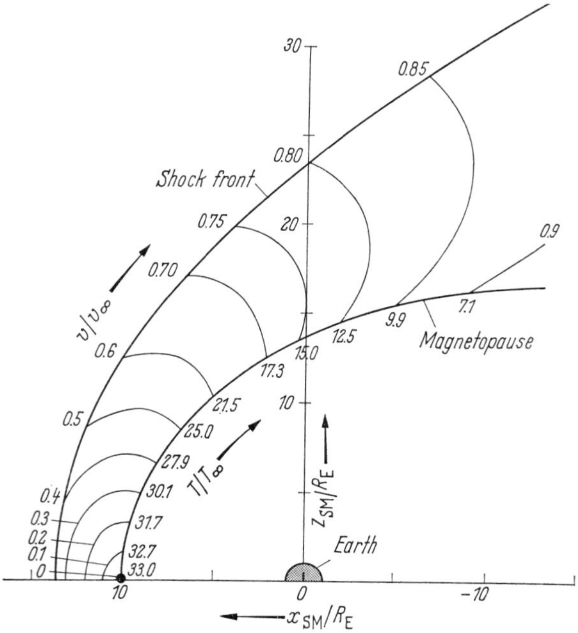

Fig. 13. Plasma velocities and temperatures in the magnetosheath[4] according to hydrodynamic theory with $M=8$ and $\gamma=2$. The ratios of velocities and temperatures to those of the unperturbed solar wind (v_∞, T_∞) are noted.

[4] J. R. Spreiter, A. L. Summers, and A. Y. Alksne[3], Figs. 5 (slightly modified) and 7.

Continuity is required of the normal mass flux density, ϱv_n, the normal momentum flux density plus pressure $\varrho v_n^2 + p$, and the normal energy flux density $\varrho(\frac{1}{2}v^2 + c_p T)v_n$. The entropy of a parcel of gas jumps in passing through a strong shock. The common equation for an adiabatic (or "isentropic") change of state is consequently not usable at the strong shock.

The density jump and the velocity jump *at the shock front* are derived from the continuity postulates [cf. Eq. (29) in [3]]. They depend on the data of the solar wind, in particular on its Mach number and on the ratio of specific heats γ. In the limit of an infinite Mach number a density increase of $(\gamma+1)/(\gamma-1)$ is obtained with a relative velocity decrease reciprocal to this. For $\gamma=2$ these ratios are 3 and $\frac{1}{3}$. They are almost reached in the subsolar point in Figs. 12 and 13.

Within the magnetosheath the sound velocity has to be increased so much that the stream velocity becomes subsonic. At a particular location the stream velocity may be v; in the original solar wind it was v_∞. The energy relationship for a plasma (or gas) parcel is then

$$\tfrac{1}{2}(v_\infty^2 - v^2) = c_p \Delta T. \tag{15.1}$$

Sound velocity V_s and absolute temperature are related by

$$\gamma(c_p - c_v)\,T = V_s^2 \tag{15.2}$$

or

$$c_p T = \frac{1}{\gamma-1} V_s^2. \tag{15.3}$$

Eqs. (15.1) and (15.3) yield

$$\frac{\Delta T}{T_\infty} = \frac{\gamma-1}{2}\left(1-\frac{v^2}{v_\infty^2}\right) M^2 \tag{15.4}$$

with the Mach number of the arriving solar wind

$$M = v_\infty / V_{s\infty}. \tag{15.5}$$

Eq. (15.4) [which is Eq. (28) in [3]] shows the temperature increase resulting from a reduction of the stream velocity. Velocity ratios v/v_∞ and temperatures derived from hydrodynamic theory[3] for $M=8$ are represented in Fig. 13.

For the sound velocity at some location, V_s, it follows from Eqs. (15.1) and (15.3) that

$$V_s^2 - V_{s\infty}^2 = \frac{\gamma-1}{2}(v_\infty^2 - v^2). \tag{15.6}$$

When the sound velocity V_s becomes identical with the stream velocity v there is

$$V_s^2 = \frac{(\gamma-1)M^2+2}{(\gamma+1)M^2} v_\infty^2. \tag{15.7}$$

This condition is fulfilled on the "sonic line" of Fig. 12. With the present parameters Eq. (15.7) yields $V_s/v_\infty = 0.59$ and a temperature increase $[(\gamma-1)M^2+2]/(\gamma+1) = 22$ on the sonic line. The flow is subsonic in the middle of the magnetosheath and becomes supersonic again in crossing the sonic line.

The computed depth of the magnetosheath at the subsolar point (shown in Figs. 12 and 13) is 3.7 R_E for a magnetopause at 10 R_E, in fair agreement with observed depths, which are given as 4 to 5 R_E.

The ratio between the depth of the magnetosheath (the "stand-off distance" of the shock front) and the height of the magnetopause should for large M be nearly invariable. Very far behind the Earth the shock front should asymptotically become the Mach cone, if it is still recognizable there.

In *transfer of the hydrodynamic theory to the plasma* it is necessary to divide all temperatures by 2, because half the dissipated energy is transferred to the electrons. The sound velocity, however, remains that of the assumed neutral gas because the mean molecular weight of the proton-electron gas appears reduced to 1/2. An increase of the proton temperature by factors from 5 to 12 in crossing the shock front was, in fact, observed (see Table 14 in [*15*], Sect. 39, and Tables 7.1 and 7.2 in [*14*]).

The shock front should be identified with the *fast magnetodynamic shock*. No matter which of the two Mach numbers (acoustic or Alfvén) is higher, the faster of the two shock fronts is more readily apt to stand up against the solar wind. *Field-line pictures*, based on the purely hydrodynamic theory together with the theorem of moving field lines, were presented by SPREITER et al.[1].

γ) *Remarks on the shock front.* The hydrodynamic theory (Sect. 15β) seemed to explain the shape and position of the shock front and the magnetosheath flow field. In cases in which plasma and magnetic-field data were observed on the way through the shock front the comparison of the jumps with the Rankine-Hugoniot relations of *magnetodynamics* did not yield a quite satisfactory result[5]. In particular the temperature increase did not correspond to theory; it was frequently too high, sometimes by a factor 2 [*14*][5].

From a *microscopic viewpoint* the shock front is not too well understood [*14*][5]. The mechanism leading to dissipation of kinetic energy at the front and in the magnetosheath is not clear; it must be connected with nonlinear processes and must have to do with the magnetic field and its irregular structure in the magnetosheath. One may think of wave-like phenomena. The magnetosheath is believed to involve wave-particle interactions.

The shock front has a certain structure, which apparently is somewhat complicated [*14*][5]. A quantity sometimes used as an indicator of the position of the front is the variance of the magnetic field, which is low in interplanetary space and high in the magnetosheath. At the front the magnetic field and the plasma quantities vary drastically. The quantity whose variation is most easily observed with some accuracy is the magnetic field strength, but even the field strength is in passage through the front not a simple function of location.

The *thickness* of the front is not so well defined. It depends on the choice of the markation points within the front structure; it is found to lie between 10 and 250 km [*14*][5]. Some authors connect the thickness with a gyroradius. An alternative explanation is by means of waves fast enough to travel upstream in the solar wind. Beginning with some frequency beyond ion gyroresonance the group velocity of propagated electromagnetic waves exceeds the solar-wind velocity. The wavelength of the limiting frequency might determine the thickness of the front (cf. [6]).

Assuming *whistler* propagation parallel to the magnetic field and data of the quiet solar wind (Sect. 4), one finds $\lambda/2\pi \approx 30$ km for the limiting frequency (0.75 Hz). The value becomes smaller for higher wind velocity; it is proportional to $(Nv)^{-1}$, whereas the gyroradius is proportional to v. Occasionally it was suggested that the density jump at the shock front takes place in a smaller interval than the jump of the magnetic field [*14*].

16. Tail.

α) *Magnetic field.* The magnetosphere has a long tail, pointing away from the Sun [*12*], [*14*], [*15*], [*17*], [*18*][1]. An illustration of the tail, summarizing

[5] J. H. WOLFE and D. S. INTRILIGATOR: Space Sci. Rev. **10** (4), 511—596 (1970).

[6] C. F. KENNEL and H. E. PETSCHEK: Magnetic turbulence in shocks, in [*13*], p. 485—513.

[1] C. P. SONNETT, D. S. COLBURN, R. G. CURRIE, and J. D. MIHALOV: The geomagnetic tail: topology, reconnection and interaction with the moon, in [13], p. 461—484.

the observational facts, was presented in Fig. 11 under assumption of a tilted dipole, corresponding to northern summer. Field lines starting or ending on the Earth at polar latitudes extend to long distances in the tail. All field lines in the tail are parallel, directed toward or away from the Sun — in the upper half toward and in the lower half away. The fields in the two halves are approximately homogeneous. They are separated by the "*neutral sheet*", in which the field reverses direction. The neutral sheet begins at approximately 10 R_E from the earth's center. For an example of measured tail fields see Fig. 46 in HESS's article [*15*].

The tail is sometimes compared with the wake of an obstacle in a streaming fluid. This may merely be a paraphrase for the fact that the tail is due to the flow of the solar wind. The tail is an extension of the magnetosphere in which the field lines are stretched out by the influence of the solar wind. What kind of force the solar wind exerts on the magnetosphere or its plasma is still uncertain (see Sect. 16β).

The idea of a *temporarily* existing tail was brought up by PIDDINGTON (1960) in a tentative explanation of the main phase of geomagnetic storms (which is not held up nowadays; see Sect. 29). This was shortly before it became known that the magnetosphere is *permanently* confined by a definite boundary. Various authors *observed* distortions of the magnetic field corresponding to a tail (e.g. HEPPNER, 1963; CAHILL, 1964) or *developed theoretical models of a magnetosphere with tail and neutral sheet* (AXFORD, 1965; DESSLER and JUDAY, 1965). Systematic studies of the tail field by NESS (1965) and others followed [*12*], [*17*].

The *field* in the magnetospheric tail (or "magnetotail") *decreases with distance* as shown in Fig. 14. The decrease is, at least to some degree, due to a *field-line connection across the neutral sheet*. The field crossing the neutral sheet not too far out is purely northward, amounting to a few gamma, but varying in intensity[1,2]. Beyond 30 R_E both directions of the cross field are found, both with values between a fraction of a gamma and 10 γ, but the direction to the north is preferred[2].

Fig. 15 (after AXFORD et al.) shows a schematic *cross section of the tail* together with the current system that the change of the direction of the magnetic field requires in the neutral sheet and at the magnetopause. The field directions above and below the neutral sheet require a current flow in the sheet from dawn to

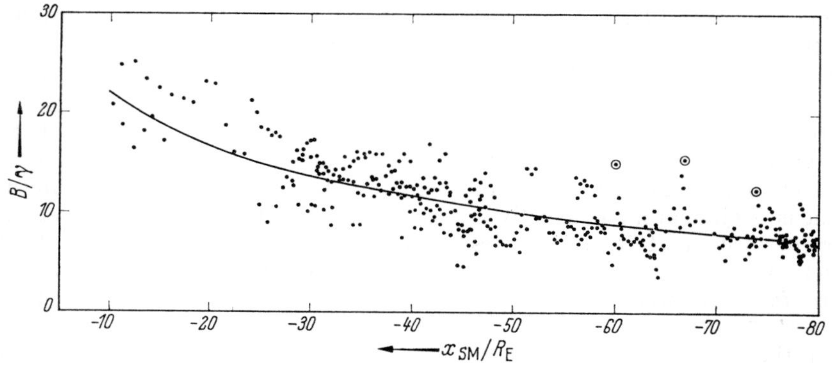

Fig. 14. Magnetic field in the tail[3]. The coordinate x_{SM} is measured from the Earth's center. The points denote hourly averages of the field magnitude in the period July to November 1966. Points in circles refer to strongly increased field magnitudes preceding an increase of K_p by one hour.

[2] J. D. MIHALOV, D. S. COLBURN, R. G. CURRIE, and C. P. SONNETT: J. Geophys. Res. **73** (3), 943—959 (1968).

[3] K. W. BEHANNON: J. Geophys. Res. **73** (3), 907—930 (1968); Fig. 23 (modified).

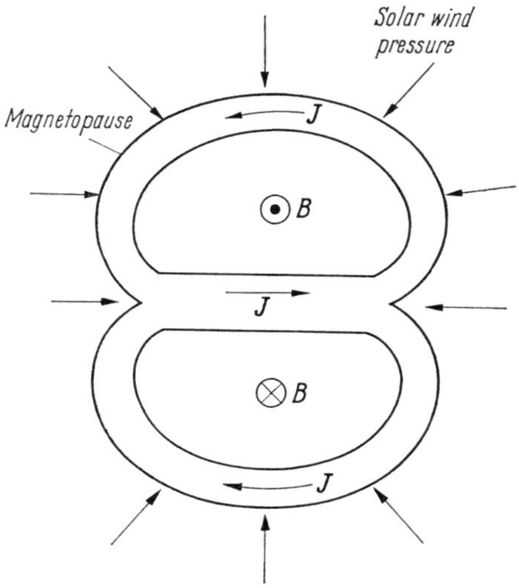

Fig. 15. Cross section of the tail[4]. Direction of view is from the Earth. The picture is schematic and is supposed to fit observations and predictions from theory.

dusk. The *thickness of the neutral sheet* was found to be between 0.1 and 1 R_E. An accurate determination was not possible because the neutral sheet appears to be almost always in motion (with velocities up to several km $\sec^{-1}$)[5]. As a consequence, *multiple crossings* of a satellite within a few hours have been observed frequently[5]. The position of the neutral sheet varies also with the tilt of the geomagnetic axis against the Sun-Earth line. This may be the cause of double crossings with a long interval between them.

The tail *extends far beyond the lunar orbit* [*12*]. If the decrease of the tail field observed between 10 and 80 R_E (Fig. 14) continues according to a power law, the field should fall below the interplanetary field near 150 R_E. Intermittent observations of a possible tail field of approximately 8 γ, however, were made by Pioneer 7 near 1000 R_E [*12*]. The tail at this distance seems less developed and may be of a filamentary structure; it is presumably near resolution (cf. Sect. 16 γ).

β) *Theoretical aspects.* The tail is obviously the result of a *force pulling field lines and plasma away from the Sun.* The numerical estimate below indicates that the solar wind is capable of such a force. It is in general believed that the solar wind sets the outer layer of the magnetospheric plasma in motion, thus stretching the field lines out. Another consequence of a pull on the outermost plasma is a circulatory convective motion throughout the magnetosphere with a current system at high latitudes in the ionosphere that manifests itself in magnetic-field variations on the Earth (Sect. 24). *Magnetodynamic waves traveling outward in the tail* were also noted as a possible cause of its large extension ([*12*], [*18*], Sect. 7 δ of [*28*]).

For the *mechanism* by which the *solar wind acts on the magnetosphere* two alternatives have mainly been considered: A viscous (or "frictional") force might arise from instabilities at the magnetopause according to a hypothesis of AXFORD

[4] W. I. AXFORD, H. E. PETSCHEK, and G. L. SISCOE: J. Geophys. Res. **70** (5), 1231—1236 (1965); Fig. 3.

[5] T. W. SPEISER and N. F. NESS: J. Geophys. Res. **72** (1), 131—141 (1967).

and HINES (Sect. 7 of the article by PARKER and FERRARO [28]) or there might be some connection of interplanetary and magnetospheric field lines, not the perfect separation that was postulated so far. This corresponds to an "open" magnetosphere (see Sect. 17). DUNGEY[6] predicted a tail length of approximately 1000 R_E on the basis of field line connection. His essential idea was that the outer ends of terrestrial field lines are moved by the solar wind with its velocity while the inner ends travel with roughly 100 msec^{-1} on a path of nearly 1 R_E (cf. Sect. 36η of [15]).

It is of interest to compare *the magnetic tension* in the tail with the *pressure of the solar wind* in impact on a surface [21]. Values of the total forces, the tension force across the tail and the solar wind pressure on the dayside magnetosphere, are given by PARKER and FERRARO in this Encyclopedia ([28] Sect. 6β, see also Sect. 28 of the present article).

The forces on the unit area may now be noted: A magnetic field of 15 γ ($=1.5 \cdot 10^{-8}$ V sec m^{-2}) in the tail exerts a tension of $B^2/2\text{u}\,\mu_0 = 0.9 \cdot 10^{-10}$ N m^{-2} ($=0.9 \cdot 10^{-9}$ dyn cm^{-2}). The pressure of the quiet solar wind (with the data of Sect. 4) is $0.9 \cdot 10^{-9}$ N m^{-2}, i.e., an order of magnitude larger than the field tension. This proves the solar-wind intensity to be sufficient to produce the tail.

PARKER [21] suggests that — in addition to other effects — the solar wind may eventually *penetrate* into the tail at large distances. It should be kept in mind that the neutral sheet of the tail is practically open at the two edges (see Fig. 15). The solar wind would readily penetrate there if the space were perfectly field-free. There is, however, the cross field in the neutral sheet, which was found weak, but irregular (Sect. 16α), in the outer area even changing its sign at some places. Disappearance of the field at these places permits the solar wind to penetrate into the neutral sheet and to push the field lines of the cross field in the neutral sheet outward.

BEARD, BIRD, and HUANG[7] (Sect. 36 below) suggested that the tail may be self-consistent, maintained by drift motions of the tail plasma. The reversal of the magnetic field across the neutral sheet represents a strong inhomogeneity of the field, in which the particles describe drift motions [Eq. (19.1)]. These drift motions (the curvature drift, in particular) lead to a current flow, according to the estimates just to the currents necessary for the existing configuration of the field (see small print below). The explanation of the tail with this hypothesis depends on finding a way for solar-wind plasma to enter the tail field, a problem studied, for example, by PARKER (see preceding paragraph) and discussed in connection with the polar cusps, which were recently discovered (see end of Sect. 16γ and Sect. 35).

In the attempt to understand the *neutral sheet* one is faced with this problem: The plasma of fairly low density (between 0.1 and 1 cm^{-3}) that is present must carry the current required in the neutral sheet (Fig. 15).

Reversal of the direction of a field of 15 γ ($=1.5 \cdot 10^{-8}$ V sec m^{-2}) requires an *areal current density* of $2c_0\sqrt{\varepsilon_0\,\mu_0}\,B/\text{u}\,\mu_0 = 2.4 \cdot 10^{-2}$ A m^{-1}. The corresponding *spatial current density* in a sheet of 2000 km thickness would be $J = 1.2 \cdot 10^{-8}$ A m^{-2}. This is a remarkable magnitude. It would empty the neutral sheet with a particle density $N = 1$ cm$^{-3} = 10^6$ m^{-3} in a time

$$l/v = l N q/J. \tag{16.1}$$

For a path length $l = 20\ R_E = 1.3 \cdot 10^8$ m (half-width of the sheet) the time is $l/v = 1.7 \cdot 10^3$ sec ≈ 30 min. The plasma obviously has to be replaced within this time.

If the magnetopause separates the magnetic field of the tail from the solar wind with the interplanetary field then a current sheet of the type shown in Fig. 15

[6] J. W. DUNGEY: J. Geophys. Res. **70** (7), 1753 (1965).

[7] D. B. BEARD, M. BIRD, and Y. H. HUANG: Planet. Space Sci. **18** (9), 1349—1355 (1970).

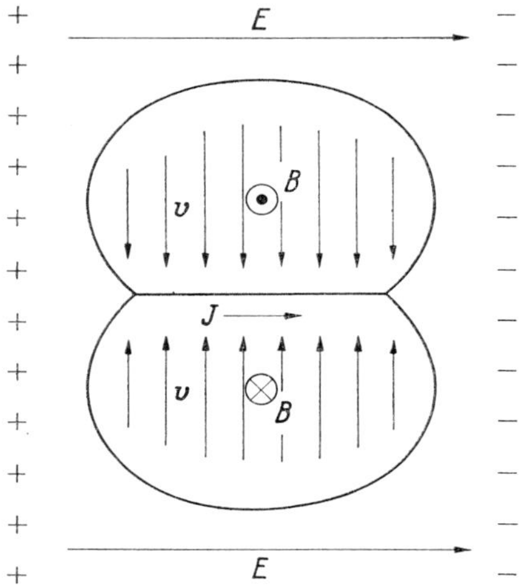

Fig. 16. Electrodynamic drift in the tail required by Alfvén's model of a neutral sheet.

exists at the magnetopause. The currents converge or diverge at the edges of the neutral sheet and must consequently flow through the neutral sheet. The current in the neutral sheet thus might be supplied from the magnetopause.

A different explanation of the current in neutral sheets (the neutral sheet of the tail or a neutral sheet at the magnetopause) was given by Alfvén[8]. Alfvén's model, adapted to the tail, is shown in Fig. 16. It is assumed there that an electric field that is present in the solar wind (cf. Sects. 17 and 30) penetrates partially into the tail, thus causing the plasma in the two halves of the tail to drift into the neutral sheet. Drift motion and electric field are, in fact, similar to those postulated by the hypothesis of convection due to friction forces (Sect. 24). Fig. 16 may as well be interpreted as demonstrating convective motion caused by friction at the magnetopause.

The direction of the electric field in Fig. 16 corresponds to a southward magnetic field component in the solar wind. How the electric field can penetrate into the tail is left open. Penetration would be made possible by a connection of interplanetary magnetic field lines with tail field lines (cf. Sect. 17γ).

A voltage of 10 **kV** over the width of the tail (approximately 40 $R_E = 2.5 \cdot 10^8$ m) would produce an electric field of $4 \cdot 10^{-5}$ V m^{-1} and a drift with $2.7 \cdot 10^3$ m sec^{-1} in the magnetic field of 15 γ ($= 1.5 \cdot 10^{-8}$ V sec m^{-2}). In the neutral sheet the *protons* move *with* $\boldsymbol{E}$ and the electrons move in the opposite direction. The velocities appear here increased in comparsion with the drift velocity by the ratio of width to thickness of the neutral sheet, which is approximately $2.5 \cdot 10^8$ m$/2 \cdot 10^6$ m $= 125$. The velocity thus is $3.4 \cdot 10^5$ m sec^{-1}. This velocity with $N = 0.2$ cm^{-3} leads approximately to the postulated current density of $1.2 \cdot 10^{-8}$ A m^{-2}. (For the general relationships see Alfvén's article[8].)

In the last description the effect of the weak magnetic field across the neutral sheet was ignored. The question remains why this magnetic field does not inhibit the current flow. On the other hand, the plasma of the neutral sheet with the above data is not a low-β plasma [cf. Eq. (8.12) and [9]]. Caution may consequently be necessary in applying some of the ideas developed in earlier sections. The theorem of frozen-in field lines together with the drift motion toward the neutral sheet demands that *in the neutral sheet* field lines and plasma move rapidly *toward the Earth.*

[8] H. Alfvén: J. Geophys. Res. **73** (13), 4379—4381 (1968).

[9] M. D. Montgomery: J. Geophys. Res. **73** (3), 871—889 (1968).

The neutral sheet resembles a special case of *tangential discontinuity* (Sect. 12β). The specialty seems that $\boldsymbol{B}$, which is tangential, preserves its magnitude, only changing its direction by 180°. However, the existence of a normal component of $\boldsymbol{B}$ (the cross field in the neutral sheet) means a significant deviation from the tangential discontinuity. If the plasma drifts from two sides toward the neutral sheet as it was hypothesized, another departure from classical magnetodynamic discontinuity surfaces becomes apparent (cf. Sect. 12ε): The normal flux density ϱv_n is not continuous at the neutral sheet. There is a flux of mass to the sheet from the two sides and a corresponding flux along or within the sheet; i.e. a nearly two-dimensional mass flow. A *current flow* along a discontinuity of the tangential magnetic-field component is commonly allowed for. But it is in magnetodynamic theory unconventional to assume a similar *mass flow* along a surface. With the low mass densities now involved such a mass flow seems well possible (see also Sect. 17γ).

γ) *Plasma flux observations.* In the magnetotail there is an area of increased plasma density, *the plasma sheet.* It is indicated in Fig. 11. The plasma sheet surrounds the neutral sheet. The neutral sheet, as a region of a weak magnetic field embedded between magnetic fields of the order of 20 γ ($=2\cdot 10^{-8}$ V sec m^{-2}) requires some plasma pressure, balancing the magnetic-field pressure outside. The earlier search for some plasma flux in the neutral sheet was a success insofar as the plasma sheet was discovered, but this sheet appears as a *wider region around the neutral sheet.* The plasma flux was, for instance, observed by means of collectors (<20 keV) and Geiger counters (>45 keV). Outside the plasma sheet there is still plasma, but of too low density to be measurable.

The plasma sheet [*12*], [*15*], [*17*], [*18*][10,11] has a width of approximately 4 to 6 R_E normal to the neutral sheet near the midnight meridian. Near the boundary of the magnetotail the width is approximately doubled. Early measurements of low-energy electrons (>200 eV) were made on Soviet space probes (by GRINGAUZ et al., 1960). ANDERSON observed tail electrons >45 keV by means of IMP 1 (1965). The Vela satellites were used in a comprehensive investigation of the plasma sheet (by BAME and HONES, by MONTGOMERY and other authors). The flux of electrons turns out to be nearly isotropic. Reported values of the flux commonly represent omnidirectional flux, obtained by integration over all directions. Most measurements in the plasma sheet were made at 17 to 32 R_E from the Earth's center. The proton flux and its energy distribution were also investigated.

Two different *populations of electrons* are discernible in the plasma sheet. The *low-energy population,* to which Fig. 17 refers, has nearly thermal character, although the peak energy is higher than in the magnetosheath and the energy spectrum has a high-energy tail. Fig. 17 shows energy spectra of magnetosheath electrons (a) and plasma sheet electrons (b), (c), (d). The peak energy of the latter varies between 0.2 and 12 keV, being in most cases around or below 1 keV. Their flux density is in general $10^8 \dots 10^9$ cm^{-2} sec^{-1} ($=10^{12}\dots 10^{13}$ m^{-2} sec^{-1}); it varies somewhat, but remains smooth as long as the satellite is located in the plasma sheet. The particle density corresponding to this flux is between 0.1 and 1 cm^{-3}. The energy spectra of *protons* in the keV range are similar to the energy spectra of the electrons.

---

[10] E. W. HONES: Review and interpretation of particle measurements made by the Vela satellites in the magnetotail, in [*13*], p. 392—408.

[11] V. M. VASYLIUNAS: Low energy particle fluxes in the geomagnetic tail, in G. SKOVLI (ed.): The Polar Ionosphere and Magnetospheric Processes, p. 25—47. New York: Gordon and Breach 1970.

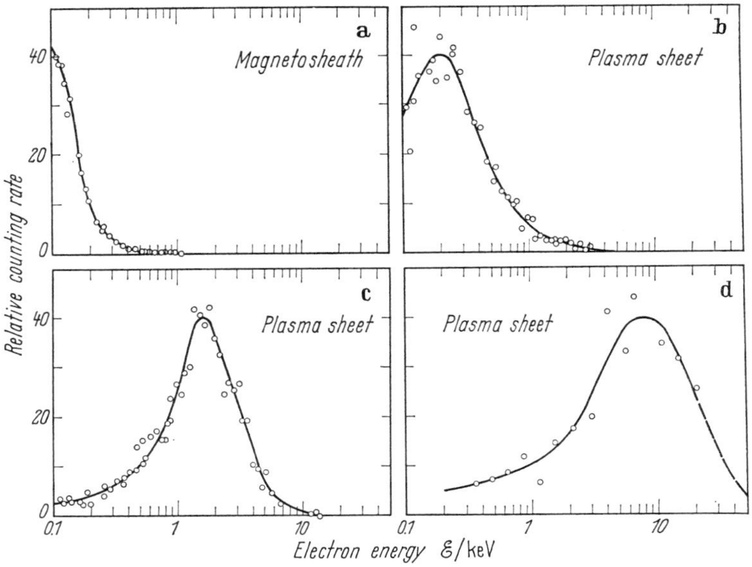

Fig. 17a—d. Energy spectrum of electrons in the magnetosheath (a) and in the plasma sheet of the tail (b, c, d)[12]. Examples measured at different times in 1964 and 1965.

A *much more variable population of electrons* is found at energies *beyond* 45 keV. Their flux density can rise to $10^7\ cm^{-2}\ sec^{-1}$ $(=10^{11}\ m^{-2}\ sec^{-1})$. They are only, but not always, present when the low-energy electron flux representative of the plasma sheet is observed. They appear in the form of outbursts with a rapid onset (i.e., a rise time of a few minutes or less) and a slow decay (from several minutes to a few hours[13, 14]). This led to the term "electron islands" and to the idea of a pulse-type injection or of a sudden acceleration followed by some relaxation. Quite generally, however, the cause of both electron populations and the nature of the plasma sheet are still unclear. The magnetic field strength appears reduced in the electron islands and, more generally, at places with an increased plasma density, in agreement with the diamagnetic behavior of plasma and with the postulate of balance between gaskinetic and magnetic pressures.

The high-energy electrons (>45 keV) have a frequency of occurrence that is highest near the dawn side of the plasma sheet and decreases toward the dusk side, whereas the low-energy electrons are distributed more equally[9,10]. There is no definite idea about the cause of the dawn-to-dusk asymmetry. Possible processes involved are local acceleration of particles in connection with instabilities and the Earth's rotation with field lines entering and leaving the tail.

The *plasma sheet* seems to be *frequently in motion*. Some motions may be due to the tilt of the magnetic dipole and to changes of the solar wind. During development of a magnetic bay the sheet becomes thinner and contains electrons of lower energy. Later during the magnetic bay it thickens and recollects electrons of higher energy[10].

---

[12] S. J. BAME, J. R. ASBRIDGE, H. E. FELTHAUSER, E. W. HONES, and I. B. STRONG: J. Geophys. Res. **72** (1), 113—129 (1967); Fig. 2.

[13] K. A. ANDERSON: J. Geophys. Res. **70** (19), 4741—4763 (1965).

[14] See, for example, Fig. 15 in [*18*] or Fig. 4 in the article by HONES[10].

The satellite Pioneer 7, which gave an indication of the magnetic field of the tail near 1000 R_E, made plasma probe measurements in the same area[15]. The ion spectrum corresponded most of the time to the solar wind, but intermittently it was found disturbed, showing less thermal contribution below 1 keV and a second peak or a more pronounced extension to higher energies. Quite likely this is a tail effect. If the well defined tail does not extend so far, it may be that the observed anomalous spectra are those of some turbulent solar wind in a wake representing the continuation of the magnetotail.

The plasma sheet with its large width does not provide the *force balance required at the neutral sheet* (see beginning of this section, Sect. 16 γ). MURAYAMA and SIMPSON[16], however, found a peak of the flux of electrons >160 keV in the neutral sheet. The peak flux is about twice the flux of adjacent regions in the tail. The observation was made in 20 neutral-sheet crossings. This high-energy flux in the neutral sheet seemed to be continuously present and it was still observable beyond 31 R_E, but it appeared highly variable in intensity. The range of magnitudes of the electron flux is similar as in the electron islands, but the spectrum may fall off toward high energies more slowly. The origin of the high-energy electrons in the neutral sheet and their relation with the plasma sheet have not yet been clarified. They may come from the radiation belt or they may have been accelerated in the neutral sheet. In any case it is likely that they are the particles that provide the balance of forces at the neutral sheet.

A new possibility for *penetration of solar-wind plasma* into the tail has been recognized at the recently discovered "*polar cusps*", which surround the high-latitude field lines extending to the hypothetical neutral points[17]. Magnetosheath plasma is found in the polar cusps. The plasma is supposed to get from the magnetosheath into the polar cusps because of field line merging (see Sect. 17 γ). It may then be moved with the polar-cusp field lines into the magnetotail, where it forms the plasma sheet. Some tail field lines become reconnected, moving with the attached plasma toward the Earth. Magnetic merging and plasma supply may occur at an increased rate during polar and magnetospheric substorms.

17. Open magnetosphere. The description of the magnetosphere in the previous sections was based on the idea of a perfect separation between the solar wind and the geomagnetic field. The separating boundary is the magnetopause. It is the theorem of frozen-in field lines that prohibits penetration of the arriving plasma into the geomagnetic field. The strict validity of this theorem, however, is a matter of debate. Some phenomena may point out a connection between interplanetary and terrestrial magnets fields. There is, for example, the replenishment of trapped particles in the outer radiation belt during the course of a geomagnetic storm, which may be due to particles streaming in on field lines (Sect. 20 β). Also the convective motion present in the high-latitude magnetosphere is possibly caused by a connection of the geomagnetic field lines with the traveling interplanetary field lines (Sect. 24).

α) *Models* of an "open magnetosphere" with a *connection of field lines across the magnetopause* were brought up by DUNGEY and in an improved version by

---

[15] J. H. WOLFE and D. D. MCKIBBIN: Review of Ames Research Center plasma-probe results from Pioneers 6 and 7, in [*13*], p. 435—460.

[16] T. MURAYAMA and J. A. SIMPSON: J. Geophys. Res. **73** (3), 891—905 (1968).

[17] Sect. 35 on p. 109, which was added in proof-reading, deals briefly with the matter of the polar cusps and gives relevant references.

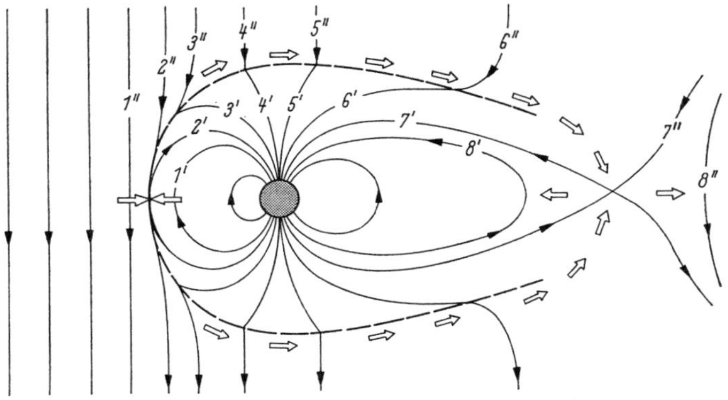

Fig. 18. Open magnetosphere wi th merging field lines[1].

LEVY, PETSCHEK, and SISCOE[2] (p. 306—317 in [*14*], [*33*]). This last model is represented in Fig. 18. It still contains a magnetopause, now marked by a kink in the magnetic field lines.

The figure shows the special case of an *interplanetary field opposite to the geomagnetic field* at the front of the magnetopause. This case was investigated in some detail. The subsolar point on the magnetopause is an X-*type neutral point* of the magnetic field (cf. Sect. 8δ). White arrows and a system of consecutive numbers in Fig. 18 demonstrate the *motion of plasma and field lines,* which will be discussed below. A second neutral point appears on the night side.

The noon-midnight meridian plane is depicted in the figure. There may be just the *two neutral points* shown in the figure [*19*], but PETSCHEK in his fundamental paper[3] implies that the configuration is nearly the same in neighboring meridian planes. The neutral points of Fig. 18 should in this case be the intersection points of *two neutral lines or rings in the equatorial plane* with the noon-midnight meridian plane. For a discussion of possible positions of the neutral points or neutral rings, also with other external fields, see DUNGEY[4]. Pictures of magnetospheric field lines for an open magnetosphere with various directions of the external field were presented by SPEISER[5].

β) Essential for the open model of the magnetosphere is the *state of motion* resulting from the progression of the solar wind with the embedded magnetic field lines. The theorem of frozen-in field lines is supposed to be valid everywhere except at the two neutral points (in the front and in the rear) and in their close vicinity. Disappearance of $\boldsymbol{B}$ and a finite conductivity both tend to make the theorem invalid. [In Eq. (10.10) the conductivity σ is finite and the scale length, being comparable with $B/|\partial/\partial\boldsymbol{r}\times\boldsymbol{B}|$, disappears with $\boldsymbol{B}$.]

The continuous progression of the solar wind leads in the model of Fig. 18 to this *state of motion*: Interplanetary and terrestrial field lines with the attached plasma approach the neutral point in a radial direction from the two sides; 1'' becomes 2'', 1' becomes 2'. In the neutral point the field lines merge and get separated so that the upper parts and lower parts depart in opposite directions—upward and downward; 2'',2' becomes 3'',3'; 3'',3' becomes 4'', 4' and so on.

[1] R. H. LEVY, H. E. PETSCHEK, and G. L. SISCOE[2]: Fig. 2 (slightly modified).

[2] R. H. LEVY, H. E. PETSCHEK, and G. L. SISCOE: AIAA (American Institute of Aeronautics and Astronautics) J. **2** (12), 2065—2076 (1964).

[3] H. E. PETSCHEK: Magnetic field annihilation, in W. N. HESS (ed.): AAS-NASA Symposium on the Physics of Solar Flares, NASA SP-50 (1964).

[4] J. W. DUNGEY: The structure of the exosphere or adventures in velocity space, in [*1*], p. 503—550, in particular p. 527—531.

[5] T. W. SPEISER: Radio Science **6** (2), 315—319 (1971); see also T. B. FORBES and T. W. SPEISER: J. Geophys. Res. **76** (31), 7542—7551 (1971).

At the rear neutral point the motions are just opposite to those at the front neutral point. In the magnetosphere the motions must, of course, form a closed circulatory system. This circulatory motion system will be discussed in Sect. 24.

The plasma stream impinging on the magnetopause from the two sides, from outside and from the magnetosphere, is led away along the magnetopause, strictly speaking in a thin transitional layer that now replaces the boundary. Continuity of the mass flux involves an acceleration of the plasma in entering the thin transitional layer, which contains a weak magnetic field only. The electric field necessary for the acceleration is that connected with the arrival of the plasma, $\boldsymbol{E} = \boldsymbol{v} \times \boldsymbol{B}/c_0\sqrt{\varepsilon_0\mu_0}$. The given electric field strength requires a higher velocity in the weaker magnetic field of the transitional layer. The acceleration occurs in entering the transitional layer. The same kind of acceleration mechanism is believed to be of importance in solar flares, where merging of oppositely directed field lines may cause ejection of material.

γ) There are mainly two *theories* that try to explain *the process of merging of field lines at a boundary* of the type of the magnetopause.

An *earlier theory*, developed by SWEET and PARKER for solar flares, assumes that "*merging*" or "*annihilation*" of oppositely directed field lines takes place *along the entire boundary surface*, which again contains a neutral point [*14*], [*33*][2]. The theorem of frozen-in field lines is in this theory supposed to be violated in a thin layer extending over the entire boundary. The relationship governing the behavior of the magnetic field and the plasma motion in the boundary layer is Eq. (10.2), which includes field variations due to motion of field lines with the plasma and to a kind of diffusion of field lines. [A velocity of diffusion $\boldsymbol{w}$ may be defined from Eq. (10.7) or in a different way.] It was attempted to apply this theory to the magnetosphere, but it yields an insignificant flow of plasma into the magnetopause only.

A *different idea of the merging process*, which was brought up by PETSCHEK[3], is the basis of Fig. 18. In PETSCHEK's theory ([*14*][2,3]) there is again a thin *boundary layer* (not indicated in Fig. 18), but the *plasma flow is of magnetodynamic nature* with the field lines frozen-in *everywhere except in the close vicinity of the neutral points* (in front and in the rear of the magnetosphere). Only in these limited areas it is necessary to assume a diffusive motion of field lines with merging. The small size of these areas allows merging at a much faster rate than in the earlier theory. Also this idea of merging aimed originally at an explanation of corpuscular emissions from solar flares and was then transferred to the magnetopause.

An observer on either side may compare the boundary with a rotational discontinuity, a front of ALFVÉN wave character (see Sect. 12γ). The variation of the field seems to represent a rotation of the tangential component by 180 degrees. The front is stationary while the plasma is seen to move into it. On both sides the field lines are slightly inclined toward the boundary and the approach velocity of the plasma equals the ALFVÉN velocity corresponding to the normal field component, $v = B_n/(\mathrm{u}\mu_0\varrho)^{\frac{1}{2}}$. An essential deviation from the rotational discontinuity described in Sect. 12γ, however, is that the plasma approaches the front from two sides, thus causing a remarkable flux in the thin transitional layer. The possibility of this kind of departure from classical magnetodynamic theory was noted in Sects. 12ε and 16β, in Sect. 16β with reference to the neutral sheet of the magnetotail.

LEVY, PETSCHEK, and SISCOE[2] derive a *maximum flow velocity* ($v_{\max}$) for the flux of plasma into the neutral point in front of the magnetosphere

$$v_{\max} = \frac{V_A}{\log\left(2\,\frac{\mathrm{u}\,\mu_0}{c_0^2\,\varepsilon_0\,\mu_0}\,\sigma\, l\,\frac{v_{\max}^2}{V_A}\right)}. \tag{17.1}$$

The various quantities are: V_A the Alfvén velocity corresponding to the total magnetic field slightly upstream of the neutral point, σ the conductivity, and l a characteristic length, which may roughly be the radius of the nearly spherical boundary surface. The flow velocity can be reduced by any resistance in the plasma flux on its way along the boundary and on return through the magnetosphere. According to a numerical estimate v_{max} is larger than 0.1 V_A, perhaps 0.2 V_A.

This velocity causes a fraction of 0.1 to 0.2 of the arriving field lines to merge with the geomagnetic field lines. The majority of the field lines, arriving near the front of the magnetopause is deflected sidewise, thus traveling around the magnetosphere.

The flow rate of magnetic field lines in normal flow through a boundary of the width d is $v B d$. A measure of the flow rate is the voltage along the boundary, which is[6]

$$U = \frac{|\boldsymbol{v} \times \boldsymbol{B}|\, d}{c_0 \sqrt{\varepsilon_0 \mu_0}} .$$

For the solar wind ($v = 320$ km/sec, $B = 5\,\gamma = 5 \cdot 10^{-9}$ V sec m^{-2}) and a width of the magnetosphere of $2 \cdot 10^5$ km the voltage becomes $U \approx 300$ kV. If one tenth of the field lines passes through the boundary the voltage across the magnetosphere is roughly 30 kV, a value agreeing in order of magnitude with other considerations (Sects. 24, 30 and 32).

18. Plasmasphere and plasmapause.

α) *General facts.* The electron and ion densities decrease on the way out from the F2-layer — sometimes quite smoothly, sometimes with a superimposed irregular structure. The charged-particle density is approximately 10^4 cm^{-3} ($= 10^{10}$ m^{-3}) near 1.5 R_E geocentric height, strictly speaking on the field line that intersects the equatorial plane at 1.5 R_E. It falls gradually to 10^2 or 10^3 cm^{-3} (10^8 or 10^9 m^{-3}) approximately. After that there is a rapid, in many observations even almost sudden drop to a value between 10^{-1} and 10^1 cm^{-3} (10^5 and 10^7 m^{-3}). This drop is during magnetically quiet periods found on a field line intersecting the equatorial plane at 6 or 7 R_E and is displaced to lower altitudes in disturbed periods. The fall of the particle concentration may be as rapid as by a factor 10 within less than 250 km.

The inner area with electron and ion concentrations well above 10 cm^{-3} (10^7 m^{-3}) is called the *plasmasphere.* The rapid decrease of the concentration appears in the literature as the "*knee*" in the distribution of thermal plasma or the "*plasmapause*". The outer area with low concentration is sometimes named the "trough".

It seems that the concentration is high or low, corresponding to the plasmasphere or trough, along *entire field lines.* It is therefore desirable to denote field lines by a symbol. The letter L, borrowed from the ideas on trapped particles (Sect. 20γ and [1]) is common for this. A field line $L = 6$, for example, is in a dipole field one that intersects the magnetic equatorial plane at 6 R_E (geocentric). In a nondipole field this is only approximately correct. The value L more strictly defines a field line shell on which a trapped fast particle remains in its drift around the Earth[1].

β) *Various methods of observation.* The first observations of the rapid decrease of the ion concentration far out were made by means of collectors on rockets (GRINGAUZ[2]).

---

[6] See the original paper by LEVY, PETSCHEK, and SISCOE[2] and the discussion of the voltage across the polar cap by HESS in this volume ([*15*], p. 193 and p. 216).

[1] Contribution to this volume by W. N. HESS [*15*] Sect. 5, p. 124.

[2] K. I. GRINGAUZ: Planet. Space Sci. **11** (3), 281—296 (1963); first published note presumably 1960.

More attention was paid to the conclusions drawn on the knee from whistler observations (CARPENTER[3,4]). These observations yielded quite an amount of material, in particular on spatial and temporal behavior of the plasmapause (knee). Also motion of whistler ducts, presumably due to plasma motion, was detected. Measurements of ion concentrations by means of mass spectrometers on OGO satellites, published since 1968[5,6], confirmed and supplemented the earlier whistler results.

The *whistlers*[7] are natural noise signals that run through some audiofrequency band, starting with a high audiofrequency and ending near 1 kHz. They originate from pulses (spherics) and have undergone a dispersion during propagation along magnetic field lines from one hemisphere to the other. The guidance along field lines results partly from the behavior of the ray in the anisotropic medium, partly from ducting by ionized columns on individual field lines. The existence of discrete ducts leads to a fine structure of a whistler: The whistler as seen in a spectrogram (frequency-vs.-time curve) consists of several traces. Each trace tells something about the electron concentration in the traversed area and the maximum height (minimum magnetic field) of the path. The minimum magnetic field shows up in the "nose frequency" of "nose whistlers". Nose whistlers with an ascending and a descending branch in the spectrogram are the type of whistler received at high latitudes.

Whistler investigations in the Antarctic sometimes gave nose whistlers with two different sets of traces[3,4]. One set had a lower nose frequency and shorter travel time, corresponding to propagation in the trough, while the other had a higher nose frequency and longer delay, indicating propagation in the plasmasphere. This type of observations in the Antarctic, where the L values were from 2.5 to 9, permitted detailed studies of the plasmapause.

The fine structure of a whistler is preserved over many successive occurrences of whistlers. It is thus possible to observe the same traces longer time and to recognize their gradual change due to displacement of the ducting ionization tube. The motion of the ionization tubes, on the other hand, is likely to represent the plasma motion[3,8].

γ) *Observational results.* Fig. 19 shows the position of the *plasmapause* in the equatorial plane according to whistler observations, most of which originated from two stations in the Antarctic. The solid curve is an average on days with moderate geomagnetic agitation (K_p from 2 to 4) in July and August 1963. The dotted curve was obtained during increasing magnetic agitation, the dashed curve corresponds to decreasing agitation. Typical is the asymmetric shape with a bulge on the evening side. Because whistlers respond to the electron concentration on a larger part of their path it may be inferred that the plasmapause is a shell of magnetic field lines or follows such a shell fairly closely.

Although the shape of the plasmapause is quite generally similar to that in Fig. 19, its extension is highly variable with magnetic activity. This becomes evident from the dotted and dashed curves of Fig. 19, which were obtained during varying activity. Fig. 20 represents the *proton concentration* measured by a mass spectrometer versus L on four days with different magnetic activity.

[3] D. L. CARPENTER: J. Geophys. Res. **71** (3), 693—709 (1966).
[4] R. A. HELLIWELL: Whistlers and VLF emissions, in [*13*], p. 106—146.
[5] H. A. TAYLOR, H. C. BRINTON, and M. W. PHARO: J. Geophys. Res. **73** (3), 961—968 (1968).
[6] C. R. CHAPPELL, K. K. HARRIS, and G. W. SHARP: The reaction of the plasmapause to varying magnetic activity, in [*16a*], p. 148—153.
[7] See contribution to Vol. 49/3 of this Encyclopedia by R. GENDRIN, p. 461—525.
[8] D. L. CARPENTER: J. Geophys. Res. **75** (19), 3837—3847 (1970).

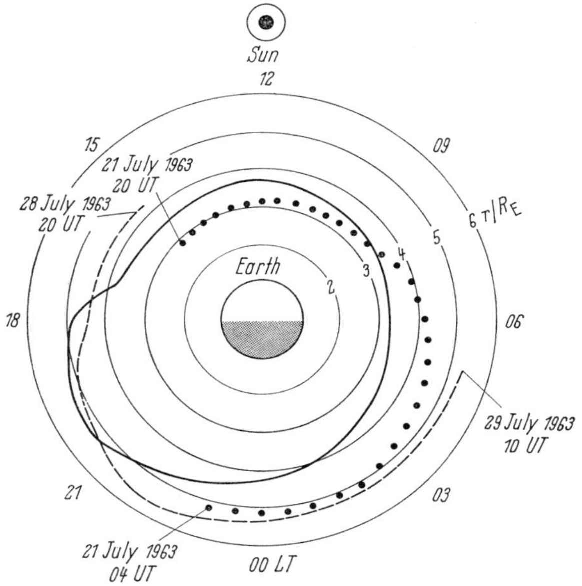

Fig. 19. The plasmapause[9]. The figure shows the position of the plasmapause in the equatorial plane as determined from whistler observations. The solid curve is an average during moderate geomagnetic activity ($K_p = 2 \ldots 4$); the dotted curve was obtained during increasing magnetic agitation, and the dashed curve was obtained during recovery from increased agitation.

The horizontal bars in the figure mark the L interval in which a concentration of 10 cm^{-3} (10^7 m^{-3}) was found for a particular value of K_p during a series of measurements. The position of the pasmapause is *not* determined by the *instantaneous* value of K_p, but by the *preceding activity*[3,6]. The best fit seems to be with the K_p of the preceding 6 hours. This past K_p was consequently chosen in Fig. 20.

The azimuthal position of the *evening bulge of the plasmapause* also depends on magnetic activity[8]. A frequently found position is 18 ... 21 h local time as shown in Fig. 19. The bulge itself and its steep slope at the western end seem to be permanent features. A study of the westward end of the bulge showed a dependence of its position on the magnetospheric substorm activity during the preceding 10 hours or so, which is measured by the auroral electrojet index (AE). During very quiet periods the bulge is rotated toward midnight or even beyond, while after heavy substorm activity it is seen in the local afternoon.

Ion temperatures were measured by a retarding potential analyzer on OGO 5 [10]. A temperature jump at the plasmapause to high values outside was frequently seen, amounting to factors from 5 to 10 or even more. The ion temperature outside (in the trough) is high, $\geqq 10^5$ °K.

The study of *motions*[3,8] of the whistler ducts indicate a corotation of the plasma with the Earth. That means that the plasma moves along the plasmapause most of the time. It turns out, however, that it does not follow the boundary of the bulge. Near the bulge it continues the circular path, not penetrating into the bulge. Radial plasma motions (cross-L drifts) have been observed in connection with magnetospheric substorms[8]. The question whether shrinkage and expansion of the plasmasphere mean an equivalent plasma motion, or erosion (or refilling) of an outer part of the plasmasphere, does not seem to have been definitely answered (cf. CARPENTER[8] and Sect. 27).

[9] D. L. CARPENTER: Fig. 6 of [3].

[10] G. P. SERBU and E. J. R. MAIER: J. Geophys. Res. **75** (31), 6102—6113 (1970).

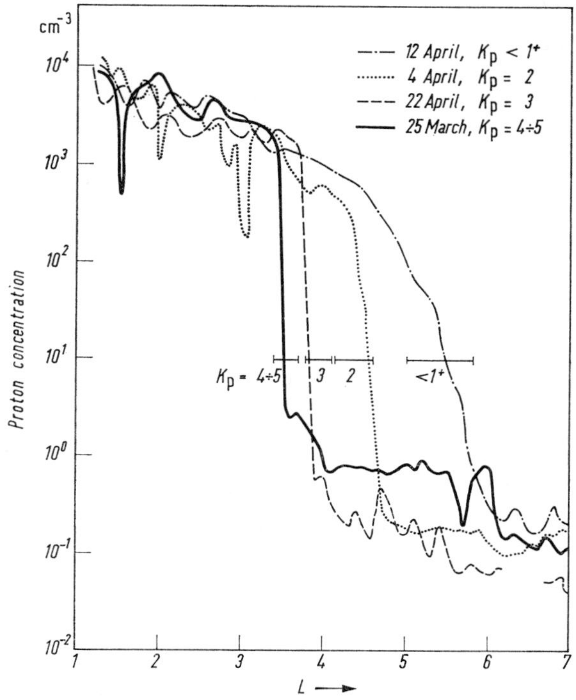

Fig. 20. Proton concentration as function of L, measured by a mass spectrometer in the time from midnight to 04.00 hrs[11]. Curves obtained on four flights in March and April 1968 are shown. The values of K_p are averages of the preceding 6 hours. The error bars indicate spreading in series of measurements.

δ) *Connection with other phenomena.* It is generally believed that the plasmasphere is the space in which the plasma corotates with the Earth, describing closed paths around the Earth. Outside the plasmapause the plasma is supposed to follow long paths coming from the magnetotail and leading back to it. Because of the distribution over a wide space the plasma in the outer area, the trough, has a very low density. In terms of moving field lines it may be said: The field lines in the plasmasphere retain the form of a loop while they rotate with the Earth; in the trough they get stretched out into the tail during their motion, appearing some time loop-shaped, some time "open". Corresponding models of motion will be discussed in Sects. 22β and 24. The plasma on open field lines may move out very rapidly in accordance with the theoretical ideas on a supersonic "polar wind" that is somewhat analogous to the solar wind (Sect. 24α).

Very sharply bounded troughs in the upper ionosphere were detected in satellite observations, in particular in topside soundings at night[12]. They are observable down to F2-peak height. Apparently troughs are situated on the lower ends of the field lines that determine the plasmapause. This may indicate that the ionospheric troughs are a consequence of the same process of plasma removal. During magnetic disturbances the ionospheric troughs move toward the equator, similarly to the ionospheric ends of the plasmapause, but correlated studies of the behavior of both do not seem to have been published.

[11] C. R. Chappell, K. K. Harris, and G. W. Sharp: l.c. [6], Fig. 4.

[12] H. Rishbeth and O. K. Garriot: Introduction to Ionospheric Physics, p. 188—189 and 251. New York and London: Academic Press 1969.

The *auroral oval* is also shifted toward the equator during disturbances (Sect. 32). Auroral activity is believed to be another effect resulting from the transition between closed and open field lines in the magnetosphere. This transition might thus be responsible not only for the plasmapause but also for several other phenomena; a relationship of all these phenomena with magnetic activity is apparent.

It may be that plasmapause, auroral oval, and geomagnetic activity depend on the *convective motion system,* which is caused by a *magnetospheric electric field* (Sect. 24). The electric field is supposed to result from the solar wind and to vary with solar-wind parameters. VASYLIUNAS evaluated some empirical relationships under this aspect (Sect. 30).

Magnetic micropulsations with periods between 10 sec and 2.5 min (types **pc 3** and **pc 4**) are according to some theories resonance oscillations in the magnetosphere. It has been found that their period depends on the state of the magnetosphere. Presumably the period is related with the plasmapause, more specifically with the height of the plasmapause, which varies with local time and with the degree of magnetic activity (see Sect. 33 and references given there).

E. Fast particles.

19. Slow and fast particles.

α) *Distinction between thermal plasma and fast particles.* The magnetosphere was characterized as the space permeated by the geomagnetic field and filled with the highly conducting atmospheric plasma. This plasma is in many respects to be considered a thermal plasma. In the ionospheric F2 layer the ion temperature is of the order 1000 °K; from roughly 750 km up the ion and electron temperatures are expected to be equal[1]. The plasma temperature increases on the way up to a value of the order 10^4 °K in the outer plasmasphere[2]. The value varies very much with time of day. Outside the plasmasphere a temperature $\geqq 10^5$ °K has been observed[2]. The temperature reached in the solar plasma of the magnetosheath, of the order of 10^6 °K, may place an upper limit on the temperatures possible in the magnetosphere.

A temperature of 10^4 °K corresponds to a *particle energy* of 1.29 eV. The *thermal particles* of the magnetosphere thus attain no energies higher than 100 or 200 eV. Geiger counters, carried up on rockets and satellites, record only much faster particles, >40 keV. The typical energies of *fast particles trapped in the radiation belt* are in the interval 40 keV to nearly 1000 MeV. Particles with energies of *a few* keV are observed in the *plasma sheet of the magnetotail* (Sect. 16 γ) and in varying concentration in the *radiation belt* (Sect. 21 α). Their concentration in the radiation belt is particularly high during magnetic storms (Sects. 21 α and 29), when they affect the geomagnetic field. In *auroras* one finds precipitation of electrons or protons of roughly 10 keV. Energies of *particles coming from the Sun,* for comparison, are around 600 eV (protons of the quiet solar wind with 340 km sec^{-1}), a few keV (solar protons causing auroral disturbances) or much higher, $\geqq 1$ MeV, occasionally more than 1 GeV, in the infrequent events of emissions from solar flares that lead to perturbations of the polar ionosphere ("polar black-out").

The *density* of thermal particles is in general much higher than that of trapped fast particles. Immediately below the plasmapause there are 100 cm^{-3} (10^8 m^{-3})

[1] W. B. HANSON: Structure of the ionosphere, in [2], p. 21—49.

[2] G. P. SERBU and E. J. R. MAIER: J. Geophys. Res. **75** (31), 6102—6113 (1970).

or more thermal particles (Sect. 18), whereas concentrations derived from observed fast-particle fluxes are $10^{-1} \ldots 10^{-4}$ cm^{-3} ($10^5 \ldots 10^2$ m^{-3}) in most part of the radiation belt. In processes involving the entire medium such as magnetodynamic (or "hydromagnetic") processes one considers the thermal plasma as the medium and tends to ignore the fast particles. Particles between 1 and 10 keV in the radiation belt, however, can reach respectable concentrations in geomagnetic storms; 1000 cm^{-3} (10^9 m^{-3}) protons are occasionally observed in this energy range ([*14*] p. 472; also see Sect. 21α).

The thermal plasma can be conceived of as a kind of gas with the peculiar attitudes following from magnetodynamic theory. The thermal particles undergo *collisions*, thus remaining confined to some plasma parcel within a finite period of time. Beyond the vertex of the F2 layer collisions between charged particles are the predominant collisions (see Sect. 7β). For the outer plasmasphere, where a charged-particle concentration of 200 cm^{-3} ($2 \cdot 10^8$ m^{-3}) and a temperature of $2 \cdot 10^4$ °K may be assumed, the mean time between proton-proton collisions is derived[3] as $7 \cdot 10^3$ sec or 2 h. That means that collisions are at this height still essential in processes of diurnal or semidiurnal character. The collision frequency due to Coulomb interaction[3] is nearly proportional to $T^{-\frac{3}{2}}$. The collision frequencies of charged particles of 0.34 eV (corresponding to 2600 °K) and 1 keV (corresponding to $7.7 \cdot 10^6$ °K) should at a given density be[3] in the ratio $10^5:1$ (approximately). Trapped fast particles, remaining above a few hundred km, consequently have nearly no collisions.

Thermal protons in the *outer plasmasphere*, even though subject to collisions, may with their thermal velocity of approximately 20 km sec^{-1} travel on *helical paths along field lines* between the two hemispheres and collide at a somewhat lower altitude. *Outside the plasmasphere*, in the trough, the thermal particles might like trapped particles travel back and forth along field lines many times without collisions, being reflected lower down. If electrons and ions cover different path lengths on a field line, an electric field parallel to the magnetic field arises. This might make magnetodynamic theory invalid [see the remarks in connection with Eq. (10.17)].

An essential difference between thermal and fast particles is in their drift motions. Only the thermal particles are thought of as being attached to an individual magnetic field line. The drift motions will be discussed now.

β) *Drift motion of slow and fast particles.* Slow particles move in an electric field with the drift velocity of Eq. (7.6),

$$\boldsymbol{v} = \frac{c_0 \sqrt{\varepsilon_0 \mu_0}}{B^2} \boldsymbol{E} \times \boldsymbol{B}.$$

In Sect. 7α it was already noted that inhomogeneity of the magnetic field causes additional drift motions, which are not obtained from the linearized macroscopic treatment of plasma motion. The expression for the drift velocity $\boldsymbol{w}$, including both types of drift motion, is (according to [*15*] Sect. 2, [*16*], [*28*] Eq. (20.6), [*31*], [*32*])

$$\frac{\boldsymbol{w}}{c_0 \sqrt{\varepsilon_0 \mu_0}} = \frac{\boldsymbol{E} \times \boldsymbol{B}}{B^2} + \frac{1}{2} m\, v_\perp^2 \frac{\boldsymbol{B} \times \dfrac{\partial}{\partial \boldsymbol{r}} B}{\pm q B^3} + m v_\parallel^2 \frac{\boldsymbol{B} \times \boldsymbol{R}}{\pm q B^2 R^2}. \tag{19.1}$$

The velocities $v_\perp$ and $v_\parallel$ are the particle velocities normal and parallel to the magnetic field; $v_\perp$ is the velocity of the gyromotion. The two signs of the charge refer to positive

[3] SPITZER's formula for the "self-collision time", i.e. the reciprocal collision frequency, Eq. (5-26) in [*32*], has been used.

and negative particles. The radius of curvature of the field lines, $\boldsymbol{R}$, is supposed to point to the center of curvature. The fields are considered temporally constant.

The last two terms in Eq. (19.1) represent the drift due to the inhomogeneity of the magnetic field. One of them is proportional to $\frac{\partial}{\partial \boldsymbol{r}} B$; it is derivable from Eq. (7.4a) with the assumption that the gyrofrequency ω_B varies on the path of the particle. The velocity $\boldsymbol{u}$ postulated by Eq. (7.4a) consists in this case of two parts, one expressing circling in the magnetic field and the other expressing the drift caused by $\frac{\partial}{\partial \boldsymbol{r}} \omega_B$ or $\frac{\partial}{\partial \boldsymbol{r}} B$, the gradient of the magnitude of $\boldsymbol{\omega}_B$ or $\boldsymbol{B}$.

The last term of Eq. (19.1) results from the centrifugal force that acts on a particle moving with velocity $v_\parallel$ on the curved field line. The drift velocity $\boldsymbol{w}_c$ given by this term is such that the force $\pm q(\boldsymbol{w}_c \times \boldsymbol{B})$ balances the centrifugal force. In linearized theory the centrifugal force does not appear because it is brought in by the nonlinear term $\varrho \boldsymbol{v} \cdot \frac{\partial}{\partial \boldsymbol{r}} \boldsymbol{v}$ of the equation of motion [Eq. (7.13)].

In reference to the inhomogeneity of the magnetic field the length

$$l = \frac{B^2}{\left| \boldsymbol{B} \times \frac{\partial}{\partial \boldsymbol{r}} B \right|} \tag{19.2}$$

may be defined as scale length. The second term of Eq. (19.1), referring to the gradient drift $\boldsymbol{w}_g$, has the magnitude

$$\frac{w_g}{c_0 \sqrt{\varepsilon_0 \mu_0}} = \frac{1}{2} \frac{m v_\perp^2}{q B l}. \tag{19.3}$$

For the curvature drift there is

$$\frac{w_c}{c_0 \sqrt{\varepsilon_0 \mu_0}} = \frac{m v_\parallel^2}{q B R}. \tag{19.4}$$

Eq. (19.3) may alternately be written

$$\frac{w_g}{c_0 \sqrt{\varepsilon_0 \mu_0}} = \frac{1}{2} v_\perp \frac{R_g}{l} \tag{19.5}$$

with the gyroradius R_g. In the dipole field of the Earth both drifts are in the same direction, toward west for positive particles and toward east for negative particles.

Of some interest is the ratio of the two drift velocities w_g and w_c to the electric-field drift velocity w_E which in an electric field normal to the magnetic field is given by

$$\frac{w_E}{c_0 \sqrt{\varepsilon_0 \mu_0}} = \frac{E}{B}. \tag{19.6}$$

The ratios are

$$\frac{w_g}{w_E} = \frac{1}{2} \frac{m v_\perp^2}{q E l} \tag{19.7}$$

and

$$\frac{w_c}{w_E} = \frac{m v_\parallel^2}{q E R}. \tag{19.8}$$

Which drift predominates depends apparently on the ratios of the particle energies measured in electron volts, $\frac{1}{2} m v_\perp^2 / q$ and $\frac{1}{2} m v_\parallel^2 / q$, to the voltage El or ER. The electric-field drift is predominant at low particle energies whereas the drifts w_g and w_c in the inhomogeneous magnetic fields predominate at high energies.

The ratios w_g/w_E and w_c/w_E may serve as criteria for the distinction between slow and fast particles.

The voltage across the magnetosphere is under regular conditions estimated at 30 kV (see Sects. 17γ and 24α). The lengths l and R are only a certain fraction of the dimension of the magnetosphere. (In the equatorial plane of the dipole field either length is $\frac{1}{3}$ of the geocentric height.) An estimate based on these figures yields roughly 5 keV as the transitional particle energy. For particles well below 5 keV the electric-field drift is decisive while the rest of Eq. (19.1) represents a minor perturbation. For fast particles, far beyond 5 keV, the electric-field drift is negligible compared with the drift in the inhomogeneous magnetic field [corresponding to the last two terms of Eq. (19.1)]. This separation at approximately 5 keV might be used to define "slow" and "fast" particles. Thermal particles are slow in this sense, the common trapped particles (> 40 keV) are fast.

The theorems of magnetodynamics rest on the postulate of disappearance of $\boldsymbol{E}+\boldsymbol{v}\times\boldsymbol{B}/c_0\sqrt{\varepsilon_0\mu_0}$. They are consequently applicable only to the thermal plasma, whose drift motion corresponds to this postulate. In particular the theorem of magnetic field lines moving with the plasma holds only for thermal plasma or slow particles. Fast particles appear to move with respect to the field lines with the drift velocity $\boldsymbol{w}_g+\boldsymbol{w}_c$ according to Eq. (19.1). Numerical examples of corresponding travel times around the Earth will be given in Sect. 20γ.

20. Trapped fast particles. Charged particles move in the magnetic field on helical paths as the result of the gyration around a field line and an additional motion parallel to the field line. Fast particles with their low collision frequencies describe long helical paths following the curved field line over quite a distance.

The fate of the particle depends on the minimum height of the path. At a low enough height the particle will collide, most likely with the consequence of no return to the original height. The other alternative is a reflection in the more intense magnetic field lower down. The thread of the helix tightens in progression into the stronger field. Defining two velocity components with reference to the magnetic field in the center of the particle orbit one derives two forces acting in the magnetic field of the orbit itself. The force

$$\pm q(\boldsymbol{v}_{\parallel}\times\boldsymbol{B})/c_0\sqrt{\varepsilon_0\mu_0}$$

leads on convergent field lines to an increase of $v_{\perp}$ while $v_{\parallel}$ decreases (cf. [*15*], Sect. 2). If $v_{\parallel}$ disappears ultimately, the particle is reflected and travels back along the same field line. Trapped particles are particles traveling back and forth between the two reflection points on a field line many times. The radiation belt is a wide region in the magnetosphere characterized by a remarkable flux of trapped particles, in particular "fast" particles with energies above 40 keV.

The following Sect. 20α is a short survey of observed facts on trapped fast particles. Source and loss processes are noted in Sect. 20β. Sect. 20γ deals with the motions of trapped particles and the adiabatic invariants of the motion. A *more extensive treatment* of the entire subject has been given by HESS *in this volume* [*15*] and by various authors [*20*], [*21a*], [*31a*][1].

α) *Trapped particles in the radiation belt.* Fig. 21 is the classical picture showing the radiation belt area as it appeared in the first measurements, which were made 1958 by means of shielded Geiger counters on Explorer IV and Pioneer III The first publications, by VAN ALLEN and FRANK, appeared 1959. The numbers in the figure are count rates in $\sec^{-1}$ on Pioneer III; the paths represent the trajectory of Pioneer III projected on a meridian plane. In preceding attempts to observe fast particels the counters became saturated (Explorer I and III in 1958) or only the bottom of the radiation belt was indicated (Sputnik II in 1957) [*14*].

[1] J. A. VAN ALLEN: Charged particles in the magnetosphere, in [*22*], p. 233—255.

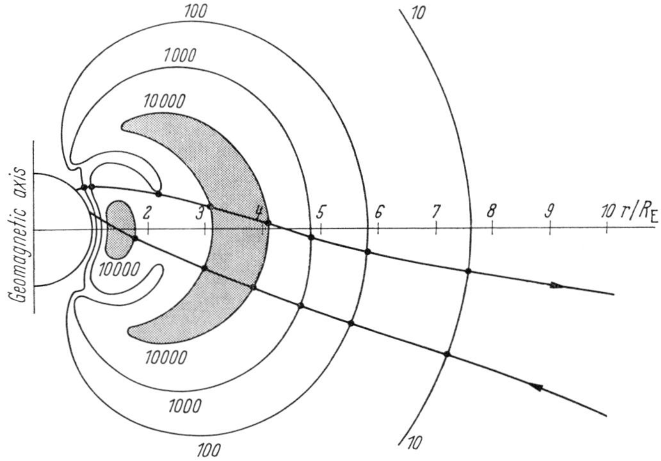

Fig. 21. The radiation belt. This is the classical picture obtained by VAN ALLEN and FRANK from Geiger counter measurements on Explorer IV and Pioneer III[2]. The numbers are counting rates. The outward and inward going lines are the projection of the Pioneer-III trajectory on a magnetic meridian plane.

On the basis of the first measurements two radiation belts were distinguished. This separation, however, was caused by the heavy shielding of the Geiger counters ($1\ \mathrm{g\ cm^{-2}}$), which let only protons >30 MeV and electrons >1.6 MeV penetrate. The first maximum of the count rate, at approximately 1.5 R_E (geocentric, equatorial) or 3000 km above ground, is according to later observations due to fast protons, while the outer maximum, between 3 and 4 R_E, has been attributed to electrons >1.6 MeV. With inclusion of slower particles in the measurements the two regions merge into one, now frequently called "the radiation belt", but characteristic differences between the two regions remain.

The *inner region* (1 to 2 R_E in the equatorial plane) contains quite *energetic protons*. An energy spectrum extending from 20 MeV upward and fading out at 800 MeV was obtained with a stack of nuclear emulsions; it is shown as Fig. 4 in [*15*]. The omnidirectional flux of protons >50 MeV has its maximum near 1.5 R_E (equatorial) with a few times $10^3\ \mathrm{cm^{-2}\,sec^{-1}}$ ($10^7\ \mathrm{m^{-2}\,sec^{-1}}$). Protons >400 keV do not show a maximum in the inner region. Up to 2 R_E their omnidirectional flux increases to approximately $10^7\ \mathrm{cm^{-2}\,sec^{-1}}$ ($10^{11}\ \mathrm{m^{-2}\,sec^{-1}}$). Figures similar to Fig. 21, but showing lines of definite proton and electron fluxes in a meridian plane were given in a paper by SCHARDT and OPP[3] (Figs. 13, 17, and 18); a corresponding representation for protons between 40 and 110 MeV is Fig. 7 in [*15*]. A second peak of the proton flux appearing in some of these figures at 2.2 R_E was not observed by some later satellites.

The natural *electron flux* in the inner region is not so well known. In the early experiments no attempt was made to discriminate between electrons and protons. In the inner region thus mainly the protons were measured. The Starfish explosion at 400 km altitude in July 1962 filled the inner region with a huge amount of electrons, which decayed only slowly, about a factor 2 per year in the interior of the inner region (cf. Sections 14 and 16 of [*15*] and p. 177 of [*14*]). The flux of natural electrons in the inner region is estimated at $3 \cdot 10^7\ \mathrm{cm^{-2}\,sec^{-1}}$

[2] J. A. VAN ALLEN and L. A. FRANK: Nature **183** (No. 4659), 430 (1959); Fig. 5 and Nature **184** (No. 4682), 219 (1959); Fig. 2.

[3] A. W. SCHARDT and A. G. OPP: Rev. Geophysics **7** (4), 799—849 (1969).

($3 \cdot 10^{11}$ m^{-2} sec^{-1}) for energies >40 keV and $2 \cdot 10^6$ cm^{-2} sec^{-1} ($2 \cdot 10^{10}$ m^{-2} sec^{-1}) for energies >580 keV [*14*], [*15*]. The electron energies in the inner region are essentially below 1 MeV.

Typical of the inner region is the *constancy of the proton flux*. The time constant is of the order of one year. The decay of 55 MeV protons could be observed after the Starfish explosion (Fig. 6 in [*15*]). It should be noted, however, that the nuclear explosion did not produce new protons, but apparently displaced some of the protons present. The proton flux thus is very smooth. The time constant of the inner-region electrons is also long: months or a year.

In the *outer region* there are no measurable protons with energies of several hundred MeV. The energy spectra of protons and electrons there are commonly represented by

$$\exp(-\mathscr{E}/\mathscr{E}_0)\; d\mathscr{E}$$

with an energy constant $\mathscr{E}_0$ that decreases with height. The energy $\mathscr{E}_0$ varies for protons from roughly 1 MeV at 2 R_E to 50 keV at 6 R_E (see Fig. 24 in [*15*] and Fig. 6.5 in [*14*]). For electrons the range of $\mathscr{E}_0$ is 250 to 40 keV between 2 and 6 R_E ([*15*] Sect. 25). Reported values of the omnidirectional fluxes near the peak of the outer region are: roughly 10^8 cm^{-2} sec^{-1} (10^{12} m^{-2} sec^{-1}) for protons >400 keV (see Fig. 17 in [3]) and 10^7 to nearly 10^9 (frequently $10^7 \cdots 10^8$) cm^{-2} sec^{-1} ($10^{11} \cdots 10^{12}$ m^{-2} sec^{-1}) for electrons >40 keV [*14*], [*15*][1,4].

Although the life times of both the protons and the electrons in the outer region appear to be in the order of days, the *proton flux* is *fairly stable*; its fluctuations are small-scale. The *electron flux*, however, is *quite variable*; it may change by more than an order of magnitude even within hours. Fig. 22, due to VAN ALLEN, demonstrates the variation of the electron flux throughout the outer region (also see Fig. 37 in [*15*]).

Geomagnetic storms and geomagnetic activity influence the electron flux in the outer region very much. During a storm the high-energy electron flux decreases first and later rises to values much higher than before the storm. The flux of energetic protons can be affected by intense storms (Fig. 28 in [*15*]; [*20*]). There is a positive correlation between K_p and the electron flux in the outer region beyond $L=4$ (Fig. 39 in [*15*] and the present Fig. 22).

β) *Sources and loss of particles*. One of the serious questions is: How are the fast particles fed into their orbits? If collisions bring them in, collisions at the other end of the helical path would similarly throw them out again. Particles can, however, be trapped quite readily when they arrive as neutral particles and become charged on their flight at the right instant, i.e., while location and direction of flight correspond to a trapped-particle path. It thus seems reasonable to assume that the energetic protons of the inner region originate from the *decay of albedo neutrons*, which are produced in the atmosphere at lower altitudes by galactic cosmic rays ([*14*] p. 71—74 and 128—129; [*15*] Sects. 6, 7γ, and 7δ). Neutrons are known to decay into a proton, an electron and an antineutrino. The neutrons would be free to move to all places, while some of the protons that result from their decay would be trapped. The energy spectrum of the protons is explainable as the energy spectrum of the neutrons, but the efficiency of the neutron source seems to be too low. However, an additional radial diffusion of protons (p. 76), caused by field fluctuations, lets the neutron decay process appear effective enough for the inner region[4a].

[4] S. N. VERNOV, E. V. GORCHAKOV, S. N. KUZNETSOV, Y. I. LOGACHEV, E. N. SOSNOVETS, and V. G. STOLPOVSKI: Particle fluxes in the outer geomagnetic field, in [*22*], p. 257—280.

[4a] M. WALT: Space Sci. Rev. **12** (4), 446—485 (1971).

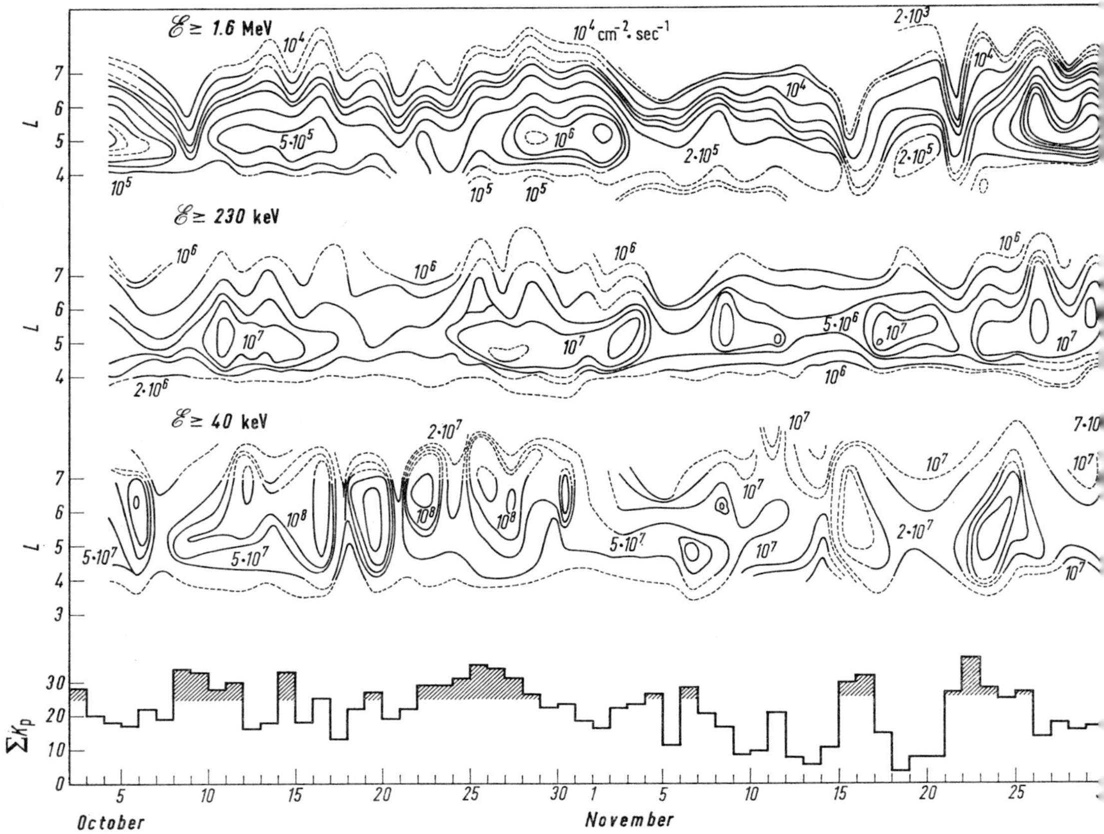

Fig. 22. Contours of constant omnidirectional intensities of electrons with energies $\mathscr{E}$ above some limit during October and November 1962 [5]. Geomagnetic activity is indicated by ΣK_p with values >25 marked by hatching. A precise definition of L is found in the contribution by W. N. HESS to this volume ([15], p. 125). See also Subsect. γ, p. 69.

For all the less energetic particles (the protons of the outer region and the electrons) it is likely that they *enter the magnetosphere from outside* in connection with some kind of perturbation. They are in this case solar-wind particles, but they must have undergone some acceleration (see Sect. 22). Solar-wind particles may possibly enter through the boundary (magnetopause) of an open magnetosphere or they may come in via the tail, eventually far behind the earth (Sect. 27; remarks and references in [1] and [19]). The tail, in addition, offers possibilities of acceleration of particles in approach toward the Earth (e.g., Fermi-acceleration, Sect. 22γ). It should also be recalled that irregularities or instabilities may prevent a perfect separation between the solar wind and the magnetosphere. Neutral points of the magnetic field also present a chance for particles to enter the magnetosphere.

Loss of trapped particles is due to a variety of processes. *Collisions* near the lower ends of the helical path may prevent a return to the opposite hemisphere because of a change of the direction of flight or because of a considerable reduction of the velocity (in a number of collisions). *Perturbations*, involving a transient

[5] H. D. OWENS and L. A. FRANK: J. Geophys. Res. **73** (1), 199—208 (1968), Fig. 3; Fig. 8 of VAN ALLEN's paper[1].

electromagnetic field, can bring the reflection points of the helix down into the denser atmosphere where the particle finally gets lost. The immediate effect of a perturbation may be a change of the pitch angle ("pitch angle scattering") or a diffusion of particles to other (lower) field lines, processes possible on a particle path at any height. A brief list of loss mechanisms will follow later in this section.

Measurements at a certain altitude (e.g., 1000 km) and at appropriate latitudes show that there is a continuous precipitation of electrons from the outer region. Observation of pitch angles sufficiently different from 90° indicates that the particles will precipitate (see Fig. 33 in [*15*]). The *continuous presence* of the trapped particles during unperturbed periods proves that there is a *balance* between supply and loss. *Rapid decay and replenishment* of the particle fluxes *in disturbances*, in particular in geomagnetic storms, were already mentioned (in Sect. 20α).

The *loss mechanisms mainly considered* for the various particles in the two regions of the radiation belt will now be listed (see [*14*], [*15*], [*20*][3,6,7]). *Inner-region protons* >100 keV are gradually slowed down in a large number of collisions. Once below 100 keV they are lost by charge exchange in collisions with slow hydrogen atoms. Beyond 200 MeV inelastic nuclear collisions lead to loss of protons. *Electrons in the inner region* are deflected in collisions with charged particles ("Coulomb scattering").

In the *outer region* diffusion across field lines is expected to bring particles down, with simultaneous acceleration (see Sect. 22γ) and with the possible effect of precipitation. Diffusion of *protons* may be caused by fluctuations of the magnetic field in numerous small disturbances coming in with the solar wind (see Sect. 22γ).

For the *electrons in the outer region*, which without doubt are also subject to diffusion, the more effective loss mechanism is perhaps pitch angle scattering (also called "pitch angle diffusion") with consequent submerging in the denser atmosphere. Pitch angle scattering in the outer region might preferably result from a resonance between an oscillatory electromagnetic field and one of the periods of the particle motion — i.e. synchrotron resonance or resonance with the bounce period of the helical path. The necessary electromagnetic field may be some noise (e.g. whistler waves) coming from other regions or may be generated by the particles themselves ([*15*], Sect. 29).

All these processes — spatial diffusion and pitch angle scattering — involve a violation of at least one of the adiabatic invariants, which are constants of the motion of trapped particles in fields varying slowly in space and time (see the following Sect. 20γ). The absence of energetic protons in the outer region may be due partly to some of the noted loss mechanisms, partly to the too large gyroradius that would be required.

γ) *Three nearly periodic motions.* The trapped particles perform three nearly periodic motions in superposition. Only two of them were mentioned so far: *the gyration* and *the motion back and forth on a helical path* that extends along a field line. This path, however, is also in motion; it is not stationary. It was seen (in Sect. 19β) that the inhomogeneity of the magnetic field causes a *drift motion* of the particles. The entire helical path consequently moves to the west (for positive particles) or east (for negative particles). The drift velocity, represented by Eq. (19.1) with neglect of the $\boldsymbol{E}\times\boldsymbol{B}$ contribution, consists of the two

[6] B. A. Tverskoy: Main mechanisms in the formation of the Earth's radiation belts, in [*22*], p. 219—231.

[7] C. S. Roberts: Pitch-angle diffusion of electrons in the magnetosphere, in [*22*], p. 305—337.

parts due to the gradient $\frac{\partial}{\partial \boldsymbol{r}} B$ and the curvature $\boldsymbol{R}$ of the field lines. The path traveling with this velocity circles around the Earth, thus covering a *field line shell.* The path returns exactly to the initial position when the field stays constant in time.

The three motions may at a first glance appear periodic, but they are not perfectly so. The gyration is perturbed by the gradual displacement of the gyration circle (or its center, the "guiding center") to a different magnetic field, the bounce motion is perturbed by the east-west drift, and all three motions are perturbed by temporal changes of the entire magnetic field. The periods of these three motions are of different orders of magnitude [*14*][3]. The gyration has periods from microseconds to milliseconds for electrons and from milliseconds to approximately one second for protons. The bounce periods for 100 keV particles crossing the equatorial plane at 4 R_E with 30° pitch angle, for example, are 0.6 sec for electrons and 24 sec for protons[3]. The same particles need approximately 135 min for a drift around the Earth. More energetic particles drift much faster around the Earth (e.g., 10 MeV particles in 1 to 5 min approximately).

In the case of a slightly perturbed periodic motion it is quite generally possible to define a quantity that remains constant during the proceeding perturbation. Such a quantity is called an "*adiabatic invariant*". An adiabatic invariant can be given for each of the three nearly periodic motions of trapped fast particles. Each invariant is constant only under the condition that the variation of the magnetic field in space and time is sufficiently slow — insignificant for the particular period of motion. For more extensive treatments of the adiabatic invariants in application to magnetospheric particles see Roederer [*21a*][8]; also see [*31a*].

The *first adiabatic invariant,* which is supposed to be a constant for the gyration, is at nonrelativistic velocities the magnetic moment[9] M. At relativistic velocities this invariant becomes

$$\left(1-\frac{v_\perp^2}{c_0^2}\right)^{-\frac{1}{2}} \mathsf{M} = \frac{p_\perp^2}{2 m_0 B} \tag{20.1}$$

with $p_\perp$ denoting the component of the momentum normal to the magnetic field and m_0 being the rest mass. For the *application to the helical paths of particles* see next page and Hess ([*15*] Sect. 2).

The *second adiabatic invariant,* referring to the motion back and forth along a field line, is (in relativistic form)

$$\mathcal{J} = \oint p_\parallel \, \mathrm{d}s, \tag{20.2}$$

[8] J. G. Roederer: J. Geophys. Res. **72** (3), 981—992 (1967).

[9] There is no general agreement about the definition of a *magnetic moment* (see Introductory Remarks, p. 1 of this volume). Here and in the later Sects. 22β and 28 the magnetic moment M is defined so that it is *in V,A,m,kg,sec-units* the product of a ring-current intensity and the enclosed area (or corresponds to such a product), however writing equations in a form usable with various systems of units. The present definition makes the magnetic moment *in each system of units* an energy (e.g., the energy of a gyrating nonrelativistic particle) divided by the magnetic flux density B, in accordance with a recommendation by IUPAP (International Union of Pure and Applied Physics). At *relativistic energies* the *first adiabatic invariant* is $\left(1-\frac{v_\perp^2}{c_0^2}\right)^{-\frac{1}{2}} \mathsf{M}$ if the present definition of M is retained (see, for example, the relativistic formulations in [*29*], p. 20 and 29). Some authors, however, define the magnetic moment of a relativistic particle as the present M divided by $\sqrt{1-\frac{v_\perp^2}{c_0^2}}$, thus identifying it with the first adiabatic invariant.

when $p_{\|}$ is the component of the momentum parallel to the magnetic field and the integration is extended along the field line over a complete bounce oscillation. *The third adiabatic invariant* is connected with the drift of the particle path on a field line shell. The invariant is the magnetic flux entering the field line shell from one side and leaving it on the other side.

The second adiabatic invariant is nontrivial when the field is not axially symmetric. In this case the helical path does not stay at constant altitude; the drift motion has a varying radial component. The first two invariants may be used to determine the successive helical paths, which in combination represent the field line shell on which the particle stays. The third invariant is applicable when the entire field changes slowly so that the particle path drifts around the Earth several times until the field variation becomes remarkable. It should be emphasized that the *"invariants" in the above formulation* are invariant in fields *varying slowly in space and time.*

The adiabatic invariants — provided, they are really invariant — together with some initial values are sufficient to determine the particle motion in a magnetic field that is known through space and time. The phases of the nearly periodic motions, being of little interest, can be ignored. From the invariants one obtains the reflection points on a field line, the family of field lines making up for the field line shell on which the helical path drifts, and the eventual change of the field line shell. The invariants don't leave an ambiguity. In particular, the paths (i.e., their characteristic parameters) become the same when the same field configuration is re-established. A similar statement is not possible for slow particles, which according to Sect. 19β experience an $\boldsymbol{E}\times\boldsymbol{B}$ drift only. They are not subject to the three adiabatic invariants and they may get displaced in return to the same field situation or even in an invariable field.

In a *time-independent magnetic field* without electric field the total momentum p of a particle is constant. Constancy of the first adiabatic invariant [Eq. (20.1)] can in this case be expressed by

$$\frac{p_{\perp}^2}{B} = \frac{p^2}{B_{\mathrm{m}}} \tag{20.3}$$

with B_{m} denoting the magnetic field intensity in the reflection or "mirror" point (a constant for the particle) [*14*], [*15*], [*20*] [*21a*]. The second invariant in the present case is after division by $2p$ defined as

$$\mathcal{I} = \frac{\mathcal{J}}{2p} = \int \left(1 - \frac{B(s)}{B_{\mathrm{m}}}\right)^{\frac{1}{2}} \mathrm{d}s, \tag{20.4}$$

the integral extended once over the path from one reflection point to the other.

The quantities $\mathcal{I}$ and B_{m}, being constants with definite values for a trapped particle, determine all the field lines to which the particle seems attached while it drifts around the Earth and the reflection points on the field lines. The two quantities thus can be assigned to the surface or "field line shell" covered by the drifting particle. It is, however, more common to use the quantities B_{m} and L instead (omitting the subscript m in general)[10]. The L value is a function of B_{m} and $\mathcal{I}$ for a given geomagnetic-field configuration. It is defined so that in an exact dipole field it represents the equatorial distance of the field line from the Earth's center in Earth radii. In the realistic geomagnetic field L is approximately this distance and, like the quantity $\mathcal{I}$, it denotes a field line shell.

Particles at some instant on the same field line, but reflected at a different B_{m}, will move on different L shells. This difference, though, is nearly negligible

[10] See contribution by W. N. HESS, [*15*], Sect. 5. — L was introduced by MCILWAIN.

for L values up to 4 or 5. Up to this limit consequently an L value can be assigned to some field line shell without specification of the trapped particles. Beyond $L=5$ the discrepancy in field line shells with identical L values and different B_m becomes larger; it is referred to as "L-shell splitting" [*14*], [*15*][8].

A quantity sometimes used[10] in place of L is the invariant latitude Λ. It is defined so that in a dipole field it is the geomagnetic latitude of the footpoints of a field line:

$$L \cos^2 \Lambda = 1. \tag{20.5}$$

21. Supplement on trapping and quasitrapping.

α) *Trapped low-energy particles.* The investigation of geomagnetic storms at all heights led to the conclusion that a ring current at geocentric heights roughly from 3 to 5 R_E (measured in the equatorial plane) is the cause of the main phase of geomagnetic storms ([*28*] Chap. C, also see Sects. 28 and 29). Satellite observations showed the depression of the field up to these heights. It is a reasonable idea to ascribe the ring current to trapped particles at these heights, but the total energy of all fast particles (>40 keV) seems too low.

FRANK (in the years 1965 to 1967) made measurements of protons and electrons with energies from 200 eV to 50 keV through the magnetosphere, using electrostatic analyzers with some collectors [*14*][1,2]. In particular the flux of protons in this energy range was found remarkable. The peak at quiet times was near 6 R_E and amounted to roughly 10^7 cm^{-2} sr^{-1} sec^{-1} (measured unidirectionally) (10^{11} m^{-2} sr^{-1} sec^{-1}). During the main phase of geomagnetic storms the height of the peak was lowered to approximately 4 R_E and both the proton flux and the electron flux turned out to be quite high: The proton flux, e.g., rose to $7 \cdot 10^7$ cm^{-2} sr^{-1} sec^{-1} ($7 \cdot 10^{11}$ m^{-2} sr^{-1} sec^{-1}) in a moderate magnetic storm with 50 γ ($=5 \cdot 10^{-8}$ V sec m^{-2}) field depression. The energy of all these fairly slow trapped protons appeared sufficient for the generation of the main phase.

An explanation of the enormous increase of the particle fluxes in the keV range in connection with magnetic storms might give a clue to the origin of the fast trapped particles, but so far not much is known about the history of these slow particles either. The descent of the maximum of the low-energy particle flux during storms is possibly connected with the descent of the plasmapause[1]. The prevalent presence in the outer area and the descent in disturbances may indicate that the particles come in from far out.

Magnetosheath plasma seems to enter the "polar cusps", which were recently discovered (Sects. 16γ and 35). From there it might convect into the tail (for references see Sect. 35) and further on, during disturbances, into the radiation belt.

β) *Regions adjacent to the radiation belt.* The radiation belt was defined as the region in which particles are trapped on magnetic field lines, describing helical paths and drifting at the same time around the Earth. The radiation belt in this definition does not extend to the magnetopause. Not all particles on field lines that intersect the equatorial plane beyond approximately 8 R_E can drift around the Earth. Depending on the reflection points of their paths they may drift around the Earth or they may reach the magnetopause in the drift motion [*14*], [*15*][3]. The existence of various shells on which the paths drift is the phenomenon called L-shell splitting in Sect. 20γ.

Electrons between 45 *and* 250 keV are observed in the region following the radiation belt near 8 R_E[4,5], which is sometimes called the distant radiation zone. These electrons are mirroring on helical paths, but not all of them drift around

---

[1] J. A. VAN ALLEN: Charged particles in the magnetosphere, in [*22*], p. 233—255.
[2] K. I. GRINGAUZ: Low-energy plasma in the Earth's magnetosphere, in [*22*], p. 339—378.
[3] J. G. ROEDERER: J. Geophys. Res. **72** (3), 981—992 (1967).
[4] K. A. ANDERSON: J. Geophys. Res. **70** (19), 4741—4763 (1965).
[5] A. W. SCHARDT and A. G. OPP: Rev. Geophysics **7** (4), 799—849 (1969).

the Earth. It is not sure whether in the outermost region any of them complete a drift path around the Earth. The electrons temporarily trapped on helical paths, but lost during their drift motion are called quasitrapped or pseudo-trapped[3,4,5]. The flux of electrons in the distant radiation zone is lower than in the radiation belt, decreasing outward, and it is much more *irregular*, largely because electrons are lost before they circle around the Earth (the drift motion may carry them out of the magnetosphere)[4,5].

The distant radiation zone, the region of quasitrapping, is subdivided into the *skirt* and the *cusp* [*20*][4,5]. Both are not too well defined. The *skirt* contains an irregular electron flux >45 keV as described in the last paragraph. It is contiguous to the radiation belt and has been observed at latitudes up to $\pm 35°$ and at all azimuths except the interval $\pm 60°$ from midnight. On the night side the *cusp* is adjacent to the radiation belt.

The *cusp*, extending roughly from 8 to 16 R_E in the equatorial plane and being confined to the area near the ecliptic plane, also contains an irregular electron flux >45 keV. Differently from the skirt the cusp plays a transitional role between the radiation belt and the plasma sheet of the tail (see the schematic, not necessarily quite correct representation of the cusp in Fig. 11). The field lines there still have the form of loops, while outside they are stretched out into the tail. In the irregular particle flux one recognizes more intense electron islands with a rapid rise and slower decay, just as in the plasma sheet of the tail (see Sect. 16γ). Doubtless there is some particle exchange between cusp and tail. Both cusp and skirt may stay in exchange with the radiation belt, in particular, during perturbations. The magnetopause, which confines the skirt region on the outside, does not seem to represent a sharp cut-off for the electron flux[4].

22. Acceleration of particles. The two most severe questions that arise in connection with the trapped particles are: 1) How do particles get into the trapped-particle orbits? 2) How are the particles accelerated to energies from a few keV to many MeV? As for the source of the particles various possibilities are seen (see Sect. 20β), all of which, however, mean a departure from the extremely simplified magnetodynamic concept. Acceleration of particles seems to be harder to understand; it has presumably to do with particular processes beyond the scope of classical magnetodynamic theory.

Acceleration of thermal or solar wind particles is supposed to yield low-energy trapped particles and auroral particles, both in the keV range. The corresponding acceleration mechanism seems to be triggered or facilitated by perturbations, which presumably arrive with the solar wind. Acceleration to a few keV might be achieved in a single process. Acceleration from some keV to many MeV, however, must be the result of many acceleration steps or of a continuous violation of some rule, e.g. of the postulate of constancy of the third adiabatic invariant.

α) *Idealizations without energy change.* The different types of drift motions were in Sect. 19β used as a criterion for the distinction of slow and fast particles. Slow particles drift with the velocity $\boldsymbol{w_E} = c_0 \sqrt{\varepsilon_0 \mu_0}\, \boldsymbol{E} \times \boldsymbol{B}/B^2$, nearly unaffected by inhomogeneity of the magnetic field. For fast particles this electric-field drift is insignificant. Their drift motion is essentially due to the inhomogeneity of the magnetic field $\left(\text{i.e., due to } \frac{\partial}{\partial \boldsymbol{r}} B \text{ and the curvature of the field lines}\right)$. The motion of fast particles is under appropriate conditions controlled by the three adiabatic invariants.

A quick view at the prerequisits of the two classes of particles lets one expect no energy gain in many processes that might be considered. At *slow* particles the force $\pm q\boldsymbol{E}$ together with the velocity $\boldsymbol{w}_{\boldsymbol{E}}$ normal to $\boldsymbol{E}$ does not lead to a transfer of power. For *fast* particles the preservation of the three adiabatic invariants seems to prevent a permanent energy gain. In a stationary situation the fast particles ought to return to the same bounce path periodically. In the distorted dipole field of the Earth a particle should after a complete drift around the Earth bounce back and forth on the same field line and between the same reflection points as at the start. In a nonstationary situation the original state of motion should be re-established in return to the initial situation. This means that after a geomagnetic storm nothing should have changed, contrary to the observation of a replenishment of the radiation belt. The present aim is to show why these statements can not be held up and energy gain is possible under realistic conditions.

β) *Energy gain of slow particles.* The statement that slow particles drifting in superimposed electric and magnetic fields do not gain energy is not generally valid as an example may demonstrate. The example, elaborated by ALFVÉN [*29*], deals with *electrons streaming into the geomagnetic field.*

It is assumed that a homogeneous stream of electrons coming from a weak homogeneous magnetic field in interplanetary space enters the geomagnetic dipole field. The electric field due to the motion in the interplanetary magnetic field is supposed to continue through the entire space, thus being homogeneous. Only motions in the equatorial plane, the x,y-plane, are considered. The magnetic field is everywhere assumed in the z-direction. The initial *drift* velocity is in the $-y$- direction, the electric field is in the x-direction. A *gyration* in the x, y-plane is assumed, initially with the velocity v_0 and later, in the stronger magnetic field, with a velocity v.

Fig. 23 represents the paths of electrons obtained for a magnetic field

$$B = B_z = B_0 + a/r^3. \tag{22.1}$$

The spatial coordinate r is the distance from the Earth's center. The field of Eq. (22.1) is in the vicinity of the earth (at small r) the dipole field of the equatorial plane and becomes far out in a gradual transition the weak interplanetary field B_0. The parameter a expressed in terms of the Earth's magnetic moment M_{E} is[1]

$$a = \frac{\mathrm{u}\,\mu_0}{4\pi}\,\mathsf{M}_{\mathrm{E}}.$$

The deflection in the direction of $-\boldsymbol{E}$ means an energization. The electrons are seen to approach the dipole only up to a certain minimum distance $(-x_{\mathrm{m}})$. The interior of the dashed loop is a zone remaining forbidden to the electrons of the given initial gyration velocity.

A quick estimate of the minimum distance $|x_{\mathrm{m}}|$ is obtained from the statement that the magnetic moment of the particles, $\frac{1}{2}mv^2/B$, stays constant. Adiabatic invariance of the magnetic moment in a slowly varying magnetic field as noted in Sect. 20γ is now implied.

[1] In V,A,m,kg,sec-units there is $\mathrm{u} = 1$, while in absolute electromagnetic units (Gauss etc.) the entire factor is $\frac{\mathrm{u}\,\mu_0}{4\pi} = 1$. The equation of the geomagnetic dipole field is

$$B = \frac{\mathrm{u}\,\mu_0}{4\pi}\,\frac{\mathsf{M}_{\mathrm{E}}}{r^3}\,(1 + 3\sin^2\varphi)^{\frac{1}{2}}.$$

About the definition of the magnetic moment see footnote [9] of the above Sect. 20γ and Introductory Remarks, p. 1.

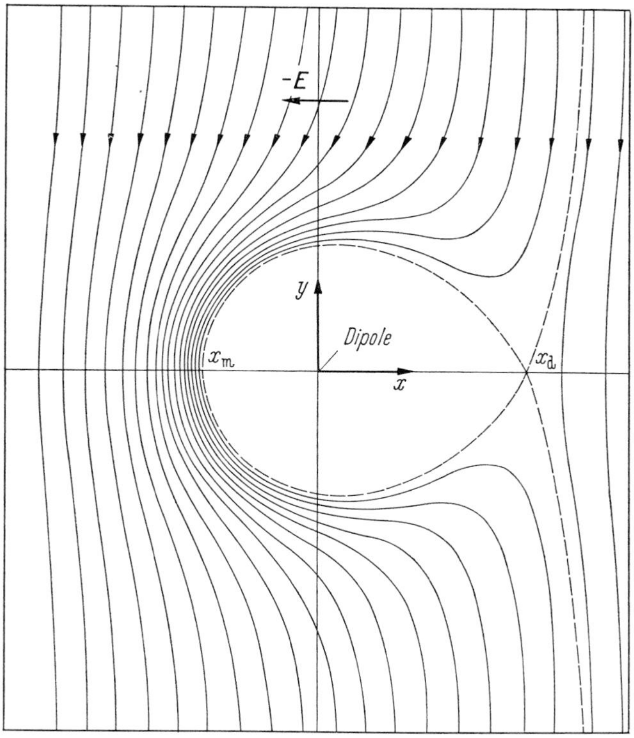

Fig. 23. Electrodynamic drift of electrons in a superposition of an unconfined magnetic field and a homogeneous electric field[2]. The magnetic field is normal to the plane of drawing and corresponds to the equatorial field of an open magnetosphere. The plane of drawing is considered to be the equatorial plane.

In the closest approach of a particle the external part of the magnetic field, B_0, can be neglected and the particle energy can be set equal to qEl for a particle deflected by the length l to the left (if the initial kinetic energy is neglected).

The particle energy is then approximately

$$qEl = \frac{1}{2} m v^2 \tag{22.2}$$

or

$$qEl = \frac{1}{2} m v_0^2 \frac{B}{B_0} \tag{22.3}$$

for an initial velocity v_0. At the closest approach one may set $l \approx |x| = r$. This is exact only for paths coming from $x=0$, though many paths reach nearly the same $|x|$ in the closest approach as Fig. 23 shows. Thus there is at the closest approach approximately

$$qEl = \frac{1}{2} m v_0^2 \frac{a}{B_0 l^3} \tag{22.4}$$

or

$$l^4 = \frac{\frac{1}{2} m v_0^2}{B_0} \frac{a}{qE}\,. \tag{22.5}$$

[2] Fig. 2.10 in [*29*].

The length l is the approximate minimum distance for an electron starting at $x=0$. The absolute minimum distance $|x_m|$ of Fig. 23 is somewhat smaller. According to Alfvén it is

$$-x_m = 0.74l, \tag{22.6}$$

whereas the largest extension of the forbidden zone is

$$x_d = 1.32l. \tag{22.7}$$

The energy acquired by the electron while traveling a distance l to the left is according to Eqs. (22.4) and (22.5)

$$qEl = \left(\frac{1}{2} m v_0^2\right)^{\frac{1}{4}} (qEr_0)^{\frac{3}{4}} \tag{22.8}$$

when r_0 denotes the distance at which the two parts of the magnetic field are equal. The idea now is that the potential energy available within the dipole magnetic field is of the order qEr_0. This is much greater than the initial particle energy $\frac{1}{2}mv_0^2$. The kinetic energy of electrons at a close approach (near x_m) is roughly qEl as given by Eq. (22.8).

With an initial kinetic energy of 100 eV and a potential energy $qEr_0=25$ kV the kinetic energy qEl becomes according to Eq. (22.8) approximately 6 keV.

After the previous statement that slow particles, drifting with the velocity $c_0\sqrt{\varepsilon_0\mu_0}\,\boldsymbol{E}\times\boldsymbol{B}/B^2$, should not gain energy the present result seems surprising. The fact is that a drift due to $\frac{\partial}{\partial \boldsymbol{r}} B$ and field-line curvature, though small, permits the force $-qE$ to supply power. With increasing kinetic energy the velocity of this drift increases, thus leading to an accordingly larger power supply. When the distance from the dipole is of the order l the kinetic energy is of the order qEl, corresponding to particles transitional between "slow" and "fast" in the sense of Sect. 19β. For these transitional particles the electric-field drift and the drift in the inhomogeneous magnetic field are equally important. The present example indicates that the classification of particles as slow and fast, depending on their predominant drift motion may not be kept up during some motion process. Particles may get into the transitional energy range and even end up in the other category. Another general conclusion is that particles can utilize a large part of the potential energy available in the inhomogeneous magnetic field.

Alfvén's motion model, as described, was originally thought of as relating to *particles (electrons) coming from the Sun* and entering the geomagnetic field. The perfect or imperfect confinement of the magnetosphere, however, does not allow unimpeded entrance of particles from the Sun. In recent years the model has found a *different application:* Particles are believed to be supplied to the magnetosphere through the *magnetotail*. The outer magnetic field in the model is interpreted as corresponding to the tail field. The electric field leading to the particle drift is a continuation of the electric field in the solar wind (cf. Sect. 16β). The present model, though not fitting closely, indicates that particles may reach certain field lines on which they are precipitated. This is a hypothetical explanation of the precipitation in auroras (Sect. 32). Trapped low-energy particles, which are considered the cause of the main phase of geomagnetic storms (Sects. 21α and 29) may have arrived in the same manner, but must have become trapped in the region of the ring current by some process. A brief discussion of particle supply from outer areas with acceleration is given by Obayashi and Nishida [*19*].

Another example of energization of slow particles — representing a thermal plasma — was discussed by HINES[3]. An inhomogeneous magnetic field, constant in time, is assumed. An electric field causes *drift motions normal to* $\boldsymbol{B}$. Plasma moving into a stronger magnetic field is being compressed in accordance with the theorem of moving magnetic field lines. The product of B and the cross section of a field line tube filled with plasma has to stay constant or

$$B\mathcal{V} = \text{const}. \tag{22.9}$$

when $\mathcal{V}$ is the specific volume of the plasma. The thermodynamic relationship for an adiabatic change of state postulates

$$T\mathcal{V}^{\gamma-1} = \text{const}. \tag{22.10}$$

Without collisions only two degrees of freedom, corresponding to the two components of motion normal to the magnetic field, have to be considered. The ratio of specific heats for two degrees of freedom is $\gamma = 2$. Eq. (22.10) with $\gamma = 2$ would also have been obtained from the postulate of constancy of the magnetic moment.

The last remark may remove any doubt about the validity of Eq. (22.10), though the energization seems to be the result of the *electric-field drift* which is normal to $\boldsymbol{E}$. HINES[3] shows, however, that the energy increase is correctly derived from the *force* $-q\boldsymbol{E}$ in conjunction with the *drift due to the gradient* $\frac{\partial}{\partial \boldsymbol{r}} B$, which plays the role of a minor correction of the total drift. For further discussions of energization of a thermal plasma HINES's paper is referred to.

From both examples of this section it is concluded that thermal particles can be energized by an electric field. It appears even likely that in a simple arrangement of stationary electric and magnetic fields some particles travel through a major part of the electric field, thus accumulating an essential fraction of the energy available in the potential field.

γ) *Acceleration of fast particles.* An acceleration mechanism that permits acceleration of plasma particles in interstellar space was proposed by FERMI in the attempt to explain cosmic rays [*14*], [*29*], [*32*]. Fig. 24 shows a relatively simple situation for the FERMI *process*. The magnetic field is assumed to be homogeneous over quite a distance, becoming more intense in some plasma

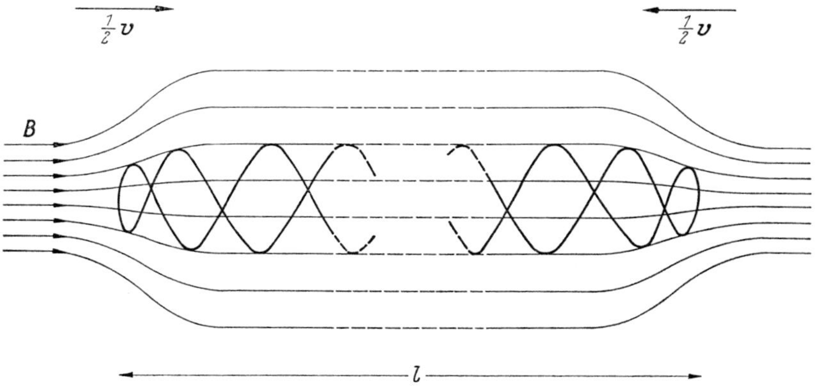

Fig. 24. Fermi acceleration. Two plasma clouds carrying the magnetic field lines with them approach each other with the relative velocity v.

[3] C. O. HINES: Planet. Space Sci. **10**, 239—246 (1963).

clouds far out. Charged particles are trapped between the clouds, where they are reflected by the stronger field. Now the two clouds with the embedded stronger field approach each other with a relative velocity $\boldsymbol{v}$. Because the particles are still reflected by the increasing magnetic fields, their path along the field lines shortens. At the same time their velocity along the field lines, $v_{\|}$, increases. The reflection at a mirror moving with $\frac{1}{2}\boldsymbol{v}$ yields a velocity increase v. The number of reflections per second is $v_{\|}/l$. The velocity increase per second is consequently

$$\frac{\mathrm{d}v_{\|}}{\mathrm{d}t} = \frac{v_{\|}}{l}\,v = -\frac{v_{\|}}{l}\,\frac{\mathrm{d}l}{\mathrm{d}t} \tag{22.11}$$

(see [*32*]).

Eq. (22.11) would also have been obtained from the second adiabatic invariant, which is in the present case $2mv_{\|}\,l$. The energization can be understood as the result of the electric field caused by $\partial\boldsymbol{B}/\partial t$ at the places of the moving clouds, a kind of *betatron effect*.

The Fermi process does not necessitate moving plasma clouds. A length reduction of the path of a trapped particle with energization of the particle can also result from the displacement to a different field line in an inhomogeneous field. In the magnetosphere a displacement to a lower field line, corresponding to a smaller L, involves this type of energization. The displacement may occur as a temporary, reversible phenomenon in cases of perturbations or during drift through a non-axially-symmetric field. A lasting displacement is only obtained when at least one of the adiabatic invariants is violated. The third adiabatic invariant is easily violated because it relates to the period of drift around the Earth — of the order of an hour or shorter (Sect. 20γ). Processes with comparable or shorter time constants are easy to find.

Radial diffusion ("*cross L diffusion*") in combination with a variation of the energy is in the outer region of the radiation belt expected from "pumping" by magnetic fluctuations [*14*], [*15*][4]. Disturbances of the type of sudden impulses or sudden commencements have rise times of a few minutes and decay times of a few hours. The third adiabatic invariant will be preserved during the slow decay, but not during the fast rise. During the *rise* the drift motions due to the inhomogeneity of the magnetic field are negligible; the duration is simply too short. This means that the electric-field drift predominates with the consequence that the particles move with the field lines like *thermal* particles. The particle displacement during the rise is not reversed during the *slow decay*. The *resultant* displacement depends on the initial position of the particle on its drift path around the Earth and may be upward or downward. A downward shift means energization. The occurrence rate of the magnetic disturbances seems, however, somewhat too low to explain the observed cross L drift in this way. Some reasons for a possible underestimate of the drift can be given [*14*], [*15*].

Other processes in which one of the adiabatic invariants is violated and an acceleration may result have been considered [*14*], [*15*]. A peak in the electron spectrum that was occasionally observed in the lower region was explained by acceleration with downward diffusion due to *resonance between fluctuations of the magnetic field and particle drift around the Earth.* The phenomenon was named "geosynchrotron".

A violation of the *first* adiabatic invariant by *resonance between electromagnetic waves and gyromotion* is also possible. The waves can be excited somewhere else or can be generated by the gyrating particles themselves in some instability. This resonance is supposed to change the pitch angles of particles, presumably

[4] M. P. Nakada and G. D. Mead: J. Geophys. Res. **70** (19), 4777—4791 (1965).

with the main result of precipitation of particles (Sect. 20β). The pitch-angle variation or "pitch-angle scattering" may be an elastic process. In connection with L-shell splitting it could lead to radial diffusion without energy change[5]. If outward diffusion with constant energy and the earlier described inward diffusion with energization occur several times, there may be a large increase of energy[5].

ALFVÉN's model of motion of slow particles in an inhomogeneous magnetic field superposed by an electric field (Sect. 22β, Fig. 23) made clear that particles can collect as much energy as is found in the form of potential energy on a scale length of the inhomogeneous magnetic field. One may infer that in statistically irregular fields (varying in space and time) particles also collect quite some energy from the electric field. STURROCK has investigated this *stochastic acceleration process* for electric fields parallel and normal to the magnetic field [*14*]. The process is deemed possible in the magnetosheath with its turbulent structure.

Acceleration of particles is also possible around *neutral points* of the magnetic field with merging of field lines[6]. The theory of adiabatic invariants breaks down in this area because the variation of the magnetic field on a length of a gyroradius is there remarkable.

F. Motions in the magnetosphere.

23. Magnetodynamic convection. Magnetodynamic or "hydromagnetic" convection is the common name for *large-scale motions* underlying the simple *laws of magnetodynamics* (or "hydromagnetic theory"), in particular being governed by the approximative relationship

$$\boldsymbol{E} + \boldsymbol{v} \times \frac{\boldsymbol{B}}{c_0 \sqrt{\varepsilon_0 \mu_0}} = 0 \tag{23.1}$$

(see Sect. 7, [*16*][1]). The particles following these laws are supposed to be thermal (see Sect. 19β), but ALFVÉN'S example of a state of motion (Sect. 22β) showed that acceleration to higher energies may occur even under conditions under which magnetodynamics with Eq. (23.1) is a valid approach in the study of motions.

In the lower magnetosphere the geomagnetic field does not vary a great deal. This makes the theorem of *field lines moving with the plasma* (Sects. 8α and β) particularly useful. Implying that the *magnetic field stays constant*, one postulates that the joint motion of plasma and field lines must not change $\boldsymbol{B}$ in magnitude or direction at any place (see Sect. 8γ). This is a condition for the possible models of motion. The magnetic field can in certain problems be identified with the dipole field.

The idea of moving field lines in the magnetosphere seems to contradict the fact that the field lines are anchored in the Earth. As GOLD[2] has emphasized, however, "moving field lines" can not be imagined in the insulating space between Earth and ionosphere. The field lines lose their identity in this space though continuity conditions for the field at the bottom and top of the insulating space are valid (the divergence $\frac{\partial}{\partial \boldsymbol{r}} \cdot \boldsymbol{B}$ necessarily disappears).

[5] J. G. ROEDERER: Shell splitting and radial diffusion of geomagnetically trapped particles, in B. M. MCCORMAC (ed.): Earth's Particles and Fields, p. 193—208. New York: Reinhold Book Corporation 1968.

[6] W. I. AXFORD: Magnetospheric convection, in [*22*], p. 421—459.

[1] W. I. AXFORD: Magnetospheric convection, in [*22*], p. 421—459.

[2] T. GOLD: J. Geophys. Res. **64**, 1219 (1959).

Boundary conditions at the "bottom" of the magnetosphere result from the conductivity of the ionosphere with contribution of the conductivity of the Earth. The height of the lower ionosphere may appear negligible in many problems. A perfectly conducting ionosphere would suppress any electric field, i.e. any potential difference between magnetic field lines, thus prohibiting motions. The ionosphere, however, has a finite HALL conductivity and a finite PEDERSEN conductivity. The combination of both leads to not so trivial boundary conditions. The ratio of the two height-integrated conductivities is even variable with time of day, since high Hall conductivity is concentrated around 110 km height, whereas the Pedersen conductivity is noteworthy through most of the ionosphere [*2*].

Attention must also be paid to the boundary conditions at the magnetopause. In excitation of motions in the ionosphere the magnetopause as outer boundary may restrict the motions in some way.

The postulate of disappearance of $E_{\parallel}$ (parallel to $\boldsymbol{B}$), which is included in Eq. (23.1) and is essential for the simple magnetodynamic approach (Sect. 10β), has been questioned occasionally. A current density component $J_{\parallel}$, which in earlier times was not considered by many authors, does not affect the basic principles so badly, but will not leave the magnetic field unchanged. In recent years the possible effects of both an $E_{\parallel}$ and a $J_{\parallel}$ were discussed ([*19*], also see Sects. 10β and 32). In the plasmasphere the conductivity may be high enough to suppress an $E_{\parallel}$; outside, where the collision frequency is very low, an $E_{\parallel}$ might be imagined [*19*]. About field-aligned currents in geomagnetic disturbances see Sect. 34.

Magnetodynamic convection in the magnetosphere includes *motions of various origins* [*16*][1]. In the lower ionosphere there is a state of motion, which is responsible for the dynamo-electric current system with the resulting variations of the geomagnetic field. The electric field associated with this motion is continued upward — magnetic field lines are equipotential — with the consequence of motions in the magnetosphere (Sects. 25 and 37). The rotary motion of the atmosphere with the lower-ionospheric plasma entails rotation of entire magnetic field lines with the attached plasma (Sect. 25).

Motions may also be generated at the magnetopause in various ways. Arrival of a more intense solar wind leads to increased compression of the magnetosphere with some transitory state of motion (Sect. 27). The solar wind seems to pull the magnetospheric plasma at the magnetopause with it, thus causing a circulation within the magnetosphere (Fig. 25 and Sect. 24). The fixed shape of the magnetosphere in conjunction with participation of the plasma in the Earth's rotation enforces a system of motions even without a particular pull or perturbing forces (Sect. 25).

Exchange of plasma at different altitudes by means of drift motion normal to the field lines is another type of magnetodynamic convection, comparable to vertical exchange in the neutral atmosphere that leads to adiabatic variation of temperature and density with height. In the exchange by magnetodynamic convection the height variation of the magnetic field is prescribed. The height variation of plasma pressure and temperature turn out different, depending on whether collisions are significant or not ([*16*], also see Sect. 22β). In observations of the plasma density and temperature as functions of height the predicted behavior is not evident; the salient feature is the plasmapause.

Fig. 25, an *example of magnetodynamic convection* given by HINES, shows motions in the *polar cap* (from 60° latitude to the pole) at low (i.e., ionospheric)

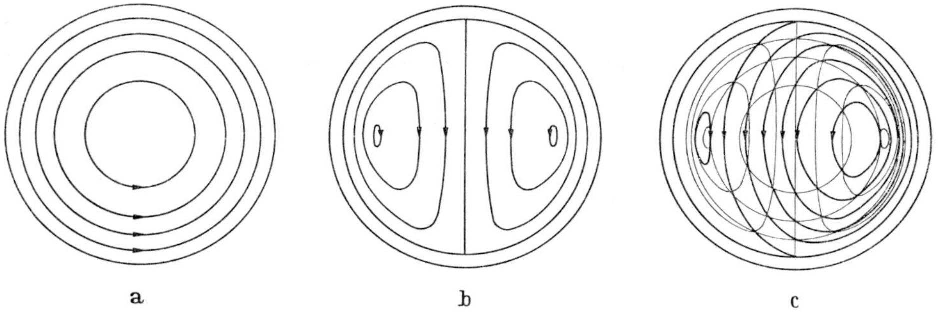

Fig. 25a–c. Simplified model of motions in the polar cap[3]. (a) Rotation, (b) drift motion generated by friction or field line merging at the magnetopause, (c) superposition of both motions.

height. In Fig. 25a the motion is *corotation* with the Earth. In Fig. 25b it is the *circulation* due to the *solar-wind pull* at the magnetopause. The Sun was assumed above the figure. The geomagnetic field is considered as nearly undistorted by the solar wind. A view at the N-S meridian plane (Fig. 9 or Fig. 18) makes clear that polar field lines are moved away from the Sun. The return flow must be found on field lines reaching the Earth at somewhat lower latitudes (60 to 70° in Fig. 25). Fig. 25c represents the *superposition* of the two motions, which is an idealization of the realistic flow pattern (Sect. 24).

Eq. (23.1) indicates that in a gradient electric field (corresponding to $\partial \boldsymbol{B}/\partial t = 0$) the stream lines and the magnetic field lines are equipotential. Thus the stream lines of Fig. 25a and b can be interpreted as equipotential lines. Addition of the two potentials yields the equipotential lines or stream lines of the composed motion of Fig. 25c.

In the lower ionosphere, where the Hall conductivity is essential, only the electrons follow the magnetodynamic drift according to Eq. (23.1); the ions are impeded by collisions. The consequence is a horizontal *Hall current opposite to the convective motion higher up* — opposite to the arrows in Fig. 25c.

24. Convective motion generated at the magnetopause.

α) *Theoretical model.* A *current system* consisting of two cells similar to the convective cells of Fig. 25c is known to exist in the ionosphere in varying intensity. It causes the quiet-time variation of the magnetic field in polar regions, S_q^p, and it appears with higher intensity at times of magnetic disturbances [*19*], [*27*]. In some recent articles the magnetic field variation corresponding to such a current system is separated from the total variation as the "**DP2** disturbance", which is present all the time, but increases during more disturbed periods, and fluctuates sometimes with time constants of the order of an hour [*19*]. The **DP2** current system is strongest in the polar regions, though it extends over the entire Earth.

The **DP2** disturbance and other polar magnetic disturbances suggest that there is a circulatory convective motion system in the magnetosphere which at 110 km height or so shows up as electron motions or currents according to the remarks at the end of last section. It is an obvious idea that the solar wind enforces the plasma immediately below the magnetopause to move with it. Thus the polar field lines with their plasma move away from the Sun. The circulation to be expected is in rough outlines that of Fig. 25c.

[3] Fig. 2 of [*16*].

Various authors (in particular AXFORD and HINES, NISHIDA, BRICE) developed motion models of this type [*19*][1,2] in order to explain polar-disturbance current systems, auroras (if possible), and the plasmapause. The solar wind is in all models supposed to cause the motion.

What the *force* is that links the terrestrial plasma with the solar wind can only be conjectured. It is not even certain whether the force causing the magnetotail (Sect. 16β) is the same as the one needed for the convective motion, which has to penetrate more deeply (Sect. 7δ of [*28*]). A regular viscous force is impossible because collisions are too rare. Interlocking may be due to instabilities with a surface structure or there may be a field line connection across the magnetopause such as shown in Fig. 18. The motions indicated in this figure agree with the present models of motion.

Fig. 26 shows the *streamlines in the equatorial plane* according to NISHIDA [*19*]. The streamlines are potential lines at the same time [see Sect. 23 and Eq. (23.1)]. Their potentials in kV are noted in the figure.

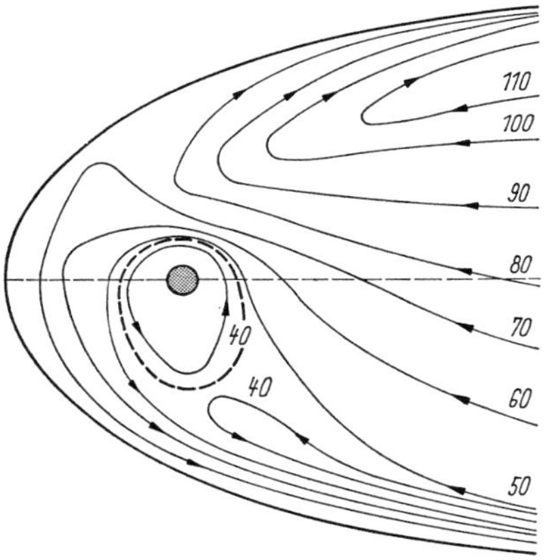

Fig. 26. Plasma flow and electric potentials in the equatorial plane[3]. The streamlines shown are equipotential lines. The numbers are potentials in kV. The dashed line is the intersection of the plasmapause with the equatorial plane.

The streamlines are obtained in this way [*19*]: The **DP 2** *current system in the ionosphere* is given and is interpreted as including Hall currents only. It has been conjectured that Pedersen currents do not contribute a great deal to the magnetic field variations (**DP 1** and **DP 2**) and are consequently not found from them. The electric potential surfaces for the Hall currents are defined by the current lines and the magnetic field lines. *Streamlines (potential lines) in the equatorial plane* are thus derived under assumption of a (closed) model of the magnetosphere with tail. These streamlines refer to a coordinate system rotating with Earth and atmosphere. A *rotary motion* finally has to be superimposed. The *resulting motion*, seen by an observer not rotating with the Earth, is represented in Fig. 26.

The dotted line in Fig. 26 separates closed streamlines from streamlines leading way back into the magnetotail. In space the dotted line means a field

[1] N. BRICE: Magnetospheric and high latitude ionospheric disturbance phenomena, in [*13*], p. 563—585.

[2] W. I. AXFORD: Magnetospheric convection, in [*22*], p. 421—459.

[3] [*19*], part of Fig. 14.

line shell. Plasma and field lines at one time inside this shell will remain inside. Plasma and field lines *outside* will drift far out. The result is quite some *reduction of the plasma density* outside. It is hypothesized that this is the explanation of the plasmapause and the plasma trough outside. Removal of plasma along extended polar field lines may proceed as a supersonic "polar wind", whose generation is due to the counteraction of the restraining gravity force and the spatial expansion, just like the generation of the solar wind[2,4]. The polar wind, though, is supposed to embrace light-weight ions (H^+ and He^+) only, each reaching its individual velocity. Charged particles may also *come in* along the extensive tail field lines (Sects. 20β and 27).

The different velocities of the various types of ions are a consequence of the electric field tending to balance the ion and electron stratifications. For a discussion of this and other difficulties and for a comparison with the *evaporative theory of ion departure* see the recent review paper by DONAHUE[5]. Hydrodynamic theory, referring to the supersonic polar wind, and evaporative theory are possibly equivalent, if the correct electric fields are accounted for and boundary conditions are chosen the same[5].

If the auroral oval (described in Sect. 32) is related to the plasmapause, the present motion model should also explain the precipitation of particles leading to auroras.

Despite the resemblance of the present motion model with Alfvén's model (Fig. 23) the two show essential differences. In Alfvén's model an acceleration is connected with the motion and the paths of particles depend on their initial velocity. Either model may be used in the attempt to explain the auroral precipitation. A way of energization in the tail may easily be found also in the present model (see Sects. 22, 27, and 32β).

Motion models of the magnetosphere have also been discussed in connection with the recent observations of magnetosheath plasma that has penetrated into the "polar cusps" — two bands extending along field lines to the postulated neutral points (Sect. 35). This plasma may be transported by the convective motion system to other parts of the magnetosphere, in particular to the tail.

β) *Observations.* Motions of the tenuous plasma in the magnetosphere above 1000 km have in most cases *not* been *observed directly*. Observational evidence of the motions has mainly come from *motions in the ionosphere* which are interpreted as an $\boldsymbol{E}\times\boldsymbol{B}$ drift with the magnetic field lines being equipotential, from *observations of* $\boldsymbol{E}$, from investigations of the *plasmapause* by means of *whistlers*, and from the *motion of artificial ion clouds* that were released by rockets or satellites. With the assumption of Eq. (23.1) all observational results can be expressed in terms of the velocity $\boldsymbol{v}$ or the electric field strength $\boldsymbol{E}$ corresponding to the drift motion. A brief review on magnetospheric electric fields with emphasis on observations was given by VÖLK and HAERENDEL[6] (see the following paragraphs and remarks on drift motions in connection with auroras, Sect. 32α).

In the context with effects on the magnetopause the motions in the outer magnetosphere are of particular interest. They were consequently the subject of a number of investigations. It has to be noticed, however, that the motions coming in at low altitudes (at the "bottom" of the magnetosphere), which will be discussed in the next section, may be included in any observational data.

The observations of *whistlers* which led to a model of the plasmasphere with the plasmapause also indicate a *motion of the whistler paths*. In general it is believed that the whistler observations confirm a motion model such as Fig. 26,

---

[4] P. M. BANKS and T. E. HOLZER: J. Geophys. Res. **74** (26), 6317—6332 (1969).

[5] T. M. DONAHUE: Rev. Geophys. and Space Phys. **9** (1), 1—9 (1971).

[6] H. VÖLK and G. HAERENDEL: Magnetospheric electric fields, in V. MANNO and D. E. PAGE (ed.): Intercorrelated Satellite Observations Related to Solar Events, p. 280—296. Dordrecht, Holland: D. Reidel Publishing Co. 1970.

although the motion of whistler paths does *not* seem to follow the boundary of the *bulge* of the plasmasphere (see Sects. 18γ and δ).

Electric-field measurements by GURNETT[6,7] at relatively low altitudes (677 to 2528 km) showed a change of the electric field in traversal of the plasmapause in the sense of a corotation inside and a convective motion outside.

Particle measurements made by means of a retarding potential analyzer on a synchronous satellite (FREEMAN et al.[6]) led to observations of a *directional ion flux* in the energy range 0...50 eV during periods of enhanced compression of the magnetosphere. The flux was *toward* the Sun when the magnetopause was not too near and *away* from it close to the magnetopause, in qualitative agreement with the expected convective motion.

Balloon measurements of the horizontal electric-field component at heights up to a little more than 40 km were made by MOZER and SERLIN[6,8] in the belief that this field component corresponds to the potential difference between magnetic field lines higher up. The observations were made at high latitude so that potential differences and electric fields in the outer magnetosphere could be derived. The electric field in the equatorial plane seems to correspond to some convective motion system which permits interesting comparisons with existing theories. The motion is found to be radially inward near midnight and outward in the afternoon and evening hours. This is, in fact, what theoretical models (e.g., Fig. 26) postulate.

Artificial Barium ion clouds are supposed to follow the motion of the surrounding plasma. Barium releases in the ionosphere at temperate latitudes confirmed the motions derived from the dynamo-electric theory of magnetic variations. Barium releases at auroral latitudes were arranged in order to study the motions accompanying high-latitude magnetic disturbances and auroral phenomena[6]. Quite a number of releases took place in the F2 layer at 200 to 400 km height, one was in the outer magnetosphere at 12.5 R_E geocentric (see Sect. 32α, in particular the references given there). The acceleration time of the cloud and the question whether a cloud follows the motion of the tenuous plasma at high altitudes have been discussed in the literature. The motions observed in the *auroral ionosphere* permit an extrapolation to motions in the equatorial plane of the magnetosphere, which seem to agree roughly with other results — as far as a comparison is possible in view of the limited observation times (twilight hours had to be chosen).

25. Motions initiated at low altitude. Drift motions in the magnetosphere may be caused by *electric fields at the bottom* as well as by electric fields at the outer boundary, the magnetopause. At the "bottom", between 90 and 150 km height, an *electric field accompanies the dynamo-electric currents.* The current system has been computed from the quiet-day variations of the geomagnetic field. A computation of the driving electric field, which itself is due to a circulatory motion system in the atmosphere (the "tidal wind system") requires the knowledge of the ionospheric conductivities (Hall and Pedersen conductivities).

Various authors mapped the electric field of the dynamo-electric theory into the magnetosphere under the assumption of equipotential magnetic field lines (see, for example, MATSUSHITA and TARPLEY[1]). In the equatorial plane the drift motions corresponding to this electric field are outward (toward the Sun) around

[7] D. A. GURNETT: Satellite measurements of DC electric fields in the ionosphere, in [*16a*], p. 239—246.

[8] F. S. MOZER and R. SERLIN: J. Geophys. Res. **74** (19), 4739—4754 (1969).

[1] S. MATSUSHITA and J. D. TARPLEY: J. Geophys. Res. **75** (28), 5433—5443 (1970).

noon and inward around midnight. At $L=4$ the maximum velocity is approximately 800 m/sec. The associated electric field amounts to approximately (1 mV/m). The vertical motion observed in the topside ionosphere in the equatorial region has an equivalent diurnal period, thus supporting the conclusions on magnetospheric motions.

The rotation of the atmosphere must entail still another motion system[2]. The atmosphere corotates with the Earth. The plasma has to rotate with the neutral gas in the lower ionosphere, where collisions are frequent. The theorem of moving magnetic field lines under this condition postulates rotation of all plasma on a field line. The consequent *rotation of the entire magnetosphere*, whose radial extension varies remarkably from day to night, *involves additional motions*, leading to a reduction or increase of the rotary velocity and to a compression or expansion of a section of the magnetosphere, depending on the time of day.

An *extremely simplified model magnetosphere* and a *simulating mechanical device* (Fig. 27) may help to clarify the expected motion system. Fig. 27a represents a cylindrical magnetosphere with field lines in the direction of the axis (dots in the figure). The magnetic field is supposed to be not affected by plasma motion, in accordance with the low β of the magnetosphere [Eq. (8.12)]. Any motion system, though connected with motion of field lines, has to leave the field unvaried. The inner cylinder (the Earth) rotates and enforces a rotary motion on plasma and field lines in the surrounding space (the magnetosphere). The outer boundary (magnetopause) is an eccentric cylinder. The rotary motion in the space of variable radial depth has to preserve the density of the dots (field lines). East-west motions must consequently be superimposed on the rotary motion. The rotary velocity must be increased at the place of small depth (noon) and decreased at the place of large depth (midnight). Radial motions are necessary to fill the entire interval with dots.

The mechanical device with some slides (Fig. 27b) will let the gas in the chambers between slides rotate with the inner cylinder. The rotary velocity here is kept constant, but radial velocities ensue in distributing the gas over the chambers whose size alternately increases and decreases during a full rotation.

Similar motions are to be expected in the magnetosphere. It should be noticed, however, that in Fig. 27a the tacit assumption was made that field lines can not leave or enter through the boundaries of the space. This is, in fact, valid at the magnetopause if the magnetosphere is closed, but at the bottom boundary (the lower ionosphere) field lines can depart or come in because the adjacent medium is not (or nearly not) conductive (see Sect. 23). At the bottom, thus, a radial velocity is not excluded. The boundary conditions at the bottom follow from the conductivities of the lower ionosphere (Hall, Pedersen, and Cowling conductivities).

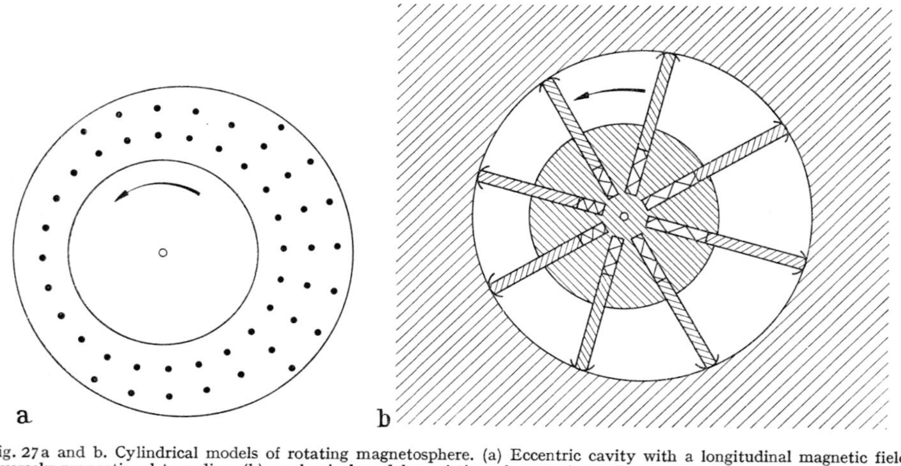

Fig. 27a and b. Cylindrical models of rotating magnetosphere. (a) Eccentric cavity with a longitudinal magnetic field inversely proportional to radius, (b) mechanical model consisting of a rotating cylinder with slides in an eccentric bore.

[2] H. Poeverlein: J. Atmos. Terr. Phys. **28**, 1111—1123 (1966).

The motion under consideration represents one type of electric-field drift. *Other electric-field drifts* (such as described in Sects. 23 and 24) are *superimposed* on it. A motion model that includes the pulling effects at the magnetopause and the rotation takes the deformation in the above sense into account if the proper magnetic-field configuration is assumed. The pictures of the resulting motions such as Fig. 26 are supposed to show the hypothesized motions in the outer regions correctly. The present consideration of a rotation with deformation, however, not marred by additional complications, makes clear that radial (vertical) motions and east-west motions throughout the magnetosphere result from rotation in connection with the given magnetic-field configuration.

One may try to *separate east-west motions and motions in the meridian planes* of a dipole field (the magnetospheric field up to several Earth radii). East-west-motions may let field lines of one meridian plane replace those of a neighboring meridian plane, while vertical motions mean exchange of field lines within a meridian plane. From this concept it follows that east-west motions are proportional to geocentric height r and vertical motions or drift motions in the meridian planes are proportional to r^2. An estimate based on the shape of the magnetosphere and on the rotation period of the Earth yields east-west velocities of the order 100 m/sec and vertical (radial) velocities of the order 10 m/sec at ionospheric levels[3].

Motions of plasma along field lines may result as a *secondary phenomenon* accompanying electric-field drift and may entail density variations of the plasma[2]. It can be expected that the total motion at ionospheric altitudes tends to be horizontal[2]. The vertical components of drift motions and of motions along field lines thus may balance in some regions, though not everywhere.

An additional difficulty arises in a model of a corotating closed magnetosphere. The magnetopause of the *closed* magnetosphere (e.g., the model of MEAD and BEARD, Fig. 9) has two *neutral points*. The magnetic-field configuration has to be retained during the rotation despite the participation of field lines in the motion. A somewhat complicated motion model has been proposed with assumption of a rotation around the polar field line. In the present author's opinion the rotation has to take place around the field lines leading to the neutral points, which themselves have to remain stationary with their plasma[2]. Otherwise infinite velocities would be obtained in approach to the neutral points. The shape of the *plasmapause* and its position in long quiet periods seem to agree roughly with the idea of a rotation accompanied by the radial motions that are imposed on the plasma by the confinement of the magnetosphere[3].

26. Effects of the inclined magnetic axis. In the past sections no attention was paid to the fact that the magnetic axis of the Earth is usually not normal to the ecliptic plane. The inclination of the Earth's rotational axis toward the normal of the ecliptic plane and the deviation of the magnetic axis from the rotational axis introduce some *asymmetry* into the configuration of the magnetosphere (see Fig. 11).

It is hypothesized that symmetry with respect to the incidence of the solar wind leads to increased magnetic activity. This is believed to be the reason for increased activity around the equinoxes (according to the "equinoctial hypothesis" [*11*]) and for certain variations of activity with universal time observed at solstices and equinoxes [*11*]. A consequence of asymmetry between the two

[3] H. POEVERLEIN, in W. H. CAMPBELL and S. MATSUSHITA (ed.): Proceedings of Upper Atmospheric Currents and Electric Field Symposium, Earth Sciences Laboratories, Boulder, Colorado, 1970, p. 72—74. — See also Sect. 37 at the end of this article.

hemispheres may eventually be plasma flux along a magnetic field line (see below) or currents flowing along field lines from one hemisphere to the other and linking the two ionospheric current systems[1].

The *neutral sheet* of the tail has been found to *move up and down* with respect to the ecliptic plane because of the rotation of the geomagnetic axis on a cone around the Earth's rotational axis [*17*], [*18*]. The neutral sheet seems to be approximately parallel to the solar-magnetospheric equatorial plane (x_{SM},y_{SM}-plane of Fig. 10), but located above or below it when the magnetic axis is not normal to the Sun-Earth line (x_{SM}-axis).

The effect of the solar wind should be strongest in the subsolar point of the magnetopause. It may be expected that this strong effect extends along the field lines in the magnetic meridian plane of the subsolar point, the x_{SM}, z_{SM} plane. An inclination of the magnetic meridian plane toward the geographic meridian plane, corresponding to geomagnetic declination, thus causes an *advancement or retardation in the effect of the solar wind at higher latitudes* (at 53° latitude the maximum is 1 hour)[2]. It looks as if some influence of the geomagnetic declination on ionospheric electron concentrations is of this sense, but it has to be noticed that any phenomenon arising at low latitude and being propagated along magnetic field lines is subject to the same advancement or retardation.

If the magnetosphere were radially symmetric and if rotational and geomagnetic axes coincided, the *corotation* of the magnetosphere with the Earth would be quite easy to describe. The rotary motion would be a pure drift motion normal to the magnetic field, connected with an $\boldsymbol{E}$ in the meridian plane. The complications arising from lack of radial symmetry were briefly discussed in Sect. 25. The deviation of the magnetic axis from the rotational axis poses another problem[2]. As Fig. 28 demonstrates the velocity vector of the *rotary motion* is

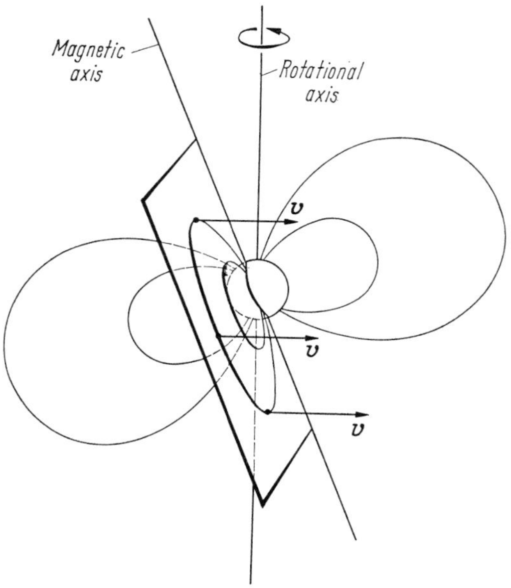

Fig. 28. Rotation of the magnetosphere under the assumption that magnetic and rotational axes do not coincide.

[1] D. van Sabben: J. Atmos. Terr. Phys. **28**, 965—981 (1966).
[2] H. Poeverlein: Kleinheubacher Berichte **13**, 199—201 (1969).

in general *not normal to the magnetic field*. The velocity comprehends a normal component and a parallel component. The normal component involves an electric field again and the extension of the electric field from ionospheric height to the outer part of a magnetic field line tends to move the plasma there in the appropriate way normal to the field line.

The velocity component parallel to the magnetic field must be provided by a gradient of the plasma pressure along the field lines. Plasma will accumulate at one end of the field line, thus forming the necessary pressure gradient. A crude numerical estimate[2] under assumption of a dipole magnetic field showed that the plasma density gradient along a magnetic field line may be noteworthy at intermediate L values, e.g. at $L=4$, corresponding to a geocentric height of 4 R_E in the equatorial plane.

27. Perturbations. The solar wind is highly variable. As Fig. 1 shows, its particle density, velocity, and embedded magnetic field vary a great deal. Also the temperature and the α-particle content vary[1]. Most of these variations affect the magnetosphere in a certain way. The position of the magnetopause is according to the simplified theory (e.g. MEAD's and BEARD's theory, Sect. 13) determined by the quantity Nmv^2, which represents the pressure of the wind. A variation of the pressure can be gradual or sudden. Sudden increases, corresponding to shock fronts, have been observed and are considered to be the cause of geomagnetic storms with sudden commencement[1,2] (see also Sects. 12δ, 12ε, and 29).

The question in the present context is for the *motions that may arise in the magnetosphere from an increase of the solar-wind pressure*. A simple model of the magnetosphere (Fig. 9, Sect. 13), in which the unperturbed solar wind is at the magnetopause in balance with the pressure of the solar wind, just requires a change in scale to get adjusted to a higher solar-wind pressure.

The magnetic pressure is proportional to B^2 or to r^{-6}. Occasionally, during a magnetic storm, the magnetopause was found depressed to a height below 6.6 R_E geocentric (the height of a synchronous satellite), i.e. to approximately $\frac{2}{3}$ of its normal height. The necessary intensification of the solar-wind pressure is by a factor $(\frac{3}{2})^6 \approx 11$. This corresponds, in fact, to solar-wind data observed during the particular storm[3].

The transition of the magnetosphere from the normal state to a state of increased compression involves some motion. At first sight it seems that plasma and field lines in the entire magnetosphere have to be moved downward, toward the Earth. This idea, however, can be disproved. A lowering of plasma and field lines from 10 to 6.6 R_E on all sides of the Earth would increase the magnetic flux through an equatorial circle at 6.6 R_E too much in comparison with the occurring increase of the magnetic field (50 to 60 γ on the ground in the reported case). The descending motion on the solar side thus necessitates an outward motion on the night side. The line integral of $\boldsymbol{E}$ around the Earth must be kept small, nearly vanishing.

Fig. 29, based on some observed facts, shows how *the magnetosphere with tail varies when the solar-wind pressure increases*. The first picture refers to the normal state. In the second picture the more intense wind has lowered the magnetopause; fewer field lines appear on the dayside. The magnetotail obtains a smaller cross section, but contains a higher magnetic flux (larger number of field lines) in each half [*18*] ([*28*] Sect. 12ε). An increase of the tail field magnitude by a factor 2 during the main phase of a sudden commencement storm is considered typical.

---

[1] K. W. OGILVIE and L. F. BURLAGA: Hydromagnetic observations in the solar wind, in [*16a*], p. 82—94.

[2] A. J. HUNDHAUSEN: Shock waves in the solar wind, in [*16a*], p. 79—81.

[3] W. D. CUMMINGS and P. J. COLEMAN: J. Geophys. Res. **73** (17), 5699—5718 (1968).

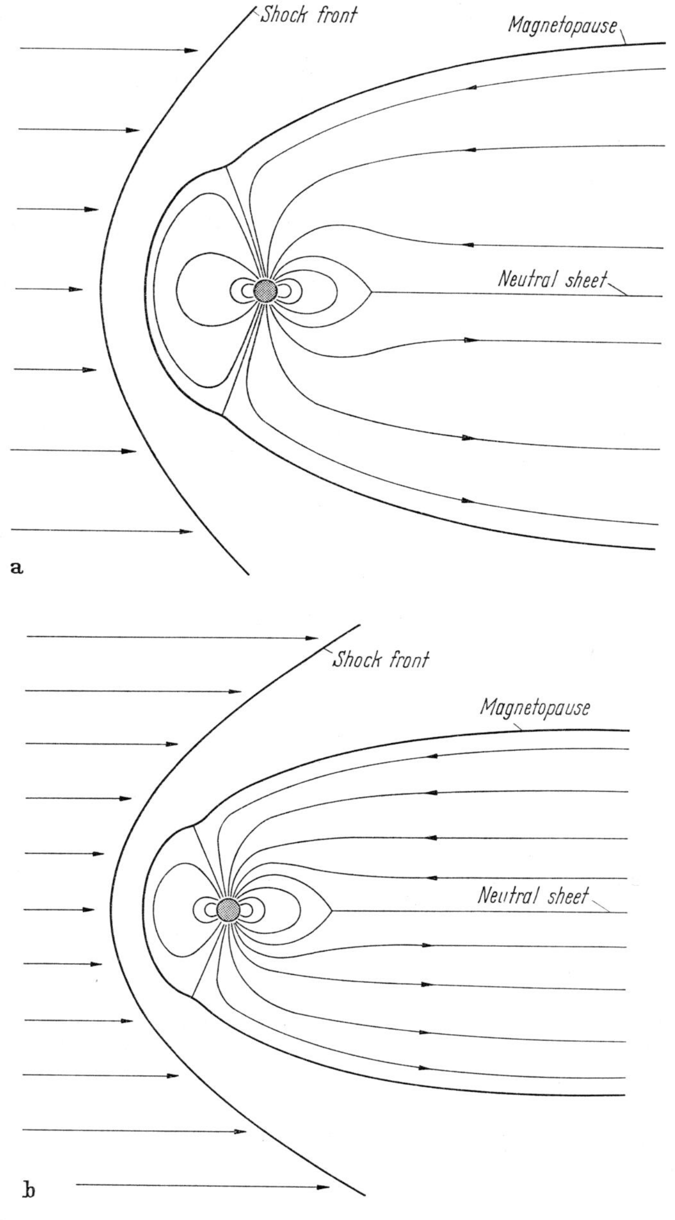

Fig. 29a and b. The magnetosphere, (a) in a normal state[4] and (b) in a state of enhanced compression by an intensified solar wind.

If the rule $p \propto B^2 \propto r^{-6}$ is transferred to the tail one finds the following proportionalities for the magnitudes of the magnetic field and the magnetic flux

[4] A. J. Dessler and B. J. O'Brien: Penetrating particle radiation, in [2], p. 53—92; Fig. 3-7. Fig. 29b is a slight modification of this figure.

in dependence on the solar-wind pressure:

$$B \propto p^{\frac{1}{2}}, \tag{27.1}$$

$$\Phi \propto B r^2 \propto p^{\frac{1}{6}}. \tag{27.2}$$

A factor 2 in B would require a pressure increase by a factor 4, which corresponds to a shift of the magnetopause by a factor $4^{\frac{1}{6}} = 1.26$, i.e. from 10 to 8 R_E, a reasonable shift in a magnetic storm.

The implication in the above relationships, Eqs. (27.1) and (27.2), is that also the pressures producing the tail are proportional to the solar-wind pressure. This is not necessarily so.

In transition from the first to the second picture of Fig. 29 some additional field lines have to become tail field lines. They have to be stretched out a long way. The connection of the two halves of a field line in the tail is presumably somewhere in the neutral sheet far out. Plasma will be moved out with the field line. This is the process (or one of the processes) emptying the outer part of the plasmasphere. It has been known that the plasmapause is lowered during magnetic perturbations (Sect. 18γ), but a general descent of all plasma was in the above discussion found incompatible with the small magnetic-field variation.

In recovery from the state of increased compression the field lines with the attached plasma are brought from the magnetotail into the magnetosphere proper. If solar-wind plasma has entered the tail it will now refill the magnetosphere or plasmasphere. The shortening of field lines snapping back from the tail may lead to Fermi acceleration of particles (Sect. 22γ). This is one possible cause of the replenishment of trapped particles and of the supply of keV-particles in geomagnetic storms (Sects. 20β and 21α).

If the tail field lines are open, the drawing-in requires a merging of field lines in the neutral sheet. Some inward-motion in the tail, presumably connected with particle acceleration, is supposed to be part of the general magnetospheric convection (Figs. 25 and 26). The described recovery process seems to cause a remarkable intensification of this motion in the tail. It has been hypothesized that auroral particles come from the tail in such motion processes and may have been accelerated in the tail ([*28*] Sect. 8δ)[5,6].

The *convective motion system* that is generated by friction or field-line merging at the magnetopause (Sect. 24, Fig. 26) may also be noted as a perturbation effect. Although it is supposed to be a permanent feature its intensity is *quite variable*, presumably strongly dependent on the solar-wind intensity. The **DP2** current system, which results from the convectic motion, is influenced by the magnetic field in the solar wind. This becomes apparent in some variations of the geomagnetic field (Sect. 30).

Perturbations in smaller dimensions, connected with auroras, also involve some motions, commonly reported as motions of auroral forms (see references in Sect. 32).

28. Forces and energies. In view of the dynamic processes in the magnetosphere some values of energies, energy flux, and forces should be of interest. The predominant energy in the magnetosphere is the magnetic energy (Sect. 8γ). The height at which the thermal energy density of the gas $\frac{1}{2}\varrho v^2$ and the magnetic

[5] T. OBAYASHI: The interaction of the solar wind with the geomagnetic field during disturbed conditions, in [*3*], p. 107—167 (in particular p. 158—161).

[6] W. I. AXFORD: Magnetospheric convection, in [*22*], p. 421—459.

energy density $\frac{1}{2} B^2/\mathrm{u}\mu_0$ are equal is approximately 130 km[1]. The condition of equal energy densities can also be formulated

$$V_s^2 = \frac{\gamma(\gamma-1)}{2} V_A^2 \frac{\varrho_P}{\varrho} \tag{28.1}$$

with the sound velocity V_s, the Alfvén velocity V_A, the plasma density ϱ_P, and the total density ϱ. The numerical factor is $\frac{5}{9}$ in an atomic gas.

The *energy of the geomagnetic field around the Earth*, conceived of as an infinitely extended dipole field, is derived as $8.4 \cdot 10^{17}$ J ($=8.4 \cdot 10^{24}$ erg). The equatorial field on the ground has been assumed to be 0.312 Gs $=3.12 \cdot 10^{-5}$ Vsec m^{-2}. The corresponding magnetic moment of the Earth[2] is $\mathsf{M}_E = 8.06 \cdot 10^{22}$ Am^2.

An increase of the geomagnetic field due to an external effect (e.g., increased solar-wind pressure) means an energy increase

$$\Delta W = \tfrac{1}{2} \mathsf{M}_E \Delta B \tag{28.2}$$

when M_E is the Earth's magnetic moment[2], and ΔB the increment of the flux density in the axial direction near the Earth[3,4]. Eq. (28.2) is valid only for transient field variations, which do not penetrate into the conducting Earth. In a not too strong initial phase of a magnetic storm there may be $\Delta B = 10\,\gamma = 10^{-8}$ V sec m^{-2}, corresponding to $\Delta W = 4.0 \cdot 10^{14}$ J (Joule).

The energy deposited in an aurora by the incoming electrons is computed from the auroral luminescence[5] at approximately 2 to $4 \cdot 10^{14}$ J during a period of 1 to 2 h.

The energy flux of the solar wind through a circular disk of 20 R_E radius, being of the size of the cross section of the magnetosphere, is $7 \cdot 10^{12}$ J/sec under quiet conditions. The solar wind thus is capable of supplying $4 \cdot 10^{14}$ J in one minute.

The *total energy of fast protons* (>100 keV) in the radiation belt was estimated[6] at $7 \cdot 10^{14}$ J. This is comparable with the above energy change for a 10 γ-perturbation. The perturbation that has to be explained in particular is the main phase of the geomagnetic storm ([*28*] and Sect. 29). In a severe storm the horizontal field component may be depressed by 200 γ [*28*]. The energy decrease for a field variation of 100 γ parallel to the magnetic axis would be $4 \cdot 10^{15}$ J, more than the trapped *fast* particles can resume. The trapped *low-energy* particles that Frank has investigated, however, turn out to be the more likely cause of the field depression in the main phase. These particles, in the energy range 200 eV to 50 keV, appear in high concentrations just during magnetic storms (Sect. 21 α). Their energy content in a storm with 50 γ field depression was given by Frank as approximately $2 \cdot 10^{15}$ J $=2 \cdot 10^{22}$ erg [*14*]. According to the above values this energy would suffice for a ΔB of 50 γ in axial direction. How the particles produce the postulated ring current is noted in Sect. 29 and treated in detail by Parker and Ferraro [*28*].

---

[1] H. Poeverlein: J. Geophys. Res. **72** (1), 251—256 (1967).

[2] About the definition of a magnetic moment see Sect. 20γ, footnote [9]. The magnetic flux density B of the geomagnetic dipole field was in footnote [1] of Sect. 22β given as a function of the Earth's magnetic moment. In the IUPAP definition a magnetic moment is an energy divided by a magnetic flux density ([9] Sect. 20γ).

[3] J. J. Maguire and R. L. Carovillano: J. Geophys. Res. **71** (23), 5533—5539 (1966).

[4] R. L. Carovillano and J. J. Maguire: Magnetic energy relationships in the magnetosphere, in [*13*], p. 290—300.

[5] [*23*] p. 222. — Ya. I. Feldstein and G. V. Starkov: J. Atmos. Terr. Phys. **33** (2), 197—203 (1971).

[6] R. A. Hoffman and P. A. Bracken: J. Geophys. Res. **70** (15), 3541—3556 (1965).

An estimate of the *energy in the magnetotail* may be made assuming a homogeneous field of 10 γ in a cylinder of 20 R_E radius and 200 R_E length. The energy of the field in this space is $2.6 \cdot 10^{15}$ J.

This would be significant in comparison with the data of the main phase, but it seems doubtful whether a large part of it is available for effects in the magnetosphere. May be, it is available after transformation into particle energy, which appears with FRANK's low-energy particles in the radiation belt. From Eqs. (27.1) and (27.2) it follows that the tail field energy is increased by a factor $4^{\frac{2}{3}} = 2.5$ when the solar-wind pressure is enhanced four times and the tail length remains unvaried. Recovery is connected with the corresponding energy decrease.

The *force that the solar wind exerts on the magnetopause* is estimated by PARKER and FERRARO ([*28*] Sects. 6β and 12ε) at $6 \cdot 10^7$ N (Newton) $= 6 \cdot 10^{12}$ dyn. The force on a *circular disk* with radius 20 R_E exposed to the *quiet-time* solar wind is computed at $4 \cdot 10^7$ N. The *tension force in the quiet tail not too far out* (with 15 γ and 20 R_E radius) is $7 \cdot 10^6$ N (see [*28*] and the values of the tail tensile stress and solar-wind pressure in the above Sect. 16β).

The tension of the tail should reduce the geomagnetic field — this was PIDDINGTON's idea when he proposed the existence of a tail during the main phase of storms. The field reduction, however, turns out to be too small. The pressure of the quiet solar wind raises the surface magnetic field, according to some computation[7] by roughly 25 γ. The effect of the tail should be smaller because of the smaller force with which it acts on the bulk of the magnetosphere. The tail thus does not seem to be the immediate cause of the main phase of storms (see Sect. 29).

At last a remark on the *kinetic energy density* of a plasma in *drift motion* may follow. It is assumed that positive and negative particles move according to

$$\boldsymbol{E} + \boldsymbol{v} \times \frac{\boldsymbol{B}}{c_0 \sqrt{\varepsilon_0 \mu_0}} = 0 \,. \tag{28.3}$$

The particle velocity $\boldsymbol{v}$ corresponds to an accelerating force $\pm q\boldsymbol{E}$ during a time

$$1/\omega_B = \frac{m}{q} \, \frac{c_0 \sqrt{\varepsilon_0 \mu_0}}{B}$$

with ω_B denoting the angular gyrofrequency. The kinetic energy density of the plasma is in a good approximation the kinetic energy density of the ions, which is

$$\frac{1}{2} N m_{\mathrm{i}} v^2 = \frac{c_0^2 \varepsilon_0 \mu_0}{2} N m_{\mathrm{i}} \left(\frac{E}{B}\right)^2 \tag{28.4}$$

or

$$\frac{1}{2} N m_{\mathrm{i}} v^2 = \frac{1}{2} \left(\frac{\omega_{N\mathrm{i}}}{\omega_{B\mathrm{i}}}\right)^2 \frac{\varepsilon_0}{\mathrm{u}} E^2 \tag{28.5}$$

when $\omega_{N\mathrm{i}}$ and $\omega_{B\mathrm{i}}$ are the (angular) plasma frequency and gyrofrequency of the ions. Since $[1 + (\omega_{N\mathrm{i}}/\omega_{B\mathrm{i}})^2]\,\varepsilon_0$ is the direct transverse dielectric constant ($\varepsilon_\perp$), the equation may be written

$$\frac{1}{2} N m_{\mathrm{i}} v^2 + \frac{1}{2\mathrm{u}} \varepsilon_0 E^2 = \frac{1}{2\mathrm{u}} \varepsilon_\perp E^2 \,. \tag{28.6}$$

The energy density $\frac{1}{2\mathrm{u}} \varepsilon_\perp E^2$ apparently resides in the moving plasma except of the small contribution $\frac{1}{2\mathrm{u}} \varepsilon_0 E^2$.

G. Solar wind and terrestrial phenomena.

29. Geomagnetic storms. Magnetic storms principally are periods of high magnetic activity. The horizontal component of the magnetic field has a typical

[7] G. D. MEAD: J. Geophys. Res. **69** (7), 1181—1195 (1964).

behavior in many magnetic storms. In this section a brief description of the typical geomagnetic storm will be given together with a view at related phenomena in the solar wind and magnetosphere. More material, observational and theoretical, is presented in Vol. 49/3 of this Encyclopedia by NAGATA and FUKUSHIMA [*27*] and PARKER and FERRARO [*28*]. In the literature one finds surveys emphasizing the relationship with the interplanetary magnetic field[1] and with solar-flare phenomena[2].

The *typical magnetic storm* (Fig. 18a in [*27*], Fig. 1 in [*28*]) starts with a sudden increase of the horizontal field component ($\mathcal{H}$), the "*sudden commencement*". An increase of 20 to 30 γ is typical; increases of 50 γ or more are observed in severe storms. The *initial phase* of the storm with the increased $\mathcal{H}$ lasts 3 to 8 hours. The *main phase*, following the initial phase, represents a depression of $\mathcal{H}$ below normal, amounting at middle latitudes to 50 to 200 γ or even more and lasting typically 12 to 24 h. The *bay disturbance*, a positive or negative field excursion of one or a few hours, appears *repeatedly* during the main phase. The bay disturbances and the development of the main phase seem to be in some internal relationship. The *recovery phase* of a storm may last a day or longer.

The duration of the initial and main phases is shorter in the strongest storms than in more moderate storms. Individual storms may deviate remarkably from the typical pattern. In particular the initial phase and the main phase seem to be pretty independent. There are storms without sudden commencement or with nearly no initial phase and a notable main phase and other storms that show a good initial phase and no perceptible main phase. Initial phase and main phase apparently are quite distinct phenomena that have different histories though both are due to a perturbation of the solar wind by a solar flare.

A *solar flare* leading to a geomagnetic storm *ejects some plasma* with a velocity beyond the regular solar-wind velocity. At the front of the ejected plasma cloud a *shock front* evolves, which is propagated in the solar wind that is already on its way off the Sun. The increase of material velocity and density will be communicated to all plasma overtaken by the front. The front has the character of a fast magnetodynamic shock (Sects. 12δ and ε). When the front reaches the magnetosheath or magnetopause of the Earth a sudden compression of the magnetosphere sets in. The compression with the accompanying magnetic-field increase represents the initial phase of a storm. Propagation of the compression through the magnetosphere to low altitude proceeds also as a magnetodynamic perturbation with Alfvén velocity ([*28*] Sects. 10β and 17γ), requiring one or two minutes (for a discussion of the time delay see [*28*] Sect. 10β). The observed time differences between locations on the Earth are noted in [*27*] Sect. 26.

Quite a number of shock fronts in the solar wind have been observed. Also the coincidence of their arrival with sudden commencements of storms has been established in some cases[3,4]. Shock fronts may also be associated with long-lasting solar plasma streams of high velocity that cause recurrent magnetic storms[3].

The *main phase* is not simply a relaxation effect during recovery from the increased solar-wind pressure. If it were, there could be no depression of the

---

[1] J. HIRSHBERG and D. S. COLBURN: Planet. Space Sci. **17** (6), 1183—1206 (1970).

[2] W. I. AXFORD: A survey of interplanetary and terrestrial phenomena associated with solar flares, in V. MANNO and D. E. PAGE (ed.): Intercorrelated Satellite Observations Related to Solar Events, p. 7—22. Dordrecht, Holland: D. Reidel Publ. Co. 1970.

[3] A. J. HUNDHAUSEN: Solar wind disturbances associated with solar activity, in V. MANNO and D. E. PAGE (ed.): Intercorrelated Satellite Observations Related to Solar Events, p. 111—129. Dordrecht, Holland: D. Reidel Publ. Co. 1970.

[4] A. J. HUNDHAUSEN: Ann. Géophys. **26** (2), 427—442 (1970).

field below normal. The depression of $\mathcal{H}$ below normal is in general even larger than the raise during the preceding initial phase.

PIDDINGTON's earlier idea that the tension force of the magnetotail on the magnetosphere causes the main-phase depression of the field had to be given up because the tension force is too small (Sect. 28, [*28*] Sect. 12 ε). A depression of 200 γ or more (in severe storms) has to originate at heights where the field is at least of this magnitude.

DESSLER and PARKER developed a theory of the main phase on the basis of an idea that SINGER proposed. The main phase in their theory is considered the effect of trapped particles carrying a sufficient amount of energy ([*28*] Sects. 11, 12, 18 to 25). In Sect. 20γ it was noted that trapped particles, while bouncing back and forth on field lines, drift on field line shells around the Earth, positive particles to the west, negative particles to the east. The resulting ring current causes a diminution of the field in the region below the ring current. The field outside and inside the particle shell is varied so that the gradient of the magnetic-field pressure balances both the centrifugal force on the particles following curved field lines and the force on the particles gyrating in the inhomogeneous field.

The field variation at the Earth $\delta B_{♁}$ relative to the equatorial field strength $B_{♁0}$, has in the case of axial symmetry been derived as

$$\frac{\delta B_{♁}}{B_{♁0}} = -\frac{2}{3}\,\frac{W_{\mathrm{p}}}{W_{\mathrm{m}}} \tag{29.1}$$

when W_{p} is the total kinetic energy of the trapped particles and W_{m} is the total energy of the magnetic field ([*28*] Eq. (22.14)). The amounts of energies (Sect. 28) indicate that keV particles, the particles between 200 eV and 50 keV that FRANK found during magnetic storms in vast numbers, must be the ones carrying the ring current of the main phase. The particles were observed between 3 and 5 R_{E} geocentric. Magnetic-field observations on satellites approve this approximate height interval as the seat of the ring current. The magnetosphere is said to be "inflated" by the outward pull of the particles.

Various possible sources of the magnetic-storm particles have been considered ([*28*] Sect. 12). It is likely that the particles come in from far out in the course of convective motions during the storm. The magnetotail offers the best chance for this (see Sect. 27). The particles may be solar-wind particles that have entered the tail.

The *magnetic bays during the main phase* are believed to be individual events during which some plasma is transported into the ring-current region [2]. A magnetic bay, also called "*polar magnetic substorm*", is a polar magnetic disturbance that is worldwide, although being produced around auroral field lines (Sect. 32). Its immediate cause may be an instability somewhere in the magnetosphere (or ionosphere) or an increased ionospheric conductivity in an auroral perturbation.

It has been found that a field variation corresponding to a bay on the Earth's surface appears in the tail region of the magnetosphere with some delay (approximately 10 or 15 min) [5]. This supports the idea of an ionospheric or magnetospheric origin. On the other hand, a striking coincidence between enhanced variability of the magnetic field in the solar wind ("interplanetary storms") and activity in main phases was recognized [1]. The solar wind showing a more intense and variable field was thought of as the flare-ejected plasma that arrives many hours after the shock front. This plasma is also characterized by an increased α-particle content. If there is some instantaneous influence of solar-wind parameters on

[5] J. P. HEPPNER, M. SUGIURA, T. L. SKILLMAN, B. G. LEDLEY, and M. CAMPBELL: J. Geophys. Res. **72** (21), 5417—5471 (1967).

the main phase of geomagnetic storms, it might be a triggering of magnetospheric processes.

30. Geomagnetic activity. The simultaneity of intense, perturbed interplanetary fields and high geomagnetic activity in the main phase of geomagnetic storms was noted in the last paragraph of the preceding section. A connection between solar wind and interplanetary magnetic field on the one side and variations of the geomagnetic field on the other has been found in a number of investigations[1]. Though the observational facts are well established, only sketchy and tentative theoretical explanations have been given till now.

A common measure of geomagnetic activity is the planetary geomagnetic activity index taken for 3 hours, K_p [*27*][2]. Frequently the eight values of a day are added up and used under the notation ΣK_p. An evaluation of ΣK_p in the period in which Figs. 3 to 5 were obtained yielded a clear *dependence of* ΣK_p *on the position in a sector of the solar wind.* Fig. 30 shows the average behavior of ΣK_p in a sector according to WILCOX and NESS. Only sectors covering $\frac{2}{7}$ of a circle were used in the figure (see Fig. 3). Circles denote averages of seven sectors passing by the Earth, + signs are averages of the four sectors with a field away from the Sun, and — signs are averages of the three sectors with a field toward the Sun, just as in Fig. 5. There is a striking similarity of the solar-wind velocity, the magnitude of the interplanetary field, and ΣK_p as functions of position in a sector (Figs. 4, 5, and 30).

The *correlation of* K_p *with the interplanetary magnetic field* has also been investigated on the basis of the IMP 3 measurements that led to the above results[3]. A good correlation between *field magnitude and* K_p has been found, confirming the relationship recognizable from Figs. 5 and 30. The magnetic activity index K_p appears correlated also with the angle of the interplanetary field toward the ecliptic plane as Fig. 31 shows. The magnetic activity is *higher* for an *interplanetary field with a southern component* than for one with a northern component.

A difference of magnetic activity depending on whether the interplanetary magnetic field is *away from* or *toward the Sun* was found, but this turned out to be due to predominant observations on the northern hemisphere. SIEBERT investigating separate activity indices for the two hemispheres found that the sign of their difference is at midnight hours correlated with the field direction in the solar-wind sector[4].

The effect of the field component normal to the ecliptic plane is by many authors ascribed to a *merging of interplanetary and terrestrial field lines* at the magnetopause (see Sect. 17 on the open magnetosphere). Merging is believed to occur at the highest rate with opposite field directions in the subsolar area of the magnetopause (see Fig. 18). Magnetic field variations may originate in merging or — more likely — may be brought in by the solar wind and enter the magnetosphere on merging field lines. This is managed by the electric field in the solar wind, $\boldsymbol{E} = -\boldsymbol{v} \times \boldsymbol{B}/c_0 \sqrt{\varepsilon_0 \mu_0}$, which penetrates along magnetic field lines into the magnetosphere. The interaction of the *magnetotail* with the solar wind may be strong when the field lines are well linked at the boundary, i.e. on the side of the tail where the two field directions are opposed (cf. SIEBERT's observation and his discussion[4]).

---

[1] See contribution by M. SIEBERT, Vol. 49/3 of this Encyclopedia, p. 272—275.
[2] M. SIEBERT: Vol. 49/3 of this Encyclopedia, p. 208—210, 222—225.
[3] K. H. SCHATTEN and J. M. WILCOX: J. Geophys. Res. **72** (21), 5185—5191 (1967).
[4] M. SIEBERT: Vol. 49/3 of this Encyclopedia, p. 269—270; J. Geophys. Res. **73** (9), 3049—3052 (1968).

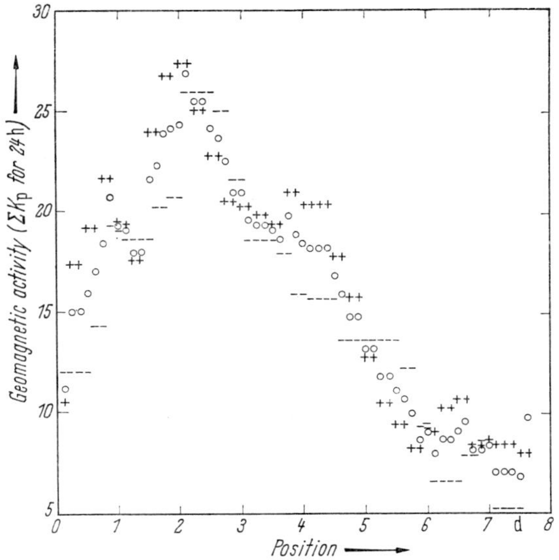

Fig. 30. Geomagnetic activity as a function of position within a solar-wind sector[5]. Compare with Figs. 4 and 5. Average values of several away-sectors and several toward-sectors are marked by + and —, while averages over all sectors of 8 d duration are indicated by ○. Superposed epoch analysis with 24-h epochs is used.

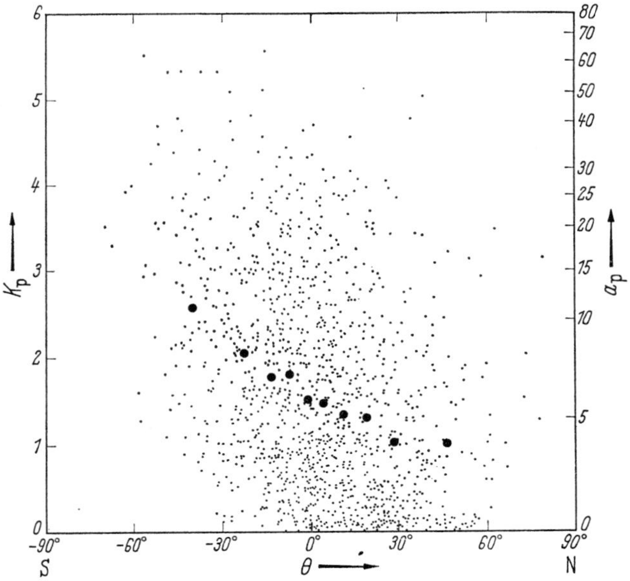

Fig. 31. Geomagnetic activity as a function of the angle θ between the interplanetary field and the ecliptic[6]. Each small dot represents an observed K_p and the corresponding 3-h average of the observed θ. The dots are slightly displaced in a vertical direction to avoid coincidence. Each big dot represents the average of all the small dots falling within a column that contains one tenth of the total data.

[5] J. M. WILCOX and N. F. NESS: J. Geophys. Res. **70**, 5793—5805 (1965); Fig. 11.
[6] K. H. SCHATTEN and J. M. WILCOX: [3] Fig. 4.

The magnetic activity index is found to be *correlated with the velocity of the solar wind*. This may indicate a relationship between the magnetospheric electric field and the solar-wind velocity as VASYLIUNAS[7] has shown.

VASYLIUNAS[7] used an empirical formula for the equatorial height of the plasmapause as a function of K_p to derive the *magnetospheric electric field* required in a model of convective motion (e.g. Fig. 26) on the dusk side. The electric field turns out to be approximately proportional to $(1+\frac{1}{5}K_p)$ for not too large K_p. Some authors had represented the correlation of K_p with the *solar-wind velocity* by a formula that approximately is

$$v = A(1+\tfrac{1}{5}K_p) \tag{30.1}$$

with $A = 262$ km sec^{-1} (according to one group of authors). The magnetospheric electric field E and the velocity v are seen to vary with the index K_p in the same way. Thus there is for not too large K_p (according to VASYLIUNAS)

$$E = Cv. \tag{30.2}$$

The constant, which is obtained as

$$C = 1.5 \cdot 10^{-9}\ \mathrm{V\,sec\,m^{-2}}, \tag{30.3}$$

suggests that the electric field is a certain fraction of the electric field $\boldsymbol{E} = -\boldsymbol{v}\times\boldsymbol{B}/c_0\sqrt{\varepsilon_0\mu_0}$ in the solar wind, where $B \approx 5\cdot 10^{-9}$ V sec m^{-2} ($=5\,\gamma$) under quiet conditions (with $c_0\sqrt{\varepsilon_0\,\mu_0} = 1$ in the present units).

It has to be kept in mind that the solar-wind velocity itself is connected with other parameters of the solar wind[1]. A good correlation between *transverse magnetic fluctuations in the solar wind* and K_p is demonstrated in Fig. 20 of SIEBERT'S article[1].

In the attempt to understand the relationship of the magnetic activity on the Earth with various data of the solar wind one has to think of the simultaneous change of these data in a perturbation of the solar wind, particularly when solar-flare plasma arrives during a magnetic storm[8]. The perturbed solar wind at these events contains a stronger magnetic field, deviating more from the ecliptic plane than the regular field. The field is also highly variant and the wind velocity is higher than normal. One may be led to the conclusion that the increased magnetic activity is already present in the arriving solar wind. The term "interplanetary storm" was introduced by HIRSHBERG and COLBURN[8] with reference to the solar-wind phenomena concurring with the main phase of geomagnetic storms (Sect. 29).

NISHIDA observed *sequences of fluctuations in the interplanetary field direction and in the geomagnetic field strength* that are *strikingly similar* [*19*]. The correspondence was evident in recordings at the geomagnetic equator and in the polar region. Fig. 32 is an example of good correspondence between the field at Huancayo and the angle of the interplanetary field toward the ecliptic plane. A time shift of 11 min for the interplanetary field was taken into account to compensate for a delay between the two fields. The period of the fluctuations is of the order of one hour, in Fig. 32 somewhat shorter. The ground field at auroral latitudes shows different fluctuations. The equatorial region is suited best for the comparison because of the enhancement of field variations due to the high Cowling conductivity of the lower ionosphere at the magnetic equator.

[7] V. M. VASYLIUNAS: J. Geophys. Res. **73** (7), 2529—2530 (1968).
[8] J. HIRSHBERG and D. S. COLBURN: Planet. Space Sci. **17** (6), 1183—1206 (1969).

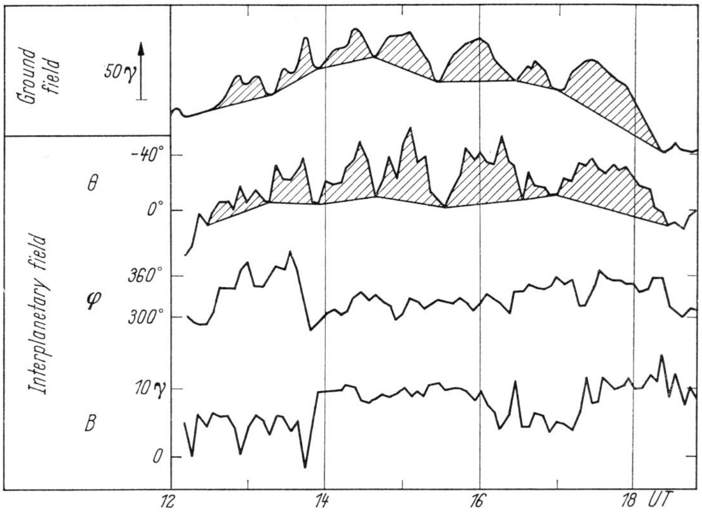

Fig. 32. Geomagnetic field at Huancayo (Peru), a station on the magnetic dip equator, and interplanetary magnetic field observed by IMP 1 near the noon meridian at 30 R_E on 2 Dec. 1963[9]. The data of the interplanetary field are the angle toward the ecliptic, θ, the longitude in the ecliptic plane counted from the direction toward the sun, φ (identical with ϕ in Fig. 2), and the magnetic flux density B. The interplanetary records are shifted by 11 min to the right to account for a time difference in the fluctuations.

NISHIDA [*19*] assumes that the **DP2** *current system,* which is present in variable intensity all the time and has a worldwide character (referred to in Sect. 24α), causes the fluctuations of the ground field. The **DP 2** current system is supposed to fluctuate accordingly. The good correspondence with variations of the magnetic field direction in interplanetary space may indicate that the interplanetary electric field penetrates into the magnetosphere or a peculiar friction force may couple the magnetosphere with the solar wind [*19*].

31. Storm effects in the ionosphere. The ionospheric perturbations that are most detrimental to radio wave propagation are sudden ionospheric disturbances (**SID**) and cases of polar cap absorption (**PCA**). Both are characterized by very high electron concentrations in the D- and lower E-regions, leading to complete or high absorption of short waves in the affected area. **SID**'s are caused by strong emissions of solar X-rays in the wavelength range 0.2 ··· 1 nm (2 ··· 10 Å). **PCA** events are due to energetic solar protons, frequently of energies from 1 to 20 MeV, occasionally reaching several hundred MeV. Both types of events have no direct relationship with magnetospheric processes, though the X-rays and energetic protons come from intense solar flares, which may also emit keV protons, causing auroral perturbations, and shock waves, initiating geomagnetic storms.

The perturbations of particular interest now are the *variations of the F2 layer in connection with magnetic storms.* These and other ionospheric perturbations are extensively treated in the literature[1–3] (for further material see the references given in the reviews referred to[1–3]).

[9] A. NISHIDA: J. Geophys. Res. **73** (17), 5549—5559 (1968); Fig. 2.

[1] K. RAWER and K. SUCHY: Radio-observations of the ionosphere. See this Encyclopedia, Vol. 49/2, p. 1—546, in particular p. 361—377.

[2] H. RISHBETH and O. K. GARRIOT: Introduction to Ionospheric Physics, in particular p. 261—270. New York: Academic Press 1969.

[3] T. OBAYASHI: J. Geomag. Geoelectr. **16** (1), 1—30 (1964).

In magnetic storms the *maximum electron concentration of the F2 layer is varied*, in middle latitudes frequently increased for a few hours and then decreased for two or more days. The variation of $N_m F2$ is best demonstrated as a function of storm time, placing zero time at the beginning of the storm (perhaps a sudden commencement). Fig. 33 is such a representation of average stormtime ($\mathbf{D}_{st}$) variations of $N_m F2$ for various geomagnetic latitudes and different seasons. The predominant effect is a depression, but it is preceded by an increase, which lasts longer than the increase of the magnetic field in the initial phase of the geomagnetic storm. At nearly equatorial latitudes and in winter also at slightly higher latitudes there is only an increase (or a short decrease with a succeeding increase).

Satellite and moon-echo observations have shown that the *total electron content decreases* in storms and at times of an increased K_p. The decrease of $N_m F2$ in storms thus is not the consequence of a redistribution of electrons and ions only. The observed *F2 layer height*, more specifically the minimum virtual heights of F and F2 in ionograms are found to increase in disturbed situations; but this increase is not so easy to interpret because it may partly be a group retardation. RÜSTER in an elaborate study[4], based on work by MARTYN and KOHL, derived the upward motion of F2 ionization in substorms, in qualitative agreement with observations. Height increases between 70 and 140 km were observed at low latitude.

F2 layer disturbances associated with increased magnetic activity appear to be delayed by roughly one day (according to J. W. KING[2]). This is interesting in comparison with the delay of some hours that the plasmapause shows in getting affected by magnetic activity (Sect. 18γ).

The *F2 layer variation* during a *magnetic storm* is of a *complex nature*. This becomes already apparent from the facts that the deviation of $N_m F2$ is partly positive, partly negative (Fig. 33) and the transition from the positive phase to the negative phase is not coincident with the transition from the initial phase to the main phase of the magnetic storm. The primary cause of a magnetic storm is a stronger compression of the magnetosphere by an increase of the solar-wind pressure (Sect. 29). During the process of increasing compression some transitory electrodynamic drift motions are going on (Sect. 27). In the state of higher compression the convective motion system that is believed to exist all the time (Sect. 24, Fig. 26) is more intense. An obvious idea is that these motions may somehow produce variations of the plasma density in the F2 layer.

The effect of *electrodynamic drift motions* on the plasma density has been investigated[2]. Density variations and magnetic-field variations in combination should arise from convergence or divergence of drift motions, but drift motions alone do not seem sufficient for the required compression or expansion of the plasma. It has to be noticed that the relative change of the magnetic field strength is much smaller than that of the plasma density. Plasma motions along field lines are more apt to generate large density variations.

Another possibility is *stoppage of motions* — electrodynamic drift or motion along field lines — in progression downward. This would cause severe density variations in the height interval of the stoppage. In the consideration of electrodynamic drift one should, however, not expect an ascent (or descent) all around the Earth. This would require a line integral $\oint \boldsymbol{E} \cdot d\mathbf{s}$ of a too large value on a path around the Earth (see Sect. 27). An ascent on the nightside should be accompanied by a descent on the dayside.

[4] R. RÜSTER: J. Atmos. Terr. Phys. **27** (11/12), 1229—1245 (1965).

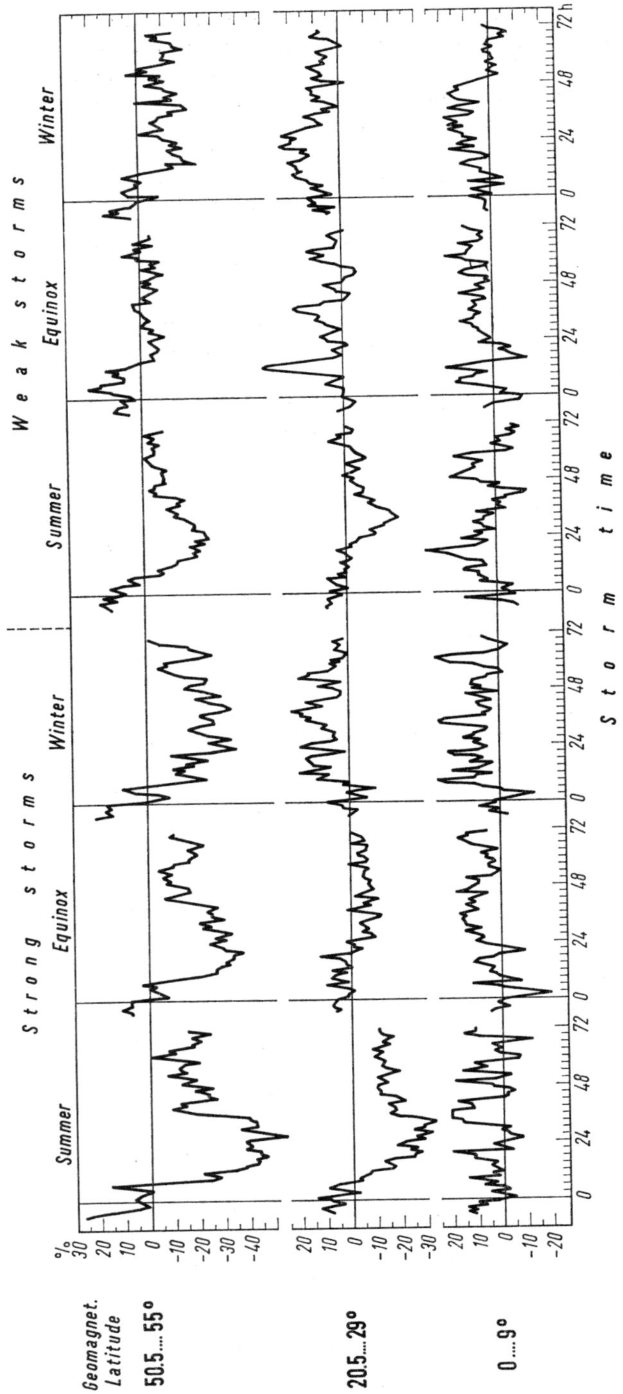

Fig. 33. Storm time variations of the electron concentration at F2 vertex in different intervals of geomagnetic latitude and in different times of the year[5].

[5] S. MATSUSHITA: J. Geophys. Res. **64**, 305—321 (1959); Fig. 3.

The different sign of the density variations at intermediate latitude in summer and winter might indicate that *motions along field lines* are involved. A motion of plasma extending along an entire field line would remove plasma from one hemisphere and supply plasma to the other. This should take place in asymmetric situations, i.e., in summer and winter. Motions along field lines can *arise from electrodynamic motions* by means of *coupling*. A study of wave modes characterized by the different motions demonstrates this coupling effect (Sect. 11γ)[6].

Other processes, which affect the electron density of the F2 layer, have also to be considered. The loss rate depends on temperature and on composition of the air. Heating of the outer atmosphere during storms has been observed by means of satellite drag studies. The composition changes with the upper height limit for turbulent mixing, e.g. of O and O_2 (ref.[2] above). Molecular gas has the higher loss rate. Height variations of the layer also influence the loss rate. Thermal expansion of the layer was occasionally noted as a possible cause of reduction of the electron concentration. The observed decrease of the total electron content up to the height of some satellite, however, seems to disprove this as the exclusive explanation.

It is likely that various effects act together[2]. Motions should in no case be disregarded. Attention may be called to the magnitude of magnetic field variations in storms, which is as large as or even larger than that of regular diurnal variations. The horizontal wind velocities required for these have an amplitude of the order of 100 m sec^{-1}. Comparable horizontal velocities in the F2 layer should thus be expected in conjunction with perturbations too.

Particularly heavy perturbations of the F2 layer occur *in the auroral zone* in some events[1]. The F2 layer can nearly disappear in consequence of the strong decrease of the electron concentration. Strong Es (sporadic E) appears in these auroral-zone perturbations.

32. Auroras. The visual phenomena of auroras have been studied for decades and are fairly well understood[1,2]. New concepts on the spatial and temporal arrangement and on the conjunction with other disturbance phenomena, however, have evolved during the last decade [*14*], [*15*], [*23*], [*24*][1,2]. Rocket and satellite observations, coordinated observations by various means, and the knowledge about the magnetosphere stimulated corrections of old oversimplified ideas.

Despite the regular nightly appearance in the auroral zone auroras are to be considered a disturbance phenomenon. Intense displays with a shift to lower latitudes follow strong solar flares within 20 to 40 h. They are caused by a faster, more intense solar wind (Sect. 4). It is, however, still undecided what the processes involved are, in particular, how the energy of solar-wind disturbances enters the magnetosphere and in what form it is propagated to the perturbed regions of magnetosphere and ionosphere.

α) *Observational facts.* The region of maximum occurrence of auroras is the *auroral zone*, a nearly circular ring around 67° geomagnetic latitude (in the south slightly more poleward). In an individual case auroras are not necessarily concentrated on the auroral zone. It has been found that auroral arcs preferably lie in a narrow belt, which encloses the magnetic dipole but is *highly eccentric and varies in position*. It is called the *auroral oval* (Fig. 34). The idea of an auroral

[6] H. Poeverlein: Phys. Rev. **136** (6A), A 1605—1613 (1964).

[1] S.-I. Akasofu: The aurora, in this Encyclopedia, Vol. 49/1. Berlin-Heidelberg-New York: Springer 1966.

[2] B. M. McCormac and A. Omholt (ed.): Atmospheric Emissions. New York: Van Nostrand Reinhold Co. 1969.

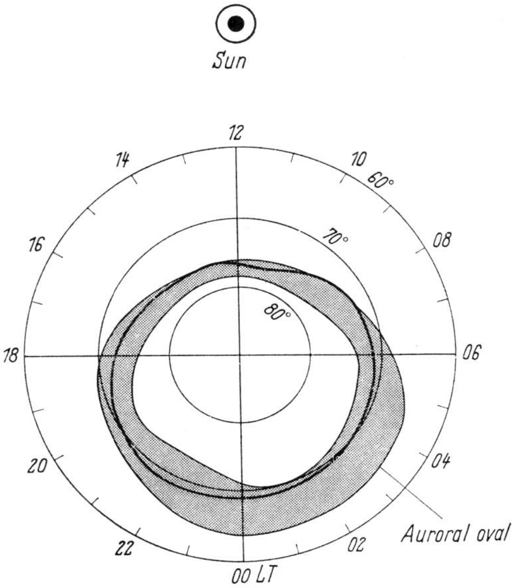

Fig. 34. Auroral oval and high-latitude boundary of trapped electrons shown in a polar map[3]. The dotted area is the auroral oval (after Feldstein). The strong line is an iso-intensity line of the trapped-electron flux plotted in invariant latitude [Eq. (20.5)]. The line corresponds to an omnidirectional flux of 10^8 m^{-2} sec^{-1} (10^4 cm^{-2} sec^{-1}) for $\mathscr{E} > 40$ keV (after Frank, van Allen, and Craven). The flux decreases rapidly inside the line toward the pole.

oval that varies its position with the degree of a disturbance was introduced by Feldstein and has been further justified by Akasofu [*23*], [*24*].

The *auroral oval* is *shifted equatorward in magnetically disturbed periods* (periods with a high K_p) and it is shifted toward the pole from its normal position in extended quiet periods (e.g. 24 h of very low K_p). On the dayside it is at higher latitude than at night (see also Sect. 35). The auroral *zone* is largely determined by the predominant location of the *midnight section of the oval*, where intense auroras are frequent.

The auroral oval lies approximately on the *magnetic field lines that represent the outer boundary of the trapping region*. The strong line in Fig. 34 refers to the field lines at which the flux of trapped electrons (>40 keV) was 10^4 cm^{-2} sec^{-1} (10^8 m^{-2} sec^{-1}), a value in the middle of a rapid decrease toward polar field lines. This and the displacement of the oval with varying K_p, analogous to the displacement of the plasmapause, seem to indicate that the field lines of the oval are at or near the transition between closed field lines and open field lines, extending far out into the tail from where particles may be supplied.

The behavior of active auroras (i.e., all auroras different from quiet arcs) is described by a superposition of a dependence on local time and an occurrence of active periods, named *auroral substorms* by Akasofu [*23*], [*24*]. The auroral substorms, starting often abruptly, growing rapidly, and dying down more gradually, are a universal-time phenomenon. Quiet auroras (quiet arcs) are common throughout the auroral oval in quiet periods. An auroral substorm starts near the equatorward edge of the oval around midnight with a sudden increase of brigthness, followed by active auroral forms and expansion in all directions, in particular poleward; the morning and evening auroras get activated and the

[3] S.-I. Akasofu: Planet. Space Sci. **14** (7), 587–595 (1966); Fig. 1.

aurora extends beyond the oval — far poleward and also equatorward. In the later course the auroral substorm subsides slowly. The entire duration is from 1 to 3 h.

Several auroral substorms are observed in a moderately disturbed period. They can be separated by intermissions, during which the quiet arc remains. In more complicated cases substorms are overlapping.

An auroral substorm is associated with a polar magnetic substorm (positive or negative bay, depending on local time) [*23*], [*24*]. This means that certain currents in the ionosphere belong to an auroral event. The disturbance field in *magnetic storms* and the corresponding current system in the lower ionosphere are usually decomposed into a part dependent on stormtime $\mathbf{D}_{st}$, in which the variation with geomagnetic longitude is averaged out, and an additional part, $\mathbf{D}_S$, that varies with both longitude and storm-time [*27*]. The variation of this last part with the time of day is demonstrated by arranging the $\mathbf{D}_S$ fields or currents according to local time and taking appropriate averages. This way one obtains a model of the field or a current system surrounding the Earth, which is denoted $\mathbf{S}_D$ (or **SD**) and is usually presented separately for the first, second, and (perhaps) third days of magnetic storms. AKASOFU [*23*] points out the similarity between the $\mathbf{S}_D$ current system and the current system of an individual polar magnetic substorm. It seems that the polar magnetic substorms, which are linked with auroral phenomena, are an essential constituent of geomagnetic storms (cf. Sect. 29).

Typical of the $\mathbf{S}_D$ current system is the *auroral* (or "polar") *electrojet*, an intense current toward the west in the lower ionosphere from late evening till late morning. The $\mathbf{S}_D$ disturbance is worldwide. In connection with local auroral forms there are, however, localized strong currents. To demonstrate the intensity of the disturbance it may be noted that the magnetic-field variation amounts to several hundred γ in strong storms and occasionally reaches 2000 γ ($=2 \cdot 10^{-6}$ V sec m^{-2}).

It is nowadays believed that the *current circuits* for the auroral electrojet and for the more local current strip are not closed whithin the lower ionosphere, but extend along magnetic field lines to high altitudes in the magnetosphere, where they are closed (see FUKUSHIMA's discussion of the three-dimensional current system in Sect. 34 of this article). In the case of the auroral electrojet the outer part, on the dayside of the magnetosphere, is called the *asymmetric ring current*. The closed ionospheric current system derived from the magnetic field on the ground takes the role of an equivalent current system, which would lead to the correct magnetic-field variation, but does not exist in this form.

Magnetic substorms appear in succession and with high intensity in the main phase of magnetic storms (Sect. 29). They seem to be essential for the development of the main phase. The combined appearance of auroral substorms and polar magnetic substorms justifies the name "*magnetospheric substorm*", which also points out their magnetospheric nature [*23*].

Auroras have been known to result from *precipitation of* keV *electrons*. X-*rays* released by the *electrons* have been observed by means of balloon detectors. X-ray bursts and electron bursts have been detected during auroral substorms. Certain quiet auroras or parts of them are due to *proton* precipitation.

The ionospheric currents in auroral events may be Hall and Pedersen currents. The electric field driving the Hall current causes drift motion in the opposite direction at a higher level (Sect. 23). The *ionospheric and magnetospheric drift motions* were investigated by means of release of Ba ion clouds in and beyond the auroral zone (Sect. 24β)[4,5]. Positive and negative perturbations of the horizontal magnetic field component were found to be connected respectively with westward and eastward drifts in the auroral-zone ionosphere, in accordance with the idea of Hall currents. The drift motions observed in the *outer* magnetosphere (in the case of a negative bay) and in the polar-cap ionosphere, however, are such that Hall currents can not fully explain the magnetic perturbation vectors of a wider area. Field-aligned currents may have to be postulated (Sect. 34)[4,5].

A study of ground magnetograms and magnetometer data of the outer magnetosphere in conjunction with results of Ba ion cloud experiments[6], published while this review was in press, lends support to the idea of field-aligned currents and gives more specific information on them.

β) *Theoretical aspects.* The explanation of all the observations on auroras, polar magnetic substorms, and related phenomena poses a difficult theoretical problem ([*15*] Sect. 45), [*23*], [*24*]. It is clearly established that particle emissions at some solar flares lead to intense auroras and geomagnetic disturbances, but there are various hypotheses for the stages intermediary between the solar wind and the observed effects.

An entrance of solar-wind particles into the magnetosphere is possible via merging field lines, via neutral points of the magnetosphere or at the far end of the tail. But, in addition, particles have to be accelerated. In other words, energy has to be transferred to particles somewhere in the magnetosphere.

The old idea that charged particles travel directly from the Sun to 100 km height where they produce auroras had to be abandoned. ALFVÉN's *model of drift motion* of incoming particles with acceleration to a few keV (Fig. 23) did not take the boundary of the magnetosphere into account. The application of this model to particles approaching from the tail of the magnetosphere is more promising. The particles would reach a lowest field line shell that corresponds to the auroral oval.

Not much different is the particle motion in *convective motion models* (Fig. 26). Particles moving with magnetic field lines from the tail inward would stick to field lines that are gradually shortened. The consequence is Fermi (or betatron) acceleration (Sect. 22γ), possibly just to the energy of auroral particles.

The theories assuming *plasma from the tail* to produce auroras are supported by the fact that the auroral oval seems to be connected to the transitional field lines between tail and closed shells and is in disturbed periods displaced accordingly. This leaves the questions open how the particles get into the tail and whether energy can be stored in the tail plasma or in the tail magnetic field. An estimate of the tail magnetic energy (Sect. 28) yielded an amount that could produce an auroral substorm. The energy carried by the auroral electrons in a substorm is estimated at $6 \cdot 10^{14}$ J ($=6 \cdot 10^{21}$ erg) [*23*]. Half of this energy may be deposited in precipitation of electrons in the aurora (Sect. 28). A magnetic storm, for comparison, needs more than 10^{15} J (Sect. 28). It should be recalled that the magnetic energy of the tail is increased in a state of higher compression (Sect. 28) and comes down to its old value in recovery.

Observations of magnetic substorms at high altitudes have shown that the substorm begins near the outer edge of the trapping region (cf. Sect. 29, [*23*]). This region may consequently be considered as the *region of origin* of a substorm. In the magnetotail the magnetic disturbance arrives later.

[4] G. HAERENDEL and R. LÜST: Electric fields in the ionosphere and magnetosphere, in [*16a*], p. 213—228.

[5] E. M. WESCOTT, J. D. STOLARIK, and J. P. HEPPNER: Auroral and polar cap electric fields from Barium releases, in [*16a*], p. 229—238.

[6] G. HAERENDEL, P. C. HEDGECOCK, and S.-I. AKASOFU: J. Geophys. Res. **76** (10), 2382—2395 (1971).

If the energy had been stored in the magnetosphere as magnetic energy, the aurora would be the result of *conversion of magnetic energy into plasma or particle energy*. Such a conversion has been known from the ideas on a subsolar neutral point (Sect. 17) and from the emission of protons in solar flares.

After discovery of the radiation belt the idea arose that the auroral particles might be particles dumped from the outer region of the radiation belt. It turned out, however, that the amount and the energy of the trapped particles are too low. The energy requirement of *one* substorm, of the order of 10^{15} J, corresponds roughly just to the energy of all trapped fast particles (Sect. 28). Low-energy particles were noted as carrying somewhat more, but an entire bright aurora may need much more [*14*], [*15*]. The conjecture at present is that *refilling of the outer radiation belt and auroral precipitation in magnetic storms both are consequences of the same process*. O'Brien found that the fluxes of auroral and trapped electrons rise frequently at the same time [*14*], (Sects. 28 and 44β of [*15*]).

The remaining question is that for the *initiation of a substorm*. There may be a driving electric field, current or plasma instability. Field line merging or some effect in interaction of the solar wind with the magnetosphere could be involved.

Electric fields may be carried in by the solar wind. A voltage of the order of 50 kV, which is supposed to exist across the magnetosphere (Sect. 17γ and Fig. 26) would be sufficient in any acceleration mechanism. Another possibility would be some charge separation when electrons and positive particles move differently. In earlier theories a different penetration depth of electrons and ions incident from outside was considered. Fejer [*23*][7] developed a theory of the $\mathbf{D}_S$ or $\mathbf{S}_D$ current system, in which charge separation results from the drift motion of keV-protons in the inhomogeneous geomagnetic field that is distorted by the solar wind. The keV-protons are the ring current particles (Sect. 21α); they appear to be fast particles in the sense of Sect. 19β, being subject to the drift in the inhomogeneous magnetic field — in contrast to the electrons, who have much lower energies.

What prohibits an understanding of the dynamics of auroras seems to be a lack of knowledge on processes in the magnetosphere. A clarification may be expected from further exploration of magnetic and electric fields, of mass motion and particle fluxes, and of the coordination of phenomena at various places on the Earth and in the magnetosphere.

H. Remarks on plasmapause and current systems.

By N. Fukushima.

The following two sections are a postscript to the contribution by Nagata and Fukushima to Vol. 49/3 of this Encyclopedia [*27*] with particular reference to Sects. 69 and 70 of that contribution, but they may also serve as a supplement to the present article on the magnetosphere. It should be noticed that in recent years additional information on particles and fields in the magnetosphere is coming up and the morphology of magnetic disturbances is discussed in view of the results of space investigations. Akasofu in a monograph [*23*] treated magnetospheric substorms and related phenomena. Interesting observational material

[7] J. A. Fejer: J. Geophys. Res. **69** (1), 123—137 (1964).

obtained in the magnetosphere is presented in several volumes of Proceedings of a Summer Advanced Study Institute [*16a*][1-3].

33. Plasmapause and resonant magnetodynamic[4] oscillation. From the analysis of geomagnetic pulsations at different latitudes we had expected the presence of some discontinuity layer in the magnetosphere that permits an effective trapping of energy in magnetodynamic oscillations[5]. Observations of whistlers, in fact, proved the existence of a discontinuity layer of the plasma density, which has been named plasmapause (Sect. 18 of this article; references are given there). The geocentric height of the plasmapause depends on both local time and geomagnetic activity. Diurnal variation of the height of the plasmapause seems to explain the diurnal variation of the period of geomagnetic pulsations **pc 4** and **pc 3**. The morphology of geomagnetic pulsations is summarized in a recent review paper by SAITO[6] and a book by JACOBS[7].

34. The equivalent current-system in the ionosphere and the three-dimensional current-system. Throughout the article of NAGATA and FUKUSHIMA [*27*] a two-dimensional equivalent current-system in the ionosphere was for convenience used to describe a world-wide geomagnetic disturbance. However, from the ground magnetic observations alone it is impossible to obtain a true current distribution in the space surrounding the Earth. A part of the two-dimensional equivalent current-system in the ionosphere may, only be a substitute for a *three-dimensional* current flow in the ionosphere and magnetosphere that produces the same geomagnetic effect on the ground. It is even possible to imagine a current flow in the ionosphere and magnetosphere whose magnetic field is observable only in and above the ionosphere and not on the ground. Fig. 35 shows that on the ground the magnetic effect of a current flowing vertically in from infinity is cancelled by the magnetic effect of the overhead spreading current in the plane ionosphere, if the electric conductivity of the thin ionosphere is uniform. Therefore, unless the ionospheric conductivity is non-uniform, the Pedersen current (parallel to the electric field) in the ionosphere does not seem to play an important role for geomagnetic disturbances at high latitudes, if it is connected with a current along a magnetic field line, a "field-aligned current". On the other hand, the Hall current (perpendicular to the electric field) in the ionosphere always produces a geomagnetic variation on the ground.

Two alternative electric current-systems have been postulated for the world geomagnetic variation during polar magnetic storms. One of them includes field-aligned currents in the magnetosphere and an east-west current-segment in the ionosphere. This model was first suggested by BIRKELAND[8] and later advocated by ALFVÉN[9]. The other, confined to the ionosphere, was originally depicted by

[1] B. M. McCORMAC (ed.): Radiation Trapped in the Earth's Magnetic Field. Dordrecht, Holland: D. Reidel Publishing Company 1967.

[2] B. M. McCORMAC (ed.): Earth's Particles and Fields. New York: Reinhold Book Corp. 1968.

[3] B. M. McCORMAC and A. OMHOLT (ed.): Atmospheric Emissions. New York: Van Nostrand Reinhold Co. 1969.

[4] The word "magnetodynamic" is used in the same sense as "hydromagnetic" (in much of the literature) or "magnetogasdynamic" and "gasmagnetic" (in Vol. 49/3 and in this volume of the Encyclopedia), see Introductory Remarks p. 1.

[5] T. HIRASAWA and T. NAGATA: Pure Appl. Geophys. **65**, 133 (1966).

[6] T. SAITO: Geomagnetic pulsations. Space Sci. Rev. **10**, 319—412 (1969).

[7] J. A. JACOBS: Geomagnetic Micropulsations. Berlin-Heidelberg-NewYork: Springer 1970.

[8] Ref. [*29*] in [*27*].

[9] H. ALFVÉN: Kungl. Sv. Vetensk.-Akad. Handl. Ser. III, **18**, No. 3 and No. 9 (1939, 1940).

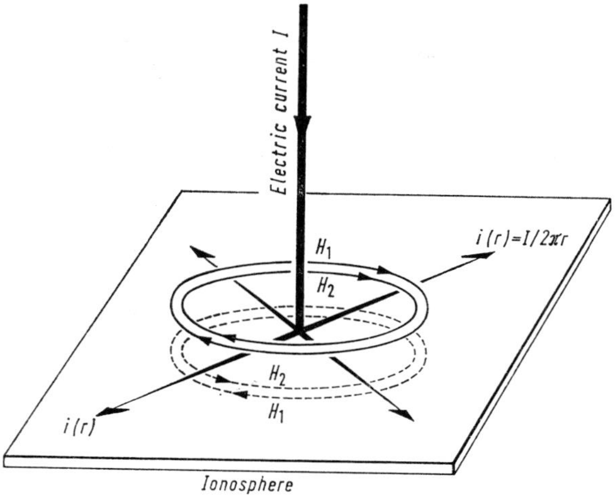

Fig. 35. Vertical current in the magnetosphere, horizontal spreading current in the ionosphere, and their magnetic effect above and below the ionosphere. Below the ionosphere $\boldsymbol{H}_1+\boldsymbol{H}_2=0$, while $\boldsymbol{H}_1+\boldsymbol{H}_2=2\boldsymbol{H}_1=2\boldsymbol{H}_2$ above the ionosphere; $H_1=H_2=(\mathrm{u}/c_0\sqrt{\varepsilon_0\mu_0})\, i/2=(\mathrm{u}/c_0\sqrt{\varepsilon_0\mu_0})\, I/4\pi r$.

CHAPMAN[10] and by VESTINE and CHAPMAN[11], and upheld by a number of investigators. For a typical minor disturbance at high latitudes, the former consists of a westward electrojet along the auroral zone in the ionosphere and field-aligned currents as shown on the left-hand side of Fig. 36, while the latter is a westward electrojet with return currents flowing in the ionosphere as shown in the right-hand picture of Fig. 36.

It is assumed here for simplicity that: (1) the geomagnetic effect at a point of observation on the Earth's surface is due entirely to the electric current flowing nearby, (2) the curvature of the Earth's surface and of magnetic field-lines can be ignored, and (3) field-aligned currents flow in or out vertically. Then the BIRKELAND-ALFVÉN current model is given by A) of Fig. 37, which is the superposition of B) and C). Current pattern B) is produced when two electric poles with different potentials make contact with the plane of uniform electric conductivity that represents the ionosphere, while pattern C) is similar to the CHAPMAN-VESTINE current-system. Pattern B) is subdivided into B1) and B2). As indicated in Fig. 35, no magnetic effect below the ionosphere results from the current patterns B1) and B2). Therefore, the magnetic effect on the ground of the current system A) is exactly the same as that produced by the current pattern C)[12]. The ground magnetic effect of the BIRKELAND-ALFVÉN current-system and of the CHAPMAN-VESTINE current-system can also be proved equivalent in the case of a spherical ionosphere[13], if the field-aligned currents are assumed to flow radially inward and outward. However, the curvature of geomagnetic field-lines must be taken into account for a more precise discussion.

The current system A) of Fig. 37 is produced if all current lines in the ionosphere are concentrated in a narrow band along the auroral zone. Even when the

[10] Ref. [*18*] in [*27*].

[11] Ref. [*17*] in [*27*].

[12] N. FUKUSHIMA: Rep. Ionos. Space Res. Japan **23**, 219 (1969); Radio Science **6**, 269—275 (1971).

[13] V. M. VASYLIUNAS: unpublished; refer to the same author: Mathematical models of magnetospheric convection and its coupling to the ionosphere, in [*16a*], p. 60—71.

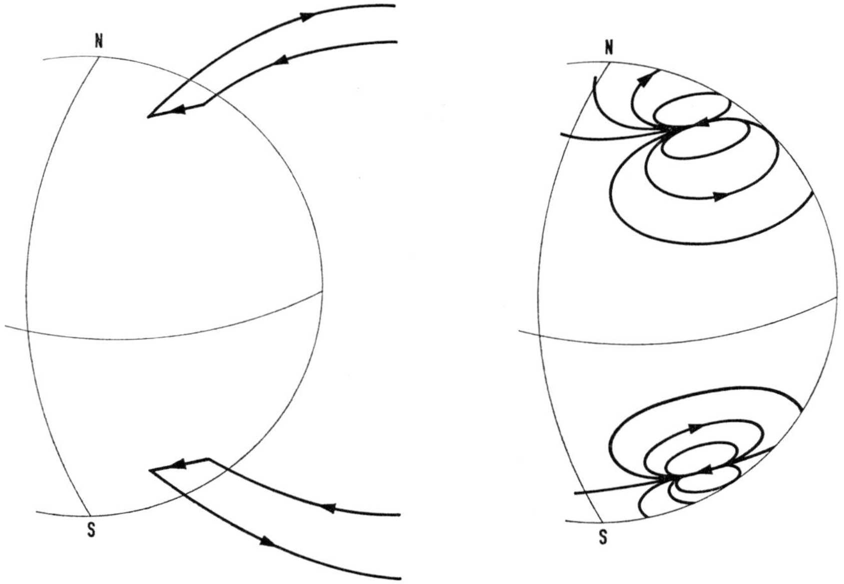

Fig. 36. Schematic illustration of the BIRKELAND-ALFVÉN current-system (left) and the CHAPMAN-VESTINE current-system (right) for polar magnetic substorm.

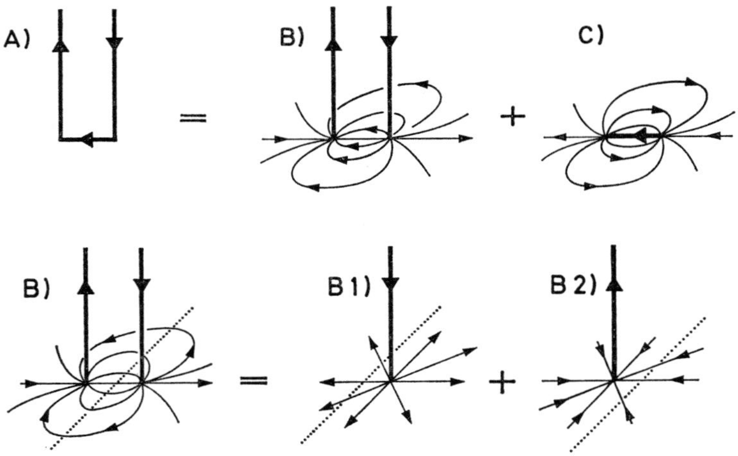

Fig. 37. The BIRKELAND-ALFVÉN current-system A) is considered to consist of B) and C), where C) is the CHAPMAN-VESTINE current-system in the ionosphere. B) produces no magnetic effect on the ground.

ionospheric conductivity along the auroral zone is much higher than that in the surrounding region, however, not all of the current lines concentrate along the auroral zone, but some of the current must spread into the surrounding region in the form of current system B). Therefore, the current system A) is always accompanied by a current system B) and the resultant current flow in the ionosphere depends on the electric-conductivity distribution in the ionosphere.

An *electric field* in the auroral zone with a *strip of enhanced conductivity* leads to *charge accumulations* as a consequence of the Hall conductivity. With

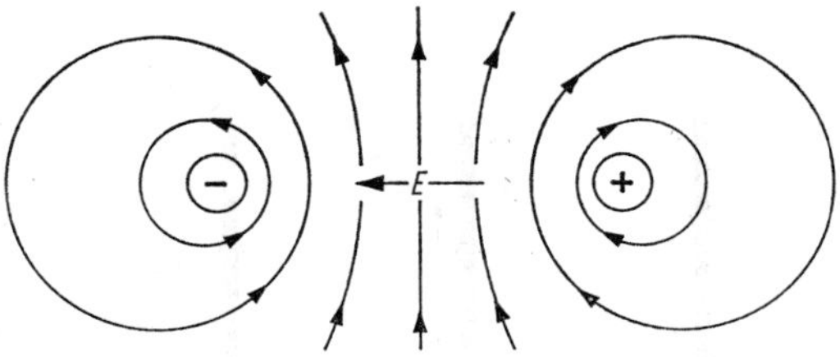

Primary Hall current with no enhancement of electric conductivity.

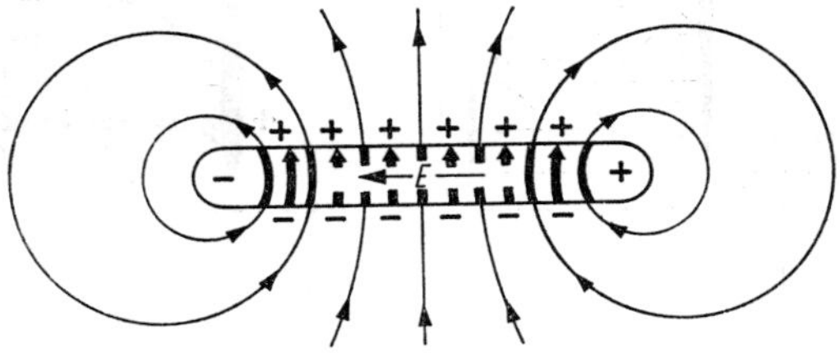

Primary Hall current and charge accumulation around the region of enhanced conductivity.

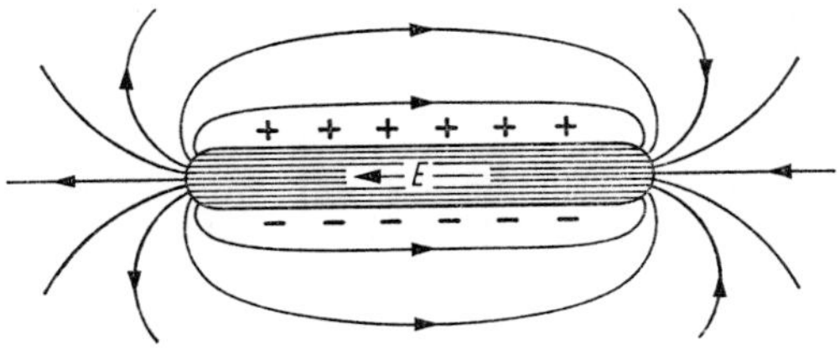

Secondary Hall current due to charge accumulation around high-conducting region.

Fig. 38. Hall current in the ionosphere under a westward electric field (view from above the ionosphere in the northern hemisphere). The **E**-vector indicates the primary electric field.

a westward electric field in the northern auroral zone there is an accumulation of positive and negative charges respectively at the northern and southern boundaries of the region of enhanced conductivity (Fig. 38). It may be noted that the region of enhanced conductivity is not necessarily located at the place of the strongest electric field in east-west direction.

Fig. 38 suggests that the westward auroral electrojet and its return currents are produced in the auroral-zone ionosphere in a primary electric field toward west with an enhanced conductivity. A *current circuit in the meridian plane* with field-aligned currents arises then as a by-product; the field-aligned current flows outward on the high-latitude side of the auroral zone (from $+++$) and inward on the low-latitude side ($---$). The right-hand picture of Fig. 39 demonstrates this. The field-aligned currents together with the ionospheric current from south to north do not produce an appreciable magnetic field on the ground, if there are field-aligned current *sheets* extending over a wide range of longitudes.

A westward electrojet along the auroral zone may, however, be generated in a different way, as a *primary Hall current* under an *intense equatorward electric field* applied from the *magnetosphere* to the ionosphere. This case is shown on the left-hand side of Fig. 39. Electric conductivity in the auroral-zone ionosphere

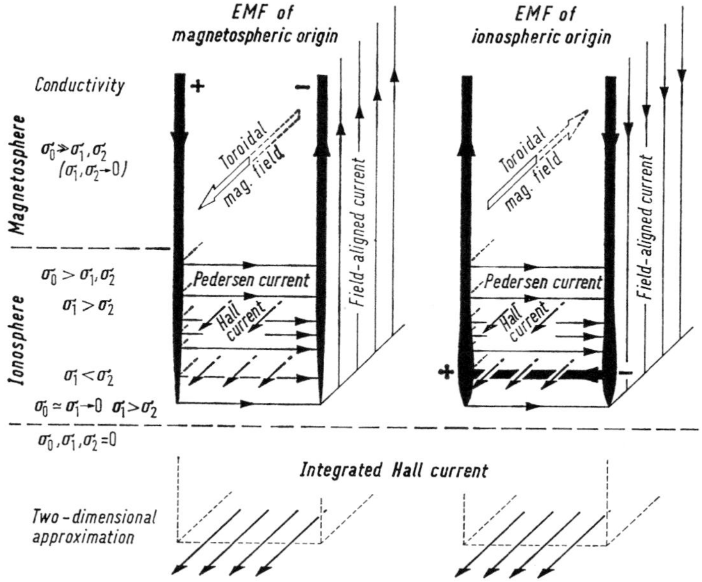

Fig. 39. Meridional cross-section of the three-dimensional current flow and toroidal magnetic field when the electromotive force is of magnetospheric origin (left) and of ionospheric origin (right). Bottom diagram shows the height-integrated Hall current. (View in the northern hemisphere; north is to the left.) At the left the three conductivities are compared: the parallel conductivity σ_0, the Pedersen conductivity σ_1, and the Hall conductivity σ_2.

is not necessarily enhanced in this case. Comparing the left and right diagrams of Fig. 39, we see that the toroidal magnetic field above the ionosphere (not observed on the ground if the longitudinal extent of the current-sheet is infinite) is just opposite in the two cases, though the ground magnetic effect is the same.

ZMUDA et al.[14] detected a *local east-west magnetic field over the auroral-zone ionosphere* (at the altitude of 1100 km) by means of the polar orbit satellite 1963 38 C. Though they attributed this magnetic field to a transverse hydromagnetic disturbance propagated from the magnetosphere down to the ionosphere, the local east-west magnetic field might be indicative of a toroidal magnetic field due to the field-aligned current-system in the meridional plane. In their recent paper, ARMSTRONG and ZMUDA[15] reported that the left-side case of Fig. 39 was detected at 0851 local time on Nov. 1, 1968, over the central part of North America at geomagnetic latitudes 64 to 66°. The westward toroidal magnetic field was about 800 γ ($=8\cdot 10^{-7}$ V sec m^{-2}). On the other hand, CLOUTIER et al.[16] reported that their rocket experiment at Fort Churchill, Canada, 2000 LT on Feb. 26, 1969, showed the right-side case of Fig. 39. The observed toroidal magnetic field was eastward, amounting to approximately 330 γ.

From these results we surmise that the field-aligned current configuration in the meridional plane may depend on local time. Further satellite and rocket experiments are desirable for the study of the three-dimensional current distribution in and above the ionosphere. Precise measurements of the magnetic field in and above the ionosphere are indispensable for a decision on the distribution of the

---

[14] A. J. ZMUDA, J. H. MARTIN, and F. T. HEURING: J. Geophys. Res. **71**, 5033 (1966).

[15] J. C. ARMSTRONG and A. J. ZMUDA: J. Geophys. Res. **75**, 7122 (1970).

[16] P. A. CLOUTIER, H. R. ANDERSON, R. J. PARK, R. R. VONDRAK, R. J. SPIGER, and B. R. SANDEL: J. Geophys. Res. **75**, 2595 (1970).

electric currents. This has been discussed by many research workers[17-23]. Electric field measurements in and above the ionosphere are essential for an estimate of the contribution of Pedersen and Hall currents. If the electric field measured in the ionosphere would not correspond to the equivalent ionospheric currents derived from the ground geomagnetic variations — especially outside the auroral zone — this would prove that field-aligned currents in the magnetosphere play an important role[24].

The above discussion dealt only with the electric current system of a stationary state. However, the polar magnetic substorm is a transient phenomenon and the field-aligned currents as well as the currents in the ionosphere change rapidly during the course of substorms. The longitudinal extent of the auroral zone with the enhanced ionospheric conductivity varies also during the development of polar substorms. The toroidal magnetic field produced by the field-aligned currents in the meridional plane may leak out to the ground from the eastern and western edges of the region of enhanced conductivity. Such an effect may be detectable through a precise analysis of the progressive change in the ground geomagnetic field variation at high latitudes. Other observational information, e.g., on auroral emissions and ionospheric conditions, has also to be examined in a thorough study of polar substorm phenomena.

I. Recent findings.

This chapter, dealing with observational facts and theoretical ideas that became known recently, was added in proof-reading.

35. Polar cusps. Satellite measurements of proton and electron fluxes in the polar magnetosphere pointed out the existence of a region on the dayside where *protons and electrons* with energies of roughly 100 eV to 1 keV are present in *great numbers*[1-3]. This region, named "polar cusp"[2], is of a wide east-west extension and appears to be *adjacent to the last closed field lines*, which represent the end of the radiation belt. The observations were made in the northern hemisphere; a southern polar cusp has also to be expected.

The energy distributions of the protons and electrons resemble those of the magnetosheath particles. This fact and the location of the polar cusps seem to indicate that they are something like wide field line tubes centered on the *field lines that should reach the neutral points of a closed magnetosphere* (Fig. 9) and the observed particles are *magnetosheath plasma* entering the polar cusp at the top. During magnetic quiescence the polar cusp is on the noon meridian at low altitude found around 78° geomagnetic latitude. It is displaced equatorward by several degrees in magnetically disturbed periods. Its latitudinal width is approximately 2 to 5°; its longitudinal width corresponds to 8 hours or more in local time.

---

[17] R. BOSTRÖM: J. Geophys. Res. **69**, 4983 (1964); — Ann. Géophys. **24**, 681 (1968).
[18] B. BONNEVIER, R. BOSTRÖM, and G. ROSTOKER: J. Geophys. Res. **75**, 107 (1970).
[19] W. D. CUMMINGS and A. J. DESSLER: J. Geophys. Res. **72**, 1007 (1967).
[20] E. SEILER and W. KERTZ: Z. Geophys. **33**, 371 (1967).
[21] S.-I. AKASOFU and C.-I. MENG: J. Geophys. Res. **74**, 293 (1969).
[22] C.-I. MENG and S.-I. AKASOFU: J. Geophys. Res. **74**, 4035 (1969).
[23] J. W. KERN: J. Geomag. Geoelectr. **18**, 125 (1966).
[24] Some evidence of field-aligned currents in magnetic substorms has been obtained from the combination of Ba ion cloud experiments with magnetic measurements (Sect. 32α).
[1] W. J. HEIKKILA and J. D. WINNINGHAM: J. Geophys. Res. **76** (4), 883—891 (1971).
[2] L. A. FRANK: J. Geophys. Res. **76** (22), 5202—5219 (1971).
[3] C. T. RUSSELL, C. R. CHAPPELL, M. D. MONTGOMERY, M. NEUGEBAUER, and F. L. SCARF: J. Geophys. Res. **76** (28), 6743—6764 (1971).

Fig. 40 may give a rough idea of position and extension of the polar cusp as determined from the particle measurements. Its shape is, however, not so well known. The particle measurements were made at low magnetospheric heights by HEIKKILA and WINNINGHAM[1] and in the outer magnetosphere on Imp 5 by FRANK[2]. Magnetic-field measurements on Imp 5 specified the outer cusp as a region of a weak and strongly fluctuating magnetic field[4]. The magnetopause was not detectable at the top of the polar cusp. A field line connection at the neutral points had been postulated[5–8] in order to allow entrance of magnetosheath plasma and to provide for force balance at disappearing $\boldsymbol{B}$. Several authors propose cusp geometry of the magnetic field around the neutral points. On the other hand, the irregularity of the field and the absence of a magnetopause in the polar cusp may indicate a more unusual behavior of the field with field line connection through the top of the cusp (blank area of Fig. 40).

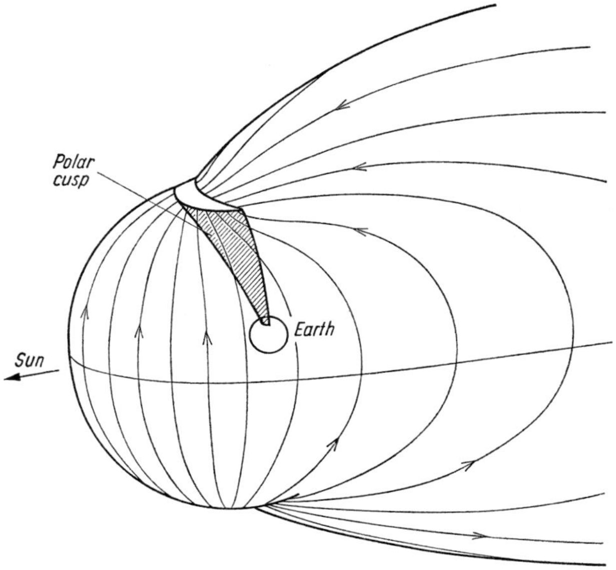

Fig. 40. Tentative sketch of the magnetopause and the northern polar cusp. Thin lines on the magnetopause are magnetic field lines and the equatorial line.

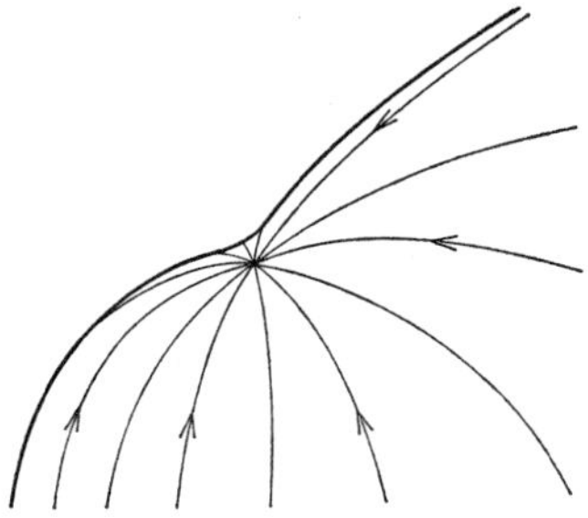

Fig. 41. Conventional concept of a neutral point on the magnetopause.

[4] D. H. FAIRFIELD and N. F. NESS: J. Geophys. Res. **77** (4), 611—623 (1972).
[5] A. I. ERSHKOVICH, V. D. PLETNEV, and G. A. SKURIDIN: J. Atmos. Terr. Phys. **29** (4), 367—376 (1967).
[6] J. R. SPREITER and A. L. SUMMERS: Planet. Space Sci. **15** (4), 787—798 (1967).
[7] M. I. PUDOVKIN, O. I. SHUMILOV, and S. A. ZAITZEVA: Planet. Space Sci. **16** (7), 881—890 (1968).
[8] M. I. PUDOVKIN and V. M. BARSUKOV: Planet. Space Sci. **19** (5), 525—531 (1971).

The conventional picture of magnetopause field lines around a neutral point is for comparison shown in Fig. 41.

The *convective motion* of the magnetosphere, i.e. a large-scale motion of field lines with plasma (Sects. 23 and 24), may move some plasma from the polar cusps to the plasma sheet (Sect. 16γ). In a circulatory motion system the reverse flux from the plasma sheet back to the polar cusps, will also occur. The various authors suggest different motion models [1–3,9]. A clear kinematic model of the moving field lines, taking the postulated merging process into account, is still missing. The magnetic field of the solar wind might affect the two polar cusps in a different way.

It is possible that the polar cusps are the sources of the particles precipitated in the *daylight aurora*. Several authors believe that the dayside and nightside parts of the auroral oval, situated at different geomagnetic latitudes, are neither causally nor spatially in a direct connection [7,8,10,11]. They may be caused by different particle populations [8,10,11] (particles from the polar cusps and from the plasma sheet).

36. Theory of the tail. The basic problem of the tail (Sect. 16β) is clearly stated in the question "What balances the magnetic tension of the tail". The magnetic field in the two halves is approximately homogeneous and of opposite direction. The total force due to the magnetic tension tends to contract the tail unless there is an equal force of a different nature pulling out. An alternative view is to consider the force density $\boldsymbol{J}\times\boldsymbol{B}$ in the (narrow) space between the two homogeneous fields, the "neutral sheet", and to integrate it. If all field lines are connected across the neutral sheet, the result is the same contracting force. A process that sustains the neutral-sheet current must involve a pulling-out force of the right size to keep the situation stationary.

According to Beard, Bird, and Huang [1] the neutral-sheet current may result from the *drift motion of particles in the tail field*, which is *not homogeneous*. In traversal of the neutral sheet the magnetic field reverses its direction. This means a gradient $\frac{\partial}{\partial \boldsymbol{r}} B$ and a strong curvature of the field lines connected through the neutral sheet. Both the gradient and the curvature lead to drift motions, which appear as currents [see Eq. (19.1)]. The curvature drift (ignored in the original paper [1]) turns out to be predominant and to yield a current of the right direction and approximately the right magnitude [2,3]. This drift is due to the centrifugal force on the particles that are trapped on tail field lines and follow them through the neutral sheet. The centrifugal force thus seems to be a pulling-out force of the required size. The concept of a smooth neutral sheet of the assumed transitory character can of course be only a kind of average, in which remarkable irregularities have been ignored.

37. Magnetospheric motion and ionospheric currents. A *strong coupling* between currents in the lower ionosphere and motions in the magnetosphere results from

---

[9] L. A. Frank and D. A. Gurnett: J. Geophys. Res. **76** (28), 6829—6846 (1971).

[10] W. Riedler: Auroral particle precipitation patterns, in [*40*]. — W. Riedler and H. Borg: Über die Morphologie der Elektronen- und Protonenausfällung im keV-Bereich, in Kleinheubacher Berichte **15** (1972), p. 167—176.

[11] R. H. Eather and S. B. Mende: J. Geophys. Res. **77** (4), 660—673 (1972).

[1] D. B. Beard, M. Bird, and Y. H. Huang: Planet. Space Sci. **18** (9), 1349—1355 (1970).

[2] H. Poeverlein: Driftbewegung in der neutralen Schicht des Magnetotails, in Kleinheubacher Berichte **15** (1972), p. 1—8.

[3] M. K. Bird: to be published (private communication 1972).

the constancy of the electric potential on magnetic field lines (see Sects. 23 and 25). However, *two cases* have to be distinguished. An *electric field induced* at ionospheric heights by motions of the atmosphere may *produce currents* (dynamo-electric effect). The accompanying electrostatic field continues upward and causes magnetospheric drift motions (Sect. 25). The other possibility is a *primary state of motion in the magnetosphere* with an electric field extending down into the ionosphere, where it causes currents.

MATSUSHITA[1] *derived magnetospheric motion systems from the* $\mathbf{S}_q$ *current system* (the current system of the diurnal magnetic variation on quiet days) for both cases. The first case corresponds to the conventional idea of a dynamo-electric effect. In the second case no motion of the neutral atmosphere is involved and $\mathbf{S}_q$ currents appear to be a consequence of magnetospheric phenomena. Although the motions in this case do not seem to agree with any known magnetospheric motion model, it becomes obvious that velocities originating in the magnetosphere on quiet days (Sect. 25) and in disturbances (Sects. 24 and 27) are sufficiently high to generate geomagnetic variations via ionospheric currents. Thus rotary motion in the absence of spherical symmetry (Sect. 25) may contribute to the diurnal magnetic field variation.

General references.

Comprehensive literature.

[*1*] DEWITT, C. (ed.): Geophysics, The Earth's Environment. New York: Gordon and Breach 1962.

[*2*] JOHNSON, F. S. (ed.): Satellite Environment Handbook, 2nd ed. Stanford, Calif.: Stanford University Press 1965.

[*3*] KING, J. W., and W. S. NEWMAN (ed.): Solar-Terrestrial Physics. London-New York: Academic Press 1967.

[*4*] ODISHAW, H. (ed.): Research in Geophysics, Vol. 1: Sun, Upper Atmosphere, and Space. Cambridge, Mass.: The M. I. T. Press 1964.

Solar wind.

[*5*] AXFORD, W. I.: Observations of the interplanetary plasma. Space Sci. Rev. **8** (3), 331—365 (1968).

[*6*] COLBURN, D. S., and C. P. SONETT: Discontinuities in the solar wind. Space Sci. Rev. **5** (4), 439—506 (1966).

[*7*] DESSLER, A. J.: Solar wind and interplanetary magnetic field. Rev. Geophysics **5** (1), 1—41 (1967).

[*8*] LÜST, R.: The properties of interplanetary space, in [*3*], p. 1—44.

[*9*] MACKIN, R. J., and M. NEUGEBAUER (ed.): The Solar Wind. Oxford: Pergamon Press 1966.

[*10*] PARKER, E. N.: Interplanetary Dynamical Processes. New York-London: Interscience Publishers 1963.

[*11*] WILCOX, J. M.: The interplanetary magnetic field, solar origin and terrestrial effects. Space Sci. Rev. **8** (2), 258—328 (1968).

Magnetosphere.

[*12*] BEHANNON, K. W., and N. F. NESS: Satellite studies of the Earth's magnetic tail, in [*13*], p. 409—434.

[*13*] CAROVILLANO, R. L., J. F. MCCLAY, and H. R. RADOSKI (ed.): Physics of the Magnetosphere. Dordrecht, Holland: D. Reidel Publishing Co. 1968.

[*14*] HESS, W. N.: The Radiation Belt and Magnetosphere. Waltham, Mass.-Toronto-London: Blaisdell Publ. Co. 1968.

[*15*] — The Earth's radiation belt, in this volume, p. 115—230.

[*16*] HINES, C. O.: Hydromagnetic motions in the magnetosphere. Space Sci. Rev. **3**, 342—379 (1964).

[1] S. MATSUSHITA: Radio Science **6** (2), 279—294 (1971).—See also A. HRUŠKA and S. MATSUSHITA: Planet. Space Sci. **19** (6), 651—657 (1971).

[16a] McCormac, B. M. (ed.): Particles and Fields in the Magnetosphere. Dordrecht, Holland: D. Reidel Publishing Co. 1970.
[17] Ness, N. F.: Observations of the interaction of the solar wind with the geomagnetic field during quiet conditions, in [3], p. 57—89.
[18] — The geomagnetic tail, in [22], p. 97—127.
[19] Obayashi, T., and A. Nishida: Large-scale electric field in the magnetosphere. Space Sci. Rev. **8** (1), 3—31 (1968).
[20] O'Brien, B. J.: Interrelations of energetic charged particles in the magnetosphere, in [3], p. 169—211.
[21] Parker, E. N.: Dynamical properties of the magnetosphere, in [13], p. 3—64.
[21a] Roederer, J. G.: Dynamics of Geomagnetically Trapped Radiation. Berlin-Heidelberg-New York: Springer 1970.
[22] Williams, D. J., and G. D. Mead (ed.): Magnetospheric Physics, invited papers presented at The International Symposium on the Physics of the Magnetosphere, Washington, D. C., 1968. Rev. Geophysics **7**, No. 1 and 2 (1969).

Geomagnetic phenomena.

[23] Akasofu, S.-I.: Polar and Magnetospheric Substorms. Dordrecht, Holland: D. Reidel Publishing Co. 1968.
[24] —, and S. Chapman: Geomagnetic storms and auroras, in [26], Vol. 2, p. 1113—1151.
[25] Chapman, S.: The Earth's Magnetism, 2nd ed. London: Methuen & Co. 1951.
[26] Matsushita, S., and W. H. Campbell (ed.): Physics of Geomagnetic Phenomena, Vols. 1 and 2. New York and London: Academic Press 1967.
[27] Nagata, T., and N. Fukushima: Morphology of magnetic disturbance. See Vol. 49/3 of this Encyclopedia, p. 5—130.
[28] Parker, E. N., and V. C. A. Ferraro: Theoretical aspects of the worldwide magnetic storm phenomena. See Vol. 49/3 of this Encyclopedia, p. 131—205.

Plasma physics and magnetodynamic phenomena.

[29] Alfvén, H., and C.-G. Fälthammar: Cosmical Electrodynamics, 2nd ed. Oxford: Clarendon Press 1963.
[30] Landau, L. D., and E. M. Lifšic: Elektrodinamika splošnyh sred, Moskva: Gosudarstvennoe Izdatel'stvo fiziko-matematičeskoj Literatury 1957. English translation: L. D. Landau and E. M. Lifshitz: Electrodynamics of Continuous Media, Oxford: Pergamon Press 1960, p. 224—233. German translation: L. D. Landau u. E. M. Lifschitz: Lehrbuch der theoretischen Physik, Bd. VIII. Berlin-Heidelberg-New York: Akademie-Verlag 1967, p. 262—275.
[31] Longmire, C. L.: Elementary Plasma Physics. New York-London: Interscience Publishers 1963.
[31a] Rossi, B., and S. Olbert: Introduction to the Physics of Space. New York: McGraw-Hill Book Co. 1970.
[32] Spitzer, L.: Physics of Fully Ionized Gases, 2nd ed. New York-London: Interscience Publishers 1962.
[33] Stern, D. P.: The motion of magnetic field lines. Space Sci. Rev. **6** (2), 147—173 (1966).
[34] Sturrock, P. A. (ed.): Plasma Astrophysics. Proc. International School of Physics "Enrico Fermi", Course 39. New York-London: Academic Press 1967.

Magnetodynamic waves.

[35] Ginzburg, V. L., and A. A. Ruhadze: Waves and resonances in magneto-active plasma. See this volume, p. 395—560.
[36] Hultqvist, B.: Plasma waves in the frequency range 0.001—10 cps in the Earth's magnetosphere and ionosphere. Space Sci. Rev. **5** (5), 599—695 (1966).

Additional references, added in proof-reading.

[37] Brandt, J. C.: Introduction to the Solar Wind. San Francisco: W. H. Freeman and Co. 1970.
[38] Piddington, J. H.: Cosmic Electrodynamics. New York-London-Sydney-Toronto: John Wiley & Sons 1969.
[39] Dyer, E. R. (ed.): Solar-Terrestrial Physics/1970. Dordrecht, Holland: D. Reidel Publishing Co. 1972.
[40] McCormac, B. M. (ed.): Earth's Magnetospheric Processes. To be published by D. Reidel Publishing Co. 1972.
[41] Kertz, W.: Einführung in die Geophysik, Bd. 2: Obere Atmosphäre und Magnetosphäre. B. I. Hochschultaschenbücher 535/a/b. Mannheim-Wien-Zürich: Bibliographisches Institut 1971.

The Earth's Radiation Belt.

By

W. N. Hess*.

With 60 Figures.

A. Introduction.

1. Historical review.

α) Before the advent of rockets and satellites we had a very incomplete and in some cases wrong idea what the *environment of the Earth* was a close as 300 km. One would then have described the Earth's magnetic field as being roughly dipolar rather like a bar magnet as far away from the Earth as you cared to look. As for the particle environment of the Earth, one would have claimed the atmosphere ended for practical purposes at may be 300 km altitude and above that one had galactic cosmic rays and a few atoms per cm^3 of hydrogen throughout space and not much else. Using satellites we have found that things are very different from this. But even before satellites there were a few important clues to what the environment was really like. It had been known for a long time that violent eruptions on the Sun — solar flares — produced disturbances at the Earth. A day or so after the flare a magnetic storm frequently occurred accompanied by troubles with radio propagation and bright auroral displays. The fact that the transit time from the Sun was a day led CHAPMAN and FERRARO[1] in 1930 to the idea that the Sun emitted a blast of plasma at the time of a flare which travelled to the Earth with a velocity of about 1000 km/sec and when it arrived pushed in the geomagnetic field to produce the magnetic storm. We now know this picture of the initial phase of a magnetic storm is essentially correct but that the solar plasma, instead of occurring in bursts only at flares, is emitted all the time and continually distorts and encloses the geomagnetic field in the solar wind.

β) People knew that solar flares also produced bright aurorae a day or so later. It took until 1957 using rockets to find that aurorae were caused by energetic particles bombarding the atmosphere. However, well before this some people suspected that energetic particles were involved in aurorae. STØRMER[2] studied how *energetic particles* from the Sun might enter the Earth's magnetic field and what their orbits would look like in a dipole magnetic field in attempts to explain aurorae. In connection with this work, STØRMER calculated the orbit of a particle trapped in the Earth's magnetic field. However, the meaning of this orbit which contained all the elements of a radiation belt was apparently never considered seriously.

γ) In 1957, before the discovery of the natural radiation belt, it was suggested by CHRISTOFOLIS[3] by analogy to laboratory size magnetic mirror machines that

* Manuscript received January 1967.

[1] S. CHAPMAN, and V. C. A. FERRARO: Terr. Magn. Atmosph. Electr. **36**, 77—97, 171—186 (1931).

[2] C. STØRMER: Arch. Sci. Phys. Nat. **24**, 317 (1907), and The Polar Aurora.

[3] N. C. CHRISTOFOLIS: J. Geophys. Res. **64**, 869 (1959).

the Earth's magnetic field should be able to trap charged particles. This idea was tested by the "Argus" high altitude nuclear explosions of 1958 which did produce *artificial radiation belts*. But before the "Argus" explosions were carried out, VAN ALLEN discovered the *natural radiation belt*[4]. The GEIGER counter that he placed on the first U.S. Satellite "Explorer 1" in 1958 was intended to study how the cosmic ray flux varied with altitude and latitude. This satellite did not carry a tape recorder so only short periods of data near ground receiving stations were received. Some of these passes showed the expected cosmic ray counting rate but some showed the surprising result of zero counts for an entire pass. There was a question about whether the instrument was malfunctioning. A similar GEIGER counter was flown on "Explorer 3" in 1958. This satellite carried a tape recorder and when one complete orbit of data was obtained the puzzle was solved[5]. The detectors count rate was the cosmic ray rate for a while and then rapidly rose to a very high count rate and then stopped counting altogether. Later it turned on again at a high counting rate and eventually returned to cosmic ray rate. When a geiger counter is subjected to a very high radiation flux it just stops counting. The pulses out become small and no counts will be registered at all. The counters on "Explorer 1" and "Explorer 3" had entered a zone of large energetic particle fluxes on part of their orbits and the counters saturated and stopped counting. This was the discovery of the radiation belt.

We have had a considerable period of time to map the radiation belt and have reasonably good ideas of the fluxes, spectra and time variability of the various particle populations. However, we still have rather incomplete and untested ideas about particle sources and acceleration mechanisms and loss processes.

Let us first consider how particles moving in a magnetic field can build up a radiation belt and then study what the characteristics of the terrestrial radiation belt are.

B. Particle motion in a magnetic field.

2. The Guiding center approximation.

α) In a uniform constant magnetic field the *motion of a charged particle* is a helix. This can be considered as a combination of circular motion around a field line and linear motion along the field line. This same description of the motion is very useful when the field varies slowly in space and time. In this case the motion can still be considered composed of a nearly-circular motion around a field line and a linear motion along a field line and also generally a drift across field lines. This motion is described by the "guiding center" of the circle [*1*]. The instantaneous position of the particle $\boldsymbol{r}$ is broken down into the circular motion of radius $\boldsymbol{\rho}$ and the motion of the guiding center whose location is $\boldsymbol{R}$ where (see Fig. 1 a)

$$\boldsymbol{r} = \boldsymbol{R} + \boldsymbol{\rho}.$$

Starting with the equation of motion of a charged particle of mass m and charge q

$$m \frac{\mathrm{d}^2 \boldsymbol{r}}{\mathrm{d} t^2} = m \boldsymbol{g} + \frac{q}{c_0 \sqrt{\varepsilon_0 \mu_0}} \frac{\mathrm{d} \boldsymbol{r}}{\mathrm{d} t} \times \boldsymbol{B} + q \boldsymbol{E}, \tag{2.1}$$

---

[4] J. A. VAN ALLEN: The first public lecture on the Discovery of the Geomagnetically Trapped Radiation. State Univ. of Iowa, Report 60—13 (1960).

[5] J. A. VAN ALLEN, G. H. LUDWIG, E. C. RAY, and C. E. MCILWAIN: Observation of High Intensity Radiation by Satellites. 1958 α and γ. Jet Propulsion **28**, 588 (1958).

where $\boldsymbol{g}$ is the acceleration of gravity, q is the electronic charge, $\boldsymbol{B}$ is the magnetic field, and $\boldsymbol{E}$ is the electric field. NORTHROP [2] has obtained a general expression for the motion of the particles' guiding center by substituting in $\boldsymbol{r}=\boldsymbol{R}+\boldsymbol{\rho}$ and averaging over one gyration to get

$$\frac{d^2 \boldsymbol{R}}{dt^2}=\boldsymbol{g}+\frac{q}{m}\left[\boldsymbol{E}+\frac{1}{c_0\sqrt{\varepsilon_0\mu_0}}\frac{d\boldsymbol{R}}{dt}\times\boldsymbol{B}\right]-\frac{|\boldsymbol{\mu}|}{m}\nabla B+0\left(\frac{\varrho}{x}\right), \tag{2.2}$$

$\boldsymbol{\mu}$ being the magnetic moment*, see Eq. (2.8).

The function 0 means terms of the order of its argument; B is the absolute value of vector $\boldsymbol{B}$.

The expansion parameter here is ϱ/x, the cyclotron radius divided by a distance x in which the magnetic field changes appreciably. The distance x is of the order of R the distance from the center of the Earth. The expansion parameter ϱ/x should be a small number in order for the guiding center approximation to hold well. In this case the higher order terms $0(\varrho/x)$ are omitted in Eq. (2.2). As ϱ becomes smaller and smaller the motion of a charged particle becomes closer and closer to the helical motion of a charged particle in a uniform $\boldsymbol{B}$ field[1].

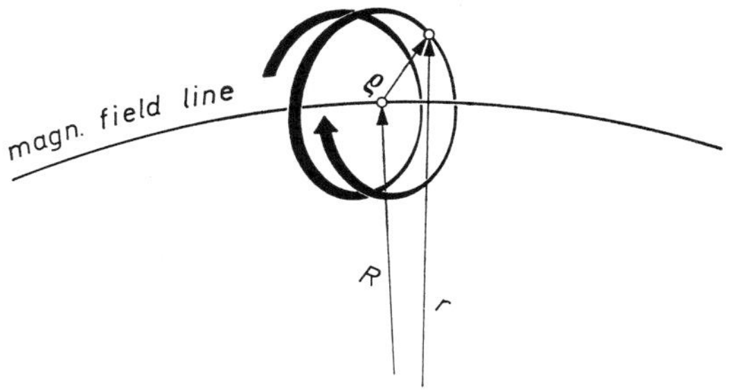

Fig. 1 a

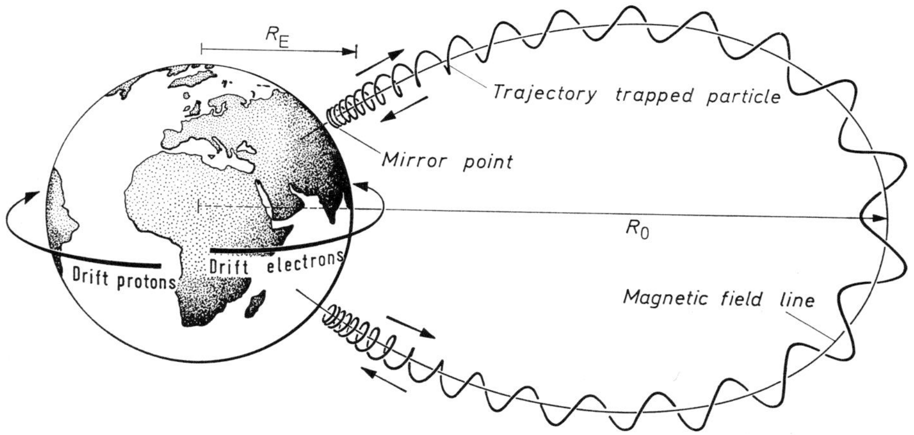

Fig. 1 b

Fig. 1 a and b. Motion of a charged particle in a dipole field. For low energy particles the motion can be broken up in three components as shown in Fig. 1 b (1) spiraling along field lines (detailed drawing with denominations in Fig. 1 a), (2) bouncing back and forth along field lines from one hemisphere to the other, and (3) drifting in longitude around the Earth with electrons drifting east and protons west. Fig. 1 a explains guiding center motion ($\boldsymbol{R}$) and particle motion ($\boldsymbol{R}+\boldsymbol{\rho}=\boldsymbol{r}$). Fig. 1 b describes mirroring (where pitch angle becomes 90°), drifting and the L-coordinate ($L\approx R_0/R_E$), see Sect. 5.

* $\boldsymbol{\mu}$ (not to be confounded with μ_0) is used for magnetic moments of particles; we use $\mathbf{M}$ for the magnetic dipole moment of Earth, see Eq. (3.1).

[1] For full theory of particle motion in a dipole field see [35—38], for a summary of these theories see [39] and Chap. VI A, p. 84 of contribution by A. B. MEINEL, S.-I. AKASOFU and S. CHAPMAN in this Encyclopedia, Vol. 49/1.

β) In the *guiding center approximation* the motion of a charged particle in a dipole field is broken down into three components as shown in Fig. 1b. The particle *gyrates* rapidly around a field line. It *bounces* back and forth along a line between its two mirror points. The particle slowly *drifts* in longitude around the earth. The speeds of these three motions are so different that they can be separated.

The period of cyclotron rotation or gyration about a field line is

$$T_c = c_0 \sqrt{\varepsilon_0 \mu_0}\, \frac{2\pi m}{B q}. \tag{2.3}$$

For non-relativistic electrons in a field of 0.1 Gs (= Gauss) we find $T_c = 3.5$ μsec, and for a non-relativistic proton in the same field $T_c = 6.3$ msec.

Charged particles (with $\alpha \neq 90°$) will move along a line at the same time that they rotate rapidly around the line. The pitch angle α, which is the angle between the particles' velocity vector and $\boldsymbol{B}$, changes in such a way that the particle may turn around and not hit the Earth. This can be seen from the following analysis.

γ) A static magnetic field does no work on a particle. This means the magnetic flux Φ linking the orbit of a particle rotating about a field line is constant, since if $\partial B/\partial t \neq 0$ the *particle's energy* would change. Therefore

$$\Phi_m = B\,\pi\,R_c^2 = \text{constant} \tag{2.4}$$

where the gyro radius is

$$R_c = c_0 \sqrt{\varepsilon_0 \mu_0}\, \frac{m}{q B^2}\, |\boldsymbol{v} \times \boldsymbol{B}| = c_0 \sqrt{\varepsilon_0 \mu_0}\, \frac{m v_\perp}{q B}. \tag{2.5}$$

We can write for the particle's kinetic energy of gyration

$$\mathscr{E}_\perp = \frac{m v_\perp^2}{2} = \frac{m v^2 \sin^2\alpha}{2} \tag{2.6}$$

substituting in Eq. (2.4) we get

$$\Phi_m = \text{const} = c_0^2\, \varepsilon_0\, \mu_0\, \frac{2\pi m \mathscr{E}_\perp}{q^2 B}. \tag{2.7}$$

Therefore

$$\frac{\mathscr{E}_\perp}{B} = \text{const} = |\boldsymbol{\mu}|. \tag{2.8}$$

It is easy to show that the constant in Eq. (2.8) is the magnetic moment of the particle's gyration around the field line.

The magnetic moment of a ring of current $i = \frac{q v_\perp}{2\pi R_c}$ and area A is given by

$$|\boldsymbol{\mu}| = i A = \frac{q v_\perp R_c}{2} = \frac{\mathscr{E}_\perp}{B}. \tag{2.9}$$

Eq. (2.8) shows that the magnetic moment* is constant along the particle's orbit. Actually, it is only an adiabatic constant. We will consider this more later.

δ) We can now see how the *particle's pitch angle* α varies with position.

We find from Eqs. (2.8), (2.9), assuming the total kinetic energy $\mathscr{E}$ to be constant:

$$\frac{\sin^2\alpha_1}{B_1} = \frac{\sin^2\alpha_2}{B_2} = \text{const}. \tag{2.10}$$

---

* According to the IUPAP recommendations, the magnetic moment, in the international system, has Am^2 as unit.

Consider a particle moving along a field line down towards the earth. It moves into a region of increasing B and therefore $\sin\alpha$ increases. At the point where $\sin\alpha = 1$, the particle turns around and starts moving up along the line. This point is called the *mirror point*.

This process is repeated at the other end of the field line where the same value of B, namely B_m, exists. The particle turns around at this point also and this is called the *conjugate point* of the original mirror point. The particle's motion between mirror points is shown in Fig. 1b. The particle is trapped between the two magnetic mirrors and can stay oscillating in space above the Earth for long periods of time. A particle with an equatorial pitch angle α_0 will mirror at a place where the magnetic field is B_M given by

$$\sin^2\alpha_0 = \frac{B_0}{B_M} \tag{2.11}$$

where B_0 is the equatorial field on this field line.

ε) The time a particle takes to go from one mirror point to the other one and back again is called the *bounce period* and is obtained by

$$T_b = 2\int_{l_1}^{l_2} \frac{dl}{v_\parallel} = \frac{4\,r_0}{v}\,\mathrm{T}(\alpha_0)\,. \tag{2.12}$$

l_1 and l_2 are the two mirror points and dl is taken along the field line and $v_\parallel = v\cos\alpha$ and the function $\mathrm{T}(\alpha_0)$ varies from 0.75 for $\alpha_0 = 90°$ to 1.4 for $\alpha_0 = 0°$ and is given approximately[2] by

$$\mathrm{T}(\alpha_0) = 1.30 - 0.56\sin\alpha_0\,.$$

The fact that this integral varies only slightly with equatorial pitch angle α_0 shows what the bounce period does not change much with the amplitude of the particle's oscillation. The period of an ordinary pendulum is to first order independent of amplitude so this seems reasonable. The bounce period for relativistic particles for field lines having various equatorial radii $r_0 \equiv R_0/R_E$ is given below* for $\mathrm{T}(\alpha_0) = 1$

R_0/R_E	1.5	2	3	4	6
T_b/sec	0.13	0.17	0.26	0.34	0.52

ζ) When a force acts on a charged particle in a magnetic field in a direction perpendicular to the field, the particle will undergo a *drift across field lines*.

The general expression for the drift velocity can be determined [2] from Eq. (2.2) by taking $\frac{1}{B}\left(\boldsymbol{B}\times\frac{d^2\boldsymbol{R}}{dt^2}\right)$.

The resultant equation can be solved to give the drift velocity perpendicular to $\boldsymbol{B}$ as

$$\begin{aligned}\frac{d\boldsymbol{R}_\perp}{dt} &= \frac{d\boldsymbol{R}}{dt}\times\frac{\boldsymbol{B}}{B} = c_0\sqrt{\varepsilon_0\mu_0}\,\frac{\boldsymbol{E}\times\boldsymbol{B}}{B^2} + \frac{|\mu|}{q}\,c_0\sqrt{\varepsilon_0\mu_0}\,\frac{\boldsymbol{B}\times\nabla\boldsymbol{B}}{B^2}\\ &\quad + \frac{m}{q}\,c_0\sqrt{\varepsilon_0\mu_0}\,\frac{\left(\boldsymbol{g}-\frac{d^2\boldsymbol{R}}{dt^2}\right)\times\boldsymbol{B}}{B^2} + 0\left(\frac{\varrho^2}{x^2}\right)\end{aligned} \tag{2.12a}$$

* $R_E = 6370$ km $\approx 6.4\cdot 10^6$ m is the radius of Earth which is a convenient unit of length for describing the magnetosphere.

[2] D. A. Hamlin, R. Karplus, R. C. Vik, and K. M. Watson: J. Geophys. Res. **66**, 1 (1961).

we need to eliminate $\frac{d^2\boldsymbol{R}}{dt^2}$ from (2.12a) which can be done by differentiating (2.12a) itself and keeping only the zeroth order terms to give

$$\frac{d^2\boldsymbol{R}_\perp}{dt^2} = \frac{d}{dt}\left[c_0\sqrt{\varepsilon_0\mu_0}\left(\frac{\boldsymbol{E}\times\boldsymbol{B}}{B^2}\right)\right]. \tag{2.12b}$$

Substituting Eq. (2.12b) into Eq. (2.12a) and keeping only first order terms in ϱ/x gives the expression for the drift velocity

$$\boldsymbol{v}_d = \frac{\boldsymbol{B}}{B^2}\times\left\{-\boldsymbol{E} + \frac{|\mu|}{q}\nabla B + \frac{m}{q}\left[-\boldsymbol{g} + v_\parallel\frac{d}{dt}\left(\frac{\boldsymbol{B}}{B}\right) + \frac{d}{dt}\left(\frac{\boldsymbol{E}\times\boldsymbol{B}}{B^2}\right)\right]\right\}. \tag{2.13}$$

The first term is the drift due to the electric field. The second term is due to the gradient of the magnetic field. The third term is due to gravity. The fourth term contains effects of time-varying fields and also implicitly a $\boldsymbol{v}_\parallel^2\frac{\partial B}{\partial l}$ term which is the curvature of field line effect. The last term is due to electric fields.

In the absence of an electric field $\boldsymbol{E}$ and time-varying magnetic fields and neglecting gravity, we can take for the total drift velocity

$$\boldsymbol{v}_d = c_0\sqrt{\varepsilon_0\mu_0}\left\{\frac{m v_\perp^2}{2q B^3}\boldsymbol{B}\times\nabla B + \frac{m v_\parallel^2}{q B^3}\boldsymbol{B}\times\frac{\partial\boldsymbol{B}}{\partial l}\right\}, \tag{2.14}$$

which can be written

$$v_d = c_0\sqrt{\varepsilon_0\mu_0}\,\frac{m}{qB R_l}\left(\frac{v_\perp^2}{2} + v_\parallel^2\right) \tag{2.15}$$

where R_l is the radius of curvature of the field line.

But rather than this we frequently want the rate of change of longitude Λ

$$\frac{d\Lambda}{dt} = \frac{v_d}{r\cos\phi} = \frac{v_d}{r_0\cos^3\phi}. \tag{2.16}$$

LEW[3] averaged this in latitude and then took the reciprocal to get the *drift period*.

$$T_d = \frac{2\pi}{\left\langle\frac{d\Lambda}{dt}\right\rangle}, \tag{2.17}$$

$$T_D/\text{min} = \frac{172}{\eta}\left(\frac{m_e}{m}\right)\left(\frac{1+\eta}{2+\eta}\right)\frac{R_E}{r_0}\,F(\phi) \tag{2.17a}$$

where $F(\phi) = 1$ for $\phi_m = 0°$; $F(\phi) = 1.5$ for $\phi_m = 90°$ and where m_e is the mass of the electron and m is the mass of the particle being studied and

$$\eta = \frac{\mathscr{E}}{m c_0^2} = \frac{\text{particles' kinetic energy}}{m c_0^2}. \tag{2.18}$$

This reduces to $T_d/\text{min} = \frac{44\text{ MeV}}{\mathscr{E}}\left(\frac{r_0}{R_E}\right)$ for $\phi_m = 0°$ for non-relativistic particles.

An electron and proton of the same kinetic energy have nearly the same drift periods since the factor $m\eta$ in the denominator is the same for both.

3. Earth's magnetic field.

α) The magnetic field of Earth is a moderately good *dipole field* so it is important to understand the properties of a dipole.

---

[3] J. A. LEW: J. Geophys. Res. 66, 2681 (1961).

The strength of a dipole magnetic field is given by

$$B(r, \phi) = \frac{\mathrm{u}}{4\pi}\, \mu_0 \frac{\mathsf{M}_\mathrm{E}}{r^3} \left[1 + 3 \sin^2 \phi\right]^{\frac{1}{2}} \tag{3.1}$$

where r = distance from the center of Earth,

ϕ = magnetic latitude,

M_E = Earth's dipole moment*.

This field has components

$$\left.\begin{aligned} B_r &= \frac{\mathrm{u}}{4\pi}\, \mu_0 \frac{\mathsf{M}_\mathrm{E}}{r^3}\, 2 \sin \phi, \\ B_\phi &= \frac{\mathrm{u}}{4\pi}\, \mu_0 \frac{\mathsf{M}_\mathrm{E}}{r^3} \cos \phi. \end{aligned}\right\} \tag{3.2}$$

From this we can find the equation of a field line by

$$\frac{r\,\mathrm{d}\phi}{\mathrm{d}r} = \frac{B_\phi}{B_r} = \frac{\cos\phi}{2\sin\phi} \tag{3.3}$$

which integrates to give the equation of a dipole field line

$$r = r_0 \cos^2 \phi. \tag{3.4}$$

The field along one field line is in the case of Earth**

$$B(r_0, \phi)/\mathrm{Gs} = 0.32 \left(\frac{R_\mathrm{E}}{r_0}\right)^3 \frac{\left[1 + 3\sin^2\phi\right]^{\frac{1}{2}}}{\cos^6 \phi}. \tag{3.5}$$

Values of the surface field of Earth given by Eq. (3.1) are in error by as much as 30 percent, see Fig. 2. A better approximation to Earth's field is to take a dipole displaced about 400 km towards the western Pacific from the center of Earth. This gives the surface field to about 10 percent, but this is not nearly good enough for current needs.

β) A *multipole expansion* of the field is used frequently now[1,2]. The higher order terms in the multipole expansion fall off radially faster than the dipole term and at several earth radii it is generally accurate enough[3] to use the dipole field description. The most widely used expansion for the geomagnetic field now is that of Jensen and Cain[2] using data for 1960. This expands the magnetic potential V into spherical harmonic terms given by

$$V = R_\mathrm{E} \sum_{n=1}^{6} \left(\frac{R_\mathrm{E}}{r}\right)^{n+1} \sum_{m=0}^{n} \left(g_n^m \cos m\Lambda + h_n^m \sin m\Lambda\right) \mathrm{P}_n^m(\cos\phi), \tag{3.6}$$

where R_E = radius of Earth. Λ is longitude, ϕ is latitude and $\mathrm{P}_n^m(\cos\phi)$ = associated Legendre polynomial.

This fit gives the Earth surface field of ~40000 γ to an accuracy of roughly ± 200 γ. After the IQSY program to map Earth's field, an expansion of more terms probably should be used. The Jensen and Cain[2] (1960) fit to the surface field of Earth is shown in Fig. 2. Some interesting features appear here. The region of low field in the South Atlantic near Brazil has important consequences in radiation belt problems. This feature of the field has been called the "Brazilian anomaly" although the use of the word anomaly is questionable here. The

---

* According to IUPAP the unit of M in the international system is A m².

** The SI-unit of B is Tesla ($1\,\mathrm{T} \equiv 1\,\mathrm{kg\,sec^{-2}\,A^{-1}} = 1\,\mathrm{Vs\,m^{-2}}$). In geomagnetism the Gaussian unit is quite generally used: 1 Gauss (Gs) $\equiv 1\,\Gamma = 10^{-4}$ T. Small field changes and fields are indicated in gamma: 1 gamma $\equiv 1\,\gamma = 10^{-5}\,\Gamma = 10^{-9}$ T.

[1] H. F. Finch, and B. R. Leaton: The Earth's Main Magnetic Field — Epoch 1955.0. Monthly Notices Roy. Astron. Soc. (Geophys. Suppl.), **7**, 314 (1957).

[2] D. C. Jensen, and J. C. Cain: J. Geophys. Res. **67**, 3568 (1962).

[3] C. Y. Fan, P. Meyer, and J. A. Simpson: J. Geophys. Res. **66**, 2607 (1961).

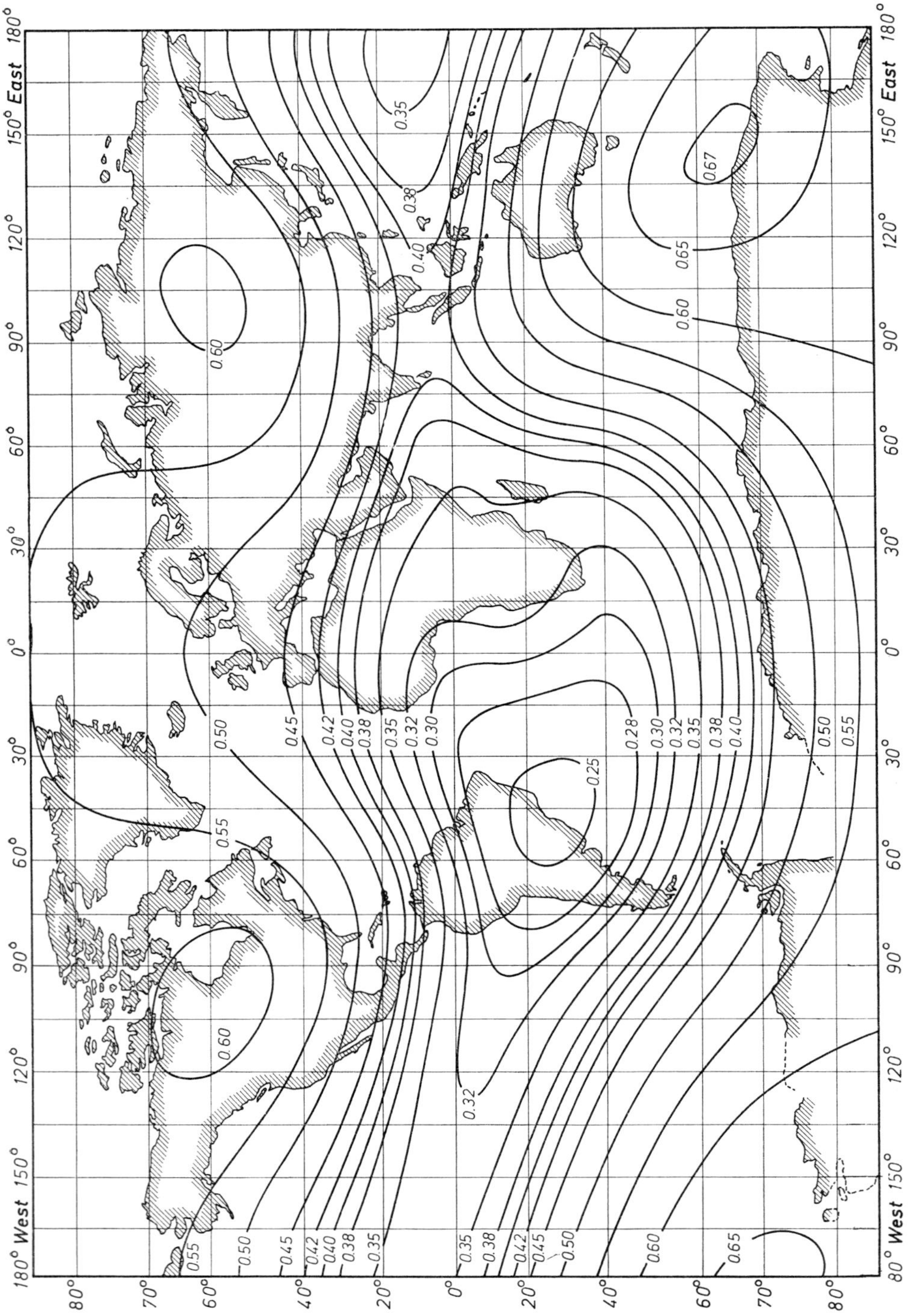

Fig. 2. Calculated constant field contours at the surface of Earth using the Jensen-Cain 48 term expansion of the magnetic field for 1960.

general features of the offset dipole model of Earth's field require it to be weak here and the field in the western Pacific to be high. Non-dipolar features might more properly be called anomalies. Using this definition, the most significant anomaly — the Capetown Anomaly — does not show up very clearly on Fig. 2.

γ) As to *its origin*, Earth's magnetic field is produced mostly by currents in the Earth's core but significant fields are also produced by minerals in the Earth's crust, by currents in the ionosphere, by trapped particles in the field producing so-called ring currents and by effects of the solar wind distorting the geomagnetic field. The geomagnetic field is time varying both with long time constants changing the main field by about 100 γ per year in many locations but also with short time constants when magnetic storms of solar origin change the surface field by as much as 1000 γ for a matter of days.

When dealing with the trapped particles in the radiation belt for most purposes it is sufficient to describe the geomagnetic field in terms of a constant field ignoring time variations and using a spherical harmonic expansion. This may not be accurate enough at large distances from the Earth — roughly past $6 R_{\mathrm{E}}$ or it may not be accurate enough closer in than this at storm times.

The magnetic field is substantially altered at high altitudes. The solar wind blowing on the terrestrial field distorts it and restricts it to a cavity in interplanetary space that is vaguely tear drop shaped. This means the field is not azimuthally symmetric. The magnetic field in the terrestrial cavity has been calculated on various assumptions concerning the solar wind[4,5]. We will return to this when discussing the outer edge of the radiation belt.

4. Adiabatic invariants. We are familiar with mechanical systems where there are constants of the motion such as energy or linear or angular momentum. There are approximate constants of the motion for the three components of motion of a charged particle in a magnetic field also, but these are adiabatic constants, that is, they are only constant for very slow changes of the variables involved.

Adiabatic invariants appear in systems having periodic motion or multiple periodic motion. They are connected with slow perturbations of the Hamiltonian which maintain the basic periodic character of the motion and do not resonate with it[1].

α) We remarked earlier that $|\boldsymbol{\mu}|$, the *magnetic moment* of a particle *rotating about a line of force*, was an adiabatic invariant of the motion. By this is meant $|\boldsymbol{\mu}|$ is a constant if changes in the magnetic field take place slowly enough. We have demonstrated that $|\boldsymbol{\mu}|$ is constant in a static $\boldsymbol{B}$ field. There is an exchange of energy between $\mathscr{E}_{\perp}$ and $\mathscr{E}_{\parallel}$ as the particle bounces between mirror points but $|\boldsymbol{\mu}| = \mathscr{E}_{\perp}/B$ remains constant. It is necessary for $\partial B/\partial t$ to be constant over the cyclotron radius for longer than one gyration period in order that $|\boldsymbol{\mu}|$ be a constant.

β) There is a second adiabatic invariant that can normally be used to describe the motion of particles in a magnetic field. This is the longitudinal invariant $\mathscr{J}$ first described by ROSENBLUTH. $\mathscr{J}$ is defined by

$$\mathscr{J} = 2 \int_{l_1}^{l_2} \mathrm{d}l\, m\, v_{\parallel} \tag{4.1}$$

where the integral is taken between the two mirror points, with coordinate l_1 and l_2, respectively[1a].

[4] D. B. BEARD: Rev. Geophys. **2**, 335 (1964).

[5] G. D. MEAD: J. Geophys. Res. **69**, 1181 (1964).

[1] M. BORN: The Mechanics of the Atom (translation). New York: Frederic Ungor Publ. Co. 1960.

[1a] ROSENBLUTH's invariant has J sec as dimension in SI-units—like PLANCK's constant (the "Wirkungsquantum").

γ) A version of the longitudinal invariant $\mathcal{J}$ used when v is a constant is obtained from Eq. (4.1) by

$$\mathcal{I}=\frac{\mathcal{J}}{2\,m\,v}=\int_{l_1}^{l_2}\mathrm{d}l\sqrt{1-\frac{B}{B_{\mathrm{M}}}}\,. \tag{4.2}$$

We see that $\mathcal{I}$, called the *integral invariant,* is the length of the field line between mirror points weighed by a function of the magnetic field along the line.

For undisturbed particle motion in a magnetic field with no collisions or radial drift, v is a constant so $\mathcal{I}$ is an adiabatic invariant. For this case the particles bouncing motion is constrained. If there were no drift in longitude the particle would have to bounce back and forth along the same field line because a change in field line for the same mirror field B_{M} would mean a change in $\mathcal{I}$. But the particle does drift in longitude and the constancy of $\mathcal{I}$ constrains the drift motion.

By using the magnetic moment invariant $|\boldsymbol{\mu}|$ and the integral invariant $\mathcal{I}$, we can describe the surface on which a particle drifts around the earth. The particle bounces back and forth between mirror points of constant B_{M}. As the particle drifts in longitude the value of $\mathcal{I}$ is conserved so there is only one set of field lines on which the particle can drift. This set of lines has approximately uniform length. Because of these constraints the particle returns to the field line it started on at the end of a drift period. The surface on which a particle drifts around the Earth resembles a cored apple. The altitude of the mirror points will change from point to point because the surface field of Earth is not constant with position.

δ) There is a third adiabatic constant of the motion of a charged particle in a magnetic field associated with the particle's drift in longitude[2]. This states that the *magnetic flux* Φ_{m} linked in the drift orbit is a constant of the motion. If magnetic field changes take place in times short compared to the drift period this invariant will be violated. For example, the sudden commencement at the start of a magnetic storm which takes place in a few minutes might alter the particle's Φ_{m}.

5. Coordinate systems.

α) The best system of coordinates for describing the radiation belt uses MCILWAIN's magnetic shell parameter L and magnetic field strength B. The magnetic shell parameter, L, is a length which reduces to the equatorial radius of a field line r_0 in the case of a dipole field.

L is defined in the following way [3]. Let us first consider a dipole field. The integral invariant $\mathcal{I}$ can be written functionally from Eq. (4.2) as

$$\mathcal{I}=r_0\,\mathrm{f}_1(\phi) \tag{5.1}$$

and the magnetic field along a field line is given from Eq. (3.2) by

$$\frac{r_0^3\,B}{\mu_0\,\mathrm{M_E}}=\mathrm{f}_2(\phi) \tag{5.2}$$

combining Eqs. (5.1) and (5.2) we get

$$\mathcal{I}^3\frac{B}{\mu_0\,\mathrm{M_E}}=r_0^3\frac{B}{\mu_0\,\mathrm{M_E}}\mathrm{f}_1^3(\phi)=\mathrm{f}_2(\phi)\cdot\mathrm{f}_1^3(\phi)\,. \tag{5.2a}$$

There is a functional relationship between $\mathrm{f}_1(\phi)$ and $\mathrm{f}_2(\phi)$ which can be expressed

$$\mathrm{f}_1^3(\phi)=\mathrm{f}_3[\mathrm{f}_2(\phi)]=\mathrm{f}_3\left(\frac{r_0^3\,B}{\mu_0\,\mathrm{M_E}}\right). \tag{5.2b}$$

[2] T. NORTHRUP, and E. TELLER: Phys. Rev. **117**, 215 (1960).

Now from Eq. (5.2a) we can write

$$\mathscr{I}^3 \frac{B}{\mu_0 \mathsf{M}_{\mathrm{E}}} = r_0^3 \frac{B}{\mu_0 \mathsf{M}_{\mathrm{E}}} \mathrm{f}_3\left(\frac{r_0^3 B}{\mu_0 \mathsf{M}_{\mathrm{E}}}\right) = \mathrm{f}_4\left(\frac{r_0^3 B}{\mu_0 \mathsf{M}_{\mathrm{E}}}\right) \tag{5.3}$$

or by inverting, this can be written as

$$\frac{r_0^3 B}{\mu_0 \mathsf{M}_{\mathrm{E}}} = \mathrm{f}_5\left(\frac{\mathscr{I}^3 B}{\mu_0 \mathsf{M}_{\mathrm{E}}}\right). \tag{5.4}$$

So far we have been considering a dipole field. Now let us generalize and consider the Earth's field and actual values of $\mathscr{I}$. We will now define L by

$$(L R_{\mathrm{E}})^3 \frac{B}{\mu_0 \mathsf{M}_{\mathrm{E}}} = \mathrm{f}_5\left(\frac{\mathscr{I}^3 B}{\mu_0 \mathsf{M}_{\mathrm{E}}}\right) \tag{5.5}$$

where f_5 is the function defined by Eq. (5.4). In calculating values of L from Eq. (5.5) values of B and $\mathscr{I}$ for the *real field of Earth* are used.

We see that for a dipole field $L = r_0$ and for a non-dipole field L is a generalization of the equatorial distance that turns out to be very nearly, but not exactly, constant along a line of force. Along most field lines L varies by less than one percent. A set of values of B and $\mathscr{I}$ corresponds uniquely to one set of value of B and L by Eq. (5.5) so the same information is contained in both sets of coordinates.

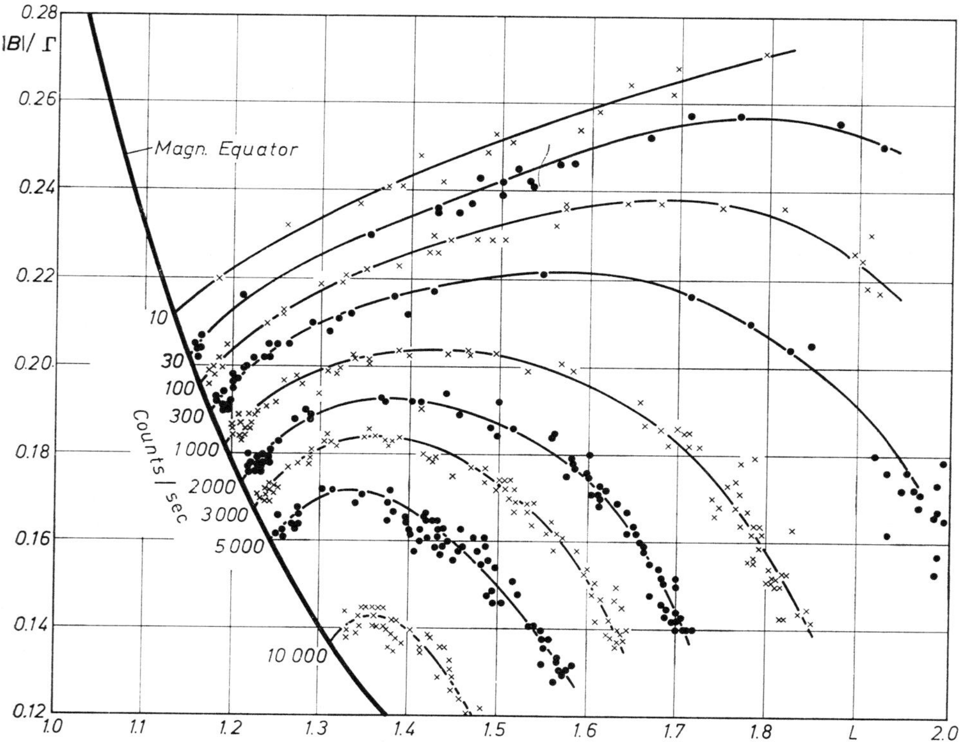

Fig. 3. The first map particle fluxes plotted in B, L coordinates. McIlwain used the Explorer 4 data on protons of $\mathscr{E} > 30$ MeV to show the usefulness of the B, L coordinates. The instrument was an unshielded Geiger-counter. Parameter of the individual curves is count per sec. [3].

It is convenient to compile experimental data in terms of L since one value of L corresponds roughly to one set of field lines making up a field shell*. Because of the close relationship between $\mathscr{I}$ and L, data can be combined in longitude by B, L-coordinates as easily as by $\mathscr{I}$, B-coordinates.

Fig. 3 is an example of this. A large amount of data from a GEIGER-MÜLLER counter on Explorer 4 is combined here and forms a single consistent pattern. This same data shown in geographic coordinates [*4*] is confusing and shows little understandable order. In Fig. 3 is shown a B, L map of inner zone protons. Coming down a line of constant L is equivalent to moving out from the earth along one particular field line.

β) Another set of coordinates used sometimes is $\mathscr{R}$ and Φ defined by

$$\mathscr{R} \equiv \frac{R}{R_E} = L \cos^2 \Phi \qquad (5.6)$$

and

$$B = \mu_0 \frac{M_E}{R^3} \sqrt{4 - 3\frac{\mathscr{R}}{L}} \qquad (5.7)$$

where Φ is a *generalized magnetic latitude*. This is a better form of the magnetic latitude to use than the dip latitude Ψ. It is also better than a geomagnetic (dipole) latitude ϕ.

Using $\mathscr{R} = 1$ in Eq. (5.6) we get a special magnetic latitude called the *invariant latitude* Λ. It is defined by

$$\cos^2 \Lambda = \frac{1}{L}. \qquad (5.8)$$

This *invariant latitude*, uniquely related to L, is commonly used now to discuss processes near the surface of Earth such as aurorae and particle precipitation. The *invariant latitude* is defined at $\mathscr{R} = 1$ which is *not* really the surface of Earth. However, the change in magnetic latitude from $\mathscr{R} = 1$ to the surface at $r = R_E$ is small enough to generally be unimportant.

γ) Some care should be exercised in using B, L-coordinates. L is really not constant along a field line so particles that are on one field line at one longitude having different pitch angles may be separated onto different but nearby lines at another longitude.

There is a fundamental reason why particles originally on one field line will drift onto different field lines at *other longitudes*. The integral invariant $\mathscr{I}$ is defined in terms of a particle's mirror point and therefore exists only for particles mirroring at the point of observation. If a particle flux is measured with an omnidirectional instrument, this flux includes particles mirroring at the observation point and also other particles mirroring lower. The omnidirectional flux contains particles of various values of $\mathscr{I}$ but we lump them together. Therefore, it is not surprising that they separate and move to different field lines at other latitudes. There is no reason for them to stay together as they drift. If this is the case, then a shell of lines of force defined by a value of L where L is uniquely related to $\mathscr{I}$ can only approximate the location of particles originally on one field line. The B, L-coordinate system is the best one available for organizing data, however it has certain limitations.

It is now known that for low altitude electrons there are variations in the flux on a B, L-ring at different longitudes [14]. The reason for this is that substantial changes take place for these particles in less than one revolution around Earth. Therefore the data at different longitudes cannot be organized into one set of data, accurately. When combining data in longitude we implicitly assume that the particules live a long time and drift many revolutions. When particles drift eastward down into the Brazilian magnetic anomaly, the low altitude particles get lost. As the remaining particles move up out of the east edge of the anomaly new particles are scattered down into the vacant region which had just been depopulated. The variation of flux with longitude due to the magnetic anomaly has been named the windshield wiper effect (see Sect. 17). This variation of $\mathscr{J}(B, L)$ with longitude has been observed on a low altitude satellite [*1*]. This longitudinal variation cannot be handled by the B, L-system but demands an additional variable such as longitude.

* A map of Earth with lines of constant L at 400 km above ground is found in this Encyclopedia, Vol. 49/2 (K. RAWER and K. SUCHY), Fig. 333, p. 523.

δ) Similarly, *local time variations* in trapped particle fluxes show the need for something other than the B, L-system.

The Earth's magnetic field is distorted by the solar wind as shown in Fig. 41, Sect. 34γ. This produces asymmetries in the field at large L values that depend on local time. The location of flux contours are known to move with local time. Data from Explorer XIV show this effect[1]. For $L>6$ a new definition of L is needed which includes local time effects and also solar wind strength effects.

C. The inner zone.

6. High energy protons.

α) The first experiment performed in the radiation belt that unambiguously identified the *particles* which were counted involved flying a stack of nuclear emulsions on an Atlas rocket[1].

The emulsion stack was recovered and developed and the nuclear tracks were read. The range and ionization of the particles were measured, the particles identified, and their energies determined. Protons of $\mathscr{E}>75$ MeV and electrons of $\mathscr{E}>12$ MeV could get through the 6 g/cm^2 shielding into these first nuclear emulsions. No electrons were found, but a large number of protons were found. The energy spectrum of protons measured by FREDEN and WHITE[2] in a similar way on a later flight is shown in Fig. 4. Other experiments[3-6] have shown very similar energy spectra and intensities of protons and have extended the data down to lower energies.

This population of protons is quite stable in time. One experiment performed shortly after a large solar flare measured an essentially identical flux and spectrum of protons as an earlier experiment that had been performed after a lengthy period of solar quiet.

It is commonly thought that a considerable part, if not all, of these protons are produced by cosmic rays producing neutrons in the Earth's atmosphere. Some of these neutrons emerging from the Earth's atmosphere and decaying in the magnetosphere produce trapped protons. This is called the *cosmic ray albedo neutron decay* (CRAND) source. We can get a quantitative picture of the flux and energy spectrum of the inner-radiation zone protons produced by neutron decay by considering the particles continuity equation[7,8] in energy space

$$\frac{\mathrm{d}N(\mathscr{E})}{\mathrm{d}t}=\frac{\mathrm{d}}{\mathrm{d}\mathscr{E}}\left[J(\mathscr{E})\right]+\mathscr{S}_{\mathrm{p}}(\mathscr{E})-\mathscr{L}(\mathscr{E}) \tag{6.1}$$

where $N(\mathscr{E})$ is the equilibrium proton-density energy spectrum, $\mathscr{S}_{\mathrm{p}}(\mathscr{E})$ is the source of protons, $\mathscr{L}(\mathscr{E})$ is the loss term, and $J(\mathscr{E})$ is the current in energy space $J(\mathscr{E})=N(\mathscr{E})\,\mathrm{d}\mathscr{E}/\mathrm{d}t$. For equilibrium $\mathrm{d}N(\mathscr{E})/\mathrm{d}t=0$. Let us now consider two special cases of the equation for equilibrium [5].

β) Case A: $L(\mathscr{E})=0$.

For protons between 5 and 100 MeV the dominant loss process is slowing down by exciting and ionizing electrons by distant coulomb collisions, and we

[1] L. A. FRANK, J. A. VAN ALLEN, and E. MACAGNO: J. Geophys. Res. **68**, 3543 (1963).
[1] S. C. FREDEN, and R. S. WHITE: Phys. Rev. Letters **3**, 9 (1959).
[2] S. C. FREDEN, and R. S. WHITE: J. Geophys. Res. **67**, 25 (1962).
[3] S. C. FREDEN, and R. S. WHITE: J. Geophys. Res. **65**, 1377 (1960).
[4] A. H. ARMSTRONG, F. B. HARRISON, H. H. HECKMANN, and L. ROSEN: J. Geophys. Res. **66**, 351 (1961).
[5] J. E. NAUGLE, and D. A. KNIFFEN: Phys. Rev. Letters **7**, 3 (1961).
[6] H. H. HECKMANN, and A. H. ARMSTRONG: J. Geophys. Res. **67**, 1255 (1962).
[7] F. S. SINGER: Phys. Rev. Letters **1**, 181 (1958).
[8] W. N. HESS: Phys. Rev. Letters **3**, 11 (1959).

can ignore other losses. This slowing down contributes to the energy-current term and is not considered here to be part of $\mathscr{L}(\mathscr{E})$; Eq. (6.1) now becomes

$$\mathscr{S}_{\rm p}(\mathscr{E}) = \frac{\mathrm{d}}{\mathrm{d}\mathscr{E}}\left[N(\mathscr{E})\,\frac{\mathrm{d}\mathscr{E}}{\mathrm{d}t}\right] = \frac{\mathrm{d}}{\mathrm{d}\mathscr{E}}\left[N(\mathscr{E})\,\frac{\mathrm{d}\mathscr{E}}{\mathrm{d}x}\,v\right]. \tag{6.2}$$

The neutron flux $J_{\rm n}$ outside the atmosphere* has been determined[9]

$$J_{\rm n} \equiv n_{\rm n}\, v_{\rm n} = 0.8\cdot 10^{-4}\left(\frac{\mathscr{E}}{\mathrm{MeV}}\right)^{-2.0}\mathrm{m^{-2}\,sec^{-1}\,MeV^{-1}}.$$

The proton-source term is given by the neutron's decay density

$$\mathscr{S}_{\rm p}(\mathscr{E}) = \eta\,\frac{\mathrm{d}\,n_{\rm n}(\mathscr{E})}{\mathrm{d}\,V} = \frac{\eta\,0.8\cdot 10^{-4}\,\mathscr{E}^{-2.0}}{v\,\tau_{\rm n}} \tag{6.3}$$

where $\tau_{\rm n}$ is the neutron mean lifetime of about 1000 sec. and $n_{\rm n}$ is the number of neutrons decaying in volumn V per MeV.

This expression is valid at low latitudes and close to the Earth. The injection coefficient η is put in here because not all of the neutrons that decay form protons that are trapped[10,11]. Some of the protons made by neutron decay have pitch angles that are so small that they will hit the earth before they mirror. These protons will not form part of the trapped radiation.

We can solve Eq. (6.3) approximately for the energy range 10 to 80 MeV by approximating

$$\left.\begin{aligned}\frac{\mathrm{d}\mathscr{E}}{\mathrm{d}x} &= 6.2\cdot 10^{-21}\,\frac{n}{\mathrm{cm^{-3}}}\left(\frac{\mathscr{E}}{\mathrm{MeV}}\right)^{-0.79}\frac{\mathrm{MeV}}{\mathrm{cm}} \equiv 6.2\cdot 10^{-25}\left(\frac{n}{m^{-3}}\right)\left(\frac{\mathscr{E}}{\mathrm{MeV}}\right)^{-0.79}\frac{\mathrm{MeV}}{\mathrm{m}}\\ v &= 1.45\cdot 10^{9}\left(\frac{\mathscr{E}}{\mathrm{MeV}}\right)^{+0.477}\mathrm{cm\,sec^{-1}} = 1.45\cdot 10^{7}\left(\frac{\mathscr{E}}{\mathrm{MeV}}\right)^{+0.477}\mathrm{m\,sec^{-1}}\end{aligned}\right\} \tag{6.4}$$

where n is the numerical air density. Assuming a solution of the form $N(\mathscr{E}) = k\left(\frac{\mathscr{E}}{\mathrm{MeV}}\right)^{-\nu}$ substituting in Eq. (6.3) we get

$$\left.\begin{aligned}J(\mathscr{E}) = v\,N(\mathscr{E}) &= \frac{1.1\cdot 10^{7}}{n/\mathrm{cm^{-3}}}\left(\frac{\mathscr{E}}{\mathrm{MeV}}\right)^{-0.72}\mathrm{cm^{-2}\,sec^{-1}\,MeV^{-1}} \equiv\\ &\equiv \frac{1.1\cdot 10^{17}}{n/\mathrm{m^{-3}}}\left(\frac{\mathscr{E}}{\mathrm{MeV}}\right)^{-0.72}\mathrm{m^{-2}\,sec^{-1}\,MeV^{-1}}.\end{aligned}\right\} \tag{6.5}$$

If a time-averaged density of $1.0\cdot 10^{5}\ \mathrm{cm^{-3}}$ atoms of O is used, then $J(\mathscr{E}) = 110\left(\frac{\mathscr{E}}{\mathrm{MeV}}\right)^{-0.72}$ $\mathrm{cm^{-2}\,sec^{-1}\,MeV^{-1}}$. This is the low energy portion of the solid curve in Fig. 4.

The average lifetime $\tau_{\rm p}$ of these protons can be obtained by using the "leaky bucket" equation. We have for the leaky bucket model shown in Fig. 5 for equilibrium

$$\mathrm{Input} = \mathrm{Output} = \frac{\mathrm{Contents}}{\tau_{\rm p}} \tag{6.6}$$

which gives, using Eq. (6.2), a particle lifetime

$$\left.\begin{aligned}\tau_{\rm p} &= \frac{\mathrm{Contents}}{\mathrm{Input}} = \frac{N(\mathscr{E})}{\mathscr{S}(\mathscr{E})}\\ &= \frac{1.1\cdot 10^{11}\,\mathrm{cm^{-3}}}{n\,(\nu+0.313)}\left(\frac{\mathscr{E}}{\mathrm{MeV}}\right)^{1.313}\mathrm{sec} \equiv \frac{1.1\cdot 10^{17}\,\mathrm{m^{-3}}}{n\,(\nu+0.313)}\left(\frac{\mathscr{E}}{\mathrm{MeV}}\right)^{1.313}\mathrm{sec}.\end{aligned}\right\} \tag{6.7}$$

---

* We will use J for omnidirectional particle fluxes and j for directional fluxes. The units of J will be either $\mathrm{m^{-2}\,sec^{-1}}$ or $\mathrm{m^{-2}\,sec^{-1}\,MeV^{-1}}$ and the units of j will be either $\mathrm{m^{-2}\,sec^{-1}\,sr^{-1}}$ or $\mathrm{m^{-2}\,sec^{-1}\,sr^{-1}\,MeV^{-1}}$.

[9] W. N. HESS, E. H. CANFIELD, and R. E. LINGENFELTER: J. Geophys. Res. **66**, 665 (1961).

[10] F. S. SINGER: Phys. Rev. Letters **5**, 300 (1960).

[11] A. M. LENCHEK, and S. F. SINGER: J. Geophys. Res. **67**, 1263 (1962).

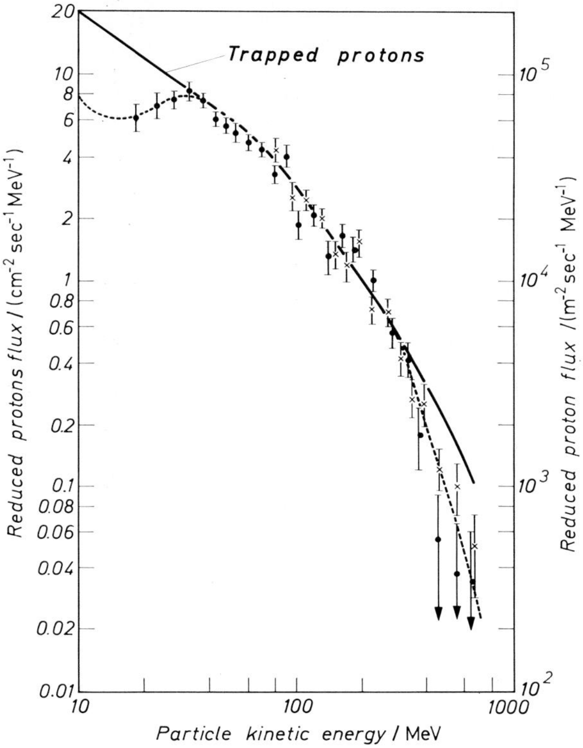

Fig. 4. Energy distribution of high energy protons in the middle of the inner zone of the radiation belt. The solid curve is the theoretical spectrum for trapped protons expected from the cosmic ray neutron albedo decay theory. Courtesy FREDEN and WHITE (ref. [3], Sect. 6).

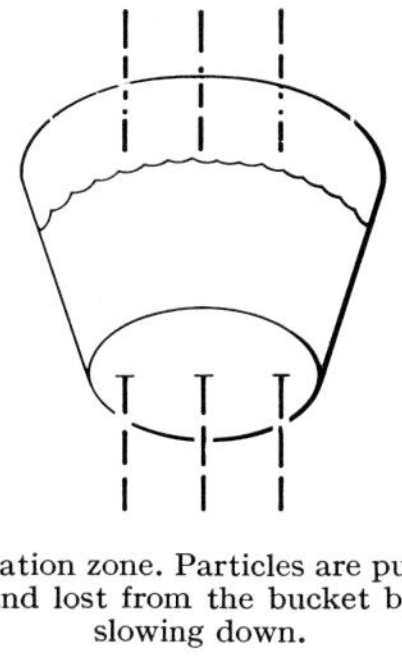

Fig. 5. The leaky bucket model of the inner radiation zone. Particles are put into the bucket (radiation belt) from a source such as cosmic ray neutron albedo decay are and lost from the bucket by processes like inelastic nuclear collisions and slowing down.

Using Eq. (6.7) and assuming $\nu = 0.5$ this gives $\tau_p = 7.0 \cdot 10^5 \left(\frac{\mathscr{E}}{\text{MeV}}\right)^{1.313}$ sec in the energy range $10\ \text{MeV} < \mathscr{E} < 80\ \text{MeV}$ for 1100 km altitude. It is important to note that a particle's lifetime is an average concept and is less fundamental and significant than a particle's time history.

γ) Case B: $d\mathscr{E}/dx = 0$.

If some *other loss process* occurs considerably faster than slowing down, then the protons will have essentially constant energy; slowing down can then be neglected. This situation is approximated for protons of $\mathscr{E} > 300$ MeV. For these energies, the protons almost all have nuclear collisions before they slow down. The cross section for an inelastic collision of a high-energy proton with an oxygen nuclei is $\sigma = 3.0 \cdot 10^{-25}\,\mathrm{cm}^2$. This gives an effective mean free path for nuclear interactions* of $\lambda_{\mathrm{eff}} = \frac{\varrho}{n\,\sigma} = 74\,\mathrm{g\,cm}^{-2}$ where ϱ is the chemical density of the absorber usually given in $\mathrm{g\,cm}^{-2}$. The range of a 300 MeV proton in oxygen is 135 $\mathrm{g\,cm}^{-2}$; therefore, these high-energy protons will usually have nuclear collisions before slowing down much.

In this case, Eq. (6.1) becomes

$$\mathscr{S}(\mathscr{E}) = \mathscr{L}(\mathscr{E}) \tag{6.8}$$

substituting

$$J(\mathscr{E}) = v\,N(\mathscr{E}) = 4.2 \cdot 10^6 \left(\frac{\mathscr{E}}{\mathrm{MeV}}\right)^{-2.54} \mathrm{cm}^{-2}\,\mathrm{sec}^{-1}\,\mathrm{MeV}^{-1}. \tag{6.9}$$

This is an asymptotic expression for $N(\mathscr{E})$ at high energies where slowing down is not important. Above 300 MeV it holds quite well but there slowing down also is important.

This high energy proton population is usually very constant in time but on occasions it has changed. At the time of the "Starfish" high altitude nuclear explosion** the high energy inner zone protons were disturbed probably by the magnetic pulse produced by the explosion. The low altitude proton fluxes were increased by a factor of 5 or more by the explosion.

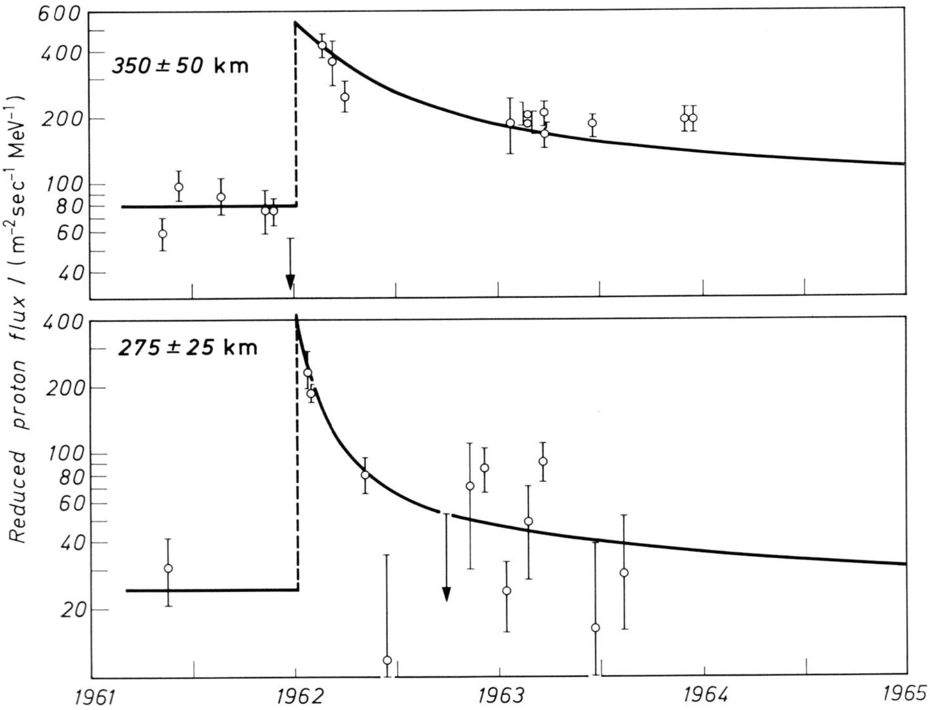

Fig. 6. Time variation of $\mathscr{E} = 55$ MeV protons at two altitudes in the inner zone showing the transient produced by the 'Starfish' nuclear explosion in July of 1962. The solid curves show the calculated decay curves for the transient as computed with the relevant atmospheric data. Courtesy: FILZ and HOLEMAN (ref. [12], Sect. 6).

* As usual in particle physics we define as "effective free path" an absorber density.

** "Starfish" was a 1.4 megaton nuclear explosion fired at 400 km altitude over Johnson Island in the Pacific Ocean in July 1962. For details see Table 2, Sect. 12α.

FILZ and HOLEMAN[12] measured the decay of this transient pulse by studying 55 MeV protons using emulsions on polar orbiting satellites for the period Aug. 1961 to June 1964.

Assuming the protons are lost by slowing down by interaction with the atmosphere, the decay of the proton flux can be studied[13] by using Eq. (6.2) with $\mathscr{S}=0$ or

$$J_0\left(\frac{\mathrm{d}\mathscr{E}}{\mathrm{d}x}\right)_0 = J\,\frac{\mathrm{d}\mathscr{E}}{\mathrm{d}x} \tag{6.10}$$

and integrating $\mathrm{d}\mathscr{E}/\mathrm{d}t$ from Eq. (6.4)

$$\left(\frac{\mathscr{E}}{\mathrm{MeV}}\right)_0^{1.313} - \left(\frac{\mathscr{E}}{\mathrm{MeV}}\right)^{1.313} = 1.18\cdot 10^{-11}\, n t, \tag{6.11}$$

n being the atomic number density usually given in cm^{-3} and t time (in sec). Which using Eqs. (6.4), (6.10) gives

$$\frac{J(\mathscr{E},t)}{J_0(\mathscr{E},0)} = \left(\frac{\mathscr{E}}{\mathscr{E}_0}\right)^{0.79} = \left(1 + 1.18\cdot 10^{-11}\, n t \left(\frac{\mathscr{E}}{\mathrm{MeV}}\right)^{-1.313}\right)^{-0.99}. \tag{6.12}$$

The experimental data on the proton decay in Fig. 6 agrees well with this expression and reasonable values of the average density n.

The spatial distribution of the high energy inner zone protons shown as a B, L-map in Fig. 3 is not complete.

More recent data of MCILWAIN's[14] from Explorer 15 shown as an $\mathscr{R}$, ϕ map (see Sect. 5 β) in Fig. 7 gives a complete picture of protons of 40 MeV $<\mathscr{E}_\mathrm{p}<$ 110 MeV.

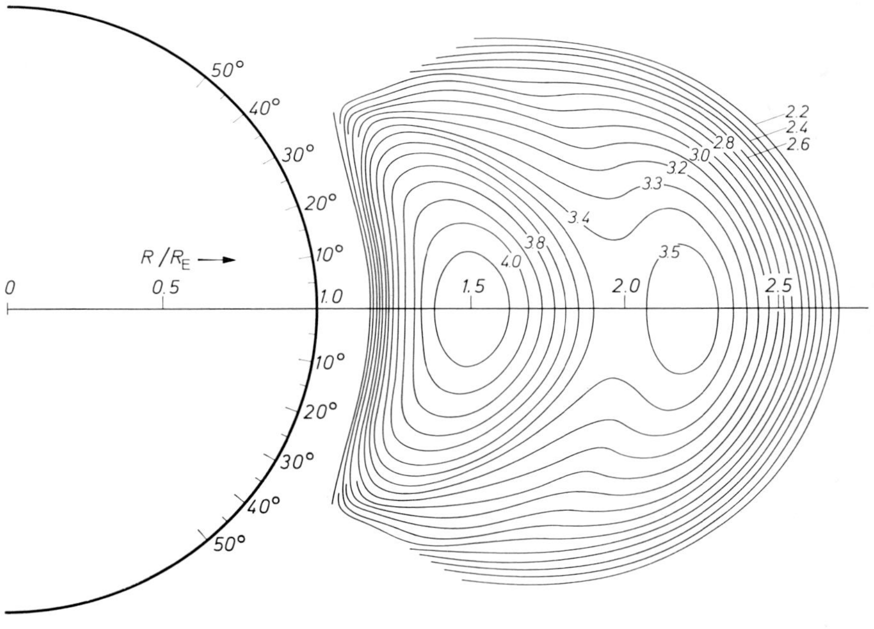

Fig. 7. An $\mathscr{R}$, ϕ map of experimentally measured proton fluxes of 40 MeV $<\mathscr{E}<$ 110 MeV as measured by Explorer 15. Parameter of the curves is $^{10}\log(1.4\, J_0)$. This shows the second peak of the proton flux after MCILWAIN (ref. [19], Sect. 6).

[12] R. C. FILZ, and E. HOLEMAN: J. Geophys. Res. **70**, 5807 (1965).
[13] E. D. RAY: J. Geophys. Res. **65**, 1125 (1960).
[14] C. E. MCILWAIN: Science **142**, 355 (1963).

Very interesting structure shows up here. There is a second peak in the radial distribution at $L \sim 2.5$. This second peak is narrow enough in its latitudinal width so that it was not observed by the detectors on the earlier Pioneer 3 and 4 flights. The inner peak in Fig. 7 is essentially the same as that seen by Explorer 4 in Fig. 3. The outer peak in Fig. 7 is not easy to understand in terms of neutron decay, either from galactic or polar cap protons. CRAND protons may be responsible for the inner peak. But neither CRAND nor polar cap neutrons will make protons at larger L values having the characteristics of the outer peak. We need another source for these outer peak protons. We will return to this in Chap. E.

We should also consider the reasons why there are not high energy protons in the outer zone. At $L=3$ the neutron source strength $\mathcal{S}(\mathcal{E})$ is down about a factor of 5 from the neutron source at the heart of the inner zone. The atmospheric density is also decreased at $L=3$ so the steady state trapped proton flux from neutron decay here should be about 1/3 what it is at $L=1.5$. But from Explorer 6 data FAN, MEYER and SIMPSON[15] say that at 40000 km altitude the trapped proton flux of $\mathcal{E} > 75$ MeV is less than the cosmic ray proton flux. This means the lifetime of a proton at $L=3$ must be at least a factor of 1000 less than the lifetime at $L=1.5$. It seems quite probable that non-adiabatic process due to magnetic waves disturb the protons motion and can shorten the protons lifetime. This probably is a problem of pitch angle scattering[16] such as is discussed in Chap. D for electrons.

7. Protons from polar cap neutrons.

α) *Protons* fairly often arrive at the Earth from the Sun where they were made *in connection with solar flares.* These protons have energies up to a few hundred MeV and can only reach the Earth in the polar regions. The terrestrial magnetic field excludes them from the equator.

Neutrons are made when these protons interact with the polar atmosphere. The decay of these solar proton-produced neutrons is an additional source of energetic protons for the inner radiation zone[1]. These neutrons have two different characteristics from those made by galactic cosmic ray protons. Because they are made by lower energy protons these neutrons are of lower average energy. Since the solar protons only hit the polar caps ($\Phi > 60°$)* the neutron source geometry is quite different from that for galactic cosmic ray protons.

β) NAUGLE and KNIFFEN[2] measured the trapped proton energy spectrum using nuclear emulsions for $1.47 < L < 1.79$. For $L < 1.6$ the *energy spectrum* is very similar to FREDEN and WHITE's spectrum in Fig. 4. For $L > 1.6$ a second lower energy component appears. A large flux of $\mathcal{E} < 30$ MeV protons appears at these larger latitudes. NAUGLE and KNIFFEN considered it probable that this polar cap neutron source was responsible for the low energy component at $L > 1.6$. (This source due to solar proton albedo neutron decay is called SPAND by DRAGT et al.[3].

LENCHEK[4], to try to explain the low energy protons found by NAUGLE and KNIFFEN, calculated the proton flux expected from polar cap neutrons. Using fairly extreme values for the source strength and proton lifetime he could get agreement with the measured proton fluxes. Most opinion now (including LENCHEK's) is that solar protons do *not* make the low energy trapped protons observed by NAUGLE and KNIFFEN and that some other new source is needed.

γ) A quite exact way of treating the *neutron decay injection* that does not use the approximate injection coefficient η but instead keeps track of the proton pitch angle at birth has been developed by DRAGT, AUSTIN and WHITE[3] and by HESS and KILLEEN[5]. These calculated

* See Sect. 5β for definition of the 'generalized magnetic latitude' Φ.

15 C. Y. FAN, P. MEYER, and J. A. SIMPSON: J. Geophys. Res. **66**, 2697 (1961).

16 A. J. DRAGT: J. Geophys. Res. **66**, 1641 (1961).

1 A. H. ARMSTRONG, F. B. HARRISON, H. H. HECKMANN, and L. ROSEN: J. Geophys. Res. **66**, 351 (1961).

2 J. E. NAUGLE, and D. A. KNIFFEN: Phys. Rev. Letters **7**, 3 (1961).

3 J. A. DRAGT, M. M. AUSTIN, and R. S. WHITE: J. Geophys. Res. **71**, 1293 (1966).

4 A. M. LENCHEK: J. Geophys. Res. **67**, 2145 (1962).

5 W. N. HESS, and J. KILLEEN: J. Geophys. Res. **71**, 2799 (1966).

proton fluxes are much lower than LENCHEK's values[4] although they start essentially the same solar proton flux incident on the polar cap. These calculations make it seem unlikely that the SPAND source produces the NAUGLE and KNIFFEN low energy protons.

The quantitative calculations[3,5] also cast suspicion on the origin of the higher energy low-latitude inner zone protons. It has been normally assumed that CRAND is the source of these high energy trapped protons. However, it seems difficult to make as many protons as are observed using this source. Off equator at $B/B_0 \sim 3$ it seems impossible to get agreement. Near equator it is marginal. Solar neutrons have also been considered[5] but they do not help much. This business must be considered undecided now. The fact that the calculated energy spectrum in Fig. 4 agrees well with the shape of the measured spectrum argues for the CRAND source but the calculation of the proton *flux* makes CRAND look marginal. Off-equator, clearly something else is needed besides CRAND plus atmospheric losses.

δ) FREDEN et al.[6] have measured the energy spectrum of protons from 10 to 200 Mev through all of the inner zone on satellite 1964-45 A. Their results are shown in Fig. 8. The large fluxes of $\mathscr{E}_p < 30$ MeV protons that are clearly present

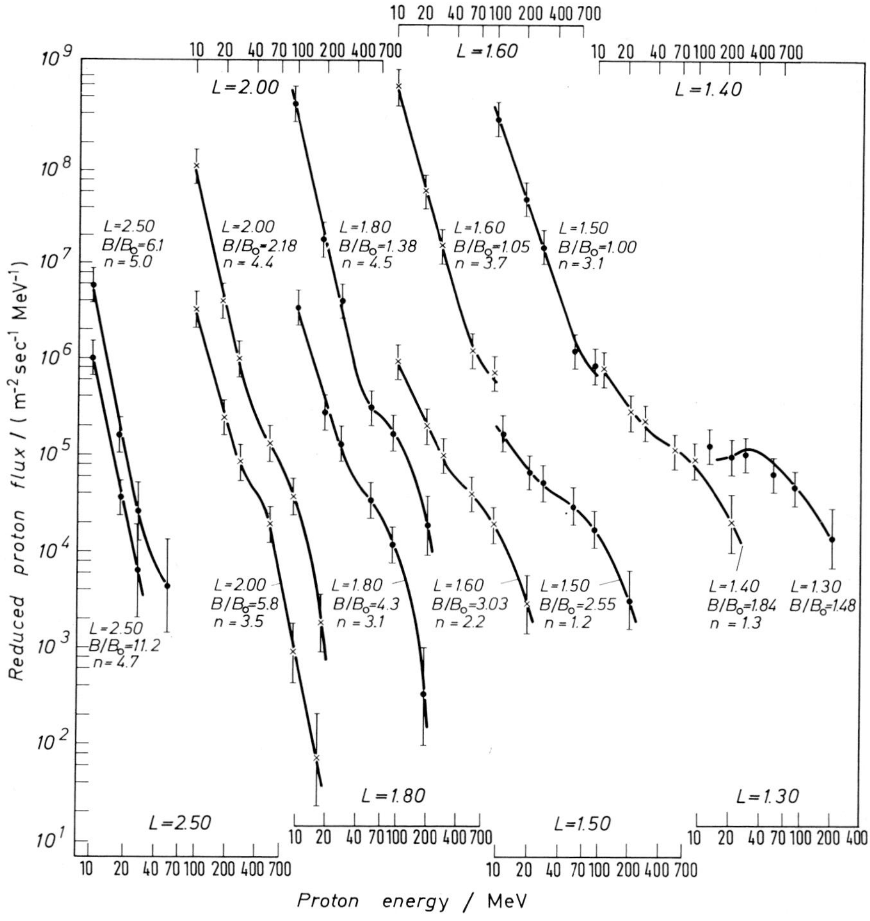

Fig. 8. Measured proton energy spectra at several different locations in the inner zone as indicated by L and B/B_0. (FREDEN et al., ref. [6], Sect. 7).

[6] S. C. FREDEN, J. B. BLAKE, and G. A. PAULIKAS: J. Geophys. Res. **70**, 3113 (1965).

cannot be explained by CRAND or SPAND and some additional source is needed that is currently not understood.

8. Solar cycle changes. One of the features of the high energy proton component of the radiation belt is the time constancy, but slow changes in proton population are expected due to changes in the galactic cosmic ray flux during the solar cycle and more importantly due to changes in the upper atmospheric density during the solar cycle.

To discuss the expected changes in proton populations for $L < 1.6$, BLANCHARD and HESS[1] wrote the time dependent form of the continuity equation, see Eqs. (6.1),

$$\frac{\mathrm{d}N(\mathscr{E}\,t)}{\mathrm{d}t} = \mathscr{S}_{\mathrm{p}}(\mathscr{E},t) - \mathscr{L}(\mathscr{E},t) + \frac{\mathrm{d}}{\mathrm{d}\mathscr{E}}\left[N(\mathscr{E},t)\,\frac{\mathrm{d}\mathscr{E}}{\mathrm{d}x}(t)\,v\right]. \tag{8.1}$$

The CRAND source is assumed to vary 25 percent during the solar cycle. The atmospheric density at ~500 km altitude varies by about a factor of 50 during the solar cycle by direct measurement[2]. Both effects make the trapped proton flux larger at solar minimum.

Eq. (8.1) was integrated to build up to an oscillating proton population which is the same from one solar cycle to the next. After achieving this condition the proton energy spectrum varies during one solar cycle as shown in Fig. 9. The dotted curves top and bottom in these figures are what the proton's spectrum would be if steady state conditions were achieved at solar maximum and solar minimum. Steady state clearly is not achieved for high energy protons or for high altitudes.

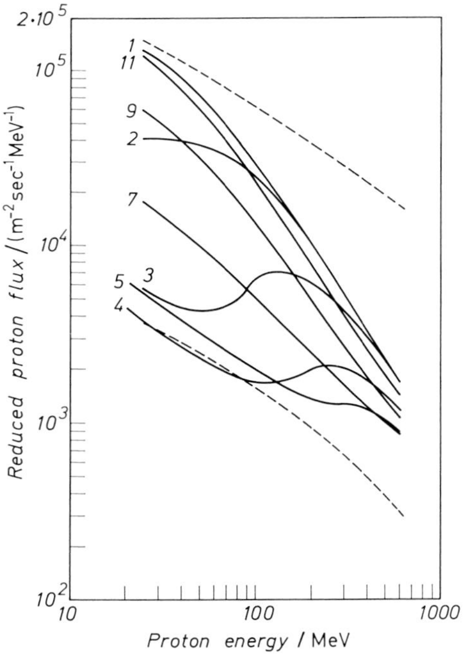

Fig. 9. Calculated variation of the proton energy spectrum for a minimum height of 330 km, $L = 1.4$ and $B = 0.225\,\Gamma$, at different times in the solar cycle (parameter; year 0 is solar minimum). The dotted curves are the steady state spectra for solar minimum and maximum atmospheric conditions. (BLANCHARD and HESS, ref. [1], Sect. 14).

[1] R. C. BLANCHARD, and W. N. HESS: J. Geophys. Res. **69**, 3927 (1964).
[2] D. G. KING HELE: Nature **203**, 959 (1964), and [*12*] 1132 (1965).

Time changes have been observed in the proton population at low altitudes that agree at least qualitatively with the predictions. It is uncertain whether the results agree quantitatively or not. If other processes are important at low altitudes besides atmospheric loss they may be found by studying these solar cycle changes.

9. The East-West effect. Another demonstration that the trapped proton flux is limited at low altitudes by the atmosphere was produced by HECKMAN and NAKANO[1]. They flew emulsions in an oriented polar satellite at about 400 km. At this altitude the trapped protons are encountered only in a quite limited region in the South Atlantic. In analyzing the tracks in the emulsions they found the 2.3 times as many protons of $\mathscr{E} > 57$ MeV entered the stack from the West than from the East.

The explanation of this, due to LENCHEK[2], is straight-forward and is due to the fact that in this region the protons moving from West to the East, j_E, have their guiding centers above the satellite and the protons moving East to West, j_W, have the guiding centers below the satellite. Assuming that the proton flux is inversely proportional to the atmospheric density at the guiding center the East-West ratio is given by

$$\frac{j_E}{j_W} = \exp\left(2\,\frac{R_c}{H}\cos\psi\right) \tag{9.1}$$

where ψ is the dip angle of the field line and H is the atmospheric scale height and R_c the gyroradius.

Solving Eq. (9.1) for H using $j_E/j_W = 1.57$ at 60 MeV gives $H \cong 60$ km. This is very similar to the atmospheric scale height measured by satellite drag at this altitude and therefore demonstrates that the proton lifetime is inversely proportional to the mirror atmospheric density.

10. Electrons.

α) There is only fragmentary data on the natural electron fluxes for $L < 2$. The reason for this is two-fold. First, nearly all early experiments in the inner zone did not discriminate between electrons and protons and the detector count rates were dominated by the penetrating proton component. Secondly, experiments conducted after the summer of 1962 in the inner zone that could measure electrons uniquely did not observe natural electrons but rather the artificial electrons from nuclear explosion "Starfish" (see Sect. 6γ).

But there are a few measurements which do help define the natural *inner zone* electron population. There have been three experiments[1-3] that measure the energy spectrum of the natural electrons. These all use magnetic analysis to separate electrons from protons. The agreement is not too good but the following seem true:

(1) The measured spectrum is reasonably flat for $100\ \text{keV} < \mathscr{E}_e < 400\ \text{keV}$.

(2) It is uncertain whether there are any electrons present of $\mathscr{E} > 800$ keV or not.

β) KELLOGG calculated the equilibrium trapped electron energy spectrum that would be made by the CRAND source considering that the electrons are lost by small angle scattering. The source spectrum is weighted by $\mathscr{E}^2$, which is the energy dependence of the scattering loss process to get the equilibrium energy spectrum[4]. Comparing this calculated spectrum with the measured spectra shows rather clearly that at low energies there are more electrons observed than we could expect from neutron decay.

---

1 H. H. HECKMAN, and G. H. NAKANO: J. Geophys. Res. **68**, 2117 (1963).
2 A. M. LENCHEK, and S. F. SINGER: J. Geophys. Res. **67**, 4073 (1962).
1 F. E. HOLLY, L. ALLEN, and R. G. JOHNSON: J. Geophys. Res. **66**, 1627 (1961).
2 W. L. IMHOF, R. V. SMITH, and P. C. FISHER: [*10*], 438 (1963).
3 L. G. MANN, S. D. BLOOM, and H. I. WEST: [*10*], 447 (1963).
4 P. J. KELLOGG: J. Geophys. Res. **65**, 2705 (1960).

It is uncertain from the experimental data whether there are electrons present of $\mathcal{E}_e > 780$ keV or not. This energy is the upper end of the neutron betha-decay spectrum except that there will be a few electrons above this upper energy limit due to the fact that the neutron that decays is in motion. A 150 MeV neutron produces decay electrons of energies up to 1.5 MeV. This yields an electron flux of $10^2 \ldots 10^3$ cm^{-2} sec^{-1} of $\mathcal{E} > 1.5$ MeV[5]. This decay in flight makes a modest number of high energy electrons but less than the spectrum measured by MANN et al.[3].

What *flux of electrons* is expected from neutron decay from the CRAND source? The neutron decay density in the center of the inner zone is about 10^{-11} cm^{-3} sec^{-1}. We know lifetimes of electrons from the "Starfish" explosion (see Chap. D).

At $L \sim 1.5$ the electron lifetime is found experimentally to be about one year. This gives an equilibrium CRAND electron flux of

$$J \sim 10^{-11} (3 \cdot 10^{10}) (3 \cdot 10^{7}) \sim 10^{7} \text{ cm}^{-2} \text{ sec}^{-1}.$$

FRANK and VAN ALLEN[6] using "Injun I" data measured the electron flux at the inner edge of the inner belt to be $J \sim 10^7$ cm^{-2} sec^{-1} at $L = 1.22$ at the equator. Crudely extrapolating this to the middle of the inner belt inversely as the atmospheric density gives $J \sim 3 \cdot 10^7$ cm^{-2} sec^{-1}.

γ) The flux calculated from the CRAND source is close enough to the extrapolated experimental flux in the center of the inner zone so that there can be little doubt that neutron decay is an important source of inner zone electrons. But from the spectral measurements there clearly are lots of *lower energy electrons* of $\mathcal{E}_e < 200$ keV present in the inner belt. These are not expected on the basis of neutron decay. There must be a second source to explain these low energy electrons. Electrons are often observed precipitating in the atmosphere in the region of the inner belt.

If low energy electrons such as those precipitated can be found beneath the inner radiation belt it is not surprising that they are found trapped in it too.

11. Other particles. If the Sun were the source of particles in the inner radiation belt we would expect to find not only protons but other heavier particles such as deuterons, tritons, and He^3 and He^4 nuclei. The Sun contains about 15 percent He nuclei and solar cosmic rays contain about 5 percent He nuclei. The lifetime of a He^4 nuclei in the inner belt would be about five times less than that of a proton the same energy because the rate of slowing down is faster for (nuclear charge) $Z = 2$. On this basis, we would expect about one percent He^4 in the inner belt. But experimentally not one $Z = 2$ track has been found in nuclear emulsion. The total number of particles measured in three experiments is given in Table 1.

Table 1. *Measurements on trapped heavy particles.*

Experiment	No. of protons	No. of deuterons	No. of tritons	No. of alphas
FREDEN and WHITE May 1959 (Ref. [1], Sect. 6)	243	0	3	0
ARMSTRONG, et al. July 1959 (Ref. [4], Sect. 6)	477	5	0	0
HECKMAN and ARMSTRONG October 1960 (Ref. [6], Sect. 6)	301	0	0	0
Total	1021	5	3	0

[5] M. P. NAKADA: J. Geophys. Res. **68**, 47 (1963).
[6] L. A. FRANK, and J. A. VAN ALLEN: J. Geophys. Res. **68**, 1203 (1963).

An upper limit[1] of the alpha-flux is 0.1 ± 0.1 percent of the proton flux in the energy interval 125 ... 185 MeV. This quite clearly shows that the sun contributes few, if any, of the heavy particles in the inner belt.

A few deuterons and tritons (roughly one-half percent each) were found in the emulsion experiments as is shown in Table 1. These particle fluxes can be explained[2] as being the result of nuclear collisions of trapped protons with O and N nuclei in the very thin atmosphere present at radiation belt altitudes. Such collisions will result in spallation products such as deuterons and tritons. No heavy particles have been observed in the inner belt that cannot be understood by the neutron-decay source.

D. Artificial radiation belts.

12. How they are made.

α) There have been seven artificial radiation belts made by the explosion of *high altitude nuclear bombs* since 1958. These artificial belts result from the release of energetic charged particles, mostly electrons, from the nuclear explosions. These seven explosions are:

Table 2. *Nuclear explosions producing artificial belts.*

Explosion	Locale	Date	Yield	Altitude	L of burst
Argus I	South Atlantic	Aug. 27, 1958	1 kt	200 km	1.7
Argus II	South Atlantic	Aug. 30, 1958	1 kt	250 km	2.1
Argus III	South Atlantic	Sept. 6, 1958	1 kt	500 km	2.0
Starfish	Johnson Island Pacific Ocean	July 9, 1962	1.4 Mt	400 km	1.12
USSR	Siberia	Oct. 22, 1962	several hundred kt	?	1.9
USSR	Siberia	Oct. 28, 1962	submegaton	?	2.0
USSR	Siberia	Nov. 1, 1962	megaton	?	1.8

β) What is there about a nuclear explosion that makes an artificial radiation belt? There are two kinds of nuclear explosions, fission and fusion. The basic element of a *fission reaction* is the capture of a neutron by a heavy element such as U^{235}, which then fissions into two lighter nuclei or fission fragments. In this process two or three neutrons are given off, of which about one percent may escape from the fissioning system. The neutron can produce trapped particles by decaying into a proton and an electron.

The fission fragments produced are unstable (they are neutron rich) and they decay by emitting electrons to become stable. One fission fragment emits about six electrons. These electrons are the most important source of all the artificial belts produced. They have energies up to about 8 MeV with an average of about 1 MeV. They can be released a long distance from the bomb because the fission fragment decay process is relatively slow. Electrons are still being given off minutes to hours after the explosion.

γ) A *fusion* bomb works by burning hydrogen to make helium. The end products are not radioactive but some intermediate steps in the reaction produce charged particles that can be trapped.

---

[1] H. H. Heckmann, and A. H. Armstrong: J. Geophys. Res. **67**, 1255 (1962).
[2] S. C. Freden, and R. S. White: J. Geophys. Res. **65**, 1377 (1960).

A fusion bomb explosion will produce a quite insignificant artificial radiation belt compared to fision or atom bomb explosion of the same yield. Incidentally, a "hydrogen" bomb may contain a considerable amount of fissionable material and thus may generate a substantial fission fragment betha-decay type radiation belt. The electrons from fission fragment betha-decay are the most important source of particles for all artificial radiation belts made to date.

δ) The *particles* from a nuclear explosion are emitted at one point in space. In a few seconds they are spread out along a field line and in less than an hour they drift around the earth and spread out in longitude to form a blanket around the earth. The thickness of the blanket depends on the initial dimensions of the particle source.

About 10^{23} neutrons are given off by a one kiloton explosion. The neutron mean life τ_n is about 1000 sec. The fraction which will decay inside the magnetosphere is about one percent.

The decay protons have energies nearly equal to the energy of the parent neutrons. This means that the protons from fission neutrons will be about 1 MeV. The neutron decay proton flux from "Starfish" was probably about 10^6 cm^{-2} sec^{-1} of $\mathscr{E} > 1$ MeV[1]. The natural proton fluxes are considerably larger than this in most regions of space so we can ignore the neutron decay protons. However, the neutron decay *electron* flux is considerably higher than this. If we have a neutron source above the atmosphere we must consider not only neutrons travelling upwards away from the explosion but partly thermalized albedo neutrons from the top of the atmosphere. Actually more total decays will result from the initially downward-moving neutrons than from the upward-moving ones.

Electron fluxes up to about 10^7 cm^{-2} sec^{-1} are expected from neutron decay from "Starfish"[1]. These are initial fluxes and will, of course, decay with time. This electron flux is not negligible but it is concealed by the considerably larger flux of fission electrons from "Starfish". Some effects on very low frequency (vlf) propagation observed shortly after "Starfish" have been interpreted in terms of neutron decay producing electrons which were precipitated into the atmosphere and changed the properties of the ionosphere[2,3].

13. Argus. The "Argus" explosions of 1958 were carried out to study the trapping of energetic particles by the earth's magnetic field[1].

The planning for "Argus" was well underway before the discovery by VAN ALLEN of the natural radiation belt. In the "Argus" planning sessions it had been suggested that a natural belt might exist around the earth. Such a belt was discovered, of course, by the Explorer I and Explorer III satellites. After each of the "Argus" explosions, trapped particles were observed by VAN ALLEN on the Explorer 4 satellite[2]. The most interesting result of the "Argus" experiment was that it demonstrated that a shell of particles put in to the magnetosphere near $L = 2$ was stable for several weeks.

14. Starfish.

α) The most interesting high altitude nuclear explosion from a particle standpoint was "Starfish". This was a 1.4 megaton bomb exploded at 400 km over Johnson Island in the central Pacific on July 9, 1962. There was quite complete *experimental coverage* of the event.

---

[1] J. KILLEEN, W. N. HESS, and R. E. LINGENFELTER: J. Geophys. Res. **68**, 4637 (1963).
[2] A. J. ZMUDA, B. W. SHAW, and C. R. HAAVE: J. Geophys. Res. **68**, 745 (1963).
[3] J. F. KENNERY, and H. R. WILLARD: J. Geophys. Res. **68**, 4645 (1963).
[1] N. C. CHRISTOFOLIS: J. Geophys. Res. **64**, 869 (1959).
[2] J. A. VAN ALLEN, C. E. MCILWAIN, and G. H. LUDWIG: J. Geophys. Res. **64**, 877 (1959).

Magnetic and electromagnetic signals and a whistler radiation by the explosion were observed at several places[1–3]. These may interact with particles in the natural belt and either change their energy or scatter them and change their pitch angle.

Just seconds after the explosion artificial aurora were observed in New Zealand[4]. These are produced by the electrons and other particles from the explosion that are not trapped. Many of these particles are injected with mirror points below the atmosphere.

On July 10, 1962 there were five satellites in orbit that had electron detectors on board and which gave useful information on the newly trapped particles, see Table 3.

Table 3. *Observing satellites during "Starfish" experiment.*

Satellite	Apogee	Perigee	Inclination	Detectors
ARIEL 1	1209 km	393 km	54°	shielded GM counter $\mathcal{E}_e > 4.7$ MeV
INJUN I	1010 km	890 km	67°	shielded GM counter, counting several MeV electrons by bremsstrahlung
TELSTAR	5630 km	955 km	44.7°	4-channel solid state detector $\mathcal{E}_e > 0.2$ MeV
TRAAC	1110 km	951 km	32.4°	shielded GM counter $\mathcal{E}_e > 1.6$ MeV
COSMOS 5	1512 km	204 km	49°	shielded GM counter $\mathcal{E}_e > 7$ MeV also probably counting bremsstrahlung

The US-UK satellite "Ariel 1"[5] showed that high energy electrons resulting from the bomb appeared very shortly after the explosion at high latitudes — up to $L=5$ or more. "Cosmos 5" showed the electron flux at $L=2.2$ increased by a factor of 100 in the first orbit after Starfish[6].

By comparing the measurements of the several different detectors having different energy responses, the energy spectrum of the new particles was determined[7]. At about 1000 km the spectrum closely resembled an equilibrium fission energy spectrum, thus identifying the decay of fission fragments as the major particle source in this region.

The "Injun I" counters mapped out the new belt at 1000 km altitude and produced the first flux contour picture of the Starfish electrons[8]. The "Telstar" satellite produced all of the information above 1000 km for the first three months after "Starfish"[9]. When the experimental data from "Injun I" and "Telstar" for a short period after "Starfish" were compared they did not seem to agree. The highest electron flux for both "Telstar" and "Injun" was about 10^9 cm^{-2} sec^{-1}. But the $J=10^7$ cm^{-2} sec^{-1} contour for "Injun I" seemed to extend out to about $L=2$ while for "Telstar" the 10^7 cm^{-2} sec^{-1} contour extended past $L=3$. The "Injun I" flux distribution seemed to be much more compressed than is "Telstar's". This difference is reasonably well understood now. It is due to the difference in the artificial electron energy spectrum at different L values.

---

[1] G. M. Allcock, C. K. Branigan, J. C. Mountjob, and R. A. Helliwell: J. Geophys. Res. **68**, 735 (1963).
[2] C. R. Wilson, and M. Sugiura: J. Geophys. Res. **68**, 3149 (1963).
[3] G. M. Crook, E. W. Greenstadt, and G. T. Inonye: J. Geophys. Res. **68**, 1781 (1963).
[4] J. B. Gregory: Nature **196**, 508 (1962).
[5] A. C. Durney, H. Elliott, R. J. Hynds, and J. J. Quenby: Nature **195**, 1245 (1962).
[6] Yu. I. Galperin, and A. D. Bulyunova: [*12*], 446—457 (1965).
[7] W. N. Hess: J. Geophys. Res. **68**, 667 (1963).
[8] B. J. O'Brien, C. D. Laughlin, and J. A. van Allen: Nature **195**, 939 (1962).
[9] W. L. Brown, and J. D. Gabbe: J. Geophys. Res. **68**, 607 (1963).

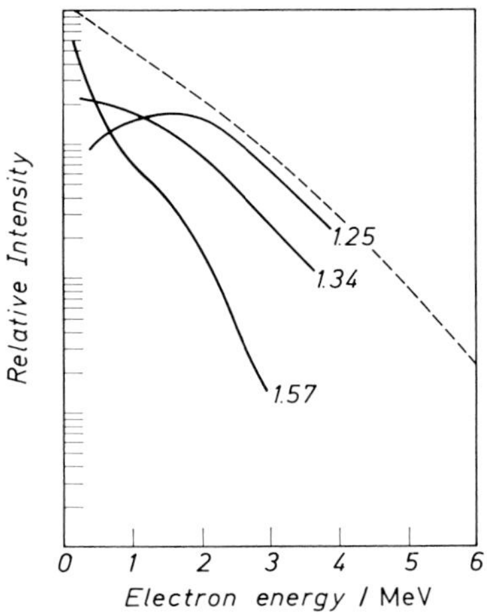

Fig. 10. Various electron energy spectra for the 'Starfish' electrons. The broken curve is the equilibrium fission energy spectrum; full curves are:
(i) the spectrum measured at $L=1.25$ on Dec. 8, 1962;
(ii) the spectrum measured at $L=1.34$ on Dec. 8, 1962;
(iii) the spectrum measured at $L=1.57$ on Dec. 8, 1962. (L is indicated as parameter.) Courtesy: WEST (ref. [10], Sect. 14).

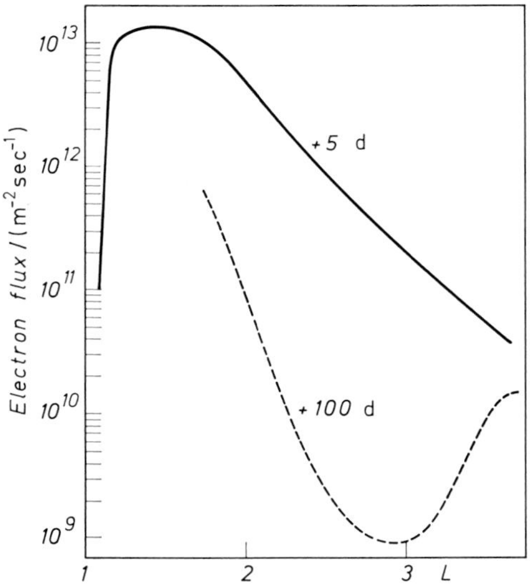

Fig. 11. Approximate omnidirectional electron fluxes of $\mathscr{E}_e > 0.5$ MeV, as function of L along the equator at $+5$ and $+100$ days after 'Starfish'.

β) A magnetic spectrometer flown on satellite 1962 β K measured the *electron energy spectra* from "Starfish" at different L values[10]. Such spectra for Dec. 8, 1962 are shown in Fig. 10. At $L=1.25$ the energy spectrum looks like the fission spectrum except that there are fewer low energy electrons. These low energy electrons probably have been lost by Coulomb scattering and energy loss between July and December. The measured spectrum at $L=1.57$ is softer. It has consid-

[10] H. I. WEST, L. G. MANN, and S. D. BLOOM: [*12*], 423 (1965). — H. I. WEST: [*16*], 634 (1966), and [*16*], 663 (1966).

erably fewer high energy electrons than a fission spectrum. Not as much decay should have taken place here, so this spectrum should be rather like the initial spectrum on July 9, 1962 at $L=1.57$.

The "Injun" and "Telstar" contours are two different pictures of the same thing. The "Injun I" picture showed the spatial distribution of electrons of $\mathscr{E}\sim 2$ MeV. The "Telstar" picture showed the spatial distribution of $\mathscr{E}\sim\frac{1}{2}$ MeV electrons and probably gave a better estimate of the total artificial belt electron population than the "Injun I" estimate because "Injun" does not include the large number of low energy electrons present. The energy spectra in Fig. 10 show why the two sets of contours had different spatial extents. Using this information, the approximate spatial distribution of all electrons from "Starfish" has been constructed for a few days after the explosion and is shown in Fig. 11.

γ) Photographs taken of the debris[11] from the "Starfish" explosion give considerable information about the *debris motion* and help in understanding where the electrons went to. A photograph from Christmas Island taken at +60 sec (shown in Fig. 12) shows some of the debris following the field lines but part of it clearly is moving upwards across field lines. This upward jet seems to move straight along the direction of the field line at the explosion site. Zinn et al. say:

"It is apparent that a part of the jet was undeflected by the bending field lines, but coasted in a straight line across the field for several hundred kilometers (until it passed out of the field of view of the camera). As it travelled it lost material into the field at a relatively constant rate, resulting in a high but thin curtain-like formation."

It is not known how far the jet went upwards but it clearly could be responsible for all of the high altitude electrons observed after the explosion. It seems likely that the low energy electrons observed after "Starfish" at large L values are made by the expansion and adiabatic cooling[12] of the upwards-moving debris observed in the Christmas Island photograph.

δ) Attempts were made to try to observe *synchrotron radiation* from the natural Van Allen belt before the "Starfish" explosion[13] by the Jicamarca Radio Observatory in Peru.

Calculations had shown that the noise from the natural belt would be less than the normal noise background and would be hard to measure[14]. By special techniques looking for the polarization of the synchrotron radiation, measurements made at Peru showed that the antenna temperature due to the natural radiation belt was less than 30° K (in the presence of a background antenna temperature of greater than 3000° K). But after "Starfish" the synchrotron radiation was easily found. Measurements in Peru showed the "Starfish" synchrotron radiation was nearly linearly polarized.

The excess antenna temperature a few hours after the explosion averaged out in longitude was about 10000 °K at 50 MHz. This radiation decayed slowly for several months following a law

$$I(t)=\frac{I_0}{1+\frac{t}{60\,\mathrm{d}}}. \tag{14.1}$$

Because most of the synchrotron radiation comes from low altitudes, this decay is biased towards low altitudes and is not representative of the whole artificial belt.

Measurements were made at several other sites of the synchrotron radiation from the July 9 explosion[15]. The radiation was only observed up to $\phi\sim 25°$. This is due to the beaming of the synchrotron radiation strongly perpendicular to the field lines. Field line normals do mostly hit the earth close to the equator. The

[11] John Zinn, H. Hoerlin, and A. G. Petschek: [*16*], 671 (1966).
[12] W. N. Hess: The Effects of High Altitude Explosions, in: [*17*], 573 (1964).
[13] G. R. Ochs, D. T. Farley jr., K. L. Bowles, and P. Bandyopadhay: J. Geophys. Res. **68**, 701 (1963).
[14] R. B. Dyce, and M. P. Nakada: J. Geophys. Res. **64**, 1163 (1959).
[15] R. B. Dyce, and S. Horowitz: J. Geophys. Res. **68**, 713 (1963).

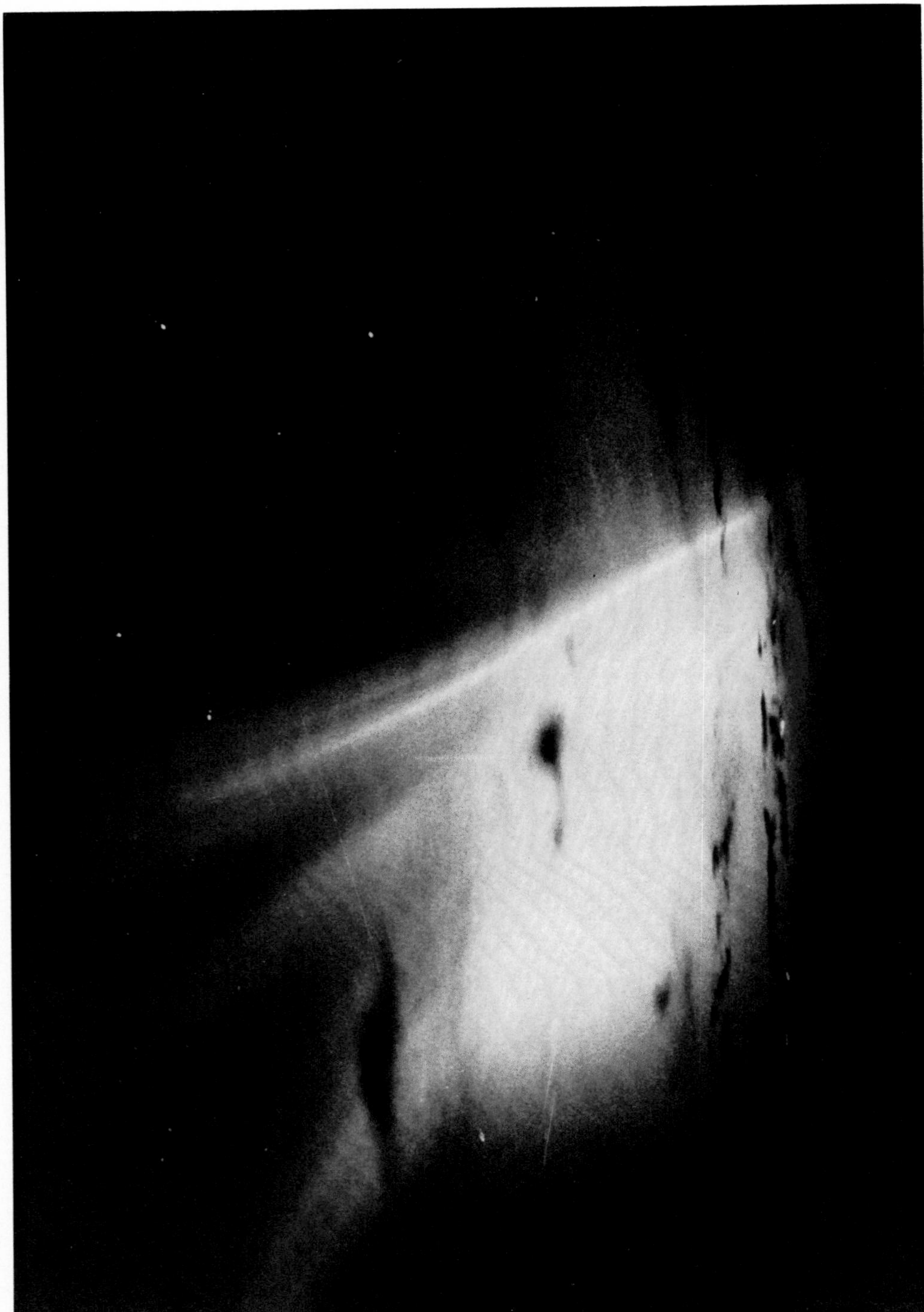

Fig. 12. Photograph of the Starfish debris taken from Christmas Island at +60 sec after the explosion. The streamers at the upper right are clearly moving up across field lines. Courtesy: Zinn (ref. [11], Sect. 14).

observed synchrotron radiation agreed with what was expected from the measured trapped particles[16, 17].

15. The Soviet high altitude explosions. On October 27, 1962, NASA launched the Explorer 15 satellite to study the artificial radiation belt. But before it got in orbit there were two artificial belts and by the time it was up for a day there was a third belt. The Soviets conducted high altitude explosions on October 22 and October 28, and then a third one on November 1, 1962. Fig. 13 shows an $\mathcal{R}$, ϕ map (see Sect. 5β) of the electron omnidirectional flux measured by the magnetic spectrometer on 1962 βK 7.3 h after the October 28 explosion[1]. The double-peaked distribution shown here is due to the superposition of the electron populations from two of the Soviet explosions. The inner peak at $L\sim 1.8$ was due to the October 22 USSR bomb and the outer peak at $L\sim 2.0$ was produced by the

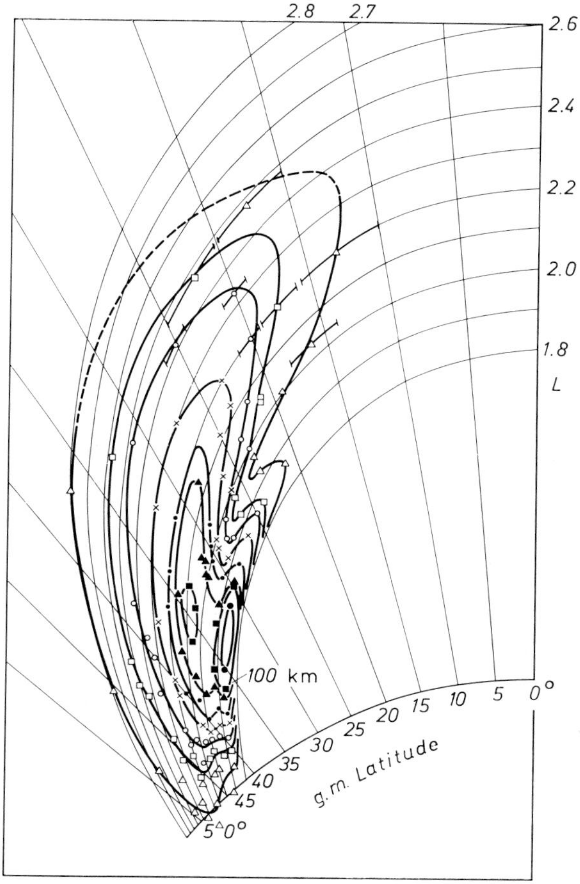

Fig. 13. Spatial distribution of the electrons from the two October 1962 USSR explosions in an $\mathcal{R}$, ϕ map as measured by West on satellite 1962 β K. Clearly the electrons were injected at quite low altitude. Count rates are as follows: ● 10^4; ■ $8\cdot 10^3$; ▲ $6\cdot 10^3$; · $4\cdot 10^3$; × $2\cdot 10^3$; ○ 10^3; □ 500; △ 100 (ref. [1], Sect. 15)

[16] A. M. Peterson, and G. L. Hower: J. Geophys. Res. **68**, 723 (1963).

[17] M. P. Nakada: J. Geophys. Res. **68**, 4079 (1963).

[1] H. I. West, L. G. Mann, and S. D. Bloom: [*12*], 423 (1965). — H. I. West: [*16*], 634 (1966), and [*16*], 663 (1966).

second USSR explosion on October 28. Clearly both explosion sites were at relatively low altitude. It seems likely that the electrons at high L values in Fig. 13 were made in a similar manner to the electrons at large L from "Starfish".

16. Decay of the electron belts.

α) Instruments on several satellites have observed the "Starfish" *particle population for several years* and certain general features of the decay have been found[1–4]. At low altitudes of a few hundred kilometers the decay is rapid due to interactions with the atmosphere. At high altitudes the decay is quite fast too, but for a different reason. The atmosphere clearly is not responsible for this loss. In a region in between these two rapid decay zones for $L < 1.7$ there was a rapid initial decay at low altitudes followed by a slow decay afterwards due to coulomb scattering with atmospheric atoms. VAN ALLEN[1] measured the decay of the artificial electrons at 1000 km with the "Injun" satellites as shown in Fig. 14. After the initial rapid decay which was larger for large B values (or low altitude) the decay slowed down and was nearly constant for all B values.

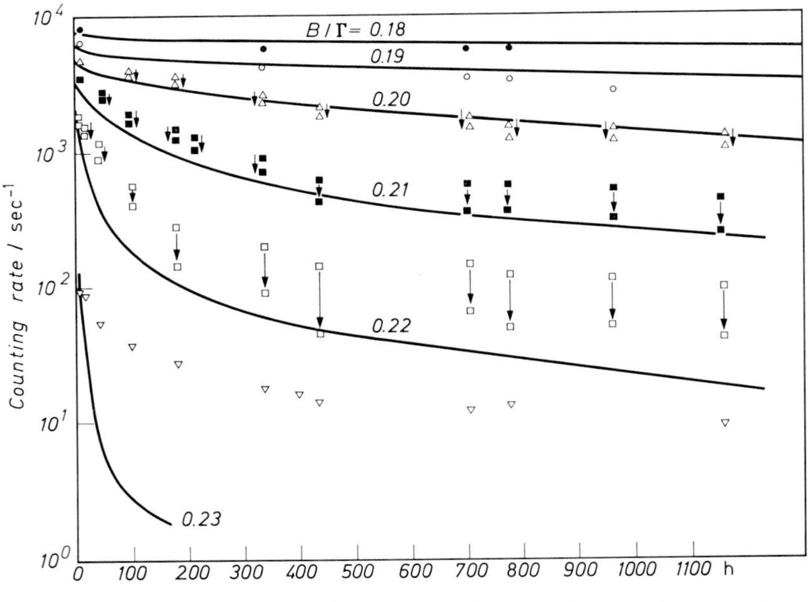

Fig. 14. Comparison of theoretical and experimental values for the variation of the trapped electron flux resulting from the 'Starfish' high-altitude nuclear explosion with time after the event (abscissa). Experimental points shown are from VAN ALLEN (ref. [1], Sect. 16) ordered by magnetic field B. The identification of individual points is in B/Γ: ● 0.18; ○ 0.19; △ 0.20; ■ 0.21; □ 0.22; ▽ 0.23. The solid lines are theoretical values based on the atmospheric scattering theory for $L = 1.25$. The lower set of points for $B = 0.20$, 0.21, and 0.22 have been corrected for the enhanced proton background observed by FILZ and HOLEMAN. Courtesy: WALT (ref. [13], Sect. 16).

β) The loss of geomagnetically trapped electrons by *atmospheric interactions* has been studied for some time. A complete FOKKER-PLANCK formulation including

[1] J. A. VAN ALLEN: [*16*], 575 (1966).
[2] C. U. BOSTROM, and D. J. WILLIAMS: J. Geophys. Res. **70**, 240 (1965).
[3] I. B. MCDIARMID, J. R. BURROWS, E. E. BUDZINSKI, and D. C. ROSE: Can. J. Phys. **41**, 1332 (1963).
[4] C. E. MCILWAIN: Science **142**, 355 (1963).

scattering and energy loss has been derived[5–7]. The atmospheric interaction is composed of many electron-atom or electron-ion collisions. Large-angle scattering has been ignored. A small-angle scattering event can take place anywhere along the field line and will in general result in either a raising or a lowering of the electron's mirror point. The result of a large number of scattering events will be a diffusion of an electron along a field line and a gradual loss of energy from the electrons. This transport phenomenon is approximated with a FOKKER-PLANCK expansion in powers of the mean scattering angle, which is small.

Scattering sufficient to cause complete loss of an electron by changes in mirror-point field B will only cause a change in L on the order of a gyroradius (~ 1 km), so populations on different L shells can be calculated independently. Diffusion in B will keep the mirror points within a tube of constant magnetic flux Φ_m, i.e.,

$$B \cdot A = \Phi_m = \text{const.},$$

where A is the area perpendicular to $\boldsymbol{B}^\circ$. Consequently, the distribution function U is defined by[8]

$$U(L, B, \mathscr{E}, t)\, \mathrm{d}\Phi_m\, \mathrm{d}B\, \mathrm{d}\mathscr{E} = \mathrm{d}N \tag{16.1}$$

where $\mathrm{d}N$ is the number of particles in the tube of magnetic flux $\mathrm{d}\Phi_m$ mirroring between B and $(B + \mathrm{d}B)$, having energies from $\mathscr{E}$ to $\mathscr{E} + \mathrm{d}\mathscr{E}$ at time t and on the magnetic shell L. To derive the FOKKER-PLANCK equation we must consider how U evolves with time. The distribution function U at $(t + \Delta t)$ is given by

$$U(B, \mathscr{E}, t + \Delta t) = \int_0^\infty \int_0^\infty U(B - \beta, \mathscr{E} - \varepsilon, t)\, \Pi(B - \beta, \mathscr{E} - \varepsilon; \beta, \varepsilon, \Delta t)\, \mathrm{d}\beta\, \mathrm{d}\varepsilon \tag{16.2}$$

where

$$\Pi(B - \beta, \mathscr{E} - \varepsilon; \beta, \varepsilon, \Delta t)\, \mathrm{d}\beta\, \mathrm{d}\varepsilon$$

is the probability that coulomb scattering and ionization will change the mirror point field by β into the interval $\mathrm{d}\beta$ and change the particle energy by ε into the interval $\mathrm{d}\varepsilon$ in a time Δt.

Making a Taylor series expansion of U and Π around $(B, \mathscr{E}, t)$ gives

$$\left.\begin{aligned} \frac{\partial U}{\partial t} &= \frac{U(B, \mathscr{E}, t + \Delta t) - U(B, \mathscr{E}, t)}{\Delta t} = -\frac{\partial}{\partial B}[U\langle\beta\rangle] + \\ &+ \frac{1}{2}\frac{\partial^2}{\partial B^2}[U\langle\beta^2\rangle] - \frac{\partial}{\partial \mathscr{E}}[U\langle\varepsilon\rangle]. \end{aligned}\right\} \tag{16.3}$$

This is the form of the FOKKER-PLANCK equation commonly used to describe Coulomb scattering of fast electrons by the thermal atmospheric atoms. The coefficients are given by

$$\langle\beta\rangle = \frac{1}{\Delta t}\int_0^\infty \int_0^\infty \mathrm{d}\beta\, \mathrm{d}\varepsilon\, \Pi(B, \mathscr{E}; \beta, \varepsilon, \Delta t)\, \beta \tag{16.4}$$

and are the average value per unit time of the variables which can be evaluated directly from considering Coulomb scattering.

γ) The *results of calculations* by WELCH et al.[8] are shown in Fig. 15 for $L = 1.25$ and $\mathscr{E} = 3.75$ MeV. Several features appear here.

[5] J. A. WELCH, and W. A. WHITAKER: J. Geophys. Res. **64**, 909 (1959).
[6] R. C. WENTWORTH, W. M. McDONALD, and S. F. SINGER: Phys. Fluids **2**, 499 (1959).
[7] W. M. McDONALD, and M. WALT: Ann. Phys. (N.Y.) **15**, 44 (1961).
[8] J. A. WELCH, R. L. KAUFMANN, and W. N. HESS: J. Geophys. Res. **68**, 685 (1963).

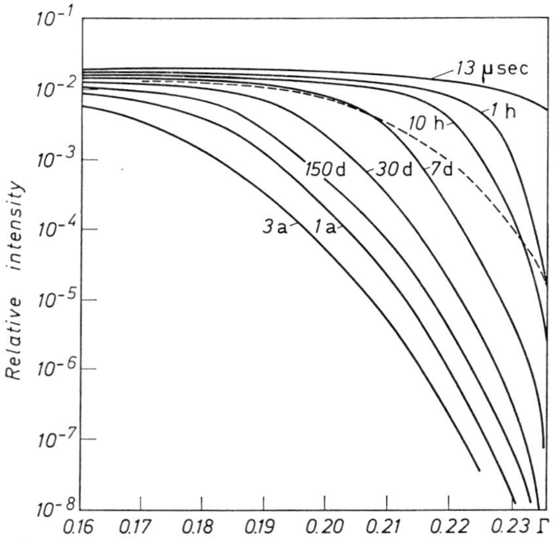

Fig. 15. Calculated changes in a trapped electron population due to Coulomb scattering for $L = 1.25$ and $\mathcal{E}_e = 3.75$ MeV. Time after the start of scattering is from top to bottom: $1.3 \cdot 10^{-5}$ sec, 1 hour, 10 hours, 1 week, 1 month, 5 months, 1 year, 3 years. Shown for comparison is data from Van Allen taken about one week after Starfish (broken curve). If the experimental data was corrected for proton contamination (ref. [13], Sect. 16) for $B > 0.21\ \Gamma$ the agreement with the calculated data would be improved considerably. (WELCH et al., ref. [8], Sect. 16.)

(1) The electrons at large B values decay rapidly at first due to the dense atmosphere there while the electrons near the equator show almost no initial decay.

(2) The initially flat distribution quite rapidly evolves into a characteristic shaped one which is in scattering equilibrium, that is, scattering up along the line just balances scattering down along the line.

(3) The decay for early times can be expressed approximately as

$$J(B, t) = J(B, 0) \frac{\tau(B)}{t + \tau(B)} \tag{16.5}$$

where $\tau(B)$ is a characteristic decay time. This is like the decay law observed by studying synchrotron radiation[9] and also using emulsions on low altitude satellites[10].

(4) After achieving the characteristic shape $U(B)$ the particles decay slowly exponentially being controlled by the rate of scattering at the *equator* and therefore by the equatorial atmospheric density. For $L = 1.25$ the calculated mean lifetime during exponential decay is $\tau \sim 1.3 \cdot 10^8$ sec.

(5) For large B and long times

$$U(B)\, n(B) \approx \text{const.}, \tag{16.6}$$

n being atmospheric density.

WALT and MCDONALD[11] [*18*] have also used a FOKKER-PLANCK equation to study Coulomb scattering of fast electrons. They used the equatorial pitch angle α_0 and electron energy as variables and obtained an approximate analytic solution by expanding the distribution function in terms of a series of orthonormal functions $G_\Phi(\cos\alpha_0)$.

[9] G. R. OCHS, D. T. FARLEY jr., K. L. BOWLES, and P. BANDYOPADHAY: J. Geophys. Res. **68**, 701 (1963).

[10] E. E. GAINES, and R. A. GLASS: J. Geophys. Res. **69**, 1271 (1964).

[11] M. WALT, and W. M. MACDONALD: J. Geophys. Res. **67**, 5013 (1962).

WALT[12] has made extensive calculations using the FOKKER-PLANCK equation to compare calculated Coulomb scattering decay curves with the experimental data of VAN ALLEN from the "Injun" satellites. Their decay curves compared with VAN ALLEN data are shown on Fig. 14. The agreement is quite good especially after the experimental data is corrected for proton contamination[13]. This proton correction also improves the agreement with the data at high B values in Fig. 15.

During the process of atmospheric scattering, the electron energy spectrum changes. The lower energy electrons are more easily scattered and therefore lost first. Because of this in general the fission energy spectrum hardens with time until an equilibrium spectrum is developed with a peak at about 2 MeV rather like that measured in Fig. 10, curve 1.25.

17. The wind shield wiper effect.

α) It is now well established that for low altitude B, L rings the electron flux is not constant along the ring but depends on *longitude*[1,2]. This is reasonable because if loss rates are comparable to drift rates we would expect to observe change in particle population during one drift around the earth.

ROEDERER and WELCH[3] showed that the usual FOKKER-PLANCK equation needs modification for particles drifting in longitude at low enough altitude so that the flux varies considerably with longitude. A FOKKER-PLANCK equation

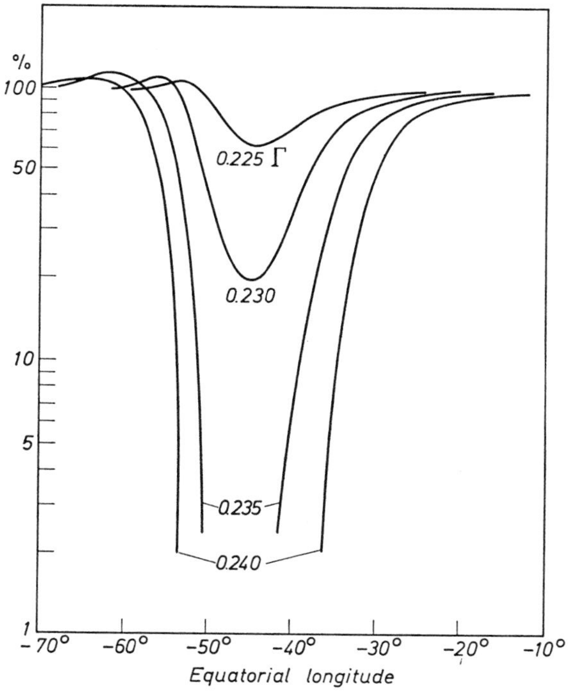

Fig. 16. The calculated longitudinal variation of the relative perpendicular flux $j_\perp(B)$ of 300 keV electrons at $L=1.25$ for several B values (parameter). As these electrons drift through the anomaly they are "wiped off" by the atmosphere and then replenished by scattering as they emerge. The curves are normalized to 100 percent far away from the anomaly. (ROEDERER and WELCH, ref. [3], Sect. 17.)

[12] M. WALT: J. Geophys. Res. **69**, 3947 (1964).

[13] M. WALT, and L. L. NEWKIRK: J. Geophys. Res. **71**, 3265 (1966).

[1] W. L. IMHOF, and R. V. SMITH: J. Geophys. Res. **70**, 569 (1965).

[2] G. A. PAULIKAS, and S. C. FREDEN: J. Geophys. Res. **69**, 1239 (1964).

[3] J. G. ROEDERER, and J. A. WELCH: [*13*], 148 (1966).

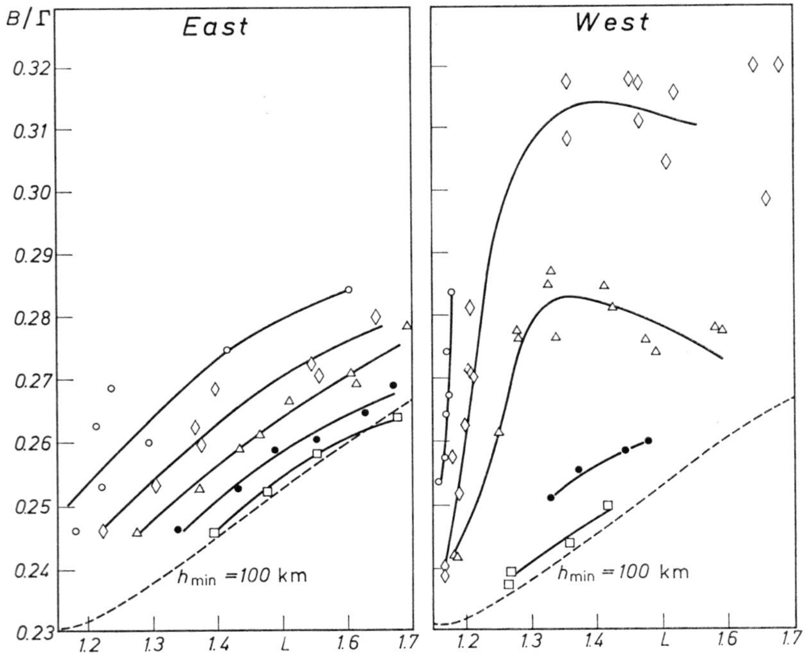

Fig. 17. Isoflux contour maps at two different longitudes for electrons in the energy interval 0.28 MeV $< \mathscr{E}_e <$ 0.71 MeV for Oct.—Nov. 1963. The windshield wiper effect of the anomaly is seen here in that the fluxes east of the anomaly (left side) are lower than West of it (right side). Flux intensities $j/\mathrm{m}^{-2}\,\mathrm{sec}^{-1}$ are as follows: □ $3\cdot10^7$; • 10^7; △ $3\cdot10^6$; ◇ 10^6; ○ $3\cdot10^5$. (IMHOF and SMITH, ref. [1], Sect. 17.)

can be set up with a new longitudinal variable $\mathcal{X}$. Calling $\mathscr{S}$ the source distribution of electrons, we have:

$$\frac{\partial U}{\partial t}+\frac{\partial}{\partial \mathcal{X}}[U\dot{\mathcal{X}}]=-\frac{\partial}{\partial \mathscr{E}}[U\langle\varepsilon\rangle]-\frac{\partial}{\partial B}[U\langle\beta\rangle]+\frac{1}{2}\frac{\partial^2}{\partial B^2}[U\langle\beta^2\rangle]+\mathscr{S}. \quad (17.1)$$

The probability function Π that appears in the Fokker-Planck coefficients is now longitude dependent:

ROEDERER and WELCH showed that this low-altitude longitude-dependent form of the FOKKER-PLANCK equation does *not* go over in a simple way to the commonly used form at higher altitude, which is longitude independent. Because of this it is not surprising that the calculated values for $B=0.23$ in Fig. 14 and Fig. 15 do not agree well with the observations. The longitude dependent form of the calculation should be used here.

ROEDERER and WELCH[3] have integrated the longitude dependent FOKKER-PLANCK equation and the mirror point flow equation for the stationary case, for low L shells.

Fig. 16 shows the calculated longitude dependence of the directional flux of 300 keV electrons for $L=1.25$ and various B values, for low solar activity and 0400 LT in the anomaly. All curves are normalized for a common level well east of the anomaly. Fig. 16 clearly shows the "windshield wiper effect", especially for those B values for which the B, L rings dip below the 100 km layer.

β) These theoretical computations for longitude dependence are in general agreement with *flux measurements* in the region near the anomaly — they, however, do *not* explain the observed continuing replenishment at high B values which occurs well west of the anomaly where the effect of multiple Coulomb scattering is small. IMHOF and SMITH[1] using detectors on low altitude polar orbiting satellites found the electron flux of 0.28 MeV $<\mathscr{E}_e<$ 0.71 MeV varied a great deal from west to east of the anomaly as shown in Fig. 17. West of the anomaly they find electrons at $L=1.4$ and $B=0.31$. These electrons are going to get wet as they drift east. ROEDERER and WELCH show that the Coulomb scattering effect is limited to a fairly narrow band of longitude (see Fig. 16). Replenishment observed to occur well west of the anomaly, where multiple Coulomb scattering should be negligible, must be explained by some other pitch angle scattering mechanism, such as interactions with electromagnetic waves or large angle Coulomb scattering.

18. Electron loss outside the inner zone ($L > 1.7$).

α) The electron decay rate is quite different for $L>1.7$. BROWN and GABBE[1] measured on "Telstar" the decay of the transient electron population introduced into space by the "Starfish" explosion. They found the decay curves in Fig. 18 for $L>1.7$. At $L=1.7$ the decay constant τ is many months but by $L=2.3$ it is about one week. It is impossible for this rapid decay to be due to the atmosphere.

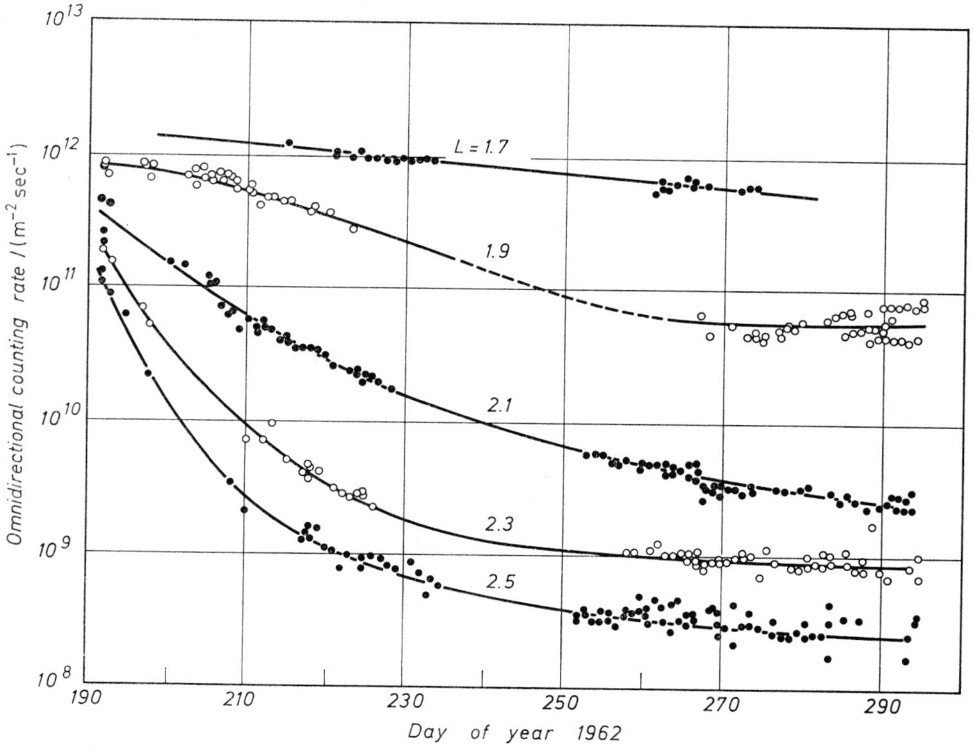

Fig. 18. Decay curves of the 'Starfish' electrons measured by BROWN and GABBE (ref. [1], Sect. 18) on satellite Telstar for $L>1.7 \ldots 25$ (parameter).

Measurements made after the three USSR explosions for $L>1.7$ give decay constants quite similar to the earlier "Telstar data". MCILWAIN[2] on "Explorer XV" found the decay was quite steady and monotonic for $L\sim 3$ after November 1. But for $L=3.8$ the decay is quite different as is shown in Fig. 19. There are several stepwise decreases in the flux apparently related to magnetic storms. On Dec. 18 a large increase in the electron flux occurred at the time of a new magnetic storm. The increase of the $\mathscr{E}>5$ MeV electrons at this time was more than a factor of 10 and made the flux roughly what it was shortly after the USSR explosions. This increase looks suspiciously like particle acceleration.

[1] W. L. BROWN, J. D. GABBE, and W. ROSENSWEIG: Bell System Tech. J. (July 1963). W. L. BROWN: [*16*], 610 (1966). — W. L. BROWN: [*19*], 189 (1965).

[2] C. E. MCILWAIN: Science **142**, 355 (1963).

β) The process going on at $L > 1.7$ causing rapid decay probably is due to some kind of magnetic disturbance which violates one or more of the adiabatic invariants. DUNGEY[3] has considered the pitch angle scattering due to a resonant *interaction of a whistler wave* on a charged particle in a $\boldsymbol{B}$ field as schematically shown in Fig. 20.

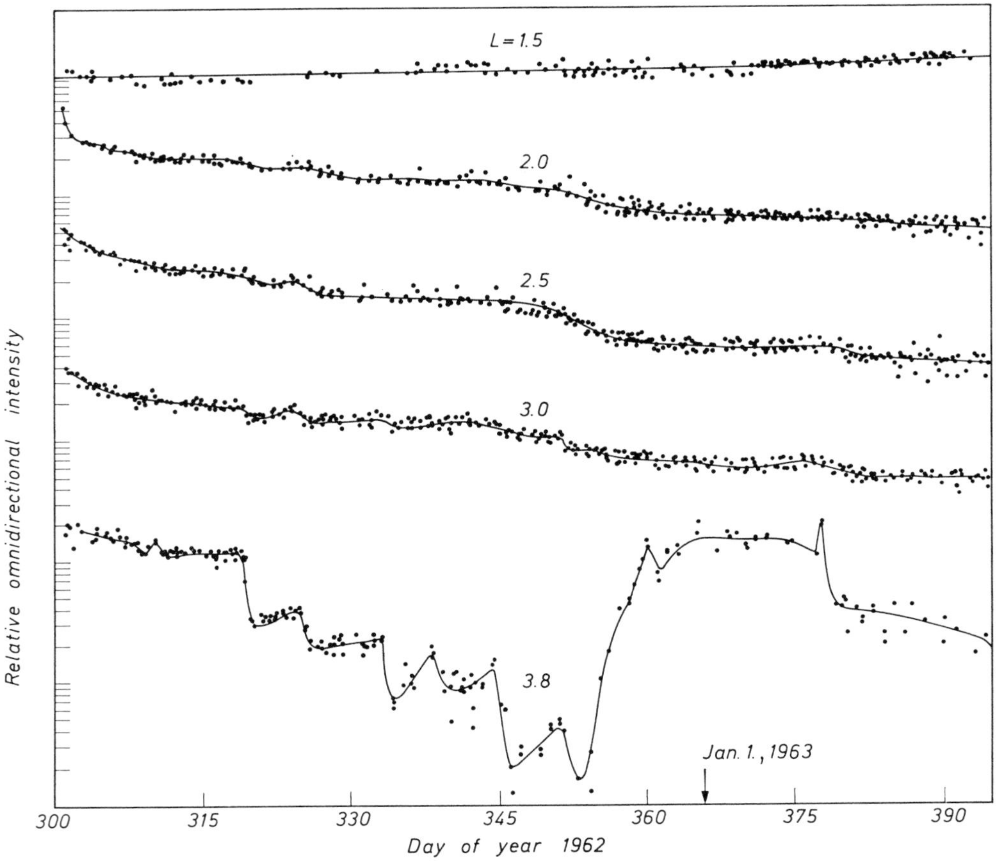

Fig. 19. Decay curves of equatorial artificial belt electrons of $\mathcal{E}_e > 5$ MeV measured by McILWAIN (ref. [2], Sect. 18) on Explorer 15 at various L values (parameter).

Taking the undisturbed particle orbit a helix

$$\left.\begin{aligned} x &= a \sin \Omega t, \\ y &= a \sin \Omega t, \\ z &= v_{\parallel}\, t \end{aligned}\right\} \tag{18.1}$$

and going to a frame of reference where the wave is at rest, DUNGEY calculated the change in $v_{\parallel}$ for one gyration of the particle from the equation of motion

$$\frac{m}{q}\,\frac{\mathrm{d}v_{\parallel}}{\mathrm{d}t} = E_z - \frac{v_{\perp}}{c_0\sqrt{\varepsilon_0\,\mu_0}}\,(B_y \sin \Omega t + B_x \cos \Omega t)\,. \tag{18.2}$$

[3] J. W. DUNGEY: J. Fluid Mech. **15**, 74 (1963); — Planetary Space Sci. **11**, 591 (1963).

The wave has been taken with the wave normal $\boldsymbol{k}$ along the $\boldsymbol{x}^0$ axis with components

$$\left.\begin{aligned} E_z &= C\cos(\alpha+\vartheta), \\ B_x &= b_x\cos(\beta_x+\vartheta), \\ B_y &= b_y\cos(\beta_y+\vartheta) \end{aligned}\right\} \tag{18.3}$$

where α, β_x, β_y are phase angles and $\vartheta = ky\sin\Theta + kz\cos\Theta$ where Θ is the angle between the wave normal, $\boldsymbol{k}$, and $\boldsymbol{B}$. Ignoring the effect of E_z as small compared to that of $\boldsymbol{B}$, the change in $v_\parallel$ per gyration of the particle is

$$\frac{m}{q}\,\delta v_\parallel = \frac{v_\perp}{c_0\sqrt{\varepsilon_0\,\mu_0}}\int\limits_0^{2\pi/\Omega} \mathrm{d}t\,[b_y\cos(\beta_y+\vartheta)\sin\Omega t + b_x\cos(\beta_x+\vartheta)\cos\Omega t]. \tag{18.4}$$

γ) It is clear that in order to disturb the particle there must be a relationship between $\boldsymbol{k}$ and Ω, otherwise the average value of the integral would be zero and there would be no change in $v_\parallel$. The *condition for* the *resonance* to occur so that $\delta v_\parallel \neq 0$ is

$$k(v_\parallel + V_\omega) = \mathrm{N}\Omega \qquad (\mathrm{N}\ \text{integer}) \tag{18.5}$$

where V_ω is the wave velocity.

But usually $V_\omega \ll v_\parallel$ and taking the first resonance $\mathrm{N}=1$ gives

$$\Omega = k v_\parallel. \tag{18.6}$$

This says the *electron should see the waves' frequency doppler-shifted up to electron's gyrofrequency* in order that an energy transfer take place.

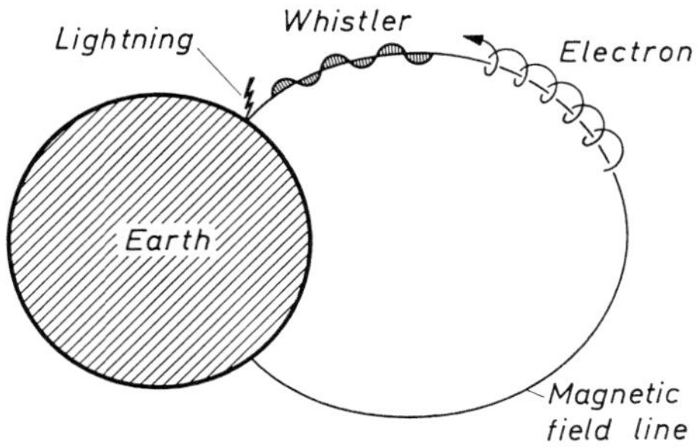

Fig. 20. A whistler, a circularly polarized very low frequency wave generated by lightning, propagates along a field line and can resonately interact with a charged particle moving along the line. (For information about whistlers see K. Rawer and K. Suchy, this Encyclopedia Vol. 49/2, Sect. 57; see also for theory contribution by A. A. Ruhadse and V. L. Ginzburg in this Vol., Sect. 25 ζ, p. 477.)

For a whole whistler whose length is r Dungey got the total change in pitch angle as

$$\delta\alpha = \frac{b}{B}\cos\vartheta\sqrt{k r} \tag{18.7}$$

where ϑ is the phase between the electron's helix and $\boldsymbol{b}$. The change in pitch angle goes as $r^{\frac{1}{2}}$, not linearly as r because the resonant condition isn't really satisfied along the whole whistler. k is the wave number.

Now saying the whistlers amplitude goes down as the wave spreads $b \propto r^{-\frac{1}{2}}$ and using

$$\left.\begin{aligned} B &= 0.3\,L^{-3}\ \mathrm{Gs}, \\ \Omega &= 5.5\cdot 10^6\,L^{-3}\ \mathrm{Hz}, \\ r &= R_\mathrm{E} \end{aligned}\right\} \tag{18.8}$$

gives for the electrons interaction with one whistler

$$\delta\alpha = 10^{-2} L^{\frac{3}{2}} \frac{b}{\gamma}. \tag{18.9}$$

The electron interacts with whistlers with a random phase so after encountering M whistlers a random walk process has changed α by

$$\Delta\alpha = \sqrt{M(\delta\alpha)^2}. \tag{18.10}$$

It will take $\Delta\alpha = 1$ for the electron to escape. The "escape time" τ_{es} is thus given by

$$\nu N b^2 L^3 10^{-4} \tau_{es} = 1 \tag{18.11}$$

where there are N whistlers per day and the electron encounters each whistler ν times, so $M = \nu N t$.

Using $\nu = 10$ and $N b^2 = 100\,\gamma^2$/day from "Vanguard III" satellite observations[4], Dungey got

$$\tau_{es} = 100\, L^{-3}\ \mathrm{d} \tag{18.12}$$

so at $L = 2.5$ the electrons lifetime τ_{es} is about five days which seems to agree with the data.

δ) This whistler scattering process seems to have many desirable features but it really is only a semi-quantitative theory and is untested.

Roberts[5] has pointed out that there are *certain problems* about Dungey's whistler-electron interaction mechanism. The most prominent one is that this whistler scattering process cannot make the observed electron angular distribution. The resonance condition cannot be satisfied if $v_{\parallel} = 0$. In this case for $L = 2$ at the equator $\Omega/2\pi \sim 100$ kHz but $f_{max} \sim 20$ kHz for a whistler so the two frequencies cannot resonate. This means that equatorial particles of $\alpha_0 = 90°$ will not be disturbed much by whistlers so that the electron pitch angle distribution should become strongly peaked at $\alpha_0 = 90°$. This is not observed to occur.

In order to find what kind of pitch angle scattering process would predict angular distributions of the kind observed Roberts has studied a process for which

$$\langle \Delta \cos\alpha \rangle = 0 \quad \text{and} \quad \langle (\Delta \cos\alpha)^2 \rangle = \text{const}$$

(where these are local pitch angles, not equatorial values). This would correspond to a process in which changes in pitch angle occur in random direction all along the electrons' trajectory. Whistlers don't work this way since the resonance condition changes along the field line.

ε) Roberts[5] has shown that pitch angle *scattering due to* some sort of *wide band noise* seems to work. This noise need not be in the whistler mode. It can be in any plasma mode. We don't have good enough data on the occurrance of such noise in the magnetosphere to evaluate this process quantitatively but the idea seems very promising.

E. Outer belt particles.

I. Protons and α-particles.

19. Protons: general features.

α) On the Explorer 12 satellite, Davis and Williamson[1] studied the *population* of outer belt protons in the energy range 0.1 MeV $< \mathscr{E}_p <$ 5 MeV. The spatial distribution of this population extends from the magnetopause in to where it merges with the inner belt (and where, usually, the detector is swamped by background). The energy spectra of these protons are quite closely exponential over a considerable range of energy and space as shown in Fig. 21. This population of

[4] J. C. Cain, I. R. Shapiro, J. D. Stolarik, and J. P. Heppner: J. Geophys. Res. **66**, 2677 (1961); **67**, 5055 (1962).

[5] C. S. Roberts: [*16*], 403 (1966).

[1] L. R. Davis, and J. M. Williamson: [*8*], 365 (1963).

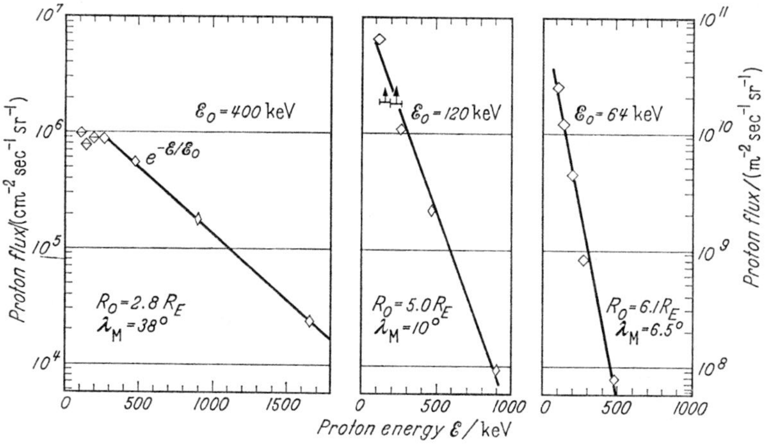

Fig. 21. Energy spectra of outer zone protons obtained on Aug. 8, 1961 by satellite Explorer 12. Measurements have been made at three different radial distances $R_0/R_E = 2.8$, 5.0 and 6.1; corresponding geomagnetic latitudes (read Φ instead of λ) are 38°, 10° and 6.5°. These spectra are nicely fit by a curve of the form $j(\mathscr{E}) = \text{const.}\exp(-\mathscr{E}/\mathscr{E}_0)$. Experimentally determined values of $\mathscr{E}_0$ are indicated. (Davis and Williamson, ref. [1], Sect. 19.)

particles is very stable in time. Outer belt proton measurements on Explorer 14 with the same instrument by Davis, Hoffman, and Williamson[2] show proton fluxes very similar to the measurements a year earlier on Explorer 12. These protons do not show large flux changes at times of magnetic storms except that some time change are found for the high energy particles for $L > 5$.

β) The *properties* of these protons form an interesting pattern: (i) large fluxes, (ii) quite stable in time, (iii) strongly limited to the equatorial regions with most of the particles having large pitch angles (outer belt electrons are not nearly as strongly peaked in pitch angle as the protons are), and (iv) exponential spectra of the directional flux given by

$$j(>\mathscr{E}) = \text{const.}\exp(-\mathscr{E}/\mathscr{E}_0). \tag{19.1}$$

γ) The *characteristic energy* $\mathscr{E}_0$ varies strongly with L and weakly with α. Empirically,

$$\mathscr{E}_0 \propto L^{-3} \quad \text{for} \quad \alpha = 90°. \tag{19.2}$$

This relationship is very suggestive. Since $B \propto L^{-3}$ then $\mathscr{E}_0 \propto B$. An energy increasing linearly with B sounds very like betatron acceleration. If the particles drift radially in the magnetic field conserving their magnetic moment $|\boldsymbol{\mu}|$ as they go the single particles energy changes as can be seen by

$$\left.\begin{aligned} |\boldsymbol{\mu}| &= \frac{\mathscr{E}}{B} = \text{const.} \quad \text{for} \quad \alpha = 90° \\ \text{or} \qquad & \\ \mathrm{E} &\propto L^{-3}. \end{aligned}\right\} \tag{19.3}$$

Kellogg[3] first considered the possibility of radial drift of particles. He suggested that outer belt electrons might be drifted in from outside conserving $|\boldsymbol{\mu}|$ and $\mathscr{J}$ but he found that the diffusion times were so long that one could not hope to make inner zone electrons this way. However, the production of outer zone electrons this way was not impossible.

[2] L. R. Davis, R. S. Hoffman, and J. M. Williamson: (Abstract), Trans. Am. Geophys. Union **45**, 84 (1964). L. R. Davis, and J. M. Williamson: [*16*], 215 (1966).

[3] P. J. Kellogg: Nature **183**, 1295 (1959).

Let us consider how radial drift of particles can be accomplished and what properties a drifting population would have.

20. The drift process.

α) If a particle in a magnetic field moves, conserving the three adiabatic invariants $|\boldsymbol{\mu}|$, $\mathcal{J}$ and Φ_m (see Sect. 4), it will not drift radially. In order to move radially at least the third *invariant* must be *violated*. The magnetic flux Φ_m linked by the orbit will change with radial position. Other invariants may also be violated in the process.

β) Only certain classes of invariant violation are useful. First consider a magnetic disturbance where a change in the particle's momentum $\Delta \boldsymbol{p}$ occurs in an *arbitrary direction*. This will probably violate all the invariants. An arbitrarily directed $\Delta \boldsymbol{p}$ will change both L and α.

From Eq. (2.4) the cross field motion is given by

$$\Delta \boldsymbol{R} = \frac{c_0 \sqrt{\varepsilon_0 \mu_0}}{q B^2} (\Delta \boldsymbol{p} \times \boldsymbol{B}) = R_c \frac{\Delta \boldsymbol{p}_\perp}{p_\perp}. \tag{20.1}$$

The particle's pitch angle α is given by

$$\cos \alpha = \frac{p_\parallel}{p} \tag{20.2}$$

and so the pitch angle is changed by an impulse $\Delta p_\parallel$ by

$$\Delta \alpha \cdot \sin \alpha = \frac{\Delta p_\parallel}{p} \quad \text{or} \quad \Delta \alpha = \frac{\Delta p_\parallel}{p_\perp}. \tag{20.3}$$

For the particle to move one gyroradius across the magnetic field

$$\frac{\Delta R}{R_c} = 1 = \frac{\Delta p_\perp}{p_\perp}. \tag{20.4}$$

If $\Delta p_\perp \approx \Delta p_\parallel$ for each impulse then we find that the pitch angle will be changed by

$$\Delta \alpha = \frac{\Delta p_\parallel}{p_\perp} \approx 1 \text{ rad}. \tag{20.5}$$

This shows the particle will be scattered into the atmosphere in the same time it diffuses radially one gyroradius. This type of diffusion obviously doesn't work well to move the particles radially.

γ) We need a process which conserves some of the adiabatic invariants in order to get radial diffusion. PARKER[1] has shown that violating the second invariant also produces particle loss rather than radial diffusion. It would seem the only reasonable processes to achieve radial diffusion are those in which $|\boldsymbol{\mu}|$ and $\mathcal{J}$ are conserved but the *third invariant* Φ_m *only is violated*.

If the first and second adiabatic invariants of trapped particles are maintained during motion in L-space, changes in both the energy and equatorial pitch angle can be calculated. The non-relativistic first invariant is:

$$|\boldsymbol{\mu}| = \frac{\mathscr{E} \sin^2 \alpha_0}{B_0} = \frac{\mathscr{E} L^3 \sin^2 \alpha_0}{0.312\, \Gamma} \tag{20.6}$$

where B_0 is the equatorial magnetic field.

[1] E. N. PARKER: J. Geophys. Res. **66**, 693 (1961).

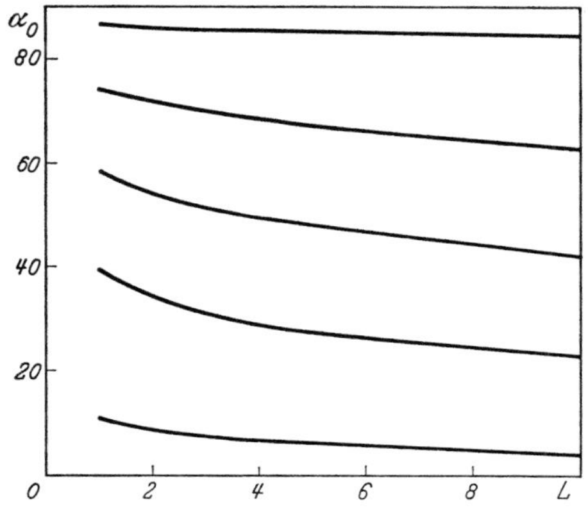

Fig. 22. Calculated variation of equatorial pitch angle α_0 for radial diffusion conserving $|\mu|$ and $\mathcal{J}$, using Eq. (21.2). Inward drifting particles move closer to the equator. (NAKADA, DUNGEY and HESS, ref. [2], Sect. 20).

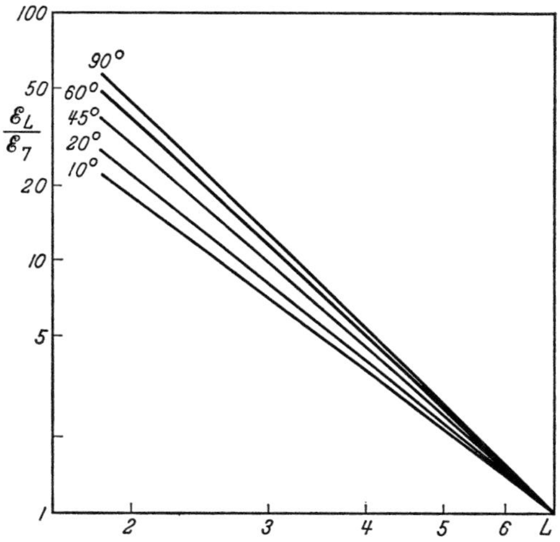

Fig. 23. Calculated change in particle energy during radial drift conserving $|\mu|$ and $\mathcal{J}$. Various initial pitch angles at $L=7$ have been adopted (parameter). Ordinate is the energy ratio compared with that at $L=7$, thus $\mathcal{E}_L/\mathcal{E}_7$. (NAKADA, DUNGEY and HESS, ref. [2], Sect. 20).

The second invariant is from Eq. (4.1)

$$\mathcal{J} = m \int \mathrm{d}l\, v \cos\alpha = m v L\, R_{\mathrm{E}} \mathrm{F}(\alpha_0) \tag{20.7}$$

where α the local pitch angle and $\mathrm{d}l$ is along the guiding center.

Since $|\mu|$ and $\mathcal{J}$ are constants, Eq. (20.6) may be divided by the square of Eq. (20.7) to give [with $\mathcal{E} \equiv \frac{1}{2} m v^2$]

$$L \left[\frac{\sin\alpha_0}{\mathrm{F}(\alpha_0)}\right]^2 = \text{const.} \tag{20.8}$$

δ) From this, the changes in α_0 with L were evaluated by NAKADA, DUNGEY and HESS[2] and are shown in Fig. 22. Two features of these results are worthy of note:

[2] J. W. DUNGEY, W. N. HESS, and M. P. NAKADA: [*12*], 399 (1965). — J. Geophys. Res. **70**, 3529 (1965).

(i) Changes in α_0 with L are relatively small for $L > 2.5$ and as DAVIS and CHANG[3] have indicated, particles diffusing inwards assume flatter helices and are therefore not lost easily;

(ii) Changes in α_0 with L are independent of energy for non-relativistic particles.

These changes in α_0 with L and Eq. (20.6) may be used to find the variation in energy with L and α_0. Results are shown in Fig. 23 for protons having α_0 values at $L=7$ as indicated on the curves. Reduced energies are given in the figure, relative to energies at $L=7$.

21. Comparison with experimental results.

α) The spectrum of protons that one would expect at some L and α_0 depends on the location and nature of the source and sink and on the energy dependence of motion in L-space. When the invariant violation mechanism involves electric fields, the velocity of L-space motion is proportional to $\boldsymbol{E}\times\boldsymbol{B}$ and does not depend on particle energy. Motion in L-space for asymmetric distortions of the geomagnetic field such as occur with sudden commencements and sudden impulses[1] depend on the guiding center of particles following magnetic field lines during rapid changes in the field and is also independent of energy.

Assuming motion in L-space is *independent of energy*, if the injection spectrum is power law, the spectrum remains power law with the same exponent during the L drift. If the injection spectrum has an exponential form $\exp(-\mathscr{E}/\mathscr{E}_0)$, the spectrum remains exponential after L-space motion and $\mathscr{E}_0$ varies in the same way with L and α_0 as has been calculated for a single particle in the previous section. This can be seen by transforming from L_1 to L_2 by

$$N(\mathscr{E}) = k_1 \exp(-\mathscr{E}_1/\mathscr{E}_{01}) \rightarrow k_2 \exp(-\mathscr{E}_2/\mathscr{E}_{02}) \tag{21.1}$$

where we know from Fig. 23 $\mathscr{E}_2$ is related to $\mathscr{E}_1$ by an expression

$$\mathscr{E}_2 = \mathrm{q}(L,\alpha)\,\mathscr{E}_1. \tag{21.2}$$

The new spectrum is then

$$N(\mathscr{E}) = k_2 \exp\left(-\frac{\mathrm{q}(L,\alpha)\,\mathscr{E}_1}{\mathscr{E}_{02}}\right) \tag{21.3}$$

or from Eq. (21.3) we get

$$\mathscr{E}_{01} = \mathrm{q}(L,\alpha)\,\mathscr{E}_{02}.$$

β) Two *predictions* of the model may be *compared with experiment*. The first prediction, that the spectrum retains its exponential form after motion in L-space is in agreement with experiment (see Fig. 21). To test the second prediction, measured[2] $\mathscr{E}_0$ values have been plotted in Fig. 24 as a function of L. The labels on the curves refer to α_0 values at $L=7$. The dashed curves in Fig. 24 are taken from Fig. 23 for corresponding changes in $\mathscr{E}$ and L and α_0. The comparison, in Fig. 24, shows good agreement between the model and experimental results.

γ) Thus far, only the spectra of particles has been compared with the model. We can also compare the *fluxes* in L-space with predicitions of LIOUVILLE's Theorem. The Liouville theorem states that f, the density of particles in phase space, is constant along a particle's orbit for a lossless process. The phase space

[3] L. DAVIS jr., and D. B. CHANG: J. Geophys. Res. **67**, 2169 (1962).

[1] E. N. PARKER: J. Geophys. Res. **65**, 3117 (1960).

[2] L. R. DAVIS, R. S. HOFFMAN, and J. M. WILLIAMSON: (Abstract), Trans. Am. Geophys. Union **45**, 84 (1964). L. R. DAVIS, and J. M. WILLIAMSON: [*16*], 215 (1966).

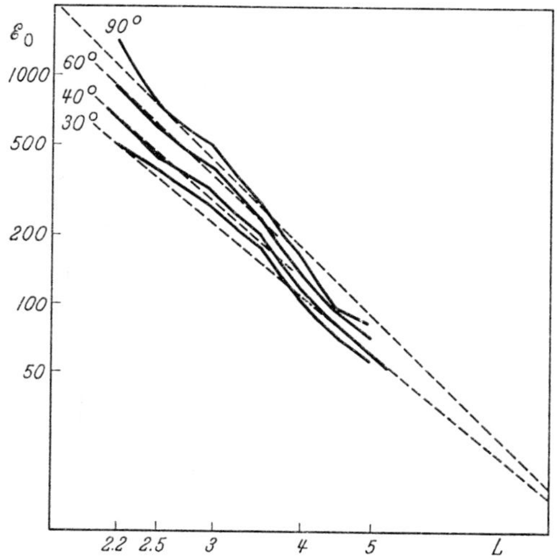

Fig. 24. Comparison on measured energies $\mathscr{E}_0$ from data of DAVIS and WILLIAMSON, ref. [1], Sect. 19 (solid curves) with calculated energy variation of protons during radial drift (broken curves) conserving $|\boldsymbol{\mu}|$ and $\mathscr{J}$ for various initial pitch angles α_0 (parameter) at $L=7$. (NAKADA, DUNGEY and HESS, ref. [2], Sect. 20.)

density f is defined by

$$\mathrm{d}n = f(v, \alpha, L)\, 2\pi \sin\alpha \,\mathrm{d}\alpha\, v^2 \,\mathrm{d}v\, \mathrm{d}V \tag{21.4}$$

where $\mathrm{d}n$ is the number of particles in volume $\mathrm{d}V$.

But also from the measured directional flux j we can write

$$\mathrm{d}n = \left[\frac{j(\mathscr{E}, \alpha, L)}{v}\right] 2\pi \sin\alpha\, \mathrm{d}\alpha\, \mathrm{d}\mathscr{E}\, \mathrm{d}V \tag{21.5}$$

using $\mathrm{d}\mathscr{E} = m v\, \mathrm{d}v$ we get

$$\frac{m^2 j(\mathscr{E}, \alpha, L)}{2\mathscr{E}} = f(\mathscr{E}, \alpha, L). \tag{21.6}$$

This shows that we can get the density of particles in phase space directly from the measured data. We can test LIOUVILLE's theorem by asking subject to the constraints that $|\boldsymbol{\mu}|$ and $\mathscr{J}$ are conserved, whether

$$f[\mathscr{E}_1(|\boldsymbol{\mu}_1|, \mathscr{J}_1), \alpha, (|\boldsymbol{\mu}_1|, \mathscr{J}_1), L_1] = f[\mathscr{E}_2(|\boldsymbol{\mu}_1|, \mathscr{J}_1), \alpha_2(|\boldsymbol{\mu}_1|, \mathscr{J}_1), L_2]. \tag{21.7}$$

δ) The results of such a test[3] for several pairs of $|\boldsymbol{\mu}|$ and $\mathscr{J}$ show that f is *not* constant along the L drift.

The value of f probably would be constant along one particular particle trajectory but the trajectories get all wound up like a ball of string and cannot be identified. While f may be constant microscopically, the measured macroscopic values of f are not constant because of the multiplicity of particle trajectories involved. From the test of the Liouville theorem we find that $j/\mathscr{E}$ increases monotonically with L. This indicates that the particle source is at large L, very probably at the magnetopause.

MIHALOV and WHITE[4] have measured the outer belt proton energy spectrum on satellite 1964—45A using a nine-channel spectrometer extending from 0.17 to 3.4 MeV. Distributions in B, L space of protons of several energies are shown in Fig. 25. This data agrees with data two years earlier of DAVIS and WILLIAMSON where they can be compared. MIHALOV and WHITE[4] say that the data supports inward radial diffusion and that the general features

[3] J. W. DUNGEY, W. N. HESS, and M. P. NAKADA: [*12*], 399 (1965). — J. Geophys. Res. **70**, 3529 (1965).
[4] J. D. MIHALOV, and R. S. WHITE: J. Geophys. Res. **71**, 2207 (1966).

agree with the calculations of NAKADA and MEAD[5]. Their theory gives j versus L plots along the equator that agree generally with the data. The measured fluxes stay large in to somewhat lower L values than given by the calculation which seem to indicate a faster diffusion rate than NAKADA and MEAD used. But the fluxes drop off abruptly at $L \approx 2$ as expected by the theory. MIHALOV and WHITE[4] say that it seems impossible to diffuse protons into the inner zone to explain the low energy protons found there.

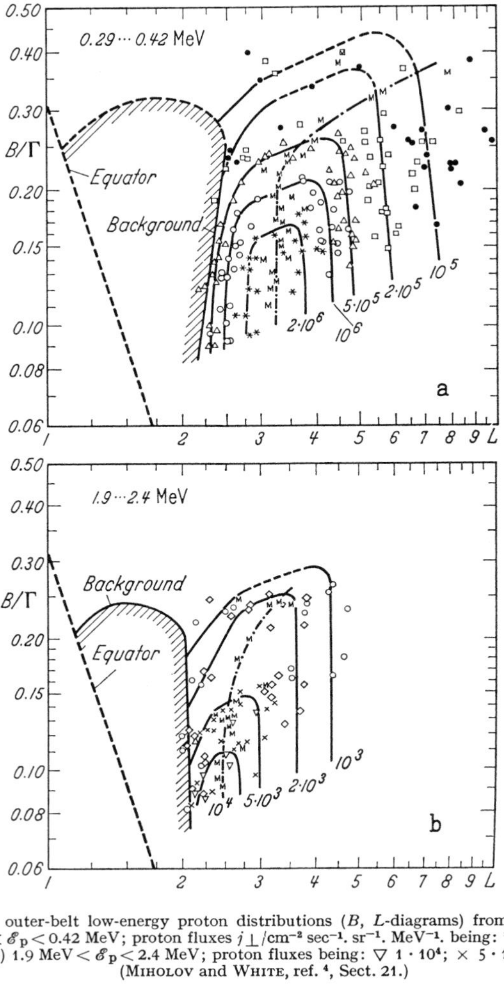

Fig. 25 a and b. Measured outer-belt low-energy proton distributions (B, L-diagrams) from a spectrometer on satellite 1964—45a. (a) 0.29 MeV $< \mathscr{E}_p <$ 0.42 MeV; proton fluxes $j_\perp$/cm^{-2} sec^{-1}. sr^{-1}. MeV^{-1}. being: * $2 \cdot 10^6$; ○ $1 \cdot 10^6$; △ $5 \cdot 10^5$; □ $2 \cdot 10^5$; ● $1 \cdot 10^5$. (b) 1.9 MeV $< \mathscr{E}_p <$ 2.4 MeV; proton fluxes being: ▽ $1 \cdot 10^4$; × $5 \cdot 10^3$; ◇ $2 \cdot 10^3$; ○ $1 \cdot 10^3$. (MIHOLOV and WHITE, ref. [4], Sect. 21.)

22. The pumping mechanism.

α) We want a *process* that *violates the third invariant only* without violating the first two to produce motion in L-space.

This means we want magnetic disturbances with times

$$T_b < \tau < T_d,$$

where T_b is the bounce period and T_d is the drift period (see Sect. 2ε and 2ζ).

[5] M. P. NAKADA, and G. D. MEAD: J. Geophys. Res. **70**, 4777 (1965).

Magnetic disturbances with ~1 min are observed regularly on sea level magnetometers. Sudden impulses and sudden commencements are of this class and occur roughly two times per day with an average amplitude of 2γ and two per month with an amplitude of $20\,\gamma$. These sudden impulses at the earth surface may be produced by sudden changes in the solar wind pressure and resulting compressions or expansions of the geomagnetic field. PARKER[1] studied what the results on trapped particles would be as a result of such a pumping on the magnetopause.

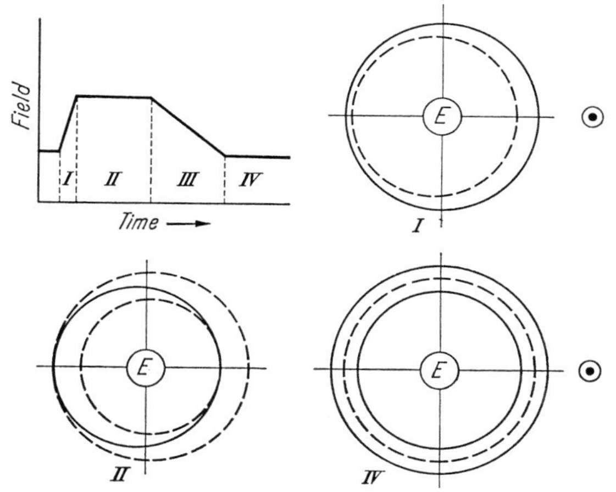

Fig. 26. The pumping mechanism for radial diffusion. A sudden commencement magnetic disturbance (**SC**) is schematically shown in the top, lefthand diagram. The way particles respond is shown in I, II and IV (Earth at center, Sun at right side). In I particles on a drift shell (solid curve) before the **SC** are moved to the dotted curve by the **SC**. In II this displaced shell of particles (solid curve) spreads and forms a broad region between both dotted circles because some particles move inwards and some outwards depending on the field change ΔB encountered by each particle. In IV the original drift shell (shown dotted) is compared with the region occupied by particles (between solid curves) after the magnetic event is over (NAKADA and MEAD, ref. [5], Sect. 21).

The basic processes moving the particles are shown in Fig. 26. Consider a ring of particles that have $\mathcal{J}=0$ and are a δ-function in L as in the solid curve of Fig. 26/I. These particles drift on a surface of constant B. When the field increases during period I (Fig. 26 top lefthand) the particles move inwards following field lines to the broken line positions in Fig. 26/I. The field compression is assymmetrical in space. The field is pushed in more on the day side than on the night side. From MEAD[2] a dipole field is distorted by the solar wind to a field whose value at the equator is

$$B_{\text{eq}}=\frac{\mathsf{M}}{r^3}+\frac{a_1}{r_{\text{b}}^3}+\frac{a_2\, r}{r_{\text{b}}^4}\cos\Lambda \tag{22.1}$$

with

$$a_1=0.25 \text{ Gs} \quad\text{and}\quad a_2=0.210 \text{ Gs};$$

r_{b} is the subsolar boundary distance from Earth divided by R_{E} (i.e. in Earth radii) and Λ is the local longitude measured from the local noon meridian.

Due to the last term in B involving Λ the particles move in farther on the day side. The particles are now on different constant B drift paths. Following this, during the period II in Fig. 26 when the field is constant, the particles drift off the distorted ring. Particles that were initially in the rear drift outwards and particles that were in the front drift inwards as in the dashed curves of Fig. 26/II. For $\mathcal{J}\neq 0$ the drifts are similar but not identical to this. Then as the field relaxes slowly back to normal during period III the particles drift rings expand to the positions shown by the solid curves in Fig. 26/IV. The dotted ring here is the

[1] E. N. PARKER: J. Geophys. Res. **65**, 3117 (1960).
[2] G. D. MEAD: J. Geophys. Res. **69**, 1181 (1964).

initial ring from Fig. 26/I. From this discussion we see that as a result of *pumping* the field some particles are moved in and some out. If this process is repeated many times the result will resemble diffusion.

β) In the *earlier attempts* to obtain a steady state distribution for particles undergoing such a redistribution, PARKER[1] used a diffusion equation. DAVIS and CHANG[3] pointed out that the FOKKER-PLANCK equation was a more correct formalism to use since the motion of the particles was like a random walk. They obtained solutions for the first two terms of the Fokker-Planck equation and demonstrated that the mean displacement coefficient $\langle\Delta r\rangle$ was as important as the mean squared displacement coefficient $\langle(\Delta r)^2\rangle$. The earlier treatments had not considered $\langle\Delta r\rangle$.

By considerung the process as akin to stirring, DUNGEY[4] demonstrated that a carefully formulated diffusion equation was equivalent to the Fokker-Planck equation (since the Fokker-Planck coefficients were related in a particular way). FÄLTHAMMAR[5] has recently shown for equatorial particles that the two Fokker-Planck coefficients are uniquely related for any process where motion in L-space is produced without violation of the magnetic moment and longitudinal invariants.

γ) Because of DUNGEY's and FÄLTHAMMAR's contributions, it is possible to use a diffusion equation with just a mean square *displacement coefficient* that is relatively easy to evaluate. The mean displacement coefficient $\langle\Delta r\rangle$ is usually difficult to obtain from the details of the motion. In a manner completely similar to the Coulomb scattering analysis in Sect. 16β the FOKKER-PLANCK equation can be derived for this case

$$\frac{\partial U}{\partial t}\,\mathrm{d}t = -\frac{\partial}{\partial r}\left[U\langle\Delta r\rangle\right] + \frac{1}{2}\,\frac{\partial^2}{\partial r^2}\left[U\langle(\Delta r)^2\rangle\right] + \mathcal{S} - \mathcal{L} \tag{22.2}$$

where $\mathcal{S}$ = particle sources and $\mathcal{L}$ = particle losses and where U is the total number of particles in the ring of width $\mathrm{d}r$ at r.

It is assumed that the onset of the magnetic disturbance or storm is fast enough so that the particle does not drift off the field line and is energized as it moves inwards due to the electric field produced by $\mathrm{d}B/\mathrm{d}t$. Then as the field relaxes adiabatically to its initial value later the particle drifts outwards.

In order to follow the particle motion appropriately we must identify the field line it is on and how this field line moves.

δ) DAVIS and CHANG[3] used an image dipole field of the CHAPMAN-FERRARO type with l the distance to the neutral plane between the two dipoles and evaluated the Fokker-Planck coefficients $\langle\Delta r\rangle$ and $\langle(\Delta r)^2\rangle$. The *steady state* FOKKER-PLANCK *equation without sources or losses* is

$$\frac{\partial}{\partial r}\left[U\langle\Delta r\rangle\right] = \frac{1}{2}\,\frac{\partial^2}{\partial r^2}\left[U\langle(\Delta r)^2\rangle\right]. \tag{22.3}$$

This can be solved directly now. Assuming a source of density n_1 at r_1, which we can consider to be the magnetopause, and also assuming that a particle sink exists at r_0 where the atmosphere or some other loss process effectively stops the diffusion, DAVIS and CHANG[3] got the solution in terms of particle density $n(r)$ where

$$U(r) = \text{const.}\; n(r)\, r^2. \tag{22.4}$$

The density is

$$n(r) = \left(\frac{r}{r_1}\right)^{-10}\left[\frac{r^7 - r_0^7}{r_1^7 - r_0^7}\right] n(r_1). \tag{22.5}$$

If we define the energy spectrum at r to be

$$N(r,\mathcal{E})\,\mathrm{d}\mathcal{E}$$

then

$$n(r) = \int \mathrm{d}\mathcal{E}\, N(r,\mathcal{E}) \tag{22.6}$$

---

[3] L. DAVIS jr., and D. B. CHANG: J. Geophys. Res. **67**, 2169 (1962).
[4] J. W. DUNGEY: [*12*], 183 (1965).
[5] C. G. FÄLTHAMMAR: J. Geophys. Res. **71**, 1487 (1966).

and, non-relativistically

$$\mathscr{E}(r)\, r^3 = \text{const.} \tag{22.7}$$

so

$$\mathrm{d}\mathscr{E}(r) = \left(\frac{r_1}{r}\right)^3 \mathrm{d}\mathscr{E}, \tag{22.8}$$

which shows how particles spread out in energy.

Therefore, the *energy spectrum* is

$$N(r, \mathscr{E}) = \left(\frac{r}{r_1}\right)^{-7} \left[\frac{r^7 - r_0^7}{r_1^7 - r_0^7}\right] N_1\left(r_1, \frac{\mathscr{E} r^3}{r_1^3}\right). \tag{22.9}$$

ε) NAKADA and MEAD[6] have extended the calculations of DAVIS and CHANG using

(i) the MEAD model of the geomagnetic field[2] which is a better approximation than the image dipole model to the real field,

(ii) the quiet time magnetopause placed at 10 R_E instead of at ∞,

(iii) the actual rate of diffusion calculated using observed geomagnetic sudden impulse and sudden commencement data and most important,

(iv) they included loss processes in the analysis—both slowing down, and charge exchange. TVERSKOY[7] was the first to introduce slowing down of particles into the diffusion problem. He used distribution functions that were dependent on magnetic moments to include the particle energy.

ζ) The FOKKER-PLANCK *equation with loss* is

$$\frac{\partial y}{\partial t} = -\frac{\partial}{\partial r}\left[y \frac{\langle \Delta r\rangle}{\Delta t}\right] + \frac{1}{2}\frac{\partial^2}{\partial r^2}\left[y \frac{\langle (\Delta r)^2\rangle}{\Delta t}\right] - \frac{\partial}{\partial|\boldsymbol{\mu}|}\left[y \frac{\langle \Delta|\boldsymbol{\mu}|\rangle}{\Delta t}\right] - \frac{y}{\tau_{\mathrm{cx}}} \tag{22.10}$$

where y is now the number of particles in dr, at r and in the range of magnetic moments d$|\boldsymbol{\mu}|$ ($|\boldsymbol{\mu}|$ replaces $\mathscr{E}$ as the energy variable). The last term in Eq. (22.10) introduces loss by the charge exchange lifetime τ_{cx}.

Looking at magnetograms for sudden impulses and sudden commencements will clearly underestimate $\langle(\Delta r^2)\rangle$. FÄLTHAMMAR[8] has shown that time-varying electric fields in the magnetosphere will also produce radial diffusion when $|\boldsymbol{\mu}|$ and $\mathscr{J}$ are conserved. The FOKKER-PLANCK coefficients should be increased by a factor of about ten to allow for this more realistic diffusion rate, changes the calculated results in Fig. 27 so that they are essentially indistinguishable from the experimental results.

NAKADA and MEAD[6], assuming the solar wind was the source of the protons, found that about 10^{-6} of the solar wind protons incident on the magnetopause were needed to supply the outer belt protons. They found about one percent of all trapped protons were lost per day mostly by coulomb collision slowing down at large L.

The results of this calculation of MEAD and NAKADA show quite convincingly that *cross-field diffusion* is the dominant process in controlling outer belt protons. They use a complete description of the diffusion process and find they get quantitative agreement with the observations for very reasonable values of the parameters used in the theory. The only real limitation of their answer is that they apply to equatorial particles with $\alpha_0 = 90°$.

η) There is more *experimental* proton *data* that points to radial diffusion in the outer belt. McILWAIN[9] found that near the equator high energy protons of 40 MeV $< \mathscr{E} <$ 110 MeV show a double-peaked structure as shown in Fig. 7, Sect. 6. It seems likely that the inner peak here is made by the CRAND source (Sect. 6) but it seems that the outer peak must be from another source. I believe that the McILWAIN second bump protons have diffused in from the magnetopause. I feel that most if not all of the outer belt protons of $\mathscr{E} >$ 100 keV have diffused in from the magnetopause. This becomes harder to demonstrate looking off-

[6] M. P. NAKADA, and G. D. MEAD: J. Geophys. Res. **70**, 4777 (1965).

[7] B. A. TVERSKOY: Geomagnetizm i Aeronomija **5**, 224 and 436 (1964) [engl. transl.: Geomagn. and Aeronomy **5**, 174, 351 (1964)].

[8] C. G. FÄLTHAMMAR: J. Geophys. Res. **70**, 2503 (1965).

[9] C. E. McILWAIN: [*12*], 374 (1965); — Relay 1 Trapped Radiation Measurements. Final Report on the Relay 1 Program NASA SP-76 1965.

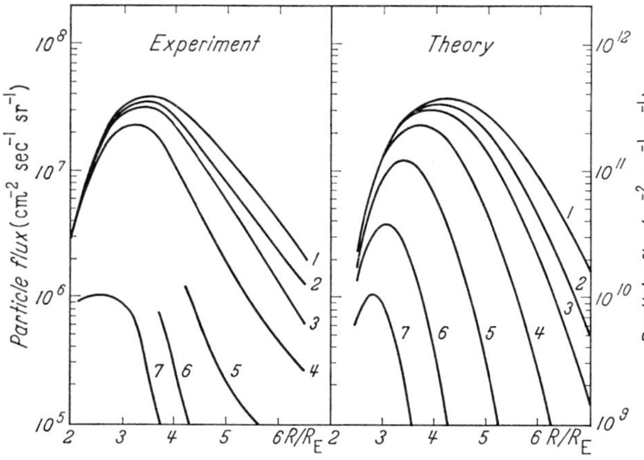

Fig. 27. Comparison of integral fluxes as functions of L with $\mathscr{E}_p$ as parameter, measured by DAVIS et al. (1964) shown at left side, and those calculated by NAKADA and MEAD (ref. [5], Sect. 21) which are seen at right side. The calculated curves are normalized to the same peak flux for the lowest energy threshold. Parameter Figs. 1 ... 7 identify the threshold energy, viz. $\mathscr{E}_p \geqq 98$; 134; 168; 268; 498; 988 and 1690 keV, respectively. The incomplete curves for energies 5 and 6 are due to detector saturation.

equator some distance. Comparing off-equator data of FILLIUS[10] with what is expected from radial diffusion conserving $|\mu|$ and $\mathscr{J}$, HESS[11] found that there were many more protons observed at $L \approx 3$ and $B/B_0 \approx 2$ than were expected from transforming data from larger L values. It is not esay to understand where these off-equator protons come from. Maybe pitch angle scattering can move high energy protons away from the equator fast enough to explain the larger fluxes observed at large B values. Pitch angle scattering cannot be a very fast process or the protons would never have as narrow a pitch angle distribution as they do. Maybe diffusion of CRAND (Sect. 6) protons outwards from the inner belt can explain the off-equator protons.

It may be that the observed narrow pitch angle distribution of protons can be understood in terms of quasi-trapped particles. ROEDERER[12] has calculated that certain regions of the magnetosphere cannot contain mirroring permanently trapped particles see Fig. 49 (Sect. 39) because they do not have complete drift paths around the Earth. At $R = 7R_E$ trapped particles can mirror only out to a magnetic latitude $\Phi_m \approx 30°$ or $B_m/B_0 \approx 3.1$. This angular distribution is quite narrow and should stay narrow if particles obey Eq. (20.8) during the inward drift qualitatively in agreement with observations.

23. Storm variations. A characteristic of the protons in the outer belt is that the proton flux is quite stable in time. The proton flux contours observed by DAVIS on Explorer 14 could be superimposed on his Explorer 12 flux contours quite accurately. But some time changes are seen in the outer belt protons.

α) MCILWAIN[1], using detectors on Relay 1, observed a significant *change in the flux of protons* of $\mathscr{E}_p > 34$ MeV on September 23, 1963. On this day there occurred the largest magnetic disturbance for two years. Below $L = 2.1$ there were not any significant changes. At $L = 2.5$ the proton flux decreased about a factor of 10. One or more adiabatic invariants were violated here. The fluxes of 18 MeV $< \mathscr{E}_p < 35$ MeV protons also decreased considerably but for the flux of $\mathscr{E}_p > 5$ MeV protons there was no large effect. It may be possible to explain the

[10] R. W. FILLIUS, and C. E. MCILWAIN: Phys. Rev. Letters **12**, 609 (1964). — R. W. FILLIUS: J. Geophys. Res. **71**, 97 (1966).

[11] W. N. HESS: [*16*], 352 (1966).

[12] J. G. ROEDERER: J. Geophys. Res. **72**, 981 (1967).

[1] C. E. MCILWAIN: [*12*], 374 (1965); — Relay 1 Trapped Radiation Measurements. Final Report on the Relay 1 Program NASA SP-76 1965.

observed flux changes in terms of violation of the third invariant only, but it seems likely that other invariants may be violated too.

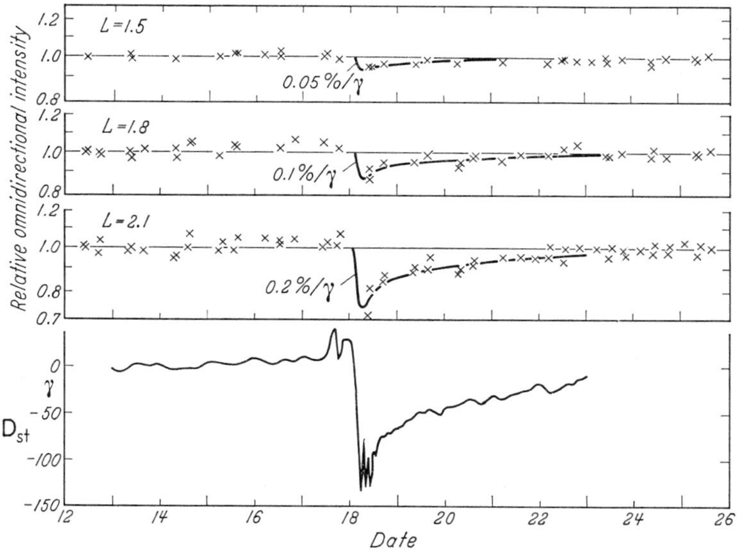

Fig. 28. Proton flux variations with time of 40 MeV $< \mathcal{E}_p <$ 110 MeV protons near the equator ($B/B_0 = 1.0 \ldots 1.2$) for three values of L in April 1965. The magnetic storm of Apr. 17 ... 20 produced the storm-time variation $\mathbf{D_{st}}$ shown in the bottom diagram. The corresponding variations in the three fluxes are shown as curves in the upper diagrams. (McIlwain, ref. [2], Sect. 23.)

β) A magnetic storm occurred during the period 17—20 April 1965 that also *affected* the *high energy trapped protons.* McIlwain[2] observed that the 40 MeV $< \mathcal{E}_p <$ 110 MeV protons seemed to behave adiabatically out to $L \approx 2.2$. That is, the proton flux in this region decreased in a way that could be attributed to adiabatic decelleration due to the ring current magnetic field decrease. As the storm decayed the proton flux returned to the pre-storm value as shown in Fig. 28. The observed flux changes are consistent with this expected adiabatic behavior.

γ) McIlwain had also observed that the shape of the second bump in Fig. 7 (Sect. 6) has changed over the period 1963—1965. The bump has become flatter and seems to have moved in. Schiff[3] has suggested that this change in the second bump is very likely due to radial diffusion associated with the breakdown of the Φ invariant as discussed in Sect. 22.

24. *L*-Diffusion of alpha-particles. The process described for the origin of outer belt protons will work equally well for other particles such as α-particles and electrons if they have the proper energies.

α) The criteria for the applicability of the *drift mechanism* is that the rise time (or decay time) of a magnetic disturbance τ be of the kind

$$T_b < \tau < T_d,$$

T_b being the bounce period and T_d the drift period (see Sect. 2). If this is true, then the disturbance will cause diffusion.

[2] C. E. McIlwain: J. Geophys. Res. **71**, 3623 (1966).
[3] M. L. Schiff: Trans. Am. Geophys. Union **47**, 136 (1966).

As an example of this, consider α-particles. It has been reported that there are low energy α-particles in the outer zone[1]. It seems very likely that they have diffused in from the magnetopause as it appears the protons have. If they have the same velocity as the protons do, then they should have a drift period of one-quarter the T_D of a proton of the same energy. This means that there may be fewer magnetic disturbance that have the appropriate values of the time constant τ to cause diffusion and so the diffusion velocity for α-particles will be slower. If we limit ourselves to the region of L where losses are not important, the energy spectrum of α-particles is related to the spectrum of the protons. This can be seen from the steady state FOKKER-PLANCK equation, Eq. (22.4). For an ensemble of storms comparing protons and α-particles we can write

$$\frac{\langle\Delta r_\alpha\rangle}{\langle\Delta r_p\rangle} = \frac{\langle(\Delta r_\alpha)^2\rangle}{\langle(\Delta r_p)^2\rangle} = \xi. \tag{24.1}$$

and this coefficient ξ cancels out of the steady state FOKKER-PLANCK equation.

β) So the *steady state energy spectrum* for α-particles is from Eq. (22.10)

$$N(r, \mathscr{E}_\alpha) = \left(\frac{r}{r_1}\right)^{-7} \left[\frac{r^7 - r_0^7}{r_1^7 - r_0^7}\right] N_1\left(r_1, \frac{\mathscr{E}_\alpha r^3}{r_1^3}\right). \tag{24.2}$$

If we now assume that the source characteristics are such that the number of α-particles of energy $\mathscr{E}$ is a *constant fraction* f_α of the number of protons of the same energy

$$N_{1\alpha}(\mathscr{E}) = f_\alpha N_{1p}(\mathscr{E}). \tag{24.3}$$

If we fit both energy spectra by an esponential

$$N_\alpha(r, \mathscr{E}) = k(r) \exp(-\mathscr{E}/\mathscr{E}_0). \tag{24.4}$$

We have at any point (neglecting losses)

$$N_\alpha(r, \mathscr{E}) = f_\alpha N_p(r, \mathscr{E}) \tag{24.5}$$

and

$$\mathscr{E}_{0\alpha}(r) = \mathscr{E}_{0p}(r). \tag{24.6}$$

II. Electrons.

25. Electrons, general features.

α) Protons in the outer belt are characterized by their stability. The proton data of DAVIS[1] from Explorer 12 and from Explorer 14 a year later could be fitted together well. Electrons in the outer zone are characterized by their *time variability*. The fluxes can change by orders of magnitude in hours.

FRANK, VAN ALLEN, and HILL[2] studied outer belt electrons on Explorer 14 and found "1. Typical intensities of electrons in the central part of the outer radiation zone ($L \approx 4.0$ near the geomagnetic equator) are given in Table 4:

Table 4. *Electron fluxes*

$J(\mathscr{E}_e > 40$ keV$)$	$\approx 3 \cdot 10^7$ cm^{-2} sec^{-1} $= 3 \cdot 10^{11}$ m^{-2} sec^{-1}
$J(\mathscr{E}_e > 230$ keV$)$	$\approx 3 \cdot 10^6$ cm^{-2} sec^{-1} $= 3 \cdot 10^{10}$ m^{-2} sec^{-1}
$J(\mathscr{E}_e > 1.6$ MeV$)$	$\approx 3 \cdot 10^5$ cm^{-2} sec^{-1} $= 3 \cdot 10^9$ m^{-2} sec^{-1}

[1] J. A. VAN ALLEN, and S. M. KRIMIGIS: Amer. Geophys. Union **46**, 140 (1965).
[1] L. R. DAVIS, and J. M. WILLIAMSON: [8], 365 (1963).
[2] L. A. FRANK, J. A. VAN ALLEN, and H. K. HILLS: J. Geophys. Res. **69**, 2171 (1964).

2. The gross temporal variations of electron intensities in the outer radiation zone during the five-month period (Oct. 1962 through Feb. 1963) were greater by a factor of

100 for electrons $\mathscr{E}_e > 40$ keV,
10 for electrons $\mathscr{E}_e > 230$ keV,
100 for electrons $\mathscr{E}_e > 1.6$ MeV.

There was no evident trend in average intensities during this period ... Temporal variations of the electron intensity increased markedly as L increased from 2.8 to 4.8."

β) MIHALOV and WHITE[3] have measured *differential energy spectra* of trapped electrons using a 10-channel directional spectrometer on satellite 1964—45 A. This was launched in Aug. 1964 into a near polar orbit with apogee 3765 km and perigee 270 km. Typical quiet time data are shown in Fig. 29. In the inner zone for $L < 1.7$ the measured spectra agree quite well with the calculated spectra[4] for the Starfish electrons (see Sect. 14) modified by Coulomb scattering by the atmosphere. There may also be some low energy electrons with $\mathscr{E}_e < 0.3$ MeV from some other source.

In the outer zone for $\mathscr{E} > 0.5$ MeV, exponential energy spectra of 0.2 MeV $< \mathscr{E}_0 <$ 0.6 MeV were observed. The spectrum becomes generally softer with increasing L. Also electrons of a particular energy $\mathscr{E}$ peak at lower L on $\mathscr{E}$ increase. These two facts would tend to indicate that inwards diffusion violating the third adiabatic invariant may be a dominant source of these quiet time electrons. However, the variation of $\mathscr{E}$ with L is such that the first two invariants $|\mu|$ and $\mathscr{J}$ do not seem to be conserved. This is perhaps not surprising because we know that electron precipitation takes place underneath all of the outer zone (but not for protons). This means that pitch angle diffusion seems important for electrons. This violates $|\mu|$ and may explain why the electron energies do not change as fast as they should if $|\mu|$ and $\mathscr{J}$ were conserved.

γ) PIZZELLA et al.[5] measured *low energy* outer belt *electrons* for 5 keV $< \mathscr{E}_e <$ 160 keV in the range $2 < L < 10$ using a scintillator on Explorer 14. For a two week period of the data they found are reproduced in Table 5.

Table 5. *Peak fluxes of electrons.*

$\bar{\mathscr{E}}$/keV	peak flux J/cm^{-2} sec^{-1}	location of peak flux
50	$8 \cdot 10^8$	$L \sim 5$
10	$1.6 \cdot 10^9$	$L \geqq 8$

These fluxes varied by about one order of magnitude during the two weeks. Using three absorbers they got a rough energy spectrum of the electrons which they fitted by an exponential distribution and found that the e-folding energy $\mathscr{E}_0$ decreased systematically with L. This is also indicative of radial diffusion. They found average values of $\mathscr{E}_0$ given in Table 6. because $\mathscr{E}_0 L^3$ is not constant with L it seems here also that $|\mu|$ and $\mathscr{J}$ are not constant during the radial diffusion.

Table 6. *Electron energy.*

L	$\mathscr{E}_0$/keV	$\mathscr{E}_0\, L^3$/keV
2	250	20
3	150	41
4	80	51
5	55	69
6	40	86
7	30	103

[3] J. D. MIHALOV, and R. S. WHITE: J. Geophys. Res. **71**, 2217 (1966).
[4] M. WALT: J. Geophys. Res. **69**, 3948 (1964).
[5] G. PIZZELLA, L. R. DAVIS, and J. M. WILLIAMSON: J. Geophys. Res. **71**, 5495 (1966).

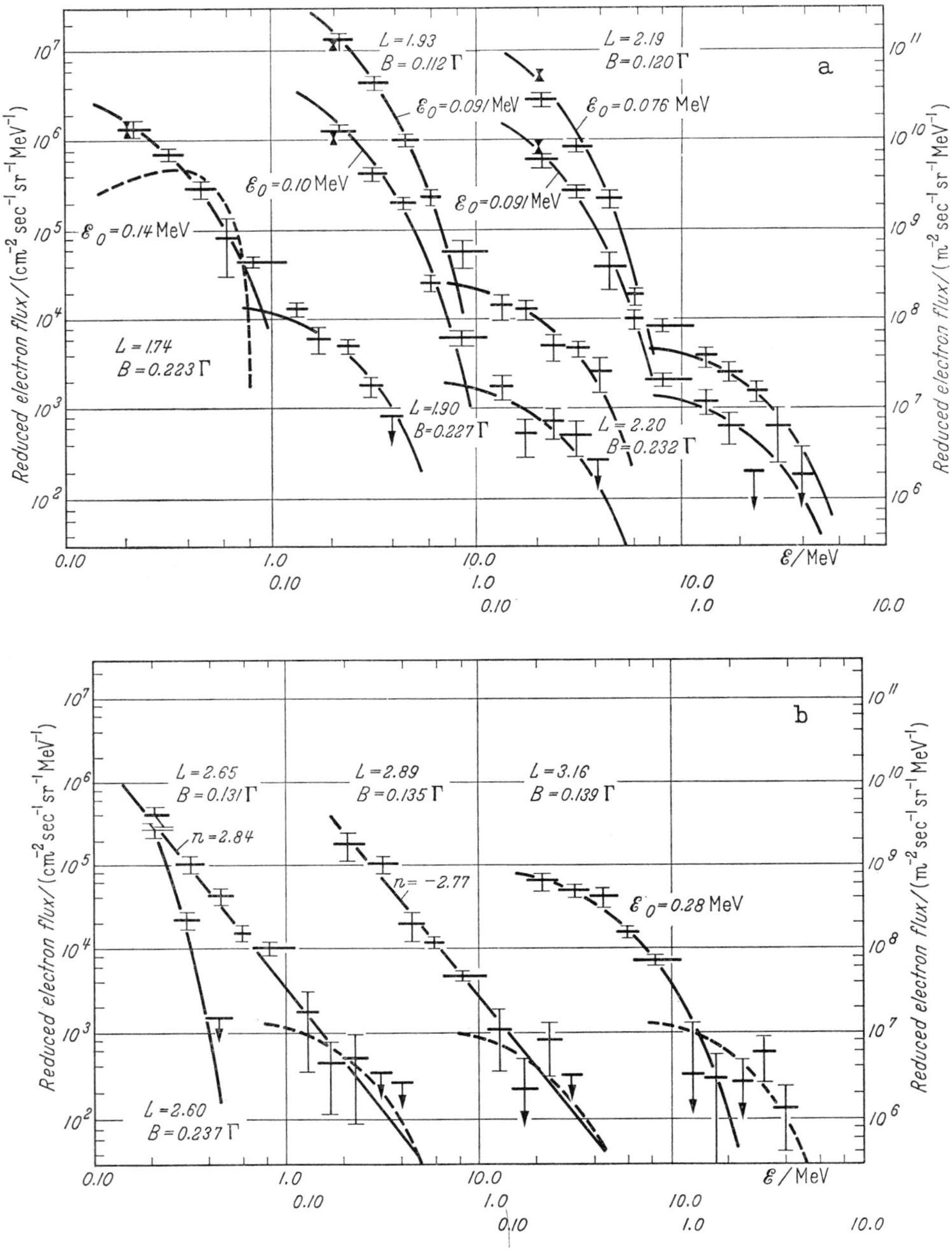

Fig. 29 a and b.

Fig. 29. a—d. Typical differential electron energy spectra measured on Satellite 1964—45a on August 15, 1964. (Mihalov and White, ref. [3], Sect. 25). (a) For $1.7 < L < 2.2$. Exponential distributions are fitted to the soft component and an equilibrium fission data spectrum (see Fig. 10, Sect. 14β) is fitted to the hard component. The neutron beta-decay spectrum is given as a broken line with the data of $L = 1.74$. (b) For $2.4 < L < 3.2$. Power laws are fitted to two sets of data. The data at $L = 3.16$ are better fitted by an exponential system. Fission beta-spectra fitting the high energy points are given as broken lines. (c) For $3.4 < L < 5.0$. Exponential spectra fit the data above 0.5 MeV. (d) For $5.9 < L < 8.2$. Exponential spectra fit the data given.

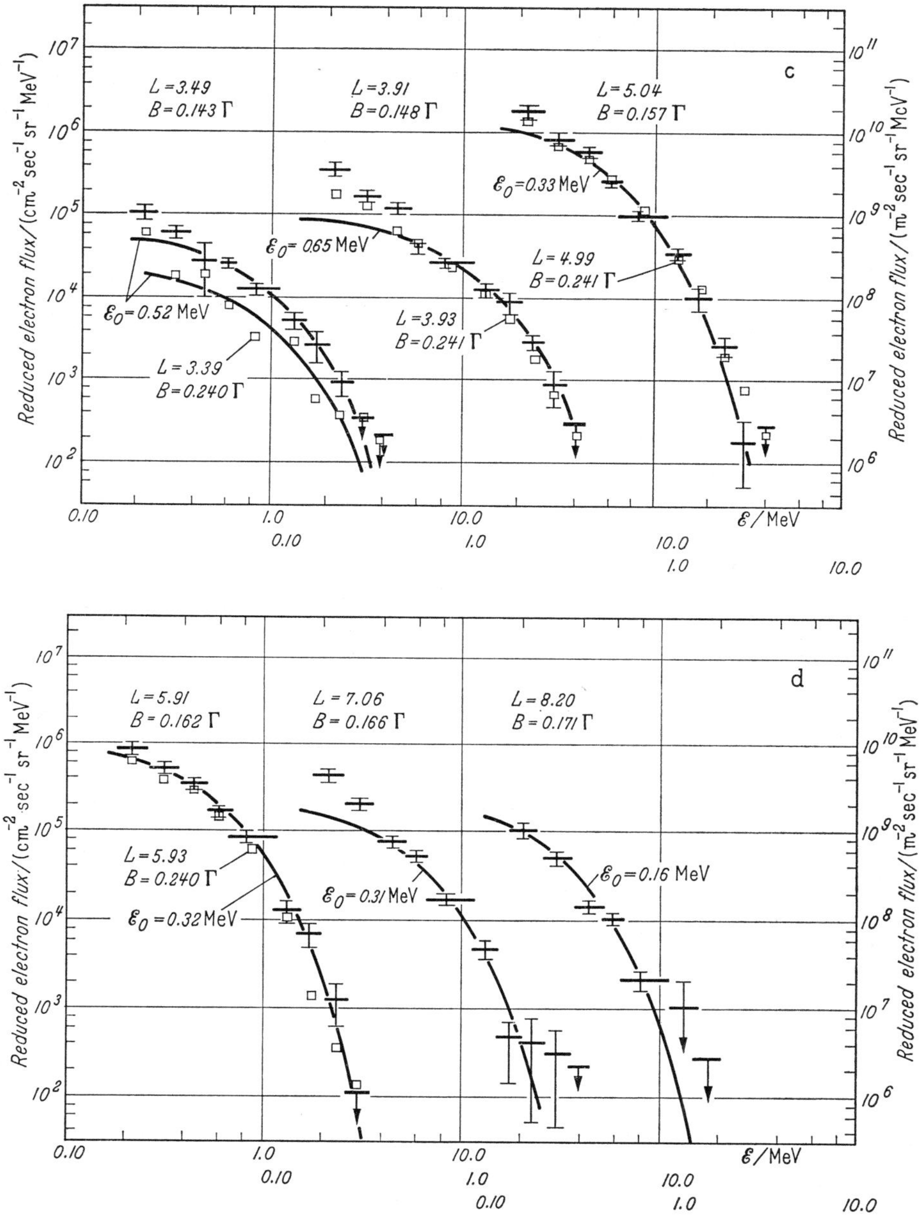

Fig. 29c and d.

26. Radial diffusion of electrons.

Frank et al.[1] observed temporal variations of $\mathscr{E}_e > 1.6$ MeV electrons on Explorer 14 that strongly suggest radial diffusion of the electrons. Following a period of geomagnetic activity in Dec. 1962 an inward moving wave of electrons was observed, see Fig. 30. The

[1] L. A. Frank, J. A. van Allen, and H. K. Hills: J. Geophys. Res. **69**, 2171 (1964).

radial motion is apparently present during magnetic quiet times and the electron pulse was probably injected during the magnetic disturbance.

Several such radial diffusion events were seen on Explorer 14. Analyzing one of these events FRANK[2] found that the radial diffusion velocity v_r (defined as the motion of the logarithmic half-maximum) was $v_r \approx 0.4\, R_E/\mathrm{d} \approx 30\ \mathrm{m\,sec^{-1}}$ at $L = 4.7$, and $0.03\, R_E/\mathrm{d} \approx 2\ \mathrm{msec^{-1}}$ at $L = 3.4$.

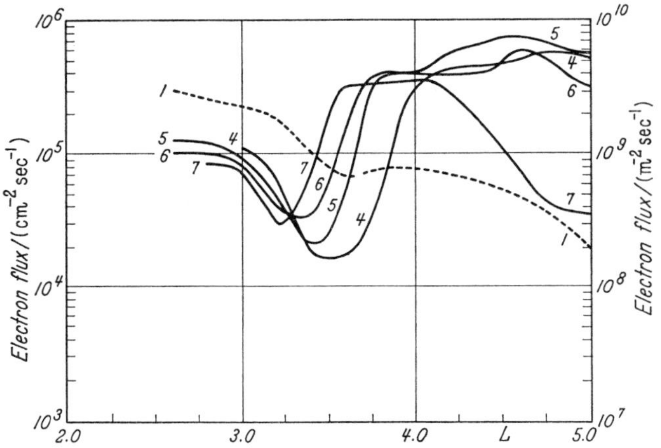

Fig. 30. Measured radial distribution of equatorial intensities of $\mathscr{E}_e > 1.6$ MeV electrons in the outer radiation zone following the magnetic storm of Dec. 1962. Parameter Figs. 1 and 4 ... 7 identify dates as Dec. 7 and 20, 23, 29, Jan. 8, 1963 Electrons are apparently drifting inwards towards Earth during this period. (FRANK, ref. [2], Sect. 26.)

The *radial dependence of* the measured '*diffusion*' *velocity* is very nearly

$$v_r = 1.6 \cdot 10^{-6}\, L^8\, R_E/\mathrm{d} = 1.2 \cdot 10^{-4}\ \mathrm{m\,sec^{-1}}\, L^8. \tag{26.1}$$

Frank found the inward diffusion velocities for two separate events were the same at the same L values indicating the process is a fairly usual one.

This process has most of the proper characteristics to be explained by the L diffusion process associated with violation of the third adiabatic invariant. The discussion of this process dealt with protons but it does not depend strongly on mass. The characteristic of the process is the drift time T_d (see Sect. 2). Eq. (2.17) shows that this is nearly the same for protons and electrons of the same energy.

27. The geosynchrotron.

α) It appears from several recent measurements that there is a form of *synchronous acceleration* of particles going on in the radiation belt.

The first reported instance of this was by IMHOF and SMITH[1] who found on Oct. 30, 1963 a nearly monoenergetic group of electrons of $\mathscr{E} = 1.35$ MeV at $L = 1.15$ below the region where such large fluxes of electrons normally exist, see Fig. 31. Calculations showed that the electrons which decayed in times of a few days were lost by coulomb scattering with the neutral atmosphere as expected. There was apparently a transient source of electrons that produced this event. IMHOF and SMITH observed a second similar event on Nov. 2, 1963 when a group of 0.75 MeV electrons appeared at low altitudes and subsequently decayed in a few days.

β) CLADIS[2] and IMHOF et al.[3] in analyzing the possible sources concluded that the electrons were produced by *resonant acceleration*. The process they considered

[2] L. A. FRANK: J. Geophys. Res. **70**, 3533 (1965).
[1] W. L. IMHOF, and R. V. SMITH: Phys. Rev. Letters **14**, 885 (1965).
[2] J. B. CLADIS: J. Geophys. Res. **71**, 5019 (1966).
[3] W. L. IMHOF, J. B. CLADIS, and R. V. SMITH: Planetary Space Sci. **14**, 569 (1966).

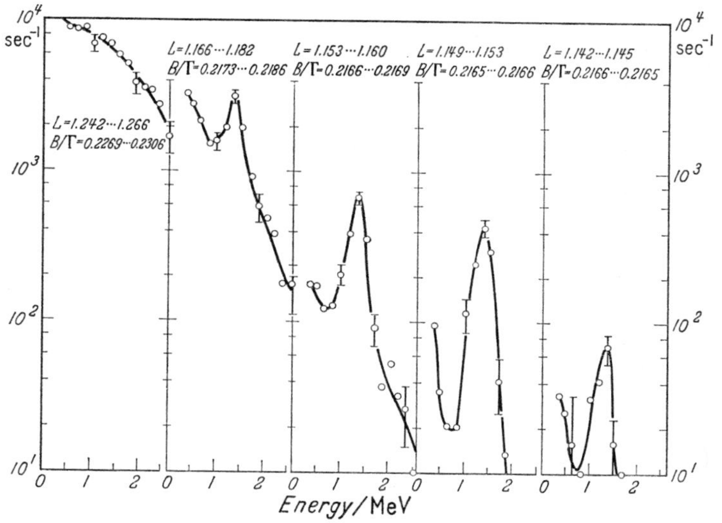

Fig. 31. Measured electron energy spectra near the inner edge of the radiation belt for several B, L values on Oct. 30, 1963. The satellite was at geogr. longitude $\lambda = 299$ °E and latitudes $\varphi = 22° \ldots 35$ °S, time 1,733 ... 1738 UT. A nearly monol-energetic group of electrons of $\mathscr{E}_e \approx 1.3$ MeV can be seen below most of the trapped particles. This peak decayed away in a few days. (IMHOF and SMITH, ref. [1], Sect. 27).

was a resonant acceleration involving time-varying magnetic fields. If some process could accelerate certain particles conserving $|\boldsymbol{\mu}|$ and $\mathscr{J}$ then these particles would be moved to a smaller L shell. It seems that the important feature of this process is not the gain in energy but the radial motion in this case. Large fluxes of electrons of $\mathscr{E}_e \sim 1$ MeV exist at $L \sim 1.20$. An acceleration will produce a radial motion given by

$$\frac{\Delta\mathscr{E}}{\mathscr{E}} = \frac{\Delta B}{B} = -3\,\frac{\Delta L}{L}\,. \tag{27.1}$$

CLADIS[2] suggests that $\Delta L = 0.013$ is what is needed to move enough electrons from a region of high flux downwards to agree with the observations. This corresponds to an acceleration of a 1 Mev electron of

$$\Delta\mathscr{E} = 3\,\frac{\Delta L}{L}\,\mathscr{E} \cong 30\ \text{keV}. \tag{27.2}$$

γ) What is the *driving mechanism* for the acceleration?

CLADIS showed by studying magnetograms that there was a world-wide magnetic oscillation with an amplitude of about 10γ (except in South America where there was $\sim 100\,\gamma$) in phase as shown in Fig. 32, at just about the time that the transient population was produced. This was probably due to intensity variations in the solar wind. The oscillations in Fig. 32 (p. 170) for Oct. 29, 1963 have a period of about 35 minutes which is the drift period at $L = 1.2$ for $\Phi = 25°$ electrons of $\mathscr{E}_e \sim 1.5$ MeV. On Nov. 2, 1963 the oscillations have a period of about 65 min which for the same L and Φ is the drift period of electrons of $\mathscr{E} \sim 0.7$ MeV. It seems very likely that a synchronous acceleration took place produced by the magnetic field fluctuations.

The energy gain for a particle of charge q in a time varying field is

$$\Delta\mathscr{E} = \int \mathrm{d}t \left(q\,\boldsymbol{v}_{\mathrm{D}} \cdot \boldsymbol{E} + \boldsymbol{\mu} \cdot \frac{\partial \boldsymbol{b}}{\partial t} \right) \tag{27.3}$$

where $\boldsymbol{b}$ is the varying magnetic field

$\boldsymbol{\mu}$ is the relativistic magnetic moment,

$\boldsymbol{E}$ is the induced electric field.

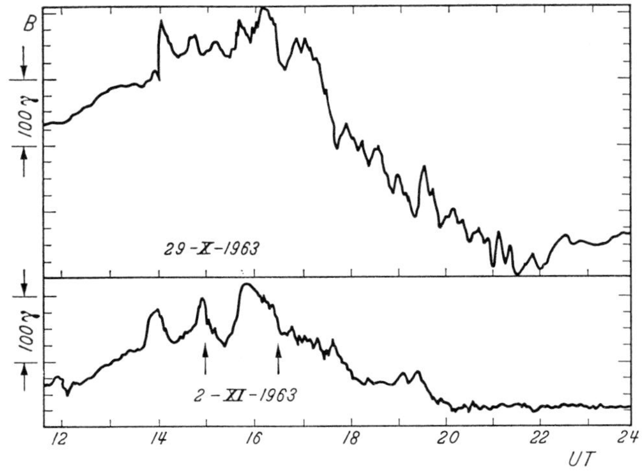

Fig. 32. A ground level magnetogram from Huancayo (Peru) at the magnetic equator showing quasi-periodic disturbance of Earth's magnetic field on Oct. 29 and Nov. 2, 1963, which is related to the electron peak in Fig. 31. (CLADIS, ref. [2], Sect. 27.)

$\boldsymbol{b}$ is given by Eq. (22.2) and can be written in the form

$$|\boldsymbol{b}| = S(t) + A(t)\, r \cos\delta \tag{27.4}$$

where δ is local time defined as the angle from noon clockwise to the point of observation.

FÄLTHAMMAR[4] obtained for the induced electric field here

$$E_{\mathrm{b}} = \frac{r}{2}\frac{\mathrm{d}S(t)}{\mathrm{d}t} + 4r^2 \frac{\mathrm{d}A(t)}{\mathrm{d}t}\cos\delta. \tag{27.5}$$

Eq. (27.3) can now be written:

$$\Delta\mathscr{E} = \int \mathrm{d}t\left[C_1 \frac{\mathrm{d}S(t)}{\mathrm{d}t} + C_2 \frac{\mathrm{d}A(t)}{\mathrm{d}t}\cos\delta\right]. \tag{27.6}$$

δ) Assuming that the *fluctuating* $\boldsymbol{b}$ *field* is *sinusoidal* and taking $\delta = \delta_0 + \omega_{\mathrm{D}} t$ the energy gain per N turns is

$$\Delta\mathscr{E} = \int_0^{N T_{\mathrm{d}}} \mathrm{d}t\left[C_1 S_0 \omega\cos\omega t + C_2 A_0 \omega \cos\omega t \cos(\omega_{\mathrm{D}} t + \delta_0)\right] \tag{27.7}$$

we can see that the S_0 symmetric term averages out to zero and the A_0 term is non-zero only if $\omega = \omega_{\mathrm{D}}$ in which case $\Delta\mathscr{E} = \frac{C_2 A_0}{2}$ per turn. Only the *asymmetric part* of the field variation plays a role in *accelerating* the particles.

If this resonant condition exists such that a particle drifts around the earth with the period of the natural magnetic disturbance, particles at the proper phase will be accelerated (particles with other phases will be decellerated). This process

[4] C. G. FÄLTHAMMAR: J. Geophys. Res. 71, 1487 (1966).

is very similar to the operation of a *synchrotron*[5]. In a *synchrotron* a particle is accelerated once per turn by an electric field. The particle gain energy and the magnetic field increase keeping the particle at constant radius in the machine. In the *geosynchrotron* the particles resonantly gain energy and move to smaller L to move into a stronger B field keeping $|\mu|$ and $\mathcal{J}$ constant.

ε) CLADIS[2] has calculated the *magnitude* of $\Delta\mathcal{E}$ for the magnetic event shown in Fig. 33 for two cases.

(1) ΔB assumed due to variations in pressure of the solar wind.

(2) ΔB assumed due to variations in the intensity of the equatorial electrojet.

In case (1) the magnetopause location was assumed to vary in time so that ΔB was obtained from MEAD's distortion field given by Eq. (22.2) and then Eq. (27.5) was used for the induced electric field and the energy gain per turn found to be $\Delta\mathcal{E}\sim 0.1$ keV. This is much too small to be interesting. For case (2) it was assumed that the electrons encountered the time varying fields only for the period when they were within a 100° wide band of longitude near Huancayo. The ΔB observed at Huancayo were stronger than at other longitudes. CLADIS feels this is due to the enhanced ionospheric conductivity here due to the large particle precipitation near the South Atlantic magnetic anomaly. This would lead to a large equatorial electrojet current as the source of ΔB. Using this model CLADIS obtained $\Delta\mathcal{E}\sim 5$ keV for the Nov. 2 event and $\Delta\mathcal{E}\sim 25$ keV for the Oct. 29 event. The ΔL obtained this way is reasonable and the energy spectrum in the displaced peak agrees well with the measured spectra when allowance is made for atmospheric scattering.

CLADIS suggests that this kind of process of resonant acceleration may be fairly common. PAI and SARABHAI[6] found magnetic fluctuations of the sort shown in Fig. 32 frequently occur during storms and found periods in the range 40...50 min frequently.

28. Electron precipitation. It is apparent that radial diffusion works for electrons as well as protons during magnetic quiet periods at least. But quite clearly there are other processes that operate on electrons. It has been observed that there is a rather steady flux of electrons precipitating into the atmosphere over large sections of the world. No comparable loss of protons has been observed. Particle precipitation is the process that generates aurorae. This was demonstrated by direct observation on rockets in 1958. But precipitation goes on more or less continuously deep inside the magnetosphere as well as at higher latitudes where the aurora occur.

α) KRASSOVSKIJ et al.[1] first observed particle precipitation using a detector on the third USSR artificial satellite. They found downward moving fluxes up to 1 erg $\text{cm}^{-2}\,\text{sec}^{-1}$. O'BRIEN[2] found direct evidence of particle precipitating on Injun 1. The angular distribution of electrons shown in Fig. 33 sometimes was narrow and symmetrical as Fig. 33a and indicated trapped particles. On other passes the satellite detected particles at pitch angles such that they would be moving downward directly into the dense atmosphere as in Fig. 33b. The upwards moving particles did not have as large a number of particles at small pitch angles as were found in the downwards moving particles. Some upwards moving particles are made at small pitch angles by backscattering from the atmosphere. Dumping was observed on about half the Injun 1 passes.

---

[5] M. S. LIVINGSTON, and J. P. BLEWETT: Particle Accelerators. New York: McGraw-Hill 1962.

[6] G. L. PAI, and V. A. SARABHAI: Planetary Space Sci. **12**, 855 (1964).

[1] V. I. KRASSOVSKIJ, I. S. ŠKLOVSKIJ, JU. I. GAL'PERIN, E. M. SVETLITSKIJ, JU. M. KUŠNIR, and G. A. BORDOVSKIJ: Iskusstv. Sputniki zemli **6**, 113 (1961) (engl. transl.: Artificial Earth Satellites **6**, 137, Plenum Press, 1961); also Planetary Space Sci. **9**, 27 (1962).

[2] B. J. O'BRIEN: J. Geophys. Res. **67**, 1227 (1962).

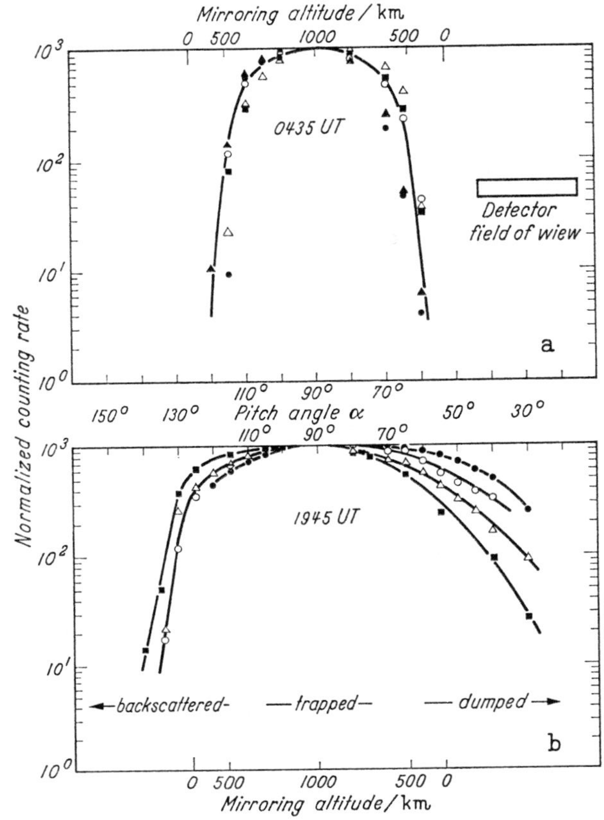

Fig. 33 a and b. Experimentally determined pitch angle distributions of $\mathscr{E}_e > 40$ keV electrons at 1000 km altitude on Jun. 30, 1961 from satellite Injun 1. (O'Brien, ref. [2], Sect. 28.) (a) A narrow symmetric distribution typical of trapped particles observed part of the time. (b) a broad and skewed distribution indicating particle precipitation also observed part of the time.

a	B/Γ	L	height/km
■	0.38	5.1	1011
○	0.38	4.7	1010
△	0.37	3.6	1007
●	0.37	3.3	1005
▲	0.35	2.4	998

b	B/Γ	L	height/km
■	0.37	3.3	974
○	0.38	3.7	977
△	0.38	4.2	980
●	0.38	4.9	983

O'Brien[3] determined the average flux of precipitating electrons observed on Injun 1 to be

$$\text{at } \Lambda = 45°: j = 10^3 \text{ cm}^{-2} \text{ sec}^{-1} = 10^7 \text{ m}^{-2} \text{ sec}^{-1},$$

$$\text{at } \Lambda = 65°: j = 10^5 \text{ cm}^{-2} \text{ sec}^{-1} = 10^9 \text{ m}^{-2} \text{ sec}^{-1}.$$

β) Using this data a *time to empty the outer belt* can be obtained. Summing over all α_0, or equivalently all Λ, we get all the particles in the tube of force of unit area at Earth's surface. Dividing by the precipitating flux we get the time to empty the belt, τ. Values of τ calculated by O'Brien are given in Table 7.

[3] B. J. O'Brien: J. Geophys. Res. **67**, 3687 (1962).

Table 7. *Average* characteristics** of electrons with energy $\mathcal{E} < 40$ keV.*

Range of Л	Equatorial plane flux-cm^{-2} sec^{-1}	1000-km Flux of patricles/cm^{-2} sec^{-2} sr^{-1}		'Lifetime' τ, order of magn.	
		Trapped	Dumped	τ/sec	τ/h
45°—50°	10^7	$2\cdot10^5$	$4\cdot10^3$	$6\cdot10^3$	2
50°—55°	$2\cdot10^7$	$2\cdot10^5$	$8\cdot10^3$	$2\cdot10^4$	6
55°—60°	$5\cdot10^6$	$1\cdot10^5$	$4\cdot10^4$	$2\cdot10^3$	0.6
60°—65°	$5\cdot10^6$	$2\cdot10^5$	$1\cdot10^5$	$3\cdot10^3$	0.8
65°—70°	$5\cdot10^6$	$1\cdot10^5$	$7\cdot10^4$	$2\cdot10^4$	6
70°—75°	$2\cdot10^6$	$6\cdot10^4$	$6\cdot10^4$	$1\cdot10^4$	3

* The variations of electron flux with time are so large that we think these average characteristics should be treated as order-of-magnitude estimates only.

** Multiply numerical flux values by 10^4 in order to get flux/m^{-2} sec^{-1}.

These lifetimes are so short that it is quite clear the trapped radiation cannot be responsible for the precipitation. The outer belt would be drained dry in short order.

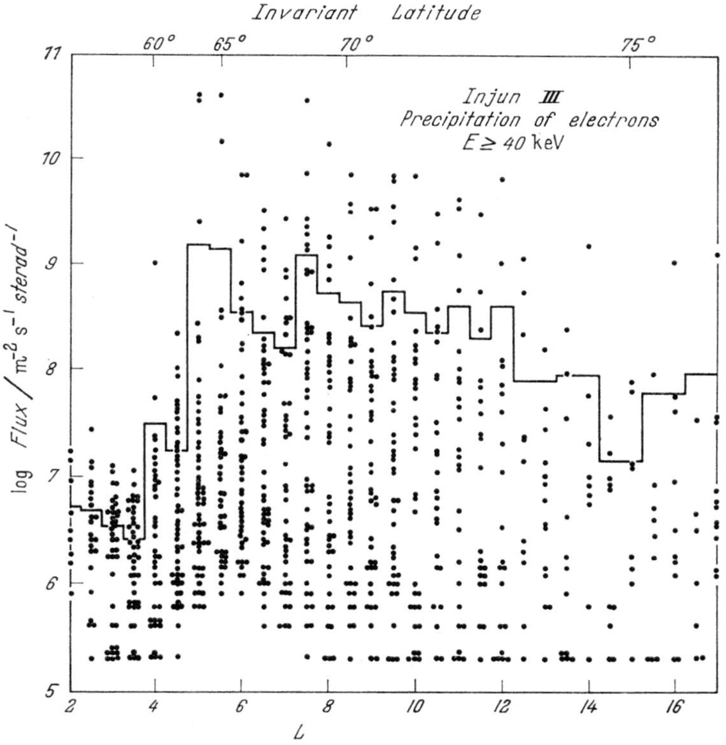

Fig. 34. Samples of precipitated electron fluxes of $\mathcal{E}_e > 40$ keV versus L measured over North America in Jan. 1963 by the Injun 3 satellite (O'Brien, ref. [9], Sect. 28).

γ) Using the magnetically oriented satellite Injun 3 at 1000 km altitude, O'Brien[4] made a detailed study concerning the *origin of electron precipitation.* The latitude distribution of the precipitating electron flux measured on Injun 3

[4] B. J. O'Brien: J. Geophys. Res. **69**, 13 (1964).

is shown in Fig. 34. It is seen that the flux is very variable but that there is always some precipitation in the auroral zone. From Injun 1 studies O'BRIEN had suggested that precipitation occurs principally during the acceleration process and that the precipitated particles are fresh. He also said the outer belt should be regarded not as a leaky bucket (see Fig. 5) that occasionally spills out particles to cause aurorae but rather as a bucket that catches a little of the splash from the acceleration process. The splash catcher shown in Fig. 35 is compared with the leaky bucket in Table 8.

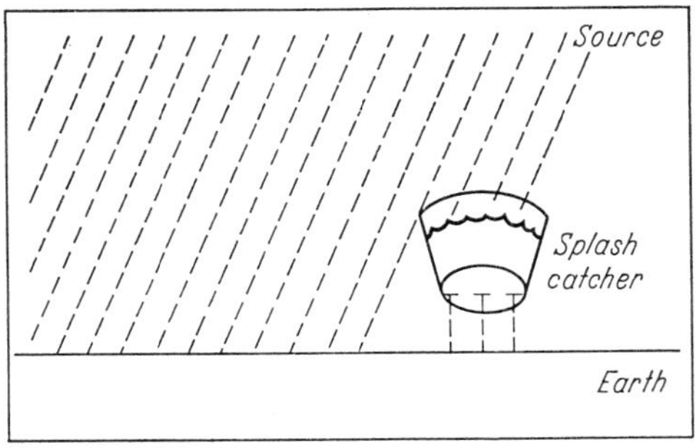

Fig. 35. The splash catcher model of the outer radiation zone. In this model one source produces both ground level effects such as electron precipitation and aurorae and concurrently it generates the trapped radiation. In this case the trapped flux increases when the precipitated flux increases as is commonly observed (after O'BRIEN).

Table 8. *Simplified comparison of two models.*

Leaky-Bucket model	Splash-catcher model
Two acceleration mechanisms which (1) create energetic trapped electrons, and (2) dump them later with no energy change and cause auroras.	One acceleration mechanism which creates energetic electrons, some of which are trapped but most of which are precipitated to cause auroras, etc.
Implication: Outer-zone electrons are the immediate cause of auroras.	*Implication*: Outer-zone electrons are generated (in part) by whatever is the immediate cause of auroras.
Test: Outer-zone intensity decreases when aurora occurs.	*Test*: Outer-zone intensity increases when aurora occurs.

δ) The *test* of the splash catcher model is to observe whether the flux of trapped electrons increases when electrons are precipitated. Injun 1 seemed to indicate this but Injun 3 really decided it. In Sect. 40 on the aurora in Fig. 53 is shown data from one pass of Injun 3 through the auroral zone during a precipitation event. Clearly the trapped *flux* does *increase where precipitation* occurs. O'BRIEN says this is the normal situation. The test has only been performed at 1000 km so it is not known whether equatorial fluxes behave in the same way or not. The $\mathscr{E} > 40$ keV electron flux tends to become quite isotropic at 1000 km during a splash.

Precipitation is largest at times of large planetary three-hour index, K_p[5]. Fluxes of precipitating electrons go up roughly a factor of 1000 for K_p going from 0 to 6. Comparison with data from Explorer 12 shows that the omnidirectional flux in the equatorial plane goes up only about a factor of 10 for the same change. If a common acceleration mechanism is responsible for both changes the acceleration must be preferentially along $\boldsymbol{B}$.

[5] Sect. 15 of contribution by T. NAGATA and N. FUKUSHIMA in this Encyclopedia, Vol. 49/3.

ε) *Precipitation* is *energy dependent*. In one splash where the $\mathscr{E} > 40$ keV precipitating flux increased four times the $\mathscr{E} > 250$ keV precipitated flux roughly doubled and the omnidirectional flux of $\mathscr{E} > 1.5$ MeV did not change by more than 10 percent. In general no significant changes in the omnidirection flux of $\mathscr{E} > 1.5$ MeV occurred when a splash occurred. Using another detector on Injun 3 it could be shown that the directional flux of trapped $\mathscr{E} > 1.3$ MeV electrons did not change in the splashes studied.

This says that $\mathscr{E} > 1$ MeV electrons can remain trapped and unchanged on field lines where fresh electrons of $\mathscr{E} > 40$ keV are being precipitated.

If precipitation were caused by a *blob* of plasma[6] pushed into the field from the solar wind and disturbing trapped particles and/or introducing new particles from the blob then we would expect that this would dustirb particles from the blob then we would expect that this would dustirb particles of all energies. Since the $\mathscr{E} > 1$ MeV electrons were not changed by splashes this *blob theory* would *not* seem *reasonable*.

McDiarmid and Burrows[7] have found that electrons up to $10^9\ \text{cm}^{-2}\ \text{sec}^{-1}\ \text{sr}^{-1}$ ($\equiv 10^{13}\ \text{m}^{-2}\ \text{sec}^{-1}\ \text{sr}^{-1}$) are occasionally found in spikes at high latitudes outside the normal trapping region at 1000 km or less altitude. They suggest that the high latitude electron islands are accelerated in the geomagnetic tail and propagate inwards to low altitudes and may form an important source of electrons for the outer radiation zone. The electrons in the *spikes* have a differential *energy spectrum* of

$$N(\mathscr{E}_e) = k_{sp}\,\mathscr{E}_e^{-n} \tag{28.1}$$

where $5 < n < 7$ usually for $40\ \text{keV} < \mathscr{E}_e < 250\ \text{keV}$. For trapped outer zone electrons $2 < n < 4$ in this energy range.

ζ) Kellogg[8] has considered the possibility of electron acceleration in a fluctuating electric field to try to explain electron precipitation. He used a simple model of a sinusoidal $\boldsymbol{E}$ field with $\boldsymbol{E} \perp \boldsymbol{B}$. He concluded that this process is marginal to produce the normally precipitated electrons from the outer zone and it is clearly not rapid enough to produce the intense dumping occasionally seen.

Parker[9] considered how a breakdown of the integral invariant $\mathscr{I}$ can accelerate particles and cause their precipitation. If the magnetic field at a particle's mirror point changes in a time short compared to a bounce period, then the integral invariant $\mathscr{I}$ is not a constant of the motion. The interesting feature here is that as the particle gains energy its mirror point is systematically lowered because all the energy gain is in $\mathscr{E}_{\parallel}$ and therefore the pitch angle decreases and the particle is lost into the atmosphere. There is no information on how rapidly this process occurs.

η) *Precipitation of protons* is also known[10]. The first measurements were earth based studies of doppler-shifted Balmer lines[11] in aurora. This showed that proton fluxes of $10^7 \ldots 10^8\ \text{cm}^{-2}\ \text{sec}^{-1}$ ($\equiv 10^{11} \ldots 10^{12}\ \text{m}^{-2}\ \text{sec}^{-1}$) with energies of order 1 keV sometimes bombard the auroral zone. Rocket measurements by McIlwain[12] and Davis[13] showed the presence of $\mathscr{E}_p > 100$ keV protons in aurora. Proton precipitation has only been observed near the auroral zone.

29. Self-excited pitch angle diffusion.

α) Kennel and Petchek[1] have studied another process that will produce electron precipitation and has many of the characteristics required of the splash

[6] J. W. Chamberlain: Astrophys. J. **134**, 401 (1961b).
[7] I. B. McDiarmid, and J. R. Burrows: J. Geophys. Res. **70**, 3031 (1965).
[8] P. J. Kellogg: Planetary Space Sci. **10**, 165 (1963).
[9] E. N. Parker: J. Geophys. Res. **66**, 693 (1961).
[10] B. J. O'Brien: [*16*], 321 (1966).
[11] J. W. Chamberlain: Physics of the Aurora and Airglow. New York: Academic Press 1961; see also: S.-I. Akasofu, S. Chapman, and A. B. Meinel: This Encyclopedia, Vol. 49/1, Sect. 64.
[12] C. E. McIlwain: J. Geophys. Res. **65**, 2727 (1960).
[13] L. R. Davis, O. E. Berg, and L. H. Meredith: [*8*], 721 (1960).
[1] C. F. Kennel, and H. E. Petchek: J. Geophys. Res. **71**, 1 (1966).

catcher model of the outer belt. They have considered pitch angle diffusion of particles in a manner somewhat similar to DUNGEY[2] and CORNWALL[3] but with one large difference. DUNGEY and CORNWALL used externally generated whistlers made by lightning. KENNEL and PETCHEK use self-generated waves. In a situation where the pitch angle distribution of trapped particles is anisotropic enough, the system will be unstable and will generate whistler waves. These *whistlers* can *scatter* the *trapped particles* thereby producing particle precipitation and as a result limiting the value of the trapped flux. The scattering process here is, as with DUNGEY's analysis, the resonant interaction where the Doppler-shifted wave frequency equals the particles' gyrofrequency.

$$v_{\text{res}} = \frac{\omega - \Omega}{|\boldsymbol{k}|} \tag{29.1}$$

where ω and $\boldsymbol{k}$ are the pulsation and wave number vector of the whistler and Ω is the gyropulsation of the electron.

β) A sufficient *condition for instability* of waves resonant with electrons whose kinetic energies are larger than the resonant energy $\mathscr{E}_{\text{res}} = \frac{1}{2} m v_{\text{res}}^2$ is that $\partial f/\partial \alpha$ be positive everywhere. So for a pitch angle distribution decreasing monotonically towards the loss cone the system will be unstable for particles above $\mathscr{E}_{\text{res}}$.

KENNEL and PETCHEK developed a dispersion relation from which they obtained the growth rate γ of the whistler waves in terms of the anisotropy and magnitude of the trapped electron flux.

γ) KENNEL and PETCHEK studied the *steady-state condition* where the whistler mode growth rate approaches a constant value. Some electron source strength $\mathscr{S}$ with whistler scattering here determines the electron's lifetime in the radiation belt and the magnitude of the precipitated flux. But the source strenght does not determine the maximum omnidirectional trapped flux J_{Max}. This is controlled by the self-excited pitch angle diffusion in this model. The larger the trapped flux the larger the loss rate in a non-linear fashion so there is a maximum allowed flux. After the maximum trapped flux J_{Max} has been achieved increasing the source strength $\mathscr{S}$ just increases the precipitation.

For a steady state situation the whistler growth rate γ due to the instability will be just balanced by the loss in wave amplitude occurring at reflection from the ionosphere, that is,

$$\exp(\gamma T_\omega) = \frac{1}{|\Re|} \tag{29.2}$$

where $\Re$ is the wave reflection coefficient and T_ω is the time for the wave to go from equator to ionosphere. The maximum trapped flux can be found from this.

These theoretical maximum omnidirectional fluxes are compared with Explorer 14 data in Fig. 36. For $L > 4$ the highest observed fluxes lie quite near the calculated upper limit and for $L > 4$ there are observed quite large precipitation fluxes as expected theoretically. For $L < 4$ the trapped fluxes don't reach the upper bound and there should be little if any precipitation of particles.

δ) There are several interesting *properties of maximum trapped flux*, J_{Max}:

(i) The calculated value of $J_{\text{Max}}(>\mathscr{E}_{\text{res}})$ is independent of energy at high energies so the upper limited spectrum should be flat. Since the measured spectrum is not flat, above some transition energy the upper limit must not be met. This means that there should not be much particle precipitation at high energies and indeed precipitation infrequently occurs for $\mathscr{E} > 1.6$ MeV. In general, the *precipi-*

[2] J. W. DUNGEY: Planetary Space Sci. **11**, 591 (1963).
[3] JOHN M. CARNWALL: J. Geophys. Res. **69**, 1251 (1964).

tation energy *spectrum* should be *softer* than the trapped particle spectrum as is observed.

(ii) In regions where there is continuous particle acceleration the flux should consistently lie near the upper bound as for $5<L<8$ in Fig. 36.

(iii) We might expect larger *time fluctuations in* the *trapped* electron *flux* for high energy electrons which do not reach the upper bound flux. These electrons could respond to changes in the source while self-limited fluxes of lower energy electrons could not change much.

(iv) When electron acceleration is continuous, more and more of the electron distribution should reach the limiting flux therefore the energy *spectrum* should *become progressively harder* with time.

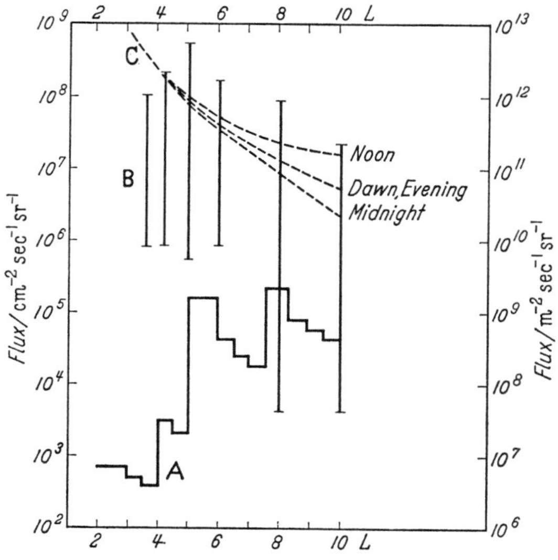

Fig. 36. The calculated limitation on trapped $\mathscr{E}_e > 40$ keV electron fluxes. The theoretical limiting flux (dotted curves on top) is compared with Explorer 14 equatorial trapped fluxes (marked **B**) as a function of the equatorial radial distance, L. The largest observed trapped fluxes are close to the theoretical upper limit. Also is shown the distribution with L shell of Injun 3 precipitated electrons (curve **A**). Strong precipitation occurs only where trapped fluxes can be comparable with the calculated limiting flux. Curves **B** and **C** both describe omnidirectional fluxes in m^{-2} sec^{-1} while curve **A** is a directional flux in m^{-2} sec^{-1} ster^{-1}. (KENNEL and PETCHEK, ref. [1], Sect. 29.)

ε) This theory of self-excited pitch angle diffusion holds for *protons* as well as electrons. The waves involved for protons are *ion cyclotron waves* rather than whistlers. KENNEL and PETCHEK calculated J_{Max} for protons too, but in this case the limiting fluxes were high enough so that the magnetic field limit was approached. For the field to control the particles' motion we need an energy density

$$\frac{B^2}{2\,\mathrm{u}\,\mu_0} > \int_0^\infty d\mathscr{E}'\,\frac{\mathscr{E}'}{v}\,\frac{dJ(\mathscr{E}<\mathscr{E}')}{d\mathscr{E}'}\,. \tag{29.3}$$

The protons approach quite close to this limit[4] for $L>5$.

It is not obvious that pitch angle diffusion affects protons much. In fact, if it were a significant effect we would not expect the radial diffusion to work as well as it seems to. In particular the variation of $\mathscr{E}_0$ with α seen in Fig. 24 should be hidden by pitch angle diffusion

[4] L. R. DAVIS, and J. M. WILLIAMSON: [8], 365 (1963).

and the peaked pitch angle distribution should be broadened. The fact that proton populations are very stable with time and that low latitude proton precipitation is unknown also rules against proton pitch angle diffusion.

III. Time variations.

Outer belt electron fluxes vary a great deal with time while the outer belt proton fluxes do not. Forbush et al.[1] studying the variations of $\mathcal{E}_e > 1.1$ MeV electrons observed on Explorer 7 found changes of one hundred on occasion. Frank, Van Allen, and Hills[2] have made the most detailed study of electron time variations using data from Explorer 14 for $\mathcal{E}_e > 40$ keV, $\mathcal{E}_e > 230$ keV, and $\mathcal{E}_e > 1.6$ MeV. Fig. 37 shows their results. The protons of $\mathcal{E}_p > 500$ keV don't change appreciably but the electrons wander all over, especially at large L values.

As more measurements have been made on the outer belt electrons, several different kinds of time variations have been sorted out[3]. These include (but are not necessarily limited to) local time variations, 27-day variations, solar cycle variations, variations related to magnetic disturbances including magnetic storm changes, and quiet time decay. Let us consider these separately.

30. Local time variations.

α) Inside the shell $L = 5$ there are no measured local time variations in the trapped particle fluxes. Beyond this, there have been several observations of variations of the electron flux. The first measurement of a local time variation in trapped electron flux was by O'Brien[4] using $\mathcal{E}_e > 40$ keV data from Injun 1. More complete data from Injun 3[5,6] is shown in Fig. 38. It is obvious that there is much more trapped flux at local noon than at local night

O'Brien suggested that this might be due to

(**a**) motion of trapped particles in the distorted geomagnetic field which could produce asymmetries.

(**b**) The possible presence of a steady-state source of electrons accelerating them within a few degrees of $\boldsymbol{B}_{\delta}$ in the equatorial plane on the sunlit side of the Earth.

(**c**) Auroral type acceleration mechanisms which might occur on only closed field lines.

O'Brien preferred (**b**) on the basis of the existance of daytime precipitation of electrons but considered the question unsolved.

β) *Other data* due to Williams and Palmer[7] from satellite 1963-38c show that the $\mathcal{E}_e > 0.28$ MeV electrons have local time variations also. But these higher energy electrons have a significantly smaller diurnal latitude shift during periods of magnetic quiet than the $\mathcal{E}_e > 40$ keV electrons do. Williams and Mead[8] have considered the motion of particles in a distorted magnetic field including a geomagnetic tail and find that if the adiabatic invariants are conserved the diurnal flux variation for $\mathcal{E}_e > 0.28$ MeV can be explained just by the geomagnetic field distortion for reasonable assumptions about the field.

Frank[9] has analyzed data from Explorer 14 for $\mathcal{E}_e > 1.6$ MeV electrons and found that the local time variations of these particles observed near the equatorial plane for $L > 6$ are

[1] S. W. Forbush, G. Pizzella, and D. Venkatesan: J. Geophys. Res. **67**, 3651 (1962).
[2] L. A. Frank, J. A. van Allen, and H. K. Hills: J. Geophys. Res. **69**, 2171 (1964).
[3] D. J. Williams: [*16*], 263 (1966).
[4] B. J. O'Brien: J. Geophys. Res. **68**, 989 (1963).
[5] L. A. Frank, J. A. van Allen, and J. D. Craven: J. Geophys. Res. **69**, 3155 (1964).
[6] T. Armstrong: J. Geophys. Res. **70**, 2077 (1965).
[7] D. J. Williams, and W. F. Palmer: J. Geophys. Res. **70**, 557 (1965).
[8] D. J. Williams, and G. D. Mead: J. Geophys. Res. **70**, 3017 (1965).
[9] L. A. Frank: J. Geophys. Res. **70**, 4131 (1965).

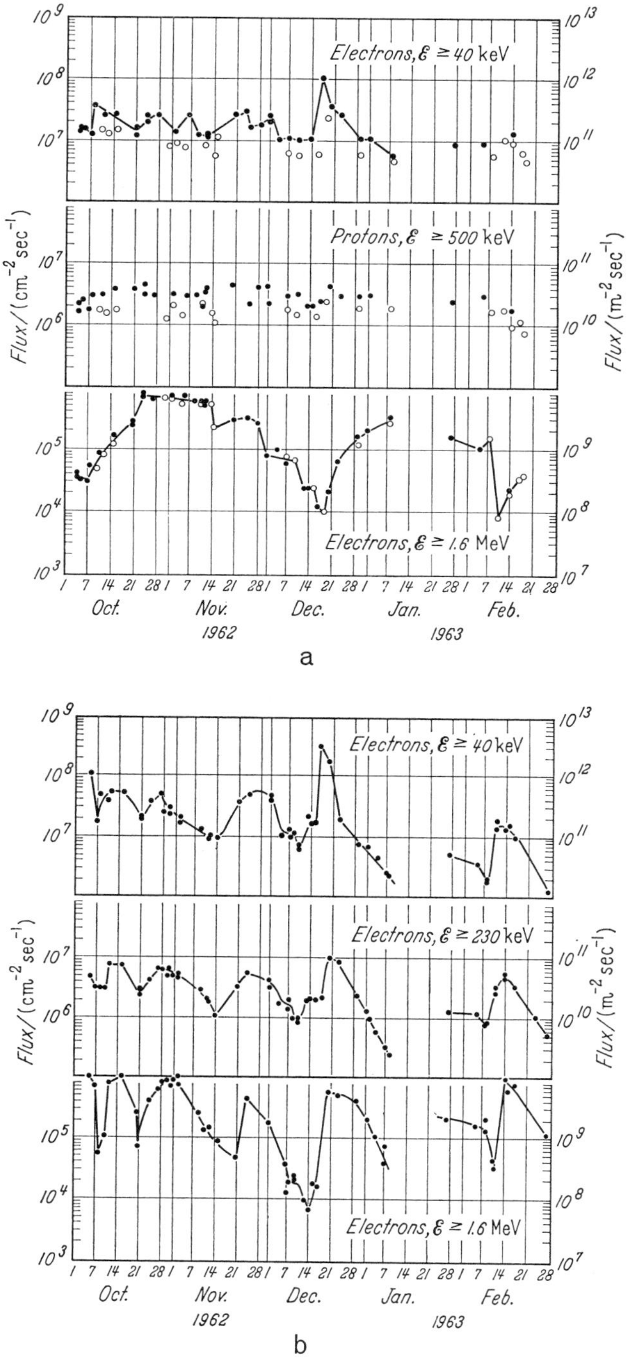

Fig. 37 a and b. Measured temporal variations of the omnidirectional intensity of electrons and protons near the geomagnetic equatorial plane at (a) $L=3.6$; $B/B_0=1.0\ldots1.9$ ●, $=1.9\ldots3.0$ ○; (b) $L=4.8$; $B/B_0=1.0\ldots1.9$ ●. Measured on Explorer 14 in 1962—3. (FRANK et al., ref. [2], Sect. 30.)

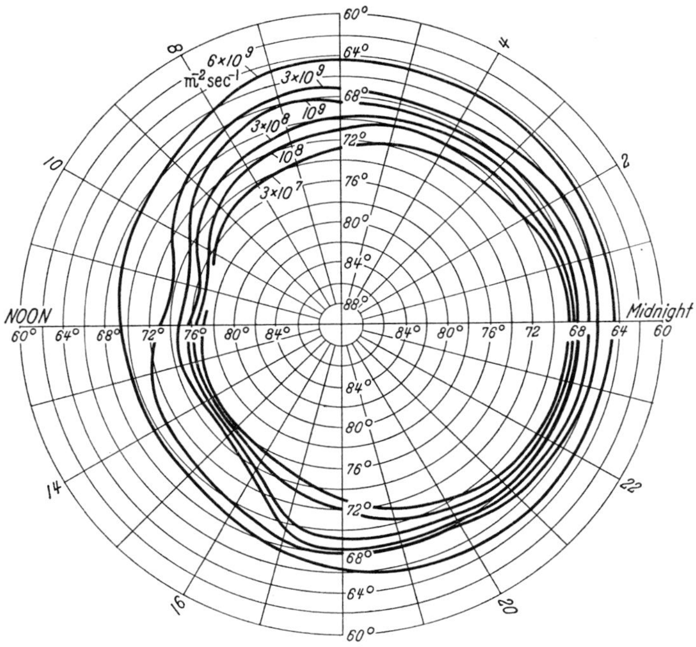

Fig. 38. The median omnidirectional intensities of trapped $\mathscr{E}_e > 40$ keV electrons measured at low altitudes with satellite Injun 3 displayed in the invariant latitude Л versus local time coordinate system. Parameter of the curves is particle flux/m⁻² sec⁻¹. (FRANK et al., ref. 5, Sect. 30.)

consistent with the expected distortion of the geomagnetic field in these regions due to the solar wind.

γ) It looks as if most of the local time variations are due to the *distortion of the geomagnetic field by the solar wind* and the presence of the geomagnetic tail (see Chap. E) but probably not all. The fact that the $\mathscr{E}_e > 40$ keV electrons show a larger effect than the higher energy electrons coupled with the daytime enhanced precipitation of these low energy electrons argues for strong electric fields, or a daytime source of $\mathscr{E}_e > 40$ keV electrons as well as field distortion.

31. Variations depending on solar activity.

α) WILLIAMS[1] has shown the presence of *27-day periodicities* in electron fluxes in the outer belt for electrons of $\mathscr{E}_e > 0.28$ MeV and $\mathscr{E}_e > 1.2$ MeV. These measurements were made on the satellite 1963—38c in a 1100 km altitude polar orbit and are shown in Figs. 39. These periodic effects were seen for $L > 3.5$ for four solar rotations in late 1963. The effect is clearly related to the sun. The electron flux increases when a boundary between two of the *sectors* found by NESS and WILCOX [20] passes the Earth. The change in electron flux is large when the magnetic field polarity changes from 'minus' (towards the Sun) to 'plus' (away from the Sun). This occurs twice per solar rotation. For the other two sector boundaries with the opposite polarity change the flux change is small. If the coupling of solar wind energy into the magnetosphere was via changes in the interplanetary magnetic field[2] then one would expect to see similar effects at all

[1] D. J. WILLIAMS: J. Geophys. Res. **71**, 1815 (1966).
[2] A. J. DESSLER, and G. K. WALTERS: Planetary Space Sci. **12**, 227 (1964).

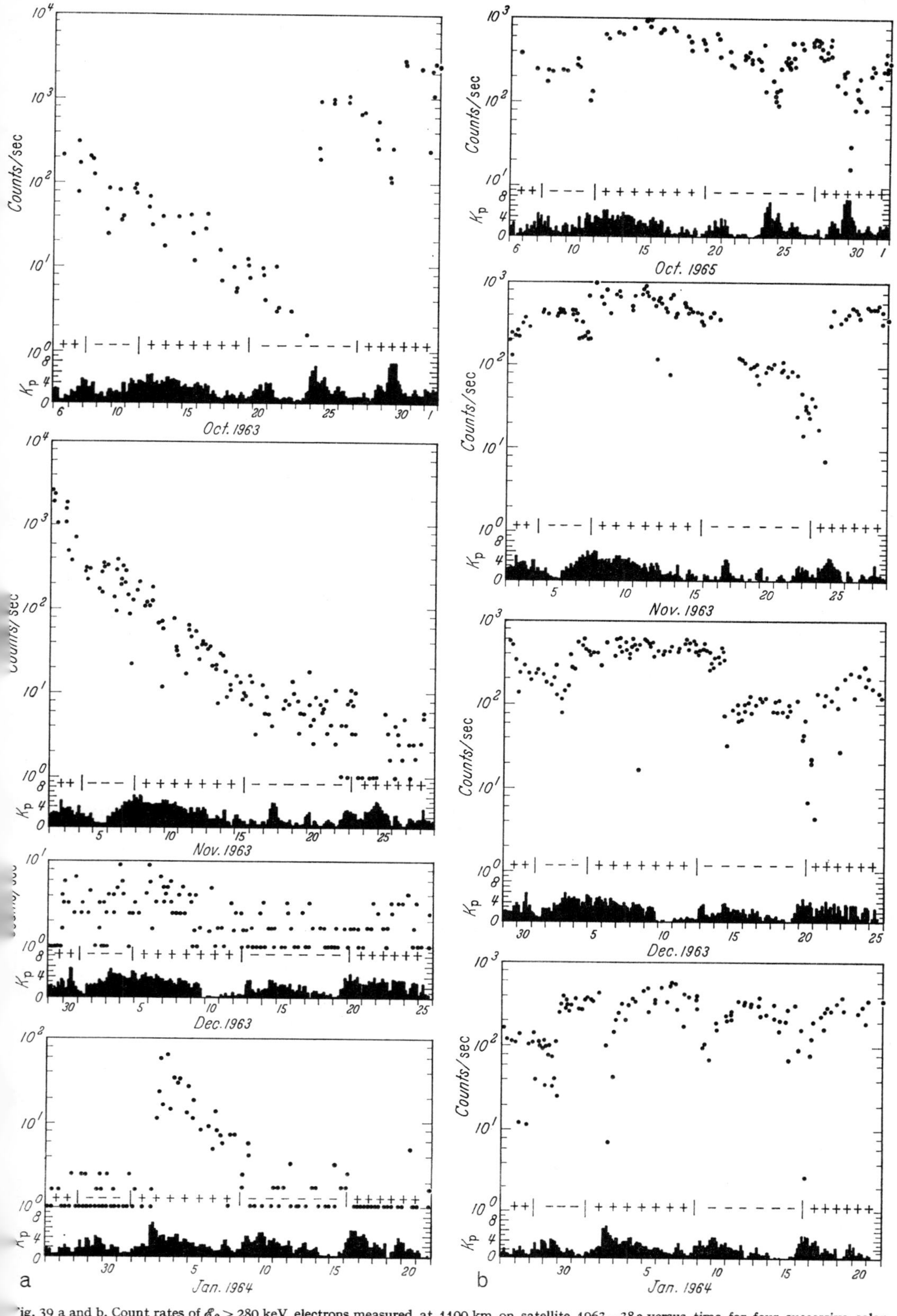

Fig. 39 a and b. Count rates of $\mathscr{E}_e > 280$ keV electrons measured at 1100 km on satellite 1963–38c versus time for four successive solar 27 d-rotations. Shown for comparison are the planetary magnetic three-hour index K_p (black pattern at bottom of each diagram), and the sector structure of the interplanetary magnetic field as measured by NESS ++ and –– indications. (a) for $L=3$; (b) for $L=5$. (WILLIAMS and SMITH, ref. [2], Sect. 32.)

sector boundaries which is not observed. Williams[3] has suggested that the coupling mechanism may be related to the Alfven-Mach-number

$$M_A = \frac{v_w}{V_A} \tag{31.1}$$

with V_A = Alven wave velocity, see Eq. (39.1), v_w = solar wind velocity.

β) *Solar cycle changes* have also been observed. Frank and Van Allen[4] have examined the structure of the outer radiation zone using $\mathscr{E}_e \sim 1$ MeV electron data from Explorer 4, 7, 12, 14, Injun 1, 3, and 4 and, OGO 1 to see if solar cycle changes can be identified. They find that the *inner edge* of the outer radiation zone —the *slot* — has *moved* systematically *outward* from 1958 to 1965.

At solar maximum in 1958 the slot was at $L \approx 2.2$ and it has moved monotonically out to $L \approx 3.0$. A similar movement of the position of maximum flux $= L_{\text{Max}}$ has been noted. In 1960 the median position[5] had $L_{\text{Max}} \approx 3.5$. In 1963 Injun 3 data[6] gave $L_{\text{Max}} \approx 4.0$ and in late 1964 OGO 1 data placed it at $L_{\text{Max}} \approx 4.5$.

Frank and Van Allen[4] suggested that this outward movement of the outer radiation zone during declining solar activity represents a change in the radial diffusion process and that the rate of radial diffusion extends deeper into the magnetosphere at solar maximum.

The outer zone fluxes do not seem to have a substantial solar cycle variation.

The flux in the slot has not changed much and Frank and Van Allen[4] say "It may be noted that the typical magnitude of the peak intensity of energetic electrons ($\mathscr{E}_e > 1$ MeV) in the outer zone near the geomagnetic equator has not displayed a marked change during the decreasing portion of the solar activity cycle perhaps indicating a 'saturated' magnetosphere with regard to the population of ~ 1 MeV electrons".

32. Time variations under quiet and disturbed conditions.

α) Because the outer belt electrons show such large temporal changes it is difficult to study their quiet *time decay*. Brown and Gabbe[1] measured decay curves for the Starfish electrons (see Fig. 18) that give their lifetime τ for $L < 3.0$ but above this L we have to rely on natural electrons.

Williams and Smith[2] studying the time dependance of the electron flux as shown in Fig. 39 decided that during magnetically quiet times such as the period 1—15 October the electron flux was just steadily decaying. This allowed them to derive lifetimes for the trapped particles as shown in Table 9.

Table 9. *Quiet lifetimes of trapped electrons.*

	$L=2$	$L=4$	$L=6$	$L=8$
$\mathscr{E}_e > 280$ keV	4 d	6 d	3 d	0.5 d
$\mathscr{E}_e > 1.6$ MeV	7 d	6 d	2 d	0.5 d

These authors feel that the region $L > 3$ would be soon emptied of trapped electrons if it were not for the continual magnetic agitation present in the magnetosphere. They feel this *continual replenishment* of particles is but one manifestation of the turbulence present at the solar wind-magnetosphere interface.

[3] D. J. Williams, and G. D. Mead: J. Geophys. Res. **70**, 3017 (1965).
[4] L. A. Frank, and J. A. van Allen: J. Geophys. Res. **71**, 2697 (1966).
[5] S. E. Forbush, G. Pizzella, and D. Venkatesan: J. Geophys. Res. **67**, 3651 (1962).
[6] T. Armstrong: J. Geophys. Res. **70**, 2077 (1965).
[1] W. L. Brown, and J. D. Gabbe: J. Geophys. Res. **68**, 607 (1963).
[2] D. J. Williams, and A. M. Smith: J. Geophys. Res. **70**, 541 (1965).

FRANK et al.[3] also calculated electron lifetimes based on the observed decrease of electron fluxes after the magnetic storm of Dec. 20, 1962. The period after the storm was quite quiet. They found lifetimes given in Table 10.

Table 10. *Lifetimes of trapped electrons after a magnetic storm.*

	$L=3.0$	$L=4.2$	$L=4.8$
$\mathcal{E}_e > 40$ keV	—	7 d	6 d
$\mathcal{E}_e > 0.28$ MeV	—	10 d	5 d
$\mathcal{E}_e > 1.6$ MeV	10 d	15 d	5 d

β) *Variation with magnetic disturbance* has been studied by FRANK et al.[3] using data from Explorer 14. They showed a direct correlation of outer zone electron flux and planetary three-hour-index K_p as shown in Fig. 39. They say:

"Detailed study of the temporal variations of different parts of the electron spectrum at various L values < 5.0 within the outer zone shows a certain orderly sequence of phenomena: (a) During the *early phase* of a magnetic storm the intensity of electrons $\mathcal{E}_e > 40$ keV *increases promptly* (within a time of less than one day after the onset of the storm), the effect is larger and extends to lower L values for larger values of K_p, (b) After a *time delay* of the order of several days, the intensity of more energetic *electrons* $\mathcal{E}_e > 1.6$ MeV also increases, and by a similar factor, the increase occurring first at the higher L values. (c) The *inner edge* of the radial profile of intensity of electrons $\mathcal{E}_e > 1.6$ MeV then *drifts inward* to $L \approx 3.0$ at a rate of $\sim 0.02\ R_E/\mathrm{d}$*. (d) The intensity of electrons of $\mathcal{E}_e > 40$ keV diminishes from its peak value to its prestorm value within about a week, but the intensity of the more energetic electrons $\mathcal{E}_e > 1.6$ MeV returns to its prestorm value considerably more slowly.

(e) To within the finite time resolution (~ 1 day) of the observations, it appears that the time history of the intensity of electrons $\mathcal{E}_e > 230$ keV is intermediate between those of the lower- and higher-energy components, though the occurrence of successive storms during the period under study was such as to result in a lesser relative variation of intensity of the intermediate energy component".

These authors say the protons of $\mathcal{E}_p > 500$ keV in the range $2.8 < L < 3.6$ didn't vary in intensity by as much as a factor of two (the experimental error) during the period of observation. It appears that the *protons* are markedly *less sensitive* to geomagnetic disturbances than are electrons.

WILLIAMS and SMITH[2], using data on electrons of $\mathcal{E}_e > 0.28$ MeV and $\mathcal{E}_e > 1.2$ MeV taken on satellite 1963—38c, showed a distinct correlation of outer zone electron flux at 1100 km with the planetary magnetic three-hour index K_p.

For $L=5$ in Fig. 39 (center) the count rates to up and down in clear relation to K_p, but at $L=2.8$ at left in Fig. 39 the period of moderate magnetic activity Oct. 10...17 produced no enhancement of the electron fluxes. It took the $K_p > 6$ period of Oct. 24 to produce a response at $L=2.8$ and then it was only the $\mathcal{E}_e > 0.28$ MeV electron flux that changed. Experimenters on Explorer 6 and Explorer 12 found qualitatively similar results to these[4–7].

There is a great variability in storm types and characteristics, and while there are certain characteristics of the trapped radiation that occur for many storms, there is by no means a unique picture of the changes that occur to the trapped particles. The initial phase of a magnetic storm is normally thought to be due to

* i.e. 1.5 m sec^{-1}, see Eq. (26.1).

[3] L. A. FRANK, J. A. VAN ALLEN, and H. K. HILLS: J. Geophys. Res. **69**, 2171 (1964).

[4] P. L. ARNOLDY, R. A. HOFFMAN, and J. R. WINCKLER: J. Geophys. Res. **65**, 1361 (1960).

[5] C. Y. FAN, P. MEYER, and J. A. SIMPSON: J. Geophys. Res. **66**, 2607 (1961).

[6] A. ROSEN, and T. A. FARLEY: J. Geophys. Res. **66**, 2013 (1962).

[7] J. W. FREEMAN jr.: J. Geophys. Res. **69**, 1691 (1964).

the impact of a cloud of enhanced-strength solar wind on the magnetosphere. The processes governing the later phases of a magnetic storm are not so well understood (see contribution by E. N. PARKER and V. C. A. FERRARO in volume 49/3, in particular Chap. C).

33. Low energy particles. There are trapped particles of considerably lower energy than are usually considered in discussions of the radiation belt.

α) For *very low energy particles* the gravitational forces F_g will be larger than the magnetic force F_m and the particle will fall to earth. We might consider a *criterion of trapping* to be given by

$$F_m > F_g$$

assuming a low altitude particle is near its mirror point

$$F_m = \mu \cdot \frac{\partial \boldsymbol{B}}{\partial l} \approx \frac{\mathscr{E}}{B}\frac{\partial B}{\partial R} \approx \frac{3\mathscr{E}}{R_E} > mg. \tag{33.1}$$

Therefore, trapped particles must have energies of

$$\mathscr{E} > \tfrac{1}{3} mg R_E \approx \begin{Bmatrix} 0.2 \text{ eV for protons} \\ 10^{-4} \text{ eV for electrons.} \end{Bmatrix} \tag{33.2}$$

This shows very low energy particles can actually be trapped but when we put in collisional processes they may not stay trapped very long. Low energy particles have not been well mapped yet but there are indirect evidences of their presence in the magnetosphere such as the time history of magnetic storms.

β) Quite frequently *magnetic disturbances* are observed on the surface of the Earth. They have several common features shown in Fig. 40.

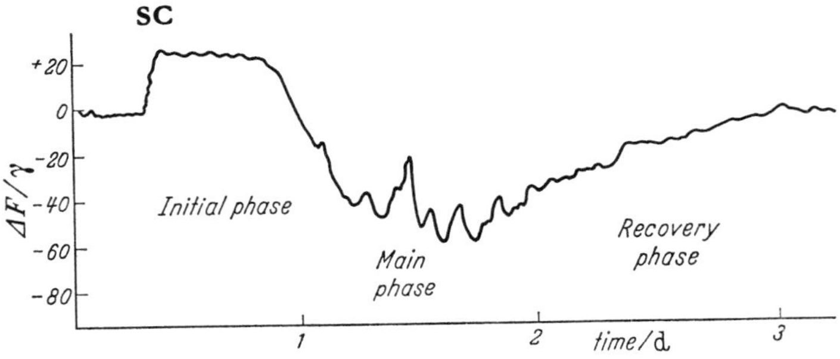

Fig. 40. Coarse variation of magnetic field change (ordinate) during a typical magnetic storm at medium latitude. (See for more details contribution by T. NAGATA and N. FUKUSHIMA in this Vol., Chap. F.)

This might be how an average magnetic storm at the equator would look, measuring the horizontal component of the $\boldsymbol{B}_{\delta}$ field [*21*]. Individual storms might be lacking some of the usual features. (See for details contributions by T. NAGATA and N. FUKUSHIMA, and by E. N. PARKER and V. C. A. FERRARO in volume 49/3.)

(1) Sudden Commencement (**SC**).
A fairly sharp increase in the magnetic field of from 10 to 30 γ in about a minute.

(2) Initial Phase.
A period of several hours when $\Delta B > 0$.

(3) Main Phase.

A period of up to a day or so when the field is *decreased* by 50—100 γ and is usually disturbed. During this time the field curvature is upwards.

(4) Recovery Phase.

A period of up to days when $\Delta B < 0$ and the field is decaying exponentially to roughly its original value. Curvature is downwards.

γ) First consider the *initial phase* of a storm[1—3]. An increase of the solar wind pressure will increase $|\boldsymbol{B}_{\delta}|$. There is a rather well established correlation between the occurrance of a severe magnetic storm and the occurrance about two days earlier of a solar flare. The flare produces a stream of higher velocity plasma than usual, which in about two days hits the magnetopause. In fact, the plasma probe on Mariner observed an increase in wind strength at the time of a magnetic storm on Earth.

If the solar wind pressure change is sudden (and it should be, because the magnetic field in space should keep a steep front on the solar wind pulse moving from the sun) then the magnetopause will be perturbed in a super-alfvenic way and we may expect a magneto-dynamic wave to propagate inwards to the surface of Earth with a velocity*

$$v \sim V_{\mathrm{A}} = \sqrt{\frac{B^2}{\mathrm{u}\mu_0 \varrho}}\,. \tag{33.3}$$

The wave will take a minute or so to reach Earth. This could produce the **SC**.

δ) What can produce the *main phase* of the magnetic storm? It has been suggested by Piddington [22] that it might be due to the stretching of field lines outwards on the dark side of the Earth by the solar wind but this does not seem to work quantitatively. The only good explanations for the main phase now involve putting trapped particles into the field. These particles behave diamagnetically and tend to reduce the field (see Sect. 12 of Parker and Ferraro's contribution to volume 49/3). This idea is referred to as a ring current and was suggested in 1932 by Chapmann and Ferraro[4]. Singer [23] in 1957 suggested that the ring current might result from the longitudinal drift of particles trapped in Earth's magnetic field.

Explorer 12 observations[5] at middle latitudes during a magnetic storm in 1961 indicated that the *ring current* must be below 4 R_{E}. Soviet high-latitude measurements[6] on Electron 2 showing consistent depression of field magnitude at distances at less than 6 R_{E}. A search for the charged particles that consitute a ring current has also been rather unsuccessful. The energy density of the trapped particle belts, so far observed seems to be inadequate to produce the magnetic effects seen at ground level.

Cahill[7], using a magnetometer on Explorer 26, followed the time history of a large magnetic storm in April 1965. All evidence indicated that the main phase of this storm resulted from a body of charged particles introduced or locally accelerated in an *evening* sector of the magnetosphere. Ground evidence of the symmetry of the main phase suggests that the particles after being initially introduced between 1800 and 2400 local time drifted west in longitude. Low energy *protons* could therefore be principally responsible for the main phase. The mean energy of the protons could be estimated from the drift speed. The particles took

---

* $\mathrm{u} \equiv 1$ in rationalized, $\equiv 4\pi$ in non-rationalized systems of units. See "Introductory Remarks" on p. 1. The international system of units (SI) is rationalized, the Gaussian system is not.

[1] M. Sugiura, and J. P. Heppner: [*21*], 5 (1965).

[2] A. J. Dessler, and E. N. Parker: J. Geophys. Res. **64**, 2239 (1959); see also E. N. Parker, and V. C. A. Ferraro: This volume, p. 164.

[3] A. J. Dessler, W. B. Hanson, and E. N. Parker: J. Geophys. Res. **66**, 3631 (1961); see also E. N. Parker, and V. C. A. Ferraro: This volume, p. 165.

[4] S. Chapman, and V. C. A. Ferraro: Terr. Magn. Atmosph. Electr. **37**, 147, 421 (1932).

[5] L. J. Cahill, and D. H. Bailey: Distortions of the Geomagnetic Field within the Inner-Magnetosphere (abstract). Trans. Am. Geophys. Union **46**, 5055 (1962). L. J. Cahill: J. Geophys. Res. **71**, 4505 (1966).

[6] Š. Š. Dolginov, Je. G. Jeroženko, and L. N. Žuzgov: [*13*], 790 (1966).

[7] L. J. Cahill jr.: J. Geophys. Res. **71**, 4505 (1966).

about four hours to drift 180° west. This corresponds to protons of energy 15...20 keV. At $L \approx 4$ these protons should be lost by charge exchange in a few days.

ε) One of the most interesting radiation belt discoveries is the finding of a *large flux of low energy protons* at 1000 km. FREEMAN[8], using a CdS detector on Injun, measured a heavy ion energy flux of about 50 erg cm^{-2} sec^{-1} sr^{-1} ($\equiv 0.05$ J m^{-2} sec^{-1} sr^{-1}) at 1000 km in the energy range of 0.5 keV to 1 MeV. These particles are most likely protons. If an average proton has an energy of 100 keV there is a proton flux of

$$J_{\mathrm{p}} \approx \frac{(50\ \mathrm{erg\ cm^{-2}\ sec^{-1}\ sr^{-1}})\ (\sim 1\ \mathrm{sr})}{(0.1\ \mathrm{MeV})\ (1.6\cdot 10^{6}\ \mathrm{erg/Mev})} \approx 3\cdot 10^{8}\ \mathrm{cm^{-2}\ sec^{-1}} \equiv 3\cdot 10^{12}\ \mathrm{m^{-2}\ sec^{-1}}. \qquad (33.4)$$

This is a quite large flux and may play a part in magnetic storms. The spatial distribution of these low energy protons found by FREEMAN seems very similar to the distribution of high energy inner belt protons.

ζ) In order to try to understand the main phase of magnetic storms, AKASOFU[9] suggested that the solar wind might not be completely ionized especially at the times of solar flares. The *neutral hydrogen atoms* in the solar wind could enter the magnetosphere freely and charge exchange with the thermal protons inside to make the ring current. AKASOFU[9] has computed that a flux of $\sim 10^{9}$ cm^{-2} sec^{-1} $\equiv 10^{13}$ m^{-2} sec^{-1} of 4 keV atoms is necessary to make a main phase decrease of $\Delta B = 40\ \gamma$. This means the solar wind must be roughly 50 percent ionized.

This seems like a very attractive idea to get large numbers of low energy particles deep inside the magnetosphere to make a main phase storm but it doesn't seem to work well. It seems impossible to get the fraction of neutral hydrogen up to even one percent for peculiar situations[10].

BRANDT and HUNTEN[11] state that if the neutral cloud did get started at the sun it would be ionized by solar radiation and swept-up coronal electrons so that at the Earth it would be < 0.1 percent neutral.

HUNTEN[11] has pointed out that there should be 7 kR of H_α optical emissions in twilight if the postulated neutral hydrogen flow exists*. Only occasionally in the auroral zone does the H_α flux get up to 2 kR and at low latitudes it doesn't get nearly this high. Further, there should be ~ 100 kR of 3914 Å emission from the neutral hydrogen. The observed values are closer to 100 R and show little or no relation to magnetic activity.

Using the neutral hydrogen model a magnetic storm main phase would not occur at the right time or right place. The **SC** occurs when the enhanced ionized solar wind strikes the magnetosphere. The neutral hydrogen should arrive at the same time as the ionized hydrogen if it has the same history. Therefore the main phase should start at the same time as the **SC** but it does not. The main phase should be asymmetric on this model. There should be a larger ΔB on the day side of Earth than on the night side. The observed ΔB on the night side is frequently larger than ΔB day side — just the opposite of what would be expected from neutral hydrogen.

F. The outer edge.

34. The magnetopause.

α) It is now well known that the geomagnetic field does not extend to infinity as it would if Earth existed in a vacuum. The terrestrial *field* is *constrained* to exist inside a cavity called the magnetosphere by the continuous streaming of particles outward from the sun[1,2].

---

* 1 R = 1 Rayleigh designates a photon flux of 10^{6} cm^{-2} sec^{-1} $= 10^{10}$ m^{-2} sec^{-1}. A detailed explanation can be found in Sect. 1γ of the contribution by A. and E. VASSY in Vol. 49/4 of this Encyclopedia.

[8] J. W. FREEMAN: J. Geophys. Res. **67**, 921 (1962).

[9] S.-I. AKASOFU: Planetary Space Sci. **12**, 905 (1964).

[10] P. A. CLOUTIER: A Comment on "The Neutral Hydrogen Flux in the Solar Plasma Flow", by S.-I. AKASOFU. Planetary Space Sci. **14**, 809 (1966).

[11] J. C. BRANDT, and D. M. HUNTEN: Planetary Space Sci. **14**, 95 (1966).

[1] S. CHAPMAN, and V. C. A. FERRARO: Terr. Magn. Atm. osph. Electr. **36**, 77, 171 (1931).

[2] F. S. JOHNSON: J. Geophys. Res. **65**, 3049 (1960).

The location of the boundary (called the magnetopause) between the solar wind and the geomagnetic field can be determined by the hydromagnetic pressure balance

$$p + \frac{B^2}{2\mathrm{u}\mu_0} = \text{const.} \tag{34.1}$$

The solar wind is a good conductor. It pushes Earth's field ahead of it until the pressure p of the solar wind is balanced by the magnetic field compression. Assuming the solar wind is a field-free plasma that strikes the boundary and is specularly reflected we can write for the transfer of momentum per particle per collision

$$2mv\cos\psi$$

where ψ is the angle of incidence.

The number of particles striking unit area of the surface per second is $Nv\cos\psi$. This gives the total pressure of the particles perpendicular to the surface

$$p_0 = 2Nmv^2\cos^2\psi. \tag{34.2}$$

The normal component of $\boldsymbol{B}$ is continuous across the surface and we are assuming the value outside the surface is $\boldsymbol{B}_0 = 0$ and with $p_{\text{int}} = 0$ Eq. (34.2) reduces to

$$p_0 = \frac{B_{\text{int}}^2}{2\mathrm{u}\mu_0} \tag{34.3}$$

which gives for the tangential magnetic field just inside the boundary

$$B_{\text{int}} = \sqrt{4\mathrm{u}\mu_0 Nmv^2}\cos\psi \equiv \sqrt{4\mathrm{u}\mu_0 Nm}\, v_\perp. \tag{34.4}$$

Taking $N = 5\ \text{cm}^{-3}$ and $\frac{mv^2}{2} = 10^{-9}\ \text{erg} \equiv 10^{-16}\ \text{J}$ (corresponding to a 600 eV proton) we find that at the subsolar point

$$p_0 = 10^{-8}\ \text{dyn cm}^{-2} \equiv 10^{-9}\ \text{Nm}^{-2}, \quad \text{and} \quad B_{\text{int}} = 50\gamma \equiv 5\cdot 10^{-8}\,\text{T}.$$

Considering that the undistorted dipole field is half of this value, we can find the location of the boundary by comparison with a hypothetic dipole field, $\boldsymbol{B}_{\text{dp}}$

$$\frac{1}{2}B_{\text{int}} \equiv \frac{1}{2}\,50\,\gamma = B_{\text{dp}}(R) = B_{\text{♁}0}\left(\frac{R_{\text{E}}}{R}\right)^3 = \frac{0.32\cdot 10^5\,\gamma}{(R/R_{\text{E}})^3},$$

which leads to

$$R_{\text{b}} = 11\,R_{\text{E}}$$

where $B_{\text{♁}0}$ is the magnetic field at Earth's surface at the equator and R_{E} is the radius of Earth. This is roughly where the magnetopause is found on a quiet day. The value of R_{b} is not very sensitive to the parameters involved.

$$\mathcal{R}_{\text{b}}^6 \equiv (R_{\text{b}}/R_{\text{E}})^6 = \left(\frac{2B_{\text{♁}0}}{B_{\text{int}}}\right)^2 = \frac{B_{\text{♁}0}^2}{\mathrm{u}\mu_0 Nmv^2}. \tag{34.5}$$

To move the boundary from 11 R_{E} to 8 R_{E} where it is sometimes found requires increasing the solar wind energy density $\frac{1}{2}Nmv^2$ by a factor of 7.

β) Protons and electrons in the solar wind will penetrate into the boundary layer approximately one cyclotron radius, be bent in opposite directions by the magnetic field and create a *surface current*. It is this surface current which produces the magnetic field change across the boundary. If the magnetopause were a plane these surface currents would just double the magnetic field inside the boundary and make it zero outside.

Several people[3–6] have solved the problem of the magnetosphere shape using the specular reflection model, a geomagnetic dipole field and a zero temperature solar wind. The tangential magnetic field $\boldsymbol{B}_{\text{int}}$ at a point just inside the magnetopause is made up of the geomagnetic dipole field at that point, $\boldsymbol{B}_{\text{g}}$, and the field due to the surface currents. MEAD and BEARD[7] broke this up into two parts. They first took a contribution that would result from a current on a *plane* surface tangent to the magnetopause at the point of observation $\boldsymbol{B}_{\text{p}}$ and second a relatively small field due to the curvature of the surface, $\boldsymbol{B}_{\text{c}}$. The total field inside the boundary is

$$\boldsymbol{B}_{\text{int}} = \boldsymbol{B}_{\text{g}} + \boldsymbol{B}^{\text{p}} + \boldsymbol{B}_{\text{c}}. \tag{34.6}$$

Outside the boundary the field is

$$\boldsymbol{B}_0 = \boldsymbol{B}_{\text{g}} - \boldsymbol{B}_{\text{p}} + \boldsymbol{B}_{\text{c}} \equiv 0. \tag{34.7}$$

Therefore

$$\boldsymbol{B}_{\text{int}} = 2(\boldsymbol{B}_{\text{g}} + \boldsymbol{B}_{\text{c}}). \tag{34.8}$$

A problem exists at this point in solving for the field because the curvature field $\boldsymbol{B}_{\text{c}}$ cannot be found until the shape of the boundary is known but the surface shape depends on the total field $\boldsymbol{B}_{\text{int}}$ which contains $\boldsymbol{B}_{\text{c}}$. A reasonable first approximation can be obtained by taking $\boldsymbol{B}_{\text{c}} = 0$ and solving for the surface shape. Then using this shape values for $\boldsymbol{B}_{\text{c}}$ can be calculated and by iteration a self-consistent surface can be found.

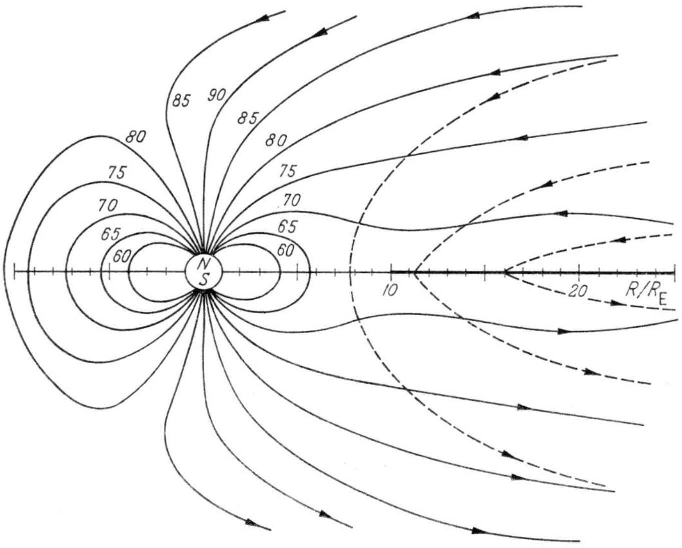

Fig. 41. The WILLIAMS-MEAD model of the geomagnetic field using a dipole field $\boldsymbol{B}_{\text{d}}$ and surface currents producing a magnetopause $\boldsymbol{B}_{\text{s}}$ and a current sheet in the tail producing a field $\boldsymbol{B}_{\text{cs}}$. Solid curves = total field ($\boldsymbol{B}_{\text{d}} + \boldsymbol{B}_{\text{s}} + \boldsymbol{B}_{\text{cs}}$); broken curves = $\boldsymbol{B}_{\text{cs}}$ only (ref. [8], Sect. 34).

γ) The *surface* found this way[8] is shown in Fig. 41. The field model shown in Fig. 41 also has a tail current added consistent with the geomagnetic tail discussed later.

[3] D. B. BEARD: J. Geophys. Res. **65**, 3559 (1960).
[4] J. R. SPREITER, and B. R. BRIGGS: J. Geophys. Res. **67**, 2983 (1962).
[5] J. R. MIDGLEY, and L. DAVIS: J. Geophys. Res. **68**, 5111 (1963).
[6] D. B. BEARD: Rev. Geophys. **2**, 335 (1964).
[7] G. D. MEAD, and D. B. BEARD: J. Geophys. Res. **69**, 1169 (1964).
[8] D. J. WILLIAMS, and G. D. MEAD: J. Geophys. Res. **70**, 3017 (1965).

MEAD[9] obtained the following field description for $\boldsymbol{B}_{\text{int}}$ without including tail current effects

$$\left.\begin{aligned} B_\phi/\Gamma &\equiv \mathcal{X}/\Gamma = \frac{0{,}31}{\mathcal{R}^3}\sin\theta + \frac{0.25}{\mathcal{R}_{\mathrm{b}}^3}\sin\theta + \frac{0.21\,\mathcal{R}}{\mathcal{R}_{\mathrm{b}}^4}(2\cos^2\theta - 1)\cos\alpha, \\ B_A/\Gamma &\equiv \mathcal{Y}/\Gamma = \frac{0.21\,\mathcal{R}}{\mathcal{R}_{\mathrm{b}}^4}\cos\theta\sin\alpha \\ -B_r/\Gamma &\equiv Z/\Gamma = \frac{0.62}{\mathcal{R}^3}\cos\theta + \frac{0.25}{\mathcal{R}_{\mathrm{b}}^3}\cos\theta + \frac{0.41\,\mathcal{R}}{\mathcal{R}_{\mathrm{b}}^4}\sin\theta\cos\theta\cos\alpha \end{aligned}\right\} \quad (34.9)$$

where $\mathcal{R} \equiv R/R_{\mathrm{E}}$ (i.e. distances reduced to Earth's radius) and $\mathcal{R}_{\mathrm{b}}$ is the reduced distance to the magnetopause at the stagnation point from Eq. (34.5)

$$\mathcal{R}_{\mathrm{b}} = 1.26\left(\frac{(0.32\,\Gamma)^2}{\mathrm{u}\mu_0\,N m v^2}\right)^{\frac{1}{6}}. \quad (34.10)$$

This expansion of the field using Eq. (34.9) is good to about 10 percent of MEAD's calculated field at all points out to R_{b}.

Using this description of the distorted field for $\mathcal{R}_{\mathrm{b}} = 10$ at the subsolar point, gives $\mathcal{X} = 77\,\gamma$ of which $46\,\gamma$ or 60 percent is due to the surface currents. Field lines for $\phi > 83°$ are blown over into the back side of Earth. There are two null points in the field at high latitudes on the front hemisphere of Fig. 41.

δ) Earth's magnetic field is compressed by the solar wind more on the *day side* of the Earth than the *night side*. This means that trapped particles will drift on non-circular orbits.

Field lines having $Л < 60°$ or L4 are not distorted much. At $Л = 65°$ or $L = 5.6$ some compression of field lines is observed and by $Л = 70°$ there is a noticeable difference in compression at noon and midnight. This would be the $L = 8.5$ shell but with this large distortion the definition of L breaks down. At about $L = 6$ the definition of L must be changed because of the distortion.

FRANK, VAN ALLEN, and MACAGNO[10] have organized electron data from Explorer 14 to show constant count rate contours of $\mathcal{E}_{\mathrm{e}} > 40$ keV electrons in the equatorial plane. The $5 \cdot 10^7\,\mathrm{cm^{-2}\,sec^{-1}} \equiv 5 \cdot 10^{11}\,\mathrm{m^{-2}\,sec^{-1}}$ flux contour at about $R = 7\,R_{\mathrm{E}}$ is very nearly circular but the $10^7\,\mathrm{cm^{-2}\,sec^{-1}} \equiv 10^{11}\,\mathrm{m^{-2}\,sec^{-1}}$ flux contour is at about $R = 9.5\,R_{\mathrm{E}}$ on the Sun side and at about $R = 7.3\,R_{\mathrm{E}}$ away from the Sun. This compares quite well what is expected from the MEAD model.

Considering the special class of trapped particles that have the integral invariant $\mathcal{I} = 0$ we can understand how they will drift around Earth: they will stay on curves of constant B. MEAD's model gives the data on the field distortion for $\mathcal{R}_{\mathrm{b}} = 10$ as presented in Table 11.

Table 11. *Distorted field after* MEAD ($r_{\mathrm{b}} = 10$).

Distance towards sun $\mathcal{R}_1 \equiv R_1/R_{\mathrm{E}} =$	B/γ	Distance away from sun to same B $\mathcal{R}_2 \equiv R_2/R_{\mathrm{E}} =$	$\Delta\mathcal{R} \equiv (R_1 - R_2)/R_{\mathrm{E}}$
10	77	7.7	2.3
9	87	7.4	1.6
8	102	7.0	1.0
6	181	5.7	0.3
4	480	3.96	0.04

ε) This model of the magnetopause we have considered so far involving single particle motion and specular motion is clearly too simple. The presence of an

[9] G. D. MEAD: J. Geophys. Res. **69**, 1181 (1964).
[10] L. A. FRANK, J. A. VAN ALLEN, and E. MACAGNO: J. Geophys. Res. **68**, 3543 (1963).

interplanetary magnetic field certainly means the particles do not move independently. The solar plasma must act as a *fluid* for large-scale processes such as *flow around the magnetosphere.*

Lees[11] first solved this problem of fluid flow. He used a hydromagnetic flow model around the magnetopause and used a solar magnetic field parallel to the solar wind, solved for the magnetopause shape and found it not much different from the shape calculated using the single particle specular reflection model.

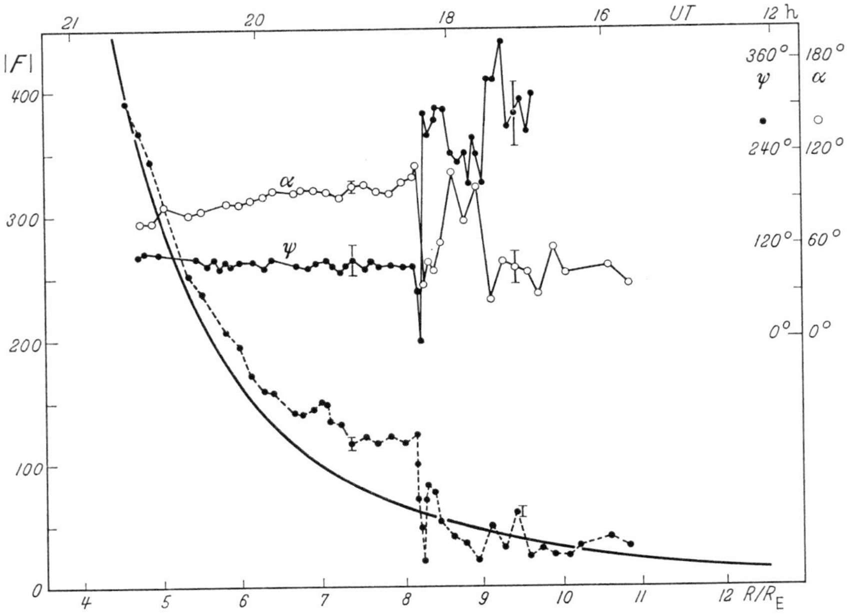

Fig. 42. A magnetometer record of the geomagnetic field measured on the inbound pass of Explorer 12 on Sept. 13, 1961. (Cahill and Amazeen, ref. [12], Sect. 34.)

ζ) The first clear *identification of the magnetopause* was made by Cahill and Amazeen[12] with a magnetometer on the Explorer 12 satellite. One radial pass of this experiment near the subsolar point is shown in Fig. 42. Inside $R=8\ R_E$ the field is generally what is expected for a dipole but at $R=8.2\ R_E$ an abrupt change occurs. Outside this the field is variable in both magnitude and direction. This clearly is the magnetopause. Just inside this boundary the field is roughly double what the dipole field would be as expected from Eq. (34.8).

The *thickness of* the *boundary* layer as measured by the time to change from steady to disturbed field[12,13] and using the radial velocity of the satellite was about 100 km or just about the proton cyclotron radius as expected. This would probably be a lower limit on the real boundary thickness since if the boundary were in motion it would seem thinner. The field just outside the boundary was typically 30 to 40 γ and it was frequently directed nearly southward.

η) Ness's[14] magnetometers on the IMP-1 satellite had the ability to measure fields of less than $\pm 0.5\ \gamma$. The IMP-1 satellite had a very eccentric orbit and went out to $R=32\ R_E$. This magnetometer experiment has provided good data about the interplanetary magnetic

[11] L. Lees: AIAA Journal (Amer. Inst. Aeronautics and Astronautics) **2**, 1576 (1965).
[12] L. J. Cahill, and P. G. Amazeen: J. Geophys. Res. **68**, 1835 (1963).
[13] R. E. Holzer, M. G. McLeod, and E. J. Smith: J. Geophys. Res. **71**, 1481 (1966).
[14] N. F. Ness, C. S. Scearce, and J. B. Seek: J. Geophys. Res. **69**, 3531 (1964).

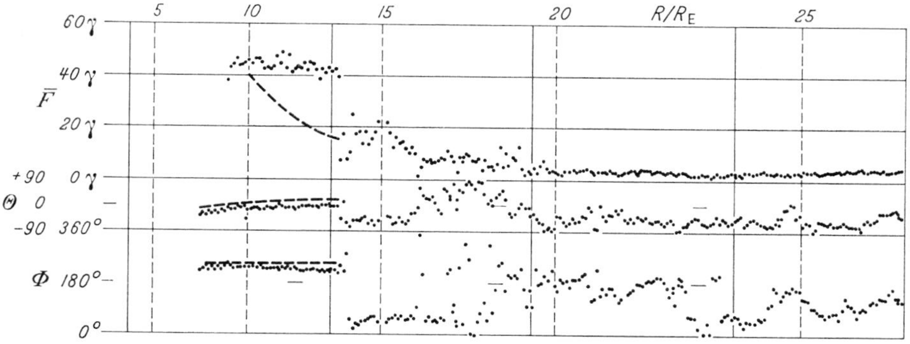

Fig. 43. Geomagnetic field measurements on outbound orbit 11 from satellite IMP-1 on Jan. 5, 1964. Record gives total field $\mathcal{F}$ and angular coordinates. The magnetosphere boundary is observed at a distance of $R = 13.6\ R_E$ and the shock wave at a distance of $R = 19.7\ R_E$ (solid vertical lines). The satellite at this time was approximately on the sunrise terminator position with respect to Earth's sunline. (NESS et al., ref. [14], Sect. 34.)

field, bow shock wave and transition zone magnetic field. Data from one pass of IMP-1 is shown in Fig. 43. Moving outwards from Earth at $R = 13.6\ R_E$, the magnitude and direction of the field abruptly changed and the field was more variable. Inside this boundary, which is the magnetopause, the field direction was appropriate for the dipole field and the magnitude was about double the expected dipole value as we expect near the magnetopause. Proceeding outwards the field appeared irregular and turbulent with a value of about $15\ \gamma$ and then at $R = 19.7\ R_E$ a second change occurred. The field became somewhat lower in magnitude and quite quiet. The values of the variance (root mean square change in B from 12 measurements in an interval of 5.45 min) are usually less than $1\ \gamma$ from the IMP-1 data, and consistent with zero outside $R = 19.7\ R_E$ while inside this transition the variance is several γ. The field outside this second transition was quite constant at about 4γ for 12 hours.

This change at $R = 19.7\ R_E$ is quite clearly a shock wave. Such a *bow shock* ahead of the magnetopause had been predicted by analogy to supersonic aerodynamics and had been discovered experimentally by FREEMAN[15]. The magnetopause locations obtained by IMP are shown in Fig. 44. Shown for comparison

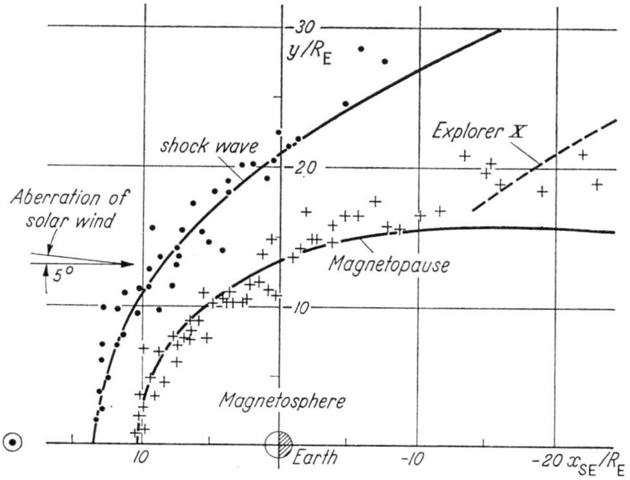

Fig. 44. Location of the shock wave and magnetopause boundary crossings measured on satellite IMP-1. The dotted line at the right side is earlier data on the magnetopause location from satellite Explorer 10. The solid curves are the calculated locations of the shock wave and magnetopause after SPREITER and JONES, with the stand-off ratio adjusted to match the observed measurements. (NESS et al., ref. [14], Sect. 34.)

[15] J. W. FREEMAN: J. Geophys. Res. **69**, 1691 (1964).

is the theoretical boundary shape calculated on the single particle specular reflection model[4]. This theoretical boundary has been rotated by 5° to allow for the aberration of the solar wind due to the motion of Earth in its orbit. The theory gives the shape and location of the front hemisphere of the magnetopause quite well.

35. The magnetosphere — open or closed?

α) DUNGEY [*24*][1] has proposed a quite different model for the magnetosphere to that discussed so far. It is based on having the magnetic field of the sun being perpendicular to the ecliptic and directed southward. Data from satellites Explorer 12[2] and IMP-1[3] do indicate a field component of this type in space sometimes but the solar field in the transition zone outside the magnetopause is confused and turbulent and the existence of a net southward component is questionable. If such a solar field does exist, then a *magnetic neutral point* might exist near the sub-solar point and a second one on the back side of the magnetosphere as shown in Fig. 45. A considerable fraction of the solar field lines that impinge on the magnetosphere may join to the terrestrial field near the front neutral point[4,5] (lines labelled 2″, 3″, 4″ in Fig. 45) and are then swept back over the poles and disconnect from the terrestrial field near the back neutral point (lines labelled 5″, 6″, 7″ in Fig. 45) while other solar field lines pass aside from the terrestrial magnetosphere (lines labelled 1″ and 8″ in Fig. 45). A plasma flow takes place in the region of the neutral points as shown in Fig. 45. Particles can reach the auroral zone and polar cap since solar field lines connect to them and particles can enter the magnetosphere through the back neutral point. Thereby particles from the Sun can reach the Earth but only near the polar regions. The major part of the terrestrial field is hidden and solar particles do not penetrate it. There is no well defined magnetopause in this model.

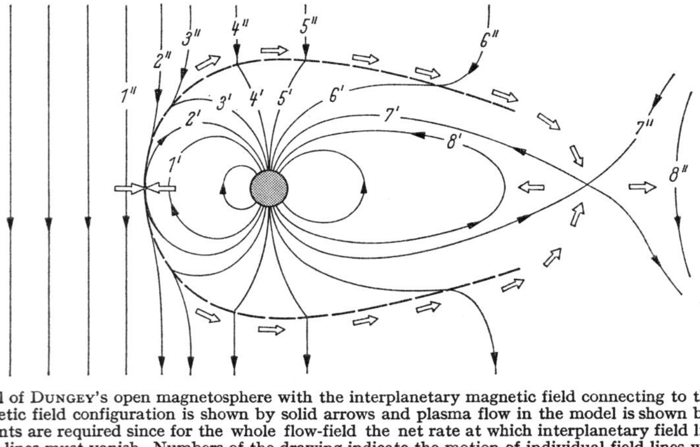

Fig. 45. A model of DUNGEY's open magnetosphere with the interplanetary magnetic field connecting to the geomagnetic field. The magnetic field configuration is shown by solid arrows and plasma flow in the model is shown by open arrows. Two neutral points are required since for the whole flow-field the net rate at which interplanetary field lines become attached to dipole lines must vanish. Numbers of the drawing indicate the motion of individual field lines with the motion progressing towards higher numbers. For simplicity the bow shock is not shown. The shaded fan in the tail represents the *neutral sheet* where the field is very low. This is a region in which plasma flow should occur in this model as indicated. (LEVY et al., ref. [5], Sect. 35.)

[1] J. W. DUNGEY: Phys. Rev. Letters **6**, 47 (1961).

[2] L. J. CAHILL, and P. G. AMAZEEN: J. Geophys. Res. **68**, 1835 (1963).

[3] N. F. NESS, C. S. SCEARCE, and J. B. SEEK: J. Geophys. Res. **69**, 3531 (1964).

[4] H. E. PETSCHEK: Magnetic Field Annihilation. NASA SP-50, Proceedings of AAS-NASA Symposium on the Physics of Solar Flares, ed. by W. N. HESS, p. 425 (1964).

[5] R. H. LEVY, H. E. PETSCHEK, and G. L. SISCOE: Aerodynamic Aspects of the Magnetospheric Flow. Avco-Everett Research Laboratory, Research Report 170 (1963).

β) This model is called *open* because geomagnetic field lines join to solar field lines. In the *closed* model the geomagnetic field is contained in a cavity with no connections across the boundary. In the open model an electric field forms at the neutral point due to the motion of the solar wind into the null point [*24*][1].

The magnitude of the potential difference along the horizontal line through the neutral point on DUNGEY's model can be obtained from considering the flow of plasma across the polar cap. Typical ionospheric wind velocities here are 100 m sec^{-1} so for an observer on Earth there is an induced electric field

$$\frac{1}{c_0\sqrt{\varepsilon_0\mu_0}}\,\boldsymbol{v}\times\boldsymbol{B}$$

to which corresponds a potential difference U_D across the polar cap of

$$U_D \approx \frac{1}{c_0\sqrt{\varepsilon_0\mu_0}}\,l\,v\,B \tag{35.1}$$

where l is a characteristic length across the polar cap. With $l \approx 6000$ km and $v = 100$ msec^{-1} we find $U_D \approx 30$ kV.

If field lines are equipotentials as seems reasonable, then this same voltage should appear along a line through the neutral point. Near the X-type neutral point in front of the magnetosphere there must be dissipation to allow solar field lines to connect to the geomagnetic field lines.

The best way to decide between the open and closed magnetosphere models would be to study the topology of high latitude field lines. Do the null points at about 80° actually exist? Is there a magnetopause over the polar cap with $B_\perp = 0$? Are there conjugate polar phenomena? For more discussion see the contribution by E. N. PARKER and V. C. A. FERRARO in this Encyclopedia, Vol. 49/3 in particular Sect. 7ε, p. 149...151 and the contribution by W. POEVERLEIN in this volume.

36. The geomagnetic tail.

α) In 1960 PIDDINGTON [*22*] predicted the existence of a geomagnetic tail in trying to explain magnetic storms. He assumed that polar cap field lines were pulled back to form a tail by the solar plasma. He thought the decreased field at Earth due to the tail could explain the main phase of a storm. We now think this is not a real explanation of a storm main phase but the existance of the tail as put forward by PIDDINGTON has now been confirmed not only at storm times as he suggested, but continuously.

When satellite Explorer 10 first found the magnetopause[1,2] in 1961 the field just inside the boundary at $R = 22\,R_E$ was quiet and well ordered, and quite clearly the geomagnetic field, but it pointed almost exactly away from the sun and it has a strength of about $30\,\gamma \equiv 3\cdot 10^{-8}$ T. At this distance the dipole field should be

$$B(R = 22\,R_E) = \frac{0.32\,\Gamma}{(22)^3} = 3\,\gamma \equiv 3\cdot 10^{-9}\,\mathrm{T}$$

so the measured field was larger than expected by a factor of 10.

NESS' magnetometer on satellite IMP-1 mapped the geomagnetic field out to $R = 32\,R_E$ on the back side of Earth and found an elongated *tail* out to this point clearly extending well past this distance[3]. At $R = 30\,R_E$ where the expected dipole field is $1\,\gamma$ the measured field is about 10...20 γ.

---

[1] J. P. HEPPENER, N. F. NESS, T. L. SKILLMAN, and C. S. SEARCE: J. Geophys. Res. **68**, 1—46 (1963).

[2] A. BONETTI, H. S. BRIDGE, A. J. LAZARUS, B. ROSSI, and F. SCHERB: J. Geophys. Res. **68**, 4017 (1963).

[3] N. F. NESS: J. Geophys. Res. **70**, 2989 (1965).

β) The *direction* of the field in this region below the ecliptic plane was consistently almost exactly antisolar — not what is expected for a dipole field. Above the ecliptic plane a few Earth radii the field generally pointed towards the Sun. The transition between these two field directions as measured on satellite IMP-1 was a *neutral sheet* or *current sheet*. As the satellite approached this point, the field decreased to a very small value and then changed its direction abruptly. This neutral plane geometry was clearly observed on almost all of the IMP-1 orbits in the magnetic tail and is therefore a normal feature of the tail. This magnetic tail geometry must be included in any complete description of the geomagnetic field.

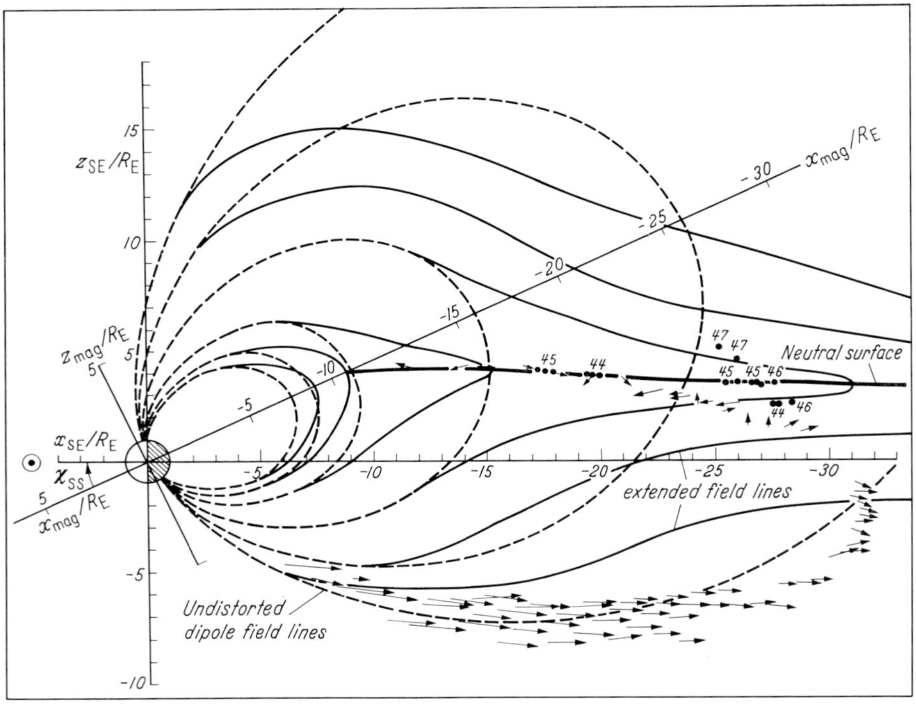

Fig. 46. The geomagnetic tail configuration (solid field lines) in the noon-midnight plane for the case of a large tilt of the dipole axis to the solar wind direction. Dipole field lines are shown dotted for comparision. The neutral surface lies in the antisolar direction (Sun is at left). It appears to be rooted to the dipole field lines at about $R=10\ R_E$. Arrows give measured data. (SPEISER and NESS, ref. [14], Sect. 39.)

WILLIAMS and MEAD[4] built it into their field model shown in Fig. 41. The tail field $\boldsymbol{B}_{cs}$ is assumed to be due to a current sheet that starts on the back side of Earth at about $R=10\ R_E$ and extends backwards to some large distance. They used a current sheet producing a 40 γ field and extending from $R=10\ R_E$ to 40 R_E and the values of $\Delta Л$ calculated this way agree well with the observed values.

The closed field lines on the back side extend out only to $R=7\ R_E$ and intersect Earth's surface at $Л=69°$. This field description is somewhat arbitrary and equally good agreement with the data can be obtained using other $\boldsymbol{B}_{cs}$ fields, for example, a 23 γ sheet extending from $R=8\ R_E$ to 100 R_E does as well.

[4] D. J. WILLIAMS, and G. D. MEAD: J. Geophys. Res. **70**, 3017 (1965).

γ) The width of the tail as measured by the IMP magnetometer is about 40 R_E. Assuming the tail is a cylinder with radius $\varrho = 22\ R_E$ and field induction $B_T = 20\gamma$ we can conserve magnetic flux if the *polar cap flux* for $\text{Л} > \text{Л}_c$ is *pulled back into* the *tail*. This gives the condition

$$\Phi_T \equiv \pi \varrho^2 B_T \equiv \pi (22\ R_E)^2 (0.0002\ \Gamma) = 2\pi (R_E \cos \text{Л}_c)^2 (0.6\ \Gamma) \equiv \Phi_c \qquad (36.1)$$

which gives $\text{Л}_c = 75°$.

The MEAD model field used to explain the high latitude limit of trapping can accommodate this total polar field pulled back into the tail and still have the front neutral point at about $\text{Л} \approx 80°$ by having an asymmetric ring around the pole for the polar cap. The ring would be at $\text{Л} \approx 80°$ in front and at $\text{Л} \approx 67°$ on the sides and back side. The best picture of the geomagnetic tail from IMP data is shown in Fig. 46.

Several authors[5–10] have suggested that the neutral sheet in the tail may be unstable. However, the "neutral" sheet doesn't seem to be really neutral but instead has a field perpendicular to the Earth-Sun line of roughly 1 γ. This should make the tail stable.

δ) ANDERSON[11] using a GEIGER-MÜLLER counter on satellite IMP-1 that counted electrons of $\mathscr{E}_e > 45$ keV found that near the subsolar point electron populations in the radiation belt were quite stable out to the magnetopause. Outside this boundary in the transition zone and tail the electron fluxes were strongly variable. *Spikes* or islands of electrons were found here especially at times of magnetic disturbances. Out to $R = 8.6\ R_E$ in the antisolar direction stable trapped particles were found. Beyond that distance in the tail *islands of electron fluxes* up to $10^7\ \text{cm}^{-2}\,\text{sec}^{-1}$ ($10^{11}\ \text{m}^{-2}\,\text{sec}^{-1}$) were consistently found for low magnetic latitudes. The fact that there is a rapid rise time and slow decay time in the flux on both inbound and outbound orbits of IMP shows that the islands are temporal not spatial features of the tail and move rapidly compared to the satellite. The occurrence of these islands of electrons in the tail is found to be anticorrelated[12] with the strength of the magnetic field in the tail, indicating that they occur in blobs with enough particle pressure to hold the field out.

ANDERSON[11, 13] also observed electrons of $\mathscr{E}_e > 45$ keV on IMP-1 in the magnetosheath. Fig. 47 shows electron *spikes* in the magnetosheath and also outside the bow shock two days after the start of a magnetic storm.

FRANK[14] found from analyzing Explorer 14 data that there is a tail of $\mathscr{E}_e > 40$ keV electrons in the antisolar direction showing large temporal variations and appearing very similar to those ANDERSON found. There was a marked dawn-dusk assymetry in the occurrance of the islands seen on the 'Vela' satellites. Many more events were observed near the dawn meridian.

The differential energy spectra of the electrons measured on Vela could be fitted quite well by a power law

$$J(\mathscr{E}) = \text{const.}\ \mathscr{E}_e^{-m} \quad \text{with} \quad 2.2 < m < 3.2 \qquad (36.2)$$

for 50 keV $< \mathscr{E}_e <$ 325 keV.

[5] R. K. JAGGI: J. Geophys. Res. **68**, 4429 (1963).

[6] D. B. CHANG, L.D. PEARLSTEIN, and M. N. ROSENBLUTH: J. Geophys. Res. **70**, 3085 (1965).

[7] H. FURTH, J. KILLEEN, and M. N. ROSENBLUTH: Phys. Fluids **6**, 459 (1963).

[8] F. C. HOH: [*16*], 547 (1966).

[9] B. COPPI, G. LAVAL, and R. PELLAT: Phys. Rev. Letters **16**, 1207 (1966).

[10] H. FURTH, in: Advanced Plasma Theory (ed. M. N. ROSENBLUTH). New York: Academic Press 1964.

[11] K. A. ANDERSON, H. K. HARRIS, and R. J. PAOLI: J. Geophys. Res. **70**, 1039 (1965).

[12] K. A. ANDERSON, and N. F. NESS: J. Geophys. Res. **71**, 3705 (1966).

[13] K. A. ANDERSON: J. Geophys. Res. **70**, 4741 (1965).

[14] L. A. FRANK: J. Geophys. Res. **70**, 1593 (1965).

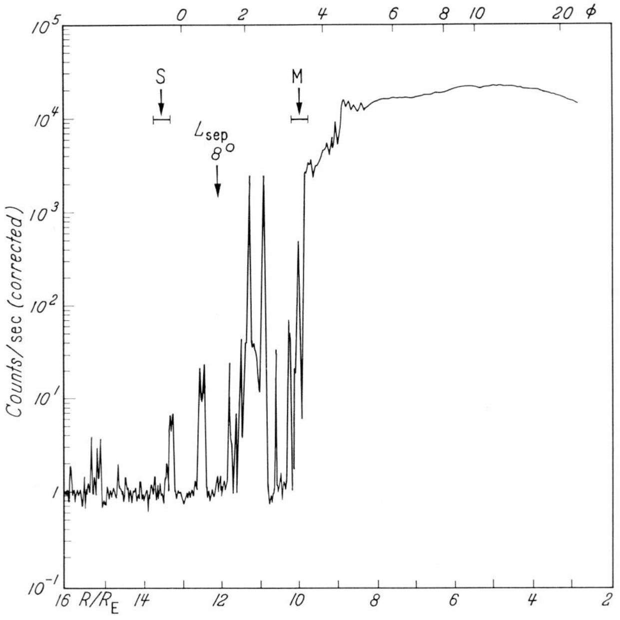

Fig. 47. The behaviour of the $\mathscr{E}_e > 45$ keV electron population in and beyond Earth's magnetosphere at small Sun-Earth probe angles. Abscissa (right to left): $\mathscr{R}$ = distance from Earth/R_E; geomagnetic latitude ϕ is indicated on top. (M = magnetopause; S = shock location; L Sep means Sun-Earth-probe angle.) On this 'in' orbit two electron spikes are evident in the magnetosheath well inside the shock location determined by the magnetometer on the same satellite. (4 Dec. 1963.) (ANDERSON et al., ref. [11], Sect. 36.)

ε) Lower energy particles have also been observed in the region of the neutral sheet in the geomagnetic tail. GRINGAUZ[15] measured a flux of $J \approx 2 \cdot 10^8$ cm^{-2} sec^{-1} ($\equiv 2 \cdot 10^{12}$ m^{-2} sec^{-1}) of $\mathscr{E}_e > 200$ eV electrons using an ion trap on space probe Lunik 2 at a Sun-Earth-probe angle of about 145° in the region from 55000 km to 75000 km.

There was some confusion for a while about whether these particles were trapped or not[16, 17] but now it is agreed they are in the tail outside the boundary of trapped particles. FREEMAN[18] with a cadmium sulfide detector on satellite Explorer 12 found fluxes of $J \approx 10^8 \ldots 10^9$ cm^{-2} sec^{-1} ($10^{12} \ldots 10^{13}$ m^{-2} sec^{-1}) of 200 eV $< \mathscr{E}_e <$ 40 keV electrons near the tail. BAME et al.[19] have measured the flux of electrons in the tail and also their energy spectra. These electrons typically have an energy of a few keV. There is a great deal of similarity between the electrons and protons observed in these islands in the tail and in the aurora.

ζ) Across the tail of the magnetosphere there must be a pressure balance in order for the system to be in equilibrium. Assuming low particle fluxes in the tail (symbol T) field the *pressure balance across the neutral sheet* (symbol nS) is*

$$\frac{B_T^2}{2\,\mathrm{u}\,\mu_0} = \frac{1}{2}\,(n m v^2)_{\mathrm{nS}}\,. \tag{36.3}$$

* u ≡ 1 for rationalized, ≡ 4π for non-rationalized systems of units. See "Introductory Remarks" on p. 1.

15 K. I. GRINGAUZ: [*9*], 574 (1963).

16 K. I. GRINGAUZ: J. Geophys. Res. **69**, 1007 (1964).

17 J. A. VAN ALLEN: J. Geophys. Res. **69**, 1011 (1964).

18 J. W. FREEMAN, quoted in: [*25*], 161 (1964). [Earlier values reported for this flux were apparently in error.]

19 S. J. BAME, J. R. ASBRIDGE, H. E. FELTHAUSER, R. A. OLSON, and I. B. STRONG: Phys. Rev. Letters **16**, 138 (1966).

Using $B_T = 20\,\gamma$ let us compare with measured particle fluxes in the neutral sheet in Table 12.

Table 12. *Comparison of energy densities in tail* (T) *and neutral sheet* (nS).

Experimenter	Particle flux/cm^{-2} sec^{-1}	Energy range/keV	Assumed average energy/keV	nS $\frac{1}{2} n m v^2$/erg cm^{-3}	$\frac{B_T^2}{2 u \mu_0}$/erg cm^{-3}
ANDERSON	$1 \cdot 10^7$	>40	60	$0.1 \cdot 10^{-9}$	$1.6 \cdot 10^{-9}$
GRINGAUZ	$2 \cdot 10^8$	>0.2	10	$0.5 \cdot 10^{-9}$	
FREEMAN	$\sim 10^9$	0.2 ... 40	5	$\sim 2.0 \cdot 10^{-9}$	
BAME	$3 \cdot 10^8$	0.35 ... 20	5	$0.7 \cdot 10^{-9}$	

The lower energy particles BAME[19] and FREEMAN[18] measured in the tail seem to roughly balance the pressure.

η) There have been various estimates of the *length of* the geomagnetic *tail*. The tail has been observed[20] to be longer than 80 R_E.

DESSLER[21] has suggested that the tail may be 20...50 a.u. long* as a result of magnetogasdynamic waves generated at the front of the magnetosphere and moving into the tail which should set the plasma in the tail in motion away from Earth. This plasma should keep the tail open until the solar wind is stopped at the edge of the 'heliosphere' at 20...50 a.u.

The Mariner 4 spacecraft passed behind Earth within 1° of the Earth-Sun line at a distance of about 3300 $R_E = 0.14$ a.u. The magnetometer on board found no detectable wake[22] and VAN ALLEN's GM counter[23] found no measurable flux of $\mathscr{E}_e > 40$ keV electrons.

VAN ALLEN considers that the wake of Earth has pretty well disappeared at this distance[24]. However, DESSLER[25] has argued that the tests for tail detectability that VAN ALLEN has used are inconclusive.

DUNGEY's open model of the magnetosphere predicts that the tail length is given by the time it takes a solar field line to disconnect from the geomagnetic field[26]. The field line moves across the polar cap with a velocity of the ionospheric winds of about 100 m sec^{-1}. It moves a distance of 2/3 R_E from front auroral zone to near auroral zone before disconnecting. The same field line far from Earth is moving at about 30 km sec^{-1} so it moves a distance of about 1000 R_E before it disconnects on this model. Therefore, according to DUNGEY, the tail will disappear at about 1000 R_E.

If the magnetic field is really not zero but instead there is a $B_\perp \approx 1\,\gamma$ at the "neutral" sheet, as seems to be the case, we can estimate the tail length from this. The magnetic flux from one half of the tail (symbol T) must cross the neutral sheet (symbol nS) so

$$\Phi_T = \tfrac{1}{2}\pi \varrho_T^2 B_T = 2\varrho_T L_T B_{nS} = 2\Phi_{nS} \tag{36.4}$$

* 1 astronomical unit (a.u.) $= 1.5 \cdot 10^8$ km $= 1.5 \cdot 10^{11}$ m.

[20] N. F. NESS, K. W. BEHANNON, S. C. CANTARANO, and C. S. SCEARCE: Observations of the Earth's Magnetic Tail and Neutral Sheet at 510,000 km by Explorer 33. Goddard Space Flight Center Report X 612-66-529, Nov. 1966.
[21] A. J. DESSLER: J. Geophys. Res. **69**, 3913 (1964).
[22] P. J. COLEMAN, L. DAVIS, D. E. JONES, and E. J. SMITH: Trans. Am. Geophys. Union **46**, 533 (1965).
[23] J. A. VAN ALLEN: J. Geophys. Res. **70**, 4731 (1965).
[24] J. A. VAN ALLEN: J. Geophys. Res. **71**, 2406 (1966).
[25] A. J. DESSLER: J. Geophys. Res. **71**, 2408 (1966).
[26] J. W. DUNGEY: J. Geophys. Res. **70**, 1753 (1965).

which, using $\varrho_T = 20\ R_E$, $B_T = 15\ \gamma$ and $B_{nS} = 1\ \gamma$ gives the tail length of

$$L_T \approx 300\ R_E.$$

ϑ) The *moon* may also have a magnetic tail. On Dec. 14, 1963, the magnetometer on satellite IMP-1 detected an unusually strong interplanetary field of 14 γ. The field was also more disturbed than usual. It was on this day that the moon nearly exlipsed the satellite. It missed by just 8 R_E at a distance of about 40 R_E. This puts the satellite within 10° of the Sun-Moon line and therefore it should have been inside the MACH-cone. The measured disturbed field may have been the wake of the moon[27]. It has only been observed once so it is uncertain. It may also have been due[28] to a class 1 + solar flare that occurred on December 13.

The moon should represent an obstacle to the flow of the solar wind. The lunar magnetic field is small. Space probe Lunik 2[29] showed that the field a few hundred km above the lunar surface is about 15...30 γ. But even if the field was initially very small, GOLD[30] has suggested it should build up to about 50 γ. Solar magnetic field lines will penetrate the Moon as a result of the solar wind striking the surface. This will build up a field of about 50 γ until the solar wind will no longer strike the surface. The magnetic field will try to diffuse through the moon but if the lunar electrical conductivity σ is similar to that of terrestrial rocks the diffusion time for the field to leak through the Moon will be the order of days to years and the solar field will build up in the Moon to the stand-off value of about 50 γ. So, independent of the nature of the solar wind interaction with the Moon, the Moon may be an obstacle to the solar magnetic field flow and therefore may have a wake. Whether the Moon will have a bow shock or not is uncertain. If the bow shock is about a cyclotron radius thick this is $R_c \approx 100$ km. The radius of the Moon is $R_☾ \approx 2000$ km so the condition for development of a bow shock that $R_c \ll R_☾$ is not really fulfilled. But it seems from measurements on satellites IMP-1 and OGO 1[31] that the shock thickness is in the range of nearly 10 to 100 km in which case the shock may be able to be fully developed. Magnetic measurements on the A-IMP satellite in orbit around the moon at altitudes of about 800 km found no discontinuities that could be considered a bow shock[32].

37. The high latitude limit of trapping.

α) Measurements on several satellites have shown that the trapped electron fluxes at high latitudes show a diurnal variation. The *trapped fluxes* normally *stop* fairly *abruptly* at some particular latitude. O'BRIEN[1], using detectors on satellite Injun 1, MCDIARMID[2] using detectors on Alouette, and FRANK et al.[3] on Injun 3, all at 1000 km showed that the $\mathscr{E}_e > 40$ keV trapped electrons had an upper latitude limit of trapping of $\Lambda_D \equiv \arccos L^{\frac{1}{2}} \approx 76°$ at local noon and $\Lambda_N \approx 69°$ at local night. WILLIAMS[4] with counters on satellite 1963—38c found that for electrons of $\mathscr{E}_e > 0.28$ MeV the trapping limits are $\Lambda_D = 72°$ and $\Lambda_N = 69°$. Data on the latitude shift for $\mathscr{E}_e > 1.2$ MeV electrons was essentially the same as for the 0.28 MeV electrons on 1963—38c. All of this data is for periods of relative magnetic quiet; usually the planetary three-hour index K_p was < 4. The fact that the diurnal variation of the latitude limit of trapping for $\mathscr{E}_e > 40$ keV is larger than for higher energies may be explained by electric fild effects.

---

[27] N. F. NESS: J. Geophys. Res. **70**, 517 (1965).
[28] E. W. GREENSTADT: J. Geophys. Res. **70**, 5451 (1965).
[29] S. H. DOLGINOV, E. G. EROŠENKO, L. N. ŽAGOV, N. V. PUŠKOV, and L. O. TYNRMINA: Geomagnetism i Aeronomija **1**, 21 (1961).
[30] T. GOLD: [*26*], 381 (1966).
[31] R. E. HOLZER, M. G. MCLEOD, and E. J. SMITH: J. Geophys. Res. **71**, 1481 (1966).
[32] N. F. NESS, K. W. BEHANNON, C. S. SCEARCE, and S. C. CANTARANO: J. Geophys. Res. **72**, 5769 (1967).
[1] B. J. O'BRIEN: J. Geophys. Res. **68**, 989 (1963).
[2] I. B. MCDIARMID, and J. R. BURROWS: Can. J. Phys. **42**, 606 (1964).
[3] L. A. FRANK, J. A. VAN ALLEN, and J. D. CRAVEN: J. Geophys. Res. **69**, 3155 (1964).
[4] D. J. WILLIAMS, and W. F. PALMER: J. Geophys. Res. **70**, 557 (1965).

β) If has been suggested by several authors[5,6] that this diurnal variation of the latitude limit of the trapping might be due to the *distortion of the magnetosphere* by the solar wind. WILLIAMS and MEAD[7] assumed that the electrons drift in longitude in the distorted magnetic field conserving the adiabatic invariants $|\mu|$ and $\mathscr{I}$ as they drift. Conserving

$$\mathscr{I} = \int_{-l_m}^{+l_m} \mathrm{d}l \sqrt{1 - B/B_M} \qquad [4.2]$$

is nearly like constraining the particle to drift on a set of field lines of constant length. In the distorted geomagnetic field with the lines pushed in on the front side and pulled out in back the particle must to lower latitude at night to keep $\mathscr{I}$ constant. Therefore, to calculate ($ЛI_D - ЛI_N$) the latitudes of constant $\mathscr{I}$ surfaces must be found. The field model (see Fig. 41) used by WILLIAMS and MEAD[7] is a dipole field $\boldsymbol{B}_{dp}$ with surface currents giving the proper front surface $\boldsymbol{B}_s$ and also a tail field to agree with the findings of NESS[8] on IMP 1.

The results of this analysis shows that when a geomagnetic tail type nightside magnetic field is used the values of Δ ЛI obtained agree with observations. Further, the tail current sheets that can be used to give agreement have fields associated with them that are reasonably close to NESS's observations of a tail field.

γ) The *diurnal variation* of the upper latitude limit of trapping for $\mathscr{E}_e > 0.28$ MeV electrons and $\mathscr{E}_e > 1.2$ MeV electrons can be understood in terms of the adiabatic motion of trapped particles. But for the $\mathscr{E}_e > 40$ keV electrons this does not seem to be the case. Those seem to be found at considerably higher latitudes than the $\mathscr{E}_e > 280$ keV but this effect seems to be mainly on the day side as shown by Table 13.

Table 13. *Observed high-latitude limit night and day.*

$\mathscr{E}_e$	$ЛI_N$	$ЛI_D$
> 40 keV	69°...71°	75°...77°
> 280 keV	69°	72°

O'BRIEN[1] suggested that there might be a more or less continuous source of $\mathscr{E}_e > 40$ keV electrons on the day side of the magnetosphere which would produce the large values of $ЛI_D$. It might also be due to electric fields.

FRITZ and GURNETT[9] using data from an open end electron multiplier tube on satellite Injun 3 detected transient pulses of $\mathscr{E}_e > 10$ keV electrons being precipitated into the atmosphere. They found that substantial fluxes of electrons with $j > 2 \cdot 10^7$ cm^{-2} sec^{-1} sr^{-1} ($\equiv 2 \cdot 10^{11}$ m^{-2} sec^{-1}. sr^{-1}.) occurred *only* at local night in the time period 1700 to 0700 h, and only for latitudes of $58° < ЛI < 76°$. The fluxes were larger when the planetary three-hour index K_p was large. They noted that a change in the electron energy spectrum occurred at or near the boundary of trapping. Inside the boundary the spectrum from 10 to 40 keV could be represented by a power law [Eq. (36.2)] with spectral index $2.2 < m < 2.8$ while outside the boundary at higher latitude the spectrum was softer and $6.8 < m < 7.3$. FRITZ and GURNETT[9] feel these electrons are related to those

[5] J. A. MALVILLE: J. Geophys. Res. **65**, 3008 (1960).
[6] D. H. FAIRFIELD: J. Geophys. Res. **69**, 3919 (1964).
[7] D. J. WILLIAMS, and G. D. MEAD: J. Geophys. Res. **70**, 3017 (1965).
[8] N. F. NESS, C. S. SCEARCE, and J. B. SEEK: J. Geophys. Res. **69**, 3531 (1964).
[9] T. A. FRITZ, and D. A. GURNETT: J. Geophys. Res. **70**, 2485 (1965).

observed by FREEMAN[10] and GRINGAUZ[11] at large altitudes on the back side of the Earth.

MCDIARMID et al.[12] detected occasional large spikes of $\mathscr{E}_e > 40$ keV electrons with fluxes up to $j = 10^9$ cm^{-2} sec^{-1} sr^{-1} ($= 10^{13}$ m^{-2} sec^{-1} sr^{-1}) *above* the trapping region and only on the night side of the Earth. Both of these experiments on particle precipitation near or beyond the upper latitude limit of trapping seem to be related to particles back in the geomagnetic tail.

38. The outer radius limit of trapping.

α) What determines the outer limit of the trapped radiation? We are using the word "trapped" here to mean "able to drift around the Earth more than once", not merely to bounce between mirror points a few times. Is the edge of the trapped radiation really the edge of the geomagnetic field? To understand this we must consider the adiabatic motion of trapped particles. Particles with pitch angle $\alpha_0 = 90°$ or equivalently with $\mathscr{J} = 0$ [Eq. (4.1)] follow a contour of $B = $const. as they drift around Earth.

MEAD[1] has shown that something quite different is the case for particles with *small pitch angles*. These particles will mirror at roughly constant altitude as they drift around Earth. The mirror altitude is controlled by Earth's magnetic field and not affected appreciably by boundary or tail currents. For these particles, assuming constant energy the second invariant requires that $\mathscr{J} = $const, [Eq. (4.2)]. For high latitude field lines the value of the integrand is very nearly unity over most of the integration path so that

$$\mathscr{J} \approx s = \text{field line length} \approx \text{const.} \tag{38.1}$$

Small pitch angle particles at high latitudes drift on a set of field lines of nearly constant length s between mirror points. Inspection of the field lines in Fig. 41 shows that these high-latitude low-altitude mirroring particles must move *out* on the night side at the equator to keep s constant.

β) This says the drift paths of particles on high-latitude field lines depend strongly on their equatorial pitch angle. This means the *L shell splitting* is very large here and

$$\Delta L \sim 5\ R_E \text{ can occur.}$$

A result of this L shell splitting will be that at $R \approx 9\ R_E$ on the day side the angular distribution of particles will be quite peaked at $\alpha_0 = 90°$ while on the night side the angular distribution will have a dip at $\alpha_0 = 90°$. Such a change in angular distribution has been found by SERLIMITSOS[2] studying ~100 keV electrons observed on satellite Explorer 14. He found that outside $R = 7 \ldots 9\ R_E$ near the equatorial plane on the back of Earth the electron angular distribution changed from being essentially isotropic to having a maximum roughly *along* the field line. The region of trapping near the equator seemed to extend to beyond $R = 12\ R_E$ on quiet days. The extent and morphology of the trapping region depended strongly on geomagnetic activity.

γ) Another result of these different drift paths will be to affect the high latitude limit of trapping measured at *low altitude* on the day side. Suppose a particle became trapped with α_0 almost 0° on the far out field lines on the day side labelled

[10] J. W. FREEMAN, quoted in: [*25*], 161 (1964). [Early values reported for this flux were apparently in error.]

[11] K. I. GRINGAUZ: [*9*], 574 (1963).

[12] I. B. MCDIARMID, J. R. BURROWS, E. E. BUDZINSKI, and M. D. WILSON: Can. J. Phys. **41**, 2064 (1963).

[1] G. C. MEAD: [*16*], 481 (1966).

[2] P. SERLEMITSOS: J. Geophys. Res. **71**, 61 (1966).

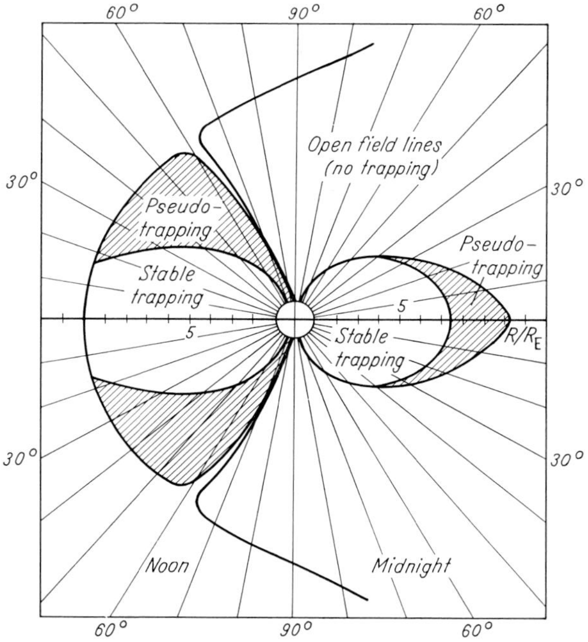

Fig. 48. Calculated locations of the regions of quasi-trapping in the magnetosphere. Particles mirroring inside those regions are unable to complete a 180° drift around Earth. Those injected into the left side will be lost into the tail; those injected into the right portions will abandon the magnetosphere through the boundary on the day side. (ROEDERER, ref. [3], Sect. 38.)

80° in Fig. 41. When it drifted to the night side in order to keep $s = const$ it would drift out to $R \approx 15\ R_E$ and very likely be lost into the geomagnetic tail on open field lines. This means that in the MEAD model magnetosphere there would be no $\alpha_0 \sim 0°$ trapped particles on the dayside 80° line since they cannot complete drift paths around Earth. We might, however, expect transient populations here related to injection of new particles. The upper limit of trapping on the day side is determined here by requiring that the particle have a complete drift path and not be lost in the tail. On the night side the last closed field line will determine the upper limit of trapping for these $\alpha_0 \sim 0°$ particles.

δ) ROEDERER[3] has made a quantitative *model calculation* of the L shell splitting near the outer edge of the magnetosphere. He has used a MEAD type magnetosphere with a dipole field normal to the solar wind flow and used

$$R_b = \text{distance to magnetopause at the subsolar point} = 10\ R_E,$$

$$R_{min} = \text{distance to start of tail current} = 8\ R_E,$$

$$R_{max} = \text{distance to end of tail current} = 200\ R_E,$$

$$B_T = \text{tail magnetic field away from neutral sheet} = 15\ \gamma.$$

Roederer calculated that particles mirroring inside the regions shown in Fig. 48 are *quasi-trapped*. They can not drift all the way around Earth even once. Those injected into the day-side shaded region will be lost into the tail and those injected into the night side shaded region will be lost through the day side magnetopause.

[3] J. G. ROEDERER: J. Geophys. Res. **72**, 981 (1967).

39. The bow shock and magnetosheath.

α) The existence of a bow shock wave upstream of the magnetosphere was predicted before it was observed. ŽIGULEV[1], KELLOGG[2], and AXFORD[3] considered the fact that the *magnetosphere* should represent an *obstacle* to the flow of the solar wind and a shock should result. A bullet in a supersonic flow produces a shock wave as it moves in air. If the speed is sufficiently high the shock wave becomes datached and lies upstream of the bullet, see Fig. 49. The sound speed in the interplanetary medium is $V_s \approx 50\,\mathrm{km\,sec^{-1}}$. If the solar wind velocity v is $400\,\mathrm{km\,sec^{-1}}$ the flow has a MACH number of $M \approx 8$. But the more significant speed for our case is the ALFVÉN speed*

$$V_A = \frac{B}{\sqrt{\mathrm{u}\,\mu_0\,\varrho}} \tag{39.1}$$

which is about $50\,\mathrm{km\,sec^{-1}}$ also. So the ALFVÉN number is $M_A \approx 8$ too. Since both M and M_A indicate supersonic flow one might expect that a shock wave would exist upstream of the magnetopause.

But the analogy with the bullet is not very good. The shock wave upstream of the bullet is produced by collisions between air atoms. The collision mean free path is given by

$$\lambda_f = \frac{1}{n\,\sigma_\times} \tag{39.2}$$

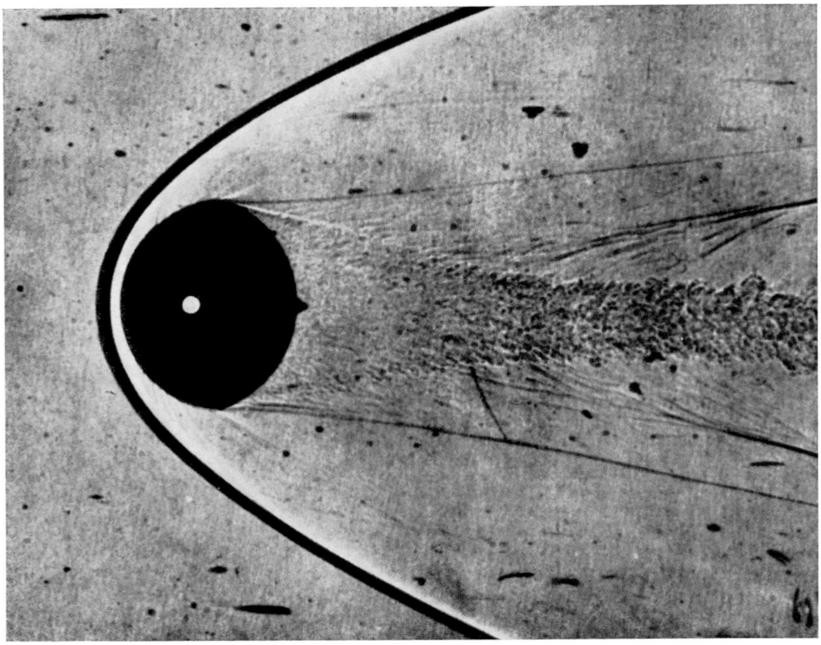

Fig. 49. A shadow-graph of a model magnetosphere in flight in Argon at a speed of Mach 4.5. The (parabolic) bow-shock is clearly visible. (SPREITER and HYETT, ref. [15], Sect. 39.)

* u = 1 for rationalized, = 4π for non-rationalized systems of units. See "Introductory Remarks" on p. 1.

[1] V. N. ŽIGULEV: Dokl. Acad. Nauk **4**, 521 (1959) [engl. transl.: Soviet Phys. „Doklady" **4**, 514 (1959)].

[2] P. J. KELLOGG: J. Geophys. Res. **67**, 3805 (1962).

[3] W. I. AXFORD: J. Geophys. Res. **67**, 3791 (1962).

where n is particle density and $\sigma_\times$ effective collision cross section. Using $n \approx 10\ \text{cm}^{-3}$ and taking a coulomb cross section of $10^{-15}\ \text{cm}^2$ corresponding to a 10 eV proton scattering we get $\lambda_f \approx 10^{14}\ \text{cm} = 1.5 \cdot 10^5\ R_E \approx 6$ a.u. This distance is so much larger than the size of the magnetosphere that we can forget about collisions.

β) A shock is a rapid transition between two different states of matter. If the collisions can't produce the rapid transition here, what can? The magnetic field clearly takes over the job of making the *plasma* act *like a fluid*. The cyclotron radius is the distance in which things can change; for a 1 keV proton at a distance of 1 a.u. ($= 1.5 \cdot 10^8$ km) this is about 1000 km.

The cyclotron radius is so much smaller than the magnetosphere that the solar wind will flow around the boundary like a fluid. The particles on one field line will move together during the flow. If the flow velocity is subsonic the flow can be laminar with no bow shock. But if the flow velocity is faster than the wave velocities (sound V_s or ALFVÉN V_A, or other) in the system there must be a bow shock. The incoming plasma must change direction to flow around the obstacle but if a signal cannot propagate upstream to tell the plasma to turn, then there must be a shock. In the case of the solar wind incident on the magnetosphere it is a *collisionless shock*. (In the contribution by V. L. GINZBURG and A. A. RUHADZE in this volume conditions in collisionless plasmas are extensively discussed.)

KELLOGG[2] and AXFORD[3] used supersonic aerodynamics to describe the bow shock. This is illegal as they stated because this assumes a collision-dominated medium. But it may be expected to give a moderately good description of the shock location. The reason for this is that conservation of mass, momentum and energy are the major factors in determining the shock location. Detailed structure of the shock cannot be determined by using supersonic aerodynamics. HIDA[4] calculated the stand-off distance and curvature of supersonic shocks in front of spherical obstacles. Applying it to the magnetosphere KELLOGG and AXFORD suggested a stand-off distance and curvature of supersonic shocks in front of spherical obstacles. Applying it to the magnetosphere KELLOGG and AXFORD suggested a stand-off distance of about 4 R_E.

γ) Using a *aerodynamical* model one can find the change in density ϱ, velocity u, pressure p and temperature T across a shock by using conservation of mass, momentum and energy. This gives (with indices 1 and 2) identifying both sides of the shock-front:

$$\varrho_1 u_1 = \varrho_2 u_2 \tag{39.3}$$

and

$$p_1 + \varrho_1 u_1^2 = p_2 + \varrho_2 u_2^2 \tag{39.4}$$

and

$$\frac{p_1}{\varrho_1} + \frac{1}{2} u_1^2 + e_1 = \frac{p_2}{\varrho_2} + \frac{1}{2} u_2^2 + e_2, \tag{39.5}$$

e being internal energy per mass unit.

These RANKINE-HUGONIOT equations can be solved to give the change of variables across the shock. Using the perfect gas expression for internal energy

$$e = \frac{p}{\varrho}\left(\frac{1}{\gamma - 1}\right) \tag{39.6}$$

where γ is the ratio of specific heats (monoatomic gas with three degrees of freedom) some manipulation gives for a strong shock where

$$\frac{p_2}{p_1} \gg 1$$

$$\frac{\varrho_2}{\varrho_1} = \frac{\gamma + 1}{\gamma - 1}, \tag{39.7}$$

[4] K. HIDA: J. Phys. Soc. Japan **8**, 740—745 (1953).

$$\frac{T_1}{T_2} = \frac{p_2}{p_1}\left(\frac{\gamma-1}{\gamma+1}\right), \tag{39.8}$$

$$u_1^2 = \frac{p_2}{2\varrho_1}(\gamma+1), \tag{39.9}$$

$$u_2^2 = \frac{p_2}{2\varrho_1}\,\frac{(\gamma-1)^2}{\gamma+1}. \tag{39.10}$$

Let us illegally use these equations with $\gamma = \frac{5}{3}$ and $M \approx 8$ and compare with the bow shock. From equation we get $p_2/p_1 \approx 80$, so it is safe to use the strong shock approximations to give

$$\frac{\varrho_2}{\varrho_1} \approx 4 \quad \text{and} \quad \frac{T_2}{T_1} \approx 20.$$

The shock stand-off distance Δ in aerodynamic flow for a spherical obstacle is given by

$$\frac{\Delta}{D} \cong 1.1\,\frac{\varrho_1}{\varrho_2} \tag{39.11}$$

where $D = 13\,R_{\mathrm{E}}$ is the effective radius of the magnetosphere. Using $\gamma = \frac{5}{3}$ so $\Delta = 3.3\,R_{\mathrm{E}}$, $\Delta/D \approx 0.25$ which seems about right.

For the case of a collisionless shock wave we cannot be really sure that these jump conditions are valid. If there is no process by which entropy can be increased across the shock then there may not really be a definite state behind the shock but just wave trains which may eventually decay and return the plasma to the original state. It appears likely, however, that various electromagnetic processes like LANDAU damping or non-linear wave-wave interactions (see contribution by GINZBURG and RUHADZE in this volume) may produce a change in entropy and a definite new state behind the shock. In this case it should be all right to use the jump conditions Eqs. (39.7, 10) even for the collisionless case.

Even though a number of people have worked on the problem, there is no good picture of the structure of the bow shock or of the detailed processes that occur to produce the turbulent transition zone behind it. Whatever the processes are, we know collisions are not important.

δ) The bow shock was first discovered by studying the energetic electrons outside the magnetopause. FREEMAN[5] analyzing the data from a cadmium sulfide detector on satellite Explorer 12 found that there was an *enhanced flux* of $1\,\mathrm{keV} < \mathscr{E}_{\mathrm{e}} < 10\,\mathrm{keV}$ electrons *outside the magnetopause*. Energy fluxes of about $30\,\mathrm{erg\,cm^{-2}\,sec^{-1}\,sr^{-1}}$ were measured commonly for a distance of about $2\ldots3\,R_{\mathrm{E}}$ outside the magnetopause and then the flux suddenly decreased. This sharp outer transition FREEMAN[5] interpreted as being the bow shock wave predicted by ŽIGULEV[1], KELLOGG[2], and AXFORD[3].

We mentioned earlier the magnetometer on satellite IMP-1 detected two transitions as it moved out radially away from Earth[6]. The inner transition was the magnetopause and then several Earth radii further out a second change was detected. This second change is the shock. On IMP-1 orbit 11 shown in Fig. 43 the shock was detected at $R = 19.7\,R_{\mathrm{E}}$. Inside this shock the magnetic field is $10\ldots20\,\gamma$ and is *very disturbed*. The *field variance* or root mean square deviation of the field components

$$\delta B = \sqrt{\frac{1}{N}\sum_{\nu=1}^{N}(B_\nu - \overline{B})^2} \tag{39.12}$$

[5] J. W. FREEMAN: J. Geophys. Res. **69**, 1691 (1964).
[6] N. F. NESS, C. S. SCEARCE, and J. B. SEEK: J. Geophys. Res. **69**, 3531 (1964).

is high in the magnetosheath between shock and magnetopause. Usual values of $\delta B/B$ measured on IMP-1 over a 5 min period in this region were in the range 0.5 to 2.0. For individual measurements over one spin of the satellite $\delta B/B \approx 0.1$ was found. This regime might be called turbulent although the use of the word here is not precise.

ε) *Outside this shock wave* we are in the undisturbed interplanetary medium. In this region the field was usually 4 to 7 γ about as expected for the magnitude of the solar magnetic field at Earth, occasionally being as high as 10 γ or low as 1 γ. The field was really quite stable in magnitude for long periods of time and moderately stable in direction.

Typically according to NESS[6] $\delta B/B$ is about 0.1 here. The best indicator of the shock location in the magnetometer record is a sudden and marked change in the variance. On Fig. 43 the location of the bow shock measured by the IMP-1 magnetometer is shown. Also shown there is the predicted shock location of SPREITER and JONES[7] who calculated the shape of the magnetopause bow shock using the supersonic aerodynamics theory we have given above.

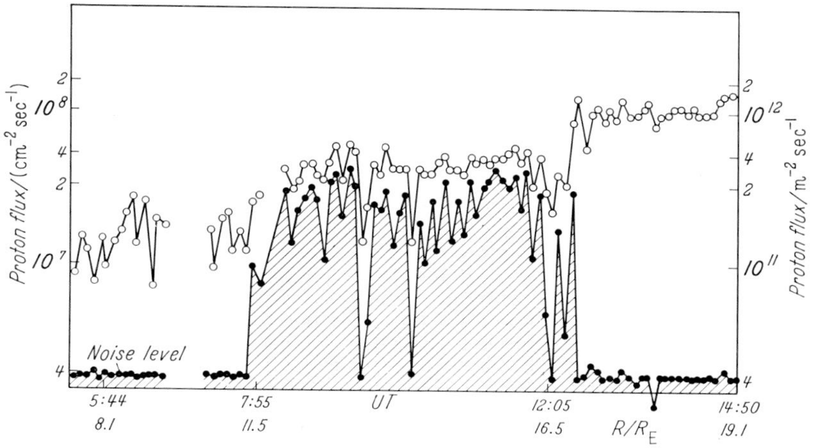

Fig. 50. Results from the MIT Faraday cup 220 ... 640 eV plasma probe showing proton flux measurements on the thermalized plasma by satellite IMP-1 during outbound orbit No. 1. Maximum and minimum proton flux curves (as seen from spin-modulation) are given. Minimum flux on noise level measn clearly directed flux .The positions of the boundaries are evidenced by the distinct spin modulation between $R = 11.3$ and 16.8 R_E (BRIDGE et al., ref. [8], Sect. 39).

At the same time that the magnetometer on IMP-1 indicated that the shock was traversed the plasma probes in this satellite showed a sudden and dramatic change. Outside the shock in interplanetary space the solar wind has a narrow spread of energies and is a highly directed stream flowing directly away from the sun. Just as the shock is passed, this flow changes to a fairly isotropic and substantially thermalized plasma. The record of the plasma probe[8] for orbit No. 1 of IMP-1 is shown in Fig. 50. This shows the maximum and minimum proton flux for each spin of the spacecraft in the energy range 220 eV $< \mathcal{E}_p <$ 640 eV as the satellite moved outwards. The magnetopause is passed at $R \approx 11.5\ R_E$ and at about $R = 17\ R_E$ the bow shock is traversed. In the magnetosheath from $R = 11.5\ R_E$ to 17 R_E the plasma shows little roll modification. Outside the shock at $R = 17\ R_E$ the plasma is very directional as the solar wind is known to be.

[7] J. R. SPREITER, and W. P. JONES: J. Geophys. Res. **68**, 3555 (1965).
[8] H. BRIDGE, A. EGIDI, A. LAZARUS, E. LYON, and L. JACOBSON: [*12*], 969 (1965).

ζ) WOLFE[9] working with the data from the 'Ames' plasma probe on IMP-1 has used the following criteria to distinguish the *three regions of space environment.*

(i) *Magnetospheric boundary*: Defined as corresponding to the appearance or disappearance of plasma (within the sensitivity of the instrument) depending upon whether the spacecraft is outbound or inbound.

(ii) *Magnetosheath* or *Transition region*: Characterized by the appearance of a wide energy spectrum by large fluctuations in flux, and by large fluctuations in the angle of incidence of the plasma stream.

(iii) *Shock wave*: The outermost boundary, tentatively identified as the bow shock, is defined simply by a sudden change from the characteristics of the transition region to one of smoother flow, narrow energy spectrum, and significant flux only in the sector containing the Sun.

WOLFE[10] also had plasma probes on OGO-1 and IMP-2. A comparison with data on the Vela satellite gave four energy spectra at different places in the magnetosheath. The significant feature here is that the spectra are quite similar at quite different Sun-Earth-probe angles. The temperatures are not measurably different. WOLFE has given data on jump conditions across the shock summarized in Table 14.

Table 14. *Observed jump conditions.*
4–6 October 1964; plasma probes on satellites IMP-2 and OGO-I.

Quantity	Preshock	Postshock
u_0/km sec^{-1}	712 ± 13	434 ± 20
n/cm^{-3}	2.46 ± 0.5	4.14 ± 1.0
T/°K	$(2.0 \pm 0.5) \cdot 10^5$	$(1.0 \pm 0.2) \cdot 10^6$*
$n m u_0^2$/erg cm^{-3}	$2.1 \cdot 10^{-8}$	$1.3 \cdot 10^{-8}$
$n k T$/erg cm^{-3}	$6.8 \cdot 10^{-11}$	$5.7 \cdot 10^{-10}$
$\frac{B^2}{2 u \mu_0}$/erg cm^{-3}	$1.0 \cdot 10^{-10}$	—
V_A/km sec^{-1}	70	
He-content	2.0%	

* 23% of ions in nonthermal tail: $n(\mathscr{E})\, d\mathscr{E} \sim \mathscr{E}^{-2.6}\, d\mathscr{E}$.

Perhaps the most interesting feature of these jump conditions given by WOLFE is that the flow velocity only goes down a factor of two across the shock. SPREITER's results[11] giving conditions in the transition zone for the aerodynamic calculation are shown in Fig. 51. For $\gamma = \frac{5}{3}$ the temperature jump across the shock at the subsolar point is 25:1 but if we take half the energy available and put it into the electrons then the temperature jump would be 12:1. Going around to Sun-Earth-probe angle by 90° the temperature jump would go from 12:1 down to 5:1. We see here that only modest changes in these parameters are expected.

FREEMAN, VAN ALLEN and CAHILL[12] reported the measurement on the Explorer 12 satellite of large fluxes of electrons of 1 keV $< \mathscr{E}_e <$ 10 keV in the magnetosheath. A cadmium sulfide detector measured energy fluxes of about 30 erg cm^{-2} sec^{-1} sr^{-1} which correspond to electron fluxes of $J \approx 3 \cdot 10^{10}$ cm^{-2} sec^{-1} ($= 3.10^{14}$ m^{-2} sec^{-1}), or a density of $n_e \sim 5$ cm^{-3}.

It seems very likely that equipartition of energy will take place behind the shock. Turbulent processes will transfer energy from protons to electrons. There-

[9] J. H. WOLFE, R. W. SILVA, and M. A. MYERS: J. Geophys. Res. **71**, 1319 (1966).
[10] J. H. WOLFE, R. W. SILVA, and M. A. MYERS: [*13*], 680 (1966).
[11] J. R. SPREITER, A. L. SUMMERS, and A. Y. ALKSNE: Planetary Space Sci. **14**, 223 (1966).
[12] J. W. FREEMAN, J. E. VAN ALLEN, and L. J. CAHILL: J. Geophys. Res. **68**, 2121 (1963).

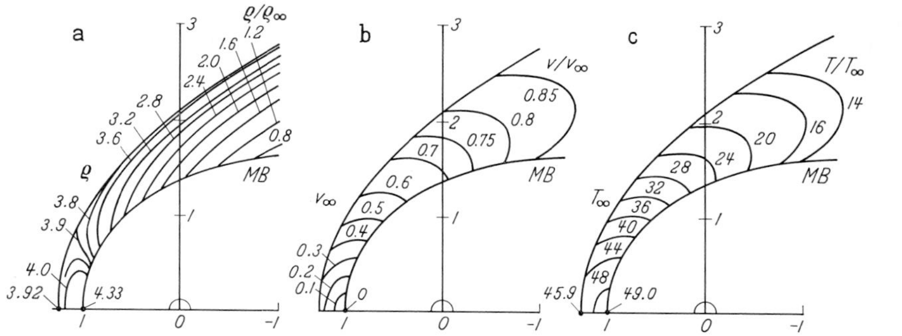

Fig. 51a—c. Calculated variation of **a**: density, **b**: velocity, and **c**: temperature through the magnetosheath for $\gamma = \frac{5}{3}$ and Mach Number $M = 12$. (MB = boundary of magnetosphere). The Earth-magnetopause distance in Solar direction is taken as unit for the cartesian coordinates. Designation of velocity is v [instead of u in Eqs. (39.3 ... 10)]. (SPREITER et al., ref. [11], Sect. 39.)

fore the average proton energy and the average electron energy should be about the same behind the shock. The average proton energy is about 2 keV so $\bar{\mathscr{E}}_p \sim \bar{\mathscr{E}}_e$ to within a factor of 3. The measured electron density seems quite sensible when we consider that the measured proton density in this region in the magnetosheath is $n_p = 4$ to $50\ \text{cm}^{-3}$, and we must have $n_e = n_p$ for charge neutrality.

These large fluxes of 1 ... 10 keV electrons were detected just outside the boundary of trapped particles. They were seen on many passes of Explorer 12 into the transition zone. The large fluxes extended out about $R = 24\ R_E$ and then often suddenly decreased. FREEMAN[5] interpreted this outer boundary as the shock front. JOKIPII[13] suggests that FERMI acceleration probably produces the energetic electrons. Some particles upstream from the shock may be FERMI accelerated between the shock, where the magnetic field induction B usually increases[14, 15].

G. Aurorae.

40. General features.

α) *Aurorae*[1] are *displays* of visible and invisible (ir and uv) light seen only at night and most often near the polar regions in the zone of invariant latitudes $65° < \text{Л} < 75°$. In the winter hemisphere they can sometimes be observed from one station (such as Byrd Station in Antarctica) at all local times. From such observations we know that the auroral zone, the region where aurorae are observed every night, are not lines of constant magnetic latitude but are distorted. At local noon the auroral zone is at invariant latitude [Eq. (5.8)] $73° < \text{Л} < 78°$ and at local midnight it is at $65° < \text{Л} < 75°$. During periods of large magnetic activity and following large solar flares aurorae are often observed at 'subauroral latitudes', i.e. considerably lower latitudes than the auroral zone. They have been seen exceptionally even at Singapore. Aurorae are seldom seen near the pole and at times of great solar and magnetic activity it is less likely than usual to see an aurora at the highest latitudes. The region of auroral occurrance moves in towards the equator at such times.

β) Before the discovery of the radiation belt, rocket studies had been made to investigate the possibility that aurorae were made by energetic particles[2]. It had

[13] J. R. JOKIPII: Astrophys. J. **143**, 961 (1966).
[14] T. W. SPEISER, and N. F. NESS: J. Geophys. Res. **72**, 131 (1967).
[15] J. R. SPREITER, and B. J. HYETT: J. Geophys. Res. **68**, 1631 (1963).
[1] S.-I. AKASOFU, S. CHAPMAN, and A. B. MEINEL: This Encyclopedia, Vol. 49/1, 1—158.
[2] J. A. VAN ALLEN: Proc. Nat. Acad. Sci. U.S. **43**, 57 (1965).

been known for some time that auroral rays were quite closely aligned with geomagnetic field lines [*27*]. This strongly implied that *energetic charged particles* were involved in the emission. Aurorae commonly occur at about 100 km. It takes about 10 keV electrons or 500 keV protons to penetrate this far into the atmosphere. The gyroradius for electrons of $\mathscr{E}_e = 10$ keV here is $R_c = 5$ m and for $\mathscr{E}_p = 500$ keV is $R_c = 2$ km. Rayed structures are often seen with thicknesses of less than one kilometer. Therefore, these thin structures must be made by electrons not by protons at 100 km altitude.

The first firm evidence of energetic particles causing aurora came in 1950 when Meinel [*28*] observed doppler-shifted hydrogen Balmer-lines in an aurora. When energetic protons of $\mathscr{E}_p \sim 10$ keV come down a field line into the auroral zone they lose some of their energy by ionization but also they may pick up an electron to become an excited neutral hydrogen atom. The optical emissions, due to transitions to the ground state of the neutral atom, will then be from an atom moving generally down the field line. A ground observer will see the emissions doppler-shifted towards the blue.

The first direct measurements of charged particle influx into the auroral zone came in 1953 when Van Allen[2] flew geiger counters on 'rockoons' (rockets launched from balloons) in the auroral zones. He sometimes found unusually large particle fluxes but only in the auroral zone. These were daytime flights and no correlation with auroral events were made.

Rocket experiments carrying particle detectors into auroral displays have shown clearly that energetic particles, both protons and electrons, are responsible for many if not all aurorae[3–5].

γ) *Direct experimenting* has shown that charged particles can generate aurora. Following several high altitude nuclear explosions[6–9] at Johnson Island in the Central Pacific aurora were observed near Apia, Samoa which is near the magnetically conjugate point to the explosions. Aurora were also observed in New Zealand.

These aurora were clearly due to the fission fragment beta-decay electrons propagating along field lines. The aurora at Apia appeared violet which is not usual but may be due to the fact that the electrons involved were of $\mathscr{E}_e \sim 1$ MeV which is considerably higher energy than for usual aurorae[10] and could therefore penetrate deeper into the atmosphere to deposit their energy and therefore involve different exitation reactions.

δ) It also has been known for some time that the *Sun* was at least partly *responsible* for the aurora. Large auroral displays are known to be closely related with magnetic storms, these in turn with solar activity. A day or so after a very sizable eruption on the Sun one quite often observes the beginning of a magnetic storm which is accompanied by aurorae in both auroral zones. This time delay indicates that the energy for these big auroral events is carried by the solar wind to Earth. That the energy for large auroral displays comes from the Sun seems quite certain. But this nearly ends our understanding of the matter. There are not many energetic particles in the solar wind. The solar wind protons are of about 1 keV and the electrons of roughly 10 eV, considerably below auroral energies.

---

[3] C. E. McIlwain: J. Geophys. Res. **65**, 2727 (1960).
[4] L. R. Davis, E. Berg, and L. H. Meredith: [*8*], 721 (1960).
[5] I. B. McDiarmid, D. C. Rose, and E. Budzinski: Can. J. Phys. **39**, 1888 (1961).
[6] P. Newman: J. Geophys. Res. **64**, 923 (1959).
[7] W. R. Steiger, and S. Matsushita: J. Geophys. Res. **65**, 545 (1960).
[8] B. A. Tinsley: Can. J. Phys. **42**, 779 (1964).
[9] J. B. Gregory: Nature **196**, 508 (1962).
[10] J. M. Malville: J. Geophys. Res. **64**, 2267 (1959).

The auroral energetic particles apparently do not come directly from the Sun. Are the energetic particles produced in the magnetosheath inside the bow shock wave or are they accelerated near Earth inside the magnetosphere? What is the coupling mechanism that produces the precipitated particles that make the aurora from the energy in the solar wind? How do the energetic particles get deep into the magnetosphere? Were the precipitated particles residents of the magnetosphere before the event or were they carried from the Sun by the solar wind?

ε) First let us consider the *energetics* of aurorae. The visible brightness of auroras varies from below the visual threshold to a brightness which produces an illumination on the ground equivalent to full moon light. Forms are rated in brightness [*29*] according to the International Brightness Coefficient (IBC) as indicated in Table 15.

Table 15. *International brightness coefficient.*

IBC	I	II	III	IV	Unit*
Visual Equivalent Brightness	Milky Way	Thin moonlit cirrus clouds	moonlit cumulus clouds	full moonlight	
Intensity OI 5577 Å	1	10	100	1000	kR
Intensity N^+ (1st. Neg.) 3914 Å	1	10	100	1000	kR
Energy Deposition	.003	.03	.3	3	$J\ m^{-2}\ sec^{-1}$

* 1 R = Rayleigh means 10^6 photons per cm^2 of source per second (i.e. $10^{10}\ m^{-2}\ sec^{-1}$).

O'Brien [*30*] has studied auroral emissions using the magnetically-oriented Injun III satellite. Photometers looking up and down field lines measured the emission at 5577 Å from atomic oxygen and 3914 Å from N_2^+ with a passband of 50 Å. Three geiger counters counting electrons of $\mathscr{E}_e > 40$ keV measured the angular distribution of these particles. Several detectors on the satellite showed that *electrons*, not protons, were the *dominant particles* in all the auroral events observed. It may be that there are fewer protons of a certain energy than electrons or just that the protons are less penetrating.

ζ) *Auroral data* for moonless nights for fifty passes in early 1963 are shown in Fig. 52. There is always some 3914 Å emission in the auroral zone and always some $\mathscr{E}_e > 40$ keV electron precipitation too (see Fig. 34). The average intensity of 3914 Å light was about 2 kR and the average electron flux being precipitated was about 4 erg $cm^{-2}\ sec^{-1}$ of $\mathscr{E}_e > 40$ keV. That there always was an aurora in the auroral zone had not been appreciated earlier because the aurora is sometimes subvisual. The process generating aurorae is a continuous one — not only associated with special events such as solar flares. The maximum intensity of 3914 Å light occurred at invariant latitude [Eq. (5.8)] $\Lambda = 69° \pm 1°$ or $L = 7.8 \pm 0.7$ somewhat higher latitude than what was earlier considered the maximum visual auroral iosphote.

One pass of Injun III is shown in Fig. 53. Before the satellite reached the auroral zone there was some 3914 Å emission — probably the night air glow. In this low latitude region more trapped electrons were seen by the geiger counter looking perpendicular to the field than dumped ones seen by the counter looking along the field. When the satellite got to the auroral zone the electron flux and light emission rose sharply and the electron flux became isotropic in the upper hemisphere — equal fluxes of dumped and trapped electrons. O'Brien says that

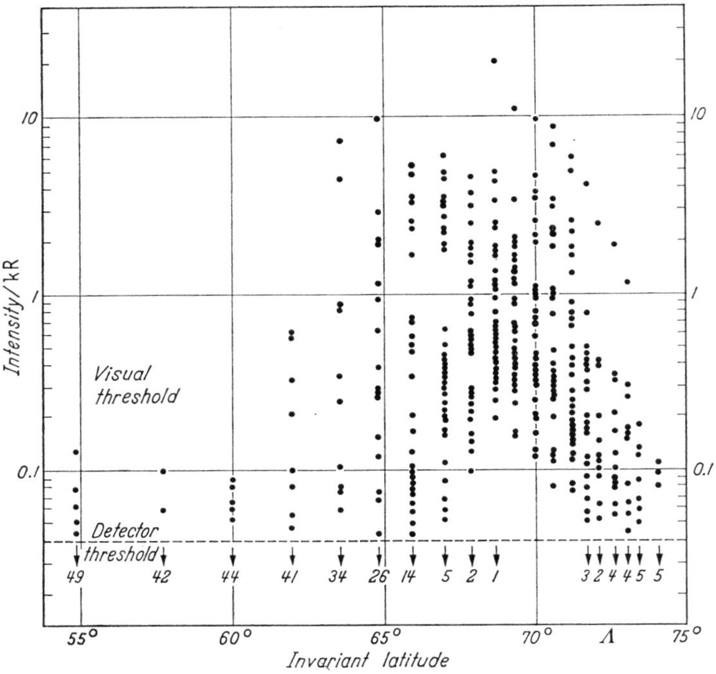

Fig. 52. Intensity (in kilo Rayleigh) of the N_2^+ 3914 Å auroral light measured by the Injun 3 satellite during approximately 50 passes early in 1963, ordered according to invariant latitude on abscissa [labelled Λ in this Fig. instead of Л in our text — see Eq. (5.8)]. (O'BRIEN [*30*].)

this is the usual situation in auroral events. At 1000 km atmospheric scattering will not yet have disturbed 40 keV electrons much so the isotropic flux during an auroral event must be considered to be related to the electron source and particle propagation down field lines.

On low altitude passes at 250 km the upward-looking photometer on Injun III saw very little if any light. This shows the flux of 10 eV electrons (which would stop above 250 km) is not larger than the flux of 10 keV electrons and therefore the energy carried by the 10 eV electrons is quite small compared to the energy in 10 keV electrons.

41. X-rays.

α) It is generally accepted that most of the energy brought into the auroral zone is carried by electrons of $\mathscr{E}_e < 30$ keV. But this does not mean that there are not lots of electrons above this energy present also. One method of studying these *higher energy electrons* is by observing X-rays by counters flown on balloons. Electrons of 100 keV will stop at an atmospheric depth of about 10^{-2} g cm^{-2} of air but the bremsstrahlung photons they emit of energies above 30 keV will penetrate several g cm^{-2} of air. Balloons float at depths of 3 to 10 g cm^{-2}, so they can measure the X-rays. WINCKLER[1] summarized the findings of the Minnesota group studying X-rays by flying balloons at Minneapolis ($L = 3.5$) as follows:

(i) At the geomagnetic latitude of 56° X-ray fluxes appear during strong negative bay disturbance or during the main phase of vigorous magnetic storms.

[1] J. G. WINCKLER: Atmospheric Phenomena, Energetic Electrons and the Geomagnetic Field. Univ. of Minn. Tech. Report CR-40 July 1961. — J. G. WINCKLER, P. D. BHAVSAR, and K. A. ANDERSON: J. Geophys. Res. **67**, 3717 (1962).

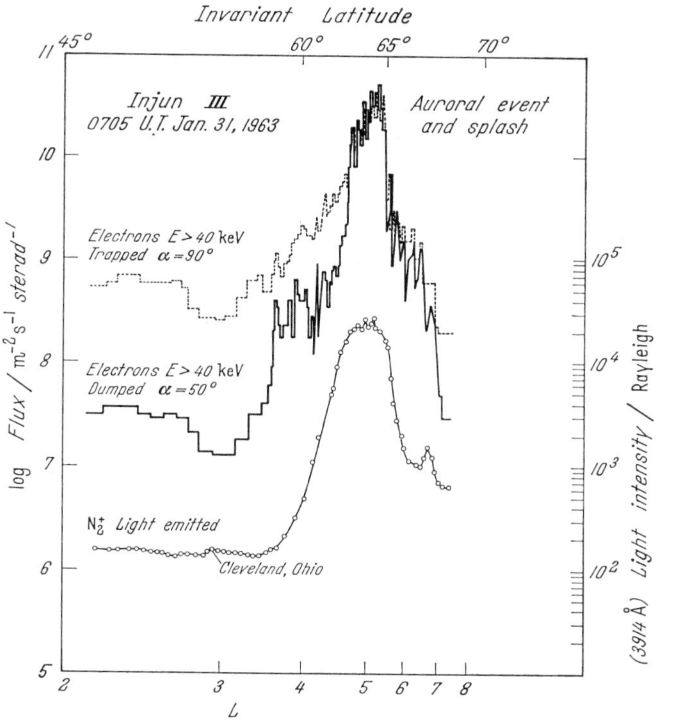

Fig. 53. Measurements of trapped and precipitated particles and 3914 Å auroral emission from a pass of the Injun 3 satellite ordered according to invariant latitude on abscissa [labelled Λ in this Fig. instead of Л in our text — see Eq. (5.8)]. It shows the simultaneous increases of trapped and precipitating particle fluxes to be compared with emitted light (lowest curve) variation. (O'BRIEN [30].)

(ii) Whenever observations of visible auroras could be made it was found that the X-rays were closely correlated with auroras above the balloon.

The correlation was particularly strong for active, rayed structures. The correlations held to the limit of accuracy in the measurements which was approximately one minute.

(iii) The X-ray bursts, magnetic disturbances and auroras occurred predominantly during hours of local nighttime.

(iv) The integrated flux of electrons producing the X-ray bursts was as high as $10^9 \ldots 10^{12}$ cm^{-2} ($10^{13} \ldots 10^{16}$ m^{-2}) during typical bursts having durations of approximately one hour. The peak of intensity reached 10^9 cm^{-2}sec^{-1} (10^{13} m^{-2}sec^{-1}) during strong auroras.

(v) The electron spectra which best fit the observed photons at balloon altitude are very steep and can be represented by power law spectra with exponent of minus four or five for the differential energy of the electrons.

β) ANDERSON[2,3] and BROWN[4] flying balloons in the *auroral zone* found that detectable X-ray fluxes were present over the auroral zone about 50 percent of the time even during periods of magnetic quiet. Measured energy spectra were very steep from 25 keV to 200 keV and few photons of $\mathcal{E}_{ph} > 200$ keV were

[2] K. A. ANDERSON: J. Geophys. Res. **65**, 3521 (1960).
[3] K. A. ANDERSON, and D. C. ENEMARK: J. Geophys. Res. **65**, 3521 (1960).
[4] R. R. BROWN: J. Geophys. Res. **66**, 1379 (1961).

observed. Anderson found that the electron precipitation required in the auroral zone to produce the observed average X-ray flux was $10^{10} \ldots 10^{11}\,\mathrm{cm^{-2}\,d^{-1}}$ ($\approx 10^{9} \ldots 10^{10}\,\mathrm{m^{-2}\,sec^{-1}}$) of $\mathscr{E}_e > 30$ keV. The bremsstrahlung efficiency (i.e. number of photons divided by number of electrons producing them) has been taken here as $\varepsilon \approx 10^{-5}$.

γ) A very interesting feature of auroral zone X-rays is the occurrance of large time *fluctuations*. WINCKLER[1] found an event at Minneapolis that showed peaks in a spectral analysis of periods of 0.8 sec and 1.6 sec. If the 0.8 sec time is the bounce period of the electrons, they would have $\mathscr{E}_e \sim 60$ keV just about right to produce the X-rays. Using high resolving time equipment, ANDERSON[5] found shorter period "microbursts" that frequently occurred in clusters. These have half-widths of about 0.2 sec. These microbursts must result from some currently not understood dynamic process occurring on auroral lines of force.

42. Particle fluxes and spectra.

α) Using all available spectral data, an average *auroral electron spectrum* has been constructed in Fig. 54. Like typical weather, this spectrum will probably never occur. Auroral spectra are clearly of many varieties and change substantially even during one event. But Fig. 54 is consistent with all the electron information given above. Most of the energy flux is at low energies — 35 percent is below 10 keV and 80 percent is below 20 keV. The total energy flux precipitated in Fig. 54 is 1.5 erg $\mathrm{cm^{-2}\,sec^{-1}}$ ($= 1.5 \cdot \mathrm{mW\,m^{-2}}$). There must be a sharp peak in the electron energy spectrum at about 5 keV to fit the available data. This is similar to one energy spectrum measured by MCILWAIN[1] on rockets in 1958. The electron energy spectrum above 100 keV is not well known yet and is therefore shown dotted on Fig. 54. There is not enough data yet to say how similar the auroral proton energy spectrum is to Fig. 54. It may be roughly the same.

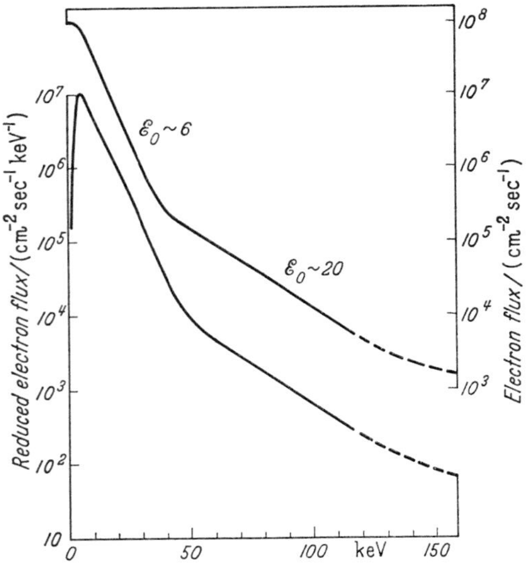

Fig. 54. Typical electron energy spectra of precipitating auroral electrons based on various experimental data. The lower curve and left hand ordinate is for a differential spectrum while the upper curve and right hand ordinate refers to an integral or energy spectrum. The high energy region shown dotted is not well known yet.

[5] K. A. ANDERSON, and D. W. MILTON: J. Geophys. Res. **69**, 4457 (1964).
[1] C. E. MCILWAIN: J. Geophys. Res. **65**, 2727 (1960).

Investigator	Particles	Intensity* $j(>\mathscr{E})$/m^{-2}sec^{-1}sr^{-1}	$\mathscr{E}$/keV	Energy range/keV	Spectrum parameters** $\mathscr{E}_0$/keV	Spectrum parameters** m***
DAVIS et al. (1960)	e	$2 \ldots 6 \cdot 10^{11}$	10	5…50		1
	p	$3 \cdot 10^7$ to $1.2 \cdot 10^9$	100	100…800		1…3.3
MCILWAIN (1960)	e	$2.1 \cdot 10^{12}$	3	3..,30	5	
	e	$5 \cdot 10^{14}$			****	****
	p	$2 \cdot 10^9$	80	80…200	30	
MCDIARMID et al. (1961)	e	$2 \cdot 10^{10}$ □	30	30…110	22	
ANDERSON and ENEMARK (1960)	e	$2 \cdot 10^9$ □	25	25..,200	28	
BHAVSAR (1962)	e	10^{13} □	22	22…100		5
ANDERSON and DEWITT (1963)	e	$5 \cdot 10^{11}$ □	25	25…200		4
KRASSOVSKIJ et al. (1962)	e				14 eq.	
O'BRIEN and LAUGHLIN (1962)	e	$6 \cdot 10^{11}$	40	>1		5 di
		10^{14}	10	>1		
	e	$3 \cdot 10^{10}$	40	>1	6	
		$6 \cdot 10^{14}$	10	>1		
O'BRIEN et al. (1962)	e	$3 \cdot 10^{10}$	40	50…90		3 di
	e	10^{12}	5	1, 50, and 90		3 di
STILWELL (1963)	e	10^{12}	10	>10	4	
MCDIARMID et al. (1963) (after HULTQVIST)		Average fluxes ϕ				
	e	$3 \cdot 10^8$ 65°	40 ($K_p < 4-$)	$40+250$	41	2.8 di
	e	$1.9 \cdot 10^7$ 60°	250 ($K_p < 4-$)	$40+250$	41	2.8 di
	e	$3 \cdot 10^9$ 65°	40 ($K_p > 4+$)	$40+250$	30	3.9
	e	$2.6 \cdot 10^7$ 60°	250 ($K_p > 4+$)	$40+250$	30	3.9
O'BRIEN (1964) O'BRIEN and TAYLOR (1964)	e	Average $4 \cdot 10^9$ □	40	>1 >40	5.7	2.2
SHARP et al. (1964)	e	10^{12} and $8 \cdot 10^{13}$	2 2	>2 >28 >2 >28	8 4…9	
(Private communication to HULTQVIST, 1964)				>2 >28	3…5	
	e			$1.5+21$	4.6, 4.1, 20, 95	
MCDIARMID and BUDZINSKI (1964)	e	$\sim 5 \cdot 10^{10}$	1	1…80	12	
EVANS (1965)	e	$3 \cdot 10^{12}$	20	$10+25$	20…35	

* Symbol □ means omnidirectional flux/m^{-2} sec^{-1}.

** Description either by exponential law, Eq. (19.1) with parameter $\mathscr{E}_0$, or by power law, with exponent m.

*** Refers to integral distribution except for cases where figure is followed by di = differential.

**** Approximately monoenergetic: 6 keV.

The Lockheed group has found[2] that the proton energy flux in aurora is not negligible. Proton aurora are more spread and diffuse than electron aurorae. Even if the protons start to come toward Earth in a narrow beam, charge exchange will spread them into a broad pattern. DAVIDSON[3] has shown that charge exchange operating on an isotropic distribution of precipitating protons will spread them out to cover a width of about 600 km. Integrating across the auroral zone the proton energy influx of $\mathscr{E}_p > 4$ keV was measured[2] to be about 15 percent of the electron flux of $\mathscr{E}_e > 80$ eV.

β) A *summary*[4] of all these auroral intensity and spectral measurements is given in Table 16.

γ) ANDERSON [*31*] has estimated average auroral electron fluxes of $\mathscr{E}_e > 30$ keV to be 10^6 cm^{-2} sec^{-1} which with an auroral area of 10000 km times 10 km or 10^{15} cm^2 gives an *energy flux* of about 10^{18} erg sec^{-1}, i.e. 10^{11} W. This is a crude estimate of the normal total particle energy influx. O'BRIEN [*32*] using an efficiency value of one percent, for converting electron kinetic energy to light, gives the same estimate for an IBS III aurora. Also O'BRIEN [*30*] has estimated that it takes about 10^{11} W to sustain the continuous quiet time electron precipitation observed throughout the auroral zone on Injun I and III.

On Feb. 11, 1958 a very bright aurora occurred over large areas of the Earth. It emitted 10^5 kR over parts of the U.S.A. for hours. KRASSOVSKIJ [*33*] estimates that it extended over more than 10 percent of Earth with an average intensity of about 10^4 kR and therefore (using an efficiency $\varepsilon = 1\%$) required about 10^{21} erg sec^{-1} i.e. 10^{14} W of particle energy to sustain it. In a three-hour period the total energy input must have been about 10^{25} erg $= 10^{18}$ J $\approx 3 \cdot 10^{11}$ kWh.

This energy must be supplied by the source of the auroral electrons. This is a problem when considering the relationship between aurorae and the trapped radiation belt particles as we will see shortly.

43. Related phenomena.

α) O'BRIEN [*32*] has reviewed all of the phenomena that occur with aurorae and gives *quantitative estimates* of the phenomena occurring during a bright aurora (class IBC III of Table 15). We repeat his estimates in updated form in Table 17.

Data from satellites Injun I and II [*30*] show that auroral electrons are accelerated above 1000 km. They are observed coming downwards at this point. Few events have even been seen with more upwards-moving particles than downward moving ones. Usually the flux was essentially isotropic during auroral events. The detectors on Injun III measuring the trapped flux and precipitated flux gave similar values during aurorae. It would be nice to know if the acceleration process takes place at the equator or not. If so, then we would expect that electrons would go down a field line both north and south and we would have conjugate aurorae.

Attempts have been made to study the conjugacy of aurorae in the Northern and Southern Hemisphere. This can be done only near the equinoxes when it is dark at both locations. Data[1] from Farewell, Alaska, and Campbell Island which are conjugate to about 100 km show the general behavior of aurorae are very similar with auroral breakup and intensity variations occurring together but it was not possible to compare individual detailed auroral forms to see if they were conjugate.

β) If we have an auroral electron flux of 10^{12} cm^{-2} sec^{-1} (10^{16} m^{-2} sec^{-1}) of particle energy 10 keV, can these particles be confined inside the geomagnetic

[2] J. E. EVANS, R. G. JOHNSON, E. G. JOKI, and R. D. SHARP: [*13*], 773 (1966).
[3] G. T. DAVIDSON: J. Geophys. Res. **70**, 1061 (1965).
[4] ANDREI KONRADI: Electron and Proton Fluxes in the Tail of the Magnetosphere. Goddard Space Flight Center X-611-65-465, November 1965.
[1] R. N. DEWITT: J. Geophys. Res. **67**, 1347 (1962).

Table 17. *Approximate quantitative estimates of phenomena for a bright aurora* (IBC III)

Phenomenon	Estimate
Solar Wind	$v \sim 700$ km sec^{-1}; particle density $n \sim 10$ cm$^{-3} = 10^{-5}$ m^{-3}.
Distorted magnetosphere	Sunward radial distance to magnetopause: 8...10 R_E (50000...65000 km).
Particle precipitation	Localized: 400 erg cm^{-2} sec$^{-1} = 0.4$ W m^{-2}; worldwide: 10^{18} erg sec$^{-1} = 10^{11}$ W.
VAN ALLEN radiation	General increase in low-energy electrons and protons at $R > 5 R_E$.
Very low frequency emissions	10^{-6} erg cm^{-2} sec$^{-1} = 10^{-9}$ W m^{-2} over 1...10 kHz.
BALMER emissions	Photon flux: 10^{10} cm^{-2} sec$^{-1} = 10^{-14}$ m^{-2} sec^{-1} at 6563 Å; $3 \cdot 10^9$ cm^{-2} sec$^{-1} = 3 \cdot 10^{13}$ m^{-2} sec^{-1} at 4861 Å.
Auroral light	Photon flux at 5577 Å (green line): 10^{11} cm^{-2} sec^{-1} = 10^{15} m^{-2} sec^{-1} or 0.36 erg cm^{-2} sec$^{-1} = 3.6 \cdot 10^{-4}$ W m^{-2}; Total spectrum: 20 erg cm^{-2} sec$^{-1} = 0.02$ W m^{-2}.
Heating	0.2 °K sec^{-1} heating flux: 60 erg cm^{-2} sec$^{-1} = 0.06$ W m^{-2}.
Sound	1...10 dyn cm^{-2} = 0.1...1 N m^{-2} at periods 10...100 sec.
X-Rays at balloon altitudes	Photon flux: x $\cdot$ 10 cm^{-2} sec^{-1} = x $\cdot$ 10^5 m^{-2} sec^{-1} with energy y $\cdot$ 10 keV $\sim$z 10^{-7} erg cm^{-2} sec^{-1} = z 10^{-10} W sec^{-1} (x, y, z very dependent on electron energy spectrum, mostly between 1 and 10).
Ionization	Ion pairs: $7 \cdot 10^{12}$ cm^{-2} sec$^{-1} = 7 \cdot 10^{16}$ m^{-2} sec^{-1} giving maximum electron densities about $5 \cdot 10^6$ cm$^{-3} = 5 \cdot 10^{12}$ m^{-3} (i.e. plasma frequency 20 MHz).
Cosmic-noise absorption	Several dB at $\sim$30 MHz (very dependent on electron energy spectrum).
Magnetic disturbance	$1000\,\gamma = 10^{-2}\,\Gamma$ $\sim$ about 1% change in the surface field below the aurora.
Micropulsations	$1\,\gamma = 10^{-5}\,\Gamma$ with periods 5...150 sec (**pc 2···4**).
Electrojets	10^5 A in localized patterns with about 10 km cross-section at altitudes of about 100 km.
Earth currents	Electric potential differences of about 1 V km^{-1} = 1 mV/m.

field? The condition for *particle containment* by the field is that the particle pressure be less than the magnetic pressure which statement can be formulated as a comparison of energy densities*:

$$\tfrac{1}{2} n m v^2 < B^2/2\mathrm{u}\mu_0. \tag{43.1}$$

For an IBC 4 aurora corresponding to 3000 erg cm^{-2} sec^{-1} = 3 W m^{-2} (see Table 15 in Sect. 40) we have

$$\tfrac{1}{2} n m v^2 \sim 10^{-6}\,\text{erg cm}^{-3} = 10^{-7}\,\text{J m}^{-3},$$

and the field energy density near the equator for $L = 6$ where $B = 140\,\gamma$ is

$$B^2/2\mathrm{u}\mu_0 \sim 10^{-7}\,\text{erg cm}^{-3} = 10^{-8}\,\text{J m}^{-3}.$$

So the particle flux cannot be contained by the magnetic field near the equator in this case.

We have assumed here that the particle pressure stays constant as we move up along the line. This is correct for an isotropic particle distribution as was observed frequently by the Injun satellites. However, if the particles are streaming down the field lines with $v_\perp = 0$ then the particles exert no pressure perpendicular

* $\mathrm{u} \equiv 1$ in rationalized, $\equiv 4\pi$ in non-rationalized systems of units. See "Introductory Remarks" on p. 1.

to the field lines and the particles could be contained. But if this were the case then we would expect that the beam might develop micro-instabilities due to the anisotropic pressure. The kink instability and mirror instability and the two-stream instability will tend to make the particle pressure isotropic. The question of containment must be considered unsettled.

44. Relation to the radiation belt.

α) When the radiation belt was found it was obvious to suggest that it was related to aurorae. After space probes Pioneer III and IV showed the existence of a time variable outer zone of the belt it was frequently suggested that the aurora was produced by particles dumped from the outer zone. It is now quite apparent that this is not the case. The *energy stored* in the trapped particles is not adequate to supply the aurora.

The bright aurora of 11. Feb. 1958 discussed by KRASSOVSKIJ [*32*] must have deposited a total energy flux of about 10^{21} erg sec^{-1} = 10^{14} W. It lasted several hours so the total energy required* was about 10^{25} erg = 10^{18} J ≈ 3 · 10^8 MWh. DESSLER[1] and VESTINE have shown that the total energy in the radiation belt must be less than 10^{23} erg = 10^{16} J ≈ 3 · 10^6 MWh, otherwise Earth's magnetic field would be distorted by an amount measurable at the surface, which it is not. How much energy is stored in the tube of force above 1 m^2 of the top of the atmosphere? The volume of the tube of force is 3 · 10^{10} m^3 above 1 m^2 at $\Lambda = 68°$. Assuming a trapped particle (electron) flux of 10^{12} m^{-2} sec^{-1} of average individual energy $\bar{\mathscr{E}}_e = 60$ keV = 10^{-7} erg = 10^{-14} J, the energy flux is 0.01 W m^{-2}. With an average particle velocity of 10^8 m sec^{-1} the total energy in this tube is 3 · 10^{-4} J. This means the tube will be drained in a time of about 10^{-2} sec to make the aurora.

This shows this aurora could not have been supplied from the previously trapped particles, as a leady bucket model of the aurora would imply.

β) O'BRIEN, using Injun III data[2] [*30*], has demonstrated an interesting relation between *auroral electrons and trapped electrons*. When an auroral event is observed and the precipitated electron flux rises, the trapped electron flux also frequently *rises* — not decreases, see Fig. 53. The $\mathscr{E}_e > 40$ keV trapped electron flux increases at this time but the $\mathscr{E}_e > 1$ MeV flux is normally not changed much. O'BRIEN has described this process by the splash catcher model shown in Fig. 35 in contrast to the leaky bucket model. The splash catcher model assumes that some 'helpful' process accelerates the auroral electrons and also the newly-trapped electrons. The VAN ALLEN belt catches some splashes of the precipitated auroral electrons. The auroral acceleration mechanism is one of the sources of the outer part of the VAN ALLEN radiation belt on this model.[3]

If the energy for the aurora does not come from the trapped particles, where does it come from? The time averaged auroral energy influx is $10^{17} \ldots 10^{18}$ erg sec^{-1} = $10^{10} \ldots 10^{11}$ W. The solar wind consists of a proton flux of about 10^8 cm^{-2} sec^{-1} (= 10^{12} m^2 sec^{-1}) of roughly 1 keV energy. This gives an energy flux of about 0.1 erg cm^{-2} sec^{-1} = 10^{-4} W m^{-2}, so that the total energy influx incident on the front face of the magnetosphere is (about 10^{10} km^2) would be 10^{19} erg sec^{-1} = 10^{12} W. Therefore if there exists a mechanism that will take a few percent of the energy of the solar wind and put it into energetic particles it would make the aurora.

γ) It appears that the energy for the aurora comes from the Sun but the energetic particles do not seem to come directly from the Sun. A solar storm that produced a large change in the trapped radiation [*1*] seen by satellite Explorer 7 was not observed by an identical detector[4] on space probe Pioneer V far from

* It appears that, economically, nature in producing an important aurora supplies an amount of energy which would be sold on Earth for some 10000000 $.

[1] A. J. DESSLER, and E. H. VESTINE: J. Geophys. Res. **65**, 1069 (1960).

[2] J. VAN ALLEN, and W. LIN: J. Geophys. Res. **65**, 2998 (1960).

[3] R. L. ARNOLDY, R. A. HOFFMAN, and J. R. WINCKLER: J. Geophys. Res. **65**, 3004 (1960).

[4] J. W. CHAMBERLAIN: Astrophys. J. **134**, 401—424 (1961).

Earth. This indicates that *local acceleration* of particles already present in Earth's magnetic field or maybe in the transition zone is the process responsible for the new particles in the radiation belt. Probably the same holds true for the precipitated auroral electrons.

The very rapid time variations of X-ray flux observed by ANDERSON[5] using scintillators on balloons in the auroral zone is strongly indicative of local acceleration. These X-rays are undoubtedly generated by bremsstrahlung of electrons striking the atmosphere. It is very difficult to believe that time variations with a time scale of a sec result from particles carried all the way from the Sun. Small scale local effects must produce these bursts.

45. Theories of the aurora.

α) There is no fully satisfactory theory of the aurora now.[1] There have been a variety of theories put forward in the past, in particular those started by FITZGERALD, BIRKELAND, and a few others at the end of last century. These considered the motion of one charge in the magnetic field of Earth, see [*35*—*37*] and [*38*] which is a summary. We will survey only those theories that are still somewhat in contention. All of them do not consider isolated charges but a plasma consisting of many charges of both signs. The CHAPMAN-FERRARO theory of magnetic storms contained the concept of a magnetosphere produced by an ionized but over-all neutral *stream of matter* approaching Earth from the sun. By assuming acceleration processes going on near the magnetopause the authors proposed to produce auroral particles. (For details see contribution by E. N. PARKER and V. C. A. FERRARO in Vol. 49/3, Chapt. F, p. 199.) Various varieties on this theme are still in use.

β) ALFVÉN [*1*] in 1939 suggested that there might be charge separation and a resultant *polarization electric field* in the neutral stream approaching Earth due to the presence of a solar magnetic field. Particles in the geomagnetic field would be accelerated by this electric field. Most people now doubt that such an electric field exists in the solar wind. Because of the large energy density of the particles of the wind, the particles move the field around at will and there should be no $\boldsymbol{v}\times\boldsymbol{B}$ electric field for particles moving with the wind velocity and therefore no charge separation.

CHAMBERLAIN[2] and KERN[3] had considered situations where charge separation occurs in the geomagnetic field and electric fields are produced parallel to magnetic field lines. To accomplish this they invoke longitudinal variations in fields or particle fluxes which may or may not exist. The parallel electric field can accelerate the particles to form an aurora.

γ) More recently CHAMBERLAIN[4] has suggested than an *instability* may produce aurora. It is assumed in this model that particles that are already inside the magnetosphere are triggered by the instability and dumped to form an aurora. Chamberlain considered a microstability which required

(i) a density gradient increasing outwards with a scale length $l \ll \frac{1}{3} R_{\mathrm{E}}$ which is the scale length of the magnetic field and

(ii) the average ion energy much be larger than the average electron energy.

If these conditions were met an instability could grow that would dump both electrons and protons.

---

[5] K. A. ANDERSON, and D. W. MILTON: J. Geophys. Res. **69**, 4457 (1964).
[1] S.-I. AKASOFU, S. CHAPMAN, and A. B. MEINEL: This Encyclopedia, Vol. 49/1, 124—140.
[2] J. W. CHAMBERLAIN: Astrophys. J. **134**, 401—424 (1961).
[3] J. W. KERN: J. Geophys. Res. **67**, 2649 (1962).
[4] J. W. CHAMBERLAIN: J. Geophys. Res. **68**, 5667 (1963).

δ) AKASOFU and CHAPMAN[5] used an X type null but had it result from a *ring current* near $R = 6\ R_E$ sufficiently large to produce a reversal of the direction of the field. Particles moving into the null can change their pitch angles and be precipitated into the atmosphere. This field geometry probably would not be stable and anyway satellite observations show that such a ring current probably doesn't exist.

CHAMBERLAIN[4], KERN[3], AKASOFU and CHAPMAN[5] invoke acceleration processes interior to the magnetosphere. They start with particles already inside and by some process inside increase their energies and dump them.

ε) The idea of using *trapped particles* to produce aurora was suggested shortly after the radiation belt was discovered but the source requirements for large aurorae cannot be met for very long by the trapped radiation. Also O'BRIEN's observation on satellite Injun that there are *larger* fluxes of trapped particles after an aurora than before, pretty well rule out these two step processes that depend on the presence of trapped particles to start with.

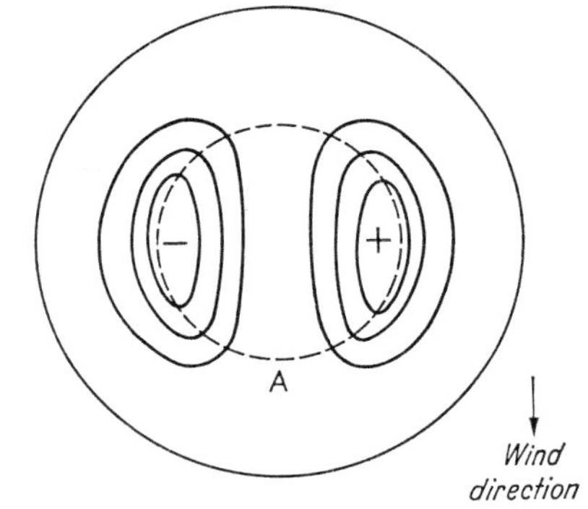

Fig. 55. The electric potential distribution in the northern hemisphere expected on the basis of the open model of the magnetosphere. (DUNGEY, ref. [6], Sect. 45.)

ζ) DUNGEY's[6] open magnetosphere model also can produce aurorae. Assuming a southward directed solar magnetic field an X type magnetic null occurs at the subsolar point where solar field lines connect to geomagnetic lines according to this model, see Fig. 45. The solar field lines blow over the polar caps and disconnect at the X-type null. To an observer on Earth there is a $\boldsymbol{v} \times \boldsymbol{B}$ electric field produced by the field lines being blown past by the solar wind of velocity $\boldsymbol{v}$. All field lines are equipotentials in this model. Looking at the polar cap the equipotentials are shown in Fig. 55. These look very like the $\mathbf{S}_D$ current system or auroral electrojets sometimes observed in the polar regions. (See for details contribution by T. NAGATA and N. FUKUSHIMA in Vol. 49/3, Sect. 38, p. 64. See also [*21*].) This would be the case if the $\mathbf{S}_D$ currents are HALL currents produced by this potential distribution which would seem to be reasonable because the HALL electrical conductivity seems to dominate in the ionosphere. A typical energy $\mathscr{E}$ that a single particle can get here is given by using an observed value of an ionospheric wind velocity of 100 m sec^{-1} and an appropriate acceleration distance l, to give

$$\mathscr{E} = q\,|\boldsymbol{E}|\,l = q\,\frac{|\boldsymbol{v}| \cdot |\boldsymbol{B}_{\circ}|}{c_0 \sqrt{\varepsilon_0\,\mu_0}}\,. \tag{45.1}$$

[5] S.-I. AKASOFU, and S. CHAPMAN: Phil. Trans. Roy. Soc. London Ser. A **253**, 359—406 (1961).
[6] J. W. DUNGEY: Phys. Rev. Letters **6**, 47 (1961).

Introducing numerical values $v = 100$ m sec^{-1}, $B_{\circ} = 0.3$ Gs and $l = \frac{1}{3} R_E$ we find $\mathscr{E} \approx 40$ keV. The voltage across the polar cap here is 40 kV. At times of major disturbances, when the solar wind velocity may be as large as 2000 km sec^{-1}, this voltage may become more than 100 kV. This theory of DUNGEY's, which uses a more realistic electric field than ALFVÉN, seems reasonable provided the southward solar field and fast merging of field lines exists.

η) AXORD and HINES [*34*] model of the aurora started with low energy *particles* back in the *magnetospheric tail* and connects them inwards gaining energy as they *move adiabatically* as in Fig. 56. This convective motion in the magnetosphere results in their model from a viscous interaction at the magnetopause which drives the motion. Field line exchange suggested by GOLD[7] allows the motion to proceed with velocity $\boldsymbol{u}$. The particles, by following the field lines in the convective motion, are adiabatically energized as they move inwards. Using field lines that are equipotentials, the $\boldsymbol{u} \times \boldsymbol{B}$ electric fields produced here when mapped onto the polar cap produce a distribution somewhat similar to the polar cap potential distribution in DUNGEY's model (see Fig. 55) and the $\mathbf{S}_D$ currents occur here too, but they are distorted by magnetospheric rotation. The convection is assumed to go in to about $R = 4.5\ R_E$ and particle precipitation and aurorae would occur into this depth or to $\phi = 62°$. Sharp narrow auroral structures are not explained here and there does not appear a well defined auroral zone in this model. (For further details see the contribution by E. N. PARKER and V. C. A. FERRARO in Vol. 49/3, Sect. 9, p. 155 ... 158.)

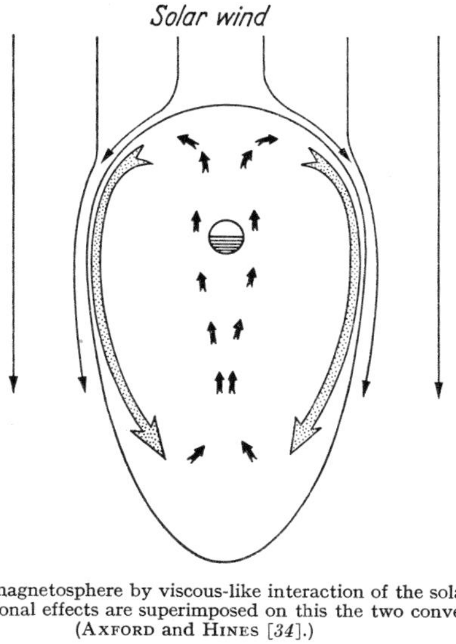

Fig. 56. The motion impressed on the magnetosphere by viscous-like interaction of the solar wind according to the theory of AXFORD and HINES. When co-rotational effects are superimposed on this the two convective cells become asymmetric. (AXFORD and HINES [*34*].)

46. Motion of particles near a magnetic null.

α) The magnetic neutral sheet in the tail of Earth's magnetic field may have a substantial voltage across it. DUNGEY's model[1] of the potential distribution responsible for the $\mathbf{S}_D$ current system in Fig. 55 has about 40 keV across the polar cap. If a *voltage* exists across the polar cap it should exist *across the magnetic tail*

[7] T. GOLD: J. Geophys. Rev. **64**, 1219 (1959).
[1] J. W. DUNGEY: Phys. Rev. Letters **6**, 47 (1961).

too. The direct conductivity σ_0 along field lines is very high so field lines should be essentially equipotentials. DUNGEY[1] has suggested that this potential will accelerate auroral particles and has given a rough idea of the flow pattern. SPEISER[2] has taken this field model and worked out how particles move in it, get accelerated and appear in the auroral zone.

The neutral sheet with a voltage across it constitutes a system in which particles can be accelerated. SPEISER has used an electric field across the tail (symbol T)

$$\boldsymbol{E}_{\mathrm{T}} = \frac{\boldsymbol{u} \times \boldsymbol{B}_{\mathrm{T}}}{c_0 \sqrt{\varepsilon_0 \mu_0}} \tag{46.1}$$

of 0.3 V km^{-1}, constant across the tail $\boldsymbol{E}_{\mathrm{T}}$ is directed along the z-axis, along the tail. This implies a flow velocity into the neutral sheet $u = 15$ km sec^{-1}. In a simple plane tail field geometry (neglecting the 'dipole field' contribution), if the particle got into the neutral sheet, it would move in the sheet indefinitely gaining energy as it changed z. SPEISER described this neutral sheet by a field geometry given by

$$\left.\begin{aligned} \boldsymbol{B} &= -b\left(\frac{x}{d}\right)\boldsymbol{y}^0 \quad \text{for} \quad |x| < d, \\ |\boldsymbol{B}| &= b \qquad\qquad\; \text{for} \quad |x| > d, \\ \boldsymbol{E} &= -a\,\boldsymbol{z}^0 \end{aligned}\right\} \tag{46.2}$$

where $\boldsymbol{x}^0$, $\boldsymbol{y}^0$, $\boldsymbol{z}^0$ are unit vectors in directions NS (perpendicular to neutral sheet), morning-evening (in neutral sheet) and Sun-Earth (along tail), respectively, where b is the strength of the magnetic field outside the neutral sheet of thickness $2d$ and a is the strength of the electric field.

β) For a charged particle in the neutral sheet *acceleration* appears by its equation of motion

$$m\,\ddot{\boldsymbol{r}} = q\left(\boldsymbol{E}_{\mathrm{T}} + \boldsymbol{v} \times \boldsymbol{B}_{\mathrm{T}}/c_0 \sqrt{\varepsilon_0 \mu_0}\right) \tag{46.3}$$

which can be written

$$\left.\begin{aligned} \ddot{x} &= c_1\,\dot{z}\,x, &&\text{(a)} \\ \ddot{y} &= 0, &&\text{(b)} \\ \ddot{z} &= -C_3 - C_1\,x\dot{x} &&\text{(c)} \end{aligned}\right\} \tag{46.4}$$

where $C_1 = \frac{q}{m}\frac{b}{d}$ and $C_3 = \frac{q}{m}a$.

Substituting the first integral of Eq. (46.4c) into Eq. (46.4a) gives

$$\ddot{x} = -C_1\,x\{-\dot{z}_0 + \tfrac{1}{2}C_1(x^2 - x_0^2) + C_3\,t\} \tag{46.5}$$

approximating for large times and solving

$$x \approx A\,t^{-\frac{1}{4}} \sin\left[\frac{2}{3}\frac{q}{m}\left(\frac{b\,a}{d}\right)^{\frac{1}{2}} t^{\frac{3}{2}} + \delta\right] \tag{46.6}$$

where A and δ are constants depending on initial conditions. The result of this model is that particles of either sign execute damped oscillations around $x = 0$ staying in the neutral sheet with their energy growing continuously. Moving across the $x = 0$ neutral plane the particle's radius of curvature changes sign and the particle starts bending back towards the $x = 0$ plane moving in z as it does so. This snake-like motion in z results in the particle being accelerated by the $\boldsymbol{E}_{\mathrm{T}}$ field. This model is too simple to be realistic. To make it better, SPEISER has

[2] T. W. SPEISER: J. Geophys. Res. **72**, 3919 (1967).

added a small magnetic field $\boldsymbol{B}_{\text{dp}} \equiv (B_x, 0, 0)$ normal to the neutral sheet to represent the Earth's dipole field.

The dipole field eventually makes the particle escape from the neutral sheet. Typical data for the particle motion with $B_x = 1\ \gamma$ is given in Table 18.

Table 18. *Particle acceleration in the neutral sheet of magnetospheric tail.*

Particle entering neutral sheet at	Proton energy at emergence	Electron energy at emergence	Distance protons travel towards Earth in neutral sheet	Distance electrons travel towards Earth in neutral sheet	Maximum pitch angle at emergence
$y = 50\ R_E$	30 keV	16 eV	24 R_E	$1/80\ R_E = 80$ km	0.3°...4°
$y = 150\ R_E$	50 keV	12 keV	120 R_E	$10\ R_E = 64000$ km	0.02°...0.2°

These particles emerge from the neutral sheet in a narrow strip with small pitch angles. A typical particle trajectory is shown in Fig. 57.

Putting in an inner boundary on the tail field as MEAD has done at $R = 8\ R_E$ this high gain strip projects down onto Earth in a narrow strip at $65° < \text{Л} < 70°$ for a 20 γ tail field. This is highly suggestive of the auroral zone.

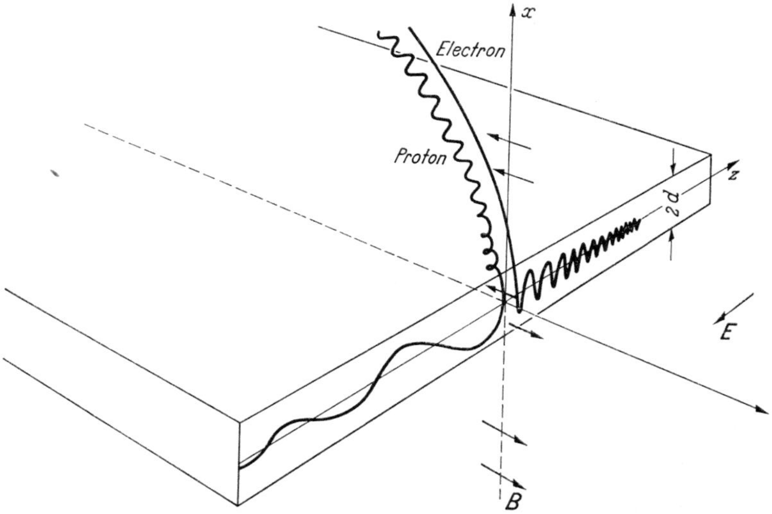

Fig. 57. A sketch of particle trajectories computed by SPEISER using a geomagnetic-tail neutral sheet, a cross-tail electric field, and a small component of the magnetic field perpendicular to the neutral sheet such as might be furnished by the Earth's dipole field. Both protons and electrons oscillate about the sheet, accelerating in opposite directions, and are turned in the same direction toward the Earth by the small magnetic-field component perpendicular to the sheet. When the particles are turned 90°, they are ejected from the neutral sheet. The dimensions shown are illustrative and are not drawn to scale (SPEISER, ref. [2], Sect. 46).

γ) TAYLOR and HONES[3] have also calculated the motion of particles in Earth's magnetic field using the *electric field* they have derived *from the* $\mathbf{S}_D$ *current system* together with a corrotational $\boldsymbol{E}$ field. The magnetic field model used was an image dipole model plus an added tail field that was uniform and in opposite directions above and below the magnetic equatorial plane to agree with the data of satellite IMP-1. They assume field lines are equipotentials and, further, $|\mu|$ and $\mathscr{I}$, the adiabatic constant (see Sect. 4α, γ) are conserved and that the total energy $W = \mathscr{E} + qU$ is a constant of the motion.

[3] H. E. TAYLOR, and E. H. HONES: J. Geophys. Res. **70**, 3605 (1965).

When the field geometry is known, specifying W, $|\boldsymbol{\mu}|$ and $\mathcal{I}$ determines a surface on which one particle must move. For different positions on this surface W, B_m and pitch angle α can be determined. The electric fields in this model are such as to prohibit entry of solar wind protons into the *tail* of the magnetosphere while permitting entry and subsequent energization of solar-wind electrons along the evening-side of the tail. The electrons, drifting from evening to dawn through the magnetosphere, constitute a current which establishes and controls the length of the tail in this model.

Their model provides the following description of auroral processes:

(i) Auroral particles originate in the solar wind with energies < 1 keV.

(ii) They are trapped near the surface of the magnetosphere and as they drift their energies are increased by the electrostatic field to 1...35 keV. They will not remain in trapped orbits (unless carried into them by nonconservative electric fields) but will either precipitate into the atmosphere and be absorbed, or will drift back out of the magnetosphere again.

(iii) There will be a definite boundary between the region where solar wind electrons may be found and the region where solar wind protons may be found, the protons confined south of the electrons. The width of this boundary should be 100...500 km. The maximum widths should occur during the day, especially near the dawn-dusk meridian and the minimum widths should occur near midnight. The location of this boundary should be in the region of auroral activity.

(iv) Electron spectra should be harder south of the boundary than north of it.

(v) Electron precipitation will occur primarily in a narrow range of latitudes on the night-side of Earth. Protons will be precipitated over a large area in the afternoon and in a narrow region south of the electrons throughout the night.

(vi) A fairly sharp boundary of the afternoon proton precipitation should occur at the magnetic noon meridian; no proton precipitation will occur at the magnetic noon meridian; no proton precipitation will occur in late morning hours.

δ) It seems likely that both of these models of auroral particle acceleration of Speiser[2] and Taylor and Hones[3] are at least partly true. *Partile injection* in the tail and on the front face of the magnetopause may both be important. There is one problem that may cause troubles with either or both of them. When a very bright aurora occurs we need almost 10^3 erg cm^{-2} sec^{-1} = 1 W m^{-2} deposited on the atmosphere. Will the magnetic field be able to contain this energy density at large distances from Earth. If the flux is isotropic as Taylor and Hones' is it seems difficult. But if the flux is highly directional as Speiser's is it may be all right.

Results of most recent observations are described by H. Poeverlein in his "Final remarks" at the end of his contribution to this volume.

H. Belts on other planets.

47. Generalities. We know something about radiation belts on several other planets. Jupiter has a substantial trapped radiation zone and Venus and Mars have faint or no belts.

What does it take for a planet to have a radiation belt? We need

(i) a planetary magnetic field,

(ii) a source of energetic particles.

The magnetic field must be large enough to make a magnetosphere somewhat larger than the planet. For the Earth the minimum planetary field to stand off the solar wind is about 70 γ and at the times of large magnetic storms it increases to about 200 γ. This minimum planetary field will decrease with the inverse distance from the Sun. This is tabulated for the planets in Table 18 using $B_{♁ \min} = 100\ \gamma$ as a basis for comparison and assuming dipolar fields.

Table 19. *Magnetic (surface) field needed to hold a magnetosphere.*

Planet		Rotation period	Mass ratio $M/M_♁$	Distance* from sun/a.u.	$B_{\min}/\gamma$	Measured B/γ
Mercury	☿	54 d	0.05	0.39	260	—
Venus	♀	240 d	0.8	0.72	140	$3 \cdot 10^3$
Earth	♁	24 h	1.0	1	100	$3 \cdot 10^4$
Moon	☾	28 d	0.0123	1	100	100
Mars	♂	24.5 h	0.11	1.52	56	100
Jupiter	♃	10 h	318	5.2	20	10^6
Saturn	♄	10 h	95	9.6	10	—
Uranus	♅	11 h	15	19.1	5	—

* 1 a.u. = astronomical unit = $1.5 \cdot 10^8$ km = 150 Gm (1 Gm = 10^6 m).

Rapid rotation and large mass are thought to be the principle ingredients necessary for the production of a planetary magnetic field so Jupiter, Saturn and Uranus might have good-sized magnetic fields.

48. Mars and Venus.

α) Space probe Mariner 2 passed about 40 Mm* away from *Venus* on the sunward side. At this distance there was no evidence for a planetary magnetic field[1] nor any evidence of trapped particles from ANDERSON's GM counter[2] or from VAN ALLEN's GM counter[3]** both detecting $\mathscr{E}_e > 40$ keV electrons. This means that Mariner 2 stayed outside any Venusian magnetosphere or bow shock. This data has been interpreted to say that the Venus surface field must be $B_♀ < 0.03$ Gs.

β) Space probe Mariner 4 flight past *Mars* came within 14 Mm from the center of the planet and then past along the flank and nearly behind the planet. There was no evidence of a planetary magnetic field[4] observed along the whole pass. There was also no observed flux of energetic particles[5,6] that could be attributed to the planet. VAN ALLEN[5] has interpreted this in the way shown in Fig. 58 to mean that the spacecraft orbit stayed outside any planetary magnetosphere or bow shock and therefore the surface field (assumed dipolar) was very likely $B_♂ < 100\ \gamma$[4, 5].

γ) This means from Table 18 above that it is very difficult imagine Mars to have any *trapped radiation*. For Venus the case is more marginal but it is clear that Venus does not have much of a radiation belt if it has any. The Moon also cannot easily have any trapped radiation. However, all these objects may

* 1 Mm = 10^6 m = 10^3 km.
** GM means GEIGER-MÜLLER.
[1] P. J. COLEMAN, L. DAVIS, E. J. SMITH, and C. P. SONETT: Science **138**, 1099 (1962).
[2] H. R. ANDERSON: J. Geophys. Res. **69**, 2651 (1964).
[3] J. A. VAN ALLEN, and L. A. FRANK: Science **138**, 1097 (1962).
[4] E. J. SMITH, L. DAVIS, P. J. COLEMAN, and D. E. JONES: Science **149**, 1232 (1965).
[5] J. A. VAN ALLEN, A. A. FRANK, S. M. KRIMIGIS, and H. K. HILLS: Science **149**, 1228 (1965).
[6] J. J. O'GALLAGHER, and J. A. SIMPSON: Science **149**, 1233 (1965).

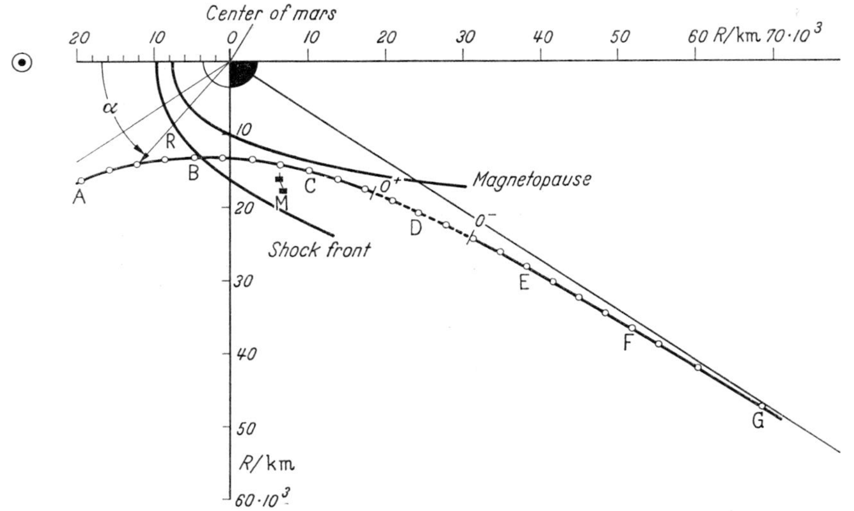

Fig. 58. A polar plot of the trajectory of space probe "Mariner 4" as it went past Mars superimposed on a presumed Martian magnetosphere for the case where the Mars dipole moment is $M_♂ = 0.001\ M_♁$. For this case Mariner 4 should have gone through the Martian magnetosheath but there was no evidence that it did so. Therefore, $M_♂$ must be $< 10^{-3}\ M_♁$. Distances from the planet's center for orbit points A ... G have been: A 25.2, B 14.0, C 18.2, D 32.1, E 47.6, F 63.5, G 83.5 Mm (1 Mm = 10^3 km). (VAN ALLEN et al., ref. [5], Sect. 48.)

have regions of energetic particles near them. If they have substantial internal electrical conductivities they will build up a magnetic field by capture of solar field in the way that GOLD[7] suggested the moon would. In this way they might produce magnetopauses at or near their surfaces. By analogy with Earth we might expect they would have bow shocks and transition zones outside this magnetopause. This would mean that energetic particles would be present near the planet but they would not be trapped particles.

49. Jupiter radio waves.

α) We have a considerable amount of information about the radiation belts of Jupiter. This information is indirect and involves *synchrotron radiation* but the interpretation appears to be quite unique. Jupiter is a strong radio source and has two different forms of radio emission[1]. In the 10 cm (3 GHz) range there is a strong radio emission[2] (decimeter radiation) that does not vary much with time. The radiation in this wave length range is linearly polarized[3]. This is considered to be synchrotron radiation from trapped electrons.

Radio emission at about 10 m wave length (30 MHz)[4] (decameter radiation) is also prominent from Jupiter. This is circularly polarized and appears mostly at certain phases of rotation of Jupiter. This decameter radiation seems to be related with the position of the moons of this planet. We will now only consider the decimeter radiation from Jupiter.

From measurements[2] made at 10...70 cm (3...0.43 GHz) it became obvious that Jupiter had an appreciable non-thermal microwave emission. The effective disk temperature at 70 cm (0.43 GHz) is about 50000 °K. DRAKE and HVATUM[5]

[7] T. GOLD: [*26*], 381 (1966).
[1] J. A. ROBERTS: Planetary Space Sci. **11**, 221—259 (1963).
[2] R. M. SLOANAKER: Astrophys. J. **64**, 346 (1949).
[3] V. RADHAKRISHNON, and J. A. ROBERTS: Phys. Rev. Letters **4**, 493 (1960).
[4] B. F. BURKE, and K. L. FRANKLIN: J. Geophys. Res. **60**, 213 (1955).
[5] F. D. DRAKE, and S. HVATUM: Non-Thermal Microwave Radiation from Jupiter. Astrophys. J. **64**, 329 (1959).

suggested that this radio noise might be synchrotron radiation from a trapped radiation belt. RADHAKRISHNON and ROBERTS[3] and later MORRIS and BERGE[6] used two 90-foot (27 m) dish antennas of Cal. Inst. Techn. with a variable separation as an interferometer and showed that the radio source was about three planet diameters in width along the equator but only about one diameter north-south. This size is nearly constant from in the above wavelength range. This source geometry including regions well off the disc of the planet strongly indicates emission from a VAN ALLEN belt.

β) Influenced by the prediction of FIELD[7] that the radiation from VAN ALLEN Belt electrons might be strongly polarized, RADHAKRISHNON and ROBERTS[3] made measurements of *polarization* at 31 cm (970 MHz). These authors found a component of linear polarization amounting to approximately 0.3 of the total emission, with the electric vector parallel to the equatorial plane of Jupiter (within $\pm 12°$). The circularly polarized component was shown to be less than 0.06 of the total emission.

By monitoring the polarization at 22 cm (1.36 GHz) for 8 h per night over a period of 7 days, MORRIS and BERGE[6] found that the direction of the electric vector rocks through $\pm 10°$ with a period of rotation of the planet. As the polarization is presumably related to the magnetic field, they inferred from this that the magnetic axis of Jupiter is inclined at 10° to the axis of rotation.

γ) This interesting discovery about the planet also provides the means of determining what is presumably the *rotation* period of the solid planet. Measurements of the variation of the direction of polarization show that the Jupiter radiation belt is not symmetric[8]. The polarization curves are not sinusoidal.

It does not seem reasonable to attribute this to an off-center dipole (suggested by WARWICK[9] to explain the decameter radiation) since the planetary eclipsing amounts to only about 6 percent for a radiation belt of the size Jupiter has. WARWICK has suggested that an offset quadrupole field might produce this asymmetry but this effect is currently unexplained. ROBERTS and EHERS[10] measured the location of the radio source with respect to the planet. They found that they are concentric to within 0.15 Jupiter radii so that the offset-dipole model of the magnetic field suggested by WARWICK doesn't seem very workable.

Several time variations of the decimeter radiation have been reported but only one type seems well established. There is about 10 percent variation in the total intensity of emission occurring *twice* per revolution of the planet[6]. The maxima occur when the tilt of the plane of polarization relative to the equator is greatest.

δ) The variation of the intensity of the radio signal with rotation of the planet is quite certainly due to the *tilt of* the *magnetic axis* of Jupiter and the variation of the radiated intensity with magnetic latitude. The dependence of emission on latitude depends on the pitch angle distribution[11]. For distributions peaked at a pitch angle $\alpha_0 = 90$ the emission is maximum in the plane of the effective magnetic equator (generalized Jovian magnetic latitude $\phi_{24} = 0$).

This means the intensity received at Earth should be maximum when the polarization vector is at the maximum 10° angle. The reason for this is that for this case the equatorial mirroring particles at all longitudes are radiating directly toward Earth. For the case when the polarization vector is a 0° the dipole axis of Jupiter is tipped either toward or away from Earth and the equatorial mirroring particles near the Jupiter-Earth meridian plane are beaming their radiation 10° away from Earth either up or down.

---

[6] D. MORRIS, and G. L. BERGE: Astrophys. J. **136**, 276 (1962).
[7] G. B. FIELD: J. Geophys. Res. **65**, 1611 (1960).
[8] J. A. ROBERTS, and M. M. KOMESAROFF: Icarus **4**, 127 (1965).
[9] J. W. WARWICK: Astrophys. J. **137**, 41—60 (1963).
[10] J. A. ROBERTS: Radio Science **69** D, 1543 (1965).
[11] K. S. THORNE: Astrophys. J., Suppl. **8**, 1—30 (1963).

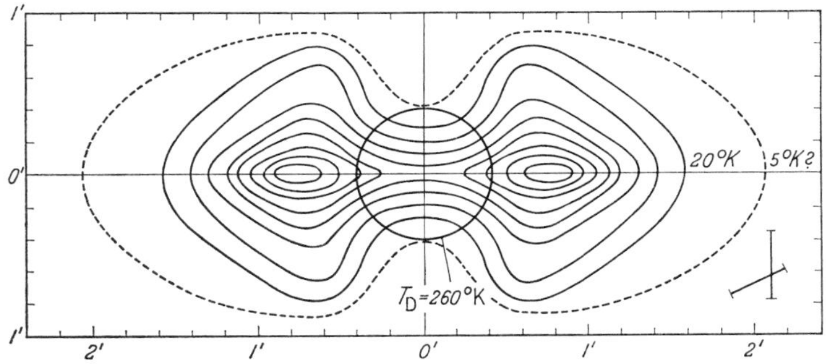

Fig. 59. The 10.4 cm (2.9 GHz) radio brightness distribution of Jupiter measured by a two 27 m ⌀ antenna interferometer. Contour interval is 20 °K. The resolution obtainable with the interferometer is shown at the lower right. Abscissa and ordinate are given in minutes of arc. (BERGE, ref. [12], Sect. 49.)

ε) More recently BERGE[12] has measured the radio source at 10.4 cm wave length (2.9 GHz). This is shown in Fig. 59. This certainly looks like a *radiation belt*. The resolution attainable by the two 90-foot (27 m) Owens Valley paraboloids used for these measurements is quite good as shown by the bars in the lower right side of Fig. 59.

An occultation of Jupiter by the Moon on April 1, 1962 was observed[8] at a wavelength of 74 cm (400 MHz). The radio disappearance occupied about twice the time of the optical disappearance and was approximately centered on it. This also suggests an extended radio source like a radiation belt.

50. Origin of Jupiter's non-thermal radiation. Since the discovery that Jovian radiation has a strong linearly polarized component and arises in a region several times the size of the planet, attention has focussed on the suggestion[1] that Jupiter is surrounded by a radiation belt similar to the VAN ALLEN Belt surrounding Earth. This presupposed a Jovian magnetic field to trap the particles, and the radio emission is assumed to be produced by the acceleration of the electrons as they spiral in the field.

α) It is now generally considered that the source of the microwave non-thermal emission is *synchrotron radiation* from electrons of several MeV. CHANG and DAVIS[2] have calculated the synchrotron radiation expected from a group of electrons sprialing in the Jupiter radiation belt. Using $2.8 \cdot 10^{16}$ erg sec^{-1} = $2.8 \cdot 10^{9}$W for the total synchrotron radiation power emitted on frequencies <10 GHz (wavelength >3 cm) they find radiation data as given in Table 20.

These values seem reasonable for the electrons. It is of interest that the lifetime to go from $\mathscr{E}_c$ to $\frac{1}{2}\mathscr{E}_c$ for a 1 Γ field is only one year.

CHANG and DAVIS[2] had suggested that if Jupiter's field was dipolar a large fraction of the radiation electrons had to be in flat helixes with $\alpha_0 \sim 90°$ in order to give the observed polarization. In Fig. 60 left are shown arrows indicating the direction of the plane of polarization for electrons mirroring at several points and also showing by the length of the arrow the intensity of radiation. It is obvious that there cannot be very many electrons of $\sin\alpha_0 = 0.15$ or we would not observe the polarization vector parallel to the equatorial plane.

[12] G. L. BERGE: Radio Science **69** D, 1552 (1965).

[1] F. D. DRAKE, and S. HVATUM: Non-Thermal Microwave Radiation from Jupiter. Astrophys. J. **64**, 329 (1959).

[2] D. B. CHANG, and L. DAVIS: Astrophys. J. **136**, 567—581 (1962).

Table 20. *Properties of the synchrotron radiation of electrons in a uniform magnetic field.*

	Magnetic field B/Γ		
	0.1	1	10
Critical energy* $\mathscr{E}_c$/MeV for radiation at wavelengths of:			
10 GHz (3 cm)	75	25	7.5
1 GHz (30 cm)	25	8	2.5
100 MHz (300 cm)	8	2.5	0.8
Time/a for energy on 1 GHz (30 cm) to decrease by 50% (in years)	30	1	$\frac{1}{30}$
Gyroradius R_c/km equivalent to $\mathscr{E}_c$ on 1 GHz (30 cm)	10	$\frac{1}{3}$	10^{-2}
Electron particle density/m^{-3} if total volume is $10\,V_{♃}$**	$2 \cdot 10^4$	2000	200
Electron energy density/J m^{-3}	$8 \cdot 10^{-8}$	$2.5 \cdot 10^{-9}$	$8 \cdot 10^{-11}$
Magnetic energy density/J m^{-3}	$4 \cdot 10^{-5}$	$4 \cdot 10^{-3}$	0.4

* The critical energy, $\mathscr{E}_c(\lambda)$, for electrons which are to emit radiation of wavelength λ is the energy for which SCHWINGER's critical frequency is equal to c_0/λ and is 0.55 times the energy at which the power radiated per unit frequency interval has its maximum at λ.

** Number required to give the observed emission of Jupiter in the decimetric range.

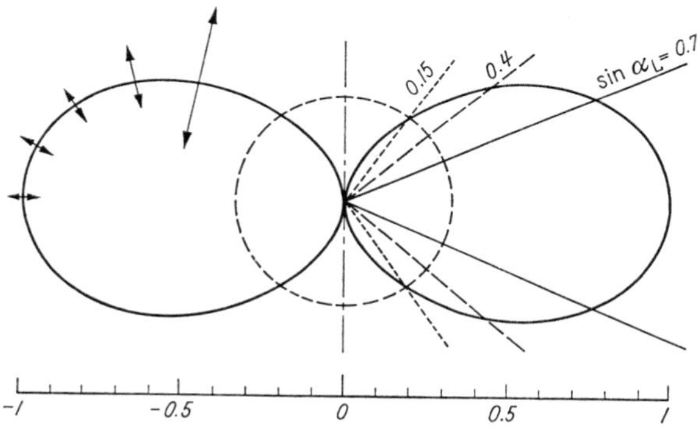

Fig. 60. A section through a dipole field showing a planet, its dipole axis, and a field shell whose equatorial radius is three times that of the planet. On the left the arrows show the planes of polarization of the synchrotron radiation emitted at the corresponding points. On the right, the straight lines show the mirror points for electrons whose helix angles in the equatorial plane are α_L. Abscissa scale gives l/r_0 where $r_0 = 3\,R_J$, l being the distance from the planet's center. (CHANG and DAVIS, ref. [2], Sect. 50.)

β) CHANG and DAVIS[2] *calculated* the *emission* from a thin cylindrical shell of electrons concentric with the dipole axis of the planet and at different distances l from the planet.

Chang and DAVIS integrated over ten thin cylindrical shells of $l/r_0 = 0.1, 0.2 \ldots 1.0$ where $r_0 = 3\,R_{♃}$ was used and evaluated the total intensity radiated and the degree of polarization as shown in Fig. 60. One can see here that the polarization is too low or even in the wrong direction for $\sin\alpha_L = 0.4$ so the electrons do really have to have large pitch angles to explain the observations.

CHANG and DAVIS[2] and THORNE[3] showed that for a thin shell of electrons having an energy spectrum

$$N(\mathscr{E})\,d\mathscr{E} = \text{const.}\,\mathscr{E}^{-(m+1)}\,d\mathscr{E} \tag{50.1}$$

the emitted radiation has a frequency spectrum

$$N(f) \propto f^{-m/2}. \tag{50.2}$$

[3] K. S. THORNE: Astrophys. J., Suppl. 8, 1—30 (1963).

Experimentally the spectrum is very flat so m = 0 and therefore

$$N(\mathscr{E})\,\mathrm{d}\mathscr{E} = \text{const.}\ \mathscr{E}^{-1}\,\mathrm{d}\mathscr{E} \tag{50.3}$$

at least over the range given by the heuristic limits:

$$1 < \frac{\mathscr{E}/\mathrm{MeV}}{\sqrt{B_0/\mathrm{Gs}}} < 100,$$

where B_0 is the magnetic field in the region of the trapped particles.

γ) Some information on the Jupiter *magnetic field* can be obtained from the decameter radiation. There is no well-based concept to the origin of the *decameter radiation* but most ideas would have it related to gyration of electrons in the magnetic field so the characteristic frequency would be the cyclotron frequency. This radiation is circularly polarized and has about 7 m wavelength, i.e. frequency of about 40 MHz. If this is a gyrofrequency f_B, then

$$\left.\begin{aligned} 2\pi f_B &= \frac{q_e}{m_e}\,\frac{B}{c_0\sqrt{\varepsilon_0\,\mu_0}}\,, \\ f_B &= 2.8\,\frac{B}{\mathrm{Gs}}\,\mathrm{MHz}. \end{aligned}\right\} \tag{50.4}$$

So that $B_{♃} \approx 10$ Gs. It seems likely that this radiation, whatever its origin, originates at low altitudes in Jupiter's ionosphere. This means at $R = 2 \ldots 3\,R_{♃}$ the magnetic field should be about 1.0 Gs. Berge[4] has measured a small circularly polarized component in Jupiter's decimeter radiation amounting to a few percent. Roberts and Komesaroff[5] showed that on the radiated frequency, f_r, the degree of circular polarization C_p was related to the magnetic field B_r in the radiating region. For maximum magnetic declination of Jupiter at Earth the following relation has been indicated:

$$C_p \approx 3.5 \cdot \sqrt{\sin\theta\,\frac{B_r/\mathrm{Gs}}{f_r/\mathrm{MHz}}}\,. \tag{50.5}$$

Taking $C_p = 0.1$ for $\lambda_r = 21$ cm ($f_r = 1.43$ GHz) gives $B_r \approx 1$ Gs. This is equivalent to a surface field of about 10...20 Gs. Using this value of $B_r = 1$ Gs from Table 20 we get an electron flux of $J \sim 10^8\ \mathrm{cm^{-2}\,sec^{-1}}$ ($= 10^{12}\ \mathrm{m^{-2}\,sec^{-1}}$) required to generate the radiation.

δ) The *origin* of the Jupiter radiation belt is not understood. Chang and Davis say:

"The primary source of the electrons appears to be the sun. Field has shown that enhanced radiation from primary cosmic-ray electrons is not sufficient. Electrons from neutron albedo seem inadequate. In the equatorial region 2 radii from the center of Jupiter, albedo electrons are in steep helices; furthermore, the number or primary cosmic-ray particles striking the surface of the planet per unit latitude range at the latitude where the shell of equatorial radius equal to 3 planetary radii intersects the planet, is smaller by a factor 1/200 at Jupiter than at the Earth if Jupiter's magnetic moment is such as to make the field at 3 Jovian radii equal to that at the surface of the Earth. On the other hand, the percent estimates of solar-wind density coupled with the observations of the influence of solar activity on the decametric radiation from Jupiter suggest that there is an ample supply of low energy electrons and energy for local acceleration transported to Jupiter from the Sun. It appears plausible that betatron acceleration leading to flat helices might result from the diffusion of the magnetic field into bubbles of solar plasma as the bubbles penetrate into the planetary field."

They are of the opinion that the radial diffusion model is not fast enough to supply the electrons. They estimate the time scale to get the electrons in to $3\,R_{♃}$

[4] G. L. Berge: Radio Science **69** D, 1552 (1965).

[5] J. A. Roberts, and M. M. Komesaroff: Icarus **4**, 127 (1965).

is of the order of a million years and is therefore much larger than the lifetime of the electron against synchrotron radiation. If this is the case, then L diffusion will not work but the velocity of L diffusion on Jupiter is not at all well known so this is still an open question.

ε) It is interesting that Saturn has no appreciable non-thermal radio emission[6]. The planets Jupiter and Saturn are similar in size and rotation period and might be expected to have about the same magnetic field. Saturn is of smaller angular diameter than Jupiter and therefore would be weaker by a factor of ten and harder to measure. Only one measurement[7] has indicated any measurable radiation and that is marginal. It has a substantially less intense radiation belt than Jupiter. This could be explained by either

(i) the planet having a small magnetic field so it can't trap particles efficiently or

(ii) by no source of particles to populate the belt.

Saturn is about the same size and rotates with about the same speed as Jupiter. On this basis from the dynamo theory we might expect that the magnetic field would be similar for the two planets. It may be that the solar wind, which is the most obvious source of radiation belts, does not reach out as far* as 8 a.u. and that Saturn is sitting there with an empty magnetosphere. It may also be that Saturn's rings inhibit the formation of a radiation belt.

General references.

[1] ALFVÉN, H.: Cosmical Electrodynamics. Oxford: Oxford Press 1950.
[2] NORTHROP, T. G.: The Adiabatic Motion of Charged Particles. New York: Interscience 1963.
[3] MCILWAIN, C. E.: Coordinated for Mapping the Distribution of Magnetically Trapped Particles. J. Geophys. Res. **66**, 3681 (1961).
[4] VAN ALLEN, J. A., C. E. MCILWAIN, and G. H. LUDWIG: Radiation Observations with Satellite 1958, ε. J. Geophys. Res. **64**, 271 (1959).
[5] HESS, W. N., Energetic Particles in the Inner Van Allen Belt. Space Sci. Rev. **1**, 278 (1962).
[6] VAN ALLEN, J. A., and L. A. FRANK: Radiation Around the Earth to a Radial Distance of 107,400 km. Nature **183**, 430 (1959).
[7] LENCHEK, A. M.: Origin and Loss of Inner Zone Protons, Radiation Trapped in the Earth's Magnetic Field. Dordrecht: D. Reidel 1966.
[8] KALLMANN-BIJL, H. (ed.): Space Research (Proc. 1st intern. Space Sci. Sympos., Nice 1960). Amsterdam: North-Holland Publ. Co. 1960.
[9] HULST, H. C. VAN DE, D. DE JAGER, and A. F. MOORE (eds.): Space Research II (Proc. 2nd Intern. Space Sci. Sympos., Florence 1961). Amsterdam: North-Holland Publ. Co. 1963.
[10] PRIESTER, W. (ed.): Space Research III (Proc. 3rd Intern. Space Sci. Sympos., Washington, D.C. 1962). Amsterdam: North-Holland Publ. Co. 1963.
[11] MULLER, P. (ed.): Space Research IV (Proc. 4th Intern. Space Sci. Sympos., Warsaw 1963). Amsterdam: North-Holland Publ. Co. 1964.
[12] KING-HELE, D. G., P. MULLER, and G. RIGHINI (eds.): Space Research V (Proc. 5th Intern. Space Sci. Sympos., Florence 1964). Amsterdam: North-Holland Publ. Co. 1965.
[13] SMITH-ROSE, R. L. (ed.): Space Research VI (Proc. 6th Intern. Space Sci. Sympos., Mar del Plata 1965). New York: Spartan Books 1966.
[14] —, and J. W. KING (eds.): Space Research VII (Proc. 7th Intern. Space Sci. Sympos., Vienna 1966). Amsterdam: North-Holland Publ. Co. 1967.
[15] MITRA, A. P., L. G. JACCHIA, and W. S. NEWMAN (eds.): Space Research VIII (Proc. 10th Plenary Meeting of COSPAR, London 1967). Amsterdam: North-Holland Publ. Co. 1968.

---

* 1 a.u. = astronomical unit = $1.5 \cdot 10^8$ km = 150 Gm (1 Gm $= 10^9$ m).

[6] J. A. ROBERTS: Planetary Space Sci. **11**, 221—259 (1963).

[7] F. D. DRAKE: Nature **195**, 893 (1962).

[16] Proceedings of the Advanced Study Institute on Radiation Trapped in the Earth's Magnetic Field (Bergen, Norway, August 16 through Sept. 3, 1965). Dontrecht: D. Reidel Publ. Co. 1966.
[17] Le Galley, D. P., and A. Rosen: Space Physics. New York: John Wiley 1964.
[18] Walt, M., and W. M. MacDonald: The Influence of the Earth's Atmosphere on Geomagnetically Trapped Particles. Rev. Geophys. **2**, 543 (1964).
[19] Chang, C. C., and S. S. Huang: Proceedings of the Plasma Space Science Symposium. (Washington, D. C., June 1963.) Dontrecht: D. Reidel Publ. Co. 1965.
[20] Ness, N. F., and J. M. Wilcox: Sector Structure of the Quiet Interplanetary Magnetic Field. Science **148**, 1592 (1965).
[21] Sugiura, M., and J. P. Heppner: The Earth's Magnetic Field. In: Introduction to Space Science. New York: Gordon and Breech 1965.
[22] Piddington, J. H.: Geomagnetic Storm Theory. J. Geophys. Res. **65**, 93 (1960).
[23] Singer, S. F.: A New Model of Magnetic Storms and Aurorae. Trans. Am. Geophys. Union **38**, 175 (1957).
[24] Dungey, J. W.: Cosmic Electrodynamics. Cambridge: Cambridge Univ. Press 1958.
[25] Frank, L. A., and J. A. van Allen: A Survey of Magnetospheric Boundary Phenomena. In: Research in Geophysics, Vol. 1. Cambridge: MIT Press 1964.
[26] Mackin, R. J., and M. Neugebauer: The Solar Wind (Proceedings on a Conference at Cal. Tech., April 1964). London: Pergamon Press 1966.
[27] Chapman, S., and J. Bartels: Geomagnetism. Oxford: Clarendon Press 1940.
[28] Meinel, A. B.: Evidence for the Entry into the Upper Atmosphere of High-Speed Protons during Aurora Activity. Science **112**, 590 (1950).
[29] Davis, T. N.: The Aurora. In: Introduction to Space Science. New York: Gordon and Breach 1965.
[30] O'Brien, B. J.: High Latitude Geophysical Studies with Satellite Injun III, Parts I to IV. J. Geophys. Res. **69**, 1—101 (1964).
[31] Anderson, K. A.: Balloon Measurements of X-Rays in the Auroral Zone. In: Auroral Phenomena. Stanford: Stanford Univ. Press 1965.
[32] O'Brien, B. J.: Auroral Phenomena. Science **148**, 449 (1965).
[33] Krassovskij, V. I.: Polar Auroras. Space Sci. Rev. **3**, 232 (1964).
[34] Axford, W. I., and C. O. Hines: A Unifying Theory of High Latitude Geophysical Phenomena and Geomagnetic Storms. Canad. J. Phys. **39**, 1443 (1961).
[35] Störmer, C.: Arch. Sci. Genève **24**, 5, 113, 221, 317 (1907); **32**, 33, 163, (1911); **35**, 483 (1913).
[36] Lemaître, G., and M. S. Vallarta: Phys. Rev. **43**, 87 (1933).
[37] —, et L. Bossy: Sur un cas limite du problème de Störmer. Bull. Cl. Sci. Acad. Roy. Belg. (5e sér.) **31**, 357 (1946).
[38] Bossy, L.: Le problème de Störmer et le mouvement des particles dans les ceintures de radiation. Ann. Géophys. **18**, 198 (1962).
[39] Ferraro, V. C. A.: The Origin of Magnetic Storms and Aurorae. Ann. Géophys. **11**, 284 (1955).

Variations rapides du champ magnétique terrestre.

Par

E. SELZER*.

Avec 75 figures.

A. Introduction. Définitions.

1. Domaine de fréquences étudié. Pour des raisons essentiellement «professionnelles» (techniques, appareillages, personnels, etc.) l'étude des *variations du champ magnétique terrestre* s'est développée pendant longtemps suivant des normes traditionnelles, particulières. C'est ainsi qu'elle est restée coupée du développement plus récent de celle des *variations électromagnétiques*, avec leur développement en radio-électricité et en radio-astronomie. Les phénomènes de base sont bien les mêmes, mais les fréquences et les longueurs d'onde en jeu sont, en général, extrêmement différentes, et l'on sait combien les questions d'échelle revêtent d'importance dans les phénomènes physiques, au point de supplanter parfois les questions de principe.

Cependant, le joint entre les deux domaines s'est accompli récemment: les radio-électriciens s'intéressent de plus en plus à la propagation des ondes électromagnétiques d'ultra basses fréquences [ainsi qu'en témoignent les Symposiums de Bad Homburg (1963), de Boulder (1964), de Paris (1966)]. Simultanément, les magnéticiens se sont attachés à l'enregistrement et à l'étude de variations magnétiques de plus en plus rapides. A l'heure actuelle, la communauté d'intérêt entre les spécialistes des deux disciplines est assurée, non seulement sur le plan expérimental par un recouvrement sur une large bande de fréquence, mais aussi sur celui de la théorie, depuis que l'on a observé que certains événements ne se produisaient pas dans les deux domaines sans une certaine dépendance[1–3]. Nous arrêterons, conventionnellement, notre définition des variations rapides du champ magnétique terrestre aux environs de 50 Hz ce qui, de toute façon, est un «no man's land» inexplorable correctement pour les phénomènes naturels étant donné la densité des réseaux de distribution d'énergie électrique autour de cette fréquence (plus précisément, entre 50 et 60 Hz).

Du côté des *basses fréquences*, nous porterons principalement notre attention sur les phénomènes plus rapides que quelques minutes, mais nous aurons souvent l'occasion de montrer leurs relations étroites avec des variations beaucoup plus lentes. Nous serons alors conduits à les présenter comme constituant la «structure fine» (ou «micro-structure») de ces dernières. Parfois même (cas des «baies magnétiques») ces microstructures se sépareront en cascades et c'est ainsi que

* Manuscrit reçu en Mars 1971.

[1] R. GENDRIN et J. VIGNERON: C.R. Acad. Sci. **263**, sér. B, 1147 (1966).

[2] R. BENOIT, A. HOURI et A. ROUILLON: C.R. Acad. Sci. **265**, 1270—1272 (1967).

[3] L. HARANG: Planet. Space Sci. **16**, 1081—1094 (1968).

nous serons conduits à parler de «microstructures *primaires*» et de «microstructures *secondaires*». Par contre, nous éviterons de recourir au terme de «micropulsations» qui n'aurait guère sa raison d'être dans ce chapitre spécialisé.

2. Séparation entre les éléments organisés et les éléments aléatoires entrant dans la constitution des variations magnétiques rapides.

α) Pour les magnéticiens de profession le caractère *organisé* des variations géomagnétiques, même pour celles les plus rapides, est peu contesté. Par contre, le physicien qui veut s'en préoccuper, par exemple pour des questions de mesure, a tendance à considérer ces variations comme un «bruit» à caractère plus ou moins aléatoire. Il n'y a pas une contradiction absolue entre ces deux points de vue et, pour une bonne part, cette opposition apparente est le fait de l'usage d'échelles de temps différentes pour juger des mêmes phénomènes: le magnéticien, par le fait même de son étude, concentre son attention sur des temps de plus en plus courts au fur et à mesure que la rapidité des phénomènes qu'il étudie augmente. Il est ainsi porté à leur trouver toujours une certaine *organisation*, ne serait-ce que celle qu'il a introduite lui-même comme procédé d'étude. De là toutes ces classifications, cette nomenclature, conduisant à une morphologie qui ne reste complexe que pour le profane. Le physicien, quant à lui, s'inquiète de l'influence des mêmes variations magnétiques sur d'autres phénomènes comportant des échelles de temps qui leur sont propres, ce qui fait qu'au-dessous d'un certain seuil fixe toutes ces variations lui apparaîtront aléatoires. Le temps n'est d'ailleurs pas seul à intervenir à ce sujet: les coordonnées d'espace jouent un rôle analogue. Le lecteur intéressé par cet aspect particulier de notre sujet pourra se reporter au chapitre «Hypothèse probabiliste du champ terrestre» de la Thèse de G. BONNET (1962)[1].

Indépendamment, cependant, de ces questions de méthode, on peut chercher à séparer d'une façon aussi objective que possible, parmi les variations géomagnétiques, celles qui se rangent plus ou moins nettement parmi des phénomènes organisés, ou aléatoires. Nous limiterons cet examen à celui des variations rapides, non sans avoir remarqué que, en ce qui a trait aux variations plus lentes, ce n'est que par rapport à des temps très longs qu'elles pourraient paraître se comporter d'une façon aléatoire.

β) Dans notre domaine des variations rapides, le caractère aléatoire de ces variations — ou, si l'on veut, leur manifestation sous la forme d'un «*bruit*» — peut se manifester de plusieurs façons: Il pourra s'agir, par exemple, d'un manque de cohérence de «signaux» d'un certain type, relativement bien organisés individuellement ou par petits groupes, mais dont la distribution en phase ou en amplitude ne répond plus à aucune loi simple quand on augmente la durée d'observation. Nous verrons que ce sera souvent le cas de certains types de pulsations dont les unes ont été déjà rangées, morphologiquement, parmi les «irrégulières» (nomenclature **pi** de la classification internationale), d'autres appartenant au groupe des «pulsations continues» (**pc**), dont les formes d'ondes sont souvent assez régulières.

Mais ce qui constitue, plus strictement, le «bruit magnétique», apparaît plutôt comme un «résidu», celui que l'on obtient quand, ayant défalqué d'un enregistrement «total» des variations magnétiques (dans une certaine bande de fréquences) toutes celles dont on a reconnu les «structures», on recueille l'énergie magnétique naturelle qui restait sur les enregistrements. Bien entendu, c'est là une opération délicate, compliquée par le fait que les bruits d'appareillages

[1] G. BONNET: Thèse Fac. Sci. Grenoble (1961); — Ann. Géophys. **18**, 1, 62—91 (1962).

atteignent souvent des niveaux (en amplitudes ou en énergies) comparables ou supérieurs à ceux de ces résidus naturels. Aussi, ce genre d'analyse n'a pu donner des résultats nets, positifs, que très récemment[2]. On a pu alors constater que le résidu de bruit ainsi trouvé n'existait pas seulement du côté des variations les plus rapides (au-delà de la fréquence 1 Hz, zone dans laquelle son amplitude — si l'on fait abstraction des oscillations de SCHUMANN qui sont partiellement structurées — ne dépasse guère* $10^{-3}\gamma$) mais qu'il était également présent du côté des variations lentes où il était alimenté, probablement, par toutes les irrégularités des variations structurées que l'on y rencontre.

γ) Ces études de bruit ne peuvent être menées à bien qu'à partir d'enregistrements ayant des normes convenables, se prêtant, en particulier, aux techniques analogiques de l'analyse spectrale électro-acoustique (tels que des enregistrements sur bandes magnétiques). Les études de F. GLANGEAUD[2] sur ce sujet ont abouti à l'établissement d'une «courbe en S» (Fig. 1 et 2**) qui donne la répartition du bruit géomagnétique naturel dans un large intervalle de fréquences (donc, de périodes). En dehors d'applications particulières, l'intérêt général d'une telle courbe tient en ce qu'elle indique la même répartition spectrale que celle qui a été trouvée par plusieurs auteurs[3–5] pour des paramètres caractéristiques de conditions interplanétaires (tels que le *vent solaire*) mesurés en satellites. Ceci ne veut pas dire que *tout* le bruit géomagnétique détecté au sol vienne de l'espace interplanétaire (ou de l'interaction de cet espace avec la magnétosphère). Il est au contraire très probable qu'une autre partie de ce bruit a son origine en d'autres domaines de l'environnement terrestre situés plus près de nous: magnétosphère elle-même, ionosphère, cavité terre-ionosphère (en dehors des fréquences privilégiées de SCHUMANN).

De même, si nous nous soucions de l'origine de la portion «structurée» des variations magnétiques — ou, plus exactement, si nous recherchons l'origine de leurs «structures» — nous devrons nous demander dans quelle mesure l'espace interplanétaire et les milieux plus proches du sol terrestre doivent se partager leur intervention. Répondre à ces questions c'est, en partie, faire la théorie de ces variations. Aussi, y reviendrons nous dans cet exposé. En tout cas, tout examen morphologique ne devra pas se limiter aux seuls cas des variations structurées et devra également se préoccuper de la «morphologie du bruit».

3. Les variations magnétiques rapides dans leurs relations avec le champ magnétique terrestre principal. Il est admis actuellement que la partie principale du champ magnétique terrestre a sa source dans un système de courants qui circulent dans le noyau de notre Globe. A l'extrémité opposée de son domaine, ce champ ne peut s'étendre jusqu'à l'infini par suite de l'existence du vent solaire et se trouve limité, comme on sait[1], à l'intérieur de la cavité formant la magnétosphère.

* $1\,\gamma = 10^{-5}$ Gauss $= 10^{-9}$ Tesla, unité couramment utilisée en magnétisme terrestre.

** N.B. Toutes les figures numérotées en chiffres arabes sont rassemblées en une sorte d'Atlas, à la fin de l'article, p. 331 et suivantes. Les quelques figures numérotées en chiffres romains sont, au contraire, placées au plus près des portions du texte auxquelles elles se rapportent.

[2] F. GLANGEAUD: C.R. Acad. Sci. **264**, 553—556 (1967).

[3] C. P. SONNETT, D. L. JUDGE, A. R. SIMS, and J. M. KELSO: J. Geophys. Res. **65**, 55 (1960).

[4] R. E. HOLZER, M. G. MCLEOD, and E. J. SMITH: J. Geophys. Res. **71**, 1481 (1966).

[5] P. J. COLEMAN: J. Geophys. Res. **71**, 5509 (1966).

[1] Voir article par E. N. PARKER et V. C. A. FERRARO au volume 49/3, p. 131 de cette encyclopédie.

Sans reprendre la description des différentes parties du domaine occupé par ce champ, remarquons que la continuité de sa manifestation à travers un espace aussi hétérogène n'est valable qu'en ce qui concerne ses propriétés et son influence *statiques*. D'une façon plus imagée, disons que seules ses lignes de forces «statiques» relient sans discontinuités le noyau terrestre à l'espace interplanétaire. Dès que ces lignes de forces sont soumises à des perturbations à partir de leurs positions moyennes, ceci dans une quelconque des parties de l'espace où elles se manifestent, ces perturbations ne sont capables de se transmettre aux parties voisines que suivant certaines normes qui dépendent d'une façon précise des propriétés physiques des milieux concernés et de leur rapidité plus ou moins grande. Il n'y a donc pas de continuité dynamique du champ magnétique terrestre superposable à sa continuité statique, bien que l'on tende vers une telle situation au fur et à mesure que des variations de plus en plus lentes entrent en jeu.

Cette entité que nous intitulons «variations magnétiques rapides du champ terrestre» a donc une existence spatiale multiple et brisée. En particulier, il n'est pas dit que, indépendamment des questions d'origine, ou d'excitation, évoquées au paragraphe précédent, la transmission de ces phénomènes puisse se faire sans des atténuations importantes (fonction de la fréquence) à travers les diverses régions de l'environnement terrestre. Inversement, même si le champ terrestre statique reste confiné à l'intérieur de la magnétosphère, ses variations sont couplées avec d'autres actions énergétiques (y compris d'autres champs magnétiques) se produisant dans l'espace interplanétaire, et elles peuvent ainsi franchir les frontières correspondantes.

Ainsi, alors qu'on peut concevoir une étude unitaire de la partie statique du champ terrestre, celle de ses variations rapides peut être considérée comme ayant trait à une grandeur indépendante. On remarquera à ce sujet, que le problème théorique de l'addition (vectorielle) de deux champs magnétiques n'a pas de signification physique indépendamment de la considération des systèmes physiques responsables de l'existence de ces champs [*1*]. Cette question conserve son ambiguïté même lorsqu'un des deux champs devient très petit par rapport à l'autre. Inversement, la séparation d'un champ variable en sa partie statique et sa partie variable reste un problème physiquement insoluble. Tout au plus pouvons nous lui donner des solutions pratiques approchées, ceci dans des limites, définies d'amplitudes et de fréquences.

Ces questions interviennent de façon particulièrement intéressante en ce qui concerne les champs magnétiques tourbillonnaires qui se manifestent dans l'épaisseur de l'interface «onde de choc statique — magnétopause» à travers laquelle notre environnement terrestre se trouve soumis aux influences planétaires.

D'une façon plus générale, au lieu de se préoccuper seulement de ces relations entre les variations magnétiques et le champ principal, on peut se demander quels sont les liens entre les variations magnétiques de tous ordres qui paraissent entrer unitairement dans la composition d'un phénomène global tel, par exemple, une baie aurorale, c'est-à-dire — en utilisant une nomenclature introduite plus haut — entre le phénomène «principal» que constitue cette baie, et ses «microstructures».

C'est ainsi que dans l'exemple que nous venons d'invoquer, les théories généralement admises qui expliquent les $\mathcal{H}$ négatifs ou positifs de ces baies[2] par des systèmes de courants circulant dans l'ionosphère (courants auroraux avec leurs «courants de fermeture») sont assez satisfaisantes en ce qui concerne la

[2] Voir article par T. NAGATA et N. FUKUSHIMA au volume 49/3, p. 5 de cette encyclopédie.

formation «macroscopique» de la baie proprement dite, mais butent sur une compréhension acceptable pour les microstructures primaires et secondaires de ces mêmes baies. Faut-il considérer ces microstructures comme des fluctuations du phénomène principal (donc des courants ionosphériques précédemment invoqués), ce qui conduit à de grosses difficultés, ou comme un autre phénomène qui ne lui serait pas lié directement mais lui serait simplement ajouté tout en étant issu d'une même cause? Ce genre de question n'apparaîtrait pas aussi capital si l'observation ne nous avait appris que le domaine spatial de cohérence de ces microstructures était souvent plus étendu que celui de la perturbation principale (macroscopique). Nous sommes ainsi conduits, parfois, à accepter un mécanisme par «courants ionosphériques» pour expliquer le phénomène principal, et un mécanisme différent — en général d'ordre magnéto-dynamique[3] — pour rendre compte de sa microstructure. Il y a là une attitude peu satisfaisante pour l'esprit tant qu'une théorie synthétique ne peut être formulée.

4. Méthode d'étude proposée. Les phénomènes dont nous occupons s'étant imposés à l'attention des magnéticiens au fur et à mesure des progrès qui étaient faits dans les procédés de détection et d'enregistrement, il serait difficile de nous familiariser avec eux sans avoir une brève connaissance des moyens expérimentaux grâce auxquels ils ont pu se manifester. Ce sera l'objet d'un court chapitre qui, évitant d'être technique — ce qui nous amènerait à des développements inconsidérés —, essaiera de présenter les principales «formes» de la pratique expérimentale en cause. Nous étendrons cet aperçu aux méthodes d'analyse et d'étude dont dépend en grande partie la mise en valeur des phénomènes.

Dans la Section suivante nous pourrons alors, sans que cela paraisse trop artificiel, donner une idée de la classification morphologique sur laquelle les spécialistes s'entendent, avec plus ou moins de rigueur, pour la nomenclature des variations magnétiques rapides. Nous ferons largement appel à une suite d'illustrations dont l'ensemble forme comme une sorte d'Atlas de ces variations. La variable «temps» sera celle que nous aurons à considérer principalement mais sans pour cela négliger l'influence des paramètres d'«espace» jouant, non seulement le long du sol de notre Globe, mais aussi dans toute l'étendue de l'environnement terrestre.

C'est en fonction de cette vue d'ensemble spatio-temporelle que nous pourrons essayer alors de dégager les grandes lignes d'une conception plus synthétique des phénomènes et — à défaut d'une théorie générale dont l'heure n'est pas encore venue — de montrer les voies principales suivant lesquelles quelques synthèses partielles sont possibles.

Les résultats précédents seront-ils capables d'ouvrir l'accès à quelques applications? De quel bénéfice peuvent-ils être pour une meilleure détermination de certains des paramètres qui déterminent les caractéristiques et le comportement de notre environnement? Ce sera l'objet de notre tentative de conclusion.

B. Méthodes d'observation, d'enregistrement et d'analyse*.

5. Limite des possibilités classiques. Nous appelons «possibilités classiques» celles qui découlent de l'usage d'un observatoire magnétique tel qu'on le concevait, disons, il y a une trentaine d'années[1]. C'était par exemple le cas de Chambon-la-Foret dont la mise en service remonte au 1er janvier 1936. Qu'y trouvait-on?:

* On pourra consulter aussi les exposés généraux [2]—[6].

[3] Voir «Introductory Remarks» où l'on explique pourquoi cette terminologie paraît préférable à l'usage d'ALFVEN «magnéto-hydrodynamique».

[1] Voir l'article de V. LAURSEN et J. OLSEN au volume 49/3, p. 276 de cette encyclopédie.

— de bons variomètres et enregistreurs LA COUR «normaux»;

— un enregistreur LA COUR «rapide» (mais sans variomètre spécialement adapté);

— un ancien enregistreur MASCART (encore utile);

— de bonnes mesures absolues, mais à 5 ou 10 gammas près, par les méthodes de GAUSS au théodolite.

Un grand cadre (ou boucle) horizontal avait été essayé au Val-Joyeux (station ayant passé son relais à Chambon) mais sans grand succès, car il était mal défendu centre le vent, ainsi que deux lignes telluriques qui avaient permis de déceler quelques «variations rapides».

L'étude des orages était donc limitée à leurs «grands mouvements» (mais même un **SSC** ne pouvait être déterminé qu'à $^1/_2$ minute près dans le temps), ou à des études statistiques où les variations rapides ne pouvaient trouver leur place. Les «caractères K» commençaient à être déterminés, mais les essais qui allaient en être faits pour étudier l'évolution géographique de certaines variations rapides (telle l'«agitation de jour» des zones polaires) ne pouvaient aboutir.

Cependant, nombre d'observateurs magnéticiens, curieux par métier, savaient que, lors des grands orages, quand les «aiguilles s'affolaient» elles devaient être soumises à des fluctuations magnétiques aussi — ou plus — rapides que la seconde (mais qu'ils croyaient toujours désordonnées). L'enregistreur rapide LA COUR, malgré la sensibilité insuffisante des variomètres avec lesquels il était utilisé, avait permis quelques investigations dans le domaine qui nous intéresse, lors des campagnes pour l'Année Polaire 1932—33, mais sans aucune généralité. En résumé, disons qu'on avait *entrevu* la plupart des phénomènes qui font l'objet de cet exposé, ils avaient même été parfois fort bien décrits pour un instant déterminé (telles les «ondes élémentaires» d'ESCHENHAGUEN[2], mais, par manque de continuité, et d'homogénéité géographique, de ces observations, les déductions capitales (telle que leur caractère de semi-mondialité) n'avaient pu être explicitées.

On ne s'étonnera pas s'il a fallu, par exemple, attendre 1949 pour que l'existence commune d'une micro-structure *pulsée* des «baies» soit bien établie (par GIACOMO[3], au moyen d'un cadre établi sur un terrain avoisinant l'observatoire solaire du Schauinsland (près de Fribourg), ou si les prévisions théoriques de SCHUMANN[4] sur les oscillations de résonance de la cavité Terre-Ionosphère aient aussi attendu plus de dix ans (de 1948 à 1960) pour être bien confirmées.

Sans nous appesantir davantage sur ce bref rappel tirons en une conclusion primordiale: les possibilités anciennes n'ont pas tellement souffert de performances techniques insuffisantes, mais surtout d'une *méthodologie* inefficace car elle n'assurait pas la *continuité géophysique* des observations. Des appareillages bien conçus pour l'étude des variations magnétiques rapides doivent donc bénéficier, non seulement de taux de performances élevés, mais aussi d'une logistique d'emploi suffisamment économique pour pouvoir être en action 24 heures sur 24 à longueurs d'années. Il y a des exceptions à la règle ci-dessus: soit occasionnelles, soit par nécessité (observations en satellites par exemple). Remarquons que dans ce dernier cas apparaît ainsi une des limitations qui résulterait d'un emploi exclusif d'observations «spatiales» et la nécessité d'une bonne et complète collaboration entre ces dernières et les observations au sol.

---

[2] M. ESCHENHAGEN: Terr. Magn **2**, 105—114 (1897).

[3] P. GIACOMO: Ann. Geophys. **5**, 171—173 (1949).

[4] W. O. SCHUMANN: Elektrische Wellen. München: Hanser 1948, 324 p.

6. Choix des méthodes modernes. Les méthodes dites «modernes»[1] demandent à être définies aussi bien que celles que nous avions dénommées «classiques».

Des principes anciens peuvent être utilisés: *variomètres à aimants*. Il faut augmenter considérablement leur sensibilité, leur rapidité. On y arrive en prenant des barreaux très courts, en matériaux modernes (alliage Platine-Cobalt par exemple, de champ coercitif pouvant atteindre 4000 oersteds), et en disposant le variomètre dans une boucle de contre-réaction (avec amplification par double cellule photo-résistante montée dans un pont type Wheatstone, et réinjection d'une partie du signal dans un couple de bobines de Helmholtz entourant l'équipage à aimant). On peut signaler à ce titre les réalisations de DÜRSCHNER[2] à Chambon-la-Forêt et celles de BRUNELLI[3] pour un certain nombre d'observatoires et stations soviétiques. Bien entendu, en intervenant sur la boucle de contre-réaction on peut introduire dans les réponses tout effet de filtrage désiré ce qui nous achemine vers les méthodes d'enregistrement sélectif. C'est d'ailleurs là une préoccupation constante: faut-il enregistrer les phénomènes *en large bande*, ou *en bande très sélective* (avec tous les intermédiaires possibles)? Sans trancher d'une façon absolue cette question nous remarquerons qu'il n'est pas indifférent — même si le résultat définitif auquel on s'intéresse est la répartition de l'énergie des variations magnétiques en bandes étroites de fréquences, c'est-à-dire la détermination d'un spectre pur de cette répartition — de suivre l'une ou l'autre de ces méthodes: Enregistrer directement en bande étroite paraît conduire plus directement au résultat cherché, mais ceci d'une façon non modifiable ultérieurement tout au moins pour le phénomène déterminé, «historique», que l'on se proposait d'étudier. Au contraire, un enregistrement initial en large bande, avec séparation ultérieure au laboratoire en bandes de plus en plus étroites, permet d'éviter toute «impasse» qui serait commise, à priori, sur le genre d'analyse spectrale souhaitée. C'est donc, en principe, celui des deux processus extrêmes qui nous paraît le plus justifié. Bien entendu, il suppose que l'on puisse mettre en mémoire — sous une forme facilement analysable par la suite — l'enregistrement initial en large bande, et aussi que des difficultés particulières, telle que l'élimination nécessaire, initiale, de tel ou tel phénomène perturbateur — par exemple: les décharges électromagnétiques orageuses (éclairs), notamment au voisinage des régions tropicales, susceptibles de saturer certains des étages d'entrée du dispositif sensible d'enregistrement — n'obligent pas à recourir à diverses restrictions sur la largeur spectrale des bandes de fréquences ouvertes à l'enregistrement.

Nous retiendrons de la discussion précédente, l'avantage des dispositifs d'enregistrement à «mémoire intégrale», c'est-à-dire, en définitive, ceux qui permettent une reproduction électrique, ou magnétique, du phénomène initial. Ceci étant, voici, très brièvement, un exposé des divers procédés actuellement utilisés; nous allons rencontrer:

les *dispositifs utilisant les lois de l'induction électromagnétique*, l'organe sensible récepteur étant un cadre, ou une bobine, *soit «à air»* (c'est-à-dire *sans noyau d'alliage métallique à perméabilité élevée*), *soit comportant justement un tel noyau*. A ces dispositifs on peut, assez logiquement, rattacher ceux qui utilisent le sol lui-même du lieu où l'on se trouve en tant qu'organe récepteur, ce qui — du point de vue pratique — revient à procéder à des mesures, ou enregistrements, de courants (électro-)telluriques;

---

[1] Voir l'article de H. SCHMIDT et V. AUSTER au volume 49/3, p. 323 de cette encyclopédie.
[2] H. DÜRSCHNER: Ann. Geophys. **7**, 4, 199—207 (1951).
[3] B. E. BRUNELLI et D. A. NIZAIEV: Izv. Akad. Nauk SSSR **8**, 1064—1068 (1957).

les *magnétomètres comportant des sondes à saturation* du type «fluxgate»[1];

les *magnétomètres à résonance*, soit protonique, soit électronique. Nous laisserons de côté d'autres dispositifs intéressants (tels que ceux utilisant des sondes à *magnétostriction*, ou à *effet Hall*), dont l'utilisation, bien que comportant certaines applications particulières où elle peut être avantageuse, ne s'est pas généralisée[1].

7. Dispositifs à induction électromagnétique — Enregistrements telluriques. Les lois de l'induction électromagnétique, résumées par l'équation de MAXWELL[1]

$$\frac{\partial}{\partial \boldsymbol{r}} \times \boldsymbol{E} \equiv \operatorname{rot} \boldsymbol{E} = -\frac{1}{\boldsymbol{c}_0 \sqrt{\varepsilon_0 \mu_0}} \frac{d\boldsymbol{B}}{dt} \tag{7.1}$$

rappellent que les magnétomètres fondés sur ce principe ramènent une mesure d'un champ magnétique variable à celle d'un champ électrique. Dans le domaine des fréquences TBF (ou VLF dans la terminologie anglo-saxonne), c'est-à-dire de l'ordre du kHz (kilo-Hertz), on mesure effectivement ce champ électrique dans l'air au moyen d'une antenne radio d'un type plus ou moins classique. Par analogie on peut dire que, même dans le domaine des UBF (ELF dans la terminologie anglo-saxonne) s'étendant de quelques centaines de Hz à quelques Hz, et dans celui des variations magnétiques «rapides» qui prolonge ce domaine vers les fréquences plus basses, les mesures de courants telluriques qui s'y pratiquent couramment reviennent également à celles du champ électrique dans le sol, bien qu'il y ait là matière à discussion théorique et qu'il soit plus correct de dire que dans ce cas les mesures, telles qu'elles sont faites — c'est-à-dire en mesurant la différence de potentiel entre deux «prises de terre» enfoncées dans le sol à des distances données (comprises entre quelques dizaines de mètres et plusieurs dizaines de kilomètres), ne sont en fait que celles des densités des courants électriques circulant dans le sol entre ces électrodes (ramenées à une moyenne mal définie si la distance est grande). Quoi qu'il en soit de cette question de principe, ces mesures telluriques demandent pour être interprétables entre plusieurs sites, la connaissance des propriétés électriques (résistivité *intégrée*) du sous-sol à l'emplacement de ces sites. Pour cette raison, malgré sa facilité de mise en place, et sa sensibilité particulièrement efficace pour les domaines de variations les plus rapides (conséquence directe du deuxième membre de l'équation de l'induction), la *méthode tellurique* est considérée comme convenant à une première mise en action rapide, mais qui doit être complétée, dès que cela devient possible, par la mise en œuvre de procédés plus rigoureux.

En Union Soviétique, où sous l'impulsion de V. A. TROITSKAJA la méthode tellurique a connu, notamment en vue de l'Année Géophysique 1957/58, un développement considérable[2,3] [2], elle y est doublée maintenant par ces autres méthodes. On peut en dire de même du réseau français des Terres Australes, où les enregistrements concernant les variations les plus rapides du champ magnétique ont d'abord été réalisés sous la forme tellurique avant de pouvoir être équipé plus complètement. En dehors de la facilité, et du faible coût, de cette méthode, les expérimentateurs qui s'intéressaient à la mise en évidence de variations magnétiques de plus en plus rapides ont aussi tablé essentiellement sur le fait qu'il était alors de plus en plus avantageux d'enregistrer des grandeurs liées aux dérivées par rapport au temps de ces variations, plutôt que ces variations elles-mêmes. Cela permettait de compenser le fait d'expérience que les amplitudes des variations décroissent fortement, en moyenne, quand leur rapidité augmente, alors que leurs dérivées par rapport au temps conservent des ordres de grandeur du même ordre [compris en général — en laissant de côté les cas ex-

---

[1] En ce qui concerne l'écriture des équations nous utilisons des unités non-spécifiées. Voir «Introductory Remarks», p. 1.

[2] V. A. TROITSKAJA: Izv. Akad. Nauk. SSSR, Ser. Geofiz. **9**, 13 é 1 (1960).

[3] V. A. TROITSKAJA: J. Geophys. Res. **66**, No. 1, 11 (1961).

trêmes — entre $^1/_{100}$ et 1 ou 2γ (gamma) par sec]. En dehors du fait que les autres dispositifs à induction (voir Sect. 8) peuvent (si l'on n'intègre pas leurs réponses) assurer les mêmes avantages — et ceci en fonction de caractéristiques mieux connues — il y a lieu de remarquer que les progrès considérables faits dans la technique de l'amplification électronique à très basse fréquence (allant jusqu'au continu) au cours de la dernière décade, a fait passer au second plan les considérations de sensibilités maintenant facilement résolues sauf dans des cas spéciaux (tel que les enregistrements en satellite, pour lesquels la mesure directe du champ électrique ambiant reste aussi un problème délicat).

Par contre, d'autres soucis, dont certains peuvent jouer en faveur, et d'autres en défaveur, de la méthode tellurique, restent toujours pleinement valables. Sans pouvoir les examiner systématiquement ici, nous donnerons deux exemples contraires:

Le premier est relatif à la protection contre les effets perturbateurs des courants *vagabonds* (d'origine industrielle) circulant dans le sol: courants de «retour» ou de «pertes» des lignes de traction électrique, notamment en «continu»; pertes dans le sol d'installations importantes utilisant le courant continu, tels que les fours électriques utilisés en métallurgie, pertes — avec redressement partiel par le sol — des mises à la terre de lignes de transport d'énergie électrique à très haute tension, ceci dans un rayon géographique beaucoup plus restreint que dans les cas précédents, etc. Dans ce cas les enregistrements telluriques sont beaucoup plus affectés, en un site donné, que les enregistrements magnétiques correspondants; (y compris ceux à induction).

Le second concerne les perturbations magnétiques directes créées au voisinage de la station d'enregistrement par le déplacement de masses métalliques magnétiques (véhicules en particulier) qui affectent, jusqu'à des distances de l'ordre de la centaine de mètres (suivant l'importance de la masse perturbatrice et son degré d'aimantation, permanente ou induite) les enregistrements magnétiques, mais en aucune manière leurs analogues telluriques.

Dans aucun de ces deux exemples il ne faut voir une contradiction avec les relations magnéto-telluriques[4]: ces relations supposent essentiellement que le phénomène excitateur soit — ainsi d'ailleurs que le terrain entourant le site choisi — *tabulaire*. Cette condition n'est aucunement remplie dans un cas comme dans l'autre: *dans le premier*, ce sont des courants de conduction circulant dans le sol qui amènent le phénomène perturbateur au voisinage du site perturbé; il en résulte, certes, des perturbations magnétiques (par *effet Ampére* et interaction *locale* le long de la discontinuité air-sol) mais le rapport des champs magnétiques et electriques reste ici plus petit que celui que donnerait la correspondance magnéto-tellurique appliquée au sol entourant le site; *dans le second*, le domaine atteint par la perturbation magnétique est trop restreint pour que des effets d'induction appréciable soient produits dans le sol; la perturbation n'apparaîtra donc que sur les seuls enregistrements magnétiques. Remarquons incidemment que cette différence de comportement apporte une justification du maintien, à une même station, des deux types d'enregistrement magnétique et tellurique: le diagnostic des causes perturbatrices s'en trouve très efficacement facilité. Enfin, il est peut-être nécessaire de préciser que nous ne traitons aucunement, dans cet article, de la méthode magnéto-tellurique — destinée à l'exploration du sous-sol dans une gamme étendue de profondeurs — dont l'emploi requiert, évidemment, de disposer au site concerné d'enregistrements simultanés magnétiques et telluriques pour des phénomènes satisfaisant aux conditions de validité de cette méthode. Remarquons simplement que, dans ce cas, un étalonnage très rigoureux des normes d'enregistrement est, plus que jamais, nécessaire, alors qu'un enregistrement tellurique que ne servirait qu'à des fins morphologiques n'aurait pas toujours besoin d'une telle rigueur.

8. Usage de sondes («capteurs» ou «antennes» magnétiques). La méthode d'induction proprement dite — avec ou sans noyau perméable — est probablement celle qui est le plus utilisée pour les études sur les variations magnétiques les plus rapides. Sans avoir à revenir sur les principes généraux qui la régissent, car ils sont bien connus, nous examinerons quelques points particuliers, importants, concernant sa mise en application.

Tout d'abord, que dire de la controverse: bobine *avec* ou *sans* noyau? — Les grands cadres sans noyau ont des avantages théoriques certains: ils sont facilement calculables, leurs coefficients de self-induction restent faibles et — dans certains cas — offrent la possibilité d'atteindre de grandes sensibilités. Ce dernier point est particuliérement vrai pour les grandes boucles horizontales, déployées

[4] L. Cagniard: Ann. Géophys. **9**, No. 2, 95—125 (1953); — Geophysics **18**, No. 3, 605—635 (1953).

sur le sol, destinées à l'enregistrement de la composante verticale. On peut, en effet, leur faire couvrir des surfaces théoriquement aussi vastes qu'on le désire. On ne rencontre en cela aucune difficulté technique sérieuse et les incidences budgétaires de ce genre d'installation restent en général raisonnables. Il convient de veiller à un parfait isolement des câbles constituant la boucle et d'assurer leur parfaite immobilité (vis-à-vis du vent, notamment). Le Tableau 1 (p. 242) donne les principales caractéristiques des plus grands cadres horizontaux réalisés à diverses époques.

Il est intéressant, si l'on réalise le projet d'une telle installation, de tenir compte de quelques règles très simples qui, en principe, conditionnent l'utilisation la plus rationnelle du terrain et du budget dont on dispose. Nous le ferons sous une forme schématisée à l'extrême, ce qui est justifié par le fait que nous ne discutons ici que de règles très générales, adaptables à chaque cas. C'est ainsi que nous supposerons que le terrain dont on dispose a des limites circulaires, le budget se traduisant lui-même par la quantité totale de cuivre constituant le conducteur qui va former le cadre. D'autre part, l'indice de performance du cadre sera stylisé par le rapport Φ/R du flux magnétique Φ qu'il collecte dans un champ déterminé (ou des variations de ce flux consécutives à une variation de ce champ), à sa résistance ohmique R. Dans ces conditions, on déduit immédiatement (à partir des expressions de la surface du cadre, de sa longueur et de sa résistance électrique):

1) Qu'il y a toujours intérêt à choisir un diamètre de section pour le câble conducteur tel qu'il permette d'utiliser tout le terrain *en un seul tour*: c'est-à-dire, ne pas mettre plusieurs tours en série;

2) Que mettre plusieurs tours en parallèle (chaque tour occupant toute la surface disponible), revenant à une augmentation de la section du câble, n'est pas contraire à la règle précédente.

Les cadres ne devraient donc comprendre, en principe, que des spires mises en parallèle.

Dans le cas où la superficie du terrain utilisable n'est pas limitée, le problème est un peu différent: pour une quantité de cuivre donnée, on vérifie que tous les cadres circulaires, à une seule spire, obtenus en changeant simplement la section, donnent la même valeur pour Φ/R (puisque, par exemple, pour une surface de section dix fois plus petite, la surface collectrice du cadre, et sa résistance, seraient toutes les deux multipliées par cent.

Bien entendu, ces règles ne peuvent être appliquées d'une façon extrême ... par exemple à des cadres très petits mais faits d'un conducteur auquel on aurait donné une section énorme, donc de résistance nulle! Elles sont aussi souvent transgressées dans la pratique, par exemple pour des questions d'adaptation d'impédance à l'ensemble d'une installation; ou parce que d'autres impératifs que ceux retenus dans notre schéma entrent en jeu: notamment, la nécessité de disposer d'un seuil minimum pour les forces électromotrices (f. e. m.) des signaux captés et à amplifier, ce qui est le cas chaque fois que le terrain dont on dispose est exigu.

La dernière de ces conditions est celle qui s'applique, prioritairement, aux *cadres verticaux* destinés aux enregistrements des composantes horizontales: à moins de disposer de grandes étendues murales verticales on est conduit à racheter par un grand nombre de tours la faiblesse relative de la surface dont on dispose: citons les cadres circulaires (2 m de diamètre, 16000 tours) qui équipent les stations du réseau de W. H. CAMPBELL[1], et ceux qui ont été employés par DUFFUS

[1] W. H. CAMPBELL: Proc. I.E.E. **51**, 1337—1342 (1963).

et SHAND (6 m de diamètre, 1100 tours) [3]. Remarquons que le fait de pouvoir disposer maintenant d'amplificateurs à très haute impédance permet justement de ne pas rechercher systématiquement des capteurs ayant les coefficients Φ/R optima.

Avec l'introduction d'un noyau de haute perméabilité (en général *mumétal*, ou alliage équivalent) les problèmes de sensibilités (grande *surface équivalente*) et de dimensions sont résolus plus facilement, mais on peut craindre l'introduction de distorsions (en amplitudes et en phases) dues aux propriétés complexes de l'alliage utilisé. Ces inconvénients ont été souvent surestimés, ceci, principalement parce qu'une attention insuffisante a été portée aux effets du champ démagnétisant s'exerçant dans ces noyaux. En particulier, l'assimilation abusive (souvent pratiquée) entre le *cylindre allongé* constituant le noyau et un *ellipsoïde de révolution* ayant même rapport d'allongement, qui paraît permettre le calcul, par les formules simples applicables aux ellipsoïdes, des valeurs de ce champ démagnétisant à l'intérieur du noyau, conduit à sous-estimer considérablement son influence (parfois, dans un rapport de 1 à 10 même). L'explication de ce paradoxe, aussi simple soit-elle, échappe facilement à l'attention: il faut remarquer que l'ellipsoïde inscrit à l'intérieur du cylindre y serait aussi pointu «que la pointe d'une épée»[2]: la répartition des «masses magnétiques libres» sur sa surface serait donc totalement différente de celle sur le cylindre. Des mesures précises faites sur des cylindres de mumétal de longueur 1 m et de diamètre 1 cm, placés dans un champ externe uniforme de l'ordre de 0,2 Gs (Gauss), ont établi que le champ démagnétisant annulait les 19/20 de ce champ extérieur, même au centre du cylindre[3], alors que la formule applicable à l'ellipsoïde conduirait à un effet démagnétisant peu important pour un tel rapport d'allongement. Il s'en suit qu'un tel noyau fonctionne dans des conditions d'*auto-asservissement*[4] qui linéarisent considérablement ses propriétés amplificatrices (ou de concentration du flux magnétique). Il revient par exemple au même qu'il soit placé dans un champ extérieur de 0,2 ou de 0,5 Gs. Pour un rapport d'allongement donné, la linéarisation est ainsi assurée par la haute perméabilité intrinsèque de l'alliage et toute augmentation de cette perméabilité permettra de maintenir le même degré de linéarisation en pratiquant des rapports d'allongements plus élevés. Bien entendu, comme dans un amplificateur électronique à contre-réaction, le processus précédent accuse la valeur du rapport de réduction quand on compare les perméabilités *intrinsèques* (ou en circuit magnétique fermé) et *apparentes* du noyau utilisé. Mais, comme pour un alliage du type mumétal la perméabilité (relative) intrinsèque (*initiale*, car ce qui précède prouve que c'est elle seule qui intervient) est considérable (par exemple 40000), on pourra bénéficier quand même d'une perméabilité apparente — ou «efficace» — de l'ordre de 1000 ou plus.

A titre d'exemple on donne comme suit les caractéristiques d'une telle sonde, constituant la «barre» d'un montage du type «Barre-Fluxmètre» [4]: *alliage*, mumétal au molybdène, de perméabilité (relative) initiale comprise entre 20000 et 40000 (selon la nature exacte du traitement thermique); *dimensions*, cylindre de révolution de longueur 2 m et diamètre 3 cm; *enroulement*, en fil de cuivre sous double émail, de 1 mm ∅, 13000 spires inscrites à l'intérieur d'une surface à méridienne de forme parabolique; *caractéristiques magnétiques* d'utilisation, perméabilité (relative) moyenne apparente 1000, surface équivalente (à un cadre à air) de l'ordre de 10^8 cm^2; *caractéristiques électriques*, résistance ohmique 25 Ω, self induction de l'odre de 90 H; *masse*, environ 25 kg venant pour moitié du noyau et pour moitié du cuivre de l'enroulement. Nous n'insistons pas sur les protections externes (bain d'huile, étui, étanchéité des prises, etc. ...) qui varient avec les conditions matérielles d'emploi.

---

[2] Cette comparaison est due à G. DUPOUY.

[3] E. SELZER: Ann. Géophys. **12**, No. 2, 144—146 (1956).

[4] E. SELZER: J. Phys. Radium **21**, 79 (1970).

Tableau 1. *Exemples de grands cadres horizontaux destinés a l'enregistrement de variations magnétiques rapides.*

Avec quelques dispositifs à noyaux perméables donnés pour comparaison.

Designation	Surface efficace S/m^2	Resistance R/Ω	Longueur L du conducteur L/m	$\frac{S/L}{m}$	$\frac{S/R}{m^2/\Omega}$
Cadre de l'observatoire de Kakioka . .	$8{,}25 \cdot 10^4$	20	1156	71	$4{,}12 \cdot 10^3$
Cadre de RÖSSIGER	$0{,}35 \cdot 10^4$	3	600	8	$1{,}2 \cdot 10^3$
Cadre de BENIOFF	$0{,}22 \cdot 10^4$				
Cadre de l'observatoire de MOSCOU (1949)	$29{,}8 \cdot 10^4$		12000	25	
Cadre de HERRENCK à UCCLE	$1{,}0 \cdot 10^4$	8	400	25	$1{,}25 \cdot 10^3$
Cadre de BERTHOLD et al. (1960) . . .	$6700 \cdot 10^4$				
Avec noyaux perméables					
Appareil de NAGAOKA	$5{,}0 \cdot 10^4$	123,7	10700	4,7	$0{,}4 \cdot 10^3$
Sonde de KATO et UTASHIRO à Onigawa (1949) (6230 tours sur noyau de «sendust»)		34			
Sonde d'un montage «Barre-Fluxmetre» (1955)	$1{,}0 \cdot 10^4$	25	2000	5	$0{,}4 \cdot 10^3$

Revenant sur le principe même de l'emploi d'un noyau soumis — de par sa perméabilité intrinsèque considérable — à un champ démagnétisant intense, il peut être intéressant de signaler que, par l'introduction d'une «contre-réaction de flux» convenable (ce qui suppose, évidemment, une amplification électronique dans le montage) on peut encore augmenter ainsi, artificiellement, l'effet de ce champ démagnétisant. Des montages de ce genre sont effectivement utilisés.

9. Sondes à saturation du type «fluxgate». Le principe de ces sondes est bien connu[1]: étant donné un petit barreau cylindrique (ou même un fil) fait d'un alliage très perméable (mumétal par exemple) on détermine le champ magnétique externe dans lequel il se trouve plongé, en superposant à ce champ l'effet d'un champ alternatif (à une fréquence de l'ordre du Kilohertz) choisi d'une intensité telle qu'à chaque alternance il sature largement ce barreau. Cette saturation périodique a pour effet de bloquer et débloquer l'effet du champ à mesurer, à une fréquence double de la précédente. Un enroulement primaire impose le champ alternatif (de fréquence f), un enroulement secondaire recueillant la somme des effets (dûs aux variations du flux magnétique dans le barreau) se manifestant ainsi sur les fréquences f et $2f$. Les diverses réalisations de magnétomètres de ce type diffèrent par la méthode qu'elles mettent en œuvre pour séparer et enregistrer le *signal utile* sur la fréquence $2f$ du *signal de commande* sur la fréquence f (par analyse spectrale, filtrage, montages symétriques à deux éléments tels que dans le secondaire les effets d'induction dûs au champ de commande soient en opposition, ceux dûs au champ à mesurer s'additionnant, etc.). On peut aussi rapprocher ces magnétomètres des amplificateurs magnétiques, ou encore, considérer que leur élément principal constitue une bobine de self avec noyau dont la perméabilité est fonction du champ pseudo-continu à mesurer. Dans ce cas, l'usage du champ alternatif auxiliaire rentre dans le cadre des méthodes classiques des mesures de self, donc du champ pseudo-continu. Il n'est plus nécessaire, alors, d'avoir recours à une saturation du noyau. On peut, par contre, monter la

[1] Voir l'article de H. SCHMIDT et V. AUSTER au volume 49/3, Sects. 11—15, p. 346 de cette encyclopédie.

self sensible en tant que partie d'un circuit oscillant et transformer alors les variations du champ pseudo-continu qui nous intéresse en variations de la fréquence propre d'un tel circuit.

Pour en revenir au montage le plus usuel maintenant, à saturation, précisons que, en l'absence du champ pseudo-continu, les signaux recueillis — avant compensation ou filtration — dans un enroulement secondaire, ne sont évidemment pas sinusoïdaux purs (ne serait-ce que par l'intervention de la loi d'aimantation propre à l'alliage perméable) mais ne comportent, dans leur développement en série de Fourier, que les harmoniques d'ordre impair (ceci pour des raisons de symétrie, comme dans tout transformateur à fer sans fuites magnétiques). L'intervention du champ pseudo-continu introduit fonctionnellement la série des harmoniques pairs, la répartition entre celui d'ordre 2 et les harmoniques supérieurs se déplaçant graduellement au profit de ces derniers quand l'intensité du champ pseudo-continu augmente.

Après avoir joué un rôle important dans des applications militaires (tel le «M.A.D.» — «Magnetic-Airborne-Detector», pour les détections de sous-marins), puis de prospection minière[2] (des M.A.D. modifiés ayant équipés les «birds» remorqués par les avions de prospection), et avoir été à la base des nouveaux compas magnétiques du type «Fluxgate», on aurait pu penser que les magnétomètres à saturation disparaîtraient devant les *magnétomètres nucléaires* [5], surtout depuis que ces derniers ont pu être réalisés sous une forme vectorielle. En fait, dans le domaine des variations magnétiques rapides qui nous occupent, ces magnétomètres ont conservé leur place dans la recherche spatiale. C'est ainsi que dans la plupart des satellites destinés à des mesures magnétiques (et quelquefois seulement pour résoudre des problèmes d'orientation ou de restitution d'«attitude»), on les trouve complétant les magnétomètres nucléaires à précession de protons, ou les magnétomètres optiques, embarqués. C'est à ce titre que nous donnerons quelques précisions concernant leurs caractéristiques et performances techniques:

La plupart de ces magnétomètres sont actuellement du type à deux noyaux en opposition, ou, mieux, à noyau toroïdal mis en action d'une façon équivalente à la précédente (deux «demi-tores» jouant le rôle des deux noyaux en opposition). Ils sont montés en détecteurs *vectoriels* (donc trois sondes disposées suivant trois axes orthogonaux). Leur réponse est en général linéarisée par un montage en contre-réaction (par réinjection, en opposition avec le champ à mesurer ou à enregistrer, d'un champ de contre-réaction proportionnel à la tension-signal de sortie). Leur vitesse de réponse peut être assez rapide (atteignant 0,1 sec), mais leur sensibilité reste nettement inférieure à celle des magnétomètres à induction proprement dits sur lesquels ils ont cependant un avantage de principe spécifique: celui de donner les valeurs statiques des champs à enregistrer. Pour que ce dernier point puisse être assuré convenablement, il convient d'ailleurs de procéder à un véritable étalonnage des sondes pour un champ nul, ou à un réglage équivalent. C'est là ce qu'il y a de plus délicat dans leur usage, la fidélité de ces réglages étant aléatoire.

10. Magnétomètres à résonance[1]. Nous ne donnerons à ce paragraphe qu'un développement très limité, ceci pour les deux raisons suivantes: l'abondance et la qualité des études — aussi bien théoriques que pratiques — qui ont été con-

[2] Voir les articles de P. H. Serson et K. Whitham (p. 384) et de J. R. Balsley (p. 395) au volume 49/3 de cette encyclopedie.

[1] Voir les articles de H. Schmidt et V. Auster (p. 323) et de P. J. Hood (p. 422) au volume 49/3 de cette encyclopédie.

sacrées à ces types de magnétomètres [*5*—*8*]; leur propriété fondamentale d'être avant tout des instruments de *mesure absolue* des champs magnétiques, plutôt que des variomètres. Cependant, bien que ce deuxième point soit justifié par les difficultés que l'on rencontre pour atteindre avec ces instruments des seuils de performance de l'ordre, par exemple, de 0,01 γ en un temps de 1 sec, ou de 0,1 γ en un temps de 0,1 sec, encore insuffisants pour enregistrer toutes les formes intéressantes de pulsations magnétiques rapides, ils ont le mérite de permettre en une seule opération — dans les limites précédentes — la mesure des variations *et* des valeurs absolues. Il conviendra, bien entendu, pour bénéficier d'une telle mesure complète, de prévoir des organes de «sortie» capables d'absorber l'amplitude dynamique considérable que cela représente (sauf si le champ à mesurer devient très faible comme c'est le cas pour les confins de l'environnement terrestre ou pour l'espace interplanétaire). Pour cette raison, ces magnétomètres qui ne donnent pas, en général, des enregistrements graphiques analogiques très satisfaisants pour une observation visuelle directe, sont particulièrement bien adaptés à des modes de sortie numériques, soit directs (sur bande magnétique par exemple) soit alimentant des télémesures (ainsi que cela se passe quand ils sont embarqués à bord de fusées ou de satellites).

Un autre point important est relatif au fait que, d'après leur principe même de fonctionnement, ces magnétomètres mesurent le *module* des champs et non leurs composantes ou leurs directions. On a pu cependant tourner cette difficulté par divers procédés, dont l'un, par exemple, consiste à annuler le champ à mesurer suivant certaines directions connues ce qui fait que la mesure du champ total qui subsiste nous donne la valeur absolue du champ dans le plan normal correspondant. On peut aussi procéder à des renversements de champs créés, connus. Dans les satellites, l'usage combiné de sondes à saturation et de magnétomètres absolus des types ci-dessus se fait souvent de la façon suivante: les sondes à saturation sont chargées de l'enregistrement vectoriel principal mais, venant s'aligner à tour de rôle parallèlement au champ total à étudier, elles sont comparées à cet instant à la mesure absolue faite par le magnétomètre absolu et recalibrées ainsi d'une façon automatique.

Nous n'entrerons pas dans le détail des divers types de magnétomètres à résonance *soit protonique* (magnétomètres à précession de protons, libre — tel que le «VARIAN»[2] — ou entretenue, par exemple par l'effet OVERHAUSER-ABRAGAM[3–8]), *soit électronique*[9,10] (magnétomètres à «pompage optique» à vapeurs de Rubidium ou de Cesium, magnétomètres à Hydrogène ou à Hélium etc.). Disons seulement que la plupart de ces magnétomètres à résonance jouent un rôle très important dans les opérations de cartographie magnétique spatiale (notamment pour la magnétosphère et ses frontières) et que le choix précis du modèle devant équiper tel ou tel satellite dépend de la mission précise qu'aura à accomplir celui-ci (et de l'ordre de grandeur moyen des champs qu'il aura à mesurer).

En résumé, il faut peut-être voir dans les grands progrès de cette *cartographie magnétique*, comparés à l'état, resté encore très précaire, de nos connaissances sur

---

[2] M. PACKARD, R. VARIAN, A. BLOOM, and D. MAUSER: Phys. Rev. **93**, Sér. II, 941 (1954).
[3] A. W. OVERHAUSER: Phys. Rev. **92**, 2, 411—415 (1953).
[4] A. ABRAGAM: Phys. Rev. **98**, 6, 1729—1735 (1955).
[5] A. ABRAGAM, J. COMBRISSON et I. SOLOMON: C.R. Acad. Sci. **245**, 157—160 (1957).
[6] I. SOLOMON: Phys. Rev. **99**, 559 (1955).
[7] J. FREYCENON et I. SOLOMON: Onde Electrique **402**, 596—601 (1960).
[8] G. BONNET: J. Phys. Radium **22**, No. 4, 204—214 (1961). Voir aussi la Note 1 de la Sect. 2.
[9] L. MALNAR et J.-P. MOSNIER: Ann. Radio. **16**, No. 63, 1—8 (1961).
[10] M. J. USHER, W. F. STUART, and S. H. HALL: J. Sci. Instr. **41**, 544—547 (1964).

la dynamique magnétique des mêmes régions, le fait que l'on dispose d'instruments bien adaptés à cette exploration, alors que — contrairement à ce qui se passe au sol — il n'existe pas encore de bons variomètres rapides et sensibles pour faire progresser aussi bien ces connaissances dynamiques. Il se joint à cette insuffisance expérimentale une difficulté de principe sur laquelle nous reviendrons, celle résultant de l'impossibilité de séparer les variables de temps et d'espace par des enregistrements faits à bord d'un seul satellite.

11. Méthodes d'analyse. L'application des diverses techniques indiquées au cours du paragraphe précédent nous procure en définitive une certaine «matière» dans laquelle se trouve rassemblée l'information à étudier. Celle-ci ne s'y trouve pas forcément à l'état brut puisque certaines opérations de mise en forme ont déjà pu être effectuées à l'enregistrement, ainsi que nous l'avons envisagé. D'autre part, le support même de cette information peut prendre des formes très variées et il est évident que le choix des méthodes d'analyse doit tenir compte, non seulement des genres de phénomènes que l'on désire étudier, mais aussi de cette «matière» même (en général — mais pas forcément dans tous les cas — des *enregistrements*) que l'on a rassemblée à cet effet.

C'est ici qu'une *logistique* appropriée doit seconder nécessairement la méthode d'analyse proprement dite. Donnons en l'exemple le plus terre-à-terre — mais non le moins important — relatif au stockage des documents qui doit, comme pour une bibliothèque, satisfaire à des normes d'*encombrement* (à moins que l'on ne s'astreigne à des disciplines d'*effacement* après une mise en mémoire de durée limitée, ou de destruction périodique, etc. ... ce qui impose des techniques d'analyse rigoureusement étudiées et appliquées), de *commodité de consultation* régulière ou épisodique, etc. C'est ainsi que certaines présentations d'enregistrements sur papier (gros rouleaux) conviennent parfaitement aux analyses méthodiques faites une fois pour toutes, mais rendent pratiquement inacceptable toute demande ultérieure de vérification épisodique d'un évènement particulier (pour répondre, par exemple, à des «checking lists» ce qui est une forme active de la collaboration internationale. (D'où la faveur des enregistrements-papier grande vitesse à «pliage accordéon» tels qu'ils ont été généralisés en de nombreuses stations.)

Passant de cette logistique à la *logique* même des méthodes d'analyse à mettre en œuvre, nous pouvons l'axer sur les trois formes suivantes:

— *l'examen direct,* soit à la vue (enregistrements sur papier, photographique ou non), soit à l'oreille (enregistrements analogiques en «direct» sur bandes magnétiques, après application de rapports de vitesse, enregistrement-lecture, convenables);

— *l'analyse analogique,* à partir d'enregistrements sur bandes magnétiques, en «direct» ou en «MF» (modulation de fréquence);

— *l'analyse numérique,* soit à partir d'enregistrements sur papier que l'on soumet à une «digitalisation» (manuelle, semi-manuelle ou automatique), soit à partir d'enregistrements déjà codés (de préférence, sur bande magnétique).

α) *Examen direct.* En dehors d'une vue d'ensemble (qu'il ne faut pas sous-estimer) cet examen comprend le relevé des phénomènes caractéristiques *et* la détermination précise de diverses grandeurs: amplitudes, instants, etc. En particulier, la détermination des **SSC** («Storm-Sudden-Commencement»), **SI** («Sudden-Impulse»), etc., à 1 sec près (ou mieux) chaque fois que le phénomène le permet (ce qui suppose aussi un rapport judicieux sensibilité/vitesse de défilement ainsi, évidemment, qu'une bonne définition du temps). Il convient de remarquer à ce dernier point de vue que la précision requise sur le temps doit être maintenant aussi parfaite dans les observatoires magnétiques que dans les observatoires sismiques. Une recommandation d'un autre ordre est — dans l'étude «à vue» d'un phénomène — de ne pas la faire sur un seul type d'enregistrement («ultra rapide» par exemple). S'aider toujours de l'ensemble des enregistrements de normes diverses dont on dispose concernant ce phénomène (notamment des enregistrements classiques dits «lents»). Dans le cas des enregistrements sur bande magnétique, un premier examen *à l'oreille* (pour diverses vitesses d'écoute utilisant au mieux nos facultés auditives) devrait toujours précéder les analyses automatiques: une oreille exercée dépasse en pouvoir séparateur harmonique la plupart des analyseurs de spectres, ceci, surtout pour les très faibles niveaux d'enregistrement et en présence de bruit.

Ce premier examen à l'oreille n'est certes pas quantitatif mais il permet une conduite plus judicieuse des analyses automatiques qui doivent lui faire suite. A noter également, que certaines variations — «perles structurées» notamment — peuvent être ainsi reproduites avec un timbre caractéristique, se rapprochant des «sons éoliens» étudiés par Lord Rayleigh[1].

β) Analyse analogique. En principe elle doit suivre l'examen direct. Il est donc souhaitable, dans le cas où l'effort principal a porté sur des enregistrements sur bande magnétique, de les accompagner quand même d'un enregistrement sur papier de caractéristiques plus ou moins voisines, par exemple au moyen d'un dispositif branché en parallèle sur l'enregistreur magnétique. (On pourra aussi s'aider des procédés de conversion: bande magnétique-papier).

Ceci étant, l'analyse analogique doit nous apporter ce qui n'est pas explicitable clairement à partir des examens directs, notamment des renseignements quantitatifs sur la constitution spectrale des variations enregistrées. Nous considèrerons à ce sujet deux grandes classes d'opérations:

— l'établissement de *spectres intégrés* (spectres de puissance), donnant la répartition de la puissance en fonction des divers composants harmoniques, ceci pour une durée d'intégration déterminée, cette durée pouvant être considérée comme un paramètre dont il conviendra de se servir judicieusement en fonction de la façon même dont on désirera découper dans le temps l'étude du phénomène global auquel on s'intéresse;

— la recherche des *évolutions spectrales en fonction du temps*, telles que celles obtenues au moyen d'appareillages analysant une bande d'enregistrement sur bande magnétique par sélection successive des différentes gammes de fréquence. Sont maintenant classiques les «Sonagraphes», «Rayspan», «Spectran», etc., la présence dans le spectre des diverses composantes n'étant alors indiquées que d'une façon relativement sommaire s'apparentant à un processus par tout ou rien.

Les deux procédés se complètent et l'on peut arriver ainsi à établir d'une façon assez précise l'évolution, en fonction des deux variables *fréquence* et *temps*, de la puissance — ou de l'intensité — spectrale des variations magnétiques analysées. En dehors des cas de structures plus ou moins complexes que nous aurons l'occasion d'examiner, une première distinction élémentaire, mais d'une grande importance, à laquelle l'emploi de ces méthodes d'analyse doit nous permettre d'accéder, est celle entre des variations *organisées* et des variations *incohérentes*. Chacun de ces termes demanderait d'ailleurs à être défini d'une façon précise (ce que nous ne ferons que pour des cas précis qui nous intéressent). En prenant ici le cas extrême d'un apport d'énergie d'une incohérence spectrale parfaite (bruit blanc) c'est dans ce cas seulement que la puissance reçue devra

[1] Lord Rayleigh: Theory of Sound, edit. 2 vol. (1926) ou 1 vol. (1945).

être trouvée proportionnelle à la largeur de la bande passante correspondant à sa mesure. La façon dont ce critère sera plus ou moins vérifié — ou transgressé — nous éclairera utilement sur la nature même des variations transportant cette énergie. A remarquer toutefois — avant d'être tenté d'appliquer ce critère sous cette forme simple — que le degré d'approximation d'incohérence que nous cherchons à atteindre n'est lui-même susceptible d'une définition précise que si l'on a fixé par une convention appropriée l'échelle des temps par rapport à laquelle on désire l'apprécier. C'est ainsi que l'ensemble des variations journalières[2] $\mathbf{S}_q$ considérées à l'échelle d'un siècle possèdera une structure totale parfaitement incohérente, alors que chacune de ces variations — ou un petit nombre d'entre elles — ont les propriétés d'un système organisé. Inversement, toute variation très rapide qui paraîtrait incohérente, par exemple à l'échelle de la seconde, pourra se révéler parfaitement organisée (ou «structurée») si on l'analyse à l'échelle du millionième de seconde. Nous avons pris là, évidemment, des cas extrêmes, mais nous rencontrerons dans notre domaine de fréquences des exemples s'apparentant à ces deux cas.

Les analyses spectrales ne sont pas le terme des études analogiques: de nombreux appareillages, disponibles maintenant sur le marché, permettent de déterminer — à partir d'enregistrements analogiques sur bandes magnétiques — les diverses fonctions de corrélation relatives soit à un signal (à une ou plusieurs composantes), soit à des signaux différents (provenant par exemple de stations d'enregistrement plus ou moins éloignées). Ne pouvant entrer ici dans les modes de fonctionnement (dont les principes sont bien connus[3]) de ces appareillages, précisons quelques unes des caractéristiques des variations magnétiques rapides que l'on peut ainsi atteindre: les degrés de cohérence dans le temps (ce qui nous ramène aux remarques déjà faites au paragraphe précédent), les degrés de cohérence dans l'espace (ce qui nous permet d'atteindre les modes de propagation), les degrés de cohérence entre les diverses composantes d'un même signal (ce qui permet d'atteindre divers effets locaux, perturbateurs ou correctifs et — dans une moindre mesure — les phénomènes de polarisation qui, à cause de la grande précision nécessaire, doivent le plus souvent être déterminés directement). Revenant sur cette dernière remarque, notons que pour atteindre une détermination des phases à 15° près dans le cas de pulsations d'une période moyenne de l'ordre de la seconde, une précision de $^1/_{24}$ sec sur les temps est nécessaire (précision en valeurs relatives ou absolues suivant la nature, elle-même relative ou absolue, des déterminations de phase recherchées). En dehors des soucis — d'une nature classique — que cela entraîne sur l'enregistrement du temps aux stations d'enregistrement utilisées, il faut veiller également à la parfaite constance des réponses en phase que cela suppose pour tous les appareillages intervenant dans les chaînes d'enregistrement, de transport et de reproduction des données. C'est là un des points les plus délicats dans la conduite des opérations que nous venons de passer en revue rapidement.

γ) *Analyse numérique.* Ce type d'analyse qui — théoriquement — peut résoudre tous les problèmes, est, cependant, relativement encore peu utilisé dans les études de variations magnétiques rapides. Nous pouvons voir deux raisons majeures à cela:

1) Les difficultés d'une digitalisation *initiale* du signal jointes aux inconvénients (complications, pertes de temps, frais, etc.) d'une digitalisation faite ultérieurement, ces inconvénients étant particulièrement importants si le signal a été enregistré initialement sur papier photographique ou sur papier ordinaire:

[2] Voir l'article de T. Nagata et N. Fukushima, Sects. 8 et 9, au volume 49/3, p. 16 de cette encyclopédie.

[3] A. Houri: Thèse de spécialité. Fac. Sci. Paris (1966). On pourra consulter aussi B. P. Lathi: Signals, Systems and Communication. New York-London-Sidney: John Wiley and Sons, Inc. 1965.

2) Les prix élevés d'une analyse complète, *à l'échelle géophysique*, de toutes les variations rapides enregistrées. (Par «à l'échelle géophysique» nous entendons les études portant sur tous les phénomènes se produisant à longueurs de journées et tous les jours de l'année. 24 heures sur 24.)

Cette étude à l'échelle géophysique n'est certes pas la seule valable et d'autres types d'études, par «ponctions» ou «échantillonnages», doivent aussi être entreprises avec l'appoint — alors possible — d'opérations digitales. Ceci pourra être le cas, par exemple, d'expériences particulières relatives à des conditions de propagation, à des relations entre variations entre points conjugués, etc. C'est ainsi que dans les études entre les stations conjuguées de Kerguelen (Océan Indien) et de Sogra (U.R.S.S.) des dispositifs digitaux doublent maintenant ceux, analogiques, exploités depuis plusieurs années. (A noter que d'autres dispositifs digitaux étaient déjà en place à Kerguelen, ainsi qu'à Port-aux-Français — Terre Adélie — pour des raisons de transmission, par «Télex», des observations entre l'Antarctique et la France, concernant les variations «lentes».)[4]

Nous ne nous étendrons pas sur la mise en application de ces méthodes digitales qui en ce qui concerne les *résultats* géophysiques n'auront pas à être différenciées fondamentalement des méthodes analogiques.

12. Remarque sur la conduite des dépouillements — Études à l'échelle individuelle — Études à l'échelle mondiale. Malgré le caractère «séculaire» des programmes d'études géomagnétiques (y compris ceux concernant les variations rapides) aucun programme de dépouillement ne doit être considéré comme figé dans un moule indéformable. Une certaine progression dans la continuité imposée aux opérations découle naturellement de la nature plus ou moins vaste du cercle des chercheurs intéressés par ces dépouillements. Laissant de côté le cas particulier des recherches individuelles, les deux stades rencontrés le plus habituellement sont ceux:

1) des recherches animées à l'échelle d'un groupe utilisant les ressources d'un laboratoire (ou d'un petit nombre de laboratoires);

2) celles définies impersonnellement à l'usage, immédiat ou futur, de la communauté mondiale.

Il est d'ailleurs fréquent que les groupes ressortissant du 1) soient les éléments constitutifs principaux de l'activité définie au 2). Dans ce cas, leurs programmes de recherche et de dépouillement devront — en plus de leurs normes particulières — s'harmoniser avec l'œuvre internationale. C'est cette dernière que nous allons maintenant résumer très schématiquement en nous limitant au cas des variations magnétiques rapides.

La raison d'agir à l'échelle mondiale dans ce domaine s'est imposée aussitôt qu'est apparue clairement que ces variations magnétiques rapides (ainsi que les variations telluriques correspondantes) n'étaient pas des manifestations purement locales, ou semi-locales (telles, par exemple, que les manifestations météorologiques usuelles), mais restaient souvent comparables (avec des degrés de corrélation d'ailleurs très variables) sur de très grandes étendues de notre Globe. Que ce soit pour étudier leur degré de corrélation — ou de non-corrélation, dans certains cas — les études synoptiques à l'échelle mondiale devenaient nécessaires.

L'organisation de ces études a pris corps au cours de réunions successives organisées principalement par le «Comité des Variations Magnétiques et Telluriques rapides» de l'Association Internationale de Géomagnétisme et d'Aéronomie («AIGA» ou «IAGA») de l'Union Géodésique et Géophysique Internationale («UGGI»), soit spécialement — Symposium d'Utrecht en 1957, de Copenhague en 1959 — soit à l'occasion des Assemblées Générales AIGA-UGGI (notamment à Toronto en 1957, Helsinki en 1960, Berkeley en 1963, Saint-Gall en 1967 et Madrid en 1969). Depuis l'Assemblée de Berkeley ce Comité a pris la forme d'un «Groupe de Travail» de la Commission IV de l'AIGA («Variations et Perturbations Magnétiques») mais, en fait, la plus grande partie de l'activité de cette Commission — ainsi que celle de la Commission V du même organisme («Relations entre le Soleil et la Magnétosphère») — concerne directement notre sujet.

---

[4] R. SCHLICH: Cahiers IPG (Institut de Physique du Globe) Paris, No. 25 (1967).

Sans insister autrement sur la structure de cette organisation (sur laquelle on pourra se documenter en consultant les rapports «IAGA»[1]) dégageons les méthodes de travail proposées à cette échelle:

Les premiers modes d'approche ont été essentiellement *morphologiques* ce qui est naturel. On a commencé par se mettre d'accord sur les propriétés morphologiques les plus communes qui se sont trouvées — par suite d'une déformation résultant des circonstances — être celles qui avaient pu être observées, enregistrées et définies, dans les observatoires, relativement nombreux, des régions de moyennes latitudes. Or, malgré le caractère mondial que nous avons signalé, une telle morphologie ne peut prétendre être représentatrice d'une morphologie à l'échelle de tout le Globe, y compris, notamment, les régions polaires et équatoriales. Grâce à l'exploitation du réseau mis en place à l'occasion de l'Année Géophysique Internationale (AGI) — 1957/58 avec continuation en 1959 — l'élargissement nécessaire a été fait: on connaît bien maintenant les types morphologiques caractéristiques, plus spécialement, des régions soit des hautes, soit des basses latitudes. Cependant, l'«Atlas des Variations Magnétiques rapides» [*9*] patronné par la Commission ci-dessus mentionnée (et qui n'a pu, d'ailleurs, qu'être imparfaitement réalisé, sous la forme d'une sorte de schématisation des réalités) propose une nomenclature mondiale unique mais qui englobe et tient compte de tous les cas observés. On conserve aussi à cette classification morphologique son caractère strictement objectif en évitant d'y introduire aucun élément qui serait lié à une loi ou à des observations particulières quelconques. Bien entendu chaque observateur, dans l'usage qu'il fait de ces structures pour classer tel ou tel phénomène, doit tenir compte de tous les éléments dont il dispose. En fait on lui demande — s'il appartient à la centaine environ de stations ou d'observatoires coopérants — de remplir mensuellement des tableaux dans lesquels cette classification morphologique (complétée par des données quantitatives mesurables directement sur les enregistrements) peut être présentée suivant des normes bien établie et suivies. Un collationnement critique est fait de l'ensemble des données ainsi reçues.

Ce travail est assuré sous la responsabilité de l'*Observatoire de l'Ebro* — Roquetas (Tarragone) — Espagne, en coordination avec le «Service International des Indices Géomagnétiques» («Int. Service of Geomagn. Indices, Kon. Ned. Meteor. Instituut, De Bilt, Netherlands») s'appuyant lui-même sur des déterminations faites à l'Institut de Géophysique de Göttingen («Institut für Geophysik, Postfach 876, 34 Göttingen, Germany»). Remarquons que le Service des Indices s'occupe avant tout des conditions d'«agitation magnétique» (Indices K — Caractères C entre autres) mais ces conditions sont étroitement liées par certains de leurs aspects aux variations rapides qui nous intéressent.

On peut regretter que tout ce travail — dont l'ampleur est grande — se fasse sur des données conventionnellement «chiffrées» (par des lettres en général) comportant pour une bonne part une information subjective. Pour la morphologie pure, une comparaison directe de tous les enregistrements analogiques mondiaux serait évidemment souhaitable, mais elle ne peut être réalisée qu'en passant par les «Centres Mondiaux de Données» («World Data Centers» — «WDC») qui disposent d'un grand nombre de ces enregistrements sous le forme de microfilms. Quant aux études quantitatives elles nécessitent une digitalisation des données qui n'a pu être mise en place qu'à un petit nombre d'observatoires (une quinzaine en 1969) et ceci uniquement pour les enregistrements dits «normaux» (ou «classiques» ou «lents»). Il y a là des problèmes d'organisation et de financement qui dépassent de beaucoup la pure technique ... et les possibilités de la communauté scientifique internationale. Remarquons enfin, que la nécessité d'étendre ces comparaisons et ces études (aussi bien morphologiques que quantitatives) aux résultats obtenus en satellites, nous incite à améliorer nos méthodes de dépouillement et de présentation des résultats, ceci dans le sens des nécessités exposées ci-dessus.

---

[1] On pourra consulter en particulier: Transactions of the General Scientific Assembly, IAGA Bulletin No. 27, 71—87, Madrid (1969).

Tableau 2. *Charactéristiques des pulsations géomagnétiques observées a Chambon-la Forêt («CF»).*
(A comparer avec le tableau équivalent — pris comme modèle — pour Kakioka, voir tome 49/3, p. 119, contribution de NAGATA et FUKUSHIMA.)

IAGA notation	Période moyenne sec	Forme d'onde	Durée	Variation diurne en		Autres remarques
				Activité	Période	
pc 5	200 ... 400	Sinusoïdal	$\frac{1}{2}$ ... 2 h	Le matin	Trop peu de cas pour conclure	Phénomène exceptionnel à CF sous sa forme pure. Relations possibles avec la «petite agitation du matin» (très irrégulière en général)
pc 4	50 ... 110	Sinusoïdal	Plusieurs heures	Principalement en fins de nuit; possibilité à toute heure sauf lère partie nuit	?	En période de minimum d'activité solaire, prennent souvent la place des **pc 3** pendant la journée
pc 3	20 ... 30	Sinusoïdal	Plusieurs heures	Durant le jour, entre 5 et 17 h TLM	Susceptible de variations brusques lors de changements de séquences; allongement progressif au cours de la journée lors de séquences stables	De toutes les classes de pulsations ce sont celles qui ont le «coefficient de présence» le plus élevé
pc 2	5 ... 7	Sinusoïdal	Une à deux heures	très tôt le matin, entre 3 et 6 h TLM	?	Sont les plus fréquentes et intenses à la suite d'une agitation magnétique
pi 2	50 ... 110	Irrégulières parfois, diminutions exponentielle des amplitudes	Par trains successifs de chacun 5 ... 12 min	La nuit, principalement entre 21 et 2 h TLM	?	Correspondent à l'ancienne nomenclature «pt». On peut continuer à désigner par pt le phénomène général, y compris les **pi 1** qui y accompagnent souvent les **pi 2**
pi 1	2 ... 5	Irrégulières se présentent parfois comme une suite d'impulsions	Par «nuages» successifs	La nuit, entre 21 et 1 h TLM	?	Peuvent être rattachées en général à un train de pt (donc à des **pi 2**, mais il arrive fréquemment dans le cas de trains successifs que les **pi 1** recouvrent l'ensemble de ces trains
pc 1	$\frac{1}{2}$ ... 2 s	Structurées en général	20 min ... 1 à 2 h	La nuit	?	Il existe des cas non structurés; les structures les plus régulières donnent lieu aux «perles»

C. Classifications et connaissances morphologiques déduites des observations au sol*.

I. Généralités.

13. Classifications et nomenclatures internationales. Les variations magnétiques rapides sont des phénomènes complexes dont les magnéticiens n'ont pu prendre conscience que peu à peu et dans des conditions dépendant beaucoup de facteurs extérieurs à leur nature même, en particulier des conditions d'enregistrement et d'analyse, ainsi que nous l'avons expliqué à la Section précédente. On comprend, dans ces conditions, que les deux aspects du sujet dont nous allons nous occuper maintenant: les problèmes de classification et de nomenclature d'une part, les connaissances morphologiques d'autre part, se soient développés parallèlement. En ce qui concerne notre méthode d'exposition, nous allons d'abord faire appel au chapitre «Morphology of Magnetic Disturbance»[1] dans lequel les auteurs exposent une vue plus générale (non limitée aux variations rapides) de ces questions. Nous supposerons donc connues du lecteur les connaissances morphologiques fondamentales qu'il trouvera dans ce chapitre. Ceci est d'autant plus souhaitable que les interactions entre les diverses formes de variations ou de perturbations magnétiques jouent un rôle important — bien que souvent complexe et encore peu compris — dans les mécanismes qui les conditionnent et qu'il est ainsi nécessaire, même si l'on ne s'intéresse qu'à un type déterminé de variations, d'être bien au courant de l'ensemble du complexe géomagnétique dans lequel ces variations ont pris place. (Pour la même raison il est recommandé d'accompagner l'examen d'un enregistrement rapide, de toute la série des enregistrements faits avec d'autres caractéristiques, au même lieu et au même moment.)

Dans l'exposé de NAGATA et FUKUSHIMA nous porterons notre attention tout spécialement sur les deux points suivants:

1) La description morphologique — principalement à une échelle «macroscopique» — des perturbations magnétiques les plus caractéristiques (orages, baies, en particulier);

2) une introduction — avec rappel de la classification et de la nomenclature internationales en vigueur actuellement — sur les «Pulsations Magnétiques».

Relativement au premier de ces points, nous nous laisserons guider par la description macroscopique pour montrer comment certaines de ses évolutions peuvent comporter une «microstructure» plus ou moins fine dont la liaison avec l'évolution principale sera intéressante à dégager.

En ce qui concerne le second point, son rappel nous permettra d'entrer directement dans le vif des études morphologiques (appliquées non seulement aux formes d'ondes mais aussi aux spectres de toute nature) en bénéficiant de la clarté d'une nomenclature déjà en place, mais sans en être prisonnier. A ce titre, nous pourrons aussi montrer l'intérêt que présentent parfois des nomenclatures plus anciennes, ou certaines nomenclatures particulières utilisées pour des recherches spécialisées.

Etant donné son intérêt pratique pour notre exposé, nous commencerons par la partie formelle de ce second point: les «pulsations magnétiques» ne sont qu'une variété particulière des variations magnétiques rapides, mais on peut les con-

* Un certain nombre d'Études Générales, notamment [*2*], [*10*], [*11*], [*12*], [*13*], [*14*], [*15*], présentent — sous un jour plus ou moins différent — cette question.

[1] Contribution de T. NAGATA et N. FUKUSHIMA au volume 49/3, p. 5—130 de cette encyclopédie.

sidérer dans une certaine mesure comme des constituants élémentaires de toutes ces variations, ceci avec d'autant plus de facilité que l'on aura réduit les différentes entités que représentent ces pulsations à une nomenclature de tendance plus «algébrique» que «phénomènologique», ce qui est justement le cas de la nomenclature internationale actuelle.

Nous savons ainsi que toutes les pulsations sont rangées en deux catégories principales, les «continues» (ou «régulières»), notées **pc** et les «irrégulières», notées **pi**. Une division supplémentaire est faite d'après la valeur moyenne de la «période dominante»[2], de **pc1** à **pc5** et entre **pi1** et **pi2**. Pour l'instant nous ne discuterons pas la signification exacte à donner aux termes «continues» et «irrégulières» représentés par les suffixes **c** et **i**. Il suffit que, par suite d'un ensemble de conventions on n'ait guère à hésiter entre l'un ou l'autre. (Nous reviendrons sur cette question à propos de l'examen plus poussé de la morphologie.) L'usage des subdivisions est ensuite automatique. Nous donnerons cependant, dès maintenant, une idée un peu plus concrète de toutes ces pulsations, en renvoyant d'une part[3] à la table 2, p. 119, du tome 49/3 qui donne les principales caractéristiques des pulsations observées à l'observatoire japonais de Kakioka, et en permettant d'autre part la comparaison de cette table avec une table analogue que nous avons établie pour l'observatoire français de Chambon-la-Forêt (voir notre Tableau 2, p. 250 ci-devant).

Revenant maintenant au premier des deux points sur lesquels nous désirions porter notre attention, nous allons l'envisager plus en détail en prolongeant l'étude morphologique de NAGATA et FUKUSHIMA par l'examen et l'analyse des enregistrements spécialisés — rapides et ultra-rapides — relatifs aux mêmes perturbations magnétiques.

II. Microstructures des perturbations magnétiques.

Nous traiterons principalement le cas des baies et celui des orages, avec ceux de quelques phénomènes plus particuliers.

14. Microstructures des baies magnétiques et des perturbations plus complexes de même type. En accord avec l'article NAGATA et FUKUSHIMA, mais sans suivre exactement le même ordre d'exposition, nous considèrerons la baie magnétique comme la forme la plus simple (bien que susceptible de se développer à une échelle mondiale) de toute une classe de perturbations dont on s'est habitué — peut-être quelque peu arbitrairement — à mettre en relief le côte «polaire». Dans l'ordre de la complexité croissante — et de leur importance perturbatrice — nous rencontrons ainsi: les *baies simples*, les *baies autorales* et les *orages polaires élémentaires* («polar elementary storms» de BIRKELAND)[1]. Il est entendu qu'il s'agit ici des différentes intensités par laquelle ce type général de phénomène peut se manifester et qu'il est très possible qu'à un même instant le phénomène mondial apparaisse sous la forme d'un orage polaire élémentaire, ou d'une baie aurorale, en un certain lieu (principalement dans la zone aurorale) et sous la

[2] Voir le tableau de la page 119 de l'article NAGATA et FUKUSHIMA au volume 49/3, p. 5 de cette encyclopédie.

[3] Article de T. NAGATA et N. FUKUSHIMA au volume 49/3, p. 5 de cette encyclopédie.

[1] Nous éviterons, par contre, de rechercher une traduction française de l'expression «polar substorm» employée par de nombreux auteurs, le terme «substorm» — que nous pourrions traduire par «sous-orage» — ayant été utilisé avec des sens très variables, allant depuis celui attaché à une simple baie, à celui qui conviendrait aux phases les plus actives des plus grands orages mondiaux symbolisées, par exemple, par les «**DP**» de AKASOFU et CHAPMAN [*16*]. .

forme d'une simple baie, plus ou moins marquée, en d'autres régions (notamment aux latitudes modérées). Mais les microstructures que nous décrirons seront alors relatives à la forme correspondante sous laquelle se sera manifestée la perturbation *au même lieu*. (Voir Fig. 17 à 22 pour ce qui suit.)

Les *baies simples* ne présentent pas toutes une microstructure. Aux moyennes latitudes seules les baies positives (croissance de $\mathcal{H}$) nocturnes, en comportent; il s'agit alors, en premier lieu, d'une *microstructure primaire* formée de trains courts, successifs, de quelques oscillations irrégulières, dont la pseudo-période s'échelonne entre un peu moins d'une minute et deux ou trois minutes, qui dans les premières nomenclatures internationales reçurent successivement les dénominations de **psc** (pour «polar — ou pulsationnel — sudden commencement») puis de **pt** (pour «pulsationnal train») [*9*], et qui rentrent maintenant dans la catégories des **pi2**. A cette microstructure primaire vient souvent s'ajouter une *microstructure secondaire* faite, en général, de pulsations encore plus irrégulières que les précédentes et correspondant à des pseudo-périodes beaucoup plus courtes que l'on range ainsi dans la catégorie des **pi1**. On distingue quelquefois parmi elles une microstructure très fine (pseudo-periode de une à quelques secondes) et une *microstructure intermédiaire* (pseudo-période autour de 10 à 15 sec). Ces microstructures secondaires débordent souvent largement, dans le temps, des **pi2** (ou **pt**) auxquels, cependant, elles paraissent liées. C'est ainsi qu'elles ne sont pas forcément limitées à la partie «aller» de la baie. Il arrive aussi fréquemment que cette phase aller comporte plusieurs trains séparés de **pt**. Chacun d'eux dure, par exemple, de l'ordre de 5 min (s'il est formé de quelques oscillations **pi2** seulement), des intervales d'une dizaine de minutes existant entre la fin d'un train et le début du suivant, tout cet ensemble se produisant au cours de 30, 40 ou 50 min ... de durée de la phase aller de la baie. Dans le cas des *baies complexes*, que l'on peut considérer comme des baies *juxtaposées*, la deuxième ou la troisième commençant avant que la précédente ne soit terminée (baies «chevauchantes»), la répartition des différents trains de **pt** paraît obéir quelquefois à un phénomène de répétition assez régulier sans qu'il soit possible de les rattacher au développement propre de la baie complexe.

De toute façon les microstructures secondaires unissent tous ces trains entre eux, apparaissant sur les enregistrements très rapides comme des nuages d'impulsions irrégulières dont la durée totale déborde de celle du phénomène primaire. (Il arrive même parfois que le début de ce nuage précède celui de la baie). Il est fréquent aussi que des trains de **pt** — accompagnés ou non d'une microstructure d'ordre plus élevé — soient enregistrés à une station déterminée sans qu'une baie y soit simultanément décélée (ce qui a conduit le Comité International des Variations Rapides à demander des relevés indépendants pour les deux types de phénomènes). Ceci n'est en soi que d'une importance théorique mitigée — si l'on pense aux variations très importantes dans les amplitudes suivant lesquelles une même baie peut être enregistrée en différents points du Globe; c'est-à-dire que la liaison supposée entre les deux phénomènes (baie et trains de **pt**) n'est pas forcément mise en défaut de ce fait sur le plan général (mondial), mais il illustre les possibilités de propagation probablement différentes qui leur sont offertes pour leur manifestation à l'échelle mondiale.

Un problème voisin du précédent nous est posé par les baies négatives qui — continuant à envisager le cas des moyennes latitudes — se présentent en général de jour. [Le cas des *grandes baies* négatives que l'on remarque épisodiquement, vers 15 h de Temps Local Moyen (TLM) à Chambon-la-Forêt est, peut être, plus particulier et nous le laissons de côte.] Elles ne sont accompagnées qu'assez rarement d'une microstructure nette et l'on peut se demander si ce fait n'est pas en contradiction avec le caractère mondial des **pt**, analogue à celui des baies. C'est ainsi que des **pt** observés de nuit à l'observatoire d'Oginawa (Japon) ont pu presque tous être retrouvés sur les enregistrements de l'observatoire de Tamanrasset (Sahara Algérien) à des heures de jour[2]. Cette question est quelque peu éclairée par des observations plus récentes[3], d'après lesquelles ce caractère mondial des **pt** ne serait bien marqué pour l'hémisphère de jour qu'à condition de faire appel à des stations d'observation *de basses latitudes*. C'est ainsi que presque tous les **pt** observés de nuit à Chambon-la-Forêt (où ils accompagnent alors, en général, les baies nocturnes, positives, classiques), sont retrouvés nettement, au même instant, à la station de Pamataï (Tahiti: Long. 149°34′ W Lat. 17°34′ S) située alors dans l'hémisphère de jour (les baies correspondantes ne s'y trouvant pas forcément aussi bien marquées), alors que la réciproque n'est qu'assez peu souvent vérifiée, Chambon-la-Forêt étant de latitude dejà trop élévée pour bien recevoir des **pt** de jour. Ces observations nous permettent de penser que, bien qu'étant des phénomènes liés (ceci, probablement lors de leur génération), les baies des moyennes et basses latitudes, et leurs microstructures, se propagent mondialement par des voies — et des processus — nettement différents. Ainsi on ne pourrait admettre que ces microstructures soient simplement des microvariations des systèmes de courants électriques par lesquels on rend compte en général de l'extension mondiale des baies. Nous allons retrouver des considérations analogues, mais en plus complexes, en examinant maintenant les baies et perturbations des latitudes plus élevées.

15. Microstructures des baies aurorales et des orages polaires. (Voir Fig. 23a et b et 24 et les références [*10*] et [*11*].) Cette nomenclature, tout en maintenant l'accord avec celle exposée dans l'article NAGATA et FUKUSHIMA revêt une forme quelque peu différente mieux adaptée à l'étude des microstructures. Revenons donc très brièvement sur les phénomènes primaires eux-mêmes: Alors que les baies «classiques» peuvent se retrouver un peu partout sur le Globe (dans l'hémisphère nocturne surtout), il n'en est pas de même pour les baies dites «aurorales» qui constituent un phénomène surtout caractéristique des régions aurorales et sub-aurorales. Il semble même que ces baies apparaissent le plus nettement à la limite extérieure de la zone aurorale et, en fait, cette qualification d'«aurorale» leur a été donnée à la suite des observations continues faites à la station de Port-aux-Français (Kerguelen) depuis l'Année Géophysique Internationale («AGI»). Les baies ainsi observées sont, certes, à rapprocher du phénomène connu sous la dénomination plus généralement adoptée d'«orage polaire», mais il n'est pas tout à fait prouvé qu'il s'agisse là de deux phénomènes exactement de la même catégorie. A Kerguelen la distinction est faite de prime abord sur une question de «degrés»: l'orage polaire serait une baie aurorale d'exceptionnelle intensité et durée, sans toutefois qu'il puisse être assimilé à un *orage mondial*, ceci à cause de sa zone d'action géographique relativement limitée (sous sa morphologie première). En dehors de cette question le phénomène est défini par les deux propriétés fondamentales suivantes:

1) être accompagné d'une aurore se produisant au voisinage du zénith de la station;

2) présenter une structure morphologique particulière, très caractéristique, se traduisant notamment par des *indentations* beaucoup plus profondes que pour les baies ordinaires, des «bords» beaucoup plus abrupts et une plus grande analogie

[2] G. GRENET, Y. KATO, J. OSSAKA, and M. OKUDA: Sci. Rep. Tôhoku Univ., Ser. V, No. 6, 1—10 (1954—1955).

[3] F. JAMET, R. REMIOT, J. ROQUET et E. SELZER: C.R. Acad. Sci. **269**, 442—444 (1969).

entre les phases «aller» et «retour». Ces dernières ne paraissent plus être un simple «retour au calme» — c'est-à-dire un processus de relaxation — comme dans les baies classiques, et comportent des indentations et des «créneaux» aussi «actifs» que ceux portés par les phases «aller». Enfin amplitudes (jusqu'à 400 γ) et durées (jusqu'à 3 heures) sont supérieures à celles constatées dans la plupart de ces baies classiques (en dehors des grands orages mondiaux).

L'extension géographique du phénomène magnétique est donc liée au domaine d'extension d'*une même aurore*.

Suivant les observations faites à Kerguelen et en Terre Adélie durant l'AGI, les aurores qui ne sont observées qu'assez bas sur l'horizon ne sont pas accompagnées de baies aurorales mais, au maximum, de baies simples plus ou moins classiques (analogues à celles des latitudes moyennes). Cette détermination des domaines d'extension a pu être complétée par différentes expériences et observations menées entre points et zones conjugués (notamment entre Baker Lake et Little America[1,2]).

D'une façon assez générale on admet que quand, partant des régions aurorales, on descend vers celles de latitudes moins élevées, les phénomènes deviennent moins violents et la «baie aurorale» s'y traduit — si les conditions d'heures locales s'y prêtent — par une baie simple. Cependant des cas intermédiaires sont possibles et à une station telle que Chambon-la-Forêt (lat. géom. 51° N) on observe quelquefois, en synchronisme avec une baie aurorale des régions polaires, une baie à flancs raides et irréguliers, ou même parfois une perturbation présentant — à l'amplitude près — la morphologie d'un orage polaire de courte durée.

Par analogie avec la nomenclature que nous avons suivie pour les baies simples, nous pouvons considérer les indentations que l'on observe tout au long du développement d'une baie aurorale — ou d'un orage polaire — comme constituant la microstructure première du phénomène. En général ces variations irrégulières pourront être rangées dans la catégorie des **pi2** bien que certains des décrochements qu'elles comportent soient plutôt à comparer à des **SI** («Sudden Impulse», voir Fig. 9, 19 et 20) ou même, pour le premier, à un **SSC**. Il est rare que l'on puisse les faire entrer dans les **pt** des anciennes classifications car elles n'en ont ni l'organisation en «trains», ni le déroulement relativement «doux». Il semble cependant que ce soit la transmission vers les latitudes plus basses de ces variations très irrégulières qui soient responsables des microstructures, partiellement adoucies, des baies classiques (surtout pour les nocturnes), y compris les trains de **pt**.

Ce que nous venons d'exposer peut être étendu aux variations plus rapides formant les microstructures secondaires des phénomènes qui nous occupent: l'étude spectrale des baies aurorales (sur laquelle nous reviendrons) les montre en effet très riches — beaucoup plus que les baies ordinaires — en pulsations irrégulières très rapides que l'on peut ranger dans la catégorie **pi1**. (Nous verrons qu'il peut s'y adjoindre aussi, avec un certain retard, certaines variations s'apparentant davantage aux pulsations **pc1**.) L'étude générale de la propagation, vers les plus basses latitudes de l'ensemble de toutes les perturbations — ou variations — constituant une baie aurorale (ou un orage polaire) est un problème complexe qui est loin d'être résolu. Nous avons déjà remarqué que, même à l'échelle des latitudes moyennes ou basses, les conditions de propagation de la partie principale des baies (classiques) d'une part, de leurs microstructures

[1] H. J. Duffus, J. A. Shand, and Sir C. S. Wright: Can. J. Phys. **40**, 218—225 (1962).

[2] H. J. Duffus, J. Kinnear, J. A. Shand, and Sir C. S. Wright: Can. J. Phys. **40**, 1133—1152 (1962).

d'autre part, devaient présenter des différences importantes. Ceci est encore plus évident si nous partons des latitudes aurorales. Il semble que ce soient les variations les plus rapides existant dans la composition spectrale d'une baie aurorale qui puissent être transmises le plus complètement vers les régions de moindres latitudes. Il semble aussi que la largeur du «fuseau horaire d'heure locale» qui constitue, en un lieu donné, la fenêtre d'admission pour un phénomène déterminé, soit plus grande ouverte pour ces fluctuations rapides que pour la baie elle-même. Nous ignorons malheureusement à la fois les mécanismes et les milieux précis de propagation: si une propagation par le jeu de la fermeture de courants circulant (en nappes horizontales) dans l'ionosphère, a été longtemps la seule admise en ce qui concerne la partie principale de la baie (bien que d'autres mécanismes aient été plus récemment suggérés[3]), nous avons déjà dit en quoi une propagation par ce même support paraissait ne pas pouvoir répondre au cas des variations rapides constituant les microstructures. Il est possible qu'une excitation d'ensemble, magnétosphérique (faisant appel soit à des ondes hydromagnétiques, soit à des averses de particules, soit à ces deux actions réunies) puisse, au moins partiellement, remplacer les propagations imaginées. Une propagation du type hydromagnétique par divers guides d'ondes ionosphériques[4], ou magnétosphériques est aussi possible. Enfin, dans le cas où le phénomène apparaît dans les régions équatoriales, il peut être le fruit, non seulement d'une propagation à partir d'une autre région, mais aussi d'une amplification à l'échelle locale en liaison avec l'*électrojet équatorial*.

III. Microstructures plus particulières.

On peut encore relever sur les magnétogrammes normaux des régions de latitudes moyennes des périodes d'agitation de durées assez prolongées mais de natures assez mal définies: «micro-orages», «agitation par baies juxtaposées», «petite agitation du matin», etc., dont l'étude des microstructures — quand elles en comportent une — peut être intéressante, notamment parce qu'elle peut nous renseigner sur leurs degrés de dépendance de perturbations des zones aurorales. En règle assez générale, l'existence d'une microstructure active dans la constitution de ces divers phénomènes est l'indice de leur couplage plus étroit avec des perturbations aurorales correspondantes. Ceci s'applique en particulier à la «petite agitation du matin» (visible très régulièrement sur les enregistrements normaux — La Cour — de Chambon-la-Forêt, mais le phénomène peut être observé en la plupart des stations de latitudes moyennes), agitation qui peut comprendre — s'ajoutant aux fluctuations assez irrégulières, de pseudo-périodes de quelques minutes, qui sont celles que l'on peut ainsi observer couramment — d'autres fluctuations beaucoup plus rapides (et de moindre amplitude) visibles seulement sur les enregistrements spéciaux. Cette agitation paraît être l'équivalent, aux latitudes moyennes, de l'agitation «de Jour» (soit **J**[5]) des régions polaires. Cette agitation **J**, lorsqu'elle se régularise (comme cela arrive fréquemment dans les stations sub-aurorales telles que Port-aux-Français à Kerguelen, et la Station du roi Baudouin sur le continent Antarctique) donne lieu à des pulsations régulières d'assez longues périodes (de l'ordre de 5 min) et, plus exceptionnellement, de «pulsations géantes» (**«pg»**) à ranger dans la catégorie **pc 5** de la classification internationale actuelle. On aurait ainsi un exemple d'une déformation progressive d'un même phénomène rapide depuis les hautes jusqu'aux moyennes latitudes.

16. Microstructures des orages magnétiques mondiaux. Par souci de simplicité d'exposition nous examinerons d'abord ces microstructures dans l'ordre même où elles peuvent être rencontrées dans le déroulement chronologique normal d'un orage[1]. Nous aurons ainsi à nous occuper des phases successives suivantes (dont

---

[3] H. Alfvén dans: A. Egeland et J. Holtet (eds.): The Birkeland Symposium on Aurora and Magnetic Storms, p. 439—444. Paris: CNRS 1968.

[4] R. N. Manchester: J. Geophys. Res. **71**, 3749—3754 (1966).

[5] P. N. Mayaud: Thèse Fac. Sci. Paris (1955); — Ann. Géophys. **12**, 84—101 (1956).

[1] Renvoyons aux articles de T. Nagata et N. Fukushima (p. 5) et de E. N. Parker et V. C. A. Ferraro (p. 131) au volume 49/3 de cette encyclopédie.

les deux premières sont à considérer comme des «pré-phases», antérieures à l'apparition du début brusque **SSC** au sol): le «*crochet magnétique*» (ou «Solar Flare Effect» — **SFE**), conséquence directe — quand on peut l'observer — de l'éruption chromosphérique solaire qui est supposée être à l'origine de l'orage (si nous laissons de côté, pour l'instant, le cas des orages à début progressifs); la *période intermédiaire* entre le **SFE** et le **SSC** (celle au cours de laquelle se manifestent certains phénomènes spéciaux dans les régions de hautes latitudes, notamment les «évènements à protons» **PCA** («polar-cap-absorption»)[2]; le *début brusque* **SSC** proprement dit; les *trois phases* classiques de l'orage.

L'existence d'une structure fine pour le *crochet magnétique* est controversée: rarement observée à Chambon-la-Forêt — et sans aucune certitude — elle a fait l'objet d'observations répétées au Japon[3]. Il est possible qu'il faille un «front» de crochet très raide pour déclencher les oscillations (ou pulsations) ainsi rapportées. Ces oscillations accompagneraient la brusque augmentation d'ionisation produite ainsi dans les basses couches de l'ionosphère (couche D principalement). On pourrait alors rapprocher ce phénomène de celui produit par le «flare» artificiel lors de l'explosion d'une charge nucléaire à haute altitude (bien que les sources et les processus d'ionisation, et la propagation des excitations initiales, soient bien différents dans les deux cas).

L'existence d'une activité magnétique appréciable due à une éruption chromosphérique, qui serait observable au sol durant la période séparant un crochet (ou **SFE**) du **SSC** qui lui est associé, est également une question qui est loin d'être éclaircie. Signalons à ce sujet l'observation demeurée unique relative à une région polaire, faisant état d'une perturbation magnétique à l'échelle macroscopique durant un évènement à protons (**PCA**)[4]. Les observations de pulsations sont plus nombreuses, notamment celles de **pc 1** qui seraient observables plusieurs heures avant le **SSC**[5,6], mais l'incertitude réside dans l'association faite entre les deux phénomènes chaque fois que l'on se base sur des cas particuliers. Du point de vue statistique il n'apparaît pas que les premières conclusions suivant lesquelles les orages seraient précédés d'une façon assez constante par l'apparition de **pc 1** aient été vérifiées: on peut penser que ce fait se produit assez souvent (les deux phénomènes étant alors liés) pour les grands orages, mais dans l'état actuel de nos connaissances sur une telle corrélation on ne peut en faire une méthode de prévision des orages.

La structure fine du **SSC** proprement dit est, par contre, bien établie[7,8] [*12*]. Cependant la composition spectrale de la microstructure enregistrée est loin de correspondre à des normes précises: presque tous les **SSC** présentent au moins quelques oscillations plus ou moins régulières pouvant descendre jusqu'à une fraction de seconde (ce qui, à l'échelle des enregistrements rapides et très rapides, complique la distinction formelle entre les **SSC** simples et les **SSC***, mais un **SSC** dure en général trop peu de temps (deux à trois minutes) pour que des trains bien formés de pulsations (tels que, par exemple, des **pc1**, puissent être observés. Certains **SSC** présentent plutôt une structure du type **pi1** analogue à celle des microstructures secondaires des baies.

La *première phase* de l'orage (valeur de $\mathcal{H}$ supérieure à la normale) présente souvent d'assez grandes oscillations irrégulières (jusqu'à une dizaine de gammas en amplitude, pseudo-périodes de plusieurs minutes) durant les 10 à 20 min qui suivent le **SSC** par lequel elle a débuté). Ces oscillations ont cependant une morphologie nettement différente de celle des **pt** (ou **pi2**) dont on serait tenté de les rapprocher. Elles sont assez semblables à celles qui ont été observées mondialement à la suite des «**SSC** (ou **SFE**) *artificiels*» engendrés par des explosions

---

[2] D. K. Bailey: Planetary Space Sci. **12**, 495—541 (1964).

[3] Y. Kato and T. Tamao: J. Geomagnet. Geoelec. **10**, 4, 203—207 (1958).

[4] A. Lebeau, S. Cartron, G. Olivieri, M. Pick, R. Schlich et G. Weill: Ann. Géophys. **20**, 309—318 (1964).

[5] V. A. Troitskaja: J. Geophys. Res. **66**, No. 1, 11 (1961).

[6] A. Cecchini, G. Dupouy, J. Roquet et E. Selzer: C. R. Acad. Sci. **250**, 4023—4026 (1960).

[7] Y. Kato and T. Saito: Sci. Rep. Tôhoku Univ., Geophys. **9**, 99—112 (1958).

[8] E. Selzer: Ann. IPG (Institut de Physique du Globe) Paris **31**, 147—193 (1963).

nucléaires à haute altitude. (Ceci est vrai aussi pour la structure fine de la phase initiale: **SSC** naturel ou artificiel.) Il est remarquable que les microstructures précédentes restent comparables (très semblables) à de grandes distances (par exemple entre Chambon-la-Forêt et Kerguelen (voir Fig. 5a et b). On ne peut dire encore s'il s'agit là de processus locaux *d'excitation* en synchronisme mondial, d'un phénomène *d'onde stationnaire*, ou d'un phénomène *de propagation* (notamment par le guide ionosphérique convenant à la propagation des UBF). Le reste de cette première phase — qui peut durer plusieurs heures (jusqu'à dix ou plus) — ne comporte pas, en général, de microstructure.

La *deuxième phase* de l'orage — ou *phase principale* — est, de beaucoup, la plus active relativement à des formes très variées de microstructures, y compris toutes sortes de *décrochements brusques* qui, bien que relevant de la *macrostructure* de l'orage, sont difficilement dissociables de tous les phénomènes impulsationnels et pulsationnels qui les animent.

On sait que — par définition pourrait-on dire — cette phase principale est celle au cours de laquelle la valeur moyenne de la composante $\mathcal{H}$ se maintient largement au-dessous de sa valeur normale (ceci, sans montrer encore la tendance systématique à un retour vers cette valeur normale qui sera la caractéristique de la troisième phase — ou phase terminale — de l'orage). Cependant, l'activité nerveuse, rapide, qui nous intéresse, n'est pas présente — en général — tout au long de cette phase. Elle procède par «orages partiels» (que l'on peut aussi dénommer: «orages de micropulsations», «sous-orages», etc. dont les grands orages mondiaux du maximum d'activité solaire de 1957—1960, et un plus petit nombre d'orages du cycle actuel commencé vers 1966/67, nous donnent de bons exemples (voir Fig. 10, 11 et 12). Entre deux orages partiels successifs prévaut, en général, une situation relativement peu agitée et sans microstructure importante, surtout en ce qui concerne les phénomènes les plus rapides. La Fig. 13 montre à la fois un exemple et une exception relatifs à cette remarque: un calme assez complet est bien visible entre les deux orages partiels qui l'encadrent, mais il est interrompu épisodiquement par une séquence de pulsations très rapides que l'on peut considérer comme «surajoutée» à la distribution des orages partiels.

Les enregistrements donnés par ces figures montrent combien l'étude directe des variations, très rapides et très complexes, qui composent la microstructure d'un orage partiel est peu facile à partir des formes d'ondes données par les enregistrements. Certains cas remarquables font cependant exception, tel celui illustré par la Fig. 14a et b, où — grâce à une augmentation importante de la vitesse d'enregistrement — on a pu faire apparaître une «émission» presque parfaitement monochromatique (pulsations de 1,8 sec de période) au milieu même d'une des phases les plus actives d'un orage partiel. Bien que relativement peu fréquent, cet exemple n'est pas unique et il semble qu'au moment même de la phase la plus active et la plus nerveuse des plus grandes perturbations mondiales, de telles émissions quasi-monochromatiques puissent être présentes. La difficulté pratique est de réaliser les conditions d'enregistrement propres à leur mise en évidence. En dehors de l'exemple donné en Fig. 14 cela a pu aussi être fait pour deux phases de l'orage du 8 juillet 1958.

17. Analyses spectrales. Une méthode plus efficace pour détecter et pouvoir examiner les phénomènes des types précédents est de procéder à leur enregistrement sur bande magnétique dont on fait ultérieurement, au laboratoire, l'analyse spectrale. C'est ainsi qu'au cours des premiers orages importants du nouveau cycle solaire F. Glangeaud a pu mettre en évidence d'autres types voisins d'émissions monochromatiques (voir Fig. 51). Dans l'exemple ainsi donné, la différence avec les cas mentionnés précédemment consiste en la longue durée du phénomène (5 h) et en ce que sa fréquence, qui est bien à très peu près monochromatique à chaque instant, glisse de façon continue de 1 à 4 Hz au cours de cette durée.

Les analyses spectrales ont permis aussi de mieux prendre connaissance des diverses entités sous lesquelles pouvait être considéré tel ou tel type de phénomène. C'est ainsi qu'a pu être éclaircie et précisée la relation entre les orages partiels que nous venons de décrire et le phénomène qui avait été introduit il y a une dizaine d'années (avant l'intervention des méthodes d'analyse spectrale) par V. A. Troitskaja sous la dénomination d' «**IPDP**» («Interval of Pulsations of Diminishing Periods» ... interprêté aussi quelquefois dans les publications du même auteur comme des «Irregular Pulsations of Diminishing Periods».) A cette époque cette notion encore assez indéterminée cadrait assez bien avec celle des orages partiels, ainsi que l'illustre le Tableau 3 sur lequel nous avons reporté les cas principaux de coïncidences entre les deux types de phénomènes au cours du développement d'un certain nombre de grands orages mondiaux. Cependant, depuis l'introduction de l'analyse spectrale — notamment à l'occasion des études entre points conjugués Kerguelen-Sogra-une «morphologie spectrale» s'est dégagée d'une façon beaucoup plus caractéristique que l'ancienne «morphologie des ondes» («formes d'ondes») pour ce groupe de phénomènes. Les Fig. 56 et 58 donnent une idée de cette morphologie spectrale caractérisée essentiellement par un spectre étalé en fréquence, l'ensemble de la large bande ainsi obtenue se déplaçant vers les fréquences plus élevées au cours de l'évolution du phénomène dans le temps (d'une durée moyenne d'une vingtaine de minutes). La question de savoir si cette «forme spectrale brute» est le résultat global du mélange d'un grand nombre d'évolutions individuelles, dont chacune serait relative à un phénomène bien défini caractérisé par une fréquence montante qui lui serait propre, ou si elle aurait plutôt la constitution d'un bruit formé de la manifestation d'un grand nombre d'actions indépendantes (à caractère impulsif par exemple) dont la densité temporelle, croissant avec le temps, donnerait l'augmentation générale observée pour les fréquences, n'est pas résolue, tout au moins dans le cas d'un **IPDP** courant. Dans des cas plus particuliers (voir Fig. 55) la première des deux hypothèses paraît vérifiée.

Quoi qu'il en soit, les normes relativement précises avec lesquelles on a pu définir cette forme spectrale brute des **IPDP**, a conduit à attacher plus d'importance à cette forme spectrale qu'à la forme d'onde (et à tout le contexte magnétique correspondant) dans la définition actuelle du phénomène. A ce titre, cette définition s'est écartée beaucoup de celle des orages partiels présents à l'intérieur d'un grand orage mondial et la plupart des exemples donnés maintenant dans la littérature, concernant les **IPDP**, ont trait aux microstructures de phénomènes plus ou moins isolés, s'apparentant aux orages polaires ou aux baies aurorales. A ce titre on peut admettre que — dans leur définition actuelle — les **IPDP** sont des phénomènes qui abordent l'environnement immédiat du sol terrestre dans les régions de latitudes élevées. Cette remarque a été utilisée dans la mesure récente de la vitesse de propagation des ondes **UBF** au voisinage du sol terrestre. Par contre, si nous revenons au sujet même des micropulsations formant les orages partiels, nous ne saurions, à l'heure actuelle, nous faire une idée précise des régions où ces perturbations ont leur origine. Divers indices sur lesquels nous reviendrons à propos des théories des pulsations donnent à penser que ces orages partiels sont couplés directement aux «orages magnétosphériques», ceci en tout cas d'une façon beaucoup plus étroite que les **DP** («Disturbance Polar» de

Tableau 3*. *Comparaison des orages partiels (ou «sous-orages-mondiaux», «SOM») avec les IPDP de* V. A. Troitskaja, *pour 1957-58-59, et l'exemple d'un orage du nouveau cycle solaire: 25. mai 1967. (Seules les heures correspondant aux premiers IPDP ont été publiées).*

Date du début de l'orage principal	Heure T.U. du SSC	Nombre des événements		Événements successifs (Heures T.U. de débuts et fins)				
				I		II	III	IV
		IPDP	SOM	IPDP	SOM	SOM	SOM	SOM
4/IX-57	12.59	2	2	17.28 19.25	17.30 19.45	21.52 23.50		
13/IX-57	00.46		2		08.16 09.10	12.00 15.00		
29/IX-57	00.16	2	2	17.00 18.45	17.12 18.45	22.35 23.20		
11/II-58	01.26	4 (5 ?)	2	*06.57*** *07.55*	01.58 06.30	*07.00*** *07.55*		
8/VII-58	07.48	2	2	16.40 18.15	17.00*** 18.15	21.05 22.03		
4/IX-58	08.42	3 (4 ?)	4	16.30 18.48	16.25 18.40	20.40 22.00	22.13 23.50	01.03 01.30 (le 5/IX)
11/VII-59	16.25		1		23.25 00.30 (le 12)			
15/VII-59	08.03		3		17.28 19.00	19.40 21.50	22.38 22.52	
25/V-67	12.35		3		21.00 22.30	23.44 00.55 (le 26)	02.10 03.10 (le 26)	

* Préparé pour le Symposium Birkeland (1967) par l'auteur.
** Événements enchevêtrés, le 1er **IPDP** signalé correspond, évidemment, au 2e **SOM**.
*** Le début de ce premier **SOM** n'a pu être matériellement enregistré.

Akasofu et Chapman[1]), ou les «polar substorms» (sous-orages polaires[2]) avec lesquels ils avaient, à une certaine époque, été associés.

Les mêmes déterminations spectrales qui ont conduit à une meilleure définition des **IPDP** ont permis, de plus, de mettre en évidence quelques autres entités voisines. Sans entrer dans trop de détails citons à ce sujet des «**SIP**» («Short Irregular Pulsations») dont le spectre se présente comme une barre verticale plus ou moins épaisse dans une représentation fréquence/temps, toute une bande de fréquence étant excitée simultanément au moment où se produit le phénomène (voir Fig. 56).

La connaissance des compositions spectrales pourrait nous dispenser, en principe, d'essayer de ramener la structure de telle ou telle phase d'un phénomène riche en variations rapides à des composants élémentaires rentrant par exemple dans la classification conventionnelle des pulsations. En fait, si nous revenons au cas le plus complexe des grands orages, on peut toujours écrire que leur microstructure est faite de la somme de divers types de pulsations; par exemple on pourra écrire:

$$\text{pulsations d'orage} = \mathbf{pi\,1} + \mathbf{pi\,2} + \mathbf{pc\,1} + \mathbf{pc\,2} + \mathbf{pc\,3}.$$

Cela voudrait dire que, dans la microstructure de l'orage correspondant à un intervalle de temps donné, l'analyse spectrale montre bien l'existence de com-

[1] S.-I. Akasofu and S. Chapman: J. Geophys. Res. **66**, 1321—1350 (1961).
[2] S.-I. Akasofu: Radio Science **69** D, 361 (1965).

posants s'apparentant, par leur période, aux pulsations mentionnées dans le membre de droite de cette égalité. Cette manière de faire qui tendrait à enlever toute interprétation morphologique au résultat de l'analyse ne nous paraît pas recommandable dans tous les cas. Il en est d'autres, au contraire, où l'étude de l'orage permet effectivement de reconnaître une structure morphologique conforme à l'écriture donnée en exemple (ou à une autre semblable). Il est ainsi fréquent que des pulsations du type **pc3** surviennent vers la fin (ainsi qu'*après* la fin) d'un grand orage. Elles apparaissent aussi souvent après une période d'agitation plus modérée (voir Fig. 29a et b et Fig. 31a et 32). Dans le cas des pulsations très régulières, de périodes peu supérieures à la seconde, que nous avons rencontrées en Fig. 11 et 14, malgré qu'elles «rentrent» parfaitement dans la définition conventionnelle des **pc1** nous trouvons préférable de leur conserver une nomenclature particulière, par exemple «pulsations régulières d'orage». Il est en effet probable que leurs modes de génération (et de propagation) sont très différents de ceux de la majorité des pulsations rangées dans la classe des **pc1** (dont les distingue également l'absence d'une microstructure spectrale).

18. Relations entre les orages partiels (ou orages de micropulsations) et les grandes remontées soudaines de la composante $\mathcal{H}$ qui se produisent au cours de la phase principale d'un grand orage. Revenant plus strictement à la microstructure des grands orages mondiaux, nous avons vu qu'elle devenait particulièrement active lors de la phase principale (ou deuxième phase) de ces orages. En fait, le premier des orages partiels peut se produire dès le début de cette phase principale (mais rarement avant). *A cet instant* on note (à une ou deux minutes près, une vérification plus précise étant, en général, impossible) que la composante $\mathcal{H}$ — qui, par définition de la phase principale, était alors très affaiblie — remonte (par un décrochement brusque de sa trace sur les magnétogrammes normaux) d'une fraction importante de sa dépression initiale, ce qui peut se traduire par plusieurs centaines de gammas et atteindre des taux de plusieurs γ/sec). Par la suite, soit au cours de l'orage partiel, soit durant l'intervalle relativement calme qui le suit, la valeur de $\mathcal{H}$ retombe (plus progressivement) à une valeur très inférieure à la normale et le même processus reprend place avec le déclenchement du deuxième orage partiel, et ainsi de suite. Il est ainsi permis de se demander si cette relation morphologique joue un rôle essentiel dans le mécanisme de déclenchement des orages partiels, où si nous avons simplement là les aspects de deux effets d'une même cause. Pour éclairer cette question nous ferons à nouveau appel à la notion d'«orage magnétosphérique» qui s'est imposée peu à peu au cours de ces dernières années pour remplacer tout un aspect de la notion plus ancienne — mais trop ambigüe — de sous-orage polaire (polar substorm). Plus précisément, nous sommes conduits à supposer l'existence de «sous-orages magnétosphériques» (magnétospheric substorms), l'observation morphologique que nous avons rapportée suggérant alors les liens qui doivent exister entre ces sous-orages magnétosphériques et les facteurs responsables de la dépression de $\mathcal{H}$ au cours de la phase principale des orages (attribués à l'existence d'un «courant équatorial» ou «ring current», ou à tout effet équivalent). Plus généralement, on peut se demander si la production de pulsations très rapides ne serait pas la résultante quasi automatique (peut-être avec des «veto» occasionnels dûs aux actions d'autres facteurs) d'une augmentation rapide de $\mathcal{H}$ (après ou sans dépression initiale). On pourrait alors rapprocher le cas qui nous occupe spécifiquement ici de celui des pulsations de **SSC** et celui des pulsations des parties «aller» des baies positives des moyennes latitudes. Les décompressions de $\mathcal{H}$ ne produiraient pas de tels effets même quand elles se produiraient de façon abrupte.

Le manque de données à l'échelle mondiale sur ces divers effets nous empêche d'aller plus avant dans notre examen de cette question.

19. Troisième phase des orages. Nous passons enfin maintenant à la *troisième phase de l'orage* bien qu'elle soit de beaucoup la moins active à plusieurs points de vue. Tous les éléments dont on dispose s'accordent pour lui reconnaître les caractéristiques générales d'un «retour au calme» ou, en d'autres termes, d'un phénomène de relaxation. Les effets que l'on peut y observer n'auraient donc essentiellement qu'un caractère passif.

Bien que dans beaucoup de cas une telle présentation des faits paraisse correcte, il convient de signaler que cette fin de l'orage est souvent accompagnée de l'apparition de divers types de pulsations dont on ne connaît pas nettement les liens qu'elles ont avec l'orage en question. Nous ne les considérerons pas comme une microstructure propre à cette phase de l'orage, mais plutôt comme un état général — peut-être de la magnétosphère — qui, ayant été perturbée par le déroulement global de l'orage, présente de ce fait diverses particularités dynamiques, dont des pulsations. Celles observées le plus couramment sont des **pc3**, en beaux trains de plus en plus réguliers au fur et à mesure que l'agitation générale due à l'orage se calme. Ces **pc3** étaient présents — peut-être — durant les parties les plus actives de l'orage, alors couvertes par d'autres variations de plus grande amplitude. Ces trains de **pc3** peuvent se prolonger — en suivant leur rythme diurne propre — pendant de nombreux jours faisant suite à l'orage. Un cas moins fréquent que celui des **pc3** — mais particulièrement intéressant parce qu'il touche à la théorie des pulsations de très courtes périodes **pc1** et **pi1** — est l'apparition de ces dernières vers la fin d'un orage — et aussi au cours des quelques jours qui suivent cette fin — ceci, même quand les parties actives de l'orage lui-même n'en avaient pas comportées. Plus précisément, il arrive que, à la suite de **pi1** qui se sont manifestés au cours de l'orage, des pulsations régulières du type **pc1** viennent prendre la place — ou s'ajouter — à ces pulsations irrégulières. (Ces faits apparaissent clairement sur les sonagrammes de ces fins d'orages.) D'autre part — sans que l'on puisse montrer leur lien avec des orages autrement que par des considérations statistiques — il est maintenant bien établi que les cas — toujours relativement peu fréquents (en moyenne, trois ou quatre périodes de l'ordre de l'heure chacune, au cours d'un mois ... mais avec la possibilité d'écarts considérables dans cet ordre de grandeur) de **pc1** se produisent préférentiellement au cours des quelques jours qui suivent les orages.

Ce qui est difficile d'expliquer dans les faits précédents, ce n'est pas tant que l'orage puisse exciter de telles variations rapides des types pulsations **pi1** ou **pc1**, mais que ces variations soient décalées dans le temps par rapport à la cause excitatrice que l'on soupçonne. Egalement, qu'elles puissent parfois se renforcer ultérieurement, ou — encore plus simplement — ne pas montrer d'effet d'amortissement auquel on pourrait s'attendre en fonction de l'éloignement dans le temps de la cause excitatrice. Pour ces raisons on peut penser que des dispositifs de «stockage» (ou de «volant») interviennent et que les relations entre les orages et ces pulsations sont loin d'être directes.

IV. Morphologie directe, et spectrale, des pulsations magnétiques.

Nous venons d'examiner les variations rapides dans leurs rapports avec les perturbations qui se manifestent à l'échelle macroscopique. L'ordre de cet examen était donc déjà tout tracé. Par contre, avec la section présente, nous abordons une étude en soi d'un certain nombre de ces variations — déjà ou non envisagées — et il nous faut préciser comment sera conduite cette étude. Au stade actuel de notre exposé il nous paraît nécessaire de considérer comme acquis ce qui, pour diverses raisons, a été déjà expliqué, même si une logique rigoureuse semblerait devoir l'imposer à nouveau ici même. Ceci doit nous permettre de pousser plus avant, en profondeur, l'examen des phénomènes et d'en essayer une compréhension plus complète. Nous basant sur la classification déjà donnée[1], nous aborderons d'abord le cas des «pulsations irrégulières» **pi** avant de nous étendre plus longuement sur les «pulsations continues» (ou «régulières») de la série des **pc**. Dans chacun de ces cas nous commencerons par les pulsations les plus lentes, pour terminer par les plus rapides (ordre des indices descendants).

20. Pulsations irrégulières pi1 et pi2. Nous avons déjà décrit ces pulsations en tant que microstructures des baies magnétiques et de quelques autres phénomènes analogues. Nous nous bornerons donc ici à certains aspects propres de leur manifestation.

Pulsations **pi2**. Nous pouvons, presque toujours, les considérer comme les oscillations qui interviennent dans la composition des trains **pt**. Ceci nous amène à donner des précisions sur cette composition. Il est rare que les oscillations successives soient de même pseudo-période. Le plus souvent leur durée est croissante, par exemple entre 50 et 120 sec (ou plus). (Notons, en passant, que la limite conventionnelle de 150 sec pour la pseudo-période maximale des **pi2** est trop faible: des oscillations atteignant presque 300 sec ayant été observées comme constituants de certains trains de **pt**[1]). Simultanément, l'amplitude est souvent décroissante et la forme devient de plus en plus vague (alors que la première oscillation est assez souvent pseudo-sinusoïdale). Une catégorie particulière — et relativement peu fréquente, sauf durant certains mois privilégiés — est constituée par des **pt** dont les oscillations successives — alors régulières — s'inscrivent à l'intérieur d'une enveloppe à décroissance exponentielle (oscillations amorties).

D'assez nombreuses études ont été faites pour établir les degrés de corrélation entre les pseudo périodes des **pi2** constituant les **pt** et l'activité magnétique générale. ROSTOKER a montré que la composition spectrale de ces **pi2** (non inclus les **pi1** qui les accompagnent) *s'élargissait* avec l'augmentation de l'activité magnétique planétaire[1], ce qui peut expliquer que les observations d'autres auteurs — dont ŠČEPETNOV cité par TROITSKAJA[2] — sur une *diminution* de la *période dominante* de ces oscillations avec l'augmentation de l'indice K_p (planétaire) aient pu faire l'objet de controverses (une *augmentation* de période avec l'augmentation du K_p ayant été aussi observée).

Le caractère *impulsif* de l'arrivée des trains de **pt**, 'la première de ses oscillations étant parfois précédée par un petit décrochement raide, donne alors à la baie correspondante — s'il y a synchronisme dans son début — la morphologie d'une «baie à début brusque» (notée «**bs**» ou «**bsp**»: «baie à début brusque suivie de pulsations», dans la classification internationale).

Ces différentes remarques suggèrent une origine impulsationnelle — mais lointaine (probablement en provenance des régions éloignées de la magnétosphère nocturne) — pour les trains de **pt**, leur propagation à travers, et le long, de l'ionosphère, diminuant les caractères impulsifs et introduisant les amortissements et les dispersions observées. Cette propagation se ferait suivant des modes indépendants de ceux responsables de la manifestation, également mondiale, des baies elles-mêmes.

Signalons enfin les résultats généraux des études de polarisation concernant ces pulsations (dans le plan des $\mathcal{H}$, D): rotations dans le sens inverse de celui des

[1] Voir Sect. 65 de l'article par NAGATA et FUKUSHIMA au volume 49/3 (p. 5) de cette encyclopédie.

[1] G. ROSTOKER: J. Geophys. Res. **72**, 2032—2039 (1967).

[2] V. A. TROITSKAJA: Solar-Terrestrial Physics, chap. VII, p. 213—274; edit. J. W. KING and W. S. NEWMAN. New York: Academic Press 1967, citant R. V. ŠČEPETNOV paru dans Geomag. i Aeron. (1966/1967).

aiguilles d'une montre pour l'hémisphère boréal; sens opposé pour l'hémisphère austral. Ces sens seraient indépendants des heures de la journée[3]. Il y a cependant de nombreuses observations particulières contraires à ces règles[4]. De plus, quand on descend vers les latitudes de plus en plus basses, les polarisations tendent à devenir rectilignes, orientées approximativement dans la direction du méridien magnétique. Ceci, rapproché du fait que dans les zones aurorales la variabilité du phénomène est parfois très grande, même entre stations distantes de moins de 1 000 km, peut être considéré comme une confirmation de ses origines aurorales.

Pulsations **pi1**. Nous avons vu que ces pulsations accompagnaient souvent les baies, les trains **pt** (composés de **pi2**), etc. Si nous les considérons maintenant en propre nous sommes conduits à préciser diverses données morphologiques qui les concernent. Précisons toutefois que nous limiterons cet examen au cas des **pi1** ne dépassant pas une dizaine de secondes pour leur pseudo-période: le cas des **pi1** plus lents formant les «microstructures intermédiaires» de YANAGIHARA[5], qui a déjà été évoqué n'a, en effet, surtout d'intérêt qu'à ce dernier titre.

Les **pi1** «courts» apparaissent le plus souvent sur les enregistrements très rapides (de vitesse de déroulement de l'ordre de 30 à 60 mm/min et d'une sensibilité permettant d'apprécier au moins 0,01 γ), comme une suite d'impulsions, ou d'oscillations très irrégulières et *incomplètes*, c'est-à-dire ne comportant même pas une oscillation entière, dont chacune d'elle — impulsion ou pseudo-oscillation — succède à la précédente d'une façon aléatoire. Il en résulte que leur spectre de fréquence s'apparente à celui d'un bruit blanc plus ou moins parfait remplissant une certaine bande de fréquences.

Dans les régions de moyennes et de basses latitudes (dans ce dernier cas des absorptions plus importantes par l'ionosphère locale peuvent limiter vers 0,5 Hz la partie observable du spectre, sauf pour les évènements de très grande intensité) les pulsations **pi1** peuvent apparaître parfois par situation magnétique apparemment calme, se présentant alors un peu comme des émissions de **pc1** qui — pour des raisons inconnues — seraient devenues très irrégulières. Aux hautes latitudes, on peut noter quelque chose d'équivalent, bien que différent: laissant encore de côté les cas non discutables — et dont nous avons déjà parlé — où les **pi1** accompagnent des perturbations bien reconnues telles que les baies, **pt**, etc., on a trouvé que d'autres cas de **pi1** accompagnaient diverses perturbations d'un caractère plus local et paraissaient ainsi être plus liées à des situations magnétiques régionales polaires qu'à la situation magnétique générale planétaire.

Ceci a été notamment mis en évidence à l'occasion des observations entre points conjugués Byrd Station (Antarctique) — Great Whale River (Canada), comparées aussi à d'autres stations du Nord du Canada plus ou moins voisines, telle que Churchill, ainsi qu'aux stations plus éloignées de College (Alaska) et de Kiruna (Suède)[6].

Les cas les plus typiques sont ceux constitués par les «noise bursts», à débuts impulsifs et durées relativement courtes et dont les relations de conjugaison sont apparues comme peu stables (Churchill étant quelquefois mieux corrélé à Byrd que Great Whale. D'autre part, l'étude de l'évolution de ces évènements — comparée à celle des **pi2** du type **pt** — au cours d'un cycle solaire[7] semble avoir montré que, en dépit de la liaison générale entre les **pi2** et les **pi1**, leur distribution statistique n'évoluait pas de la même façon au cours de ce cycle: les **pi2** du type **pt** s'observeraient en plus grand nombre à l'époque d'un minimum d'activité

[3] G. ROSTOKER: Can. J. Phys. **45**, 1319—1335 (1967).
[4] Y. KATO, J. OSSAKA, M. OKUDA, T. WATANABE, and T. TAMAO: Sci. Rep. Tôhoku Univ., Ser. V, Geophys. **8**, 19—23 (1956).
[5] K. YANAGIHARA: Mem. Kakioka Mag. Obs. **9**, 15—74 (1960).
[6] R. G. GREEN: Trans. Am. Geophys. Union **48**, 68 (1967).
[7] K. YANAGIHARA: J. Geophys. Res. **68**, (11), 3383—3397 (1963).

solaire, alors que les **pi 1** du type «noise bursts» sont beaucoup plus fréquents au moment où le soleil est le plus actif. Une complication dans ce genre d'étude provient de la difficulté que l'on a à bien reconnaître les **pt** dans les régions polaires quand ils y sont plus ou moins masqués par une agitation générale de grande amplitude. Ceci explique peut-être que des «noise bursts» (de **pi 1**) ayant été observés isolément (et non accompagnés de **pi 2**) aux hautes latitudes, apparaissent en synchronisme avec un train de **pt** des latitudes moyennes.

V. Pulsations régulières — ou continues — (pc).

Il y a lieu, pour éviter une confusion, de rappeler que les termes «régulier» et «continu» sont employés ici avec des sens très particuliers — mais admis, par convention, à l'échelle internationale — et qu'il nous faut préciser: ces termes ont été introduits à l'origine à propos des premières pulsations de cette catégorie dont on avait remarqué l'existence (dont la période était de l'ordre de 20 à 30 sec, rentrant donc dans les **pc 3** de la classification actuelle), pulsations que l'on avait d'abord désignées par «pulsations de jour» à cause de leur manifestation essentiellement diurne. L'adjectif de «continu» était relatif au fait que ces pulsations étaient observables souvent pendant plusieurs heures sans presque d'interruptions, et qu'elles s'étendaient sur une portion importante — parfois la quasi-totalité — de la partie diurne de la journée. La qualification de «régulières» venait de ce qu'elles avaient assez souvent une forme d'onde se rapprochant sensiblement d'une sinusoïde, ce qui permettait, en particulier, de leur affecter avec une précision convenable une période définie. Par la suite on a reconnu qu'il convenait d'élargir dans les deux sens (vers les périodes plus longues et vers les périodes plus courtes) les cas à étudier. On l'a fait sans changer les qualificatifs ci-dessus, que l'on a donc appliqués peu à peu à l'ensemble de toute la gamme, depuis 1 sec jusqu'à 10 min, sans trop se soucier si toutes les pulsations que l'on incluait dans cette gamme avaient bien le droit à être ainsi qualifiées. Comme ce n'est pas toujours le cas, mais comme on ne désire pas renoncer pour autant à une nomenclature commode, on arrivera à mentionner quelquefois des pulsations «continues» **pc**, «épisodiques», ou, encore, des pulsations continues de *forme* irrégulière. Dans ce dernier cas, il est évident que l'on doit avoir des raisons pour ne pas les ranger parmi les «pulsations irrégulières» **pi 1** ou **pi 2**.

21. Pulsations pc5. Définies comme des **pc** dont la période est comprise entre 150 et 600 sec, on ne trouve dans la gamme ainsi définie qu'une seule entité morphologique bien marquée: celle des «pulsations géantes» ou «**pg**». Cependant, de même que nous n'avions pas institué une identité totale entre **pi2** et **pt**, il serait incorrect de poser comme on le fait quelquefois: **pc5** = **pg**. (Ce serait, en effet, se priver de la possibilité de classer des pulsations éventuelles qui — sans ressortir du phénomène bien défini qu'est une pulsation géante **pg**, auraient une période comprise dans la gamme 150 ... 600 sec.)

Les pulsations géantes sont connues depuis longtemps en tant que phénomène apparaissant spécifiquement sur les magnétogrammes (les «normaux» suffisent) des stations magnétiques des hautes latitudes. En dehors de leur période (en général plutôt supérieure qu'inférieure à 5 min) et de leur régularité assez grande, elles se font surtout remarquer par leur grande amplitude qui dépasse souvent plusieurs dizaines de gammas. La durée globale d'un de ces phénomènes peut varier entre une dizaine de minutes (avec alors seulement 2, 3 ou 4 oscillations) et quelques heures. Elles se présentent souvent par situation magnétique d'agitation modérée, et à des heures locales préférentielles centrées autour de 6 à 8 h TLM du matin (mais avec de très grandes fluctuations).

Géographiquement, leur action reste assez centrée autour de leur zone d'origine (jusqu'à une distance de l'ordre d'un millier de kilomètres), avec un bon effet de conjugaison dans l'hémisphère opposé. Leur fréquence d'apparition (qui reste relativement faible: quelques cas par mois au plus pour les cas bien nets) ne paraît pas liée étroitement au déroulement d'un cycle solaire, mais il n'en serait pas de même de leur période moyenne qui deviendrait plus courte lors des minimums d'activité solaire, ce qui s'expliquerait par le changement correspondant de la densité du plasma dans la magnétosphère[1]. Nous retiendrons également les observations faites — aux deux extrémités d'un tube de forces de conjugaison — sur leurs diagrammes de polarisation (dans le plan des $\mathcal{H}$, D): sens de rotation inverses aux deux extrémités, pour un même phénomène se produisant à un même moment (ce qui est conforme à l'hypothèse suivant laquelle ces pulsations reflèteraient une oscillation d'ensemble, suivant un mode toroïdal, des lignes de forces aboutissant à ces extrémités. Pour un observateur regardant dans le sens des lignes de forces (sens positif de $\mathcal{H}$) le sens de rotation serait celui de droite à gauche pour tous les cas — les plus importants — correspondant à un méridien de la matinée. A ceci correspondrait donc dans le plan des $\mathcal{H}$, D, une polarisation dans le sens inverse des aiguilles d'une montre pour l'extrémité boréale, et le sens opposé au précédent pour l'extrémité australe. Dans les cas — relativement rares et correspondant à des amplitudes plus faibles — de **pg** *du soir* (correspondant au deuxième maximum diurne d'occurence signalé par certains autours[2,3] les sens de polarisation sont, en général les inverses des précédents.

Les pulsations géantes ont toujours été considérées comme un phénomène peu commun, et cependant il est assez fréquent que l'on observe, soit en zone sub-aurorale, soit à des latitudes moins élevées, des oscillations des composantes magnétiques rentrant dans la gamme des **pc5** mais dont la régularité, ou l'amplitude, ne paraissent pas de qualité ou de valeur suffisantes pour être reconnues comme des **pg**. On peut se demander si cette manière de faire est justifiée et si certaines de ces oscillations ne devraient pas — à la qualité près — être considérées comme de même essence que les **pg**. Ceci paraît être le cas de pulsations assez régulières, de période de l'ordre de 5 min, d'amplitude moyenne en général un peu faible comparée à celle des **pg**, mais qui, très fréquentes le matin (entre 5 et 12 h TLM) en diverses stations sub-aurorales (notamment à Kerguelen et à la station du Roi Baudouin), y prennent parfois — par simple renforcement progressif de leur amplitude — toutes les caractéristiques d'indiscutables **pg**. Il y aurait donc un passage continu entre ces formes «non reconnues» de **pg** et les **pg** les plus caractérisés. Dans les stations situées au-delà des zones aurorales (stations polaires proprement dites), ce phénomène n'est plus observé, mais une forte composante spectrale dans la même bande de fréquence se manifeste comme participant à l'«agitation de jour» («agitation **J**» des régions polaires[4]), tout au moins dans l'hémisphère qui s'y prête. A l'opposé — toujours aux mêmes heures locales — on a depuis longtemps remarqué dans les stations de latitudes moyennes[5] une «petite agitation du matin» centrée aussi sur la même bande de fréquence et — en général — très irrégulière. (C'est d'ailleurs le plus souvent à l'intérieur de cette «petite agitation» qu'apparaissent — superposées à ses oscillations irrégulières, par exemple de 5 min de pseudo-période — les trains, beaucoup plus réguliers, de **pc3** formant la partie principale des «pulsations de jour».) En résumé, bien qu'il ne semble pas justifié, dans l'état actuel de nos connaissances, de considérer toutes les manifestations que nous venons de mentionner, comme la manifestation d'un même type de phénomène qui serait simplement déformé de diverses façons tout au long de l'échelle des latitudes, il convient de ne pas perdre de vue cette prédominance, sous des formes très variées, d'une composante dans la même gamme que les **pc5** des variations magnétiques apparaissant dans tout le secteur matinal de notre Globe.

---

[1] Y. KATO and T. SAITO: Rep. Ionos. Space Res. Japan **18** (1964).
[2] Y. KATO and T. TAMAO: J. Phys. Soc. Japan **17**, Suppl. A II, 39—43 (1962).
[3] T. NAGATA, S. KOKUBUN, and T. IIJIMA: J. Geophys. Res. **68**, 4621—4625 (1963).
[4] P. N. MAYAUD: Thèse Fac. Sci. Paris (1955); — Ann. Géophys. **12**, 84—101 (1956).
[5] C'est le cas, en particulier, à l'Observatoire de Chambon-la-Forêt.

22. Pulsations pc 4 (45 ... 150 sec). La portion des «pulsations continues» **pc** tombant dans cette gamme avait été presque totalement ignorée à l'époque des premières observations régulières des **pc**[1]. Il est possible que cela soit dû au fait que ces pulsations ne se manifestent que d'une façon plus modérée en période d'activité solaire telle que celle des années 1957/59. Quand, par la suite, on a observé leur manifestation parfois importante (ce dont on a tenu compte dans la classification adoptée à l'Assemblée Générale de l'UGGI de Berkeley, 1963[2]), on ne s'est pas trouvé d'accord pour leur reconnaître une entité unique, bien définie. Certains auteurs[3] voient en elles une extension — par situations magnétiques relativement calmes (quand la magnétosphère serait «décomprimée»), des types **pc2** et **pc3** (le regroupement de ces divers types s'effectuant à nouveau quand l'activité solaire — et magnétique — augmente); d'autres auteurs[4] les considèrent comme un groupe ayant, au contraire, sa spécificité propre (et un mécanisme de formation bien distinct); enfin, d'observations plus récentes[5] il semblerait résulter que certaines d'entre elles au moins ne seraient que le résultat de la propagation dans l'hémisphère de jour — à partir de l'hémisphère de nuit — des nombreux trains de **pt** (**pi2**) pratiquement toujours présents en telle ou telle portion de cet hémisphère. Ces différences de conception entraînant souvent des différences dans les classifications, on ne doit pas s'étonner si certains résultats contradictoires ont pu être publiés concernant telle ou telle de leurs propriétés (ceux, notamment, concernant l'évolution de leurs périodes avec l'activité du cycle solaire et ceux relatifs à leurs diagrammes de polarisation). Nous éviterons donc de discuter plus avant ces questions qui demandent clarification, mais nous aurons à tenir compte de l'existence d'une composante importante dans la bande spectrale qui les concerne quand nous examinerons les principaux modes qui doivent être rendus possibles dans l'environnement terrestre (soit au titre de leur génération, soit à celui des propagations et des transmissions).

23. Pulsations pc 3 (10 ... 45 sec). Avec les **pc 1** (mais depuis plus longtemps) les **pc3** ont été les plus étudiées — tout au moins au point de vue morphologique — de toutes les pulsations magnétiques. Cependant, contrairement aux **pc1**, relativement peu de conceptions théoriques vraiment précises et paraissant valables, ont été élaborées les concernant. Cela est surtout vrai en relation avec la question fondamentale par laquelle on se demande quelle est l'action précise qui déclenche spécifiquement certaines de leurs apparitions — ou disparitions — subites à l'échelle mondiale (dans les limites de modulation jour-nuit permises), ou leurs brusques discontinuités morphologiques et de périodes. Pour ces raisons, leur étude est particulièrement intéressante.

Dans la première classification de Copenhague, en 1957 [*9*], qui revenait à distinguer essentiellement des «pulsations de jour» et des «pulsations de nuit», les **pc3** constituaient en fait la presque totalité des pulsations que l'on rangeait dans la première de ces catégories sous la dénomination générale de «pulsations continues» (ou «**pc**», sans indice) — les pulsations **pt** constituant l'autre catégorie. Ce sont, en effet, celles que l'on rencontre de la façon la plus permanente,

---

[1] Colloque de Copenhague (1957); Symposium d'Utrecht (1959).

[2] J. A. Jacobs, Y. Kato, S. Matsushita, and V. A. Troitskaja: J. Geophys. Res. **69**, (1), 180 (1964).

[3] V. A. Troitskaja and A. V. Gul'elmi: Space Sci. Rev. **7**, 689—768 (1967).

[4] Y. Kato and S. Takei: JUSCO Conference, Kyoto (1969). A cette occasion ces auteurs introduisirent la notion d'*Ondes de Surface* magnéto-dynamiques qui, excitées simultanément par le vent solaire sur les côtés matin-soir de la magnétopause, y déterminent des polarisations contraires quand on les observe du sol.

[5] F. Jamet, R. Remiot, J. Roquet et E. Selzer: C.R. Acad. Sci. **269**, 442—444 (1969).

en la plupart des régions du Globe, dans l'intervalle d'Heure Locale allant de 04 ou 05 à 17 ou 18 h, ces limites n'étant elles-mêmes que très approximatives. La qualification de «continues» ne veut pas dire qu'elles remplissent sans discontinuité tout cet intervalle, mais qu'on les y trouve «communément» à une heure quelconque (avec, par exemple, une probabilité — ou fréquence d'occurence — de l'ordre de 50%). Elles y forment des «trains» de durées plus grandes, en général, que les trains de **pt**, et qui se raccordent les uns aux autres tantôt d'une façon pseudo-continue[1] (voir Fig. 28a et b), tantôt en passant par des discontinuités — telles que des sautes brusques dans la valeur de leurs périodes ou de leurs amplitudes — plus ou moins marquées (voir Fig. 29a et b), celles que nous avons tenu à signaler dès le début de cette section.

Ces pulsations présentent donc deux types très distincts d'évolution journalière:

d'une part, une modulation liée à l'action *jour-nuit* dont elles subissent l'effet, en général assez progressif, qui est sous la dépendance directe de l'Heure Locale (donc de la longitude); cette modulation agit surtout de manière décisive le matin et le soir, conditionnant ainsi d'une façon générale le *lever* et le *coucher* de ces pulsations; il est difficile de déterminer dans quelle mesure elle agit encore au cours de la journée[2] (à l'intérieur des heures locales limites) car un tel effet doit, de toute façon, être faible, et il est masqué par l'autre type d'évolution, c'est-à-dire:

d'autre part, toutes les fluctuations se produisant dans l'intervalle de temps local permis par le premier effet et qui paraissent pouvoir être rattachées à un effet mondial: un grand nombre d'observations on montré qu'elles se présentaient dans la plupart des cas en synchronisme mondial dans les diverses stations d'observation situées dans cet intervalle de temps local favorable (en dehors de cet intervalle le synchronisme ne peut pas être observé puisque le phénomène manque au moins à une des stations).

Au voisinage des heures de *lever* et de *coucher* les deux effets précédents peuvent interférer: c'est ainsi que si le phénomène mondial est absent vers l'heure locale du lever relative à une station déterminée, les **pc3** devront attendre pour se manifester à cette station que le phénomène mondial existe à nouveau. Le *lever* en Heure Locale y sera donc retardé tout en coïncident en Heure Universelle avec l'instant du *lever* de ces mêmes pulsations à une station située un peu plus à l'Ouest. Ceci explique aussi que les *levers* des pulsations puissent se produire d'une façon brusque, quand ils sont commandés par le phénomène mondial au lieu de l'être par la modulation progressive liée à l'Heure Locale.

Ces considérations, qui nous paraîssent actuellement assez claires, ont eu quelque mal à être bien comprises à l'origine. Nous sommes davantage habitués maintenant, dans le domaine des variations magnétiques, à considérer que des évènements qui se produisent — ou devraient se produire — en synchronisme

[1] Dans un grand nombre de cas il paraît même plus correct de considérer qu'il n'y a pas des «raccords» entre trains successifs, mais une modulation plus ou moins rythmée de l'amplitude des pulsations se traduisant par une enveloppe en forme de «navette», avec une succession de noeuds et de ventres. Tout porte à croire que ceci est le résultat d'un battement entre deux fréquences voisines, phénomène qui interviendrait ici d'une façon plus simple que dans le cas des pulsations **pc 1** du type «perles».

[2] Ceci nous rattache à la «variation diurne de l'amplitude» des **pc 3** ou, encore — ainsi que cela est présenté par de nombreux auteurs — à la variation diurne de la «probabilité de présence» de ces pulsations. Dans les deux cas la définition précise des normes de dépouillement à adopter est délicate et imprécise, ce qui rend les résultats incertains. C'est une des raisons pour laquelle nous donnons la préférence à la considération des «heures de lever et de coucher».

mondial, peuvent également être sous le contrôle d'une modulation (pouvant aller jusqu'à tout ou rien) commandée par l'Heure Locale. On pourrait citer comme autre exemple la modulation de l'amplitude des **SSC** en remarquant qu'ici elle n'atteint le «rien» que si l'on ne dispose pas d'une sensibilité suffisante. Inversement, on peut se demander si une grande augmentation de sensibilité ne nous permettrait pas de mettre en évidence des **pc3** *de nuit*. La réponse est plutôt négative — contrairement à ce que nous avions vu au sujet des «**pt** de jour»[3] — mais il est bien évident que ce genre de question ne peut comporter de réponse absolue ... c'est toujours une affaire de *degré*. Remarquons en outre que certaines observations de **pc3** *du soir* en des régions particulières (équatoriales ou sub-équatoriales[4]) pose un autre problème à distinguer du précédent et que l'on n'a pas encore éclairci.

24. Remarques. Nous avons insisté quelque peu sur les faits précédents car de leur compréhension dépend l'intérêt que l'on peut attacher à l'étude des **pc3** en vue d'une meilleure compréhension de la dynamique de la magnétosphère elle-même. Respectant notre plan général d'exposition, nous n'allons pas examiner tout de suite les *modes d'oscillations* pouvant expliquer la manifestation générale des **pc3** (nous renvoyons pour cela au chapitre final de notre article), mais nous allons essayer d'aller plus avant dans l'examen des modes généraux d'*excitation* et de *transmission* qui peuvent être d'accord avec les deux types de modulation décrits ci-dessus. A cette fin, il est peut-être commode de partir des deux idées premières que l'on s'était faites — jusqu'à ces dernières années — à ce sujet: l'évolution des conceptions est ici très intéressante à suivre dans la mesure où elle a été imposée par celle relative au comportement général de l'environnement terrestre face à l'action du vent solaire. *La première* idée à laquelle on s'était rattaché après s'être convaincu du synchronisme mondial des fluctuations brusques des **pc3** (à l'intérieur des intervalles d'Heure Locale permis), c'était que ces fluctuations traduisaient des discontinuités de structure (ou de vitesses) pour le vent solaire (ou toute action directe, analogue, du soleil). *La seconde*, c'était que la modulation jour-nuit par l'Heure Locale devait être l'effet d'une absorption imposée par l'ionosphère locale (ou tout autre agent, ou milieur absorbant, dont le mécanisme d'absorption serait commandé par l'Heure Locale ... cette variante non précisée permettant de pallier l'objection d'un effet d'absorption par l'ionosphère contraire à ses propriétés connues).

Or, les connaissances morphologiques actuelles sur la magnétosphère donnent très simplement une explication de l'effet jour-nuit par la rotation de la partie pseudo-sphérique de la magnétosphère correspondant à l'ensemble des lignes de forces fermées (rotation liée à celle du Globe terrestre lui-même), par rapport à la position semi-fixe (vue suivant l'axe Soleil-Terre) de toute la partie arrière éloignée (après la plasmapause) de cette magnétosphère (queue, zone neutre, etc). C'est uniquement la partie avant (côté Soleil), connectée, de la magnétosphère qui pourrait être excitée directement par le vent solaire pour donner naissance aux **pc3** (la partie arrière pouvant être soumise à d'autres types d'excitation susceptibles, par exemple, de donner lieu aux trains de **pt**). Ainsi cet effet jour-nuit ne serait plus le résultat d'une action secondaire de modulation par absorption, mais un effet direct d'excitation différenciée. De ce fait, il n'y aurait plus la même nécessité de rechercher l'explication des fluctuations en synchronisme mondial dans des discontinuités supposées du vent solaire (d'autres faits pouvant être par ailleurs invoqués pour ou contre la réalité de telles dis-

[3] F. Jamet, R. Remiot, J. Roquet et E. Selzer: C.R. Acad. Sci. **269**, 442—444 (1969).
[4] R. Hutton: Radio Science **69** D, 1169 (1965); — Nature **186** (1960).

continuités, mais ceci échappe à notre sujet). Ces fluctuations pourraient avoir leur source dans toutes sortes de discontinuités, géométriques ou physiques, pouvant prendre place à l'intérieur même de la magnétosphère, telles que des variations brusques dans les dimensions, ou dans la forme, des cavités de résonance favorisant l'amplification et la mise en forme des pulsations. Les conclusions précises de cette manière actuelle de concevoir les mécanismes généraux conditionnant la manifestation des **pc3** n'ont pas encore été dégagées mais on peut espérer, quand elles le seront, que des nombreuses — et relativement faciles — observations des **pc3**, on saura déduire des renseignements utiles sur l'évolution à chaque instant aussi bien du vent solaire que de la magnétosphère.

On peut cependant se demander si cette nouvelle conception du mécanisme des **pc3** est due uniquement à nos progrès dans la connaissance de l'environnement terrestre, ou si elle se trouve aussi étayée par certaines propriétées spécifiques de ces pulsations, ce qui lui confèrerait, évidemment, un plus grand poids. Nous pensons pouvoir répondre affirmativement sur ce dernier point, notamment par la considération des effets saisonniers appliqués aux hémisphères conjugués Nord et Sud. Ces effets présentent certaines propriétés qui seraient paradoxales si les *levers* et *couchers* des **pc3** étaient directement liés aux *levers* et *couchers* correspondants du Soleil, ceci en dépit d'une vague concordance globale. C'est ainsi qu'à Chambon-la-Forêt, station de latitude moyenne, les heures de *lever* et de *coucher* des **pc3** sont indépendantes des saisons, ainsi que le montrent des statistiques établies sur plus de 13 ans d'observations. (Certaines années, un léger effet saisonnier semble se manifester, ce qui expliquerait que des auteurs japonais[1] aient mentionné un tel effet.) Ceci veut dire que ces pulsations apparaissent bien avant le lever du Soleil en hiver, et bien après ce lever en été. Cette indépendance envers les variations des levers et couchers du Soleil ne peut qu'être renforcée quand on descend vers les latitudes plus basses (c'est ce qu'on peut appeler l'élargissement de l'effet équatorial[2]) mais elle cesse très brusquement quand on remonte, au contraire, vers les latitudes très élevées (de l'ordre de 60 à 70°). C'est ainsi que dans les régions polaires la «durée des journées de **pc3**» est, en moyenne, très longue pendant l'été correspondant et très courte pendant la nuit polaire. Ceci montre que la dépendance entre le Soleil et les **pc3** ne doit pas être considérée sur un plan local, mais qu'il convient de la faire intervenir tout au long des parties connectées des lignes de forces remplissant la demi-magnétosphère éclairée. Les relations de conjugaison qui existent entre les extrémités Nord et Sud de ces lignes de forces — et qui ont été, bien vérifiées pour les **pc3** si l'on défalque quelques influences ionosphériques locales[3] — sont bien en accord avec une telle excitation globale de tout un tube de forces qui ne dépend plus alors que de la rotation d'ensemble de toute la «portion connectée» de la magnétosphère. L'effet des saisons reprend ses droits dans les régions de hautes latitudes, justement là où la connection des lignes des forces n'est plus assurée. (Il est possible qu'une étude plus fine de l'effet saisonnier sur les **pc3**, et la détermination précise des latitudes auxquelles il tend brusquement à se manifester, nous permette d'atteindre, par des mesures au sol, la détermination des zones de connection et de déconnection que nous venons d'évoquer.)

25. Cas litigieux. Les indications générales que nous venons de donner nous permettent de ne pas nous engager dans un exposé détaillé de toutes les réparti-

[1] T. SAITO: Sci. Rep. Tôhoku Univ., Ser. V, Geophys. **14**, 81—106; — J. Geomagnet. Geoelec. **16**, 115—151 (1964).

[2] E. SELZER: Comm. Sympos. URSI-IAGA, Belgrade (1966).

[3] R. SCHLICH: Comm. Sympos. IAGA—UGGI «Points conjugués», St-Gall (1967); — Ann. Géophys. **24**, 411—429 (1968).

tions géographiques qui ont été abondamment publiées pour les **pc3**. Il convient cependant de signaler une controverse qui a pris naissance il y a quelques années avec la publication de résultats[1–3] faisant état d'une variation avec la latitude de la période d'un même groupe de certains **pc3** (ainsi que d'autres types de pulsations). En principe, les mécanismes les plus simples de vibrations que l'on puisse appliquer à tout un tube de forces de la demi-magnétosphère de jour ne prévoient pas une telle variation, mais dès qu'on imagine un mode de vibration un peu plus compliqué un tel phénomène devient possible: par exemple, des effets de torsion peuvent changer les correspondances de couplage le long du tube entre les composantes $\mathcal{H}$ et D tout en apportant des irrégularités aux figures de polarisation. La combinaison — ou la composition — des rotations de ces figures de polarisation avec celles des oscillations proprement dites peut affecter les périodes localement observées. Ceci permettrait, de plus, d'expliquer que, bien que les auteurs précédemment cités aient essentiellement relevé des allongements de période quand l'observation se fait à des latitudes plus élevées, des effets inverses aient pu aussi être signalés[4–6].

Afin de compléter cette étude des **pc3** nous ferons maintenant un certain nombre de remarques qui peuvent nous aider à mieux comprendre ce type de variations magnétiques. Elles concernent l'influence de paramètres tels que l'agitation magnétique générale (symbolisée, par exemple, par les indices K_p), l'influence du cycle solaire, etc. ...

Du fait même que nous n'avons pas rangé les **pc3** parmi les variations qui devaient être considérées comme des microstructures d'autres phénomènes nous n'avons pas envisagé de lien strict les rattachant à une agitation magnétique macroscopique bien définie. Il semble cependant que ces pulsations aient besoin d'une agitation générale légère, synchrone ou antérieure, pour se manifester avec une amplitude notable. Une agitation trop forte, au contraire, en troublant leur régularité, nuit à cette manifestation. On peut expliquer ainsi, par une simple règle de bon sens, que de nombreux auteurs aient trouvé:

1) Qu'elles apparaissaient de façon privilégiée pour des valeurs de K_p de l'ordre de 2 ou 3;

2) qu'elles suivaient statistiquement le rythme de 27 jours de la rotation solaire, ce qui n'est qu'une conséquence secondaire du 1) étant donné la périodicité mise en évidence pour les K_p eux-mêmes, tout au moins pour ceux relatifs à des agitations modérées (orages ou agitations à débuts progressifs).

Ceci n'est pas en contradiction avec le fait que des pulsations de grande amplitude apparaissent souvent dans la gamme des **pc3** au cours de certaines phases des grands orages mondiaux. Ces pulsations sont alors, en effet, presque toujours très irrégulières et doivent être considérées comme la composante, dans cette gamme, d'une agitation très générale et en très large bande, plutôt que comme les trains organisés de **pc3** qui nous intéressent ici. Il arrive cependant que cette composante de l'agitation se régularise au cours du déroulement de l'orage, donnant lieu à des **pc3** de grande amplitude mais tout à fait classiques que nous appelerons «pulsations issues d'orage». Nous les rangeons parmi les «pulsations d'orage».

---

[1] M. Siebert: Planetary Space Sci. **12**, 137 (1964).
[2] H. Voelker: Max-Planck Inst. Aeronom., Bericht Nr. **11**, (S) (1963).
[3] H. Voelker: Rep. No. 8815, NBS, Boulder, Col. [Abst. Radio Science **69** D, 1187] (1965).
[4] G. R. A. Ellis: Aust. J. Phys. **13**, 625 (1960).
[5] R. A. Duncan: J. Geophys. Res. **66**, (7), 2087 (1961).
[6] T. J. Herron and J. R. Heirtzler: Nature **210**, 361 (1966).

Enfin, on s'est demandé si l'on pouvait mettre en évidence, statistiquement, une évolution générale, soit de la période, soit de l'amplitude des **pc3** avec le cycle solaire. En dehors des influences rattachables à ce qui précède (degré d'agitation) la réponse à ces questions n'est pas sûre: il semble toutefois que la période moyenne des **pc3** soit plus grande durant la phase du minimum d'activité solaire alors que ce serait l'inverse pour les **pc4**. (Ceci pourrait d'ailleurs être la source de recouvrements entre **pc3** et **pc4** au voisinage de ces époques, leurs périodes venant réciproquement à la rencontre de l'une l'autre.)

26. Pulsations pc 2. En terminant cette section relative aux **pc 3** donnons très rapidement les quelques *propriétés caractéristiques des* **pc2** qui, mis à part leur gamme conventionnelle qui couvre l'intervalle 5 ... 10 sec, les ont fait ranger à part. En effet, si du côté des périodes plus courtes la frontière avec les **pc1** est particulièrement bien définie, il y a peu de distinctions morphologiques à faire — à part justement la valeur des périodes moyennes — entre **pc2** et **pc3**. La distinction que l'on a, cependant, tenu à faire est basée sur les deux faits suivants:

1) On observe une absence relative de pulsations régulières juste autour de 10 sec;

2) des pulsations régulières de 5 à 6 sec de période apparaissent assez souvent le matin de très bonne heure (donc, avant le lever des **pc3** conventionnelles), surtout au voisinage d'une phase de maximum d'activité d'un cycle solaire; d'après ce que nous avons vu sur l'évolution de la période des **pc3** avec le cycle solaire, ces **pc** de très courtes périodes pourraient cependant être interprétées comme une de leurs manifestations extrêmes. En fait, plusieurs auteurs (voir [*14*]) traitent comme un tout une classe de pulsations «**pc2,3**» groupant les **pc2** et les **pc3**, malgré la complication que cela entraîne dans l'étude de l'évolution diurne de ce groupe unique.

27. Pulsations pc 1 ($^1/_5$ ... 5 sec). Ces pulsations se sont dégagées peu à peu de travaux généraux portant, *les uns*, sur les limites extrêmes, en allant vers les phénomènes les plus rapides, des variations proprement géomagnétiques (à l'exclusion de celles qui — telles que les oscillations de SCHUMANN[1] — sont d'origine météorologique; *les autres*, sur des structures caractéristiques — soit des formes d'ondes, soit des compositions spectrales — qui, à certaines occasions (assez peu fréquentes il est vrai mais qui se reproduisent sans contestation possible) prennent des apparences tout à fait remarquables quand on dispose d'appareillages ayant les caractéristiques très poussées (notamment en sensibilités) requises.

La première catégorie de ces travaux, entrepris vers les années 1945/50 alors que l'on ne savait encore presque rien de la morphologie des pulsations, était axée sur des problèmes énergétiques: déceler les énergies électromagnétiques naturelles pouvant être présentes dans les régions non perturbées artificiellement du sol terrestre, au voisinage et au-delà immédiat de la fréquence 1 Hz, et déterminer la répartition spectrale de ces énergies. Les mesures portaient donc sur des niveaux énergétiques et ne s'inquiétaient pas des formes d'ondes. Un de leurs résultats les plus marquants a été d'arriver à cette séparation entre phénomènes géomagnétiques proprement dits et phénomènes météorologiques que nous avons signalée plus haut.

---

[1] Voir Sects. 55—69 du présent article.

Les travaux portant sur les structures ne se sont développés que plus tard (à partir de l'Année Géophysique Internationale 1957/58), *d'abord sur les formes d'ondes* — notamment pour les «pulsations en perles» — relevées sur les enregistrements telluriques rapides du réseau mis en place par V. A. TROITSKAJA, et — ultérieurement — sur de nombreux enregistrements analogues, magnétiques ou telluriques, *puis sur les spectres de puissance.* Nous n'insistons pas sur le caractère remarquable de ces structures et sur leur description, ce dont le lecteur pourra se rendre compte d'une façon plus efficace en se reportant aux nombreuses illustrations que nous avons consacrées à ces phénomènes (voir les Fig. 35 à 40 pour les formes d'ondes et les Fig. 46 à 54 pour les structures spectrales). Des descriptions plus détaillées de ces structures ont été faites plus récemment, notamment par HEACOCK[2,3] qui à côté de la notion de «perle» a introduit celle de «grain». Encore plus récemment, F. GLANGEAUD[4] s'est préoccupé de déterminer la correspondance exacte entre les structures des formes d'ondes et celles des spectres et a pu rattacher ainsi la distinction entre «perle» et «grain» à l'intervention de deux mécanismes foncièrement différents: les perles étant dues à des modulations d'amplitude conséquences du rapprochement, ou de l'éloignement, du milieu («paquet d'onde») générateur des pulsations, les grains résultant, eux, d'interférences entre diverses composantes spectrales présentes dans les perles (voir Fig. 35 à39 et 41, 42). Sans entrer ici dans la théorie de ces pulsations, ce que nous ferons au chapitre E, nous signalerons seulement — ce qui nous permettra une exposition morphologique plus claire — que l'on admet actuellement qu'elles sont engendrées (suivant un processus d'interaction onde — particules) en des régions privilégiées de certains tubes de forces de paramètre L de MCILWAIN[5] convenable. Les paquets d'ondes dans lesquels elles ont été engendrées les transportent le long de ces tubes suivant un mouvement de va-et-vient entre les deux hémisphères Nord et Sud, la réflexion de ces paquets d'ondes au voisinage des couches ionosphériques correspondantes y déterminant les processus d'injection et de guidage qui leur permettent de se manifester à terre. Nous retiendrons essentiellement, pour l'instant, de ce schéma, qu'il y aura lieu de se préoccuper au point de vue morphologique: des régions d'injection, des modalités de propagation pseudo-horizontale dans les guides ionosphériques, des compositions spectrales et de leur évolution, de leur polarisation, ainsi que de leurs relations de conjugaison entre les deux hémisphères. Il y a lieu aussi de préciser que la forme «en perles» n'est pas la seule: d'abord, des «perles» peuvent être très irrégulières (même sans l'intervention de «grains»); d'en d'autres cas elles peuvent manquer totalement. Les structures spectrales sont en accord avec ces diverses variétés. On peut voir, Fig. 51, un exemple de pulsations (baptisées ici «émission» en se référant au mécanisme de génération du phénomène) dont le spectre ne comporte aucune structure, étant à chaque instant quasi monochromatique; d'autres exemples peuvent être donnés (Fig. 40) dans lesquels l'émission, bien que non monochromatique, est aussi partiellement non structurée. L'ensemble de tous ces cas constituant les **pc1** — sous réserve de satisfaire aux conditions générales de régularité et d'ordre de grandeur *pour les oscillations composantes individuelles* (celles qui forment les pulsations proprement dites) — on a été conduit à distinguer les **pc1** structurées et les **pc1** non structurées. Parmi les premières, les «pulsations en perles» (ou «**PP**») forment une classe particulière. Ce sont celles dont nous allons nous occuper maintenant. Signalons qu'elles sont

---

[2] R. R. HEACOCK: J. Geophys. Res. **68**, 1871—1884 (1963).
[3] R. R. HEACOCK and V. P. HESSLER: J. Geophys. Res. **68**, 953—954 (1963).
[4] F. GLANGEAUD: Ann. Geophys. **26**, 299—312 (1970).
[5] Voir l'article de W. N. HESS dans ce volume, p. 115.

aussi abondamment mentionnées dans la littérature sous la dénomination d' «émission hydromagnétique structurée» (symbole «**HM**») bien qu'une telle dénomination puisse paraître ici prématurée.

28. Amplitude — Fréquence d'occurence — Période dominante des PP. Les deux premières de ces caractéristiques sont évidemment liées: Si le seuil pour les amplitudes décelables est de l'ordre de 0,001 γ à 1 Hz (comme il convient pour les appareillages à utiliser) on peut compter sur une moyenne de quelques cas par mois (3 ou 4 par exemple) aux latitudes moyennes (2, 3 ou 4 fois plus aux latitudes plus élevées). Les chiffres mensuels individuels peuvent s'écarter considérablement de ces valeurs (il y a des mois sans aucun cas et des mois où ils sont presque journaliers). Chaque «cas» peut comprendre de quelques minutes à quelques heures d'«émission». Ils ne sont bien observés qu'en situation magnétique générale «calme» (d'où le nom de «pulsations de temps calme» que l'on donne quelquefois à ces pulsations), mais ceci n'implique pas obligatoirement que ces pulsations ne puissent se superposer — avec ou sans lien réel — avec d'autres types de variations magnétiques. (C'est ainsi qu'elles peuvent se superposer à des **pi 1** pour entrer dans la composition de certains **IPDP**). Les circonstances précises déterminant leur occurence tel et tel jour de préférence à tel autre, sont encore très mal connues. On s'était probablement trompé en pensant qu'elles pouvaient nous aider à prévoir les orages magnétiques (bien que l'on ait observé des cas nets où elles avaient précédé un début **SSC** d'orage[1]), mais il est assez bien établi qu'elles suivent de préférence, à deux ou trois jours près, la fin des grands orages. Nous verrons qu'un des aspects satisfaisants de la théorie généralement admise à leur sujet est qu'elle suppose un processus de génération résultant de la coïncidence de deux évènements aléatoires, probablement non liés l'un à l'autre, ce qui introduit le caractère, lui-même aléatoire, de leur manifestation habituelle. D'autre part, aucune loi n'a pu être clairement dégagée concernant la dépendance éventuelle de leur apparition d'effets saisonniers ou liés au déroulement général d'un cycle solaire. A l'échelle journalière on peut être au contraire assez précis, leur variation diurne (considérée statistiquement) paraît être sous le double dépendance d'une génération primaire (magnétosphérique), maximale autour du midi local, et d'un processus d'absorption par l'ionosphère (faible aux hautes latitudes, mais croissant rapidement quand on descend vers des latitudes plus basses) qui serait au contraire le plus marqué à ce moment. Le résultat global est donc que leur variation diurne paraît inversée quand on passe des hautes aux basses (ou moyennes) latitudes: maximum de jour dans le premier cas (avec minimum non nul la nuit) et maximum la nuit (plus exactement, vers le début et vers la fin de la nuit) dans le second.

Leur période dominante est loin de remplir toute la bande (0.2 ... 5 sec) des **pc1**. La valeur la plus commune oscille entre 1 et 2 sec. L'influence, à l'échelle statistique, de la latitude a été discutée: d'après divers auteurs — et certaines considérations théoriques — leur période devrait diminuer quand on descend en latitude ... mais l'absorption plus forte de l'ionosphère pour les courtes périodes contrarierait cet effet. Cette question est compliquée par les processus de propagation horizontale (le long du sol terrestre). En fait des **PP** de courte période ont pu — en certaines occasions exceptionnelles — être observées avec netteté jusqu'à l'équateur[2,3] Ceci nous amène aux influences géographiques.

---

[1] A. Cecchini, G. Dupouy, J. Roquet et E. Selzer: C.R. Acad. Sci. **250**, 4023—4260 (1960).

[2] J. Roquet: C.R. Acad. Sci. **268**, 581 (1969).

[3] J. Roquet: Comm. Sympos. «Micropulsations», AIGA—UGGI, Madrid (1969).

29. Répartition et extension géographique des PP. La théorie des **PP** (voir la Sect. 53), si on la limitait aux processus qui se passent dans la magnétosphère, ne prévoierait leur manifestation qu'en des zones magnétiquement conjuguées, relativement étroites chacune (quelques centaines de kilomètres en longitude et en latitude). Des observations de plus en plus nombreuses[1,2] montrent qu'il n'en est pas ainsi: non seulement elle peut se produire tout au long d'une «tranche méridienne», mais, même en longitude des mêmes trains de perles et de grains ont pu être observés entre des stations séparées par plusieurs milliers de kilomètres de distance[3]. Pour tenir compte de ces faits, il convient, non seulement de mettre en jeu les propriétés de guide d'ondes des hautes couches de l'ionosphère (pour les ondes électromagnétiques dans la gamme UBF), mais aussi d'admettre que les sources magnétosphériques de **PP** sont beaucoup plus étendues — notamment en longitude — qu'il n'avait été d'abord envisagé.

Un autre aspect de la question est celui relatif à la comparaison des structures de **PP** observées en zones conjuguées. Dans le cas où ces structures ne sont pas trop compliquées — notamment quand la forme en «perles» n'est pas éclipsée par une superposition complexe de «grains» — il est possible de définir en chaque station, en dehors de la période propre de chaque oscillation, une «période de répétition» relative à la succession des perles les unes aux autres. Cette période de répétition apparaît aussi — et même plus facilement — dans les structures spectrales, ainsi que nous le verrons ultérieurement. Dans ces conditions on observe:

1) Que les périodes d'oscillation et les périodes de répétition sont à très peu près les mêmes aux deux zones conjuguées;

2) que les perles observées à une des zones (la Nord par exemple) se placent dans le temps, «juste» entre les perles observées à l'autre zone. (Le «juste» veut dire à mieux qu'une dizaine de secondes près, ceci pour des périodes de répétition de l'ordre de 100 sec.) Des observations relatives à des structures de **pi1** montrent que dans leur cas les structures aux régions conjuguées sont à peu près en coïncidence. Ces deux types d'observation devront être pris en considération dans l'examen des théories concernant les deux types de pulsations.

30. Structures spectrales des pc 1. Les analyses spectrales ont fait faire de grands progrès dans la connaissance des **pc1** et, plus spécialement, des **PP**. Les premiers résultats obtenus dans cette voie[1] ont été jugés si curieux qu'il a fallu un certain temps — et leur reproduction dans des conditions variées — pour que tous les intéressés soient persuadés que certaines de leurs particularités n'étaient pas imputables à des effets expérimentaux artificiels. Cette remarque concerne surtout les structures répétitives et périodiques des fluctuations de fréquences, que l'on sait maintenant être rattachées à la périodicité des perles elles-mêmes (observées alors sur leurs formes d'ondes). Un fait particulièrement intéressant, prolongement d'une remarque amorcée plus haut, consiste en ce que, quand la structure perlée de la forme d'onde est rendue confuse par la modulation supplémentaire introduite par les «grains», la structure du spectre conserve sa clarté. (C'est d'ailleurs grâce à cela que la nature exacte des grains a pu être déterminée.) D'autre part, en dehors de leur périodicité, l'ouverture «en éventail» des montées successives de fréquences, est une caractéristique des plus importantes de ces

[1] F. Glangeaud, J. Roquet et E. Selzer: Ann. Géophys. **24**, 871—877 (1968).
[2] F. Glangeaud, P. Gouin, J. Roquet, and E. Selzer: Bull. Geophys. Obs. Addis Ababa **12**, 67 (1968).
[3] E. Selzer et R. Stefant: Rap. National Fr. Ass. UGGI, 159—170 (1967).
[1] L. R. Tepley: J. Geophys. Res. **66**, 1651—1658 (1961).

structures spectrales car elle permet d'atteindre, par extrapolation, la fréquence limite supérieure qui pourrait théoriquement être atteinte pour les oscillations individuelles (celles des **PP** elles-mêmes), si le nombre des trajets allers-retours entre hémisphères des paquets d'ondes responsables pouvait augmenter indéfiniment et la localisation approximative du tube de forces (par son paramètre L) où doit se passer l'évènement principal.

Nous reviendrons sur cette question (voir Sect. 53 de cet article), mais pour l'instant, nous demanderons au lecteur de se familiariser, par l'examen des Fig. 46 à 54, avec les divers exemples de structures spectrales — obtenues en général au sonagraphe — rassemblées à cette intention. On pourra suivre sur ces exemples des cas plus complexes ou plus simples que le cas «typique moyen» donné par la Fig. 48, et se rendre compte que les éléments de base — sur lesquels s'appuieront les considérations théoriques — s'y retrouvent dans leur essence fondamentale.

31. Polarisations des pc 1. Du point de vue expérimental, les diagrammes de polarisation des **PP** demandent, pour être établis d'une façon correcte, des soins attentifs; sinon des erreurs de phase introduites par de légers glissements dans le temps de certaines composantes par rapport à d'autres (qu'elles appartiennent à un même enregistrement ou à des enregistrements différents) peuvent dénaturer complètement les résultats. Une précision relative dans le temps à 0,1 sec est à peine suffisante (pour des périodes d'oscillation de l'ordre de la seconde). Toutes les constantes de temps des appareils de transmission ou de reproduction doivent être éliminées (en valeurs relatives) à cette précision minimale. Ceci étant, on s'explique déjà un certain nombre de résultats contradictoires publiés sur la question. Cependant il paraît établi, notamment à la suite d'un grand nombre de comparaisons précises faites entre les stations Nord et Sud conjuguées de Sogra et de Kerguelen, que les hodographes de polarisation, en général elliptiques, relatifs aux **PP** y sont toujours décrits dans des sens opposés, en général: sens opposé à celui des aiguilles d'une montre le matin à Sogra, sens inverse l'après midi, et l'inverse — simultanément, au décalage près signalé plus haut de l'ordre d'une demi-période de répétition — à Kerguelen[1,2]. Les exceptions observées se sont produites en concordance aux deux stations, donc sans affecter l'inversion de sens entre les deux stations. Dans l'hypothèse d'une propagation par paquets d'ondes le long d'un tube de forces, à tout moment les ondes du paquet sont donc polarisées dans un sens unique mais ce sens serait susceptible de subir quelques renversements épisodiques brusques, sans compter le renversement régulier vers le milieu de la journée. Des observations récentes faites non plus entre stations conjuguées, mais entre Lerwick et Chambon-la-Forêt situés tous les deux dans l'hémisphère Nord, approximativement suivant le même méridien (latitudes géomagnétiques: $+62{,}5°$ et $+50{,}4°$, respectivement)[3,4], ont montré que les signaux (**PP**) qui arrivaient en différents points d'un plan méridien pouvaient présenter à un même instant des sens de polarisation différents, la comparaison des sens entre deux stations pouvant varier avec chaque évènement — ou même au cours de l'un d'entre-eux. Ceci conduit à penser que:

1) Le sens sous lequel on «voit», d'une station, tourner les ondes d'un paquet, est sous la dépendance — en plus du sens propre des ondes — d'un «effet de

[1] R. GENDRIN, M. GOKHBERG, S. LACOURLY et V. TROITSKAJA: C.R. Acad. Sci. **262**, 845—848 (1966).

[2] R. E. GENDRIN and V. A. TROITSKAJA: Radio Science, J. Res. Natl. Bur. Std. **69** D, 1107 (1965).

[3] F. GLANGEAUD: Ann. Géophys. **26**, 299—312 (1970).

[4] F. GLANGEAUD: Thèse Fac. Sci. Paris (1970).

perspective» (ou de tout effet analogue) déterminé par la position relative de l'onde (là où elle peut émettre de l'énergie en direction de la station) par rapport à celle-ci;

2) au cours de l'évolution d'un évènement cette position relative est susceptible de changer très rapidement, c'est-à-dire que le lieu de l'émission secondaire de l'onde captée par la station ne reste pas forcément fixe.

Indépendamment des questions de sens, en ce qui concerne *la forme* elle-même des diagrammes de polarisation (en général elliptique mais d'une excentricité plus ou moins élevée), *statistiquement* on se rapproche d'une polarisation circulaire aux stations telles que Lerwick que l'on peut considérer comme assez voisines des sources d'émission secondaire, ou de «pénétration», dans les guides d'ondes circum-terrestres, et d'une polarisation linéaire aux stations de basses latitudes. Ce résultat est bien de la nature de ce que l'on peut attendre d'une propagation dans ces guides, des hautes vers les basses latitudes.

32. Propagations des pc 1. Le problème complet de la propagation des **PP** (et, plus généralement des **pc1**, avec extension possible aux **pi1**) ne peut être envisagé qu'en fonction d'une présentation d'ensemble des mécanismes responsables de ces pulsations. Nous allons donc seulement donner ici quelques indications sur la façon dont on peut se préoccuper d'une «propagation au sol», apparente ou réelle. Des premiers résultats, obtenus soit suivant un méridien[1,2], soit perpendiculairement[3,4], on pouvait déjà déduire que, dans un cas comme dans l'autre les durées de propagation apparente n'excédaient pas quelques secondes, ceci pour des distances de l'ordre de quelques milliers de kilomètres. Depuis, des mesures plus précises ont pu être faites, soit par la méthode des «perles de culture»[5,6] (par laquelle, grâce à des opérations de filtrage, on arrive à opérer sur des fréquences pures), soit par la «méthode des grains» qui permet, en prenant un certain nombre de précautions[7,8,9], d'arriver aussi à des résultats assez précis, cohérents entre eux et en accord avec les précédents. On trouve ainsi que, pour cette propagation «au sol» (qui doit, en fait, prendre place dans un guide ionosphérique) les vitesses de groupe et de phase sont pratiquement confondues (pas de dispersion notable) et de l'ordre de 300 à 400 km sec^{-1}. L'imprécision dans ce genre de détermination vient non pas tant de l'incertitude des mesures, mais de celle sur les trajets suivis (y compris la position de la «source secondaire» d'émission).

D. Essai d'une morphologie spatiale des variations magnétiques rapides*.

Il n'a pas été facile de mettre au point de bonnes mesures magnétiques en satellite. Quand cela a été rendu possible — ne serait-ce que, en dehors de l'appareillage de mesures proprement dit, par la nécessité de disposer de satellites suffisamment amagnétiques — on s'est d'abord occupé, avec raison, de faire de la *cartographie magnétique spatiale*. Cette opération fondamentale est loin d'être terminée, ne serait-ce que par suite de diverses évolutions — plus ou moins

* Voir [*17*] et [*18*] comme référence d'ensemble.

1 L. R. Tepley, R. Heacock, and B. J. Fraser: J. Geophys. Res. **70**, 2720—2725 (1965).

2 R. N. Manchester: J. Geophys. Res. **71**, 3749—3754 (1966).

3 E. Selzer: Comm. Sympos. Belgrade (1966).

4 R. Schlich et J. Bitterly: Ann. Géophys. **23**, No. 3, 407—412 (1967).

5 R. C. Wentworth, L. Tepley, K. D. Amundsen, and R. R. Heacock: J. Geophys. Res. **71**, (5), 1492—1498 (1966).

6 L. Tepley and R. K. Landshoff: J. Geophys. Res. **71**, (5), 1499 (1966).

7 R. N. Manchester: J. Geophys. Res. **73**, 3549—3556 (1968).

8 F. Glangeaud: C.R. Acad. Sci. **268**, 113—116 (1969).

9 F. Glangeaud: Thèse, loc. cit. Sect. 31, (notamment les pages 65 à 70), (1970).

importantes en intensité, et plus ou moins rapide dans le temps — de cette cartographie. Nous ne faisons ici que la rappeler en raison de la base fondamentale que constitue pour toute étude sur des variations rapides du champ, les connaissances relatives au champ principal lui-même. C'est ainsi que Spoutnik III (1958), Lunik I et II (1959), Vanguard III (1959), Explorer VI (1960), et bien d'autres ensuite ont surtout contribué à préciser les différences entre les valeurs du *champ fondamental observées* et les *valeurs calculées* à partir de développements déduits de l'ensemble des valeurs mesurées au sol. Encore s'agissait-il là d'une mise en place — et d'une rectification — dans un cadre général plus ou moins prévu. Des informations plus inattendues furent ensuite fournies par les résultats de Pioneer V (sur le champ interplanétaire), par Explorer X et XII (sur l'existence d'une «magnétogaine» — ou «magnetosheath»), par IMP I (mesures du champ sur les frontières mêmes de la magnétosphère, premières déterminations relatives à la queue de la magnétosphère et à l'existence d'une zone neutre: «neutral sheet»), etc. ... Nous ne reviendrons sur les résultats de ces explorations que dans la mesure où ils concernent également plus directement notre sujet, ce qui est le cas — entre d'autres — des trois derniers satellites que nous venons de nommer.

33. Résultats d'Explorer X. En fait, les premiers enregistrements spatiaux satisfaisants de *variations* magnétiques ont été ceux d'Explorer X entre les 25 et 27 mars 1961. Rappelons que ce satellite avait reçu un équipement très complet. En dehors des dispositifs prévus pour les mesures de plasma, qui ne nous intéressent qu'indirectement, il était muni:

1) D'un magnétomètre à vapeur de rubidium permettant des mesures du module du champ total ainsi que de ses composantes vectorielles (ceci, grâce à l'intervention de champs auxiliaires);

2) de sondes à saturation (du genre «flux-gates»), assurant des mesures vectorielles et étalonnées périodiquement au moyen de l'équipement précédent.

Résumons l'essentiel des résultats donnés par ce satellite:

1) Les mesures transmises permirent de vérifier l'existence d'une frontière entre le domaine du champ terrestre et le domaine interplanétaire. La délimitation exacte de cette frontière resta toutefois mal déterminée, la trajectoire du satellite lui ayant été presque parallèle pendant près de 24 heures, et plusieurs «sorties» ou «entrées» ayant été ainsi probablement enregistrées. L'interprétation actuelle (depuis, notamment, les mesures assurées par Explorer XIV et Explorer XVIII, alias IMP I, en 1962/63) fait intervenir la zone de transition turbulente, «magnétogaine» ou «magnetosheath», mentionnée plus haut. Nous retiendrons essentiellement de ces explorations qu'elles précisèrent l'existence d'une zone interne à champ magnétique stable (tout au moins en l'absence de grand orage) et à faible densité de plasma, et d'une zone externe à champ instable, turbulent, et à forte densité de plasma.

2) Il est intéressant de comparer les fluctuations enregistrées par Explorer X, avec celles relevées à l'échelle mondiale par les stations magnétiques au sol. Les résultats d'une telle comparaison ont été donnés sous forme graphique par Heppner et al.[1] apportant la preuve d'une corrélation «globale»; c'est-à-dire que chaque décrochement ou variation brusque enregistrés à bord du satellite correspond à un évènement particulier observé au sol, mais sans qu'il soit possible de pousser une telle concordance jusqu'à la forme ou la durée individuelle des phénomènes.

Même sujette à une telle restriction, la remarque précédente nous apparaît comme très importante pour une théorie d'ensemble des variations magnétiques. Comme elle avait été faite, du côté des variations au sol, à partir d'enregistrements lents («normaux») et semi-rapides, il était intéressant de l'étendre aux enregistre-

[1] J. P. Heppner, M. Sugiura, T. L. Skillman, B. G. Ledley, and M. Campbell: NASA Goddard Space Flight Center-Report X-612-67-150 (1967).

ments rapides. C'est ainsi qu'il a été possible de vérifier[2] que la plupart des décrochements enregistrés par les magnétomètres embarqués sur le satellite correspondaient à des trains de pulsations **pt** sur les enregistrements de Chambon-la-Forêt, ceci quand l'heure locale de l'évènement le permettait. Citons, à titre d'exemple, le décrochement de 2203 TU du 26 mars 1961. Cette constatation ne peut cependant pas être étendue — en l'état présent de la technique des mesures spatiales — aux oscillations individuelles composant le phénomène. Elle va à l'encontre des théories suivant lesquelles les baies, ainsi que les microstructures qui les accompagnent, seraient d'origine essentiellement ionosphérique.

34. Correspondance Sol-Magnétosphère. Des correspondances analogues entre évènements brusques au sol et dans la magnétosphère — notamment entre des débuts brusques de baies (accompagnées de leurs microstructures) et les «éruption d'électrons» détectés par satellites dans la queue de la magnétosphère, au voisinage de sa zone neutre — ont été signalés par divers auteurs[1]. On a ainsi songé à rechercher l'origine des baies — et des aurores qui leur sont associées dans les régions de hautes latitudes — dans ces éruptions. Cependant, une mesure précise des instants relatifs aux débuts des deux types de phénomènes a montré que, dans la plupart des cas, l'évènement à électrons était légèrement postérieur — et non pas antérieur — au début du phénomène terrestre. De ce fait les premières interprétations proposées sont à réexaminer. D'autres auteurs[2,3] ont étudié les liens qui apparaissaient entre les **IPDP** observés au sol et les perturbations magnétosphériques suivantes: les variations de flux des particules dans la zone de Van Allen; les variations de champ magnétique enregistrées par divers satellites, notamment IMP I (1963), Electron II (1964) et Explorer XXVI (1964). Ils ont pu montrer ainsi que:

1) Les **IPDP** apparaissent au cours d'une brusque diminution du flux des électrons piégés et plus la diminution de flux est importante, plus la fréquence finale de l'**IPDP** est élevée;

2) l'apparition des **IPDP** est liée à une augmentation de l'intensité du champ total dans la queue de la magnétosphère et à une diminution du champ au voisinage du maximum de la zone de piégeage; de plus, des oscillations hydromagnétiques de grande amplitude apparaîtraient aux environs de cinq rayons terrestres.

Parmi les satellites relativement récents équipés pour des mesures magnétiques précises, la plupart ont continué l'œuvre importante d'exploration spatiale du champ fondamental («cartographie magnétique»), notamment avec la série des Explorer (N° XXIV et suivants) dans la vaste région occupée par la queue de la magnétosphère. Parmi les résultats qui intéressent plus directement notre sujet il convient de signaler la détection par Mariner IV d'un évènement temporel paraissant lié à un **SSC** enregistré aux stations terrestres[4].

La variation observée alors que le satellite se trouvait en plein espace interplanétaire (et non à l'intérieur de la cavité magnétosphérique) s'est présentée cette fois sous une forme rappelant les **SSC** enregistrés aux stations terrestres. Il n'en est pas toujours ainsi: le **SSC** de l'orage du 8 juillet 1966, enregistré par

[2] E. Selzer: Rapport Général sur la participation française aux A.I.S.C., présentation F. du Castel — Comité Français des A.I.S.C., Paris (1967).

[1] L. J. Cahill, Jr.: J. Geophys. Res. **71**, (19), 4505 (1966).

[2] R. Gendrin et S. Lacourly dans: A. Egeland et J. Holtet (eds.): The Birkeland Symposium on aurora and magnetic storms, p. 275—289. Paris: édit., CNRS 1968.

[3] R. Gendrin, S. Lacourly, V. A. Troitskaja, M. Gokhberg, and R. V. Shepetnov: Planetary Space Sci. **15**, (8), 1239—1259 (1967).

[4] J. J. O'Gallagher and J. A. Simpson: Phys. Rev. Letters **16**, 1212 (1966).

les satellites IMP III et Explorer XXXIII, qui se trouvaient alors en dehors de la magnétosphère, par le satellite OGO III, situé dans la queue magnétique à environ 30 rayons terrestres au-delà de la Terre, et enregistré également mondialement au sol (à 0902.15 TU $\pm$ 5 sec), a pu fournir d'intéressantes remarques à ce sujet; elles ont trait non seulement aux différences, relativement importantes, constatées sur les instants auxquels en chacun de ces lieux ce **SSC** s'est manifesté (près de 3 min de retard à l'extérieur de la magnétosphère par rapport au sol dans un même plan de front perpendiculaire à la direction générale Soleil-Terre de propagation, alors que ce retard n'était que de l'ordre de $^1/_2$ min pour Ogo III), mais — ce qui nous intéresse plus spécialement ici — sur le caractère beaucoup moins abrupt de ce **SSC** dans l'espace interplanétaire que dans la magnétosphère. Il y apparaît comme une simple augmentation très progressive du champ (étalée sur une durée de 5 min) alors que le **SSC** au sol se présente comme un décrochement rapide. On peut voir dans cela la transformation d'une onde progressive en une onde de choc lorsqu'on passe d'un espace libre à un espace confiné. Nous retiendrons de ces exemples — ainsi que de plusieurs autres cas que nous ne pouvons détailler ici — que la relation précise entre un **SSC** observé au sol et une modification brusque se manifestant, soit dans la magnétosphère, soit dans l'espace interplanétaire, est loin d'être bien élucidée. Par contre-coup, le rôle de l'ionosphère elle-même ne peut être encore bien connu (puiqu'on ne sait pas exactement comment lui parvient l'excitation). On voit là une infériorité manifeste — malgré leur avantage d'opérer in-situ — des enregistrements en satellite comparés à ceux des réseaux mondiaux de stations au sol: ils ne donnent, à tout instant *déterminé*, qu'une information très localisée.

35. Observations spatiales d'ondes et de pulsations magnétiques. La même remarque peut s'appliquer aux *observations spatiales d'ondes ou de pulsations magnétiques* qui soulèvent des difficultés expérimentales encore beaucoup plus grandes que celles relatives aux observations précédentes. Les deux principales de ces difficultés sont:

1) La grande sensibilité et la grande vitesse d'enregistrement requises (avec comme corollaire une finesse bien adaptée pour la transmission par télémesure au sol);

2) la nécessité de séparer — pour toute interprétation complète — les structures (périodiques ou non) dues à l'espace traversé, de celles se déroulant dans le temps. Pour séparer valablement l'influence de ces deux variables des mesures simultanées par plusieurs satellites sont nécessaires. Encore faut-il procéder à une discussion serrée de leurs positions relatives: ni trop près ni trop loin (dans ce dernier cas ils risquent de se trouver dans des régions de l'espace peu comparables). Un exemple important des difficultés rencontrées est relatif au cas des satellites qui, sans inversion de leur sens de parcours (s'éloignant, disons, de la Terre) ont traversé un grand nombre de fois les surfaces de séparation magnétosphère — magnétogaine (la magnétopause) et magnétogaine — espace interplanétaire (onde de choc stationnaire). L'explication, maintenant bien établie, est qu'ils se sont fait rattraper plusieurs fois par ces surfaces frontières, qui peuvent être animées de mouvements — dans l'un ou l'autre sens — beaucoup plus rapides que les satellites. Dans ce cas bien précis on sait actuellement reconnaître (par la forme générale des variations magnétiques enregistrées) dans quelle région se trouve le satellite et ne pas se laisser induire en erreur par ces traversées de frontières. On tiendra compte également de ces connaissances dans l'interprétation plus générale des variations temporelles observées. C'est ainsi que nous donnerons brièvement les résultats les plus probants obtenus au cours

de ces toutes dernières années sur les ondes et variations magnétiques observées dans l'espace interplanétaire, dans la magnétogaine, dans la magnétosphère (côté éclairé, côté nuit et queue).

α) *Espace interplanétaire.* Des mesures simultanées de champ et de densité de plasma ont apporté quelques éclaircissements à notre problème: en général assez calme, le champ magnétique interplanétaire (dont on admet que la partie principale émane du soleil et a la structure en spirale, liée à celle du plasma, proposée par PARKER[1]) paraît être le siège d'ondes se propageant dans les divers modes hydromagnétiques (modes couplés) dont on a relevé les caractéristiques spectrales. Les résultats les plus probants ont été obtenus avec Mariner II[2] en 1962, et avec Pioneer VI[3] en 1965. Dans ce dernier cas, des oscillations quasi sinusoïdales ont été mises en évidence ayant une *période apparente* d'environ 5 min. Ainsi que nous l'avons indiqué plus haut une interprétation plus poussée demande que l'on se préoccupe des divers référentiels susceptibles d'intervenir dans une telle observation, y compris le choix d'un d'entre-eux auquel serait plus directement lié le mécanisme du phénomène, et auquel on désirerait le rapporter pour cette raison. Mais ce choix lui-même peut répondre à diverses hypothèses. C'est ainsi que l'on peut supposer que les ondes observées se propagent «portées» par la structure fondamentale, tournante, des spirales de PARKER, mais on ne peut affirmer que — quelle que soit la distance au soleil — cette structure tourne d'une vitesse angulaire uniforme égale à celle du soleil lui-même. Remarquons à ce sujet que la critique que nous avons faite aux mesures en satellite — d'opérer dans un complexe spatio-temporel non séparable — est aussi, dans une certaine mesure, applicable aux mesures faites au sol. Mais cette critique est habituellement masquée par le fait que nos interprétations elles-mêmes sont alors également conduites dans un référentiel lié à notre Globe. Nous résumerons cette discussion en disant que dans ce genre d'expérience il faut non seulement déterminer tous les paramètres, mais aussi savoir choisir en fonction de ce que l'on cherche (si c'est possible) ceux que l'on considèrera comme fondamentaux.

β) *Magnétogaine**. Les mesures y ont été nombreuses (en tant que complément des déterminations sur les positions des frontières qui la limitent: Magnétopause; Onde de choc stationnaire) notamment au moyen des satellites Mariner IV, 1964/65[4], Ogo I, 1965[5–7] et Vela III, 1965[8]. La plupart de ces mesures ont indiqué la présence quasi permanente de fluctuations irrégulières ayant plus ou moins les caractéristiques d'un bruit et couvrant une large bande de fréquence. Certaines de ces fluctuations auraient tendance à être tourbillonnaires. D'autre part, un certain alignement parallèlement à la magnétopause se manifesterait au voisinage de celle-ci. Les mesures de Vela III se distinguent des autres par la mise en évidence, à certains moments, parmi cette agitation irrégulière, de trains de pulsations sinusoïdales pouvant durer quelquefois plusieurs minutes, ceci dans des bandes de fréquences privilégiées (correspondant notamment à des périodes de 10 ... 15,

* Traduction française du terme anglais «Magnetosheath».

[1] Voir l'article de H. POEVERLEIN dans ce volume, p. 7.

[2] P. J. COLEMAN, JR.: Rep. Inst. Geophys. Planet, Phys., Univ. Calif., No. 501 (1966); No. 504 (1966).

[3] N. F. NESS, C. S. SCEARCE, and S. CANTARANO: J. Geophys. Res. **71**, (13), 3305 (1966).

[4] J. L. SISCOE, E. J. SMITH, L. DAVIS, JR., P. J. COLEMAN, JR., and D. E. JONES: Trans. Amer. Geophys. Un. **47**, 143 (1966).

[5] R. E. HOLZER, M. G. MCLEOD, and E. J. SMITH: J. Geophys. Res. **71**, (5), 1481 (1966).

[6] J. WOLFE, R. W. SILVA, and M. A. MYERS dans: Space Research IV (R. L. SMITH-ROSE, ed.). Washington: Spartan Books 1966.

[7] F. L. SCARF, J. H. WOLFE et R. W. SILVA: Comm. Sympos. URSI-IAGA, Belgrade (1966).

[8] E. W. GREENSTADT, G. T. INOUYE, I. M. GREEN, and D. L. JUDGE: Space Sci. Lab., TRW Systems Rept, 99900-6070-R000 (1966).

20 ... 25 et 30 ... 40 sec. Bien qu'il soit tentant d'établir un lien entre ces dernières observations et les pulsations **pc3** enregistrées au sol — dont on aurait trouvé ainsi la source directe d'excitation — aucune corrélation directe de cet ordre n'a pu encore être établie. Par ailleurs, il est intéressant de noter que les spectres de répartition énergétique déduits des mesures dans la magnétogaine donnent à peu près la même loi de répartition que ceux établis pour divers paramètres (notamment les fluctuations de vitesses du vent solaire) relatifs à l'espace interplanétaire. Une loi analogue a été retrouvée pour les résidus d'agitation magnétique (ayant la structure d'un bruit) toujours présents dans les enregistrements de variations magnétiques obtenus au sol[9].

γ) *Magnétosphère.* Ainsi que pour les autres régions les premiers résultats bien établis l'ont été pour des fluctuations du champ (ou «pulsations») d'assez longue période — en général de plusieurs minutes — ceci pour des raisons expérimentales déjà exposées. D'autre part, qu'il s'agisse de la magnétosphère de jour ou de celle de nuit, quand on a cherché à retrouver sur les enregistrements au sol l'équivalent de ce que l'on avait observé dans l'espace, on a eu, en général, des résultats positifs pour les stations au sol ayant grosso modo la même position méridienne que celle qu'avait le satellite au moment de l'évènement. Ceci apparaît quelque peu en contradiction avec l'extension à prédominance mondiale, ou semi-mondiale, que nous avons relevée pour un grand nombre de variations magnétiques entre stations au sol, et laisserait supposer que les fluctuations observées dans l'espace appartiendraient essentiellement aux catégories de variations (au sol) dont l'extension en longitude est — contrairement au cas le plus répandu — assez réduite (cas des **pg** par exemple). Inversement, les variations dont l'extension mondiale au sol est la plus grande, seraient plus difficilement observables dans l'espace. En fait, c'est bien dans la gamme des **pg** qu'ont eu lieu le plus grand nombre d'observations, notamment avec la série des Pioneer[10,11] et celle des Explorer — plus particulièrement Explorer XII, 1961[12,13] et Explorer XIV, 1962/63[14]. Le résultat le plus remarquable a été l'observation de pulsations de caractère hydro-magnétique, dans un mode transversal pour des périodes de l'ordre de 200 sec et un mode longitudinal pour des périodes plus courtes (100 sec environ). Dans le cas d'Explorer XII les pulsations transversales furent observées à 55000 km du sol et furent retrouvées au sol — avec un retard de 1,5 min environ — à une des extrémités de la ligne de forces correspondante; leur amplitude n'y était que légèrement diminuée par rapport à celle observée dans l'espace (6 ... 8 γ) ce qui n'est pas surprenant pour les périodes relativement longues de ces pulsations.

Les observations faites par Ogo II[15] permirent de descendre dans l'échelle des périodes (c'est-à-dire de monter en fréquence avec une bande de 0,01 à 1000 Hz). Entre 400 et 1500 km d'altitude Ogo II a enregistré des émissions du type «éruptif» quand il passait au voisinage des zones aurorales. On peut rapprocher de ces observations celles faites au moyen du satellite «1963-38-C»[16] lancé sur

---

[9] F. GLANGEAUD: C.R. Acad. Sci. **264**, 553—556 (1967).

[10] D. L. JUDGE and P. J. COLEMAN, JR.: J. Geophys. Res. **67**, (13), 5071 (1962).

[11] C. P. SONETT, A. R. SIMS, and I. J. ABRAMS: J. Geophys. Res. **67**, (4), 1191 (1962).

[12] V. L. PATEL: Planetary Space Sci. **13**, 485 (1965).

[13] V. L. PATEL and L. J. CAHILL, JR.: Phys. Rev. Letters **12**, 213 (1964).

[14] L. J. CAHILL, JR.: in "Space Physics" (D. P. LE GALLEY et A. ROSEN, eds.), p. 301. New York: John Wiley & Sons 1964.

[15] J. C. CAIN, R. A. LANGEL, and S. J. HENDRICKS: Publication X-612-66-305 du NASA-Goddard Space Flight Center (1966).

[16] A. J. ZMUDA: Cité par H. B. LIEMOHN dans la publication D1-82-0890 — Seattle (USA): Boeing Sci. Res. Lab. p. 40 (1969).

une orbite polaire presque circulaire (altitude variant peu et de l'ordre de 1100 km): une bonne corrélation fut trouvée entre les perturbations enregistrées par le satellite (fluctuations de nature magnéto-dynamique, transversales) et les perturbations de la zone aurorale, au sol, correspondant à peu près aux mêmes lignes de forces. Les perturbations étaient observées principalement au-dessus de l'«ovale auroral» et la zone correspondante se déplaçait vers les latitudes moins élevées quand l'agitation magnétique générale (mesurée, par exemple, par l'indice K_p) augmentait, comme cela se passe au sol.

36. Résultats les plus récents et premières conclusions. Les résultats les plus récents sur l'observation dans l'espace de fluctuations magnétiques rapides ont été obtenus au moyen de sondes à induction de grande sensibilité ce qui a permis — ainsi que nous l'avions signalé pour Ogo I et Ogo II — de monter dans l'échelle des fréquences. La lacune la plus importante dans ces observations paraît concerner la gamme des **pc1** et celle des **pc3**. Un enregistrement correct, dans l'espace, de pulsations structurées du type « perles » reste un problème expérimental très difficile.

La conclusion générale de cet examen est que, pour l'instant, les données par satellite restent d'un ordre général: elles confirment que la plupart des fluctuations magnétiques rapides observées au sol peuvent être rattachées à des évènements décelables dans les régions élevées de la magnétosphère, ou même dans l'espace interplanétaire, mais il ne faut pas encore compter sur des corrélations précises qui nous permettraient de décider entre diverses hypothèses avancées pour les origines de ces phénomènes.

Par rapport aux idées admises avant ces observations en satellites, nous noterons parmi les changements les plus importants:

1) Qu'il est peu de cas où une origine purement ionosphérique (à fortiori, sous-ionosphérique) puisse être admise pour les fluctuations observées au sol. L'ionosphère intervient considérablement, mais en tant que facteur de transmission, de modulation, ou d'amplification (par exemple pour certaines catégories de pulsations équatoriales amplifiées par l'électrojet[1,2].

2) qu'il n'est pas possible de distinguer clairement, parmi les variations enregistrées au sol, celles qui ont leur source directe dans le plasma solaire (émissions corpusculaires ou électromagnétiques, solaires), et celles qui seraient dues à des actions prenant place dans les zones de transition situées entre l'espace interplanétaire et la magnétosphère.

Divers points importants restent en suspens:

a) La contribution des nappes de courants électriques circulant dans l'ionosphère. Les densités de ces courants ne sont pas mesurables facilement, bien que l'emploi de fusées puisse apporter ici un concours efficace aux satellites.

b) Le rôle joué par les zones de radiation (telles que les ceintures de Van Allen), bien que quelques résultats nouveaux reprennent cette question.

c) Celui joué par les zones de turbulence qui peuvent exister au voisinage de la «zone neutre» («neutral sheet»), à l'intérieur de la queue de la magnétosphère.

37. Observations par Ogo III et Ogo V. Alors que la rédaction de ce chapitre de notre article était terminée, des résultats préliminaires relatifs aux observations faites par les satellites Ogo III et Ogo V viennent d'être publiés[1]. Pour le première fois, répondant ainsi

[1] J. Roquet: Comm. 2ème Symp. Aéronom. Equator., Sao-Paulo, Sept. 1965. — Ann. Géophys. **22**, 508 (1966).

[2] J. Roquet: Comm. Coll. Phys. Soleil-Terre, Belgrade, Août 1966. J. Atm. Terrest. Phys. **29**, 453 (1967).

[1] J. P. Heppner, B. G. Ledley, T. L. Skillman, and M. Sugiura: Publication X-612-69-429 du NASA-Goddard Space Flight Center (1969).

à certaines des questions importantes que nous nous posions, une vue d'ensemble de la distribution des pulsations dans toute la gamme couvrant les périodes comprises entre 0,8 et 240 sec, est donnée à l'échelle de la magnétosphère. Nous allons résumer les résultats très importants déduits de ces observations, même si pour certains d'entre-eux des confirmations précises doivent être données ultérieurement:

1) La question de la distribution spatiale des **pc 3** paraît résolue. Ces pulsations ont été enregistrées à bord des satellites avec une grande netteté *à toute altitude* dans la magnétosphère côté Jour (au voisinage du plan équatorial). Elles y ont la même apparence — et paraissent suivre les mêmes lois de variation diurne — que les pulsations correspondantes au sol avec lesquelles on a pu, dans un certain nombre de cas, les comparer directement. On peut déduire de ces observations que les **pc 3** doivent — pendant toutes les durées de leur manifestation au sol — correspondre à une vibration d'ensemble de toute cette magnétosphère de Jour, depuis la magnétopause jusqu'à l'ionosphère et — après le «relai» assuré par celle-ci — jusqu'au sol. Les satellites ont observé cependant un maximum d'amplitude au voisinage de $L = 5$ (correspondant à la latitude sub-aurorale pour laquelle l'amplitude est maximale également pour les **pc 3** au sol) et c'est surtout entre la «coquille magnétique» correspondante et le sol que les **pc 3** se conserveraient très semblables à eux-mêmes. Au-dessus, et plus particulièrement quand on se rapproche de la magnétopause et — à fortiori — quand on pénètre dans la magnétogaine, la présence de ces **pc 3** continue à être détectée mais leur correspondance directe avec les **pc 3** au sol n'est pas encore bien établie. En particulier, la confirmation très nette des résultats de Vela III précédemment cités[2] sur la présence quasi habituelle de pulsations — ou de suites d'impulsions — avec une période dans la gamme des **pc 3**, qui a été ainsi confirmée dans la magnétogaine, ne nous apprend pas encore s'il faut y voir la vraie source des **pc 3** observées dans la magnétosphère et au sol ou si — au contraire — l'entretien de cet état pulsatoire de la magnétogaine ne viendrait pas de l'intérieur. Il ne serait pas contradictoire de supposer que les échanges d'énergie se font, à travers la magnétopause, dans le sens *externe vers l'interne* pour tout ce qui a trait à l'agitation erratique, aléatoire (dont la présence dans la magnétogaine a été également confirmée — s'il était besoin encore — par les observations de Ogo III et Ogo V), et dans le sens *interne vers l'externe* pour ce qui concerne les oscillations sinusoïdales régulières. Ceci permettrait de conserver le rôle des cavités résonantes de la magnétosphère — alors qu'il semble difficile d'en concevoir de convenables dans la magnétogaine — sans pour cela inverser le bilan total des échanges énergétiques (qui resterait ainsi dans le sens externe-interne).

2) En ce qui concerne les **pc 4**, également retrouvées dans la magnétosphère, elles y seraient distribuées moins uniformément que les **pc 3** avec prédominance relative des coquilles de L plus élevés (plus près de la magnétopause).

3) En ce qui concerne les pulsations irrégulières de toutes périodes, leur distribution à un instant déterminé semble n'affecter que des régions d'étendues limitées, situées, dans l'ensemble, soit du côté des instabilités en provenance de la queue de la magnétosphère (côté nuit), soit dans les parties les plus élevées de la magnétosphère côté jour.

4) Enfin, les observations relatives aux périodes les plus courtes (jusqu'à 0,8 sec), ont montré la difficulté qu'il y a à essayer de distinguer, dans l'espace, et dans cette gamme, les fluctuations régulières des fluctuations irrégulières (par exemple: les **pc 1** des **pi 1**). Toutes paraissent irrégulières (tout en étant situées au voisinage des coquilles prévues par les théories actuelles: valeurs de L élevées, maximum à l'aube ou dans la nuit). On peut se demander dans quelle mesure il n'y a pas là une vérité première (au lieu d'un effet des difficultés expérimentales) et si la distinction quasi absolue que nous faisons au sol entre les **pc 1** et les **pi 1** n'est pas simplement l'effet — en ce qui concerne les premières — d'une simple «régularisation», ou d'une mise en forme structurée particulière, qui pourrait être due aux divers processus de propagation ou de transmission. L'association fréquente de **pi 1** à des **pc 1** relevée (en synchronisme ou pas) lors de certains évènements au sol (tels que les **IPDP**) milite en faveur d'une telle interprétation.

E. Essai d'une présentation synthétique d'une théorie des pulsations magnétiques.

38. Introduction. On remarquera que le titre ci-dessus est restrictif: nous ne pensons pas pouvoir nous engager dans une théorie de l'ensemble des variations magnétiques rapides et nous laisserons de côté tout examen, en propre, de ce que nous avons appelé les «microstructures de phénomènes magnétiques macroscopiques» (orages, baies, etc. ...). Ceci parce

[2] E. W. GREENSTADT, G. T. INOUYE, I. M. GREEN, and D. L. JUDGE: Space Sci. Lab., TRW Systems Rept, 99900-6070-R000 (1966).

que sous cette forme leur théorie fait corps avec celle de la perturbation qu'ils accompagnent. Par contre, dans la mesure où ces microstructures peuvent se décomposer (ou être décomposées par une analyse adéquate) en éléments pulsationnels, il sera possible — mais non sans précautions — de leur appliquer certaines des théories avancées pour les pulsations correspondantes.

Une remarque d'un autre genre est que le champ magnétique terrestre, tel qu'il est défini dans son *état statique moyen* (celui que l'on appelle souvent le *champ principal moyen*) par sa donnée première (aimantation du noyau terrestre se traduisant globalement — en première approximation — par le moment magnétique de la Terre), et par un nombre réduit (en se limitant encore à une première approximation) de *conditions aux limites* et de *contraintes*, possède un ensemble de *propriétés dynamiques* qui lui sont propres. C'est ainsi — qu'avec un petit nombre de restrictions que nous allons préciser — on peut concevoir une dynamique de ce système ainsi stylisé et presque intégralement réduit à son entité magnétique. La restriction fondamentale — et très simple — qu'il convient de formuler à ce sujet, est l'obligation de tenir compte de l'*inertie mécanique* qui s'attache à un tel système à côté de son inertie magnétique pure. Il nous faudra donc faire état d'une certaine *densité de masse* du milieu, soit ϱ, dans la mesure où l'hypothèse du «champ figé dans la matière» d'un plasma très conducteur lie la dynamique de cette masse à la dynamique du champ magnétique. De même, nous aurons à tenir compte des apports d'énergie (sous forme cinétique, thermique, de radiations, etc. ...) qui — hors son énergie magnétique propre — pourront être considérés comme des liaisons externes de notre système magnétique. Bien entendu, il ne s'agit pas de traiter ainsi la physique complète d'une partie, ou du tout, de l'environnement terrestre, mais de présenter — à titre de premier schéma — un modèle de base. Par l'introduction de facteurs physiques nouveaux (au premier plan, les propriétés du plasma magnétosphérique) ce modèle initial pourra, ensuite, être rendu plus réel.

39. Rappel des données statiques et des contraintes définissant l'état moyen du champ magnétique terrestre. Sans revenir sur les données élémentaires (champ d'un dipôle, ou champ d'une sphère uniformément aimantée, moment magnétique total de l'ordre de $8 \cdot 10^{22}$ Am^2), ni sur les développements précis donnés par l'analyse sphérique harmonique pour le champ au sol et — en altitude — jusqu'à 3 ou 4 rayons terrestres, nous retiendrons des connaissances actuelles que le champ principal qui nous intéresse ici a son origine dans le noyau de la Terre, un milieu semi-liquide doué apparemment d'une conductibilité très élevée. Ceci nous permet de considérer les lignes de forces issues de ce milieu comme parfaitement «ancrées» — sans glissement possible, dans ce noyau.

A l'autre extrémité de l'environnement terrestre, ces mêmes lignes de forces, soumises à divers types d'interaction avec le milieu interplanétaire (dont la principale, et la plus permanente, est celle avec le vent solaire) sont, soit recourbées dans la partie semi-close de la magnétosphère, soit allongées suivant la queue magnétosphérique, avec connexion possible avec les champs magnétiques externes.

Afin de pouvoir se rendre compte — même très schématiquement — des propriétés «magnéto-mécaniques» du système ainsi défini, il importe:

1) D'examiner d'un peu plus près sa constitution interne: ses zones particulières, là où la présence de concentrations de plasma plus denses — ou, au contraire, l'absence totale de plasma — détermine des hétérogénéités dont il conviendra de discuter l'influence sur notre problème.

2) D'étudier les frontières (magnétopause) et les «zones-frontières» (telle que la magnétogaine) et les échanges d'énergie qui s'y effectuent.

La première de ces questions doit tenir compte des effets (passifs ou actifs) de l'ionosphère, des propriétés propagatives de la cavité Terre-Ionosphère, des effets de «volant énergétique» susceptibles d'être assumés par les ceintures de radiation (VAN ALLEN), par la zone neutre et les concentrations d'électrons mises en évidence dans la queue, etc. ... Elle doit comprendre également la définition de cavités résonnantes plus ou moins distinctes, ainsi que l'examen des modes de couplage s'exerçant entre ces cavités. Le long des frontières également l'étude des couplages est primordiale.

D'un autre point de vue, notre magnéto-mécanique demande à être précisée:

Dans la plus grande partie de la magnétosphère nous admettrons que les conditions de l'approximation «magnéto-dynamique» (appelée très souvent «hydro-magnétique», ou «HM» dans la littérature) sont suffisamment réalisées pour pouvoir être prises comme première base de présentation. Dans la mesure où cela s'avèrera insuffisant, ou non satisfaisant, nous introduirons les autres entités physiques indispensables (par exemple, un champ électrique; des averses de particules suivant — ou ne suivant pas — l'approximation adiabatique; etc. ...). Autrement dit, prenant acte de ce que dans l'état actuel de nos connaissances aucune théorie cohérente n'a pu encore être établie — sauf, dans une certaine mesure, pour les **pc1** — nous partirons des notions et explications les plus simples, ne les compliquant qu'en fonction des nécessités rencontrées. C'est ainsi que nous allons commencer notre examen en rappelant — sous une forme quelque peu nouvelle — les états de tension mécanique que l'on doit trouver dans un champ magnétique pur.

Tensions internes dans un champ magnétique. Les résultats sont tout à fait classiques: pour la portion de tube de forces figurée (voir Fig. I), le milieu interne (tube) exerce sur le milieu qui l'entoure une pression latérale (orientée vers l'extérieur) de valeur

$$p = \frac{B^2}{2\,\mathrm{u}\,\mu_0}$$

et une tension normale aux sections droites (orientée vers l'intérieur) de même valeur absolue.

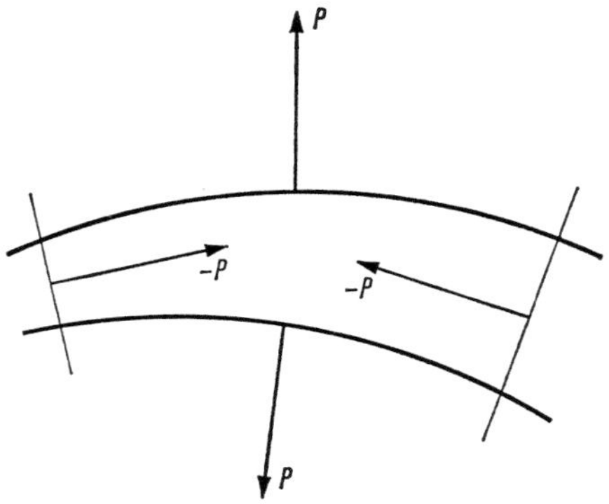

Fig. I. Actions exercées par la portion du tube considéré sur le milieu qui l'entoure.

μ_0 perméabilité du vide, B induction magnétique. $\mathrm{u} = 1$ dans un système d'unités rationalisé (p.e. SI), mais $\mathrm{u} = 4\pi$ dans un système non-rationalisé (comme p.e. celui de GAUSS).

Malgré sa banalité nous allons donner quelques explications sur l'établissement de ce résultat afin de bien préciser sa relation — et, plus généralement, la relation des forces de liaison intervenant aux frontières d'un système — avec le comportement de la densité d'énergie inclue, à tout moment, dans ce système. Les valeurs des forces de liaison sont, dans la plupart des cas, proportionnelles à celle de la densité d'énergie, mais les coefficients de proportionnalité ne sont égaux à l'unité que dans des cas particuliers, leur diversité dans les autres cas définissant les caractéristiques du tenseur auquel on peut ramener la distribution de ces forces.

40. Relations entre les densités d'énergie et les forces de liaison.

α) Nous raisonnerons dans un cas assez général: Considérons une petite portion interne de volume V d'un milieu quelconque supposé dans un certain état statique de tension (voir Fig. II). Identifions cette portion de ce milieu avec un «système interne» Σ dont nous nous proposons d'examiner les *échanges d'énergie avec le milieu extérieur*. (V aura été choisi assez petit pour que le milieu puisse être considéré comme homogène en ce qui concerne les propriétés assurant les tensions envisagées.)

Soit M un point de la surface S limitant Σ, et $\mathrm{d}S$ — représenté par son vecteur normal $\boldsymbol{n}^0 \mathrm{d}S$ — un petit élément (plan) de cette surface en M ($\boldsymbol{n}^0$ vecteur unitaire orienté vers l'extérieur).

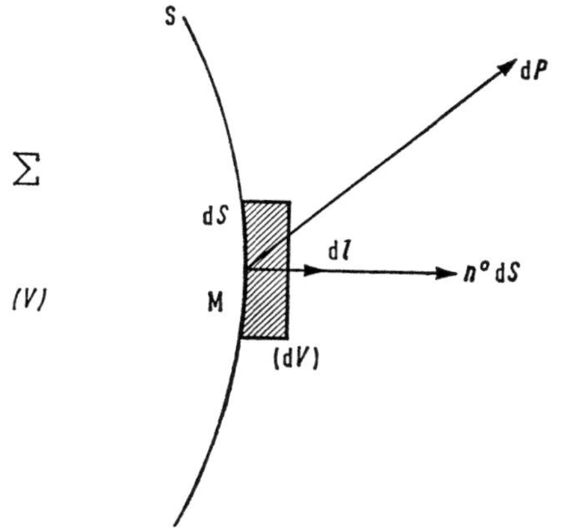

Fig. II. Relations entre les densités d'énergie et les forces de liaison (voir texte).

Soit $\mathrm{d}\boldsymbol{P}$ la force élémentaire de liaison exercée par Σ sur le milieur externe, à travers $\mathrm{d}S$. On peut écrire:

$$\mathrm{d}\boldsymbol{P} = F(\boldsymbol{n}^\circ)\ \mathrm{d}\boldsymbol{S} = F_{\boldsymbol{n}}\,\mathrm{d}\boldsymbol{S}.$$

$F(\boldsymbol{n}^\circ)$, écrit plus simplement $F_{\boldsymbol{n}}$, étant une fonction vectorielle caractéristique des propriétés du milieu en M.

Soit $\mathrm{d}\boldsymbol{l}$ un déplacement choisi suivant la direction normale $\boldsymbol{n}$

$$\mathrm{d}\boldsymbol{l} = \boldsymbol{n}^\circ\ \mathrm{d}l.$$

Dans ce cas $\mathrm{d}\boldsymbol{P}$ effectuera (contre le milieu extérieur) un travail

$$\mathrm{d}\mathscr{E} = \mathrm{d}\boldsymbol{P}\cdot\mathrm{d}\boldsymbol{l} = \mathrm{d}\boldsymbol{P}\cdot\boldsymbol{n}^\circ\ \mathrm{d}l = F_{\boldsymbol{n}}\cdot\boldsymbol{n}^0\ \mathrm{d}S\ \mathrm{d}l$$

ou, plus simplement: $\mathrm{d}\mathscr{E} = F_{\boldsymbol{n}}\,\mathrm{d}V$ en confondant $F_{\boldsymbol{n}}$ avec sa composante normale et laissant de côté le cas des déplacements tangentiels qui se traiteraient d'une façon analogue en faisant intervenir — au lieu de densités en volume — des densités d'énergie de surface.

Soit W l'énergie qui «remplissait» la surface S avant le déplacement $d\boldsymbol{l}$ et W' cette même énergie après le déplacement. Les densités d'énergie correspondantes

$$w = W/V \quad \text{et} \quad w' = W'/(V + dV)$$

s'exprimeront sous la forme:

$$w' = \frac{W - F_{\boldsymbol{n}}\, dV}{V + dV} = \frac{W}{V} \frac{\left(1 - F_{\boldsymbol{n}} \frac{dV}{W}\right)}{\left(1 + \frac{dV}{V}\right)} \tag{40.1}$$

ou

$$w' = w \frac{\left(1 - \frac{F_{\boldsymbol{n}}}{w} \frac{dV}{V}\right)}{\left(1 + \frac{dV}{V}\right)} \approx w \left[1 - \left(\frac{F_{\boldsymbol{n}}}{w} + 1\right) \frac{dV}{V}\right] \tag{40.2}$$

et, en explicitant les variations relatives

$$\frac{dw}{w} = -\left(\frac{F_{\boldsymbol{n}}}{w} + 1\right) \frac{dV}{V} \tag{40.3}$$

$F_{\boldsymbol{n}}/w$ est un nombre sans dimension soit $K_{\boldsymbol{n}}$, dépendant de $\boldsymbol{n}$ d'où:

$$\frac{dw}{w} = -(K_{\boldsymbol{n}} + 1) \frac{dV}{V}. \tag{40.4}$$

Cette relation, et le facteur $(K_{\boldsymbol{n}} + 1)$ déterminent ainsi une «loi de compression en densité d'énergie» qui sera la même que la loi de compression en $F_{\boldsymbol{n}}$ (correspondant à la même valeur de $\boldsymbol{n}$). En effet, de $F_{\boldsymbol{n}}/w = K_{\boldsymbol{n}}$ on déduit:

$$\frac{dF_{\boldsymbol{n}}}{F_{\boldsymbol{n}}} = \frac{dw}{w} = -(K_{\boldsymbol{n}} + 1) \frac{dV}{V},$$

ce qui peut aussi s'écrire en valeurs finies:

$$F_{\boldsymbol{n}} \cdot V^{(K_{\boldsymbol{n}}+1)} = \text{const.} \quad \text{et} \quad w \cdot V^{(K_{\boldsymbol{n}}+1)} = \text{const.} \tag{40.5}$$

(les deux constantes étant dans le rapport $K_{\boldsymbol{n}}$), et par suite définit la loi de compressibilité adiabatique du milieu, loi qui dépend du facteur *numérique* (scalaire) $K_{\boldsymbol{n}}$, fonction de la direction du vecteur $\boldsymbol{n}$.

β) Le *calcul direct* des valeurs de $K_{\boldsymbol{n}}$ à partir *des facteurs* définissant l'état complet du milieu considéré est en général compliqué et difficile. (Un exemple très simple est cependant fourni par les gaz parfaits monoatomiques, on trouve alors $K_{\boldsymbol{n}} = 0$ dans le cas d'une transformation isothermique et $K_{\boldsymbol{n}} = \frac{2}{3}$ dans celui d'une transformation adiabatique; on peut également suivre ce qui se passe pour les gaz polyatomiques.) Par contre, il est possible dans certains cas intéressants, de se rendre compte directement de la loi de «compressibilité en densité d'énergie» relative à diverses valeurs principales de $\boldsymbol{n}$. Si cette loi peut être posée sous une forme (anisotrope):

$$w\, V^{\gamma_{\boldsymbol{n}}} = \text{const.} \tag{40.6}$$

(par analogie avec la loi de compression — isotropique — adiabatique des gaz), $\gamma_{\boldsymbol{n}}$ étant un coefficient numérique *fonction de* $\boldsymbol{n}$, on en déduira immédiatement, non seulement la loi de compressibilité correspondante

$$F_{\boldsymbol{n}}\, V^{\gamma_{\boldsymbol{n}}} = \text{const.} \tag{40.7}$$

relative aux $F_{\boldsymbol{n}}$, mais surtout les valeurs numériques exactes $F_{\boldsymbol{n}} = (\gamma_{\boldsymbol{n}} - 1)\, w$ permettant de passer de l'expression de w à celle des $F_{\boldsymbol{n}}$ (puisque $F_{\boldsymbol{n}} = K_{\boldsymbol{n}} w$ et que $\gamma_{\boldsymbol{n}} = K_{\boldsymbol{n}} + 1$). On voit qu'en général il n'y aura pas identité, mais seulement proportionnalité entre les $F_{\boldsymbol{n}}$ et les w, le coefficient de proportionnalité dépendant de $\boldsymbol{n}$ et pouvant même s'inverser quand $\boldsymbol{n}$ varie, car les $K_{\boldsymbol{n}}$ peuvent aussi bien prendre des valeurs négatives que positives, *comme c'est effectivement le cas pour les champs magnétiques et électrostatiques.*

γ) Appliquons par exemple la méthode précédente au cas de la portion de tube de forces considéré au début de cette Section:

(i) *Actions sur la surface latérale.* Prenons $\boldsymbol{n}$ perpendiculaire aux lignes de forces. Nous supposons donc que nous donnons aux parois latérales un déplacement normal qui augmente les surfaces des sections droites. Dans ces conditions le volume V varie proportionnellement à ces surfaces et comme le flux magnétique qui les traverse $\Phi = BS$ reste constant, B varie suivant la loi:

$$B = \frac{\text{const.}}{S} = \frac{\text{const.}}{V}$$

donc pour la densité d'énergie:

$$w = \frac{B^2}{2\,\mathrm{u}\,\mu_0} = \frac{\text{const.}}{V^2} \tag{40.8}$$

ou

$$\frac{\mathrm{d}w}{w} = -2\,\frac{\mathrm{d}V}{V} = \frac{\mathrm{d}F_{\boldsymbol{n}}}{F_{\boldsymbol{n}}}\,.$$

Il en résulte que $K + 1 = 2$ d'où $K = 1$ et par suite:

$$F_{\boldsymbol{n}} = w = \frac{B^2}{2\,\mathrm{u}\,\mu_0}\,. \tag{40.9}$$

L'intérieur du tube de forces (c'est-à-dire le système Σ) exerce donc sur le milieu extérieur une pression normale dont le module est égal à la valeur même de la densité de l'énergie magnétique.

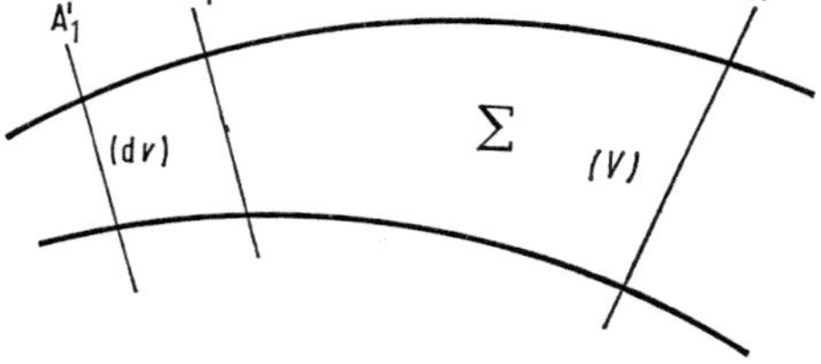

Fig. III. Détermination des actions sur les sections droites d'un tube de forces (voir texte).

(ii) *Actions sur les sections droites.* Examinons les actions qui s'exercent sur une des sections, A_1 par exemple (voir Fig. III). Supposons que nous la déplacions de A_1 en A_1', maintenant la section A immobile. Dans ces conditions le déplacement de A_1 «étire» les lignes de forces en incorporant de nouvelles portions d'énergie à l'intérieur du système Σ mais sans rien changer à la valeur de la densité d'énergie:

$$w = \frac{B^2}{2\mathrm{u}\,\mu_0} = \text{const.} \tag{40.10}$$

Nous aurons donc:

$$\frac{\mathrm{d}F_{\boldsymbol{n}}}{F_{\boldsymbol{n}}} = \frac{\mathrm{d}w}{w} = \text{const.} = 0\,\frac{\mathrm{d}V}{V}\,. \tag{40.11}$$

La «loi de compressibilité» en $F_{\boldsymbol{n}}$ se réduit donc à:

$$F_{\boldsymbol{n}} \cdot V^{(0)} = \text{const.,}\quad \text{d'où}\quad K_{\boldsymbol{n}} + 1 = 0$$

soit

$$K_{\boldsymbol{n}} = -1 \tag{40.12}$$

d'où

$$F_{\boldsymbol{n}} = -\frac{B^2}{2\mathrm{u}\,\mu_0}\,. \tag{40.13}$$

On retrouve donc pour valeur absolue de $F_{\boldsymbol{n}}$ la même valeur que précédemment, *mais cette fois avec le signe négatif* ce qui est bien conforme au résultat à vérifier.

Nous retiendrons de tout cela — en dehors de cette application spécifique à un tube de forces — que les valeurs de $K_{\boldsymbol{n}}$ applicables aux transformations de divers systèmes ont — tout en restant assez simples — une certaine diversité ($\frac{2}{3}$, $+1$, -1, etc. ...). Seule la valeur $K_{\boldsymbol{n}} = +1$ permet d'identifier quantitativement les forces de liaison (et d'interaction) à une densité d'énergie et la probabilité pour qu'il en soit ainsi reste modeste. Dans tous nos problèmes d'interaction on risque donc de commettre des erreurs — au moins numériques — si on compare les divers phénomènes ou actions en présence en se basant uniquement sur leurs densités énergétiques.

41. Superposition de plusieurs états de tension. On peut rendre notre problème plus réel en supposant que plusieurs causes de tensions soient superposées dans un même milieu. Par exemple, des causes cinétiques (thermiques) correspondant à une densité d'énergie w_1 telle que

$$w_1 = \tfrac{1}{2}\,\varrho\, v^2,$$

et des causes magnétiques correspondant à une densité w_2 telle que

$$w_2 = \frac{B^2}{2\mathrm{u}\,\mu_0}\,.$$

(On pourrait y ajouter des causes électriques, radiatives, etc. ...) En supposant d'abord qu'il n'existe pas de transformations possibles internes (à Σ) entre ces diverses formes d'énergie, il correspondra à chacune d'elle un état défini de tensions, symbolisé par l'existence de constantes $K_{\boldsymbol{n}}$ correspondantes. (Ceci pour chaque valeur de $\boldsymbol{n}$ commune). On pourra donc écrire:

$$F_{\boldsymbol{n}(1)} = K_{\boldsymbol{n}(1)} \cdot w_{(1)}$$

$$F_{\boldsymbol{n}(2)} = K_{\boldsymbol{n}(2)} \cdot w_{(2)}$$

et, faisant la somme:

$$(i,\ \text{indice courant})\quad F_{\boldsymbol{n}(\text{total})} = \text{Somme}\,(\dots K_{\boldsymbol{n}(i)} \cdot w_{(i)} \dots)\,. \tag{41.1}$$

Par exemple, en additionnant des effets de pression thermique (isotrope) p et de tensions anisotropes $f_{\boldsymbol{n}}$, on pourra écrire:

et

$$\left.\begin{aligned} F_{\boldsymbol{n}(1)} &= p \quad \text{et} \quad F_{\boldsymbol{n}(2)} = f_{\boldsymbol{n}} \\ F_{\boldsymbol{n}\,(\mathrm{total})} &= K_{(1)}\, w_{(1)} + K_{\boldsymbol{n}(2)}\, w_{(2)}. \end{aligned}\right\} \tag{41.2}$$

Remarquons que les w sont toujours isotropes; seules leurs compressions peuvent ne pas l'être; quant aux coefficients K, le premier est unique (indépendant d'une direction $\boldsymbol{n}$), le second étant au contraire fonction de la direction (ce que nous traduisons par les deux notations $K_{(1)}$ et $K_{\boldsymbol{n}(2)}$, respectivement). On pourra suivre l'évolution générale des tensions s'exerçant dans ce milieu en analysant les lois de «compressibilité» applicables séparément aux deux types de densités d'énergie qui s'y manifestent.

Dans le cas plus complexe où des transformations internes de l'une en l'autre de ces énergies seraient possibles, l'évolution des tensions dans le milieu correspondant reflèterait cette complexité. Un exemple nous en est fourni par la magnétogaine, portion de l'environnement terrestre à l'intérieur de laquelle on peut penser qu'il se produit une transformation permanente d'énergie cinétique (thermique) en énergie magnétique par un processus tourbillonnaire s'apparentant à ceux proposés par F. HOYLE[1] pour certaines atmosphères stellaires. En affectant les indices 1 et 2 aux deux densités d'énergie respectives correspondantes, la relation Eq. (41.2) est applicable avec transfert continu de la valeur de $w_{(1)}$ en celle de $w_{(2)}$, les coefficients $K_{(1)}$ et $K_{\boldsymbol{n}(2)}$ étant constants. Si la magnétogaine était isolée (au point de vue énergétique) des deux milieux qu'elle sépare (l'espace interplanétaire sur une de ses faces, la magnétosphère supérieure sur l'autre face), nous pourrions écrire de plus: $w_{(1)} + w_{(2)} = \text{const}$. En fait, il est admis qu'elle reçoit en permanence (bien qu'avec un taux variable) de l'énergie de l'espace interplanétaire (notamment du vent solaire et de toutes les sources fluctuantes d'énergie qui peuvent l'accompagner) et qu'elle transmet, de façon quasi permanente, également, de l'énergie vers la magnétosphère. Ceci donne son rôle caractéristique à cette magnétogaine qui, en définitive, transmet principalement à la magnétosphère sous la forme $w_{(2)}$ de l'énergie qu'elle a reçu de l'espace intermédiaire sous la forme $w_{(1)}$.

(A remarquer que nous faisons abstraction ici — sans les considérer comme forcément négligeables — des énergies des types (1) et (2) qui pourraient transiter par la magnétogaine sans changer de nature.)

Le processus que nous venons d'évoquer peut être envisagé comme un des schémas possibles pour la génération, aux frontières supérieures de la magnétosphère, d'une activité magnétique anisotrope plus ou moins organisée, et cette activité pourrait être considérée comme une ébauche de certaines catégories de pulsations.

Ce genre de transformation dans le sens d'une «vectorialisation» de grandeurs scalaires par le mécanisme d'un transfert entre deux formes (scalaires) d'énergie, peut prendre place en d'autres régions de l'environnement terrestre. Des transformations inverses sont aussi possibles, par exemple quand des ondes magnétodynamiques organisées déterminent un échauffement de diverses portions du plasma magnétosphérique.

[1] F. HOYLE: Chapitre "The Build-up of large magnetic fields inside stars" du livre: "Magnetohydrodynamics" (K. ROLF et M. LANDSHOFF, eds.), p. 29—35. Stanford, Calif.: Stanford University Press 1957.

Ainsi, sans pouvoir éclaicir le détail des mécanismes d'interaction qui se produisent à tout instant dans les différentes régions formant l'environnement terrestre, les processus généraux examinés ci-dessus sont en mesure de nous donner une justification d'ensemble de leur évolution.

42. Phénomènes d'ondes stationnaires et phénomènes de propagation. Nous commençons à connaître assez bien (notamment, grâce aux satellites) les formes et les dimensions des domaines où les phénomènes vibratoires que nous étudions ont leurs sources, ainsi que les modes de propagation — principalement magnétodynamiques — qui permettent à ces phénomènes de prendre forme. La confrontation de ces données nous conduit à remarquer que — sauf pour les variations magnétiques les plus rapides (celles dont les périodes sont inférieures à quelques secondes) — les longueurs d'onde en jeu atteignent des ordres de grandeur qui sont une fraction considérable des dimensions de ces domaines. Dans ces conditions, plutôt qu'à des processus de propagation, on doit faire appel à l'établissement de régimes de vibrations stationnaires pour tâcher d'expliquer d'une façon valable (qualitativement et — si possible — quantitativement) les oscillations observées. Pour chaque classe de pulsation il y aura donc lieu de déterminer quelles pourraient être les *cavités résonnantes* susceptibles de correspondre au mieux aux caractères morphologiques propres de cette classe, ainsi que les modes de vibration de ces cavités. Pourront échapper à cette façon de concevoir les processus oscillants, les mécanismes rendant compte des pulsations **pc1** et **pi1** pour lesquelles les longueurs d'ondes sont suffisamment courtes. Il résulte de cette remarque préliminaire, que — plutôt que de pouvoir présenter une théorie unique, générale — nous sommes conduits à deux types de théorie très distincts.

Cette nécessité logique nous conduit également à envisager deux modes d'approche très différents pour l'étude expérimentale des degrés de synchronisme dans les apparitions mondiales des différents types de pulsations. On pourra toujours définir — et essayer de mesurer avec la plus grande précision réalisable — des «temps de propagation *apparents* au sol», mais ces temps devront être considérés suivant les cas, *soit* comme correspondant à des propagations *réelles* (même si ces propagations représentent le plus souvent un effet global à partir d'une suite de propagations composantes correspondant aux divers milieux traversés, ce qui augmentera la difficulté de l'interprétation des mesures), *soit* comme la manifestation tangible d'une déformation permanente ou transitoire de figures — ou d'«états» — d'ondes stationnaires.

En fait, c'est seulement dans le cas des **pc1** et des **pi1** — correspondant par suite à des propagations réelles — que, comme nous l'avons vu (voir Sect. 32) des temps de propagation au sol ont pu être mesurés d'une façon suffisamment précise et cohérente. Dans le cas de pulsations de périodes plus longues on a tout juste pu vérifier, d'une façon très partielle et grossière, un pseudo-synchronisme mondial; ceci sous réserve, évidemment, que les conditions d'apparition de ces pulsations soient satisfaites, simultanément, aux diverses stations utilisées pour ces mesures.

43. Phénomènes et ondes magnétodynamiques.

(Les références générales [*19*], [*20*], [*21*] et [*22*] pourront être consultées avec profit pour cette Section et les suivantes.)

α) Sous ce titre vient se ranger un chapitre très important de physique classique dont on pourrait seulement s'étonner qu'il ait été ouvert si tardivement dans l'histoire de la physique. Le sujet dont il traite était en effet complètement inconnu des physiciens lorsque, vers 1942, H. ALFVÉN découvrit la nécessité de l'introduire dans l'étude du comportement de la plupart des *plasmas* considérés *en astrophysique*, notamment ceux que l'on rencontre dans l'étude du Soleil et de ses relations avec l'environnement terrestre. Sous cette forme ALFVÉN montrait qu'il s'agissait là de déterminer suivant quelles conditions il était possible de généraliser, à l'échelle des dimensions cosmiques, les lois de l'électromagnétisme classique déduites de la théorie de Maxwell et vérifiées principalement par des expériences de laboratoire. En même temps, il montrait comment l'application de ces lois devait conduire à l'observation de nouveaux phénomènes — et, en particulier, à celle d'un nouveau type d'ondes — au laboratoire même (à condition que certains ordres de grandeurs puissent y être réalisés). On s'étonnera moins qu'on ne serait tenté de le faire, de ce que ceci n'ait pas été mis en lumière plus tôt, si l'on remarque que ce nouveau domaine peut être considéré comme une généralisation de l'hydrodynamique classique quand on lui adjoint *deux conditions* indépendantes, aussi peu usuelles à priori l'une que l'autre (en dehors des conditions cosmiques), à savoir: opérer en présence d'un champ magnétique puissant et choisir un liquide extrêmement conducteur. En fait, ce n'est qu'après plusieurs années d'efforts que, connaissant cependant assez exactement les conditions générales à réaliser, L. LUNDQUIST réussit — en 1949 — à vérifier par une expérience de laboratoire la justesse des vues de ALFVÉN[1]. (Confirmées aussi par[2].)

β) Ne nous arrêtant pas autrement à ce rappel historique, nous constatons que ce domaine de la «magnéto-hydrodynamique» — suivant sa dénomination initiale par ALFVÉN — appelé aussi (pour simplifier) «hydro-magnétique» (d'où les notations symboliques «M-H» ou «H-M» employées par de nombreux auteurs), et qu'il paraît en définitive plus correct de ranger sous le vocable *«magnétodynamique»* (symbole «M-D») *qui est celui que nous retiendrons ici*, se rattache maintenant aux trois grands chapitres suivants de la physique:

l'étude des plasmas chauds de laboratoire (notamment celle de la fusion nucléaire);

celle des plasmas chauds intervenant en astrophysique (soleil, étoiles, etc.);

celle des plasmas froids concernant principalement la géophysique et, en particulier, la magnétosphère.

Cependant, l'étude générale des ondes dans les plasmas étant traitée dans d'autres articles[3] nous rappellerons seulement ici les quelques données les plus indispensables relatives au troisième cas.

γ) A cet effet, plutôt que d'essayer de résumer une réalité complexe, nous allons nous laisser guider par des *analogies mécaniques* qui nous permettrons de saisir une première approximation des phénomènes. Nous les avons préparées par les exposés constituant les Sects. 40 et 41. Nous partons des expressions des états de tension qui conditionnent les forces de liaison qui s'exercent entre une portion du plasma magnétosphérique, que nous assimilerons à la portion d'un tube de forces comprise entre deux sections droites S_1 et S_2, et le plasma qui l'entoure (voir Fig. IV). Nous avons vu que le plasma exerçait sur les parois latérales du tube une pression $B^2/2u\,\mu_0$, et sur les sections droites S_1 et S_2 une tension (dirigée de l'extérieur vers l'intérieur) de valeur arithmétique égale. Cependant, pour des raisons que nous développerons plus loin (voir Sect. 45), il apparaît plus avanta-

[1] S. LUNDQUIST: Phys. Rev. **76**, 1805 (1949); — Nature **164**, 145 (1949).
[2] B. LEHNERT: Phys. Rev. **94**, 815 (1954).
[3] Voir les contributions par H. POEVERLEIN (p. 7) et par V. L. GINZBURG et A. A. RUHADZE (p. 395) dans ce volume. Voir aussi Chap. A de l'article de K. RAWER et K. SUCHY an volume 49/2, p. 1 de cette encyclopédie.

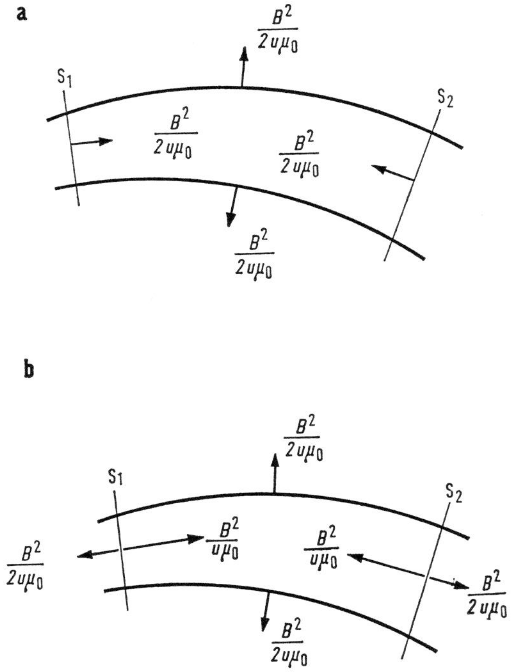

Fig. IVa et b. Illustration de deux représentations équivalentes, a et b pour l'ensemble des liaisons s'exerçant entre une portion d'un tube de forces et le milieu qui l'entoure (explication dans le texte, p. 293—294).

geux de présenter l'ensemble de ces forces de liaison en suivant une autre décomposition: on admet que la pression «latérale» puisse prendre le comportement d'une pression isotrope, c'est-à-dire que la pression $B^2/2\mathrm{u}\,\mu_0$, au lieu de s'exercer uniquement le long des faces latérales du tube de forces, s'exercerait aussi, avec le même sens — interne vers l'externe — le long des deux sections droites S_1 et S_2. Bien entendu, afin que l'on puisse aboutir à une présentation algébrique correcte de l'ensemble des forces de liaison en jeu, il importe que nous compensions l'introduction de la pression de sens interne vers l'externe supposée maintenant s'exercer le long des sections latérales, par une tension de valeur égale, et de sens inverse, soit de valeur $B^2/2\mathrm{u}\,\mu_0$, dirigée de l'extérieur vers l'intérieur. Il s'en suit que nous aurons maintenant le long de ces sections latérales, une tension globale égale à deux fois l'ancienne, soit $B^2/\mathrm{u}\,\mu_0$. Dans la mesure où nous aurons adopté la valeur $B^2/2\mathrm{u}\,\mu_0$ pour la pression isotrope régnant dans le plasma, nous serons obligés de prendre la valeur $B^2/\mathrm{u}\,\mu_0$ pour celle de la tension longitudinale.

La clef de ces analogies repose donc sur l'assimilation du plasma magnétosphérique à un milieu matériel dont la densité massique (donc l'inertie) serait bien sa densité propre, mais dont les propriétés élastiques (ou thermodynamiques) seraient à la fois celles d'un solide (verge, tige ou corde vibrante), et celles d'un gaz. Dans les deux cas, les petites déformations, ou déplacements, mis en jeu par le comportement dynamique du plasma, sont supposés se faire sans la moindre dégradation d'énergie (c'est-à-dire sans transformation d'énergie mécanique en chaleur), ce qui est justifié en première approximation par les propriétés *extrêmement conductrices* du plasma (ne donnant lieu théoriquement à aucune «chaleur de Joule»). C'est là la condition même de notre assimilation à des équivalents mécaniques *parfaitement élastiques.*

δ) On peut s'étonner que nous fassions appel à deux systèmes mécaniques aussi différents (sauf en ce qui concerne justement leurs propriétés élastiques) que ceux représentés par une tige métallique ou par un gaz. Mais on doit garder présent à l'esprit le fait que les propriétés intrinsèques du plasma sont caractérisées, en ce qui concerne ses forces de liaison internes, par une *anisotropie tensorielle.* Celle-ci peut être considérée comme la somme: d'une part, de l'anisotropie qui se manifeste dans les vibrations transversales d'une verge, d'autre part, de l'isotropie propre aux déformations des gaz. Le plasma se comporte ainsi comme ayant une double nature mécanique. Il est intéressant de remarquer du point de vue historique, que la première de ces natures (assimilation à une verge ou à une corde vibrante) a été la première à être mise en évidence par ALFVÉN lors de sa découverte des phénomènes magnéto-dynamiques.

Ce n'est que quelques années plus tard que la deuxième de ces natures fut également explicitée. C'est ainsi que furent découvertes, dans l'ordre, les «ondes transversales de Alfvén» et les «ondes longitudinales» (ces dernières, par extension des travaux de Alfvén telle qu'elle a été conduite par HERLOFSON[4], HOFFMAN et TELLER[5], et VAN DE HULST[6]. Ces deux types d'ondes constituent en définitive des cas particuliers — mais spécialement importants — du phénomène général des ondes magnéto-dynamiques et, encore plus général, des ondes électromagnétiques de toute nature susceptibles de se propager dans les plasmas.

ε) Etant donné l'état rudimentaire des théories actuelles sur les *pulsations magnétiques* de périodes supérieures à celles des **pc1** (c'est-à-dire de toutes celles qui mettent en jeu des longueurs d'onde de dimensions comparables à celles de la Terre) il est en général suffisant de faire figurer les ondes magnéto-dynamiques qui interviennent dans ces théories en les réduisant aux deux cas précédents: ondes transversales et ondes longitudinales, chacun de ces cas étant pris sous sa forme la plus schématique. Les autres cas — et les autres formes — interviennent certainement le plus souvent sans «détruire» — peut-on l'espérer — les principes mêmes des théories. Des défauts beaucoup plus graves peuvent provenir, par contre, de conditions physiques qui, dans la magnétosphère, rendraient erronées les hypothèses qui sont à la base de la théorie magnéto-dynamique. C'est ainsi que ALFVÉN lui-même a fait remarquer que l'existence d'une composante longitudinale du champ électrique (c'est-à-dire parallèle aux lignes de forces du champ magnétique principal), en permettant au plasma magnétosphérique d'échapper aux conditions de «gel» magnéto-dynamique rendrait inapplicable l'intervention sous toute forme des ondes correspondantes (schématisées ou non). En l'absence de données suffisantes sur de telles conditions nous ne retiendrons cet aspect des phénomènes que dans la discussion de certains cas particuliers (par ex., celui concernant l'extension vers les basses latitudes de pulsations rapides polaires). Pour le reste, les modes schématiques énoncés plus haut constitueront une base d'exposition utile, à laquelle il ne faudra pas cependant s'attacher aveuglément.

44. Ondes transversales (on dit quelquefois: *ondes de Alfvén pures*). Ce sont les tensions longitudinales (parallèles aux lignes de forces du champ magnétique principal) qui sont en cause. En substituant leur valeur $T = B^2/\text{u}\,\mu_0$ dans l'expression $V = \sqrt{T/\varrho}$ qui donne la vitesse des ondes transversales dans une tige (ou

[4] N. HERLOFSON: Nature **165**, 1020 (1950).

[5] F. DE HOFFMAN and E. TELLER: Phys. Rev. **80**, 692 (1950).

[6] H. C. VAN DE HULST: Proc. Sympos. on Motions of Gaseous Masses of Cosmical Dimensions, Central Air Documents Office, Dayton, Ohio (1951).

corde vibrante) on obtient la vitesse de propagation (parallèlement au champ magnétique principal B), dite vitesse de ALFVÉN V_A, de ces ondes transversales*:

$$V_A = \sqrt{\frac{B^2}{u\,\mu_0 \cdot \varrho}} = \frac{B}{\sqrt{u\,\mu_0 \cdot \varrho}}. \tag{44.1}$$

Dans cette expression ϱ doit être considéré comme représentant la masse spécifique totale du plasma, c'est-à-dire doit comprendre l'apport dû aux particules neutres, bien que l'hypothèse du «gel» ne s'applique *directement* qu'aux composantes électrisées. Ceci peut être exprimé sous une autre forme en écrivant $\varrho = N M m$, expression dans laquelle N est la densité numérique, en nombre de particules par unité de volume, M la «masse relative moléculaire moyenne» et m la masse de l'ion hydrogène. On obtient ainsi:

$$V_A = \sqrt{\frac{B^2}{u\,\mu_0\, N M m}} = \frac{B}{\sqrt{u\,\mu_0\, N M m}} \tag{44.2}$$

expression qui permet de suivre les variations de la vitesse de ALFVÉN en fonction de l'altitude à partir des valeurs mesurables N et M, et de B**.

L'interprétation précédente sera correcte tant que le plasma pourra être supposé suffisamment ionisé pour que — par chocs — le mouvement de ses particules neutres et ionisées soit, statistiquement, uniformisé. On remarquera que la vitesse V_A est celle d'une propagation sans dispersion (Vitesse de groupe = Vitesse de phase $= V_A$). Ceci n'est valable que pour des fréquences inférieures à une certaine limite déterminée par la condition que la fréquence de l'onde considérée soit très inférieure à la fréquence de gyrorésonance des ions intervenant comme constituants du plasma, condition que l'on écrit en général $\omega \ll \omega_{B\mathrm{i}}$, en introduisant les fréquences angulaires, ω pour l'onde et $\omega_{B\mathrm{i}}$, parfois dénommée ω_i tout simplement, pour les ions (principalement les protons), correspondantes. (Bien entendu la gyrofréquence $\omega_{B\mathrm{e}}$ des électrons est considérablement plus grande que les précédentes.) Quand, au contraire, ω se rapproche de $\omega_{B\mathrm{i}}$, de la dispersion apparaît. C'est ce qui se passe, comme nous le verrons (voir Sect. 53) pour certaines pulsations **pc1** du type «pulsations en perles», dans leur trajet à travers la magnétosphère, alors que leur dispersion au cours de leur propagation

* *Remarque* sur les Unités: Comme pour tout cet ouvrage («Introductory Remarks», p. 1), nous avons écrit l'expression de la vitesse de ALFVÉN, V_A, en unités non-spécifiées. Elle s'écrit $V_A = B/\sqrt{\mu_0\,\varrho}$ dans le système MKSA (Giorgi), rationalisé ($u = 1$). Cette manière de faire, adoptée par souci d'homogénéité, n'est pas usuelle en ce qui concerne cette vitesse que — probablement pour une raison historique — on continue généralement, depuis qu'elle a été donnée la première fois par ALFVÉN lui-même, à écrire dans le système u.e.m.c.g.s. (non rationalisé, $u = 4\pi$, $\mu_0 = 1$). Dans ce cas, V_A est donné par l'expression:

$$V_A = \frac{B}{\sqrt{4\pi\,\varrho}}. \tag{44.1 bis}$$

On ne s'étonnera donc pas de trouver l'écriture ci-dessus dans un grand nombre d'ouvrages, même modernes (voir, par exemple, J. A. JACOBS: Geomagnetic Micropulsations. Berlin-Heidelberg-New York: Springer 1970). Ceci va de pair avec la conservation du *gamma* (γ), sous-unité du gauss ($1\,\gamma = 10^{-5}$ gauss), en tant qu'unité usuelle de champ magnétique en ce qui concerne le champ terrestre. Ceci illustre aussi le fait que dans tous les secteurs spécialisés de la science (tel que celui constitué par le «magnétisme terrestre»), une nomenclature de «corps de métier» résiste d'une façon durable aux évolutions plus générales du «langage».

** A partir du sol jusqu'à environ 90 km d'altitude on a $M = 29$, au dessus de 100 km M diminue à peu près d'une façon linéaire jusqu'à 300 km ($M = 19$). Aux altitudes entre 300 et 800 km la diminution continue mais différement jour et nuit et dépendant de l'activité solaire; CIRA 1965 donne pour 800 km des valeurs entre 3,5 et 15,5.

ultime dans le guide sub-magnétosphérique reste négligeable. Par contre, pour la plupart des autres types de pulsations (dont les périodes sont plus longues) on pourra supposer une dispersion négligeable.

Remarquons enfin que certains auteurs considèrent le type de propagation par ondes transversales que nous venons d'examiner, comme caractéristique d'un «milieu incompressible», ceci par opposition à la propagation par ondes longitudinales que nous allons envisager dans la section suivante. Bien qu'un tel point de vue paraisse pouvoir se déduire des analogies mécaniques que nous avons données, les phénomènes de «superposition des modes» (avec ou sans couplage), qui sont toujours présents, rendent sa rigueur discutable.

45. Ondes longitudinales.

α) Les mécanismes par lesquels on rend compte de l'existence de ces ondes font appel aux propriétés d'un *milieu compressible*, isotrope, donc de la nature d'un gaz ordinaire. Nous avons vu (voir Sect. 43) que le plama magnétosphérique ne répondait pas *globalement* à ces normes, mais que nous pouvions le considérer comme la superposition de deux milieux dont l'un — basé sur la généralisation isotropique de l'effet de pression latérale $B^2/2\text{u}\,\mu_0$ — déterminait ensuite, par différence, les propriétés anisotropiques de l'autre, à savoir, une tension longitudinale de valeur $B^2/\text{u}\,\mu_0$. En fait, c'est l'utilisation de ce second résultat qui nous a permis de nous rendre compte de la possibilité de l'existence des ondes transversales. Par un retour en arrière, nous pouvons donc maintenant — au moins en principe — postuler l'existence du milieu isotrope, superposé, donc d'ondes de compressions-dilatations analogues aux ondes acoustiques qui se propagent dans un gaz (ou dans un liquide compressible et isotrope).

Une telle manière de faire peut apparaître — à juste titre — comme très artificielle. Il semble, à priori, que l'on pourrait aussi bien admettre une propagation d'ondes transversales basées sur une tension (longitudinale) qui aurait la valeur $B^2/2\text{u}\,\mu_0$ (donc moitié de la précédente), quitte à se plier aux conséquences de l'existence complémentaire d'une pression latérale, également *anisotrope* (puisqu'elle ne serait que *latérale*), dont l'effet serait d'ailleurs délicat à étudier. Plus généralement, une infinité de décompositions algébriques des types précédents pourraient être imaginées, et nous serions ainsi tentés de considérer tous les raisonnements précédents comme étant absurdes.

En fait, cette indétermination dans les décompositions possibles, loin d'être sans lien avec la réalité, illustre l'impossibilité d'une séparation en deux modes qui resteraient non couplés. Certaines analyses mathématiques ont d'ailleurs conduit à démontrer l'inexistence des modes séparés, sauf dans des cas particuliers. Ces analyses ne conduisent à l'un ou l'autre de ces modes que par suite des incorrections introduites par des approximations de calcul que l'on croit pouvoir faire sans danger, mais qui en fait postulent déjà, implicitement, l'existence de modes séparés. On voit donc que nos raisonnements par analogies mécaniques ne sont ni plus justes ni plus faux que les calculs compliqués — et inextricables si on veut qu'ils suivent la réalité — que l'on présente souvent comme leur étant très supérieurs. Le phénomène naturel reste une entité complexe qui, en fonction des cas d'espèces, peut se rapprocher avec une approximation plus ou moins bonne, des cas schématisés par les modes théoriques élémentaires couplés d'une façon convenable. La figure 45.1 essaie de donner une illustration directe de cette impossibilité foncière de séparer totalement les deux modes tranversal et longitudinal dont — pour la commodité d'une exposition — nous faisons un exposé séparé.

β) Voyons donc comment peut être développée l'analogie avec un milieu compressible, isotrope, tout en *introduisant l'effet du champ magnétique*: Partons de l'expression de la vitesse du son dans un gaz:

$$V_s = \sqrt{\frac{\gamma p}{\varrho}}, \tag{45.1}$$

γ étant le rapport des chaleurs spécifiques à pression et à volume constants $\gamma = c_p/c_V$, p la pression du gaz et ϱ sa masse spécifique. Nous allons supposer que l'on puisse donner à la pression magnétique $B^2/2\text{u}\,\mu_0$ les mêmes propriétés que celles de la pression p. Le terme ϱ conservera évidemment la même signification. Quand au rapport γ on sait que, ainsi que nous l'avons établi plus haut (voir Sect. 40), il résulte de la valeur de l'exposant à prendre dans la relation générale des compressions adiabatiques:

$$p\,V^{\gamma} = \text{const.} \tag{45.2}$$

qui lie la pression p au volume V au cours de cette compression. Dans le cas de la pression magnétique nous avons vu (voir Sect. 40) qu'il fallait lui donner la valeur 2.

En définitive, si la pression qui s'exerce dans le plasma était toute entière d'une nature magnétique, une vitesse V_B, tel que

$$V_B = \sqrt{\frac{2(B^2/2\,\mathrm{u}\,\mu_0)}{\varrho}} \equiv \sqrt{\frac{(B^2/\mathrm{u}\,\mu_0)}{\varrho}} \tag{45.3}$$

remplacerait la vitesse V_s. Comme la présence du champ magnétique ne supprime pas pour autant celle d'un gaz nanti de ses propriétés usuelles, nous avons en fait une pression totale

$$p + B^2/2\,\mathrm{u}\,\mu_0$$

pour assurer la propagation d'ondes de compressions-décompressions. Cependant, il serait incorrect d'ajouter les pressions elles-mêmes dans l'expression des vitesses; ce sont les termes du type $\gamma \cdot p$ qui doivent être ajoutés. En définitive, nous aurons l'expression suivante pour la vitesse des ondes mixtes (magnétiques-acoustiques) qui se propageront longitudinalement:

$$V = \sqrt{\frac{(B^2/\mathrm{u}\,\mu_0) + \gamma\,p}{\varrho}} \tag{45.4}$$

suivant l'importance relative des deux termes figurant au numérateur ces ondes seront d'une nature plus essentiellement magnétique ou acoustique.

Dans la relation précédente séparons plus complètement les termes correspondants aux deux pressions. Nous écrirons:

$$V = \sqrt{\frac{B^2}{\mathrm{u}\,\mu_0\,\varrho} + \frac{\gamma\,p}{\varrho}} = \sqrt{(V_{\mathrm{A}})^2 + (V_{\mathrm{s}})^2}\,. \tag{45.5}$$

C'est là une formule fondamentale qui ne préjuge en rien des valeurs relatives de V_{A} et de V_{s}.

γ) En fait, dans la majorité des cas applicables à la *magnétosphère* celle-ci a une constitution telle que la densité de l'énergie magnétique est bien supérieure à celle du gaz seul de façon que V_{A} y est très supérieur à V_{s}. On pourra alors, soit négliger totalement ce dernier terme, soit développer l'expression de V sous la forme:

$$V = V_{\mathrm{A}} \sqrt{1 + \frac{(V_{\mathrm{s}})^2}{(V_{\mathrm{A}})^2}} \approx V_{\mathrm{A}} \cdot \left(1 + \frac{1}{2}\,\frac{(V_{\mathrm{s}})^2}{(V_{\mathrm{A}})^2}\right) \tag{45.6}$$

ou encore

$$V \approx V_{\mathrm{A}} + \frac{1}{2}\,\frac{(V_{\mathrm{s}})^2}{V_{\mathrm{A}}}\,. \tag{45.7}$$

En pratique — étant donné le peu de précision avec laquelle il est possible de mener à bien des études quantitatives, théoriques, relativement à la magnétosphère — on se contente presque toujours d'adopter la valeur V_{A} pour vitesse des ondes longitudinales magnétosphériques. Ceci montre bien l'importance pratique de ce paramètre, la quasi-identité des vitesses des ondes magnéto-dynamiques dans la magnétosphère, qu'elles s'y propagent suivant le mode transversal ou le mode longitudinal, simplifiant de surcroît la nomenclature. On parlera de «vitesse de Alfvén» sans avoir à préciser le mode de propagation concerné. Cette généralisation dans l'usage de ce paramètre est, certes, commode, mais elle ne doit pas nous faire oublier qu'elle est la conséquence d'une simple coïncidence de forme et qu'elle ne s'applique en principe qu'imparfaitenent aux propagations s'effectuant suivant le mode longitudinal.

46. Directions de propagation et polarisations — Généralisation des cas les plus simples considérés précédemment.

α) Nous venons de présenter les deux cas les plus simples, fondamentaux, d'ondes magnéto-dynamiques — transversales et longitudinales — (voir Sects. 44 et 45), et, bien que les qualificatifs précédents ne prêtent pas à confusion dans le contexte d'analogies mécaniques qui a servi à cette présentation, nous pensons qu'il est nécessaire, avant toute généralisation de ces deux cas, de préciser leurs *relations avec le champ magnétique* principal $\boldsymbol{B}$ dans lequel se fait la propagation.

Le cas des ondes transversales — ou ondes de Alfvén proprement dites — est le plus clair: le vecteur magnétique de perturbation $\boldsymbol{b}$ est, de même que le déplacement de matière qu'il entraîne (hypothèse du gel), normal à $\boldsymbol{B}$. La propagation, elle, se fait le long des lignes de forces principales, donc parallèlement à $\boldsymbol{B}$ (théoriquement, dans les deux sens). La composition de $\boldsymbol{b}$ — variable, par exemple suivant une loi sinusoïdale — et de $\boldsymbol{B}$, produit au niveau de la perturbation une légère rotation du champ résultant, donc des lignes de forces, sans affecter toutefois son module. Si on considère un tube de forces étroit, axé sur une ligne de forces, son épaisseur ne change donc pas au passage de la perturbation qui s'y traduit seulement par un coude plus ou moins prononcé.

Dans le cas des ondes longitudinales, la propagation la plus simple est celle qui se fait normalement à $\boldsymbol{B}$ (pour autant que cette condition puisse être définie pour une onde ayant une certaine étendue spatiale, le champ $\boldsymbol{B}$ n'étant pas uniforme). C'est également suivant cette même direction que se font les déplacements alternatifs de matière et les compressions et décompressions correspondantes, à la fois de la matière et du champ magnétique principal. C'est ce parallélisme entre les compressions-décompressions et la propagation qui justifie le qualiquatif de «longitudinal». Par contre, le vecteur de perturbation magnétique $\boldsymbol{b}$ est dirigé nécessairement parallèlement au champ principal (ce qui assure les variations alternatives du module de ce dernier), donc normalement à la propagation. Si nous considérons à nouveau un tube de forces étroit atteint en un de ses points par la perturbation, il ne s'y coude pas (comme dans le cas des ondes transversales), mais y subit un petit étranglement, ou dilatation, transitoire. Ceci est illustré par la Fig. V.

β) Examinons maintenant les développements à attendre d'une telle perturbation si nous la considérons comme marquant, conventionnellement, un état donné, localisé. Deux *processus principaux* sont possibles:

a) La compression (ou dilatation) se transmet en tant que telle suivant son axe propre, c'est-à-dire que l'onde longitudinale «antérieure à l'état donné» poursuit sa route;

b) la compression (ou dilatation) n'ayant pas pu se faire localement sans une certaine rotation des lignes de forces dans la partie du tube de forces assurant le «raccord» entre les portions où le champ principal est inchangé et celle où il vient d'être augmenté (ou diminué), une onde transversale prend naissance à la hauteur de ce raccord et va se propager le long du tube de forces correspondant.

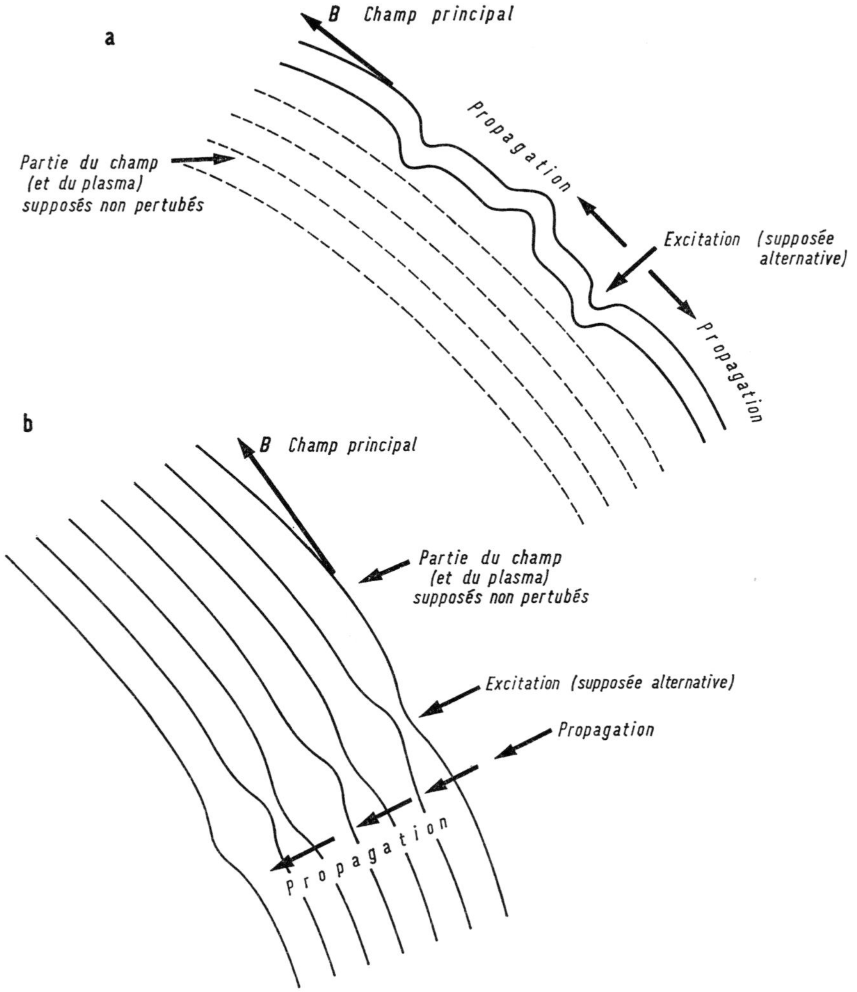

Fig. V a et b. Essais d'illustration de l'impossibilité de séparer totalement un mode transversal (*principalement*, le cas a) d'un mode longitudinal (*principalement*, le cas b).

On voit qu'il paraît impossible qu'il ne se produise aucun couplage entre les deux modes, tout au moins lors de processus élémentaires, locaux. Cependant, il ne faudrait pas en déduire qu'obligatoirement un tel couplage persiste quand on fait la somme des effets de ces perturbations élémentaires. En effet, si, au lieu d'une excitation quasi ponctuelle, nous supposons que le tube de forces considéré ci-dessus soit attaqué par une onde plane, longitudinale, tout au long de sa surface latérale (en négligeant la courbure des lignes de forces pour la portion du tube considérée), cette excitation s'y produisant en synchronisme, ce tube tout entier subira des compressions (ou dilatations) synchrones ce qui ira à l'encontre d'une propagation transversale. En fait, si le synchronisme est parfait, aucune portion du tube de forces ne présentera de «raccord» à aucun instant et ainsi il n'y aurait même pas d'excitation transversale. L'avantage de cette discussion théorique, limite, du problème, est qu'elle nous montre que les causes de couplage entre les deux modes sont essentiellement le fait des imperfections inéluctables dans l'application des surfaces d'ondes du mode longitudinal sur les tubes de forces. La courbure des lignes de forces du champ $\boldsymbol{B}$ nous donne donc une première cause, obligatoire, de couplage. Cet effet sera renforcé si l'onde longitudinale progresse dans la magnétosphère dans une direction qui s'écarte délibérément de la normale aux lignes de forces rencontrées. Ce cas se présente notamment dans la haute magnétosphère lorsqu'elle est soumise à une brusque compression de la part d'une influence extérieure (par une discontinuité dans le vent solaire, par exemple), ces conditions d'inclinaison dépendant alors essentiellement de la région de la haute magnétosphère par laquelle se transmet l'excitation. Il peut arriver — notamment dans le cas d'une perturbation du type **SSC** (voir Sect. 16) — qu'il soit impossible de reconnaître si la transmission de cette perturbation jusqu'à l'ionosphère (et le sol) s'est faite suivant l'un ou l'autre des deux modes, ou tous les deux simultanément. ... bien que dans la plupart des cas on admette que les **SSC** se transmettent par ondes longitudinales, l'excitation étant surtout équatoriale.

γ) La discussion précédente nous conduit aussi à nous intéresser à tous les cas de *propagations obliques*[1], Θ — angle entre la direction de $\boldsymbol{B}$ et la direction de propagation — ayant toute valeur comprise entre 0 et $\pi/2$. Nous ne ferons pas l'étude de ces cas, mais nous donnerons seulement quelques résultats: Des propagations obliques sont possibles pour les ondes transversales (ondes de Alfvén pures, généralisées), avec une vitesse $V = V_{\mathrm{A}} \cdot \cos\Theta$, et pour les ondes longitudinales. Mais dans ce cas apparaît aussi un autre type d'ondes beaucoup plus proche d'une onde sonore ordinaire que d'une onde magnéto-dynamique, le champ magnétique intervenant cependant pour modifier la vitesse des ondes acoustiques correspondantes. Les nomenclatures en usage relativement à ces deux types d'ondes ne sont pas unanimement suivies et donnent même lieu, parfois, à des confusions importantes. En ce qui concerne notre exposé, en dehors des ondes de Alfvén proprement dites (transversales), avec Θ nul ou non nul, nous utiliserons les dénominations suivantes pour les deux autres (inspirées de celles de Alfvén): *Onde magnéto-dynamique modifiée* pour les ondes longitudinales qui ont fait l'objet de notre examen et dans lesquelles la pression p cinétique ne joue qu'un rôle «modificateur»; *Onde acoustique modifiée* pour les ondes pseudo sonores ou pseudo acoustiques) pour lesquelles la présence du champ magnétique ne fait que modifier légèrement — en les retardant — les processus de transmissions sonores.

Désignant par V_1 et par V_2 les vitesses correspondantes, la formule suivante donne en fonction de Θ, suivant les signes choisis, ces vitesses:

$$V_{1,2}^2 = \tfrac{1}{2}\left[(V_{\mathrm{A}}^2 + V_{\mathrm{s}}^2) \pm \sqrt{(V_{\mathrm{A}}^2 + V_{\mathrm{s}}^2)^2 - 4\cdot V_{\mathrm{A}}^2 \cdot V_{\mathrm{s}}^2 \cdot \cos^2\Theta}\right]. \tag{46.1}$$

Cette relation peut être traduite par les familles de courbes représentées sur la Fig. VI. On remarquera que pour $\Theta = 0$ la représentation graphique de l'expression algébrique se décompose en les deux droites $V = V_{\mathrm{A}}$ et $V = V_{\mathrm{s}}$ suivant le signe choisi devant le radical. On serait tenté d'affecter une de ces droites à V_1 et l'autre à V_2. Cependant, l'examen des courbes pour $\Theta \neq 0$ montre qu'il ne doit pas en être ainsi: l'attribution — ou, plutôt — la distribution des deux droites entre V_1 et V_2 doit se faire par association deux à deux des quatre demi-droites dont on dispose.

---

[1] Voir l'article de V. L. Ginzburg et A. A. Ruhadze dans ce volume, en particulier Chap. E, p. 508.

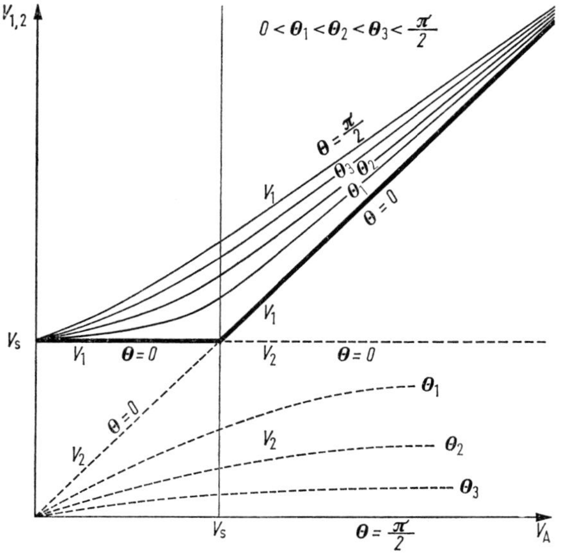

Fig. VI. Schéma destiné à illustrer la séparation des deux droites $V_{1,2}=V_A$ et $V_{1,2}=V_s$ en quatre demi-droites convenablement associées (voir texte p. 301—302). Traits continus ——V_1. Traits discontinus ---V_2.

A savoir: pour V_1 *(ondes magnéto-dynamiques modifiées)*, en prenant $V_1=V_s$ tant que V_A est inférieur à V_s; et $V_1=V_A$ ensuite.

Pour V_2 *(ondes acoustiques modifiées)* c'est l'inverse: il conviendra de prendre $V_2=V_A$ tant que V_A est inférieur à V_s et $V_2=V_s$ ensuite.

Cette manière de faire, qui n'a pas toujours été respectée dans les ouvrages traitant de ce sujet (car du point de vue algébrique elle oblige à changer le signe devant le radical en cours de représentation, au moment où V_A traverse la valeur V_s, mais ceci est une opération parfaitement correcte), s'impose pour des raisons de continuité comme on pourra s'en rendre compte en raisonnant sur des valeurs de Θ petites mais non nulles, puis les faisant tendre vers zéro.

δ) Pour *propagation à travers* le champ magnétique, $\Theta=\pi/2$, la relation se décompose en

$$\left.\begin{aligned} V_1 &= \sqrt{V_A^2+V_s^2} \\ \text{et} \qquad V_2 &= 0. \end{aligned}\right\} \tag{46.2}$$

Dans ce cas il n'y a aucune ambiguïté dans la répartition des graphiques entre V_1 et V_2. Nous retrouvons le résultat fondamental, établi précédemment, sur la valeur de la vitesse de propagation d'une onde longitudinale quand cette propagation se fait normalement au champ magnétique $\boldsymbol{B}$. Quant au fait que V_2 soit nul pour une telle propagation normale, hypothétique, elle est en accord avec le raisonnement que nous avons exposé à la Sect. 45 relativement à la contribution *globale* des deux termes de pression (de forme $\gamma\cdot p$) magnétique et acoustique. En fait, l'examen de la variation de V_2 en fonction de Θ montre que V_2 reste toujours très voisin de zéro tant que Θ n'est pas très petit, donc tant que V_1 conserve son entité caractéristique (qu'il perd pour $\Theta=0$) d'onde, partiellement au moins, longitudinale.

En résumé, on voit que, malgré la difficulté de représenter correctement par des schémas physiques et algébriques, une réalité beaucoup plus complexe — et qui ne se laisse pas décomposer en modes, ce qui constitue une opération plus ou moins artificielle ainsi que nous l'avions indiqué plus haut (voir Sect. 45) —, ces schémas paraissent valables quand ils s'appliquent aux cas bien tranchés les plus importants, tout en devenant flous et sans objet si on voulait chercher à les appliquer aux cas limites, litigieux, et sans réel intérêt. Nous arrêterons donc là leur discussion.

47. Cas où l'approximation magnéto-dynamique devient insuffisante. Nous avons déjà remarqué que l'approximation magnéto-dynamique n'était plus valable dès que la fréquence des ondes considérées devenait d'un ordre de grandeur comparable à la fréquence de giration des ions présents dans le plasma. En toute rigueur nous devrions alors retourner à la théorie générale de propagation des ondes dans les plasmas (relative aux «plasmas froids»), ce qui entraînerait une complication peu efficace tout en nécessitant une extension de notre sujet qui serait incompatible avec les limites qu'il doit garder.

α) Nous adopterons un point de vue beaucoup plus simple, limitant strictement notre exposé aux quelques indications indispensables pour comprendre les *résultats d'expériences*, laissant à ces dernières le soin de nous guider. Les voici:

1) Durant les trajets magnétosphériques — mais pas dans ceux à l'intérieur des guides ionosphériques ou, à fortiori, plus bas dans l'atmosphère neutre — les propagations ne peuvent plus être considérées comme se faisant sans dispersion: il convient maintenant de distinguer entre les différents types de vitesses, notamment entre la *vitesse de phase* et la *vitesse de groupe*.

2) Les *polarisations* de ces ondes — qui, pour celles qui interviennent dans la théorie des pulsations magnétiques **pc1** et **pi1**, s'apparentent aux ondes transversales d'Alfvén — ne peuvent plus être considérées comme linéaires, et elles sont en général de forme *elliptique* ou circulaire.

3) De telles formes de polarisation permettent de distinguer des sens, droite et gauche, de rotation, ce qui amène à classer ces ondes en *ondes à gauche* — sens définis par rapport à la direction et au sens du champ magnétique $\boldsymbol{B}$ et non par rapport à la direction et au sens de propagation — et *ondes à droite*.

β) Une propriété importante liée à ces sens de polarisation est celle du «*guidage*» éventuel de ces ondes *par le champ magnétique* $\boldsymbol{B}$. On démontre que les *ondes à gauche* sont — pour l'ordre de grandeur des fréquences qui nous intéressent, soit autour de 1 Hz — mieux guidées par le champ magnétique que les *ondes à droite*. Rappelons cependant que ce sont ces ondes à droite qui, dans une bande de fréquences plus élevées — celle du kHz, donc des TBF — assurent le guidage partiel des *atmosphériques sifflants* («whistlers»)[1,2]. Etant donné les guidages, plus ou moins rigoureux qui semblent nécessités par les interprétations les plus généralement admises pour les observations de pulsations du type **pc1** entre points — ou régions — conjugués, ce sont aux ondes à gauche auxquelles on fait appel le plus souvent pour expliquer la propagation de ces pulsations. Une telle hypothèse est loin d'être une nécessité absolue, étant donné que, si l'on a bien vérifié que la propagation des **pc1** leur permettait d'aller d'une région donnée du Globe, à la région conjuguée, on n'a pas encore démontré expérimentalement que c'étaient là les seuls trajets suivis exclusivement. Quoi qu'il en soit, elle a été prise comme point de départ par de nombreux auteurs. Parmi eux, Kitamura et Jacobs[3] ont calculé les trajectoires de «paquets d'ondes»

[1] Au sujet des «whistlers» voir les articles de R. Gendrin (p. 461—525) au volume 49/3 et de V. L. Ginzburg et A. A. Ruhadze (p. 395) dans ce volume.

[2] Voir aussi l'article de K. Rawer et K. Suchy, vol. 49/2 de cette encyclopédie, Sect. 57.

[3] T. Kitamura and J. A. Jacobs: Planetary Space Sci. **11**, 869—878 (1963).

gauches à travers la magnétosphère et trouvé que dans une magnétosphère ne présentant pas d'irrégularités (c'est-à-dire, considérée comme homogène à l'échelle locale) le guidage par le champ magnétique était très insuffisant pour ramener une onde montante au point conjugué de sa région d'émission: en s'élevant dans la magnétosphère la normale à l'onde devenait peu à normale à $\boldsymbol{B}$ et il n'y avait plus de propagation.

Reprenant ces calculs, mais en introduisant des hypothèses de non-homogénéité pour la magnétosphère — soit par l'effet de la plasmapause soit par celle de feuillets d'irrégularités d'ionisation — GLANGEAUD et LACOUME[4] ont pu montrer qu'un guidage très précis pouvait être ainsi justifié. On a donc démontré des conditions qui sont suffisantes — mais pas forcément nécessaires — pour retrouver les résultats donnés par l'observation.

En ce qui concerne les trajets dans l'espace situé entre la base de la magnétosphère et le sol terrestre, les résultats expérimentaux de GLANGEAUD[5] ont établi qu'ils n'introduisaient aucune dispersion nouvelle dans la propagation des ondes responsables des **pc1**. En d'autres termes, on peut donc admettre que les vitesses de phase et les vitesses de groupe pour ces ondes sont redevenues égales, aussi bien le long des trajets dans le guide ionosphérique (le «SMIG»[6]) que — ce qui était prévisible — dans l'atmosphère neutre. Remarquons enfin que, contrairement aux ondes auxquelles on rattache les pulsations de périodes plus longues, les longueurs d'onde ayant trait aux **pc1** ne dépassant pas une étendue de l'ordre d'un millier de km, on peut, pour ces pulsations, envisager des processus de propagation proprement dits.

γ) *Influence de l'ionosphère et de la cavité terre-ionosphère.* Les propriétés dynamiques de la magnétosphère ne peuvent être étendues sans changement profond aux milieux qui la bordent inférieurement et la séparent du sol terrestre. Cette question est à la base de l'examen du problème fondamental de la transmission jusqu'au sol des variations magnétiques. Celui-ci peut se décomposer de la façon suivante:

1) Les propriétés de *guides d'ondes* soit de l'ionosphère, soit de la cavité terre-ionosphère (Sect. 48);

2) les propriétés transductrices, passives ou actives, de l'ionosphère (Sect. 49).

48. Guides d'ondes. L'étude des deux guides d'ondes mentionnés ci-dessus au 1) est un sujet trop spécialisé pour pouvoir être abordé ici. Nous renvoyons le lecteur aux Sects. 55 ... 59 ainsi qu'aux références[1,2]. Nous allons seulement rappeler très brièvement la façon dont ces guides peuvent intervenir dans le cas des variations qui nous occupent.

α) Nous commençons par la *cavité terre-ionosphère* qui joue un rôle très important dans la propagation des ondes électromagnétiques dans la gamme T.B.F. («V.L.F.» dans la nomenclature anglaise)[3]. On pense que, par contre, son rôle dans la propagation des ondes U.B.F. (anglais: «E.L.F.») et des variations rapides qui nous intéressent, est limité. En fait, la littérature traitant de cette

[4] F. GLANGEAUD et J. L. LACOUME: Comm. Sympos. IUCSTP. Leningrad (1970).

[5] F. GLANGEAUD: Ann. Géophys. **26**, 299—312 (1970).

[6] «SMIG», pour: «Sub Magnetospheric-Ionospheric Guide». Il s'agit du guide constitué par la couche à minimum de vitesse pour les ondes de ALFVÉN, couche située dans les hautes régions de l'ionosphère presqu'à la base de la magnétosphère et dont la hauteur réelle peut varier dans d'assez larges limites, notamment en fonction de la température.

[1] R. N. MANCHESTER: J. Geophys. Res. **71**, (15), 3749—3754 (1966).

[2] C. GREIFINGER and P. S. GREIFINGER: J. Geophys. Res. **73**, (23), 7473—7490 (1968).

[3] Voir l'article de V. L. GINZBURG et A. A. RUHADZE dans ce volume, Sect. 328, p. 520.

question est contradictoire; cela tient, vraisemblablement, à ce que les données expérimentales sur la façon dont la paroi supérieure de la cavité (couche D de l'ionosphère) réagit à ces ondes électromagnétiques d'ultra basses fréquences, sont encore mal connues: dans une cavité à parois rigides leur propagation serait affectée d'un amortissement considérable, mais si cette couche D participe aux oscillations les conditions de propagation sont très différentes. Il est intéressant à ce sujet de se référer à deux phénomènes, l'un déclenché par l'action artificielle d'une explosion nucléaire à haute altitude (intra-ionosphérique), telle que celles de la série Argus de 1958[4,5], de l'expérience Starfish du 9 juillet 1962 et des expériences des 22 et 28 novembre 1962[6–11], l'autre qui — ayant pour origine les décharges électriques orageuses semi-permanentes issues principalement des régions tropicales du Globe — se traduit, à l'échelle de l'ensemble de la cavité, par les «oscillations de SCHUMANN» (sujet que nous traitons en Annexe, Sects. 55 à 59).

En se référant à la Fig. 61 on peut constater que la perturbation magnétique mondiale engendrée par Starfish (et il en serait de même dans le cas des autres expériences citées) a été constituée initialement par la mise en vibration sur 8 Hz environ (fréquence fondamentale des oscillations de SCHUMANN) de l'ensemble de la cavité. D'une façon plus précise, disons que l'on a pu constater qu'il y avait eu, *dans le diélectrique* compris entre le sol et l'ionosphère, un renforcement important durant plusieurs secondes de la résonance naturelle de la cavité, et *sur sa paroi supérieure* une modification à l'échelle mondiale des propriétés de la partie inférieure de la couche D[12], sans qu'il soit toutefois possible de déterminer dans quelle mesure cette «paroi» avait participé à des oscillations synchrones de celles qui s'étaient ainsi établies dans le diélectrique. Bien qu'elle soit en général rejetée actuellement, la possibilité d'une participation de la cavité terre-ionosphère à une propagation de certains types de variations magnétiques rapides n'est pas totalement exclue.

β) En ce qui concerne le *guide d'ondes ionosphérique*, il est dénommé suivant les auteurs soit «sur-ionosphérique», soit «sous-magnétosphérique», ou encore «SMIG» («sub-magnéto-iono-guide») dans la littérature anglo-saxonne, ce qui correspond au fait que le minimum de vitesse V_A des ondes de ALFVÉN se situe vers le sommet de l'ionosphère, soit vers 500 km d'altitude (encore que cette valeur soit variable avec un grand nombre de facteurs, l'extension de l'ionosphère vers le haut correspondant d'autre part à une notion mal définie). Cette altitude est donc très supérieure à celle (100—150 km) où l'on admet que peuvent circuler les courants électriques (supposés horizontaux) auxquels on rapporte la plupart des perturbations magnétiques observées au sol, soit conventionnellement, soit comme dernier «relais» pour les perturbations venant de plus haut.

En conséquence, chaque fois que nous sommes conduits à admettre que la manifestation au sol d'une certaine variation magnétique présente une extension

[4] E. SELZER: C.R. Acad. Sci. **249**, 1133—1135 (1959).

[5] S. ESCHENBRENNER, L. FERRIEUX, R. GODIVIER, R. LACHAUX, H. LARZILLIERE, A. LEBEAU, R. SCHLICH et E. SELZER: Ann. Géophys. **16**, (28), 264—271 (1960).

[6] J. ROQUET, R. SCHLICH et E. SELZER: C.R. Acad. Sci. **255**, 549—551 (1962).

[7] J. ROQUET, R. SCHLICH et E. SELZER: C.R. Acad. Sci. **255**, 1225—1227 (1962).

[8] J. ROQUET, R. SCHLICH et E. SELZER: J. Geophys. Res. **68**, (12), 1731—1732 (1963)

[9] E. SELZER: Chapitre de "Natural Electromagnetic Phenomena below 30 Kc/s", ed. D. F. BLEIL, p. 107—142. New York: Plenum Press 1964.

[10] H. A. BOMKE, I. A. BALTON, H. H. GROTE, and A. K. HARRIS: J. Geophys. Res. **69**, 3125—3136 (1964).

[11] B. CANER: Pub. Dom. Obs. **31**, No. 1, 1—15 (1964).

[12] R. GENDRIN et R. STEFANT: C.R. Acad. Sci. **255**, 2273—2275 (1962).

géographique beaucoup plus large que l'extension qu'elle est supposée avoir au dessus de l'ionosphère, à la verticale des lieux correspondants, nous aurons à examiner si cette «diffusion» (ou «distribution») horizontale peut être attribuée à un effet de guide d'ondes (sous la forme d'onde électromagnétique ou magnéto-dynamique' suivant celui des deux guides indiqués ci-dessus qui est supposé intervenir, ou à un «transport» par un certain «courant de retour» circulant dans l'ionosphère. Comme l'hypothèse préliminaire (sur une non-excitation générale par la magnétosphère à la verticale des lieux d'observation) ne sera pas toujours confirmée, on voit qu'un grand nombre de mécanismes seront possibles à priori, ceci d'autant plus que des processus partiels, ou mixtes, peuvent être envisagés.

49. Propriétés transductrices. Le paragraphe précédent avait trait essentiellement à des propagations horizontales. Nous nous occupons maintenant plutôt d'une transmission suivant la verticale (ou, tout au moins, qui soit transverse par rapport à l'ionosphère).

α) Plusieurs hypothèses peuvent être formulées sur la constitution même de l'*énergie incidente* et sur les transformations qu'elle peut subir au niveau de l'ionosphère (indépendamment des phénomènes d'absorption et d'amortissement proprement dits: l'énergie incidente peut être apportée par des averses de particules (électrons ou protons), ou des ondes magnéto-dynamiques, avec possibilités de couplages de ces deux actions; elle peut se transformer soit en courants électriques ionosphériques — donnant eux-mêmes naissance à des ondes électromagnétiques dans le diélectrique de la cavité terre-ionosphère — soit se transformer directement en ondes électromagnétiques. La première de ces transformations était encore la seule considérée comme possible il y a peu d'années. Cependant d'après ALFVÉN et FÄLTHAMMER[1] il n'y avait pas une différence essentielle entre la propagation d'une onde magnéto-dynamique et celle d'une onde électromagnétique — de même fréquence, ultra-basse — dans l'ionosphère même, quand on tenait compte de l'interaction entre ces ondes et les constituants ionisés du milieu.

β) Les relations qui régissent le *couplage* entre une onde magnéto-dynamique incidente et l'onde électromagnétique transmise, de part et d'autre d'une interface plane séparant un plasma (tel que le plasma ionosphérique) d'un espace vide (auquel on peut pour ce problème assimiler l'atmosphère neutre située au-dessous de l'ionosphère) ont été explicitées par divers auteurs[2,3]. Ces relations font apparaître plusieurs faits intéressants:

1) Un courant électrique est engendré dans l'interface mais il n'intervient pas directement dans les relations de couplage;

2) la transmission ne s'effectue que pour autant que l'onde incidente (qu'elle soit une onde transversale pure — anisotropique —, ou une onde de compression d'un des modes magnéto-soniques) aborde l'interface suivant une direction comprise à l'intérieur d'un cône de petit angle centré sur la normale au plan de celle-ci. Dans ces conditions l'onde électromagnétique émerge dans une direction quelconque du demi-espace inférieur (mais liée, bien entendu, à la direction exacte de l'onde incidente). Ces relations géométriques rappelent celles qui se présentent lors de la réfraction d'un rayon optique passant d'un milieu d'indice considérale à un autre d'indice faible. Elles nous indiquent également que la

[1] H. ALFVÉN and C.-G. FÄLTHAMMAR: "Cosmical Electrodynamics", 2. edit., sect. 3.4.4., p. 83—85. Oxford: Clarendon Press 1963.

[2] N. ULLAH and S. L. KAHALAS: Phys. Fluids *6*, 284 (1963).

[3] B. HULTQUIST: Space Sci. Rev. **5**, No. 5, 599—695 (1966).

direction d'arrivée au sol de l'onde électromagnétique ultime ne pourra guère nous renseigner sur la direction exacte qu'avait l'onde magnéto-dynamique dont elle est issue.

Malgré leur intérêt théorique il n'est pas sûr que les relations de couplage ainsi déterminées correspondent d'assez près à la réalité, ne serait-ce que parce que les propriétés de guidage que nous avons examinées précédemment — notamment, celles par l'ionosphère elle-même — doivent toujours intervenir dans une certaine mesure (et il suffit qu'elles interviennent très peu, du point de vue des distances de propagation, pour influer déjà considérablement sur les directions d'émergence de l'onde électromagnétique inférieure)[4]. Cette question devient encore plus complexe quand on se soucie des modifications apportées aux polarisations des ondes lors de cette traversée plus ou moins directe de l'ionosphère, voir Fig. 59a et b.

γ) Une autre façon de poser le problème est d'essayer de déterminer dans quelle mesure les *lignes de forces* du champ magnétique terrestre conservent leur connexion de part et d'autre de l'ionosphère lors de variations ou perturbations plus ou moins grandes et plus ou moins rapides. En particulier, dans le cas où ces variations se traduisent par des déplacements latéraux de ces lignes de forces, nous devons admettre que ces déplacements n'affectent pas les «points d'attache» fixes que ces lignes de forces ont avec la partie solide, conductrice, du globe terrestre (à partir d'une certaine profondeur fonction de la rapidité des déplacements), mais par contre qu'ils se produisent dans l'ionosphère, simplement plus ou moins amortis par elle. Il y a donc induction — et production de courants — à la fois dans l'ionosphère et dans les couches superficielles du sol (ceci sans couplage appréciable *direct* entre ces deux effets, pour autant que nous maintenons jusqu'au sol leur cause commune). On voit donc que les courants qui peuvent se développer dans l'ionosphère lors d'une perturbation magnétique peuvent se présenter sous des formes partielles de diverses sortes et c'est par une abstraction particulière que, dans une des formes d'exposition que nous avons rappelée plus haut, on peut être conduit à les présenter sous un terme unique auquel serait attribué tout ce qui se passe au sol (et dans le sol). D'un point de vue un peu différent, on peut voir que les questions précédentes mettent en cause les relations entre les perturbations magnétiques et les perturbations ionosphériques. Le fait que ces dernières se produisent en général à une échelle géographique plus restreinte que les premières (des centaines au lieu de milliers de kilomètres) et qu'elles soient sensibles également à d'autres facteurs que ceux purement magnétiques, ne facilite pas ce genre de comparaisons. Il est intéressant de signaler à ce sujet — malgré leurs résultats échappant aux normes prévues — les efforts qui ont été faits pour tâcher d'expliciter en deux régions conjuguées, notamment entre Kerguelen et Sogra[5] et entre l'Afrique du Sud (Hermanus) et le Sud-Ouest de la France (Hurouqué)[6], l'influence des ionosphères locales sur le comportement des divers types de pulsations qui y sont enregistrées (notamment sur les **pc1** et les **pc3**). Superposées aux relations générales de conjugaison, des différences apparaissent fréquemment — ou même systématiquement — dans ces comportements, mais il n'a pas été possible d'établir de corrélation nette entre les écarts aux relations de conjugaisons et l'évolution des caractéristiques des ionosphères locales, correspondantes. V. A. Troitskaja[7] en conclut que d'autres influences que les seules conditions ionosphériques locales doivent intervenir sur la manifestation des écarts aux relations de conjugaison (des in-

[4] F. Glangeaud: Thèse Fac. Sci., Paris (1970).

[5] V. A. Troitskaja et R. Gendrin: Colloque sur les Expériences Franco-Soviétiques en Points Conjuguées. Publication CNFRA (Comité National Français des Recherches Antarctiques) No. 25, 1—89 (1970).

[6] R. Schlich: Ann. Géophys. **24**, (2), 411—429 (1968).

[7] V. A. Troitskaja: Solar-Terrestrial Physics, Chap. VII, p. 249—251; J. W. King and W. S. Newman, eds.; London and New York: Academic Press 1967.

fluences qui, par exemple, interviendraient sur des parties médianes du trajet magnétique entre points conjugués: mais ce seraient alors des influences magnétosphériques et non plus ionosphériques, à moins que les propagations dans des guides relativement voisins du sol soient en cause).

δ) En *résumé*, le rôle que l'on a attribué à l'ionosphère dans la manifestation des variations magnétiques au sol a subi — depuis la mise en évidence de cet élément fondamental de l'environnement terrestre — l'évolution suivante:

D'abord, et pendant longtemps, on a attribué à peu près intégralement les variations au sol à l'effet direct de grandes nappes de courants électriques ionosphériques. C'était à la fois une convention (notamment pour l'altitude adoptée pour ces nappes, de l'ordre de 100 à 150 km) et une hypothèse, à laquelle on attribuait — et on les attribue toujours pour certaines classes de variations macroscopiques importantes telles que les baies, les orages, etc. — un certain degré de réalité.

Ensuite, avec la découverte des ondes magnéto-dynamiques, et leur détection (directe et indirecte) dans la magnétosphère, une partie du rôle fondamental que l'on faisait jouer ainsi à l'ionosphère a été dévolue à des milieux plus élevés. Ceci concernait surtout les variations rapides, pour lesquelles l'ionosphère gardait cependant son rôle de transducteur.

Enfin, l'importance que l'on a été conduit à donner récemment (au cours de ces quelques dernières années) aux phénomènes de guidage tels que ceux que nous avons signalés plus haut, a redonné un rôle important aux diverses couches ionosphériques. Mais, contrairement à ce qui se passait jadis (lors de la première phase), ce rôle est devenu très compliqué et il est loin d'être bien compris. Nous verrons notamment un exemple de ceci dans le cas des **pc1**.

50. Application aux pulsations continues pc 2, pc 3, pc 4, pc 5. Ainsi que nous l'avons indiqué précédemment (voir Sect. 42) il s'agira ici de vibrations donnant lieu à des phénomènes d'ondes stationnaires.

Les *modes de vibration* à envisager sont, tout au moins sous leur forme schématique, assez simples bien que, ainsi que nous l'avions remarqué (voir Sect. 46) ils aient parfois été empreints de quelque confusion.

α) Ceux, parmi ces modes, qui sont à base de vibrations transversales polarisées rectilignement (ondes de ALFVÉN pures) circulant le long de tubes de forces, pourront, à priori, présenter toute direction de polarisation dans leur plan de front. Mais ceci soulève les problèmes de *couplage entre tubes voisins*: si cette direction de polarisation est perpendiculaire au plan méridien elle pourra se transmettre telle quelle aux tubes voisins se déduisant du premier par une légère rotation de ce plan: de proche en proche on pourra arriver ainsi à la vibration d'un tore complet et l'on aura ce que l'on appelle le *mode toroïdal*. Chacun de ces tores correspond ainsi à un tube de forces initial s'étendant entre deux latitudes géomagnétiques, Nord et Sud, bien déterminées, soit $\pm\phi$, dans l'approximation du dipôle. Il s'en déduit une période de vibrations bien définie (vibrations du type «corde» ou «tige», mais ici avec une infinité de cordes ou de tiges juxtaposées latéralement).

Même sans calculs on se rend compte que la période ainsi introduite doit augmenter avec la longueur du tube de forces générateur, et par suite les pulsations correspondantes devraient présenter des périodes d'autant plus longues que la latitude ϕ augmente. Or, comme nous l'avons vu (voir Sect. 25), ce fait est loin d'être établi pour les **pc3** et les **pc4**. Seuls, certains cas particuliers (notamment, divers groupes caractéristiques de pulsations isolées, et certaines pulsations accompagnant les **SSC** d'orages) y satisfont d'une manière nette. Dans la plupart des autres cas, chaque fois que l'on peut identifier à des latitudes différentes un phénomène, ou un évènement, pulsatoire, on trouve qu'il est formé de pulsations individuelles ayant la même période aux divers lieux d'observation. Evidemment, ceci peut toujours prêter à discussion, notamment quand l'évènement considéré n'est pas formé de pulsations identiques, à moins que l'on puisse — ce qui reste assez exceptionnel — établir une correspondance oscillation par oscillation. Une autre méthode d'étude de cette question consiste à opérer par voie statistique en examinant si la distribution générale des périodes aux diverses stations de latitudes différentes présente des maximums concordants ou non (c'est-à-dire se présentant pour des valeurs identiques pour les périodes). Dans cette méthode (dont la mise en œuvre est plus facile) les résultats sont plus délicats à interprêter.

β) Quoi qu'il en soit, l'étude quantitative du problème, en dépit des données incertaines sur lesquelles elle est obligée de s'appuyer, montre que si nous admettons ce mode toroïdal, on ne peut échapper à une *variation extrêmement rapide de la période* avec la latitude. C'est là une première contradiction avec les cas les plus courants de **pc3**. De plus, les valeurs numériques obtenues ne concordent (en ordre de grandeur) avec les valeurs expérimentales (observées) que pour des choix étroits de latitudes (géomagnétiques) ϕ. Cette concordance elle-même est douteuse car elle apparaît davantage comme une simple conséquence numérique de la grande étendue de variation que nous avons signalée, que comme une propriété spécifique des latitudes correspondantes. En d'autres termes, on pourra toujours trouver une latitude pour laquelle les calculs et l'observation seront en accord, mais cela ne voudra pas dire que des pulsations correspondantes soient localisées d'une façon privilégiée sur les tubes de forces aboutissant à ces latitudes.

Pour ces différentes raisons on est arrivé à la conclusion que le mode toroïdal est peu propre à fournir un modèle valable pour les pulsations continues des catégories **pc2** et **pc3**. Il paraît convenir très bien, par contre, aux pulsations **pc5**, notamment aux pulsations géantes **pg**, dont on sait par ailleurs qu'elles se trouvent localisées d'une façon assez précise sur des tubes de forces de latitude déterminée. Nous y reviendrons (voir Sect. 52).

γ) Les considérations précédentes ne doivent pas nous faire oublier l'intérêt qui s'attache à jeter les bases d'un calcul aussi précis que possible pour les *périodes de résonance* prévues par le mode toroïdal pour chaque tube de forces déterminé (défini par exemple par la valeur ϕ_1 correspondant à ses extrémités. Il s'agit de calculer l'intégrale

$$\frac{1}{2}\,T = \int \frac{\mathrm{d}l}{V_\mathrm{A}} \tag{50.1}$$

le long d'une ligne de forces, en tenant compte de la valeur de

$$V_\mathrm{A} = \frac{B}{\sqrt{\mathrm{u}\,\mu_0 \cdot \varrho}} \tag{50.2}$$

dont on évalue au mieux les paramètres.

Pour ce calcul il est légitime — tant que l'on opère pour des latitudes pas trop élevées, et que les lignes de forces correspondantes ne sont pas trop déformées par la proximité de la plasmapause ou de la magnétopause — d'opérer dans l'approximation du dipôle.

Dans ce cas, la valeur de B ne soulève aucune difficulté. On prendra:

$$B = \mathrm{u}\,\mu_0 \frac{M}{R^3} (1 + 3 \cos^2 \phi)^{\frac{1}{2}} \qquad (50.3)$$

M: moment magnétique du dipôle; R: distance au centre; ϕ: angle polaire.

Les difficultés apparaissent avec le choix à faire pour ϱ. La diversité des solutions essayées par les divers auteurs qui se sont occupés de cette question montre à quel point une grande incertitude règne encore dans ce domaine.

Pour cette raison nous ne reproduirons pas les calculs correspondants, dont certains supposent ϱ constant dans toute l'étendue de la magnétosphère traversée par les lignes de forces considérées, d'autres adoptant des lois empiriques (les unes en $1/R^3$, les autres en $1/R^8$) pour ce paramètre. Nul doute que quand les mesures en satellites et fusées pourront constituer un réseau serré de mesures dans le temps et dans l'espace, on ne puisse adopter des valeurs plus adéquates pour la distribution de ϱ et — aussi compliquée que se présentera cette distribution — effectuer les calculs de périodes par intégrations numériques. Il est vrai qu'à ce moment là ces calculs auront perdu une grande partie de leur intérêt, à savoir: se servir de l'observation des pulsations (et des mesures de leur période) en tant que moyen de détermination des densités magnétosphériques.

N'abandonnons pas cet examen du mode toroïdal sans signaler certaines études récentes[1,2] qui mettent en évidence la grande sensibilité de ce mode — notamment au point de vue des périodes — aux discontinuités (ou aux forts gradients) de densités liés à la position exacte, à un instant déterminé, de la plasmapause quand elle est tangentée sur de longs parcours. Ce sujet va être repris différemment en δ.

δ) Examinons maintenant le cas où la *direction de polarisation* se trouve *dans le plan méridien*. Dans ces conditions il n'y a pas de couplage direct entre les vibrations du tube de forces élémentaire que nous considérons initialement et celles de tubes voisins de mêmes latitudes limites mais d'azimuts différents. Par contre, des couplages se produiront avec les tubes situés dans le plan méridien initial mais au-dessus ou au-dessous du tube considéré en premier. De proche en proche, tous les tubes constituant une «tranche méridienne» pourront être mis en vibration. C'est le *mode poloïdal.*

Comme les tubes situés à des altitudes différentes ne correspondent certainement pas à une même période (étant donné leurs dimensions géométriques qui deviennent très différentes), les couplages que nous venons de prendre en considération se font entre systèmes oscillants non accordés ce qui, *à priori*, ne doit pas favoriser la formation d'oscillations importantes. En fait, le problème doit être abordé en tant que concernant la vibration de la tranche méridienne toute entière, limitée vers le haut par la plasmapause ou par la magnétopause, et vers le bas par l'ionosphère ou par la couche à vitesse de Alfvén maximale. Les calculs numériques qui ont pu être faits à partir de ce schéma ont donné des valeurs de T tombant principalement dans la gamme des **pc3**. Leur résultat dépend essentiellement, non seulement de celle des deux hypothèses qui est retenue pour la limite inférieure de la cavité, mais surtout (bien que d'une façon moins instable que dans le cas limite que nous avions considéré en γ ci-dessus) de la position exacte de la magnétopause, ou de la plasmapause, à l'époque, ou à l'instant considérés. Un assez grand nombre d'auteurs se sont intéressés à cette question, essayant de trouver une relation entre les périodes des **pc3** ou des **pc4** — et ils ont fait de même avec les **pi2** — et les positions des frontières supérieures qui viennent d'être invoquées, telles que l'on peut les déterminer maintenant, plus ou moins régulièrement, par les explorations en satellites.

---

[1] F. GLANGEAUD et J. L. LACOUME: Comm. Sympos. IUCSTP, Leningrad (1970).

[2] F. GLANGEAUD: Ann. Géophys. 26, 299—312 (1970).

ε) On pense ainsi avoir établi que la période de ces catégories de *pulsations décroissait* quand la magnétosphère se contractait — notamment lors d'une augmentation générale du niveau de l'activité magnétique — le phénomène inverse prenant place lors des déplacements contraires.

Ces conclusions ne sont cependant pas à l'abri de plusieurs critiques: *d'une part*, il est rare que l'on ait qu'une seule période, bien définie, pour l'ensemble des **pc3** observés à un même instant en différents points du Globe. Il est alors difficile de dire que l'on a assisté à un certain instant à la variation des caractéristiques d'un système oscillant déterminé, ou si il n'y a pas eu un certain chassez-croisé dans les observations (confondant la variation d'un paramètre *d'un* système avec un changement de système). *D'autre part*, nous avons vu que l'on observe souvent, en synchronisme mondial, des sauts brusques de période pour des trains de **pc3** et, comme — dans l'hypothèse à vérifier — ces sauts devraient être causés par des déplacements importants de la magnétopause (ou de la plasmapause) il est difficile d'admettre que de tels déplacements puissent se produire aussi brusquement et se succéder parfois irrégulièrement dans les deux sens. Une hypothèse plus vraisemblable serait que une certaine couche — peut-être la plasmapause — joue le rôle de «verrou» ou de filtre à certains instants et pas à d'autres, ce qui entraînerait des changements brusques dans les cavités en cause. Cette hypothèse basée sur des phénomènes de «commutation» paraît plus plausible que l'hypothèse précédente sur les changements brusques de dimensions d'une même cavité. Bien entendu cela ne résoud pas la question des «causes premières» et il est possible que de telles causes, agissant à partir de l'espace interplanétaire, interviennent à la fois sur les dimensions des cavités et sur la plus ou moins grande «transparence» de leurs parois. Parmi les causes premières que l'on peut invoquer (sans essayer de les remonter jusqu'au Soleil lui-même) on a étudié plus particulièrement les suivantes: variations de vitesses (en valeur et en direction) du vent solaire, turbulence de ce vent, rôle de filtre actif ou passif de la magnétogaine. Les résultats les plus intéressants paraissent avoir été obtenus en comparant l'activité des **pc3** et **pc4** à l'inclinaison suivant laquelle le vent solaire prend contact avec la magnétosphère: l'activité des pulsations paraît se manifester chaque fois que le vent solaire attaque plus ou moins tangentiellement les surfaces de séparation limitant la magnétosphère.

ζ) On a étudié si on devait attribuer aux deux catégories **pc3** et **pc4** des *cavités résonnantes* différentes ou si elles pouvaient partager le même domaine. Certains indices laissent à penser que celui des **pc4** serait plus restreint que celui des **pc3** et se limiterait, dans la gamme des altitudes, entre deux et trois rayons terrestres comptés à partir du sol. En fait, la divergence la plus grande a trait aux modes stationnaires eux-mêmes. Si, s'appuyant sur des arguments du type de ceux présentés ici, on n'hésite plus à faire appel au mode poloïdal pour les **pc3**, la question reste des plus controversée pour les **pc4**, soit qu'on les rapproche des **pc3** dont ils ne représenteraient qu'une extension vers les périodes plus longues (extension se produisant particulièrement au cours des années les plus calmes des cycles solaires de onze ans), soit qu'on les considère comme un extension, dans le sens inverse, des **pc5** (dont nous allons voir qu'il convient de les rattacher à un mode toroïdal). Cette question n'est en rien résolue et les difficultés que l'on éprouve à vérifier expérimentalement l'existence, ou la non-existence, d'un effet de latitude sur la période des différents types de pulsations (ce qui pourrait aider à lever l'ambiguïté relative au mode dont sont issues les **pc4**), ne facilitent pas l'éclaicissement de cette question. De l'autre côté des **pc3**, les **pc2** ne soulèvent pas les mêmes difficultés: la plupart de leurs propriétés permet de les considérer comme des **pc3** dont la période aurait été diminuée au-delà des normes conventionnelles sous l'influence des divers facteurs susceptibles den réduire les dimensions des cavités résonnantes: augmentation des actions (vitesse ou turbulence) du vent solaire, augmentation du niveau d'agitation générale, mondiale (K_p), etc.

51. Transformation des modes. Bien qu'il apparaisse que notre conclusion soit, en ce qui concerne les **pc3**, de considérer qu'elles sont la manifestation d'ondes stationnaires se rattachant au mode poloïdal — ce qui explique, en particulier, l'indépendance de leur période d'un effet marqué de latitude pour la plupart d'entre-elles — une excitation particulière d'un faisceau plus ou moins étroit de tubes de forces traversant le domaine spatial siège de ce mode poloïdal n'est pas exclue, cette excitation donnant alors lieu à un autre mode stationnaire, du type toroïdal. Nous avons vu (voir Fig. V, p. 300) que de tels couplages étaient possibles.

Dans ces conditions, l'énergie de l'onde stationnaire principale (dans le mode poloïdal) peut être transmise plus efficacement à l'ionosphère (dans les régions où les tubes de forces ainsi excités y aboutissent) pour être retransmise au sol (sous forme d'ondes électromagnétiques ordinaires).

Quoi qu'il en soit, un problème important lié à la forme géométrique et aux frontières du système résonnant est celui posé par l'influence des saisons: la magnétosphère de jour, siège des **pc3**, n'est évidemment pas identique dans ses parties été et hiver, les différences se faisant de plus en plus marquées quand on se rapproche des zones polaires. Nous avons déjà signalé (voir Sect. 24) que cet effet des saisons ne se manifestait pas aussi rapidement que ce à quoi on aurait pu s'attendre, quand on s'éloignait de part et d'autre de l'équateur. Il redevient au contraire prépondérant quand on atteint les latitudes à partir desquelles les lignes de forces ne se ferment plus vers l'avant (côté Soleil) mais sont rejetées — en restant probablement ouvertes — vers la queue de la magnétosphère. Ces faits militent en faveur d'une onde stationnaire homogène pour les oscillations — quels que soient leur mode — responsables des **pc3** et assimilées.

Ils peuvent être compliqués par l'influence des ionosphères locales mais — à part quelques cas particuliers — il semble que cette influence ne se manifeste qu'épisodiquement et soit plus limitée que l'on ne pensait (voir Sect. 49γ).

En définitive, l'examen des points précédents nous confirme que les **pc2**, **pc3** et **pc4** sont essentiellement d'origine magnétosphérique et que «l'effet de jour» qui les caractérise est dû *directement* à la rotation de la Terre (changeant la forme des cavités résonnantes et les couplages énergétiques avec le milieu).

52. Cas des pc 5 — Pulsations géantes — Relations avec les TBF et avec les orages. Avec les **pc 5** nous avons un des exemples les plus satisfaisants de bon accord entre les connaissances morphologiques que l'on a maintenant pour cette classe de pulsations et les considérations théoriques qu'on leur applique: la vibration stationnaire d'un tube de forces assez étroit, mettant en jeu une onde de Alfvén transversale, polarisée perpendiculairement au plan méridien du lieu. On explique ainsi la netteté de la conjugaison applicable à ces pulsations, les localisations aurorales pour les extrémités de ce tube de forces, ainsi que l'effet de latitude. Plus difficile à éclaircir reste la dépendance de leur période de l'époque à laquelle on se trouve par rapport à l'évolution du cycle solaire: périodes devenant plus courtes quand on aborde la phase la plus calme du cycle solaire[1], ce qui est l'inverse des observations faites en ce qui concerne les **pc2** et **pc3**.

---

[1] Y. KATO and T. SAITO: Rep. Ionos. Space Res. Japan **18**, 183—187 (1964).

Une question délicate, soulevée récemment, serait leur lien avec les sous-orages polaires. Plus généralement, les auteurs que nous venons de citer ont pensé pouvoir établir deux triéquivalences complémentaires, à savoir:

pc5↔sous-orage polaire↔émission TBF dite «chœur de l'aube»,

pc4↔orage mondial ↔émission TBF dite «souffle».

La rigueur de ces analogies reste précaire mais elles ont l'avantage de mettre en relief certains des points cruciaux qui devraient être élucidés; notamment: — laissant de côté la distinction entre les deux types cités d'émissions TBF, ce qui dépasse notre domaine[2] — la relation entre les sous-orages polaires et les orages mondiaux et celle entre **pc5** et **pc4**. Revenant à ce dernier point, nous y retrouvons la tendance à considérer les **pc5** comme un phénomène plus localisé — et probablement plus auroral — que les **pc4**, ces derniers paraissant ne pas affecter eux-mêmes d'aussi vastes portions de la magnétosphère que les **pc3**. Il nous faut cependant remarquer qu'il paraît excessif de lier les **pc4** à des orages (mondiaux).

Ceci semble en contradiction avec la remarque de V. A. TROITSKAJA et d'autres auteurs sur le fait d'observation que ce type de pulsations se montre plus fréquemment au cours des périodes calmes du cycle solaire. On retrouve ainsi que les **pc4** sont, parmi tous les types de pulsations, celles pour lesquelles nos connaissances ont beaucoup de peine à dépasser le niveau morphologique direct: on ne peut affirmer nettement que ces pulsations soient magnétosphériques ou aurorales; qu'elles soient excitées directement — ou à retardement (mais par quel processus?) — par les orages; qu'elles soient différentes ou non des **pi2** que l'on trouve souvent simultanément dans la magnétosphère de nuit.

53. Cas des pc 1 et des pi 1. Ainsi que nous l'avons indiqué à la Sect. 42, les longueurs d'ondes relativement courtes (quelques centaines de km) que l'on peut attribuer à ces deux catégories de pulsations quand on suppose que, comme les précédentes, elles ont pour support des ondes magnétodynamiques, permettent d'envisager pour elles de véritables processus de propagation. Il importe alors de déterminer pour chacune d'elles les caractéristiques suivantes: modes d'excitation et d'entretien, modes de propagation dans la magnétosphère, guidage ou traversée ionosphérique, arrivée au sol. La discussion de ces caractéristiques ne peut cependant être faite strictement une par une. En particulier, les deux premières dépendent étroitement du mode magnétodynamique envisagé pour les ondes. C'est ce que nous allons examiner en premier lieu. Ainsi que pour les développement suivants, nous nous appuierons largement sur les bases déjà données précédemment (voir Sect. 47), ce qui nous permettra de limiter les présents développements à un essai de synthèse.

α) *Mode magnétodynamique intéressé — Propagation dans la magnétosphère.* L'adoption d'un mode déterminé dépend des conditions — reconnues expérimentalement — auxquelles doit satisfaire la propagation. Dans le cas qui nous intéresse, c'est en se basant sur les résultats des expériences entre points conjugués que le *mode gauche* (voir Sect. 47) a été adopté, ceci aussi bien pour les **pi1** que pour les **pc1**. Nous avons déjà remarqué (Sect. 47) que ces résultats ont été peut-être abusivement stylisés; c'est-à-dire que, ayant reconnu l'excellence des relations entre «points conjugués», on en a déduit que c'était là un résultat spécifique, exclusif, de la conjugaison, sans qu'une exploration expérimentale suffisante des autres possibilités ait été menée. Or, peu-à-peu, les nombreux enregistrements faits dans le monde entre *points* ou *régions* conjugués, ont montré

[2] Voir l'article de R. GENDRIN au volume 49/3, p. 461—525, de cette encyclopédie.

que la conjugaison n'était parfois définie qu'à un millier de km près; (elle l'est en moyenne à une centaine de km près — et parfois mieux — en latitude, et trois à quatre cents km en longitude). Un des problèmes qui se pose est donc de savoir dans quelle mesure cet «étalement» de la conjugaison serait dû à une imperfection dans le guidage à travers la magnétosphère, ou à des facilités de propagation isotropique dans d'autres milieux, tel que l'ionosphère.

Une autre question fondamentale a trait aux liens à établir entre les caractères morphologiques spécifiques de cette catégorie de pulsations, notamment quand elles sont structurées, — faisant intervenir, en particulier, les *périodes d'oscillations* et celles *de répétition* — et les mécanismes de base retenus pour leur théorie. On peut se rendre compte par l'énorme variété de ces derniers, tels qu'ils ont été discutés au cours des quelques années qui ont suivi l'A. G. I. («Année Géophysique Internationale» 1957/58), avant d'avoir atteint une certaine unification à partir de 1965 environ, combien la compréhension du phénomène a été hésitante, erronée et peu efficace, avant de réussir à déboucher sur le mécanisme actuellement admis dans ses grandes lignes par presque tous les auteurs.

C'est ainsi que les premières théories[1-4] ont examiné les possibilités d'une correspondance entre: période d'oscillation et aller-retour (de paquets d'électrons ou de protons) entre points conjugués, d'une part; période de répétition et dérive des particules autour du champ magnétique terrestre, d'autre part. On dut abandonner ce point de vue qui ne pouvait assurer les ordres de grandeur requis pour ces périodes qu'en faisant appel à des protons ultra rapides, d'une énergie telle qu'aucune observation ne pouvait le justifier. (Il aurait fallu, par exemple, qu'en un temps de l'ordre d'une seconde les paquets de particules aient pu faire un aller-retour entre les deux hémisphères Nord et Sud.)

Les tentatives suivantes pour échapper à ces difficultés firent appel à des mécanismes très divers (résonance entre les couches supérieures de l'ionosphère et celle à maximum de vitesse pour V_A[5]; résonance entre le sommet et la base de l'ionosphère; etc. ...). Elles eurent le mérite d'introduire une meilleure compréhension des ordres de grandeur numériques à respecter, préparant ainsi la voie à la compréhension actuelle, assez satisfaisante, du phénomène. Celle-ci fait intervenir des «paquets d'ondes» (du mode gauche, ainsi que nous l'avons vu, voir Sect. 47) dont les impacts successifs dans les «régions miroirs» Nord et Sud donnent la période de répétition (c'est-à-dire celle des «perles» et non celle des «grains», tels que nous les avons introduits, voir Sect. 27). Quant à la période d'oscillation, elle traduit les effets au sol — à partir des pertes d'énergie subies par ces paquets d'ondes lors de leur réflexion sur ces points miroirs — de la période propre des ondes en jeu, ou de celle des mouvements de révolution, autour du champ terrestre, des particules (protons principalement) accompagnant les paquets d'ondes («effet synchrotron»). Précisons que ce mécanisme, tel que nous venons de l'évoquer pour des trajets aller-retour entre points conjugués, s'applique spécifiquement aux **pc1**, un mécanisme semblable étant invoqué pour les **pi1**, mais l'aller-retour étant remplacé par deux demi-trajets allant chacun des régions équatoriales vers les points conjugués. La distinction précédente est imposée par les *oppositions de phases* présentées par les **pc1** aux points conjugués, alors qu'on observe un accord de phase (tout au moins grossier) pour les **pi1**.

---

[1] L. R. TEPLEY and R. C. WENTWORTH: J. Geophys. Res. **67**, 3317—3333 (1962).
[2] R. C. WENTWORTH and L. R. TEPLEY: J. Geophys. Res. **67**, 3335—3343 (1962).
[3] R. R. HEACOCK: J. Geophys. Res. **68**, 589—591 (1963).
[4] R. GENDRIN: Ann. Géophys. **19**, (3), 197—214 (1963).
[5] J. A. JACOBS and T. WATANABE: Planetary Space Sci. **11**, 869—878 (1963).

Le stade suivant s'attaqua à la compréhension du guidage à travers la magnétosphère, d'un point conjugué à l'autre. En effet, bien que les ondes gauches soient assez bien guidées par les lignes de forces du champ magnétique (dans le cas de ces fréquences très basses; si nous étions dans la gamme TBF ce serait l'onde droite qui bénéficierait d'un tel guidage), les calculs de trajectoires effectués par KITAMURA et JACOBS[6] ont montré l'impossibilité d'une propagation tout le long d'une ligne de forces, tout au moins dans l'hypothèse où la densité du plasma magnétosphérique ne présenterait qu'une variation progressive: Ils avaient trouvé, en effet, que, au cours de la propagation, la normale à l'onde tendait rapidement à s'orienter perpendiculairement au champ magnétique. De plus, les rayons avaient tendance à diverger vers l'extérieur des lignes de forces. Ces auteurs en avaient déduit que, de même que pour les TBF dans le cas d'une onde droite (cas des atmosphériques sifflants), l'existence, vérifiée expérimentalement, d'un guidage, devait entraîner la présence de colonnes ionisées. Reprenant ces travaux, GLANGEAUD et LACOUME[7] ont montré que les discontinuités — ou les forts gradients de densités — introduits par la présence de la plasmapause, suffisaient, par l'intervention de réflexions multiples, à maintenir un paquet d'onde dans une localisation volumétrique bien limitée et à assurer une bonne focalisation des ondes au-dessus de la région ionosphérique conjuguée de celle de départ. A partir de là, ces ondes peuvent parvenir à des latitudes très variées suivant des processus qui ont été précisés.

β) *Modes d'excitation et d'entretien.* Nous avons déjà vu (voir Sect. 27) qu'il était maintenant généralement admis que les **pc1** étaient engendrés par des processus d'interaction «particules-ondes» se produisant dans les parties les plus éloignées (zones équatoriales) de tubes de forces privilégiés (correspondant à des valeurs du paramètre L de McILWAIN — de l'ordre de 5, ce qui correspond à des latitudes de raccordement au sol de l'ordre d'une soixantaine de degrés. Les modèles proposés pour ces interactions se sont inspirés de ceux qui avaient été introduits bien avant, dans une autre bande de fréquences, celle des TBF. Il convient cependant de souligner les différences suivantes, importantes, entre les deux cas:

— dans la gamme TBF ce sont des ondes «à droite» (de même sens de rotation que les électrons autour du champ magnétique, d'où l'expression «d'ondes du type électron») dont on considère l'interaction avec les faisceaux de particules chargées; dans la gamme des **pc1** nous avons déjà indiqué les raisons pour lesquelles des ondes «à gauche» (du type «ionique»[8]) devaient être retenues comme modèle

— même en tenant compte de cette inversion des modes (sans laquelle il n'y aurait plus aucune équivalence), les deux cas nous conduisent à des résultats opposés en ce qui concerne les dispersions dont sont affectées les ondes dans leurs trajets magnétosphériques: dans la gamme TBF ce sont les fréquences les plus élevées qui présentent la vitesse de phase la plus élevée (ce qui donne aux «atmosphériques sifflants» — ou «Whistlers» — leur morphologie acoustique caractéristique allant de l'aigu au grave), alors que dans celle qui nous occupe ce sont les fréquences les plus basses qui arrivent les premières (voir Fig. 50 et 52). Rappelons à ce sujet que cette dispersion ne se manifeste dans cette gamme des «Ultra-Basses-Fréquences» («UBF») que dans la partie la plus

[6] T. KITAMURA and J. A. JACOBS: Planetary Space Sci. **16**, 863—879 (1968).
[7] F. GLANGEAUD et J. L. LACOUME: Comm. Sympos. IUCSTP Leningrad (1970).
[8] Une explication détaillée de la nomenclature des modes se trouve chez K. RAWER et K. SUCHY au volume 49/2 de cette Encyclopédie, Sect. 7, p. 42—59.

élevée de la gamme, pour les périodes plus courtes que quelques secondes — ce qui est bien le cas des **pc1** et des **pi1** — c'est-à-dire quand on se rapproche en fréquence de la gyrofréquence des ions dans les conditions magnétosphériques de la traversée.

Compte tenu des remarques précédentes, nous pouvons transposer ainsi, des TBF aux UBF, les schémas d'interaction entre une onde se propageant le long d'une ligne de forces (onde dont il nous faut supposer l'existence à priori, mais avec une amplitude qui, en théorie, pourrait être aussi minime qu'on le voudrait) et un faisceau de particules présentant aussi une composante non négligeable parallèlement à la même direction. Si ω est la fréquence (angulaire) d'une des composantes déterminée de l'onde, $\boldsymbol{k}$ son facteur d'onde (compté positivement si l'onde circule dans le sens du champ, négativement dans le cas contraire), $\boldsymbol{V}$ la composante de vitesse du faisceau de particules (compté également positivement si elle est dirigée dans le sens du champ, négativement dans le cas contraire), la fréquence ω_r sur laquelle pourra prendre place une résonance et, par suite, un transfert d'énergie du faisceau vers l'onde, s'exprimera — en tenant compte de l'effet DOPPLER, suivant lequel la fréquence de l'onde est «vue» par le faisceau dans un système de référence animé de la vitesse $\boldsymbol{V}$ — par la relation suivante

$$\omega_r = \omega - \boldsymbol{k}\cdot\boldsymbol{V}. \tag{53.1}$$

Cette relation est valable algébriquement dans tous les cas, à condition de respecter les conventions de signe qui ont été posées, et de considérer ω_r comme positif ou négatif (ω étant pris positif par hypothèse) suivant qu'il correspond ou non au même sens de rotation que celui de l'onde dans le référentiel fixe.

Pour que le transfert se fasse effectivement, il faudra que la gyrofréquence d'une des composantes du faisceau (prise également avec son signe propre, positif ou négatif, suivant qu'elle correspond à une rotation du champ magnétique qui soit de même sens, ou de sens contraire, respectivement, que celle relative à ω) soit égale à ω_r.

γ) On peut alors expliciter ainsi les *différents cas possibles*:

(i) *Onde à gauche* et *faisceau de particules à gauche* également. (Mode «ionique», c.-à.-d. rotation du champ comme celle des ions positifs dans le même champ magnétique[8]. La nomenclature «à gauche» étant toujours appliquée, conventionnellement, en prenant comme direction et sens de référence ceux donnés par le champ magnétique.) Ce sera, par exemple, l'interaction entre un faisceau de protons et une onde elle-même du «mode ionique». Dans ce cas ω_r devra avoir le même signe que ω, ce qui implique: soit que le terme $\boldsymbol{k}\cdot\boldsymbol{V}$ soit positif mais petit (mais on ne peut espérer alors qu'il se produise un transfert d'énergie appréciable), soit que $\boldsymbol{k}\cdot\boldsymbol{V}$ soit négatif — le faisceau de particules étant animé d'une vitesse *antiparallèle* par rapport à celle de l'onde — son effet venant donc s'ajouter à celui de ω. Un transfert d'énergie important peut alors se faire lors de la résonance.

(ii) *Onde à gauche* (mode «ionique»[8]) et *faisceau de particules à droite* (faisceau d'électrons). Il faut donc que la valeur de ω_r sur laquelle se fera la résonance soit «à droite», donc de signe contraire à $\boldsymbol{\kappa}$ en valeur algébrique. En explicitant sa valeur absolue, l'équation s'écrira:

$$-\omega_r = \omega - \boldsymbol{k}\cdot\boldsymbol{V}. \tag{53.2}$$

Ceci ne pourra se produire que si $\mathbf{k}\cdot\mathbf{V}$ reste positif, c'est-à-dire si le faisceau se propage dans le même sens que l'onde (propagation *parallèle* à celle de l'onde). Pour qu'un transfert important d'énergie puisse se faire, il faudra que ce terme

$\boldsymbol{k}\cdot\boldsymbol{V}$ ait une valeur suffisamment importante, ce qui ne sera réalisé que dans le cas d'électrons animés de vitesses relativistes.

(iii) *Onde à droite* (mode «électronique»[8]) et *faisceau de particules á gauche* (par exemple, action d'un faisceau de protons sur une onde «à droite»). Dans ce cas il faut que ω_r soit de signe contraire à celui de ω, donc soit négatif suivant nos conventions. En explicitant ce signe, la relation s'écrira:

$$-\omega_r = \omega - \boldsymbol{k}\cdot\boldsymbol{V}. \qquad \text{comme pour} \quad (53.2)$$

On voit qu'il faudra que le terme en $\boldsymbol{k}\cdot\boldsymbol{V}$ soit suffisamment grand (protons d'assez grande énergie) pour qu'un échange important d'énergie puisse prendre place.

(iv) *Onde à droite* (mode «électronique»[8]) et *faisceau de particules à droite* (par exemple, action d'un faisceau d'électrons sur une onde «à droite»). On devra alors avoir le même signe (donc positif) pour ω_r et pour ω et la relation s'écrira sans changement (signes explicités):

$$\omega_r = \omega - \boldsymbol{k}\cdot\boldsymbol{V}. \qquad (53.3)$$

Il faudra donc, soit que le terme $\boldsymbol{k}\cdot\boldsymbol{V}$ reste petit (mais l'échange d'énergie sera alors très faible), soit que le faisceau soit animé d'une propagation *antiparallèle* par rapport à l'onde.

(v) *Application aux cas réels.* La comparaison des déductions précédentes avec les résultats déjà connus nous a déjà conduits à rejeter le mode «à droite» (mode «électronique»[8]) (voir Sect. 47) qui — dans le cas des UBF, contraire à celui rencontré dans la gamme TBF — n'assure pas un guidage suffisant dans la magnétosphère. La résonance du mode «à gauche» (mode «ionique»[8]) avec un faisceau d'électrons nécessitant — pour être efficace — des électrons animés de vitesse relativistes (non détectés en général lors des cas de **pc1** observés) il ne nous reste plus, comme seul cas possible, que l'interaction entre un faisceau d'ions (protons) et une onde «à gauche». C'est bien là l'hypothèse la plus généralement admise.

La mise en résonance, consécutive à l'interaction, se produirait à chaque passage des paquets d'ondes au voisinage de l'équateur magnétique, assurant ainsi un mécanisme d'entretien périodique. L'étude exacte de ce mécanisme est d'ailleurs délicate et nous ne pourrons pas la faire ici.

δ) A partir des données ci-dessus il devient facile de comprendre l'ensemble des résultats expérimentaux relatifs aux **pc1**. En particulier, les phénomènes de dispersion croissante (structure des spectres «en éventail», voir Sect. 30 et Fig. 50 et 52) se comprennent aisément, sans cacher toutefois la complexité parfois très grande des cas à expliciter. Cette dispersion se produit quasi intégralement au cours des trajets d'aller et retour magnétosphériques (ce qui a permis de fixer avec quelque précision les valeurs du paramètre L ($3{,}5 < L < 5{,}5$) déterminant les tubes de forces où ils se produisent) mais — ainsi que l'a démontré expérimentalement F. GLANGEAUD[9] — elle reste fixée à une valeur limite à partir du moment où la propagation se fait le long des guides d'ondes ionosphériques et, dans l'atmosphère, au-dessous de l'ionosphère. Durant ces derniers trajets, la valeur relative des fréquences ω des ondes magnéto-dynamiques, comparée à la gyrofréquence ionique, est suffisamment basse pour que l'approximation d'une onde sans dispersion (en particulier assurant l'identité: vitesse de groupe = vitesse de phase) soit valable.

[9] F. GLANGEAUD: Ann. Géophys. **26**, 299—312 (1970).

ε) *Cas des* **pi1**. La théorie des **pi1** est calquée sur celle des **pc1** mais en introduisant dans celle-ci les modifications indispensables pour rendre compte des propriétés particulières de ces pulsations. Ces dernières peuvent se résumer ainsi:

1) Moins grande régularité que les **pc1**; souvent même, extrêmement irrégulières (en forme et en moments d'apparition);

2) pseudo-période plus longue (descendant rarement au-dessous de 2 sec) et rejoignant celle de variations plus lentes (**pc2**, en particulier);

3) synchronisme (approximatif) des arrivées de paquets d'ondes dans les régions conjuguées (au lieu de l'«alternance» propre aux **pc1**).

Les adaptations (à partir de la théorie des **pc1**) se font ainsi:

a) Valeurs plus grandes pour le paramètre L, ce qui reporte le phénomène sur des tubes de forces plus élevés, assurant un contact équatorial plus proche avec les parties de la magnétosphère de nuit extérieure à la plasmapause, donc avec les zones d'instabilité liées à la zone neutre magnétosphérique, à la queue de la magnétosphère, aux lignes de forces ouvertes, etc. ..., et, par cela même, avec les perturbations dont les **pi1** paraissent issues (baies magnétiques, **pt** — voir Sect. 20 et Fig. 17—21 — **pi2**, sous-orages,

b) Excitation des mécanismes de couplage ondes-particules, par des faisceaux de particules irréguliers et erratiques (car ils sont sous la dépendance directe des instabilités issues de la zone arrière de la magnétosphère,

c) Déclenchements simultanés de propagations dans les deux sens (à partir des zones équatoriales), ce qui peut s'expliquer quand on renonce à l'hypothèse — faite pour les **pc1** — sur l'existence d'une onde magnéto-dynamique préétablie, et que l'on adopte celle d'une génération d'ondes irrégulières et dans les deux sens par les faisceaux de particules eux-mêmes. Il convient d'ajouter que les mécanismes correspondants ne sont pas, pour autant, éclaircis.

54. Généralisation — Application à quelques cas particuliers — Conclusion. A l'issue de cet exposé faisons un bilan rapide des théories que nous avons esquissées, ceci en vue de nous faire quelque idée de leur extension possible vers une explication plus complète d'une réalité très complexe. Nous avons essentiellement essayé de dégager des mécanismes permettant de comprendre comment des manifestations pulsatoires simples du champ géomagnétique pouvaient être engendrées, entretenues, ou, même, amplifiées. Sous leur forme stylisée, ces manifestations restent, cependant, exceptionnelles. L'aspect réel des variations magnétiques rapides est, en général, plus compliqué, qu'on les examine et les étudie par la voie de leur «forme d'onde» ou par celle de leur spectre de puissance. On pourra s'en rendre compte en comparant, par exemple, les spectres simples donnés par les Fig. 41a et b, et ceux, beaucoup plus complexes, représentés par les Fig. 46a et b.

α) Une question se pose: peut-on, en géomagnétisme, considérer une variation (ou perturbation) complexe, comme une simple *somme des variations* composantes? La réponse doit être, presque toujours, négative, car des couplages existent, la plupart du temps, entre ces composantes, ce qui empêche que l'ensemble ait des propriétés linéaires. Nous allons voir ceci d'un peu plus près par l'examen d'un phénomène complexe, mais limité cependant assez étroitement dans le temps, celui introduit vers 1960 par V. A. TROITSKAJA[1–3] sous la nomenclature «IPDP», soit «Interval of Pulsations of Diminishing Periods» (voir Fig. 56 et 58).

[1] V. A. TROITSKAJA: Izv. Akad. Nauk., SSSR, Ser. Geofiz. **9**, 1321 (1960).
[2] V. A. TROITSKAJA: J. Geophys. Res., **66**, 11 (1961).
[3] V. A. TROITSKAJA et M. V. MELNIKOVA: IAGA Bull. **16**c, 135 (1961).

D'après sa représentation spectrale on pourrait considérer un **IPDP** comme une somme de pulsations simultanées dont le spectre s'élargirait dans le temps, sa valeur moyenne en fréquence étant croissante. Mais on peut aussi le considérer comme une somme de pulsations ayant chacune une période fonction croissante du temps. D'une autre façon, on peut dire que le terme «Diminishing Periods» («Périodes décroissantes») peut être appliqué, soit à un ensemble de phénomènes pulsatoires dont chaque période individuelle serait décroissante, soit à un ensemble de phénomènes dont chacun resterait fixe en fréquence, la valeur moyenne de leurs périodes augmentant par des suppressions (disparitions) des phénomènes les plus lents et par des additions (apparitions) de phénomènes plus rapides. Si, maintenant, on se réfère aux mécanismes que nous avons indiqués pour les composantes pulsées élémentaires, les deux points de vue précédents pourraient être présentés ainsi: Dans le premier cas, on envisagerait un passage progressif des conditions dynamiques propres à un tube de forces déterminé, à un tube voisin de valeur de L plus élevée. Dans le second cas, chaque tube de forces resterait fixe dans sa définition «mécanique», mais certains de ces tubes (ceux de L le plus bas) cesseraient peu à peu d'être actifs, d'autres (de L supérieur à tous ceux déjà représentés) participant progressivement à la vibration. En fait, aucune théorie d'ensemble des **IPDP** n'a pu encore être établie et plutôt que de procéder à une telle entreprise[4] on s'efforce d'enrichir le contexte expérimental qui les concerne en cherchant tous les liens qui pourraient les rattacher à telle ou telle autre manifestation naturelle. C'est ainsi qu'on a mis en évidence la relation qu'il y avait entre la vitesse de variation des périodes dans un **IPDP** et l'amplitude des variations de flux de particules à l'intérieur des zones de radiation inclues dans la magnétosphère (Zones de Van Allen)[5,6].

Une situation analogue se présentera pour l'ensemble des variations qui interviennent dans le développement d'une «baie magnétique», ou d'un «sous-orage», polaire ou magnétosphérique et, à fortiori, dans celui d'un grand orage magnétique mondial. Ceci ne veut pas dire qu'il n'existe pas de théorie sur ces phénomènes — en particulier sur les orages[7] — mais ces théories ne «descendent» pas jusqu'à la gamme des microstructures et des microvariations dont nous nous occupons ici.

β) *En conclusion générale pour ce chapitre,* nous remarquerons que, pour un ensemble de phénomènes aussi variés et aussi complexes que le sont les variations magnétiques rapides, nous ne pouvions nous attendre à une compréhension complète, satisfaisante. Dans bien des cas nous devons encore nous contenter d'une description et d'une classification morphologiques dont la nécessité reste impérative puisque c'est ce premier pas dans la connaissance qui nous permet ensuite d'examiner notre sujet en des termes précis.

La détermination des domaines spatiaux qui sont le siège de ces variations, entrée récemment dans une phase plus directe par les observations en fusées et satellites constitue le pas suivant, également primordial. Elle est en bonne voie.

L'étude des relations entre phénomènes, soit appartenant au même domaine géographique (ou spatial), soit mettant en jeu les domaines solaires et interplanétaires (sinon universels, comme ce serait le cas pour les rayons cosmiques) ouvre la voie à la détermination des filiations de *cause à effet*, détermination qui

[4] K. R. Roxburgh: Thèse Univ. British Columbia (1970).

[5] V. A. Troitskaja et R. Gendrin: Comm. Sympos. URSI-IAGA Belgrade (1966).

[6] V. A. Troitskaja, O. V. Bolšakova et E. T. Matveena: Geomagn. i Aeronomija **6**, (3) (1966). [Engl. transl.: Geomag. and Aeron. **6**, (3), 393, 1966].

[7] Voir la contribution de E. N. Parker et V. C. A. Ferraro au volume 49/3, p. 131 de cette encyclopédie.

est très importante pour toute compréhension intrinsèque (mécanismes, théories, etc.) des phénomènes. C'est elle, qui pourra permettre l'utilisation de certaines observations continues, au sol, de variations géomagnétiques (ceci dans une bande de fréquences aussi large que possible), à la détermination de diverses caractéristiques du milieu qui entoure notre Globe. C'est elle, également, qui permettra de faire des prévisions sur la manifestation au sol de divers phénomènes magnétiques dont certains (orages magnétosphériques) peuvent entraîner certaines conséquences à éviter (sur les transmissions radiotélégraphiques par exemple).

Nous rejoindrons un point de vue plus universel en remarquant que les variations géomagnétiques, par le fait même qu'elles s'exercent et qu'elles se manifestent dans un domaine spatial de grande étendue (quand on le compare, en particulier, à nos laboratoires de physique) se prêtent à un élargissement des connaissances et des conceptions concernant cette même physique. Nous en avons eu un exemple remarquable, vers 1950, avec la première découverte par ALFVÉN des phénomènes magnéto-dynamiques qui — en plus de leurs implications en astrophysique et en géophysique — ont ouvert plusieurs nouveaux chapitres de physique classique.

F. Oscillations de Schumann.

55. Bref historique.

α) La *cavité Terre-Ionosphère*, espace compris entre la surface du sol terrestre d'une part, les plus basses couches de l'ionosphère d'autre part, se comporte comme une «cavité résonnante» pour certaines ondes électromagnétiques. On peut considérer en effet — tout au moins en une première approximation — que c'est une atmosphère neutre qui remplit cet espace (à part de faibles proportions d'ions dus à des circonstances météorologiques ou artificielles), et que ses deux frontières, inférieure et supérieure, jouent à l'échelle du Globe le rôle de parois très conductrices. Cette propriété fondamentale n'a cependant été envisagée qu'à une époque relativement récente, lorsque W. O. SCHUMANN jeta les bases d'une première théorie approchée concernant les propriétés de cette cavite[1–5]. La fréquence fondamentale qu'il calcula ainsi pour la résonance (de l'ordre de 10 Hz) s'est révélée cependant par la suite — aussi bien par des calculs plus approchés que par des mesures expérimentales — être trop élevée. La valeur moyenne admise actuellement est en effet de l'ordre de 8 Hz (avec des fluctuations pour les valeurs individuelles pouvant atteindre $\pm 0{,}5$ Hz), cette fréquence fondamentale étant accompagnée de pseudo-harmoniques (dénommés plus correctement «partiels») sur les fréquences successives: 14, 20, 26, 32 ... Hz[6].

L'existence de propriétés résonnantes pour la cavité ne suffit pas, bien entendu, pour que des oscillations de résonance correspondantes s'y établissent. Il faut encore qu'une ou plusieurs causes excitatrices puissent s'exercer. Depuis les premiers travaux de SCHUMANN cités plus haut, tous les auteurs s'accordent pour voir dans les décharges orageuses des *éclairs* (soit entre nuages électrisés, soit entre nuages et le sol) la source quasi unique de ces excitations. En effet, aucune des perturbations naturelles rattachables à d'autres causes (en provenance, par exemple, de la magnétosphère ou de l'ionosphère) ne paraît jouir d'une continuité

[1] W. O. SCHUMANN: Elektrische Wellen, p. 324. München: Hanser 1948.
[2] W. O. SCHUMANN: Z. Naturforsch. **7**a, 149—154 (1952).
[3] W. O. SCHUMANN: Z. Naturforsch. **7**a, 250—252 (1952).
[4] W. O. SCHUMANN: Z. Angew. Physik **6** (1), 35—43 (1954).
[5] W. O. SCHUMANN: Z. Angew. Physik **6** (6), 267—271 (1954).
[6] Voir aussi l'article de V. L. GINZBURG et A. A. RUHADZE dans ce volume, en particulier Sect. 32, p. 520.

suffisante pour arriver à établir dans cette cavité terre-ionosphère un niveau de résonance appréciable. Il n'en serait pas de même pour les excitations provoquées par certaines interventions artificielles telles que des explosions nucléaires; dans la Sect. 58 nous reviendrons d'ailleurs brièvement sur ce sujet. Nous laissons également de côté ici les propriétés de «guide d'ondes» de la cavité qui — contrairement aux phénomènes de résonance — n'ont pas besoin pour se manifester de pouvoir étendre leur action à l'ensemble de la surface du Globe.

En collaboration avec H. L. KOENIG, SCHUMANN chercha lui-même à mettre en évidence ces oscillations propres permanentes de résonance dont il avait démontré la nécessaire existence théorique.

β) Les *travaux expérimentaux* de KOENIG ainsi entrepris[7,8] se heurtèrent aux difficultés résultant de la faiblesse des signaux de résonance recherchés, par rapport à diverses causes perturbatrices artificielles propres aux régions du Globe comportant une forte activité humaine (dans le cas considéré: région de Munich, en Allemagne, perturbée non seulement par le 50 Hz des réseaux de distribution d'énergie, mais aussi par le 1/3 de 50 Hz utilisé pour la traction électrique ferroviaire). L'élimination du 50 Hz (par des filtres appropriés) empêcha seulement l'observation des modes supérieurs, mais celle — également nécessaire — du 16,66 Hz fut une gêne considérable pour la détermination des premiers termes eux-mêmes. Cependant, KOENIG put mettre clairement en évidence la présence d'une oscillation quasi permanente sur une fréquence comprise entre 8 et 9 Hz. En opérant simultanément en plusieurs stations distantes de plus de cent kilomètres, il put établir le caractère non local de ces oscillations ainsi que leur nature d'onde électromagnétique (usage de doubles capteurs: antennes linéaires et bobinages sur noyaux perméables)[9]. Il put également — mais moins clairement — se faire une première idée de la variation diurne de ce phénomène, séparant les oscillations observées au cours d'une journée, en un fond pseudo continu de très faible amplitude, et en des évènements (ou groupes d'évènements) plus marqués, attribuables à des foyers orageux (météorologiques) particuliers.

Par la suite, H. FOURNIER, opérant dans une région moins perturbée artificiellement *, put mettre en évidence[10,11] le caractère quasi permanent d'oscillations magnétiques au voisinage de 8 Hz, tout en retrouvant les résultats de KOENIG sur les groupements particuliers, consécutifs à des recrudescences isolées d'activité, formés par certaines suites d'oscillations (se prêtant ainsi à des mesures susceptibles de donner l'ordre de grandeur des facteurs d'amortissement en jeu) (voir Fig. 64 à 68).

Peu après, BALSER et WAGNER commencèrent à publier dans une série d'articles[12,13] les résultats très complets obtenus par des analyses spectrales, digitales, statistiques, à partir de leurs observations: mise en évidence de la fréquence fondamentale sur 8 Hz et des trois pseudo-harmoniques (sur environ 14, 20, 26 et 32 Hz), Fig. 69 et 70.

Par une méthode différente (analyse spectrale analogique intégrée sur des durées de l'ordre de dix minutes) BENOIT et HOURI[14–16] retrouvèrent l'essentiel

* Celle du Centre de Recherche Géophysique de Garchy, à environ 200 km au Sud de Paris.

[7] W. O. SCHUMANN et H. L. KOENIG: Naturwissenschaften **8**, 183—184 (1954).

[8] H. L. KOENIG: Dissertation Technische Hochschule München (1958).

[9] H. L. KOENIG: Z. Angew. Physik **11**, 264—274 (1959).

[10] H. G. FOURNIER: C.R. Acad. Sci. **251**, 671—673 (1960).

[11] H. G. FOURNIER: C.R. Acad. Sci. **251**, 962—964 (1960).

[12] M. BALSER and C. A. WAGNER: Nature **188**, 638 (1960).

[13] M. BALSER and C. A. WAGNER: J. Geophys. Res. **67**, (10), 4081—4083 (1962).

[14] R. BENOIT et A. HOURI: Ann. Géophys. **17**, (4), 370—373 (1961).

[15] R. BENOIT et A. HOURI: C.R. Acad. Sci. **255**, 2496—2498 (1962).

[16] A. HOURI: Thèse 3ème cycle Fac. Sci. Paris (1966).

des résultats précédents en établissant leur validité à tout instant déterminé, et non pas seulement par voie statistique (Fig. 69 et 70).

γ) Les travaux qui suivirent, notamment ceux de GENDRIN et STEFANT[17,18], de RYCROFT[19], de CHAPMAN et JONES[20,21], s'attaquèrent à diverses formes plus évoluées des *conditions de résonance*: des valeurs moyennes, leur intérêt s'étendit à l'évolution journalière (diurne, saisonnière, etc.) des paramètres définissant le phénomène, ainsi qu'aux relations entre cette évolution et les causes présumées (foyers orageux principalement) de l'excitation de la cavité. De plus, ils se préoccupèrent — notamment à partir d'études théoriques telles que celles de GALEJS[22–24], d'améliorer les modèles représentatifs de la cavité, se donnant en particulier pour la paroi supérieure une frontière plus élaborée que celle — d'abord admise dans les premières théories — d'une simple discontinuité brutale *isolant-conducteur*. Ceci conduisit à s'intéresser, dans l'évolution observée expérimentalement des paramètres de la résonance, non seulement aux causes d'excitation, mais aussi aux normes complexes — quand on s'en préoccupe avec quelque détail — régissant la constitution de la cavité elle-même et, plus spécifiquement, de sa paroi supérieure constituée par la basse ionosphère. Pour pouvoir examiner plus avant cette question il est maintenant nécessaire de voir brièvement l'essentiel des mécanismes qui lient les paramètres définissant la constitution même de la cavité à ceux qui caractérisent ses propriétés résonnantes. Auparavant, signalons les excellentes revues d'ensemble dues à WAIT [*23*], KLEIMENOVA [*24*], BROCK-NANNESTAD [*25*] et LOKKEN, SHAND et WRIGHT [*26*].

56. Propriétés résonnantes de la cavité terre-ionosphère.

α) Prenons d'abord le cas — celui considéré par SCHUMANN — d'une ionosphère schématisée par une *paroi infiniment conductrice* (en remarquant que, en ce qui concerne le sol terrestre, l'assimilation à une surface conductrice parfaite est admise à tous les degrés d'approximation envisagés). Chaque source individuelle est supposée pouvoir se ramener à un dipôle électrique (ou doublet de Hertz) vertical (c'est-à-dire radial par rapport à la surface terrestre supposée sphérique), et le seul mode de propagation énergétique capable d'atteindre d'assez grandes distances sans subir un amortissement exagéré — c'est-à-dire le seul pouvant contribuer efficacement à une résonance de l'ensemble de la cavité — est alors du type transversal-magnétique (TM), caractérisé par une onde de composantes E_r, E_θ, H_φ dans un système de coordonnées sphériques (r, θ, φ) dont l'origine aura été prise au centre de la Terre. En particulier, on peut montrer que l'onde transversal-électrique (TE) qui serait engendrée par un doublet électrique horizontal — ou par le doublet magnétique vertical équivalent, donc de composantes H_r, H_θ, E_φ — s'affaiblirait très rapidement avec la distance à la source. Plus généralement, pour toute onde d'un mode autre que TM la fréquence critique déterminant les propriétés de filtre passe-haut de la cavité serait très supérieure aux fréquences envisagées et elle ne pourrait donc intervenir efficacement dans

[17] R. GENDRIN and R. STEFANT dans: AGARD Conference on "Propagation of radio frequencies below 300 kc/s, Munich 1962 (édit. J. BLACKBAND), p. 371—400. London: Pergamon Press 1964.

[18] R. STEFANT: Ann. Géophys. **19**, 250—283 (1963).

[19] M. J. RYCROFT: Radio Sci. **69** D, (8), 1071—1081 (1965).

[20] F. W. CHAPMAN and D. L. JONES: Nature **202**, (4933), 654—657 (1964).

[21] F. W. CHAPMAN and D. L. JONES: Radio Sci. **68** D, 1177—1185 (1964).

[22] J. GALEJS dans: Natural Electromagnetic Phenomena (édité par J. BLEIL), p. 205—258. New York: Plenum Press 1964.

[23] J. GALEJS: Radio Sci. **69** D, (5), 705—720 (1965).

[24] J. GALEJS: Radio Sci. **69** D, (8), 1043—1065 (1965).

une perturbation éventuelle du mode TM, sauf près de la source de l'excitation ce qui exclut son effet sur la résonance.

Dans ces conditions, SCHUMANN a montré que la cavité terre-ionosphère doit présenter pour ces ondes TM une suite de fréquences propres dont les valeurs sont données par l'expression: (c_0 étant la vitesse de la lumière, R_E le rayon terrestre et f_n la fréquence pour l'ordre n)

$$f_n = \frac{c_0}{2\pi R_{\mathrm{E}}} \cdot \sqrt{n(n+1)} \tag{56.1}$$

ce qui donne pour les fréquences successives: 10,6, 18,4, 26,0, 35,5 et 41,1, en se limitant aux cinq premiers modes. (Les modes suivants seraient d'ailleurs difficilement observables par suite des interférences avec les distributions d'énergie sur 50 ou 60 Hz et leurs harmoniques: ils auraient peut-être, cependant, été observés par BALSER et WAGNER[1].

β) A la suite des mesures qu'il confia à KOENIG[2], SCHUMANN reconnut que l'expérience paraissant donner des fréquences de résonance observées dont les valeurs étaient d'environ 20% plus faibles que celles qu'il avait calculées, il convenait de reprendre ces calculs à partir de modèles plus perfectionnés. C'est ce qu'il fit en adoptant pour la paroi supérieure de la cavité une *surface ionosphérique à conductibilité finie.* Des hypothèses semblables furent faites par GENDRIN et STEFANT[3] ainsi que par RYCROFT et WORMELL[4], ceci avec une ionosphère de conductibilité de l'ordre de $5 \cdot 10^{-6}\ \mathrm{Sm}^{-1} (\equiv \mathrm{AV}^{-1}\ \mathrm{m}^{-1})$, limitée abruptement à une altitude d'environ 75 km. Par la suite, GALEJS étudia de plus près les différents modèles d'ionosphère qui pourraient conduire à un accord plus parfait entre la théorie et les observations, notamment soit en tenant compte de l'anisotropie de la conductibilité ionosphérique, soit en considérant des modèles d'ionosphère à plusieurs couches, ou à variation progressive de conductibilité[5–7].

γ) *Discussion.* Ce qui rend le choix précis de ces modèles plus délicat et plus complexe que s'il ne s'agissait que de retrouver, avec une approximation suffisante, les valeurs observées pour les fréquences de résonance, c'est que d'autres caractéristiques physiques de la résonance jouent également un rôle important dans les propriétés observées, et dans les mécanismes en cause.

En premier lieu, il est nécessaire de considérer le «coefficient de qualité» Q (ou «coefficient de surtension», par analogie avec les circuits électriques) de la cavité. Il est vrai que la cavité terre-ionosphère ne constitue peut-être pas un système physique assez bien défini pout qu'un tel coefficient Q puisse lui être affecté d'une façon précise. Ceci suppose, en particulier, que l'hypothèse initiale d'une vibration d'ensemble, bien cohérente, de l'ensemble de la cavité, soit suffisamment correcte. De plus, l'ionosphère, donc une des deux parois définissant fondamentalement la cavité, n'a pas une constitution statique et ses variations dans le temps doivent entraîner certaines évolutions, soit journalières, soit saisonnières, dans les valeurs moyennes, ou globales, des coefficients Q observés. Enfin, du fait même de la constitution, certainement non parfaitement homo-

[1] M. BALSER and C. A. WAGNER: Nature **188**, 638 (1960).

[2] H. L. KOENIG: Z. Angew. Physik **11**, 264–274 (1959).

[3] R. GENDRIN and R. STEFANT dans: Propagation of radio frequencies below 300 kc/s. (AGARD Conference Munich 1962, ed. J. BLACKBAND), p. 371—400, London: Pergamon Press 1964.

[4] M. J. RYCROFT and T. W. WORMELL dans: Propagation of radio frequencies below 300 kc/s. (AGARD Conference Munich 1962, éd. J. BLACKBAND), p. 421—434. London: Pergamon Press 1964.

[5] J. GALEJS: J. Geophys. Res. **73**, (1), 339 (1968).

[6] J. GALEJS: J. Geophys. Res. **75**, (13), 2529—2539 (1970).

[7] J. GALEJS: J. Geophys. Res. **75**, (16), 3237—3250 (1970).

gène, de la cavité, il est probable que son comportement, en ce qui concerne les pertes d'énergie, doit dépendre des positions, relatives et absolues, qui — à un moment ou à une époque donnés — sont celles des plus importants centres (orageux) d'excitation, ainsi que du lieu d'observation. On conçoit que dans de telles conditions il ne puisse pas y avoir un accord immédiat, précis, dans les valeurs mesurées pour Q par divers groupes d'observateurs et qu'il ne soit pas non plus facile de déterminer, par des considérations théoriques ou empiriques, les modèles cadrant au mieux avec les observations.

Cependant, cette complexité porte en elle son intérêt, puisque c'est de son étude que doivent pouvoir être dégagée l'influence des divers paramètres — soit relatifs à l'ionosphère, soit relatifs aux sources — qui la conditionnent. C'est ce à quoi nous allons nous employer maintenant.

57. Variation des fréquences de résonance et du coefficient de qualité en fonction des diverses fluctuations affectant la constitution de la cavité ou ses conditions d'excitation et d'observation.

α) Sans revenir sur la définition des fréquences de résonance, précisons celle de la notion générale de «*coefficient de qualité*» Q, appliquée à la cavité.

On sait que dans le cas de circuits électriques ce coefficient se confond avec le facteur de surtension $L\omega/R$ du circuit (L: coefticient de self-induction; ω: pulsation correspondant à la fréquence propre du circuit; R: résistance ohmique), mais les paramètres L, ω et R n'étant pas susceptibles d'une définition immédiate, simple, dans le cas d'une cavité, l'expression précédente ne leur est pas directement applicable.

On suivra au contraire plus près la propriété qui nous intéresse maintenant dans l'étude de la cavité, si nous examinons les conditions d'amortissement, soit en amplitudes, soit en énergies, qui s'y manifestent. De ces deux formes, c'est la seconde qui s'introduit le plus naturellement en considérant la fraction d'énergie «perdue» par cycle. «Perdue» veut exprimer ici le fait qu'elle échappe aux bilans énergétiques fluctuants propres aux conditions de résonance de la cavité, sans prendre forcément pour cela une forme calorifique plus ou moins «dégradée». En fait, dans le cas de la cavité Terre-ionosphère, les pertes d'énergie se produiront:

1) *En faible proportion dans le sol* qui, bien qu'il soit supposé très conducteur, absorbera un peu d'énergie par induction suivie d'une transformation thermique (avec dégradation: chaleur de Joule) qui, cependant, ne sera pas forcément totale: possibilité de transformations chimiques, par microélectrolyse, électrophorèse, électrofiltration, ou encore de transfert d'énergie dans un guide hypothétique sous-terrestre (postulé par certains auteurs) et plus ou moins couplé au guide formé par la cavité Terre-ionosphère;

2) *en faible proportion* également *dans l'atmosphère neutre* remplissant le volume même de la cavité (avec dégradation thermique);

3) *dans une proportion beaucoup plus considérable* (et pouvant être importante en valeur absolue), *dans les couches ionosphériques inférieures* (couches D et E), soit avec dégradation (sous forme thermique), soit sous d'autres formes que nous n'examinerons pas ici. Si ΔW est l'énergie perdue par cycle, pour une énergie W disponible au début du même cycle, on définit alors le coefficient Q par la relation $2\pi\, W/\Delta W$ en vérifiant que — appliquée au cas du circuit électrique envisagé précédemment — cette expression est bien en accord avec celle donnée pour le coefficient de surtension. On comprend ainsi que le coefficient Q de la cavité ait une valeur numérique qui soit sous une dépendance directe des phénomènes de pertes — ou d'absorption — se produisant dans ces couches ionosphériques inférieures.

β) A partir de l'expression précédente on peut relier la valeur de Q à la *loi de décroissance* — supposée quasi-exponentielle — des amplitudes successives (dans le même sens), A_N, A_{N+1}, etc. ... de la grandeur variable qui, dans la cavité, est liée à l'énergie de résonance par une relation de la forme

$$W_N \propto A_N^2. \tag{57.1}$$

Posant

$$A_{N+1} = A_N \exp(-\varkappa\tau)$$

(τ étant la période de résonance considérée et $\varkappa$ le coefficient d'amortissement, $\delta = \varkappa\tau$ étant le décrément logarithmique), on a en effet:

$$Q = 2\pi \frac{W}{\Delta W} = 2\pi \frac{A_N^2}{[A_N^2 - A_{N+1}^2]} \tag{57.2}$$

ou

$$Q = 2\pi \frac{1}{1 - A_{N+1}^2/A_N^2} = 2\pi \frac{1}{[1 - \exp(-2\varkappa\tau)]} \tag{57.3}$$

ou

$$Q \approx \frac{\pi}{\varkappa\tau} = \frac{\pi}{\delta}, \tag{57.4}$$

si $\varkappa\tau$ (ou δ) sont suffisamment petits. Nous remarquerons donc que cette dernière expression ne s'identifie avec celle, fondamentale, que nous avons pris pour définition

$$Q = 2\pi \frac{W}{\Delta W}$$

que pour des valeurs de Q qui ne sont pas trop petites.

Enfin, l'analogie avec la valeur de Q définie pour une cavité et celle introduite — sous forme de coefficient de qualité, ou de surtension — pour les circuits oscillants, nous suggère une nouvelle expression de Q, soit:

$$Q = \frac{f}{\Delta f}, \tag{57.5}$$

quotient de la fréquence de résonance du circuit (ou de la cavité), à sa *largeur de bande à la résonance*. Cette expression pourrait, en principe, faire également l'objet d'une mesure sur la cavité Terre-ionosphère, s'il était possible d'exciter la cavité par une action d'ensemble dont on ferait varier graduellement la fréquence de part et d'autre des fréquences propres de la cavité. En fait, elle apparaît comme d'une réalisation utopique.

Par contre, des mesures d'amortissement (en amplitude) d'une des grandeurs liées à l'énergie stockée dans la cavité sont parfois possibles, notamment quand un groupe serré de causes excitatrices lui apporte, à un moment quasi instantané, une énergie suffisante pour que l'on puisse négliger, au cours de la durée de deux ou trois oscillations nécessaires pour la mesure du décrément logarithmique, les apports d'énergie — indépendants des premiers — en provenance d'autres sources. La Fig. 66 donne un exemple de trains d'oscillations relatifs à la résonance fondamentale sur 8 Hz environ qui ont pu être suffisamment mis en forme par H. Fournier pour permettre ainsi des mesures de Q à ce moment. Cependant une telle méthode doit, essentiellement, être pratiquée sous une forme statistique — ou, tout au moins, comporter diverses méthodes de vérification sous une telle forme — car il est difficile de se faire une idée juste du degré de rigueur avec lequel les conditions énoncées plus haut sont satisfaites pour chacun des trains de pulsations qui auront été retenus comme devant, apparemment, donner satisfaction.

γ) *Discussion.* En définitive, aucune méthode ne paraît très sûre pour mesurer, *à un moment donné,* le Q de la cavité, ceci en contraste avec les mesures des fréquences propres pour lesquelles on ne rencontre pas de difficulté particulière à partir du moment où l'on a pu enregistrer correctement — avec un bruit de fond pas trop élevé par rapport au signal — la forme d'onde des oscillations. Une difficulté supplémentaire, d'ordre théorique, provient de l'incertitude qui existe sur le mode précis suivant lequel se déroule le processus de résonance. C'est ainsi que pour l'application correcte de l'expression en $W/\Delta W$ il serait nécessaire de

Tableau 4. *Oscillations de* SCHUMANN, *valeurs calculées.*

Valeurs calculées	Mode 1		Mode 2		Mode 3		Mode 4		Mode 5	
	$\frac{f}{\text{Hz}}$	Q	$\frac{f}{\text{Hz}}$	Q	$\frac{f}{\text{Hz}}$	Q	$\frac{f}{\text{Hz}}$	Q	$\frac{f}{\text{Hz}}$	Q
Formule (56.1) (pour mémoire)	1,6	∞	18,4	∞	26,0	∞	35,5	∞	41,1	∞
GALEJS, tenant compte d'un champ magnétique radial										
de jour										
ionosphère isotropique	8,1	8,3	14,2		20,4	7,0	26,6			
ionosphère anisotropique	7,8	6,8	13,8		19,8	5,6	26,0			
de nuit										
ionosphère isotropique	8,6	8,8	15,3		21,3	9,3	27,4			
ionosphère anisotropique	7,7	1,7... ...2,8	13,8... ...15,2		20,1... ...23,2	6,6... ...4,5	26,9... ...29,5			

savoir sous quelle forme principale — magnétique, électrique, ou, simultanément, magnétique et électrique — l'énergie mise en jeu lors des oscillations propres, se trouve ainsi stockée. La plupart des auteurs estiment qu'il suffit de prendre en considération l'énergie *magnétique* de l'onde stationnaire — ceci, parce que cette énergie est beaucoup plus considérable que l'énergie électrique des mêmes ondes *dans le cas du mode Transversal-Magnétique* («TM») que l'on pense être le seul à s'imposer quand on observe la cavité assez loin de ses sources d'excitation (par exemple, dans les régions de moyennes latitudes alors que les sources sont situées dans les régions tropicales d'Asie, d'Afrique ou d'Amérique. L'existence — non toujours soupçonnée — de sources d'excitation plus rapprochées, pourrait donc fausser les résultats.

On conçoit que, dans ces conditions, les valeurs mesurées pour ce coefficient Q (par différents auteurs et à des époques et en des lieux très variés) présentent une très grande dispersion. En particulier, l'étude des relations entre Q et diverses influences temporelles (par ex., saisonnières) ou spatiales, n'a guère conduit à des résultats probants. Par contre, l'ordre de grandeur trouvé — statistiquement — pour ce coefficient de qualité de la cavité (beaucoup plus élevé que les théories très simplifiées proposées en premier ne l'avaient fait prévoir) et le fait très remarquable (que l'on peut considérer maintenant comme établi d'une façon quasi certaine) que les valeurs de Q ne sont pas les mêmes pour les fréquences propres d'ordre successif, et croissent notablement avec cet ordre (par ex., de $Q=4$ pour le fondamental sur 8 Hz, jusqu'à $Q=6$ pour l'harmonique sur 26 Hz) a obligé les théoriciens à reprendre d'une façon beaucoup plus détaillée les modèles de cavité (côté ionosphère) qu'ils avaient d'abord adoptés. La difficulté est de trouver *des* modèles (ils ne peuvent être uniques puisqu'ils doivent tenir compte d'un certain nombre d'évolutions) qui satisfassent à la fois d'une façon convenable aux valeurs à trouver pour les fréquences et à celles à trouver pour les coefficients Q. Même s'il s'agit d'illustrer ce qui se passe à un instant déterminé, on ne peut s'attendre à être conduit ainsi à un modèle unique, car il n'est pas sûr que la cavité oscille dans tout son entier suivant les lois d'un système défini et, seules, des valeurs de Q partielles ont-elles, peut-être, une certaine réalité; elles dépendraient ainsi du lieu d'observation. Nous avons résumé dans

Tableau 5. *Oscillations de* SCHUMANN, *valeurs observées.*

Auteurs et lieux d'observation	Mode 1		Mode 2		Mode 3		Mode 4	
	f / Hz	Q	f / Hz	Q	f / Hz	Q	f / Hz	Q
BALSER et WAGNER (U.S.A.) Moyennes établies sur deux jours consécutifs 27 et 28 juin 1960 (par calculs numériques)	7,8		14,1		20,3		26,4	
BENOIT et HOURI (Tunisie) Filtrage, détection et intégration sur 30 sec sur chaque fréquence étudiée — 11/12 juin 1961	non observable avec l'appareillage utilisé		14,4		20,8		26,8	
CHAPMAN et JONES (U.K.) Moyennes établies sur deux années: 1961, 1962	8,0 ± 0,1		14,1 ± 0,2		20,0 ± 0,4		26,0 ± 0,8	
STEFANT (France) Moyennes établies sur deux jours consécutifs 13 et 14 juillet 1962	7,85 ± 0,2	3 à 4	14,2 ± 0,3	4 à 5	19,9 ± 0,3	6 à 7	26,25 ± 0,3	Déterminations trop imprécises
RYCROFT (Cambridge, U.K.) Moyennes établies à partir de 5 mois des années 1962—1963	7,8 ± 0,2		14,1 ± 0,2		20,0 ± 0,2		26,0 ± 0,4	

les Tableaux 4 et 5 les principaux résultats numériques (calculés et observés) qui permettent de se rendre compte des ordres de grandeur des paramètres que nous venons d'examiner.

58. Modèles proposés pour la basse ionosphère en fonction des observations sur la résonance de la cavité.

α) Nous avons déjà dit comment une discontinuité brutale, unique, était une approximation très insuffisante pour la paroi supérieure de la cavité. L'effort des théoriciens s'est porté sur une augmentation aussi limitée que possible de la complexité d'un tel modèle, de façon à permettre — sans faire appel à un formalisme trop compliqué — une concordance satisfaisante entre les résultats susceptibles d'être déduits par *calcul du nouveau modèle adopté*, et ceux donnés par l'observation. C'est ainsi qu'ont été considérés successivement (sans que nous attachions à ce dernier terme une signification historique précise):

— une ionosphère bordée inférieurement par une «paroi» fixe mais dont la conductivité serait fonction de la fréquence des ondes qui l'abordent[1,2];

— le cas en quelque sorte inverse du précédent: conductibilité fixe mais bord inférieur fonction de la fréquence[3];

— ionosphère à deux couches[4,5], ou plus;

[1] H. R. RAEMER: J. Geophys. Res. **66**, (5), 1580—1583 (1961).
[2] H. R. RAEMER: J. Res. Nat. Bur. Std. **65** D, (6), 581—594 (1961).
[3] J. GALEJS: Radio Sc. **69** D, (8), 1043—1065 (1965), en particulier Section 1.
[4] J. R. WAIT: J. Geophys. Res. **63**, 125—135 (1958).
[5] F. W. CHAPMAN and D. L. JONES: Nature **202**, (4933), 654—657 (1964); — Radio Sci. **68** D, 1177—1185 (1964).

— ionosphère à variation continue (en général exponentielle) de la conductivité[6–9];

— ionosphère formée d'une juxtaposition de structures régionales[10]; etc. ...

β) Plutôt que d'examiner un par un les différents modèles qui ont été ainsi proposés, nous donnerons quelques *conclusions* plus générales sur le parti que l'on peut en tirer:

Les deux premiers modèles (faisant intervenir de deux façons différentes la fréquence des ondes), permettent — grâce à ce paramètre — de se rapprocher des faits mieux que ne l'avait permis le modèle de la discontinuité unique: les fréquences correctes sont retrouvées pour la résonance (aussi bien pour les harmoniques que pour le fondamental), mais les valeurs de Q sont deux fois trop faibles (par rapport aux mesures).

Parmi les modèles à deux couches, ceux proposés par CHAPMAN et JONES[5] apparaissent surtout satisfaisants quand, ainsi que l'a fait JONES[11, 12], on leur donne vers le bas le type d'extension qui a été étudié et recommandé par COLE[13] et par PIERCE et COLE[14, 15]. Il s'agit de la mise en lumière du rôle joué par ce que l'on pourrait appeler la «sous-ionosphère», zone constituée par la portion de la haute atmosphère située entre 30 et 55 km environ et dans laquelle la présence d'une quantité assez considérable (en valeur relative) d'ions lourds parmi les molécules neutres, affecte d'une façon non négligeable la propagation des ondes e. m. pour les fréquences inférieures à 100 Hz. JONES[11, 12] a pu montrer ainsi le bon accord qui pouvait être établi, en définitive, entre les résultats — apparemment quelque peu divergents — de divers auteurs.

Le travail de MADDEN et THOMPSON[10] constitue un pas supplémentaire, en ce qu'il permet — par l'introduction de ses facteurs régionaux — de tenir compte de divers effets de dissymétrie et d'anisotropie (effets jour et nuit, effets polaires et équatoriaux, etc.). La considération de l'influence qu'exerce le champ magnétique terrestre œuvre dans le même sens. On arrive ainsi, en tenant compte de tous ces facteurs, à pouvoir retrouver — et, le cas échéant, prévoir — les effets des diverses influences qui donnent aux propriétés de la cavité leur caractère fluctuant. Bien entendu, l'énergie e. m. qui se manifeste dans la cavité aux fréquences de résonance, ceci en tel ou tel lieu géographique, dépend aussi — dans une très large mesure — de la répartition, en lieux, temps et intensités, des sources d'excitation. C'est ce que nous allons examiner pour terminer.

59. Conditions d'excitation et d'observation. Nous avons déjà remarqué que les *causes* d'excitation sont bien connues (même si les *mécanismes* exacts ne le sont pas autant, bien que de grands progrès aient été accomplis au cours de ces dernières années sur les processus de décharges orageuses responsables. La *localisation* de ces causes est, dans ses grandes lignes, assez simple — régions équatoriales d'Asie, d'Afrique et d'Amérique —, mais dans ses détails, à un

[6] J. GALEJS: J. Geophys. Res. **66**, (9), 2787—2792 (1961).
[7] J. GALEJS: IRE Trans. Ant. Prop. **AP-9**, 554—562 (1961).
[8] J. GALEJS: J. Geophys. Res. **67**, (7), 2715—2728 (1962).
[9] J. GALEJS dans: Proc. International Conference on the Ionosphere, p. 467—474. London: Chapman & Hall ed. 1962.
[10] J. MADDEN and W. THOMPSON: Rev. Geophys. **3**, (2), 211—254 (1965).
[11] D. L. JONES: Thèse, Université de Londres (1964).
[12] D. L. JONES: J. Geophys. Res. **69**, (19), 4037—4046 (1964).
[13] R. K. COLE: Radio Science **69** D, (10), 1345—1345 (1965).
[14] E. T. PIERCE: J. Res. Nat. Bur. Std. **64** D, Radio Prop., (4), 383—386 (1960).
[15] E. T. PIERCE: J. Geophys. Res. **68**, (13), 4125—4127 (1963).

instant précis, plus compliquée (étant donné le grand nombre de foyers orageux individuels et la faible densité des infrastructures météorologiques dans les régions où ils sont les plus marqués).

Cependant, comme nous nous intéressons surtout ici aux états oscillatoires de l'ensemble de la cavité — et non aux propriétés spécifiques de *guide d'ondes* de telle ou telle portion de cette cavité envers la propagation des signaux e. m. émis par chacun de ces foyers orageux (ces études, qui sont aussi intéressantes, doivent être faites en plaçant chaque fois les points d'observation à des distances relativement modérées de ces foyers), nous pouvons nous contenter de considérer les trois grandes régions données ci-dessus comme schématisant suffisamment — en tenant compte également de leurs heures locales d'activité — nos sources d'excitation.

D'autre part, nous référant à la constitution TM des ondes admises à circuler (sans affaiblissement prohibitif) dans la cavité, pour y établir les régimes stationnaires de résonance, nous prévoyons la dépendance suivante entre le type — et l'orientement — des capteurs utilisés par les stations d'observation et les positions relatives de ces stations par rapport aux sources:

— *Antenne électrique verticale.* En principe, la réception est isotrope (dans le plan horizontal) et les trois foyers généraux (Asie, Afrique, Amérique) doivent être détectés (cas des mesures de BALSER et WAGNER[1]).

— *Capteurs magnétiques* (par ex. sondes à noyaux de ferrite ou de mumétal). Ces sondes étant sensibles à la composante magnétique (horizontale) de l'onde dirigée parallèlement à leur axe, une sonde orientée Sud-Nord (S-N), donc sensible à une composante S-N également, ne recevra essentiellement que des ondes ayant suivi approximativement (lors de l'établissement de la résonance) un trajet de direction Est-Ouest (E-O) — venant par ex. d'Amérique Centrale ou d'Asie pour une sonde ainsi orientée disposée en France (observatoire de Chambon-la-Forêt dans le cas des mesures de GENDRIN et STEFANT[2]); de même, une sonde orientée E-O ne captera que des signaux ayant cheminé dans une direction générale S-N, par ex. venant d'Afrique si cette sonde est installée en France.

Ce sont bien là les résultats qui ont été observés — avant même que l'on en ait eu une claire conception — ce qui est une vérification assez directe du mode de résonance seul susceptible de s'établir efficacement dans la cavité. Une étude systématique de ces faits a été conduite par la suite à la station de Chambon-la-Forêt[3,4], ce qui fait que les variations journalières régulières commencent à y être bien connues — compte tenu également des saisons — ce qui prépare la voie à des études plus spécifiquement géophysiques dans lesquelles l'observation continue des niveaux de réception de la résonance de la cavité (fondamental et harmoniques) servirait de base à une surveillance permanente, semi-globale, de la basse ionosphère («sous-ionosphère», couches D et E — les couches F ne participant pas à la résonance). C'est là un des développements les plus actuels des études sur la résonance de Schumann et qui doit être mené — s'il veut être efficace — par des mises en œuvre techniques rapides, telles que les analyses spectrales analogiques *en temps réel* mises au point par FOURNIER[5]. Ces possibilités d'étude indirecte du comportement, à chaque instant, de *l'ionosphère globale* avaient déjà fait l'objet d'un certain nombre de vérifications très intéressantes à l'occasion des perturbations apportées à l'état

[1] M. BALSER and C. A. WAGNER: Nature **188**, 638 (1960). — J. Geophys Res., **67**, (10), 4081—4083 (1962).

[2] R. GENDRIN et R. STEFANT: C.R. Acad. Sci. **255**, 2273—2275 (1962).

[3] J. STERNE et J. ETCHETO: C.R. Acad. Sci. **259**, 3584—3587 (1964).

[4] J. ETCHETO, R. GENDRIN et J. F. KARCZEWSKI: Ann. Géophys. **22**, 646—648 (1966).

[5] H. G. FOURNIER: Ann. Géophys. **21**, (3), 465—467 (1965).

général, mondial, de la basse ionosphère (en plus d'autres effets — en particulier magnétiques — apportés à la magnétosphère), lors de l'explosion nucléaire à très haute altitude (environ 320 km) du 9 juillet 1962 (expérience «Starfish»), GENDRIN et STEFANT [2], BALSER et WAGNER [6]. On verra également, en se référant à la Fig. 61, les résultats intéressants mis en évidence par LOKKEN et le «Pacific Naval Laboratory» du Canada à ce sujet.

Références générales.

[1] SELZER, E.: Oscillations propres et oscillations forcées dans la magnétosphère. J. Atm. Terrest. Phys. **29**, 339—350 (1967).
[2] TROITSKAJA, V. A., dans ODISHAW (ed.): Research in Geophysics. vol. 1, p. 485—532. Cambridge, USA: M.I.T. Press 1964.
[3] DUFFUS, H. J.: Techniques for measuring high frequency components of the geomagnetic field. From "Methods and Techniques in Geophysics", vol. 2 (1966).
[4] SELZER, E.: La méthode «barres-fluxmètres» d'enregistrement des variations magnétiques rapides. Ann. Intern. Geophys. Years **4**, part 4—7, 287—301 (1957).
[5] LE BORGNE, E., et J. LE MOUEL: Magnétomètres à protons — Magnétomètres à pompage optique. Note IPG, Nos. 1 et 2, p. 1—122 (1964).
[6] SCHLICH, R.: Enregistrement numérique des variations du champ magnétique terrestre. Note IPG No. 25, 1—16 (1967).
[7] MALNAR, L., et J.-P. MOSNIER: Un magnétomètre à pompage optique pour l'étude du champ spatial. Ann. Radioélec. **16**, No. 63, 1—8 (1961).
[8] BONNET, G.: Les possibilités nouvelles des magnétomètres à protons. Thèse Fac. Sci., Grenoble (1961). — Ann. Géophys. **18**, fasc. 1, 62—91 (1961/62).
[9] Atlas synthétique (artificiel) et Rapport du Comité Spécial No. 10 — IAGA — (Variations rapides) pour l'Assemblée Générale de Toronto (1957).
[10] CAMPBELL, W. H., and S. MATSUSHITA: Auroral zone geomagnetic micropulsations with periods of 5 to 30 seconds. J. Geophys. Res. **67**, (2), 555—573 (1962).
[11] — A review of seven studies of geomagnetic pulsations associated with auroral zone disturbance phenomena. Rep. No. 8815, N.B.S., Boulder, Col. Abstract in: Radio Science **69** D, 1187 (1965).
[12] TROITSKAJA, V. A.: Micropulsations and the State of the Magnetosphere, dans: Solar-Terrestrial Physics, ed. J. W. KING and W. S. NEWMAN, p. 213—274. London and New York: Academic Press 1967.
[13] CAMPBELL, W. H.: Geomagnetic Pulsations, dans: Physics of Geomagnetic Phenomena, ed. S. MATSUSHITA and W. H. CAMPBELL, vol. 2, p. 822—909. New York and London: Academic Press 1967.
[14] SAITO, T.: Geomagnetic Pulsations, dans: Space Science Reviews, ed. C. DE JAGER, p. 329—412. Dordrecht-Holland: D. Reidel Publishing Co. 1969.
[15] JACOBS, J. A.: Geomagnetic Micropulsations, p. 1—179. Berlin-Heidelberg-New York: Springer 1970.
[16] AKASOFU, S.-I., and S. CHAPMAN: The ring current, geomagnetic disturbance, and the Van Allen radiation belts. J. Geophys. Res. **66** (1961).
[17] LIEMOHN, H. B.: ELF Propagation and Emission in the Magnetosphere. Seattle (USA): Boeing Sci. Res. D 1-82-0890 (1969).
[18] NESS, N. F.: The geomagnetic Tail: Magnetospheric Physics, ed. D. J. WILLIAMS and G. D. MEAD, p. 97—127. Washington, D.C.: American Geophysical Union 1968/69.
[19] ALFVÉN, H.: Cosmical Electrodynamics, p. 1—237. Oxford: Clarendon Press 1950.
[20] —, and C.-G. FÄLTHAMMAR: Cosmical Electrodynamics (2ème edit.), p. 1—228. Oxford: Clarendon Press 1963.
[21] DUNGEY, J. W.: Cosmic Electrodynamics, p. 1—182. Cambridge: University Press 1958.
[22] HULTQUIST, B.: Plasma Waves in the Frequency Range 0.001—10 CPS in the Earth's Magnetosphere and Ionosphere. Space Science Reviews, ed. C. DE JAGER, p. 599—695. Dordrecht (Holland): D. Reidel Publishing Co. 1966.
[23] WAIT, J. R.: Can. J. Phys. **42**, 575—582 (1964).
[24] KLEIMENOVA, N.: Izv. Akad. Nauk SSSR, Geofiz. Ser. No. 12, 1798—1813, A.G.U. Translation No. 12, p. 1091—1100 (1963).
[25] BROCK-NANNESTAD, L.: Tech. Report No. 10, SACLANT ASW Research Center, La Spezia, Italie (1962).
[26] LOKKEN, J. E., J. A. SHAND, and C. S. WRIGHT: J. Geophys. Res. **68**, No. 3, 789—794 (1963).

[6] M. BALSER and C. A. WAGNER: J. Geophys. Res. **68**, (13), 4115—4118 (1963).

Annexe:
Atlas montrant différents types de variations observées.

Liste des figures de l'atlas.

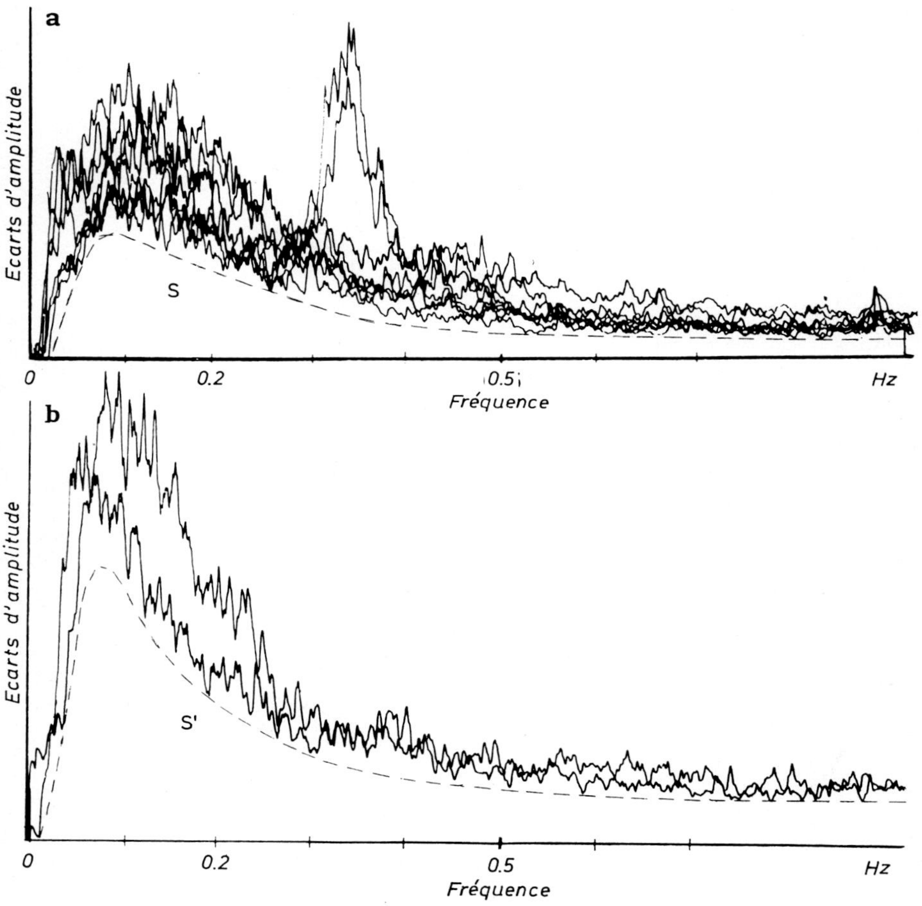

Fig. 1 a et b. Illustration du mode de détermination des «courbes en S» de F. GLANGEAUD. Les deux cas représentés sont relatifs à la même journée du 29 août 1966. Un orage à début brusque **SSC** s'est déclenché ce jour là à 13 h 15 m T.U. *La figure supérieure «a»* est relative à la *période calme* qui existait avant ce **SSC**: par la superposition de huit spectres des écarts en amplitude pris dans l'intervalle de temps 5 h 30 — 12 h 00 (en T.U.), chacun d'eux étant intégré sur $^1/_2$ heure, on a fait apparaître la courbe minimale S, unique, caractéristique des situations magnétiques calmes. Il est intéressant de remarquer que sur deux de ces spectres, ceux obtenus entre 6 et 7 h T.U., des perturbations individuelles, passagères (il s'agit de pulsations en perles pc-1) viennent se «poser» sur le bruit de fond sans modifier de façon sensible celui-ci. *La figure inférieure «b»* montre la superposition des deux spectres relatifs à l'intervalle 14 h—15 h T.U., période agitée faisant suite au **SSC**, avec accroissement général — mais sans changement de forme générale — du niveau du spectre résiduel passant de S à S'.

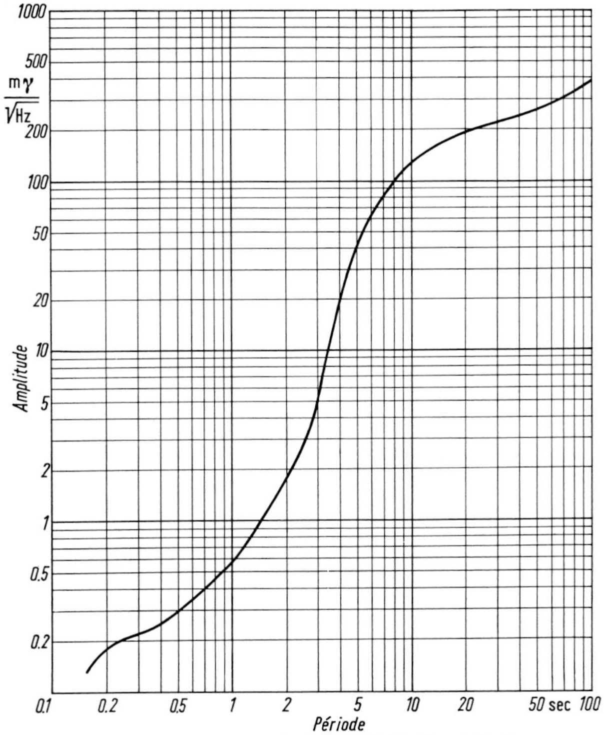

Fig. 2. Spectre du bruit géomagnétique permanent (courbe S normalisée), d'après F. GLANGEAUD. Alors que les spectres donnés aux Fig. 1 et 2 avaient été obtenus par des procédés quasi automatiques, il s'agit ici du résultat d'une synthèse construite, avec normalisation, pour toute la bande spectrale étudiée.

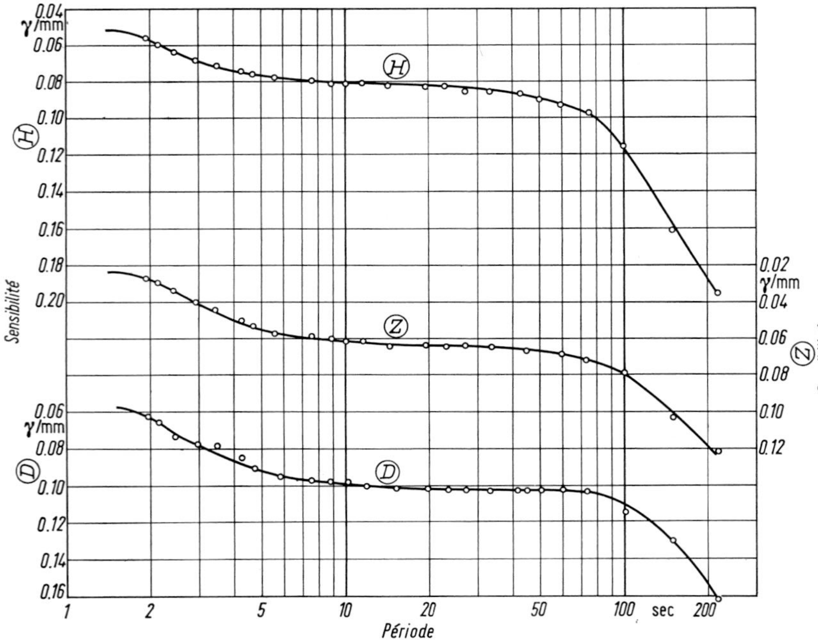

Fig. 3. Exemple d'une réponse type d'un montage « barres-fluxmètres » utilisé à la station de GARCHY (France). Ces montages sont encore en service mais notons qu'actuellement il est plus avantageux, chaque fois que cela est possible, d'enregistrer sur bande magnétique et d'appliquer alors une sensibilité « inverse de la courbe en S» de la Fig. 2.

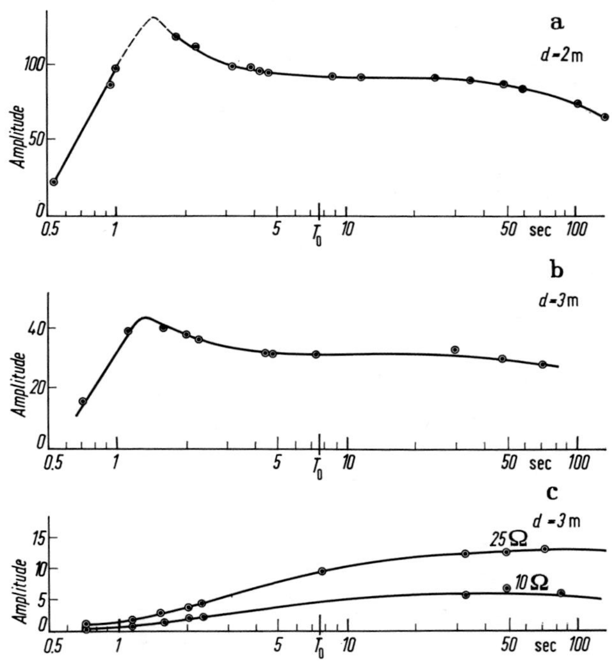

Fig. 4a—c. Réponses d'un montage « barres-fluxmètres » utilisé en Terre Adélie (station Charcot, en 1957/58).

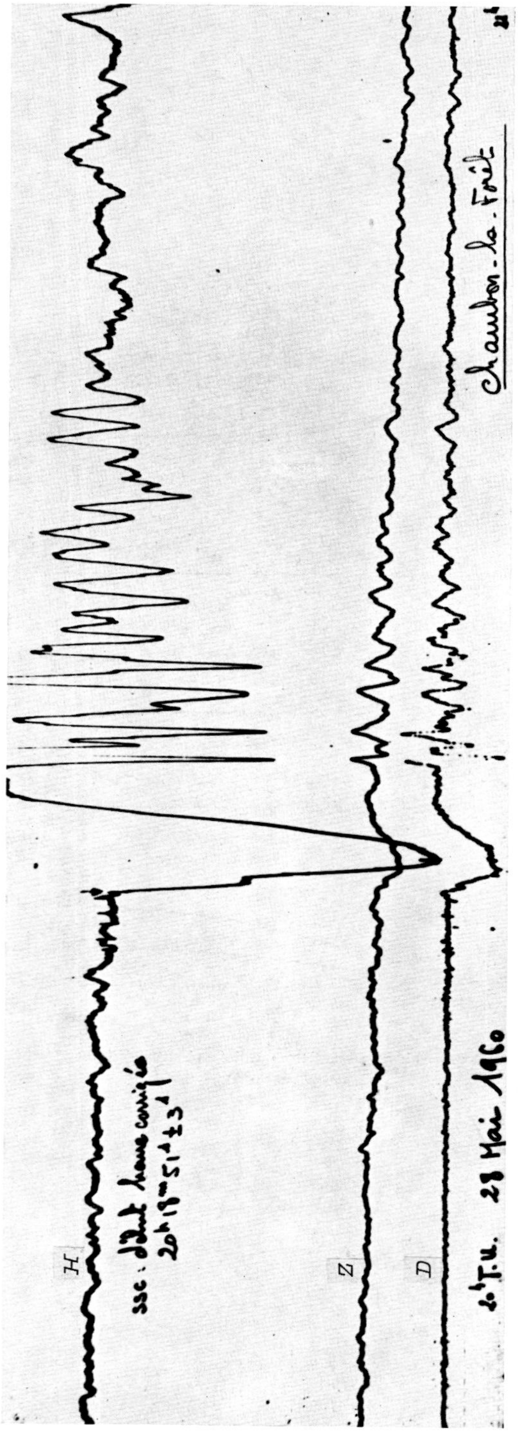

Fig. 5a.

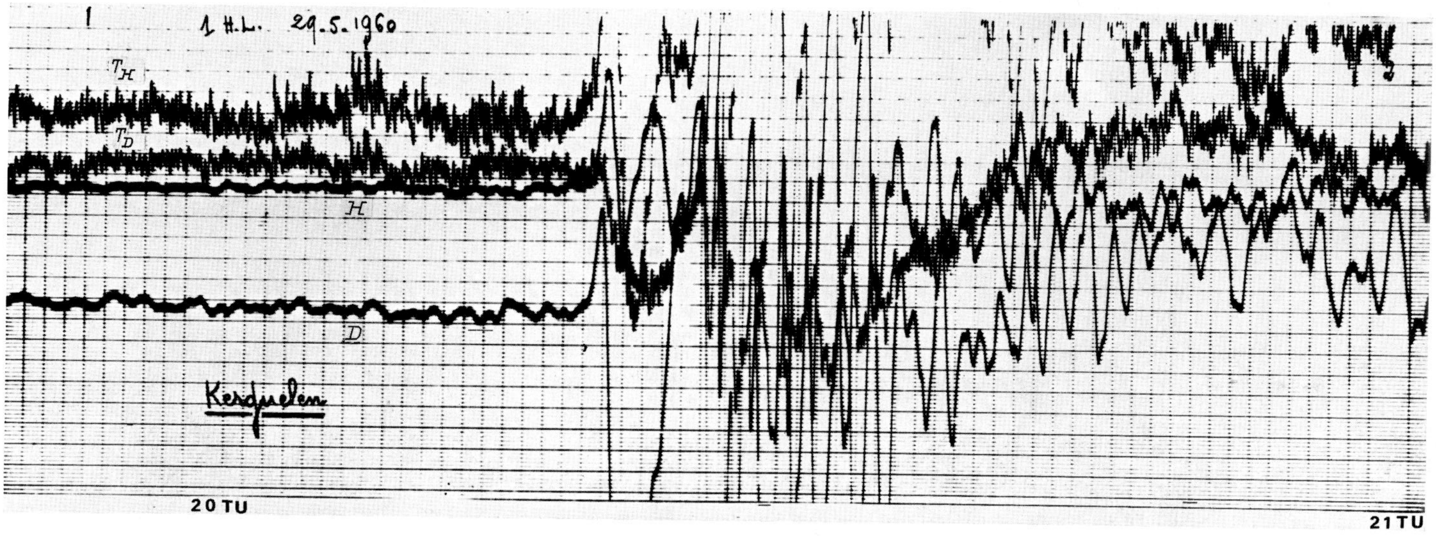

Fig. 5b.

Fig. 5a et b. Début brusque **SSC** d'un orage mondial enregistré aux stations de Chambon-la-Forêt (France) et de Port-aux-Français (Kerguelen). Les deux stations sont situées à des latitudes comparables: 48° 01′ N pour Chambon, 49° 21′ S. pour Kerguelen. Leur différence de longitude (2° 16′ E.G. pour Chambon et 70° 15′ E.G. pour Kerguelen) donne une Heure Locale en avance d'environ 5 heures à Kerguelen sur Chambon. On remarquera la similitude d'aspect de ces enregistrements, notamment du grand décrochement initial et des pulsations de l'ordre de 1 à 2 min qui le suivent. A noter que sur l'enregistrement de Kerguelen (deux composantes magnétiques $\mathcal{H}$ et $\mathcal{D}$ et deux composantes telluriques E—W et N—S) les fluctuations rapides visibles sur les traces telluriques sont artificielles (réseau local d'énergie électrique). *Heure du* **SSC**: *20. 19 T.U. le 28 Mai 1960.*

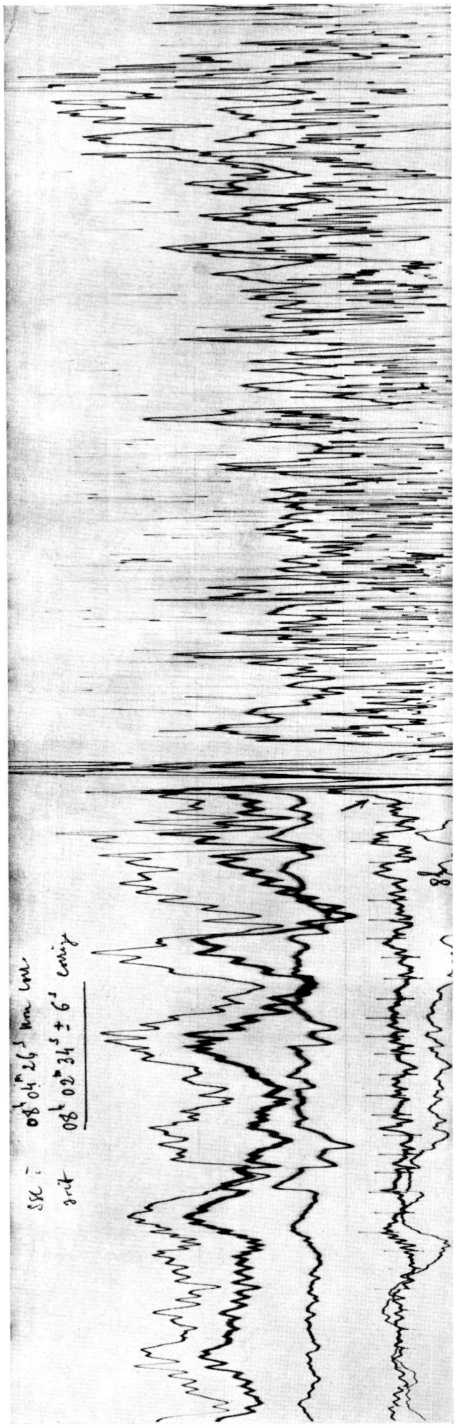

Fig. 6. Début brusque **SSC** de l'orage mondial intense du 15 juillet 1959 à 08 h 02 min 34 s ± 6 s T.U. (Enregistrement type « Barre-Fluxmètre », complété par d'autres dispositifs plus rapides (30 mm/min) en ce qui concerne la définition du temps.)

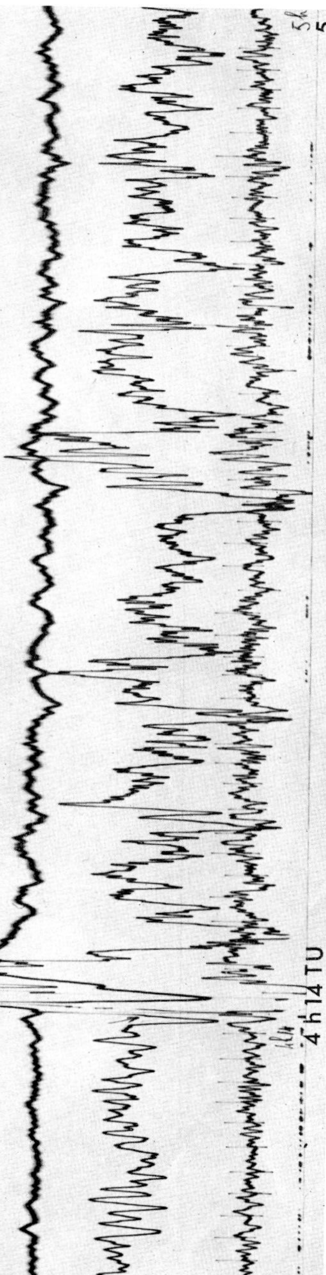

Fig. 7. Début brusque **SSC** du 20 août 1959 à 04. 14 T.U. («Barre-Fluxmètre» comme sur la Fig. 9. On remarque que, sans changer le régime des pulsations «**pc-3**», déjà existant, le **SSC** a créé d'autres pulsations de périodes plus courtes ou plus longues.

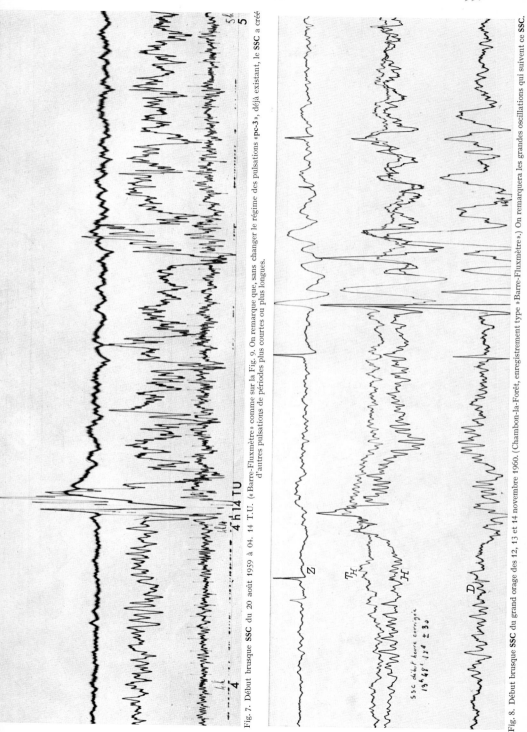

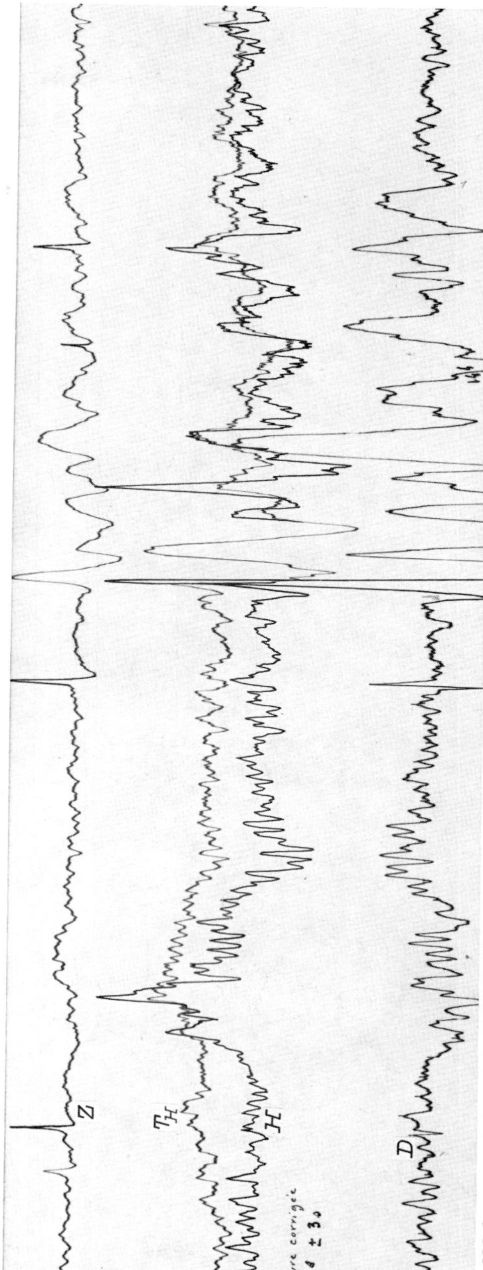

Fig. 8. Début brusque **SSC** du grand orage des 12, 13 et 14 novembre 1960. (Chambon-la-Forêt, enregistrement type »Barre-Fluxmètre«.) On remarquera les grandes oscillations qui suivent ce **SSC**.

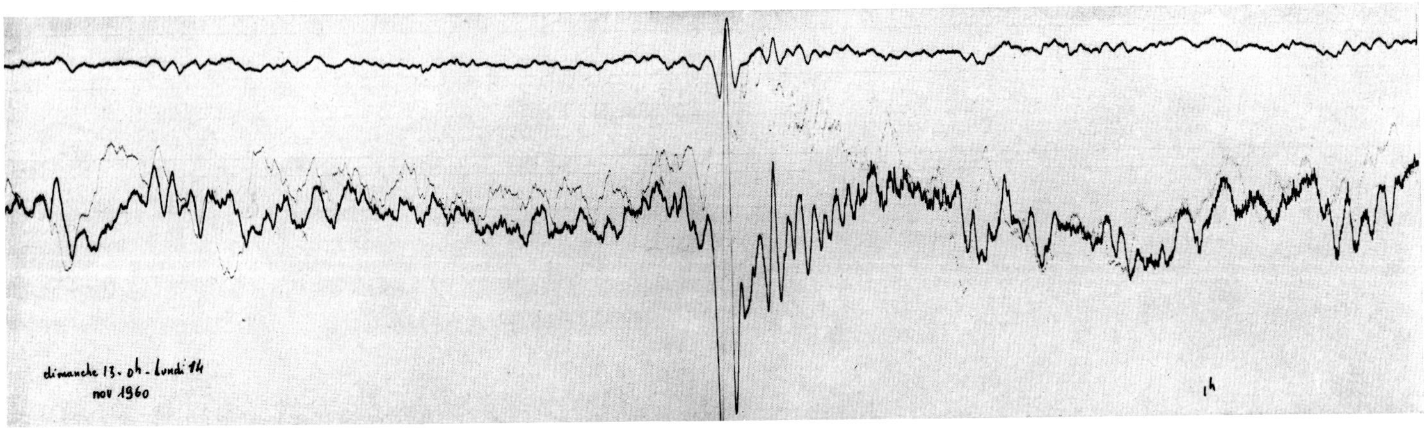

Fig. 9. Exemple de **SI** en cours d'orage. Chambon-la-Forêt, le 14 novembre 1960, à 00.33 T.U. (Enregistrement type «Barre-Fluxmètre»).

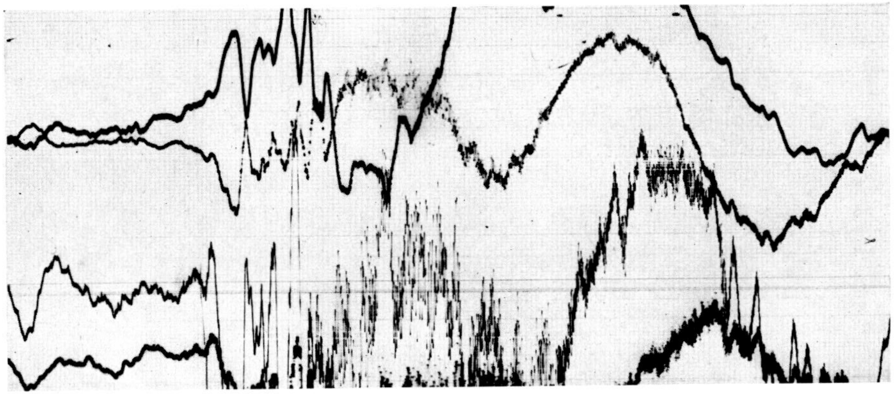

Fig. 10. Exemple d'«orage de micropulsations» (autres dénominations: «orage partiel», ou «sous-orage») *durant l'orage mondial du 4 septembre 1957*. Il s'agit du deuxième de ces orages partiels à s'être manifesté durant l'orage principal. Celui-ci a débuté à 12.59 T.U. Le premier orage partiel a commencé vers 17.30 T.U., le second vers 21.52 T.U. Indépendamment des critères très spécifiques déduits des analyses spectrales, on notera l'analogie générale qui existe entre ces orages partiels et les **IPDP** de V. A. TROITSKAJA (qui a notamment donné un tel phénomène pour 17.28 T.U. le même jour) et les **DP** de AKASOFU et CHAPMAN.

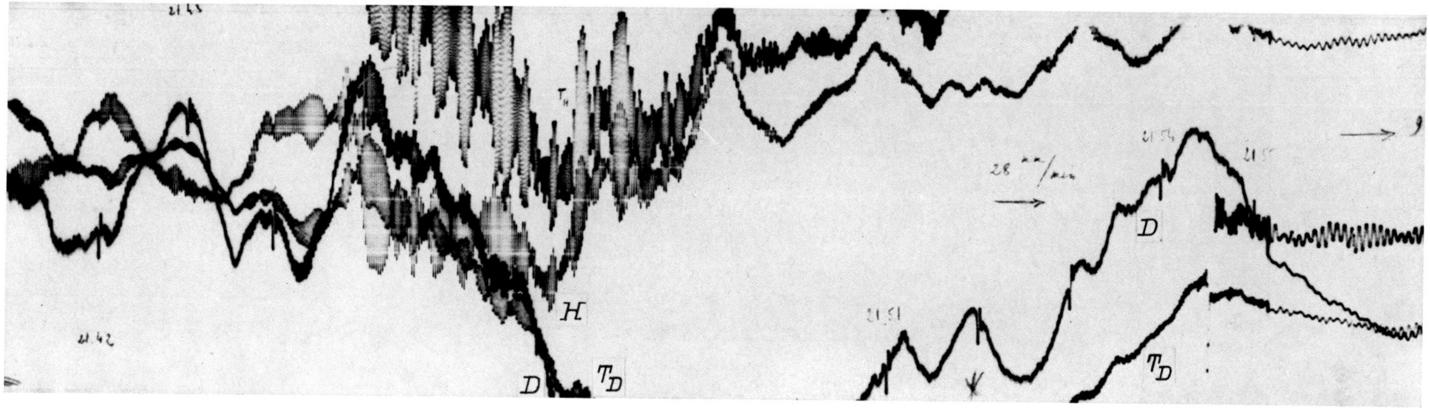

Fig. 11. Phase le plus active du deuxième orage partiel s'étant produit au cours de l'orage mondial du 8 juillet 1958. Comme complément de l'exemple donné en Fig. 10, on donne ici une «image» plus détaillée de ce qui constitue généralement la structure fine d'un orage partiel. Cette vision «grossie» n'est évidemment rendue possible que si l'enregistrement original a été fait à une vitesse suffisante. (Ce qui n'était pas le cas de l'exemple donné en Fig. 10 qui avait par contre l'avantage de permettre une vue d'ensemble du phénomène). A noter, relativement au même sujet, le changement de vitesse de défilement qui a permis, dans le cas du 8 juillet 1958, de mieux mettre en évidence, un peu après 21.54 T.U., les oscillations individuelles élémentaires — ici, régulières — composant cette phase de l'orage partiel. Remarquons encore, que la technique moderne d'enregistrement sur bande magnétique en grande largeur spectrale permet de résoudre ultérieurement, au laboratoire, toutes ces questions d'échelle. Encore faut-il que les cas «historiques» qui ont attiré notre attention sur ces phénomènes lors du maximum d'activité solaire de 1957/60, veuillent bien se reproduire durant le cycle actuel, ou les suivants.

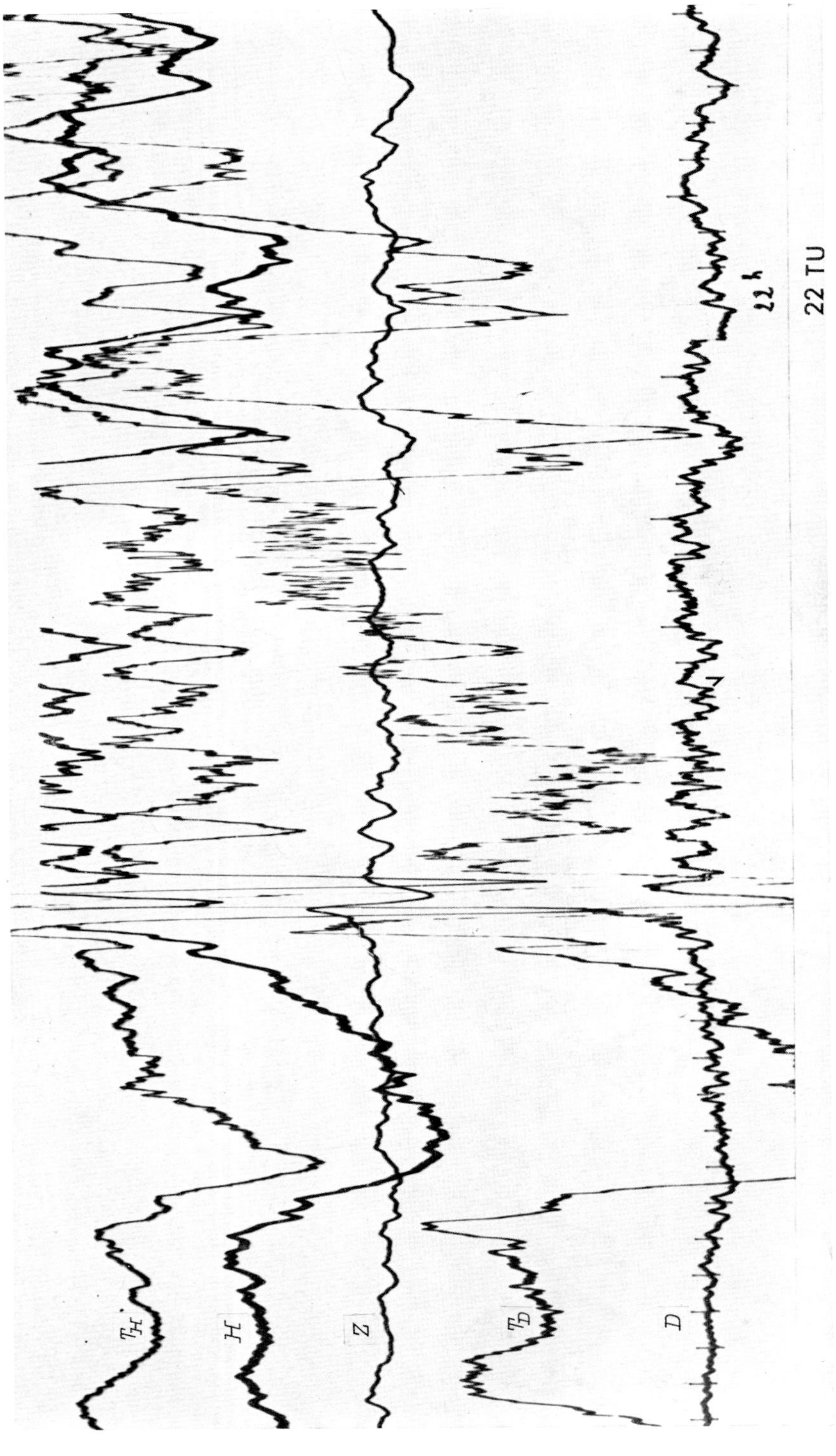

Fig. 12. Exemple de pulsation d'orage ayant tendance à se grouper en orages partiels. Chambon-la-Forêt, le 6 mai 1960. (Enregistrement type «Barre-Fluxmètre», H, Z, D, T_D: Tellurique Est-Ouest et T_H: Tellurique Nord-Sud.)

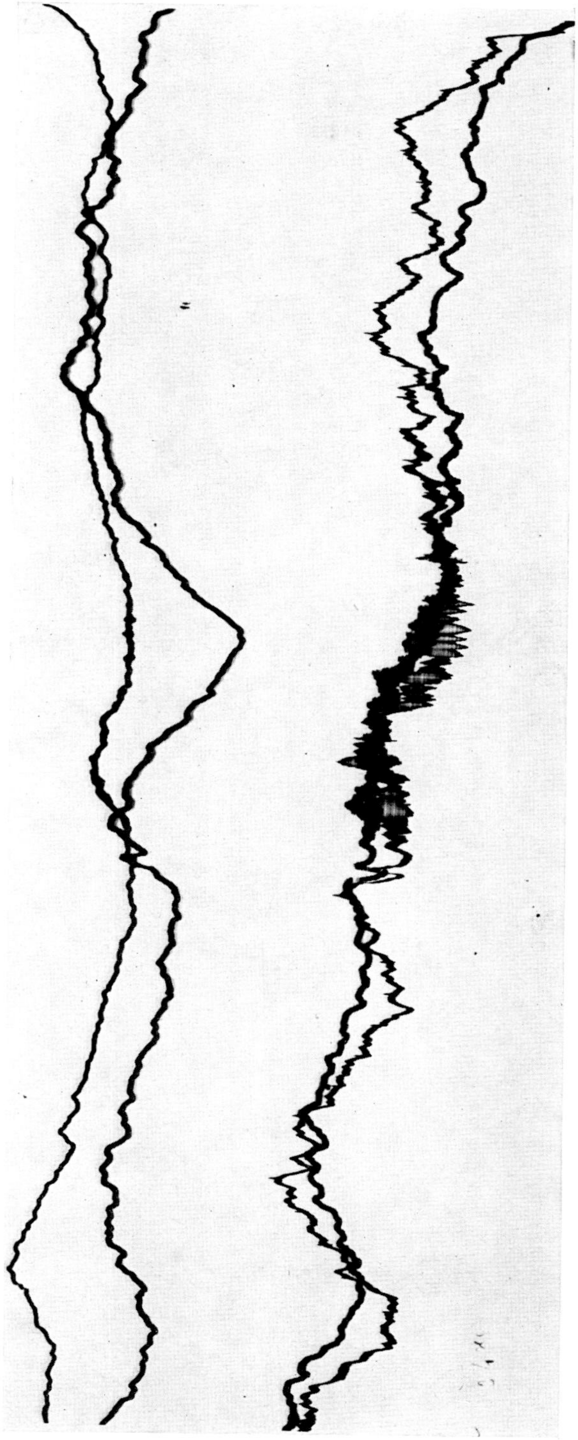

Fig. 13. Pulsations d'orage se manifestant dans un intervalle relativement calme, entre deux orages partiels. Il s'agit ici d'un phénomène se manifestant assez rarement et qui illustre le fait que, tout en étant liées — par définition pourrait-on dire — aux orages, les «pulsations d'orage» ne sont pas forcément synchrones des grands mouvements macroscopiques formant la structure principale de ces orages. T = app. 1 s. *Chambon-la-Forêt, le 4 septembre 1957.*

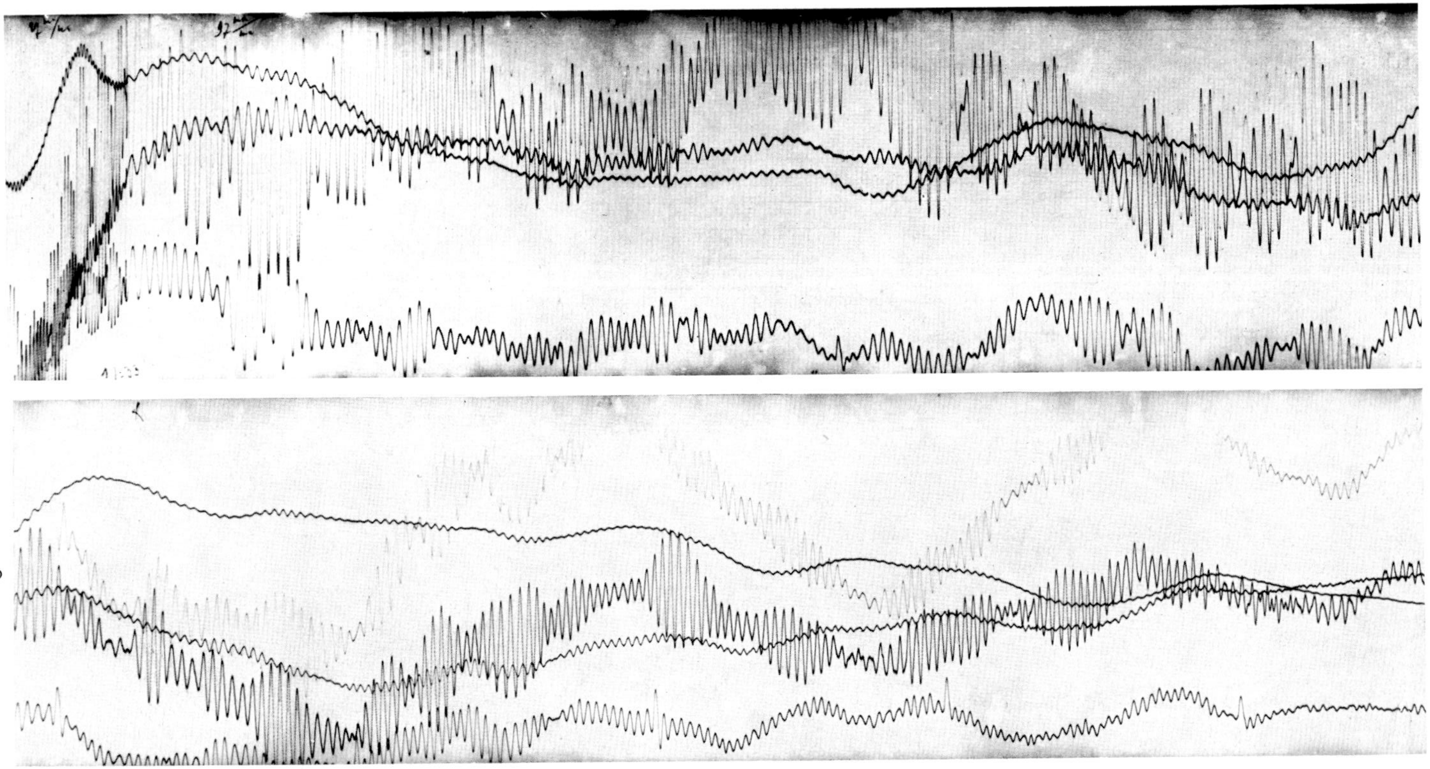

a

b

Fig. 14 a et b. Pulsations d'orage présentées à grande échelle, observées lors de la première des phases très actives de l'orage mondial du 15 juillet 1959. L'échelle adoptée permet une mesure précise de la période qui se montre constante à très peu près, et voisine de 1,8 s pendant toute leur durée (plus de dix minutes). Ceci différencie ce phénomène des IPDP de V. A. Troitskaja et aussi de *l'ensemble* des pulsations constituant un «orage partiel», bien que nous ayons vu (Fig. 15) qu'il pouvait parfois être dégagé de la complexité des composantes de ce dernier. Dans une représentation «fréquence-temps» telle qu'en donnent les analyses spectrales, ces pulsations se traduiraient par une ligne fine qui serait, ici, parallèle approximativement à l'axe des temps, mais qui en d'autres cas pourrait montrer une évolution spectrale continue en fonction de ce temps (voir Fig. 51). Une autre remarque, cette fois très spécifique relativement à cet orage du 15 juillet 1959, a trait à l'amplitude considérable de ces pulsations (pour cet ordre de grandeur de leur période), qui dépasse le gamma, ce qui est tout a fait exceptionnel pour une région de latitude modérée (observatoire de Chambon-la-Forêt).

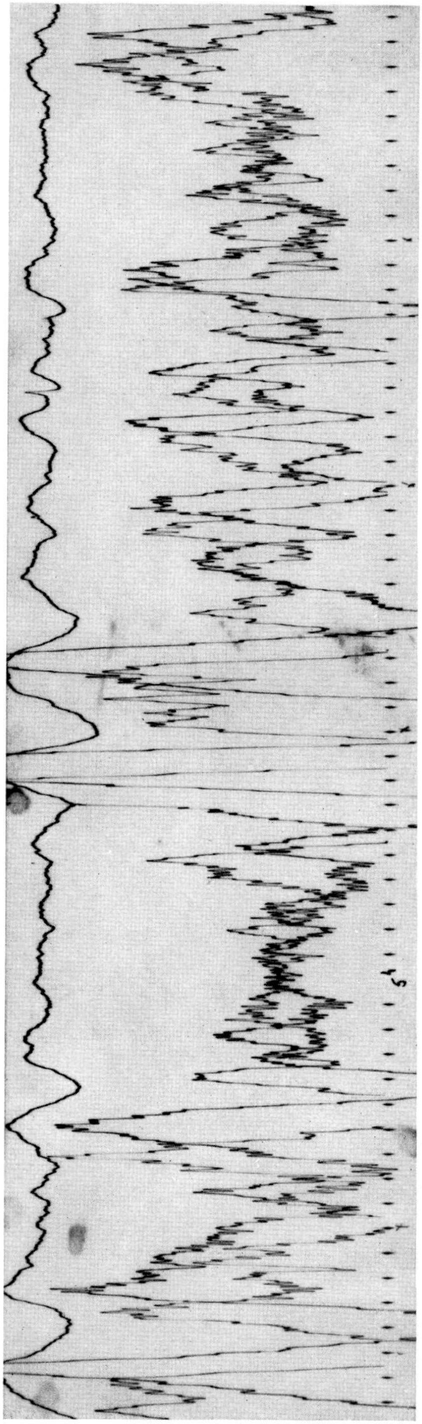

Fig. 15. «Pulsations d'orage» au cours du grand orage des 30 septembre/1er octobre 1961. Il ne s'agit plus cette fois d'un phénomène unitaire, ni de pulsations régulières (comme c'était le cas pour les phénomènes décrits par les figures précédentes), mais d'une agitation pulsée serrée engendrée par l'orage. (Chambon-la-Forêt, enregistrement type «Barre-Fluxmètre».)

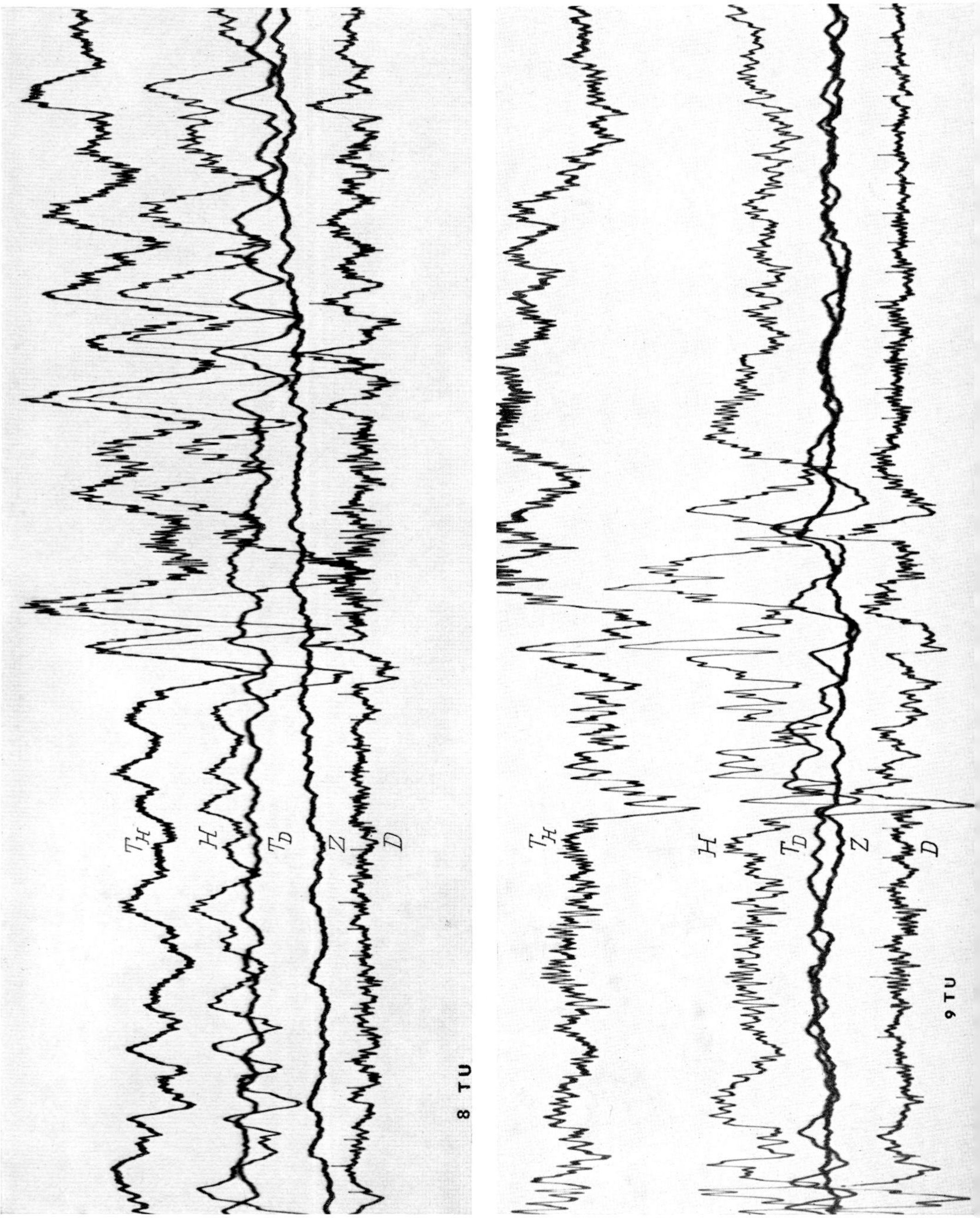

a

b

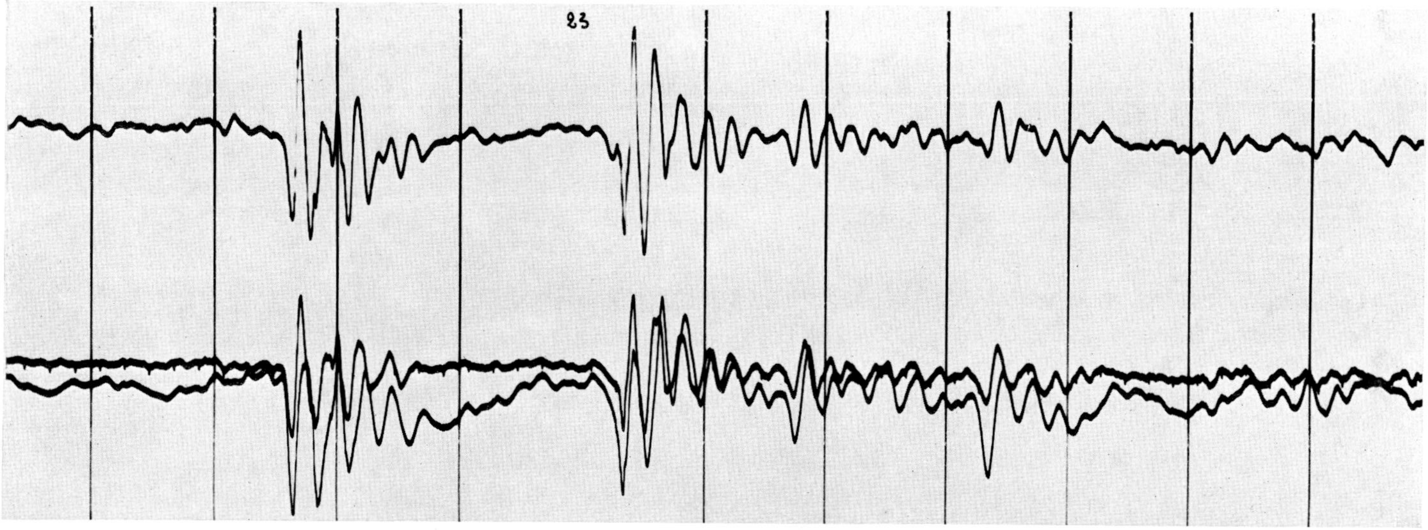

Fig. 17. Propriétés structurales de « trains de pulsations ». Par « trains de pulsations » on désigne, encore aujourd'hui, le phénomène unitaire, complexe, désigné jadis (Symposium de Copenhague en 1957, et d'Utrecht en 1959) par le symbole « **pt** » (« pulsation train »). Dès 1948 il avait été reconnu (Giacomo 1949) que ce type de pulsations accompagnait la phase « aller » d'un grand nombre de « baies ». Depuis, on leur a découvert une structure fine. La nomenclature « pt » peut continuer à désigner valablement le phénomène global (sauf celui de la baie dont il dérive), comprenant donc à la fois les pulsations des types « **pi-2** » et « **pi-1** » des nouvelles nomenclatures. L'exemple présenté ici, est remarquablement typique de la formation répétitive de ces trains et il est donné à ce titre (Enregistré à la station de Garchy, à environ 100 km au Sud de Chambon-la-Forêt, autour de 23 h T.U., date non retrouvée).

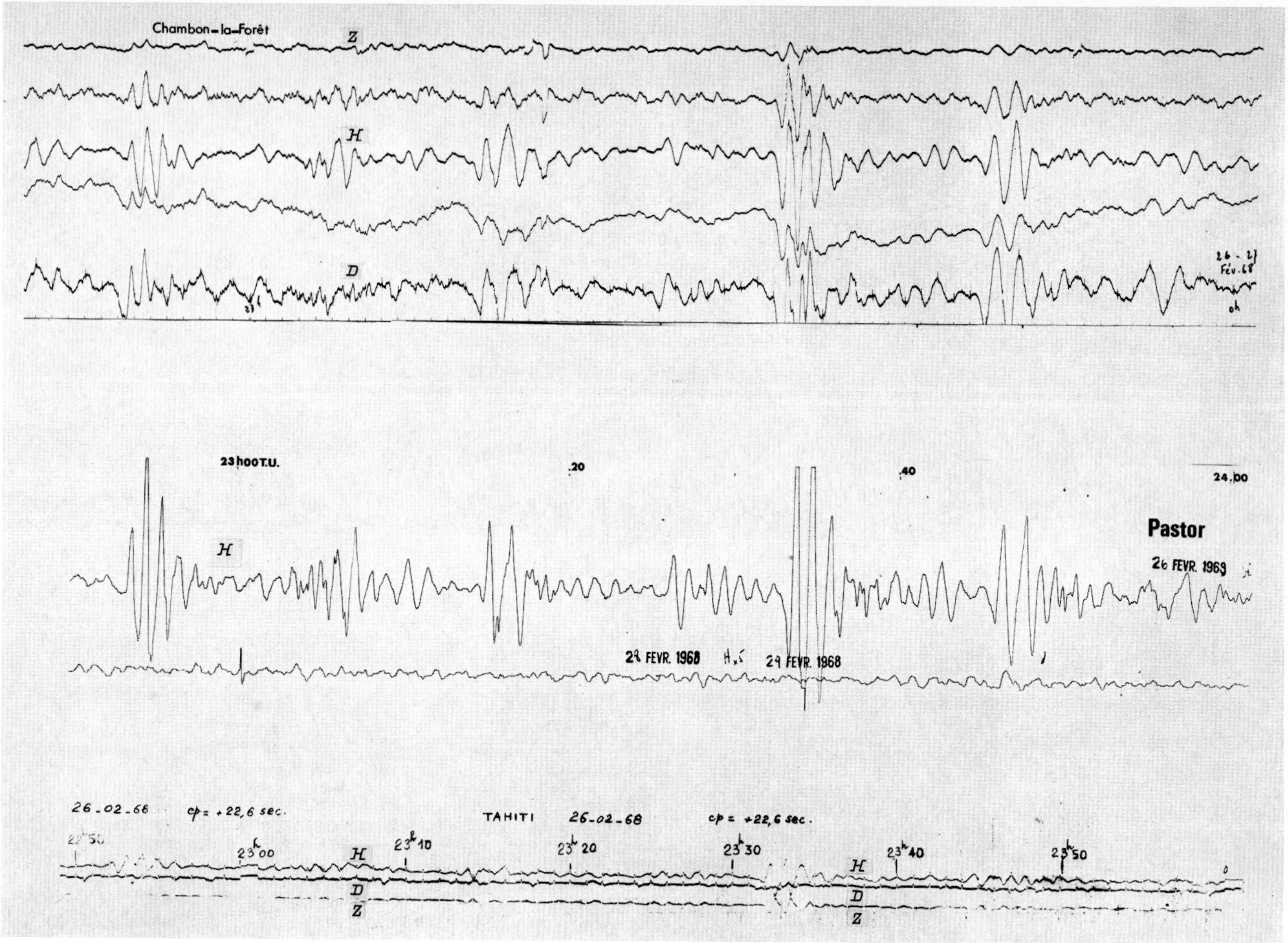

Fig. 18. Propriétés structurales de « trains de pulsations », présentation synoptique. Ces enregistrements mettent en évidence le caractère mondial des « **pt** ». Leur juxtaposition avait d'abord pour but de comparer les variations magnétiques pulsatoires entre hémisphères obscur (station de *Chambon-la-Forêt*) et éclairé (station de *Tahiti*). Au cours de cette comparaison on a reconnu qu'il y avait avantage à remplacer Chambon par *Pastor*, station d'environ même longitude mais équatoriale, donc de latitude plus comparable à celle de Tahiti. La comparaison permet de se rendre compte des deux effets, de latitude, et de longitude, sur les « **pt** ». Le premier explicite une augmentation de la polarisation en $\mathcal{H}$ (suivant le méridien magnétique) des « **pi-2** » constituant la structure primaire des « **pt** », quand la latitude tend à devenir équatoriale, (comparaison Chambon-Pastor). Le second (comparaison Pastor — Tahiti) illustre les relations Nuit-Jour (Tahiti est en « retard » d'environ 11 h, en Heure Locale, sur Pastor — et de 10 h sur Chambon) relativement au même phénomène. On peut reconnaître à Tahiti, *de jour*, la présence de pulsations synchrones des « **pt** » de nuit de Pastor, mais la forme d'onde de ces pulsations est moins nette (justifiant ainsi le terme de « pulsations de nuit » donné aux « **pt** »). De plus, les **pi-2** constituants des « **pt** » de nuit, bien que se retrouvant oscillation par oscillation sur les enregistrements rapides de Tahiti, y ont pris l'apparence générale de « **pc-4** » ce qui montre un lien entre les catégories formelles **pi-2** et **pc-4** de la classification

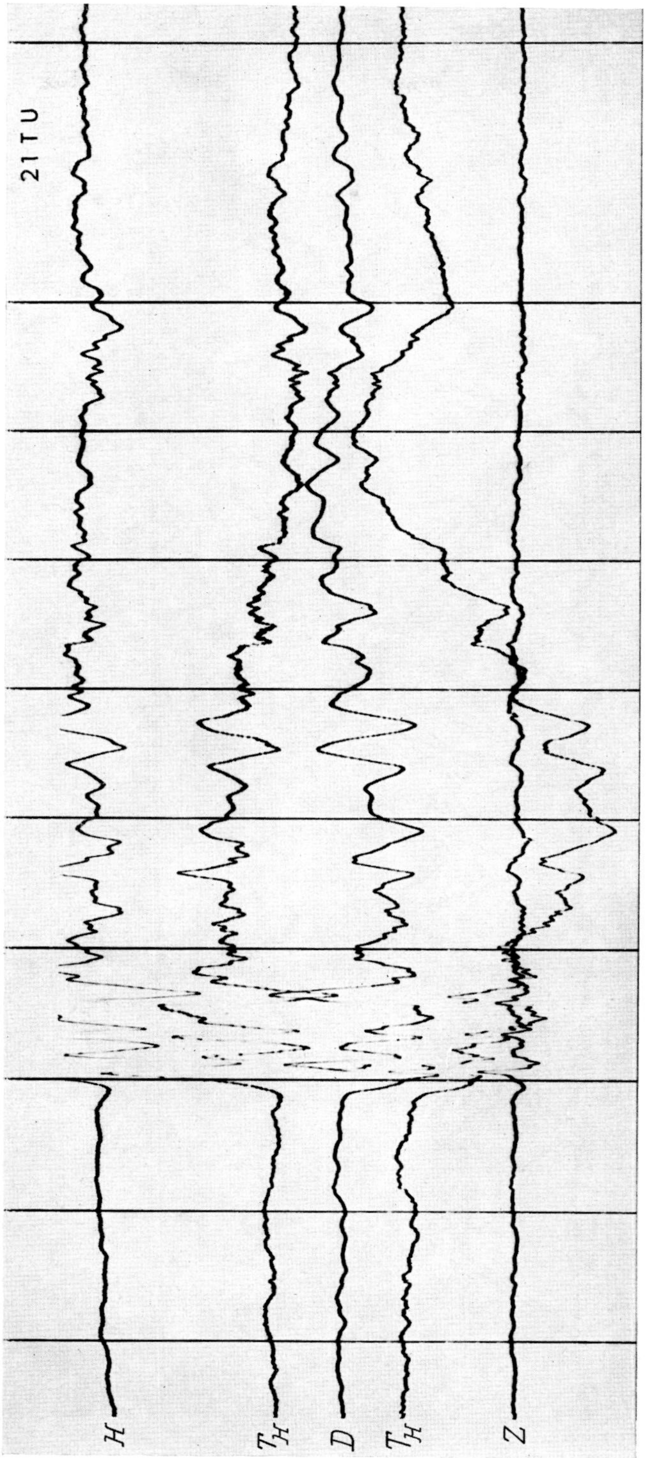

Fig. 19. Exemple d'une baie complexe accompagnée de ses microstructures **pi-2** et **pi-1**. *Le 12 avril 1965, vers 20.25.* Le mouvement général de la baie elle-même est très «filtré» par le type même des enregistrements dont on donne ici un extrait. Il apparaît cependant. *Station de Garchy*, enregistrements type «Barre-Fluxmètre», marques de temps toutes les cinq minutes (sauf aux heures rondes).

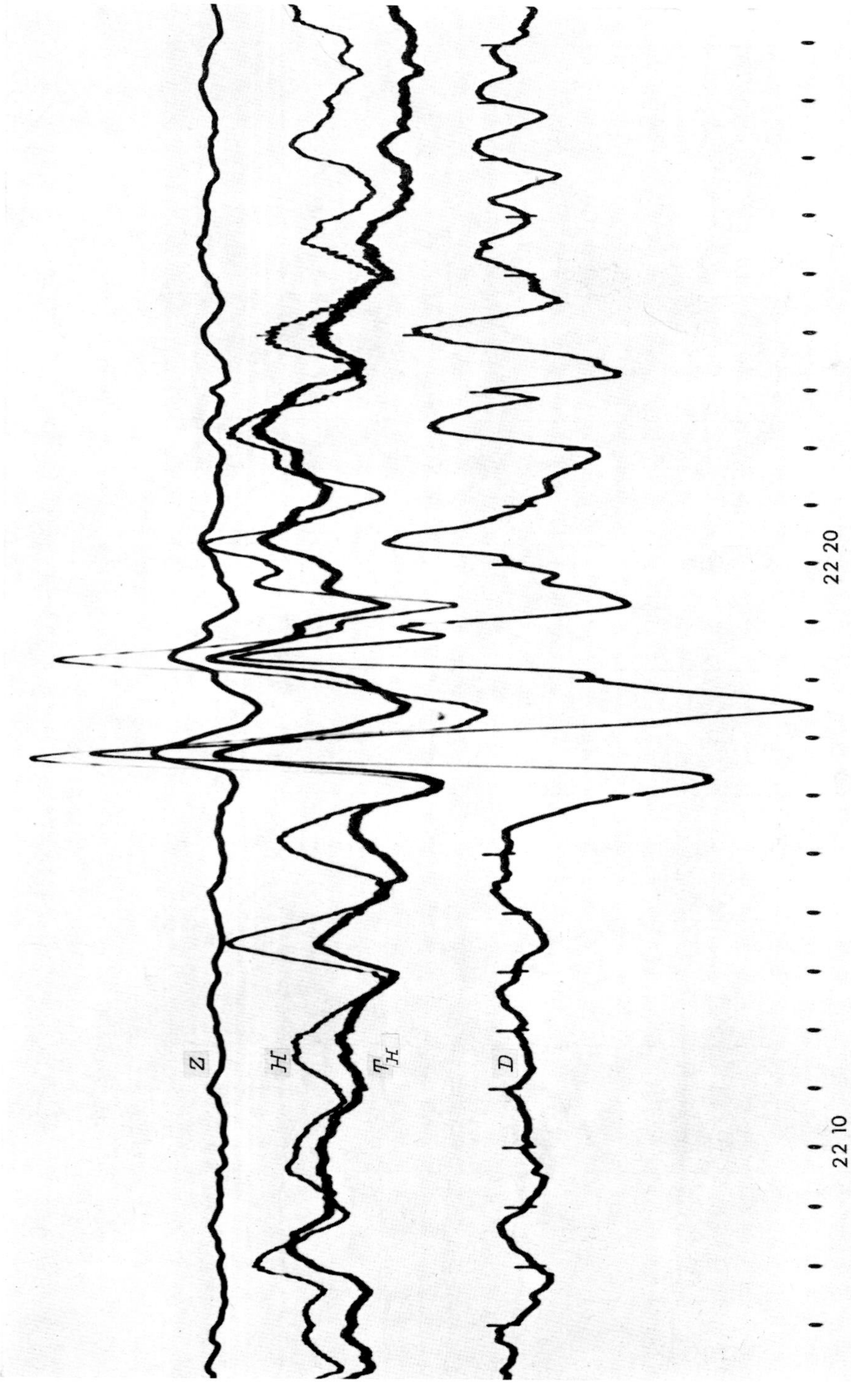

Fig. 20. Exemple d'un des trains de **pt** de grande amplitude (entre 10 et 15 gamma) accompagnant une baie complexe. *Chambon-la-Forêt*, le *26 mars 1961*. (Enregistrement de même type que le précédent.)

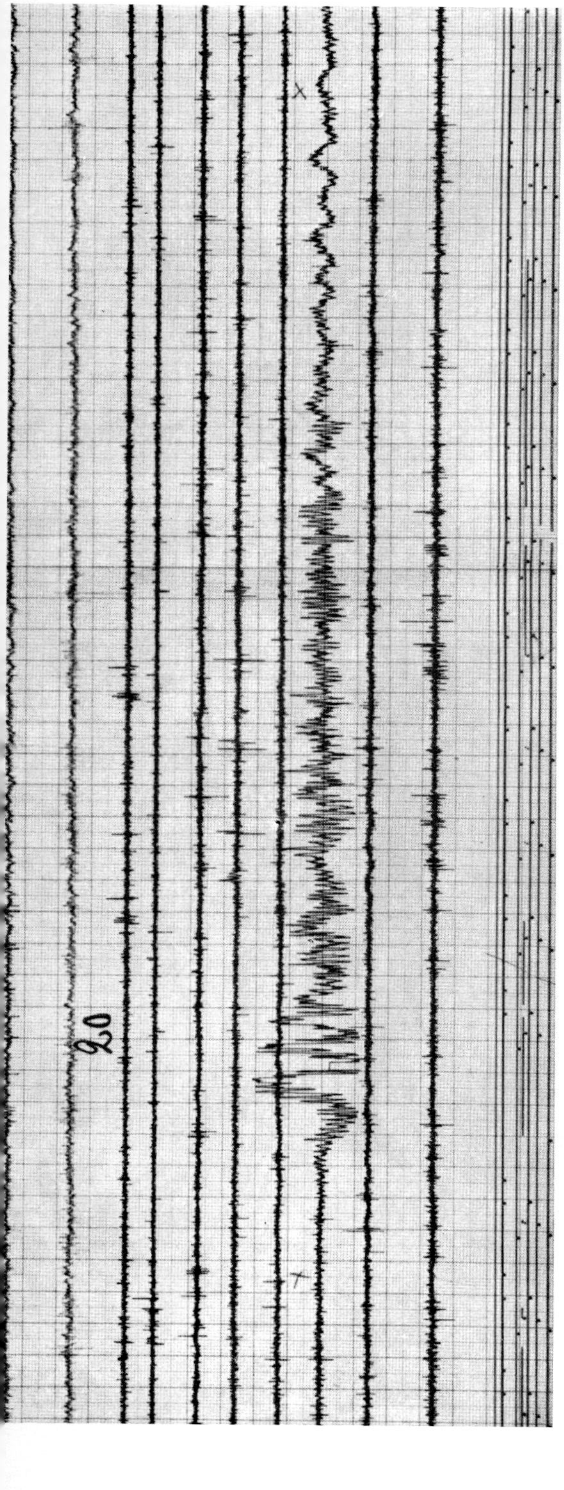

Fig. 21. Microstructures primaire (**pi-2**) et secondaire (**pi-1**) d'une baie complexe. *Garchy, 28. mars 1966.* (Enregistrement rapide, vitesse de défilement 30 mm/mn, traces multiples.)

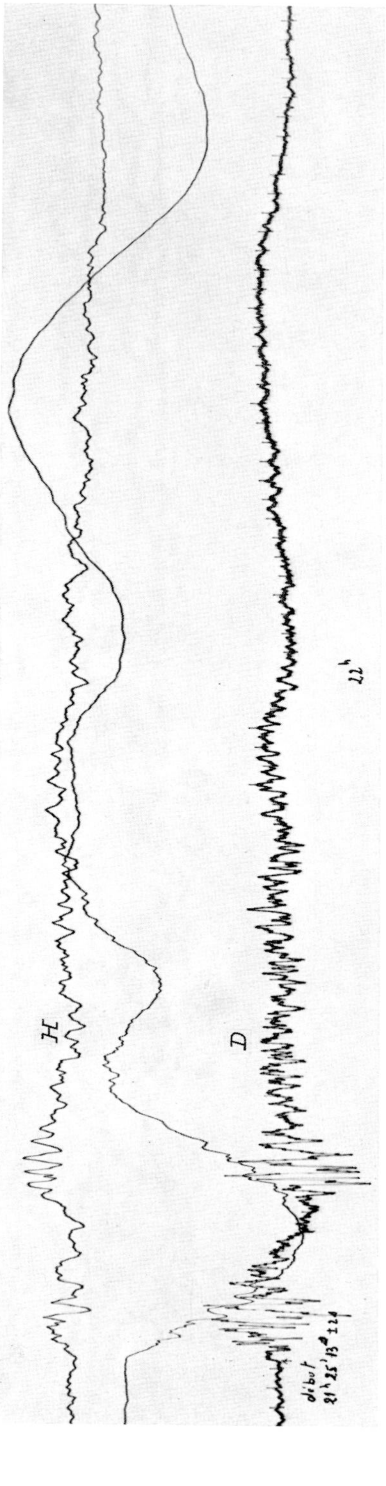

Fig. 22. Trains successifs de **pt** (**pi-2** + **pi-1**) se produisant au cours d'une baie complexe. On remarquera en particulier la grande variété des pseudo-périodes constituant les **pi-1** (Ne pas porter attention à la trace centrale de l'enregistrement qui est mal réglée). *Chambon-la-Forêt, 4 septembre 1960.* (Enregistrement «Barre-Fluxmètre».)

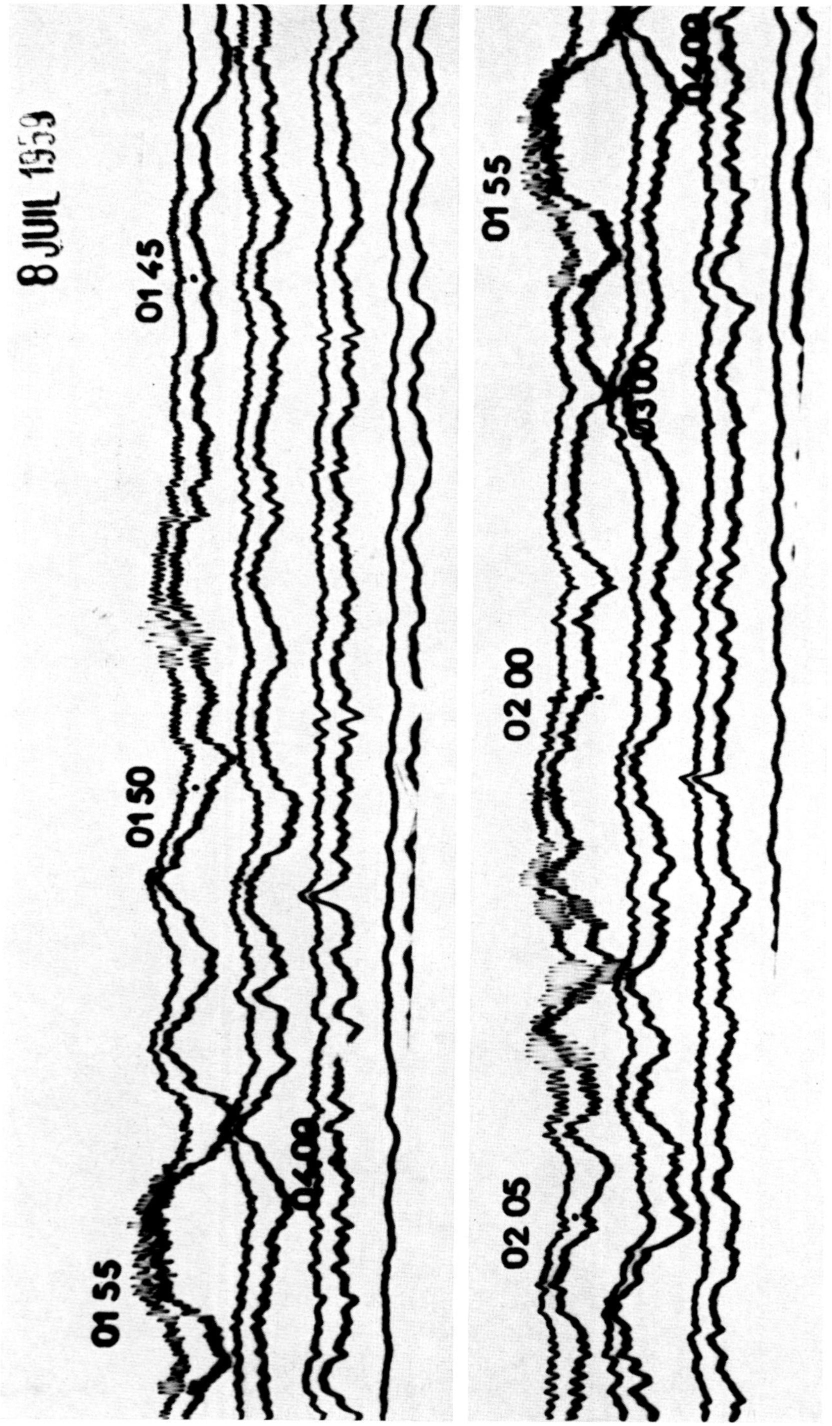

Fig. 23a.

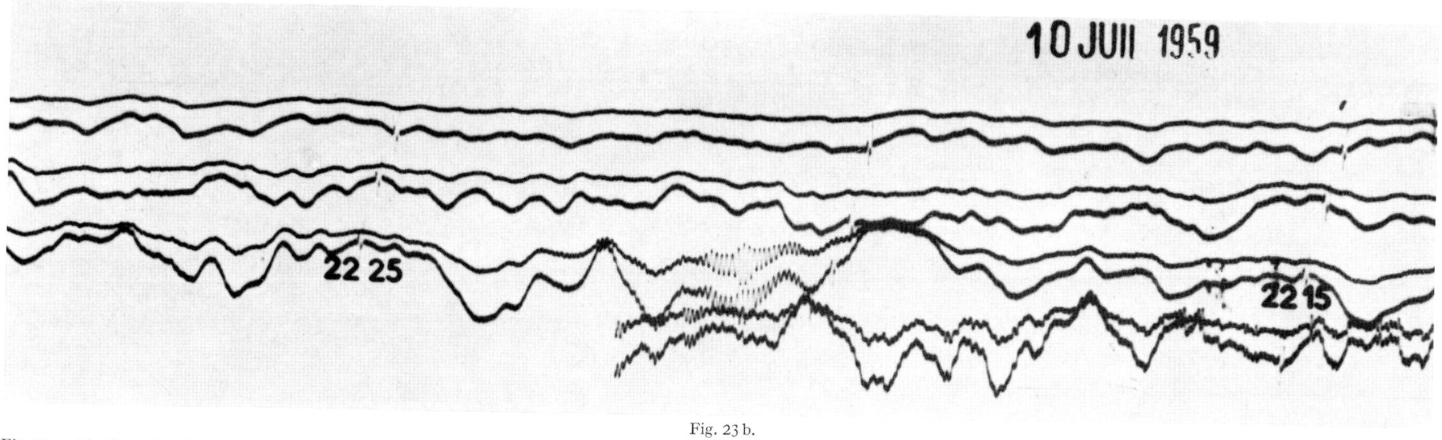

Fig. 23b.

Fig. 23a et b. *Exemples de pulsations rapides* (**pc-1** ou **pi-1**) observées à l'intérieur de la zone aurorale. Station de *Dumont d'Urville* — Terre Adélie — *8 et 10 juillet 1959*. Il s'agit des premiers types d'enregistrement (deux composantes telluriques) mis en place en Terre Adélie pour l'étude des variations très rapides. Sur l'enregistrement du 8 juillet se présentent différents groupes de pulsations aux périodes variables, descendant jusqu'à 2 s environ. Sur celui du 10 juillet le groupe qui apparaît est formé de pulsations de période d'environ 3,2 s. Heures en T.U. Enregistrements Lachaux — T.A.A.F. et Exp. Pol. Françaises.

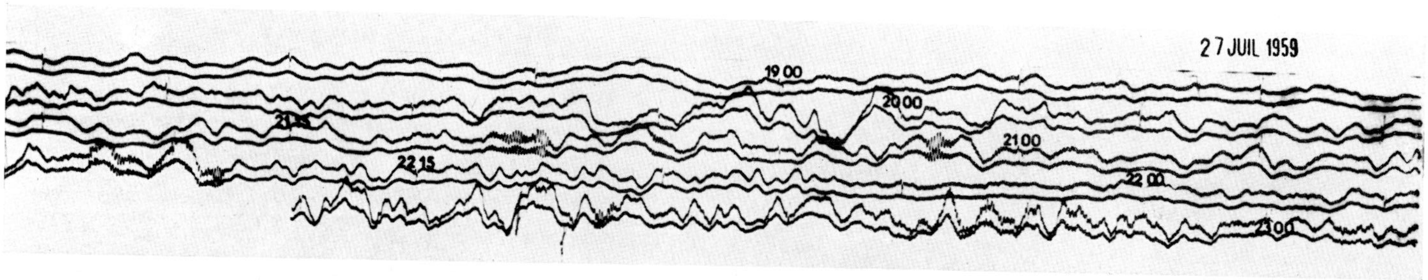

Fig. 24. Ensemble complexe de pulsations enregistrées à l'intérieur de la zone aurorale. Station de *Dumont d'Urville* — Terre Adélie — *27 juillet 1959* — Cet enregistrement est du même type que les précédents. Il correspond à une situation magnétique plus agitée. Les pulsations de périodes les plus courtes qui y apparaissent (environ 3 s) sont à ranger parmi les **pi-1**. Enregistrement Lachaux — T.A.A.F. et Exp. Pol. Françaises.

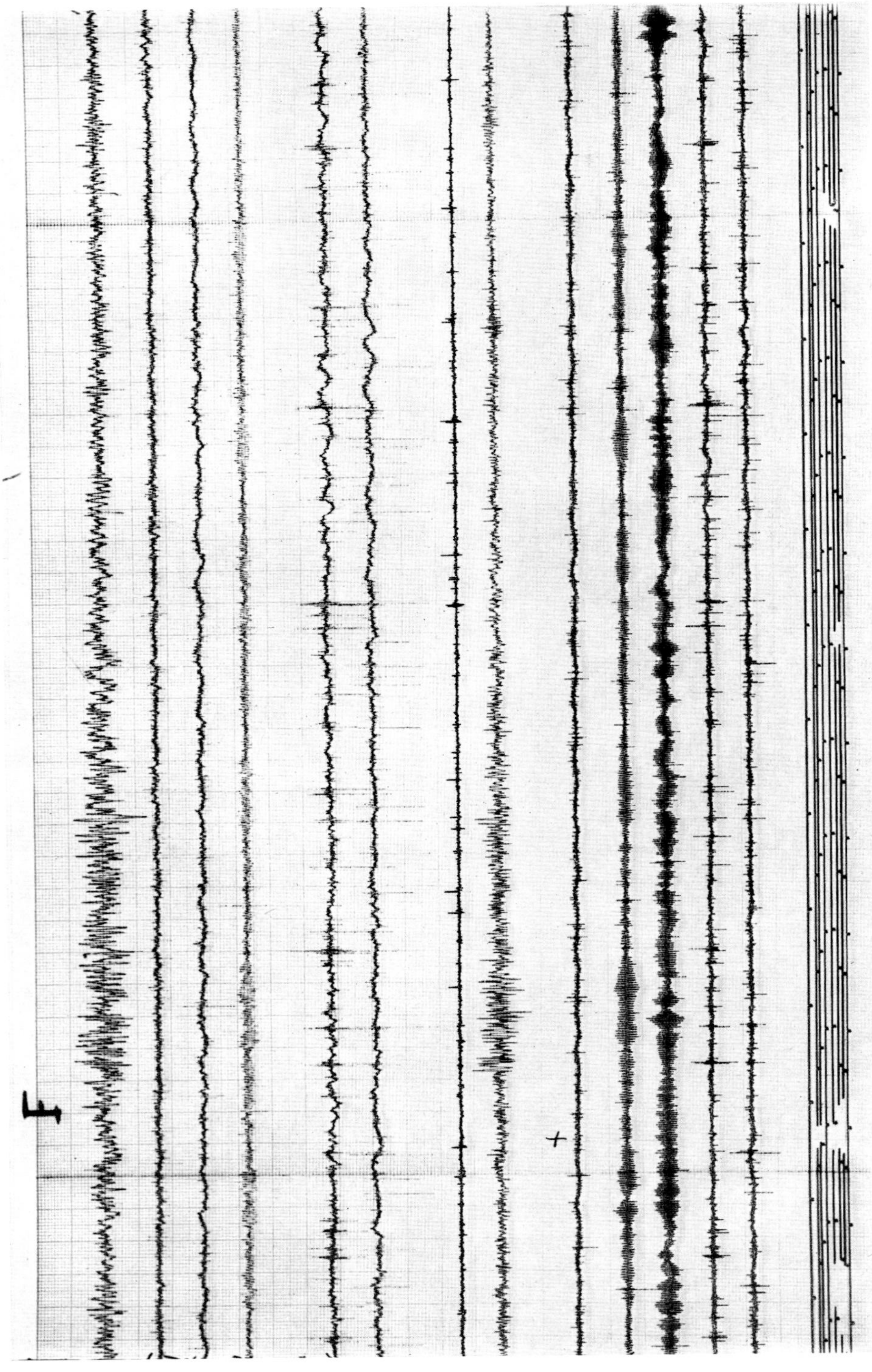

Fig. 25a.

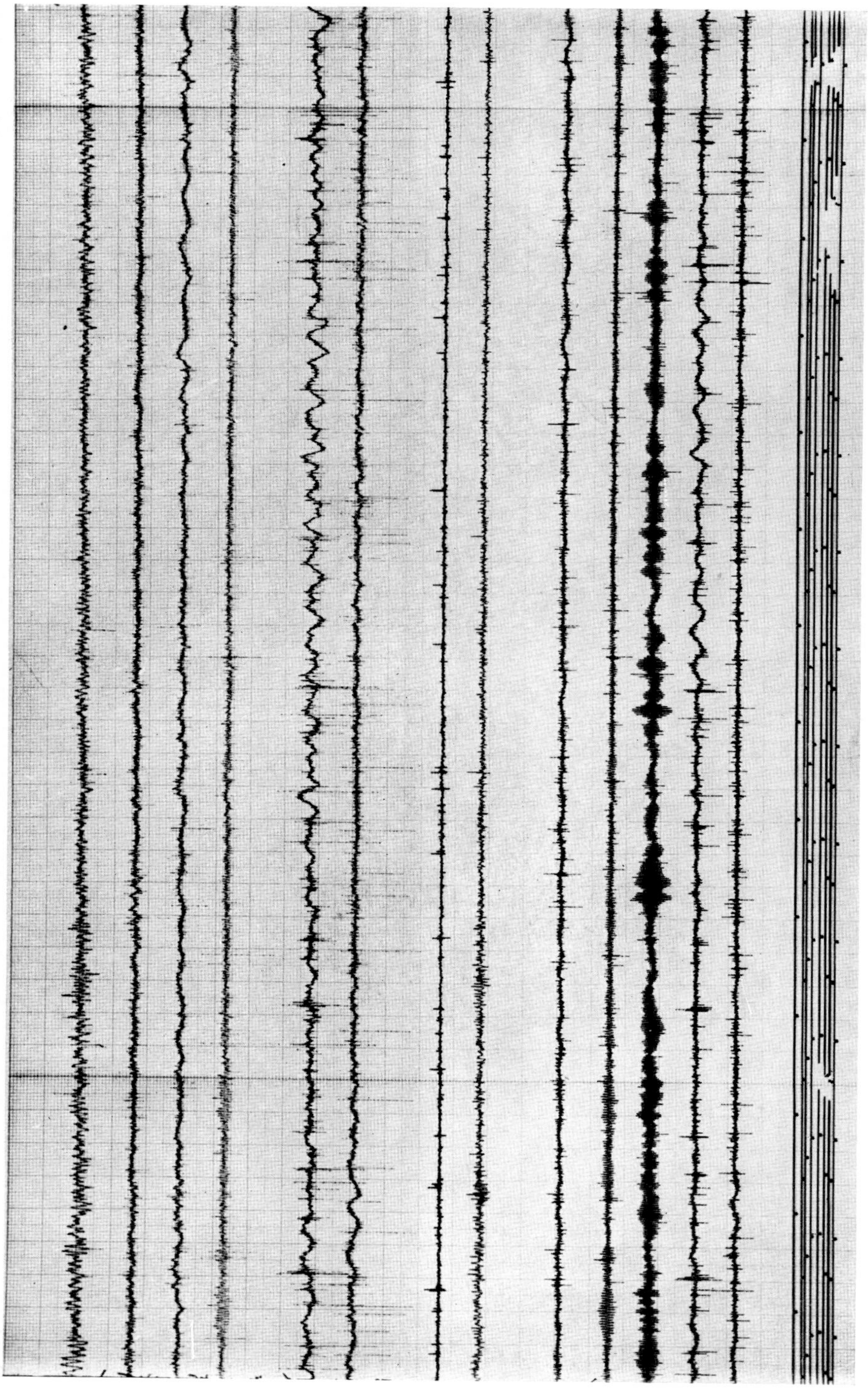

Fig. 25 b.

Fig. 25 a et b. Exemples de pulsations des types **pi-1** et **pc-1** relatives à des jours différents mais rassemblées sur une même « page » d'un enregistrement très rapide (pliage du papier en « accordéon »). Sur la ligne supérieure et médiane apparaissent des **pi-1** et sur les 4ème et 3ème lignes par le bas on remarque des pulsations en perles **pc-1**. Station de *Garchy*, *mars 1966*.

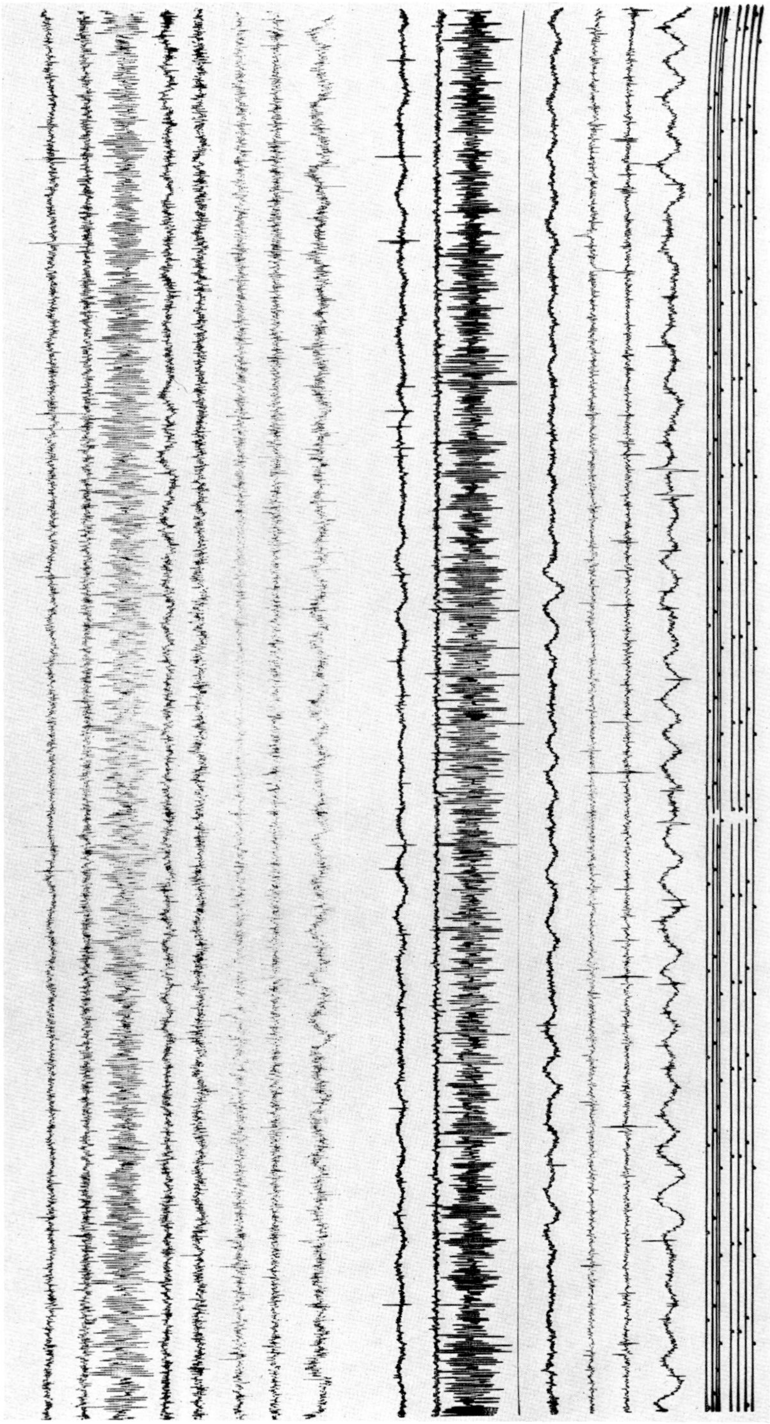

Fig. 26. Distribution typique de pulsations **pi-1**. On remarquera la distribution aléatoire de leurs amplitudes successives et de leurs groupements dans le temps, ce qui les distingue des pulsations en perles et, plus généralement, de toutes les pulsations structurées du type **pc-1**. Station de *Garchy*, enregistreur très rapide (30 mm/min sur l'original).

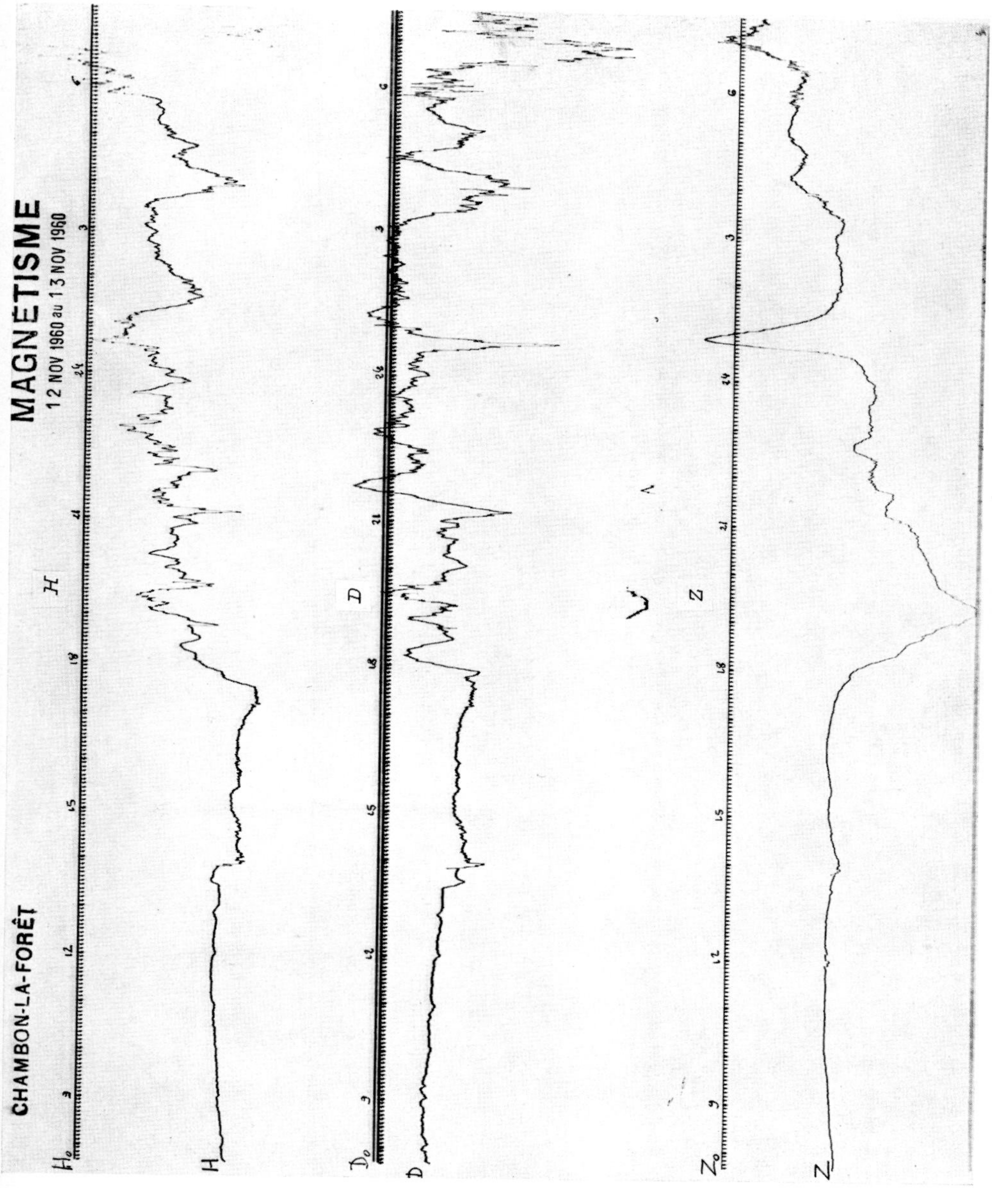

Fig. 27a.

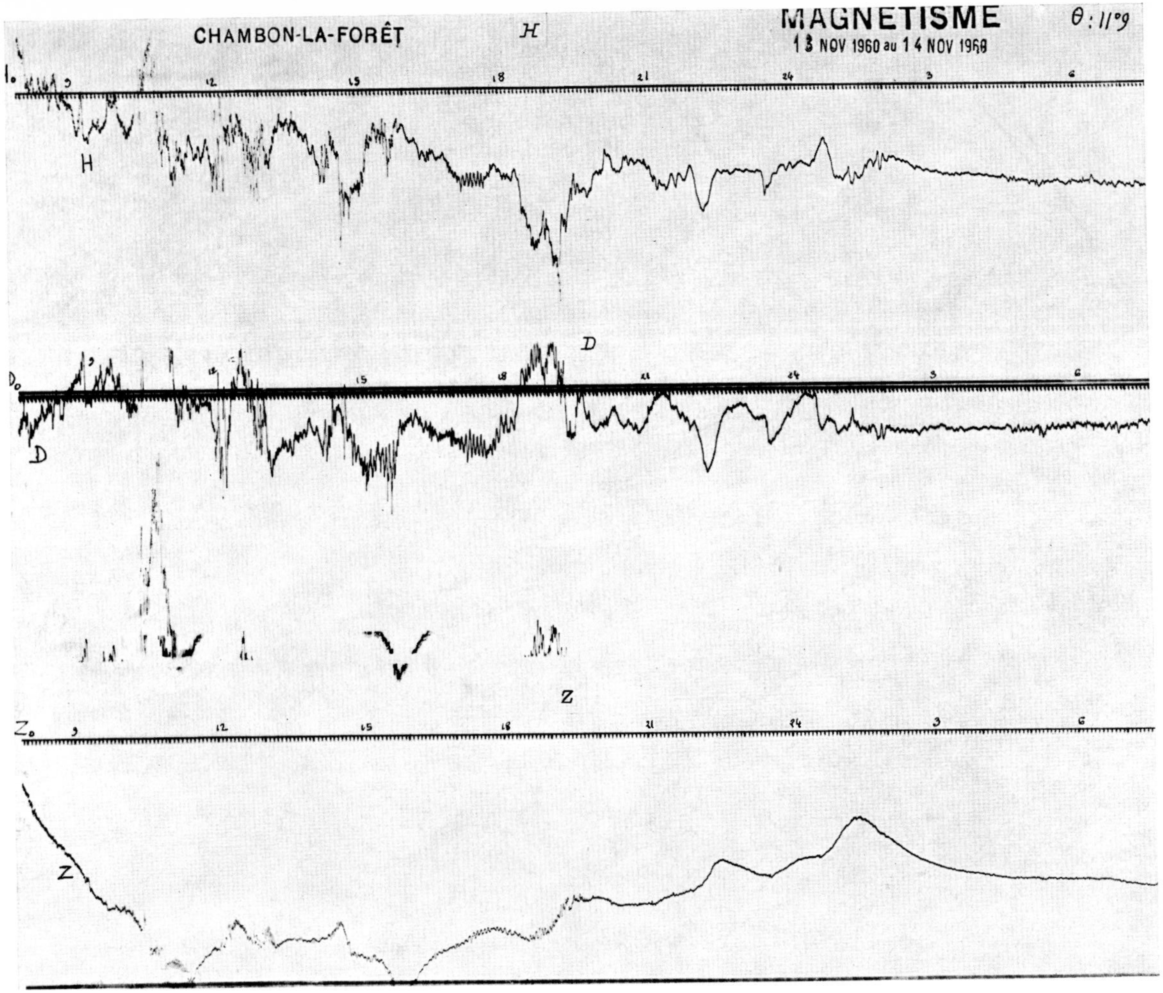

Fig. 27b.

Fig. 27a et b. Magnétogrammes normaux La Cour du 12 au 14 novembre 1960. Ces magnétogrammes sont donnés en exemple de «*pulsations géantes*» (**«pg»**) se rangeant dans la catégorie **«pc-5»** de la classification systématique internationale, qui se manifestent au cours du grand orage des *12/13 novembre 1960* (notamment un peu avant 18 h T.U. le 13 novembre)

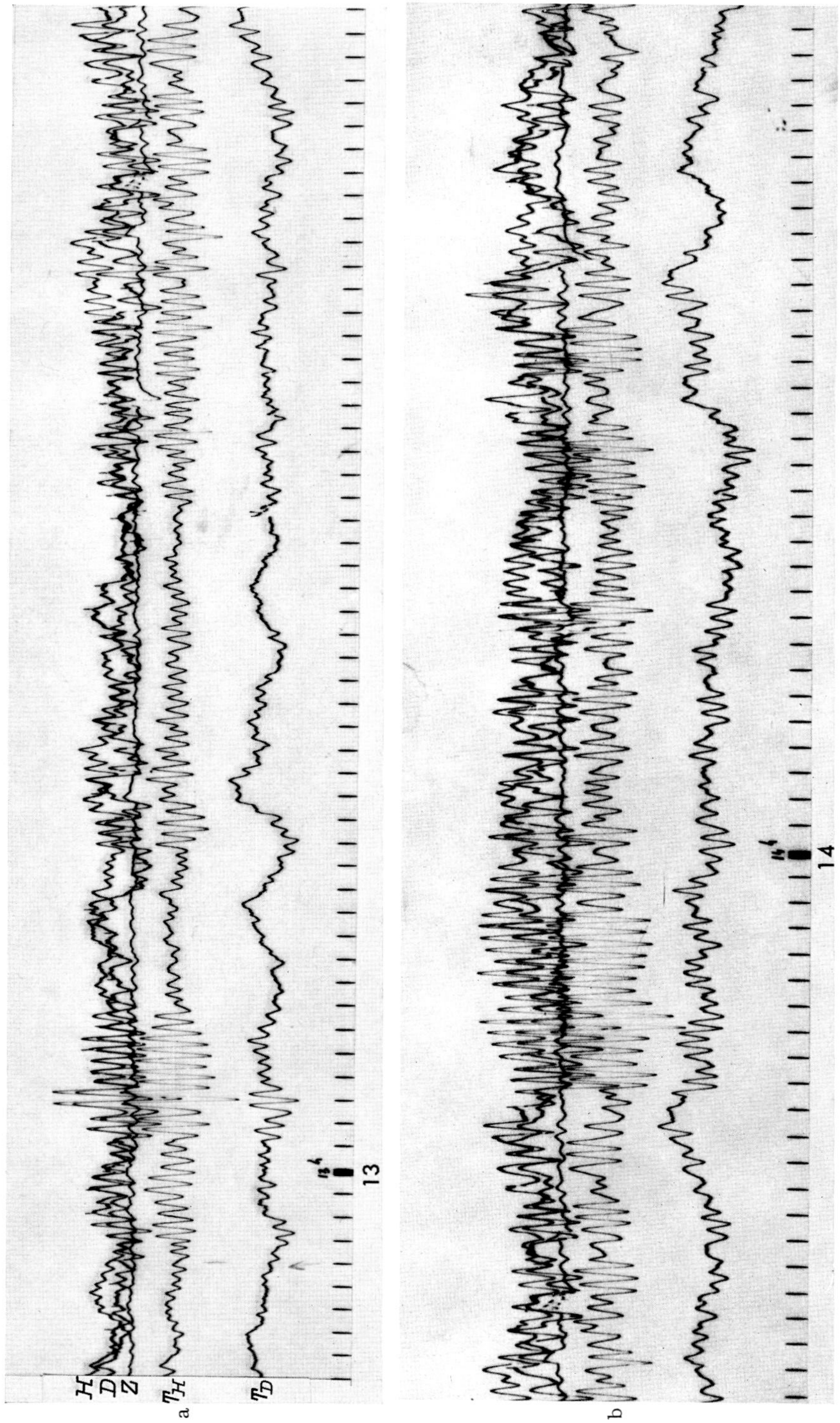

a

b

Fig. 28a et b. Exemple de pulsations continues de type **pc-3**, le 11 fév. 1963, Chambon-la-Forêt, de 12.51 à 14.32 T.U. Marques de temps toutes les minutes. Amplitude maximale de ces pulsations environ 2 gamma. Période de l'ordre de 30 s. Enregistrement type « Barre-Fluxmètre » (trois composantes magnétiques, 2 composantes telluriques).

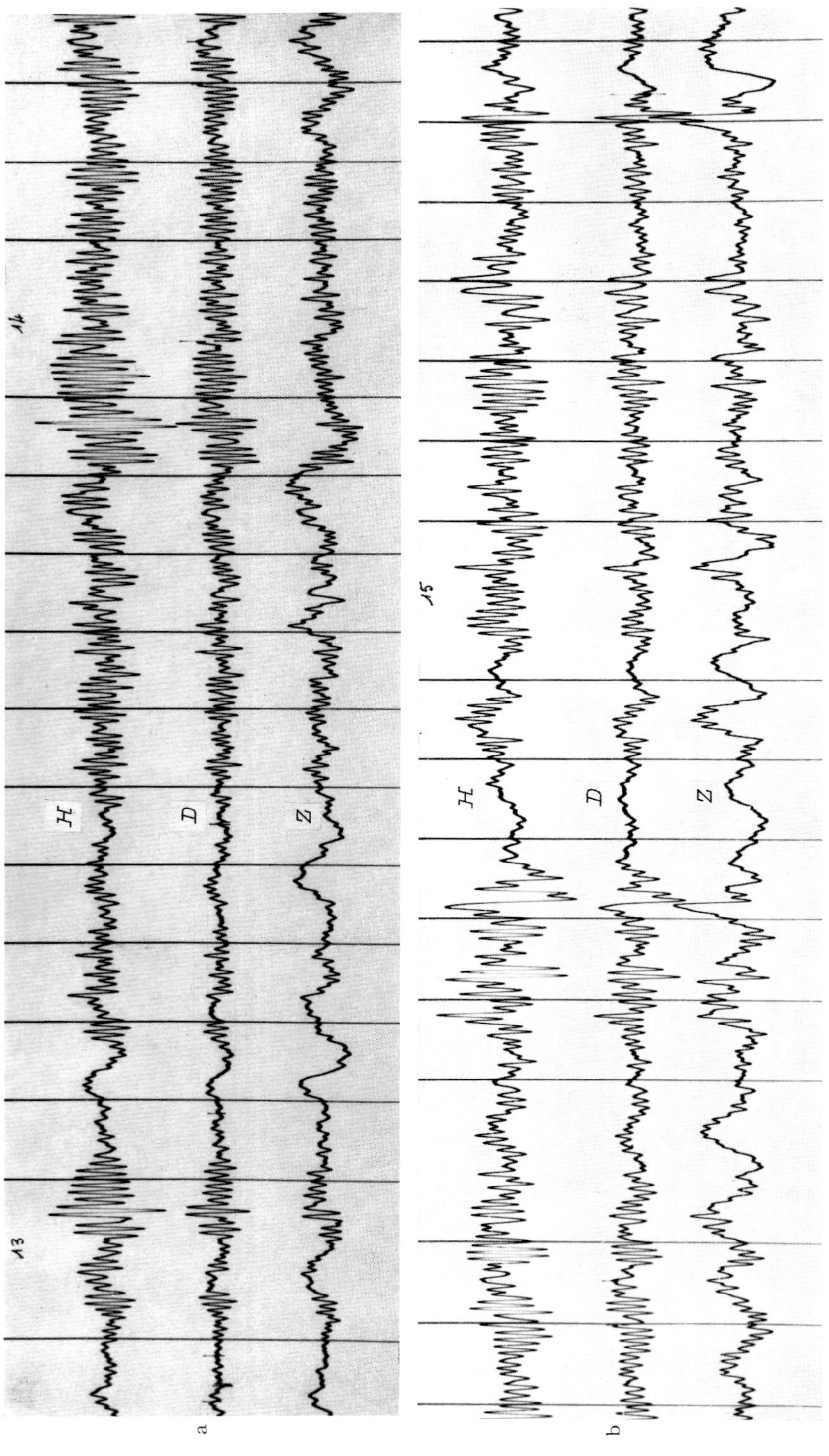

Fig. 29a et b. Pulsations continues **pc-3** par situation magnétique peu agitée, Garchy, le 11 février 1963, entre 13 h et 15 h T.U. On remarquera l'organisation en paquets d'oscillations (parfois en fuseaux) qui traduisent en général une action d'ensemble, synchrone, à l'échelle mondiale (sous réserve des effets de modulation jour-nuit, liés à l'Heure Locale. (Marques de temps toutes les 5 min, sauf aux heures rondes).

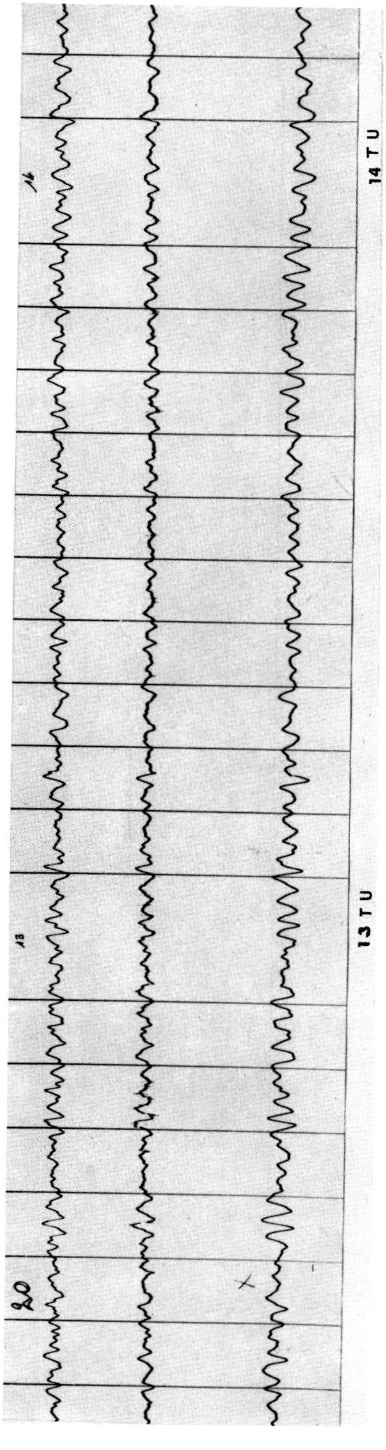

Fig. 30. Exemple de pulsations continues du type **pc-4** Garchy, le 15 mars 1963, autour de 13/14 h T.U. (Enregistrement type «Barre-Fluxmètre», marques de temps toutes les 5 min)

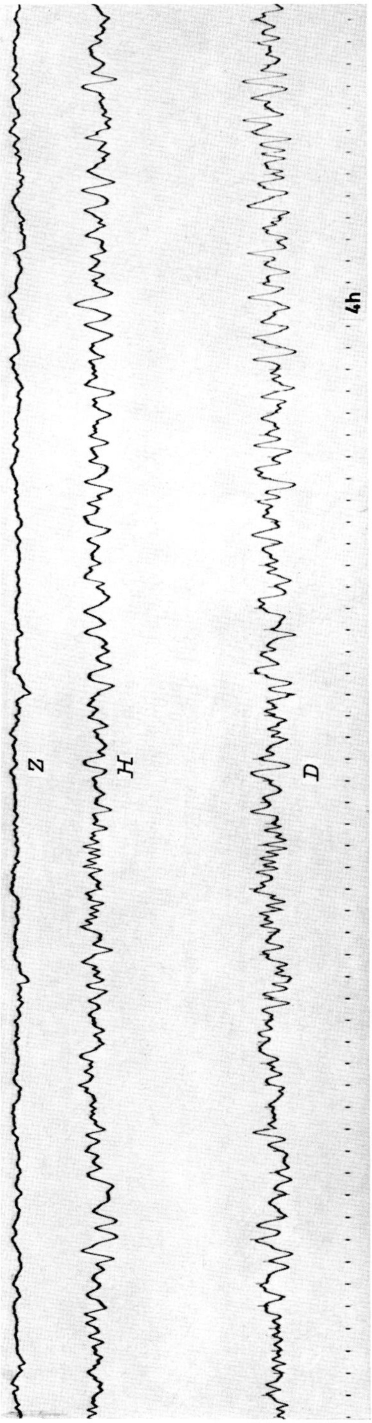

Fig. 31. Exemple «d'agitation du matin» formée ici, principalement, d'un mélange de **pc-3** et de **pc-4**. Chambon-la-Forêt, le 7 juin 1961. (Enregistrement type «Barres-Fluxmètres», marques de temps toutes les minutes).

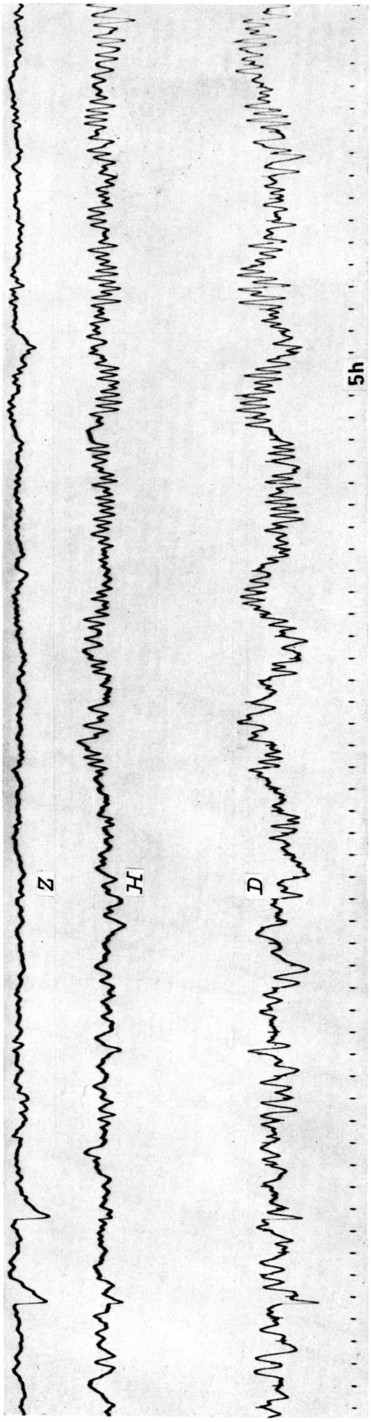

Fig. 32. Suite de l'enregistrement précédent. Faisant suite à «l'agitation du matin", les pulsations **pc-3** deviennent de plus en plus régulières et fournies. *Chambon-la-Forêt.* (Enregistrement type Barre-Fluxmètre.)

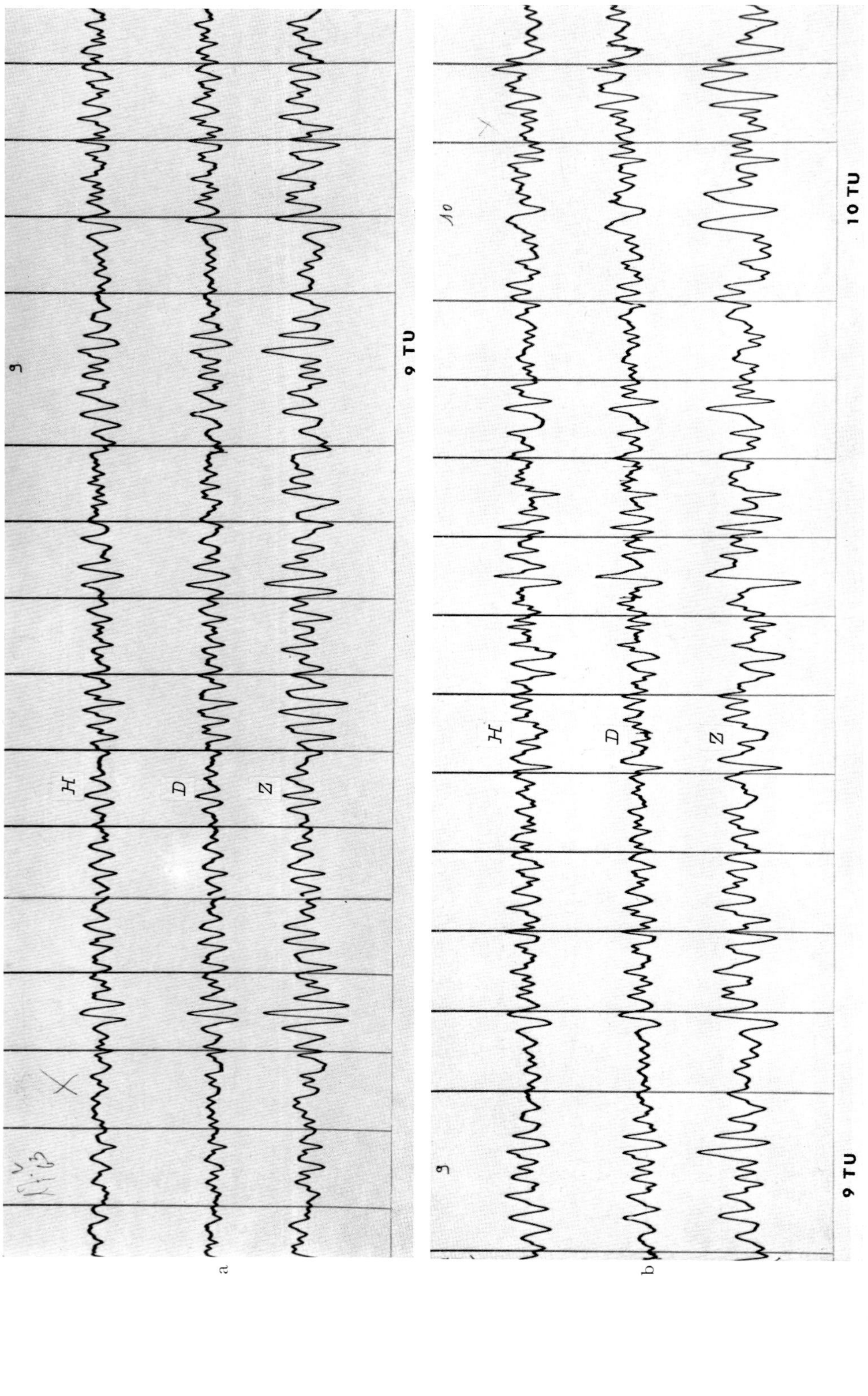

a b

Fig. 33a et b. Pulsations continues du type **pc-4** auxquelles viennent s'adjoindre peu à peu des **pc-3** (surtout après 9 h T.U.). Garchy, le 17 février 1963.

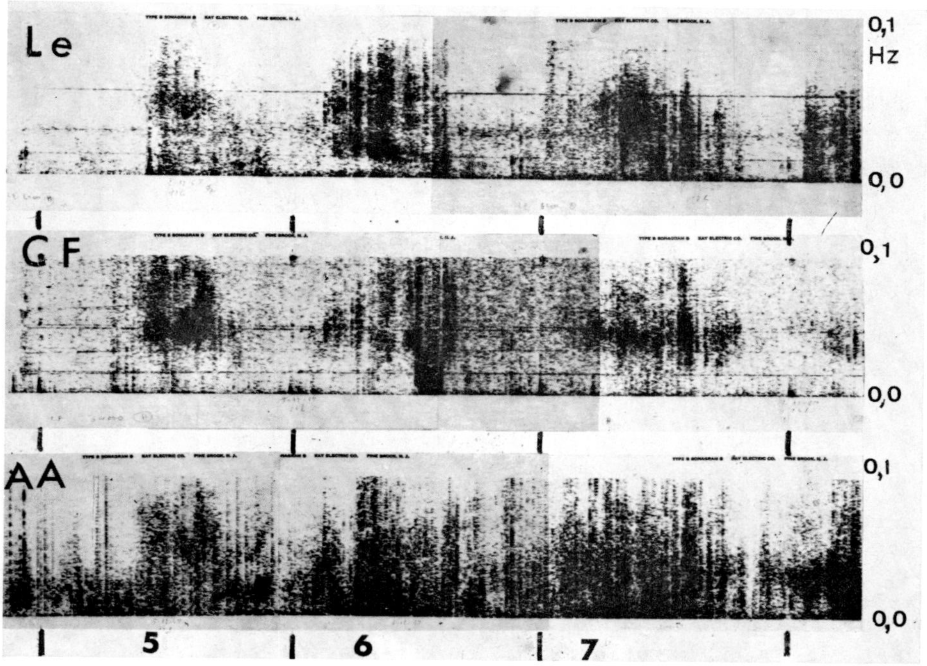

Fig. 34. Observation simultanée à Lerwick, Chambon-la-Forêt et Addis-Abéba de pulsations **pc-3** et **pc-4** et comparaison des spectres de fréquence correspondants. Cette comparaison, qui porte sur les journées des 5, 6 et 7 décembre 1967, permet — d'une façon plus synthétique que par l'examen des formes d'onde — de mettre en évidence le caractère mondial de ces pulsations (sous réserve de l'effet général de modulation par l'Heure Locale). (Cliché GLANGEAUD-ROQUET).

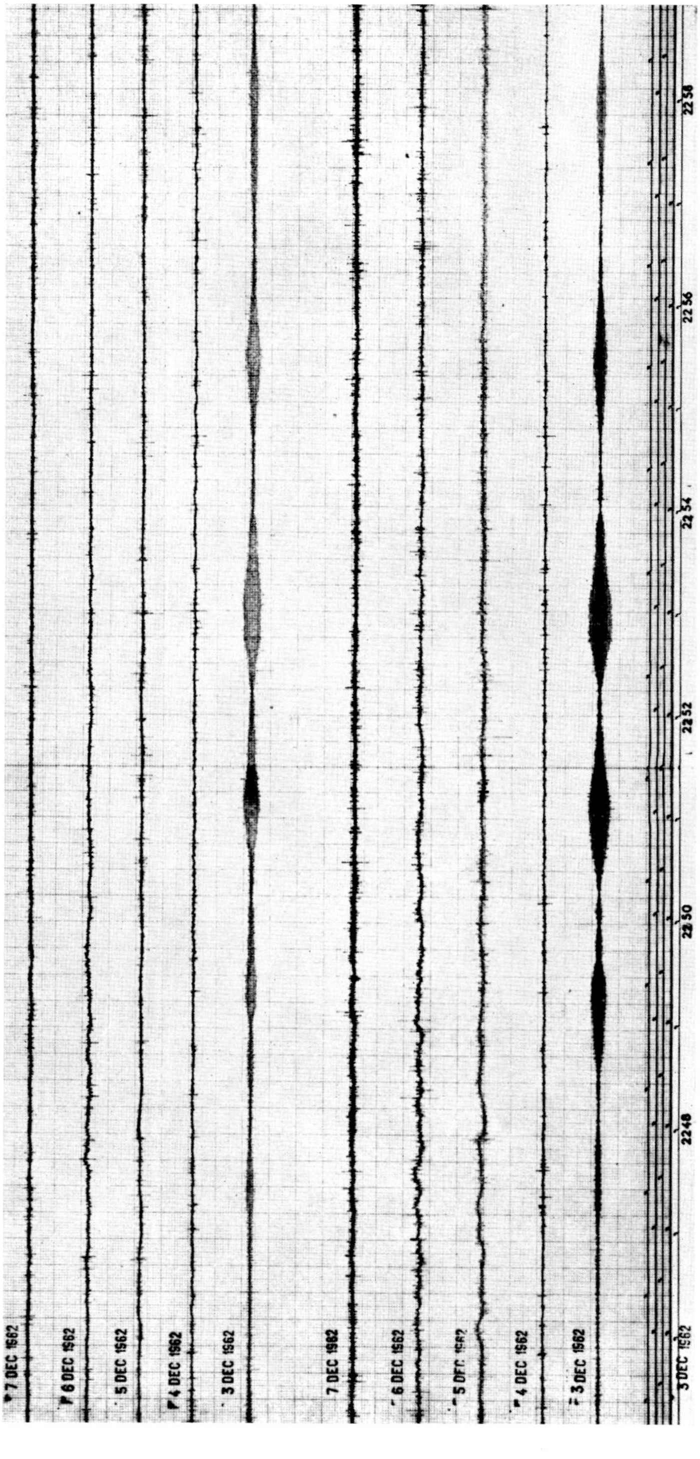

Fig. 35. Exemple de pulsations **pc-1** structurées en forme de perles simples, sans «grains». Kerguelen, 3 décembre 1962. (Cliché SCHLICH — T.A.A.F.).

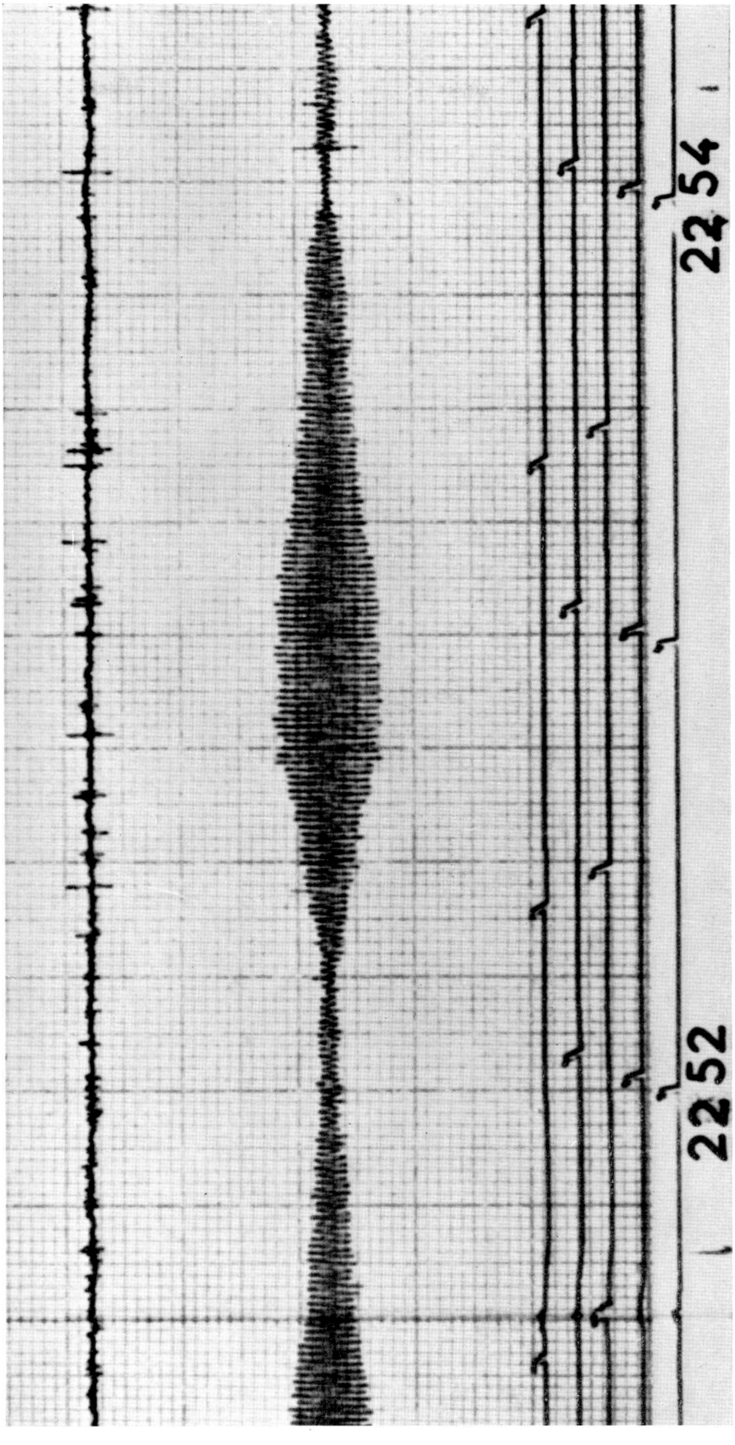

Fig. 36. Détail d'une des perles du Cliché donné en Fig. 35. Ce détail met bien en évidence l'absence de tout phénomène d'interférence au cours de l'évolution de cette «perle».

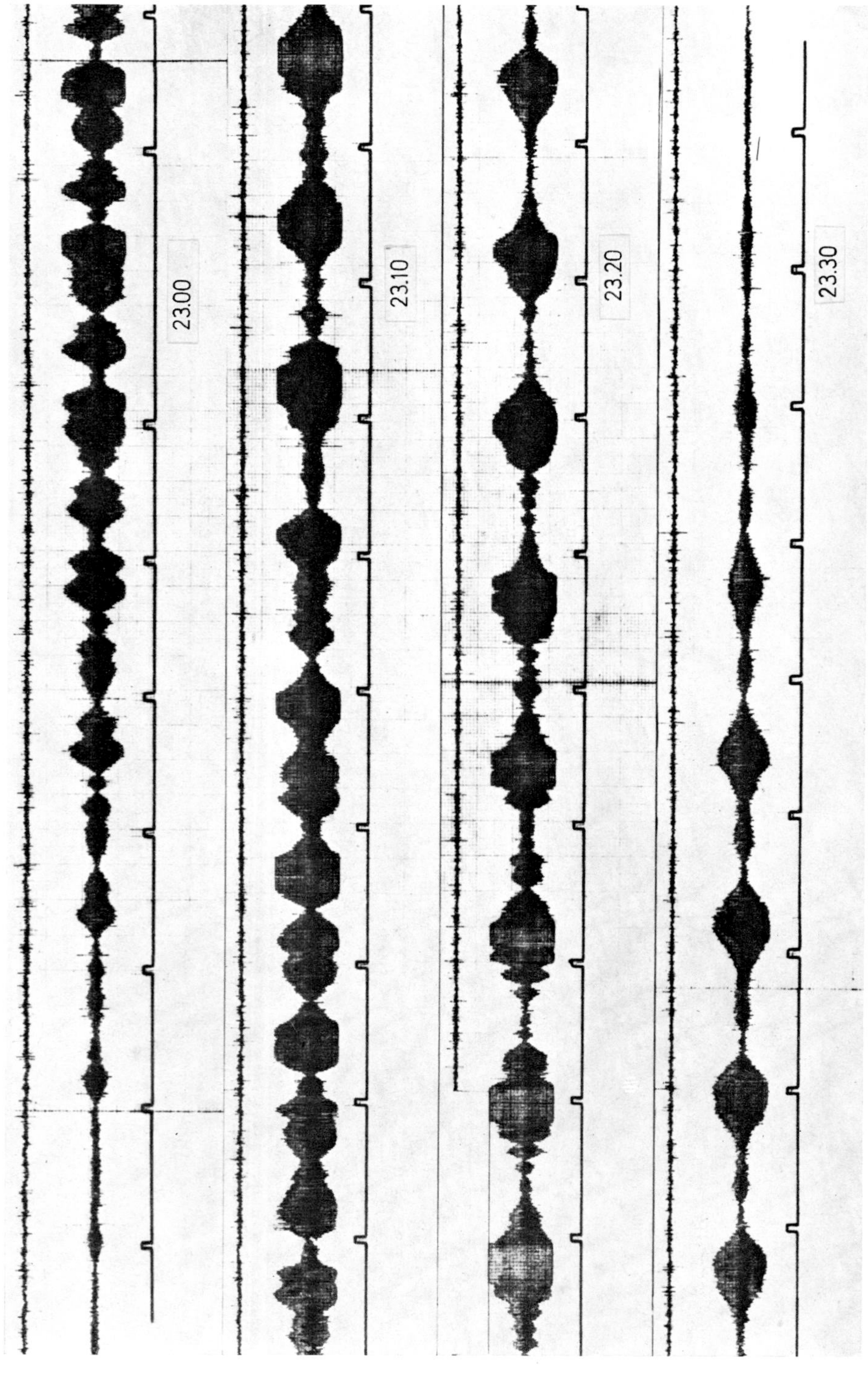

Fig. 37. Evolution d'une structure de pulsations **pc-1** entre 22.53 et 23.31 T.U., le 3 octobre 1962, à Kerguelen. On peut voir des «perles simples» se dégager progressivement d'une structure plus complexe (formée de «perles» et de «grains»). (Cliché SCHLICH — T.A.A.F.).

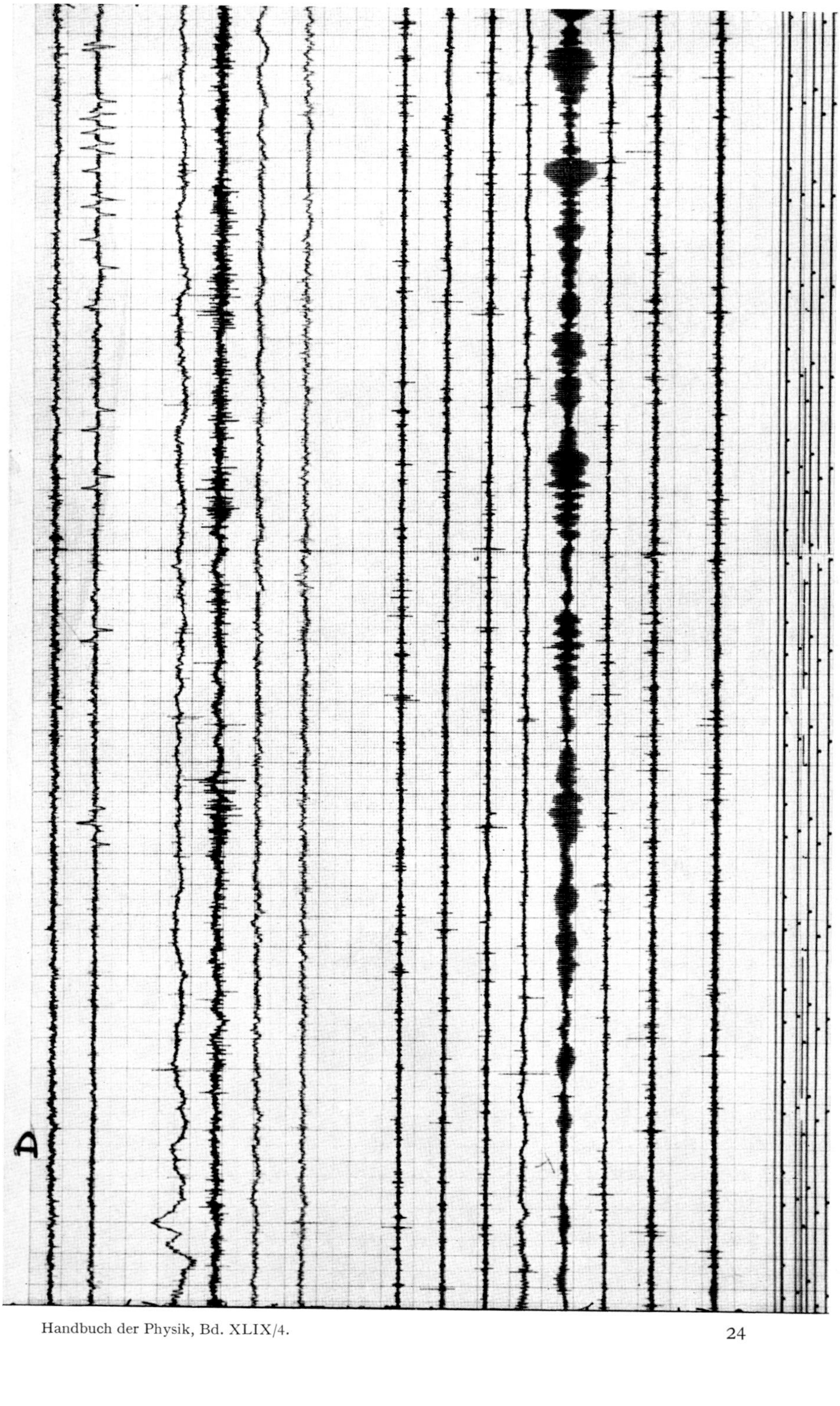

Fig. 38a.

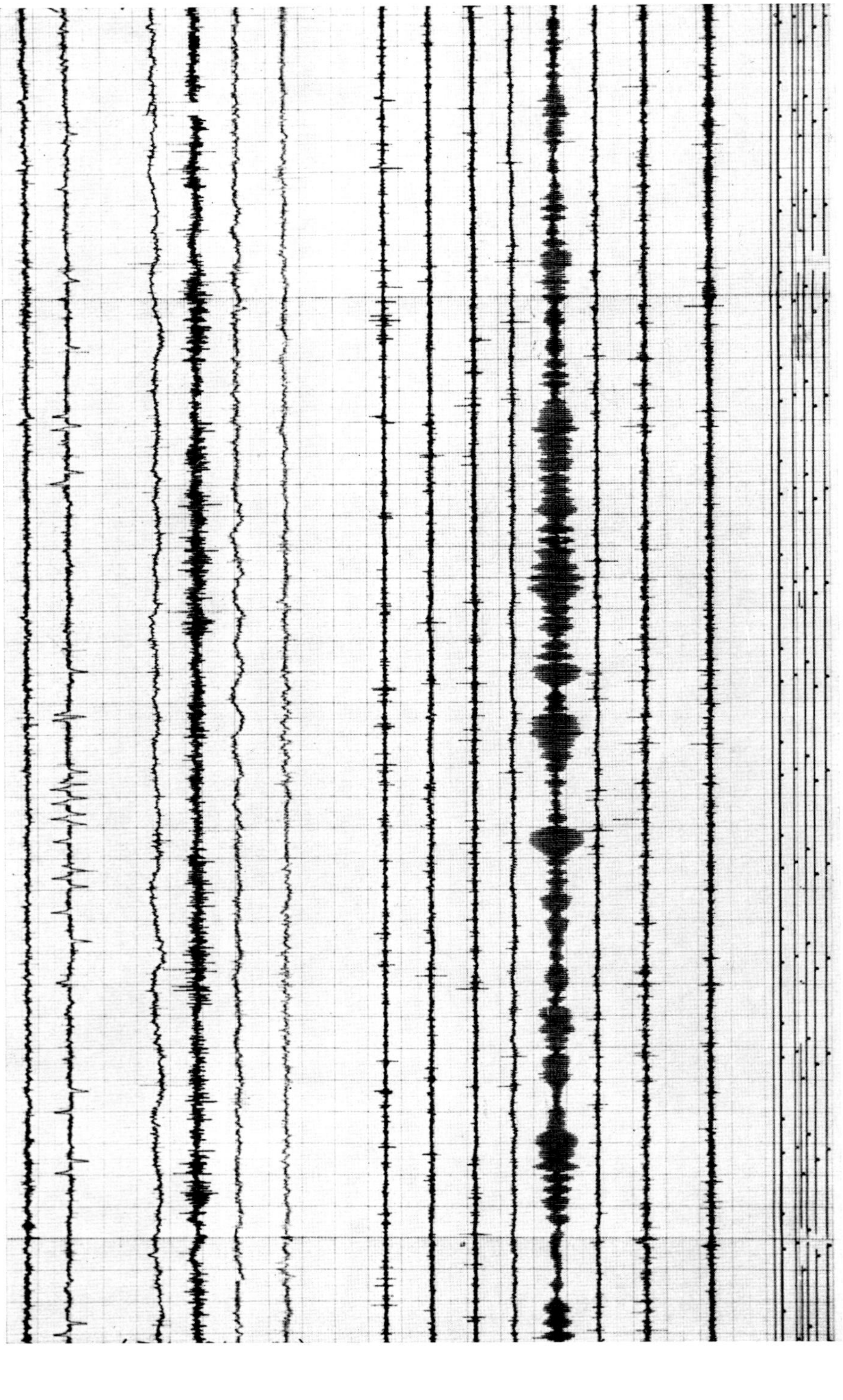

Fig. 38b.

Fig. 38a et b. Formes d'onde de «pulsations en perles» **pc-1**, sous-structurées en «grains». Garchy, le 28 mars 1966. La prédominance des «grains» empêche de bien voir la périodicité des «perles».

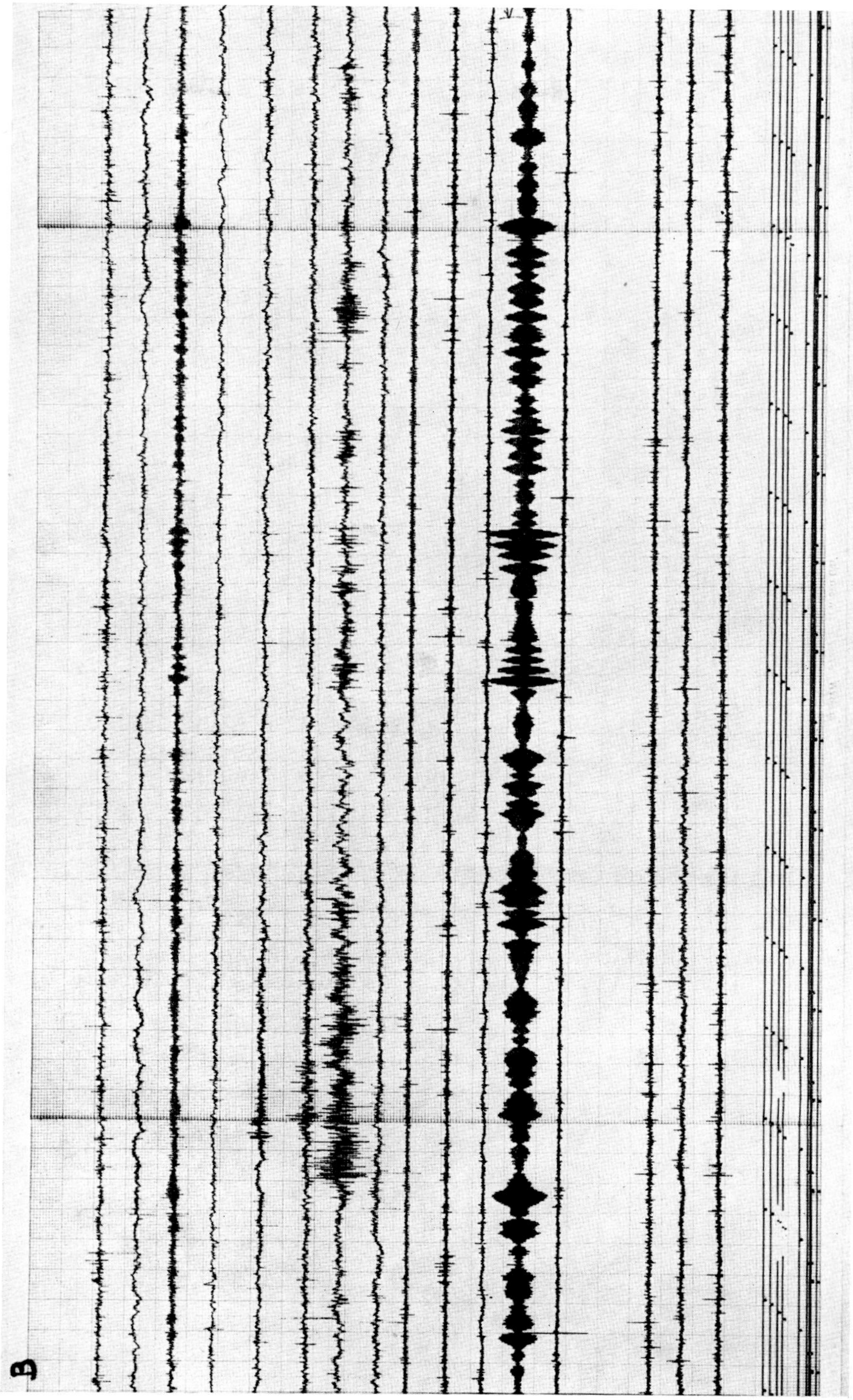

Fig. 39. Formes d'ondes de «pulsations en perles» **pc-1** faisant apparaître d'une façon particulièrement nette une sous-structure en grains. Garchy, le 4 avril 1966. (On remarque en outre, sur une autre ligne, correspondant au 8 avril, un évènement du type **pi-2** + **pi-1**).

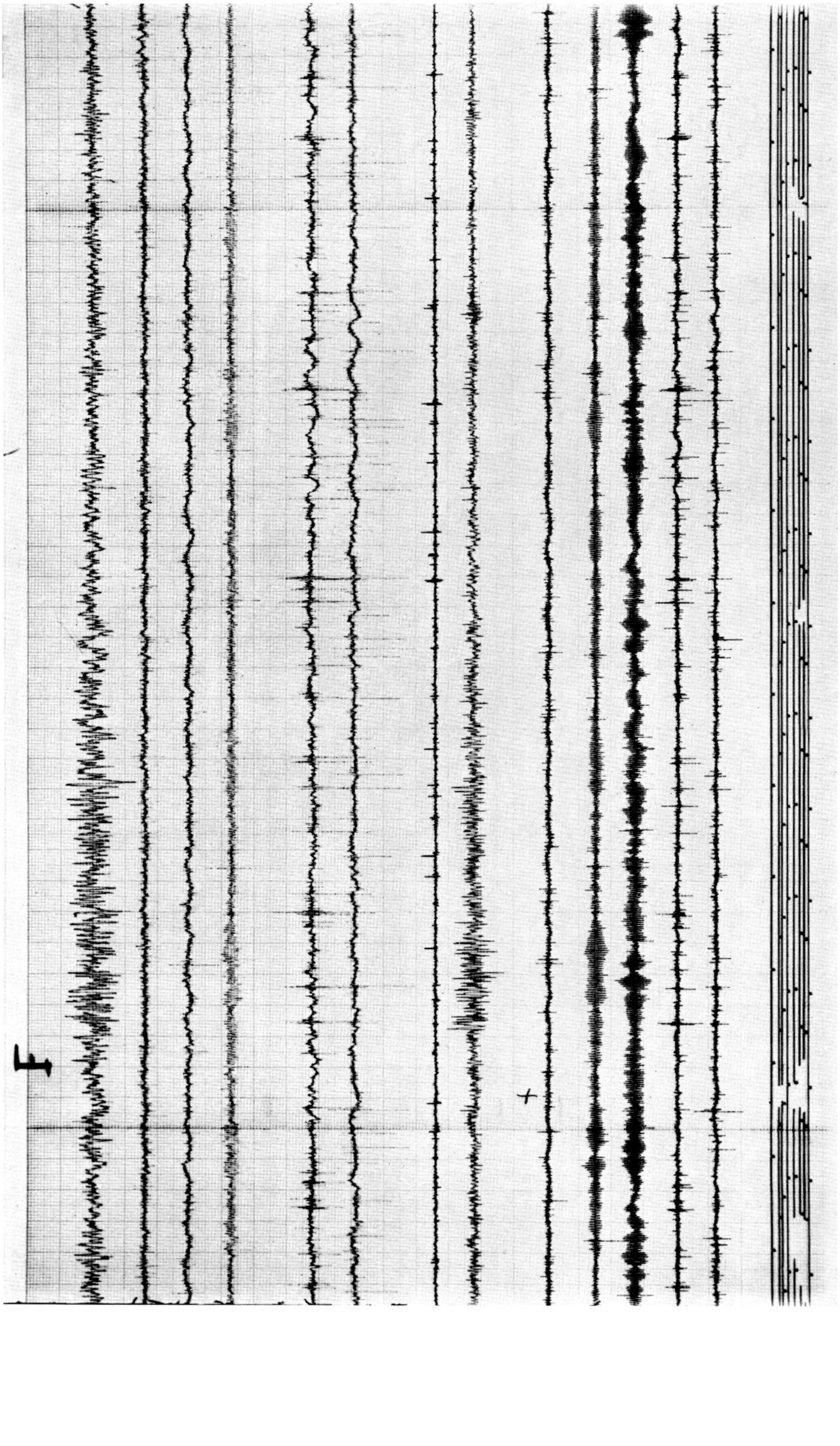

Fig. 40. Comparaison de formes d'ondes de pulsations **pi-1** et **pc-1**, Garchy, mars 1966. Sur cette «page» d'un enregistrement rapide (type «accordéon», vitesse de défilement 30 mm/min) de la station on peut voir simultanément un exemple de **pi-1**, le 14 mars (haut et milieu de la feuille, composantes telluriques E—W et N—S, respectivement) et deux exemples plus ou moins structurés en «perles», les 17 et 18 mars.

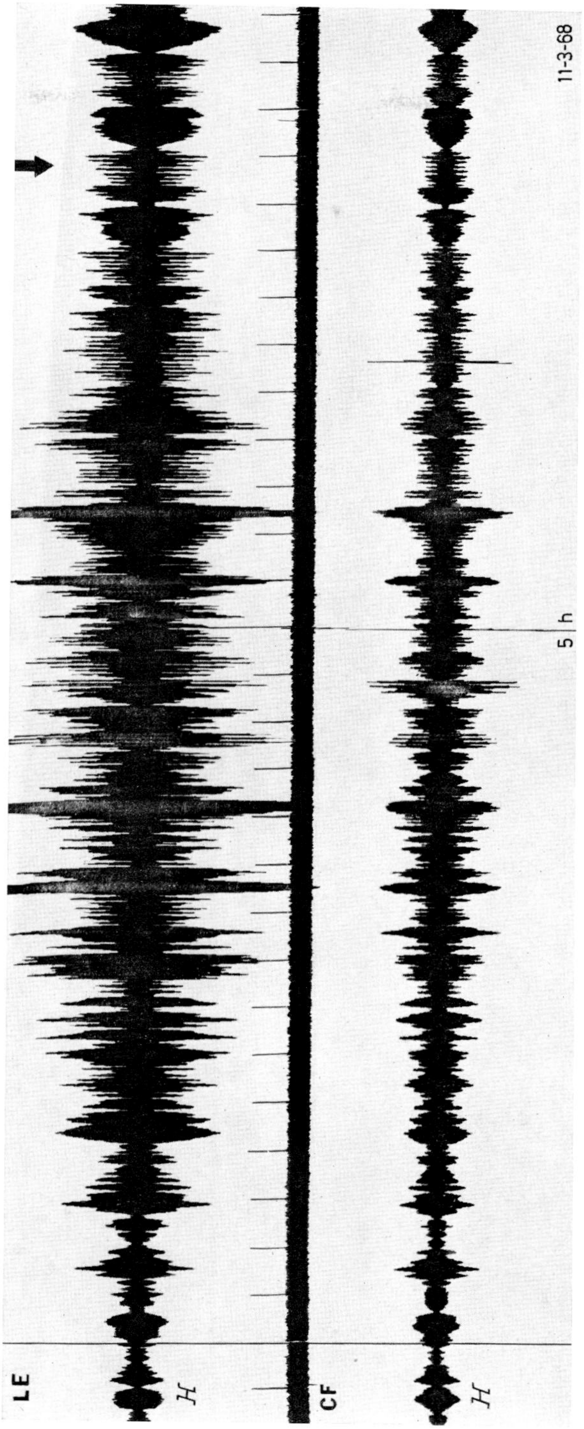

Fig. 41 a.

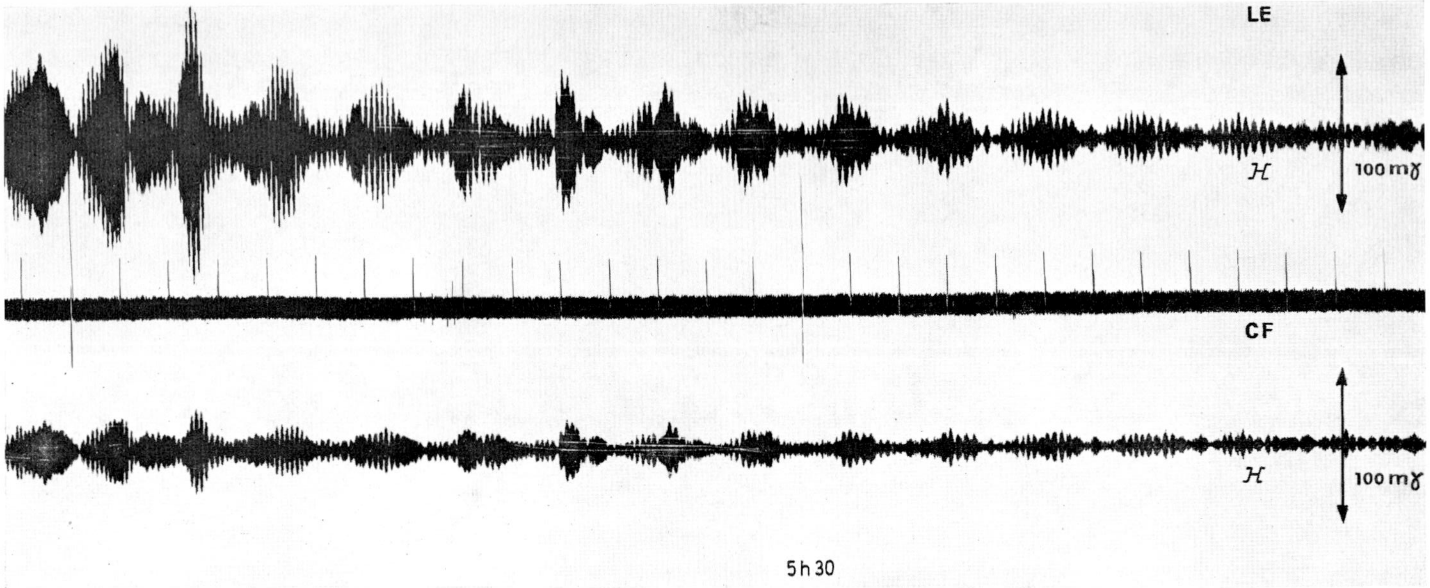

Fig. 41 b.

Fig. 41 a et b. Comparaison entre les formes d'ondes («perles» et «grains») de pulsations **pc-1** enregistrées à Lerwick et Chambon-la-Forêt, le 11 mars 1968. Les deux stations sont approximativement situées suivant le même méridien mais présentent une différence de latitude d'environ 12° (lat. de Lerwick: + 60° 08′, lat. de Chambon-la-Forêt: + 48° 01′). Sur ce document on peut suivre l'évolution relative des «grains» par rapport aux «perles» ce qui a permis de mieux comprendre la nature précise des «grains» (GLANGEAUD 1968). On voit d'autre part que cette évolution est à très peu près la même aux deux stations. (Cliché GLANGEAUD).

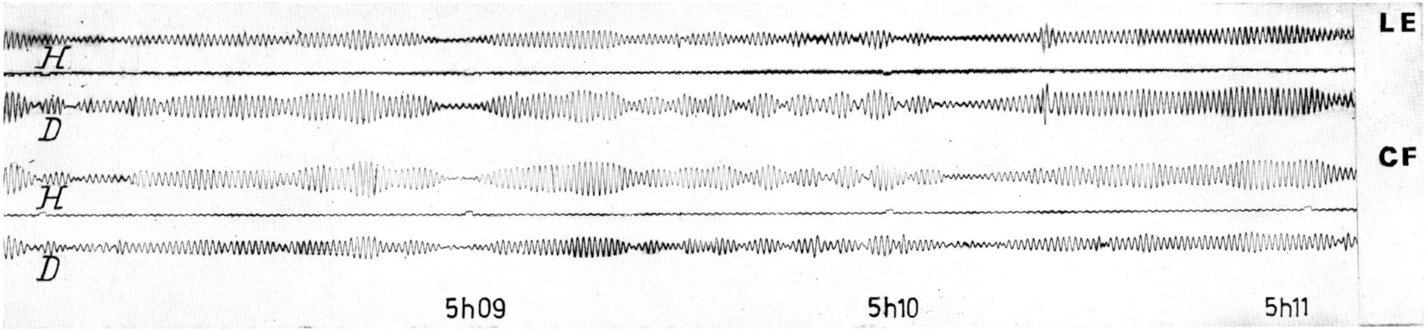

Fig. 42. Ce document présente d'une façon détaillée un épisode central de la comparaison Lerwick-Chambon qui a fait l'objet de la Fig. 41 a. Il est centré à l'aplomb fléché porté par le document précédent. Les «grains» qui apparaissaient comme de simples traits sur la Fig. 54 peuvent être examinés ici avec leur forme propre. De plus, on peut remarquer un léger retard de l'arrivée des

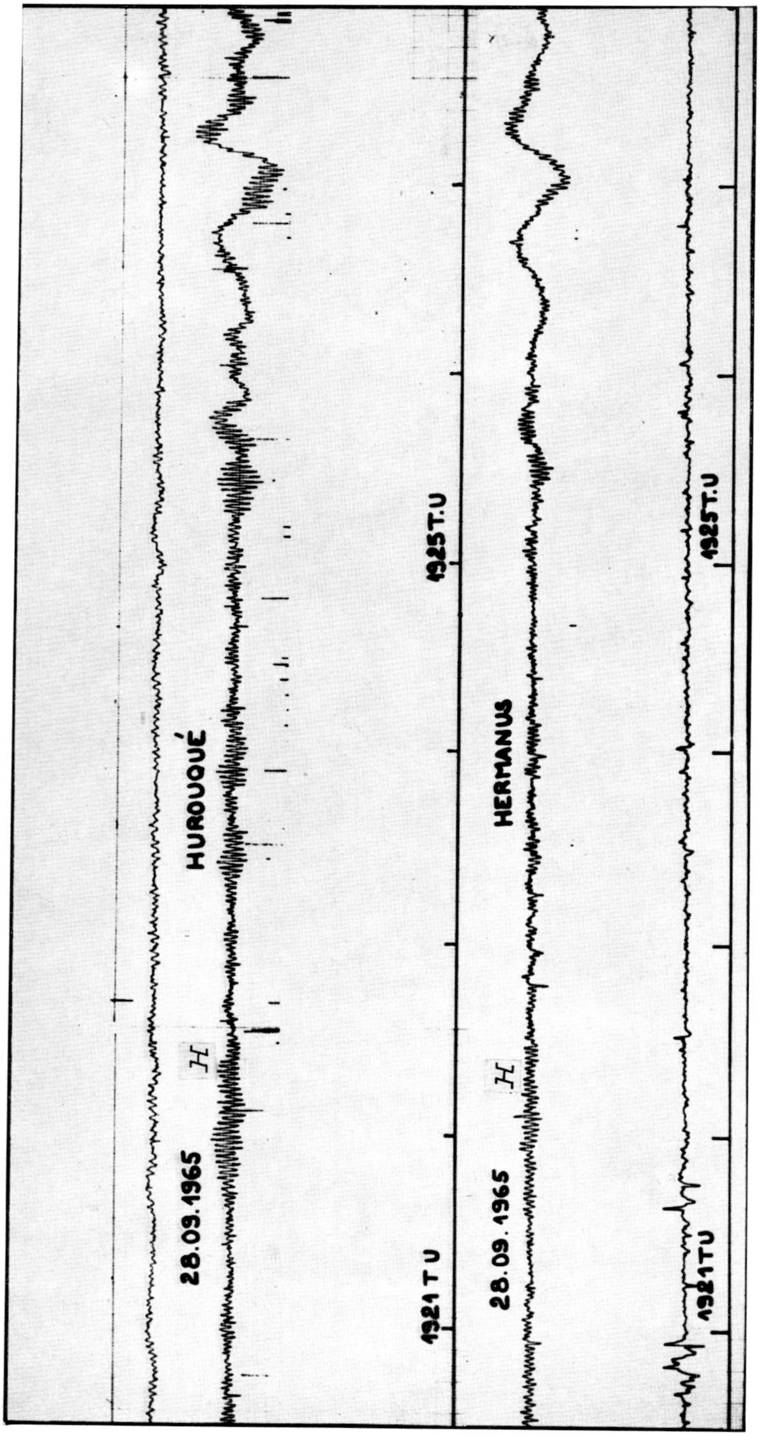

Fig. 43. Comparaison de pulsations **pi-1** en deux points conjugués (Observatoire d'Hermanus en Afrique du Sud, et station de Hurouqué dans le Sud-Ouest de la France) *des moyennes latitudes*, le *28 septembre 1965*. La précision dans le calage des temps a permis de prouver que les trains de pi-1 se manifestaient en synchronisme aux deux stations (à une approximation de l'ordre de la seconde), ainsi d'ailleurs que les «grands» mouvements de type **pi-2** qu'ils accompagnent. (Cliché Schlich).

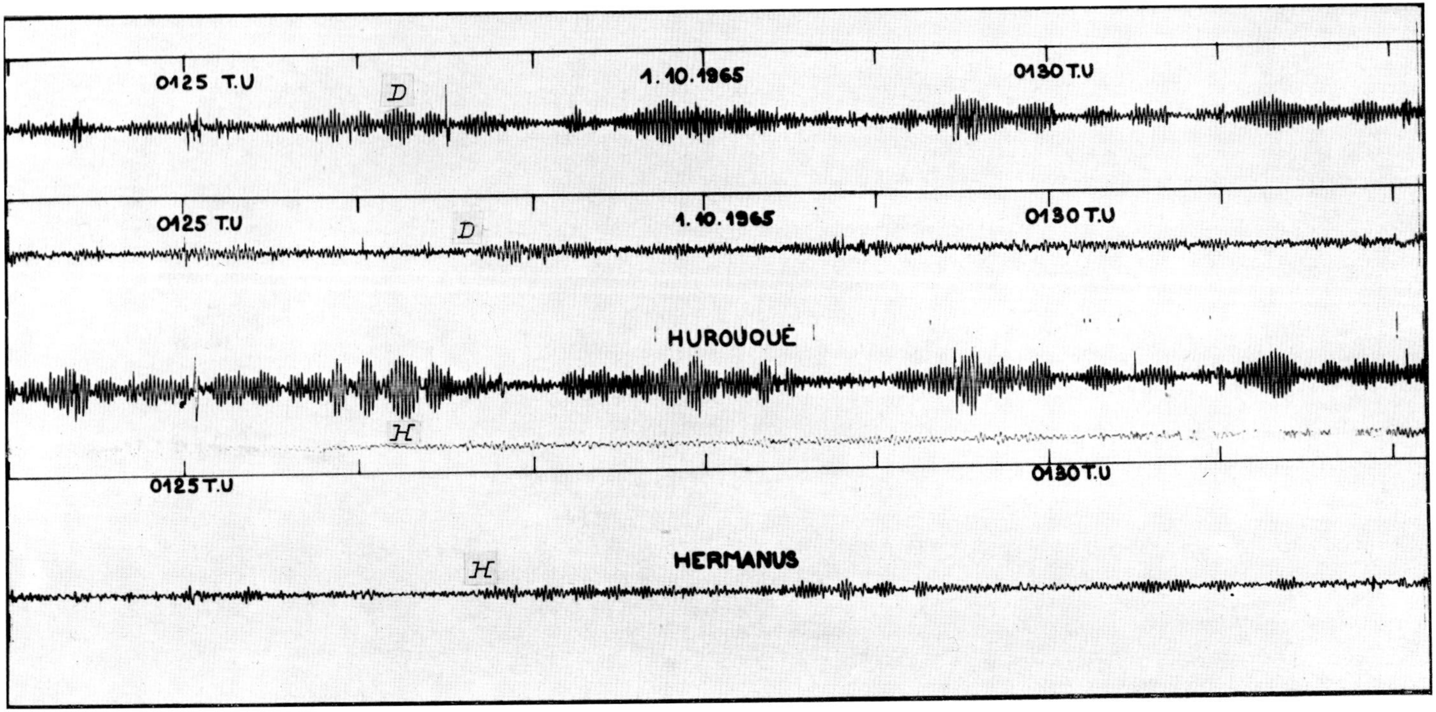

Fig. 44. Comparaison de pulsations **pc-1** en deux points conjugués (Observatoire d'Hermanus et station de Hurouqué, cf. Fig. 43), *des moyennes latitudes, le 1er octobre 1965*. Contrairement à ce qui se passe pour les **pi-1** (cf. Fig. 43) le synchronisme des «trains d'ondes» n'est pas conservé et fait place à une alternance plus ou moins rigoureuse de ces trains. (Cliché SCHLICH).

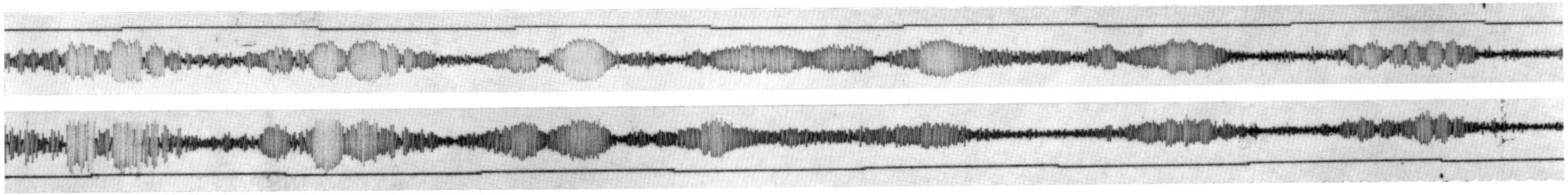

Fig. 45. Comparaison de pulsations **pc-1** en deux points approximativement conjugués (Kerguelen — trace inférieure — et Borok — trace supérieure —) *de latitudes sub-aurorales*, le *20 février 1964*. Il s'agit ici de formes d'ondes qui ont été restituées au laboratoire à partir d'enregistrements d'origine sur bandes magnétiques. On remarquera que les «perles» et «grains» que montrent ces reproductions ne sont ni en phase ni en antiphase parfaite aux deux stations mais leurs similitudes de formes sont frappantes. (Cliché GENDRIN-TROITSKAJA).

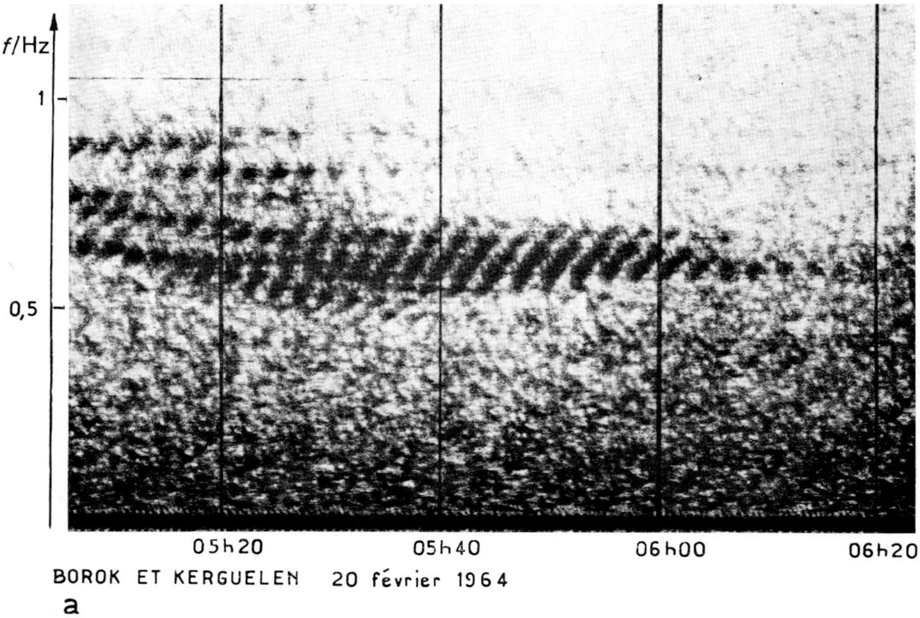

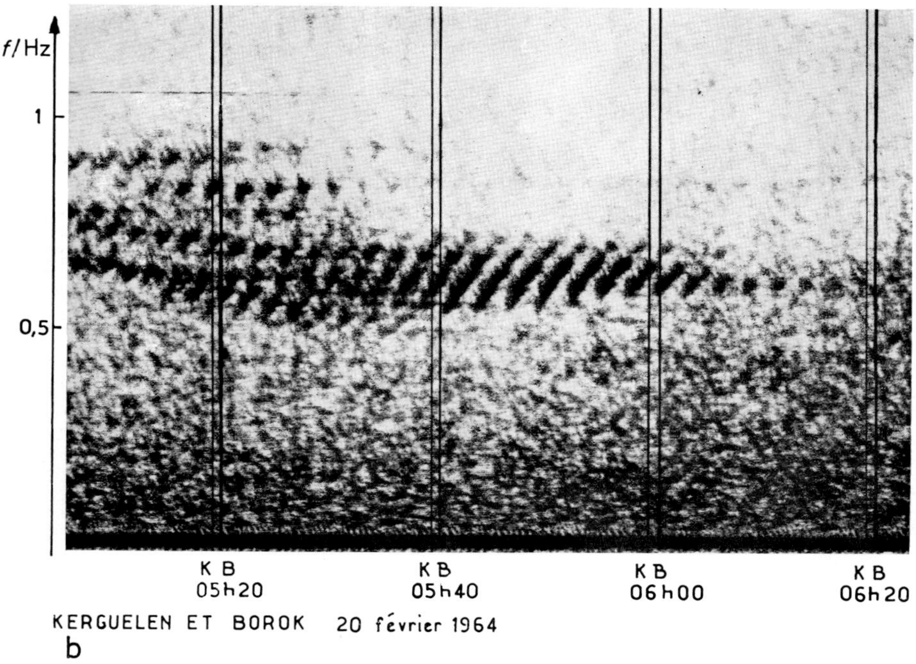

Fig. 46a et b. Comparaison des analyses spectrales (sonagrammes) faites sur des séquences d'enregistrement de Kerguelen et de Borok faisant immédiatement suite à la séquence donnée en Fig. 45. Ces sonagrammes ont servi aux auteurs. (R. Gendrin et V. A. Troitskaja) dans leurs efforts pour arriver à une détermination aussi précise que possible du décalage dans le temps des structures spectrales qui apparaissent comme très semblables aux deux stations. La précision atteinte a été de l'ordre de 20 s ce qui a limité partiellement la rigueur des conclusions que l'on a pu ainsi atteindre au sujet des *relations d'anti-phase* applicables aux formes d'ondes telles qu'elles sont observées aux deux stations. Les doubles traits verticaux de la représentation inférieure indiquent le décalage de temps qui a été reconnu le plus efficace (par approximation) pour amener les deux structures spectrales en quasi-coïncidence. (Cliché Gendrin-Troitskaja).

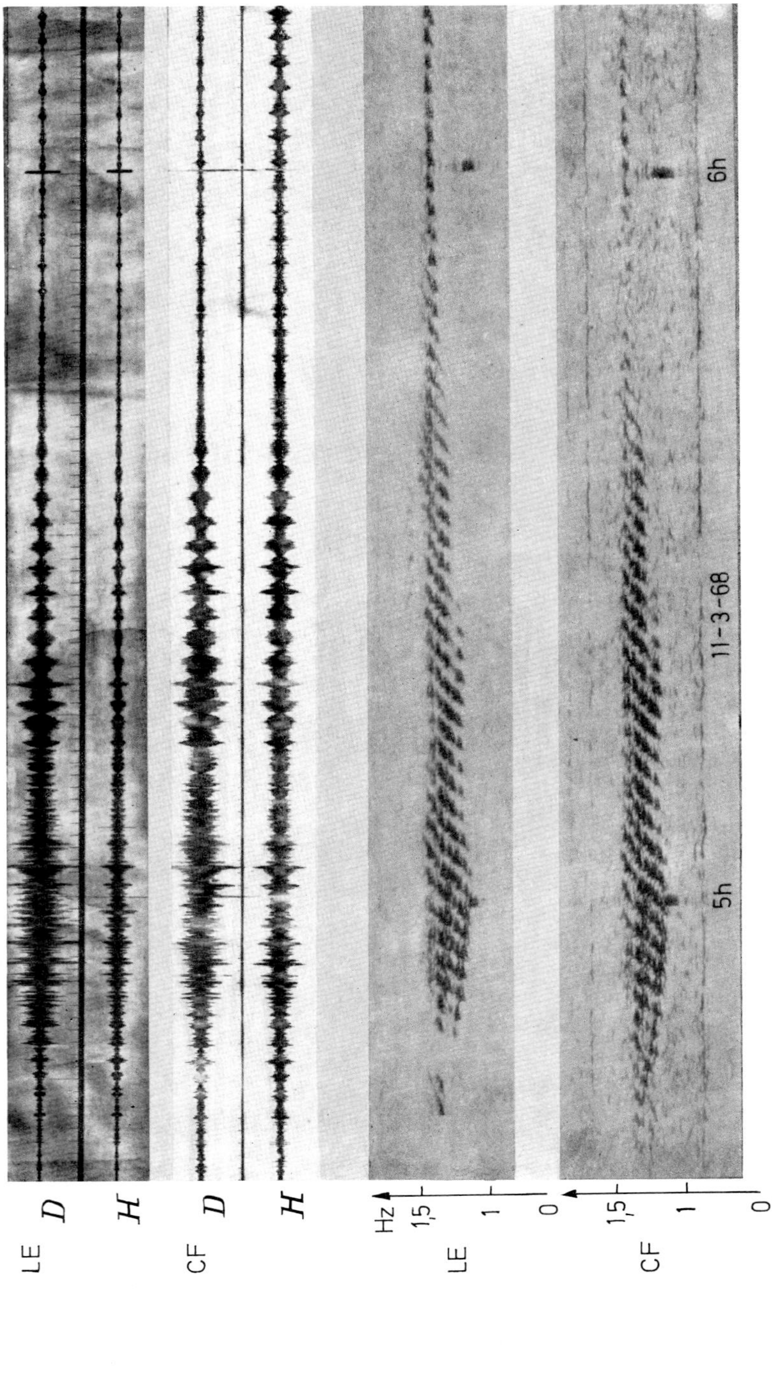

Fig. 47. Comparaisons analogues à celles résumées par les figures précédentes (cf. Fig. 45 et 46) mais les deux stations faisant l'object de la comparaison (Lerwick et Chambon-la-Forêt) ne sont pas conjuguées et sont seulement situées, approximativement, suivant un même méridien. La comparaison est étendue à deux composantes (H et D) ce qui permet de la faire porter, si on le désire, sur les diagrammes de polarisation. Elle est ici appliquée aux formes d'ondes (partie supérieure du cliché) et aux résultats d'analyse spectrale (partie inférieure). (Cliché GLANGEAUD)

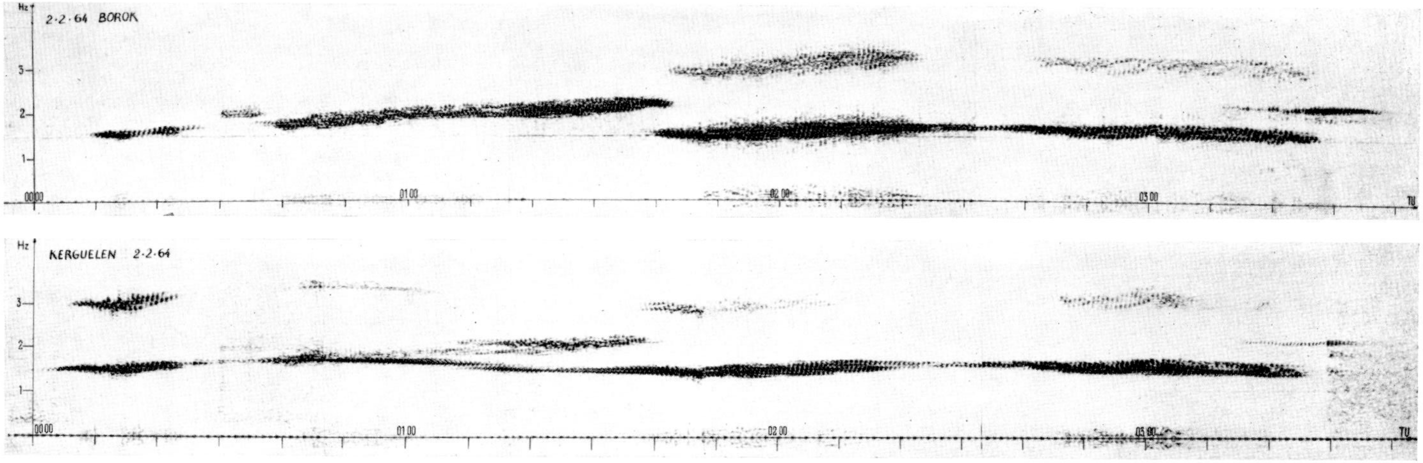

Fig. 48. Comparaison de sonagrammes Borok-Kerguelen concernant des cas de **pc-1** structurées à spectres fins (émissions quasi-monochromatiques). La similarité des deux spectres est très grande, sauf en ce qui concerne les intensités des émissions qui varient parfois de façons non parallèles. 2 février 1964. (Cliché Gendrin-Troitskaja).

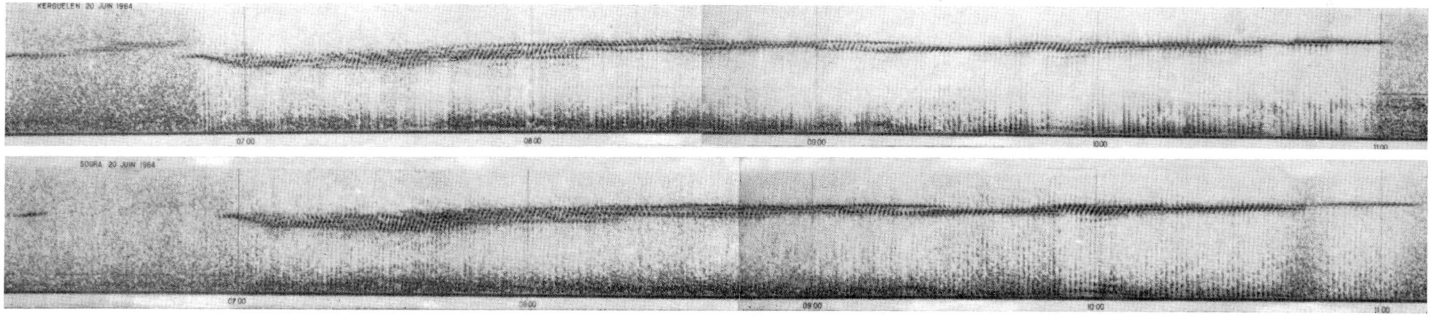

Fig. 49. Comparaison de sonagrammes analogue à la précédente (cf. Fig. 48) mais la station de Sogra ayant remplacé celle de Borok en tant que station conjuguée plus précise de la station de Kerguelen. Les deux sonagrammes ont été alignés, dans ce montage, par rapport à la quasi-identité de leurs structures, ce qui a permis de mettre en évidence, ensuite, le léger décalage des instants de temps universel T.U., correspondants. *20 juin 1964*. (Cliché Gendrin-Troitskaja).

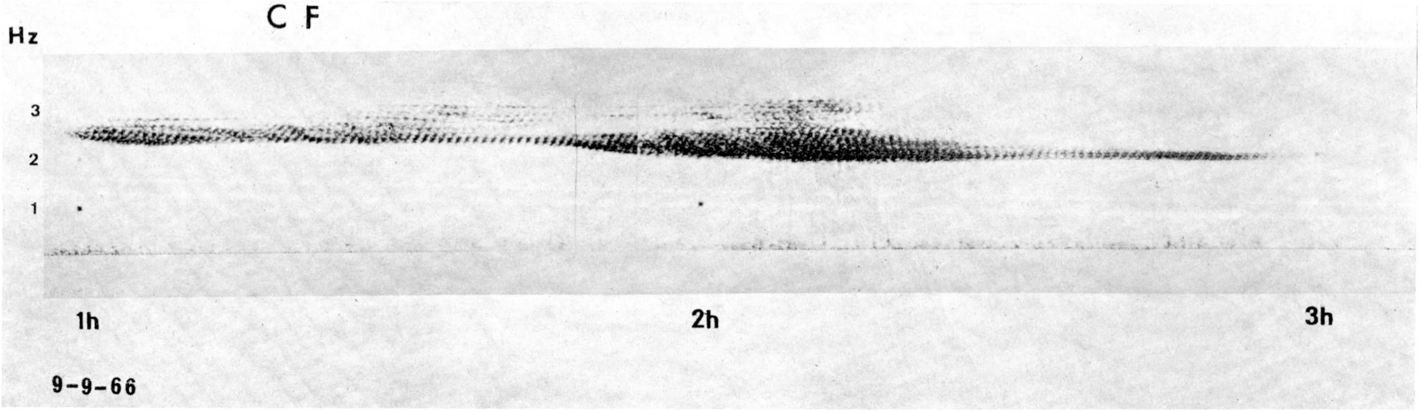

Fig. 50. Observation à Chambon-la-Forêt, du spectre de fréquences de perles structurées de 2 à 3 Hz. La frontière inférieure des pulsations est bien marquée alors que la limite supérieure semble déterminée par des phénomènes d'absorption. 9 septembre 1966. (Cliché GLANGEAUD).

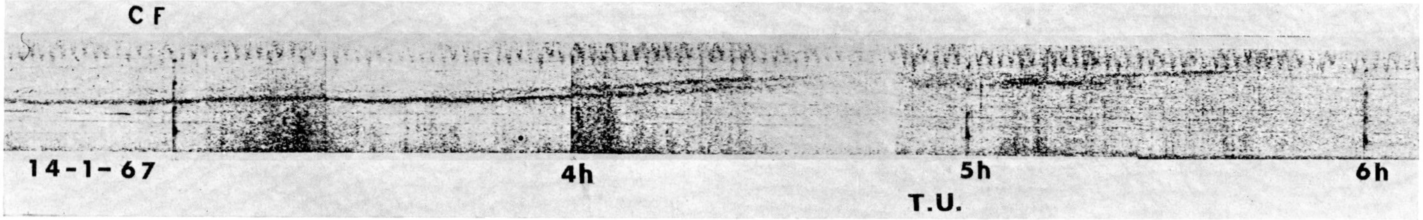

Fig. 51. Sonagramme d'une émission monochromatique, de longue durée, observée le 14 janvier 1967 à Chambon-la-Forêt. Ce phénomène se différencie nettement de ceux du type «perles». Il s'agit ici de pc-1 non structurées et se produisant au cours d'un orage (**SSC** le 13 janvier à 12.02 T.U.). Elles se sont manifestées pendant plus de cinq heures de suite (bien que leur continuité apparente puisse être décomposée en petits trains successifs presque jointifs) au cours desquelles leur fréquence s'est élevée progressivement de 1 à 4 Hz. (Cliché GLANGEAUD).

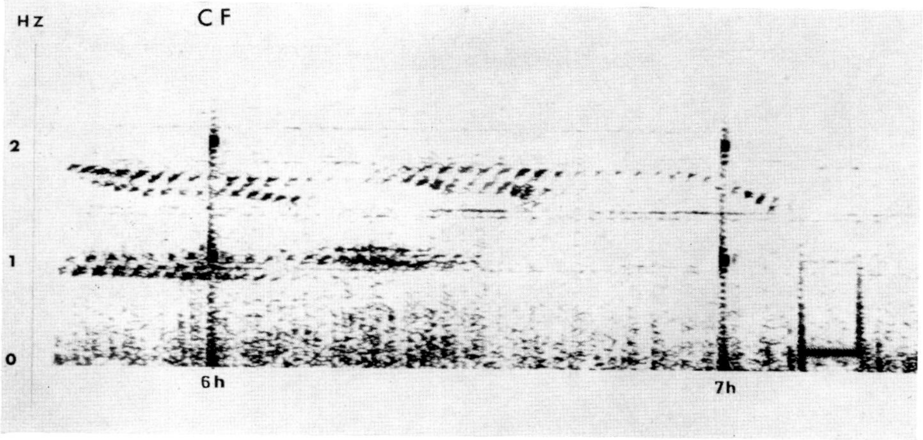

Fig. 52. Exemple d'une émission de **pc-1** structurées, le 25 janvier 1967, à Chambon-la-Forêt. (Cliché GLANGEAUD-PETIAU)

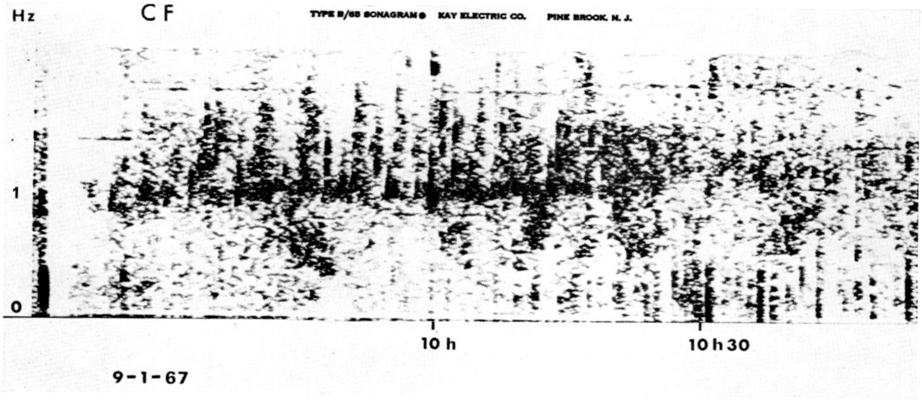

Fig. 53. Exemple peu commun d'une émission pseudo monochromatique se détachant sur un fond d'émissions impulsives. Chambon-la-Forêt, le 9 janvier 1967. (Cliché GLANGEAUD.)

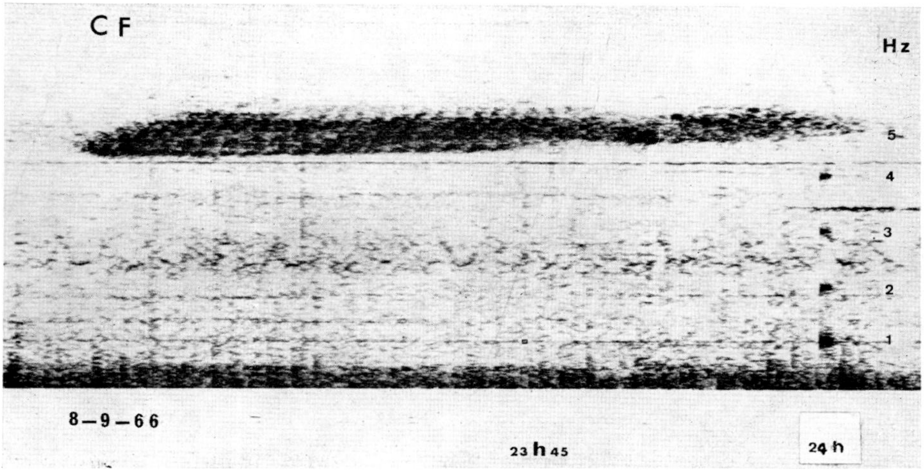

Fig. 54. Observation à Chambon-la-Forêt de «perles» (**pc-1** structurées) d'une fréquence de 5 Hz. (Cliché GLANGEAUD-PETIAU.)

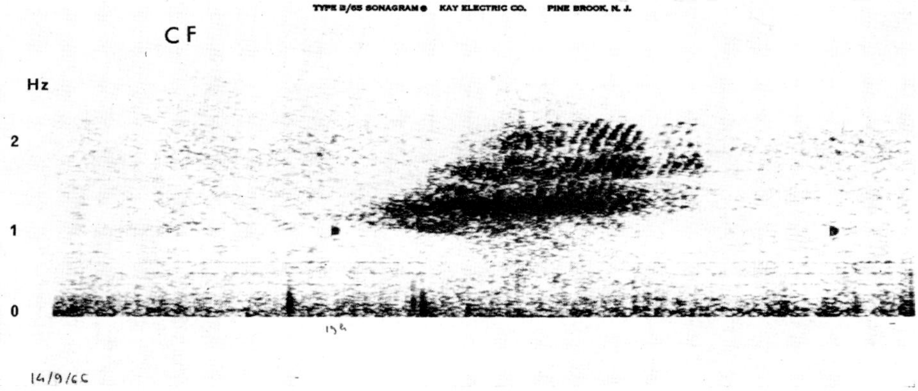

Fig. 55. Exemple d'un évènement du type «perles» (**pc-1** structurées) à émission multiple s'étageant entre 0,8 et 2,2 Hz. Chambon-la-Forêt le 14 septembre 1966. On remarquera qu'aux fréquences les plus hautes les périodes de répétition des éléments structurés sont plus grandes, mais il y a rarement continuité dans la montée du spectre entre les fréquences les plus basses et les plus hautes (cf.avec la Fig. 51). On pourra également noter la forme générale de la représentation spectrale, globale, de cet évènement qui rappelle ainsi celle des spectres des «**IPDP**» introduits par V. A. TROITSKAJA). On conçoit, ainsi qu'il puisse se présenter des cas intermédiaires pour lesquels la nature structurée, ou non structurée, d'un phénomène ne puisse être bien élucidée (soulevant alors à nouveau la question de la distinction établie entre **pc-1** et **pi-1**). (Cliché GLANGEAUD-PETIAU.)

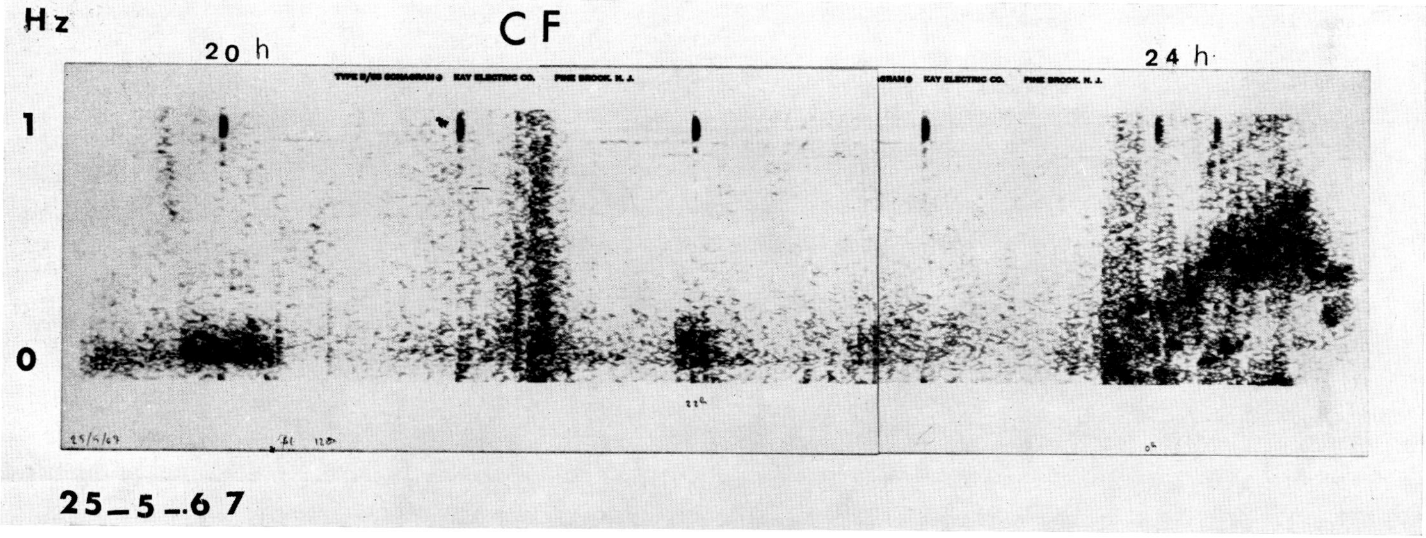

Fig. 56. Exemple de deux évènements successifs formés de pulsations irrégulières **pi-1** mais dont les spectres de fréquences ne suivent pas la même évolution générale. Le premier de ces évènements (un peu après 21 h T.U.) est du type «**S.I.P.**» (Short Irregular Pulsation) de la classification de V. A. Troitskaja, le second, autour de 24 h T.U., se rangeant parmi les «**I.P.D.P.**» (Irregular Pulsation of Diminishing Period) du même auteur. *Chambon-la-Forêt, le 25 mai 1967.* (Cliché Glangeaud.)

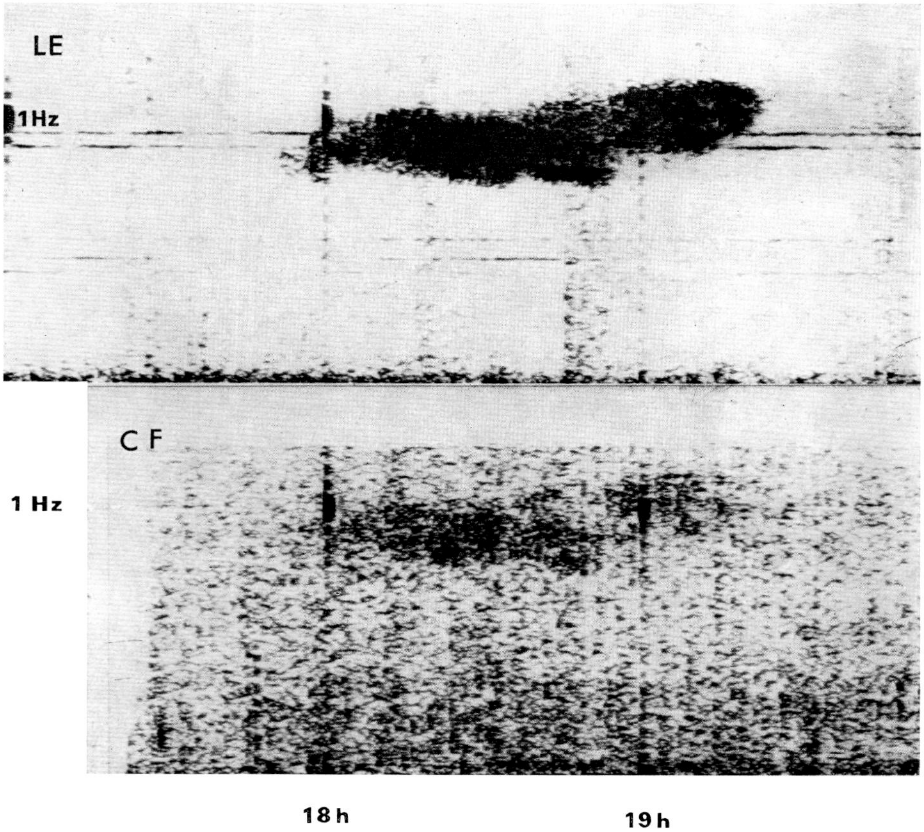

Fig. 57. Comparaison de pulsations en perles (**pc-1**) entre Lerwick et Chambon-la-Forêt. Le 11 novembre 1967, entre 18 et 19 h T.U. On remarquera qu'aux moyennes latitudes correspondant à Chambon-la-Forêt les amplitudes observées sont plus faibles (noircissement moins accentué du sonagramme). (Cliché GLANGEAUD.)

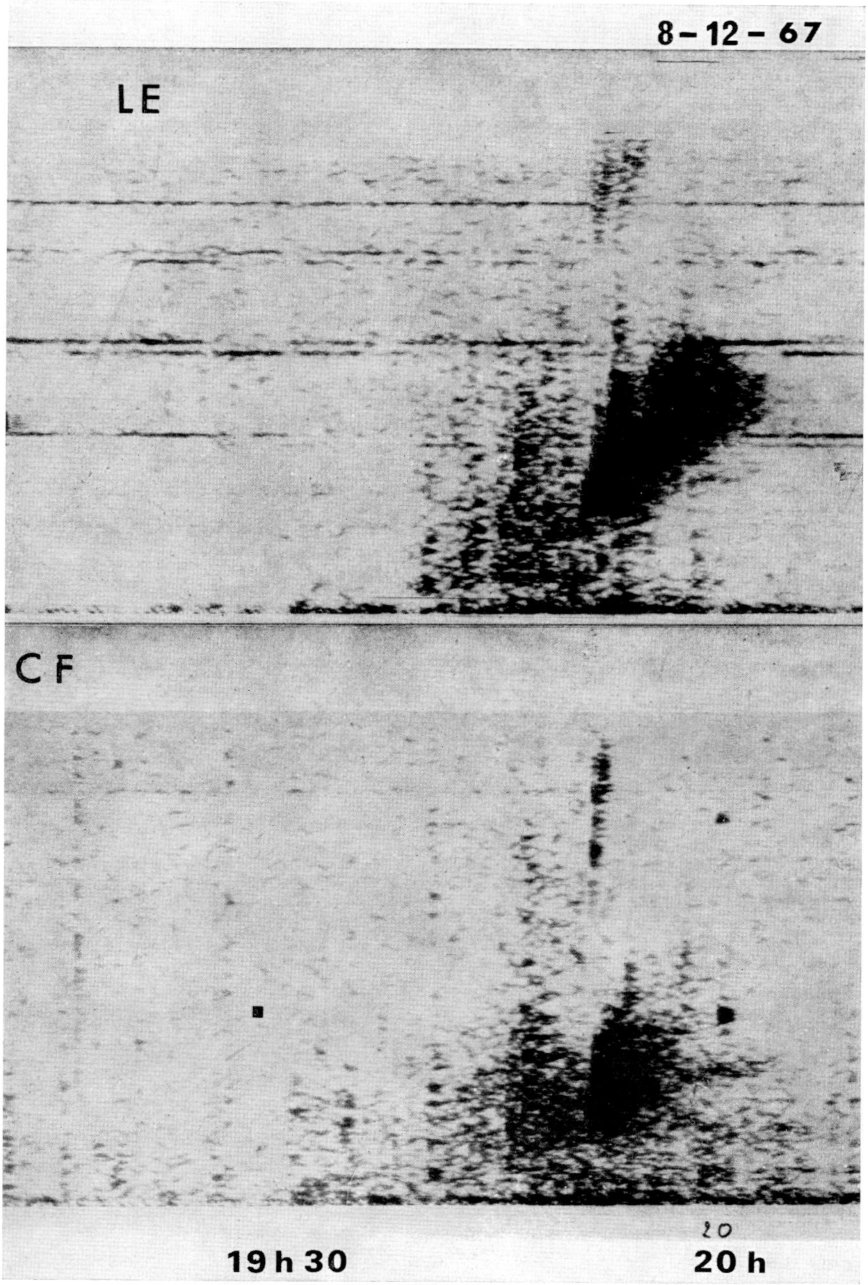

Fig. 58. Comparaison de pulsations irrégulières (**pi-1**), formant un «**IPDP**» au sens de V. A. TROITSKAJA, entre Lerwick et Chambon-la-Forêt, le 8 décembre 1967, entre 19.30 et 20.10 T.U. En première approximation, les phénomènes sont simultanés. On peut faire correspondre les divers éléments du sonagramme sans que les diverses amplitudes soient conservées (Cliché GLANGEAUD.)

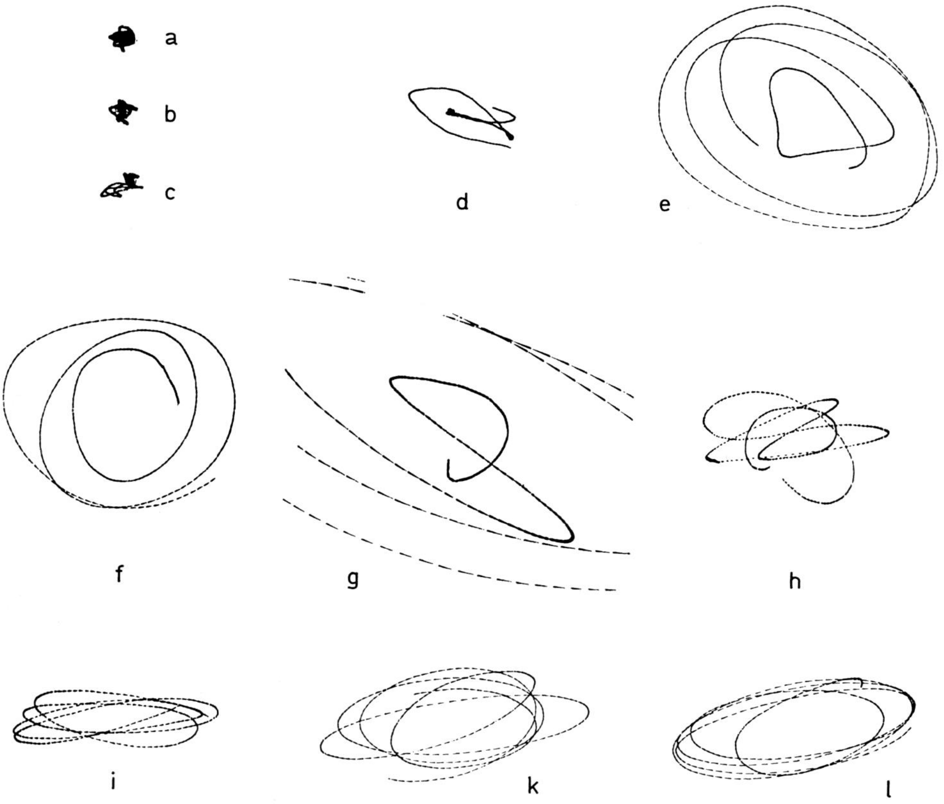

Fig. 59a.

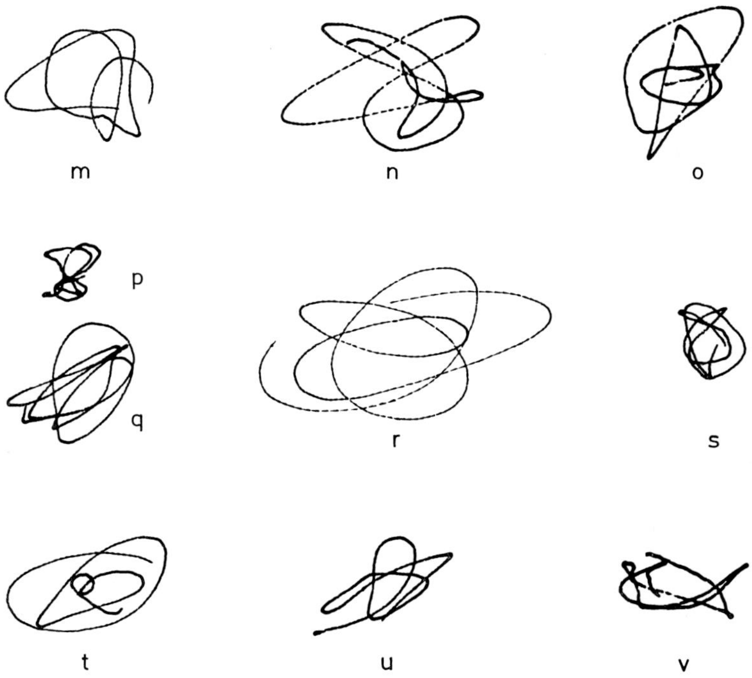

Fig. 59b.

Fig. 59a et b. Evolution de la polarisation d'un évènement du type **IPDP**, le 3 mars 1968 à Lerwick. Chaque prise de vue correspond à 6 s de pose. Elles sont faites en suivant la lecture d'un enregistrement en $\mathcal{H}$ et D sur ruban magnétique. Le déroulement à la lecture étant très accéléré un système de fléchage de la trace oscillographique (difficilement visible sur le cliché) permet de reconnaître les sens de rotation: il est ici *gauche* pour les premières vues et devient *droit* en h. Heure de Début: 18 h 07 min 50 s T.U.; Fin: 18 h 08 min 14 s T.U. Valeur d'échelle: 1 cm vaut 500 milligamma. A noter le début brusque et la variabilité de la polarisation tant en sens qu'en direction, ce qui est très différent de ce que l'on observe pour les **pc-1**. (Cliché GLANGEAUD.)

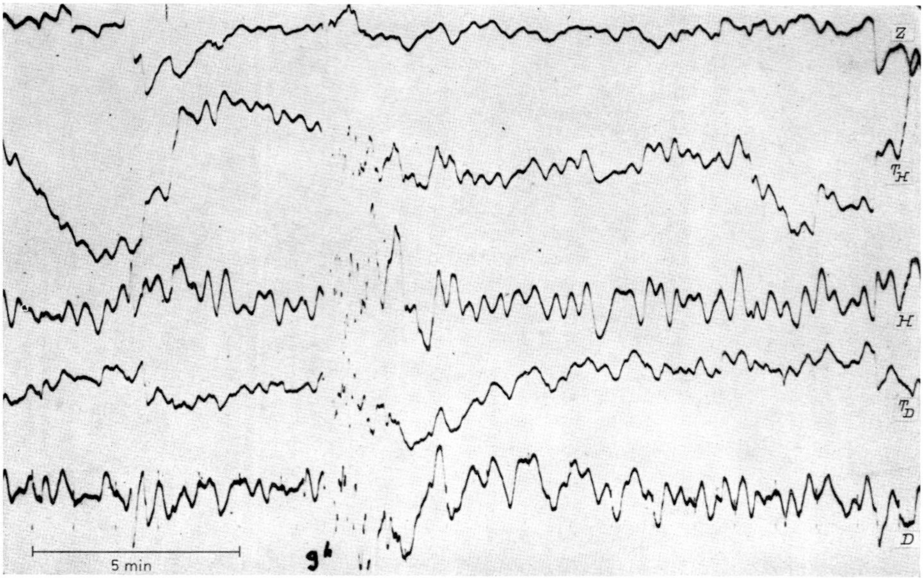

Fig. 60. Perturbation du champ magnétique terrestre, dans la gamme des variations magnétiques rapides, provoquée par l'explosion nucléaire à haute altitude du 9 juillet 1962. Il s'agit de l'expérience «Starfish», à environ 400 km au-dessus de l'ile Johnson (Ocean Pacifique), dont les effects (mondiaux) ont pu être comparés à ceux d'un petit orage magnétique (à «début brusque», rappelant un «**SSC**» assez important, mais l'«orage» ne durant qu'une dizaine de minutes), ou à ceux d'une éruption chromosphérique solaire («crochet magnétique» ou «solar-flare-effect» «**SFE**»). L'enregistrement. reproduit ici est celui obtenu à Chambon-la-forêt, sur le dispositif «Barres-Fluxmètres».

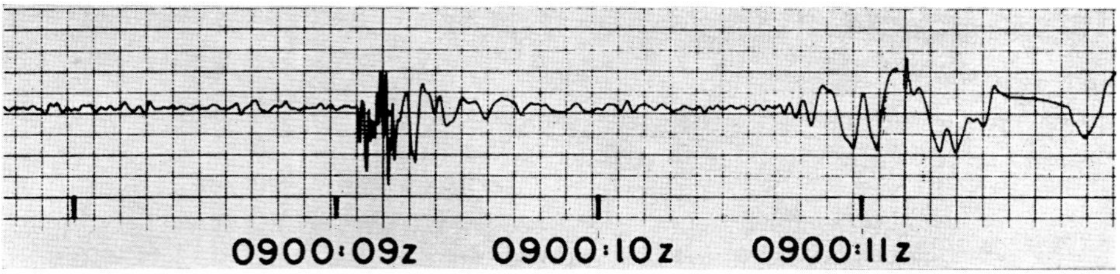

Fig. 61. Enregistrement détaillé des trois premières secondes de la perturbation causée par l'expérience «Starfish» (cf Fig. 60) obtenu dans l'Antarctique. L'intérêt de cet enregistrement est de bien mettre en évidence les deux groupes de signaux (mondiaux) engendrés par l'explosion: le premier de ces groupes débute en synchronisme presque parfait (à $^1/_{10}$ s) avec l'instant zéro, nominal, de l'explosion (09.00.09 T.U.); le deuxième se présente environ 1,7 s plus tard. Le fait le plus remarquable est que ces deux instants — donc également leur différence — se retrouvent inchangés en tout point de la surface du Globe terrestre. D'un autre point de vue, si l'explication du second signal met en jeu des phénomènes complexes, basés sur des processus magnétohydrodynamiques qui, pour cette expérience particulière, n'ont pas été totalement éclaircis, les mécanismes intervenant dans la propagation du premier signal paraissent mieux compris: il s'agit essentiellement d'une onde électromagnétique circulant dans la cavité Terre—Ionosphère, y déterminant au passage — comme on peut l'observer sur l'enregistrement — une recrudescence du mode fondamental (sur environ 8 Hz) des oscillations de SCHUMANN. (Cliché LOKKEN et P. N. L. Groupe du Canada.)

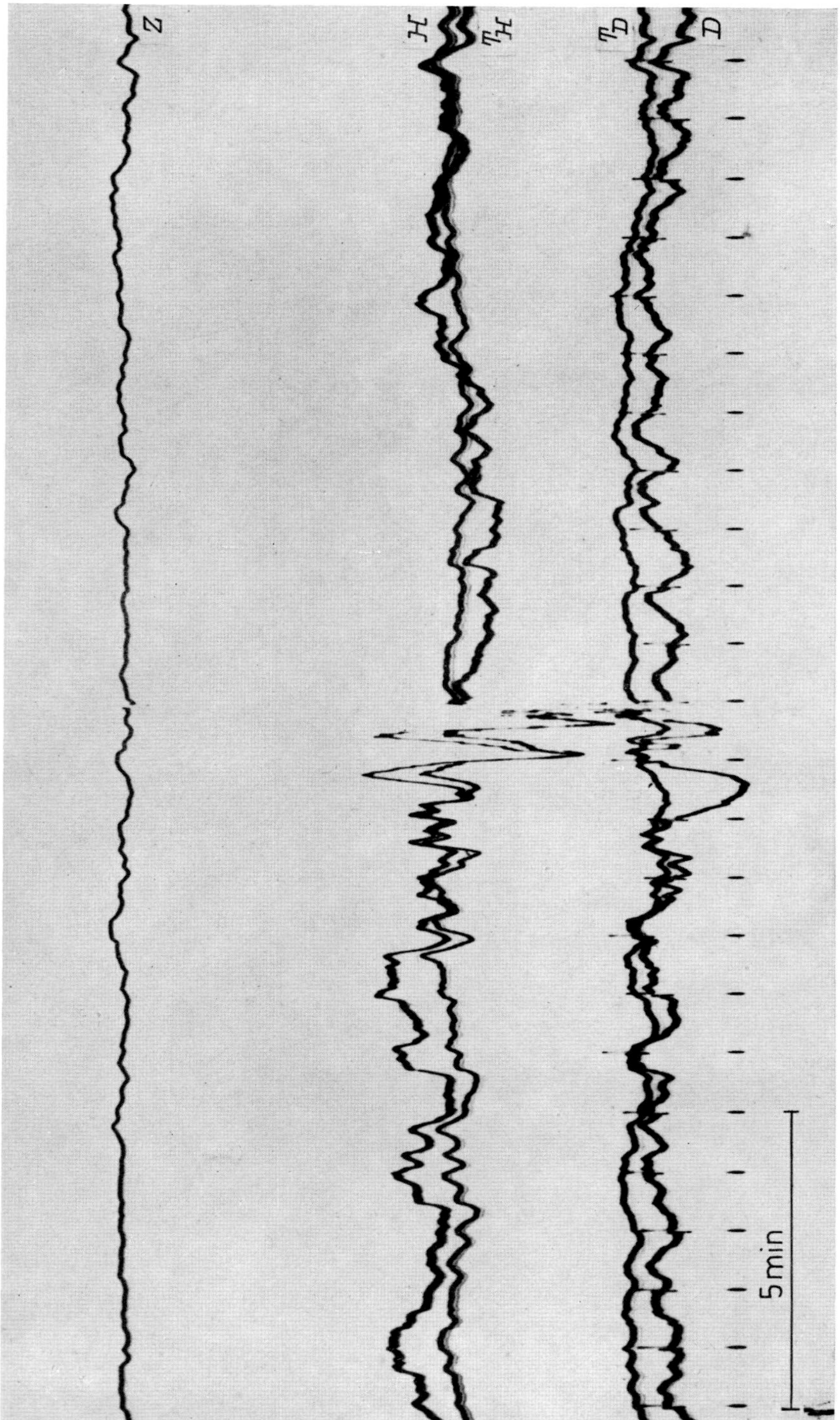

Fig. 62. Enregistrement à la station de Chambon (France) de la perturbation mondiale, dans la gamme des variations magnétiques rapides, provoquée par une explosion nucléaire à haute altitude, le 22 octobre 1962 à 03.40.46 T.U. Les faits observés ont été très comparables, dans leur généralité, à ceux relatifs au 9 juillet 1962. (Cf. Fig. 60 et 61.)

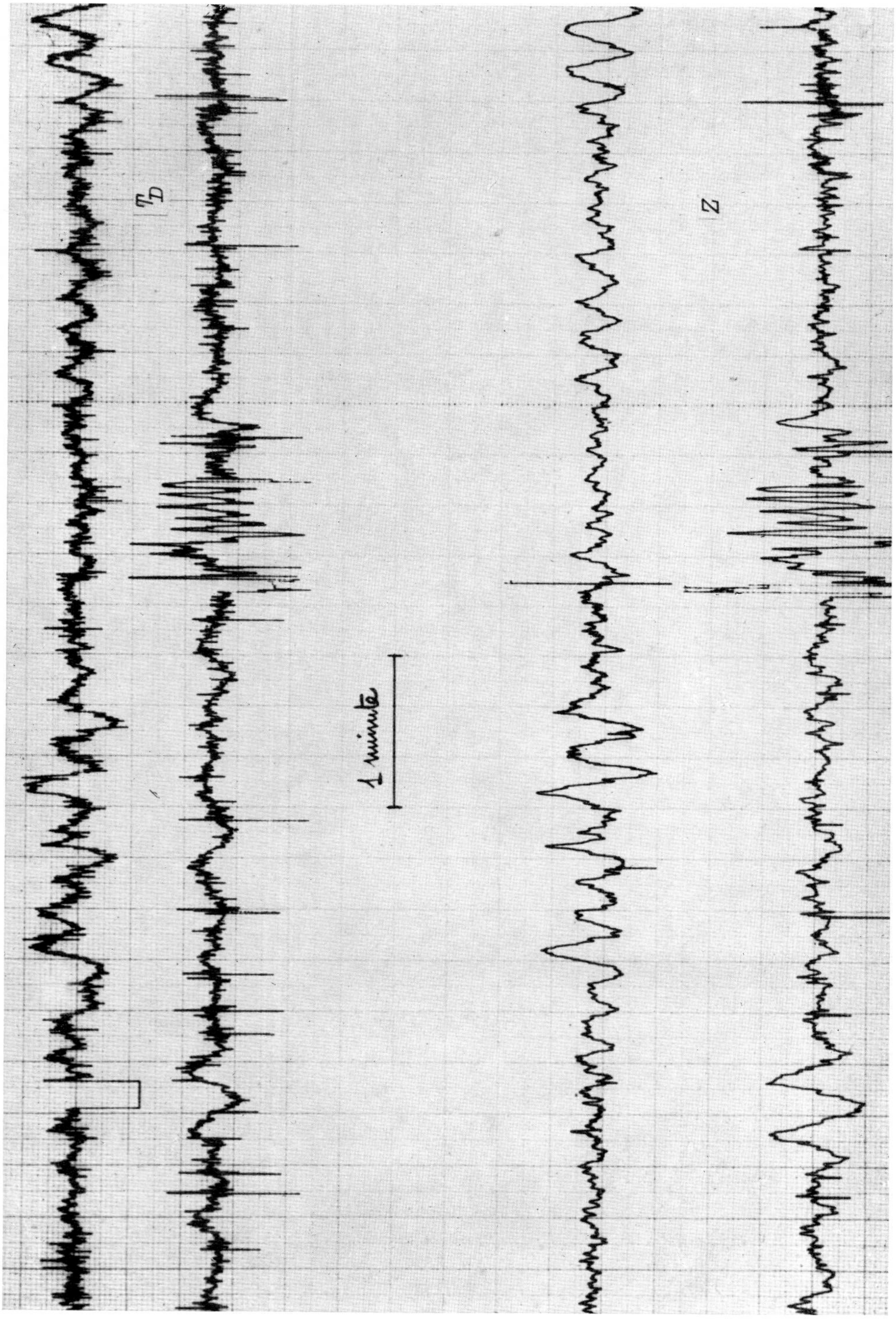

Fig. 63. Enregistrement analogue à celui donné en Fig. 62, mais relatif à un évènement semblable, le 28 octobre 1962 à 04.41.18 T.U.

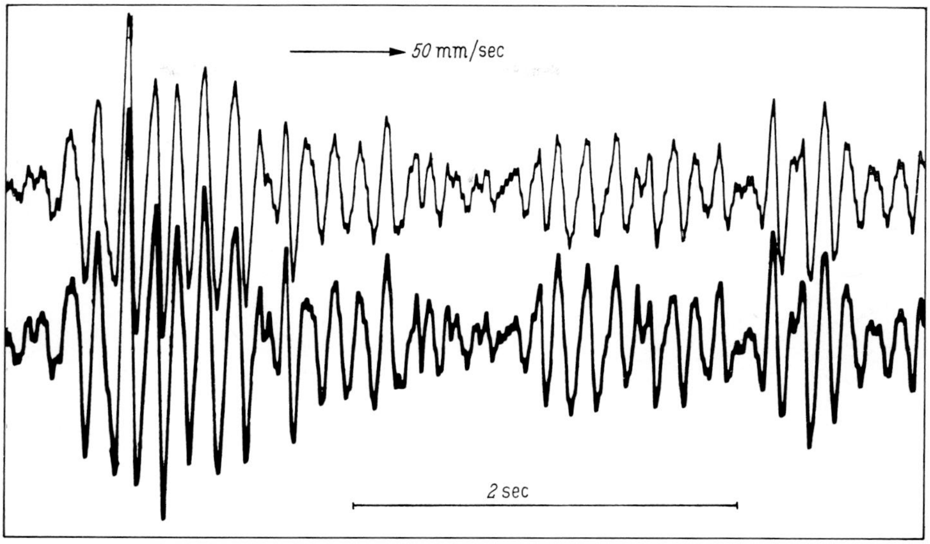

Fig. 64. Enregistrement de la forme d'onde d'oscillations de SCHUMANN, dans un cas de predominance du mode fondamenta (sur environ 8 Hz), à la station de Garchy, le 23 janvier 1962, vers 21.10 T.U., par H. FOURNIER.

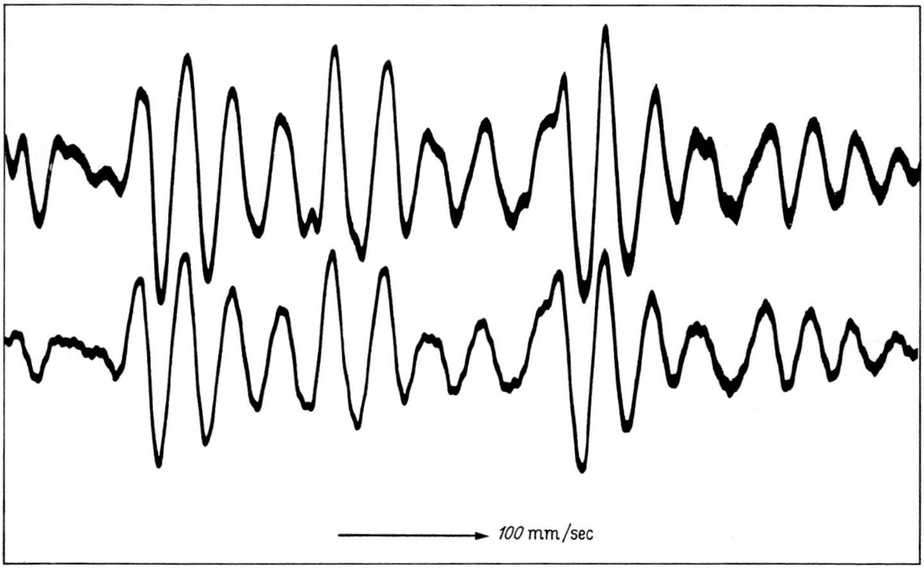

Fig. 65. Enregistrement analogue à celui présenté en Fig. 64, le 4 mars 1962, à 16.03 T.U., par H. FOURNIER. Etant donné la très grande sensibilité utilisée (0,00016 γ/mm pour la voie magnétique — trace inférieure —) on ne s'étonnera pas qu'un certain épaississement du trait soit dû à des traces du 50 Hz industriel.

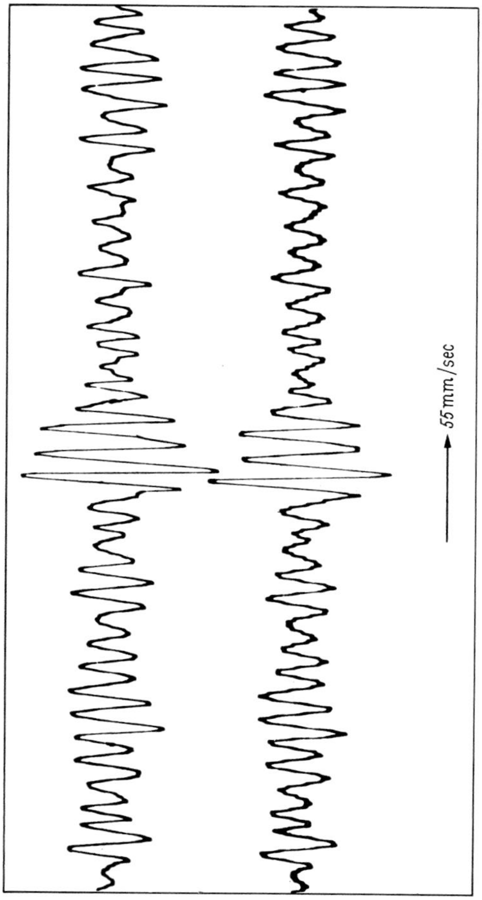

Fig. 66. Enregistrement de la forme d'onde d'oscillations de SCHUMANN dans un cas de renforcement brusque de niveau. Garchy, le 4 mars 1962 vers 10.25 T.U., par H. FOURNIER.

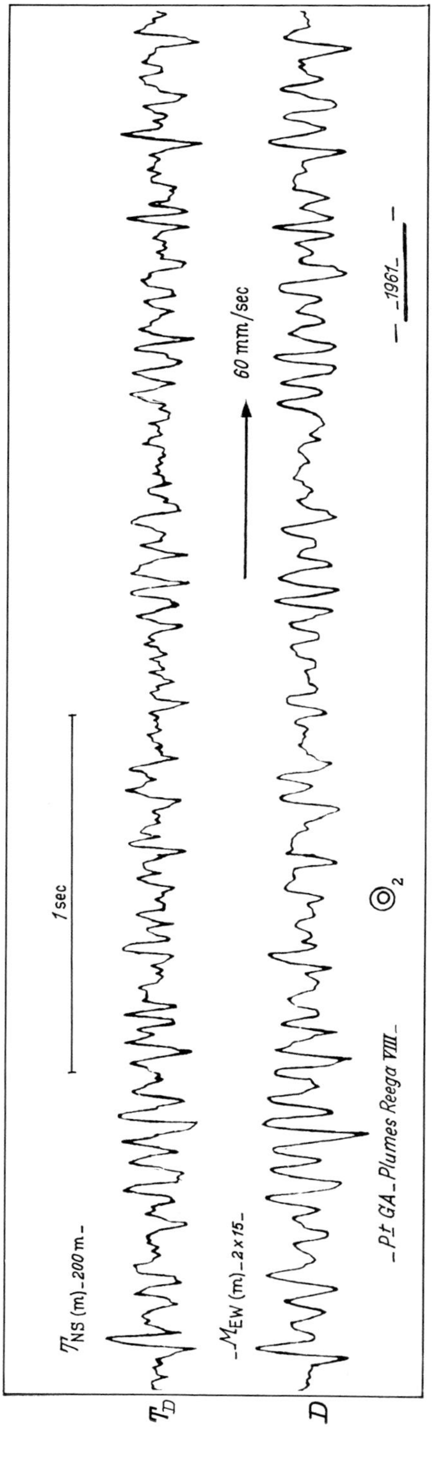

Fig. 67. Enregistrement de l'harmonique 2 d'oscillations de SCHUMANN. Garchy, FOURNIER (1961).

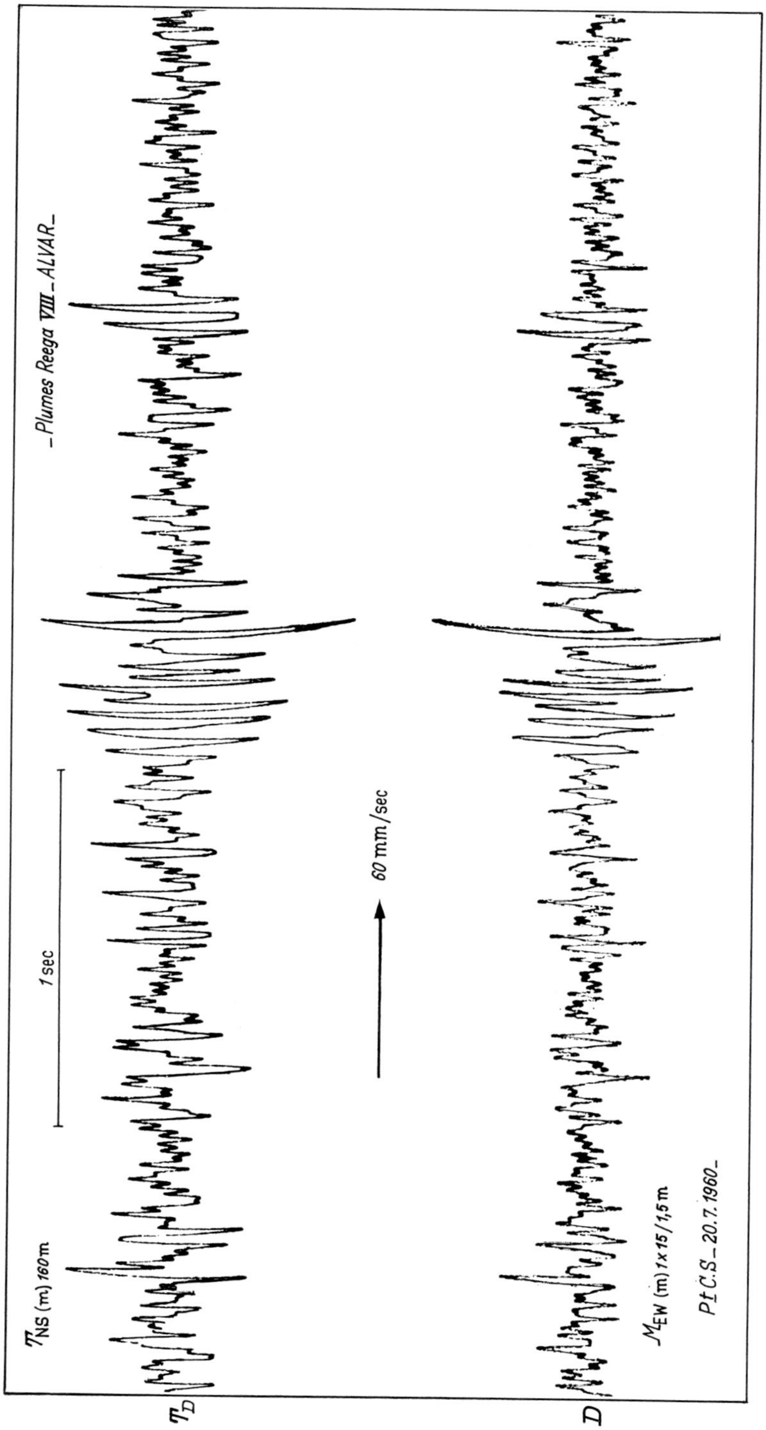

Fig. 68. Enregistrement des harmoniques 2 et 3 des oscillations de Schumann. Garchy, le 20 juillet 1960, Fournier (1960).

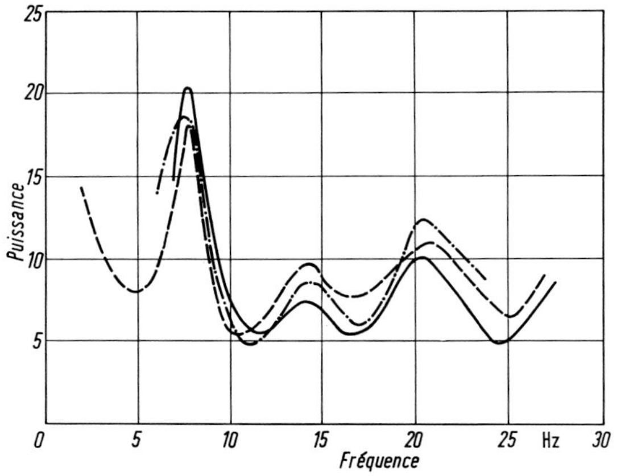

a

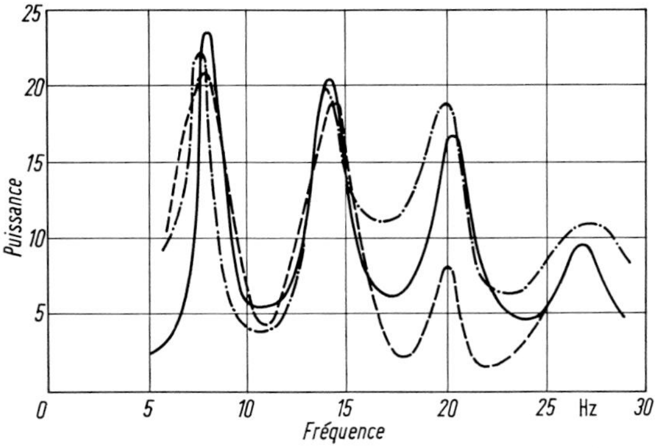

b

Fig. 69a et b. Spectres de puissance (en unités arbitraires) entre 0 et 30 Hz mettant an évidence les résultats de divers auteurs relativement, soit aux calculs, soit aux données d'expérience, concernant la répartition des oscillations de SCHUS MANN entre sa fréquence fondamentale et ses harmoniques. Les trois courbes représentent: premièrement les résultats expérimentaux de BALSER et WAGNER (— · —), ceux de BENOIT et HOURI (— — —) et ceux obtenus par calcul (—— D'après GALEJS (1964).

Waves and Resonances in Magneto-active Plasma.*

By

V. L. GINZBURG and A. A. RUHADZE.

With 12 Figures.

Introduction.

It is the aim of this paper to give a consistent theory of resonances and propagation phenomena for all types of electromagnetic waves in plasma. A great number of different problems occur at such investigation. There are not only the problems of refraction, of attenuation and, eventually, amplification of radio waves in plasma, but also those of heating and acceleration of the plasma by electromagnetic waves; there should also be considered effects due to scattering in plasma, and transition between different modes of waves. One could even say that the central problem of the theory of controlled thermonuclear fusion, namely that of the stability of a plasma, is closely related to the investigations concerning wave propagation in plasma.

The literature concerning these problems is extremely widespread. We may therefore state at once that no attempt is made in the following to review these papers since this would be a task going beyond the given frame of our work. We intend only to present and evaluate the theories concerning the propagation of electromagnetic waves in plasma, shedding some light on a few specific questions using a consistent basis of argumentation. In particular, the more recent development of the electrodynamics of plasmas will be considered; in this approach a plasma is taken as a material medium having not only dispersion with frequency but spatial dispersion also. We shall present here the theory of propagation of electromagnetic waves in plasma in the framework of such electrodynamic theory. It is our feeling that by introducing this general theory a great number of papers written on these subjects may be understood without difficulty.

When applying this general theory we shall get into touch with some specific phenomena which are common to the propagation of electromagnetic waves in the ionosphere as well as in interplanetary space. Here, again, we cannot claim to present the whole picture but shall consider only the most important phenomena. Detailed discussions will be found in [*1*], [*2*], [*3*], [*4*], [*5*], [*6*].

A. Foundations of plasma theory.

1. Plasma as a material medium with frequency and spatial dispersion.

α) We call plasma an ionized gas containing a large number of charged particles, but sometimes also neutral atoms and molecules. In spite of the presence of charged particles the plasma is "quasineutral" in most cases. This means

* Manuscript received September 1966 in Russian language. Translation by K. RAWER and G. SCHMIDTKE, revised by R. GRABOWSKI, S. BAUER and A. A. RUHADZE.

that if we consider a large enough volume and a long enough time the plasma behaves like an electrically neutral system. Specific and particular conditions are found in cases where the distribution of charged particles is inhomogeneous so that macroscopic electrical fields appear. (Such fields are concerned with distances several orders of magnitude greater than the average distance between two particles; see also Sect. 6 below.) In particular, electromagnetic fields may arise in a plasma as a results of *external* sources. It is, however, important to note that these fields influence the motion of the charged particles, inducing space charges and currents in the plasma. The latter cause electromagnetic fields and so change the total field in the system. In this manner mutually self-consistent effects of particles and fields take place in a plasma. Considering induced charges and currents the *field equations* in plasma, as in any other medium, are those of MAXWELL[1,2,3].

$$\left.\begin{aligned} &\frac{\partial}{\partial \boldsymbol{r}}\times\boldsymbol{B}\equiv \operatorname{rot}\boldsymbol{B}=\frac{\mu_0}{c_0\sqrt{\varepsilon_0\mu_0}}\left[\varepsilon_0\frac{\partial \boldsymbol{E}}{\partial t}+\mathrm{u}(\boldsymbol{J}+\boldsymbol{J}_0)\right]; \quad \frac{\partial}{\partial \boldsymbol{r}}\cdot\boldsymbol{B}\equiv \operatorname{div}\boldsymbol{B}=0\\ &\frac{\partial}{\partial \boldsymbol{r}}\times\boldsymbol{E}\equiv \operatorname{rot}\boldsymbol{E}=-\frac{1}{c_0\sqrt{\varepsilon_0\mu_0}}\frac{\partial \boldsymbol{B}}{\partial t}; \quad \frac{\partial}{\partial \boldsymbol{r}}\cdot\boldsymbol{E}\equiv \operatorname{div}\boldsymbol{E}=\frac{\mathrm{u}}{\varepsilon_0}(\varrho+\varrho_0)\,. \end{aligned}\right\} \tag{1.1}$$

Here ϱ_0 and $\boldsymbol{J}_0$ are charge and current densities of external field sources in the plasma, while ϱ and $\boldsymbol{J}$ are the corresponding parameters induced in the plasma; $\boldsymbol{E}$ and $\boldsymbol{B}$ are the electric field strength and the magnetic induction, respectively. Our writing of Eq. (1.1) may be justified by the fact that the fields $\boldsymbol{E}$ and $\boldsymbol{B}$ have immediate physical meaning; this appears from the expression for the (generalized) LORENTZ force, namely that force $\boldsymbol{F}$ which acts on a charge q moving with velocity $\boldsymbol{v}$:

$$\boldsymbol{F}=q\left\{\boldsymbol{E}+\frac{1}{c_0\sqrt{\varepsilon_0\mu_0}}\boldsymbol{v}\times\boldsymbol{B}\right\}. \tag{1.2}$$

We may further define that the fields $\boldsymbol{E}$ and $\boldsymbol{B}$ as well as the current density $\boldsymbol{J}$, occuring in Eqs. (1.1) and in the following, shall be statistical averages (i.e. averages as defined in statistical physics) of the corresponding microscopic quantities. Thus fluctuations are not considered. We may yet note that under normal circumstances the wavelength is much larger than the average distance between neighbouring particles; in this case the average statistical values of the fields can be identified with the result obtained by averaging over "infinitely small" volume or time-intervals.

β) The field equations Eq. (1.1) differ from those in vacuum by the *induced charge density* ϱ *and current density* $\boldsymbol{J}$. However, these two quantities are not independent from each other; their connection is given by the balance equation (equation of continuity):

$$\frac{\partial\varrho}{\partial t}+\frac{\partial}{\partial \boldsymbol{r}}\cdot\boldsymbol{J}\equiv\frac{\partial\varrho}{\partial t}+\operatorname{div}\boldsymbol{J}=0. \tag{1.3}$$

Thus one can give a description of induction phenomena using only one of the two quantities; we may take the density of the induced current $\boldsymbol{J}$, a vector quantity, as the linking quantity. As mentioned above, the fields $\boldsymbol{E}$ and $\boldsymbol{B}$ should be considered as being the cause of the current induced in the plasma. This

[1] We use the self-evident symbol $\partial/\partial\boldsymbol{r}$ for the Nabla-operator, ∇.

[2] $c_0\sqrt{\varepsilon_0\mu_0}=1$ in SI-units, $=c_0$ in GAUSS's cgs units.

[3] $u=1$ for rationalized, $=4\pi$ for non-rationalized systems of units (see "Introductory Remarks" on p. 1).

means that the current density $\boldsymbol{J}$ can be described as some function of the fields $\boldsymbol{E}$ and $\boldsymbol{B}$. Taking account of the above equation

$$\frac{\partial}{\partial \boldsymbol{r}} \times \boldsymbol{E} \equiv \operatorname{rot} \boldsymbol{E} = -\frac{1}{c_0 \sqrt{\varepsilon_0 \mu_0}} \frac{\partial \boldsymbol{B}}{\partial t}, \qquad [1.1]$$

connecting the field $\boldsymbol{B}$ with $\boldsymbol{E}$, we can write this functional relation in the form

$$\boldsymbol{J} = \Phi(\boldsymbol{E}). \qquad (1.4)$$

A specific determination of this functional relation is only possible after introducing some specific model for the plasma or, equivalently, a model describing the motion of the charged particles under the action of the electromagnetic field. We shall consider different models in Sects. 6 through 11 below.

γ) Meanwhile we shall discuss some premisses concerning other problems of the *electrodynamics of plasmas*, whithout involving a specific form for Eq. (1.4). We may remark first of all that one may use a generalized quantity $\boldsymbol{D}$ insead of $\boldsymbol{J}$:

$$\boldsymbol{D}(\boldsymbol{r}, t) = \varepsilon_0 \, \boldsymbol{E}(\boldsymbol{r}, t) + \mathrm{u} \int_{-\infty}^{t} \mathrm{d}t' \, \boldsymbol{J}(\boldsymbol{r}, t'). \qquad (1.5)$$

This field quantity allows us to combine the induced current density with the displacement current density in Eq. (1.1). With $\boldsymbol{D}$ the field equations, Eq. (1.1), may be rewritten in the simplified form:

$$\left.\begin{aligned} &\frac{\partial}{\partial \boldsymbol{r}} \times \boldsymbol{B} \equiv \operatorname{rot} \boldsymbol{B} = \frac{\mu_0}{c_0 \sqrt{\varepsilon_0 \mu_0}} \left[\frac{\partial \boldsymbol{D}}{\partial t} + \mathrm{u} \boldsymbol{J}_0\right]; \quad &&\frac{\partial}{\partial \boldsymbol{r}} \cdot \boldsymbol{B} \equiv \operatorname{div} \boldsymbol{B} = 0 \\ &\frac{\partial}{\partial \boldsymbol{r}} \times \boldsymbol{E} \equiv \operatorname{rot} \boldsymbol{E} = -\frac{1}{c_0 \sqrt{\varepsilon_0 \mu_0}} \frac{\partial \boldsymbol{B}}{\partial t}; \quad &&\frac{\partial}{\partial \boldsymbol{r}} \cdot \boldsymbol{D} \equiv \operatorname{div} \boldsymbol{D} = \mathrm{u}\, \varrho_0. \end{aligned}\right\} \qquad (1.6)$$

Thus there appears in a material medium a second field quantity, $\boldsymbol{D}$, which is not really needed in vacuum. Using the traditional name we shall call $\boldsymbol{D}$ the vector of dielectric displacement; the relation

$$\boldsymbol{D} = \varepsilon_0 \boldsymbol{E} + \mathrm{u} \int_{-\infty}^{t} \mathrm{d}t' \, \Phi(\boldsymbol{E}), \qquad (1.7)$$

which is derived from Eqs. (1.4), (1.5) and connect $\boldsymbol{D}$ and $\boldsymbol{E}$, is designated as the *material field equation*. Eqs. (1.6) and (1.7) together are MAXWELL's complete system of field equations in plasma, allowing us to resolve any electrodynamic problem[4].

Later we shall pay particular attention to linear electromagnetic phenomena in plasma. In this case not only the material field equation, Eq. (1.7), but also, by definition, Eq. (1.4) appear as linear relations. In a most general form, these may be written as time and space integrals[5]:

$$\left.\begin{aligned} \boldsymbol{J}(\boldsymbol{r}, t) &= \int_{-\infty}^{t} \mathrm{d}t' \int \mathrm{d}^3 \boldsymbol{r}' \, \hat{\mathsf{S}}(\boldsymbol{r}, \boldsymbol{r}'; t, t') \cdot \boldsymbol{E}(\boldsymbol{r}', t') \\ \boldsymbol{D}(\boldsymbol{r}, t) &= \int_{-\infty}^{t} \mathrm{d}t' \int \mathrm{d}^3 \boldsymbol{r}' \, \hat{\mathsf{E}}(\boldsymbol{r}, \boldsymbol{r}'; t, t') \cdot \boldsymbol{E}(\boldsymbol{r}', t'), \end{aligned}\right\} \qquad (1.8)$$

[4] Only plasmas without sharp boundaries are considered in this paper; therefore boundary problems are not relevant. Under conditions as found at sharp boundaries, for example at a boundary between plasma and vacuum, we would have to use boundary conditions, different from the field equations in the differential form as given above. These conditions can be obtained by integrating the field equations, Eq. (1.6), see [2], [7].

[5] The tensorial functions with ^ are defined as conductivity and permittivity *densities* in four-dimensional $\boldsymbol{r}'$, t' space.

which may be written in coordinates, using the summation rule (over m)

$$\left.\begin{aligned} J_l(\boldsymbol{r}, t) &= \int_{-\infty}^{t} \mathrm{d}t' \int \mathrm{d}^3\boldsymbol{r}'\, \hat{\sigma}_{lm}(\boldsymbol{r}, \boldsymbol{r}'; t, t')\, E_m(\boldsymbol{r}', t') \\ D_l(\boldsymbol{r}, t) &= \int_{-\infty}^{t} \mathrm{d}t' \int \mathrm{d}^3\boldsymbol{r}'\, \hat{\varepsilon}_{lm}(\boldsymbol{r}, \boldsymbol{r}'; t, t')\, E_m(\boldsymbol{r}', t'). \end{aligned}\right\} \tag{1.8a}$$

Thus it appears clearly that if the conditions in the plasma (where charges and currents are induced) are given until an instant t at all points $\boldsymbol{r}$, their future development at any point depends on the field, as requested by the (classical) principle of causality. The dispersion in the plasma with respect to time and space coordinates, i.e. frequency and spatial dispersion also appears. Indeed, their exists a physical relation between frequency dispersion on one side and conservation of charges and relaxation processes in the plasma on the other. As to spatial dispersion, it is so to say, connected with the possibility of transfer of the field action at one point in the plasma to another one. This possibility is due to processes of transport (in the sense of ordered motions), and to thermal motion of particles.

In order to compute spatial and frequency dispersion in more detail let us consider a plasma which is homogeneous in space and time[6]. It is evident that, in this case, the kernels of the integral relations, Eq. (1.8), appear as functions of time and coordinate differences only, i.e. they depend on $(\boldsymbol{r}-\boldsymbol{r}')$ and $(t-t')$. We shall consider such a plasma in more detail below. Using FOURIER development, the electromagnetic field in the plasma can always be represented as a sum of plane monochromatic waves each of which depends on time t and coordinates $\boldsymbol{r}$ after

$$\exp(-i\omega t + i\boldsymbol{k}\cdot\boldsymbol{r}).$$

As the field equations are linear these components can be considered individually; thus we may write[7]

$$\boldsymbol{E}(\boldsymbol{r}, t) = \tilde{\tilde{\boldsymbol{E}}}(\omega, \boldsymbol{k}) \exp(-i\omega t + i\boldsymbol{k}\cdot\boldsymbol{r}),$$

---

[6] We are speaking of homogeneity in the time average, or in the macroscopic aspect. In this context the quantities $\boldsymbol{E}$ and $\boldsymbol{B}$ (also $\boldsymbol{D}$) are revealed as to be statistical averages, as we have stated above. The hypothesis of homogeneity of the medium is, in principle, not contradictory in itself.

[7] We apply FOURIER-transform twice here, in space and time; the operations may be seen by the change of arguments $\boldsymbol{r}\to\boldsymbol{k}$, and $t\to\omega$. In the literature one often uses the same letter symbol for the original and the transformed physical quantity, distinguishing by the arguments only. In this and the following chapter we prefer to indicate the transformation by the special symbol $\tilde{\tilde{}}$, or $\tilde{}$ in cases where only one coordinate is transformed (normally $\boldsymbol{r}\to\boldsymbol{k}$). Thus, for whichever field quantity, $A(\boldsymbol{r}, t)$, we write

$$\tilde{A}(\boldsymbol{k}, t) = \int \mathrm{d}^3\boldsymbol{r}\, A(\boldsymbol{r}, t) \exp(-i\boldsymbol{k}\cdot\boldsymbol{r}),$$

and

$$\tilde{\tilde{A}}(\boldsymbol{k}, \omega) = \int \mathrm{d}^3\boldsymbol{r} \int \mathrm{d}t\, A(\boldsymbol{r}, t) \exp(i\omega t - i\boldsymbol{k}\cdot\boldsymbol{r}),$$

so that inversely:

$$A(\boldsymbol{r}, t) = \int \mathrm{d}^3\boldsymbol{k}\, \tilde{A}(\boldsymbol{k}, t) \exp(+i\boldsymbol{k}\cdot\boldsymbol{r}),$$

and

$$A(\boldsymbol{r}, t) = \int \mathrm{d}^3\boldsymbol{k} \int \mathrm{d}\omega\, \tilde{\tilde{A}}(\boldsymbol{k}, \omega) \exp(-i\omega t + i\boldsymbol{k}\cdot\boldsymbol{r}).$$

Consequently, the unit for $\tilde{A}(\boldsymbol{k}, t)$ is the volume unit times the unit of $A(\boldsymbol{r}, t)$ and that for $\tilde{\tilde{A}}(\boldsymbol{k}, \omega)$ is (volume · time) unit times that of $A(\boldsymbol{r}, t)$. For the electromagnetic field quantities

considering one component only of the summation. Similar relations are valid for the dielectric displacement $\boldsymbol{D}$ and the magnetic induction $\boldsymbol{B}$; thus we may write Eqs. (1.8) in the form:

$$\left.\begin{aligned}\tilde{\tilde{\boldsymbol{J}}}(\omega, \boldsymbol{k}) &= \tilde{\tilde{\mathsf{S}}}(\omega, \boldsymbol{k}) \cdot \tilde{\tilde{\boldsymbol{E}}}(\omega, \boldsymbol{k}) \\ \tilde{\tilde{\boldsymbol{D}}}(\omega, \boldsymbol{k}) &= \tilde{\tilde{\mathsf{E}}}(\omega, \boldsymbol{k}) \cdot \tilde{\tilde{\boldsymbol{E}}}(\omega, \boldsymbol{k}),\end{aligned}\right\} \tag{1.9}$$

in components

$$\left.\begin{aligned}\tilde{\tilde{J}}_l(\omega, \boldsymbol{k}) &= \tilde{\tilde{\sigma}}_{lm}(\omega, \boldsymbol{k})\, \tilde{\tilde{E}}_m(\omega, \boldsymbol{k}) \\ \tilde{\tilde{D}}_l(\omega, \boldsymbol{k}) &= \tilde{\tilde{\varepsilon}}_{lm}(\omega, \boldsymbol{k})\, \tilde{\tilde{E}}_m(\omega, \boldsymbol{k})\end{aligned}\right\} \tag{1.9a}$$

where

$$\left.\begin{aligned}\tilde{\tilde{\sigma}}_{lm}(\omega, \boldsymbol{k}) &= \int_0^\infty \mathrm{d}t \int \mathrm{d}^3\boldsymbol{r}\, \hat{\sigma}_{lm}(\boldsymbol{r}, t) \exp(i\,\omega\, t - i\,\boldsymbol{k}\cdot\boldsymbol{r}) \\ \tilde{\tilde{\varepsilon}}_{lm}(\omega, \boldsymbol{k}) &\equiv \int_0^\infty \mathrm{d}t \int \mathrm{d}^3\boldsymbol{r}\, \hat{\varepsilon}_{lm}(\boldsymbol{r}, t) \exp(i\,\omega\, t - i\,\boldsymbol{k}\cdot\boldsymbol{r}).\end{aligned}\right\} \tag{1.10}$$

The tensor $\tilde{\tilde{\mathsf{S}}} \equiv ((\tilde{\tilde{\sigma}}_{lm}(\omega, \boldsymbol{k})))$ is called tensor of complex conductivity while $\tilde{\tilde{\mathsf{E}}} \equiv ((\tilde{\tilde{\varepsilon}}_{lm}(\omega, \boldsymbol{k})))$ is the tensor of complex dielectric permeability. Since the general characteristics of these are investigated in [*2*], [*8*], we shall only discuss a few features here. Taking account of the definition, Eq. (1.5), one easily establishes[8] the following relation between S and E:

$$\tilde{\tilde{\mathsf{E}}}(\omega, \boldsymbol{k}) = \varepsilon_0\, \mathsf{U} + i\,\frac{\mathrm{u}}{\omega}\,\tilde{\tilde{\mathsf{S}}}, \tag{1.11}$$

$$\tilde{\tilde{\varepsilon}}_{lm}(\omega, \boldsymbol{k}) = \varepsilon_0\, \delta_{lm} + i\,\frac{\mathrm{u}}{\omega}\,\tilde{\tilde{\sigma}}_{lm}(\omega, \boldsymbol{k}). \tag{1.11a}$$

δ_{lm} being KRONECKER's symbol ($\delta_{jj}=1$, $\delta_{ij}=0$ for $i\neq j$). In the following, since Eq. (1.11) allows us to express $\tilde{\tilde{\mathsf{S}}}$ in terms of $\tilde{\tilde{\mathsf{E}}}$, we shall only discuss the tensor $\tilde{\tilde{\mathsf{E}}} \equiv ((\tilde{\tilde{\varepsilon}}_{lm}(\omega, \boldsymbol{k})))$. As indicated by Eqs. (1.8), the tensor $\hat{\mathsf{E}}(\boldsymbol{r}, t)$ apparently is a real function of the variables $\boldsymbol{r}$ and t because it connects the real field quantities $\boldsymbol{E}(\boldsymbol{r}, t)$ and $\boldsymbol{D}(\boldsymbol{r}, t)$. The tensor $\tilde{\tilde{\mathsf{E}}}(\omega, \boldsymbol{k})$, however, appears as a complex function — even in the basic variables ω and $\boldsymbol{k}$. The reality of $\hat{\mathsf{E}}(\boldsymbol{r}, t)$ gives the

---

the SI-units are:

$\boldsymbol{E}$: $\mathrm{V\,m^{-1}}$;	$\boldsymbol{D}$: $\mathrm{A\,s\,m^{-2}}$;	$\boldsymbol{B}$: $\mathrm{V\,s\,m^{-2}}$;	$\boldsymbol{J}$: $\mathrm{A\,m^{-2}}$;	σ: $\mathrm{A V^{-1} m^{-1}}$;	ε: $\mathrm{A\,s\,V^{-1}\,m^{-1}}$
$\tilde{\tilde{\boldsymbol{E}}}$: $\mathrm{V\,m^2\,s}$;	$\tilde{\tilde{\boldsymbol{D}}}$: $\mathrm{A\,s^2\,m}$;	$\tilde{\tilde{\boldsymbol{B}}}$: $\mathrm{V\,s^2\,m}$;	$\tilde{\tilde{\boldsymbol{J}}}$: $\mathrm{A\,s\,m}$;	$\tilde{\tilde{\sigma}}$: $\mathrm{A V^{-1} m^2\,s}$;	$\tilde{\tilde{\varepsilon}}$: $\mathrm{A\,s^2\,V^{-1}\,m^2}$
$\tilde{\boldsymbol{E}}$: $\mathrm{V\,m^2}$;	$\tilde{\boldsymbol{D}}$: $\mathrm{A\,s\,m}$;	$\tilde{\boldsymbol{B}}$: $\mathrm{V\,s\,m}$;	$\tilde{\boldsymbol{J}}$: $\mathrm{A\,m}$;	$\tilde{\sigma}$: $\mathrm{A V^{-1} m^2}$;	$\tilde{\varepsilon}$: $\mathrm{A\,s\,V^{-1}\,m^2}$.

In most of the literature (and in the later chapters of this contribution) the different FOURIER-transforms are not distinguished by the *special symbols* $\sim$ and $\approx$, but only by the *arguments*. One simply writes $A(\boldsymbol{k}, t)$ instead of $\tilde{A}$, and $A(\boldsymbol{k}, \omega)$ instead of $\tilde{\tilde{A}}$. Note that the dimension of A then depends on its arguments as explained above for $\tilde{A}$ and $\tilde{\tilde{A}}$.

[8] Strictly speaking this relation has the form [*2*]

$$\tilde{\tilde{\varepsilon}}_{lm}(\omega, \boldsymbol{k}) = \varepsilon_0\, \delta_{lm} + 2\pi\,\mathrm{u}\,\delta_+(\omega)\,\tilde{\tilde{\sigma}}_{lm}(\omega, \boldsymbol{k}) \quad \text{with} \quad \delta_+(\omega) = \frac{1}{2}\,\delta(\omega) + \frac{i}{2\pi}\,\frac{\mathscr{P}}{\omega}$$

the symbol $\mathscr{P}$ indicating a principal value in the case $\omega = 0$. It is easily shown that this is identical with Eq. (1.11) in the case $\omega \neq 0$.

following limitations for real and imaginary part of the FOURIER-transformed tensorof complex permittivitiy (dielectric permeability)[7], $\tilde{\tilde{E}}(\omega, \boldsymbol{k})$:

$$\left.\begin{aligned} \tilde{\tilde{E}}(\omega, \boldsymbol{k}) &= \tilde{\tilde{E}}^*(-\omega, -\boldsymbol{k}), \\ \operatorname{Re}\tilde{\tilde{\varepsilon}}_{lm}(\omega, \boldsymbol{k}) = \operatorname{Re}\tilde{\tilde{\varepsilon}}_{lm}(-\omega, -\boldsymbol{k}); &\quad \operatorname{Im}\tilde{\tilde{\varepsilon}}_{lm}(\omega, \boldsymbol{k}) = -\operatorname{Im}\tilde{\tilde{\varepsilon}}_{lm}(-\omega, -\boldsymbol{k}). \end{aligned}\right\} \quad (1.12)$$

Thus in the basic variables ω and $\boldsymbol{k}$ the real part of $\tilde{\tilde{E}}(\omega, \boldsymbol{k})$ is an even function of the frequency $\omega/2\pi$ and the imaginary part is an odd one. This property shall be used quite often in the following. The dependence of $\tilde{\tilde{E}}$ on ω characterizes the frequency dispersion of the plasma while its dependence on $\boldsymbol{k}$ takes care of spatial dispersion.

2. A few parameters of ionospheric and interplanetary plasma. The specific form after which E depends on ω and $\boldsymbol{k}$ can be obtained with a specific plasma model. It is obtained from the plasma parameters, in particular relaxation time, thermal velocity of particles, characteristic frequencies and dimensions of the plasma. The values of some parameters are given below in Tables 1, 2, 3 as function of the distance from Earth's surface for both cases, that of the ionosphere and that of interplanetary plasma, see [*9*], [*10*]. As these values are always variable with longitude, latitude, season and local time, the indicated values can only be taken as orientation. They also depend on solar activity in a very fundamental manner.

More recent measurements at heights between 200 and 300 km, i.e. in the F-region of the ionosphere, show the electron temperature to be greater than ion and neutral temperature by a factor 2 or 3. However, at lower as well as at greater heights this difference seems to be less important.

In the high-frequency range,

$$\omega \gg \omega_{N\mathrm{e}} = \sqrt{\frac{\mathrm{u}\, q^2}{\varepsilon_0\, m_\mathrm{e}}\, N_\mathrm{e}}$$

the dielectric permeability of a plasma is similar to that of vacuum

$$E \approx \varepsilon_0\, U; \qquad \varepsilon_{lm} \approx \varepsilon_0\, \delta_{lm}.$$

In this frequency range only transverse electromagnetic waves can propagate in the plasma. The refraction index is almost one (like in vacuum), but absorption is determined by the properties of plasma. It may be seen from Table 3 that in the case of the ionosphere this frequency range is roughly $\omega \gg 10^8$ Hz. We consider here mainly waves of lower frequencies, $\omega \lesssim \omega_{N\mathrm{e}}$, as the propagation of high frequency waves is discussed in another volume [*11*], see also [*2*], [*3*], [*4*], [*5*]. Apart from transverse electromagnetic waves, this frequency range admits longitudinal plasma waves, also electron and ion cyclotron waves

$$\omega \sim s\, \omega_{B\mathrm{h}}; \qquad \omega_{B\mathrm{h}} = (q/m_\mathrm{h})\,(B_{\delta}/c_0\sqrt{\varepsilon_0\, \mu_0})$$

(with $\mathrm{h} = \mathrm{e}, \mathrm{i}$ and $s = 1, 2, 3 \ldots$) and finally low frequency magneto-acoustic and magneto-gas-dynamic waves ($\omega \ll \omega_{B\mathrm{i}}$). For all these waves the relation between wave vector, $\boldsymbol{k}$, and frequency, $\omega/2\pi$, is determined by the dispersion relation which will be discussed rather thoroughly below. It is important that the relevant relations between ω and $\boldsymbol{k}$ are of the same number as are the types of waves.

The spatial attenuation of the wave due to thermal motion V_T of the particles e, i, may be neglected for the periods considered, provided that $\omega \gg |\boldsymbol{k}|\, V_{T\mathrm{e,i}}$.

Table 1. *Typical values of atmospheric parameters.*

R_E radius of Earth. N_n number density of neutral particles*, $N_{e,i}$ that of charged particles (of one sign). T temperature (in °K)*. $\overline{m}$ averaged molecular weight of neutral particles and ions*. $B_{\circ}$ magnetic field of Earth (in Gs). $\beta = \frac{N_{e,i} \cdot kT}{B_{\circ}^2/2u\,\mu_0}$ ratio of kinetic pressure of charged particles to magnetic pressure, which is identic with the ratio of the energy relevant densities.

Height	N_n / m^{-3}	$N_{e,i}$ / m^{-3}	T_n / °K	N_n / $N_{e,i}$	$\overline{m}$ / m_H	$B_{\circ}$ / Gs	β
			Ionosphere				
100 km	$(2–4) \cdot 10^{19}$	$(0.2–10) \cdot 10^{10}$	230	$10^{-10}–10^{-8}$	28	0.49	$(2–100) \cdot 10^{-8}$
200 km	$3 \cdot 10^{16}$	$(0.3–5) \cdot 10^{11}$	400–800	$10^{-5}–10^{-6}$	24	0.45	$(4–100) \cdot 10^{-7}$
300 km	$3 \cdot 10^{15}$	$(0.1–2) \cdot 10^{12}$	10^3	$10^{-3}–10^{-4}$		0.44	$(4–100) \cdot 10^{-6}$
400 km	$5 \cdot 10^{14}$	$(0.5–1.5) \cdot 10^{12}$	$1.5 \cdot 10^3$	$10^{-2}–10^{-3}$	20	0.40	$(3–10) \cdot 10^{-5}$
500 km	$5 \cdot 10^{13}$	$(0.4–1) \cdot 10^{12}$	$1.8 \cdot 10^3$	$10^{-2}–10^{-1}$		0.37	$(4–10) \cdot 10^{-5}$
700 km	$6 \cdot 10^{12}$	$(2–5) \cdot 10^{11}$	$2 \cdot 10^3$	10^{-1}	16	0.35	$(2–5) \cdot 10^{-5}$
1000 km	10^{11}	10^{11}	$3 \cdot 10^3$	1	14	0.33	$2 \cdot 10^{-5}$
3000 km	10^6	$(5–7) \cdot 10^9$	$4 \cdot 10^3$	10^4		0.16	$2 \cdot 10^{-6}$
			Magnetosphere				
(3–4) R_E		$(3–5) \cdot 10^8$	$5 \cdot 10^3$		1	0.01	$(4–20) \cdot 10^{-5}$
			Interplanetary medium				
100 R_E		10^8	10^4		1	0.001	$(3–10) \cdot 10^{-3}$

Table 2. *Typical collisional parameters in the terrestrial atmosphere.*

ν_{ei} and ν_{en} collision frequency of electrons with ions and neutral particles, respectively. $\nu_{e,(i+n)}$ total collision frequency of electrons. Corresponding definitions for ν_{ii}, ν_{in} and $\nu_{i,(i+n)}$. $l_{e,i}$ mean free path of particles e or i. V_T mean thermal velocity of particles (indicated by suffix e, i, n). (As for the definition of collision frequencies see 'Introductory Remarks', p. 1.)

Height	ν_{ei} / sec^{-1}	ν_{en} / sec^{-1}	$\nu_{e(n+i)}$ / sec^{-1}	ν_{ii} / sec^{-1}	ν_{in} / sec^{-1}	$\nu_{i(n+i)}$ / sec^{-1}	V_{Te} / m sec^{-1}	$V_{Ti,n}$ / m sec^{-1}	$l_{e\,i}$ / m
				Ionosphere					
100 km	$6 \cdot 10^2$	$2 \cdot 10^5$	$2 \cdot 10^5$	0.3	10^3	10^3	$7.2 \cdot 10^4$	$3 \cdot 10^2$	$3.6 \cdot 10^{-2}$
200 km	$2 \cdot 10^2$	$2 \cdot 10^3$	$2 \cdot 10^3$	0.1	9	10	$1.1 \cdot 10^5$	$5 \cdot 10^2$	$6 \cdot 10^1$
300 km	10^3	10^3	$3 \cdot 10^3$	7	7	15	$1.5 \cdot 10^5$	$7 \cdot 10^2$	$5 \cdot 10^1$
400 km	10^3	10^2	10^3	7	1	7	$1.2 \cdot 10^5$	$1.8 \cdot 10^2$	$1.8 \cdot 10^2$
500 km	$7 \cdot 10^2$	10	$7 \cdot 10^2$	3	0.1	3	$2.1 \cdot 10^5$	10^3	$3 \cdot 10^2$
700 km	$2 \cdot 10^2$	—	$2 \cdot 10^2$	1	—	1	$2.4 \cdot 10^5$	$1.3 \cdot 10^3$	$1.2 \cdot 10^3$
1000 km	40	—	40	0.2	—	0.2	$2.8 \cdot 10^5$	$1.6 \cdot 10^3$	$7 \cdot 10^3$
3000 km	14	—	14	0.1	—	0.1	$3 \cdot 10^5$	$2 \cdot 10^3$	$2 \cdot 10^4$
				Magnetosphere					
(3–4) R_E	10^{-2}	—	10^{-2}	$3 \cdot 10^{-4}$	—	—	$3 \cdot 10^5$	$1 \cdot 10^4$	$3 \cdot 10^7$
				Interplanetary medium					
100 R_E	$3 \cdot 10^{-3}$	—	$3 \cdot 10^{-3}$	10^{-4}	—	—	$5 \cdot 10^5$	$2 \cdot 10^4$	10^8

* Typical average values. The actual values depend largely on solar activity and are very different by day and by night, see COSPAR Internat. Reference Atmosphere (CIRA), 1972.

Table 3. *Typical plasma parameters in the terrestrial atmosphere.*

$\omega_{N\mathrm{e}} = \sqrt{\frac{u}{\varepsilon_0}\frac{q^2}{m_\mathrm{e}}N_\mathrm{e}}$ plasma pulsation, $f_{N\mathrm{e}} \equiv \omega_{N\mathrm{e}}/2\pi$ plasma (or LANGMUIR) frequency (for electrons). Similar definition for ions with $\omega_{N\mathrm{i}} = \sqrt{\frac{u}{\varepsilon_0}\frac{q^2}{m_\mathrm{i}}N_\mathrm{i}}$. $\lambda_\mathrm{D} \equiv \left(\sum_\mathrm{h} \omega_{N\mathrm{h}}^2/V_{T\mathrm{h}}^2\right)^{-\frac{1}{2}}$ DEBYE radius. $\omega_{B\mathrm{e}} = (q/m_\mathrm{e})(B_0/c_0\sqrt{\varepsilon_0\mu_0})$; $f_{B\mathrm{e}} \equiv \omega_{B\mathrm{e}}/2\pi$ gyro (or LARMOR) frequency for electrons. Similar definition for ions with $\omega_{B\mathrm{i}} = (q/m_\mathrm{i})(B_0/c_0\sqrt{\varepsilon_0\mu_0})$; $f_{B\mathrm{i}} \equiv \omega_{B\mathrm{i}}/2\pi$. $r_{B\mathrm{e}} = V_{T\mathrm{e}}/\omega_{B\mathrm{e}}$ LARMOR radius for electrons, $r_{B\mathrm{i}} = V_{T\mathrm{i}}/\omega_{B\mathrm{i}}$ for ions.

Height	$\omega_{N\mathrm{e}}$ / sec^{-1}	$\omega_{N\mathrm{i}}$ / sec^{-1}	λ_D / cm	$\omega_{B\mathrm{e}}$ / sec^{-1}	$\omega_{B\mathrm{i}}$ / sec^{-1}	$\omega_{B\mathrm{e}}/\omega_{N\mathrm{e}}$	$\omega_{B\mathrm{i}}/\omega_{N\mathrm{i}} = V_\mathrm{A}/c_0$	$r_{B\mathrm{e}}$	$r_{B\mathrm{i}}$
Ionosphere									
100 km	$(3–20)\cdot 10^6$	$(2–10)\cdot 10^4$	1	$8\cdot 10^6$		1	10^{-2}	9 mm	1.5 m
200 km	$(1–4)\cdot 10^7$	$(5–20)\cdot 10^4$	0.2–1	$7.9\cdot 10^6$		0.3	$2\cdot 10^{-3}$	1.5 cm	2.5 m
300 km	$(2–8)\cdot 10^7$	$(1–4)\cdot 10^5$	0.1–0.7	$7.7\cdot 10^6$	$2\cdot 10^2$	0.1	10^{-3}	2 cm	3.5 m
400 km	$(4–7)\cdot 10^7$	$(3–5)\cdot 10^5$	0.2–0.4	$7.3\cdot 10^6$		0.1	10^{-3}	2.5 cm	4 m
500 km	$6\cdot 10^7$	$4\cdot 10^5$	0.3–0.6	$7\cdot 10^6$		0.2	$2\cdot 10^{-3}$	3 cm	5 m
700 km	$(2–4)\cdot 10^7$	$(1–3)\cdot 10^5$	0.4–0.7	$6.4\cdot 10^6$		0.2	$2\cdot 10^{-3}$	4 cm	6.5 m
1000 km	$2\cdot 10^7$	$2\cdot 10^5$	1	$5.7\cdot 10^6$	10^2	0.3	$3\cdot 10^{-3}$	6 cm	15 m
3000 km	$5\cdot 10^6$	$5\cdot 10^4$	4	$2.8\cdot 10^6$		0.5	$5\cdot 10^{-3}$	10 cm	20 m
Magnetosphere									
(3–4) R_E	10^6	$2\cdot 10^4$	30	$(1–3)\cdot 10^5$	50	0.2	$4\cdot 10^{-3}$	2 m	200 m
Interplanetary medium									
100 R_E	$5\cdot 10^5$	10^4	50	$2\cdot 10^4$	5	$3\cdot 10^{-2}$	$6\cdot 10^{-4}$	20 m	4 km

In the presence of a magnetic field B the conditions are

$$|\omega - s\,\omega_{B\mathrm{h}}| \gg |k_\parallel|\, V_{T\mathrm{h}} \quad \text{and} \quad \omega_{B\mathrm{h}} \gg |k_\perp|\, V_{T\mathrm{h}} \qquad (\mathrm{h} = \mathrm{e}, \mathrm{i})$$

where $\boldsymbol{k}_\parallel$ and $\boldsymbol{k}_\perp$ are the components of the wave vector parallel and perpendicular to the magnetic field. Thus the spatial dispersion of the dielectric permeability of the plasma can be neglected under the above conditions, as far as it is brought about by thermal motion of particles. In this case the analysis of the spectra of plasma oscillations is rather easy, it corresponds to that used in crystal optics [7], [8]. If, however, $\omega \lesssim |k|\, V_{T\mathrm{e,i}}$ then spatial dispersion becomes essential and the thermal motion of the particles plays a decisive rôle when the spectra of plasma oscillations are investigated.

I. Principles of linear electrodynamics.

Before going over to investigate the electromagnetic properties of precise models of plasma we must discuss a few general relations of electrodynamics of such media in relation with spatial dispersion.

3. Energy considerations.

α) The first question in such a medium concerns the energy of the electromagnetic field. It is evident that the energy contained in the medium is changed by the field generators outside which provoke the field inside. This *change of energy* is due to the mutual reaction of the electromagnetic field with its generators outside, in other words to the energy needed to build up the field in the medium.

In order to compute the change of energy in the medium we multiply the first Eq. (1.6) by $\boldsymbol{E}$, the third one by $\boldsymbol{B}$; then both equations are subtracted one from the other. It results ($\partial/\partial t = \dot{}$)

$$\frac{1}{\mathrm{u}}\left(\boldsymbol{E}\cdot\dot{\boldsymbol{D}}+\frac{\boldsymbol{B}}{\mu_0}\cdot\dot{\boldsymbol{B}}\right)=-\frac{c_0\sqrt{\varepsilon_0\,\mu_0}}{\mathrm{u}}\frac{\partial}{\partial\boldsymbol{r}}\cdot\left(\boldsymbol{E}\times\frac{\boldsymbol{B}}{\mu_0}\right)-\boldsymbol{E}\cdot\boldsymbol{J}_0. \tag{3.1}$$

Integrating over a certain volume V, limited by a surface $\boldsymbol{S}$, we obtain

$$\frac{1}{\mathrm{u}}\int_V \mathrm{d}^3\boldsymbol{r}\left(\boldsymbol{E}\cdot\dot{\boldsymbol{D}}+\frac{\boldsymbol{B}}{\mu_0}\cdot\dot{\boldsymbol{B}}\right)=-\frac{c_0\sqrt{\varepsilon_0\,\mu_0}}{\mathrm{u}}\oint_S \mathrm{d}\boldsymbol{S}\cdot\left(\boldsymbol{E}\times\frac{\boldsymbol{B}}{\mu_0}\right)-\int_V \mathrm{d}^3\boldsymbol{r}\,\boldsymbol{E}\cdot\boldsymbol{J}_0. \tag{3.2}$$

We suppose that when approaching infinity the electromagnetic fields go towards zero so that with increasing volume V the contribution of the surface integral on the right side becomes negligible in Eq. (3.2). Then the left hand term (with $\boldsymbol{H}=\boldsymbol{B}/\mu_0$, the magnetic field strength)

$$\frac{1}{\mathrm{u}}\int_V \mathrm{d}^3\boldsymbol{r}\,(\boldsymbol{E}\cdot\dot{\boldsymbol{D}}+\boldsymbol{H}\cdot\dot{\boldsymbol{B}}) \tag{3.3}$$

is the total change of field energy; Eq. (3.2) says that this is the average Joule heating, i.e. the power applied by the outside generators, viz.

$$-\int_V \mathrm{d}^3\boldsymbol{r}\,\boldsymbol{E}\cdot\boldsymbol{J}_0. \tag{3.4}$$

Even in the case of a monochromatic plane wave field

$$\left.\begin{aligned}\boldsymbol{E}(\boldsymbol{r},t)&=\boldsymbol{E}\exp\left(i(\omega t-\boldsymbol{k}\cdot\boldsymbol{r})\right)+\boldsymbol{E}^*\exp\left(-i(\omega t-\boldsymbol{k}\cdot\boldsymbol{r})\right)\\ \boldsymbol{B}(\boldsymbol{r},t)&=\boldsymbol{B}\exp\left(i(\omega t-\boldsymbol{k}\cdot\boldsymbol{r})\right)+\boldsymbol{B}^*\exp\left(-i(\omega t-\boldsymbol{k}\cdot\boldsymbol{r})\right)\\ \boldsymbol{D}(\boldsymbol{r},t)&=\boldsymbol{D}\exp\left(i(\omega t-\boldsymbol{k}\cdot\boldsymbol{r})\right)+\boldsymbol{D}^*\exp\left(-i(\omega t-\boldsymbol{k}\cdot\boldsymbol{r})\right)\end{aligned}\right\} \tag{3.5}$$

(where ω and $\boldsymbol{k}$ are real quantities), the surface integral in Eq. (3.2) can be neglected for $V\to\infty$.

Note that in this case the surface integral

$$\oint_S \mathrm{d}\boldsymbol{S}\cdot(\boldsymbol{E}\times\boldsymbol{H}) \tag{3.6}$$

does not vanish, but it does not increase with increasing V, while the volume integral, Eq. (3.3) increases. Thus the contribution (3.6) can be neglected compared with the contribution (3.3).

Putting the expressions Eq. (3.5) into the relation Eq. (3.2), and averaging over a time interval considerably greater than $1/\omega$, yield the following expression for the power per unit volume supported by the external (electrical) generators of the fields:

$$\left.\begin{aligned}Q&=-\frac{1}{V}\int_V \mathrm{d}^3\boldsymbol{r}\,\boldsymbol{E}\cdot\boldsymbol{J}_0=\frac{1}{\mathrm{u}V}\int_V \mathrm{d}^3\boldsymbol{r}\,\boldsymbol{E}\cdot\dot{\boldsymbol{D}}\\ &=\frac{i\omega}{\mathrm{u}}\{\varepsilon_{ij}^*(\omega,\boldsymbol{k})-\varepsilon_{ji}(\omega,\boldsymbol{k})\}\,E_i\,E_j^*.\end{aligned}\right\} \tag{3.7}$$

The material equations, Eqs. (1.9), have been used when deriving this expression. In the special case of the monochromatic wave field, Eqs. (3.5), the situation is stationary; this means that the energy flux is not changed by time averaging. The power due to the generators outside is completely transformed into heat. Thus Q is the Joule heating produced by unit volume and unit time in the field of the plane, monochromatic wave, Eqs. (3.5). $Q\equiv 0$ if the tensor E characterizing the dielectric permeability of the medium is hermitean in the variables ω and $\boldsymbol{k}$, i.e. if

$$\varepsilon_{ij}^*(\omega,\boldsymbol{k})=\varepsilon_{ji}(\omega,\boldsymbol{k}).$$

In a medium satisfying this condition no heat is produced inside so that the wave is not absorbed. We may therefore state that absorption of waves in a medium is due to the anti-Hermitean part of the tensor of dielectric permeability.

β) The fact that the tensor $\mathsf{E}(\omega, \boldsymbol{k})$ depends upon the wave vector $\boldsymbol{k}$ has the consequence that, even in an isotropic medium, the dielectric permeability is a *tensor*. In an isotropic (not gyrotropic) medium we can indeed establish the tensor $\boldsymbol{k}\boldsymbol{k} = ((k_i k_j))$. Thus in the most general case, the dielectric permeability of an isotropic medium is to be represented [2], [8] by a tensor

$$\mathsf{E}(\omega, \boldsymbol{k}) = (\mathsf{U} - \boldsymbol{k}\boldsymbol{k}/k^2)\,\varepsilon^{\text{tr}}(\omega, \boldsymbol{k}) + (\boldsymbol{k}\boldsymbol{k}/k^2)\,\varepsilon^{\text{l}}(\omega, \boldsymbol{k}), \tag{3.8}$$

in components:

$$\varepsilon_{ij}(\omega, \boldsymbol{k}) = (\delta_{ij} - k_i k_j/k^2)\,\varepsilon^{\text{tr}}(\omega, \boldsymbol{k}) + (k_i k_j/k^2)\,\varepsilon^{\text{l}}(\omega, \boldsymbol{k}). \tag{3.8a}$$

Putting this expression into Eq. (3.7) we get, for an *isotropic medium*[1],

$$Q = \frac{2\omega}{\text{u}\,k^2}\{\text{Im}\,\varepsilon^{\text{l}}(\omega, \boldsymbol{k})\,|\boldsymbol{k}\cdot\boldsymbol{E}|^2 + \text{Im}\,\varepsilon^{\text{tr}}(\omega, \boldsymbol{k})\,|\,(\boldsymbol{k}\times\boldsymbol{E})|^2\}. \tag{3.9}$$

The first term describes the absorption (due to the medium) of the longitudinal field component, $\boldsymbol{E}^{\text{l}} \| \boldsymbol{k}$, and the second one that of the transverse component, $\boldsymbol{E}^{\text{tr}} \perp \boldsymbol{k}$. Quite generally the quantities $\varepsilon^{\text{l}}(\omega, \boldsymbol{k})$ and $\varepsilon^{\text{tr}}(\omega, \boldsymbol{k})$ determine the character of propagation by longitudinal and transverse waves in an isotropic medium, see Sects. 4 and 17. In this respect one usually designates $\varepsilon^{\text{l}}(\omega, \boldsymbol{k})$ and $\varepsilon^{\text{tr}}(\omega, \boldsymbol{k})$ as longitudinal and transverse dielectric permeabilities.

If the medium is in full thermodynamic equilibrium electromagnetic waves of any sort must be absorbed, so that always $Q > 0$. It then follows from Eq. (3.9) for an isotropic medium in thermodynamic equilibrium:

$$\text{Im}\,\varepsilon^{\text{l}}(\omega, \boldsymbol{k}) > 0 \quad \text{and} \quad \text{Im}\,\varepsilon^{\text{tr}}(\omega, \boldsymbol{k}) > 0. \tag{3.10}$$

γ) If any one of these inequalities is violated we might have $Q < 0$. This, however, means that energy can pass over from the medium into the electromagnetic field. In such a medium a field produced by fluctuations will increase with time, i.e. such a medium is in an *unstable* condition. We may note here that while Eq. (3.10) must necessarily be violated in the case of a developing instability, this condition is not sufficient. Another necessary condition for the development of an instability is that in that range of ω and $\boldsymbol{k}$ where either $\text{Im}\,\varepsilon^{\text{l}} < 0$ or $\text{Im}\,\varepsilon^{\text{tr}} < 0$, electromagnetic waves find suitable propagation conditions in the medium.

4. Waves and their behaviour in time. The question is now which waves can propagate in the medium if exterior sources are absent, or with other words: under which conditions is an expression of the type $\exp(-i\,\omega\,t + i\,\boldsymbol{k}\cdot\boldsymbol{r})$ a solution of the homogeneous system of equations valid for the field?

α) In the case where the expression for the tensor of dielectric permeability, E, is known, our question is easily answered. It is, of course, the question after the electromagnetic eigenoscillations of the medium, or, using another terminology, that after the *normal modes* (or characteristic waves) in the medium. In order to answer the question we therefore ask which non-trivial solutions of the field equations Eq. (1.6) exist if exterior sources happen to be absent.

[1] Im designates the imaginary part of a complex quantity.

In the absence of exterior sources we may write these field equations for an infinite, homogeneous medium, supposing the variation in space and time being given by $\exp(-i\,\omega\,t+i\,\boldsymbol{k}\cdot\boldsymbol{r})$, as

$$\left.\begin{aligned} \boldsymbol{k}\times\boldsymbol{B} &= \frac{-\omega\,\mu_0}{c_0\sqrt{\varepsilon_0\,\mu_0}}\,\mathsf{E}(\omega,\boldsymbol{k})\cdot\boldsymbol{E}; & \boldsymbol{k}\cdot\boldsymbol{B} &= 0 \\ \boldsymbol{k}\times\boldsymbol{E} &= \frac{\omega}{c_0\sqrt{\varepsilon_0\,\mu_0}}\,\boldsymbol{B}; & \boldsymbol{k}\cdot\mathsf{E}(\omega,\boldsymbol{k})\cdot\boldsymbol{E} &= 0. \end{aligned}\right\} \tag{4.1}$$

These equations may also be written in coordinates:

$$\left.\begin{aligned} \varepsilon_{ijk}\,\frac{\partial}{\partial x_i}\,B_k = (\boldsymbol{k}\times\boldsymbol{B})_i &= \frac{-\omega\,\mu_0}{c_0\sqrt{\varepsilon_0\,\mu_0}}\,\varepsilon_{ij}(\omega,\boldsymbol{k})\,E_j; & k_i B_i &= 0; \\ (\boldsymbol{k}\times\boldsymbol{E})_i &= \frac{\omega}{c_0\sqrt{\varepsilon_0\,\mu_0}}\,B_i; & k_i\,\varepsilon_{ij}(\omega,\boldsymbol{k})\,E_j &= 0. \end{aligned}\right\} \tag{4.1a}$$

We have introduced $\mathfrak{E}\equiv(((\varepsilon_{ijk})))$, the (antisymmetric) permutation tensor of third grade, see Sect. 6ε. The summation convention (or rule) is applied through out this contribution so that one has to sum up over each index which appears twice, e.g. $\varepsilon_{ij}\,E_j$ means $\sum_j \varepsilon_{ij}\,E_j$, etc.

This system of equations is easily reduced to the following system of three algebraic equations for the components of the electric field $\boldsymbol{E}$:

$$\left(k^2\,\mathsf{U}-\boldsymbol{k}\,\boldsymbol{k}-\frac{\omega^2\,\mu_0}{c_0^2\,\varepsilon_0\,\mu_0}\,\mathsf{E}\right)\cdot\boldsymbol{E}=0. \tag{4.2}$$

(The coefficient of the last term could als be written as k_0^2/ε_0.)

β) The condition for this system of homogeneous equations to have non-trivial solutions yield the *dispersion relation*. This relation is basic for the non-trivial solutions of the field equations, Eq. (1.6), in the medium and answers the questions we asked above. Apparently, the dispersion relation has the form[1]

$$\det\left(k^2\,\mathsf{U}-\boldsymbol{k}\,\boldsymbol{k}-\frac{\omega^2\,\mu_0}{c_0^2\,\varepsilon_0\,\mu_0}\,\mathsf{E}\right)=0. \tag{4.3}$$

In the particular case of an *isotropic medium* the system splits up into two equations (provided $\omega\neq0$)

$$\left.\begin{aligned} \varepsilon^{\mathrm{l}}(\omega,\boldsymbol{k})\,\boldsymbol{E}^{\mathrm{l}} &= 0, \\ \left[k^2-\frac{\omega^2\,\mu_0}{c_0^2\,\varepsilon_0\,\mu_0}\,\varepsilon^{\mathrm{tr}}(\omega,\boldsymbol{k})\right]\boldsymbol{E}^{\mathrm{tr}} &= 0 \end{aligned}\right\} \tag{4.4}$$

where

$$\boldsymbol{E}^{\mathrm{l}}\equiv\boldsymbol{k}\,\frac{\boldsymbol{k}\cdot\boldsymbol{E}}{k^2} \tag{4.4a}$$

is the longitudinal component of the electric field, and

$$\boldsymbol{E}^{\mathrm{tr}}\equiv\boldsymbol{E}-\boldsymbol{k}\,\frac{\boldsymbol{k}\cdot\boldsymbol{E}}{k^2} \tag{4.4b}$$

is its transverse component, both with respect to the wave vector $\boldsymbol{k}$. Of course, the dispersion relation Eq. (4.3) must also split up into two equations in this

[1] det means determinant.

case, namely

$$\left.\begin{aligned} \varepsilon^{\mathrm{l}}(\omega, \boldsymbol{k}) &= 0, \\ k^2 - \frac{\omega^2 \mu_0}{c_0^2\, \varepsilon_0\, \mu_0}\, \varepsilon^{\mathrm{tr}}(\omega, \boldsymbol{k}) &= 0. \end{aligned}\right\} \tag{4.5}$$

The first of these equations is the condition that longitudinal waves may exist in the medium, the second one is the condition for transverse waves.

γ) The dispersion relation, Eq. (4.3) connects the frequency, $f = \omega/2\pi$ (pulsation ω), of the wave with the wave vector $\boldsymbol{k}$, for electromagnetic normal eigenmodes. With this relation we can determine the value of the frequency as function of $\boldsymbol{k}$ for given parameters of the medium (material constants) and this means *establish the spectrum of eigen-oscillations* as one usually says. On the other hand for a given value of the frequency, $\omega/2\pi$, and a given direction we can find the corresponding projection of the wave vector $\boldsymbol{k}(\omega)$. These two approaches correspond to two different formulations of a problem when solving the integro-differential equations of the field. The integral character of these equations is due to the material equation, Eq. (1.8). One either speakes of an *initial value problem or a boundary value problem.* We shall now consider these problems in more detail.

We suppose now an infinite and homogeneous medium, in which at initial time ($t = 0$) an electromagnetic field happened to be produced. This may result from fluctuations, may be it was provoked by sources outside of the medium. Let us consider the variation of this field in the medium, in the absence of outer sources. In order to resolve such a problem it is not sufficient to know the initial values $\boldsymbol{E}(\boldsymbol{r}, 0)$ and $\boldsymbol{B}(\boldsymbol{r}, 0)$ only, but one also must know the initial value of the electric displacement vector $\boldsymbol{D}(\boldsymbol{r}, 0)$. It is determined by the "prehistory" of the field $\boldsymbol{E}(\boldsymbol{r}, t)$ until the instant $t = 0$:

$$\boldsymbol{D}(\boldsymbol{r}, 0) = \int_{-\infty}^{0} \mathrm{d}t' \int \mathrm{d}^3\boldsymbol{r}'\, \hat{\varepsilon}(\boldsymbol{r} - \boldsymbol{r}', -t') \cdot \boldsymbol{E}(\boldsymbol{r}', t'), \tag{4.6}$$

in components

$$D_i(\boldsymbol{r}, 0) = \int_{-\infty}^{0} \mathrm{d}t' \int \mathrm{d}^3\boldsymbol{r}'\, \hat{\varepsilon}_{ij}(\boldsymbol{r} - \boldsymbol{r}', -t')\, E_j(\boldsymbol{r}', t'). \tag{4.6a}$$

The physical reason of this necessity is related with the inertia of electrons and ions and the relaxation processes in the medium, or, in other words, with dispersion in time, i.e. frequency dispersion. The knowledge of the "prehistory" would not be necessary if, apart from the initial values of the fields $\boldsymbol{E}$ and $\boldsymbol{B}$, we also knew the parameters of all individual particles in the medium at initial time $t = 0$. In that case we could *compute* $\boldsymbol{D}(\boldsymbol{r}, 0)$ without any difficulty. This formulation of the problem, however, would not be realistic. Therefore we need the knowledge of $\boldsymbol{E}(\boldsymbol{r}, t)$ in principle for the whole "prehistory", i.e. for each $t < 0$.

δ) In order to resolve the problem with the initial conditions formulated above we use a FOURIER-LAPLACE *development* [12][2]:

$$\boldsymbol{E}(\boldsymbol{r}, t) = \frac{1}{(2\pi)^4} \int \mathrm{d}^3\boldsymbol{k}\, \mathrm{e}^{i\boldsymbol{k}\cdot\boldsymbol{r}} \int_{-\infty+i\sigma}^{+\infty+i\sigma} \mathrm{d}\omega\, \mathrm{e}^{-i\omega t}\, \tilde{\tilde{\boldsymbol{E}}}(\boldsymbol{k}, \omega), \quad (t \geqq 0), \tag{4.7a}$$

$$\tilde{\tilde{\boldsymbol{E}}}(\boldsymbol{k}, \omega) = \int \mathrm{d}^3\boldsymbol{r}\, \mathrm{e}^{-i\boldsymbol{k}\cdot\boldsymbol{r}} \int_{0}^{\infty} \mathrm{d}t\, \mathrm{e}^{i\omega t}\, \boldsymbol{E}(\boldsymbol{r}, t), \quad (\operatorname{Im}\omega \equiv \sigma \geqq 0). \tag{4.7b}$$

---

[2] FOURIER-transformation is indicated by the symbols $\tilde{}$ or $\tilde{\tilde{}}$. According to their definition by Eqs. (4.7) the unit for $\tilde{\tilde{\boldsymbol{E}}}(\boldsymbol{k}, \omega)$ (in SI-units) is $\mathrm{V\,s\,m^2}$, as $\boldsymbol{E}(\boldsymbol{r}, t)$ has $\mathrm{V\,m^{-1}}$ for unit. Similarly, the FOURIER-transforms $\tilde{\tilde{\boldsymbol{D}}}$, $\tilde{\tilde{\boldsymbol{B}}}$ have $\mathrm{A\,s^2\,m}$ and $\mathrm{V\,s^2\,m}$ resp. for SI-units, see footnote 7 of (Sect. 1).

Similar forms can be written down for the electric and magnetic induction vectors, $\boldsymbol{D}(\boldsymbol{r}, t)$ and $\boldsymbol{B}(\boldsymbol{r}, t)$. In this way the field equations, Eq. (1.6), can be transformed into the following expressions:

$$\left.\begin{aligned} &\frac{\omega\,\mu_0}{c_0\sqrt{\varepsilon_0\,\mu_0}}\,\tilde{\tilde{\boldsymbol{D}}}(\boldsymbol{k},\omega)+\boldsymbol{k}\times\tilde{\tilde{\boldsymbol{B}}}(\boldsymbol{k},\omega)=i\,\frac{\mu_0}{c_0\sqrt{\varepsilon_0\,\mu_0}}\,\tilde{\boldsymbol{D}}(\boldsymbol{k},t=0); \quad \boldsymbol{k}\cdot\tilde{\tilde{\boldsymbol{B}}}(\boldsymbol{k},\omega)=0\\ &\frac{\omega}{c_0\sqrt{\varepsilon_0\,\mu_0}}\,\tilde{\tilde{\boldsymbol{B}}}(\boldsymbol{k},\omega)-\boldsymbol{k}\times\tilde{\tilde{\boldsymbol{E}}}(\boldsymbol{k},\omega)=i\,\frac{1}{c_0\sqrt{\varepsilon_0\,\mu_0}}\,\tilde{\boldsymbol{B}}(\boldsymbol{k},t=0); \quad \boldsymbol{k}\cdot\tilde{\tilde{\boldsymbol{D}}}(\boldsymbol{k},\omega)=0 \end{aligned}\right\} \tag{4.8}$$

$\boldsymbol{D}(\boldsymbol{k}, t=0)$ and $\boldsymbol{B}(\boldsymbol{k}, t=0)$ are FOURIER representations after the space coordinates $\boldsymbol{r}$ of the initial values $\boldsymbol{D}(\boldsymbol{r}, 0)$ and $\boldsymbol{B}(\boldsymbol{r}, 0)$. Using Eq. (1.9), Eq. (4.8), can be written as:

$$\left.\begin{aligned} \left\{k^2\,\mathsf{U}-\boldsymbol{k}\,\boldsymbol{k}-\frac{\omega^2\,\mu_0}{c_0^2\,\varepsilon_0\,\mu_0}\,\mathsf{E}(\boldsymbol{k},\omega)\right\}\cdot\tilde{\tilde{\boldsymbol{E}}}(\boldsymbol{k},\omega)&=i\,\frac{\omega\,\mu_0}{c_0^2\,\varepsilon_0\,\mu_0}\,\tilde{\boldsymbol{D}}(\boldsymbol{k},t=0)+\\ &\quad+i\,\frac{1}{c_0\sqrt{\varepsilon_0\,\mu_0}}\,\boldsymbol{k}\times\tilde{\boldsymbol{B}}(\boldsymbol{k},t=0), \end{aligned}\right\} \tag{4.9}$$

which gives three algebraic equations:

$$\left.\begin{aligned} \left\{k^2\,\delta_{ij}-k_i\,k_j-\frac{\omega^2\,\mu_0}{c_0^2\,\varepsilon_0\,\mu_0}\,\varepsilon_{ij}(\boldsymbol{k},\omega)\right\}\tilde{\tilde{E}}_j(\boldsymbol{k},\omega)&=i\,\frac{\omega\,\mu_0}{c_0^2\,\varepsilon_0\,\mu_0}\,\tilde{D}_i(\boldsymbol{k},t=0)+\\ &\quad+i\,\frac{1}{c_0\sqrt{\varepsilon_0\,\mu_0}}\,[\boldsymbol{k}\times\tilde{\boldsymbol{B}}(\boldsymbol{k},t=0)]_i. \end{aligned}\right\} \tag{4.9a}$$

Different from Eq. (4.2) this is an inhomogeneous system. Its solution may formally be written as

$$\tilde{\tilde{\boldsymbol{E}}}(\boldsymbol{k},\omega)=\frac{1}{\mathscr{D}(\boldsymbol{k},\omega)}\,\boldsymbol{\varDelta}(\boldsymbol{k},\omega); \qquad \tilde{\tilde{E}}_i(\boldsymbol{k},\omega)=\frac{\varDelta_i(\boldsymbol{k},\omega)}{\mathscr{D}(\boldsymbol{k},\omega)}. \tag{4.10}$$

$\mathscr{D}(\boldsymbol{k}, \omega)$ being the determinant of the system, see Eq. (4.3):

$$\mathscr{D}(\boldsymbol{k},\omega)\equiv\det\left\{k^2\,\delta_{ij}-k_i\,k_j-\frac{\omega^2\,\mu_0}{c_0^2\,\varepsilon_0\,\mu_0}\,\varepsilon_{ij}(\omega,\boldsymbol{k})\right\} \tag{4.11}$$

and $\boldsymbol{\varDelta}(\boldsymbol{k}, \omega)$ is the algebraic complement depending on the right side of the system Eq. (4.9). This means $\boldsymbol{\varDelta}$ depends on the initial values $\boldsymbol{D}(\boldsymbol{r}, 0)$ and $\boldsymbol{B}(\boldsymbol{r}, 0)$.

Now, introducing the formal solution, Eq. (4.10), into the FOURIER-LAPLACE development we get the field for arbitrary initial values. We write the initial values as a FOURIER development[3] with the FOURIER component $\tilde{\boldsymbol{E}}(\boldsymbol{k}, t)$:

$$\tilde{\boldsymbol{E}}(\boldsymbol{k},t)=\frac{1}{2\pi}\int\limits_{-\infty+i\sigma}^{+\infty+i\sigma}\mathrm{d}\omega\,\mathrm{e}^{-i\omega t}\,\frac{\boldsymbol{\varDelta}(\boldsymbol{k},\omega)}{\mathscr{D}(\boldsymbol{k},\omega)}, \tag{4.12}$$

$$\tilde{E}_i(\boldsymbol{k},t)=\frac{1}{2\pi}\int\limits_{-\infty+i\sigma}^{+\infty+i\sigma}\mathrm{d}\omega\,\mathrm{e}^{-i\omega t}\,\frac{\varDelta_i(\boldsymbol{k},\omega)}{\mathscr{D}(\boldsymbol{k},\omega)}. \tag{4.12a}$$

ε) Usually, this integral is computed using the *theory of residues*; it follows from an integration path which is closed by a half-circle of infinite radius. The value of the integral is mainly determined by the poles of the integrand lying inside this contour, but branch points and other characteristic points also have

[3] The FOURIER-component $\tilde{\boldsymbol{E}}(\boldsymbol{k}, t)$ has $\mathrm{V\,m^2}$ for SI-unit, see above and footnote 7 in Sect. 1.

to be considered. Details related to this integration will not be discussed here as they have been in [2], [13]. We consider only the contributions to the integral which are due to the poles of the integrand caused by a vanishing determinant of the system Eq. (4.9):

$$\det\{\ldots\}(\boldsymbol{k},\omega) \equiv \mathscr{D}(\boldsymbol{k},\omega) = 0. \tag{4.13}$$

Let $\omega_n(\boldsymbol{k})$ be the roots of this equations [which is identical with the dispersion equation, Eq. (4.3)]. With the sum of residues of the poles[4] we then obtain the following expression for the variation of the field in time:

$$\widetilde{\boldsymbol{E}}(\boldsymbol{k},t) \sim \sum_n e^{-i\omega_n(\boldsymbol{k})t}. \tag{4.14}$$

If $\mathrm{Im}\,\omega_n(\boldsymbol{k}) < 0$ for all roots inside the integration contour then all members of the sum decrease exponentially with increasing t; for great values of t only the member with the smallest decrement must be considered which may be

$$\gamma_n \equiv \mathrm{Min}[\mathrm{Im}\,\omega_n(\boldsymbol{k})].$$

If, however, we have a root with $\mathrm{Im}\,\omega_n(\boldsymbol{k}) = 0$ then the corresponding term of the sum Eq. (4.14) does not decrease with time and we are in the presence of a stationary oscillation in the medium; these are called *stationary normal modes*. Finally, if there was at least one root with $\mathrm{Im}\,\omega_n(\boldsymbol{k}) > 0$, then the oscillations of the corresponding mode would be increasing with time. This can not be a stationary case but is a case of *instability*. The increase of oscillation amplitude is described (instead of a decrement) by an *increment* $\gamma_n = \mathrm{Im}\,\omega_n(\boldsymbol{k})$.

We thus have shown that the roots of the dispersion equation, Eq. (4.3), determine the development with time of initial fluctuations in the medium. These fluctuations themselves are caused by the particularities of the medium and not by the "chance" for such initial conditions to be produced[5]. This shows how important the dispersion equation is for the whole linear electrodynamics of dense media.

5. Waves and their behaviour in space.

α) The dispersion equation, Eq. (4.3), does not only determine the variations in time of the electromagnetic field in the medium but also its *variations in space*. In particular the dispersion equation is extremely important in the case where an electromagnetic wave hits the border of a medium; it determines the manner how the wave may enter into the medium and how it propagates inside.

On the other hand the dielectric permeability tensor $\boldsymbol{\mathsf{E}}(\boldsymbol{k},\omega)$ appearing in Eq. (4.3) is, strictly speaking, only defined in an infinite and homogeneous medium. Though we meet here a certain inconsistency the considerations given above may be applied if the dimensions of the medium are considerably larger than the wavelength (in this medium) of the oscillations. The dispersion equation, Eq. (4.3), then correctly describes the space variation of electromagnetic waves at distances large compared to the wavelength. At such distances the spatial variation of the field is determined by the properties of the medium, not by specific conditions at any boundary.

When resolving the initial value problem we have to look for complex solutions $\omega_n(\boldsymbol{k})$ of the dispersion equation for predetermined values of $\boldsymbol{k}$. Usually when resolving a problem of this kind the complex projection $k_1(\omega)$ of the

[4] We note that the branch points of the function $\Delta(\boldsymbol{k},\omega)$ do not lead to similar time variations, see [2].

[5] While all other contributions to the integral Eq. (4.12) depend on the particularities of the function $\Delta(\boldsymbol{k},\omega)$ and therefore depend considerably on the specific initial fluctuation present.

vector $\boldsymbol{k}(\omega)$ into a predetermined direction is looked for, supposing that ω and two other orthogonal projections $k_{2,3}(\omega)$ exist and are real. The spatial variation of the field is then given by an expression of the form:

$$\boldsymbol{E}(\boldsymbol{r}, t) \sim \sum_{\mathrm{n}} \mathrm{e}^{i\boldsymbol{k}_{\mathrm{n}}(\omega)\cdot\boldsymbol{r} - i\omega t} \tag{5.1}$$

where $\boldsymbol{k}_{\mathrm{n}}(\omega)$ satisfies the dispersion relation, Eq. (4.3).

In the general case the $\boldsymbol{k}_{\mathrm{n}}(\omega)$ are complex quantities. Let ϑ be the angle between the predetermined direction and the wave vector $\boldsymbol{k}_{\mathrm{n}}(\omega)$. If now $\mathrm{Im}\, k_{\mathrm{n}\vartheta}(\omega) > 0$ then the wave must decrease in the given direction; in the opposite case it will increase.

It is, however, not allowed to draw conclusions concerning the admittance of the medium from the sign of $\mathrm{Im}\, k_{\mathrm{n}\vartheta}(\omega)$. In order to do this either the problem of limits must be accurately resolved, or one has to make complicated considerations [*3*] based on the form of the dispersion relation, Eq. (4.3). The problem of the stability of the medium can be analyzed much easier by considering the initial value problem, i.e. by resolving the dispersion relation, Eq. (4.3), as function of ω. If it is known that the plasma is stable (i.e. $\mathrm{Im}\,\omega < 0$, see Sect. 4) then $\mathrm{Im}\, k_{\vartheta}(\omega)$ characterizes the spatial decrease of a given mode of oscillations in the given direction. On the other hand in an instable case $\mathrm{Im}\, k_{\vartheta}(\omega)$ characterizes the spatial amplification of a wave with the given frequency $\omega/2\pi$.

β) When describing the propagation of electromagnetic waves in a medium we often use the ***index vector n*** instead of the wave vector $\boldsymbol{k}$:

$$\boldsymbol{k} = \frac{\omega}{c_0}\boldsymbol{n}. \tag{5.2}$$

The modulus of the index vector is called the complex refraction index. The dispersion relation, Eq. (4.3), can be written with $\boldsymbol{n}$ in the form

$$\det\left\{n^2\,\mathsf{U} - \boldsymbol{n}\boldsymbol{n} - \frac{1}{\varepsilon_0}\,\mathsf{E}\left(\omega, \frac{\omega}{c_0}\boldsymbol{n}\right)\right\} = 0, \tag{5.3}$$

in components:

$$\det\left\{n^2\,\delta_{ij} - n_i n_j - \frac{1}{\varepsilon_0}\,\varepsilon_{ij}\left(\omega, \frac{\omega}{c_0}\boldsymbol{n}\right)\right\} = 0. \tag{5.3a}$$

From this equation the projection of the vector $\boldsymbol{n}(\omega)$ for a given direction is easily determined

$$n(\omega, \vartheta) = \mu(\omega, \vartheta) + i\,\chi(\omega, \vartheta). \tag{5.4}$$

μ being the real refraction index and χ the extinction index of the wave.

γ) In the general case of complex $\boldsymbol{k}(\omega)$ a wave described by

$$\sim \exp(-i\omega t + i\boldsymbol{k}\cdot\boldsymbol{r}) \qquad [5.1]$$

can be called a plane wave in a broad sense only. In this case the planes of constant phase, i.e. planes perpendicular to the vector $\mathrm{Re}\,\boldsymbol{k}(\omega)$, do not coincide with the planes of constant amplitude[1], i.e. planes perpendicular to the vector $\mathrm{Im}\,\boldsymbol{k}(\omega)$. Therefore we better designate these waves as *inhomogeneous plane waves* distinguishing them from homogeneous plane waves for which both systems of planes coincide. Even if this is not so but if the decrement remains small, i.e. $|\mathrm{Im}\,\boldsymbol{k}(\omega)| \ll |\mathrm{Re}\,\boldsymbol{k}(\omega)|$ one may use the approximation of vanishing

[1] H. Arzélies: C. R. Acad. Sci. Paris **235**, 1619–1621 (1952); see also [*2*] Sect. I.6.

extinction; one then finds with very high accuracy that the phase- and group-velocity of the real wave have the same values as those found with the approximation $\mathrm{Im}\,\boldsymbol{k}=0$. This is, for example, the case of a transparent medium of weak attenuation[2]. For a transparent medium with vanishing extinction the *phase velocity* of the wave is

$$\boldsymbol{V}_{\mathrm{p}} \equiv \frac{\omega}{k^2}\,\boldsymbol{k} = \frac{c_0}{n}\,\frac{\boldsymbol{n}}{n} = \frac{c_0}{n}\,\boldsymbol{n}^0, \tag{5.5}$$

$\boldsymbol{n}^0$ being the unit vector in the direction of $\boldsymbol{n}$. V_{p} characterizes the propagation velocity of the *surfaces of constant phase,*

$$\boldsymbol{k}\cdot\boldsymbol{r} - \omega t = const.$$

It is therefore called *phase velocity.*

The vector quantity

$$\boldsymbol{V}_{\mathrm{g}} \equiv \frac{\partial\omega}{\partial\boldsymbol{k}} \tag{5.6}$$

however, characterizes the transport velocity of the wave amplitude, consequently of its energy, and is called the *group velocity*[3].

We stated above that for general waves propagating in a medium the quantities ω and $\boldsymbol{k}$ are not independent but appear to be related by the dispersion relation, Eq. (4.3) for example. There arises the question: in which manner should the dielectric permeability tensor $\mathsf{E}(\omega, \boldsymbol{k})$ be considered to depend on independent variables ω and $\boldsymbol{k}$? Apparently, the answer is that (in the absence of exterior sources) the normal modes or the eigenoscillations of the medium coincide with the solutions of the field equations, Eq. (1.6), if no macroscopic currents and space charges are present at the beginning, i.e. $\boldsymbol{J}_0=0$ and $\varrho_0=0$. With arbitrary values of both, $\boldsymbol{J}_0$ and ϱ_0, the quantities ω and $\boldsymbol{k}$ are no longer connected by the dispersion relation; in that case the tensor $\mathsf{E}(\omega, \boldsymbol{k})$ describes the linear "answer" of the medium given to an electromagnetic "excitation field" which has the form

$$\boldsymbol{E} = \boldsymbol{E}_0 \exp(i\,\boldsymbol{k}\cdot\boldsymbol{r} - i\,\omega\,t)$$

with mutually independent ω and $\boldsymbol{k}$.

δ) Finally we want to discuss briefly the problem of *decrease* (or *increase*) in media with weak attenuation (or amplification) in relation to the initial and boundary value problems and the connection between these problems. This is a very important question for some considerations given below; it must therefore be reconsidered several times in following sections.

In a medium with weak attenuation the anti-Hermitean part of the tensor of permittivity [which describes wave absorption, see Eq. (3.7)] is small compared with the hermitean part. This means that in the dispersion relation, Eq. (4.3), the imaginary parts of the summands are small compared with the real ones, i.e.

$$\mathrm{Im}\,\mathscr{D}(\boldsymbol{k}, \omega) \ll \mathrm{Re}\,\mathscr{D}(\boldsymbol{k}, \omega)\,.$$

If now we look for the behaviour of the wave field in time (initial value problem), an approximate solution of Eq. (4.3) can be written in the form[4]

$$\omega \to \omega + i\gamma \tag{5.7}$$

[2] A medium without attenuation could well be non-transparent, for example in a case of internal total reflection, see Sect. 32 and [*1*] Sect. II.7. In such a case the wave amplitude decreases exponentially in the given direction.

[3] See K. RAWER and K. SUCHY: [*11*], Sect. 10 ε; see also [*8*], Sect. I.3.

[4] We do not use different designations for a complex frequency and for a real one. In cases were ω is complex we also write ω for $\mathrm{Re}\,\omega$ and γ for $\mathrm{Im}\,\omega$. Thus in the following

$$\omega \leftrightarrow \omega + i\gamma$$

according to conditions.

where $\omega(\boldsymbol{k})$ designates the real roots of

$$\operatorname{Re}\mathscr{D}(\boldsymbol{k},\omega)=0, \tag{5.8}$$

an equation which characterizes the frequency spectrum of the oscillations, see Sect. 4. In this approach the real quantity

$$\gamma(\boldsymbol{k})=-\frac{\operatorname{Im}\mathscr{D}(\boldsymbol{k},\omega)}{\frac{\partial}{\partial\omega}\operatorname{Re}\mathscr{D}(\boldsymbol{k},\omega)} \tag{5.9}$$

is the decrement of attenuation (or the increment of amplification) of the oscillations. If $\gamma(\boldsymbol{k})<0$ the wave energy is dissipated and the energy of the medium increases. If, however, $\gamma(\boldsymbol{k})>0$ the medium delivers energy to the wave, therefore the oscillation is amplified. It can be seen from Eq. (5.9) that the sign of $\gamma(\boldsymbol{k})$ depends fundamentally on the relative signs of the quantities $\operatorname{Im}\mathscr{D}(\boldsymbol{k},\omega)$ and $\partial/\partial\omega \operatorname{Re}\mathscr{D}(\boldsymbol{k},\omega)$.

ε) In order to analyze the *spatial decrease* of a wave in a given direction we have to solve the dispersion relation, Eq. (4.3), for $k_\vartheta(\omega)$:

$$k_\vartheta(\omega)=\operatorname{Re}k_\vartheta(\omega)+i\operatorname{Im}k_\vartheta(\omega). \tag{5.10}$$

Here $\operatorname{Re}k_\vartheta(\omega)$ designates the real roots of Eq. (5.8), and the quantity

$$\operatorname{Im}k_\vartheta(\omega)=-\frac{\operatorname{Im}\mathscr{D}(\boldsymbol{k},\omega)}{\frac{\partial}{\partial k_\vartheta}\operatorname{Re}\mathscr{D}(\boldsymbol{k},\omega)} \tag{5.11}$$

characterizes the spatial decrease (or increase) of the wave.

ζ) From Eqs. (5.9) and (5.11) one easily obtains a relation between time and space decrease:

$$\operatorname{Im}k_\vartheta=\frac{\gamma(\vartheta)}{V_g(\vartheta)};\qquad \gamma(\vartheta)=V_g(\vartheta)\cdot\operatorname{Im}k_\vartheta=\frac{\omega\chi}{\frac{\partial}{\partial\omega}(\omega\,\mu_\vartheta(\omega))} \tag{5.12}$$

where

$$V_g=\frac{\partial\omega}{\partial(\operatorname{Re}k_\vartheta)}$$

is the group velocity of the wave *in the direction* given by the angle ϑ. Eqs. (5.8) ... (5.11) can very directly be interpreted in the case of purely longitudinal waves in an isotropic medium, where we may write after Eq. (4.5):

$$\mathscr{D}(\boldsymbol{k},\omega)\equiv\varepsilon^{1}(\omega,\boldsymbol{k}).$$

For an isotropic medium, in the case of stability,

$$\operatorname{Im}\varepsilon^{1}(\omega,\boldsymbol{k})>0,$$

and the waves are evanescent in space as well as with increasing time, thus $\gamma<0$ and $\operatorname{Im}k_\vartheta>0$. A change of the sign of $\operatorname{Im}\varepsilon^{1}(\omega,\boldsymbol{k})$ however, may bring about a change of sign of γ and $\operatorname{Im}k_\vartheta$ and this means excitation of the wave or, equivalently, instability of the medium. Thus, we arrive again to the conclusion that the *characteristic condition for instability* of a medium is

$$\operatorname{Im}\varepsilon^{1}(\omega,\boldsymbol{k})<0. \tag{5.13}$$

This condition is only a necessary one. According to Eq. (5.9) a sufficient condition for instability must also take account of the sign of $\partial/\partial\omega \operatorname{Re}\mathscr{D}$.

II. Different descriptions of plasma.

Having discussed the general foundations of linear electrodynamics of material media of any kind we may now specify and investigate the properties of plasma. It is the precise aim of such investigations to obtain specific material equations so that Eqs. (1.8) are exact. We could also say that we try to obtain an expression for the dielectric permeability tensor $\mathsf{E}(\omega, \boldsymbol{k})$. In this chapter we shall — with one exception — discuss a few different models as are used to describe plasmas, and indicate the relevant ranges of validity.

6. Motion of individual particles.

α) In the simplest description of plasma one uses the hypothesis of individual *charged particles* moving freely in *fields of exterior origin*. For a particle with mass m and charge q the equations of motion are (in the non-relativistic approximation):

$$\left.\begin{aligned} \frac{\partial \boldsymbol{r}}{\partial t} &= \boldsymbol{v} \\ \frac{\partial \boldsymbol{v}}{\partial t} &= \frac{q}{m}\left\{\boldsymbol{E} + \frac{1}{c_0\sqrt{\varepsilon_0\mu_0}}\,\boldsymbol{v}\times\boldsymbol{B}\right\} + \boldsymbol{g} \end{aligned}\right\} \tag{6.1}$$

where $\boldsymbol{g}$ is the acceleration produced by forces of non-electromagnetic origin, such as the gravitational force. The integration of the system Eqs. (6.1) in the case of fields[1] $\boldsymbol{E}$, $\boldsymbol{B}$, which are arbitrarily inhomogeneous (in space) and variable (in time) presents a mathematically extremely difficult task. In particular, the inhomogeneity in space is a great difficulty. The work of H. ALFVÉN [*14*] describes general methods of approximate integration in the case of strong magnetic and weak electric fields (see also the survey given by D. V. SIVUHIN in [*6*]). In the literature this approach has been designated as the "drift approximation". In the following we shall give its basic ideas in a simple form for application.

β) We suppose that the field vectors $\boldsymbol{E}, \boldsymbol{B}$ and $\boldsymbol{g}$ are invariable in space and with time, $\boldsymbol{E}$ and $\boldsymbol{g}$ being both *transverse fields*, i.e. perpendicular to the magnetic field $\boldsymbol{B}$. Then the Eqs. (6.1) admit the following stationary solution for the particle velocity $\boldsymbol{v}$:

$$\boldsymbol{v} = \frac{c_0\sqrt{\varepsilon_0\mu_0}}{B^2}\left[\left(\boldsymbol{E} + \frac{m}{q}\,\boldsymbol{g}\right)\times\boldsymbol{B}\right]. \tag{6.2}$$

From the form of the right-hand side we see that under the influence of fields $\boldsymbol{E}$ and $\boldsymbol{g}$ (perpendicular to the magnetic field $\boldsymbol{B}$) the particles make a drift motion across the magnetic field $\boldsymbol{B}$. The velocity of the electric drift motion,

$$\boldsymbol{v}_{\boldsymbol{E}} = \frac{c_0\sqrt{\varepsilon_0\mu_0}}{B^2}\,\boldsymbol{E}\times\boldsymbol{B} \tag{6.2a}$$

is independent of the particle mass and charge. This means that under the influence of such a field electrons and ions describe a "bulk-motion" both drifting with the same velocity and in the same direction. We have vanishing electric current in this case.

On the other hand, under the influence of a non-electromagnetic force $\boldsymbol{g}$, where the corresponding field is independent of the sign of the particle charge (as in the case of a gravitational force) electrons and ions drift in opposite directions which gives rise to an electric current in the plasma. We shall see below that such a current may cause an instability in the plasma.

[1] As we do not introduce the magnetic field vector $\boldsymbol{H}$ we may use simplified language and call in the following the magnetic induction $\boldsymbol{B}$ simply the "magnetic field".

In the ionosphere the presence of the field $\boldsymbol{g}$ may be caused by the curvature of the magnetic field lines of Earth. Free motion of particles along field lines which have a non-vanishing curvature results in a centrifugal force with an equivalent acceleration field

$$g = \frac{V_T^2}{R_0}$$

where V_T is the thermal velocity of the particles and R_0 the radius of curvature of the magnetic field lines of Earth. (It is of the order of the Earth's radius, R_{E}.)

There may be another possible cause for a field $\boldsymbol{g}$ to appear in the ionospheric plasma, namely a non-vanishing radial pressure gradient. It produces a radial force equivalent to an acceleration field

$$g \approx V_T^2 / L_0 .$$

L_0 being a length characterizing the radial inhomogeneity of the plasma. We shall discuss these questions in more detail when considering the stability of an inhomogeneous plasma, see Sects. 36 and 37.

γ) A differential drift of charged particles of different signs in the plasma may also be caused by the action of a constant *electric field parallel* to the magnetic field $\boldsymbol{B}$. If non-electromagnetic forces are neglected ($\boldsymbol{g} \equiv 0$) Eqs. (6.1) lead to a speed $\boldsymbol{v}$ parallel to the field and increasing linearly with time[2]:

$$\boldsymbol{v}_{\parallel} = \frac{q}{m} \boldsymbol{E}\, t. \tag{6.3}$$

If, however, the charged particle undergoes a friction-force during its motion in the plasma then the average result of collisions with particles of another kind, for example with neutral ones, can be described by an average "field" caused by friction

$$\boldsymbol{g} = -\bar{\nu}\, \boldsymbol{v}$$

where $\bar{\nu}$ is an averaged collision frequency[3]. The drift speed then reaches a limiting value which is stationary:

$$\boldsymbol{v}_{\parallel} = \frac{q \boldsymbol{E}}{m \bar{\nu}} . \tag{6.4}$$

δ) Using Eqs. (6.1) for electrons and for ions separately one can also determine the current which is induced in a plasma by the action of an *alternating electromagnetic* (high frequency) *field*. This procedure yields the tensor of dielectric premeability. The integration of the system Eqs. (6.1) is possible in this case due to the weakness of the electromagnetic wave field and its slow spatial changes (in the case of a high frequency). Looking for a solution in terms of a plane wave with wave-number vector $\boldsymbol{k}$ and frequency $\omega / 2\pi$:

$$\exp(i \boldsymbol{k} \cdot \boldsymbol{r} - i \omega t)$$

supposing $\omega \gg \bar{\nu}$, one may neglect the dependence of the wave fields $\boldsymbol{E}$ and $\boldsymbol{B}$ on the spatial coordinates, linearize the system of Eqs. (6.1), and have a stationary solution in the following form giving the complex amplitude of the wave motion:

$$\left.\begin{aligned} \boldsymbol{v} &= \frac{q}{m} \frac{(\bar{\nu} - i\omega)}{\omega_B^2 + (\bar{\nu} - i\omega)^2} \times \\ &\times \left\{ \boldsymbol{E} + \frac{\omega_B}{(\bar{\nu} - i\omega)} \left(\frac{1}{B_{♁}} \boldsymbol{E} \times \boldsymbol{B}_{♁} + \frac{\omega_B}{(\bar{\nu} - i\omega)} \frac{1}{B_{♁}^2} (\boldsymbol{E} \cdot \boldsymbol{B}_{♁})\, \boldsymbol{B}_{♁} \right) \right\} . \end{aligned}\right\} \tag{6.5}$$

---

[2] Our considerations are only concerned with non-relativistic particles here.

[3] K. Rawer and K. Suchy: this Encyclopedia, vol. 49/2, Sects. 1 and 3.

Here $\boldsymbol{B}_{\circ}$ is the constant exterior magnetic field, $\boldsymbol{E}$ the electric wave field and

$$\omega_B = \frac{q\, B_{\circ}}{m\, c_0 \sqrt{\varepsilon_0\, \mu_0}} \tag{6.6}$$

is the gyro-pulsation.

An eventually present exterior (constant) electric field does not influence the oscillatory movement of the particles with frequency $\omega/2\pi$. When deriving Eq. (6.5) we have supposed that $\boldsymbol{g} = -\bar{\nu}\, \boldsymbol{v}$, approximately, i.e., we introduced an averaged effective friction force acting on charged particles of a given kind, for example due to collisions with neutral particles[3].

ε) The current density $\boldsymbol{J}$ induced in the plasma by the motion of charged particles of kinds e (electrons) and i (ions) is

$$\boldsymbol{J} = \sum_{\mathrm{h=e,\,i}} q_{\mathrm{h}}\, N_{\mathrm{h}}\, \boldsymbol{v}_{\mathrm{h}}. \tag{6.7}$$

N_{h} being the number density of charged particles of kind h. The summation goes over all kinds of charged particles which are present. Putting Eq. (6.5) into Eq. (6.7), using Eqs. (1.9) and (1.11) one finds the *conductivity* and the *permittivity (dielectric permeability) of the plasma for a wave with frequency* $\omega/2\pi$. The result is[4] the dielectric permeability tensor

$$\left.\begin{aligned} \frac{1}{\varepsilon_0}\mathsf{E} \equiv \frac{1}{\varepsilon_0}((\varepsilon_{lm})) = \mathsf{U} - \sum_{\mathrm{h=e,\,i}} \frac{\omega_{N\mathrm{h}}^2(\omega + i\bar{\nu}_{\mathrm{h}})}{\omega[(\omega + i\bar{\nu}_{\mathrm{h}})^2 - \omega_{B\mathrm{h}}^2]} \times \\ \times \left\{\mathsf{U} + \frac{i\,\omega_{B\mathrm{h}}}{(\omega + i\bar{\nu}_{\mathrm{h}})}\left[\mathfrak{E}\cdot\boldsymbol{B}_{\circ}^0 + \frac{i\,\omega_{B\mathrm{h}}}{(\omega + i\bar{\nu}_{\mathrm{h}})}\boldsymbol{B}_{\circ}^0\,\boldsymbol{B}_{\circ}^0\right]\right\}, \end{aligned}\right\} \tag{6.8}$$

$$\left.\begin{aligned} \frac{1}{\varepsilon_0}\varepsilon_{lm}(\omega, \boldsymbol{k}) = \delta_{lm} - \sum_{\mathrm{h=e,\,i}} \frac{\omega_{N\mathrm{h}}^2(\omega + i\bar{\nu}_{\mathrm{h}})}{\omega[(\omega + i\bar{\nu}_{\mathrm{h}})^2 - \omega_{B\mathrm{h}}^2]} \times \\ \times \left\{\delta_{lm} + \frac{i\,\omega_{B\mathrm{h}}}{(\omega + i\bar{\nu}_{\mathrm{h}})}\left[\sum_n \varepsilon_{lmn}\, B_{\circ n}/B_{\circ} + \frac{i\,\omega_{B\mathrm{h}}}{(\omega + i\bar{\nu}_{\mathrm{h}})} B_{\circ l}\, B_{\circ m}/B_{\circ}^2\right]\right\}. \end{aligned}\right\} \tag{6.8a}$$

Here the following notations have been used:

$$\omega_{B\mathrm{h}} = \frac{q_{\mathrm{h}}}{m_{\mathrm{h}}\, c_0 \sqrt{\varepsilon_0\, \mu_0}}\, B_{\circ}. \qquad [6.6]$$

$\omega_{B\mathrm{h}}/2\pi$ being the gyro- or LARMOR *frequency* of particles of kind h;

$$\omega_{N\mathrm{h}} = \sqrt{\frac{\mathrm{u}\, q_{\mathrm{h}}^2\, N_{\mathrm{h}}}{\varepsilon_0\, m_{\mathrm{h}}}}. \tag{6.9}$$

$\omega_{N\mathrm{h}}/2\pi$ being the LANGMUIR *frequency* of particles of kind h (in the particular case of electrons, h = e, the LANGMUIR frequency is also called plasma-frequency)[5]. Finally, $B_{\circ l}$, $B_{\circ m}$ and $B_{\circ n}$ are the cartesian components of the magnetic field in an arbitrary righthanded frame.

$$\mathfrak{E} \equiv (((\varepsilon_{lmn}))) \tag{6.10}$$

[4] K. RAWER and K. SUCHY: this Encyclopedia, vol. 49/2, Sects. 5 and 6, in particular Eqs. (5.3) ... (5.5) and (6.16) ... (6.18).

[5] u ≡ 1 for rationalized, ≡ 4π for non-rationalized systems of units (see "Introductory Remarks" on p. 1).

is the usual *permutation tensor* of third grade: ε_{lmn} equals ± 1 according as $l\,m\,n$ is an even or odd permutation, respectively, on 1 2 3, and zero otherwise.

If we take the z^0-axis of a system of cartesian coordinates in the direction $\boldsymbol{B}^0_\delta$ of the magnetic field, then the tensor, Eq. (6.8), simplifies to:

$$\mathsf{E} \equiv ((\varepsilon_{lm})) = \left(\begin{pmatrix} \varepsilon_{xx} & \varepsilon_{xy} & 0 \\ \varepsilon_{yx} & \varepsilon_{yy} & 0 \\ 0 & 0 & \varepsilon_{zz} \end{pmatrix}\right) \tag{6.11}$$

with

$$\left.\begin{aligned} \frac{1}{\varepsilon_0}\,\varepsilon_{xx} &= \frac{1}{\varepsilon_0}\,\varepsilon_{yy} = 1 - \sum_{\mathrm{h}} \frac{\omega^2_{N\mathrm{h}}(\omega + i\,\bar{\nu}_{\mathrm{h}})}{\omega\,[(\omega + i\,\bar{\nu}_{\mathrm{h}})^2 - \omega^2_{B\mathrm{h}}]}\,, \\ \frac{1}{\varepsilon_0}\,\varepsilon_{xy} &= -\frac{1}{\varepsilon_0}\,\varepsilon_{yx} = -i \sum_{\mathrm{h}} \frac{\omega^2_{N\mathrm{h}}\,\omega_{B\mathrm{h}}}{\omega\,[(\omega + i\,\bar{\nu}_{\mathrm{h}})^2 - \omega^2_{B\mathrm{h}}]}\,, \\ \frac{1}{\varepsilon_0}\,\varepsilon_{zz} &= 1 - \sum_{\mathrm{h}} \frac{\omega^2_{N\mathrm{h}}}{\omega\,(\omega + i\,\bar{\nu}_{\mathrm{h}})}\,. \end{aligned}\right\} \tag{6.12}$$

ζ) When discussing the kinetic plasma model in Sect. 8 below we shall classify more strictly the *conditions for applying* the simple system of Eqs. (6.8) ... (6.12), and for using the basic approximation of "individual particles", too. We may be allowed to state here only that, when deriving these formulae from Eqs (6.1), we have neglected spatial changes of the wave fields $\boldsymbol{E}$ and $\boldsymbol{B}$ "in the localization range of charged particles".

We should yet explain what that means. The charged particles of a plasma, except from the regular (periodic) motion induced by the wave-field, are subject to (apparently irregular) thermal motions, too. As in almost all applications (and certainly in our approximation) the wave fields are small, the average (absolute) speed values due to thermal motions is much greater than that due to the regular one. Therefore, the "localization" of a particle is not given by the amplitude of the regular oscillation but by the path-length obtained from thermal motion during a time-element $1/\omega$, thus for motion along the magnetic field $\boldsymbol{B}_\delta$:

$$L_\parallel \sim V_{T\mathrm{h}}/\omega\,, \tag{6.13a}$$

$V_{T\mathrm{h}}$ being the average thermal velocity of particles of kind h. The situation is completely different in the transverse direction, for which the LARMOR radius determines a (generally better) localization with radius

$$L_\perp \sim V_{T\mathrm{h}}/\omega_{B\mathrm{h}} \equiv r_{B\mathrm{h}}\,. \tag{6.13b}$$

In order that the inhomogeneity of the wave field be negligible we have to satisfy the necessary conditions that $L_\parallel$ as well as $L_\perp$ are small compared to the wavelength λ in the plasma. Using wave-numbers we come to the conditions

$$\frac{k_\parallel V_{T\mathrm{h}}}{\omega} \ll 1\,; \qquad \frac{k_\perp V_{T\mathrm{h}}}{\omega_{B\mathrm{h}}} \ll 1\,. \tag{6.14a, b}$$

This signifies that Eqs. (6.8) ... (6.12) can only be applied for waves the phase velocity of which is considerably greater than the thermal velocity of the relevant particles, and that the wavelength is much greater than the LARMOR radii. It can be seen from Tables 1 ... 3 that, for example in the case of ionospheric plasma the conditions (6.14a, b) are quite generally satisfied with waves the wavelength of which is longer than a few cm, the frequency being greater than 10^5 Hz[6].

Another limitation of the applicability of our Eqs. (6.8) ... (6.12) is connected with the introduction of the effective collision frequency $\bar{\nu}_h$, which is closely related with the collision frequency of the charged particles of kind h with those of other kinds for example with neutral ones. The "individual particle approach" does not take account of the interrelations between charged particles of the same kind. The approximation given by a unique "effective collision frequency" and an effective frictional force, $-\bar{\nu}\,\boldsymbol{v}$, is not very good for collisions between charged particles of different kinds, either[7]. Strictly speaking, the present approximation is satisfying only if, apart from Eqs. (6.14a, b) also the following conditions are valid

$$\bar{\nu}_{en} \gg \begin{cases} \bar{\nu}_{ei} \\ \bar{\nu}_{ee} \end{cases}; \quad \bar{\nu}_{in} \gg \begin{cases} \bar{\nu}_{ie} \\ \bar{\nu}_{ii} \end{cases} \tag{6.15a, b}$$

i.e. if the collisions with neutrals give the largest effects. In the case of the ionospheric plasma these conditions are only satisfied up to 200 km. The "individual particle model" is, therefore, only a low altitude approximation[8]. With some precautions it may be applied similarly to altitudes up to 700 km, where electron-ion collisions are predominant.

7. Fluid-dynamic plasma model.

α) The approach just opposite to the "individual particle approximation" is a "quasi-hydrodynamic" plasma model. This model starts with the hypothesis that the correlations between the behaviour of charged particles among themselves, and the neutrals is so strong that the plasma can be considered as behaving like a conducting fluid. In the general case of *magnetofluid dynamics*, combining the equations of electrodynamics with the equations of motion in a compressible, conductive fluid, one has the following system of basic equations, see [1], [17]:

$$\left.\begin{aligned}
&\frac{\partial \boldsymbol{B}}{\partial t} = \frac{\partial}{\partial \boldsymbol{r}} \times \left(\boldsymbol{v} \times \boldsymbol{B} - \frac{c_0^2\,\varepsilon_0\,\mu_0}{\mathrm{u}\,\mu_0\,\sigma}\,\frac{\partial}{\partial \boldsymbol{r}} \times \boldsymbol{B}\right),\\
&\frac{\partial}{\partial \boldsymbol{r}} \cdot \boldsymbol{B} = 0, \quad \frac{\partial \bar{\varrho}}{\partial t} + \frac{\partial}{\partial \boldsymbol{r}} \cdot (\bar{\varrho}\,\boldsymbol{v}) = 0,\\
&\bar{\varrho}\left(\frac{\partial \boldsymbol{v}}{\partial t} + \left(\boldsymbol{v} \cdot \frac{\partial}{\partial \boldsymbol{r}}\right)\boldsymbol{v}\right) = -\frac{\partial}{\partial \boldsymbol{r}} P - \frac{1}{\mathrm{u}\,\mu_0}\,\boldsymbol{B} \times \left(\frac{\partial}{\partial \boldsymbol{r}} \times \boldsymbol{B}\right) +\\
&\quad + \bar{\varrho}\,\boldsymbol{g} + \eta\left(\frac{\partial}{\partial \boldsymbol{r}}\right)^2 \boldsymbol{v} + \left(\xi + \frac{1}{3}\,\eta\right)\frac{\partial}{\partial \boldsymbol{r}}\left(\frac{\partial}{\partial \boldsymbol{r}} \cdot \boldsymbol{v}\right).
\end{aligned}\right\} \tag{7.1}$$

In the last equation the term in parenthesis on the left side means "substantial differentiation", D/Dt.

Here $\bar{\varrho}$ is the density of the fluid, σ its conductivity, η and ξ are two viscosity coefficients. All exterior non-electrodynamic forces are summarized in the term $\bar{\varrho}\,\boldsymbol{g}$.

The system Eqs. (7.1) must be completed by two more equations: the equation of state of the fluid giving a relation between pressure, P, density $\bar{\varrho}$, and

[6] A few (very rare) exceptions are indicated by K. RAWER and K. SUCHY: this encyclopedia, vol. 49/2, in Sect. 8.

[7] A quite general discussion of collisional effects is found in K. SUCHY: Ergeb. Exact. Naturw. **35**, 103—294 (1964).

[8] Collisions between charged particles of different kinds can approximately be taken account of by a frictional force if some precautions are respected, see K. RAWER and K. SUCHY: this Encyclopedia, vol. 49/2, Sect. 3 in particular Subject δ and ε. See [1] for more details.

temperature, T, and the equation of heat (and energy) transport:

$$P = P(\bar{\varrho}, T), \tag{7.2}$$

$$\left.\begin{aligned} \bar{\varrho}\, T\left(\frac{\partial S}{\partial t} + \left(\boldsymbol{v}\cdot\frac{\partial}{\partial \boldsymbol{r}}\right) S\right) &= \frac{\partial}{\partial \boldsymbol{r}}\cdot\left(\varkappa\,\frac{\partial}{\partial \boldsymbol{r}}\, T\right) + \frac{c_0^2\,\varepsilon_0\,\mu_0}{\mathrm{u}\,\mu_0\,\sigma}\left|\frac{\partial}{\partial \boldsymbol{r}}\times\boldsymbol{B}\right|^2 \\ &\quad + \mathsf{T}\cdot\cdot\left(\frac{\partial}{\partial \boldsymbol{r}}\,\boldsymbol{v}\right), \end{aligned}\right\} \tag{7.3}$$

where $\mathsf{T} \equiv ((\vartheta_{jk}))$ is a tensor describing frictional effects, S is the entropy per mass unit, and $\varkappa$ is the heat conductivity of the fluid. The last term in Eq. (7.3) describes heating by viscosity because the coefficients ϑ_{jk} are that of the viscosity tensor T[1]:

$$\mathsf{T} \equiv ((\vartheta_{jk})) = \eta\left[\left(\frac{\partial}{\partial \boldsymbol{r}}\,\boldsymbol{v}\right) + \left(\frac{\partial}{\partial \boldsymbol{r}}\,\boldsymbol{v}\right)^{\mathrm{T}}\right] + \left(\xi - \frac{2}{3}\,\eta\right)\left(\frac{\partial}{\partial \boldsymbol{r}}\cdot\boldsymbol{v}\right)\mathsf{U}, \tag{7.4}$$

i.e.

$$\vartheta_{jk} = \eta\left(\frac{\partial v_j}{\partial r_k} + \frac{\partial v_k}{\partial r_j} - \frac{2}{3}\,\delta_{jk}\left(\frac{\partial}{\partial \boldsymbol{r}}\cdot\boldsymbol{v}\right)\right) + \xi\,\delta_{jk}\left(\frac{\partial}{\partial \boldsymbol{r}}\cdot\boldsymbol{v}\right). \tag{7.4a}$$

β) The system of Eqs. (7.1) ... (7.3), which form the basic equations of magneto-fluid-dynamics, has some *deficiencies* when applied to plasmas. First the coefficients of transport, namely σ, ξ, η, $\varkappa$, and the tensor T, can only give a phenomenological description. In order to determine these 13 coefficients the microscopic structure of the fluid should be studied, and the kinetic equations be solved (see [*15*]). Second, there is a serious difference between a plasma and a "classical fluid", for example mercury, where a really microscopic description is hardly possible but is not really needed either, because reliable phenomenological coefficients can be applied without difficulty. The situation is very different in the case of a plasma where the fluid model neither describes the motions of the individual components — electrons, ions and neutrals —, nor provides an easy description with phenomenological coefficients. Therefore, one has often used a generalized dynamic system with two or three fluids. We shall discuss this approximation below in Sect. 15, after introducing the kinetic equation with self-consistent field in Sect. 8, and the collisional integral of the charged particles in Sect. 12.

Finally, the most annoying deficiency of the system of Eqs. (7.1) ... (7.3) is that it is not applicable for describing a real plasma; in particular it does not allow for propagation of electromagnetic waves in plasma. The reason is that the system is only valid for "high density", i.e. in cases where any free path is small compared to any characteristic length. In the frequency notation of plasma physics this condition requires collisional frequencies to be much larger than any characteristic frequencies, such that[2]

$$\omega \ll \bar{\nu}_{\mathrm{h}} \quad \text{and} \quad \omega_{B\mathrm{h}} \ll \bar{\nu}_{\mathrm{h}}, \qquad (\mathrm{h} = \mathrm{e}, \mathrm{i}). \tag{7.5}$$

It may be seen from Tables 1 ... 3 that the second condition is *not* satisfied for electrons in the ionosphere, and certainly not in the magnetosphere.

There are, on the other hand, quite a few cases where the results obtained with magneto-fluid-dynamics appear to be acceptable in a rarified plasma, even if the conditions of Eqs. (7.5) appear not to be satisfied. In any specific case the reasons for this behaviour are easily understood. We shall describe in Sect. 11 below which results of the strict kinetic theory can also be obtained in the framework of magneto-fluid-dynamics.

[1] Upper index ^{T} designates the transposed tensor.

[2] In reduced quantities X, Y, Z (see Sect. 25γ) this means

$$Z_{\mathrm{h}} \gg 1 \quad \text{and} \quad Y_{\mathrm{h}} \ll Z_{\mathrm{h}}. \quad X \equiv (\omega_N/\omega)^2; \quad Y \equiv \omega_B/\omega; \quad Z \equiv \bar{\nu}/\omega.$$

8. Kinetic equations with self-consistent field.

α) A plasma being essentially a gas containing charged particles, the most general description can be expected from appropriate kinetic gas theory. This method uses a description in terms of *probability distribution* applied to particles exerting some interaction amongst themselves. Applying the methods of statistical physics, see [*16*], we introduce a distribution function

$$f(t, \ldots \boldsymbol{r}_{\mathrm{s}} \ldots, \ldots p_{\mathrm{s}} \ldots)$$

characterizing the distribution of all particles in the plasma in position (coordinates $\boldsymbol{r}_{\mathrm{s}}$) and momentum ($\boldsymbol{p}_{\mathrm{s}}$). The function f thus depends on a very large number of variables. In spite of this difficulty, we can obtain a convenient kinetic description because the interaction is quite weak in geophysical and cosmical plasmas.

β) This permits a "*thin gas approximation*", i.e. a low density approximation. In all plasmas which we have to consider here, the average energy due to COULOMB interaction of charged particles is quite small compared with the kinetic energy $\mathscr{E}_{\mathrm{kin}}$ of the particle. We use $\vartheta_{\mathrm{h}} = k\,T_{\mathrm{h}}$ as a measure for the average thermal energy of a particle[1].

$$\frac{\mathrm{u}}{4\pi\,\varepsilon_0}\,\frac{q_{\mathrm{h}}^2}{\bar{r}_{\mathrm{h}}} \approx \frac{\mathrm{u}}{4\pi\,\varepsilon_0}\,q_{\mathrm{h}}^2\,N_{\mathrm{h}}^{\frac{1}{3}} \ll \mathscr{E}_{\mathrm{kin\,h}} \approx \vartheta_{\mathrm{h}}\,. \tag{8.1}$$

Here $\bar{r}_{\mathrm{h}}$ is the average distance between one of the (charged) particles considered, and one of opposite sign, so that $\bar{r}_{\mathrm{h}}^{-1} \approx N_{\mathrm{h}}^{\frac{1}{3}}$ for neutrality. If this condition is satisfied the motion of individual particles is nearly independent. Therefore we may consider a development after the parameter

$$q_{\mathrm{h}}^2\,N_{\mathrm{h}}^{\frac{1}{3}}/\vartheta_{\mathrm{h}}$$

and find in zero order approximation the distribution function as a product of "individual particle factors",

$$f(t, \ldots \boldsymbol{r}_{\mathrm{w}} \ldots, \ldots \boldsymbol{p}_{\mathrm{w}} \ldots) \approx \prod_{\mathrm{w}} f(t, \boldsymbol{r}_{\mathrm{w}}, \boldsymbol{p}_{\mathrm{w}})\,, \tag{8.2}$$

$f(t, \boldsymbol{r}_{\mathrm{w}}, \boldsymbol{p}_{\mathrm{w}})$ being the probability for the particle w to be at the moment t in the spatial cell $\boldsymbol{r}_{\mathrm{w}}$ and momentum cell $\boldsymbol{p}_{\mathrm{w}}$.

As spatial variations are less important for our considerations we define the distribution function f in a way that the integral over the momentum space gives the numerical particle density, N, thus[2]

$$\int \mathrm{d}^3 p\, f = N. \tag{8.3}$$

γ) The conservation of the total number of particles[3] leads to LIOUVILLE's *equation*[4] (dropping the index w):

$$\frac{\mathrm{d}f}{\mathrm{d}t} = \frac{\partial f}{\partial t} + \frac{\partial f}{\partial \boldsymbol{r}}\cdot\frac{\mathrm{d}\boldsymbol{r}}{\mathrm{d}t} + \frac{\partial f}{\partial \boldsymbol{p}}\cdot\frac{\mathrm{d}\boldsymbol{p}}{\mathrm{d}t} = 0\,. \tag{8.4}$$

---

[1] $k = 1.38 \cdot 10^{-23}$ J/°K is the Boltzmann constant.

[2] The function f has $(\mathrm{Js})^{-3} \equiv \mathrm{m}^{-3}(\mathrm{Ns})^{-3} \equiv \mathrm{kg}^{-3}\,\mathrm{m}^{-6}\,\mathrm{sec}^{+3}$ for SI-unit and $\boldsymbol{p}$ has $(\mathrm{Ns}) \equiv \mathrm{kg\,m\,sec}^{-1}$, such that $\int \mathrm{d}^3\boldsymbol{p}\, f$ has m^{-3}, $\int \mathrm{d}^3\boldsymbol{r}\, f$ has $(\mathrm{Ns})^{-3}$ as unit, and $\int \mathrm{d}^3\boldsymbol{r} \int \mathrm{d}^3\boldsymbol{p}\, f$ is dimensionless.

[3] A more detailed discussion concerning the "conservation of phase volume" is found in [*17*].

[4] K. RAWER and K. SUCHY: this Encyclopedia, vol. 49/2, Sect. 2, use the individual particle velocity $\boldsymbol{c}$, instead of momentum $\boldsymbol{p}$.

Of course, $\boldsymbol{v}=\mathrm{d}\boldsymbol{r}/\mathrm{d}t$ is the particle velocity (for non-relativistic particles $\boldsymbol{p}=m\,\boldsymbol{v}$), and $\mathrm{d}\boldsymbol{p}/\mathrm{d}t$ describes the effect of forces; for charged particles

$$\frac{\mathrm{d}\boldsymbol{p}}{\mathrm{d}t}=\boldsymbol{F}=q\left[\boldsymbol{E}+\frac{1}{c_0\sqrt{\varepsilon_0\,\mu_0}}\,\boldsymbol{v}\times\boldsymbol{B}\right], \qquad [1.2]$$

the generalized LORENTZ force. We obtain therefore from Eq. (8.4)

$$\frac{\partial f}{\partial t}+\boldsymbol{v}\cdot\frac{\partial f}{\partial \boldsymbol{r}}+q\left[\boldsymbol{E}+\frac{1}{c_0\sqrt{\varepsilon_0\,\mu_0}}\,\boldsymbol{v}\times\boldsymbol{B}\right]\cdot\frac{\partial f}{\partial \boldsymbol{p}}=0. \qquad (8.5)$$

One equation of this kind has to be written down for each kind of particles, h, which are present in the plasma.

δ) The *current density*, $\boldsymbol{J}$, and *charge density*, ϱ, can be obtained from the statistical distribution function by adequate integration, and summation over all kinds of particles:

$$\boldsymbol{J}=\sum_{\mathrm{h}} q_{\mathrm{h}}\int \mathrm{d}^3\boldsymbol{p}\,\boldsymbol{v}\,f_{\mathrm{h}}, \qquad (8.6)$$

$$\varrho=\sum_{\mathrm{h}} q_{\mathrm{h}}\int \mathrm{d}^3\boldsymbol{p}\,f_{\mathrm{h}}. \qquad (8.7)$$

Finally, we have the normalization integral

$$N_{\mathrm{h}}(\boldsymbol{r},t)=\int \mathrm{d}^3\boldsymbol{p}\,f_{\mathrm{h}} \qquad [8.3]$$

which describes local particle densities.

The kinetic Eq. (8.5), together with Eqs. (8.3), (8.6), (8.7) and MAXWELL'S equations, Eqs. (1.1), constitute a complete system of equations describing the motion of the plasma particles as being coordinated with the forces produced by the motion itself. In this context we call Eq. (8.5) the kinetic equation with self-consistent field. The utility of this formulation for describing the electromagnetic properties of plasmas was first clarified by VLASOV [*18*] and by LANDAU [*19*].

ε) Strictly speaking, our introduction of the kinetic equation with self-consistent field is not precise enough — it is not even completely consistent. A precise and consistent deduction may be found quite generally in BOGOLJUBOV'S statistical theory [*20*], and in more special monographs [*13*], [*21*]. One may also find there how, in a higher approximation, the effect of correlation between neighbouring particles, i.e. that of interaction by collisions can be introduced. When these correlations between the motions of different particles are introduced there appears, instead of zero, a *collision sum* (or integral) on the right side of Eq. (8.5). We obtain

$$\frac{\partial f_{\mathrm{h}}}{\partial t}+\boldsymbol{v}\cdot\frac{\partial f_{\mathrm{h}}}{\partial \boldsymbol{r}}+q_{\mathrm{h}}\left[\boldsymbol{E}+\frac{1}{c_0\sqrt{\varepsilon_0\,\mu_0}}\,\boldsymbol{v}_{\mathrm{h}}\times\boldsymbol{B}\right]\cdot\frac{\partial f_{\mathrm{h}}}{\partial \boldsymbol{p}}=\left(\frac{\delta f_{\mathrm{h}}}{\delta t}\right)_{\mathrm{col}}=\sum_{\mathrm{j}}\left(\frac{\delta f_{\mathrm{h}}}{\delta t}\right)_{\mathrm{col}}^{(\mathrm{hj})}. \qquad (8.8)$$

The expression on the right side, called collision integral after BOLTZMANN, will be discussed in Chap. B. II.

ζ) We may only note here that

$$\left(\frac{\delta f_{\mathrm{h}}}{\delta t}\right)_{\mathrm{col}}^{(\mathrm{h\,j})}\propto \bar{\nu}_{\mathrm{h\,j}}\,f_{\mathrm{h}} \qquad (8.9)$$

where $\bar{\nu}_{hj}$ is the average collision frequency[5] of particles of kind h with those of kind j. We may therefore state that the zero order equation, i.e. the kinetic one with self-consistent field, Eq. (8.5), is only useful for describing quick processes developing in a time which is shorter than the free flight time of a particle, i.e. the time between two subsequent collisions. In terms of frequencies, the condition for the present theory is

$$\omega \gg \bar{\nu}_h \approx \sum \bar{\nu}_{hj}, \quad \text{i.e.} \quad Z_h \ll 1. \tag{8.10}$$

Under these condition the plasma is, roughly speaking, free from collisions so that the computation of the collision integral gives only very small contributions in Eq. (8.8). This is an approximation for very thin plasma.

In more detailed deductions of Eqs. (8.5) or (8.8) the question of the effective field $\boldsymbol{E}_{\text{eff}}$ is discussed[6]. The result is that in plasma the effective field $\boldsymbol{E}_{\text{eff}}$ is identical with the statistical average field, $\boldsymbol{E}$. Otherwise we would have had to introduce a corrected field $\boldsymbol{E}_{\text{eff}}$ in Eqs. (8.5), (8.8) as, for example, is required for the case of dielectric material with polarizable molecules. There are different ways to prove that this is not so in plasmas.

9. Quasi-gasdynamic approximation for a collisionless plasma.

α) We shall now demonstrate that under certain conditions, even if collisions are completely absent, i.e. in the case of Eq. (8.10), a quasi-gasdynamic description of plasma may be acceptable. We shall start with the kinetic equation with self-consistent field, Eq. (8.5), which is valid in the absence of collisional effects. We are now looking for a *closed solution* in which only macroscopic quantities are present; these appear as weighted average over the momentum space, such as

$$N_h(\boldsymbol{r}, t) = \int d^3\boldsymbol{p}\, f_h \tag{[8.2]}$$

and

$$N_h(\boldsymbol{r}, t)\, \boldsymbol{v}_h(\boldsymbol{r}, t) = \int d^3\boldsymbol{p}\, \boldsymbol{v}\, f_h. \tag{9.1}$$

These quantities are particle density and particle flux, respectively.

Provided one has a chance to obtain a solution in closed form, the quasi-gas-dynamic equations have a few advantages against the kinetic equations. In fact, the kinetic treatment is considerably more involved than fluid mechanics. Note that the gas-dynamic quantities $N_h(\boldsymbol{r}, t)$ and $\boldsymbol{v}_h(\boldsymbol{r}, t)$ depend only on the two variables $\boldsymbol{r}$ and t, while the distribution function $f_h(\boldsymbol{r}, \boldsymbol{p}, t)$ depends on three.

To avoid confusion we should clarify the terminology. Equations in which the fluid is characterized not only by a velocity field $\boldsymbol{v}(\boldsymbol{r}, t)$ as well as by a variable density field $\bar{\varrho}(\boldsymbol{r}, t)$ are called fluid-dynamic [see Eq. (7.1)] or, better, gasdynamic equations. Quasi-gasdynamic means that velocity and density fields exist seperately for each sort of particles. We may, in the following, omit the word "quasi".

β) By integrating the kinetic equation, Eq. (8.5), over the momentum space, one easily obtains the *gasdynamic equations* for electrons (h=e) and ions (h=i):

$$\frac{\partial N_h}{\partial t} + \frac{\partial}{\partial \boldsymbol{r}} \cdot (N_h \boldsymbol{v}_h) = 0, \tag{9.2}$$

$$\frac{\partial}{\partial t}(N_h \boldsymbol{v}_h) + \frac{\partial}{\partial \boldsymbol{r}} \cdot \mathsf{V}_h = \frac{q_h N_h}{m_h}\left[\boldsymbol{E} + \frac{1}{c_0\sqrt{\varepsilon_0 \mu_0}}\, \boldsymbol{v}_h \times \boldsymbol{B}\right] \tag{9.3}$$

with the velocity tensor[1]

$$\mathsf{V}_h(\boldsymbol{r}, t) = \int d^3\boldsymbol{p}\, \boldsymbol{v}\boldsymbol{v}\, f_h = \frac{1}{m_h^2}\int d^3\boldsymbol{p}\boldsymbol{p}\boldsymbol{p}\, f_h. \tag{9.4}$$

---

[5] A detailed discussion of this notion is found in K. SUCHY: Ergeb. Exakt. Naturw. **35**, 103—294 (1964); also K. SUCHY and K. RAWER: J. Atmosph. Terr. Phys. **33**, 1853—1868 (1971). For an abstract see K. RAWER and K. SUCHY: this Encyclopedia, vol. 49/2, Sect. 4, and the "Introductory Remarks" in this volume, p. 1.

[6] See [1] and the literature indicated there.

[1] V has $m^{-1}\,sec^{-2}$ for SI-unit.

Eq. (9.2) is the continuity equation while Eq. (9.3) is the dynamic equation. However, a "closed gasdynamic solution" can not be obtained yet because the velocity tensor V_h is not determined by the gasdynamic quantities number density, Eq. (8.2), and particle flux, Eq. (9.1). In the general case it appears to be extremely difficult to determine the tensor V_h. One knows, however, two particular cases where this tensor can be computed and closed solutions of the gasdynamic equations can therefore be obtained. These cases will be discussed in Sects. 10 and 11.

10. Cold plasma approximation.

α) We are now looking for processes where the characteristic velocity is large compared to the thermal velocity of electrons and ions, i.e.

$$\frac{L_k}{\tau} \approx \frac{\omega}{|\boldsymbol{k}|} \gg V_{T\mathrm{h}} \tag{10.1}$$

where[1]

$$V_{T\mathrm{h}} = \sqrt{\frac{\vartheta_\mathrm{h}}{m_\mathrm{h}}}, \quad (\vartheta_\mathrm{h} = k\,T), \tag{10.2}$$

is the average thermal velocity of particles of kind h. The characteristic velocity of the process may be given as ratio of a characteristic length L_k (which may be given by the reciprocal value of a wave number $\boldsymbol{k}$), and a characteristic time τ (eventually given by $1/\omega$). If Eq. (10.1) is valid then the thermal velocity can be neglected. This is the precise meaning of "cold plasma".

In this case the distribution function f is particularly simple, namely it is a Dirac distribution δ:

$$f_\mathrm{h} \sim \delta(\boldsymbol{v} - \boldsymbol{v}_\mathrm{h}). \tag{10.3}$$

δ(α) is defined by the transition to the limit of a series of functions so that

$$\lim_{\alpha \neq 0} \delta(\alpha) = 0 \quad \text{but} \quad \int_{-\infty}^{+\infty} \mathrm{d}\alpha\, \delta(\alpha)$$

remains constant throughout.

Consequently, the tensor $V_\mathrm{h} = N_\mathrm{h}\,\boldsymbol{v}_\mathrm{h}\,\boldsymbol{v}_\mathrm{h}$, Eq. (9.4), is extremely simple and can easily be combined with the dynamic Eq. (9.3) in the form

$$\frac{\partial \boldsymbol{v}_\mathrm{h}}{\partial t} + \left(\boldsymbol{v}_\mathrm{h} \cdot \frac{\partial}{\partial \boldsymbol{r}}\right) \boldsymbol{v}_\mathrm{h} = \frac{q_\mathrm{h}}{m_\mathrm{h}} \left[\boldsymbol{E} + \frac{1}{c_0 \sqrt{\varepsilon_0 \mu_0}}\, \boldsymbol{v}_\mathrm{h} \times \boldsymbol{B}_\mathrm{h}\right]. \tag{10.4}$$

The dynamic equation in this form, together with the continuity equation, Eq. (9.2) for both, electrons and ions and Maxwell's equations Eq. (1.1) form a complete system for the cold, collisionless plasma.

Current-density, $\boldsymbol{J}$, and charge-density, ϱ, are found in the way given by Eqs. (8.5) ... (8.7) and (9.1):

$$\boldsymbol{J} = \sum_{\mathrm{h=e,i}} q_\mathrm{h} N_\mathrm{h} \boldsymbol{v}_\mathrm{h}, \tag*{[6.7]}$$

$$\varrho = \sum_{\mathrm{h=e,i}} q_\mathrm{h} N_\mathrm{h}. \tag{10.5}$$

As we sum up over electrons and ions of the plasma we may call this procedure "dynamics of two fluids" — or, better, "gasdynamics with two components" — of a "cold" plasma in the absence of collisons.

[1] We use $\vartheta_\mathrm{h} = kT_\mathrm{h}$ as a measure for the average thermal energy of a particle, see Sect. 8β. $k = 1.38 \cdot 10^{-23}$ J/°K is the Bolzmann-constant, while k means $|\boldsymbol{k}|$, the wave number.

β) One may linearize the gasdynamic equations (9.3) and (10.4), and then adopt harmonic time-dependence and "plane wave" coordinate-dependence after

$$\exp(i\boldsymbol{k}\cdot\boldsymbol{r}-i\omega t).$$

In this manner the differential equations are replaced by algebraic ones, so that it is then easy to obtain the tensorial expressions for conductivity and dielectric permittivity of the plasma. For a plasma without exterior electric field and ordered currents the equations of two-fluid dynamics lead to a dielectric permittivity tensor which is identical with Eqs. (6.8) ... (6.12) for $\bar{\nu}_h=0$. Condition Eq. (8.10) is, of course, satisfied.

Strictly speaking, a few more conditions must be satisfied if two fluid dynamics is to be applied to a collisionless "cold" plasma in the presence of a magnetic field. The full set of necessary conditions is, instead of:

$$\omega\gg\bar{\nu}_h\approx\sum_j\bar{\nu}_{hj},\quad \text{i.e.}\quad Z_h\ll 1, \qquad [8.10]$$

and

$$\frac{\omega}{|\boldsymbol{k}|}\gg V_{Th}=\left(\frac{\vartheta_h}{m_h}\right)^{\frac{1}{2}}, \qquad [10.1, 2]$$

now more precisely:

$$|\omega\pm\omega_{Bh}|\gg\begin{cases}k_\parallel V_{Th}\\ \bar{\nu}_h\end{cases} \qquad (10.6)$$

and

$$\omega_{Bh}\gg k_\perp V_{Th}. \qquad (10.7)$$

$k_\parallel$ and $k_\perp$ being the components of the wave vector $\boldsymbol{k}$ parallel and transverse to the magnetic field $\boldsymbol{B}_{\overset{+}{0}}$.

11. Non-isothermal "hot" plasma approximation with $T_e\gg T_i$.

α) As KLIMONTOVIČ and SILIN [2] have shown a second range of parameter values for which the quasi-gasdynamic description applies to a collisionless plasma is valid under the following conditions [see Eq. (10.1)]:

$$\left(\frac{\vartheta_i}{m_i}\right)^{\frac{1}{2}}=V_{Ti}\ll\frac{L_{\boldsymbol{k}}}{\tau}\approx\frac{\omega}{|\boldsymbol{k}|}\ll V_{Te}=\left(\frac{\vartheta_e}{m_e}\right)^{\frac{1}{2}}. \qquad (11.1)$$

From these conditions it follows that the ions must be considered as "cold" so that the continuity equation, Eq. (9.2), and the dynamic equation, Eq. (10.4), can be taken over from Sect. 10, insofar as ions are concerned. Consequently, these equations characterize the "bulk motions", $\boldsymbol{v}_p$ of the plasma. In order to describe the motion of the ions like that of a fluid the effect of an electric field must in this equation be expressed by fluid quantities. As for electrons, we must, of course use the kinetic equation, Eq. (8.5). The above condition, Eq. (11.1), means that time-derivatives can be neglected in this equation. We can therefore write down the solution for the *distribution function f* as a MAXWELLdistribution, f^M, the argument of which contains the electric potential which characterizes the exterior electric field $\boldsymbol{E}=-\frac{\partial}{\partial\boldsymbol{r}}\phi$;

$$f_e=f^M\left(\frac{m_e}{2\vartheta_e}|\boldsymbol{v}_p-\boldsymbol{v}_e|^2+\frac{q_e\phi}{\vartheta_e}\right). \qquad (11.2)$$

$\boldsymbol{v}_e$ is, of course, the average directed electron velocity from Eq. (9.1). The corresponding current density is

$$\boldsymbol{J}_e=q_e N_e\boldsymbol{v}_e=-q N_e\boldsymbol{v}_e. \qquad (11.3)$$

Putting the expression Eq. (11.2) into the gasdynamic equation for electrons, Eq. (9.3), and taking account of $\boldsymbol{v}_e \gg V_{Te}$ we find

$$\boldsymbol{E} = -\frac{\vartheta_e}{q N_e} \frac{\partial}{\partial \boldsymbol{r}} N_e - \frac{1}{c_0 \sqrt{\varepsilon_0 \mu_0}} \boldsymbol{v}_e \times \boldsymbol{B}. \tag{11.4}$$

β) In this case we can *eliminate the electric field* from the dynamic equation of the ions. In fact we have the condition of neutrality of the plasma, and we know the total current density

$$\boldsymbol{J} \equiv \boldsymbol{J}_e + \boldsymbol{J}_i = -q N_e \boldsymbol{v}_e + q_i N_i \boldsymbol{v}_i \tag{11.5}$$

so that we finally get from Eq. (9.3):

$$\frac{\mathrm{D}\boldsymbol{v}_i}{\mathrm{D}t} \equiv \frac{\partial \boldsymbol{v}_i}{\partial t} + \left(\boldsymbol{v}_i \cdot \frac{\partial}{\partial \boldsymbol{r}}\right) \boldsymbol{v}_i = -\frac{V_s^2}{N_i} \frac{\partial}{\partial \boldsymbol{r}} N_i + \frac{1}{c_0 \sqrt{\varepsilon_0 \mu_0}} \frac{1}{N_i m_i} \boldsymbol{J} \times \boldsymbol{B}, \tag{11.6}$$

where[1]

$$V_s = \sqrt{\left|\frac{q_i}{q}\right| \frac{\vartheta_e}{M}} \tag{11.7}$$

is the *velocity of sound* in the plasma, M being the effective particle mass for sound propagation, $M \approx \overline{m}_i$. We may also simplify MAXWELL's field equations, Eqs. (1.1), first of all by neglecting displacement currents, which is justified if changes are slow:

$$\frac{1}{\tau} \approx \omega \ll \omega_{Ni}, \quad \text{i.e.} \quad X_i^{\frac{1}{2}} \gg 1. \tag{11.8}$$

Furtheron, we may also eliminate the electric field from MAXWELL's equations. Taking account of the fact that in the second of these equations

$$\frac{\partial}{\partial \boldsymbol{r}} \times \boldsymbol{E} = -\frac{1}{c_0 \sqrt{\varepsilon_0 \mu_0}} \frac{\partial \boldsymbol{B}}{\partial t} \tag*{[1.1b]}$$

only the component of the electric field $\boldsymbol{E}$ perpendicular to $\dot{\boldsymbol{B}} = -\mathrm{i}\,\omega\,\boldsymbol{B}$ gives a significant contribution we may concentrate upon this component, $\boldsymbol{E}_\perp$. Under our hypothesis of slow changes the left-hand side of EULER's equation for the ions, Eq. (10.4), is almost zero. Neglecting it, Eq. (10.4) just gives:

$$\boldsymbol{E}_\perp = -\frac{1}{c_0 \sqrt{\varepsilon_0 \mu_0}} \boldsymbol{v}_i \times \boldsymbol{B} \tag{11.9}$$

so that the perpendicular component of the electric field $\boldsymbol{E}$ is known under the condition

$$\frac{1}{\tau} \approx \omega \ll \omega_{Bi}, \quad \text{i.e.} \quad Y_i \gg 1. \tag{11.10}$$

Then, putting Eq. (11.9) into

$$\frac{\partial}{\partial \boldsymbol{r}} \times \boldsymbol{E} = -\frac{1}{c_0 \sqrt{\varepsilon_0 \mu_0}} \frac{\partial \boldsymbol{B}}{\partial t} \tag*{[1.1b]}$$

we get the required relation

$$\frac{\partial}{\partial \boldsymbol{r}} \times (\boldsymbol{v}_i \times \boldsymbol{B}) = \frac{\partial \boldsymbol{B}}{\partial t}. \tag{11.11}$$

[1] In the following our considerations will be restricted to ions with one elementary charge only, so that $|q_i| = q$. Formulas, may, however, easily be generalized for ions with more than one elementary charge.

γ) Now we combine the above equations. Neglecting the electronic mass m_e compared with the ion mass m_i we obtain $\boldsymbol{v} \approx \boldsymbol{v}_i$. Further, in a *fully ionized plasma* (without negative ions) the mass density is

$$\bar{\varrho} = m_i N_i \tag{11.12}$$

and now we may write the system of gasdynamic equations for a collisionless plasma, Eqs. (7.1), as

$$\left.\begin{aligned} &\frac{\partial \boldsymbol{B}}{\partial t} = \frac{\partial}{\partial \boldsymbol{r}} \times (\boldsymbol{v} \times \boldsymbol{B}), \\ &\frac{\partial}{\partial \boldsymbol{r}} \cdot \boldsymbol{B} = 0, \quad \frac{\partial \bar{\varrho}}{\partial t} + \frac{\partial}{\partial \boldsymbol{r}} \cdot (\bar{\varrho}\, \boldsymbol{v}) = 0, \\ &\bar{\varrho}\left(\frac{\partial \boldsymbol{v}}{\partial t} + \left(\boldsymbol{v} \cdot \frac{\partial}{\partial \boldsymbol{r}}\right) \boldsymbol{v}\right) = -V_s^2 \frac{\partial}{\partial \boldsymbol{r}} \bar{\varrho} - \frac{1}{\mathrm{u}\,\mu_0} \boldsymbol{B} \times \left(\frac{\partial}{\partial \boldsymbol{r}} \times \boldsymbol{B}\right). \end{aligned}\right\} \tag{11.13}$$

Comparing with Eqs. (7.1) we see that the present Eqs. (11.13) are obtained from those of the gasdynamic model by omitting external non-electromagnetic forces and all frictional (dissipative) terms, and by adopting infinite conductivity. The equation of state of the plasma (assuming neutrality) now gives the following expressions for the pressure:

$$P = V_s^2 \bar{\varrho} = \left|\frac{q_i}{q}\right| N_i \vartheta_e = N_e \vartheta_e. \tag{11.14}$$

This equation shows that the plasma pressure is determined (as the ions are "cold") by the electronic temperature alone; these conditions are only possible in non-isothermal plasmas with

$$\vartheta_e \gg \vartheta_i.$$

This condition stems from the following reasoning: The characteristic "flux velocity" is that of sound waves which is $\approx (P/\bar{\varrho})^{\frac{1}{2}}$ (see for example [7], [22]). After the equation of state, Eq. (11.14), we have in our plasma $V_s \approx (\vartheta_e/M)^{\frac{1}{2}}$, and this relation is the reason why one designates V_s as the velocity of sound in a collisionless plasma. On the other hand the condition Eq. (11.1) says that it must be much larger than the thermal velocity of ions, $V_s \gg V_{Ti}$. Apparently, in thermal equilibrium we should have $V_{si} = V_{Ti}$ such that the condition, Eq. (11.1), is only acceptable if $\vartheta_e \gg \vartheta_i$, i.e. in a non-thermalized plasma.

δ) Finally, we repeat the *conditions* under which a collisionless plasma may be described by gasdynamic equations:

$$V_{Ti} \ll \frac{\omega}{|\boldsymbol{k}|} \ll V_{Te} \tag*{[11.1]}$$

and

$$\omega \ll \omega_{Bi} \ll \omega_{Ni} \quad \text{i.e.} \quad 1 \ll Y_i \ll X_i^{\frac{1}{2}}. \tag*{[11.10] (11.15)}$$

Strictly speaking, the first of these conditions must be satisfied only for the parallel projection of the wave vector, $k_{\|}$, provided a strong magnetic field is present (see Sect. 26). The condition $\omega_{Bi} \ll \omega_{Ni}$ $[Y_i \ll X_i^{\frac{1}{2}}]$ (11.15) is necessary to have quasi-neutrality in the case $\omega \ll \omega_{Bi}$ $[Y_i \gg 1]$ (11.10).

This does not mean, however, that one could not go from Eqs. (11.13) to the case of an isotropic plasma by putting $\boldsymbol{B}_{\mathfrak{z}} = 0$. With a vanishing magnetic field the continuity equation for ions, together with Eq. (11.6) forms a complete system of equations, and there is no necessity to simplify MAXWELL's second equation, Eq. (1.1 b). Quasi-neutrality is existing as long as $\omega \ll \omega_{Ni}$.

Collisions have entirely been neglected when deducing the system of Eqs. (11.13). But this does not mean that the condition, Eq. (8.10), both for electrons and ions, is necessary. When analyzing the kinetic equations for ions (see Sect. 10) we could neglect collisions if

$$\omega \gg \bar{\nu}_{\mathrm{i}} \approx \sum_{\mathrm{h}} \bar{\nu}_{\mathrm{h\,i}},$$

(8.10). Thus one may neglect collisions in the kinetic equation for electrons only if

$$\bar{\nu}_{\mathrm{e}} \ll |\boldsymbol{k}|\, V_{T\mathrm{e}}, \quad \text{i.e.} \quad Z_{\mathrm{e}} \ll \frac{|\boldsymbol{k}|\, V_{T\mathrm{e}}}{\omega}. \tag{11.16}$$

$\bar{\nu}_{\mathrm{e}}$ being also defined by Eq. (8.10). Therefore ω could be greater or smaller than $\bar{\nu}_{\mathrm{e}}$, and the system of Eqs. (11.13) remains valid only if

$$\omega \gg \bar{\nu}_{\mathrm{i}}, \quad \text{i.e.} \quad Z_{\mathrm{i}} \ll 1, \qquad [8.10]$$

is satisfied.

Also, the partial identity of the system of Eqs. (11.13) with that obtained by dynamics of an ideal fluid, Eqs. (7.1), shows that the gasdynamic description of the plasma is valid in a considerably larger range than would be expected from the conditions of Eq. (7.5).

B. Particle collisions in plasma.

We shall now consider in more detail collisions of charged particles in a plasma. This means we compute the distance-correlation in their motion. There is an abundant literature (an important bibliography can be found in the monographs [*13*], [*21*]) presenting great advances in this field. It has become possible to compute the collision integral not only with COULOMB interaction but also including emission and absorption of electro-magnetic waves by the plasma particles; this latter effect is also called the polarization effect in a plasma.

Strictly speaking we should start with LIOUVILLE's equation, Eq. (8.4), for the statistical distribution function f of the N particles, taking account of the correlation between particles (method of BOGOLJUBOV [*20*], and other methods he has inspired [*13*], [*21*]). We shall proceed otherwise, starting with the most general expression for the collision integral as given by BOLTZMANN, see [*16*], [*17*], and then simplify it for the case of a plasma. In this way the expression for the case of pure COULOMB interaction was first obtained by LANDAU [*23*]. This method, somewhat less strict, but simple and clear, has also been used for computing the collision integral taking account of the polarization of the plasma by V. P. SILIN [*24*].

12. Collision-integral in a completely ionized plasma.

α) In this case, if the thin gas approximation is valid, i.e. if the condition

$$\frac{\mathrm{u}}{4\pi\,\varepsilon_0}\, q_{\mathrm{h}}^2\, N_{\mathrm{h}}^{\frac{1}{3}} \ll \vartheta_{\mathrm{h}} \qquad [8.1]$$

is satisfied, two-body collisions are by far more frequent than such with three or four partners. Under similar conditions we may state that the distribution function of a given, individual particle must be exchangable as far as collisions with any other particle are concerned. Let $\boldsymbol{p}, \boldsymbol{p}'$ and $\boldsymbol{p}_1, \boldsymbol{p}_1'$ be the momenta before and after collision of both colliding particles h, j, and let W be the probability of the reaction

$$\boldsymbol{p} + \boldsymbol{p}' \rightarrow \boldsymbol{p}_1 + \boldsymbol{p}_1'. \tag{12.1}$$

The change of the distribution function f_h of h-particles resulting from collisions with particles of kind j, see Eq. (8.8), can be written in the following form:

$$\left(\frac{\delta f_h}{\delta t}\right)_{\text{col}}^{(hj)} = \int d^3p'\, d^3p_1\, d^3p_1'\, W(p, p'; p_1, p_1')\,[f_h(p)\, f_j(p') - f_h(p_1)\, f_j(p_1')]\,. \quad (12.2)$$

This is the expression for BOLTZMANN'S *collision integral*. The fundamental task of the theory is the computation of $W(p, p'; p_1 p_1')$, the reaction probability.

β) We shall use the *quantum-mechanical methods* as these are more illustrative than classical ones, (see any course of quantum mechanics, for example [*25*]). We suppose that the interaction energy of two particles is a function of the mutual distance only, and write

$$\left.\begin{aligned} U(|r_h - r_j|) &= \int d^3k\, \tilde{U}(k) \exp(i\, k \cdot (r_h - r_j)) \\ \tilde{U}(k) &= \frac{1}{(2\pi)^3} \int d^3r\, U(|r|) \exp(i\, k \cdot r)\,. \end{aligned}\right\} \quad (12.3)$$

$\tilde{U}(k)$ is the FOURIER component of the interaction potential, a function of the vector k. In our case the direction does not interfere so that $\tilde{U}(k) = \tilde{U}(|k|) = \tilde{U}(k)$. One shows in quantum mechanics that the wave function of a free particle with momentum p and energy $\mathcal{E} = p^2/2m$ (in a homogeneous medium) has the form of a plane wave

$$\left.\begin{aligned} \psi(r, t) &= A \exp\left(-i \frac{\mathcal{E}}{\hbar} t + i \frac{p \cdot r}{\hbar}\right) \\ &= A \exp\left(-\frac{i}{\hbar} (\mathcal{E} t - p \cdot r)\right) \end{aligned}\right\} \quad (12.4)$$

where $A = (2\pi\hbar)^{-3/2}$ is a normalization constant and $2\pi\hbar$ is PLANCK'S constant $= 6.6251 \cdot 10^{-34}$ Js. Under the influence of the change $U(r)$ the particles proceed to transitions between different states of Eq. (12.4), i.e. they are "scattered". The "*scattering probability*" with interaction of particles h and j after the expression Eq. (12.4) is then

$$W(p, p'; p_1, p_1') = \frac{2\pi}{\hbar} |U_{p, p'; p_1, p_1'}|^2, \quad (12.5)$$

where

$$U_{p, p'; p_1, p_2'} \equiv \int d^3k\, \tilde{U}(k) \langle p_1 | e^{ik \cdot r_h} | p \rangle \langle p_1' | e^{-ik \cdot r_j} | p' \rangle\,. \quad (12.6)$$

Apparently, this is BORN'S approximation of collision theory. The matrix elements $\langle \ldots \rangle$ under the integral are of course related with the wave-function, Eq. (12.4); at their computation we have to take account of the free motion of the particle which signifies a systematic change of its location, r_h, with time t:

$$r_h = r_{h0} + v_h t\,. \quad (12.7)$$

With the DIRAC-function δ the *matrix element* can be written as

$$\langle p_1 | e^{ik \cdot r_h} | p \rangle = \delta(p_1 - p - \hbar k)\, \delta\left(\frac{p_1^2}{2m_h} - \frac{p^2}{2m_h} - \hbar k \cdot v_h\right) \quad (12.8)$$

and similarly for the element with p_1', p'. It can easily be seen that Eq. (12.8) takes account of the conservation laws of energy and of momentum; this is feasable for any transition $p \to p_1$ because there is yet the "interaction quantum", a wave quantum $\hbar k$ which is emitted at the occasion of the "scattering process".

This quantum is then absorbed by the second particle, j, and brings it from the state with momentum $\boldsymbol{p}'$ to that with $\boldsymbol{p}_1'$.

γ) Putting the expressions Eqs. (12.5), (12.6), (12.8) into Eq. (12.2) we obtain (with the DIRAC-function $\delta[\ldots]$, see Sect. 10α)

$$\left.\begin{aligned}\left(\frac{\delta f_{\mathrm{h}}}{\delta t}\right)_{\mathrm{col}}^{(\mathrm{hj})} = \frac{2\pi}{\hbar}\int \mathrm{d}^3\boldsymbol{p}'\,\mathrm{d}^3\boldsymbol{k}\,|\tilde{U}(k)|^2\,\delta\left[\frac{(\boldsymbol{p}-\hbar\boldsymbol{k})^2}{2m_{\mathrm{h}}} + \frac{(\boldsymbol{p}'+\hbar\boldsymbol{k})^2}{2m_{\mathrm{j}}} - \frac{\boldsymbol{p}^2}{2m_{\mathrm{h}}} - \frac{\boldsymbol{p}'^2}{2m_{\mathrm{j}}}\right]\times \\ \times\,\{f_{\mathrm{h}}(\boldsymbol{p})\,f_{\mathrm{j}}(\boldsymbol{p}') - f_{\mathrm{h}}(\boldsymbol{p}-\hbar\boldsymbol{k})\,f_{\mathrm{j}}(\boldsymbol{p}'+\hbar\boldsymbol{k})\}.\end{aligned}\right\}\quad(12.9)$$

The conservation laws of energy and momentum are apparently satisfied in this expression. In the framework of BORN's *approximation* the expression Eq. (12.9) is a general one because it has been received without any limitations concerning the interaction law, except for its central symmetry.

δ) Now we go over to the (simpler) limiting case where the *interchanged quantum* makes only a *small* contribution:

$$|\hbar\boldsymbol{k}| \ll \begin{cases}|\boldsymbol{p}|\\|\boldsymbol{p}'|\end{cases}. \quad (12.10)$$

We develop the integrand of Eq. (12.9) after increasing orders of $\hbar\boldsymbol{k}$, and, taking account of the central symmetry of U (which only depends on $|\boldsymbol{k}|$), we obtain after some easy computations:

$$\left(\frac{\delta f_{\mathrm{h}}}{\delta t}\right)_{\mathrm{col}}^{(\mathrm{hj})} = \frac{\partial}{\partial\boldsymbol{p}_{\mathrm{h}}}\cdot\int \mathrm{d}^3\boldsymbol{p}'\,\mathsf{I}^{\mathrm{hj}}(\boldsymbol{p},\boldsymbol{p}')\cdot\left[\frac{\partial f_{\mathrm{h}}(\boldsymbol{p})}{\partial\boldsymbol{p}}f_{\mathrm{j}}(\boldsymbol{p}') - \frac{\partial f_{\mathrm{j}}(\boldsymbol{p}')}{\partial\boldsymbol{p}'}f_{\mathrm{h}}(\boldsymbol{p})\right] \quad (12.11)$$

where the tensor I^{hj} is

$$\mathsf{I}^{\mathrm{hj}} \equiv ((I_{lm}^{\mathrm{hj}})) \equiv \pi\int \mathrm{d}^3\boldsymbol{k}\,|\tilde{U}(k)|^2\,\boldsymbol{k}\,\boldsymbol{k}\,\delta(\boldsymbol{k}\cdot\boldsymbol{v}_{\mathrm{h}} - \boldsymbol{k}\cdot\boldsymbol{v}_{\mathrm{j}}) \quad (12.12)$$

i.e.

$$I_{lm}^{\mathrm{hj}}(\boldsymbol{p},\boldsymbol{p}') = \pi\int \mathrm{d}^3\boldsymbol{k}\,|\tilde{U}(k)|^2\,k_l\,k_m\,\delta(\boldsymbol{k}\cdot(\boldsymbol{v}_{\mathrm{h}} - \boldsymbol{v}_{\mathrm{j}})). \quad (12.12\mathrm{a})$$

With the expression Eqs. (12.11) and (12.12) we may go into the kinetic equations, Eq. (8.8), for the distribution function f_{h}:

$$\left.\begin{aligned}\frac{\partial f_{\mathrm{h}}}{\partial t} + \boldsymbol{v}\cdot\frac{\partial f_{\mathrm{h}}}{\partial\boldsymbol{r}} + q_{\mathrm{h}}\left[\boldsymbol{E} + \frac{1}{c_0\sqrt{\varepsilon_0\,\mu_0}}\,\boldsymbol{v}\times\boldsymbol{B}\right]\frac{\partial f_{\mathrm{h}}}{\partial\boldsymbol{p}} \\ = \left(\frac{\delta f_{\mathrm{h}}}{\delta t}\right)_{\mathrm{col}}^{(\mathrm{hj})} \equiv \frac{\partial}{\partial\boldsymbol{p}}\cdot\left\{\mathsf{D}\cdot\frac{\partial}{\partial\boldsymbol{p}} - \boldsymbol{A}\right\}f_{\mathrm{h}}\end{aligned}\right\}\quad(12.13)$$

tensor D and vector $\boldsymbol{A}$ being defined by

$$\mathsf{D} \equiv \sum_{\mathrm{j}}\int \mathrm{d}^3\boldsymbol{p}'\,\mathsf{I}^{\mathrm{hj}}(\boldsymbol{p},\boldsymbol{p}')\,f_{\mathrm{j}}(\boldsymbol{p}');\qquad \boldsymbol{A} \equiv \sum_{\mathrm{j}}\int \mathrm{d}^3\boldsymbol{p}'\,\mathsf{I}^{\mathrm{hj}}(\boldsymbol{p},\boldsymbol{p}')\cdot\frac{\partial f_{\mathrm{j}}(\boldsymbol{p}')}{\partial\boldsymbol{p}'} \quad (12.14)$$

i.e.

$$D_{lm} = \sum_{\mathrm{j}}\int \mathrm{d}^3\boldsymbol{p}'\,I_{lm}^{\mathrm{hj}}(\boldsymbol{p},\boldsymbol{p}')\,f_{\mathrm{j}}(\boldsymbol{p}'), \quad (12.14\mathrm{a})$$

$$A_l = \sum_{\mathrm{j}}\int \mathrm{d}^3\boldsymbol{p}'\,I_{lm}^{\mathrm{hj}}(\boldsymbol{p},\boldsymbol{p}')\,\frac{\partial f_{\mathrm{j}}(\boldsymbol{p}')}{\partial p_m'}. \quad (12.14\mathrm{b})$$

D is the *diffusion tensor* while vector $\boldsymbol{A}$ identifies the *frictional effect* (in the velocity space).

The summation in Eqs. (12.14) covers all kinds of particles, j, which may collide with the particles of kind h. The Eqs. (12.13) after their form are FOKKER-PLANCK equations.

The equation for the collision integral, Eq. (12.11), and the kinetic equation, Eq. (12.13), are valid for any kinds of gases insofar as the interaction-law of the particles $\tilde{U}(k)$, or $U(r)$, is not yet specified.

ε) We may now come to the *completely ionized plasma*. In order to find out the interaction energy between two particles in plasma we have to compute the potential of a charged particle in steady motion with velocity $\boldsymbol{v}_\mathrm{h}$, i.e., $\boldsymbol{r}_\mathrm{h} = \boldsymbol{v}_\mathrm{h}\, t$. From the simplified field equations, Eqs. (1.6), we obtain

$$\operatorname{div} \boldsymbol{D} \equiv \frac{\partial}{\partial \boldsymbol{r}} \cdot \boldsymbol{D} = \mathrm{u}\, q_\mathrm{h}\, N_\mathrm{h}\, \delta(\boldsymbol{r} - \boldsymbol{v}_\mathrm{h}\, t)\,. \tag{12.15}$$

We now make a FOURIER development and, using the material equation, Eq. (1.9), obtain the potential due to *one* considered particle as

$$\phi(\boldsymbol{r}) = \frac{\mathrm{u}\, q_\mathrm{h}}{(2\pi)^3} \int \mathrm{d}^3\boldsymbol{k}\, \frac{\exp(i\,\boldsymbol{k}\cdot\boldsymbol{r})}{\boldsymbol{k}\cdot\mathsf{E}\cdot\boldsymbol{k}} \tag{12.16}$$

so that the electric field is

$$\boldsymbol{E} = -\frac{\partial}{\partial \boldsymbol{r}}\,\phi\,.$$

E is the dielectric permittivity tensor, see Sect. 6, Eqs. (6.8) and (6.11). In the special case of vacuum E goes over into the unite tensor U and Eq. (12.16) is then identical with the well-known expression for the COULOMB-field.

Apparently the interaction-energy of particle h with particle i must be

$$U(\boldsymbol{r}) = q_\mathrm{j}\, \phi(\boldsymbol{r})\,, \tag{12.17}$$

ϕ being the electrostatic potential. Therefore the required quantity is:

$$\tilde{U}(k) = \frac{u\, q_\mathrm{h}\, q_\mathrm{j}}{(2\pi)^3}\, \frac{1}{\boldsymbol{k}\cdot\mathsf{E}\cdot\boldsymbol{k}}\,. \tag{12.18}$$

Finally, putting Eq. (12.18) into Eq. (12.12) we get the tensor I^hj for a completely ionized plasma.

$$\mathsf{I}^\mathrm{hj}(\boldsymbol{p},\boldsymbol{p}') = \pi\, \frac{(\mathrm{u}\, q_\mathrm{h}\, q_\mathrm{i})^2}{(2\pi)^3} \int \mathrm{d}^3\boldsymbol{k}\, \frac{\boldsymbol{k}\boldsymbol{k}\, \delta(\boldsymbol{k}\cdot(\boldsymbol{v}_\mathrm{h} - \boldsymbol{v}_\mathrm{j}))}{|\boldsymbol{k}\cdot\mathsf{E}\cdot\boldsymbol{k}|^2} \tag{12.19}$$

with elements:

$$I^\mathrm{hj}_{lm}(\boldsymbol{p},\boldsymbol{p}') = \pi\, \frac{(\mathrm{u}\, q_\mathrm{h}\, q_\mathrm{j})^2}{(2\pi)^3} \int \mathrm{d}^3\boldsymbol{k}\, \frac{k_l\, k_m\, \delta(\boldsymbol{k}\cdot\boldsymbol{v}_\mathrm{h} - \boldsymbol{k}\cdot\boldsymbol{v}_\mathrm{j})}{|k_l\, k_m\, \varepsilon_{lm}|^2}\,. \tag{12.19a}$$

The expression Eq. (12.19) takes account of the polarization of the plasma; it has first been derived in recent papers by A. LENARD, V. P. SILIN and R. BALESKU, see [*13*], [*21*], [*24*] for a rather complete bibliography.

ζ) Replacing the anisotropic permittivity tensor E in Eq. (12.19) by the unit tensor U, and cutting off the range of integration in the $\boldsymbol{k}$-space by a lower limit k_min and an upper limit k_max, we obtain the *Landau formula* [*28*]

$$\mathsf{I}^\mathrm{hj}(\boldsymbol{p},\boldsymbol{p}') = 2\pi \left(\frac{\mathrm{u}\, q_\mathrm{h}\, q_\mathrm{j}}{\varepsilon_0\, 4\pi}\right)^2 \mathscr{L}\, \frac{u^2\, \mathsf{U} - \boldsymbol{u}\boldsymbol{u}}{u^3}\,, \tag{12.20}$$

where $\boldsymbol{u} = \boldsymbol{v}_\mathrm{j} - \boldsymbol{v}_\mathrm{h}$ is the relative velocity of the colliding particles and[1]

$$\mathscr{L} = \log\left(\frac{k_\mathrm{max}}{k_\mathrm{min}}\right) \tag{12.20a}$$

[1] log designates logarithm with basis e (natural logarithm).

is called COULOMB's logarithm. The "cutoff" of the COULOMB potential stems from the hypothesis that the interaction is of COULOMB character only in a limited distance range, $r_{\min} < r < r_{\max}$, but not at very small and very large distances. $r_{\min} \approx 1/k_{\max}$ is determined by the fact that BORN's approximation has been used above, thus thermal vibrations give a limitation at small distances; therefore[2]

$$\frac{\mathrm{u}}{4\pi\,\varepsilon_0}\,\frac{q^2}{r_{\min}} = \frac{3}{2}\,\vartheta. \tag{12.21}$$

As to the large distance cutoff $r_{\max} \approx 1/k_{\min}$ it is due to the fields of other particles which must be considered at such distances[3].

The order of magnitude is the DEBYE radius which has been chosen by LANDAU with

$$r_{\max}^2 \approx \lambda_{\mathrm{D}}^2 = \sum_{\mathrm{h}} \frac{V_{T\mathrm{h}}^2}{\omega_{N\mathrm{h}}^2} \approx \frac{2\,\varepsilon_0}{\mathrm{u}\,q^2}\,\frac{\vartheta_{\mathrm{e}}}{N_{\mathrm{e}}}. \tag{12.22}$$

The cutoff parameters $r_{\min}$ and $r_{\max}$ come in through a logarithmic term. Therefore a certain inaccuracy in these values is not very important at the determination of $\mathscr{L}$.

The integral in Eq. (12.19) which takes account of the polarization effect contains an automatic cutoff at $k_{\min} \approx 1/\lambda_{\mathrm{D}}$. This is the justification for LANDAU's cutoff method[4].

Ionospheric plasmas (see Tables 1 ... 3), and also most cases of magnetospheric plasmas, are not so far from thermal equilibrium that one may use LANDAU's expression as a reasonable approximation. One has to take here:

$$\left.\begin{aligned} \mathscr{L} &= \log\left(\frac{3}{2}\,\frac{2\pi}{q^2\,\mathrm{u}/\varepsilon_0}\,\vartheta_{\mathrm{e}}\sqrt{\frac{2\vartheta_{\mathrm{e}}}{N_{\mathrm{e}}\,q^2\,\mathrm{u}/\varepsilon_0}}\right) \\ &\approx \frac{3}{2}\log\left(2.2\cdot 10^4\,\frac{T_{\mathrm{e}}/{}^\circ\mathrm{K}}{(N_{\mathrm{e}}/\mathrm{m}^{-3})^{\frac{1}{3}}}\right). \end{aligned}\right\} \tag{12.23}$$

According to Table 1, the order of magnitude of $\mathscr{L}$ is 10 ... 12 in the inner ionosphere, 12 ... 16 in the outer ionosphere, 16 ... 19 in the magnetosphere and about 20 in innerplanetary space.

13. Quasi-gasdynamic approximation for the completely ionized plasma.

α) Knowing the kinetic equations for the completely ionized plasma with COULOMB interaction we come now to a quasi-gasdynamic description including the effect of collisions. For this model apart from the fluid-dynamic parameters N_{h}, Eq. (8.7) and $\boldsymbol{v}_{\mathrm{h}}$, Eq. (9.1), we must still know the temperature for each kind of particle. The *equation of state* is then

$$P_{\mathrm{h}}(\boldsymbol{r},t) = N_{\mathrm{h}}(\boldsymbol{r},t)\,\vartheta_{\mathrm{h}}(\boldsymbol{r},t) = \tfrac{1}{3} m_{\mathrm{h}} \int \mathrm{d}^3\boldsymbol{p}\,|\boldsymbol{v}-\boldsymbol{v}_{\mathrm{h}}|^2 f_{\mathrm{h}}. \tag{13.1}$$

This equation of state, defining the temperature, should be attributed only to "non-pathological" distribution functions, i.e. functions which resemble the

[2] In the limiting case of quantum mechanics, if $q^2\,\mathrm{u}/\varepsilon_0 \ll \hbar v$ one has to take $r_{\min} \sim \lambda_B = \hbar/m\,v$, the de Broglie-wavelength, see [*1*], [*13*].

[3] See also K. RAWER and K. SUCHY: this Encyclopedia, vol. 49/2, Sect. 3 δ. See also [*1*] for a more detailed discussion.

[4] If the plasma is far from thermodynamic equilibrium, LANDAU's collision integral can lead to wrong results. This has been shown in [*26*] for a non-thermal plasma under the condition $T_{\mathrm{e}} < 10^2\,T_{\mathrm{i}}$.

MAXWELL-BOLTZMANN distribution rather well. We have found in Sect. 11 that for a collisionless plasma the temperature is constant in time and space.

β) The equations for the *macroscopic* (gasdynamic) *quantities* are also known as "transport equations". They are obtained from the kinetic equation after multiplying by 1, or $m_{\rm h}\boldsymbol{v}$, or $\frac{1}{2}m_{\rm h}\boldsymbol{v}^2$, and integration over the momentum space. Considering LANDAU's collision integral, in which only elastic collisions are accounted for, we take the particle density (which is the "zero order momentum" in the $\boldsymbol{p}$-space) as invariable such that there is no change in the continuity equation

$$\frac{\partial N_{\rm h}}{\partial t} + \frac{\partial}{\partial \boldsymbol{r}} \cdot (N_{\rm h}\boldsymbol{v}_{\rm h}) = 0. \qquad [9.2]$$

With the "first order momentum" we have [see Sect. 9β, Eqs. (9.3), (9.4)] the dynamic equation (force densities are found on the right side)

$$\left.\begin{aligned} m_{\rm h} N_{\rm h}\left[\frac{\partial \boldsymbol{v}_{\rm h}}{\partial t} + \left(\boldsymbol{v}_{\rm h}\cdot\frac{\partial}{\partial \boldsymbol{r}}\right)\boldsymbol{v}_{\rm h}\right] = -\frac{\partial}{\partial \boldsymbol{r}}P - \frac{\partial}{\partial \boldsymbol{r}}\cdot\Pi_{\rm h} + \\ + q_{\rm h}N_{\rm h}\left[\boldsymbol{E} + \frac{1}{c_0\sqrt{\varepsilon_0\mu_0}}\boldsymbol{v}_{\rm h}\times\boldsymbol{B}\right] + \boldsymbol{R}_{\rm h}, \end{aligned}\right\} \qquad (13.2)$$

with the tensor average

$$\Pi_{\rm h} = m_{\rm h}\int d^3\boldsymbol{p}\{(\boldsymbol{v}-\boldsymbol{v}_{\rm h})(\boldsymbol{v}-\boldsymbol{v}_{\rm h}) - \tfrac{1}{3}U|\boldsymbol{v}-\boldsymbol{v}_{\rm h}|^2\}f_{\rm h}, \qquad (13.3)$$

in components $[\Pi \equiv ((\Pi_{jk}))]$

$$\Pi_{jk} = m_{\rm h}\int d^3\boldsymbol{p}\{(v_j - v_{{\rm h}j})(v_k - v_{{\rm h}k}) - \tfrac{1}{3}\delta_{jk}|\boldsymbol{v}-\boldsymbol{v}_{\rm h}|^2\}f_{\rm h}, \qquad (13.3\,\text{a})$$

and the vector average

$$\boldsymbol{R}_{\rm h} = m_{\rm h}\int d^3\boldsymbol{p}\,\boldsymbol{v}\left(\frac{\delta f_{\rm h}}{\delta t}\right)_{\rm col}. \qquad (13.4)$$

$\Pi_{\rm h}$ is also called the "*viscosity tension*" tensor; the vector $\boldsymbol{R}$ characterizing the momentum change of h-particles resulting (per unit volume) from collisions with all other particles, is the "*friction-force*" vector.

With the "second momentum" we find the energy balance equation as[1]

$$\frac{3}{2}N_{\rm h}\left[\frac{\partial\vartheta_{\rm h}}{\partial t} + \left(\boldsymbol{v}_{\rm h}\cdot\frac{\partial}{\partial \boldsymbol{r}}\right)\vartheta_{\rm h}\right] + P_{\rm h}\frac{\partial}{\partial \boldsymbol{r}}\cdot\boldsymbol{v}_{\rm h} + \Pi_{\rm h}\cdot\cdot\left(\frac{\partial}{\partial \boldsymbol{r}}\boldsymbol{v}_{\rm h}\right) = \mathcal{Q}_{\rm h} - \frac{\partial}{\partial \boldsymbol{r}}\cdot\boldsymbol{q}_{\rm h} \qquad (13.5)$$

with

$$\boldsymbol{q}_{\rm h} = \tfrac{1}{2}m_{\rm h}\int d^3\boldsymbol{p}\,\boldsymbol{v}|\boldsymbol{v}-\boldsymbol{v}_{\rm h}|^2 f_{\rm h} \qquad (13.6)$$

and

$$\mathcal{Q}_{\rm h} = \frac{1}{2}m_{\rm h}\int d^3\boldsymbol{p}|\boldsymbol{v}-\boldsymbol{v}_{\rm h}|^2\left(\frac{\partial f_{\rm h}}{\partial t}\right)_{\rm col}. \qquad (13.7)$$

$\boldsymbol{q}_{\rm h}$ is the flux-vector of heat energy as transported by particles of kind h; $\mathcal{Q}$ characterizes the heat transfer by h-particles resulting from collisions with all other kinds of particles. For a completely ionized plasma[2]

$$\boldsymbol{R}_{\rm e} = -\boldsymbol{R}_{\rm i}, \qquad (13.8)$$

$$\mathcal{Q}_{\rm e} + \mathcal{Q}_{\rm i} = -\boldsymbol{R}_{\rm e}\cdot\boldsymbol{v}_{\rm e} - \boldsymbol{R}_{\rm i}\cdot\boldsymbol{v}_{\rm i} = -\boldsymbol{R}_{\rm e}(\boldsymbol{v}_{\rm e} - \boldsymbol{v}_{\rm i}). \qquad (13.9$$

[1] The third member reads in components $\Pi_{{\rm h}jk}\dfrac{\partial v_{{\rm h}j}}{\partial r_k}$.

[2] Our considerations are limited here to the case of a simple plasma consisting of electrons and one kind of ions only.

These relations express the laws of conservation of momentum and energy for COULOMB collisions.

γ) In order to complete the equations of transport we needed *relations between* the *macroscopic quantities*, namely "viscosity tension" tensor, Π_h, the heat flux vector, $\boldsymbol{q}_h$, the friction force vector, $\boldsymbol{R}_h$, and the heat transfer by collisions, $\mathcal{Q}_h$, with the gasdynamic quantities, namely number density, N_h, average velocity, $\boldsymbol{v}_h$, and temperature T_h (i.e. ϑ_h in our energy measure of temperature). These relations can also be found as *approximate solutions of the kinetic equation*, expressing the distribution function in terms of N_h, v_h and ϑ_h, and then using Eqs. (13.4) and (13.6). Such local approximate solution can, at least in principle, only be obtained in cases where the variations of the distribution function f are slow enough in space and time. Therefore, the acceptability of a local solution is related with the relaxation processes; these are due to the collisions between particles which have a tendency to bring the distribution function into maxwellian shape.

MAXWELL's distribution function is an exact solution of the kinetic equation under the particular conditions that space and time derivatives are zero, i.e. in case of homogeneity and stationary behaviour. Now, if the derivatives are not exactly zero, but small enough anyhow, then the strict solution of the kinetic equation must be quite near to a local maxwellian distribution f_h^M; therefore, for each kind of particles, we may take a maxwellian distribution as first approximation, and write:

$$f_h = f_h^M + f_h^{(1)} \equiv \frac{N_h}{(2\pi\, m_h\, \vartheta_h)^{\frac{3}{2}}} \exp\left(-\frac{m_h |\boldsymbol{v}-\boldsymbol{v}_h|^2}{2\vartheta_h}\right) + f_h^{(1)}, \tag{13.10}$$

and

$$f_h^{(1)} \ll f_h^M.$$

The correction $f_h^{(1)}$ depends on the quantities which cause derivations from the homogeneous and stationary distribution. Apparently $f_h^{(1)}$ is proportional to the space- and time-gradients. The quantities Π_h, $\boldsymbol{q}_h$, $\boldsymbol{R}_h$ and $\mathcal{Q}_h$ can be shown to be linear functions of these gradients the determination of which is the main task of transport theory in plasma.

δ) Different *approximation methods* have been developed in order to determine these quantities. The best known ones are CHAPMAN and ENSKOG's method[3] and that of GRAD[4]. We shall not go into details of these approximation methods to the kinetic equation. They are all based on a development of the general distribution function after some set of special functions, the coefficients of the development being then determined. We limit our description to a summary of results obtained by BRAGINSKIJ[5] for Π_h, $\boldsymbol{q}_h$, $\boldsymbol{R}_h$ and $\mathcal{Q}_h$ in the case of a "simple plasma"[6].

ε) The total *friction force* $\boldsymbol{R}_h$ consists of two contributions, the friction force due to the relative motion ($\boldsymbol{u}$) and that friction which is due to thermal motion (index T):

$$\boldsymbol{R}_h = \boldsymbol{R}_{h\boldsymbol{u}} + \boldsymbol{R}_{hT}, \tag{13.11}$$

$$\boldsymbol{R}_{e\boldsymbol{u}} = -m\,\bar{\nu}_e N_e (0.51\, \boldsymbol{u}_\parallel + \boldsymbol{u}_\perp) = -q_e N_e \left(\frac{J_\parallel}{\sigma_\parallel} + \frac{J_\perp}{\sigma_\perp}\right), \tag{13.12a}$$

$$\boldsymbol{R}_{eT} = -0.71\, N_e \frac{\partial \vartheta_e}{\partial \boldsymbol{r}_\parallel} + \frac{3}{2}\frac{N_e\,\bar{\nu}_e}{\omega_{Be}}\left[\boldsymbol{B}^0_{\circ} \times \frac{\partial \vartheta_e}{\partial \boldsymbol{r}}\right]. \tag{13.12b}$$

---

[3] Explained in detail in [*15*].
[4] See [*27*], where this method is developed.
[5] S. I. BRAGINSKIJ: Contribution to [*6*].
[6] Different from the conventions used in [*6*] we shall converse here the sign of the particle charge, i.e. $q_e = -q = -1.601 \cdot 10^{-19}$ A s.

Here

$$\boldsymbol{u} \equiv \boldsymbol{v}_e - \boldsymbol{v}_i \quad \text{and} \quad \boldsymbol{J} = q_e N \boldsymbol{u} \tag{13.13}$$

is the total current in the plasma ($q_i = -q_e$) and $\bar{\nu}_e$ is the relevant average transport collision frequency (see "Introductory Remarks", p. 1). We distinguish components parallel and transverse to the direction of the magnetic field, $\boldsymbol{B}_0$, for example:

$$\boldsymbol{u}_\parallel = (\boldsymbol{u} \cdot \boldsymbol{B}^0)\, \boldsymbol{B}^0; \qquad \boldsymbol{u}_\perp = \boldsymbol{B}^0 \times (\boldsymbol{u} \times \boldsymbol{B}^0). \tag{13.13a}$$

$\boldsymbol{B}^0 \equiv \boldsymbol{B}/B$ is the unit vector in the magnetic field direction. The electric conductivity[7] being of dyadic character, different values apply for directions parallel with and transverse to the magnetic field:

$$\sigma_\perp = \frac{q^2 N_e}{m_e \bar{\nu}_e}; \qquad \sigma_\parallel = 1.96\, \sigma_\perp. \tag{13.14}$$

The friction force for ions is

$$\boldsymbol{R}_i = -\boldsymbol{R}_e. \tag*{[13.8]}$$

ζ) The electron *heat flux* contains also two contributions, one due to the relative (macroscopic) motion, $\boldsymbol{u}$, another one due to thermal motion[8]:

$$\boldsymbol{q}_e = \boldsymbol{q}_{e\boldsymbol{u}} + \boldsymbol{q}_{eT}, \tag{13.15}$$

$$\boldsymbol{q}_{e\boldsymbol{u}} = 0.71\, N_e \vartheta_e \boldsymbol{u}_\parallel - \frac{3}{2} \frac{N_e \vartheta_e \bar{\nu}_e}{\omega_{Be}} \boldsymbol{B}^0 \times \boldsymbol{u}, \tag{13.16}$$

$$\boldsymbol{q}_{eT} = -\varkappa_{e\parallel} \frac{\partial \vartheta_e}{\partial \boldsymbol{r}} - \varkappa_{e\perp} \frac{\partial \vartheta_e}{\partial \boldsymbol{r}_\perp} + \frac{5}{2} \frac{N_e \vartheta_e}{m_e \omega_{Be}} \boldsymbol{B}^0 \times \frac{\partial \vartheta_e}{\partial \boldsymbol{r}}. \tag{13.17}$$

$\boldsymbol{B}^0$ is the unit vector in the direction of the magnetic field. In Eq. (13.17) appear the coefficients of electron heat conduction[9]:

$$\varkappa_{e\parallel} = 3.16 \frac{N_e \vartheta_e}{m \bar{\nu}_e}; \qquad \varkappa_{e\perp} = 4.66 \frac{N_e \vartheta_e \bar{\nu}_e}{m_e \omega_{Be}^2}. \tag{13.18}$$

The ion heat flux is

$$\boldsymbol{q}_i = -\varkappa_{i\parallel} \frac{\partial \vartheta_i}{\partial \boldsymbol{r}_\parallel} - \varkappa_{i\perp} \frac{\partial \vartheta_i}{\partial \boldsymbol{r}_\perp} - \frac{5}{2} \frac{N_i \vartheta_i}{m_i \omega_{Bi}} \boldsymbol{B}^0 \times \frac{\partial \vartheta_i}{\partial \boldsymbol{r}} \tag{13.19}$$

with the coefficients of ion heat conduction:

$$\varkappa_{i\parallel} = 3.9 \frac{N_i \vartheta_i}{m_i \bar{\nu}_i}; \qquad \varkappa_{i\perp} = 2 \frac{N_i \vartheta_i \bar{\nu}_i}{m_i \omega_{Bi}^2}. \tag{13.20}$$

Finally, through collisions, the ions transfer heat to the electrons:

$$Q_i = 3 \frac{m_e}{m_i} \bar{\nu}_e N_e (\vartheta_e - \vartheta_i) \tag{13.21}$$

and the heat transferred to the electrons is

$$Q_e = -Q_i - \boldsymbol{R}_e \cdot \boldsymbol{u}. \tag{13.22}$$

---

[7] The SI-unit for σ is A/V m.

[8] The SI-unit for $\boldsymbol{q}$ is J m^{-2} s^{-1} ≡ W m^{-2}.

[9] The SI-unit for $\varkappa$ is m^{-1} sec^{-1}.

η) The tensor of viscous tension, Π_h, Eq. (13.3), can be expressed in the usual manner using a displacement tensor, W_h, and a certain number of *viscosity coefficients*, $\eta_{h\lambda}$ ($\lambda = 0, 1, 2, 3, 4$). The displacement tensor is, of course ($h = e, i$)[10]:

$$W_h = \frac{\partial}{\partial \boldsymbol{r}} \boldsymbol{v}_h + \left(\frac{\partial}{\partial \boldsymbol{r}} \boldsymbol{v}_h\right)^T - \frac{2}{3} U \left(\frac{\partial}{\partial \boldsymbol{r}} \cdot \boldsymbol{v}_h\right), \tag{13.23}$$

in components:

$$\mathscr{W}_{h\,ij} = \frac{\partial v_{h\,i}}{\partial \boldsymbol{r}_j} + \frac{\partial v_{h\,j}}{\partial r_i} - \frac{2}{3} \delta_{ij} \frac{\partial}{\partial \boldsymbol{r}} \cdot \boldsymbol{v}_h \tag{13.23a}$$

and we have:

$$\Pi_h = -\sum_{\lambda=0}^{4} \eta_{h\lambda} W_h^{(\lambda)}, \quad \text{i.e.} \quad \Pi_{h\,ij} = -\sum_{\lambda=1}^{4} \eta_{h\lambda} \mathscr{W}_{h\,ij}^{(\lambda)}, \tag{13.24}$$

where

$$\left.\begin{aligned} &\eta_{e0} = 0.73 \frac{N_e \vartheta_e}{\bar{\nu}_e}; \quad \eta_{e1} = 0.51 \frac{N_e \vartheta_e \bar{\nu}_e}{\omega_{Be}^2}; \quad \eta_{e2} = 4\eta_{e1}; \\ &\eta_{e3} = \frac{1}{2} \frac{N_e \vartheta_e}{\omega_{Be}}; \quad \eta_{e4} = 2\eta_{e3}, \end{aligned}\right\} \tag{13.25}$$

and

$$\left.\begin{aligned} &\eta_{i0} = 0.96 \frac{N_i \vartheta_i}{\bar{\nu}_i}; \quad \eta_{i1} = \frac{N_i \vartheta_i \bar{\nu}_i}{\omega_{Bi}^2}; \quad \eta_{i2} = 4\eta_{i1}; \\ &\eta_{i3} = \frac{1}{2} \frac{N_i \vartheta_i}{\omega_{Bi}}; \quad \eta_{i4} = 2\eta_{i3}. \end{aligned}\right\} \tag{13.26}$$

The five (second order) displacement tensors $W_h^{(\lambda)} \equiv ((\mathscr{W}_h^{(\lambda)}))$ are given by the following equations:

$$\left.\begin{aligned} W_h^{(0)} &= \tfrac{3}{2} (\boldsymbol{B}^0 \boldsymbol{B}^0 - \tfrac{1}{3} U) [(\boldsymbol{B}^0 \boldsymbol{B}^0 - \tfrac{1}{3} U) \cdot\cdot\, W_h] \\ W_h^{(1)} &= (U^\perp \cdot W_h) \cdot U^\perp + \tfrac{1}{2} U^\perp (\boldsymbol{B}^0 \boldsymbol{B}^0 \cdot\cdot\, W_h) \\ W_h^{(2)} &= (U^\perp \cdot W_h) \cdot \boldsymbol{B}^0 \boldsymbol{B}^0 + (\boldsymbol{B}^0 \boldsymbol{B}^0 \cdot W_h) \cdot U^\perp \\ W_h^{(3)} &= -\tfrac{1}{2} (U^\perp \cdot W_h) \cdot B^0 + \tfrac{1}{2} B^0 \cdot (W_h \cdot U^\perp) \\ W_h^{(4)} &= -(\boldsymbol{B}^0 \boldsymbol{B}^0 \cdot W_h) \cdot B^0 + B^0 \cdot (W_h \cdot \boldsymbol{B}^0 \boldsymbol{B}^0) \end{aligned}\right\} \tag{13.27}$$

where

$$B^0 \equiv \begin{pmatrix} 0 & -B_3^0 & +B_2^0 \\ +B_3^0 & 0 & -B_1^0 \\ -B_2^0 & +B_1^0 & 0 \end{pmatrix} \equiv \begin{pmatrix} 0 & -\cos(\boldsymbol{B}, 3) & +\cos(\boldsymbol{B}, 2) \\ +\cos(\boldsymbol{B}, 3) & 0 & -\cos(\boldsymbol{B}, 1) \\ -\cos(\boldsymbol{B}, 2) & +\cos(\boldsymbol{B}, 1) & 0 \end{pmatrix}$$

is an antisymmetric tensor of second order which has the directional cosine of the magnetic field vector for components; and

$$U^\perp \equiv U - \boldsymbol{B}^0 \boldsymbol{B}^0 \tag{13.28}$$

is a symmetric tensor of second order in which products of such directional cosines appear as components.

B^0 can also be written in a more sophisticated form using the antisymmetric unit tensor of third order (see Sect. 6 ε),

$$\mathfrak{E} \equiv (((\varepsilon_{\mu\varkappa\nu}))).$$

Its components, $\varepsilon_{\mu\varkappa\nu}$, equal $+1$ or -1 according to $\mu\varkappa\nu$ being an even or odd permutation of 1 2 3, and are zero otherwise. Thus the tird order tensor $\mathfrak{E}$ has only six non-vanishing

[10] Upper index T designates the transposed tensor.

components. When applied to a vector $\boldsymbol{B}$ for $\mu \neq \nu$:

$$\sum_{\varkappa} \varepsilon_{\mu\varkappa\nu} B_{\varkappa} = \pm B_{\varkappa} \qquad (\mu \neq \varkappa \neq \nu),$$

but for $\mu = \nu$:

$$\sum_{\varkappa} \varepsilon_{\mu\varkappa\nu} B_{\varkappa} = 0.$$

With this in mind we may write the second order tensor B^0 as

$$\mathsf{B}^0 = \left(\left(\sum_{\varkappa} \varepsilon_{\mu\varkappa\nu} B_{\varkappa}^0\right)\right) = \mathfrak{E} \cdot \boldsymbol{B}^0.$$

For convenience we write the preceding expressions also in components, using [in analogy with Eq. (13.28)]

$$\delta_{jl}^{\perp} = \delta_{jl} - B_j^0 B_l^0, \tag{13.28a}$$

$$\left.\begin{aligned}
\mathscr{W}_{\mathrm{h}jl}^{(0)} &= \tfrac{3}{2}\left(B_j^0 B_l^0 - \tfrac{1}{3}\delta_{jl}\right) \sum_{\mu}\sum_{\nu} \left(B_\mu^0 B_\nu^0 - \tfrac{1}{3}\delta_{\mu\nu}\right) \mathscr{W}_{\mathrm{h}\mu\nu}\\
\mathscr{W}_{\mathrm{h}jl}^{(1)} &= \sum_{\mu}\sum_{\nu} \left(\delta_{j\mu}^{\perp}\delta_{l\nu}^{\perp} + \tfrac{1}{2}\delta_{jl}^{\perp} B_\mu^0 B_\nu^0\right) \mathscr{W}_{\mathrm{h}\mu\nu}\\
\mathscr{W}_{\mathrm{h}jl}^{(2)} &= \sum_{\mu}\sum_{\nu} \left(\delta_{j\mu}^{\perp} B_l^0 B_\nu^0 + \delta_{l\nu}^{\perp} B_j^0 B_\mu^0\right) \mathscr{W}_{\mathrm{h}\mu\nu}\\
\mathscr{W}_{\mathrm{h}jl}^{(3)} &= \tfrac{1}{2}\sum_{\mu}\sum_{\nu}\sum_{\varkappa} \left(\delta_{j\mu}^{\perp}\varepsilon_{l\varkappa\nu} + \delta_{l\nu}^{\perp}\varepsilon_{j\varkappa\mu}\right) B_\varkappa^0 \mathscr{W}_{\mathrm{h}\mu\nu}\\
\mathscr{W}_{\mathrm{h}jl}^{(4)} &= \sum_{\mu}\sum_{\nu}\sum_{\varkappa} \left(B_j^0 B_\mu^0 \varepsilon_{l\varkappa\nu} + B_l^0 B_\nu^0 \varepsilon_{l\varkappa\mu}\right) B_\varkappa^0 \mathscr{W}_{\mathrm{h}\mu\nu}.
\end{aligned}\right\} \tag{13.27a}$$

ϑ) The expressions given in Subsects. ε through η refer to ions with one unit charge ($q_{\mathrm{i}} = -q_{\mathrm{e}}$), in the case of a rather strong magnetic field, i.e. when the effect of collisions is negligible in the transverse direction:

$$\omega_{B\,\mathrm{i}} \gg \bar{\nu}_{\mathrm{i}} \quad \text{and} \quad \omega_{B\,\mathrm{e}} \gg \bar{\nu}_{\mathrm{e}}. \tag{13.29}$$

As expressions for the *average transport collision frequencies* (see "Introductory Remarks", p. 1) we may write:

$$\bar{\nu}_{\mathrm{e}} = \nu_{\mathrm{eff}} = \frac{1}{3}(2\pi)^{-\frac{3}{2}} \left(\frac{\mathrm{u}}{\varepsilon_0}\right)^2 q_{\mathrm{e}}^2 q_{\mathrm{i}}^2 N_{\mathrm{e}} \mathscr{L}/\vartheta_{\mathrm{i}}^{\frac{3}{2}}, \tag{13.30}$$

$$\bar{\nu}_{\mathrm{i}} = \bar{\nu}_{\mathrm{ii}} = \frac{1}{3\sqrt{2}}(2\pi)^{-\frac{3}{2}} \left(\frac{\mathrm{u}}{\varepsilon_0}\right)^2 q_{\mathrm{i}}^2 q_{\mathrm{i}}^2 N_{\mathrm{i}} \mathscr{L}/\vartheta_{\mathrm{i}}^{\frac{3}{2}}. \tag{13.30a}$$

In the ionosphere the inequality, Eq. (13.29), is valid above a level of about 200 ... 300 km height. Beginning with these heights collisions with neutrals are almost negligible, so that the plasma behaves like a completely ionized one.

Besides, transport coefficients in the case of vanishing magnetic field are identical with those given above for the direction parallel to B^0 *. As for the case of ions with multiple charge, $|q_{\mathrm{i}}/q_{\mathrm{e}}| > 1$, the necessary changes in the coefficients of our formulae are found in Table 4. This table suppose, of course, the same charge for all ions, not a mixture of ions with different charges.

* For cases where the inequality, Eq. (13.29) is not valid, S. I. BRAGINSKIJ [6] gives formulae for arbitrary values of $\omega_{B\mathrm{h}}$ and $\bar{\nu}_{\mathrm{h}}$.

Table 4. *Numerical values of coefficients needed in the equations for transport phenomena.*

Multiplicity of ionic charge	Equations: (13.12a)	(13.12b) and (13.16)	(13.18)	(13.18)
	for quantity: $\boldsymbol{R}_{e\boldsymbol{u}}$	$\boldsymbol{R}_{eT}$ and $\boldsymbol{q}_{e\boldsymbol{u}}$	$\varkappa_{e\parallel}$	$\varkappa_{e\perp}$
1	0.51	0.71	3.16	4.66
2	0.44	0.9	4.9	4.0
3	0.40	1.0	6.1	3.7
4	0.38	1.1	6.9	3.6
∞	0.29	1.5	12.5	3.2

ι) We may take Eqs. (13.2) through (13.30), together with MAXWELL's equations, Eqs. (1.1), as the complete set of equations for a "two-fluid model" of a completely ionized, uniform plasma. More precisely, we may speak of a *magneto-gas-dynamic system with two components*. As indicated above such description may be used in case when the spatial and time derivatives of all gasdynamic quantities are small enough, i.e. when for each kind of particles ($h = e, i$) the collision frequency is great enough:

$$\bar{\nu}_h \gg (\omega, k V_{Th}, k v_h). \tag{13.31}$$

In the presence of a magnetic field this condition "splits up" into two, for the parallel and on the transverse direction with respect to $\boldsymbol{B}^0$:

$$\left.\begin{aligned} \bar{\nu}_h &\gg (\omega, k_\parallel V_{Th}, k_\parallel v_h) \\ \omega_{Bh} &\gg (k_\perp V_{Th}, k_\perp v_h). \end{aligned}\right\} \tag{13.31a}$$

The difference between parallel and perpendicular conditions is caused by the difference of free particle orbits in these characteristic cases. In particular for parallel motion there is no influence of the magnetic field upon the orbit such that the mean free path is reasonably taken as characteristic length. Quite different, for perpendicular motion the orbit is circular and the LARMOR radius must be identified with the characteristic length. In the general (oblique) case the orbit is a helix.

14. Collisions between particles in a weakly ionized plasma. We come now to the case of a weakly ionized plasma for which collisions with neutral particles play a determining rôle. This is the case for the ionospheric plasma below a height of, roughly, 200 km.

α) In such a plasma, except for elastic scattering (elastic collisions) the collisions with the neutrals may produce quite a few *non-elastic processes* like excitation of neutrals, ionization of these, charge transfer and so on[1]. In the BOLTZMANN equation, Eq. (8.8), the collision integral, $\delta f/\delta t$, is intended to take account of all these processes. However, BOLTZMANN when writing the collision integral in Eq. (12.2) had only elastic collisions in mind. For non-elastic processes the collision integral is extremely complicated. Therefore, we shall restrict our considerations here to purely elastic collisions. These play a dominant rôle in investigations of the propagation of weak electromagnetic waves in plasma, i.e. in the linear approximation.

[1] As for the elementary processes, see S. BRAUN [*28*].

β) Up to now we do not yet have a completely satisfying expression of BOLTZMANN'S collision integral for weakly ionized plasmas, even in the case where *elastic collisions* are *predominant*. In this context approximate model-like representations of the collisional term in the kinetic equations are often used for practical computations. We shall say a few words about one of these model-like expressions for the collision integral which have reached rather great importance during the last years.

The term „model-like" indicates that one does not obtain the relevant expression logically, from the accurate expression for the collision integral by some procedure of approximation. One simply builds up an expression which satisfies a certain number of general requirements.

γ) The expression for the collision integral given by BATNAGAR, GROSS and KROOK [29] is designated by the abbreviation *GBK-integral*. In order to obtain a model-like expression in the case of elastic collisions, these authors start with the general conservation laws of number of particles, momentum and energy. These laws are written as follows, for collisions between particles of kind g with those of kind h:

$$\int d\boldsymbol{p}\left(\frac{\delta f_g}{\delta t}\right)_{col}^{(gh)}=0, \tag{14.1}$$

$$m_g\int d\boldsymbol{p}\,\boldsymbol{v}\left(\frac{\delta f_g}{\delta t}\right)_{col}^{(hg)}+m_h\int d\boldsymbol{p}\,\boldsymbol{v}\left(\frac{\delta f_h}{\delta t}\right)_{col}^{(gh)}=0, \tag{14.2}$$

$$m_g\int d\boldsymbol{p}\,v^2\left(\frac{\delta f_g}{\delta t}\right)_{col}^{(hg)}+m_h\int d\boldsymbol{p}\,\boldsymbol{v}^2\left(\frac{\partial f_h}{\partial t}\right)_{col}^{(gh)}=0. \tag{14.3}$$

Then, when introducing MAXWELL'S distribution for all kinds of particles, the collision-integral should go to zero. This appears to be a logical consequence of BOLTZMANN'S "H-Theorem" [13], [15], [16], which states that under arbitrary initial conditions the distribution function for the particles of a gas with collisions tends towards the Maxwellian distribution. This process of approach by collisions may be called a relaxation process. The *relaxation* time is of the order of the effective time of free travel of a particle; thus, fundamentally this is a determination of an *effective collision frequency*. The BGK collisional integral has the properties which are required for this transition; it can be written in the form:

$$\left(\frac{\delta f_g}{\delta t}\right)_{col}^{(gh)}=-\nu_{gh}[f_g-N_g\,\phi_{gh}] \tag{14.4}$$

where ν_{gh} is considered as to be a constant quantity having the meaning of an effective collision frequency, and the function ϕ_{gh} is defined by

$$\phi_{gh}=\frac{1}{(2\pi\,m_g\,\vartheta_{gh})^{\frac{3}{2}}}\exp\left(-\frac{m_g|\boldsymbol{v}-\boldsymbol{v}_g|^2}{2\vartheta_{gh}}\right) \tag{14.5}$$

with

$$\left.\begin{array}{ll}\text{particle density} & N_g\equiv\int d\boldsymbol{p}\,f_g,\\ \text{average velocity} & \boldsymbol{v}_g\equiv\frac{1}{N_g}\int d\boldsymbol{p}\,\boldsymbol{v}\,f_g\end{array}\right\} \qquad [9.1]$$

$$\text{effective temperature}\quad \vartheta_g=\frac{P_g}{N_g}=\frac{m_g}{3N_g}\int d\boldsymbol{p}\,|\boldsymbol{v}-\boldsymbol{v}_g|^2 f_g \qquad [13.1]$$

and

$$\vartheta_{gh}\equiv\frac{m_g\,\vartheta_g+m_h\,\vartheta_h}{m_h+m_h}. \tag{14.6}$$

These definitions of average velocity vector and effective temperature of the particles agree with the hypotheses made above. Putting the expression Eq. (14.4) into Eqs. (14.1) ... (14.3) one finds that the laws of conservation of energy and momentum are satisfied under the condition

$$m_{\mathrm{g}} N_{\mathrm{g}} \nu_{\mathrm{gh}} = m_{\mathrm{h}} N_{\mathrm{h}} \nu_{\mathrm{hg}}. \tag{14.7}$$

At that time there was no relation known in the theory of the model BGK collisional integral from which one could determine the collision frequencies $\bar{\nu}_{\mathrm{gh}}$ themselves. They had to be found from supplementary considerations, for example from simple molecular-kinetic reasonnings [1].

15. Quasi-gasdynamic approximation of weakly ionized plasmas.

α) With the kinetic equations, using the BGK collisional integral we are now able to build up the gasdynamics of a weakly ionized plasma as we have done in Sect. 13 for the completely ionized plasma. Apparently, a weakly ionized plasma is a mixture of three components, electrons (e), ions (i) and neutrals (n). There is no difficulty in computing the quantities: viscosity tension tensor Π_{h}, heat flux vector $\boldsymbol{q}_{\mathrm{h}}$, friction force vector B_{h} and heat transfer Q_{h} for the different components ($\mathrm{h} = \mathrm{e}, \mathrm{i}, \mathrm{n}$). In fact the kinetic equation admits an exact solution in case of a small deviation from the local Maxwellian distribution, Eq. (13.10). One finally obtains the following complete system of equations for the *dynamics of a three-component gas* as description of a weakly ionized plasma:

$$\frac{\partial N_{\mathrm{h}}}{\partial t} + \frac{\partial}{\partial \boldsymbol{r}} \cdot (N_{\mathrm{h}} \boldsymbol{v}_{\mathrm{h}}) = 0, \qquad [9.2]$$

$$\left.\begin{aligned} & m_{\mathrm{h}} N_{\mathrm{h}} \left[\frac{\partial \boldsymbol{v}_{\mathrm{h}}}{\partial t} + \left(\boldsymbol{v}_{\mathrm{h}} \cdot \frac{\partial}{\partial \boldsymbol{r}}\right) \boldsymbol{v}_{\mathrm{h}}\right] \\ & = q_{\mathrm{h}} N_{\mathrm{h}} \left\{\boldsymbol{E} + \frac{1}{c_0 \sqrt{\varepsilon_0 \mu_0}} \boldsymbol{v}_{\mathrm{h}} \times \boldsymbol{B}\right\} - \frac{\partial P_{\mathrm{h}}}{\partial \boldsymbol{r}} - \frac{\partial}{\partial \boldsymbol{r}} \cdot \Pi_{\mathrm{h}} - m_{\mathrm{h}} N_{\mathrm{h}} \sum_{\mathrm{g}} \nu_{\mathrm{hg}} (\boldsymbol{v}_{\mathrm{h}} - \boldsymbol{v}_{\mathrm{g}}), \end{aligned}\right\} \tag{15.1}$$

$$\left.\begin{aligned} & \frac{3}{2} N_{\mathrm{h}} \left[\frac{\partial \vartheta_{\mathrm{h}}}{\partial t} + \left(\boldsymbol{v}_{\mathrm{h}} \cdot \frac{\partial}{\partial \boldsymbol{r}}\right) \vartheta_{\mathrm{h}}\right] + P_{\mathrm{h}} \frac{\partial}{\partial \boldsymbol{r}} \cdot \boldsymbol{v}_{\mathrm{h}} + \frac{\partial}{\partial \boldsymbol{r}} \cdot \boldsymbol{q}_{\mathrm{h}} + \Pi_{\mathrm{h}} \cdot\cdot \left(\frac{\partial}{\partial \boldsymbol{r}} \cdot \boldsymbol{v}_{\mathrm{h}}\right) \\ & = \frac{1}{2} m_{\mathrm{h}} N_{\mathrm{h}} \sum_{\mathrm{g}} \nu_{\mathrm{hg}} |\boldsymbol{v}_{\mathrm{h}} - \boldsymbol{v}_{\mathrm{g}}|^2 - \frac{3}{2} \sum_{\mathrm{g}} \frac{m_{\mathrm{h}} N_{\mathrm{h}} \nu_{\mathrm{hg}}}{(m_{\mathrm{h}} + m_{\mathrm{g}})} (\vartheta_{\mathrm{h}} - \vartheta_{\mathrm{g}}) \end{aligned}\right\} \tag{15.2}$$

where $\mathrm{g}, \mathrm{h} = \mathrm{e}, \mathrm{i}, \mathrm{n}$ and $q_n \equiv 0$. The viscosity tension tensor, Π_{h}, is now given by the expression:

$$\left.\begin{aligned} \Pi_{\mathrm{h}} = & - \frac{P_{\mathrm{h}}}{\sum_{\mathrm{g}} \nu_{\mathrm{hg}}} \left\{\boldsymbol{v}_{\mathrm{h}} \frac{\partial}{\partial \boldsymbol{r}} + \frac{\partial}{\partial \boldsymbol{r}} \boldsymbol{v}_{\mathrm{h}} - \frac{2}{3} \mathsf{U} \left(\frac{\partial}{\partial \boldsymbol{r}} \cdot \boldsymbol{v}_{\mathrm{h}}\right)\right\} + \\ & + \frac{m_{\mathrm{h}} N_{\mathrm{h}}}{\sum_{\mathrm{g}} \nu_{\mathrm{hg}}} \sum_{\mathrm{g}} \nu_{\mathrm{hg}} \left\{(\boldsymbol{v}_{\mathrm{h}} - \boldsymbol{v}_{\mathrm{g}}) (\boldsymbol{v}_{\mathrm{h}} - \boldsymbol{v}_{\mathrm{g}}) - \frac{1}{3} \mathsf{U} |\boldsymbol{v}_{\mathrm{h}} - \boldsymbol{v}_{\mathrm{g}}|^2\right\}, \end{aligned}\right\} \tag{15.3}$$

i.e. in components:

$$\left.\begin{aligned} \Pi_{\mathrm{h}jk} = & - \frac{P_{\mathrm{h}}}{\sum_{\mathrm{g}} \nu_{\mathrm{hg}}} \left\{\frac{\partial v_{\mathrm{h}j}}{\partial r_k} + \frac{\partial v_{\mathrm{h}k}}{\partial r_j} - \frac{2}{3} \delta_{jk} \operatorname{div} \boldsymbol{v}_{\mathrm{h}}\right\} + \\ & + \frac{m_{\mathrm{h}} N_{\mathrm{h}}}{\sum_{\mathrm{g}} \nu_{\mathrm{hg}}} \sum_{\mathrm{g}} \nu_{\mathrm{hg}} \left\{(\boldsymbol{v}_{\mathrm{h}} - \boldsymbol{v}_{\mathrm{g}})_j (\boldsymbol{v}_{\mathrm{h}} - \boldsymbol{v}_{\mathrm{g}})_k - \frac{1}{3} \delta_{jk} |\boldsymbol{v}_{\mathrm{h}} - \boldsymbol{v}_{\mathrm{g}}|^2\right\}, \end{aligned}\right\} \tag{15.3a}$$

and the heat flux vector $\boldsymbol{q}_\mathrm{h}$ by:

$$\left.\begin{aligned}\boldsymbol{q}_\mathrm{h} = &-\frac{5}{2}\frac{P_\mathrm{h}}{m_\mathrm{h}\sum\limits_\mathrm{g}\nu_\mathrm{hg}}\frac{\partial}{\partial\boldsymbol{r}}\vartheta_\mathrm{h} + \\ &+\frac{1}{2}\frac{m_\mathrm{h}N_\mathrm{h}}{\sum\limits_\mathrm{g}\nu_\mathrm{hg}}\sum_\mathrm{g}\nu_\mathrm{hg}\left\{(\boldsymbol{v}_\mathrm{h}-\boldsymbol{v}_\mathrm{g})\left[5\frac{\vartheta_\mathrm{h}-\vartheta_\mathrm{g}}{m_\mathrm{h}+m_\mathrm{g}} - |\boldsymbol{v}_\mathrm{h}-\boldsymbol{v}_\mathrm{g}|^2\right]\right\}.\end{aligned}\right\} \quad (15.4)$$

β) The *conditions of validity* of Eqs. (15.2), 15.3) correspond to those of Eqs. (13.2) ... (13.20) for the completely ionized plasma, namely Eqs. (13.31), which must, of course, be valid here for each of the three kinds of particles.

We note that in §§ 13 and 14 of [*1*] the dynamics of the weakly ionized plasma is described by equations of motion which do not take account of viscous tensions wherefore an Eq. (15.1) without the terms containing Π appears in that monograph. Such a system is only closed in the case of constant temperature for all components of the plasma. If this is not true then the energy balance has to be established, i.e. Eq. (15.2). We admit that the terms containing Π, i.e. viscous tensions, are small under the conditions of validity of the present approximations, namely Eqs. (13.31). However, in certain cases these terms just determine the attenuation (or amplification) of low frequency waves in the plasma. They must therefore be taken into account when considering drift oscillations of the inhomogeneous plasma (see part E). For such cases as considered in [*1*] the Π-terms are in fact negligible.

The system of Eqs. (15.1), (15.2) takes account of all kinds of collisions of the charged particles, with the neutrals as well as amongst themselves. This has been done because this system, together with the BGK model of the collisional integral has recently been successfully applied to the completely ionized plasma, recently[1]. The results so obtained were found to be in qualitative agreement with those obtained with LANDAU's accurate collisional integral[2]. Here, however, we shall apply the BGK model of the collisional integral only for the weakly ionized plasma where collisions of charged particles amongst themselves are negligible[3].

γ) Finally we state that the system Eqs. (15.1), (15.2), in the case of a weakly ionized plasma leads to results which are, even quantitatively, correct. In order to demonstrate this we may consider the behaviour of a *plasma in a constant external electric field*, $\boldsymbol{E}_0$. For simplicity a magnetic field should not be present. As a consequence of the electric field a drift of electrons is produced in the plasma (also a drift of ions, however this is negligible against the motion of the electrons). The drifting electrons produce Joule heating. The case has been considered in [*31*] using an accurate collisional integral; see also [*1*]. The system of Eqs. (15.1), (15.2) for drifting electrons leads to expressions for drift velocity

$$\boldsymbol{v}_\mathrm{e} = \frac{q_\mathrm{e}\boldsymbol{E}_0}{m_\mathrm{e}\nu_\mathrm{en}}, \quad (15.5)$$

and effective electron temperature

$$\vartheta_\mathrm{e} = \vartheta_\mathrm{n} + \frac{2}{3}\frac{q_\mathrm{e}^2 E_0^2}{m_\mathrm{e}\delta\nu_\mathrm{en}^2} \quad (15.6)$$

[1] See, for example, [*54*] and B. B. KADOMTSEV and D. P. POGUTZE: [*6*] vol. 5, 209 (1963). See also: A. B. MIHAILOVSKIJ and O. P. POGUTZE: Usp. Fiz. Nauk. **20** ... (engl. transl.: Sov. Phys. JETP **20**, 630 (1965); german transl.: Techn. Physik **11**, 153 (1966). The subject is also considered in: B. COPPI: Phys. ZS. **12**, 213 (1964) and **14**, 172 (1964).

[2] This agreement could, of course, be checked only in cases where the latter formula could be resolved.

[3] In the ionosphere, this hypothesis is not true near the peak of the F2 region, and above it. See K. RAWER and K. SUCHY, this Encyclopedia, vol. 49/2, Sect. 3.

where

$$\delta = 2\,\frac{m_e}{m_n} \tag{15.7}$$

is the energy transfer coefficient of electrons with neutrals in elastic collisions. These result fully agree with those of the exact theory [*31*], [*1*].

16. Limits of application of the linear approximation. The models and equations considered in Sects. 13 through 15 for describing a plasma apply to any motion of the plasma, to non-linear ones as well, and this with and without collisions. In the following we shall restrict our considerations, however, to "linear motions", i.e. to cases where the equations of motion can be linearized. Further, as stated in the introduction, only oscillations and waves in the plasma shall be considered. There remains, of course, the problem of applicability of the linear approximation.

α) We may begin by stating that the question is not trivial at all, and not completely resolved up to now. Looking upon plasma electrodynamics from a sufficiently general viewpoint one could say that *non-linear phenomena* occur more often such that linear phenomena are exceptional and not the rule.

In this context, the complete system of equations of plasma electrodynamics which includes the kinetic equation (the equations of fluid- or gas-dynamic) together with the field equations, is non-linear. Generally speaking these equations can be linearized, however not rarely the range of applicability of the linear approximation is not very large. This range depends fundamentally on the shape of the statistical distribution function (or, as one may call it, on the equilibrium function of distribution of particles), also on density, temperature and collision frequency of particles. One important observation is the fact that the presence of a rather weak electromagnetic field produces non-linearity, and this is a characteristic peculiarity of plasmas as compared with other states of matter.

Under the influence of an electromagnetic field the waves change the macroscopie plasma parameters like density, temperature and oriented (overall) velocity as well as the microscopic ones i.e. the statistical distribution function; for example in a small range of velocity around the phase velocity of the wave resonance interaction occurs between particles and the wave. The limits of applicability of the linear approximation depend essentially on the smallness of these changes.

β) In the case of a plasma which is thermodynamically in equilibrium, and homogeneous in space sufficiently small perturbations decrease when time goes on. Such a plasma is *stable* against small perturbations. Strictly speaking only in this case limits of applicability of the linear approximation can be determined. We shall see below in Sects. 28, 29, 37 and 38 that in an *instable* plasma linearity is only valid for a very short time after a sufficiently small perturbation, namely as long as the inhomogeneity or the anisotropy of the velocity distribution (or any other perturbation parameter) is yet small. However, under conditions of instability even an extremely small perturbation increases with time, so that it becomes strong after a short delay; then, of course, the linear approximation ceases to be applicable. This time as an order of magnitude is given by the reciprocal of the logarithmic increment characterizing the increasing development of the perturbation. It is a measure of the time in which the homogeneous state of the plasma is destroyed by the development of non-linear motions.

When determining limits of applicability of the linear approximation in a stable plasma we distinguish two limiting cases, namely plasma with and without collisions. The first case is given by the condition that the process in question needs considerably more time than the free flight time of a particle, the second case stems from the opposite condition.

γ) In a *plasma with collision* the non-linear effect of wave propagation is mainly heating; it is facilitated by the slowness of energy transfer from the electrons to the heavier particles, namely ions and neutrals. This condition is due to the smallness of the collisional coefficient

$$\delta = 2\frac{m_e}{m_{i,n}} \tag{16.1}$$

which is the average of the specific energy transfer from an electron (e) when colliding wish a heavy particle (i or n). The electrons get a certain amount of energy from the field by the process described by Eq. (15.5), namely acceleration in a field[1]. This means heating of the electronic component. The heating is not important, according to Eq. (15.6), as long as

$$\frac{q_e^2 E_0^2}{m_e \nu_{en}^2}\frac{1}{\vartheta_0} = \frac{v_e^2}{V_{Te}^2} \ll \delta \ll 1. \tag{16.2}$$

In cases where this inequality is valid, even the drift of electrons in the wave field may, thermodynamically, be neglected. Condition Eq. (16.2) describes the range of applicability of the linear approximation in a plasma with collisions. In the special case of a completely ionized plasma, ν_{en} in Eq. (16.2) has to be replaced by ν_{ei}, the collision frequency of electrons with ions (see[*1*] for details).

δ) The picture is considerably more involved in the case of a *collisionless plasma*, i.e. when the characteristic time of the considered process is small against the free-flight time of particles. Non-linear phenomena at the propagation of electromagnetic waves in such a plasma are mainly due to density changes, and to the truncation of the distribution function of the particles due to resonance with the phase velocity of the wave.

The first of these effects, density change, arises under the action of the averaged high frequency force[2] [*32*], [*1*]:

$$\boldsymbol{F} = \frac{q^2}{u\, m_e\, \omega^2}\frac{\partial}{\partial \boldsymbol{r}}|\boldsymbol{E}_0|^2 \tag{16.3}$$

acting upon each particle. This force produce a redistribution of the electron density in the plasma. Apparently, the effect remains unimportant when the potential corresponding to this force is small compared with the average thermal energy of a particle, thus when

$$\frac{q_e^2|\boldsymbol{E}_0|^2}{m_e\,\omega^2} \ll \vartheta_e, \quad \text{i.e.} \quad |\boldsymbol{v}_e|^2 = \frac{q_e^2|\boldsymbol{E}_0|^2}{m_e^2\,\omega^2} \ll V_{Te}^2, \tag{16.4}$$

$$\boldsymbol{v}_e = \frac{q_e}{m_e\,\omega}\boldsymbol{E}_0 \tag{16.5}$$

is the velocity which is reached by the electrons after being one halfperiod in the high-frequency field.

The other phenomenon, viz. truncation of the distribution function of particles, is a consequence of resonance interaction with an electromagnetic wave.

---

[1] Eq. (15.5) may also be applied to an alternative field, taking $\boldsymbol{E}_0$ as its amplitude, provided $\omega \ll \nu_{en}$.

[2] This expression for the force is easily obtained from the equation of motion in a collisionless plasma by averaging over the time, see Sect. 6.

The phenomenon is also known as "quasi-nonlinear effect", and occurs quite generally in a collisionless plasma. However, truncation occurs quite slowly provided the energy density of the electromagnetic field is small, against the thermal energy of the particles, i.e. their average kinetic energy:

$$\mathscr{E} \ll N_e \vartheta_e. \tag{16.6}$$

The characteristic time for truncation is

$$\tau \approx \left(\frac{q_e |\boldsymbol{E}_0|}{m_e} |\boldsymbol{k}|\right)^{-\frac{1}{2}}. \tag{16.7}$$

$\boldsymbol{k}$ being the wave number vector of the wave. Thus, in order to have a range of linear behaviour one can always indicate conditions concerning the amplitude $\boldsymbol{E}_0$ of the wave, provided the truncation of the particle distribution function is negligible against the time of development of the linear process considered[3]. Considering particle collisions one establishes the equilibrium distribution function; the truncation of the actual distribution function remains negligible under the condition

$$\nu \gg \tau^{-1} \approx \left(\frac{q_e |\boldsymbol{E}_0|}{m_e} |\boldsymbol{k}|\right)^{\frac{1}{2}}. \tag{16.8}$$

ε) The inequalities Eqs. (16.2), (16.4) and (16.8) indicate the limits of applicability of the linear approximation to the propagation of electromagnetic waves in a plasma. If these conditions are satisfied, non-linear effects may be neglected but are potentially present with small amplitude. Non-linear effects accompany any case of wave propagation in a plasma but their importance increases with increasing wave amplitude.

C. Waves in plasma.

In the preceeding chapters we considered — with different emphasis — the different model approaches which are used to describe the conditions in a plasma. The most general model, of course, is one using the kinetic equations. This is the description we shall apply in most cases in this chapter. Only occasionally the quasi-gasdynamic equations will be used.

I. Homogeneous and isotropic plasma.

17. Permittivity (dielectric permeability) of a collisionless plasma in the absence of external fields. We begin with this simplest case supposing the plasma to be unlimited in space, homogeneous and isotropic. Under these conditions the distribution function of the particles can be formulated as to depend on one parameter only, namely the absolute value of the momentum. First suppose the distribution function[1] for each kind of particles (h) to be Maxwellian:

$$f_{0\mathrm{h}}(\boldsymbol{v}) = \frac{N_\mathrm{h}}{(2\pi m_\mathrm{h} \vartheta_\mathrm{h})^{\frac{3}{2}}} \exp\left(-\frac{m_\mathrm{h}|\boldsymbol{v}|^2}{2\vartheta_\mathrm{h}}\right). \tag{17.1}$$

This expression, apparently, satisfies the stationary and homogeneous kinetic equation with selfconsistent field, see Sect. 8, in particular Eq. (8.4).

[3] We do not give such conditions in detail here. The reader is referred to the paper by A. A. Vedenov in [6].

[1] We define the distribution function f in the momentum ($\boldsymbol{p}$) space so that $\int d^3\boldsymbol{p}\, f_\mathrm{h} = N_\mathrm{h}$. The SI-unit for f is therefore $(\mathrm{sec/kg\ m^2})^3 \equiv (\mathrm{J\ sec})^{-3}$.

α) In order to obtain the (dielectric) permittivity of the plasma we have to consider a small *deviation* of the distribution function *from the equilibrium* distribution, Eq. (17.1). Thus, we suppose

$$f_{\mathrm{h}} = f_{0\mathrm{h}} + \delta f_{\mathrm{h}} \tag{17.2}$$

where the deviation δf_{h} is due to fluctuations of the fields $\boldsymbol{E}$ and $\boldsymbol{B}$. Linearizing the kinetic equation, Eq. (8.4), and supposing δf, $\boldsymbol{E}$ and $\boldsymbol{B}$ to be small of first order, we have:

$$\frac{\partial \delta f_{\mathrm{h}}}{\partial t} + \boldsymbol{v} \cdot \frac{\partial \delta f_{\mathrm{h}}}{\partial \boldsymbol{r}} + q_{\mathrm{h}} \boldsymbol{E} \cdot \frac{\partial f_{0\mathrm{h}}}{\partial \boldsymbol{p}} = 0. \tag{17.3}$$

In agreement with the assumptions made above we have omitted here constant external fields, $\boldsymbol{E}_0$ and $\boldsymbol{B}_0$. We should, however, remark that this assumption is not equivalent with using the Maxwellian distribution, Eq. (17.1), as zero order approximation. It is enough to state that even in the presence of a homogeneous and constant magnetic field, $\boldsymbol{B}_0$, Eq. (17.1) satisfies the zero order kinetic equation, see Sects. 22 and 23.

As the field equations, Eqs. (1.6) are linear and we use the linearized kinetic equation, Eq. (17.3), we are allowed to express the time and space function for all variable quantities as a FOURIER-development composed of harmonic waves of the form

$$\exp(-i\omega t + i\boldsymbol{k}\cdot\boldsymbol{r}). \tag{17.4}$$

The corresponding solution of Eq. (17.3) yields

$$\delta f_{\mathrm{h}} = \frac{i q_{\mathrm{h}}}{\omega - \boldsymbol{k}\cdot\boldsymbol{v}} \boldsymbol{E} \cdot \frac{\partial f_{0\mathrm{h}}}{\partial \boldsymbol{p}}. \tag{17.5}$$

β) We shall now *compute the current* which is induced in the plasma by changes of $\boldsymbol{E}$ or $\boldsymbol{B}$. This procedure gives us the conductivity of the plasma, and so its (dielectric) permittivity. In correspondance with Eqs. (8.5), (8.6) we have to write here for the current density:

$$\left.\begin{aligned} \boldsymbol{J} = \mathsf{S}\cdot\boldsymbol{E} &= -i\frac{\omega}{\mathrm{u}}(\mathsf{E} - \varepsilon_0 \mathsf{U})\cdot\boldsymbol{E} \\ &= -i\sum_{\mathrm{h}} q_{\mathrm{h}}^2 \int \mathrm{d}^3\boldsymbol{p}\, \frac{\boldsymbol{v}\,\boldsymbol{E}}{\omega - \boldsymbol{k}\cdot\boldsymbol{v}} \cdot \frac{\partial f_{0\mathrm{h}}}{\partial \boldsymbol{p}} \end{aligned}\right\} \tag{17.6}$$

which reads in components[2]:

$$\left.\begin{aligned} J_l = \sigma_{lm} E_m &= -i\frac{\omega}{\mathrm{u}}(\varepsilon_{lm} - \varepsilon_0 \delta_{lm}) E_m \\ &= -i\sum_{\mathrm{h}} q_{\mathrm{h}}^2 \int \mathrm{d}^3\boldsymbol{p}\, \frac{v_l E_m}{\omega - k_n v_n} \frac{\partial f_{0\mathrm{h}}}{\partial p_m}. \end{aligned}\right\} \tag{17.6a}$$

Here $\mathsf{S} \equiv ((\sigma_{lm}))$ is the tensor describing the complex conductivity of the plasma:

$$\mathsf{S}(\omega, \boldsymbol{k}) = -i\sum_{\mathrm{h}} q_{\mathrm{h}}^2 \int \mathrm{d}^3\boldsymbol{p}\, \frac{\boldsymbol{v}\,\dfrac{\partial f_{0\mathrm{h}}}{\partial \boldsymbol{p}}}{(\omega - \boldsymbol{k}\cdot\boldsymbol{v})} \tag{17.7}$$

which reads in components:

$$\sigma_{lm}(\omega, \boldsymbol{k}) = -i\sum_{\mathrm{h}} q_{\mathrm{h}}^2 \int \mathrm{d}^3\boldsymbol{p}\, \frac{v_l\,\dfrac{\partial f_{0\mathrm{h}}}{\partial p_m}}{(\omega - k_n v_n)}. \tag{17.7a}$$

---

[2] When writing tensorial equations in components we use the summation rule. Thus, for example,

$$k_l v_l \quad \text{means} \quad \sum_l k_l v_l, \quad \text{etc.}$$

Using Eq. (1.11) or, more directly, Eq. (17.6) from the description given by Eq. (17.7) which admits a complex conductivity we may go over to one admitting a complex (dielectric) permittivity, viz.

$$\frac{1}{\varepsilon_0}\,\mathsf{E}(\omega,\boldsymbol{k}) = \mathsf{U} + \sum_{\mathrm{h}} \frac{\mathrm{u}\,q_{\mathrm{h}}^2}{\varepsilon_0\,\omega} \int \mathrm{d}^3\boldsymbol{p}\, \frac{1}{(\omega-\boldsymbol{k}\cdot\boldsymbol{v})}\,\boldsymbol{v}\,\frac{\partial f_{0\mathrm{h}}}{\partial \boldsymbol{p}} \tag{17.8}$$

which reads in components:

$$\frac{1}{\varepsilon_0}\,\varepsilon_{lm}(\omega,\boldsymbol{k}) = \delta_{lm} + \sum_{\mathrm{h}} \frac{\mathrm{u}\,q_{\mathrm{h}}^2}{\varepsilon_0\,\omega} \int \mathrm{d}^3\boldsymbol{p}\, \frac{v_l}{(\omega-k_n v_n)}\,\frac{\partial f_{0\mathrm{h}}}{\partial p_m}\,. \tag{17.8a}$$

In Eqs. (17.6) ... (17.8) we have to sum over all kinds (h) of charged particles in the plasma. If conditions are such that the plasma is collisionless, the neutral particles ($q_{\mathrm{n}}=0$) apparently do not participate in influencing electromagnetic phenomena. We should, however, state here that the „collisionless" approximation does not really mean a plasma where no collisions occur for ever, but one which has attained by collisions a certain statistical distribution function and then only undergoes a process which is so rapid that collisions do not play a role in it. Thus, we even suppose the problem with collisions to be resolved before we are in a position to admit a „collisionless" process.

γ) Our next question shall be concerned with the denominator $(\omega-\boldsymbol{k}\cdot\boldsymbol{v})$ in Eqs. (17.5) ... (17.8). It results in a *pole of the integrand* at $\omega=\boldsymbol{k}\cdot\boldsymbol{v}$. The question is, how to deal with it. The answer can be formulated in different manners.

One method has been used in [*11*][3] taking the position that FOURIER-analysis is adequate and thus ω must rigorously be a real quantity. On the other hand, $\boldsymbol{k}$ represents a wave component and under real physical condition there is always either attenuation (under stable conditions) or amplification (in unstable conditions). Thus, $\boldsymbol{k}$ contains a (possibly very small) imaginary part. The consequence is then that $\boldsymbol{k}\cdot\boldsymbol{v}$ is not attainable with a real ω; with other words, the pole is not exactly on the ω-integration path and the integral is certainly finite.

We prefer another argumentation considering LAPLACE analysis as to be more adequate here. Looking upon the distribution function as being defined for all values of t, we meet a difficulty with past time, far away from the instant considered. This difficulty disappears if we assume that $\delta f_{\mathrm{h}}\to 0$ for $t\to-\infty$; the meaning is, of course, that equilibrium existed at the beginning[4] so that no waves at all were existing. Adopting this reasoning we admit strictly real wave components, thus $\boldsymbol{k}$ real, but we have to admit now an imaginary part in ω; it is positive with the signs chosen in Eq. (17.4). This part could certainly be extremely small. We shall see below in Sect. 18 that collisions between particles quite naturally introduce a small, positive imaginary component to ω in Eq. (17.5).

The result of this reasonning is that the expressions Eq. (17.7) for S and (17.8) for E are to be considered as limiting cases of functions of a complex variable ω; the functions are certainly defined in the upper half of the complex ω-plane. The pole ω' lies on the real axis[5] (see Fig. 1) but a realistic integration path has to take account of the (extremely small) imaginary part in ω, therefore it avoids the pole[6], as shown by the conntur C in Fig. 1. We shall meet integrals of this type more often and always itoetgrae along a contour of this type.

[3] K. RAWER and K. SUCHY: this Encyclopedia, Vol. 49/2, Sect. 9ε use the eikonal formula and admit one complex coordinate. See in particular their Fig. 37. The procedure has first been used by O. E. H. RYDBECK: Trans. Chalmers Univ., Gothenburg, Nr. 101 (1951).

[4] One often speaks of including the field "adiabatically" into an infinitely far past.

[5] With the reasonning given (in small print) above the pole would not be on the real axis, but just below (in case of attenuation), or above (in case of amplification). The integration path would then be along the real axis.

[6] See [*2*] and [*13*] for details.

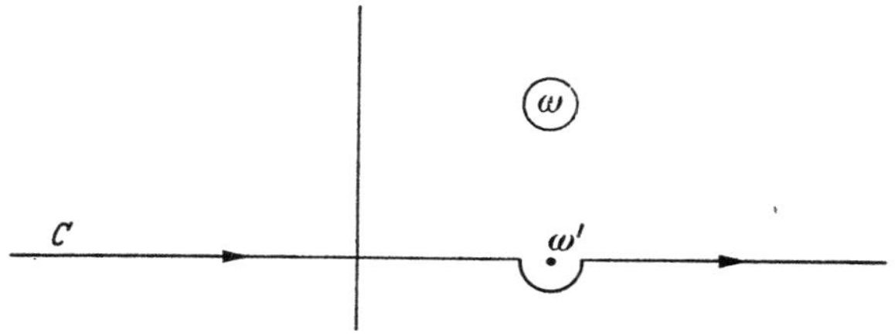

Fig. 1. Integration in the complex ω-plane.

δ) Taking account of these reasonnings and using the well known relation[7]

$$\lim_{\alpha\to+0}\frac{1}{x+i\alpha}=\frac{P}{x}-i\pi\,\delta(x)$$

we may rewrite the dielectric susceptibility tensor of the plasma. Eq. (17.8), in the following form

$$\frac{1}{\varepsilon_0}\,\mathsf{E}(\omega,\boldsymbol{k})=\mathsf{U}+\sum_{\mathrm{h}}\frac{\mathrm{u}\,q_{\mathrm{h}}^2}{\varepsilon_0\,\omega}\int \mathrm{d}^3\boldsymbol{p}\,\boldsymbol{v}\,\frac{\partial f_{0\mathrm{h}}}{\partial\boldsymbol{p}}\left\{\frac{P}{(\omega-\boldsymbol{k}\cdot\boldsymbol{v})}-i\pi\,\delta(\omega-\boldsymbol{k}\cdot\boldsymbol{v})\right\} \tag{17.9}$$

which reads in components

$$\frac{1}{\varepsilon_0}\,\varepsilon_{lm}(\omega,\boldsymbol{k})=\delta_{lm}+\sum_{\mathrm{h}}\frac{\mathrm{u}\,q_{\mathrm{h}}^2}{\varepsilon_0\,\omega}\int \mathrm{d}^3\boldsymbol{p}\,v_l\,\frac{\partial f_{0\mathrm{h}}}{\partial p_m}\left\{\frac{P}{(\omega-k_n v_n)}-i\pi\,\delta(\omega-\boldsymbol{k}\cdot\boldsymbol{v})\right\} \tag{17.9a}$$

and δ depends on $(\omega-\boldsymbol{k}\cdot\boldsymbol{v})$.

This form shows that only particles with velocity $\boldsymbol{v}$ satisfying $\omega\approx\boldsymbol{k}\cdot\boldsymbol{v}$ contribute to the imaginary part of the (dielectric) permittivity. Just this imaginary part is responsible for the absorption of electromagnetic energy in the plasma, see Chap. A.II. The condition may be written as

$$\frac{\omega}{|\boldsymbol{k}|}\equiv V_{\mathrm{ph}}\approx|\boldsymbol{v}|\cos\theta, \tag{17.10}$$

V_{ph} being the phase velocity of the electromagnetic wave and θ the angle between the wave vector $\boldsymbol{k}$ and particle velocity $\boldsymbol{v}$. Eq. (17.10) is known as emission condition for ČERENKOV radiation. The same relation apparently holds for the inverse process, which is ČERENKOV *absorption*. We see that in a collisionless, isotropic plasma there exists nevertheless a dissipative process, namely ČERENKOV absorption due to those particles of the plasma which have adequate velocity. The result is easily understood: in the absence of collisions and of a magnetic field the plasma particles have straight lines as orbits, and they travel with constant velocity. Emission and absorption of electromagnetic waves occurs then as CERENKOV emission and absorption.

ε) From Eqs. (17.7) ... (17.9) one easily proves that for an isotropic plasma the tensor of (dielectric) permittivity has only two independent components [see Eq. (3.8)], namely a transverse and a longitudinal one[8]:

$$\frac{1}{\varepsilon_0}\,\varepsilon^{\mathrm{tr}}(\omega,\boldsymbol{k})=1+\sum_{\mathrm{h}}\frac{\mathrm{u}\,q_{\mathrm{h}}^2}{2\varepsilon_0\,\omega\,k^2}\int \mathrm{d}^3\boldsymbol{p}\,\frac{(\boldsymbol{k}\times\boldsymbol{v})^2}{(\omega-\boldsymbol{k}\cdot\boldsymbol{v})}\,\frac{\partial f_{0\mathrm{h}}}{\partial\mathscr{E}_{\mathrm{h}}} \tag{17.11a}$$

---

[7] P means the principal value characterizing the conditions at $x=0$.

[8] k always means the absolute value of the wave (number) vector $\boldsymbol{k}$; not to be confounded with the BOLTZMANN constant k.

and

$$\frac{1}{\varepsilon_0}\,\varepsilon^{\mathrm{l}}(\omega, \boldsymbol{k}) = 1 + \sum_{\mathrm{h}} \frac{\mathrm{u}\, q_{\mathrm{h}}^2}{\varepsilon_0\, \omega\, k^2} \int \mathrm{d}^3 \boldsymbol{p}\, \frac{(\boldsymbol{k}\cdot\boldsymbol{v})^2}{(\omega - \boldsymbol{k}\cdot\boldsymbol{v})}\, \frac{\partial f_{0\mathrm{h}}}{\partial \mathscr{E}_{\mathrm{h}}}, \tag{17.11b}$$

where

$$\mathscr{E}_{\mathrm{h}} = \frac{|\boldsymbol{p}_{\mathrm{h}}|^2}{2 m_{\mathrm{h}}} \tag{17.12}$$

is the kinetic energy of particles of kind h. The integration must, again, follow the contour C. For the Maxwellian distribution, Eq. (17.1), the Eqs. (17.11) are identic with [2]:

$$\frac{1}{\varepsilon_0}\,\varepsilon^{\mathrm{tr}}(\omega, \boldsymbol{k}) = 1 - \sum_{\mathrm{h}} \frac{\omega_{N\mathrm{h}}^2}{\omega^2}\, \mathrm{J}_+\left(\frac{\omega}{k V_{T\mathrm{h}}}\right), \tag{17.13a}$$

$$\frac{1}{\varepsilon_0}\,\varepsilon^{\mathrm{l}}(\omega, \boldsymbol{k}) = 1 + \sum_{\mathrm{h}} \frac{\omega_{N\mathrm{h}}^2}{k^2 V_{T\mathrm{h}}^2}\left[1 - \mathrm{J}_+\left(\frac{\omega}{k V_{T\mathrm{h}}}\right)\right] \tag{17.13b}$$

with

$$\mathrm{J}_+(x) \equiv x \cdot \mathrm{e}^{-\frac{1}{2}x^2} \int_{-i\infty}^{x} \mathrm{d}x\, \mathrm{e}^{\frac{1}{2}x^2} = -i\sqrt{\frac{\pi}{2}}\, x\, \mathrm{W}\left(\frac{x}{\sqrt{2}}\right). \tag{17.14}$$

The special function $\mathrm{W}(x)$ has been discussed and tabulated in [33]. Eqs. (17.13) give the transverse and longitudinal dielectric susceptibility of a non-relativistic plasma for a real wave vektor $\boldsymbol{k}$; they are valid in the whole complex plane of ω [9].

In the following, we shall need asymptotic expansions of the function $\mathrm{J}_+(x)$. These are:

$$\left.\begin{aligned}
&\mathrm{J}_+(x) \approx \left(1 + \frac{1}{x^2} + \frac{3}{x^4} + \cdots\right) - i\sqrt{\frac{\pi}{2}}\, x\, \mathrm{e}^{-\frac{1}{2}x^2} \\
&\qquad\qquad \text{for } |x| \gg 1,\ |\mathrm{Re}\, x| \gg |\mathrm{Im}\, x|; \\
&\mathrm{J}_+(x) \approx -i\sqrt{\frac{\pi}{2}}\, x \quad \text{for } |x| \ll 1; \\
&\mathrm{J}_+(x) \approx -i\sqrt{2\pi}\, x\, \mathrm{e}^{-\frac{1}{2}x^2} \quad \text{for } |x| \gg 1,\ |\mathrm{Im}\, x| \gg |\mathrm{Re}\, x|,\ \mathrm{Im}\, x < 0.
\end{aligned}\right\} \tag{17.15}$$

18. Permittivity (dielectric permeability) of a plasma with collisions. We shall now admit collisions and compute their influence. Unfortunately, LANDAU's kinetic equation, Eqs. (12.13, 14) and (12.20), can not be resolved in the general case; even after linearization this equation remains an integral equation.

For this reason we are unable to give a general expression for the tensor of dielectric susceptibility in a completely ionized plasma with COULOMB collisions which would be valid for all frequencies and wave numbers. Fortunately, the ranges in ω and $\boldsymbol{k}$ for which the dielectricity tensor $\mathsf{E}(\omega, \boldsymbol{k})$ can be given cover sufficiently well the cases of plasma oscillations. This is in particular so for the ionospheric plasma when the magnetic field of Earth can be neglected, i.e. when $\omega \gg \omega_{B\mathrm{e}}$, and when $\omega \gg (\nu_{\mathrm{e}}, \nu_{\mathrm{i}})$, see Tables 1, 2 and 3 in Sect. 2.

α) In the limiting case of a *small collision frequency*, $\omega \gg \nu_{\mathrm{h}}$ the collision integral in the kinetic equation, Eq. (12.13), is a small term so that solutions of this equation can be obtained by successive approximation, developing after powers of the deviation from the undisturbed distribution function. In the zero order approximation we have, of course, Eq. (17.5) as solution. In the first order we have to admit a correction of the distribution function admitting only first order derivatives of this function; we must take account of all kinds of collisions with

---

[9] We note that Eqs. (17.13) could also be considered as referring to a complex wave vector $\boldsymbol{k}$ and a real frequency $\omega/2\pi$, see Subsect. α (small print). With this interpretation by the action of collisions between particles the vector $\boldsymbol{k}$ gets a small negative imaginary part.

the particles considered, including collisions between particles of the same kind[1]. The result is the following correction for the solution δf_{h} of Eq. (17.5):

$$\left.\begin{aligned}\delta f_{\mathrm{h}}^{(1)} = \frac{i}{(\omega - \boldsymbol{k}\cdot\boldsymbol{v})} \sum_{\mathrm{g}} \frac{\partial}{\partial \boldsymbol{p}} \cdot \int \mathrm{d}\boldsymbol{p}'\, I^{(\mathrm{h},\mathrm{g})}(\boldsymbol{p},\boldsymbol{p}') \cdot \Big\{ \frac{\partial f_{0\mathrm{h}}(\boldsymbol{p})}{\partial \boldsymbol{p}}\, \delta f_{\mathrm{g}}(\boldsymbol{p}') + \\ + \frac{\partial \delta f_{\mathrm{h}}(\boldsymbol{p})}{\partial \boldsymbol{p}} f_{0\mathrm{g}}(\boldsymbol{p}') - f_{0\mathrm{h}}(\boldsymbol{p}) \frac{\partial \delta f_{\mathrm{g}}(\boldsymbol{p}')}{\partial \boldsymbol{p}'} - \delta f_{\mathrm{h}}(\boldsymbol{p}) \frac{\partial f_{0\mathrm{g}}(\boldsymbol{p}')}{\partial \boldsymbol{p}'} \Big\},\end{aligned}\right\} \tag{18.1}$$

or in components (l, m):

$$\left.\begin{aligned}\delta f_{\mathrm{h}}^{(1)} = \frac{1}{(\omega - k_n v_n)} \sum_{\mathrm{g}} \frac{\partial}{\partial p_l} \int \mathrm{d}\boldsymbol{p}'\, I_{lm}^{(\mathrm{h},\mathrm{g})}(\boldsymbol{p},\boldsymbol{p}') \Big\{ \frac{\partial f_{0\mathrm{h}}(\boldsymbol{p})}{\partial p_m}\, \delta f_{\mathrm{g}}(\boldsymbol{p}') + \\ + \frac{\partial \delta f_{\mathrm{h}}(\boldsymbol{p})}{\partial p_m} f_{0\mathrm{g}}(\boldsymbol{p}') - f_{0\mathrm{h}}(\boldsymbol{p}) \frac{\partial \delta f_{\mathrm{g}}(\boldsymbol{p}')}{\partial p'_m} - \delta f_{\mathrm{h}}(\boldsymbol{p}) \frac{\partial f_{0\mathrm{g}}(\boldsymbol{p}')}{\partial p'_m} \Big\}.\end{aligned}\right\} \tag{18.1a}$$

The current tensor $I^{(\mathrm{h},\mathrm{g})}(\boldsymbol{p},\boldsymbol{p}')$, depending on the values of $\boldsymbol{p}$ and $\boldsymbol{p}'$, is given by Eq. (12.20). The summation over g covers all kinds of particles (including $\mathrm{g}=\mathrm{h}$). Eq. (18.1) produces an additional term to the induced current density in the plasma; this, in turn, produces additional members to the (dielectric) permittivity tensor as given formerly (without collisions) in Eqs. (17.9) and (17.11). In the case of a rather cold plasma, where

$$\omega \gg (k\, V_{T\mathrm{e}},\, k\, V_{T\mathrm{i}})$$

only the additional electronic current is of importance; in a completely ionized plasma the correction is only due to collisions of electrons with ions, not amongst electrons[1]. The additional term in the dielectricity tensor E is

$$\delta \mathsf{E} = i\, \varepsilon_0 \frac{\omega_{N\mathrm{e}}^2}{\omega^3}\, \nu_{\mathrm{eff}}\, \mathsf{U}, \tag{18.2}$$

$$\delta E_{lm} = i\, \varepsilon_0 \frac{\omega_{N\mathrm{e}}^2}{\omega^3}\, \nu_{\mathrm{eff}}\, \delta_{lm} \tag{18.2a}$$

The effective collision frequency has been derived in Sect. 13:

$$\bar{\nu}_{\mathrm{e}} = \nu_{\mathrm{eff}} = \frac{1}{3} (2\pi)^{-\frac{3}{2}} \left(\frac{\mathrm{u}}{\varepsilon_0}\right)^2 q_{\mathrm{e}}^2\, q_{\mathrm{i}}^2\, N_{\mathrm{e}}\, \mathscr{L} / \vartheta_{\mathrm{e}}^{\frac{3}{2}}. \qquad [13.30\mathrm{a}]$$

With Eq. (18.2) our former expressions for the transverse and longitudinal dielectric susceptibility, Eqs. (17.13a, b), under the conditions

$$\omega \gg \bar{\nu}_{\mathrm{eff}}; \qquad \omega \gg k\, V_{T\mathrm{e}}; \qquad \omega \gg k\, V_{T\mathrm{i}},$$

may be written:

$$\frac{1}{\varepsilon_0}\, \varepsilon^{\mathrm{tr}} = 1 - \frac{\omega_{N\mathrm{e}}^2}{\omega^2} \left[1 - i \sqrt{\frac{\pi}{2}}\, \frac{\omega}{k\, V_{T\mathrm{e}}} \exp\left(\frac{-\omega^2}{2k^2 V_{T\mathrm{e}}^2}\right) - i\, \frac{\nu_{\mathrm{eff}}}{\omega}\right], \tag{18.3a}$$

$$\frac{1}{\varepsilon_0}\, \varepsilon^{\mathrm{l}} = 1 - \frac{\omega_{N\mathrm{e}}^2}{\omega^2} \left[1 + 3\, \frac{k^2 V_{T\mathrm{e}}^2}{\omega^2} - i \sqrt{\frac{\pi}{2}}\, \frac{\omega^3}{k^3 V_{T\mathrm{e}}^3} \exp\left(\frac{-\omega^2}{2k^2 V_{T\mathrm{e}}^2}\right) - i\, \frac{\nu_{\mathrm{eff}}}{\omega}\right]. \tag{18.3b}$$

In these equations the collisional influence appears only in a few additional terms all of which are proportional to the effective collision frequency, ν_{eff}. These terms describe the influence of wave absorption in a plasma, but only if the other imaginary terms in the Eqs. (18.3a), (18.3b) are small enough in the fre-

[1] These collisions play an important role to obtain conditions of statistical equilibrium, or to approach towards these. The reason why they do not appear in certain formulae of wave propagation is simply that they do not change the electric current density, see [*11*].

quency range considered. These terms describe collisionless absorption which is also called "spatial dispersion"[2]. As they are of exponential form, the condition is easily satisfied under most circumstances, particularly if $|\boldsymbol{k}|$ is small. In the long wave limit $|\boldsymbol{k}|\to 0$, spatial dispersion disappears completely; and one has instead of Eqs. (18.3a), (18.3b):

$$\frac{1}{\varepsilon_0}\,\varepsilon^{l}(\omega,0)=\frac{1}{\varepsilon_0}\,\varepsilon^{tr}(\omega,0)=1-\frac{\omega_{Ne}^2}{\omega^2}\left(1-i\,\frac{\nu_{\text{eff}}}{\omega}\right), \tag{18.4}$$

the well-known SELLMEIER formula[3].

We shall prove in Sect. 20 that the absorption of weakly attenuated electromagnetic *transverse* waves in a plasma is only determined by collisions; the reason is that the phase velocity, V_{ph}, of such waves is greater than the light velocity in vacuum, c_0; the imaginary part of ε_{tr} due to ČERENKOV absorption is exactly zero in ranges where ω and $\boldsymbol{k}$ are both real.

β) Under circumstances where $k\,V_{Te}\gg(\bar\nu_e,\omega)$ and $k\,V_{Ti}\gg(\bar\nu_i,\omega)$ the collisional influence is small. In that case the *collisionless* ČERENKOV *absorption* is the dominant absorption influence in the plasma. Therefore Eqs. (17.13a), (17.13b) are valid, which can be written as

$$\frac{1}{\varepsilon_0}\,\varepsilon^{tr}=1+i\sqrt{\frac{\pi}{2}}\sum_h\frac{\omega_{Nh}^2}{\omega k\,V_{Th}}\approx 1+i\sqrt{\frac{\pi}{2}}\,\frac{\omega_{Ne}^2}{\omega k\,V_{Te}}\quad\text{for}\quad k\,V_{Te}\gg(\bar\nu_e,\omega), \tag{18.5a}$$

and

$$\frac{1}{\varepsilon_0}\,\varepsilon^{l}=1+\sum_h\frac{\omega_{Nh}^2}{k^2V_{Th}^2}\left(1+i\sqrt{\frac{\pi}{2}}\,\frac{\omega}{k\,V_{Th}}\right)\approx 1+\sum_h\frac{1}{k^2\lambda_{Dh}^2}\equiv 1+\frac{1}{k^2\lambda_D^2} \tag{18.5b}$$

where we have introduced the DEBYE length[4] of the different particles

$$\lambda_{Dh}=\left(\frac{\varepsilon_0\,\vartheta_h}{\mathrm{u}\,q_h^2\,N_h}\right)^{\frac{1}{2}},\quad\text{and}\quad \lambda_D^2\equiv\frac{1}{\sum\limits_h\frac{1}{\lambda_{Dh}^2}} \tag{18.6}$$

which may be called the DEBYE length of the plasma as whole.

γ) Finally, under circumstances where

$$k\,V_{Te}\gg(\bar\nu_e,\omega),\quad\text{but}\quad \omega\gg(\bar\nu_i,k\,V_{Ti})$$

the electronic plasma component at least can be taken as collisionless. This means that the collision integral for electrons is negligible in the kinetic equation. In that case only collisions amongst ions must be taken into account. We have then the following expressions for transverse and longitudinal (dielectric) permittivity:

$$\frac{1}{\varepsilon_0}\,\varepsilon^{tr}=1+i\sqrt{\frac{\pi}{2}}\,\frac{\omega_{Ne}^2}{\omega\,k\,V_{Te}}\quad\text{for}\quad k\,V_{Ti}\gg(\bar\nu_e,\omega), \tag{18.7a}$$

$$\left.\begin{aligned}\frac{1}{\varepsilon_0}\,\varepsilon^{l}=1+\frac{\omega_{Ne}^2}{k^2V_{Te}^2}\left(1+i\sqrt{\frac{\pi}{2}}\,\frac{\omega}{k\,V_{Te}}\right)-\frac{\omega_{Ni}^2}{\omega^2}\times\\ \times\left[1+3\,\frac{k^2V_{Ti}^2}{\omega^2}-i\sqrt{\frac{\pi}{2}}\,\frac{\omega^3}{k^3V_{Ti}^3}\exp\left(\frac{-\omega^2}{2k^2V_{Ti}^2}\right)-i\,\frac{8}{5}\,\frac{\bar\nu_{ii}\,k^2}{\omega^3}\,V_{Ti}^2\right]\end{aligned}\right\} \tag{18.7b}$$

with [see Eq. (13.30a)]:

$$\bar\nu_{ii}=\frac{1}{3\sqrt{2}}\,(2\pi)^{-\frac{3}{2}}\left(\frac{\mathrm{u}}{\varepsilon_0}\right)^2 q_i^2\,q_i^2\,N_i\,\mathscr{L}/\vartheta_i^{\frac{3}{2}}. \qquad [13.30\text{b}]$$

---

[2] See [*1*], Sect. 8.

[3] K. RAWER and K. SUCHY: this Encyclopedia, Vol. 49/2, Sect. 6, Eq. (6.41).

[4] K. RAWER and K. SUCHY: this Encyclopedia, Vol. 49/2, Sect. 56, Eq. (56.20).

We just remark that the contribution of ion terms to the transverse (dielectric) permittivity is negligible in the frequency range we consider. On the other side, the ion terms may become important, in particular if we have a non-isotropic plasma with hot electrons, $T_e \gg T_i$. This case happens to exist in the middle of the ionosphere.

δ) We come now to the case of a *weakly ionized plasma*. In order to deduce the (dielectric) permittivity we use the model description of the collision integral given by BATNAGAR, GROSS, and KROOK [*29*]. It is the essential advantage of this representation of the collision integral that it allows the kinetic equation, Eqs. (8.8) and (14.4), to be resolved without any restriction concerning the frequency range or the wavelength. In fact if we suppose that the density of the neutrals remains homogeneous, and then linearize the kinetic equation using the deviation δf_h from the equilibrium distribution which is given by Eq. (17.1), we get

$$\frac{\partial\,\delta f_h}{\partial t} + \boldsymbol{v}\cdot\frac{\partial\,\delta f_h}{\partial \boldsymbol{r}} + q_h\,\boldsymbol{E}\cdot\frac{\partial f_{0h}}{\partial \boldsymbol{p}} = -\nu_{hn}\left(\delta f_h - \eta_h\,N_h\,\Phi^{(0)}_{hn}\right) \tag{18.8}$$

where ν_{hn} is the collision frequency of charged particles of kind (h) with the neutrals (n),

relative density variation

$$\eta_h \equiv \frac{1}{N_h}\int d\boldsymbol{p}\,\delta f_h, \tag{18.9}$$

distribution function

$$\Phi^{(0)}_{hn} \equiv (2\pi\, m_h\,\vartheta_{hn})^{-\frac{3}{2}} \exp\left(\frac{m_h\,|\boldsymbol{v}|^2}{2\vartheta_{hn}}\right) \tag{18.10}$$

and reduced average temperature

$$\vartheta_{hn} = \frac{m_h\,\vartheta_n + m_n\,\vartheta_h}{m_h + m_n}. \tag{18.11}$$

When deducing Eq. (18.8) the temperature variation of the charged particles has been neglected. This is the "isotropic model" [*22*]. If the charged particles are light compared to the neutrals, $m_h \ll m_n$, then, quite independently of the value of ϑ_n, one has $\vartheta_{hn} \approx \vartheta_h$. This is very accurately so for electrons (h = e) as the relative difference is of the order of m_h/m_n. For ions however (h = i), this is not so, such that $\vartheta_{hn} \neq \vartheta_h$ except for the case where mass and temperature of ions and neutrals are equal: $m_h = m_n$ and $\vartheta_h = \vartheta_n$. This might happen in some parts of the ionosphere, but certainly not in all parts of it[5]. In this case we have

$$\Phi^{(0)}_{hn}\,N_h = f_{0h}. \tag{18.12}$$

In these circumstances the kinetic equation, Eq. (18.8), can now be written for a monochromatic, plane wave

$$\left.\begin{matrix}\boldsymbol{E}\\ \delta f_h\end{matrix}\right\} \propto \exp(-i\omega t + i\boldsymbol{k}\cdot\boldsymbol{r}) \tag*{[17.2]}$$

in the following form:

$$\delta f_h = i\,\frac{q_h}{\vartheta_h}\,\frac{(\boldsymbol{v}\cdot\boldsymbol{E})\,f_{0h}}{(\omega + i\,\bar{\nu}_{hn} - \boldsymbol{k}\cdot\boldsymbol{v})} + i\,\frac{\bar{\nu}_{hn}\,\eta_h\,f_{0h}}{(\omega + i\,\bar{\nu}_{hn} - \boldsymbol{k}\cdot\boldsymbol{v})}. \tag{18.13}$$

[5] In many regions of the ionosphere $m_h \neq m_n$ because of quick charge transfer processes. See S. BAUER, this Encyclopedia, Vol. 49/5.

η_h is easily determined by integrating over the momentum space, then applying Eq. (18.9). It is, however, easier to express η_h by the current density $\boldsymbol{j}_h$ using the equation for the (dielectric) permittivity of particles of kind (h) only, resulting in

$$\eta_h = \frac{\boldsymbol{k} \cdot \boldsymbol{j}_h}{q_h N_h \omega}. \tag{18.14}$$

Putting this expression into Eq. (18.13) we can find the current density which is induced in the plasma. After some simple computation[6] we find for the transverse and longitudinal (dielectric) permittivity in a weakly ionized plasma

$$\frac{1}{\varepsilon_0} \varepsilon^{\mathrm{tr}}(\omega, \boldsymbol{k}) = 1 - \sum_h \frac{\omega_{Nh}^2}{\omega(\omega + i\nu_{hn})} \mathrm{J}_+\left(\frac{\omega + i\bar{\nu}_{hn}}{k V_{Th}}\right), \tag{18.15a}$$

$$\frac{1}{\varepsilon_0} \varepsilon^{\mathrm{l}}(\omega, \boldsymbol{k}) = 1 + \sum_h \frac{\omega_{Nh}^2}{k^2 V_{Th}^2} \frac{1 + \mathrm{J}_+\left(\frac{\omega + i\bar{\nu}_{hn}}{k V_{Th}}\right)}{1 - i\frac{\bar{\nu}_{hn}}{(\omega + i\bar{\nu}_{hn})} \mathrm{J}_+\left(\frac{\omega + i\bar{\nu}_{hn}}{k V_{Th}}\right)}. \tag{18.15b}$$

Here the summation goes only over the charged plasma particles. Eqs. (18.15a), (18.15b) are valid for any relations between $\bar{\nu}_{hn}$ and $k V_{Th}$.

ε) It should be remarked that in the long-wave limit ($\omega \to 0$) the longitudinal permittivity gets the form

$$\frac{1}{\varepsilon_0} \varepsilon^{\mathrm{l}}(0, \boldsymbol{k}) = 1 + \frac{1}{k^2 \lambda_D^2} \tag{18.16}$$

and this is valid for the collisionless plasma ($\omega \ll k V_{Th}$) as well as for one with collisions ($\omega \bar{\nu}_h \ll k^2 V_{Th}^2$). The result is that the static potential in an isotropic plasma is always of DEBYE form, namely

$$\phi(\boldsymbol{r}) = \frac{q_h}{r} \exp(-r/\lambda_D). \tag{18.17}$$

This follows immediately from Eq. (12.16) in the static limit, i.e. for $v_h \to 0$, using the expressions Eq. (18.15). This is a hint that a "screening" or "truncating" effect occurs in the COULOMB interaction for distances larger than the DEBYE length λ_D, (see Sect. 12).

19. Waves in a homogeneous and isotropic plasma. We proceed now to investigate the propagation of electromagnetic waves in a homogeneous and isotropic plasma, in the absence of external fields. The (dielectric) permittivity tensor for this case has been obtained in Sect. 17 for collisionless conditions and in Sect. 18 for a plasma with collisions. We intend to use the same order in the following; thus we shall always begin with the collisionless case and consider collisional effects later. The general dispersion equation for electromagnetic waves, Eq. (4.3), splits up in an anisotropic medium, as has been shown in Sect. 4. One has different equations, namely

$$\varepsilon^{\mathrm{l}}(\omega, \boldsymbol{k}) = 0 \tag{19.1}$$

for longitudinal waves, and

$$k^2 - \frac{\omega^2}{c_0^2} \frac{1}{\varepsilon_0} \varepsilon^{\mathrm{tr}}(\omega, \boldsymbol{k}) = 0 \tag{19.2}$$

for transverse waves.

The first kind shall be discussed in the following Sect. 20, the second in Sect. 21.

---

[6] $\mathrm{J}_+(x) \equiv x \exp(-\tfrac{1}{2}x^2) \int_{i\infty}^{x} d\tau\, e^{\frac{1}{2}\tau^2}$

20. Longitudinal waves in a homogeneous and isotropic plasma.

α) In the *collisionless* limiting *case*

$$\omega \gg (\bar{\nu}_e, \bar{\nu}_i) \tag{20.1}$$

the equation for longitudinal plasma oscillations becomes

$$1 + \sum_h \frac{\omega_{Nh}^2}{k^2 V_{Th}^2}\left[1 - J_+\left(\frac{\omega}{k V_{Th}}\right)\right] = 0 \tag{20.2}$$

with h = e, i.

In the *high frequency range*

$$\omega \gg (k V_{Te}, k V_{Ti}) \tag{20.3}$$

with the additional condition $\mathrm{Re}\,\omega \gg \mathrm{Im}\,\omega$, the ion terms in Eq. (20.2) can be neglected relative to the electron terms. Eq. (20.2) can then approximately be written in the following form [see Eq. (17.15)]:

$$1 - \frac{\omega_{Ne}^2}{\omega^2}\left(1 + 3\frac{k^2 V_{Te}^2}{\omega^2}\right) + i\sqrt{\frac{\pi}{2}}\,\frac{\omega\,\omega_{Ne}^2}{k^3 V_{Te}^3}\exp\left(\frac{-\omega^2}{2k^2 V_{Te}^2}\right) = 0. \tag{20.4}$$

This case is called the "plasma with electrons only"[1], or simply "electron plasma".

The imaginary term in Eq. (20.4) is due to spatial dispersion and is small compared with the real part. Therefore, at the resolution of the equation the same methods can be used which have been introduced in Sect. 5, when determining the spectrum of weakly attenuated waves.

The final result is[2]:

$$\omega \to \omega + i\gamma; \qquad \omega^2 \approx \omega_{Ne}^2 + 3k^2 V_{Te}^2 = \omega_{Ne}^2(1 + 3k^2 \lambda_{De}^2) \tag{20.5}$$

with

$$\gamma = -\sqrt{\frac{\pi}{8}}\,\frac{\omega_{Ne}}{k^3 \lambda_D^3}\exp\left(-\frac{1}{2k^2 \lambda_{De}^2} - \frac{3}{2}\right) \tag{20.6}$$

and the DEBYE length for electrons:

$$\lambda_{De} = \frac{\omega_{Ne}}{V_{Te}} = \left(\frac{\varepsilon_0\,\vartheta_e}{u\, q_e^2 N_e}\right)^{\frac{1}{2}}. \tag{18.6}$$

From the "high-frequency" condition, Eq. (20.3), with Eq. (20.5) follows then that

$$\omega^2 \approx \omega_{Ne}^2 \quad\text{and}\quad k^2 \lambda_{De}^2 \ll 1. \tag{20.7}$$

This means a significant negative argument in the exponential of Eq. (20.6) for γ, and therefore an exponential decrease of the oscillation.

β) These plasma oscillation are called high-frequency LANGMUIR *electron oscillations*. The corresponding dispersion relation between the frequency, $\omega/2\pi$, and the wave vector, $\boldsymbol{k}$, is given by the upper curve in Fig. 2.

Above in Sect. 17γ we adopted the position that the wave vector, $\boldsymbol{k}$, is to be considered as real, such that the dispersion relation, Eq. (20.4), is to be used to determine a complex frequency $\omega \to \omega + i\gamma$ (see Sect. 17γ). The resolution implies a boundary value problem but its resolution is not difficult, even for the inverse task where from a real frequency ω, given at the boundary, the corresponding complex $\boldsymbol{k}$ is to be found.

---

[1] A. SOMMERFELD used the expression "verschmierte Ionen" ("smeared-out ions") in order to indicate that the ions are only used in this model to neutralise the over-all charge of the electrons.

[2] We refer to Sect. 5 and Eq. (5.7) to explain the introduction of a complex frequency, $\omega \to \omega + i\gamma$.

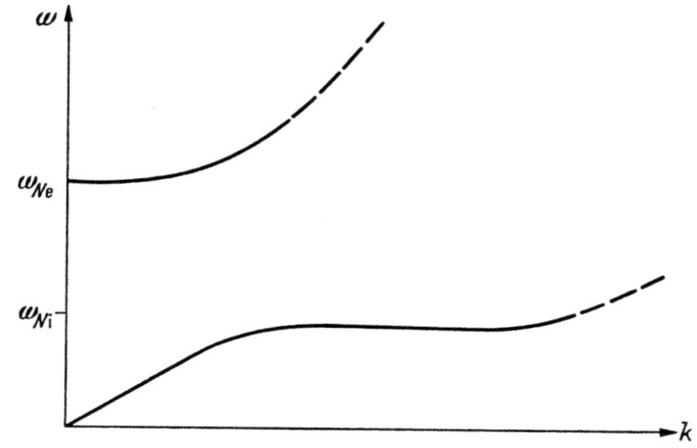

Fig. 2. Dispersion diagram for longitudinal waves.

One may immediately use Eq. (5.12) which gives a relation between the evanescence of the wave with time and in space. However, one has to take account of the fact that after Eq. (20.5) the group velocity of LANGMUIR waves is smaller than the thermal velocity $V_{T\mathrm{e}}$ of the electrons:

$$V_{\mathrm{gr}} = \frac{\partial\omega}{\partial k} \approx 3\,\frac{k\,V_{T\mathrm{e}}}{\omega_{N\mathrm{e}}}\,V_{T\mathrm{e}} \ll V_{T\mathrm{e}}. \tag{20.8}$$

At the same time their phase velocity is greater than the thermal velocity:

$$V_{\mathrm{ph}} = \frac{\omega}{k} \approx \frac{\omega_{N\mathrm{e}}}{k} \gg V_{T\mathrm{e}}. \tag{20.9}$$

As result we obtain refraction and absorption indices for LANGMUIR waves; with[3]

$$k = \frac{\omega}{c_0}\,n = \frac{\omega}{c_0}\,(\mu + i\,\chi) \tag{20.10}$$

we receive

$$\mu^2 = \frac{1 - \omega_{N\mathrm{e}}^2/\omega^2}{3V_{T\mathrm{e}}^2/c_0^2} \tag{20.11a}$$

and

$$\chi = \frac{c_0^5}{6\mu^4\,V_{T\mathrm{e}}^5}\sqrt{\frac{\pi}{2}}\,\exp\left(-\frac{c_0^2}{2\mu^2\,V_{T\mathrm{e}}^2}\right). \tag{20.11b}$$

With these relations we obtain

$$V_{\mathrm{gr}} \approx 3\mu\,\frac{V_{T\mathrm{e}}^2}{c_0} = \left(3\left(1 - \frac{\omega_{N\mathrm{e}}^2}{\omega^2}\right)\right)^{\frac{1}{2}} V_{T\mathrm{e}}, \tag{20.12a}$$

$$V_{\mathrm{ph}} = \frac{c_0}{\mu} = \left(\frac{3}{1 - \omega_{N\mathrm{e}}^2/\omega^2}\right)^{\frac{1}{2}} V_{T\mathrm{e}}, \tag{20.12b}$$

so that

$$V_{\mathrm{gr}} \approx (1 - \omega_{N\mathrm{e}}^2/\omega^2)\,V_{\mathrm{ph}}. \tag{20.12c}$$

γ) It can be seen from Eqs. (20.5), (20.6) that LANGMUIR oscillations decrease more rapidely with increasing wave number k, i.e. with decreasing wavelength. However, these equations are not adequate to describe high frequency oscilla-

[3] As to the definition of such indices, see, for example, K. RAWER and K. SUCHY: this Encyclopedia, Vol. 49/2, Sect. 5 δ.

tions where $k \geqq 1/\lambda_{De}$. In this wavelength range LANGMUIR oscillations are strongly evanescent.

We see this when considering the limiting case where $\text{Im}\,\omega \gg \text{Re}\,\omega$, supposing $|\omega| \gg k V_{Te}$. Using the asymptotic description of Eq. (17.15) we can rewrite the dispersion equation in the form

$$1 + i\sqrt{2\pi}\,\frac{\omega}{k\,V_{Te}}\,\frac{1}{k^2\,\lambda_{De}^2}\exp\left(\frac{-\omega^2}{2k^2\,V_{Te}^2}\right) = 0 \tag{20.13}$$

and replace ω by $-ik\,V_{Te}$. Then, using that $\text{Re}\,\xi \gg \text{Im}\,\xi$ ($\text{Re}\,\xi \gg 1$) we obtain from Eq. (20.11)

$$k^2\,\lambda_{De}^2 + \sqrt{2\pi}\,\xi\,e^{\frac{1}{2}\xi^2} = 0. \tag{20.14}$$

Separating in this equation real and imaginary parts one easily shows that the only solution which satisfies the conditions $\text{Re}\,\xi \gg 1$; $\text{Im}\,\xi \ll 1$ is the following (log = log nat):

$$\text{Im}\,\xi = \frac{\pi}{\text{Re}\,\xi}; \qquad \text{Re}\,\xi \approx \sqrt{\log(k^2\,\lambda_{De}^2)},$$

therefore

$$\frac{\omega}{k\,V_{Te}} \approx \frac{\pi}{\sqrt{\log k^2\,\lambda_{De}^2}} - i\sqrt{\log k^2\,\lambda_{De}^2}. \tag{20.15}$$

From the condition $|\omega| \gg k\,V_{Te}$, Eq. (20.3), follows then

$$k^2\,\lambda_{De}^2 \gg 1;$$

this implies that, different from the long-wave case considered above, the expression Eq. (20.12) is justified in the limiting case of high frequency, i.e. short wavelength. The dispersion relation or spectrum given by Eq. (20.12) is therefore the continuation of the LANGMUIR oscillations, Eq. (20.5), towards shorter wavelengths. However, these oscillations are strongly evanescent so that one may not be inclined to call them „oscillations" at all. For this reason this part of the spectrum of the high-frequency LANGMUIR waves is only indicated by a broken curve in Fig. 2.

δ) The collisionless attenuation of LANGMUIR oscillations in the electronic plasma occurs in the range of long wavelengths as well as for short ones. It has been detected by LANDAU [19] and is therefore called LANDAU *damping*. Its nature can be described by saying that it is attenuation by ČERENKOV absorption of longitudinal electronic waves in the plasma (see Sect. 17). In the range of long waves the phase velocity of such oscillations is considerably greater than the thermal velocity of the electrons, $V_{ph} \gg V_{Te}$. Therefore the damping is weak for such long waves. Only a very small part of all electrons contributes to this damping, viz. those of the tail of the Maxwellian distribution with energy great compared to the average thermal energy. On the other side an important part of the electron population contributes in the range of short wavelengths where, after Eqs. (20.15)

$$\text{Re}\,\omega/k \sim V_{Te}/\sqrt{\log k^2\,\lambda_{De}^2} \ll V_{Te}$$

so that an important part of the electrons is able to take part in the LANDAU process. This is the reason why LANGMUIR oscillations of higher frequency are quicker attenuated in a plasma.

ε) Let us now consider the intermediate range of *rather low frequencies*, where

$$k\,V_{Ti} \ll \omega \ll k\,V_{Te}. \tag{20.16}$$

Supposing $\text{Re}\,\omega \gg \text{Im}\,\omega$ the dispersion relation Eq. (20.2) takes the form

$$\left.\begin{aligned} &1 + \frac{\omega_{Ne}^2}{k^2\,V_{Te}^2}\left(1 + i\sqrt{\frac{\pi}{2}}\,\frac{\omega}{k\,V_{Te}}\right) - \\ &\quad - \frac{\omega_{Ni}^2}{\omega^2}\left[1 + 3\,\frac{k^2\,V_{Ti}^2}{\omega^2} - i\sqrt{\frac{\pi}{2}}\,\frac{\omega^3}{k^3\,V_{Ti}^3}\exp\left(\frac{-\omega^2}{2k^2\,V_{Ti}^2}\right)\right] = 0\end{aligned}\right\} \tag{20.17}$$

in this frequency range. In this equation the imaginary terms are small compared to the real ones. We therefore can write the solution in the following form, replacing the complex ω by $\omega + i\gamma$ (see Sect. 17γ)

$$\omega^2 = \omega_{Ni}^2 \frac{1 + 3k^2 \lambda_{Di}^2 (1 + k^{-2} \lambda_{De}^{-2})}{1 + k^{-2} \lambda_{De}^{-2}}, \tag{20.18a}$$

$$\gamma = -\sqrt{\frac{\pi}{8} \frac{m_i}{m_e}} \left|\frac{q_e}{q_i}\right| \frac{\omega^4}{k^3 V_{Ti}^3} \left[1 + \left|\frac{q_i}{q_e}\right| \sqrt{\frac{m_i}{m_e}} \left(\frac{\vartheta_e}{\vartheta_i}\right)^{\frac{3}{2}} \exp\left(\frac{-\omega^2}{2k^2 V_{Ti}^2}\right)\right]. \tag{20.18b}$$

The spectrum of the plasma oscillations which are considered here is shown by the lower curve in Fig. 2. It is important to remark[4] that it follows from the very condition $\omega \gg k V_{Ti}$ (20.16) that such oscillations with rather small attenuation ($\omega \gg \gamma$) can only exist in a non-isothermal plasma where $T_e \gg T_i$, and in the wavelength range defined by $k^2 \lambda_{Di}^2 \ll 1$.

Exactly as in the case of Langmuir oscillation of electrons one shows that for $k^2 \lambda_{Di}^2 \gg 1$ the spectrum Eqs. (20.18a), (20.18b) corresponds to strongly attenuated oscillations. With $\operatorname{Im} \omega \gg \operatorname{Re} \omega$ and the present condition

$$k T_{Ti} \ll \omega \ll k V_{Te}$$

(20.16), using Eq. (20.13), the spectrum is described by

$$1 + \frac{\omega_{Ne}^2}{k^2 V_{Te}^2} + i\sqrt{2\pi} \frac{\omega\, \omega_{Ni}^2}{k^3 V_{Ti}^3} \exp\left(\frac{-\omega^2}{2k^2 V_{Ti}^2}\right) = 0. \tag{20.19}$$

Similarly to the case of Eq. (20.15) we obtain the following solution (log = log nat):

$$\frac{\omega}{k V_{Ti}} \approx \frac{\pi}{\sqrt{\log (k^2 \lambda_{Di}^2 + |q_e/q_i|\, (\vartheta_i/\vartheta_e)}} - i \sqrt{\log (k^2 \lambda_{Di}^2 + |q_e/q_i|\, (\vartheta_i/\vartheta_e)}. \tag{20.20}$$

The part of the ion spectrum that corresponds to this solution is indicated by a dotted curve in Fig. 2.

In the range of longer waves, defined by

$$\frac{1}{\lambda_{De}} \gtrsim k \ll \frac{1}{\lambda_{Di}}, \tag{20.21}$$

the oscillations of the non-isothermal plasma with the spectrum of Eq. (20.18) appear as weakly attenuated, on a frequency which is near to Langmuir's ion frequency, $\omega \approx \omega_{Ni}$. This is the case of the horizontal part of the lower curve in Fig. 2.

ζ) In the range of very short waves finally, when

$$k \lambda_{De} \gg 1 \tag{20.22}$$

the spectrum described by Eqs. (20.18a), (20.18b) goes over into that of *ion sound waves*:

$$\omega^2 = k^2 \left|\frac{q_i}{q_e}\right| \frac{\vartheta_e}{m_i} \left(1 + 3\left|\frac{q_e}{q_i}\right| \frac{\vartheta_i}{\vartheta_e}\right), \tag{20.23a}$$

$$\gamma = -\sqrt{\frac{\pi}{8}}\, \omega \sqrt{\left|\frac{q_i}{q_e}\right| \frac{m_e}{m_i}} \left[1 + \left|\frac{q_i}{q_e}\right| \sqrt{\frac{m_i}{m_e}} \left(\frac{\vartheta_e}{\vartheta_i}\right)^{\frac{3}{2}} \exp\left(-\frac{1}{2}\left|\frac{q_i}{q_e}\right| \frac{\vartheta_e}{\vartheta_i} - \frac{3}{2}\right)\right]. \tag{20.23b}$$

[4] As ϑ is defined by $\vartheta = kT$ (k Boltzmann-constant), ϑ_e/ϑ_i may be replaced by T_e/T_i.

Such oscillations of the non-isothermal plasma are called ion sound waves because the spectrum described by Eq. (20.23 a) compares with that of ordinary sound oscillations in a gas or liquid [22], the phase velocity being

$$V_s = \sqrt{\left|\frac{q_i}{q_e}\right| \frac{\vartheta_e}{m_i}\left(1 + 3\left|\frac{q_e}{q_i}\right|\frac{\vartheta_i}{\vartheta_e}\right)}. \tag{20.24}$$

The spectrum Eq. (20.23 a) can also be obtained from the gasdynamic equations of a collisionless, non-isothermal plasma, i.e. from Eq. (11.13), in the absence of a magnetic field, of course. This is a hint that the oscillations considered are in a way similar to sound waves in a liquid.

η) We now introduce expressions for *refraction and absorption indices* of attenuated ion waves in a non-isothermal plasma. These can either be obtained from Eqs. (5.12) or, directly, from the dispersion relation, Eq. (20.17). Taking account of

$$k = \frac{\omega}{c_0} n = \frac{\omega}{c_0}(\mu + i\chi), \quad \text{and} \quad |\chi| \gg |\mu| \qquad [20.10]$$

we find

$$\mu^2 = \frac{\omega_{Ne}^2 c_0^2/(\omega^2 V_{Te}^2)}{\frac{\omega_{Ni}^2}{\omega^2}\left(1 + \frac{\omega_{Ne}^2}{\omega^2}\frac{m_e}{m_i}\frac{\vartheta_i}{\vartheta_e}\frac{3}{(\omega_{Ni}^2/\omega^2 - 1)}\right) - 1} \tag{20.25 a}$$

$$\chi = \sqrt{\frac{\pi}{8}}\,\frac{\omega_{Ni}^2}{\omega^2}\,\frac{c_0^3}{V_{Ti}^3}\,\frac{1}{\mu^2}\,\frac{\left\{\left|\frac{q_e}{q_i}\right|\sqrt{\frac{m_e}{m_i}}\left(\frac{\vartheta_i}{\vartheta_e}\right)^{\frac{3}{2}} + \exp\left(\frac{-c_0^2}{2\mu^2 V_{Ti}^2}\right)\right\}}{\left(\frac{\omega_{Ni}^2}{\omega^2}\left(1 + 3\,\frac{\mu^2 V_{Ti}^2}{c_0^2}\right) - 1\right)}. \tag{20.25 b}$$

In the limiting case of very low frequencies, $\omega \ll \omega_{Ni}$ (which corresponds to the ion sound oscillations of the plasma) the Eqs. (20.25 a), (20.25 b) simplify to[5]

$$\mu^2 = \frac{c_0^2}{V_s^2}, \tag{20.26a}$$

$$\chi = \sqrt{\frac{\pi}{8}}\,\frac{c_0}{V_s}\left(\frac{\vartheta_e}{\vartheta_i}\right)^{\frac{3}{2}}\left[\left|\frac{q_e}{q_i}\right|\sqrt{\frac{m_e}{m_i}}\left(\frac{\vartheta_i}{\vartheta_e}\right)^{\frac{3}{2}} + \exp\left(\frac{-c_0^2}{2\mu^2 V_{Ti}^2}\right)\right]. \tag{20.26b}$$

Apparently these expressions do not depend on the frequency such that in the low frequency limiting case there is no dispersion, neither for the refraction nor for the absorption index.

ϑ) We have repeatedly indicated above that oscillations with the spectrum described by Eq. (20.18a) are only possible in a non-isothermal plasma where $T_e \gg T_i$. We may ask two questions now: How can this necessary condition be explained? Can such oscillations appear with weak attenuation in the ionospheric or exospheric plasma? The answer may be found from the condition for weak attenuation, $\gamma \ll \omega$, from which follows from Eq. (20.15):

$$1 \gg \sqrt{\frac{\pi}{2}}\, x^2\, e^{-\frac{1}{2}x^2} \tag{20.27}$$

with

$$x^2 = \frac{\omega^2}{k^2 V_{Ti}^2} \gg 1, \tag{20.28}$$

In spite of the fact that the asymptotic function $J_+(x)$ defined above in Sect. 17 by Eq. (17.15) ceases to be a good approximation for $x^2 > 4$, we may conclude that the oscillations which we consider here should have weak attenuation for $T_e/T_i > x^2 > 6$. This a is rather important

[5] Of course, $\vartheta_e/\vartheta_i \equiv T_e/T_i$.

deviation from isothermal conditions which, apparently, does not occur at ionospheric heights. We have noted in Sect. 1 that according to Table 3 the greatest deviation from temperature equilibrium in the ionosphere occurs between 200 and 300 km of height, and that at these heights

$$T_e/T_i \approx 2 \dots 3$$

in the middle ionosphere.

Therefore one should not exspect weakly attenuated ion sound (oscillations) waves in the ionospheric plasma nor in the exosphere. Conditions in the magnetosphere are not well enough known to make a statement.

On the other hand high frequency electron oscillations of the LANGMUIR type can certainly exist in the ionospheric plasma. In a collisionless plasma such oscillations are weakly attenuated ($\gamma \ll \omega$), if $k^2 \lambda_{De}^2 \ll \frac{1}{6}$. This means that their wavelength in the ionospheric plasma should be greater than a few cm (see Sect. 2, Table 3).

ι) Finally in the range of *lowest frequencies*

$$\omega \ll k V_{Ti} \tag{20.29}$$

the longitudinal (dielectric) permittivity of a collisionless plasma is given by Eqs. (18.15a), (18.15b); it is, in general, independent of the frequency. The explanation is that the longitudinal field can not penetrate into the plasma because *screening* occurs. An effective screening radius may be determined from the wavenumber value k which gives zero refraction index in the dispersion relation; in the present case:

$$1 + \sum_h \frac{\omega_{Nh}^2}{k^2 V_{Th}^2} \to 1 + \frac{1}{k^2 \lambda_D^2} = 0. \tag{20.30}$$

Resolving it for k gives an imaginary value. The screening radius, r_{sc}, is then

$$r_{sc} \approx \frac{1}{\operatorname{Im} k} = \lambda_D \tag{20.30a}$$

and this is independent of the frequency. This screening is not dissipative, and is due to the polarization of the plasma.

An similar case of screening may occur even for not so low frequencies, when

$$k V_{Ti} \ll \omega \ll k V_{Te} \qquad [20.16]$$

provided that

$$\omega \gg \omega_{Ni}.$$

In this case, too, the dispersion relation Eq. (20.17) does not depend on the frequency and one obtains as screening radius

$$r_{sc} \approx \frac{1}{\operatorname{Im} k} = \lambda_{De}. \tag{20.30b}$$

$\varkappa$) *Collisions between particles* have been neglected in the preceding discussion. We shall now take account of their influence in the spectrum of longitudinal oscillations in the particular case of an isotropic plasma. This means that we do not admit gyrotropic (magneto-active) conditions.

We have shown in Sect. 18 that at least in the ionospheric plasma the influence of the Earth's magnetic field in this spectrum is negligible in the frequency range

$$\omega \gg \omega_{Be}.$$

There is namely

$$\omega_{Be} \gg (\bar{\nu}_e, \bar{\nu}_i),$$

and therefore our condition means also

$$\omega \gg (\bar{\nu}_e, \bar{\nu}_i) . \qquad [20.1]$$

We shall consider a particular case of this type below in this section.

Considering first the limiting case of high frequencies

$$\omega \gg k V_{Te} \quad \text{and} \quad \omega \gg \bar{\nu}_e \qquad (20.31)$$

where electron oscillations of the LANGMUIR type can exist. It follows in this case from Eqs. (18.3) and (18.14) that there appears a small additional term on the left side of the dispersion relation for longitudinal waves, Eq. (20.4), which is caused by collisions. It is, in general, imaginary and reads[6]

$$i \frac{\omega_{Ne}^2 \bar{\nu}_e}{\omega^3} . \qquad (20.32)$$

This term takes account of the collisional absorption of electron oscillations of the LANGMUIR type. The attenuation decrement, Eq. (20.6), is, of course, influenced and reads now:

$$\gamma = -\sqrt{\frac{\pi}{8}} \frac{\omega_{Ne}}{k^3 \lambda_{De}^3} \exp\left(-\frac{1}{2k^2 \lambda_{De}^2} - \frac{3}{2}\right) - \frac{1}{2} \bar{\nu}_e \qquad (20.33)$$

but the frequency spectrum itself, Eq. (20.5), is not influenced (in a first order approximation at least).

It follows from Eq. (20.32) that under the condition

$$\bar{\nu}_e > \sqrt{\frac{\pi}{8}} \frac{\omega_{Ne}}{k^3 \lambda_{De}^3} \exp\left(-\frac{1}{2k^2 \lambda_{De}^2} - \frac{3}{2}\right) \qquad (20.34)$$

it is the influence of collisions (of the electrons) which mainly causes dissipation of longitudinal waves in a plasma.

In the opposite limiting case of short wavelength LANDAU damping makes the decisive contribution. In particular, in the limiting case

$$k^2 \lambda_{De}^2 \gg 1 , \qquad (20.35)$$

$$\omega \gg k V_{Te} \qquad [20.9]$$

collisions can entirely be neglected and the spectrum of (heavily attenuated) longitudinal oscillations is given by Eq. (20.15).

The dispersion relation Eq. (20.17) for longitudinal waves in the intermediate frequency range, determined by

$$k V_{Ti} \ll \omega \ll k V_{Te} , \qquad [20.16]$$

is also influenced when account is taken of collisions. It appears then that under the conditions[7]

$$\omega \gg \bar{\nu}_i \quad \text{and} \quad k V_{Te} \gg \bar{\nu}_e \qquad (20.36)$$

[6] We have to use $\bar{\nu}_e = \nu_{eff}$ in the case of a completely ionized plasma, but $\bar{\nu}_e = \nu_{en}$ in the case of a weakly ionized plasma. For a detailed discussion see K. SUCHY: Ergebn. exakt. Naturw. 35, 103—294 (1964), or the shorter text in K. RAWER and K. SUCHY: this Encyclopedia. Vol. 49/2, Sect. 3, p. 11. For a summary with definitions of the different notions see "Introductory Remarks", p. 1.

[7] We remark that no additional condition concerning $\bar{\nu}_e/\omega$ is needed for this consideration.

ion collisions with ions and/or neutrals play the decisive role. Taking account of this influence we obtain an additional term to the left hand side of Eq. (20.17) which is for a *completely ionized plasma*

$$i\frac{8}{5}\frac{\omega_{Ni}^2}{\omega^2}\frac{\bar{\nu}_{ii}k^2V_{Ti}^2}{\omega^3}, \tag{20.37}$$

and for a *weakly ionized plasma*

$$i\frac{\omega_{Ni}^2\bar{\nu}_{in}}{\omega^3}. \tag{20.38}$$

These are small additions to the imaginary part on the left side of Eq. (20.17), so that the frequency spectra of oscillations as given by Eqs. (20.18a) and (20.23a) remain unchanged (in a first order approximation). However, to the attenuation decrement γ we have an additional term $\Delta\gamma$ which, in a *completely ionized plasma*; is

$$\Delta\gamma = -\frac{4}{5}\bar{\nu}_{ii}\frac{k^2V_{Ti}^2}{\omega^2} \tag{20.39}$$

but in a weakly ionized plasma:

$$\Delta\gamma = -\tfrac{1}{2}\bar{\nu}_{in}. \tag{20.40}$$

As stated above such oscillations are almost excluded in the ionospheric exospheric and magnetospheric plasmas because the deviation from thermal equilibrium, through existing, is too small, see Subsect. ϑ. The conclusions there obtained do not change when allowing for collisions.

One final remark concerning the terms "completely ionized" and "weakly ionized" should yet be made. These limiting cases are determined by the additional terms due to collisions in the expressions for the (dielectric) permittivity of the plasma and/or for the attenuation decrement of the oscillations. We have to consider collisions with neutrals as well as amongst charged particles. This means that for *electron* (LANGMUIR) *oscillations* a plasma can be taken as *completely ionized* when

$$\nu_{\text{eff}} \gg \bar{\nu}_{\text{en}} \tag{20.41 a}$$

and for *ionic oscillations*, when

$$\bar{\nu}_{ii}\frac{k^2V_{Ti}^2}{\omega^2} \approx \bar{\nu}_{ii}\frac{\vartheta_i}{\vartheta_e} = \bar{\nu}_{ii}\frac{T_i}{T_e} \gg \bar{\nu}_{in}. \tag{20.41 b}$$

Under opposite conditions, the plasma is to be taken as weakly ionized.

21. Transverse waves. The general Eq. (19.2) is now to be used to describe transverse waves in an isotropic plasma.

α) In the *collisionless* (limiting) *case*

$$\omega \gg (\bar{\nu}_e, \bar{\nu}_i) \tag*{[20.1]}$$

this equation takes the form

$$k^2c_0^2 - \omega^2 + \sum_h \omega_{Nh}^2\,\mathrm{J}_+\left(\frac{\omega}{k\,V_{Th}}\right) = 0 \tag{21.1}$$

the function J_+ being defined by Eq. (17.14).

In the *high-frequency range* where

$$\omega \gg (k\,V_{Te},\ k\,V_{Ti}) \tag*{[20.3]}$$

we find the following spectrum of transverse electro-magnetic oscillations in the plasma:

$$\omega^2 = \omega_{Ne}^2 + k^2c_0^2 \tag{21.2}$$

as shown in Fig. 3. Different from the spectrum of LANGMUIR oscillations, Eq. (20.5), the expression Eq. (21.2) remains valid in the limiting case of arbitrarily short wavelength. In the case of extremely small electron density,

$$k^2 c_0^2 \gg \omega_{Ne}^2 \tag{21.3}$$

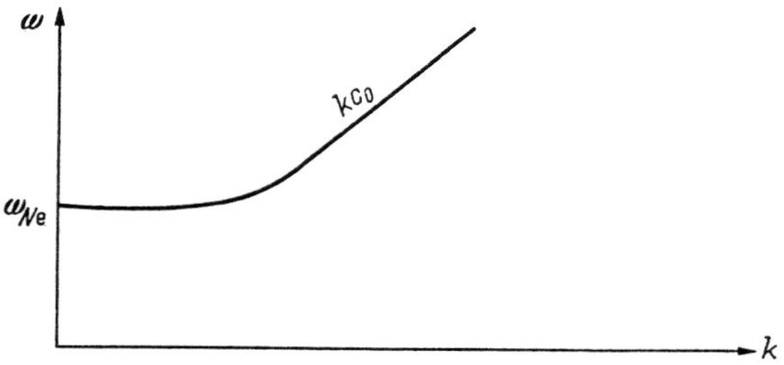

Fig. 3. Dispersion diagram for high-frequency, transverse waves.

the spectrum given by Eq. (21.2) corresponds to that of (transverse) electromagnetic waves in free space.

When solving the dispersion equation, Eq. (21.1), and determing the spectrum Eq. (21.2) we have completely neglected the imaginary terms in this equation which are caused by ČERENKOV absorption of the waves. This has been done consciously and is completely justified, because the phase velocity of high-frequency transverse waves is greater than the velocity of light in free space:

$$V_{\mathrm{ph}} \equiv \frac{\omega}{k} > c_0 \tag{21.4}$$

as can be seen from Eq. (21.2). Therefore such waves in the plasma cannot suffer from ČERENKOV absorption, such that for a collisionless plasma in the frequency range given by Eq. (21.4) we have

$$\mathrm{Im}\, \varepsilon_{\mathrm{tr}} \equiv 0.$$

Strictly speaking there always remains a small imaginary part in Eq. (17.13a), and, therefore, a very small imaginary term in Eq. (21.1) appears even under the above conditions. This is due to the fact that a Maxwellian distribution, Eq. (17.1), has been used when deriving the (dielectric) permittivity of a collisionless plasma. Now, such a distribution makes allowance for particles of arbitrarily high velocity, at least in principle. Apparently, such particles with extremely high velocity could produce ČERENKOV absorption of waves, even in the case $\omega > k c_0$. In order to describe such absorption, it is, however, necessary to use relativistic considerations [2] and these show that the imaginary part of $\varepsilon_{\mathrm{tr}}$ is strictly zero, such that there is no absorption at all in a collisionless plasma when $\omega > k c_0$. Besides, the spectrum given by Eq. (21.2) remains unchanged by these considerations.

We should yet note that for transversal waves in an isotropic plasma the group velocity V_{gr}, different from the phase velocity V_{ph}, is always smaller than the velocity of light in free space. Indeed one easily finds from Eq. (21.2)

$$V_{\mathrm{gr}} \equiv \frac{\partial \omega}{\partial k} = c_0 \frac{k c_0}{\omega} = \frac{c_0}{\sqrt{1 + \omega_{Ne}^2/(k^2 c_0^2)}} < c_0. \tag{21.5}$$

β) *Collisions between particles* are the only reason for absorption of *high-frequency transverse waves* in a plasma. Taking account of such collisions the dispersion equation in the case

$$\omega \gg (\bar{\nu}_e, k V_{Te}) \qquad [20.31]$$

becomes

$$k^2 c_0^2 - \omega^2 + \omega_{Ne}^2 \left(1 - i \frac{\bar{\nu}_e}{\omega}\right) = 0, \tag{21.6}$$

see Eqs. (18.3) and (18.4). $\bar{\nu}_e$ equals ν_{eff} in the case of a completely ionized plasma, but $\bar{\nu}_e = \nu_{en}$ in that of a weakly ionized one. The spectrum of frequencies determined by this equation coincides with that of Eq. (21.2), the attenuation decrement of the oscillations being

$$\gamma = -\frac{1}{2} \bar{\nu}_e \frac{\omega_{Ne}^2}{k^2 c_0^2 + \omega_{Ne}^2}. \tag{21.7}$$

From Eqs. (21.2) and (21.7) it follows that one is justified in considering the plasma as being purely electronic, as far as high-frequency transverse oscillations are concerned. This approximation is valid up to terms of the order m_e/m_i, for the collisionless as well as for a plasma where collisions are rather frequent.

γ) Detailed analysis shows that the motion of the ions may be neglected for transverse waves, even in the *low-frequency range* where

$$\omega \ll k V_{Te}. \tag{21.8}$$

In this frequency range from the general dispersion equation, Eq. (21.1), we obtain for a collisionless plasma:

$$\omega = -i \sqrt{\frac{2}{\omega}} \frac{k^3 c_0^2 V_{Te}}{\omega_{Ne}^2}. \tag{21.9}$$

The fact that the frequency of transverse waves appears as purely imaginary in this equation means, of course, that low-frequency transverse oscillations in a plasma decrease aperiodically with time.

δ) We note that Eq. (21.9) continues to be valid also in the *range of lowest-frequencies* where

$$\omega \ll k V_{Ti} \tag*{[20.29]}$$

Furthermore, it even remains valid in the presence of a few particle collisions if only

$$\bar{\nu}_e \ll k V_{Te}. \tag{21.10}$$

Under this condition Čerenkov absorption due to electrons is the main reason for aperiodic decrease of oscillations in the plasma.

If, however, the collision frequency is rather high, such that

$$\bar{\nu}_e \gg k V_{Te} \tag{21.11}$$

then collisional absorption prevails in the case of low-frequency waves, i.e. for

$$\omega \ll \bar{\nu}_e. \tag{21.12}$$

It follows then from the general dispersion equation, Eq. (19.2), that

$$\omega = -i \frac{\varepsilon_0}{\mathrm{u}} \frac{k^2 c_0^2}{\sigma} \tag{21.13}$$

where σ is here the electrostatic conductivity of the plasma, viz. for a completely ionized plasma:

$$\sigma = \sigma_{\|} = 1.96 \frac{\varepsilon_0}{\mathrm{u}} \frac{\omega_{Ne}^2}{\nu_{\text{eff}}} \tag{21.14}$$

but for a weakly ionized plasma

$$\sigma = \frac{\varepsilon_0}{\mathrm{u}} \frac{\omega_{Ne}^2}{\bar{\nu}_{en}}. \tag{21.15}$$

Eq. (21.14) can be obtained from the gasdynamic equations for two fluids see Sects. 15/16; Eq. (21.15) follows from Eq. (18.14).

ε) Having in mind the possible development of small oscillations and the corresponding initial value problem we may refer to the method we used in the preceding to resolve the dispersion equation with respect to the frequency, $\omega/2\pi$. There is no difficulty, however, to obtain from these solutions a complex wave number function $\boldsymbol{k}(\omega)$, viz. a wave vector characterizing *a wave propagation in space*. This is a problem connected with the values at the limits. In an isotropic plasma the absolute value of the vector $\boldsymbol{k}$ must be independent of the direction of wave propagation. Therefore, from Eqs. (21.2) through (21.7) we may obtain the refraction index $\mu(\omega)$ as well as the absorption index $\chi(\omega)$ of high-frequency transverse waves in an isotropic plasma. Considering Eqs. (5.7) and (5.12) one finds:

$$\mu^2 = 1 - \frac{\omega_{Ne}^2}{\omega^2}, \tag{21.16a}$$

$$\chi = \frac{\bar{\nu}_e}{2\mu} \frac{\omega_{Ne}^2}{\omega^3} = \frac{\bar{\nu}_e}{2\omega\mu}(1-\mu^2). \tag{21.16b}$$

We realize at this occasion that the waves considered here can only propagate in a plasma when the condition

$$\omega^2 \geqq \omega_{Ne}^2 \tag{21.17}$$

is satisfied. Otherwise one had $\mu^2 < 0$, meaning total reflection of the wave at the limits; for details see for example [1]. The depth to which field penetrates into a plasma with $\mu^2 < 0$ is

$$\delta \approx \frac{1}{\mathrm{Im}|\boldsymbol{k}|} \approx \frac{c_0}{\omega_{Ne}}. \tag{21.18}$$

Table 5. *Electromagnetic waves*

Type of oscillation	Frequency spectrum	Attennation decrement
1. Longitudinal electronic LANGMUIR oscillations	$\omega^2 = \omega_{Ne}^2(1 + 3k^2\lambda_{De}^2)$	$\gamma = -\frac{1}{2}\bar{\nu}_e - \sqrt{\frac{\pi}{8}}\frac{\omega_{Ne}}{k^2\lambda_{De}^2} \times$ $\times \exp\left(-\frac{3}{2} - \frac{1}{2k^2\lambda_{De}^2}\right)$
2. Longitudinal ion waves	$\omega^2 = \frac{\omega_{Ni}^2}{1 + k^{-2}\lambda_{De}^{-2}}$	$\gamma = -\frac{1}{2}\bar{\nu}_i - \sqrt{\frac{\pi}{8}}\left\lvert\frac{q_e}{q_i}\right\rvert \frac{m_i}{m_e}\frac{\omega^4}{k^3 V_{Te}^3} \times$ $\times \left[1 + \left\lvert\frac{q_i}{q_e}\right\rvert \sqrt{\frac{m_i}{m_e}}\left(\frac{T_e}{T_i}\right)^{\frac{3}{2}} \exp\left(\frac{-\omega^2}{2k^2 V_{Ti}^2}\right)\right]$
3. Low-frequency ion sound waves	$\omega^2 = k^2 V_s^2$ $= k^2 \left\lvert\frac{q_i}{q_e}\right\rvert \frac{\vartheta_e}{m_i}\left(1 + 3\left\lvert\frac{q_e}{q_i}\right\rvert\frac{T_i}{T_e}\right)$	$\gamma = -\frac{1}{2}\bar{\nu}_i - \sqrt{\frac{\pi}{8}}\,\omega \sqrt{\left\lvert\frac{q_i}{q_e}\right\rvert \frac{m_e}{m_i}} \times$ $\times \left[1 + \left\lvert\frac{q_i}{q_e}\right\rvert \sqrt{\frac{m_i}{m_e}}\left(\frac{T_e}{T_i}\right)^{\frac{3}{2}} \times\right.$ $\left.\times \exp\left(-\frac{3}{2} - \frac{1}{2}\left\lvert\frac{q_i}{q_e}\right\rvert\frac{T_e}{T_i}\right)\right]$
4. Transverse electromagnetic waves	$\omega^2 = \omega_{Ne}^2 + k^2 c_0^2$	$\gamma = -\frac{1}{2}\bar{\nu}_e \frac{\omega_{Ne}^2}{k^2 c_0^2 + \omega_{Ne}^2}$

ζ) Completely different is the behaviour for low-frequency waves under the condition

$$\omega \ll k V_{Te}. \qquad [21.8]$$

Resolving Eq. (21.9) after $\boldsymbol{k}(\omega)$ we find the following expression for the *penetration of low-frequency transverse waves* into a collisionless plasma:

$$\delta \approx \frac{1}{\operatorname{Im}|\boldsymbol{k}|} \approx 2 \sqrt[6]{\frac{2}{\pi}} \left(\frac{c_0^2 V_{Te}}{\omega\, \omega_{Ne}^2} \right)^{\frac{1}{3}}. \qquad (21.19)$$

Here the limitation is not given by internal total reflection but by absorption. The mechanism of penetration in this case is related with the ČERENKOV effect produced by thermal electrons. The penetration depth increases with decreasing frequency, $\omega/2\pi$; in the limiting case $\omega \to 0$ the penetration depth after Eq. (21.19) becomes virtually infinite. This consideration shows that an almost static field, for example a magnetic one, easily penetrates into a plasma in a transverse direction. In this respect the case of longitudinal penetration is quite different; we have remarked in Sect. 18 that then, due to a current curl screening occurs; in the limit of an almost static transverse field, the screening depth is given by the DEBYE length, λ_D. Eq. (21.19) is also known as formula of the *abnormal skin effect* in a collisionless plasma.

η) This is quite different from the *normal skin effect* characterizing the penetration of low-frequency transverse waves into a plasma under prevailing influence of collisions. It exists, besides, for conductors of any kind, not only plasmas [7]. In this case the penetration depth follows from Eq. (21.13) as

$$\delta \approx \frac{1}{\operatorname{Im}|\boldsymbol{k}|} = \sqrt{\frac{\varepsilon_0}{u}} \frac{c_0}{\sqrt{\frac{1}{2}\sigma\omega}}. \qquad (21.20)$$

in a homogeneous and isotropic plasma

Refractive index	Absorption index	Condition of existence
$\mu^2 = \frac{1 - \omega_{Ne}^2/\omega^2}{3 V_{Te}^2/c_0^2}$	$\chi = \frac{1}{6} \frac{c_0^2}{V_{Te}^2} \frac{1}{\mu \omega_{Ne}} \left[\overline{\nu}_e + \sqrt{\frac{\pi}{2}} \frac{c_0^3 \omega_{Ne}}{V_{Te}^3 \mu^3} \times \exp\left(\frac{-c_0^2}{2\mu^2 V_{Te}^2} \right) \right]$	$\omega \gg (k V_{Te}, \overline{\nu}_e)$ $k^2 \lambda_{De}^2 \ll 1$ $\mu \frac{V_{Te}}{c_0} \ll 1$
$\mu^2 = \frac{\omega_{Ne}^2}{\omega^2} \frac{c_0^2}{V_{Te}^2} \left(\frac{\omega_{Ni}^2}{\omega^2} - 1 \right)^{-1}$	$\chi = -\frac{\gamma}{\omega} \frac{c_0}{V_s} \frac{\omega_{Ni}^3}{\omega^3} \left(\frac{\omega_{Ni}^2}{\omega^2} - 1 \right)^{-\frac{3}{2}}$	$V_{Ti} \ll \frac{\omega}{k} = \frac{c_0}{\mu} \ll V_{Te}$ $k^2 \lambda_{Di}^2 \ll 1; \quad T_e \gg T_i$ $\omega \gg \overline{\nu}_i; \quad k V_{Te} \gg \overline{\nu}_e$
$\mu^2 = \frac{c_0^2}{V_s^2}$	$\chi = -\frac{c_0}{V_s} \frac{\gamma}{\omega} = \frac{c_0}{V_s} \left\{ \frac{1}{2} \frac{\overline{\nu}_i}{\omega} + \sqrt{\frac{\pi}{8} \left| \frac{q_i}{q_e} \right| \frac{m_e}{m_i}} \left[1 + \left| \frac{q_i}{q_e} \right| \sqrt{\frac{m_i}{m_e}} \left(\frac{T_e}{T_i} \right)^{\frac{3}{2}} \times \exp\left(-\frac{3}{2} - \frac{1}{2} \left| \frac{q_i}{q_e} \right| \frac{T_e}{T_i} \right) \right] \right\}$	$V_{Ti} \ll \frac{\omega}{k} = \frac{c_0}{\mu} \ll V_{Te}$ $k^2 \lambda_{De}^2 \ll 1; \quad \overline{\nu} \gg k V_{Te}$ $\omega_{Ni} \gg \omega \gg \overline{\nu}_i$
$\mu^2 = 1 - \frac{\omega_{Ne}^2}{\omega^2}$	$\chi = \frac{\overline{\nu}_e}{2\mu} \frac{\omega_{Ne}^2}{\omega^3}$	$\omega \gtrsim k c_0 \gg k V_{Te}$ $\omega \gg \overline{\nu}_e$

ϑ) A *summary* of the results of this and the preceding section is presented in Table 5 giving the characteristic parameters of electro-magnetic waves in a homogeneous, isotropic plasma for the two limiting cases of large and small relative plasma density. For the *completely ionized* plasma we have

$$\bar{\nu}_{\mathrm{e}} = \nu_{\mathrm{eff}} \tag{21.21a}$$

and

$$\bar{\nu}_{\mathrm{i}} = \frac{8}{5}\,\nu_{\mathrm{i\,i}}\,\frac{k^2 V_{T\mathrm{i}}^2}{\omega^2} = \frac{8}{5}\,\nu_{\mathrm{i\,i}}\,\mu^2\,\frac{V_{T\mathrm{i}}^2}{c_0^2}\,. \tag{21.21b}$$

On the other side we may write for a *weakly ionized* plasma

$$\bar{\nu}_{\mathrm{e}} = \nu_{\mathrm{en}} \tag{21.22a}$$

and

$$\bar{\nu}_{\mathrm{i}} = \nu_{\mathrm{in}}\,. \tag{21.22b}$$

II. Homogeneous, gyrotropic (magneto-active) plasma.

The analysis we made in Chap. I, in particular Sects. 19 through 21, has shown that in the absence of outer fields there exist in an isotropic plasma only three kinds (branches) of weakly attenuated oscillations. Two of these correspond to longitudinal oscillations of the field, one to a transverse oscillation. These are high-frequency longitudinal LANGMUIR oscillations, low-frequency longitudinal ion-sound-waves[1] and high-frequency transverse electro-magnetic waves.

In the presence of a strong magnetic field the features occuring in a plasma become considerably more involved, even with weak electromagnetic oscillations. There appear several new branches of possible oscillations, on high as well as on low frequencies. A complete analysis of all thinkable branches of oscillations in a magneto-active (gyrotropic) plasma would be beyond the scope of this contribution. We shall, therefore, discuss only such waves which could propagate in ionospheric or magnetospheric plasmas, with emphasis on weakly attenuated waves. The special problems of propagation theory in a gyrotropic plasma are discussed in the relevant literature [*1*] ... [*6*][2].

22. Dielectric permeability of a collisionless plasma in the presence of a magnetic field. Before analyzing the spectra of possible oscillations in a gyrotropic (magneto-active) plasma, we need (as we did in Sect. 17) the tensor of dielectric permeability[1]. For the most simple model, the model of independent particles, this tensor has been written down in Sect. 6, Eqs. (6.8) through (6.12). We must now determine the limits of applicability of these simple formulae which are often called the "equations of elementary theory", and which are widely used when investigating electromagnetic wave propagation in the ionosphere, see for example [*1*] and [*9*]. A critical analysis should be starting with the equation of kinetic statistics, or, where this is acceptable, with the gasdynamic approach of two fluids.

α) First let us consider a collisionless plasma assuming, under undisturbed conditions, a Maxwellian distribution of particle velocities, $f_{0\mathrm{h}}$, after Eq. (17.1). We use cylindrical coordinates in the velocity space, $v_\perp$, φ, v_z with the z^0 axis

[1] This particular type is only possible in a non-isothermal plasma where $T_{\mathrm{e}} > 6\,T_{\mathrm{i}}$, see Sect. 19.

[2] See in particular K. RAWER and K. SUCHY: this Encyclopedia, Vol. 49/2, Sects. 6 through 10.

[1] The expression "tensor" where used in this chapter generally means a tensor of second order, i.e. a dyade. It is remembered that a tensor of first order is a vector, one of zero order a scalar.

parallel to the external magnetic field, $\boldsymbol{B}_{\bar{0}}$. The relations with corresponding cartesian coordinates are

$$v_x = v_\perp \cos\varphi; \qquad v_y = v_\perp \sin\varphi; \qquad v_z = v_z.$$

The *equation of kinetic statistics* for particles of species h with the statistical distribution function $f_{\rm h}$, when introducing a harmonic variation of the fluctuating parameters after

$$\exp(-i\,\omega\,t + i\,\boldsymbol{k}\cdot\boldsymbol{r}),$$

reads[2]:

$$-i(\omega - \boldsymbol{k}\cdot\boldsymbol{v})\,\delta f_{\rm h} - \omega_{B\,\rm h}\frac{\partial\,\delta f_{\rm h}}{\partial\varphi} + q_{\rm h}\boldsymbol{E}\cdot\frac{\partial f_{0\,\rm h}}{\partial\boldsymbol{p}} = 0. \tag{22.1}$$

Eq. (22.1) is a linearized kinetic equation for a perturbation of the distribution function; it admits a solution which is periodic in φ (with period 2π) which reads as follows[3]:

$$\begin{aligned}\delta f_{\rm h} &= \frac{q_{\rm h}}{m_{\rm h}\,\omega_{B\,\rm h}}\int\limits_{\infty}^{\varphi} d\varphi_1\left(\boldsymbol{E}\cdot\frac{\partial f_{0\,\rm h}}{\partial\boldsymbol{v}}\right)\exp\left[\frac{i}{\omega_{B\,\rm h}}\int\limits_{\varphi}^{\varphi_1} d\varphi_2(\omega - \boldsymbol{k}\cdot\boldsymbol{v})\right]\\ &= \frac{i\,q_{\rm h}}{\vartheta_{\rm h}} f_{0\,\rm h}\sum_{s=-\infty}^{+\infty}\frac{\exp(-i\,s\,\varphi + i\,k_\perp v_\perp \sin\varphi/\omega_{B\,\rm h})}{\omega - k_z v_z - s\,\omega_{B\,\rm h}}\times\\ &\quad\times\left\{\frac{s\,\omega_{B\,\rm h}}{k_\perp}\mathrm{J}_s\left(\frac{\boldsymbol{k}_\perp\cdot\boldsymbol{v}_\perp}{\omega_{B\,\rm h}}\right)E_x + i\,v_\perp \mathrm{J}_s'\left(\frac{\boldsymbol{k}_\perp\cdot\boldsymbol{v}_\perp}{\omega_{B\,\rm h}}\right)E_y + v_z \mathrm{J}_s\left(\frac{\boldsymbol{k}_\perp\cdot\boldsymbol{v}_\perp}{\omega_{B\,\rm h}}\right)E_z\right\}.\end{aligned} \tag{22.2}$$

We have supposed the wave vector $\boldsymbol{k}$ to lie in the $\boldsymbol{x}^0$, $\boldsymbol{z}^0$ plane, i.e. $\boldsymbol{k} = (k_\perp, 0, k_z)$, which is compatible with an arbitrary direction of $\boldsymbol{k}$ as the coordinates can yet be rotated in the $\boldsymbol{x}^0$, $\boldsymbol{y}^0$ plane.

β) Substituting now the expression Eq. (22.2) into that of the current density, induced in the plasma, Eqs. (8.5), (8.6), we first obtain the conductivity tensor, and then the *tensor of permittivity* (*dielectric permeability*), E, in a collisionless, gyrotropic (magneto-active) plasma:

$$\begin{aligned}&\frac{1}{\varepsilon_0}\mathsf{E}(\omega,\boldsymbol{k})\\ &\quad= U + \sum_{\rm h} i\frac{(\mathrm{u}/\varepsilon_0)\,q_{\rm h}^2}{m_{\rm h}\,\omega\,\omega_{B\,\rm h}}\int d\boldsymbol{p}\,\boldsymbol{v}\int\limits_{\infty}^{\varphi} d\varphi_1\frac{\partial f_{0\,\rm h}}{\partial\boldsymbol{v}}\exp\left(\frac{i}{\omega_{B\,\rm h}}\int\limits_{\varphi}^{\varphi_1} d\varphi_2(\omega - \boldsymbol{k}\cdot\boldsymbol{v})\right)\\ &\quad= U + \sum_{\rm h}\frac{(\mathrm{u}/\varepsilon_0)\,q_{\rm h}^2}{\omega}\int d\boldsymbol{p}\frac{\partial f_{0\,\rm h}}{\partial\mathscr{E}_{\rm h}}\sum_s\frac{\boldsymbol{F}_{\rm h}(s)\,\boldsymbol{F}_{\rm h}^*(s)}{(\omega - k_z v_z - s\,\omega_{B\,\rm h})},\end{aligned} \tag{22.3}$$

which reads in components:

$$\begin{aligned}&\frac{1}{\varepsilon_0}\varepsilon_{lm}(\omega,\boldsymbol{k})\\ &\quad= \delta_{lm} + \sum_{\rm h} i\frac{(\mathrm{u}/\varepsilon_0)\,q_{\rm h}^2}{m_{\rm h}\omega\omega_{B\,\rm h}}\int d\boldsymbol{p}\,v_l\int\limits_{\infty}^{\varphi} d\varphi_1\frac{\partial f_{0\,\rm h}}{\partial v_m}\exp\left(\frac{i}{\omega_{B\,\rm h}}\int\limits_{\varphi}^{\varphi_1} d\varphi_2(\omega - \boldsymbol{k}\cdot\boldsymbol{v})\right)\\ &\quad= \delta_{lm} + \sum_{\rm h}\frac{(\mathrm{u}/\varepsilon_0)\,q_{\rm h}^2}{\omega}\int d\boldsymbol{p}\frac{\partial f_{0\,\rm h}}{\partial\mathscr{E}_{\rm h}}\sum_s\frac{F_{l\rm h}(s)\,F_{m h}^*(s)}{(\omega - k_z v_z - s\,\omega_{B\,\rm h})}.\end{aligned} \tag{22.3a}$$

We made use here of the particle energy

$$\mathscr{E}_{\rm h} = \tfrac{1}{2} m_{\rm h}\, v_{\rm h}^2 \tag{22.4}$$

---

[2] Compare K. Rawer, and K. Suchy: this Encyclopedia, Vol. 49/2, Sect. 1.

[3] J_s is the Bessel-function of order s, J_s' its derivative.

and of the special vector function[3]

$$\boldsymbol{F}_{\mathrm{h}}(s) \equiv \left(\frac{s\omega_{B\mathrm{h}}}{k_\perp} \mathrm{J}_s\left(\frac{\boldsymbol{k}_\perp \cdot \boldsymbol{v}_\perp}{\omega_{B\mathrm{h}}}\right);\ -i\, v_\perp \mathrm{J}_s'\left(\frac{\boldsymbol{k}_\perp \cdot \boldsymbol{v}_\perp}{\omega_{B\mathrm{h}}}\right);\ v_z \mathrm{J}_s\left(\frac{\boldsymbol{k}_\perp \cdot \boldsymbol{v}_\perp}{\omega_{B\mathrm{h}}}\right)\right). \tag{22.5}$$

The integration in Eq. (22.3) has to be performed by an integration path encircling (from below) the pole at $\omega = k_z v_z + s\,\omega_{B\mathrm{h}}$, see Sect. 17. It is equivalent with this procedure to use the following interpretation of the integrand:

$$\frac{1}{(\omega - k_z v_z - s\omega_{B\mathrm{h}})} = \frac{\mathscr{P}}{(\omega - k_z v_z - s\omega_{B\mathrm{h}})} - i\,\pi\,\delta(\omega - k_z v_z - s\,\omega_{B\mathrm{h}}), \tag{22.6}$$

where $\mathscr{P}$ is the principal value, δ being the DIRAC function.

γ) This presentation of the dielectricity tensor of a gyrotropic plasma clearly shows the *cyclotron resonances*; these are related with any zero of the denominator, $\omega - k_z v_z - s\,\omega_{B\mathrm{h}}$. The terms corresponding to the principal value of the integrand (symbol $\mathscr{P}$) determine that part of the tensor $\mathsf{E}(\omega, \boldsymbol{k})$ which is hermitean, while those with the δ-function give the anti-Hermitean contribution to the same tensor. The latter is responsible for wave absorption. Thus, in a collisionless plasma, only those particles give a contribution to wave attenuation for which the condition

$$\omega - k_z v_z - s\,\omega_{B\mathrm{h}} = 0$$

is fulfilled.

The nature of such absorption is easily understood, taking into account that the free motion of a charged particle in a magnetic field is a helical orbit composed by a circular motion around this field with the gyropulsation

$$\omega_{B\mathrm{h}} = q_{\mathrm{h}}\, B_{\bar{0}} / (m_{\mathrm{h}}\, c_0 \sqrt{\varepsilon_0\, \mu_0}),$$

and a translatorial motion with constant velocity v_z parallel to the magnetic field. Now, particles in a non-linear orbit radiate electromagnetic waves as it is the case with other kinds of acceleration, for example linear deceleration in "Bremsstrahlung". A particle in a circular orbit radiates magneto-Bremsstrahlung, i.e. cyclotron-radiation. The condition to be fulfilled is the same as for the effect found by VAVILOV and ČERENKOV. It is shown in [1] that the frequencies of waves radiated in this context are given by the conditions:

$$\omega = s\,\omega_{B\mathrm{h}} + k_z v_z \qquad (s = 0, \pm 1, \pm 2, \ldots). \tag{22.7}$$

For $s=0$ this condition corresponds to ČERENKOV emission, and for $s \neq 0$ to *magneto-Bremsstrahlung*.

Of course, a particle which is able to emit radiation (at any frequency $\omega/2\pi$) may also absorb radiation on the same frequency. This is essentially the explanation of collisionless absorption of electromagnetic waves by particles in a plasma.

It must be emphasized that for $k_z = 0$, i.e. for waves propagating perpendicularly to the magnetic field $\boldsymbol{B}_{\bar{0}}$, the anti-hermitean part of the tensor $\mathsf{E}(\omega, \boldsymbol{k})$ is vanishing so that such waves suffer from no absorption in a collisionless plasma. In fact, it can be seen from the expression Eq. (22.3), taking account of Eq. (22.6), that with $k_z = 0$ the contribution of the terms containing the δ-function vanishes so that the (dielectric) permittivity of the plasma is real[4]. This statement is, however, only true when (for any kind of particle, h)

$$\omega \lesssim \omega_{B\mathrm{h}}.$$

δ) In the limiting case of *high-frequencies*, $\omega \gg \omega_{B\mathrm{h}}$, the influence of the magnetic field $\boldsymbol{B}_{\bar{0}}$ is negligible. Therefore, in the high-frequency case absorption must

[4] However, when admitting for relativistic effects (the transverse DOPPLER effect) absorption appears also with $k_z = 0$, see [1], [2].

be possible independently on the direction of $\boldsymbol{k}$, thus also for $k_z=0$; this follows when considering the situation with vanishing magnetic field, $\boldsymbol{B}_{\overset{+}{0}}=0$. The transition towards this situation in Eqs. (22.2), (22.3) is not trivial at all; it is related with the complicated mathematical problem how BESSEL functions can be asymptotically approximated for large values of argument and index, see [34], [35]. For $\omega_{B\mathrm{h}}\to 0$ the arguments of the BESSEL functions are large in Eqs. (22.2), (22.3). Then, the terms with[5]

$$|s| < s_{\max} = \boldsymbol{k}_\perp \cdot \boldsymbol{v}_\perp / \omega_{B\mathrm{h}}$$

give contributions of similar order, while those of terms with large $|s| > s_{\max}$ decrease exponentially. Therefore the summation in Eqs. (22.2), (22.3) must be extended until at least $|s| = s_{\max}$, when considering the transition $\omega_{B\mathrm{h}}\to 0$. Thus the denominator takes the form

$$\omega - s_{\max}\,\omega_{B\mathrm{h}} = \omega - \boldsymbol{k}_\perp \cdot \boldsymbol{v}_\perp = \omega - \boldsymbol{k}\cdot\boldsymbol{v}$$

and collisionless absorption i.e. LANDAU damping must appear at the transition. Adopting a Maxwellian distribution $f_{\mathrm{h}}^{\mathrm{M}}$, the dielectric tensor $\mathsf{E}(\omega, \boldsymbol{k})$ can be fully expressed with tabulated integrals as follows, see [33], [34]:

$$\left.\begin{aligned}
\frac{1}{\varepsilon_0}\,\varepsilon_{xx} &= 1 - \sum_{\mathrm{h}}\sum_{s} \frac{s^2\omega_{N\mathrm{h}}^2}{\omega(\omega - s\,\omega_{B\mathrm{h}})}\,\frac{\mathscr{A}_s(z_{\mathrm{h}})}{z_{\mathrm{h}}}\,\mathrm{J}_+(\beta_{s\mathrm{h}}),\\
\frac{1}{\varepsilon_0}\,\varepsilon_{yy} &= \frac{1}{\varepsilon_0}\,\varepsilon_{xx} + 2\sum_{\mathrm{h}}\sum_{s} \frac{z_{\mathrm{h}}\omega_{N\mathrm{h}}^2}{\omega(\omega - s\,\omega_{B\mathrm{h}})}\,\mathscr{A}_s'(z_{\mathrm{h}})\,\mathrm{J}_+(\beta_{s\mathrm{h}}),\\
\frac{1}{\varepsilon_0}\,\varepsilon_{xy} &= -\frac{1}{\varepsilon_0}\,\varepsilon_{yx} = -i\sum_{\mathrm{h}}\sum_{s} \frac{s\,\omega_{N\mathrm{h}}^2}{\omega(\omega - s\,\omega_{B\mathrm{h}})}\,\mathscr{A}_s'(z_{\mathrm{h}})\,\mathrm{J}_+(\beta_{s\mathrm{h}}),\\
\frac{1}{\varepsilon_0}\,\varepsilon_{xz} &= \frac{1}{\varepsilon_0}\,\varepsilon_{zx} = \sum_{\mathrm{h}}\sum_{s} \frac{s\,\omega_{N\mathrm{h}}^2}{\omega\,\omega_{B\mathrm{h}}}\,\frac{k_\perp}{k_z}\,\frac{\mathscr{A}_s(z_{\mathrm{h}})}{z_{\mathrm{h}}}\,[1 - \mathrm{J}_+(\beta_{s\mathrm{h}})],\\
\frac{1}{\varepsilon_0}\,\varepsilon_{yz} &= -\frac{1}{\varepsilon_0}\,\varepsilon_{zy} = -i\sum_{\mathrm{h}}\sum_{s} \frac{\omega_{N\mathrm{h}}^2}{\omega\,\omega_{B\mathrm{h}}}\,\frac{k_\perp}{k_z}\,\mathscr{A}_s'(z_{\mathrm{h}})\,[1 - \mathrm{J}_+(\beta_{s\mathrm{h}})],\\
\frac{1}{\varepsilon_0}\,\varepsilon_{zz} &= 1 + \sum_{\mathrm{h}}\sum_{s} \frac{\omega_{N\mathrm{h}}^2(\omega - s\,\omega_{B\mathrm{h}})}{\omega k_z^2 V_{T\mathrm{h}}^2}\,\mathscr{A}_s(z_{\mathrm{h}})\,[1 - \mathrm{J}_+(\beta_{s\mathrm{h}})].
\end{aligned}\right\} \tag{22.8}$$

Here

$$\mathscr{A}_s(z_{\mathrm{h}}) \equiv \mathrm{e}^{-z_{\mathrm{h}}}\,\mathrm{I}_s(z_{\mathrm{h}}), \quad \mathscr{A}'(z_{\mathrm{h}}) \equiv \frac{\mathrm{d}}{\mathrm{d}z_{\mathrm{h}}}\,\mathscr{A}, \tag{22.9}$$

and $\mathrm{I}_y(z_{\mathrm{h}})$ means the BESSEL function with imaginary argument,

$$z_{\mathrm{h}} \equiv k_\perp^2\, V_{T\mathrm{h}}^2/\omega_{B\mathrm{h}}^2, \tag{22.10}$$

further

$$\beta_{s\mathrm{h}} \equiv \frac{\omega - s\,\omega_{B\mathrm{h}}}{|k_z|\,V_{T\mathrm{h}}} \tag{22.11}$$

and $\mathrm{J}_+(x)$ is the BESSEL function defined in Eq. (17.14). Going over to the low temperature limit as we have done in Sect. 6 with

$$k_\parallel V_{T\mathrm{h}} \ll \omega; \qquad k_\perp V_{T\mathrm{h}} \ll \omega_{B\mathrm{h}}, \tag{6.14}$$

which means here

$$z_{\mathrm{h}} \equiv \frac{k_\perp^2\, V_{T\mathrm{h}}^2}{\omega_{B\mathrm{h}}^2} \ll 1; \qquad \beta_{s\mathrm{h}} \equiv \frac{\omega - s\,\omega_{B\mathrm{h}}}{|k_z|\,V_{T\mathrm{h}}} \gg 1; \qquad \frac{k_\parallel V_{T\mathrm{h}}}{\omega} \ll 1, \tag{22.12}$$

[5] Under our particular assumptions $\boldsymbol{k}_\perp \cdot \boldsymbol{v}_\perp = k_\perp v_\perp$.

and now find[6] the dielectric tensor $E^{(0)}$:

$$\left.\begin{aligned}
&\frac{1}{\varepsilon_0}\,\varepsilon^{(0)}_{xx}=\frac{1}{\varepsilon_0}\,\varepsilon^{(0)}_{yy}\equiv\frac{1}{\varepsilon_0}\,\varepsilon^{(0)}_{\perp}=1-\sum_{\mathrm{h}}\frac{\omega^2_{N\mathrm{h}}}{\omega^2-\omega^2_{B\mathrm{h}}}\,,\\
&\frac{1}{\varepsilon_0}\,\varepsilon^{(0)}_{xy}=\frac{1}{\varepsilon_0}\,\varepsilon^{(0)}_{yx}\equiv i\,\frac{1}{\varepsilon_0}\,g^{(0)}=-i\sum_{\mathrm{h}}\frac{\omega^2_{N\mathrm{h}}\omega_{B\mathrm{h}}}{\omega\,(\omega^2-\omega^2_{B\mathrm{h}})}\,,\\
&\frac{1}{\varepsilon_0}\,\varepsilon^{(0)}_{zz}\equiv\frac{1}{\varepsilon_0}\,\varepsilon^{(0)}_{\parallel}=1-\sum_{\mathrm{h}}\frac{\omega^2_{N\mathrm{h}}}{\omega^2}\approx 1-\frac{\omega^2_{N\mathrm{e}}}{\omega^2}\,,\\
&\varepsilon^{(0)}_{xz}=\varepsilon^{(0)}_{zx}=\varepsilon^{(0)}_{yz}=\varepsilon^{(0)}_{zy}=0\,.
\end{aligned}\right\}\tag{22.13}$$

This tensor, Eq. (22.13) coincides with that of Eq. (6.12) which has been obtained with the simplest model of independent particles, for $\bar{\nu}_{\mathrm{h}}=0$. We shall call this a collisionless "cold" plasma because it is equivalent to the gasdynamic model of two fluids as considered in Eq. (10.4) where we considered a collisionless cold plasma. Such quasi-gasdynamic considerations lead also to Eqs. (22.13).

ε) We now consider the *transition towards gasdynamics of one fluid* only, see Eq. (11.13). Taking account of the fact that the mass of any ion is great against the electronic mass, m_{e}, the transition is obtained by

$$\omega\ll\omega_{B\,\mathrm{i}}\ll\omega_{N\,\mathrm{i}}\,;\qquad \frac{k^2_{\perp}\,V^2_{T\mathrm{h}}}{\omega^2_{B\mathrm{h}}}\ll 1\,;\qquad V_{T\,\mathrm{i}}\ll\frac{\omega}{|k_z|}\ll V_{T\,\mathrm{e}}\,,\tag{22.14}$$

see Sect. 11 δ.

Taking account of these inequalities we find the components of the tensor of dielectric permeability of a magneto-active (gyrotropic) plasma from Eq. (22.8):

$$\left.\begin{aligned}
&\frac{1}{\varepsilon_0}\,\varepsilon_{xx}=\frac{\omega^2_{N\mathrm{i}}}{\omega^2_{B\mathrm{i}}}=\frac{c^2_0}{V^2_{\mathrm{A}}}\\
&\frac{1}{\varepsilon_0}\,\varepsilon_{yy}=\frac{\omega^2_{N\mathrm{i}}}{\omega^2_{B\mathrm{i}}}+i\sqrt{2\pi}\,\frac{\omega^2_{N\mathrm{e}}}{\omega^2_{B\mathrm{e}}}\,\frac{k^2_{\perp}\,V_{T\mathrm{e}}}{|k_z|\,\omega}\,,\\
&\frac{1}{\varepsilon_0}\,\varepsilon_{yz}=-\frac{1}{\varepsilon_0}\,\varepsilon_{zy}=i\,\frac{\omega^2_{N\mathrm{e}}}{\omega\,\omega_{B\,\mathrm{e}}}\,\frac{k_{\perp}}{k_z}\left(1+i\sqrt{\frac{\pi}{2}}\,\frac{\omega}{|k_z|\,V_{T\,\mathrm{e}}}\right),\\
&\frac{1}{\varepsilon_0}\,\varepsilon_{zz}=-\frac{\omega^2_{N\,\mathrm{i}}}{\omega^2}+\frac{\omega^2_{N\mathrm{e}}}{k^2_z\,V^2_{T\mathrm{e}}}\left(1+i\sqrt{\frac{\pi}{2}}\,\frac{\omega}{|k_z|\,V_{T\,\mathrm{e}}}\right),\\
&\varepsilon_{xz}=\varepsilon_{zx}=\varepsilon_{xy}=\varepsilon_{yx}=0
\end{aligned}\right\}\tag{22.15}$$

with the ALFVEN velocity V_{A} as given by

$$V^2_{\mathrm{A}}=c^2_0\,\frac{\omega^2_{B\,\mathrm{i}}}{\omega^2_{N\,\mathrm{i}}}=\frac{B^2_{\circ}}{\mathrm{u}\,\mu_0\,N_{\mathrm{i}}\,m_{\mathrm{i}}}\,.\tag{22.16}$$

The tensor Eq. (22.15) gives us some more information than is obtainable from the linearized system of equations of magneto-gasdynamics, Eqs. (11.13). These latter equations do not take account of collisionless dissipation as produced by ČERENKOV and cyclotron absorption[7]. The expressions Eqs. (22.15) take account of such dissipation which is in fact produced only by the electrons of the plasma. We do not consider here such attenuation due to ions because it remains extremely small under the conditions we have in mind.

[6] Upper index $^{(0)}$ designates the Hermitean part.

[7] We note that it would have been possible to take account of collisionless dissipation when deducing Eqs. (11.13). However, when computing the dissipation terms we had restricted ourselves to the linear approximation. This would be less general than our present treatment.

23. Collisional absorption. It is not difficult to take account of dissipation due to collisions in Eqs. (22.13), (22.15). This is also described as "Bremsabsorption" of waves.

α) In the high-frequency limit we may make use of

$$\omega \gg \bar{\nu}_{\mathrm{h}}; \qquad |\omega \pm \omega_{B\mathrm{h}}| \gg \bar{\nu}_{\mathrm{h}} \tag{23.1}$$

when computing the contribution by collisions to the dielectricity tensor $\boldsymbol{E}$. *Successive approximations* can then be introduced, building up the solution of the kinetic equation in successive steps according to different terms of development of the collision integral[1]. As result we obtain another collisional term to be added to the statistical distribution function f^{M} and this is a contribution, $\delta f_{\mathrm{h}}^{(1)}$, which is additional to the δf_{h} already described in Eq. (22.2)[2].

$$\left.\begin{aligned}\delta f_{\mathrm{h}}^{(1)} &= \sum_s \frac{i}{(\omega - k_z v_z - s\omega_{B\mathrm{h}})} \exp\left(-i s\varphi + i\frac{k_\perp v_\perp}{\omega_{B\mathrm{h}}}\sin\varphi\right)\frac{1}{2\pi}\times\\ &\times\int_{-\pi}^{+\pi} \mathrm{d}\varphi_1 \exp\left(i s\varphi_1 - i\frac{k_\perp v_\perp}{\omega_{B\mathrm{h}}}\sin\varphi_1\right)\sum_{\mathrm{g}}\frac{\partial}{\partial \boldsymbol{p}}\cdot I^{\mathrm{hg}}(\boldsymbol{p},\boldsymbol{p}')\times\\ &\times\left[\frac{\partial f_{0\mathrm{h}}}{\partial \boldsymbol{p}}\delta f_{\mathrm{g}}(\boldsymbol{p}') + \frac{\partial \delta f_{\mathrm{h}}}{\partial \boldsymbol{p}} f_{0\mathrm{h}}(\boldsymbol{p}') - \frac{\partial \delta f_{\mathrm{g}}}{\partial \boldsymbol{p}'} f_{0\mathrm{h}}(\boldsymbol{p}) - \frac{\partial f_{0\mathrm{g}}}{\partial \boldsymbol{p}'}\delta f_{\mathrm{h}}(\boldsymbol{p})\right].\end{aligned}\right\} \tag{23.2}$$

In a completely ionized, "cold" plasma the anti-hermitean contribution to the dielectricity tensor Eq. (22.12) has the form[3]

$$\left.\begin{aligned}\frac{1}{\varepsilon_0}\varepsilon_{xx}^{(\mathrm{a})} &= i\sqrt{\frac{\pi}{8}}\sum_{\mathrm{h}}\frac{\omega_{N\mathrm{h}}^2}{\omega|k_z|V_{T\mathrm{h}}}\left[\exp\left(\frac{-(\omega-\omega_{B\mathrm{h}})^2}{2k_z^2V_{T\mathrm{h}}^2}\right)\right.\\ &\quad\left.+\exp\left(\frac{-(\omega+\omega_{B\mathrm{h}})^2}{2k_z^2V_{T\mathrm{h}}^2}\right)\right]+\frac{1}{\varepsilon_0}\delta\varepsilon_\perp^{(\mathrm{a})}\\ \frac{1}{\varepsilon_0}\varepsilon_{yy}^{(\mathrm{a})} &= \frac{1}{\varepsilon_0}\varepsilon_{xx}^{(\mathrm{a})} + i\sqrt{2\pi}\sum_{\mathrm{h}}\frac{\omega_{N\mathrm{h}}^2}{\omega_{B\mathrm{h}}^2}\frac{k_\perp^2V_{T\mathrm{h}}}{|k_z|\,\omega}\exp\left(\frac{-\omega^2}{2k_z^2V_{T\mathrm{h}}^2}\right),\\ \frac{1}{\varepsilon_0}\varepsilon_{xy}^{(\mathrm{a})} &= -\frac{1}{\varepsilon_0}\varepsilon_{yx}^{(\mathrm{a})} \equiv i\frac{1}{\varepsilon_0}g^{(\mathrm{a})} = -\sqrt{\frac{\pi}{8}}\sum_{\mathrm{h}}\frac{\omega_{N\mathrm{h}}^2}{\omega|k_z|V_{T\mathrm{h}}}\left[\exp\left(\frac{-(\omega-\omega_{B\mathrm{h}})^2}{2k_z^2V_{T\mathrm{h}}^2}\right)-\right.\\ &\quad\left.-\exp\left(\frac{-(\omega+\omega_{B\mathrm{h}})^2}{2k_z^2V_{T\mathrm{h}}^2}\right)\right]+\frac{1}{\varepsilon_0}\delta_g^{(\mathrm{a})}\\ \frac{1}{\varepsilon_0}\varepsilon_{zz}^{(\mathrm{a})} &= \frac{1}{\varepsilon_0}\varepsilon_\parallel^{(\mathrm{a})} = i\sqrt{\frac{\pi}{2}}\sum_{\mathrm{h}}\frac{\omega\,\omega_{N\mathrm{h}}^2}{|k_z|^3V_{T\mathrm{h}}^3}\exp\left(\frac{-\omega^2}{2k_z^2V_{T\mathrm{h}}^2}\right)+\frac{1}{\varepsilon_0}\delta\varepsilon_\parallel.\end{aligned}\right\} \tag{23.3}$$

The first terms in these expressions take account of collisionless absorption of waves by plasma particles; they are exponentially small. The second terms are due to collisions between particles, reading:

$$\left.\begin{aligned}\frac{1}{\varepsilon_0}\delta\varepsilon_\perp^{(\mathrm{a})} &= i\sum_{\mathrm{h}}\frac{\bar{\nu}_{\mathrm{h}}\omega_{N\mathrm{h}}^2(\omega^2+\omega_{B\mathrm{h}}^2)}{\omega(\omega^2-\omega_{B\mathrm{h}}^2)^2},\\ \frac{1}{\varepsilon_0}\delta\varepsilon_\parallel^{(\mathrm{a})} &= i\sum_{\mathrm{h}}\frac{\bar{\nu}_{\mathrm{h}}\omega_{Nh}^2}{\omega^3},\\ \frac{1}{\varepsilon_0}\delta g^{(\mathrm{a})} &= \sum_{\mathrm{h}}\frac{2\bar{\nu}_{\mathrm{h}}\omega_{N\mathrm{h}}^2\omega_{B\mathrm{h}}}{(\omega^2-\omega_{B\mathrm{h}}^2)^2}.\end{aligned}\right\} \tag{23.4}$$

[1] See Sect. 18 where a similar method has been applied to an isotropic plasma.

[2] Using LANDAU's collision integral we neglect the influence of the magnetic field on the collisional process of particles. This is justified when $\omega_{N\mathrm{e}} \gg \omega_{B\mathrm{e}}$, a condition satisfied in the ionospheric plasma, but not everywhere in the magnetosphere.

[3] Upper index $^{(\mathrm{a})}$ identifies the antihermitean part.

The summation goes over electrons and ions (h = e, i) with

$$\bar{\nu}_e = \nu_{\text{eff}} \quad \text{and} \quad \bar{\nu}_i = \frac{m_e}{m_i}\,\nu_{\text{eff}}, \tag{23.5}$$

see Eq. (18.2) for definition of ν_{eff}.[1]

β) The one-fluid gasdynamic model may be applied, provided the inequalities Eqs. (22.14) are satisfied. *Collisions of electrons* can then be *neglected* if

$$\bar{\nu}_e \ll (k_z V_{Te}, \omega_{Be}). \tag{23.6}$$

In the ionosphere and, a fortiori, in the magnetosphere the second condition is always satisfied, see Tables 1 through 3 (Sect. 2). On frequencies $\omega \gg \bar{\nu}_i$ only ion-ion collisions contribute to collisional dissipation. One may take account of the latter by successive approximations (as explained above under α); in this method the solution of the kinetic equation is essentially built up by developing the collision integral for ions. We finally obtain the following supplement to the dielectricity tensor as given by Eq. (22.15):

$$\left.\begin{aligned}
\frac{1}{\varepsilon_0}\,\delta\varepsilon_{xx} &= i\,\frac{7}{10}\,\frac{\omega_{Ni}^2 \nu_{ii} k_\perp^2 V_{Ti}^2}{\omega\,\omega_{Bi}^4} + i\,\frac{\omega_{Ne}^2 \nu_{\text{eff}}}{\omega\,\omega_{Be}^2},\\
\frac{1}{\varepsilon_0}\,\delta\varepsilon_{yy} &= i\,\frac{4}{5}\,\frac{\omega_{Ni}^2 \nu_{ii} k_\perp^2 V_{Ti}^2}{\omega^3\,\omega_{Bi}^2}\\
\frac{1}{\varepsilon_0}\,\delta\varepsilon_{yz} &= -\frac{1}{\varepsilon_0}\,\delta\varepsilon_{zy} = -\frac{4}{5}\,\frac{\omega_{Ni}^2 \nu_{ii} k_\perp k_z V_{Ti}^2}{\omega^4\,\omega_{Bi}},\\
\frac{1}{\varepsilon_0}\,\delta\varepsilon_{zz} &= i\,\frac{8}{5}\,\frac{\omega_{Ni}^2 \nu_{ii} k_z^2 V_{Ti}^2}{\omega^5}.
\end{aligned}\right\} \tag{23.7}$$

It must be noted that any arbitrary relation between ω and $\bar{\nu}_e$ could be admitted here.

γ) In other frequency and wavelength ranges the oscillations in a magneto-active (gyrotropic) plasma are, generally speaking, strongly attenuated, i.e. evanescent. This is in particular so if

$$\bar{\nu}_h \gg (\omega, k_z V_{Th}). \tag{23.8}$$

There is just the exception of cyclotron oscillations which will be discussed in detail in Sect. 24. Therefore, we shall not consider other particular cases of the dielectricity tensor of a completely ionized gyrotropic plasma but proceed to a weakly ionized plasma which will be considered in the following.

In this case the BATNAGAR-GROSS-KROOK (BGK) model for the collision integral allows to derive a general expression of the dielectricity tensor, independent on frequency and wavelength of the oscillations. In fact a linearized kinetic equation for particles of kind h can be written in the following form for a *weakly ionized gyrotropic plasma* [see Eq. (18.8)]:

$$-i(\omega - \boldsymbol{k}\cdot\boldsymbol{v})\,\delta f_h + q_h \boldsymbol{E}\cdot\frac{\partial f_{0h}}{\partial \boldsymbol{p}} - \omega_{Bh}\frac{\partial\,\delta f_h}{\partial\varphi} = -\bar{\nu}_{hn}(\delta f_h - \eta_h f_{0h}). \tag{23.9}$$

When writing down this equation we have used the isothermal model of the collision integral, putting $\exp(-i\omega t + i\boldsymbol{k}\cdot\boldsymbol{r})$ to describe the dependence on time and space coordinates for all variable quantities, i.e. supposing a plane wave. The general solution of Eq. (23.9) which is periodical in the angle φ, is

[1] We note, that Eqs. (23.4) are incorrect in the region $\omega \ll \omega_{Bi}$, where

$$\frac{1}{\varepsilon_0}\,\delta\varepsilon_\perp^{(a)} = i\,\frac{\omega_{Ni}^2 \nu_i \omega}{\omega_{Bi}^4}, \qquad \frac{1}{\varepsilon_0}\,\delta g^{(a)} = 2\,\frac{\omega_{Ni}^2 \bar{\nu}_i \omega^2}{\omega_{Bi}^5}. \tag{23.4'}$$

easily found. It reads [see Eq. (22.2)]:

$$\left.\begin{aligned}\delta f_{\mathrm{h}} &= \frac{1}{\omega_{B\mathrm{h}}}\int\limits_{\infty}^{\varphi} \mathrm{d}\varphi_1 \left(\frac{q_{\mathrm{h}}}{m_{\mathrm{h}}}\boldsymbol{E}\cdot\frac{\partial f_{0\mathrm{h}}}{\partial\boldsymbol{v}} - \nu_{\mathrm{hn}}\eta_{\mathrm{h}} f_{0\mathrm{h}}\right)\exp\left(i\int\limits_{\varphi}^{\varphi_1}\mathrm{d}\varphi_2\frac{\omega-\boldsymbol{k}\cdot\boldsymbol{v}+i\bar{\nu}_{\mathrm{hn}}}{\omega_{B\mathrm{h}}}\right)\\
&= i\frac{q_{\mathrm{h}}}{\vartheta_{\mathrm{h}}} f_{0\mathrm{h}}\sum_s \frac{1}{(\omega+i\bar{\nu}_{\mathrm{hn}}-k_z v_z - s\omega_{B\mathrm{h}})}\exp\left(-i\,s\,\varphi + i\frac{k_\perp v_\perp}{\omega_{B\mathrm{h}}}\sin\varphi\right)\times\\
&\times\left\{\frac{s\omega_{B\mathrm{h}}}{k_\perp}\mathrm{J}_s\left(\frac{\boldsymbol{k}_\perp\cdot\boldsymbol{v}_\perp}{\omega_{B\mathrm{h}}}\right)E_x + i v_\perp \mathrm{J}_s'\left(\frac{\boldsymbol{k}_\perp\cdot\boldsymbol{v}_\perp}{\omega_{B\mathrm{h}}}\right)E_y+\right.\\
&\left.+\left(v_z E_z + \bar{\nu}_{\mathrm{hn}}\eta_{\mathrm{h}}\frac{\vartheta_{\mathrm{h}}}{q_{\mathrm{h}}}\right)\mathrm{J}_s\left(\frac{\boldsymbol{k}_\perp\cdot\boldsymbol{v}_\perp}{\omega_{B\mathrm{h}}}\right)\right\}.\end{aligned}\right\}\quad(23.10)$$

Putting this expression into the equation of induced current density, Eq. (8.5), we get after some easy algebraic computation [using Eq. (18.14)] the dielectricity tensor E for a weakly ionized gyrotropic plasma as:

$$\left.\begin{aligned}\frac{1}{\varepsilon_0}\mathsf{E}(\omega,\boldsymbol{k}) &= \mathsf{U} + \sum_{\mathrm{h}}\left(\mathsf{U} + i\frac{\bar{\nu}_{\mathrm{hn}}\boldsymbol{G}_{\mathrm{h}}\boldsymbol{k}}{\omega - i\bar{\nu}_{\mathrm{hn}}\boldsymbol{G}_{\mathrm{h}}\cdot\boldsymbol{k}}\right)\left(1+i\frac{\bar{\nu}_{\mathrm{hn}}}{\omega}\right)\times\\
&\times\left[\mathsf{E}^{(\mathrm{h})}(\omega+i\bar{\nu}_{\mathrm{hn}},\boldsymbol{k}) - \mathsf{U}\right],\end{aligned}\right\}\quad(23.11)$$

in components:

$$\left.\begin{aligned}\frac{1}{\varepsilon_0}\varepsilon_{lm}(\omega,\boldsymbol{k}) &= \delta_{lm} + \sum_{\mathrm{h}}\left(\delta_{lm} + i\frac{\bar{\nu}_{\mathrm{hn}} G_{\mathrm{h}l} k_m}{\omega - i\bar{\nu}_{\mathrm{hn}}\boldsymbol{G}_{\mathrm{h}}\cdot\boldsymbol{k}}\right)\left(1+i\frac{\bar{\nu}_{\mathrm{hn}}}{\omega}\right)\times\\
&\left[\times\,\varepsilon_{nm}^{(\mathrm{h})}(\omega+i\bar{\nu}_{\mathrm{hn}},\boldsymbol{k}) - \delta_{nm}\right],\end{aligned}\right\}\quad(23.11\,\mathrm{a})$$

where the vector $\boldsymbol{G}_{\mathrm{h}}$ is defined as:

$$\boldsymbol{G}_{\mathrm{h}} = \left\{\begin{aligned}&\frac{\omega_{B\mathrm{h}}}{k_\perp}\sum_s\frac{s\,\mathscr{A}_s(z_{\mathrm{h}})}{(\omega+i\bar{\nu}_{\mathrm{hn}}-s\omega_{B\mathrm{h}})}\mathrm{J}_+\left(\frac{\omega+i\bar{\nu}_{\mathrm{hn}}-s\omega_{B\mathrm{h}}}{|k_z|\,V_{T\mathrm{h}}}\right)\\
&-i\frac{\omega_{B\mathrm{h}}}{k_\perp}\sum_s\frac{z_{\mathrm{h}}\mathscr{A}_s'(z_{\mathrm{h}})}{(\omega+i\bar{\nu}_{\mathrm{hn}}-s\omega_{B\mathrm{h}})}\mathrm{J}_+\left(\frac{\omega+i\bar{\nu}_{\mathrm{hn}}-s\omega_{B\mathrm{h}}}{|k_z|\,V_{T\mathrm{h}}}\right)\\
&-\frac{1}{k_z}\sum_s\mathscr{A}_s(z_{\mathrm{h}})\left[1-\mathrm{J}_+\left(\frac{\omega+i\bar{\nu}_{\mathrm{hn}}-s\omega_{B\mathrm{h}}}{|k_z|\,V_{T\mathrm{h}}}\right)\right].\end{aligned}\right\}\quad(23.12)$$

$\mathsf{E}^{(\mathrm{h})}(\omega - i\bar{\nu}_{\mathrm{hn}},\boldsymbol{k})$ is a tensor which is identic with that described by Eq. (22.8) for h particles, if we replace ω by $(\omega+i\bar{\nu}_{\mathrm{hn}})$. The summation in Eq. (23.11) goes over all kinds of charged particles in the plasma. z_{h} is defined by Eq. (22.10).

δ) We note again that Eq. (23.11) for the weakly ionized plasma is valid, whichever relation between ω, $\omega_{B\mathrm{h}}$, $\bar{\nu}_{\mathrm{hn}}$, $k_\perp$, $V_{T\mathrm{h}}$ and $k_z V_{T\mathrm{h}}$ is existing. In the limiting case of a *cold plasma* the conditions Eqs. (22.12) are satisfied and the dielectricity tensor Eq. (23.11) coincides with that of Eqs. (6.11), (6.12), valid for the simplest model of independent particles (with $\bar{\nu}_{\mathrm{h}}=\bar{\nu}_{\mathrm{hn}}$). Under the conditions Eqs. (22.14) the one fluid gasdynamic model is able to describe a collisionless plasma. If yet

$$\bar{\nu}_{\mathrm{en}} \ll (k_z V_{T\mathrm{e}},\ \omega_{B\mathrm{e}})\quad(23.13)$$

is true, we get from Eq. (23.11) [see Eqs. (22.15) and (23.7)]:

$$\left.\begin{aligned}
\frac{1}{\varepsilon_0}\,\varepsilon_{xx} &= 1 - \frac{\omega_{N\mathrm{i}}^2(\omega + i\bar{\nu}_{\mathrm{in}})}{\omega\,[(\omega + i\bar{\nu}_{\mathrm{in}})^2 - \omega_{B\mathrm{i}}^2]}\,,\\
\frac{1}{\varepsilon_0}\,\varepsilon_{yy} &= 1 - \frac{\omega_{N\mathrm{i}}^2(\omega + i\bar{\nu}_{\mathrm{in}})}{\omega\,[(\omega + i\bar{\nu}_{\mathrm{in}})^2 - \omega_{B\mathrm{i}}^2]} + i\sqrt{2\pi}\,\frac{\omega_{N\mathrm{e}}^2}{\omega_{B\mathrm{e}}^2}\,\frac{k_\perp^2\,V_{T\mathrm{e}}}{|k_z|\,\omega}\,,\\
\frac{1}{\varepsilon_0}\,\varepsilon_{yz} &= -\frac{1}{\varepsilon_0}\,\varepsilon_{zy} = i\,\frac{\omega_{N\mathrm{e}}^2}{\omega\,\omega_{B\mathrm{e}}}\,\frac{k_\perp}{k_z}\left(1 + i\sqrt{\frac{\pi}{2}}\,\frac{\omega}{|k_z|\,V_{T\mathrm{e}}}\right),\\
\frac{1}{\varepsilon_0}\,\varepsilon_{xy} &= -\frac{1}{\varepsilon_0}\,\varepsilon_{yx} = -i\left(\frac{\omega_{N\mathrm{i}}^2\,\omega_{B\mathrm{i}}}{\omega\,[(\omega + i\bar{\nu}_{\mathrm{in}})^2 - \omega_{B\mathrm{i}}^2]} + \frac{\omega_{N\mathrm{e}}^2}{\omega\,\omega_{B\mathrm{e}}}\right),\\
\frac{1}{\varepsilon_0}\,\varepsilon_{zz} &= 1 - \frac{\omega_{N\mathrm{i}}^2}{\omega\,(\omega + i\bar{\nu}_{\mathrm{in}})} + \frac{\omega_{N\mathrm{e}}^2}{k_z^2\,V_{T\mathrm{e}}^2}\left(1 + i\sqrt{\frac{\pi}{2}}\,\frac{\omega}{|k_z|\,V_{T\mathrm{e}}}\right).\\
\varepsilon_{xz} &= \varepsilon_{zx} = 0.
\end{aligned}\right\} \tag{23.14}$$

ε) Finally we define a "*longitudinal (dielectric) permititvity*" of a gyrotropic plasma (a scalar), by

$$\varepsilon(\omega, \boldsymbol{k}) = \boldsymbol{k}^0 \cdot \mathsf{E}(\omega, \boldsymbol{k}) \cdot \boldsymbol{k}^0 = \frac{1}{k^2}\,\boldsymbol{k} \cdot \mathsf{E}(\omega, \boldsymbol{k}) \cdot \boldsymbol{k}\,, \tag{23.15}$$

in components[4]

$$\varepsilon(\omega, \boldsymbol{k}) = \frac{k_l\,k_m}{k^2}\,\varepsilon_{lm}(\omega, \boldsymbol{k})\,. \tag{23.15a}$$

Eq. (12.16) shows that this quantity characterizes the potential of the electric field which is provoked in the plasma by the presence of space charges. For a collisionless plasma we obtain from Eqs. (22.8):

$$\frac{1}{\varepsilon_0}\,\varepsilon(\omega, \boldsymbol{k}) = 1 + \sum_{\mathrm{h}} \frac{\omega_{N\mathrm{h}}^2}{k^2\,V_{T\mathrm{h}}^2}\left[1 - \sum_s \frac{\omega}{(\omega - s\,\omega_{B\mathrm{h}})}\,\mathscr{A}_s(z_{\mathrm{h}})\,\mathrm{J}_+\left(\frac{\omega - s\,\omega_{B\mathrm{h}}}{|k_z|\,V_{T\mathrm{h}}}\right)\right]. \tag{23.16}$$

We may also take account of collisions in the expression for the longitudinal (dielectric) permittivity. In a weakly ionized plasma, from Eq. (23.11) we obtain, including collisions:

$$\frac{1}{\varepsilon_0}\,\varepsilon(\omega, \boldsymbol{k}) = 1 + \sum_{\mathrm{h}} \frac{\omega_{N\mathrm{h}}^2}{k^2\,V_{T\mathrm{h}}^2}\;\frac{1 - \sum\limits_s \dfrac{\omega + i\bar{\nu}_{\mathrm{hn}}}{\omega + i\bar{\nu}_{\mathrm{hn}} - s\,\omega_{B\mathrm{h}}}\,\mathscr{A}_s(z_{\mathrm{h}})\,\mathrm{J}_+\left(\dfrac{\omega + i\bar{\nu}_{\mathrm{hn}} - s\,\omega_{B\mathrm{h}}}{|k_z|\,V_{T\mathrm{h}}}\right)}{1 - \sum\limits_s \dfrac{i\nu_{\mathrm{hn}}}{\omega + i\bar{\nu}_{\mathrm{hn}} - s\,\omega_{B\mathrm{h}}}\,\mathscr{A}_s(z_{\mathrm{h}})\,\mathrm{J}_+\left(\dfrac{\omega + i\bar{\nu}_{\mathrm{hn}} - s\,\omega_{B\mathrm{h}}}{|k_z|\,V_{T\mathrm{h}}}\right)}\,. \tag{23.17}$$

For vanishing magnetic field, $\boldsymbol{B}_{\overset{+}{0}} \to 0$, Eqs. (23.16), (23.17) go over into the corresponding equations for the longitudinal (dielectric) permittivity of an isotropic plasma, see Sect. 17.

ζ) In the *limit of static conditions*, defined more precisely by

$$\omega \ll k_z\,V_{T\mathrm{h}}; \qquad \omega\,\bar{\nu}_{\mathrm{h}} \ll k_z^2\,V_{T\mathrm{h}}^2, \tag{23.18}$$

the expressions Eqs. (23.16), (23.17) come down to

$$\frac{1}{\varepsilon_0}\,\varepsilon(0, \boldsymbol{k}) = 1 + \frac{1}{k^2\,\lambda_{\mathrm{D}}^2} \tag{23.19}$$

[4] The summation rule is applied throughout this paper.

which is identic with Eq. (18.15). It follows that the potential o a statie charge in a gyrotropic plasma appears as a DEBYE potential, Eq. (18.16), as it is the case in an isotropic plasma in the absence of outer fields. One shows that this conclusion remains valid for a completely ionized plasma with collisions. Thus the COULOMB interaction in a gyrotropic (magneto-active) plasma remains limited by the DEBYE length λ_D, as it was used when deriving the collision integral after LANDAU, see Sect. 16.

24. Waves in homogeneous gyrotropic (magneto-active) plasmas. As we have remarked above, there is a large number of different branches of possible waves in a gyrotropic plasma, so that a detailed analysis of all of these will not be given here. Such detailed description can be found in [*1*] ... [*6*], [*11*], [*36*]. We shall limit our present investigation to the characteristic features of wave propagation in such a plasma.

α) In a gyrotropic plasma, quite generally, waves can not simply be described as being either longitudinal or transverse[1]. We have to take account of the symmetry of the dielectricity tensor of the plasma and analyse the *general dispersion equation* which we have derived from it in the form

$$A\,k^4 + B\frac{\omega^2}{c_0^2}k^2 + C\frac{\omega^4}{c_0^4} = 0 \qquad [4.3]$$

where

$$\varepsilon_0 A \equiv \frac{\boldsymbol{k}\boldsymbol{k}}{k^2}\cdot\cdot\,\mathsf{E}(\omega,\boldsymbol{k}) = \frac{1}{k^2}\left(k_\perp^2\,\varepsilon_{xx} + k_z^2\,\varepsilon_{zz} + 2k_\perp k_z\,\varepsilon_{xz}\right), \tag{24.1}$$

$$\left.\begin{aligned} \varepsilon_0^2 B \equiv{}& \frac{\boldsymbol{k}\boldsymbol{k}}{k^2}\cdot\cdot\left(\mathsf{E}(\omega,\boldsymbol{k})\cdot\mathsf{E}(\omega,\boldsymbol{k})\right) - A\,\mathrm{Tr}\left(\mathsf{E}(\omega,\boldsymbol{k})\right)\\ ={}& -\varepsilon_{xx}\,\varepsilon_{zz} + \varepsilon_{xz}^2 - \frac{k_z^2}{k^2}\left(\varepsilon_{yy}\,\varepsilon_{zz} + \varepsilon_{yz}^2\right) - \\ &-\frac{k_\perp^2}{k^2}\left(\varepsilon_{xx}\,\varepsilon_{yy} + \varepsilon_{xy}^2\right) + 2\frac{k_\perp k_z}{k^2}\left(\varepsilon_{xy}\,\varepsilon_{yz} - \varepsilon_{xz}\,\varepsilon_{zy}\right), \end{aligned}\right\} \tag{24.2}$$

$$\begin{aligned} \varepsilon_0^3 C \equiv{}& \det\left|\mathsf{E}(\omega,\boldsymbol{k})\right|\\ ={}& \varepsilon_{zz}\left(\varepsilon_{xx}\,\varepsilon_{yy} + \varepsilon_{xy}^2\right) + \left(\varepsilon_{xx}\,\varepsilon_{yz}^2 - \varepsilon_{yy}\,\varepsilon_{xz}^2\right) + 2\,\varepsilon_{xy}\,\varepsilon_{xz}\,\varepsilon_{yz}. \end{aligned} \tag{24.3}$$

It is easily seen that A is identical with the "longitudinal dielectric (relative) permeability".

β) For any particular case the conditions can be specified under which the wave field in a plasma appears as essentially longitudinal. The condition is that the longitudinal electric field component $|\boldsymbol{E}_l|$ is much greater than the transverse component $|\boldsymbol{E}_{\mathrm{tr}}|$, with

$$\left.\begin{aligned} \boldsymbol{E}_l &\equiv \boldsymbol{k}^0(\boldsymbol{k}^0\cdot\boldsymbol{E}) = \frac{1}{k^2}\,\boldsymbol{k}(\boldsymbol{k}\cdot\boldsymbol{E})\\ \boldsymbol{E}_{\mathrm{tr}} &\equiv \boldsymbol{k}^0\times(\boldsymbol{E}\times\boldsymbol{k}^0) = \frac{1}{k^2}\left(\boldsymbol{k}\times(\boldsymbol{E}\times\boldsymbol{k})\right). \end{aligned}\right\} \tag{24.4}$$

In fact by scalar multiplication of Eq. (4.2) by $\boldsymbol{k}$ we obtain

$$\boldsymbol{E}_l = -\boldsymbol{k}\,\frac{\boldsymbol{k}\cdot\mathsf{E}(\omega,\boldsymbol{k})\cdot\boldsymbol{E}_{\mathrm{tr}}}{k^2\,\varepsilon(\omega,\boldsymbol{k})}, \tag{24.5}$$

$$= -\boldsymbol{k}\,\frac{k_i\,\varepsilon_{ij}(\omega,\boldsymbol{k})\,E_{j\,\mathrm{tr}}}{k^2\,\varepsilon(\omega,\boldsymbol{k})}. \tag{24.5a}$$

[1] We use the terms in the strict sense, i.e. relative to the wave vector $\boldsymbol{k}$. In a purely longitudinal wave the electric wave field $\boldsymbol{E}$ is parallel with $\boldsymbol{k}$, while it is perpendicular to it in a purely transverse wave, see Eq. (24.4) below. Unfortunately, in the literature the same terms are often used relative to the magnetic field. We use the expressions "parallel" and "perpendicular" in this latter context.

It is easily seen from this equation that under the condition

$$\varepsilon(\omega, \boldsymbol{k}) = \frac{1}{k^2} \boldsymbol{k} \cdot \mathsf{E}(\omega, \boldsymbol{k}) \cdot \boldsymbol{k} = 0, \tag{24.6}$$

in components

$$\varepsilon(\omega, \boldsymbol{k}) = \frac{k_i k_j}{k^2} \varepsilon_{ij}(\omega, \boldsymbol{k}) = 0, \tag{24.6a}$$

the electric field of the wave is purely longitudinal. One may also designate such a field as a "potential field", because in a purely longitudinal wave

and

$$\left.\begin{aligned} \boldsymbol{E} = \boldsymbol{E}_{\mathrm{l}} &= -\nabla\phi \equiv -\operatorname{grad}\phi \\ \boldsymbol{B} &= 0 = \boldsymbol{E}_{\mathrm{tr}}, \end{aligned}\right\} \tag{24.7}$$

ϕ being a scalar potential. It must, however, be noted that the electric displacement vector $\boldsymbol{D}$ of such wave may be different from zero and is not necessarily longitudinal. A detailed classification of the different types of longitudinal waves is given in [8].

Eq. (24.6) is the dispersion equation for purely longitudinal waves in a plasma. Strictly speaking, these do not appear as waves particularly related with the gyrotropic anisotropy of the medium. Therefore Eq. (24.6) is not qualified to describe oscillations in a gyrotropic plasma in a strict way, but only approximately and under certain specific conditions. These stem from the requirement that the roots of Eq. (24.6) coincide approximately with those of the full dispersion equation, Eq. (4.3). It is easily seen that this holds in the frequency range

$$\omega^2 \ll k^2 c_0^2, \tag{24.8}$$

thus if the index of refraction is great (the phase velocity of the waves is small, $V_{\mathrm{ph}} \ll c_0$); in this range the first lefthand member in Eq. (4.3) is large against the two following ones which, a fortiori, can be neglected. We shall precise below the conditions for longitudinal wave character in particular cases, see for example Sect. 26.

25. Waves in homogeneous, cold gyrotropic plasmas.

α) When discussing the propagation of electromagnetic waves we begin with the easiest case, i.e. that of a cold, collisionless plasma. Eqs. (22.13) are valid under these conditions and we have*

$$\varepsilon_0 A = \varepsilon_{\perp}^{(0)} \sin^2\Theta + \varepsilon_{\parallel}^{(0)} \cos^2\Theta, \tag{25.1}$$

$$\varepsilon_0^2 B = -\varepsilon_{\perp}^{(0)} \varepsilon_{\parallel}^{(0)} (1 + \cos^2\Theta) - (\varepsilon_{\perp}^{(0)2} - g^{(0)2}) \sin^2\Theta, \tag{25.2}$$

$$\varepsilon_0^3 C = \varepsilon_{\parallel}^{(0)} (\varepsilon_{\perp}^{(0)2} - g^{(0)2}), \tag{25.3}$$

where $\varepsilon_{\parallel}^{(0)}$, $\varepsilon_{\perp}^{(0)}$ and $g^{(0)}$ have been explained in Sect. 22δ, in Eqs. (22.13).

Θ is the angle between the wave vector, $\boldsymbol{k}$, and that of the magnetic field, $\boldsymbol{B}_{\circ}$, i.e.

$$|\boldsymbol{k}_{\perp}| = k \sin\Theta; \quad |\boldsymbol{k}_{\parallel}| \equiv k_z = k \cos\Theta. \tag{25.4}$$

With these expressions we may define a *complex refractive index*

$$n \equiv \mu + i\chi = \frac{c_0}{\omega} |\boldsymbol{k}|.$$

* Upper index (0) designates the Hermitean part.

μ and χ being functions of frequency, $\omega/2\pi$, and of the angle Θ. From Eq. (4.3) one obtains two solutions

$$n_{1,2}^2 = \frac{1}{2A}\left(-B \pm \sqrt{B^2 - 4AC}\right). \tag{25.5}$$

This shows that in the general case of a cold plasma, for given ω and Θ, two types of waves can propagate which have different indices, thus different phase velocities. These two types are usually called ordinary and extraordinary waves[1]. In the general case both waves have elliptic polarization:

$$\frac{E_{y1,2}}{E_{x1,2}} = -\frac{i g^{(0)}}{n_{1,2}^2 - \varepsilon_\perp^{(0)}}. \tag{25.6}$$

(The indices 1, 2 designate the ordinary and extraordinary wave, respectively.)

Eq. (25.5) is particularly simple in the limit of waves propagating perpendicularly to the external magnetic field ($\Theta = \frac{1}{2}\pi$), or parallel with it ($\Theta = 0$). We find for the case of *perpendicular propagation*:

$$\Theta = \frac{1}{2}\pi:\ n_1^2 = \varepsilon_\parallel^{(0)2}; \qquad n_2^2 = \frac{\varepsilon_\perp^{(0)2} - g^{(0)2}}{\varepsilon_\perp^2}, \tag{25.7}$$

and for *parallel propagation*:

$$\Theta = 0:\ n_{1,2}^2 = \varepsilon_\perp^{(0)} \pm g^{(0)}. \tag{25.8}$$

Another solution of Eq. (4.3) for parallel propagation is

$$\varepsilon_\parallel^{(0)} = 0 \tag{25.9}$$

which is the condition for purely longitudinal waves ($\boldsymbol{E} \parallel \boldsymbol{k}$) in a cold plasma.

Eqs. (25.8) describe transverse electromagnetic waves ($\boldsymbol{E} \perp \boldsymbol{k}$) propagating along the external magnetic field ($\boldsymbol{k} \parallel \boldsymbol{B}_{\male}$); they have circular polarization because Eq. (25.6), under these particular conditions, gives

$$\frac{E_{y1,2}}{E_{x1,2}} = \mp i. \tag{25.10}$$

Waves propagating perpendicularly to the magnetic field ($\boldsymbol{k} \perp \boldsymbol{B}_{\male}$) are neither purely longitudinal nor purely transverse waves, as is the case for $\boldsymbol{k} \parallel \boldsymbol{B}_{\male}$. For perpendicular propagation ($\boldsymbol{k} \perp \boldsymbol{B}_{\male}$) the ordinary wave ($n_1$) is transverse ($\boldsymbol{E} \perp \boldsymbol{k}$), but the extraordinary wave has a longitudinal as well as a transverse component of the electric wave field.

For waves propagating in an oblique direction to the magnetic field $\boldsymbol{B}_{\male}$,

$$0 \neq \Theta \neq \tfrac{1}{2}\pi,$$

the situation is considerably more involved.

β) Fig. 4 schematically shows how the squared refraction indices (which are always real under our present hypotheses depend on the frequency $f = \omega/2\pi$; the figure refers to the ordinary and extraordinary wave, supposing an intermediate Θ-value ($0 < \Theta < \frac{1}{2}\pi$). It has further been supposed that $\omega_{Ne} > \omega_{Be}$.

[1] K. Rawer and K. Suchy: this Encyclopedia, Vol. 49/2, use another nomenclature, based upon wave polarization; it is explained in Sect. 7δ there. The usual nomenclature depending on the sign of the root in Eq. (25.5) is not everywhere identic with that based on polarisation.

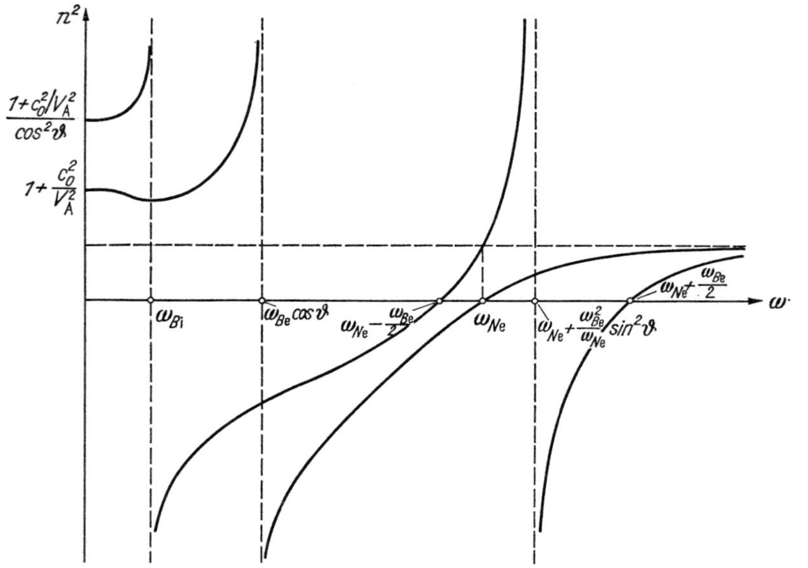

Fig. 4 Refractive index diagram in a cold gyrotropic plasma.

This inequality is satisfied in the ionospheric plasma. In the magnetospheric plasma we have the stronger relation $\omega_{Ne} \gg \omega_{Be}$, see Table 3. A more detailed discussion is found in [1], [3]; see also the survey given by V. D. SAFRANOV in [6].

In the frequency ranges where $n^2 < 0$ the plasma is no more transparent for the relevant waves. In this case, the depth of penetration of a wave of frequency $\omega/2\pi$ is

$$\delta \approx \frac{c_0}{\omega} \frac{1}{|n|}. \tag{25.11}$$

Of course, this problem must be resolved using wave theory.

In the frequency ranges where $n^2 > 0$ the plasma is transparent. These frequency ranges are related with the eigen-oscillations of the plasma at frequencies $\omega_j(\boldsymbol{k})$, $k = n\,\omega/c_0$ being a real number; the index $j = 1, 2, 3, \ldots$ designates the different branches of the oscillation modes. Looking at the upper half plane in Fig. 4, one recognizes that in a cold plasma containing electrons and one kind of ions only, there are five branches of possible oscillations. All these appear as unattenuated but only because the approximation we used here completely neglects dissipative influences. Such effects are particularly important relative to the penetration problem and with that of internal total reflection (see the lower half plane in Fig. 4).

Even in the case of a cold plasma it is, generally speaking, extremely difficult to express ω analytically as a function of $\boldsymbol{k}$. A graphic representation is, however, quite easy, as can be seen from Fig. 4. One easily realizes that in a cold gyrotropic (magneto-active) plasma there exist five branches of possible eigenoscillations, i.e. such for which $n^2(\omega) > 0$. Analytical expressions $\omega(\boldsymbol{k})$ can only be given in the limit of low and high-frequencies.

γ) We now consider waves in a cold plasma under the particular condition

$$A = \frac{1}{\varepsilon_0}\,\varepsilon(\omega, \boldsymbol{k}) \to 0. \tag{25.12}$$

It then follows from Eq. (25.5) that

$$n_1^2 = -\frac{C}{B}; \quad n_2^2 = -\frac{B}{A} \to \infty. \tag{25.13}$$

Provided $B \neq 0$, only the refractive index n_2 goes to infinity while n_1 remains finite. Now Eq. (24.6) shows $A = 0$ to be the condition for the waves to be longitudinal; therefore in a cold plasma waves with an infinite refractive index must be longitudinal ones[2]. The condition for being longitudinal can be written in the following form:

$$1 - \left(\frac{\omega_{N\mathrm{e}}^2}{\omega^2} + \frac{\omega_{N\mathrm{i}}^2}{\omega^2}\right)\cos^2\Theta - \left(\frac{\omega_{N\mathrm{e}}^2}{\omega^2 - \omega_{B\mathrm{e}}^2} + \frac{\omega_{N\mathrm{i}}^2}{\omega^2 - \omega_{B\mathrm{i}}^2}\right)\sin^2\Theta = 0, \tag{25.14}$$

which reads in the X, Y nomenclature[3] ($X \equiv \omega_N^2/\omega^2$; $Y \equiv \omega_B/\omega$; $Z \equiv \bar{\nu}/\omega$):

$$1 - (X_\mathrm{e} + X_\mathrm{i})\cos^2\Theta - \left(\frac{X_\mathrm{e}}{1 - Y_\mathrm{e}^2} + \frac{X_\mathrm{i}}{1 - Y_\mathrm{i}^2}\right)\sin^2\Theta = 0. \tag{25.14a}$$

The zeros of Eq. (25.14) define the eigenfrequencies $\omega_\mathrm{j}/2\pi$,

$$\omega_\mathrm{j}(\Theta) \qquad (\mathrm{j} = 1, 2, 3)$$

of three possible longitudinal eigenoscillations in a cold, gyrotropic plasma, viz.

$$\omega_{1,2}^2 \approx \tfrac{1}{2}(\omega_{N\mathrm{e}}^2 + \omega_{B\mathrm{e}}^2) \pm \tfrac{1}{2}\sqrt{(\omega_{N\mathrm{e}}^2 + \omega_{B\mathrm{e}}^2)^2 - 4\omega_{N\mathrm{e}}^2\,\omega_{B\mathrm{e}}^2\cos^2\Theta}, \tag{25.15}$$

and

$$\omega_3^2 \approx \omega_{B\mathrm{i}}^2\left(1 - \left|\frac{g_\mathrm{i}}{g_\mathrm{e}}\right|\frac{m_\mathrm{e}}{m_\mathrm{i}}\tan^2\Theta\right) \tag{25.16}$$

corresponding to

$$1 \approx \tfrac{1}{2}(X_\mathrm{e} + Y_\mathrm{e}^2) \pm \tfrac{1}{2}\sqrt{(X_\mathrm{e} + Y_\mathrm{e}^2)^2 - 4X_\mathrm{e}\,Y_\mathrm{e}^2\cos^2\Theta} \tag{25.15a}$$

and

$$1 \approx Y_\mathrm{i}^2\left(1 - \frac{X_\mathrm{i}}{X_\mathrm{e}}\tan^2\Theta\right) \tag{25.16a}$$

In Eqs. (25.15) X_i has been neglected against X_e, the roots ω_1 and ω_2 then being obtained from the approximation valid in a "LORENTZ gas"[4]:

$$1 - X_\mathrm{e} - Y_\mathrm{e}^2 + X_\mathrm{e}\,Y_\mathrm{e}^2\cos^2\Theta \approx 0. \tag{25.17}$$

As to Eq. (25.16) it corresponds to the ion gyro resonance defined by $Y_\mathrm{i}^2 \approx 1$, i.e. $\omega^2 \approx \omega_{B\mathrm{i}}^2$, which gives in Eq. (25.14), approximatively:

$$(1 - Y_\mathrm{i}^2)(1 - X_\mathrm{e}\cos^2\Theta) - X_\mathrm{i}\sin^2\Theta = 0. \tag{25.18}$$

The case of electron gyro resonance, $Y_\mathrm{e}^2 \approx 1$, is contained in Eq. (25.17). The approximation Eq. (25.18) is certainly valid in a neighbourhood of the gyro resonance, $\omega \approx \omega_{B\mathrm{i}}$, the width of which depends upon the value of X_e. The greater X_e the smaller is the resonance range. Eqs. (25.15) and (25.16) become impractical for values of Θ which are so near to $\pi/2$, that $\cos^2\Theta \ll (m_\mathrm{e}/m_\mathrm{i}) \ll 1$. In this particular case one better uses the first Eq. (25.15) together with

$$\omega_2^2 \approx \omega_k^2 \equiv \omega_{B\mathrm{i}}^2 + \frac{\omega_{B\mathrm{e}}^2\,\omega_{N\mathrm{i}}^2}{\omega_{N\mathrm{e}}^2 + \omega_{B\mathrm{e}}^2}, \tag{25.19}$$

---

[2] This is so for not vanishing angle Θ. The special case of „parallel" propagation, exactly along the magnetic field, $\Theta = 0$, is an exeption. In this very particular case the waves are longitudinal at any value of n.

[3] See K. RAWER and K. SUCHY: this Encyclopedia, Vol. 49/2, Sect. 1.

[4] See K. RAWER and K. SUCHY: this Encyclopedia, Vol. 49/2, Sect. 56, Eq. (56.5c).

which follows from Eq. (25.14) by suitable development, and

$$\omega_3^2 \approx \frac{\omega_{Bi}^2 \omega_{Ne}^2 \cos^2\Theta}{\omega_{Ne}^2 + \omega_{Be}^2}. \tag{25.20}$$

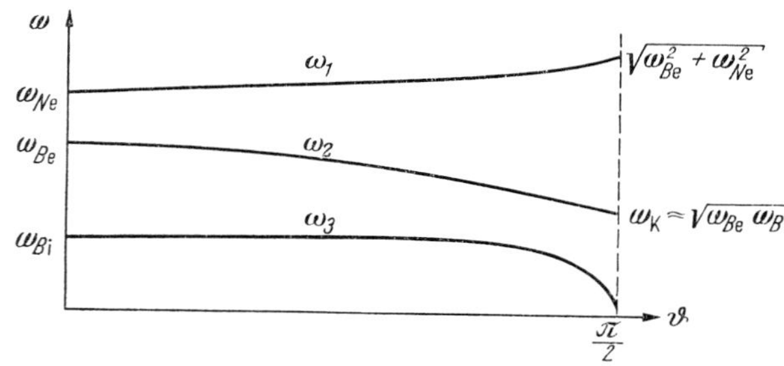

Fig. 5. Angular dependence of the three longitudinal eigen oscillations in a cold gyrotropic plasma.

The dependence of the three resonances $\omega_j(\Theta)$ on the angle is shown schematically[5] in Fig. 5. The lower curve corresponding to oscillations on the frequency ω_3 is not everywhere significant but in a very small angular intervall near $\Theta = \pi/2$; where $\omega_3 < \omega_{Bi}$. At all other angles Θ these oscillations are heavily attenuated by cyclotron absorption due to the ions in the plasma; see for example [3].

δ) We have noted above that in the limiting cases of *high- and low-frequencies* the spectra $\omega(\boldsymbol{k})$ can be described analytically for arbitrary waves. In the *high-frequency* range,

$$\omega \gg \omega_{Be}, \quad \text{i.e.} \quad Y_e \ll 1 \tag{25.21}$$

is true so that the magnetic field is of no importance. The spectra are therefore coincident with those of an isotropic (non-magnetized) plasma; these have been investigated in Sects. 20 and 21. The opposite situation of *extremely low-frequencies* is characterized by[6]

$$\omega \ll \omega_{Bi} \ll \omega_{Ne}, \quad \text{i.e.} \quad 1 \ll Y_i \ll X_e^{\frac{1}{2}}. \tag{25.22}$$

Neglecting dissipative processes in a cold plasma, the dispersion relation, Eq. (4.3), leads to the following two spectral branches:

$$\omega_1^2 = \frac{k^2 V_A^2 \cos^2\Theta}{1 + V_A^2/c_0^2} \tag{25.23}$$

and

$$\omega_2^2 = \frac{k^2 V_A^2}{1 + V_A^2/c_0^2}, \tag{25.24}$$

V_A being the ALFVÉN velocity, see Eq. (22.16). The inequality

$$V_A^2 \equiv \frac{B_0^2}{u \mu_0 N_i m_i} \ll c_0^2 \tag{25.25}$$

[5] $\omega_{Ne} > \omega_{Be}$ has been supposed, which is at least true in the ionospheric plasma.
[6] We suppose here that Θ is far enough from $\pi/2$ so that $\cos^2\Theta > (m_e/m_i)$.

is valid in most cases, and particularly under any conditions in the ionospheric plasma. If it is valid, Eqs. (25.23) and (25.24) are identical with the well-known expressions for the spectra of ALFVÉN waves, and of electron acoustic waves (i.e. fast magneto-sound waves), respectively. These have been derived using the simple conducting fluid model of magneto gas dynamics, see Eqs. (7.1) through (7.5).

ε) We have derived Eqs. (25.23), (25.24) under the assumption of a cold plasma, satisfying the inequalities, Eqs. (22.12). Speaking more generally, the *range of validity* of our equations is given by the condition that the phase velocity of the waves is larger than the thermal velocities of ions as well as electrons, i.e. for

$$\frac{\omega}{k} \approx V_{\mathrm{A}} \gg V_{T\mathrm{e}}. \tag{25.26}$$

This condition means that the magnetic energy density is by far greater than the thermal one, or

$$\beta \equiv \frac{2\mathrm{u}\,\mu_0 N_{\mathrm{i}}}{B_{\delta}^2}\,\vartheta \ll \frac{m_{\mathrm{e}}}{m_{\mathrm{i}}}, \tag{25.27}$$

with $\vartheta = kT$. The condition Eq. (25.27) is certainly satisfied in the ionosphere, but not everywhere in the magnetosphere. Table 1 indicates that if should fail above 10 R_{E} or abaot 10^4 km.

The condition Eq. (25.27) is, strictly speaking, too stringent. For the validity of Eqs (25.23), (25.24) it is not really necessary that the phase velocity of the waves, $c = V_{\mathrm{ph}} = \omega/k$ be larger than the thermal velocity of the electrons, $V_{T\mathrm{e}}$. In fact, c could be smaller, it should only be greater than the thermal velocity of the ions, $V_{T\mathrm{i}}$ (and the corresponding sound velocity V_s). It is therefore sufficient to require

$$\beta \equiv \frac{2\mathrm{u}\,\mu_0 N_{\mathrm{i}}}{B_{\delta}^2}\,\vartheta \ll 1. \tag{25.28}$$

There is no difficulty to note expressions for the refractive indices of the waves we are considering here. Using the relation

$$k \equiv \frac{\omega}{c} = \frac{\omega}{c_0}\,n = \frac{\omega}{c_0}\,(\mu + i\,\chi)$$

we find from Eqs. (25.23) and (25.24), see Fig. 4:

$$n_1^2 = \mu_1^2 = \frac{1}{\cos^2\Theta}\left(1 + \frac{c_0^2}{V_{\mathrm{A}}^2}\right), \tag{25.29}$$

$$n_2^2 = \mu_2^2 = 1 + \frac{c_0^2}{V_{\mathrm{A}}^2}. \tag{25.30}$$

Since these expressions are independent of the frequency, the corresponding waves have no dispersion.

ζ) For a cold plasma one may even indicate an analytical expression for one branch of the spectrum in an *intermediate frequency range* characterized by

$$\omega_{B\mathrm{i}} \ll \omega \ll \omega_{B\mathrm{e}}, \quad \text{i.e.} \quad Y_{\mathrm{i}} \ll 1 \ll Y_{\mathrm{e}}, \tag{25.31}$$

again for values of Θ which are not too near to $\pi/2$. Different from the low frequency case, Eqs. (25.23), (25.24), these are pure electron oscillations; they can propagate in a plasma with high enough density:

$$\omega_{N\mathrm{e}}^2 \gg \omega\,\omega_{B\mathrm{e}}, \quad \text{i.e.} \quad X_{\mathrm{e}} \gg Y_{\mathrm{e}}, \tag{25.32}$$

again for values of Θ which are not too near to $\pi/2$. Using Eqs. (4.3) and the expressions Eqs. (25.1) ... (25.3) it is easily shown that these waves have the spectrum

$$\omega_3 = |\omega_{Be} \cos\Theta| \frac{k^2 c_0^2}{\omega_{Ne}^2}. \tag{25.33}$$

The branch of waves for which this condition holds is that of the *whistler mode*. It has this name because whistling atmospherics appear on this branch[7]. Underlining the circular polarization of these waves they may also be called "spiral waves"[8].

These spiral waves as well as ALFVÉN waves and fast magneto-sound waves, considered in Subsect. δ, exist not only in a cold plasma, i.e. under the condition Eq. (22.12), that their phase velocity is greater than the thermal velocity of the electrons. They may exist as long as

provided
$$\left.\begin{aligned} &\omega \lesssim k_z V_{Te}, \\ &\omega \gg k_z V_{Ti}. \end{aligned}\right\} \tag{25.34}$$

Otherwise, the waves are strongly attenuated, see Sect. 26 below.

In our approximation (i.e. neglecting dissipative effects) the refractive index of spiral waves is given by

$$n_3^2 = \mu_3^2 = \frac{\omega_{Ne}^2}{\omega|\omega_{Be}\cos\Theta|}. \tag{25.35}$$

It appears from Fig. 4 that the spiral waves are, essentially, the continuation of the branch of fast magneto-sound waves from $\omega \ll \omega_{Bi}$ (i.e. $Y_i \gg 1$) into the intermediate range characterized by

$$\omega_{Bi} \ll \omega \ll \omega_{Be}, \quad \text{i.e.} \quad Y_i \ll 1 \ll Y_e. \qquad [25.31]$$

η) We now come to consider dissipative effects, i.e. the *absorption* of electromagnetic waves *in a cold plasma*.

In a zero order approximation we consider the plasma as to be really cold, i.e. only collisional attenuation is thought to be important; this is the case of small values of the parameters given in Eq. (22.12). Going over to a better approximation we have to consider also first order terms of these parameters. They describe heating which must be attributed to collisionless ČERENKOV magneto-Bremsstrahlung absorption. Both of these contributions appear in the anti-Hermitean part of the dielectricity tensor, Eqs. (23.3), (23.4); they are certainly small against its Hermitean part, Eq. (22.13)[9]. Therefore the real part μ of the complex refraction index $n = \mu + i\chi$ is only negligibly influenced by the absorption terms; its imaginary part, χ, however, varies as a linear function of the anti-Hermitean part of the dielectricity tensor $\mathsf{E}(\omega, \boldsymbol{k})$ such that:

$$\chi_{1,2} = i \frac{1}{2\mu_{1,2}} \frac{\mu_{1,2}^4 \delta A + \mu_{1,2}^2 \delta B + \delta C}{2\mu_{1,2}^2 A + B}. \tag{25.36}$$

The quantities A, B, C have been defined in Eqs. (25.1) ... (25.3); and $\mu_{1,2}^2$ in Eq. (25.5). The attenuative influence is characterized by the three quantities δA, δB,

[7] See contribution by R. GENDRIN in volume 49/3 of this Encyclopedia, p. 461, particularly Sects 2, 5 and 9. See also Sect. 57 in K. RAWER and K. SUCHY: this Encyclopedia, Vol. 49/2.

[8] They may be compared with torsional waves in the acoustics of solids.

[9] The dissipative corrections of the hermitean part of the dielectricity tensor are neglected for good reasons; they are negligible in most situations, except for the neighbourhood of such resonance frequencies where $\mu^2 \to \infty$. In these cases their influence „cuts off" the pole of μ^2, see [1], [3], [4].

δC:

$$\varepsilon_0\,\delta A = \varepsilon_{xx}^{(a)} \sin^2\Theta + \varepsilon_{zz}^{(a)} \cos^2\Theta, \tag{25.37}$$

$$\left.\begin{aligned}\varepsilon_0^2\,\delta B = &-(\varepsilon_{xx}^{(a)}\,\varepsilon_{\parallel}^{(0)} + \varepsilon_{zz}^{(a)}\,\varepsilon_{\perp}^{(0)})\,(1+\cos^2\Theta) - \\ &-2\,(\varepsilon_{xx}^{(a)}\,\varepsilon_{\perp}^{(0)} - g^{(a)}\,g^{(0)})\sin^2\Theta - (\varepsilon_{yy}^{(a)} - \varepsilon_{xx}^{(a)})\,\varepsilon_{\parallel}^{(0)}\cos^2\Theta - \\ &-(\varepsilon_{yy}^{(a)} - \varepsilon_{xx}^{(a)})\,\varepsilon_{\perp}^{(0)}\sin^2\Theta,\end{aligned}\right\} \tag{25.38}$$

$$\left.\begin{aligned}\varepsilon_0^3\,\delta C = &\,\varepsilon_{zz}^{(a)}(\varepsilon_{\perp}^{(0)2} - g^{(0)2}) + 2\,(\varepsilon_{xx}^{(a)}\,\varepsilon_{\perp}^{(0)} - g^{(a)}\,g^{(0)})\,\varepsilon_{\parallel}^{(0)} + \\ &+(\varepsilon_{yy}^{(a)} - \varepsilon_{xx}^{(a)})\,\varepsilon_{\parallel}^{(0)}\,\varepsilon_{\perp}^{(0)}.\end{aligned}\right\} \tag{25.39}$$

The tensor components, $\varepsilon_{ij}^{(a)}$ appearing in these expressions are given by Eqs. (23.3), (23.4). We note that Eqs. (25.36) and (25.37) ... (25.39) are also valid for a weakly ionized, plasma provided the inequalities, Eqs. (23.1) are satisfied. In this case $\bar{\nu}_h$ must be replaced by ν_{hn} in the expressions Eq. (23.4). The simplifications introduced above do not apply if the inequalities Eqs. (23.1) are not satisfied. In that case the anti-Hermitean part of the dielectricity tensor, Eqs. (6.11) and (6.12)[10] is not small against the hermitean one. Generally speaking electromagnetic waves are heavily attenuated under these conditions.

ϑ) We shall now consider in more detail wave attenuation in the *particular cases* for which an analytical description of the spectrum $\omega(\boldsymbol{k})$ can be obtained. We begin with longitudinal waves in a gyrotropic (magneto-active) plasma; in the absence of absorption, the relevant resonance frequencies have been expressed by Eqs. (25.15), (25.16) and (25.19), (25.20). Taking account of dissipative processes there appear now imaginary contributions to the real resonance frequency values obtained above; these corrections signify attenuation of longitudinal oscillations i.e. decrease of amplitude with increasing time. Transition towards dissipation is described by

$$\omega \to \omega + i\,\gamma$$

where $\gamma = \gamma_{1,2,3}$ for the three resonance frequencies $\omega_{1,2,3}$.

$$\left.\begin{aligned}\gamma_{1,2}(\Theta) = &-\frac{1}{2}\left\{\frac{\omega_{Ne}^2}{\omega_{1,2}^2}\cos^2\Theta + \frac{\omega_{Ne}^2\,\omega_{1,2}^2\sin^2\Theta}{(\omega_{1,2}^2 - \omega_{Be}^2)^2}\right\}^{-1}\times \\ &\times\left\{\left[\frac{\omega_{Ne}^2}{\omega_{1,2}^2}\,\bar{\nu}_e + \sqrt{\frac{\pi}{2}}\,\frac{\omega_{Ne}^2\,\omega_{1,2}^2}{k^3\,V_{Te}^3\,|\cos^3\Theta|}\exp\left(\frac{-\omega_{1,2}^2}{2k^2\,V_{Te}^2\cos^2\Theta}\right)\right]\times\right. \\ &\times\cos^2\Theta + \left[\frac{\omega_{Ne}^2\,(\omega_{1,2}^2 + \omega_{Be}^2)\,\bar{\nu}_e}{(\omega_{1,2}^2 - \omega_{Be}^2)^2} + \sqrt{\frac{\pi}{8}}\,\frac{\omega_{Ne}^2}{k\,V_{Te}\,|\cos\Theta|}\times\right. \\ &\left.\left.\times\left\{\exp\left(\frac{-(\omega_{1,2} - \omega_{Be})^2}{2k^2\,V_{Te}^2\cos^2\Theta}\right) + \exp\left(\frac{-(\omega_{1,2} + \omega_{Be})^2}{2k^2\,V_{Te}^2\cos^2\Theta}\right)\right\}\right]\sin^2\Theta\right\},\end{aligned}\right\} \tag{25.40a, b}$$

$$\left.\begin{aligned}\gamma_3(\Theta) = &-\frac{1}{2}\left\{\frac{\omega_{Ne}^2}{\omega_3^2}\cos^2\Theta + \frac{\omega_{Ni}^2\,\omega_3^2\sin^2\Theta}{(\omega_3^2 - \omega_{Bi}^2)^2}\right\}^{-1}\left\{\left[\frac{\omega_{Ni}^2\,(\omega_3^2 + \omega_{Bi}^2)\bar{\nu}_i}{(\omega_3^2 - \omega_{Bi}^2)^2} + \right.\right. \\ &+\sqrt{\frac{\pi}{8}}\,\frac{\omega_{Ni}^2}{k\,V_{Ti}\,|\cos\Theta|}\left\{\exp\left(\frac{-(\omega_3 - \omega_{Bi})^2}{2k^2\,V_{Ti}^2\cos^2\Theta}\right) + \right. \\ &\left.\left. + \exp\left(\frac{-(\omega_3 + \omega_{Bi})^2}{2k^2\,V_{Ti}^2\cos^2\Theta}\right)\right\}\right]\sin^2\Theta + \\ &\left. + \left[\frac{\omega_{Ne}^2\,\bar{\nu}_e}{\omega_3^2} + \sqrt{\frac{\pi}{8}}\,\frac{\omega_{Ne}^2\,\omega_3^2}{k^3\,V_{Te}^3\,|\cos^3\Theta|}\exp\left(\frac{-\omega_3^2}{2k^2\,V_{Te}^2\cos^2\Theta}\right)\right]\cos^2\Theta\right\}.\end{aligned}\right\} \tag{25.40c}$$

[10] These are the expressions giving the permittivity (dielectric permeability) of a weakly ionized, cold plasma, as shown in Sect. 23.

With $X_e \equiv \omega_{Ne}^2/\omega^2$, $Y_e \equiv \omega_{Be}/\omega$, $Z_e \equiv \bar{\nu}_e/\omega$ and $Q_{e,i} \equiv k V_{Te,i}/\omega$, Eqs. (25.40a, b and c) can be written more easily:

$$\left.\begin{aligned}
&-\frac{2}{\omega_{1,2}}\{\cos^2\Theta + \sin^2\Theta/(1-Y_{e1,2}^2)^2\}\,\gamma_{1,2}(\Theta)\\
&\quad=\left[Z_{e1,2}+\sqrt{\frac{\pi}{2}}\,Q_{e1,2}^{-3}\frac{1}{|\cos^3\Theta|}\exp\left(-\frac{1}{2}Q_{e1,2}^{-2}\frac{1}{\cos^2\Theta}\right)\right]\cos^2\Theta+\\
&\quad+\left[\frac{1+Y_{e1,2}^2}{(1-Y_{e1,2}^2)^2}Z_{e1,2}+\sqrt{\frac{\pi}{8}}\,Q_{e1,2}^{-1}\frac{1}{|\cos\Theta|}\times\right.\\
&\quad\times\left\{\exp\left(-\frac{1}{2}Q_{e1,2}^{-2}\frac{(1-Y_{e1,2})^2}{\cos^2\Theta}\right)+\right.\\
&\quad\left.\left.+\exp\left(-\frac{1}{2}Q_{e1,2}^{-2}\frac{(1+Y_{e1,2})^2}{\cos^2\Theta}\right)\right\}\right]\sin^2\Theta,
\end{aligned}\right\}\quad(25.41\,\text{a, b})$$

$$\left.\begin{aligned}
&-\frac{2}{\omega_3}\{X_{e3}\cos^2\Theta + X_{i3}\sin^2\Theta/(1-Y_{i3}^2)^2\}\cdot\gamma_3(\Theta)\\
&\quad=X_{e3}\left[Z_{e3}+\sqrt{\frac{\pi}{8}}\,Q_{e3}^{-3}\frac{1}{|\cos^3\Theta|}\exp\left(-\frac{1}{2}Q_{e3}^{-2}\frac{1}{\cos^2\Theta}\right)\right]\cos^2\Theta+\\
&\quad+X_{i3}\left[\frac{(1+Y_{i3}^2)}{(1-Y_{i3}^2)^2}Z_{i3}+\sqrt{\frac{\pi}{8}}\,Q_{i3}^{-1}\frac{1}{|\cos\Theta|}\times\right.\\
&\quad\left.\times\left\{\exp\left(-\frac{1}{2}Q_{i3}^{-2}\frac{(1-Y_{i3})^2}{\cos^2\Theta}\right)+\exp\left(-\frac{1}{2}Q_{i3}^{-2}\frac{(1-Y_{i3})^2}{\cos^2\Theta}\right)\right\}\right]\sin^2\Theta.
\end{aligned}\right\}\quad(25.41\,\text{c})$$

In the particular case where $\Theta \approx \frac{1}{2}\pi$, Eqs. (25.40) and (25.41) continue to be useful for γ_1, but not for γ_2 and γ_3, if $\cos^2\Theta \ll m_e/m_i$. This corresponds to the situation with Eqs. (25.15), (25.16) as discussed in Subsect. γ, above. One then has, instead of Eqs. (25.40), (25.41),

$$\gamma_2 = -\frac{1}{2}\frac{\omega_2^2}{\omega_{Ni}^2}\left\{\frac{\omega_{Ne}^2}{\omega_2^2}\bar{\nu}_e\cos^2\Theta + \left(\frac{\omega_{Ne}^2\bar{\nu}_e}{\omega_{Be}^2} + \frac{\omega_{Ni}^2\bar{\nu}_i}{\omega_2^2}\right)\sin^2\Theta\right\}, \quad (25.42\text{b})$$

$$\gamma_3 = -\frac{1}{2}\frac{\omega_3^2}{\omega_{Ne}^2}\left[\frac{\omega_{Ne}^2\bar{\nu}_e}{\omega_3^2} + \frac{\omega_{Ni}^2\bar{\nu}_i}{\omega_{Bi}^2}\tan^2\Theta\right] \quad (25.42\text{c})$$

and

$$\left.\begin{aligned}
-\frac{2}{\omega_2}X_{i2}\gamma_2\left(\Theta\approx\frac{1}{2}\pi\right) &= X_{e2}Z_{e2}\cos^2\Theta+\\
&+(X_{e2}Z_{e2}/Y_{e2}^2+X_{i2}Z_{i2})\sin^2\Theta,
\end{aligned}\right\}\quad(25.43\,\text{b})$$

$$-\frac{2}{\omega_3}\gamma_3\left(\Theta\approx\frac{1}{2}\pi\right) = Z_{e3} + \frac{X_{i3}}{X_{e3}}\frac{Z_{i3}}{Y_{i3}^2}\tan^2\Theta. \quad (25.43\text{c})$$

The resonance frequencies $\omega_2/2\pi$ and $\omega_3/2\pi$ must, of course, be determined by Eqs. (25.19), (25.20). Apparently, in Eqs. (25.43), (25.44) collisionless attenuation has been neglected; under realistic conditions it uses to be quite small against collisional absorption, at least for $\Theta \approx \frac{1}{2}\pi$. We further note that for a completely ionized plasma one has to put:

$$\bar{\nu}_e = \nu_{\text{eff}} \quad\text{and}\quad \bar{\nu}_i = \frac{m_e}{m_i}\nu_{\text{eff}}, \quad (25.44)$$

but for a weakly ionized plasma:

$$\bar{\nu}_e = \nu_{en} \quad\text{and}\quad \bar{\nu}_i = \nu_{in}. \quad (25.45)$$

It is not difficult to find similar expressions for the absorption decrement γ also for ALFVÉN waves and magneto-acoustic waves in a cold plasma. The resonance frequencies are then found from Eqs. (25.23), (25.24). For these waves

$$\left.\begin{aligned}\gamma_1 &= -\frac{1}{2}\frac{\omega_{Bi}^2}{\omega_{Ni}^2}\sum_h \frac{\omega_{Nh}^2 \bar{\nu}_h}{\omega_{Bh}^2}\frac{\omega_1^2}{\omega_{Bh}^2} - \\ &\quad - \sqrt{\frac{\pi}{8}}\frac{m_e}{m_i}\frac{\omega_1^6 \tan^2\Theta}{\omega_{Bi}^2 k^3 V_{Te}^3 |\cos^3\Theta|}\exp\left(\frac{-\omega_1^2}{2k^2 V_{Te}^2 \cos^2\Theta}\right),\end{aligned}\right\} \quad (25.46\text{a})$$

$$\left.\begin{aligned}\gamma_2 &= -\frac{1}{2}\frac{\omega_{Bi}^2}{\omega_{Ni}^2}\sum_h \frac{\omega_{Nh}^2 \bar{\nu}_h}{\omega_{Bh}^2}\frac{\omega_2^2}{\omega_{Bh}^2} - \\ &\quad - \sqrt{\frac{\pi}{8}}\frac{m_e}{m_i}|\omega_2|\frac{V_{Te}}{V_A}\frac{\sin^2\Theta}{|\cos\Theta|}\exp\left(\frac{-\omega_2^2}{2k^2 V_{Te}^2 \cos^2\Theta}\right),\end{aligned}\right\} \quad (25.46\text{b})$$

and with the symbols X, Y, Z and Q:

$$\left.\begin{aligned}-\frac{2}{|\omega_1|}\gamma_1 &= \frac{Y_{i1}^2}{X_{i1}}\sum_h \frac{X_{h1} Z_{h1}}{Y_{h1}^4} - \\ &\quad - \sqrt{\frac{\pi}{2}}\frac{m_e}{m_i}Y_{i1}^{-2}Q_{e1}^{-3}\frac{\tan^2\Theta}{|\cos^3\Theta|}\exp\left(-\frac{1}{2}Q_{e1}^{-2}\frac{1}{\cos^2\Theta}\right),\end{aligned}\right\} \quad (25.47\text{a})$$

$$\left.\begin{aligned}-\frac{2}{|\omega_2|}\gamma_2 &= \frac{Y_{i2}^2}{X_{i2}}\sum_h \frac{X_{h2} Z_{h2}}{Y_{h2}^4} - \\ &\quad - \sqrt{\frac{\pi}{2}}\frac{m_e}{m_i}\frac{V_{Te}}{V_A}\frac{\sin^2\Theta}{|\cos\Theta|}\exp\left(-\frac{1}{2}Q_{e2}^{-2}\frac{1}{\cos^2\Theta}\right).\end{aligned}\right\} \quad (25.47\text{b})$$

Here collisionless magneto-Bremsstrahlung absorption i.e. cyclotron attenuation of waves has been neglected. It is quite small compared to ČERENKOV electron absorption, at least for $\Theta \neq 0$[11]. Eqs. (25.46), (25.47) may be used in the limit of a completely ionized and a weakly ionized plasma, if account is taken of Eqs. (25.44) and (25.45), respectively.

Transverse waves in the Whistler-mode (so called spiral waves) will be considered in Sect. 26ε.

26. Waves in homogeneous, warm, gyrotropic (magneto-active) plasmas. In the preceding sections the limiting case of a cold plasma has been considered, assuming the conditions

$$z_h \equiv \frac{Q_{h\perp}^2}{Y_h^2} \equiv \left(\frac{k_\perp V_{Th}}{\omega_{Bh}}\right)^2 \ll 1; \quad \beta_{sh} \equiv \frac{\omega - s\omega_{Bh}}{|k_z| V_{Th}} \gg 1; \quad \frac{k_\| V_{Th}}{\omega} \equiv Q_{h\|} \ll 1; \quad [22.12]$$

according to which the phase velocity of waves greatly exceeds the thermal velocity of electrons as well as ions. Just because of this assumption we found exponentially small attenuation in collisionless plasmas. Also waves in the intermediate frequency range,

$$k_z V_{Ti} \ll \omega \ll k_z V_{Te} \quad [25.34]$$

appeared to be weakly attenuated.

α) If, apart from Eq. (22.12) the inequalities

$$\omega \ll \omega_{Bi} \ll \omega_{Ni}; \quad \left(\frac{k_\perp V_{Th}}{\omega_{Bh}}\right)^2 \ll 1; \quad V_{Ti} \ll \frac{\omega}{|k_z|} \ll V_{Te} \quad [22.14]$$

are also satisfied, we are in a position to use the expressions Eqs. (22.15) for the dielectricity tensor, Eqs. (23.7) for the contributions of collisionless and collisional

---

[11] As mentionned above, this kind of attenuation is negligible in almost all realistic cases against collisional absorption, which appears in the first term of each of the Eqs. (25.46), (25.47).

attenuation to this tensor, and, in case of a weakly ionized plasma, the simplified Eqs. (23.14) for the dielectricity tensor. As mentioned in Sects. 22 and 23, these expressions are derived from the approximation of *one fluid gasdynamics*, see Eq. (11.13). In this approximation the dispersion equation, Eq. (4.3) splits into two equations, namely:

$$\left.\begin{aligned} k_z^2\,\varepsilon_0-\left(\frac{\omega}{c_0}\right)^2\varepsilon_{xx}&=0\,,\\ \left(k^2\,\varepsilon_0-\left(\frac{\omega}{c_0}\right)^2\varepsilon_{yy}\right)\varepsilon_{zz}+\left(\frac{\omega}{c_0}\right)^2\varepsilon_{yz}^2&=0\,.\end{aligned}\right\}\tag{26.1}$$

β) The first of these equations

$$k_z^2\,\varepsilon_0-\left(\frac{\omega}{c_0}\right)^2\varepsilon_{xx}=0 \tag{26.2}$$

describes ALFVÉN *oscillations* in a magneto-active plasma, the spectrum of which is

$$\omega^2=k^2\,V_A^2\cos^2\Theta\,. \tag{26.3}$$

Its attenuation is described by an imaginary part

$$\gamma=-\frac{1}{2}\left(\frac{m_e}{m_i}\,\bar\nu_e+\bar\nu_i\right). \tag{26.4}$$

The particular values of $\bar\nu_e$ and $\bar\nu_i$ are

$$\left.\begin{aligned}\bar\nu_e=\nu_{\text{eff}}\,;\qquad \bar\nu_i&=\frac{m_e}{m_i}\,\nu_{\text{eff}}+\frac{7}{10}\,\nu_{ii}\,(k_\perp V_{Ti}/\omega_{Bi})^2\\ &=\frac{m_e}{m_i}\,\nu_{\text{eff}}+\frac{7}{10}\left(\frac{\varrho_{i\perp}}{Y_i}\right)^2\nu_{ii}\end{aligned}\right\}\tag{26.5}$$

for a completely ionized plasma, and

$$\bar\nu_e=\nu_{en}\,;\qquad \bar\nu_i=\nu_{in} \tag{26.6}$$

for a weakly ionized plasma.

The spectrum, Eq. (26.3), appears as the continuation towards lower values of phase velocity of that of ALFVÉN oscillations in a cold plasma, see Eqs. (25.23), (25.24) for attenuation see Eqs. (25.46), (25.47). The condition is

$$\omega\ll k_z V_{Te}, \tag{26.7}$$

such that ALFVÉN waves can exist in a plasma with $V_{Te}\gg V_A$, which means ($\vartheta\equiv kT$, k BOLTZMANN constant)

$$\beta_e\equiv\frac{2u\,\mu_0 N_e\vartheta_e}{B_0^2}\gg\frac{m_e}{m_i}\,. \tag{26.8}$$

The ALFVÉN velocity V_A was given in Eq. (25.25).

We note that indicating the attenuation decrement in Eq. (26.4) we have neglected collisionless absorption; under the conditions of Eq. (22.14) it is always small against collisional absorption[1]. Finally, ALFVÉN waves may exist for $\omega\gg\bar\nu_e$ as well as for $\omega\gg\bar\nu_e$. A sufficient condition for weak attenuation is

$$\left.\begin{aligned}&\omega\gg\bar\nu_i\,;\qquad \omega_{Be}\gg\bar\nu_e,\\ \text{i.e.}\qquad &Z_i\ll 1\,;\qquad Z_e\ll Y_e\,.\end{aligned}\right\}\tag{26.9}$$

This remark applies to slow ALFVÉN waves as described by Eqs. (26.3), as well as to the fast ones, which derive from Eqs. (25.23) and (25.24).

---

[1] Collisionless attenuation of ALFVÉN waves in the low-frequency range, $\omega\ll k_z V_{Te}$ (26.7) is discussed in [1], [3].

γ) From the second part of the splitted dispersion equation

$$\left(k^2 \varepsilon_0 - \left(\frac{\omega}{c_0}\right)^2 \varepsilon_{yy}\right) \varepsilon_{zz} + \left(\frac{\omega}{c_0}\right)^2 \varepsilon_{yz}^2 = 0 \tag{26.10}$$

we find the spectra for *fast and slow magneto-sound waves* as

$$\omega_\pm^2 = \tfrac{1}{2} k^2 \left[V_A^2 + V_s^2 \pm \sqrt{(V_A^2 + V_s^2)^2 - 4 V_A^2 V_s^2 \cos^2\Theta}\right] \tag{26.11}$$

with attenuation decrements

$$\gamma_\pm = -\sqrt{\frac{\pi}{8}} \sqrt{\frac{|q_i|}{|q_e|}} \sqrt{\frac{m_e}{m_i}} \frac{k V_s}{2|\cos\Theta|} \left\{1 \pm \frac{\cos^2\Theta\left(\frac{V_s^2}{V_A^2}\cos 2\Theta - 1\right)}{\sqrt{1 + \frac{V_s^4}{V_A^4} - 2\frac{V_s^2}{V_A^2}\cos 2\Theta}}\right\} + \delta\gamma_\pm . \tag{26.12}$$

Here

$$V_s = \sqrt{\left|\frac{q_i}{q_e}\right| \frac{\vartheta_e}{m_i}}$$

is the ion sound velocity, and $\delta\gamma_\pm$ is the correction due to collisions. For a completely ionized plasma it reads

$$\left.\begin{aligned}\delta\gamma_\pm = -\frac{4}{5}\nu_{ii}\frac{V_{Ti}^2}{V_s^2}\left\{\frac{1}{2} + \frac{9}{8}\frac{V_s^2}{V_A^2}\sin^2\Theta \mp \frac{1}{8}\sqrt{1 + \frac{V_s^4}{V_A^4} - 2\frac{V_s^2}{V_A^2}\cos^2\Theta} \times\right. \\ \times\left.\left[\left(1 - \frac{V_s^2}{V_A^2}\cos 2\Theta\right)\left(4 + 3\frac{V_s^2}{V_A^2}\sin^2\Theta\right) + 2\sin^2\Theta\left(2 + 3\frac{V_s^2}{V_A^2}(1 + \cos^2\Theta)\right)\right]\right\}\end{aligned}\right\}, \tag{26.13}$$

but for a weakly ionized plasma it is simply

$$\delta\gamma_\pm = -\frac{1}{2}\nu_{in}\frac{k^2 V_s^2 \cos^2\Theta}{\omega_\pm^4 - (k V_s V_A \cos\Theta)^2}\left[(\omega_\pm^2 - k^2 V_A^2) + \omega_\pm^2\left(\frac{\omega_\pm^2}{k^2 V_s^2 \cos^2\Theta} - 1\right)\right]. \tag{26.14}$$

δ) The spectra, Eqs. (26.11) ... (26.14), are particularly simple in the case of a plasma of *low pressure*, provided

$$\beta \equiv \frac{2u\,\mu_0 N_e(\vartheta_e + \vartheta_i)}{B_\delta^2} \ll 1 \tag{26.15}$$

which following definition of V_A [see for example Eq. (25.25)] means

$$V_A^2 \gg V_s^2, V_{Ti}^2 . \tag{26.16}$$

The fast magneto-sound wave now becomes purely transverse, its spectrum coincides with the spectra Eqs. (25.23) and (25.46)[2], thus indicating that the spectrum of magneto-sound waves in a cold plasma ($\omega \gg k_z V_{Te}$) continues without change into the range of low-frequencies ($\omega \gtrsim k_z V_{Te}$). They, however, can only exist in the low-frequency range if $\beta \gg m_e/m_i$. Putting $\beta \gg 1$ we obtain for the *slow magneto-sound waves*:

$$\omega_-^2 = k^2 V_s^2 \cos^2\Theta, \tag{26.17}$$

and as attenuation decrement

$$\gamma_- = -\sqrt{\frac{\pi}{8}}|\omega_-|\sqrt{\frac{q_i}{q_e}\frac{m_e}{m_i}} - \frac{1}{2}\bar{\nu}_i . \tag{26.18}$$

[2] The exponent in Eqs. (25.46), (25.47) equals one in the frequency range considered here.

For a completely ionized plasma

$$\bar{\nu}_{\mathrm{i}} = \frac{8}{5} \nu_{\mathrm{ii}} \frac{q_{\mathrm{e}}}{q_{\mathrm{i}}} \frac{\vartheta_{\mathrm{i}}}{\vartheta_{\mathrm{e}}} \equiv \frac{8}{5} \nu_{\mathrm{ii}} \frac{q_{\mathrm{e}}}{q_{\mathrm{i}}} \frac{T_{\mathrm{i}}}{T_{\mathrm{e}}}, \tag{26.19}$$

but for a weakly ionized one we just have $\bar{\nu}_{\mathrm{i}} = \nu_{\mathrm{in}}$.

The oscillations described by Eqs. (26.17), (26.18) are purely longitudinal. They correspond to ion sound waves in a non-magnetic plasma. Like these they are heavily attenuated except for cases where $T_{\mathrm{e}} > 6 T_{\mathrm{i}}$. This is the inequality we found when neglecting ČERENKOV attenuation by ions on high frequencies, viz. for $\omega \gg k_z V_{T\mathrm{i}}$. But in the frequency range $\omega \lesssim k_z V_{T\mathrm{i}}$ ČERENKOV attenuation by ions is strong so that any oscillation rapidly decreases. Let us finally note that all results found in this section. for the spectrum of oscillations in the intermediate frequency range,

$$k_z V_{T\mathrm{i}} \ll \omega \ll k_z V_{T\mathrm{e}} \tag{26.20}$$

remain valid with any relation between ω and $\bar{\nu}_{\mathrm{e}}$, provided the inequalities $\omega \gg \bar{\nu}_{\mathrm{i}}$ and $\bar{\nu}_{\mathrm{e}} \ll k_z V_{T\mathrm{e}}$ be satisfied.

Oscillations with the spectrum according to Eq. (26.17) can exist with weak attenuation in a gyrotropic plasma for the range of rather high frequencies (for ion sound waves) where $\omega \gtrsim \omega_{B\mathrm{i}} (1 \gtrsim Y_{\mathrm{i}})$. This follows from the fact that ion sound waves are able to exist in isotropic plasmas (i.e. in the absence of external fields). The proof is easy when starting from the dispersion equation for longitudinal waves in a gyrotropic plasma, Eq. (24.6). Finally, in the high-frequency limit where $\omega \gg \omega_{B\mathrm{e}}$, the effect of the magnetic field may, apparently, be neglected. In this approximation Eq. (24.6) coincides with the dispersion equation for longitudinal waves in an isotropic (non-magnetized) plasma, see Sect. 20.

The intermediate range

$$\left.\begin{aligned} &\omega_{B\mathrm{i}} \ll \omega \ll \omega_{B\mathrm{e}} \\ \text{i.e.}\qquad &Y_{\mathrm{i}} \ll 1 \ll Y_{\mathrm{e}} \end{aligned}\right\} \tag*{[25.31]}$$

is obtained either with intermediate frequencies, or with rather strong magnetic fields. Under the conditions

$$\omega \gg (\bar{\nu}_{\mathrm{i}}, k V_{T\mathrm{i}}); \qquad k_z V_{T\mathrm{e}} \gg (\bar{\nu}_{\mathrm{e}}, \omega) \tag{26.21}$$

the effect of the magnetic field upon the ions in the plasma is always negligible while that upon electrons is quite strong. We obtain from Eq. (24.6) the following spectrum for longitudinal waves:

$$\omega^2 = k^2 V_{\mathrm{s}}^2 / (1 + k^2 \lambda_{\mathrm{De}}^2) \tag{26.22}$$

where λ_{D} is the DEBYE length, Eq. (18.6). The attenuation decrement is here

$$\gamma = -\frac{1}{2} \bar{\nu}_{\mathrm{i}} - \sqrt{\frac{\pi}{8}} \frac{q_{\mathrm{e}}}{q_{\mathrm{i}}} \frac{m_{\mathrm{i}}}{m_{\mathrm{e}}} \frac{\omega^4}{k^3 V_{T\mathrm{e}}^3 |\cos\Theta|}. \tag{26.23}$$

For a completely ionized plasma one has to put

$$\bar{\nu}_{\mathrm{i}} = \frac{8}{5} \nu_{\mathrm{ii}} \left(\frac{k V_{T\mathrm{i}}}{\omega}\right)^2 \equiv \frac{8}{5} Q_{\mathrm{i}}^2 \nu_{\mathrm{ii}}, \tag{26.24}$$

but for a weakly ionized one $\bar{\nu}_{\mathrm{i}} = \nu_{\mathrm{in}}$ simply holds.

Finally on lower frequencies

$$\omega \ll \omega_{B\mathrm{i}}, \quad \text{i.e.} \quad Y_{\mathrm{i}} \gg 1, \tag{26.25}$$

and under the conditions Eqs. (26.15), (26.16) used above when deriving Eqs. (26.17), (26.18) we obtain

$$\omega^2 = \frac{(k\, V_{\rm s} \cos\Theta)^2}{1 + k^2 \lambda_{\rm De}^2} \tag{26.26}$$

where again $\lambda_{\rm D}$ is the DEBYE length, see Eq. (18.6). The attenuation decrement is then

$$\gamma = -\frac{1}{2}\bar{\nu}_{\rm i} - \sqrt{\frac{\pi}{8}\frac{q_{\rm e}}{q_{\rm i}}\frac{m_{\rm i}}{m_{\rm e}}}\frac{\omega^4}{(k\, V_{T\,\rm e}|\cos\Theta|)^3}\,. \tag{26.27}$$

Here, for a complete ionized plasma one has to put

$$\bar{\nu}_{\rm i} = \frac{8}{5}\nu_{\rm ii}\left(\frac{k_z V_{T\,\rm i}}{\omega}\right)^2 \equiv \frac{8}{5} Q_{\rm iz}^2 \nu_{\rm ii}, \tag{26.28}$$

but for a weakly ionized one $\bar{\nu}_{\rm i} = \nu_{\rm in}$.

In the limit

$$k^2 \lambda_{\rm De}^2 \ll 1 \tag{26.29}$$

the spectrum Eq. (26.25) goes over into that obtained with the gasdynamic model, Eq. (26.17). Fig. 6 shows the spectra of ion-sound waves under the condition Eq. (26.28) for a plasma which is gyrotropic and not isothermal, $T_{\rm e} > 6\, T_{\rm i}$. Fig. 6 is based upon Eq. (24.6), see [3].

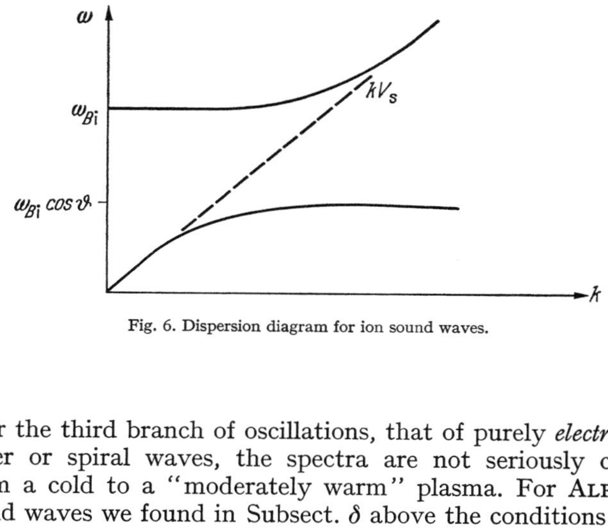

Fig. 6. Dispersion diagram for ion sound waves.

ε) Also for the third branch of oscillations, that of purely *electron waves*, also called whistler or spiral waves, the spectra are not seriously changed when changing from a cold to a "moderately warm" plasma. For ALFVÉN and fast magneto-sound waves we found in Subsect. δ above the conditions

$$\beta \equiv \frac{2\mathrm{u}\,\mu_0 N_{\rm e}(\vartheta_{\rm e} + \vartheta_{\rm i})}{B_{\delta}^2} \ll 1\,, \tag*{[26.15]}$$

and

$$k_z V_{T\,\rm i} \ll \omega \ll k_z V_{T\,\rm e}. \tag*{[26.20]}$$

The same conditions apply here. We have to start with the spectral equation of spiral waves, Eq. (25.33). Taking account of the particular conditions under which such waves can exist, namely

$$\omega_{\rm Ne}^2 \gg \omega\, \omega_{B\,\rm e}, \quad \text{i.e.} \quad X_{\rm e} \gg Y_{\rm e}, \tag{26.30}$$

and

$$\omega_{B\mathrm{i}} \ll \omega \ll \omega_{B\mathrm{e}}, \quad \text{i.e. } Y_{\mathrm{i}} \ll 1 \ll Y_{\mathrm{e}}, \tag{[25.31]}$$

and further

$$\varepsilon_{zz} \gg \varepsilon_{xy}; \quad \varepsilon_{yx} \gg (\varepsilon_{xx}, \varepsilon_{yy}); \quad \varepsilon_{xy} \gg (\varepsilon_{yz}, \varepsilon_{xz}) \tag{26.31}$$

we find from the general dispersion equation, Eq. (4.3), for the attenuation decrement of spiral waves

$$\left.\begin{aligned} \gamma_3 = &-\frac{1}{2}|\omega_3| \frac{1+\cos^2\Theta}{|\omega_{B\mathrm{e}}\cos\Theta|}\bar{\nu}_{\mathrm{e}} - \\ &-\sqrt{\frac{\pi}{8}}\,\eta\,\frac{\omega_3^3}{k V_{T\mathrm{e}}|\omega_{B\mathrm{e}}|}\tan^2\Theta \exp\left(\frac{-\omega_3^2}{2k^2 V_{T\mathrm{e}}^2\cos^2\Theta}\right), \end{aligned}\right\} \tag{26.32}$$

i.e.

$$\left.\begin{aligned} -\frac{2}{|\omega_3|}\gamma_3 = &\frac{1+\cos^2\Theta}{|Y_{\mathrm{e}}\cos\Theta|} Z_{\mathrm{e}} + \\ &+\sqrt{\frac{\pi}{2}}\,\eta\, Q_{\mathrm{e}3}^{-1}\frac{1}{|Y_{\mathrm{e}}|}\tan^2\Theta \exp\left(-\frac{1}{2}Q_{\mathrm{e}3}^{-2}\frac{1}{\cos^2\Theta}\right). \end{aligned}\right\} \tag{26.32a}$$

The function η is defined by

$$\eta = \begin{cases} \frac{1}{2}\left(\frac{\omega_3}{k V_{T\mathrm{e}}\cos\Theta}\right)^4 \equiv \frac{1}{2} Q_{\mathrm{e}}^{-4}\frac{1}{\cos^4\Theta} & \text{for } \omega_3^2 \gg k_z^2 V_{T\mathrm{e}}^2, \text{ i.e. } Q_{\mathrm{e}3}^2 \ll 1, \\ 1 & \text{for } \omega_3^2 \ll k_z^2 V_{T\mathrm{e}}^2, \text{ i.e. } Q_{\mathrm{e}3}^2 \gg 1. \end{cases} \tag{26.33}$$

Here for a completely ionized plasma one has to put $\bar{\nu}_{\mathrm{e}} = \nu_{\mathrm{eff}}$, but for a weakly ionized one $\bar{\nu}_{\mathrm{e}} = \nu_{\mathrm{en}}$. Again when the plasma becomes cold, collisionless ČERENKOV attenuation by electrons is exponentially small so that the absorption of spiral waves is mainly due to electron collisions. This can be seen from the expression Eq. (26.32) in the limit $\omega_3^2 \gg (k_z V_{T\mathrm{e}})^2$.

With decreasing phase velocity, $V_{\mathrm{ph}} \approx \omega/k_z$, collisionless attenuation increases and, for $\omega_3 \lesssim k_z V_{T\mathrm{i}}$ it becomes so important that spiral waves cease to exist, i.e. they are seriously attenuated. This is due to the effect of the ions in the plasma.

For spiral waves again, the relation between ω and $\bar{\nu}_{\mathrm{e}}$ is quite arbitrary in this context. A sufficient condition for weak attenuation is that the inequalities

$$\bar{\nu}_{\mathrm{e}} \gg \omega_{B\mathrm{e}}, \quad \text{i.e. } Z_{\mathrm{e}} \ll Y_{\mathrm{e}}, \tag{[26.9]}$$

and

$$\bar{\nu}_{\mathrm{e}} \ll k_z V_{T\mathrm{e}}, \quad \text{i.e. } Z_{\mathrm{e}} \ll Q_{\mathrm{e}} \tag{26.34}$$

are satisfied. The latter equation is only needed for small values of the phase velocity, if

$$\omega_3 \lesssim k_z V_{T\mathrm{e}}. \tag{[26.7]}$$

ζ) As shown above in Sect. 25β (Fig. 4), the number of different branches of possible oscillations is not greater than five in a cold magneto-active plasma. Taking account of thermal motion a sixth branch appears which is called *slow magneto-sound wave*[3]. Its existence depends on the non-isothermal behaviour of the plasma because $T_{\mathrm{e}} > 6\,T_{\mathrm{i}}$ must be valid, see Sect. 20. All six branches of oscillations can be described with the gasdynamic equations which we have written down in Sects. 13 and 15. For this reason these branches are often called „gasdynamic" or „hydrodynamic" branches[4]. It is, however, not so that all waves which could exist in a hot, magneto-active plasma, and their spectra were described by these six branches. In fact, the general dispersion equation, Eq. (4.3), if thermal motion is taken account of, is a transcendent equation so that the number of possible branches could be infinite in a hot, gyrotropic plasma. The large majority of these oscillations, however, decrease rapidly. Only in a few particular cases thermal motion of the particles leads to the possibility for slowly decreasing, large scale oscillations, as considered above.

[3] For $\beta \ll 1$ this is the ion sound wave.

[4] We note that longitudinal waves do not appear as independent branches in a magneto-active plasma. Any oscillation which does not derive from a potential is rather accurately longitudinal in a region where the refractive index is high.

η) Apart from these, the possible oscillations in the neighbourhood of the gyro- or cyclotron-frequencies of electrons and ions merit some attention. These "*cyclotron-oscillations*" depend upon the condition

$$\omega \approx s\,\omega_{B\,\mathrm{e,i}}, \quad \text{i.e.} \quad s\,Y_{\mathrm{e,i}} \approx 1, \qquad (s = 1, 2, \ldots). \tag{26.35}$$

They have been seriously studied during the last years, particularly because they seemed promising for applications of heating a plasma under conditions where collisional absorption (i.e. "ohmic" heating) is not very efficient (for example under conditions of high "nuclear temperature"). We do not intend to explain in detail the theory of such oscillations, referring to [*1* ... *6*] and the bibliographies given there.

In the following our considerations are restricted to such cyclotron waves which propagate along the external magnetic field. Cyclotron resonance as well as cyclotron absorption can be most clearly demonstrated with these waves[5].

We begin with the particular case of a collisionless plasma. The general dispersion equation, Eq. (4.3), for waves propagating parallel with the external magnetic field splits into three equations:

$$\left.\begin{aligned} \frac{1}{\varepsilon_0}\,\varepsilon_{zz} &= 1 + \sum_{\mathrm{h}} \left(\frac{\omega_{N\mathrm{h}}}{k\,V_{T\mathrm{h}}}\right)^2 \left[1 - \mathrm{J}_+\left(\frac{\omega}{k\,V_{T\mathrm{h}}}\right)\right] = 0 \\ &\text{and} \\ k^2 c_0^2/\omega^2 &= 1 - \sum_{\mathrm{h}} \frac{\omega_{N\mathrm{h}}^2}{\omega\,(\omega \mp \omega_{B\mathrm{h}})}\,\mathrm{J}_+\left(\frac{\omega \mp \omega_{B\mathrm{h}}}{k\,V_{T\mathrm{h}}}\right) \end{aligned}\right\} \tag{26.36}$$

i.e.

$$\left.\begin{aligned} 0 &= 1 + \sum_{\mathrm{h}} X_{\mathrm{h}}\,Q_{\mathrm{h}}^{-2}\left[1 + \mathrm{J}_+(Q_{\mathrm{h}}^{-1})\right] \\ &\text{and} \\ (k\,c_0/\omega)^2 &= 1 - \sum_{\mathrm{h}} \left(X_{\mathrm{h}}/(1 \mp Y_{\mathrm{h}})\right)\,\mathrm{J}_+\left(Q_{\mathrm{h}}^{-1}(1 \mp Y_{\mathrm{h}})\right). \end{aligned}\right\} \tag{26.36a}$$

The first equation implies longitudinal oscillations of fields in the plasma; it coincides exactly with Eq. (20.2) for the isotropic plasma. The second and third equation mean transverse, ordinary and extraordinary, waves propagating along the field. These equations are fully symmetric with respect to the sign of the cyclotron (or LARMOR) frequency, $\omega_{B\mathrm{h}}/2\pi$ or Y_{h}, respectively. It is therefore sufficient to discuss one of these.

Let us begin with the neighbourhood of the electron cyclotron frequency, $\omega \approx \omega_{B\mathrm{e}}$ (i.e. $Y_{\mathrm{e}} \approx 1$). We suppose now

$$|\omega - \omega_{B\mathrm{e}}| \gg k\,V_{T\mathrm{e}} \quad \text{i.e.} \quad |1 - Y_{\mathrm{e}}| \gg Q_{\mathrm{e}}, \tag{26.37}$$

which means, expressed in optical terms, that the frequency lies outside of the "resonance absorption line". We then obtain from the second and third of the above equations as real and imaginary part of the complex refractive index $\mu + i\chi$:

$$\mu^2 = 1 - \frac{\omega_{N\mathrm{e}}^2}{\omega\,(\omega - \omega_{B\mathrm{e}})} \to -\frac{\omega_{N\mathrm{e}}^2}{\omega\,(\omega - \omega_{B\mathrm{e}})} \tag{26.38}$$

and

$$\chi = \sqrt{\frac{\pi}{8}}\,\frac{\omega_{N\mathrm{e}}^2}{\omega^2}\,\frac{1}{\mu^2}\,\frac{c_0}{V_{T\mathrm{e}}}\,\exp\left(\frac{-(\omega - \omega_{B\mathrm{e}})^2 c_0^2}{2\mu^2\omega^2 V_{T\mathrm{e}}^2}\right), \tag{26.39}$$

i.e.

$$\mu^2 = 1 - \frac{X_{\mathrm{e}}}{1 - Y_{\mathrm{e}}} \to -\frac{X_{\mathrm{e}}}{1 - Y_{\mathrm{e}}} \tag{26.38a}$$

[5] We note that in the considered case of "parallel" propagation there are no "subharmonic" resonances; these appear only with $k_\perp \neq 0$, see [*1* ... *5*].

and

$$\chi = \sqrt{\frac{\pi}{8}}\, X_e \frac{1}{\mu^2} \frac{c_0}{V_{Te}} \exp\left(-\frac{1}{2}\left(\frac{c_0}{\mu V_{Te}}\right)^2 (1-Y_e)^2\right). \tag{26.39a}$$

These expressions show that the electronic cyclotron wave propagates only for *frequencies below the gyro-*(cyclotron-)frequency:

$$\omega < \omega_{Be}, \quad \text{i.e.} \quad Y_e > 1. \tag{26.40}$$

In the collisionless limit, outside of the absorption resonance line the attenuation is exponentially weak. If we take account of collisions between particles we obtain [under the condition Eq. (23.1)] a supplement, $\delta\chi$, to the absorption index χ:

$$\delta\chi = \frac{1}{2\mu} \frac{\omega_{Ne}^2 \bar{\nu}_e}{\omega(\omega-\omega_{Be})^2} = \frac{1}{2\mu} \frac{X_e Z_e}{(1-Y_e)^2}. \tag{26.41}$$

For a completely ionized plasma $\bar{\nu}_e = \nu_{eff}$, but for a weakly ionized one $\bar{\nu}_e = \nu_{en}$.

From Eqs. (26.38), (26.39), (26.41) expressions for the frequency of the corresponding oscillation and for the attenuation decrement can be found. They are valid for cyclotron oscillations but only outside the resonance absorption line. The *frequency of oscillation* is given by:

$$\omega \approx \omega_{Be}\left(1 - \frac{\omega_{Ne}^2}{k^2 c_0^2}\right), \tag{26.42}$$

i.e.

$$\left(\frac{\omega}{k c_0}\right)^2 \approx \frac{Y_e - 1}{X_e Y_e}, \tag{26.42a}$$

and the attenuation decrement by

$$\gamma = -\frac{1}{2}\bar{\nu}_e - \sqrt{\frac{\pi}{2}} \frac{\omega_{Be}^2 \omega_{Ne}^4}{k^5 c_0^4 V_{Te}} \exp\left(\frac{-(\omega-\omega_{Be})^2}{2k^2 V_{Te}^2}\right), \tag{26.43}$$

i.e.

$$-\frac{2}{\omega}\gamma = Z_e + \sqrt{2\pi}\left(\frac{\omega}{k c_0}\right)^5 \frac{c_0}{V_{Te}} X_e^2 Y_e^2 \exp\left(-\frac{1}{2} Q_e^{-2}(1-Y_e)^2\right). \tag{26.43a}$$

It follows accidentially that for weakly attenuated cyclotron oscillations

$$k^2 c_0^2 \gg \omega_{Ne}^2, \tag{26.44}$$

i.e.

$$\left(\frac{k c_0}{\omega}\right)^2 \gg X_e. \tag{26.44a}$$

If the frequency lies quite near to the gyro-(or cyclotron-)frequency, so that it falls on the resonance absorption line, i.e. if

$$|\omega - \omega_{Be}| \ll k V_{Te}, \quad \text{i.e.} \quad |1-Y_e| \ll Q_e, \tag{26.45}$$

then we find from Eqs. (26.38), (26.39), in the collisionless limit

$$n \equiv \mu + i\chi = \frac{1}{2}(\sqrt{3}+i)\left(\sqrt{\frac{\pi}{2}} \frac{\omega_{Ne}^2 c_0}{\omega^2 V_{Te}}\right)^{\frac{1}{3}}, \tag{26.46}$$

i.e.

$$n \equiv \mu + i\chi = \frac{\sqrt{3}+i}{2}\left(\sqrt{\frac{\pi}{2}} \frac{c_0}{V_{Te}} X_e\right)^{\frac{1}{3}}. \tag{26.46a}$$

In this particular frequency range the cyclotron waves are heavily damped because χ almost equals μ. The penetration of such cyclotron waves into the plasma is described as an *anomalous skin effect*. The characteristic penetration depth, δ, is [compare Eq. (21.19)]

$$\delta = \frac{c_0}{\chi\omega} = 2\left(\sqrt{\frac{2}{\pi}}\,\frac{c_0^2 V_{Te}}{\omega\,\omega_{Ne}^2}\right)^{\frac{1}{3}} = 2\,\frac{c_0}{\omega}\left(\sqrt{\frac{2}{\pi}}\,\frac{V_{Te}}{c_0}\Big/X_e\right)^{\frac{1}{3}}. \tag{26.47}$$

Eqs. (26.46), (26.47) remain valid if collisions occur between particles, provided

$$\bar{\nu}_e \ll k V_{Te} \sim V_{Te}\left(\frac{\omega\,\omega_{Ne}^2}{c_0^2 V_{Te}}\right)^{\frac{1}{3}} = \omega\,\frac{V_{Te}}{c_0}\left(\frac{c_0}{V_{Te}}\,X_e\right)^{\frac{1}{3}}, \tag{26.48}$$

i.e.

$$Z_e \ll Q_e \approx \frac{V_{Te}}{c_0}\left(\frac{c_0}{V_{Te}}\,X_e\right)^{\frac{1}{3}}. \tag{26.48a}$$

In the ionospheric and exospheric plasma this condition is satisfied at heights above 200 km. It breaks down near 100 km and below, but at these *lower heights* the ionospheric plasma is weakly ionized.

Supposing

$$|\omega - \omega_{Be}| \ll \nu_{en}, \quad \text{i.e.} \quad |1 - Y_e| \ll Z_e \tag{26.49}$$

and weak ionization, we find

$$n = \mu + i\chi = \frac{1+i}{\sqrt{2}}\sqrt{\frac{\omega_{Ne}^2}{\omega\nu_{en}}} = \frac{1+i}{\sqrt{2}}\sqrt{\frac{X_e}{Z_e}}. \tag{26.50}$$

This leads to

$$\delta = \sqrt{\frac{2c_0^2\,\nu_{en}}{\omega\,\omega_{Ne}^2}} = \frac{c_0}{\omega}\sqrt{2\,\frac{Z_e}{X_e}} = \sqrt{\frac{2\varepsilon_0}{\mathrm{u}}}\,\frac{c_0}{\sqrt{\omega\sigma}} \tag{26.51}$$

where

$$\sigma = \frac{\varepsilon_0}{\mathrm{u}}\,\frac{\omega_{Ne}^2}{\nu_{en}} = \frac{\varepsilon_0}{\mathrm{u}}\,\frac{X_e}{Z_e}\,\omega \tag{26.52}$$

is the "static" conductivity of the plasma parallel to the magnetic field.

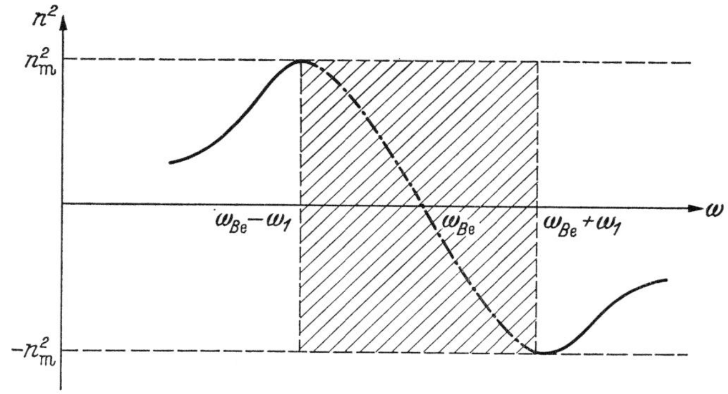

Fig. 7. Refractive index diagram for the electronic cyclotron wave near resonance absorption.

Fig. 7 shows the behaviour of n^2 for the electronic cyclotron wave in the neighbourhood of the absorption resonance line, i.e. for $Y_e \approx 1$. The hatched region means strong absorption in the plasma.

Parallel propagation of ionic cyclotron waves, $\omega \sim \omega_{Bi}$, i.e. $Y_i \sim 1$, is quite analogous in character. All formulae given above are easily rewritten replacing

Table 6. *Weakly attenuated oscillations in cold gyrotropic (magneto-active) plasmas.*

Type of oscillation	Frequency spectrum	Attenuation decrement
1. Longitudinal oscillations	$\omega_1^2 \approx \omega_{Ne}^2 + \omega_{Be}^2$	$\gamma_1 = -\frac{\bar{\nu}_e}{2}$
	$\omega_2^2 \approx \omega_{Be}^2 \cos^2\Theta$	$\gamma_2 = -\frac{\bar{\nu}_e}{2}\cos^{-2}\Theta$
	$\omega_3^2 \approx \omega_{Bi}^2\left(1 - \frac{g_i}{g_e}\frac{m_e}{m_i}\tan^2\Theta\right)$	$\gamma_3 = -\nu_i\left[1 - \frac{\bar{\nu}_e}{2\bar{\nu}_i}\frac{m_e}{m_i}\tan^2\Theta\right]$
(special)	$\omega_2^2 \approx \sqrt{\omega_{Be}^2\,\omega_{Bi}^2}$	$\gamma_2 = -\frac{\bar{\nu}_i}{2}\left[1 + \frac{\bar{\nu}_e}{\bar{\nu}_i}\left(1 + \frac{m_i}{m_e}\tan^2\Theta\right)\right]$
	$\omega_3^2 \approx \omega_{Bi}^2\cos^2\Theta$	$\gamma_3 = -\frac{\bar{\nu}_e}{2}\left[1 + \frac{\bar{\nu}_i}{\bar{\nu}_e}\frac{m_e}{m_i}\sin^2\Theta\right]$
2. ALFVÉN-waves	$\omega_1^2 = \frac{k^2 V_A^2\cos^2\Theta}{1 + V_A^2/c_0^2}$	$\frac{\gamma_1}{\lvert\omega_1\rvert} \approx -\frac{1}{2}\frac{\omega_{Bi}^2}{\omega_{Ni}^2}\sum_h \frac{\omega_{Nh}^2\bar{\nu}_h}{\omega_{Bh}^2\lvert\omega_1\rvert}\frac{\omega_1^2}{\omega_{Bh}^2}$
3. Fast magneto-sound waves	$\omega_2^2 = \frac{k^2 V_A^2}{1 + V_A^2/c_0^2}$	$\frac{\gamma_2}{\lvert\omega_2\rvert} \approx -\frac{1}{2}\frac{\omega_{Bi}^2}{\omega_{Ni}^2}\sum_h \frac{\omega_{Nh}^2\bar{\nu}_h}{\omega_{Bh}^2\lvert\omega_2\rvert}\frac{\omega_2^2}{\omega_{Bh}^2}$
4. Spiral (or whistlr waves)	$\omega_3 = \frac{k^2 c_0^2}{\omega_{Ne}^2}\lvert\omega_{Be}\cos\Theta\rvert$	$\frac{\gamma_3}{\lvert\omega_3\rvert} \approx -\frac{1}{2}\frac{\bar{\nu}_e(1+\cos^2\Theta)}{\lvert\omega_{Be}\cos\Theta\rvert}$

the electronic quantities by the ionic ones, i.e. replacing suffix e by i. It is to be remembered at that occasion that for a completely ionized plasma then

$$\bar{\nu}_i = \frac{m_e}{m_i}\nu_{\text{eff}},$$

but for a weakly ionized one $\bar{\nu}_i = \nu_{in}$.

ϑ) *Summary*: The above Tables 6 and 7 give frequency spectra and attenuation decrements for all kinds of weakly attenuated oscillations of a magneto-active plasma. Refraction and absorption indices of the corresponding waves are also noted, as well as the conditions for such oscillations to be able to occur in a given plasma.

D. The stability problem*.

In the preceding chapters we assumed a plasma with isotropically distributed (individual) particle velocities, and considered electromagnetic, magneto-acoustic and other oscillations in such a plasma. In detailed computations we supposed a Maxwellian distribution of plasma particles, and the hypothesis of thermodynamic equilibrium has been adopted. Oscillations of small amplitude were found to behave, as exspected, stable, i.e. decreasing with time; the energy contained in such oscillations is finally absorbed by the plasma, i.e. it is transformed into thermal energy, real plasma, however, is usually far from thermodynamic equilibrium; this is true for a plasma in the laboratory, in the ionosphere, in the magnetosphere, and, a fortiori, for the solar wind plasma. Real plasma,

* In this and the following chapters we normally use the word "frequency" for the number of oscillations per sec, f, while $\omega = 2\pi f$ is called "pulsation". In some cases any one of both notations may be called "frequency".

[Conditions: $T_e = 0 = T_i$; $\omega \gg (k_z V_{Te}, k_z V_{Ti})$]

Refractive index	Absorption index	Conditions of existence
—	—	$\omega_{Ne} \gg \bar{\nu}_e$; $\omega_{Ne}^2 \gg \omega_{Be}^2$
—	—	$\omega_{Ne}^2 \gg \omega_{Be}^2 \gg \bar{\nu}_e^2$; $\cos^2\Theta > \frac{m_e}{m_i}$
—	—	$\omega_{Ne}^2 \gg \omega_{Be}^2$; $\omega_{Bh} > \bar{\nu}_h$; $\cos^2\Theta > \frac{m_e}{m_i}$
—	—	$\omega_{Ne}^2 \gg \omega_{Be}^2$; $\cos^2\Theta < \frac{m_e}{m_i}$
—	—	$\omega_{Ne}^2 \gg \omega_{Be}^2$; $\omega_{Bi} > \bar{\nu}_e$; $\cos^2\Theta < \frac{m_e}{m_i}$
$\mu_1^2 = \cos^{-2}\Theta\left(1 + \frac{c_0^2}{V_A^2}\right)$	$\chi_1 \approx \frac{\mu_1}{2}\frac{\omega_{Bi}^2}{\omega_{Ni}^2}\sum_h \frac{\omega_{Nh}^2 \bar{\nu}_h}{\lvert\omega_1\rvert \omega_{Bh}^2}\frac{\omega^2}{\omega_{Bh}^2}$	$\omega_1 \ll \omega_{Bi}$
$\mu_2^2 = 1 + \frac{c_0^2}{V_A^2}$	$\chi_2 \approx \frac{\mu_2}{2}\frac{\omega_{Bi}^2}{\omega_{Ni}^2}\sum_h \frac{\omega_{Nh}^2 \bar{\nu}_h}{\lvert\omega_2\rvert \omega_{Bh}^2}\frac{\omega^2}{\omega_{Bh}^2}$	$\omega_2 \ll \omega_{Bi}$
$\mu_3^2 = \frac{\omega_{Ne}^2}{\omega\lvert\omega_{Be}\cos\Theta\rvert}$	$\chi_3 \approx \frac{\mu_3}{4}\frac{\bar{\nu}_e(1+\cos^2\Theta)}{\lvert\omega_{Be}\cos\Theta\rvert}$	$\omega_{Bi} \ll \omega_3 \ll \omega_{Be}$

normally, has some excess energy which can produce oscillations of any kind with amplitudes at least instantaneously increasing with time. In analogy with mechanical problems one speaks in these cases of *instability against oscillations.* We shall see below that even small deviations from thermodynamic equilibrium, for example an anisotropic velocity distribution, may produce such instability.

27. General stability criteria. The fundamental question related to the origin of oscillations in plasmas having an anisotropic distribution function is that of stability. This leads to the initial value problem of Sect. 4, namely that of oscillations with small amplitude. The behaviour in time depends on the roots of the dispersion equation

$$\det\left[k^2\,\mathsf{U} - \boldsymbol{k}\,\boldsymbol{k} - \left(\frac{\omega}{c_0}\right)^2 \frac{1}{\varepsilon_0}\,\mathsf{E}(\omega,\boldsymbol{k})\right] = 0$$

which reads in components [4.13]

$$\det\left[k^2\,\delta_{ij} - k_i k_j - \left(\frac{\omega}{c_0}\right)^2 \frac{1}{\varepsilon_0}\,\varepsilon_{ij}(\omega,\boldsymbol{k})\right] = 0$$

α) The *roots* of this equation

$$\omega_n(\boldsymbol{k}) \qquad (n = 1, 2, \ldots) \tag{27.1}$$

show us, how small fluctuations of the electromagnetic quantities,

$$\varphi_{\boldsymbol{k}}(t) \sim \varphi_{\boldsymbol{k}}(0)\exp(-i\,\omega_n t) \tag{27.2}$$

develop when time t progresses. If there is a root with

$$\mathrm{Im}\left(\omega_n(\boldsymbol{k})\right) > 0 \tag{27.3}$$

Table 7. *Weakly attenuated oscillations in warm gyrotropic (magneto-active) plasmas.*

Type of oscillation	Frequency spectrum	Attenuation decrement
1. Longitudinal waves	$\omega^2 = \frac{k^2 V_s^2 \cos^2\Theta}{1 + k^2 \lambda_{De}^2}$	$\gamma = -\frac{\bar{\nu}_i}{2} - \sqrt{\frac{\pi}{8}} \frac{q_e}{q_i} \frac{m_i}{m_e} \frac{\omega^4}{\|k_z\|^3 V_{Te}^3}$
Ion sound oscillations	$\omega^2 = \frac{k^2 V_s^2}{1 + k^2 \lambda_{De}^2}$	$\gamma = -\frac{\bar{\nu}_i}{2} - \sqrt{\frac{\pi}{8}} \frac{q_e}{q_i} \frac{m_i}{m_e} \frac{\omega^4}{k^2 \|k_z\| V_{Te}^3}$
2. ALFVÉN waves	$\omega^2 = k^2 V_A^2 \cos^2\Theta$	$\gamma = -\frac{1}{2}\left(\frac{m_e}{m_i}\bar{\nu}_e + \bar{\nu}_i\right)$
3. Magneto-sound waves		
3.1. Fast	$\omega_+^2 = k^2 V_A^2$	$\gamma_+ = -\frac{1}{2}\left(\frac{m_e}{m_i}\bar{\nu}_e + \bar{\nu}_i\right) - \sqrt{\frac{\pi}{8}} \frac{m_e}{m_i} \frac{k V_{Te} \sin^2\Theta}{\|\cos\Theta\|}$
3.2. Slow	$\omega_-^2 = k^2 V_s^2 \cos^2\Theta$	$\gamma_- = -\frac{\bar{\nu}_i}{2} - \sqrt{\frac{\pi}{8}} \frac{q_i}{q_e} \frac{m_e}{m_i} \|\omega_-\|$
4. Spiral (or Whistler) waves	$\omega = \frac{k^2 c_0^2}{\omega_{Ne}^2} \|\omega_{Be} \cos\Theta\|$	$\gamma = -\frac{1}{2}\|\omega_3\| \frac{\bar{\nu}_e(1+\cos^2\Theta)}{\|\omega_{Be}\cos\Theta\|} - \sqrt{\frac{\pi}{8}} \frac{\omega^3 \sin^3\Theta}{k V_{Te}\, \omega_{Be} \cos^2\Theta}$
5. Gyro- (or cyclotron-) waves [$\Theta = 0$; h = e, i]	$\omega \approx \omega_{Bh}$	$\gamma = -\frac{\bar{\nu}_h}{2} - \sqrt{\frac{\pi}{8}} \frac{\omega_{Nh}^4 \omega_{Bh}^2}{k^5 c_0^4 V_{Th}} \times \exp\left(\frac{-(\omega - \omega_{Bh})^2}{2k^2 V_{Th}^2}\right)$

then the *system* is *instable* because the fluctuation increases. If, however,

$$\operatorname{Im}\left(\omega_n(\boldsymbol{k})\right) < 0 \tag{27.4}$$

the *system* is *stable* because fluctuations once produced die out.

How could now the dielectricity tensor $\mathsf{E} \equiv ((\varepsilon_{ij}(\omega, \boldsymbol{k})))$ be determined for a plasma with anisotropic distribution of particle velocities? To this end we apparently have to choose one or another descriptive model of the plasma (see Chap. B. I). The most general model is, of course, that of the kinetic equation. However, in this Sect. we shall not use a specific model because there exists yet another method for computing the dielectricity tensor of plasmas which applies at least to a certain class of anisotropic distributions of particle velocities.

β) We limit our considerations to the investigation of such velocity distributions which may be described as *bulk-motion* in an instable, collisionless, gyrotropic plasma. The behaviour as a "bulk" means that macroscopic motions of the plasma components exist. We mention in this context: interaction between a ray of charged particles with a plasma; plasma in the presence of an external (constant and homogeneous) electric field. In such a plasma the distribution function

[Conditions: $\beta \equiv \dfrac{2\mathrm{u}\,\mu_0 (N_e \vartheta_e + N_i \vartheta_i)}{B^2_{\stackrel{+}{0}}} \ll 1;\quad k_z V_{T\,i} \ll \omega \ll k_z V_{T\,e}$]

Refractive index	Absorption index	Conditions of existence
$\mu^2 = \dfrac{c_0^2}{V_s^2}\dfrac{\omega_{N\,i}^2}{\omega^2}\Big/\left(\dfrac{\omega_{N\,i}^2}{\omega^2}\cos^2\Theta - 1\right)$	$\chi = -\dfrac{\gamma}{\omega}\dfrac{c_0}{V_s}\dfrac{\omega_{N\,i}^3}{\omega^3}\Big/\left(\dfrac{\omega_{N\,i}^2}{\omega^2}\cos^2\Theta - 1\right)^{\frac{1}{2}}$	$\begin{cases}\omega \ll (\omega_{B\,i}, \omega_{N\,i}) \\ T_e \gg T_i\end{cases}$
$\mu^2 = \dfrac{c_0^2}{V_s^2}\dfrac{\omega_{N\,i}^2}{\omega^2}\Big/\left(\dfrac{\omega_{N\,i}^2}{\omega^2} - 1\right)$	$\chi = -\dfrac{\gamma}{2}\dfrac{c_0}{V_s}\dfrac{\omega_{N\,i}^3}{\omega^3}\Big/\left(\dfrac{\omega_{N\,i}^2}{\omega^2} - 1\right)^{\frac{3}{2}}$	$\begin{cases}\omega_{B\,i} \ll \omega \ll \omega_{B\,e} \\ \omega \ll \omega_{N\,i};\ T_e \gg T_i\end{cases}$
$\mu^2 = \dfrac{c_0^2}{V_A^2}\cos^{-2}\Theta$	$\chi = \dfrac{\mu}{2}\dfrac{1}{\lvert\omega\rvert}\left(\dfrac{m_e}{m_i}\bar{\nu}_e + \bar{\nu}_i\right)$	$\omega \ll \omega_{B\,i}$
$\mu_+^2 = \dfrac{c_0^2}{V_A^2}$	$\chi_+ = -\mu_+\dfrac{\gamma_+}{\omega}$	$\omega \ll (\omega_{N\,i}, \omega_{B\,i})$
$\mu_-^2 = \dfrac{c_0^2}{V_s^2}\cos^{-2}\Theta$	$\chi_- = -\mu_-\dfrac{\gamma_-}{\omega}$	
$\mu^2 = \dfrac{\omega_{N\,e}^2}{\omega\,\lvert\omega_{B\,e}\cos\Theta\rvert}$	$\chi = -\dfrac{1}{2}\mu\dfrac{\gamma}{\omega}$	$\omega_{B\,i} \ll \omega \ll \omega_{B\,e}$
$\mu^2 \approx -\dfrac{\omega_{N\,h}^2}{\omega(\omega - \omega_{B\,h})}$	$\chi = \dfrac{1}{2\mu}\dfrac{\omega_{N\,h}^2}{\omega(\omega-\omega_{B\,h})^2}\bar{\nu}_h + \sqrt{\dfrac{\pi}{8}}\dfrac{\omega_{N\,h}^2}{\omega^2}\dfrac{c_0}{\mu^2 V_{T\,h}} \times \exp\left(\dfrac{-(\omega-\omega_{B\,h})^2 c_0^2}{2\omega^2\mu^2 V_{T\,h}^2}\right)$	$\lvert\omega - \omega_{B\,h}\rvert \gg (\bar{\nu}_h, k V_{T\,h})$

of particles of kind h has the simple form:

$$f_{0\,h} = f_{0\,h}(\boldsymbol{v} - \boldsymbol{u}_h), \tag{27.5}$$

$\boldsymbol{u}_h$ being the velocity of the systematic motion[1]. This motion is supposed to be constant and homogeneous. Instead of resolving the kinetic equation, the tensor $\mathsf{E} \equiv ((\varepsilon_{ij}))$ can be found from the equations derived in Chap. C (Sects. 17 and 22), together with the Lorentz transformation. The total current density through the plasma can, in fact, be obtained as the sum of the currents due to the individual components, so that the effective current density due to an electric field $\boldsymbol{E}$ is finally:

$$\boldsymbol{j} = \sum_h \boldsymbol{j}_h = \sum_h \mathsf{S}^{(h)}(\omega, \boldsymbol{k}) \cdot \mathsf{E}, \tag{27.6}$$

i.e.

$$j_l = \sum_h \sigma_{lm}(\omega, \boldsymbol{k})\, E_m. \tag{27.6a}$$

[1] In the case where an external magnetic field is present, $\boldsymbol{u}_h$ is supposed to be parallel to the magnetic fieldline. This is the only case of physical interest in the present context.

Here $\mathsf{S} \equiv ((\sigma_{lm}(\omega, \boldsymbol{k})))$ is that part of the conductivity tensor that is due to particles of kind h[2].

γ) In order to determine this tensor we take advantage to introduce a *frame moving with the particles* h, so that these particles are at rest in the new frame which itself moves against the laboratory frame with speed $\boldsymbol{u}_{\mathrm{h}}$. In this particular system of coordinates the current contribution reads

$$\boldsymbol{j}'_{\mathrm{h}} = \mathsf{S}^{(\mathrm{h})}(\omega'_{\mathrm{h}}, \boldsymbol{u}'_{\mathrm{h}}) \cdot \boldsymbol{E}'_{\mathrm{h}}, \tag{27.7}$$

i.e.

$$j'_{\mathrm{h}l} = \sigma^{(\mathrm{h})}_{lm}(\omega'_{\mathrm{h}}, \boldsymbol{u}'_{\mathrm{h}})\, E'_m. \tag{27.7a}$$

$\boldsymbol{j}'$ and $\boldsymbol{E}'$ being current density and electric field in the *moving* system. These quantities are connected with those in the "fixed system", $\boldsymbol{j}$ and $\boldsymbol{E}$, by the LORENTZ transformation:

$$\left.\begin{aligned} \boldsymbol{j}_{\mathrm{h}} &= \mathsf{A}(\boldsymbol{u}_{\mathrm{h}}) \cdot \boldsymbol{j}'_{\mathrm{h}} \\ \boldsymbol{E}'_{\mathrm{h}} &= \mathsf{B}(\boldsymbol{u}_{\mathrm{h}}) \cdot \boldsymbol{E} \end{aligned}\right\} \tag{27.8}$$

where

$$\mathsf{A}(\boldsymbol{u}) = \mathsf{U} + \frac{1}{\sqrt{1-u^2/c_0^2}}\left[\left(1-\sqrt{1-u^2/c_0^2}\right)\frac{\boldsymbol{u}\boldsymbol{u}}{u^2} + \frac{\boldsymbol{u}\,\boldsymbol{k}'}{\omega'}\right], \tag{27.9}$$

in components:

$$\alpha_{ij}(\boldsymbol{u}) = \delta_{ij} + \frac{1}{\sqrt{1-u^2/c_0^2}}\left[\left(1-\sqrt{1-u^2/c_0^2}\right)\frac{u_i u_j}{u^2} + \frac{u_i k'_j}{\omega'}\right], \tag{27.9a}$$

and

$$\mathsf{B}(\boldsymbol{u}) = \frac{\omega'}{\omega}\mathsf{U} + \frac{1}{\sqrt{1-u^2/c_0^2}}\left[\left(\sqrt{1-u^2/c_0^2}-1\right)\frac{\boldsymbol{u}\boldsymbol{u}}{u^2} + \frac{\boldsymbol{k}\,\boldsymbol{u}}{\omega}\right], \tag{27.10}$$

in components:

$$\beta_{ij}(\boldsymbol{u}) = \frac{\omega'}{\omega}\delta_{ij} + \frac{1}{\sqrt{1-u^2/c_0^2}}\left[\left(\sqrt{1-u^2/c_0^2}-1\right)\frac{u_i u_j}{u^2} + \frac{k_i u_j}{\omega}\right]. \tag{27.10a}$$

ω' and $\boldsymbol{k}'$ are the LORENTZ transforms of wave pulsation ω and wave vector $\boldsymbol{k}$:

$$\omega' = \frac{\omega - \boldsymbol{u}\cdot\boldsymbol{k}}{\sqrt{1-u^2/c_0^2}}, \tag{27.11}$$

$$\boldsymbol{k}' = \boldsymbol{k} + \boldsymbol{u}\,\frac{\left(1-\sqrt{1-u^2/c_0^2}\right)\boldsymbol{u}\cdot\boldsymbol{k} - \omega\, u^2/c_0^2}{u^2\sqrt{1-u^2/c_0^2}}. \tag{27.12}$$

δ) With Eqs. (27.6) through (27.12) the *tensor of dielectric permeability* for the plasma as whole, in the laboratory frame, can easily be established [37]:

$$\left.\begin{aligned} \frac{1}{\varepsilon_0}\mathsf{E}(\omega, \boldsymbol{k}) &= \mathsf{U} + i\,\frac{\mathrm{u}}{\varepsilon_0}\,\frac{1}{\omega}\sum_{\mathrm{h}} \mathsf{A}(\boldsymbol{u}_{\mathrm{h}}) \cdot \mathsf{S}^{(\mathrm{h})}(\omega'_{\mathrm{h}}, \boldsymbol{k}'_{\mathrm{h}}) \cdot \mathsf{B}(\boldsymbol{u}_{\mathrm{h}}) \\ &= \mathsf{U} + \sum_{\mathrm{h}} \frac{\omega'_{\mathrm{h}}}{\omega}\,\mathsf{A}(\boldsymbol{u}_{\mathrm{h}}) \cdot \left[\frac{1}{\varepsilon_0}\mathsf{E}^{(\mathrm{h})}(\omega'_{\mathrm{h}}, \boldsymbol{k}'_{\mathrm{h}}) - \mathsf{U}\right] \cdot \mathsf{B}(\boldsymbol{u}_{\mathrm{h}}), \end{aligned}\right\} \tag{27.13}$$

in components:

$$\left.\begin{aligned} \frac{1}{\varepsilon_0}\varepsilon_{lm}(\omega, \boldsymbol{k}) &= \delta_{lm} + i\,\frac{\mathrm{u}}{\varepsilon_0}\,\frac{1}{\omega}\sum_{\mathrm{h}} \alpha_{lp}(\boldsymbol{u}_{\mathrm{h}})\,\sigma^{(\mathrm{h})}_{pq}(\omega'_{\mathrm{h}}, \boldsymbol{k}'_{\mathrm{h}})\,\beta_{qm}(\boldsymbol{u}_{\mathrm{h}}) \\ &= \delta_{lm} + \sum_{\mathrm{h}} \frac{\omega'_{\mathrm{h}}}{\omega}\,\alpha_{lp}(\boldsymbol{u}_{\mathrm{h}})\left[\frac{1}{\varepsilon_0}\,\varepsilon_{pq}\,\omega'_{\mathrm{h}}, \boldsymbol{k}'_{\mathrm{h}}) - \delta_{pq}\right]\beta_{qm}(\boldsymbol{u}_{\mathrm{h}}). \end{aligned}\right\} \tag{27.13a}$$

---

[2] Apply the summation rule wherever two identic indices do appear.

Here $E^{(h)}(\omega_h', \boldsymbol{k}_h')$ is the dielectricity tensor for particles of kind h in "their own frame", i.e. in that where the macroscopic speed of these particles is zero. The corresponding expressions can be found in Sects. 17, 18, 22 and 23.

When using Eq. (27.13) one should have in mind that four of the parameters appearing in the expression for the dielectricity tensor, $E^{(h)}(\omega_h', \boldsymbol{k}_h')$, are to be transformed. These are: wave frequency $\omega/2\pi$ and wave vector $\boldsymbol{k}$ (both by DOPPLER-shift), but also particle density (as a consequence of volume contraction) and mass of the particles (charges remaining invariant). The final result of all these charges is, that ω_{Nh}^2 is an invariant, but

$$\omega_{Bh} \to \omega_{Bh}' = \omega_{Bh}\sqrt{1 - u_h^2/c_0^2}. \tag{27.14}$$

Eqs. (27.9) through (27.13) are generalizations of well-known relations first established by MINKOWSKI, see, for example, [7]. Our generalization is concerned with media having dispersion with frequency and/or spatial dispersion. It may, for example, be applied to a collisionless plasma.

ε) The relations become much simpler for the *non-relativistic case*, i.e. in the limit where

$$u_h = |\boldsymbol{u}_h| \ll c_0. \tag{27.15}$$

In this case we simply have

$$\omega' = \omega - \boldsymbol{u}\cdot\boldsymbol{k}; \qquad \boldsymbol{k}' = \boldsymbol{k}, \tag{27.16}$$

$$\mathsf{A}(\boldsymbol{u}) = \mathsf{U} + \frac{\boldsymbol{u}\boldsymbol{k}}{\omega'}, \tag{27.17}$$

$$\mathsf{B}(\boldsymbol{u}) = \frac{\omega'}{\omega}\mathsf{U} + \frac{\boldsymbol{k}\boldsymbol{u}}{\omega}, \tag{27.18}$$

in components:

$$\alpha_{ij}(\boldsymbol{u}) = \delta_{ij} + \frac{u_i k_j}{\omega'}, \tag{27.17a}$$

$$\beta_{ij}(\boldsymbol{u}) = \frac{\omega'}{\omega}\delta_{ij} + \frac{k_i u_j}{\omega}. \tag{27.18a}$$

ζ) We could substitute the expression Eq. (27.13) into the dispersion equation in the form of Eq. (4.13), and analyse cases at the limits of admissible conditions in order to obtain a complete theory of the interaction between a plasma and bulk motion of charged particles. This, however, would be a very extensive program which we can not consider here in detail[3].

We shall consider here only two examples of the theory of such interactions of charged particles with bulk motion with a plasma. It is intended to explain the physical behaviour of such phenomena, and at the same time illustrate by these example the efficiency of our general method.

In a first example we investigate the interaction of a beam of electrons of small density with high-frequency LANGMUIR oscillations of an electronic plasma. With this example the possibility of beam-instability has been discovered by AHIEZER and FAJNBERG [40], and by BOHM and GROSS [41]. Moreover, this example is particularly interesting for applications in the field of ionospheric physics. Occasionally the Earth is reached by intensive fluxes of charged particles, electrons as well as ions. These particles, arriving from space, interact with the magnetosphere and, finally, with the upper ionosphere, producing plasma oscillations of different kinds. We shall, later on (in Sect. 70) study the spectra of these oscillations, and the way how they can be excited by fluxes of fast electrons.

[3] A great number of publications is devoted to this problem. Bibliographies may be found in [1], [2], [3] and [4]. [38] and [39] are summaries in which the stability of a plasma is investigated.

η) The second task which we shall consider in Sect. 29 below refers to the *stability* of a plasma supporting a (strong) current under conditions where all electrons are drifting against the ions. This situation can be described by saying that there is an ionic "grid" at rest through which the plasma electrons are moving in bulk. This problem is of great practical interest for controlled nuclear fusion; a particular case is the investigation of the stability of a plasma containing an external electric field and being heated by ohmic losses.

For simplicity we limit our considerations to the investigation of longitudinal (electric potential) waves in a gyrotropic plasma. We shall further suppose that the systematic motion of the particles, $|\boldsymbol{u}_{\mathrm{h}}|$, is small against the velocity of light, c_0. This means that we shall only consider non-relativistic electrons interacting with a plasma. In the case of a field which derives from a potential, i.e. for longitudinal waves, the dispersion equation, Eq. (4.13) can be written in the form

$$\left.\begin{aligned} \varepsilon(\omega, \boldsymbol{k}) &= \boldsymbol{k}^0 \cdot \boldsymbol{E}(\omega, \boldsymbol{k}) \cdot \boldsymbol{k}^0 \equiv \frac{1}{k^2} \boldsymbol{k} \cdot \mathsf{E}(\omega, \boldsymbol{k}) \cdot \boldsymbol{k} \\ &= \varepsilon_0 + \sum_{\mathrm{h}} [\varepsilon^{(\mathrm{h})}(\omega - \boldsymbol{u}_{\mathrm{h}} \cdot \boldsymbol{k}, \boldsymbol{k}) - \varepsilon_0] = 0, \end{aligned}\right\} \tag{27.19}$$

in components:

$$\varepsilon(\omega, \boldsymbol{k}) = \frac{k_i k_j}{k^2} \varepsilon_{ij}(\omega, \boldsymbol{k}) = \varepsilon_0 + \sum_{\mathrm{h}} [\varepsilon^{(\mathrm{h})}(\omega - \boldsymbol{u}_{\mathrm{h}} \cdot \boldsymbol{k}, \boldsymbol{k}) - \varepsilon_0] = 0. \tag{27.19a}$$

Here $\varepsilon^{(\mathrm{h})}(\omega_{\mathrm{h}}', \boldsymbol{k})$ designates the longitudinal dielectric permeability of particles of kind h in their own (center of gravity) frame. The relevant expressions have been given in Sects. 17, 18, 22 and 23, see Eqs. (17.13), (17.14) and (23.16) (without summing up over h).

After these preliminary remarks we may now consider specific problems in the following sections.

28. Interaction of an electron beam of small density with high frequency oscillations of a plasma.

α) First of all we may consider the interaction between an electron beam and a plasma in the case where the external *magnetic field vanishes*. For high frequency oscillations the inequalities are certainly valid:

$$\omega' \equiv \omega - \boldsymbol{u} \cdot \boldsymbol{k} \gg k V_{T\mathrm{e}1}, \tag{28.1}$$

$$\omega \gg k V_{T\mathrm{e}0}. \tag{28.2}$$

$\boldsymbol{u}$ being the systematic (average) velocity of the electrons in the beam, $V_{T\mathrm{e}1}$ and $V_{T\mathrm{e}0}$ the average thermal velocity of electrons in the beam, and in the plasma, respectively[1]. Applying the results found in Sects. 17 and 18 we may write the *dispersion relation* Eq. (27.19), in the form

$$\left.\begin{aligned} 1 - \frac{\omega_{N1\mathrm{e}}^2}{(\omega - \boldsymbol{u}\cdot\boldsymbol{k})^2} - \frac{\omega_{N0\mathrm{e}}^2}{\omega^2} + i\sqrt{\frac{\pi}{2}} \left\{ \frac{\omega\,\omega_{N0\mathrm{e}}^2}{k^3 V_{T0\mathrm{e}}^3} \exp\left(\frac{-\omega^2}{2k^2 V_{T0\mathrm{e}}^2}\right) + \right. \\ \left. + \frac{\omega_{N1\mathrm{e}}^2(\omega - \boldsymbol{u}\cdot\boldsymbol{k})}{k^3 V_{T1\mathrm{e}}^3} \exp\left(\frac{-(\omega - \boldsymbol{u}\cdot\boldsymbol{k})^2}{2k^2 V_{T1\mathrm{e}}^2}\right) \right\} = 0. \end{aligned}\right\} \tag{28.3}$$

The imaginary terms in this equation are, of course, dissipative. They are due to ČERENKOV absorption, and to emission of waves by plasma (and beam) electrons.

[1] Index $_1$ designates quantities referring to the beam, index $_0$ those which refer to the plasma at rest.

It follows from the conditions Eqs. (28.1), (28.2) that they are small compared to the first three terms which are non-dissipative. They can be neglected in first approximation. We shall therefore look for a solution of the form

$$\omega = \boldsymbol{u}\cdot\boldsymbol{k} + i\gamma, \quad \text{where} \quad \gamma \ll \omega. \tag{28.4}$$

The result must be presented in different approximations according to the size of $\boldsymbol{u}\cdot\boldsymbol{k}$ as compared with $\omega_{N0\mathrm{e}}$.

For $(\boldsymbol{u}\cdot\boldsymbol{k})^2 \neq \omega_{N0\mathrm{e}}^2$

$$\gamma^2 = -\frac{\omega_{N1\mathrm{e}}^2}{1-\omega_{N0\mathrm{e}}^2/(\boldsymbol{u}\cdot\boldsymbol{k})^2}. \tag{28.5}$$

For $(\boldsymbol{u}\cdot\boldsymbol{k})^2 \approx \omega_{N0\mathrm{e}}^2$ splitting occurs into:

$$\left.\begin{aligned} \gamma_{1,2} &= \frac{-i\pm\sqrt{3}}{2}\left(\frac{N_{1\mathrm{e}}}{2N_{0\mathrm{e}}}\right)^{\frac{1}{3}}\boldsymbol{u}\cdot\boldsymbol{k}, \\ \text{and}\qquad \gamma_3 &= -i\left(\frac{N_{1\mathrm{e}}}{2N_{0\mathrm{e}}}\right)^{\frac{1}{3}}\boldsymbol{u}\cdot\boldsymbol{k}. \end{aligned}\right\} \tag{28.6}$$

β) We can see from these expressions that for

$$(\boldsymbol{u}\cdot\boldsymbol{k})^2 \lesssim \omega_{N0\mathrm{e}}^2 \tag{28.7}$$

the considered oscillations are *unstable*, i.e. $\operatorname{Im}\omega = \operatorname{Re}\gamma > 0$. Thus, if an electron beam of small density penetrates a plasma ($N_{1\mathrm{e}} \ll N_{0\mathrm{e}}$), LANGMUIR oscillations are excited in the plasma. The frequency of these is

$$\operatorname{Re}\omega \approx \omega_{N0\mathrm{e}} \tag{28.8}$$

while the amplitude increase can be characterized by an "oscillation increment"

$$\operatorname{Im}\omega \approx \left(\frac{N_{1\mathrm{e}}}{N_{0\mathrm{e}}}\right)^{\frac{1}{3}}\omega_{N0\mathrm{e}}. \tag{28.9}$$

This instability is, of course, independent from the dissipative processes mentioned above, as the terms corresponding to ČERENKOV absorption and to wave emission by particles have been neglected. We designate it as a *gas-dynamic beam-instability*[2].

γ) Apart from this instability, dissipative (or kinetic) instabilities may also appear. These are either due to ČERENKOV oscillations of electrons in the beam, or to LANGMUIR oscillations of electrons in the plasma at rest. Provided the contribution of the beam can be neglected in the real part of the dielectric permeability, the condition for oscillations is[3]

$$\boldsymbol{u}\cdot\boldsymbol{k} > \omega \approx \omega_{N0\mathrm{e}}. \qquad [28.7]$$

In this case Eq. (28.3) reads (ω be replaced by $\omega + i\gamma$):

$$\omega^2 = \omega_{N0\mathrm{e}}^2, \tag{28.10}$$

$$\left.\begin{aligned} \gamma = -\sqrt{\frac{\pi}{8}}\,\omega_{N0\mathrm{e}}^2\Bigg[&\frac{\omega_{N0\mathrm{e}}^2}{k^3 V_{T0\mathrm{e}}^3}\exp\left(\frac{-\omega_{N0\mathrm{e}}^2}{2k^2 V_{T0\mathrm{e}}^2}\right) + \\ &+\frac{\omega_{N1\mathrm{e}}^2}{k^3 V_{T1\mathrm{e}}^3}\left(1-\frac{\boldsymbol{u}\cdot\boldsymbol{k}}{\omega_{N0\mathrm{e}}}\right)\exp\left(\frac{-(\omega_{N0\mathrm{e}}-\boldsymbol{u}\cdot\boldsymbol{k})^2}{2k^2 V_{T1\mathrm{e}}^2}\right)\Bigg]. \end{aligned}\right\} \tag{28.11}$$

[2] Note, that Eqs. (28.5), (28.6) could also be obtained by starting with the gasdynamic equations of a cold, collisionless plasma, as derived in Chap. B. I, above.

[3] Gasdynamic instabilities are not considered in this subsection.

This shows instability $[\gamma = \mathrm{Im}\,\omega > 0]$ to occur possibly provided $\boldsymbol{u}\cdot\boldsymbol{k} > \omega_{N0e}$. If this condition is satisfied the electrons in the beam emit LANGMUIR waves, and this emission is stronger than the absorption of these waves in the plasma at rest. The consequence of this situation is developing instability in the system. It should be noted that in this case the increment which determines the rise of oscillations is exponentially small. Therefore instabilities develop exponentially slowly in the frequency range under discussion.

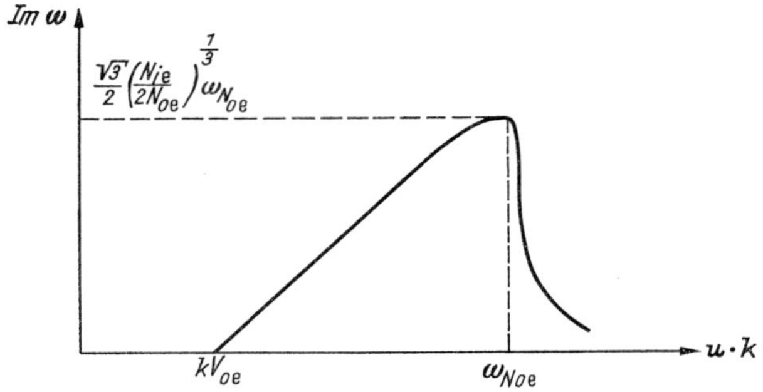

Fig. 8. Generalized dispersion diagram for the beam instability.

Fig. 8 shows the increment describing the rise of beam-instability as a function of the relevant frequency $f = \omega/2\pi$. The increment reaches a maximum at

$$\boldsymbol{k}\cdot\boldsymbol{u} \approx \omega_{N0e}. \qquad [28.8]$$

This is the condition which provokes excitation, and so determines the frequency of the developing waves.

It follows further from the condition Eq. (28.2) that such a high frequency instability can only develop with an electron beam of speed higher than the thermal electron velocity, V_{T0e}; a more detailed computation yields

$$|\boldsymbol{u}| > 2V_{T0e}. \qquad (28.12)$$

There should, of course, occur diffusion of beam electrons as a consequence of thermal motion. If now

$$|\boldsymbol{u}| < V_{T1e}\left(\frac{N_{0e}}{N_{1e}}\right)^{\frac{1}{3}}, \qquad (28.13)$$

only the slower, *kinetic instability* can develop; if, however,

$$|\boldsymbol{u}| > V_{T1e}\left(\frac{N_{0e}}{N_{1e}}\right)^{\frac{1}{3}} \qquad (28.14)$$

then the quicker, *gasdynamic instability* forms.

δ) Qualitatively the description given above for developing high-frequency beam-instabilities remains valid also in the case where a *longitudinal magnetic field* exists[4]. Furtheron, the dispersion equation for waves propagating exactly

[4] Different from earlier chapters, "longitudinal" means here: parallel to the velocity vector $\boldsymbol{u}$ characterizing the beam.

parallel to the magnetic field is identical with Eq. (28.3) which has been discussed above. We must, however, note some pecularities in conjunction with the developing of a beam instability on longitudinal (potential) waves in a magnetized plasma. These are related with the critical condition $k_\perp \neq 0$.

For illustration let us consider the most important case of an electron beam (of low density) in a cold, gyrotropic plasma. When thermal motion in the plasma as well as in the beam is negligible we have the inequalities

$$k_\perp V_{T\mathrm{e}} \ll \omega_{B\mathrm{e}} \tag{28.15}$$

and

$$|\omega - \boldsymbol{k}\cdot\boldsymbol{u} - \omega_{B\mathrm{e}}| \gg k_z V_{T\mathrm{e}}, \tag{28.16}$$

and these are valid in the plasma and in the beam as well. Using now the results of Sect. 22 we may write the dispersion equation, Eq. (27.19) [under the conditions Eqs. (28.15), (28.16)] in the following form:

$$\left.\begin{aligned} 1 - \left[\frac{\omega_{N1\mathrm{e}}^2}{(\omega - \boldsymbol{u}\cdot\boldsymbol{k})^2 - \omega_{B\mathrm{e}}^2} + \frac{\omega_{N0\mathrm{e}}^2}{\omega^2 - \omega_{B\mathrm{e}}^2}\right]\sin^2\Theta \; - \\ - \left[\frac{\omega_{N1\mathrm{e}}^2}{(\omega - \boldsymbol{u}\cdot\boldsymbol{k})^2} + \frac{\omega_{N0\mathrm{e}}^2}{\omega^2}\right]\cos^2\Theta = 0, \end{aligned}\right\} \tag{28.17}$$

i.e.

$$1 - \left[\frac{X_{1\mathrm{e}}}{\left(1 - \dfrac{\boldsymbol{u}\cdot\boldsymbol{k}}{\omega}\right)^2 - Y_\mathrm{e}^2} + \frac{X_{0\mathrm{e}}}{1 - Y_\mathrm{e}^2}\right]\sin^2\Theta - \left[\frac{X_{1\mathrm{e}}}{\left(1 - \dfrac{\boldsymbol{u}\cdot\boldsymbol{k}}{\omega}\right)^2} + X_{0\mathrm{e}}\right]\cos^2\Theta = 0.$$

Θ being the angle between the direction of wave propagation and the magnetic field.

Dissipative effects due to Čerenkov radiation and magneto-Bremsstrahlung have been neglected when deriving Eq. (28.17). Therefore, the considered case is one of strong interaction between beam and plasma. We may call this also "gasdynamic instability" in the system beam plus plasma. As the beam density is supposed to be small, $N_{1\mathrm{e}} \ll N_{0\mathrm{e}}$, we exspect strong interaction in the following two limiting cases:

$$|\omega - \boldsymbol{u}\cdot\boldsymbol{k}| \ll \omega \tag{28.18}$$

or

$$|(\omega - \boldsymbol{u}\cdot\boldsymbol{k})^2 - \omega_{B\mathrm{e}}^2| \ll |\omega^2 - \omega_{B\mathrm{e}}^2|. \tag{28.19}$$

ε) In Subsect. α above we have assumed condition Eq. (28.18), however, in the absence of a magnetic field. We called the instability a "gasdynamic beam-instability". It occurs quite similarly in a magnetized plasma. Only Eqs. (28.5), (28.6) which describe the rise of oscillations giving an increment must be slightly changed. Introducing

$$\omega = \boldsymbol{u}\cdot\boldsymbol{k} + i\gamma$$

into Eq. (28.17) we now get for $A \neq 1$:

$$\gamma^2 = -\frac{\omega_{N1\mathrm{e}}^2 \cos^2\Theta}{1 - A}, \tag{28.20}$$

but for $A \approx 1$ we find splitting:

$$\left.\begin{aligned} \gamma_{1,2} &= \frac{-i \pm \sqrt{3}}{2}\left(\frac{N_{1\mathrm{e}}}{2N_{0\mathrm{e}}}\right)^{\frac{1}{3}} \frac{\boldsymbol{u}\cdot\boldsymbol{k}}{1 - \tan^2\Theta\,(\boldsymbol{u}\cdot\boldsymbol{k})^4/[(\boldsymbol{u}\cdot\boldsymbol{k})^2 - \omega_{B\mathrm{e}}^2]^2}, \\ \gamma_3 &= -i\left(\frac{N_{1\mathrm{e}}}{2N_{0\mathrm{e}}}\right)^{\frac{1}{3}} \frac{\boldsymbol{u}\cdot\boldsymbol{k}}{1 - \tan^2\Theta\,(\boldsymbol{u}\cdot\boldsymbol{k})^4/[(\boldsymbol{u}\cdot\boldsymbol{k})^2 - \omega_{B\mathrm{e}}^2]^2} \end{aligned}\right\} \tag{28.21}$$

where

$$A \equiv \frac{\omega_{N0e}^2}{(\boldsymbol{u}\cdot\boldsymbol{k})^2}\cos^2\Theta + \frac{\omega_{N0e}^2}{(\boldsymbol{u}\cdot\boldsymbol{k})^2 - \omega_{Be}^2}\sin^2\Theta = \frac{\chi_{0e}\cos^2\Theta}{(\boldsymbol{u}\cdot\boldsymbol{k}/\omega)^2} + \frac{\chi_{0e}\sin^2\Theta}{(\boldsymbol{u}\cdot\boldsymbol{k}/\omega)^2 - Y_e^2}. \tag{28.22}$$

Eqs. (28.20), (28.21) coincide with Eqs. (28.5), (28.6) for $\Theta=0$, but also in the trivial limit $\omega_{Be}\to 0$ [i.e. $Y_e \to 0$].

Eqs. (28.20), (28.21) and (28.5), (28.6) have similar structure such that the character of the beam instability is the same with and without a magnetic field. There exists, however, a specific influence of the magnetic field upon oscillations with $k_\perp \neq 0$ in so far as the field has a stabilizing effect. It is easy to see that oscillations are unstable [$\mathrm{Im}\,\omega = \mathrm{Re}\,\gamma > 0$], for $A \gtrsim 1$.

Without a magnetic field this condition is satisfied if

$$(\boldsymbol{u}\cdot\boldsymbol{k})^2 \lesssim \omega_{N0e}^2. \tag*{[28.7]}$$

But in the presence of a magnetic field the unstable range is smaller.

ζ) In a *weakly magnetized plasma* with

$$\omega_{N0e}^2 > \omega_{Be}^2, \quad \text{i.e.} \quad X_{0e} > Y_e^2 \tag{28.23}$$

unstable oscillations occur in the ranges

$$(\boldsymbol{u}\cdot\boldsymbol{k})^2 < \omega_{Be}^2\cos^2\Theta, \quad \text{i.e.} \quad \left(\frac{\boldsymbol{u}\cdot\boldsymbol{k}}{\omega}\right)^2 < Y_e^2\cos^2\Theta, \tag{28.24}$$

and

$$\omega_{Be}^2 < (\boldsymbol{u}\cdot\boldsymbol{k})^2 < (\omega_{N0e}^2 + \omega_{Be}^2\sin^2\Theta), \quad \text{i.e.} \quad Y_e^2 < \left(\frac{\boldsymbol{u}\cdot\boldsymbol{k}}{\omega}\right)^2 < (X_{0e} + Y_e^2\sin^2\Theta) \tag{28.25}$$

only, while in the range

$$\omega_{Be}^2\cos^2\Theta < (\boldsymbol{u}\cdot\boldsymbol{k})^2 < \omega_{Be}^2, \quad \text{i.e,} \quad Y_e^2\cos^2\Theta < \left(\frac{\boldsymbol{u}\boldsymbol{k}}{\omega}\right)^2 < Y_e^2 \tag{28.26}$$

the magnetic field stabilises such that oscillations can not rise up.

η) If, on the other hand in a plasma with a *strong magnetic field*, where

$$\omega_{N0e}^2 < \omega_{Be}^2, \quad \text{i.e.} \quad X_{0e} < Y_e^2 \tag{28.27}$$

we have *instability* in the ranges

$$(\boldsymbol{u}\cdot\boldsymbol{k})^2 < \omega_{N0e}^2\cos^2\Theta, \quad \text{i.e.} \quad \left(\frac{\boldsymbol{u}\cdot\boldsymbol{k}}{\omega}\right)^2 < X_{0e}\cos^2\Theta \tag{28.28}$$

and

$$\omega_{Be}^2 < (\boldsymbol{u}\cdot\boldsymbol{k})^2 < \omega_{Be}^2 + \omega_{N0e}^2\sin^2\Theta, \quad \text{i.e.} \quad Y_e^2 < \left(\frac{\boldsymbol{u}\cdot\boldsymbol{k}}{\omega}\right)^3 < Y_e^2 + X_{0e}\sin^2\Theta, \tag{28.29}$$

but *stabilization* occurs in the range

$$\omega_{N0}^2\cos^2\Theta < (\boldsymbol{u}\cdot\boldsymbol{k})^2 < \omega_{Be}^2, \quad \text{i.e.} \quad \chi_e\cos^2\Theta < \left(\frac{\boldsymbol{u}\cdot\boldsymbol{k}}{\omega}\right)^2 < Y_e^2. \tag{28.30}$$

Fig. 9 shows the increment of rising oscillations as function of the relevant frequency (pulsation $\omega \approx \boldsymbol{u}\cdot\boldsymbol{k}$) for a gyrotropic plasma.

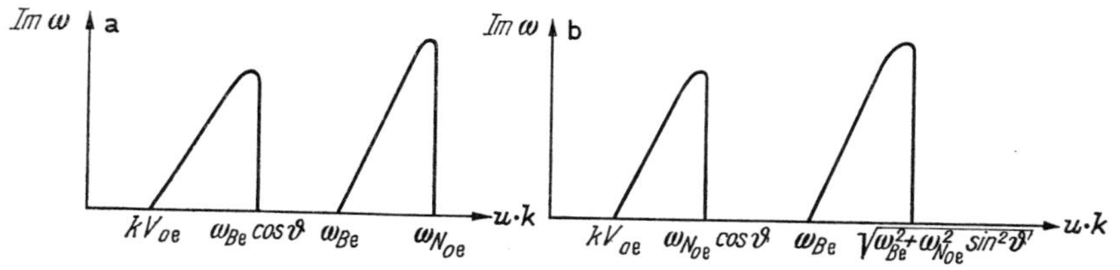

Fig. 9. Generalized dispersion diagram in the case of a strong stabilizing magnetic field.

It should not be overlooked that the inequalities

$$\left.\begin{aligned} |\boldsymbol{u}| &\gg V_{T0\,e} \\ |\boldsymbol{u}| &\gg (N_{0\,e}/N_{1\,e})^{\frac{1}{3}}\, V_{T1\,e} \end{aligned}\right\} \tag{28.31}$$

and

are necessary for the development of high frequency beam instability anyhow, even in the absence of a magnetic field. Eqs. (28.31) derive from the conditions

$$\omega \gg k\, V_{T0\,e} \qquad [28.1]$$

and

$$|\omega - \boldsymbol{u}\cdot\boldsymbol{k}| \gg k_z V_{T1\,e} \qquad [28.2]$$

when supposing plasma and beam to be cold.

ϑ) Let us now consider oscillations in the other frequency range where instability particularly occurs in the presence of a magnetic field; this is the range given by Eq. (28.19) with frequencies near the *"shifted" gyro-frequency,*

$$(\omega_{B\,e} + \boldsymbol{u}\cdot\boldsymbol{k})\,.$$

We take advantage to introduce an imaginary part of the pulsation by putting[5]

$$\omega = \boldsymbol{u}\cdot\boldsymbol{k} + \omega_{B\,e} + i\gamma \tag{28.32}$$

where $\gamma \ll |\omega|$. We then use Eq. (28.17) but under the condition

$$\frac{\omega_{N0\,e}^2}{\omega^2-\omega_{B\,e}^2}\sin^2\Theta + \frac{\omega_{N0\,e}^2}{\omega^2}\cos^2\Theta = \frac{X_{0\,e}}{1-Y_e^2}\sin^2\Theta + X_{0\,e}\cos^2\Theta \approx 1\,, \tag{28.33}$$

and find that in a magnetized plasma a new type of beam instability may appear the increment of which is given by

$$\left.\begin{aligned} \gamma^2 &= -\frac{N_{1\,e}}{N_{0\,e}}\,\frac{(1+\boldsymbol{u}\cdot\boldsymbol{k}/\omega_{B\,e})\sin^2\Theta}{\dfrac{\omega^2}{(\omega^2-\omega_{B\,e}^2)^2}\sin^2\Theta + \dfrac{1}{\omega^2}\cos^2\Theta} \\ \left(\frac{\gamma}{\omega}\right)^2 &= -\frac{N_{1\,e}}{N_{0\,e}}\,(1+\boldsymbol{u}\cdot\boldsymbol{k}/\omega_{B\,e})/[(1-Y_e^2)^{-2}+\cot^2\Theta]\,. \end{aligned}\right\} \tag{28.34}$$

It follows from this relation that instability ($\gamma > 0$) occurs for

$$1 + \boldsymbol{u}\cdot\boldsymbol{k}/\omega_{B\,e} < 0\,. \tag{28.35}$$

The increment describing the rise time of the oscillation is of the order

$$\operatorname{Im}\omega = \gamma \approx \sqrt{\frac{N_{1\,e}}{N_{0\,e}}}\,\omega \sim \sqrt{\frac{N_{1\,e}}{N_{0\,e}}}\,\omega_{B\,e}\,. \tag{28.36}$$

[5] See Sect. 5 for further explanation.

This instability depends upon the magnetic field through ω_{Be}. It is known as *cyclotron instability*, because the effective frequency of the corresponding oscillations is near to the electron gyro- or cyclotron frequency,

$$(\omega - \boldsymbol{u} \cdot \boldsymbol{k}) \approx \omega_{Be}.$$

Both, cyclotron instability and beam instability are resonance phenomena, and develop only under conditions Eqs. (28.31), i.e. if the ordered velocity of the electron beam, $\boldsymbol{u}$, is great enough.

ι) Electromagnetic oscillations and corresponding instabilities in a plasma may also develop in the presence of an *ordered* (beam) *velocity* which is *smaller than the thermal* velocities. There is in particular the case of the kinetic type of instability which is linked with ČERENKOV emission of electromagnetic waves; this type develops if the ordered velocity $|\boldsymbol{u}|$ is greater than the phase velocity of the waves. On the other side we have shown in Sects. 19, 20 and 24 ... 26 that oscillations may develop in a plasma and excite waves with a phase velocity below the thermal velocity of the electrons but above that of the ions. Consequently, the corresponding instability can develop with a beam velocity below the thermal velocity of the electrons. Finally, it has been shown that in a plasma of sufficiently large dimensions beam instabilities may be produced even with an arbitrarily small ordered velocity, well below the thermal velocity of electrons. As the discussion of these questions would go too far the interested reader is refered to special publications, see [*1* ... 4] and [*38*], [*39*] for specific references. Besides, detailed computations can easily be made using the general method explained in Sect. 27.

29. Stability of an anisotropic plasma containing electric currents. This is a second example of the general method for investigating interaction effects as explained in Sect. 27η. Consider a plasma which is exposed to an external electric field $\boldsymbol{E}_0$. The field in the plasma provokes a drift motion of the charged particles only; in the absence of an external magnetic field negative ions (in particular electrons) and positive ones must drift in opposite directions. Due to their smaller q/m ratio the ions must drift with a considerably smaller velocity than the electrons.

We simplify the following considerations by supposing that all electrons of the plasma drift with a uniform ordered velocity $\boldsymbol{u}$ relative to the ions which are supposed to be at rest[1]. Under these conditions we may identify the electrons of the plasma with one electron beam propagating through a grid of ions.

α) We use again the general method indicated in Sect. 27, at first limiting our considerations to *longitudinal* (electric potential) *oscillations*. In order to investigate the stability of such a system we make use of the dispersion equation, Eq. (27.19), in the form

$$1 + \frac{\omega_{Ne}^2}{k^2 V_{Te}^2}\left[1 - \mathrm{J}_+\left(\frac{\omega - \boldsymbol{u}\cdot\boldsymbol{k}}{k V_{Te}}\right)\right] + \frac{\omega_{Ni}^2}{k^2 V_{Ti}^2}\left[1 - \mathrm{J}_+\left(\frac{\omega}{k V_{Ti}}\right)\right] = 0, \tag{29.1}$$

i.e.

$$1 = X_e Q_e^{-2}\left[1 - \mathrm{J}_+\left(Q_e^{-1}\left(1 - \frac{\boldsymbol{u}\cdot\boldsymbol{k}}{\omega}\right)\right)\right] + X_i Q_i^{-2}\left[1 - \mathrm{J}_+(Q_i^{-1})\right] = 0. \tag{29.1a}$$

Of course, for $\boldsymbol{u} = 0$ Eq. (29.1) coincides with the dispersion equation for longitudinal waves in an isotropic plasma, as studied in Sect. 20, see Eq. (20.2). In this particular case there can not be any instability.

First of all we suppose now $\boldsymbol{u} \neq 0$, but yet

$$|\boldsymbol{u}| \ll V_{Te}. \tag{29.2}$$

---

[1] We also neglect the particular phenomenon of electron scattering and diffusion; see [*5*] for this special problem.

It can easily be shown that in this case oscillations can only be unstable in the frequency range

$$\left.\begin{aligned} &k V_{T\,\mathrm{i}} \ll \omega \ll k V_{T\,\mathrm{e}} \\ \text{i.e.} \qquad & \\ &Q_{\mathrm{i}} \ll 1 \ll Q_{\mathrm{e}}, \end{aligned}\right\} \tag{29.3}$$

where the dispersion relation Eq. (29.1) takes the form

$$1+\frac{\omega_{N\,\mathrm{e}}^2}{k^2 V_{T\,\mathrm{e}}^2}-\frac{\omega_{N\,\mathrm{i}}^2}{\omega^2}+i\sqrt{\frac{\pi}{2}}\,\frac{\omega_{N\,\mathrm{e}}^2(\omega-\boldsymbol{u}\cdot\boldsymbol{k})}{k^3 V_{T\,\mathrm{e}}^3}=0 \tag{29.4}$$

i.e.

$$1+X_{\mathrm{e}}Q_{\mathrm{e}}^{-2}-X_{\mathrm{i}}+i\sqrt{\frac{\pi}{2}}\,\frac{X_{\mathrm{e}}}{Q_{\mathrm{e}}^3}\left(1-\frac{\boldsymbol{u}\cdot\boldsymbol{k}}{\omega}\right)=0. \tag{29.4a}$$

The imaginary part being small we put $\omega\to\omega+i\gamma$ and obtain the following spectrum of the oscillations:

$$\text{pulsation:}\quad \omega^2=\frac{\omega_{N\,\mathrm{i}}^2}{1+\omega_{N\,\mathrm{e}}^2/k^2 V_{T\,\mathrm{e}}^2}, \tag{29.5}$$

with the attenuation decrement $-\gamma$:

$$-\gamma=\sqrt{\frac{\pi}{8}}\,\frac{q_{\mathrm{e}}}{q_{\mathrm{i}}}\,\frac{m_{\mathrm{i}}}{m_{\mathrm{e}}}\,\frac{\omega^4}{k^3 V_{T\,\mathrm{e}}^3}\left(1-\frac{\boldsymbol{k}\cdot\boldsymbol{u}}{\omega}\right). \tag{29.6}$$

Since $V_{\mathrm{ph}}=\omega/k$ is the phase velocity of the developing waves one could also write $(1-\cos\vartheta\, u/V_{\mathrm{ph}})$ for the parenthesis in Eq. (29.6). ϑ being the angle[2] between the ordered velocity $\boldsymbol{u}$ of the electrons and the wave normal $\boldsymbol{k}$, we have

$$\cos\vartheta=\frac{\boldsymbol{k}\cdot\boldsymbol{u}}{k\,u}. \tag{29.7}$$

For $\boldsymbol{u}=0$ the spectrum coincides with that of ion sound waves in an isotropic, non-isothermal plasma. As can be seen from Eq. (29.6) an electron velocity $\boldsymbol{u}$ different from zero makes the attenuation decrement γ increase provided $\cos\vartheta<0$, i.e. in cases where the projection of the wave vector upon the velocity direction has direction opposite to the latter. For $\cos\vartheta>0$, i.e. when the projection of $\boldsymbol{k}$ has the same direction as $\boldsymbol{u}$, the decrement can either be positive or negative. It changes sign at $u=V_{\mathrm{ph}}/\cos\vartheta$, so that the plasma is *instable* for

$$u>V_{\mathrm{ph}}\sec\vartheta. \tag{29.8}$$

The instability is due to ČERENKOV emission of ion sound waves by the electrons which move through the grid of ions. It can only occur in a non-isothermal plasma[3], i.e. for

$$T_{\mathrm{e}}\gg T_{\mathrm{i}}\quad\text{and}\quad u>V_{\mathrm{ph}}\gg V_{T\,\mathrm{i}}. \tag{29.9}$$

β) Apart from an ion sound instability a *gas-dynamic beam instability* may also develop in a plasma, provided the electron velocity exceeds the thermal velocity of the electrons. There is no dissipation of energy occuring in this context. In fact under the conditions

$$u>V_{T\,\mathrm{e}}\quad\text{and}\quad\omega\gg k V_{T\,\mathrm{e}} \tag{29.10}$$

we obtain from Eq. (29.1)

$$1-\frac{\omega_{N\mathrm{e}}^2}{(\omega-\boldsymbol{u}\cdot\boldsymbol{k})^2}-\frac{\omega_{N\,\mathrm{i}}^2}{\omega^2}\equiv 1-\frac{X_{\mathrm{e}}}{(1-\boldsymbol{u}\cdot\boldsymbol{k}/\omega)^2}-X_{\mathrm{i}}=0. \tag{29.11}$$

---

[2] Not to be confounded with ϑ_{h} (h = e, i, n...) which is a constituent temperature energy throughout this contribution ($\vartheta_{\mathrm{h}}\equiv k T_{\mathrm{h}}$).

[3] Note that JOULE heating increases preferentially the electron temperature in the plasma, so that this latter almost always happens to be non-isotherm see [1], § 38.

The solutions of this equation corresponding to unstable oscillations in case $\omega \ll \boldsymbol{u} \cdot \boldsymbol{k}$, are the following:

$$\omega^2 = \frac{\omega_{N\mathrm{i}}^2}{1 - \omega_{N\mathrm{e}}^2/(\boldsymbol{u} \cdot \boldsymbol{k})^2} \quad \text{for } (\boldsymbol{u} \cdot \boldsymbol{k})^2 \neq \omega_{N\mathrm{e}}^2 \tag{29.12}$$

but splitting occurs into

$$\omega_{1,2} = \frac{1 \pm i\sqrt{3}}{2} \left(\frac{q_\mathrm{i} m_\mathrm{e}}{2 q_\mathrm{e} m_\mathrm{i}}\right)^{\frac{1}{3}} \boldsymbol{u} \cdot \boldsymbol{k} \tag{29.13}$$

and

$$\omega_3 = \left(\frac{q_\mathrm{i} m_\mathrm{e}}{2 q_\mathrm{e} m_\mathrm{i}}\right)^{\frac{1}{3}} \boldsymbol{u} \cdot \boldsymbol{k} \tag{29.14}$$

for $(\boldsymbol{u} \cdot \boldsymbol{k})^2 \approx \omega_{N\mathrm{e}}^2$.

Evidently, solutions with $\mathrm{Im}\,\omega > 0$ exist always in the frequency range

$$(\boldsymbol{u} \cdot \boldsymbol{k})^2 \lesssim \omega_{N\mathrm{e}}^2 \tag{29.15}$$

these are unstable plasma conditions and oscillations may rise aperiodically according to the increment

$$\gamma = \mathrm{Im}\,\omega \approx \left(\frac{m_\mathrm{e}}{m_\mathrm{i}}\right)^{\frac{1}{3}} \omega_{N\mathrm{e}}. \tag{29.16}$$

This beam instability of a plasma in an external electric field has first been theoretically nvestigated by BUNEMANN [42]; it is therefore designated after this author.

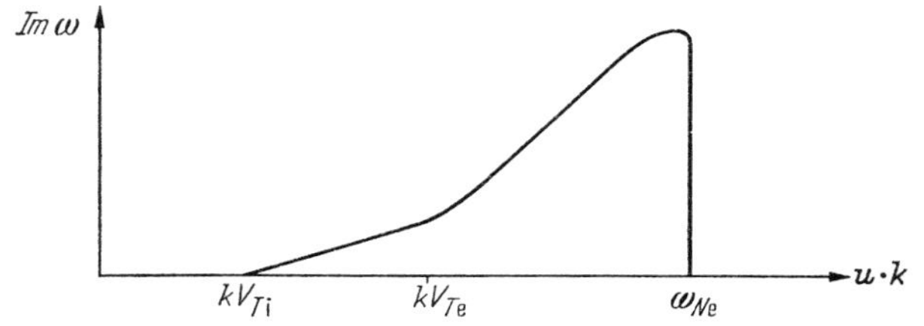

Fig. 10. Generalized dispersion diagram for the BUNEMANN-instability.

Fig. 10 shows how the increment depends on $(\boldsymbol{u} \cdot \boldsymbol{k})$ for the BUNEMANN and for the ion sound instability. The *spectrum* is such that in the range

$$\left.\begin{aligned} k V_{T\mathrm{i}} &< |\boldsymbol{u} \cdot \boldsymbol{k}| < k V_{T\mathrm{e}} \\ \text{i.e.} \quad Q_\mathrm{i} &< \left|\frac{\boldsymbol{u} \cdot \boldsymbol{k}}{\omega}\right| < Q_\mathrm{e} \end{aligned}\right\} \tag{29.17}$$

only the ion sound instability could develop in a non-isothermal plasma, while in the range

$$\left.\begin{aligned} k V_{T\mathrm{e}} &< |\boldsymbol{u} \cdot \boldsymbol{k}| < \omega_{N\mathrm{e}} \\ \text{i.e.} \quad Q_\mathrm{e} &< \left|\frac{\boldsymbol{u} \cdot \boldsymbol{k}}{\omega}\right| < X_\mathrm{e}^{\frac{1}{2}} \end{aligned}\right\} \tag{29.18}$$

the BUNEMANN instability may also exist. If so, then it has a much larger increment and oscillations are quickly increasing.

γ) We now admit an additional *external magnetic field* the direction of which shall be *parallel* to that of the electric field (which should be along the z-axis).

We shall see how this field influences the instabilities discussed above in Subsects. α and β. In the presence of a magnetic field we must use the dispersion equation, Eq. (27.19), for *longitudinal oscillations*, i.e. due to an electric potential. Considering Eq. (23.16) we may write it in the following form:

$$\left.\begin{aligned} &1+\frac{\omega_{Ne}^2}{k^2 V_{Te}^2}\left[1-\sum_s \frac{\omega-\boldsymbol{u}\cdot\boldsymbol{k}}{\omega-\boldsymbol{u}\cdot\boldsymbol{k}-s\omega_{Be}}\mathscr{A}_s\left(\frac{k_\perp^2 V_{Te}^2}{\omega_{Be}^2}\right)\mathrm{J}_+\left(\frac{\omega-\boldsymbol{u}\cdot\boldsymbol{k}-s\omega_{Be}}{|k_z|\,V_{Te}}\right)\right]+\\ &+\frac{\omega_{Ni}^2}{k^2 V_{Ti}^2}\left[1-\sum_s \frac{\omega}{\omega-\omega_{Bi}}\mathscr{A}_s\left(\frac{k_\perp^2 V_{Ti}^2}{\omega_{Bi}^2}\right)\mathrm{J}_+\left(\frac{\omega-s\omega_{Bi}}{|k_z|\,V_{Ti}}\right)\right]=0,\end{aligned}\right\} \tag{29.19}$$

i.e.

$$\left.\begin{aligned} &1+X_e Q_e^{-2}\left[1-\sum_s \frac{1-(\boldsymbol{u}\cdot\boldsymbol{k}/\omega)}{1-(\boldsymbol{u}\cdot\boldsymbol{k}/\omega)-sY_e}\mathscr{A}_s(Q_{e\perp}^2/Y_e^2)\,\mathrm{J}_+\left(\frac{1-(\boldsymbol{u}\cdot\boldsymbol{k}/\omega)-sY_e}{Q_{e\perp}}\right)\right]+\\ &+X_i Q_i^{-2}\left[1-\sum_s \frac{1}{1-Y_i}\mathscr{A}_s(Q_{i\perp}^2/Y_i^2)\,\mathrm{J}_+\left(\frac{1-sY_i}{Q_{i\perp}}\right)\right]=0\end{aligned}\right\} \tag{29.19a}$$

Note first of all that for

$$|\omega-\boldsymbol{u}\cdot\boldsymbol{k}|\gg\omega_{Be} \quad\text{and}\quad \omega\gg\omega_{Bi}, \tag{29.20}$$

i.e. for $\boldsymbol{B}_{\mathfrak{z}}\to 0$, Eq. (29.19) becomes identic with Eq. (29.1) which has been found without a magnetic field. In this limit all results which have been obtained above remain valid. This is also so for $\boldsymbol{k}_\perp=0$ because in this case Eq. (29.19) also coincides with Eq. (29.1) as found without a magnetic field.

Thus we exspect an influence of the magnetic field on a plasma instability due to an external electric field not everywhere but only at low frequencies, namely for

$$|\omega-\boldsymbol{u}\cdot\boldsymbol{k}|\ll\omega_{Be} \quad\text{and}\quad k_\perp\neq 0. \tag{29.21}$$

For simplicity we shall also suppose that

$$k_\perp^2 V_{Ti}^2\ll\omega_{Bi}^2, \quad\text{i.e.}\quad Q_i^2\ll Y_i^2. \tag{29.22}$$

Thus, we consider the transverse wavelength of the oscillations as to be large against the gyro- (or LARMOR) radius of the ions.

Let us begin with the case $|\boldsymbol{u}|<V_{Te}$. One shows that in this case only oscillations in the frequency range

$$\left.\begin{aligned} &k_z V_{Ti}\ll\omega\ll k_z V_{Te}\\ \text{i.e.}\qquad &Q_{iz}\ll 1\ll Q_{ez}\end{aligned}\right\} \tag{29.23}$$

could be instable. In this frequency range Eq. (29.19) takes the form

$$\left.\begin{aligned} &1+\frac{\omega_{Ne}^2}{k^2 V_{Te}^2}\left[1+i\sqrt{\frac{\pi}{2}}\,\frac{\omega-\boldsymbol{u}\cdot\boldsymbol{k}}{|k_z|\,V_{Te}}\right]-\frac{k_\perp^2}{k^2}\,\frac{\omega_{Ni}^2}{\omega^2-\omega_{Bi}^2}-\frac{k_z^2}{k^2}\,\frac{\omega_{Ni}^2}{\omega^2}=0\\ \text{i.e.}\qquad &1+X_e Q_e^{-2}\left[1+i\sqrt{\frac{\pi}{2}}\,Q_{ez}^{-1}(1-\boldsymbol{u}\cdot\boldsymbol{k}/\omega)\right]-\frac{k_\perp^2}{k^2}\,\frac{X_i}{1-Y_i^2}-\frac{k_z^2}{k^2}X_i=0.\end{aligned}\right\} \tag{29.24}$$

The dissipative (imaginary) part is quite small here; it is due to the ČERENKOV effect upon the electrons. Putting $\omega\to\omega+i\gamma$ we obtain the following spectrum of oscillations:

pulsation:

$$\omega^2=\frac{1}{2}\left\{\omega_{Bi}^2+\frac{\omega_{Ni}^2}{1+\omega_{Ne}^2/k^2 V_{Te}^2}\pm\sqrt{\left(\omega_{Bi}^2+\frac{\omega_{Ni}^2}{1+\dfrac{\omega_{Ne}^2}{k^2 V_{Te}^2}}\right)^2-\frac{4\omega_B^2\,\omega_N^2\cos^2\vartheta}{1+\dfrac{\omega_{Ne}^2}{k^2 V_{Te}^2}}}\right\} \tag{29.25}$$

attenuation decrement:

$$\gamma = -\sqrt{\frac{\pi}{8}}\, \frac{q_e m_i}{q_i m_e}\, \frac{\omega^4}{k^3 V_{Te}^3 |\cos^3\vartheta|}\, \frac{1-u\cos\vartheta/V_{ph}}{\cos^2\vartheta + \dfrac{\omega^4\sin^2\vartheta}{(\omega^2-\omega_{Be}^2)^2}}. \tag{29.26}$$

Here again $V_{ph} = \omega/k$ is the phase velocity and

$$\cos\vartheta = \frac{\boldsymbol{k}\cdot\boldsymbol{u}}{k\,u} = \frac{k_z}{k}. \qquad [29.7]$$

Under the particular condition $\boldsymbol{u} = 0$ the spectrum, of course, coincides with that of decreasing ($\gamma < 0$) ion sound waves in a non-isothermal ($T_e \gg T_i$) magnetized plasma. Two branches of Eq. (29.25) describe oscillations in the frequency ranges $\omega > \omega_{Bi}$ and $\omega < \omega_{Bi}$, respectively, see Sect. 26. With increasing $\boldsymbol{u}$ the decrement becomes smaller so that the oscillations decrease more slowly, and for

$$u > \sec\vartheta\, V_{ph} \gg V_{Ti} \tag{29.27}$$

the oscillations begin to increase. We then have $\gamma > 0$, i.e. instability. The condition for increasing amplitude of the ion sound waves is the same as in the absence of a magnetic field. Moreover the increment describing the increase after Eq. (9.26) is of the same order as that for a plasma without a magnetic field, see Eq. (29.6). Thus there is no remarkable influence of the magnetic field upon the ion sound instability of a plasma supporting a current, at least for $u < V_{Te}$.

δ) The situation is different for the *gasdynamic beam instability* which develops when

$$u > V_{Te}. \tag{29.28}$$

Such instability may develop in the frequency range

$$\omega \gg k_z V_{Te}, \quad \text{i.e.} \quad Q_{ez} \ll 1 \tag{29.29}$$

under the condition

$$|\omega - \boldsymbol{u}\cdot\boldsymbol{k}| \ll \omega_{Be}, \quad \text{i.e.} \quad \left|1 - \frac{\boldsymbol{u}\cdot\boldsymbol{k}}{\omega}\right| \ll Y_e. \qquad [29.20]$$

The dispersion equation Eq. (29.19) must be written as

$$1 - \frac{\omega_{Ne}^2\cos^2\vartheta}{(\omega-\boldsymbol{u}\cdot\boldsymbol{k})^2} - \frac{\omega_{Ni}^2\cos^2\vartheta}{\omega^2} - \frac{\omega_{Ni}^2\sin^2\vartheta}{\omega^2-\omega_{Bi}^2} = 0, \tag{29.30}$$

i.e.

$$1 - X_e\frac{\cos^2\vartheta}{(1-\boldsymbol{u}\cdot\boldsymbol{k}/\omega)^2} - X_i\cos^2\vartheta - X_i\frac{\sin^2\vartheta}{1-Y_i^2} = 0. \tag{29.30a}$$

Analyzing this equation for $\omega \ll \omega_{Bi}$ and for $\omega \gg \omega_{Bi}$ we find that unstable oscillations can only occur when the condition

$$\omega \ll \boldsymbol{u}\cdot\boldsymbol{k} \tag{29.31}$$

is satisfied. One then has the spectrum:

$$\omega^2 = \frac{\omega_{Ni}^2}{a - \eta\dfrac{\omega_{Ne}^2}{(\boldsymbol{u}\cdot\boldsymbol{k})^2}} \quad \text{for } (\boldsymbol{u}\cdot\boldsymbol{k})\, a \mp \eta\,\omega_{Ne}^2 \tag{29.32}$$

but splitting into

$$\omega_{1,2} = \frac{1 \pm i\sqrt{3}}{2}\left(\frac{\eta}{2}\,\frac{q_i m_e}{q_e m_i}\right)^{\frac{1}{3}} \boldsymbol{u}\cdot\boldsymbol{k} \tag{29.33}$$

and

$$\omega_3 = \left(\frac{\eta}{2}\,\frac{q_i m_e}{q_e m_i}\right)^{\frac{1}{3}} \boldsymbol{u}\cdot\boldsymbol{k} \tag{29.34}$$

for (29.33) and (29.34): $(\boldsymbol{u}\cdot\boldsymbol{k})\, a \approx \eta\,\omega_{Ne}^2$, with

$$\eta = \begin{cases} 1 & \text{for } \omega \ll \omega_{Bi} \\ \cos^2 \vartheta & \text{for } \omega \gg \omega_{Bi} \end{cases} \tag{29.35}$$

and

$$a = \begin{cases} \sec^2 \vartheta \left(1 + \dfrac{\omega_{Ni}^2}{\omega_{Bi}^2} \sin^2 \vartheta\right) & \text{for } \omega \ll \omega_{Bi} \\ 1 & \text{for } \omega \gg \omega_{Bi}. \end{cases} \tag{29.36}$$

These expressions show some analogy with Eqs. (29.12) ... (29.14). Like in the absence of a magnetic field low frequency oscillations could be unstable if

$$(\boldsymbol{u} \cdot \boldsymbol{k})^2 \, a \lesssim \eta \omega_{Ne}^2. \tag{29.37}$$

As, however, $\eta \leqq 1$ and $a \geqq 1$ we come to the conclusion that, generally speaking, the effect of a magnetic field, consists in a limitation of the range of instability. Thus the magnetic field, again, has a stabilizing effect. Consequently, in frequency ranges where the plasma remains unstable the magnetic field at least decreases the increment of developing oscillations which becomes now

$$\operatorname{Im} \omega = \gamma \lesssim \left(\frac{\eta}{2} \frac{q_i m_e}{q_e m_i}\right)^{\frac{1}{3}} \boldsymbol{u} \cdot \boldsymbol{k}. \tag{29.38}$$

The stabilizing effect of the magnetic field is only important in the frequency range

$$|\omega - \boldsymbol{u} \cdot \boldsymbol{k}| \ll \omega_{Be}, \quad \text{i.e.} \quad \left|1 - \frac{\boldsymbol{u} \cdot \boldsymbol{k}}{\omega}\right| \ll Y_e \qquad [29.20]$$

and for $k_z = 0$.

The effect is therefore only important in the presence of a rather strong field, i.e. when

$$\omega_{Be} > \omega_{Ne}, \quad \text{i.e.} \quad Y_e > X_e^{\frac{1}{2}}. \tag{29.39}$$

Fig. 11 shows the stabilizing effect of a magnetic field upon the instability of a plasma under an external electric field. Curve 1, valid for $k_z = 0$, shows no influence, but there is one visible in curve 2 for $k_z \neq 0$ and $\omega_{Be} > \omega_{Ne}$.

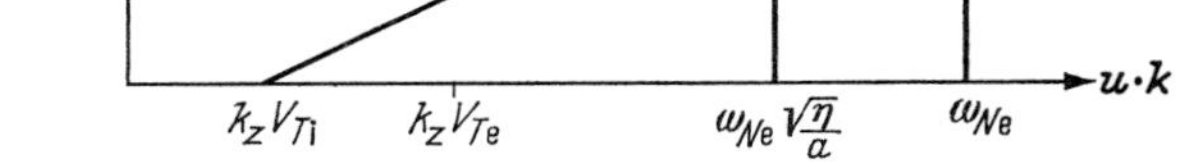

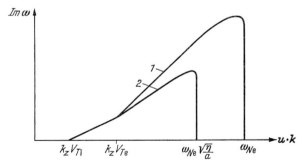

Fig. 11. Generalized dispersion diagram for supersonic velocity of the beam and with magnetic stabilization.

ε) Finally, it must be noted that in Sects. 28 and 29 only examples of instable oscillations in plasmas with anisotropic statistical distribution function have been considered. Our considerations are far from covering the whole theory of possible interactions of beams of charged particles with a plasma. Studies are now in progress concerning such general theory with applications to the explanation of *strong perturbation phenomena* like radio emission from planets, ionospheric storms, and, perhaps, magnetospheric irregularities. It can be hoped that the general method indicated in Sect. 27 for beam-plasma interactions neglecting collisions may be helpful in this context.

E. Oscillations and waves in inhomogeneous plasmas.

I. Generalities and high frequency oscillations.

In Chaps. A through D we considered homogeneous plasmas only. Real conditions in geophysics and space physics are, however, such, that, strictly speaking, only inhomogeneous plasmas occur. The relevant dimensions depend on the region which is under consideration. To begin with the magnetosphere its characteristic scale is of the order of the radius of Earth, RE, thus $\mathscr{L}_B = 10.000$ km $= 10^7$ m, and this is, of course, also the scale for the magnetic field of Earth. The temperature changes in ionosphere and exosphere have a scale of $\mathscr{L}_T =$ 500 km $= 5 \cdot 10^5$ m. The horizontal scale of ionization variations in the ionosphere is about the same, but the vertical scale of ionization is certainly smaller, $\mathscr{L}_N = 100$ km $= 10^5$ m. Finally there are more localized irregularities which are due to different causes. One could take for these a scale of about a few percent of the overall scale, thus say 10^5 m in the magnetosphere, but only 10^3 m in the ionosphere*.

30. The approximation of geometrical optics.

α) In an inhomogeneous plasma the parameters describing the behaviour of matter depend on the space coordinates, characterized by the radius vector $\boldsymbol{r}$. We are used to have a dispersion effect of matter at propagation as a function of distance strongly decreasing in importance with the distance $(\boldsymbol{r}-\boldsymbol{r}')$. In an inhomogeneous plasma, however, the material parameters are variable and no more identic at $\boldsymbol{r}$ and $\boldsymbol{r}'$. Therefore the arguments of the characteristic tensor functions, S and E, appearing in the general material equation, Eq. (1.8), are no more simple functions of the distance $(\boldsymbol{r}-\boldsymbol{r}')$ but depend on $\boldsymbol{r}$ and $\boldsymbol{r}'$, individually. Thus, the notion of a dielectricity tensor as defined in Eq. (1.10) becomes questionable for an inhomogeneous plasma. Due to this circumstance an adequate theory for the propagation of electromagnetic waves in an inhomogeneous plasma can not easily be formulated.

Of course, one obtains a simplification of this general problem when studying oscillations corresponding to wavelengths which are considerably smaller than the scale of inhomogeneity in the medium. In these cases the approximation of geometrical optics is very helpful [7][1]. The applicability of this simple approximation depends on the parameter $\lambda/\mathscr{L}_0$, where λ is the wavelength relevant to the oscillation considered (and in the direction of the material gradient) while $\mathscr{L}_0$ is the scale of inhomogeneity of the medium. The condition for applying geometrical optics is

$$\lambda \ll \mathscr{L}_0. \tag{30.1}$$

This is the condition which will be adopted through most of this chapter. If it is satisfied the plasma is nearly homogeneous, but we do better to speak of a "weakly inhomogeneous" plasma under these circumstances.

---

* See K. RAWER and K. SUCHY: this Encyclopedia, vol. 49/2, Sect. 49, p. 400. Numerical estimates for the different ionospheric layers are given there.

[1] The spatial dispersion can even be neglected rather often, namely in all cases where the argument of the tensor-functions S and E can be replaced by a delta function, i.e. if it can be "localized".

When investigating the properties of such a weakly inhomogeneous plasma we better write the material equation, Eq. (1.8), in the form[2]

$$\boldsymbol{D}(\boldsymbol{r},t)=\int_{-\infty}^{t}\mathrm{d}t'\int\mathrm{d}^3\boldsymbol{r}'\,\hat{\mathsf{E}}(t-t',\boldsymbol{r},\boldsymbol{r}-\boldsymbol{r}')\cdot\boldsymbol{E}(\boldsymbol{r}',t'),\tag{30.2}$$

in components:

$$D_l(\boldsymbol{r},t)=\int_{-\infty}^{t}\mathrm{d}t'\int\mathrm{d}^3r'\,\hat{\varepsilon}_{lm}(t-t',\boldsymbol{r},\boldsymbol{r}-\boldsymbol{r}')\,E_m(\boldsymbol{r}',t').\tag{30.2a}$$

The assumed weakness of the inhomogeneity means that the "Kernel" of the integral in Eq. (30.2) is stronger dependent upon $(\boldsymbol{r}-\boldsymbol{r}')$ than on $\boldsymbol{r}$. The reason is that the dependence from $(\boldsymbol{r}-\boldsymbol{r}')$ varies with fractions of the wavelength of the oscillation while the dependence from $\boldsymbol{r}$ goes with the scale of inhomogeneity, $\mathscr{L}_0$.

β) As usually done when applying geometrical optics we use a generalized wave function for the electromagnetic field in the medium

$$\boldsymbol{E}\sim\boldsymbol{E}(\omega)\exp\left(-i\omega t+i\Psi(\boldsymbol{r})\right),\tag{30.3}$$

$\Psi(\boldsymbol{r})$ being the *eiconal*. In a homogeneous plasma we had

$$\Psi(\boldsymbol{r})=\boldsymbol{k}\cdot\boldsymbol{r}$$

and the wave vector $\boldsymbol{k}$ was constant, i.e. independent from the coordinates. Then by substituting into the dispersion relation, Eq. (4.3), $\boldsymbol{k}$ can be determined as a function of the plasma parameters and frequency. In analogy with this procedure we can now at least define a space function $\boldsymbol{k}(\boldsymbol{r})$ by writing

$$\boldsymbol{k}(\boldsymbol{r})=\frac{\partial}{\partial\boldsymbol{r}}\Psi(\boldsymbol{r})\equiv\nabla\Psi(\boldsymbol{r}).\tag{30.4}$$

It is evident that in a weakly ionized medium $\boldsymbol{k}(\boldsymbol{r})$ is a "weakly variable function". Since the variations are caused by the variations of the characteristic properties of the medium in space we can suppose that $\boldsymbol{k}(\boldsymbol{r})$ and the characteristic material parameters have the same scale of inhomogeneity, $\mathscr{L}_0$. We then call $\boldsymbol{k}(\boldsymbol{r})$ the *wave vector* and

$$\lambda(\boldsymbol{r})=\frac{2\pi}{k(\boldsymbol{r})}\tag{30.5}$$

the wavelength of the wave with wave vector $\boldsymbol{k}(\boldsymbol{r})$.

γ) If the wave vector is being "weakly variable" with the coordinates we can proceed to construct solutions in the form of *successive approximations* after the inhomogeneity parameter $\lambda/\mathscr{L}_0$, Eq. (30.1). The method is as follows: In the zero order approximation we take the medium as homogeneous, i.e. we neglect all spatial derivatives of its characteristics, even gradients. In the first order approximation first order derivatives (gradients) are admitted; in the second order then account is taken of second order derivatives, etc.[3]. With this procedure solutions for electrodynamic problems of any kind can be approximated in successive steps; in almost all cases any desired degree of accuracy can be reached in this manner.

Below we give solutions for the electromagnetic field "in the absence of external sources" which means that we have to discuss propagation of such waves in the absence of excitation

[2] No variation with *time* is considered for the properties of the medium. The plasma is supposed to be inhomogeneous in space only. Variations with time and space are important in connection with scatter phenomena. These are not considered here.

[3] Important note: When determining the eiconal $\boldsymbol{Y}(\boldsymbol{r})$ by computing the integral of Eq. (30.4) the inhomogeneity of the wave vector, $\boldsymbol{k}(\boldsymbol{r})$, can not entirely be neglected, even for zero order calculation. The reason is, that the range of integration is so large here that it could easily be greater than the scale of inhomogeneity.

form outside. Another description of this task can be given by saying that the electromagnetic eigenoscillations of the medium are looked for.

δ) Substituting the wave field Eq. (30.3) into the general equations of electrodynamics, Eqs. (1.6), and neglecting gradients in zero order approximation we just get the approximation of geometric optics:

$$\left.\begin{aligned} \boldsymbol{k}\times\boldsymbol{B} &= -\frac{\omega}{c_0}\boldsymbol{D}, & \boldsymbol{k}\cdot\boldsymbol{D} &= 0;\\ \boldsymbol{k}\times\boldsymbol{E} &= \frac{\omega}{c_0}\boldsymbol{B}, & \boldsymbol{k}\cdot\boldsymbol{B} &= 0. \end{aligned}\right\} \tag{30.6}$$

Substituting again into the "material equation" (i.e. that characterizing the medium) we get

$$\boldsymbol{D} = \mathsf{E}(\omega, \boldsymbol{k}, \boldsymbol{r}) \cdot \mathsf{E} \exp\left(-i\omega t + i\Psi(\boldsymbol{r})\right) \tag{30.7}$$

where the *effective dielectricity tensor* E is obtained from that defined by Eq. (1.10) as the space and time average of the FOURIER transform

$$\mathsf{E}(\omega, \boldsymbol{k}, \boldsymbol{r}) = \int_0^\infty \mathrm{d}t \int \mathrm{d}^3 R\, \hat{\mathsf{E}}(t, \boldsymbol{r}, \boldsymbol{R}) \exp(i\omega t - i\boldsymbol{k}\cdot\boldsymbol{R}). \tag{30.8}$$

Eq. (30.8) is of course the analog to Eq. (1.3). It gives the equivalent expression for the tensor of dielectric permeability in a weakly ionized medium.

The system Eqs. (30.6), (30.7) can only be resolved if the determinant of the system vanishes, i.e. if the following condition holds

$$\left| k^2 \mathsf{U} - \boldsymbol{k}\boldsymbol{k} - \frac{\omega^2}{c_0^2}\mathsf{E}(\omega, \boldsymbol{k}, \boldsymbol{r}) \right| = 0 \tag{30.9}$$

in components:

$$\left| k^2 \delta_{lm} - k_l k_m - \frac{\omega^2}{c_0^2}\varepsilon_{lm}(\omega, \boldsymbol{k}, \boldsymbol{r}) \right| = 0. \tag{30.9a}$$

In all other cases the field supposed in Eq. (30.3) is not a solution of the general equations of electrodynamics. Eq. (30.3) is, of course, analog to the general dispersion relation, Eq. (4.3). There is, however, a principal difference insofar as Eq. (30.9) is not really a dispersion equation from which the spectrum of possible eigenoscillations $\omega(\boldsymbol{k})$ could be computed. Trying to resolve Eq. (30.9) after ω in an inhomogeneous medium we get a function $\omega(\boldsymbol{k}, \boldsymbol{r})$ with two arguments, thus dependent on the space variable. This is, evidently, not a description of eigenoscillations of the plasma as whole which we are now looking for. The effective spectrum of eigenoscillations of the plasma as whole should in fact not be dependent upon the space coordinates. Thus, Eq. (30.9) can not exactly be considered as to be a dispersion relation.

ε) Eq. (30.9) is analyzed in connection with *two different problems*. One of these is the determination of the *distribution of fields in space*; it will be considered in Sect. 31 in more detail. In this context one takes ω as a real given quantity and tries to determine the eiconal function, $\Psi(\boldsymbol{r})$, in space. In this case Eq. (30.9) is used as eiconal equation. This is a problem of the kind currently found in propagation theory of radio waves, and, quite generally, in geometric optics. However, in these theories one uses to neglect the spatial dispersion of the dielectric properties of the medium by putting $\mathsf{E} = \mathsf{E}(\omega, \boldsymbol{r})$ only. Eq. (30.9) also allows a generalization of the simple eiconal equation of geometrical optics by taking account of spatial dispersion when computing an effective dielectric tensor E which depends upon $\boldsymbol{k}$ instead of ω [*43*].

In Sect. 31 below we intend to discuss in more detail problems of the first kind where the eiconal $\Psi(\boldsymbol{r})$ is to be determined, for a given frequency $\omega/2\pi$[4].

[4] Neglecting spatial dispersion such problems related with the theory of radio wave propagation in plasmas are discussed in [*1*], [*5*], [*36*]. Besides, spatial dispersion does not seriously influence these methods.

As to problems of the second kind, one is, essentially, asking for the spectrum of eigenoscillations of the (inhomogeneous) medium supposing different conditions at the limits. The most important conditions are, of course, those given by nature itself, for example "trapped" oscillations in a plasma region, see Sect. 31.

Apart from the full Eq. (30.9) we shall often limit our considerations to the much simpler scalar equation ($\boldsymbol{k}^0 = \boldsymbol{k}/k$)

$$\varepsilon(\omega, \boldsymbol{k}, \boldsymbol{r}) \equiv \boldsymbol{k}^0 \cdot \mathsf{E}(\omega, \boldsymbol{k}, \boldsymbol{r}) \cdot \boldsymbol{k}^0 = 0 \tag{30.10}$$

which reads in components:

$$\varepsilon(\omega, \boldsymbol{k}, \boldsymbol{r}) \equiv \frac{k_l k_m}{k^2} \varepsilon_{lm}(\omega, \boldsymbol{k}, \boldsymbol{r}) = 0. \tag{30.10a}$$

This equation describes the space distribution of the electric potential field $\boldsymbol{E}$, submitted to the condition of vanishing rotational:

thus:

$$\left.\begin{aligned} \nabla \times \boldsymbol{E} &\equiv \frac{\partial}{\partial \boldsymbol{r}} \times \boldsymbol{E} = 0, \\ \boldsymbol{E} &= -\nabla \phi \equiv -\frac{\partial}{\partial \boldsymbol{r}} \phi. \end{aligned}\right\} \tag{30.11}$$

ϕ being the electric potential of the field $\boldsymbol{E}$.

In this context Eq. (30.10) is often called the eiconal equation for longitudinal waves, i.e. waves deriving from a potential ϕ which is variable along the propagation path. We ask now how for a spatially inhomogeneous plasma the spectrum of eigenfrequencies ω_j could be determined. This orientation has only recently been adopted in systematic studies of radio wave propagation in inhomogeneous plasmas [*43*]; it is related with the stability problem of magnetic confinement, see Sect. 37. Of course, we can only summarize the advance of knowledge which has recently been made in this field.

ζ) Let us now consider as an example a (one-dimensionally) inhomogeneous, otherwise isotropic, collisionless plasma[5]. Our problem is the determination of $\mathsf{E}(\omega, \boldsymbol{k}, \boldsymbol{r})$.

Aiming to give a more principal than numerical description in view of magnetospheric or ionospheric plasmas we may be allowed to simplify the problem by taking one axis, $\boldsymbol{x}^0$, of our cartesian coordinates along the direction of inhomogeneity and another one, $\boldsymbol{z}^0$, along that of the external magnetic field. In the case of the ionosphere $\boldsymbol{x}^0$ would be vertical, so that we consider, strictly speaking, conditions at the magnetic equator. The magnetic field may be taken as approximately homogeneous ($\mathscr{L}_B$ being $\gg \mathscr{L}_N$), its curvature being neglected.

We use a slightly simplified statistical description of the steady state distribution function, $f_{0\mathrm{h}}$, for particles of kind h, and have the equilibrium condition

$$r_{B\mathrm{h}} \cos\varphi \frac{\partial f_{0\mathrm{h}}}{\partial x} = \frac{\partial f_{0\mathrm{h}}}{\partial \varphi}, \tag{30.12a}$$

where $\varrho_B \equiv v_\perp/\omega_B$ is the gyro-radius of the spiraling motion of charges around the magnetic field line. We thus get the equivalent relation

$$v_\perp \cos\varphi \frac{\partial f_{0\mathrm{h}}}{\partial x} = \omega_{B\mathrm{h}} \frac{\partial f_{0\mathrm{h}}}{\partial \varphi}. \tag{30.12b}$$

The gyro (or LARMOR-) frequency, $\omega_{B\mathrm{h}}/2\pi$, now depends on the coordinate x, thus

$$\omega_{B\mathrm{h}}(x) = \frac{1}{c_0\sqrt{\varepsilon_0 \mu_0}} \frac{q_\mathrm{h}}{m_\mathrm{h}} B_{\dot{\mathfrak{z}}}(x). \tag{30.13}$$

[5] For inhomogeneities in two or even three dimensions the theoretical work is not yet advanced enough so that it is indicated to omit this more difficult problem here.

Eq. (30.12) means that the degree of inhomogeneity of plasma is characterized by the ratio of the gyro-radius $r_{B\mathrm{h}}$ to the scale of inhomogeneity $\mathscr{L}_0$. If this ratio is small, then the plasma is slightly inhomogeneous.

The partial differential equation, Eq. (30.12), gives as general solution an arbitrary function of two specific arguments, namely the kinetic energy:

$$\zeta_{\mathrm{h}} = \tfrac{1}{2} m_{\mathrm{h}} v^2 \tag{30.14}$$

and the expression

$$\xi_{\mathrm{h}} = v_{\perp} \sin\varphi + \int^{x} \mathrm{d}x' \omega_{B\mathrm{h}}(x'). \tag{30.15}$$

These are the "characteristics" of Eq. (30.12). The general solution is

$$f_{0\mathrm{h}} = f_{0\mathrm{h}}(\zeta_{\mathrm{h}}, \xi_{\mathrm{h}}). \tag{30.16}$$

Under real conditions the gyro-(LARMOR-) radius, $\varrho_{B\mathrm{h}}$, is small compared to the scale of inhomogeneity, $\mathscr{L}_0$, such that ($\varrho_B = v_{\perp}/\omega_B$):

$$\frac{1}{\mathscr{L}_0} \frac{v_{\perp\mathrm{h}}}{\omega_{B\mathrm{h}}} = \frac{r_{B\mathrm{h}}}{\mathscr{L}_0} \ll 1. \tag{30.17}$$

In ionospheric and magnetospheric plasmas[6] the left side is of the order of $10^{-7} \ldots 10^{-5}$ for electrons (h = e), and $10^{-4} \ldots 10^{-2}$ for ions (h = i).

The inequality, Eq. (30.17) being certainly valid we are allowed to separate the solution Eq. (30.16) and develop the factor depending upon ξ_{h} after $\varrho_B \frac{\partial}{\partial x}$ which gives

$$f_{0\mathrm{h}}(\zeta_{\mathrm{h}}, \xi_{\mathrm{h}}) \approx \left[1 + \frac{v_{\perp} \sin\varphi}{\omega_{B\mathrm{h}}} \frac{\partial}{\partial x}\right] F_{0\mathrm{h}}(\zeta_{\mathrm{h}}, x). \tag{30.18}$$

$F_{0\mathrm{h}}(\zeta_{\mathrm{h}}, x)$ is an arbitrary function of the (kinetic energy) characteristic ζ_{h} and the coordinate x. We may take the Maxwellian distribution function

$$f_{0\mathrm{h}}^{\mathrm{M}}(\zeta_{\mathrm{h}}, x) \equiv \frac{N_{\mathrm{h}}(x)}{(2\pi m_{\mathrm{h}} \vartheta_{\mathrm{h}}(x))^{\frac{3}{2}}} \exp\left(\frac{-\zeta_{\mathrm{h}}}{\vartheta_{\mathrm{h}}(x)}\right) \tag{30.19}$$

and identify f_0^{M} with $F_{0\mathrm{h}}$.

This could appear to be incorrect because we are considering an inhomogeneous plasma; however, the inhomogeneity is so weak, that the plasma can *locally* be considered as to be isotropic. Thus, the energy distribution could be Maxwellian at a given place, even though different at different x coordinates.

η) It is important to note here that the electric *current density* $\boldsymbol{J}_0$ does not vanish in an inhomogeneous plasma. It can easily be seen from Eq. (30.18) that a current is flowing perpendicularly to the inhomogeneity direction, $\boldsymbol{x}^0$, and to the magnetic field direction, $\boldsymbol{z}^0$. The current density is

$$\boldsymbol{J}_0 \equiv (0, J_{0y}, 0) \quad \text{with}^7$$

$$J_{0y} = \frac{c_0 \sqrt{\varepsilon_0 \mu_0}}{B_{\mathrm{ō}}} \sum_{\mathrm{h}} \frac{\partial}{\partial x} (N_{\mathrm{h}} \vartheta_{\mathrm{h}}). \tag{30.20}$$

We may now introduce the one of MAXWELL's equations which gives the relation between current and magnetic field

$$\nabla \times \boldsymbol{B}_{\mathrm{ō}} \equiv \frac{\partial}{\partial \boldsymbol{r}} \times \boldsymbol{B}_{\mathrm{ō}} = \frac{\mathrm{u}\mu_0}{c_0 \sqrt{\varepsilon_0 \mu_0}} \boldsymbol{J}_0$$

---

[6] See Tables 1 (p. 401), 2 (p. 401) and 3 (p. 402).

[7] Like in Sect. 13, Eq. (13.1) we use an energetic definition of the kinetic temperature $\vartheta_{\mathrm{h}} = kT_{\mathrm{h}}$ (k here: BOLTZMANN's constant).

so that finally the following equilibrium condition is obtained:

$$\frac{\partial}{\partial x}\left(\frac{1}{\mathrm{u}\mu_0}\,\frac{1}{2}\,B_{\text{♁}}^2+P_{\text{♁}}\right)=0. \tag{30.21}$$

The *total kinetic pressure*

$$P_{\text{♁}}=\sum_{\mathrm{h}} N_{\mathrm{h}}\vartheta_{\mathrm{h}}\equiv\sum_{\mathrm{h}} N_{\mathrm{h}}\,k\,T_{\mathrm{h}} \tag{30.22}$$

has been introduced here. Since the parameters in the paranthesis are energy densities, Eq. (30.21) means constant energy density along the inhomogeneity axis $\boldsymbol{x}^0$. In other words, the sum of magnetic field and particle kinetic energy remains constant in spite of the lack of homogeneity.

The current $\boldsymbol{J}_0$ is not connected with charge transport through the plasma, but is "diamagnetic" in nature. It is due to the microcurrents of the spiralling charges which did mutually compensate in a homogeneous plasma but do not exactly so in an inhomogeneous one. It is noted that in the absence of collisions the centers of the individual microcurrents are at rest. It can be derived from Fig. 12 that the remaining overall current along $\boldsymbol{y}^0$ can be approximated as follows in successive steps:

$$J_{0y}=\sum_{\mathrm{h}} q_{\mathrm{h}} N_{\mathrm{h}} V_{\mathrm{d\,h}}\sim\sum_{\mathrm{h}} q_{\mathrm{h}}\,\Delta x\,\frac{\partial}{\partial x}\,(V_{T\mathrm{h}} N_{\mathrm{h}})\sim\sum_{\mathrm{h}}\frac{q_{\mathrm{h}} N_{\mathrm{h}} V_{T\mathrm{h}}^2}{\omega_{B\mathrm{h}}\mathscr{L}_0}\,. \tag{30.23}$$

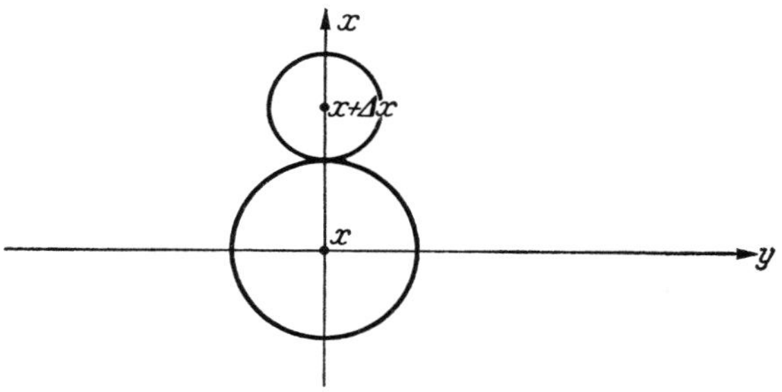

Fig. 12. Explanation of the LARMOR-drift: LARMOR-circles.

This estimation is equivalent with the expression for the current after Eq. (30.20). The characteristic quantity

$$V_{\mathrm{d\,h}}\sim\frac{1}{N_{\mathrm{h}}}\,\Delta x\,\frac{\partial}{\partial x}\,(V_{T\mathrm{h}} N_{\mathrm{h}}) \tag{30.24}$$

has the dimension of a velocity. It is a virtual drift velocity of the charges indicating which linear drift velocity were needed in order to produce the current effect of the non-compensated part of the motions along the LARMOR circles. It is now usually called "LARMOR *drift of the charged particles*".

It follows from the equilibrium condition Eq. (30.21) that the ratio of the inhomogeneity scales $\mathscr{L}$ of total pressure and magnetic field energy is of the order of

$$\mathscr{L}_P/\mathscr{L}_B\sim 2\,\frac{\mathrm{u}\mu_0 P_{\text{♁}}}{B_{\text{♁}}^2}=\beta. \tag{30.25}$$

In the ionosphere $\beta\lesssim 10^{-5}\ll 1$, see Table 1 (p. 401). In the magnetosphere $\beta=10^{-4}\ldots 10^{-3}$.

The inhomogeneity of the magnetic field is therefore in almost all cases negligible compared to that of particle density and temperature[8].

[8] The condition $\beta\ll 1$, besides, is also satisfied in most experimental plasmas produced in laboratories.

In spite of the above restriction concerning collisionless plasmas the results of this section remain valid if a weak collisional effect is introduced; only the condition

$$\bar{\nu}_{\mathrm{h}} \ll \omega_{B\mathrm{h}} \qquad (\mathrm{h}=\mathrm{e},\,\mathrm{i}) \tag{30.26}$$

must be satisfied for electrons, as well as for ions. In the ionosphere this is so at heights above 200 km (see Table 1, p. 401; 2, p. 401; 3, p. 402) and, a fortiori, in the magnetosphere also.

31. Oscillation spectra.

α) Trying now to determine the effective dielectric tensor for an inhomogeneous plasma we suppose that the *deviation of the distribution function* from complete equilibrium (i.e. homogeneity) be small. We write it in the form of a wave with frequency and wave vector $\boldsymbol{k}=(0, k_y, k_z)$

$$\delta f_{\mathrm{h}} \sim \delta f_{\mathrm{h}}(x) \exp(-i\omega t + i k_y y + i k_z z). \tag{31.1}$$

Similarly the induced variations of electric field $\boldsymbol{E}$ and magnetic field $\boldsymbol{B}$ may be described by a wave propagating in the $\boldsymbol{y}^0$, $\boldsymbol{z}^0$ plane. After BOLTZMANN-VLASSOV the kinetic equation for δf_{h} can be written as[1]

$$\left.\begin{aligned}&\left(\omega-(k_y v_y + k_z v_z)\right)\delta f_{\mathrm{h}} + i v_x \frac{\partial \delta f_{\mathrm{h}}}{\partial x} - i\omega_{B\mathrm{h}} \frac{\partial \delta f_{\mathrm{h}}}{\partial \varphi}\\ &\quad = -i q_{\mathrm{h}} \left\{\boldsymbol{E} + \frac{1}{c_0\sqrt{\varepsilon_0\mu_0}}\,\boldsymbol{v}\times\boldsymbol{B}\right\}\cdot\frac{\partial f_{0\mathrm{h}}}{\partial \boldsymbol{P}}\,.\end{aligned}\right\} \tag{31.2}$$

The left side describes motion effects while the forces are given on the right side. Eq. (31.2) has the same characteristic as Eqs. (30.14), (30.15), namely

$$v_\perp \sin\varphi + \omega_{B\mathrm{h}} x = C_{\mathrm{h}} = v_\perp \sin\varphi' + \omega_{B\mathrm{h}} x'. \tag{31.3}$$

The magnetic field has been taken as constant over the whole space of interest. This assumption is justified by the large scale of $|\boldsymbol{B}_{\delta}|$ as compared to that of N. Therefore, the general solution of Eq. (31.2) may be written as

$$\left.\begin{aligned}\delta f_{\mathrm{h}}(x) = &\frac{q_{\mathrm{h}}}{m_{\mathrm{h}}\omega_{B\mathrm{h}}} \int\limits_{\infty}^{\varphi} \mathrm{d}\varphi_1 \left\{\boldsymbol{E}(x') + \frac{1}{c_0\sqrt{\varepsilon_0\mu_0}}\,\boldsymbol{v}\times\boldsymbol{B}_{\delta}(x)\right\}\cdot\frac{\partial f_{0\mathrm{h}}(x',\varphi_1)}{\partial \boldsymbol{v}}\times\\ &\times \exp\left[\frac{i}{\omega_{B\mathrm{h}}}\int\limits_{\varphi}^{\varphi_1}\mathrm{d}\varphi_2\,(\omega - k_y v_\perp \sin\varphi_2 - k_z v_z)\right].\end{aligned}\right\} \tag{31.4}$$

In these expressions the variable x' appearing in the integrand is connected with x and φ by the relation Eq. (31.3). Now we substitute for $\vartheta f_{\mathrm{h}}(x)$ and the fields $\boldsymbol{E}(x)$ and $\boldsymbol{B}(x)$ expressions of the form

$$\exp\left(i \int^{x} \mathrm{d}x'\, k_x(x')\right) \tag{31.5}$$

limiting our considerations to the zero order approximation of geometric optics[2]. The induced current density can be determined with the expression shown in Eq. (31.4) and we finally obtain the following relation for the dielectric tensor of an inhomogeneous plasma, in the zero order approximation of geometric optics[3]:

$$\mathsf{E}(\omega, \boldsymbol{k}, x) = \mathsf{U} + \sum_{\mathrm{h}} \frac{1}{\vartheta_{\mathrm{h}}}\left(1 - \frac{k_y V_{T\mathrm{h}}^2}{\omega\,\omega_{B\mathrm{h}}}\frac{\partial}{\partial x}\right)\vartheta_{\mathrm{h}}[\mathsf{E}^{(\mathrm{h})}(\omega, \boldsymbol{k}, x) - \mathsf{U}], \tag{31.6}$$

---

[1] φ is the angle between $\boldsymbol{v}_\perp$ and the x-axis which identifies the direction of inhomogeneity, see Sect. 22α.

[2] This means: the space derivatives of $\boldsymbol{k}(x)$ are neglected.

[3] The computation is exactly analogous to that which has been made in Sect. 22 for a homogeneous plasma.

which reads in components:

$$\varepsilon_{lm}(\omega, \boldsymbol{k}, x) = \vartheta_{lm} + \sum_{\mathrm{h}} \frac{1}{\vartheta_{\mathrm{h}}}\left(1 - \frac{k_y V_{T\mathrm{h}}^2}{\omega\,\omega_{B\mathrm{h}}}\frac{\partial}{\partial x}\right)\vartheta_{\mathrm{h}}\left[\varepsilon_{lm}^{(\mathrm{h})}(\omega, \boldsymbol{k}, x) - \delta_{lm}\right] \tag{31.6a}$$

$\mathsf{E}^{(\mathrm{h})} \equiv ((\varepsilon_{lm}^{(\mathrm{h})}))$ is the dielectric tensor of the particles of kind h; it must be equivalent to the corresponding tensor for a homogeneous plasma, see Eqs. (22.8)[4].

β) We just note that the tensor Eq. (22.8) has been written in another system of coordinates, namely one with $\boldsymbol{z}^0$ along $\boldsymbol{B}_{\dot{0}}$ and $\boldsymbol{x}^0$ in the $\boldsymbol{k}$, $\boldsymbol{B}_{\dot{0}}$-plane. Different from this; Eq. (31.6) is in arbitrary coordinates so that $\boldsymbol{k} = (k_x, k_y, k_z)$. The transformation of coordinates from one system to the other is a simple rotation. However, we do not need to transform at all because Eq. (4.3)[5], which we take as eiconal equation in an inhomogeneous medium, does not change its general form at such transformation. One has only to remember that the case of an inhomogeneous plasma can be introduced in these expressions by applying the operator

$$\sum_{\mathrm{h}} \frac{1}{\vartheta_{\mathrm{h}}}\left(1 - \frac{k_y V_{T\mathrm{h}}^2}{\omega\omega_{B\mathrm{h}}}\frac{\partial}{\partial x}\right)\vartheta_{\mathrm{h}} \tag{31.7}$$

to the components ε_{lm} of the dielectric tensor E. The summation after h goes over all kinds of charged particles, and

$$k_\perp^2 = k_x^2 + k_y^2.$$

To give a simple example we take the longitudinal dielectric permeability see Eq. (23.15) which must be written for an inhomogeneous magnetized plasma in the following form (log means log nat):

$$\begin{aligned}\varepsilon(\omega, \boldsymbol{k}, x) = 1 + \sum_{\mathrm{h}} \frac{\omega_{N\mathrm{h}}^2}{k^2 V_{T\mathrm{h}}^2}\Bigg\{1 - \sum_s \frac{\omega}{\omega - s\,\omega_{B\mathrm{h}}}\Bigg(1 - \frac{k_y V_{T\mathrm{h}}^2}{\omega\omega_{B\mathrm{h}}} \times \\ \times\left[\frac{\partial \log N_{\mathrm{h}}}{\partial x} + \frac{\partial\vartheta_{\mathrm{h}}}{\partial x}\frac{\partial}{\partial\vartheta_{\mathrm{h}}}\right]\Bigg)\mathscr{A}_s\left(\frac{k_\perp^2 V_{T\mathrm{h}}^2}{\omega_{B\mathrm{h}}^2}\right)\mathrm{J}_+\left(\frac{\omega - s\,\omega_{B\mathrm{h}}}{|k_z|V_{T\mathrm{h}}}\right)\Bigg\}.\end{aligned} \tag{31.8}$$

Eqs. (31.6) and (31.8) show that one new characteristic frequency appears in an inhomogeneous plasma for each kind of particles. It is called LARMOR *drift frequency* $\omega_{\mathrm{dh}}/2\pi$:

$$\omega_{\mathrm{dh}} = k_y V_{\mathrm{dh}} \sim \frac{k_y V_{T\mathrm{h}}^2}{\omega_{B\mathrm{h}}\mathscr{L}_0}\,. \tag{31.9}$$

This is the frequency for which a wave with wave vector k_y along $\boldsymbol{y}^0$ appears in the presence of a drift with the LARMOR drift velocity along the same $\boldsymbol{y}^0$ axis. This characteristic frequency determines the range of influence of a weak inhomogeneity with given scale $\mathscr{L}_0$. For frequencies well above ω_{dh} i.e. for

$$\omega \gg \omega_{\mathrm{dh}} \tag{31.10}$$

all terms in Eqs. (31.6), (31.8) which contain derivatives in space can be neglected, so the expressions then are coincident with the corresponding expressions for a homogeneous plasma.

Moreover the expressions admitting weak collisional influence (see Sect. 23) remain also valid in this limit. A different situation appears in the frequency range

$$\omega \lesssim \omega_{\mathrm{dh}}. \tag{31.11}$$

In this frequency range the influence of an inhomogeneity is not negligible. It is therefore not a priori certain that the collisional influence can be obtained similarly as in the homo-

[4] In this case, however, N_{h} and ϑ_{h} are to be taken as depending on the coordinate $\boldsymbol{x}^\circ$.
[5] The definitions of parameters used are given in Eqs. (24.1...3).

geneous case. This question will be discussed in Sect. 36 where drifting oscillations of a plasma are investigated. In Sects. 31 through 35 we shall remain of the high frequency side, $\omega \gg \omega_{\mathrm{dh}}$, so that the formulae derived in Sects. 17, 18, 22 and 23 remain valid.

γ) We come now to the determination of the frequency *spectrum of eigen-oscillations* of an inhomogeneous plasma. We intend to discuss this problem considering a simple example, viz. a plasma in the absence of external fields, with isotropic distribution of the particle velocities. The method which shall be explained in the following will be helpful later in the more complicated cases of inhomogeneous plasmas which are also magnetized and anisotropic.

δ) We take an easy case first, that of the socalled SELLMEIER dispersion formula[6], valid in a plasma which is not exposed to a magnetic field. Let us consider high frequency transverse waves the phase velocity of which be greater than the vacuum light velocity c_0. Applying Eqs. (18.3), (18.15) and (30.9) and taking account of collisions we obtain the following eiconal equation for high frequency transverse waves

$$k^2 = \frac{\omega^2}{c_0^2}\left(1 - \frac{\omega_{N\mathrm{e}}^2}{\omega^2} + i\,\frac{\omega_{N\mathrm{e}}^2 \bar{\nu}_{\mathrm{e}}}{\omega^3}\right) = \frac{\omega^2}{c_0^2}\,(1 - X_{\mathrm{e}} + i\,X_{\mathrm{e}}\,Z_{\mathrm{e}})\,. \tag{31.12}$$

This is the well-known dispersion equation in a homogeneous, non-magnetized plasma. On the other side we use the gasdynamic equations of the "two fluid" model[7] and obtain the following differential equation for the transverse electric field $\boldsymbol{E}_{\mathrm{tr}}$

$$\left[\nabla^2 \equiv \frac{\partial}{\partial \boldsymbol{r}}\cdot\frac{\partial}{\partial \boldsymbol{r}} \equiv \Delta\right]$$

$$\frac{\partial}{\partial \boldsymbol{r}}\cdot\frac{\partial}{\partial \boldsymbol{r}}\,\boldsymbol{E}_{\mathrm{tr}} + \frac{\omega^2}{c_0^2}\left(1 - \frac{\omega_{N\mathrm{e}}^2}{\omega^2} + i\,\frac{\omega_{N\mathrm{e}}^2 \bar{\nu}_{\mathrm{e}}}{\omega^3}\right)\boldsymbol{E}_{\mathrm{tr}} = 0 \tag{31.13}$$

which should be valid in an inhomogeneous plasma as well. When comparing Eqs. (31.12) and (31.13) it becomes evident that both are equivalent so that the zero order approximation of geometric optics apparently coincides here with the eiconal equation Eq. (31.12).

ε) Eq. (31.13) has the form of a wave equation; a well-known example is the SCHRÖDINGER equation with complex potential[8] $\mathscr{V}$

$$\frac{\mathrm{d}^2\Psi}{\mathrm{d}x^2} + \mathscr{V}(\omega, x)\,\Psi = 0\,. \tag{31.14}$$

We like to know the quickly varying solutions of this equation for which the approximation of geometrical optics, Eq. (30.1), could possibly be applied. Such solutions are called "quasi-classical" in quantum mechanics; in the mathematical literature the designation "asymptotic solutions" is used. The theory of asymptotic solutions of second order differential equations of the form of Eq. (31.14) has been intensively studied and is rather well developed[9]. We only remember quickly the main theorems concerning STOKES' phenomenon and the eigenvalue spectra of Eq. (31.14). The most interesting case for practical application is that of a nearly real function $\mathscr{V}(\omega, x)$, i.e. $\mathrm{Re}\,\mathscr{V} \gg \mathrm{Im}\,\mathscr{V}$. We suppose that frequencies of eigenoscillations also satisfy the condition

$$\mathrm{Re}\,\omega^2 \gg \mathrm{Im}\,\omega^2\,.$$

[6] See K. RAWER and K. SUCHY: this Encyclopedia, Vol. 49/2, Sect. 6, p. 41.

[7] These apply because the thermal motion of the particles has only negligible influence upon the high frequency oscillations which are here considered.

[8] See, for example [*25*].

[9] See [*43*] where reference is made to special literature.

The asymptotic solutions of the SCHRÖDINGER equation, Eq. (31.14) are as follows

$$\Psi(x) = C_+ \exp\left(i \int^x \mathrm{d}x' \sqrt{\mathscr{V}(\omega, x')}\right) + C_- \exp\left(i \int^x \mathrm{d}x' \sqrt{\mathscr{V}(\omega, x')}\right). \qquad (31.15)$$

ζ) It is known that the so-called *turning points* play an important role in connection with such solutions. The turning points are defined by

$$\mathscr{V}(\omega, x) = 0. \qquad (31.16)$$

In the neighbourhood of such turning points[10] the oscillatory solutions Eq. (31.15) go over into an aperiodically attenuated function at the turning point itself[11]. $\mathscr{V}(\omega, x)$ as well as ω being almost real quantities we may suppose that the turning points themselves are lying near the real ω- or x-axis, i.e. that the corresponding imaginary part be small[12]. A range between turning points on the real ω- or x-axis where $\operatorname{Re} \mathscr{V} > 0$ is usually called a "*transparency range*"[13].

In the general case the complex x-plane can be subdivided in distinct regions according to the signs of the coefficients C_+ and C_- in Eq. (31.15). Transition from one region to another one means a discontinuous change of these coefficients. The delimitation lines between these regions are called STOKES' *lines*; on these lines

$$\operatorname{Im} \sqrt{\mathscr{V}(\omega, x)} = 0. \qquad (31.17)$$

Evidently, the turning points are situated on STOKES' lines and it can easily be shown that in the neighbourhood of the turning points the solutions of Eq. (31.14) can mostly be expressed by BESSEL or related functions[14]. STOKES' phenomenon and the asymptotic behaviour have been studied, see for example [*35*]. The coefficients are found from the asymptotic expansions of the solutions, using the connecting relations existing in the relevant family of solution functions[15]. In the particular case of an isolated turning point the coincidence of the asymptotic expansions of the BESSEL functions with the asymptotic solution Eq. (31.15) allows to determine C_+ and C_-. One so gets a "quantization rule" determining the spectrum of discrete eigenvalues of Eq. (31.14). In a configuration where only two turning points exist at x_1, x_2, the quantum rule takes the well-known form of the BOHR-SOMMERFELD integral [*44*]

$$\int_{x_1}^{x_2} \mathrm{d}x\, k_x(\omega, x) = (\mathrm{n} + \tfrac{1}{2})\,\pi, \qquad (31.18)$$

where $k_x(\omega, x) = \sqrt{\mathscr{V}(\omega, x)}$ and n is an integer, usually much greater than one[16]. The integration in Eq. (31.18) covers the transparency range, between the turning points x_1 and x_2 [17].

[10] See K. RAWER and K. SUCHY: this Encyclopedia, Vol. 49/2, Sects. 8β, p. 60; 14, p. 177; 15, p. 180. (These authors use the term "reflection point".)

[11] We do not take account of increasing solutions here as we consider a spatially unlimited, inhomogeneous plasma. In a spatially confined medium increasing solutions have to be considered.

[12] See K. RAWER and K. SUCHY: this Encyclopedia, Vol. 49/2, Sect. 15, Figs. 58 through 61.

[13] For such ranges we have an oscillatory behaviour as described by Eq. (30.2).

[14] Asymptotic solutions are not adequate in this neighbourhood. See K. RAWER and K. SUCHY: this Encyclopedia, Vol. 49/2, Sect. 15, p. 180, there a detailed discussion of different configurations is given. The function $\mathscr{V}(\omega, x)$ is usually expanded in a power serie. If the linear term does not vanish, the wave equation, Eq. (31.14) can be approximated by STOKES' differential equation, with AIRY functions as solutions.

[15] A general method is that of the "scattering matrix", see K. RAWER and K. SUCHY: this Encyclopedia, Vol. 49/2, Sect. 13, p. 162.

[16] Reason why we replace in the following $(\mathrm{n} + \frac{1}{2})$ by n.

[17] In [*45*] it has recently been shown that Eq. (31.18) is valid in the case of an arbitrary complex function $\mathscr{V}(\omega, x)$, for any pair of turning points x_μ, x_ν, provided they form a "complex of kind P". This just means that there exists no connection through infinity of x between any ranges which are disconnected by STOKES' lines.

The quantization rule, Eq. (31.18) has a simple physical interpretation. It means that an eigenfunction in the transparency range must be such that an integer number of half-waves lies between both turning points. A simple analog is an extended chord, the turning points being identified with the points of fastening[18]. In this manner the quantum rule Eq. (31.18) determines the spectrum of eigenvalues of the wave equation, Eq. (31.14). We use these notions for the sake of their analogy with the well-known "quasi-classical" quantization rules used in quantum mechanics.

η) Supposing the imaginary parts to be small for the frequency $\omega \to \omega + i\gamma$ as well as for k_x, x_1 and x_2 we easily derive from Eq. (31.18) two equations namely one for the real and one for the imaginary part[19]. These relations *determine frequency* $\omega/2\pi$ *and attenuation decrement* γ of the oscillations which may rise:
the pulsation ω follows from:

$$\int_{(x_1)}^{(x_2)} \mathrm{d}x \, \mathrm{Re}\, k_x(\omega, x) = \mathrm{n}\,\pi \qquad (\text{n integer}), \tag{31.19}$$

the attenuation decrement is given by

$$\gamma \int_{(x_1)}^{(x_2)} \mathrm{d}x \frac{\partial}{\partial\omega} \mathrm{Re}\, k_x(\omega, x) = -\int_{(x_1)}^{(x_2)} \mathrm{d}x \, \mathrm{Im}\, k_x(\omega, x). \tag{31.20}$$

Integration goes over the transparency range situated between x_1 and x_2 the end points of which are characterized by the definition that (x_1) and (x_2) are the projections of the turning points x_1, x_2 onto the real x-axis.

It is important to note that Eq. (31.20) determining the attenuation decrement γ can only be applied if the frequency $\omega/2\pi$ determined by Eq. (31.19) has a real value[20]. The sign of γ indicates whether the oscillations are increasing ($\gamma < 0$) or decreasing ($\gamma > 0$). In the last case they may be called "kinetically stable".

In cases where there are two turning points with a transparency range in between one uses the term "*trapped oscillations*". The dispersion relations Eqs. (31.19), (31.20) apply even when the transparency ranges are not very far distant on the x-axis (but anyhow distant by at least one wavelength). Thus, each pair of turning points leads to one quantization rule so that the total number of such rules equals that of the transparency ranges existing.

If, over the whole x-axis, no turning points exist at all there can be no "trapped oscillations"[21], and no quantization rule determining an eigenoscillation[22]. An apparent exception is due to spatial limitation of the plasma (in the direction of inhomogeneity), for example a plasma layer with non-dissipative conditions at the limits. A discrete spectrum as is easily shown, appears in this particular case an Eq. (31.18) may be used to determine the oscillation frequencies.

In the majority of all cases which we consider, pairs of roots $\pm k_{xs}(\omega, x)$ are found from the eiconal Eqs. (30.9) and (30.10), and the transparency ranges are sufficiently distant from each other. To each pair of roots belongs one differential equation (in the zero order approximation of geometric optics) which has the form of Eq. (31.14) with

$$\mathscr{V}(\omega, x) = k_{xs}^2(\omega, x). \tag{31.21}$$

[18] For Eq. (31.14) the amplitude of an oscillation decreases exponentially beyond a turning point, as it is the case for non-idealised fastening points of a chord.

[19] These assumptions which are rather generally valid allow to neglect that part of the integrals in Eqs. (31.19), (31.20) which would be obtained by integration along an imaginary x-axis.

[20] One speakes of "fluid-dynamic stability" in this case.

[21] In this case the solutions of Eq. (31.15) are called "unlimited" in mathematical language while they are "limited" when trapped between turning points.

[22] "Unlimited" solutions after Eq. (31.15) can then be constructed for any value of the frequency.

The corresponding "quantization rule" can be used to determine the eigenvalues of ω. This resumes the general philosophy of the theory of oscillations in an inhomogeneous plasma[23].

Thus the spectrum of oscillations of an inhomogeneous plasma is, strictly speaking, discrete. It appears, however, as well filled up with spectral lines[24] so that the discreteness is not really typical. Compared with a homogeneous plasma the main difference is that the spectrum of an inhomogeneous plasma is not determined by a local relation but by one covering a certain, rather large range of plasma (along the x-axis of inhomogeneity).

32. Propagation of radio waves.

α) Having discussed the fundamental characteristics of oscillations in an inhomogeneous plasma we now come to the problem of propagation of *waves of transverse polarization* in such a plasma. Taking the complex wave vector component k_x from Eq. (31.12), and using the quantization rule Eq. (31.18) we may specify the dispersion relations, Eqs. (31.19), (31.20) by writing

$$\int_{(x_1)}^{(x_2)} \mathrm{d}x \,\mathrm{Re}\, k_x(\omega, x) \equiv \int_{(x_1)}^{(x_2)} \mathrm{d}x \sqrt{-k_y^2 - k_z^2 + \frac{\omega^2}{c_0^2} - \frac{\omega_{N\mathrm{e}}^2}{c_0^2}} = \mathrm{n}\,\pi \qquad (\mathrm{n\ integer}), \tag{32.1}$$

and

$$\left.\begin{aligned} \gamma &= -\frac{1}{2}\int_{(x_1)}^{(x_2)} \mathrm{d}x \frac{\omega_{N\mathrm{e}}^2 \bar{\nu}_\mathrm{e}}{\omega^2} \Big/ \int_{(x_1)}^{(x_2)} \frac{\mathrm{d}x}{\mathrm{Re}\, k_x(\omega, x)}, \\ \frac{\gamma}{|\omega|} &= -\frac{1}{2}\int_{(x_1)}^{(x_2)} \mathrm{d}x\, X_\mathrm{e} Z_\mathrm{e} \Big/ \int_{(x_1)}^{(x_2)} \frac{\mathrm{d}x}{\mathrm{Re}\, k_x(\omega, x)}. \end{aligned}\right\} \tag{32.2}$$

These equations which may be used to determine the eigenvalues ω and γ; they become identical to Eqs. (20.38) and (21.1) in the limit

$$\lim_{\mathcal{L}_0 \to \infty} \frac{\mathrm{n}\,\pi}{\mathcal{L}_0} \to k_x .$$

β) Let us now analyze Eqs. (31.22), (31.23) considering radio wave propagation in the *ionosphere*. It follows from Eq. (31.22) that in an inhomogeneous plasma, like in a homogeneous one, the condition

$$\omega^2 < \omega_{N\mathrm{e}}^2, \quad \text{i.e.} \quad X_\mathrm{e} < 1 \tag{32.3}$$

must be satisfied for any eigenoscillation of transverse polarization. In an inhomogeneous plasma this condition needs to be satisfied at any point which is to be reached by a wave. At a surface (or a point) where condition Eq. (31.24) is no more valid but $\omega_{N\mathrm{e}} = \omega^2$, $(X_\mathrm{e} = 1)$ reflection of (transverse) electromagnetic waves occurs at normal incidence, i.e. when $k_y = k_z = 0$. For oblique propagation (k_x or k_y, or both $\neq 0$) reflection occurs at the *turning points*

$$\omega^2 = c_0^2 (k_y^2 + k_x^2) + \omega_{N\mathrm{e}}^2, \quad \text{i.e.} \quad X_\mathrm{e} = 1 - \frac{k_y^2 + k_z^2}{k_0^2}, \tag{32.4}$$

---

[23] An example for trapped oscillations in the ionosphere are hf-waves "excited" by satellite transmitters in the "valley" between regions E and F. These oscillations use to be kinetically stable, i.e. decreasing with time. See Sect. 32.

[24] In the approximation of geometric optics.

where

$$k_0 \equiv \frac{\omega}{c_0}.$$

In the lower ionosphere the density of charged particles as well as that of electrons increases with height x so that X_e (or ω_{Ne}^2) is a monotonuously increasing function up to a certain height level.

This is either the peak of the E-region (usually in day time) or that of the F-region (at night). Under most conditions the maximum of ω_{Ne} is reached in the F2-layer[1], between 250 and 500 km; values of more than $8 \cdot 10^7\ \text{sec}^{-1}$ for ω_{Ne} are only extremely rarely occuring anywhere on Earth. Therefore waves with pulsations $\omega > 8 \cdot 10^7\ \text{sec}^{-1}$, i.e. frequencies above 20 MHz are practically not reflected at normal incidence, anywhere in the ionosphere. Such waves are therefore "unlimited" in the ionospheric plasma, the corresponding spectrum is continuous; as there is no quantization, oscillations may occur on any frequency. On the other hand, under suitable ionospheric conditions, frequencies below this limit may be reflected by the ionosphere.

Since above the F2-peak the electron density in the ionosphere decreases with increasing x there are two turning points, i.e. two levels where at normal incidence reflection occurs, one below and one above the peak.

γ) Transparency exists below the lower turning point and above the upper one; thus we have *two ranges of transparency*. In the upper one the waves are "unlimited", as there is no second turning point existing; so in fact these waves are not "trapped". In the lower range, however, there exists another reflecting level which is the surface of Earth. This reflection is almost lossless, i.e. non-dissipative on high frequencies. Therefore for electromagnetic radiation in the lower transparency range strictly speaking, the quantization rule, Eqs. (31.19), (31.20) should be applied.

Of course, this statement is purely hypothetic on high frequencies, where the spectrum is so dense that it appears practically as a continuous one[2]. At the lowest frequencies (i.e. for rather long waves), however, quantization is an important effect.

The lowest eigenoscillations of the space between Earth's surface and a lower border in the ionosphere have been predicted by SCHUMANN[3] and detected by himself[4] and others[5], see[6] for more detailed descriptions including equations.

Thus, at the lowest frequencies we have eigen-resonances due to the vertical inhomogeneity, so that local spectra as given by Eqs. (21.2) and (21.7) are of no importance there. The reflectivity of the ionosphere is rather high on these frequencies because the vertical gradient of k_z is so steep.

δ) *With increasing frequency* the reflectivity continues to be high in the ulf-range ($f < 3$ kHz) and also in the elf-range ($3\ \text{kHz} < f < 30\ \text{kHz}$). But in the lf-range ($30\ \text{kHz} < f < 300\ \text{kHz}$) terrestrial radio waves penetrate deeper into the lower ionosphere and collisional absorption becomes important; on these frequencies quantization effects have no importance because the reflectivity of the ionosphere is too poor. Therefore the local spectra described by Eqs. (21.2) and (21.7) can be taken for granted. At even higher frequencies the reflectivity increases again but it is depending on the hour. During daylight hours D-region

[1] See K. RAWER and K. SUCHY: [*11*], Sects. 35 and 40 for region E, Sect. 37 for region F.

[2] In fact the real ionosphere is far from being so accurately homogeneous in the horizontal plane that our theory could be applied reasonably of high frequencies.

[3] W. O. SCHUMANN: Z. Naturforsch. **7**A, 149—154, 250—252 (1952).

[4] H. KÖNIG: Naturwissenschaften **41**, 183—184 (1954); Z. angew. Phys. **11**, 264—274 (1959).

[5] See W. T. BLACKBAND (ed.): Propagation of Radio Waves at Frequencies below 300 kc/s. Oxford: Pergamon 1964.

[6] See contribution by E. SELZER in this volume, p. 231.

ionization leads to very poor reflectivity in the mf-range (300 kHz $<$ f $<$ 3 MHz), while it is quite high at night in the same range; reflection occurs either from the E- or from F-region. Any way, on these frequencies the range in height is large compared to the wavelength. Therefore, the number of wavelengths over the thickness of the "subionospheric duct" is so large that the lines in the spectrum are quite dense and the always existing horizontal inhomogeneity of the ionosphere "smears them out" so that the spectrum is practically continuous.

ε) During daylight hours there usually exists a "valley" between the ionospheric regions E and F[7]. This minimum of electron density, situated roughly between 120 and 180 km of height, is rather shallow at noon but quite deep in the morning. It provokes another range of transparency which is quite limited in frequency. This *"duct inside the ionosphere"* may be excited by transmissions from low altitude satellites; it has a certain practical importance at oblique incidence on frequencies of the order of 20 MHz where some sort of trapping has been observed[8].

Of course, the wavelength is also very small compared to the thickness and the very dense line spectrum is again smeared out by inhomogeneity. On the other hand, long waves mostly can not propagate in this duct because of the condition $\omega > \omega_{N\mathrm{e}}$, the minimum in the valley of $\omega_{N\mathrm{e}}/2\pi$ being of the order of 1 MHz (vacuum wavelength $\approx$ 300 m).

ζ) For the *magnetosphere* there certainly exist electron density inhomogeneities, but not simply following stratification along a height coordinate. Field-aligned ionization should be rather frequent but it can not easily be treated with a one-dimensional model such as ours. Any way, trapping may occur in such structures and it may be important in connection with propagation of "whistlers"[9].

33. Longitudinal waves in an inhomogeneous isotropic plasma.

α) Applying the dispersion relations, Eqs. (18.3) and (18.15), we obtain the following eiconal equation for *electronic* LANGMUIR *oscillations* on high frequencies ($\omega \gg k V_{T\mathrm{e}}$)

$$1-\frac{\omega_{N\mathrm{e}}^2}{\omega^2}\left(1+3\frac{k^2 V_{T\mathrm{e}}^2}{\omega^2}\right)+i\sqrt{\frac{\pi}{2}}\frac{\omega\,\omega_{N\mathrm{e}}^2}{k^3 V_{T\mathrm{e}}^3}\exp\left(\frac{-\omega^2}{2k^2V_{T\mathrm{e}}^2}\right)+i\frac{\omega_{N\mathrm{e}}^2\bar{\nu}_{\mathrm{e}}}{\omega^3}=0 \tag{33.1}$$

i.e.

$$1-X_{\mathrm{e}}(1+3Q_{\mathrm{e}}^2)+i\sqrt{\frac{\pi}{2}}X_{\mathrm{e}}Q_{\mathrm{e}}^{-3}\exp\left(-\tfrac{1}{2}Q_{\mathrm{e}}^{-2}\right)+iX_{\mathrm{e}}Z_{\mathrm{e}}=0. \tag{33.1a}$$

Collisional as well as collisionless (ČERENKOV) absorption are accounted for in this relation. We may determine from it the complex wave number $k_x(\omega, x)$. Supposing small imaginary parts and using the quantization rule, Eqs. (31.19), (31.20), we get the following dispersion relations describing the spectrum of longitudinal LANGMUIR waves in an inhomogeneous (isotropic) plasma:

$$\int_{(x_1)}^{(x_2)} \mathrm{d}x\,\mathrm{Re}\,k(\omega,x)\equiv\int_{(x_1)}^{(x_2)}\mathrm{d}x\sqrt{-k_y^2-k_z^2+\frac{\omega^2-\omega_{N\mathrm{e}}^2}{3V_{T\mathrm{e}}^2}}=\mathrm{n}\pi \tag{33.2}$$

and

$$\gamma=-\sqrt{\frac{\pi}{8}}\,\frac{\displaystyle\int_{(x_1)}^{(x_2)}\frac{\mathrm{d}x}{\mathrm{Re}\,k_x(\omega,x)}\frac{1}{V_{T\mathrm{e}}^2}\left[\sqrt{\frac{2}{\pi}}\,\bar{\nu}_{\mathrm{e}}+\frac{\omega_{N\mathrm{e}}^2\exp\left(-\frac{3}{2}-\frac{1}{2}k^{-2}\lambda_{D\mathrm{e}}^{-2}\right)}{\omega^{-2}k^3V_{T\mathrm{e}}^3}\right]}{\displaystyle\int_{(x_1)}^{(x_2)}\frac{\mathrm{d}x}{\mathrm{Re}\,k_x(\omega,x)}V_{T\mathrm{e}}^{-2}} \tag{33.3}$$

---

[7] See K. RAWER and K. SUCHY: [*11*], Sect. 30.

[8] E. CHVOJKOVA (E. WOYK): Radio Sci. D **69**, 453—457 (1965).

[9] See the contribution by R. GENDRIN in volume 49/3 of this Encyclopedia, p. 461.

with

$$k^2 = \mathrm{Re}\, k_x^2(\omega, x) + k_y^2 + k_z^2. \tag{33.4}$$

$\mathrm{Re}\, k_x^2(\omega, x)$ being determined by Eq. (33.2). In a homogeneous plasma these relations become identic with Eqs. (20.5), (20.6) and (20.33).

Again, integration in Eqs. (33.2), (33.3) covers the transparency range of the plasma with respect to the considered oscillations. This range is defined by

$$\left.\begin{aligned} &\mathrm{Re}\, k_x > 0 \\ \text{whence} \qquad & \\ &\omega^2 > \omega_{Ne}^2, \quad \text{i.e.} \quad X_e < 1. \end{aligned}\right\} \tag{33.5}$$

On the other side we have now the conditions

$$\omega \gg k V_{Te} \tag*{[20.3]}$$

so that ω^2 can not very much be different from ω_{Ne}^2:

$$\omega^2 \gtrsim \omega_{Ne}^2, \quad \text{i.e.} \quad X_e \lesssim 1. \tag{33.6}$$

Reflexion occurs at the level where

$$\omega^2 = 3(k_y^2 + k_z^2) V_{Te}^2 + \omega_{Ne}^2, \quad \text{i.e.} \quad X_e = 1 - 3(k_y^2 + k_z^2) V_{Te}^2/\omega^2. \tag{33.7}$$

which is the condition for a turning-point.

β) In the *ionospheric plasma* for a given frequency $\omega/2\pi$ of longitudinal LANGMUIR waves there exist two turning points, and two height ranges of transparency, as we have found for transverse waves. In contrast to these, however, the *transparency ranges* are quite narrow, lying in the immediate neighbourhood of the turning points where $\omega^2 = \omega_{Ne}^2(x)$. Thus each range is limited on one side by a turning-point. (For the range which is lower in height this is the upper end; it is the lower end for the upper range.) On the other side of each of the two transparency ranges, however, there is no turning point but a range of strong absorption. This is due to the fact that the condition

$$k^2 V_{Te}^2 < \omega_{Ne}^2 \tag*{[20.7]}$$

ceases to be satisfied when the electron density decreases, and so does $\omega_{Ne}^2(x)$. We therefore exspect strong LANDAU damping at this end of the range see Eq. (20.15). Thus electron LANGMUIR waves in the ionosphere are not confined between two turning-points but between one only (at the inner side) and a range of strong absorption (at the outer side). Therefore, quantization does not apply and Eqs. (33.2), (33.3) must not be used. Instead, the local formulae, Eqs. (20.5), (20.33) can be used provided geometric optics applies, i.e. when the wavelength of the oscillation is small against the characteristic scale of inhomogeneity. We may state that it is the bell-like shape of electron density profile in the ionosphere which causes the situation of non-quantization for electron LANGMUIR waves.

γ) The longitudinal *ion sound waves* occur at low frequencies in an isotropic plasma where (see Sect. 20)

$$T_e \gg T_i.$$

Considering Eq. (20.17) together with Eqs. (20.37), (20.38) we obtain the eiconal equation for these waves in the form[1]

$$1 - \frac{\omega_{Ni}^2}{\omega^2} + \frac{\omega_{Ne}^2}{k^2 V_{Te}^2}\left(1 + i\sqrt{\frac{\pi}{2}}\frac{\omega}{k V_{Te}}\right) + i\alpha \frac{\omega_{Ni}^2 \bar{\nu}_i}{\omega^3} = 0 \tag{33.8}$$

i.e.

$$1 - X_i + X_e Q_e^{-2}\left(1 + i\sqrt{\frac{\pi}{2}} Q_e^{-1}\right) + i\alpha X_i Z_i = 0 \tag{33.8a}$$

[1] $\bar{\nu}_i$ in Eq. (33.8) means $\bar{\nu}_{in}$ for a weakly ionized plasma, but $\bar{\nu}_{ii}$ for completely ionized plasmas. Similarly, in Eqs. (31.12), (31.13), (32.1), (32.2) and (31.1 ... 3) we have to put $\bar{\nu}_e = \bar{\nu}_{en}$ or $\bar{\nu}_e = \bar{\nu}_{eff}$ in these extreme cases.

where[1]

$$\alpha = \begin{cases} 1 & \text{for weakly ionized plasmas,} \\ \frac{8}{5}\frac{k^2 V_{Ti}^2}{\omega^2} \equiv \frac{8}{5} Q_i^2 & \text{for a completely ionized plasma.} \end{cases} \tag{33.9}$$

When writing down Eq. (33.8) we neglected the thermal motion of the ions. This is justified because the plasma has to be non-isothermal in order to propagate ion sound waves at all. The relevant condition is

$$1 \gg \left(\frac{T_e}{T_i}\right)^{\frac{3}{2}} \sqrt{\frac{m_i}{m_e}} \exp\left(\frac{T_e}{2T_i}\right). \tag{33.10}$$

δ) The *spectrum* of *low-frequency longitudinal waves* in a non-isothermal, isotropic but inhomogeneous plasma follows from the relevant dispersion relation. It can be obtained from the eiconal equation, Eq. (33.8) using the quantization rules, Eqs. (31.19), (31.20), which give here:

$$\int_{(x_1)}^{(x_2)} dx \operatorname{Re} k_x(\omega, x) \equiv \int_{(x_1)}^{(x_2)} dx \sqrt{-k_y^2 - k_z^2 + \frac{\omega^2}{V_s^2} \frac{\omega_{Ni}^2}{(\omega_{Ni}^2 - \omega^2)}} = n\pi \qquad \text{(n integer)} \tag{33.11}$$

and

$$\gamma = -\frac{1}{2}\omega^2 \frac{\int_{(x_1)}^{(x_2)} \frac{dx}{\operatorname{Re} k_x(\omega, x)} \frac{\omega_{Ni}^2}{(\omega_{Ni}^2 - \omega^2)} \left[\sqrt{\frac{\pi}{2}} \frac{1}{k V_{Te} V_s^2} + \alpha \frac{k^2 \bar{\nu}_i}{\omega^4}\right]}{\int_{(x_1)}^{(x_2)} \frac{dx}{\operatorname{Re} k_x(\omega, x)} \frac{1}{V_s^2} \frac{\omega_{Ni}^4}{(\omega_{Ni}^2 - \omega^2)^2}}. \tag{33.12}$$

In this expression we have supposed that $\gamma \ll \omega$ and

$$k^2 = \operatorname{Re} k_x^2(\omega, x) + k_y^2 + k_z^2, \tag{33.13}$$

$\operatorname{Re} k_x(\omega, x)$ being given by Eq. (33.11). It follows from Eq. (33.12) that the oscillations considered can only exist on the lower frequencies, namely for:

$$\omega^2 < \omega_{Ni}^2(x), \quad \text{i.e.} \quad X_i(x) > 1. \tag{33.14}$$

In the opposite case longitudinal oscillations are impossible because DEBYE screening is so efficient that k^2 becomes < 0, see Sect. 18.

Low-frequency longitudinal waves in a non-isothermal plasma are thus reflected from ranges where $\omega_{Ni}^2 < \omega^2$. In a plasma layer the ion density distribution over x is a bell-shaped function such that the transparency range is limited toward lower ion densities on both sides; low-frequency oscillations would therefore be "trapped" inside such a layer, if at all ion sound waves could be existent there[2]. Of course, waves which are trapped in an inhomogeneous plasma must satisfy the quantization rule, i.e. they have an entire number of half-waves between successive turning points. The corresponding spectrum can be obtained from the quantization rule, Eq. (33.11).

ε) Finally, we should note that quantization is of real importance only under conditions where "wave optics" has to be applied to the waves under the particular conditions considered. In all other cases, where straightforward ray considerations are good enough local spectra give a satisfying description, at least qualitatively. These spectra can immediately be deduced from the relevant eiconal equation, which has to be considered as a dispersion relation. There are

[2] As stated above in Sects. 20 and 21, low frequency ion sound waves are not to be exspected in the terrestrial ionosphere which is not enough non-isothermal. As shown there, the condition that such waves could propagate is $T_e > 6 T_i$.

even cases where such simple considerations lead to local spectra, which are quantitatively satisfying. Examples are found in Sects. 37 and 38. Note that the condition for applying quantization essentially means that the number of half waves between *turning* points is not excessively great; the reflection level must be so clearly defined that a pattern of "standing waves" can appear.

34. Potential-depending (longitudinal) waves in an inhomogeneous magneto-active (gyrotropic) plasma. As we have stated in Sect. 25 the number of different oscillation branches is infinite in a homogeneous gyrotropic (magneto-active) plasma. The multiplicity of possible oscillation branches is yet increased if the plasma is (at least weakly) inhomogeneous. There appears then a new type of oscillations due to the LARMOR drift of the particles; these oscillations are called "drift oscillations". They shall be considered later in Chap. E.II, Sects. 36 through 38.

α) In this Section our considerations are concerned with "*higher frequencies*" only; the frequency range is given by [see Sect. 31 in particular Eq. (31.10)]

$$\omega \gg \omega_{d\,h} \sim \frac{k_y V_{T\,h}^2}{\omega_{B\,h} \mathscr{L}_0} \qquad (h = e, i), \tag{34.1}$$

i.e.

$$\mathscr{L}_0 \gg k_y V_{T\,h}^2 / Y_h, \tag{34.1 a}$$

$\omega_{d\,h}/2\pi$ being the frequency of drift oscillations. As stated in Sect. 30 on these frequencies terms containing spatial derivatives can be neglected in the dielectric tensor. Consequently, at this limit the eiconal-equation for electromagnetic waves, Eq. (30.9), plays the same rôle as the dispersion relation of a homogeneous plasma, Eq. (4.3). The only essential difference is that density and temperature of the particles (and, even the magnetic field) depend upon the spatial coordinate x, the coordinate of stratification. The eigen-frequencies of a (weakly) inhomogeneous, gyrotropic plasma are therefore determined by integral relations and not by local ones as in a homogeneous plasma (see Sect. 28). Again, these integral rules correspond to the quantization rule of BOHR and SOMMERFELD; they have the general form of Eq. (31.19) and the complex function $k_x(\omega, x)$ is given by the eiconal-equation. In all other aspects there is no difference from the homogeneous, gyrotropic plasma. This concerns the number of branches, the condition for their possible excitation etc., always with the exception of drift oscillations, i.e. when Eq. (34.1) is satisfied. Moreover, the approximation of geometric optics we have used to obtain the spectra of homogeneous plasmas can be taken as quantitatively right also in an inhomogeneous one. These are characterized as "local oscillations". We are now looking for a more general theory which could give us the characteristic spectra with quantitative accuracy.

β) We begin this analysis with the easiest case, that of *longitudinal (potential) oscillations* in a (weakly) inhomogeneous plasma. The relevant eiconal-equation is Eq. (30.10). In the limit of a cold plasma, but taking account of weak attenuation it can be written in the following form:

$$\left.\begin{aligned} & k_\perp^2 \Bigg[1 - \sum_h \frac{\omega_{N\,h}^2}{\omega^2 - \omega_{B\,h}^2}\left(1 - i\,\frac{\bar{\nu}_h}{\omega}\,\frac{\omega^2 + \omega_{B\,h}^2}{\omega^2 - \omega_{B\,h}^2}\right) + \\ & \quad + i\sqrt{\frac{\pi}{8}} \sum_h \frac{\omega_{N\,h}^2}{\omega |k_z| V_{T\,h}} \left\{\exp\left(\frac{-(\omega - \omega_{B\,h})^2}{2k_z^2 V_{T\,h}^2}\right) + \exp\left(\frac{-(\omega + \omega_{B\,h})^2}{2k_z^2 V_{T\,h}^2}\right)\right\}\Bigg] + \\ & \quad + k_z^2 \left[1 - \sum_h \frac{\omega_{N\,h}^2}{\omega^2}\left(1 - i\,\frac{\bar{\nu}_h}{\omega}\right) + i\sqrt{\frac{\pi}{2}} \sum_h \frac{\omega\,\omega_{N\,h}^2}{|k_z|^3 V_{T\,h}^3} \exp\left(\frac{-\omega^2}{2k_z^2 V_{T\,h}^2}\right)\right] = 0 \end{aligned}\right\} \tag{34.2}$$

i.e.

$$\left.\begin{aligned} &k_\perp^2 \Big[1 - \sum_{\mathrm{h}} \frac{X_{\mathrm{h}}}{1-Y_{\mathrm{h}}^2}\Big(1 - i Z_{\mathrm{h}} \frac{1+Y_{\mathrm{h}}^2}{1-Y_{\mathrm{h}}^2}\Big) + \\ &\quad + i\sqrt{\frac{\pi}{8}} \sum_{\mathrm{h}} X_{\mathrm{h}} \varrho_{\mathrm{h}z}^{-1} \Big\{\exp\Big(-\frac{1}{2}\varrho_{\mathrm{h}z}^{-2}(1-Y_{\mathrm{h}})^2\Big) + \\ &\quad + \exp\Big(-\frac{1}{2}\varrho_{\mathrm{h}z}^{-2}(1+Y_{\mathrm{h}})^2\Big)\Big\}\Big] + \\ &\quad + k_z^2\Big[1 - \sum_{\mathrm{h}} X_{\mathrm{h}}(1 - iZ_{\mathrm{h}}) + i\sqrt{\frac{\pi}{2}}\sum_{\mathrm{h}} X_{\mathrm{h}} \varrho_{\mathrm{h}z}^{-3} \exp\Big(-\frac{1}{2}\varrho_{\mathrm{h}z}^{-2}\Big)\Big] = 0. \end{aligned}\right\} \quad (34.2\mathrm{a})$$

Determining $k_x(\omega, x)$ from this equation, and applying quantization after the rule of Eq. (31.19) we obtain dispersion relations giving the possible frequencies of longitudinal oscillations in a (weakly) inhomogeneous, gyrotropic plasma, and the relevant attenuation decrements by the imaginary part. As shown in Sect. 25 there exist three branches for longitudinal oscillations in a cold, gyrotropic plasma: two of these (on high frequencies) are electron modes, the third one is an ion mode and at lower frequencies, see Eqs. (25.15 ... 20). Neglecting the dissipative (imaginary) members in Eq. (34.2) we get the following quantization rule as *dispersion equation* for determining the spectrum of *longitudinal electron oscillations* in an inhomogeneous plasma:

$$\int_{(x_1)}^{(x_2)} \mathrm{d}x \,\mathrm{Re}\, k_x(\omega, x) \equiv \int_{(x_1)}^{(x_2)} \mathrm{d}x \left\{-k_y^2 - k_z^2 \frac{(\omega^2-\omega_{N\mathrm{e}}^2)(\omega^2-\omega_{B\mathrm{e}}^2)}{\omega^2(\omega^2-\omega_{N\mathrm{e}}^2-\omega_{B\mathrm{e}}^2)}\right\}^{\frac{1}{2}} = \mathrm{n}\pi \quad (34.3)$$

(n integer).

This equation determines two branches of oscillations corresponding to the roots ω_1 and ω_2 found for a homogeneous plasma (see Sect. 25). It is easily seen that the relevant pulsation ω is determined for the *upper branch* by

$$\left.\begin{aligned} &\omega_{N\mathrm{e}}^2 \leqq \omega^2 \leqq \omega_{N\mathrm{e}}^2 + \omega_{B\mathrm{e}}^2 \\ \text{i.e.}\quad & \\ &X_{\mathrm{e}} \leqq 1 \leqq X_{\mathrm{e}} + Y_{\mathrm{e}}^2 \end{aligned}\right\} \quad (34.4)$$

and for the *lower branch* by[1]

$$\left.\begin{aligned} &\omega_{B\mathrm{e}}^2 \geqq \omega^2 \geqq |\omega_{B\mathrm{e}}\,\omega_{B\mathrm{i}}| \\ \text{i.e.}\quad & \\ &Y_{\mathrm{e}}^2 \geqq 1 \geqq |Y_{\mathrm{e}} Y_{\mathrm{i}}|. \end{aligned}\right\} \quad (34.5)$$

Longitudinal waves can not propagate outside of these conditions, but are totally reflected at the relevant limits where $\mathrm{Re}\, k_x^2(\omega, x)$ changes its sign (and becomes negative in the "non-transparent" region).

For this reason in the ionospheric plasma quantization applies to oscillations in the upper electron mode branch, but not in the lower one where the transparency range is unlimited. On the first branch the transparency range is situated in a small range above $\omega_{N\mathrm{e}}$ as given by Eq. (34.4). There are two such levels in the ionosphere, an upper and a lower one, so that we find one transparency range below the peak of the ionosphere and another one above. The magnetic field is, in fact, almost constant in the ionosphere. The situation is different in the magnetosphere where the change of $\omega_{B\mathrm{e}}$ can be more important even than that of $\omega_{N\mathrm{e}}$ such that the frequency ranges where the plasma is transparent to electromagnetic oscillations are given by the inhomogeneity of the magnetic field due to plasma irregularities with "frozen-

[1] A lower limit is not found if only electron are considered; it appears, however, when the ion terms are considered in Eq. (34.3), see Sect. 25 and Fig. 5.

in" fields. If, however, the large scale inhomogeneity of the magnetic field in the magnetosphere is considered it is found that the relevant scale is quite large, so large that the corresponding quantization of the spectrum of longitudinal electron oscillations is of no interest and Eqs. (25.15) can be taken as valid.

γ) The amount of *damping* of the oscillations can be found from the dissipative terms in the eiconal equation. These are small and therefore the *attenuation decrement* which appears as imaginary part, γ, of the pulsation, is always small against the pulsation ω. One derives for γ the expression:

$$\left.\begin{aligned}\gamma = \int\limits_{(x_1)}^{(x_2)} \frac{dx}{\operatorname{Re} k_x}\Big\{k_\perp^2 \Big\{\frac{\bar{\nu}_e\omega_{Ne}^2(\omega^2+\omega_{Be}^2)}{\omega(\omega^2-\omega_{Be}^2)^2} + \sqrt{\frac{\pi}{8}}\,\frac{\omega_{Ne}^2}{\omega|k_z|\,V_{Te}}\times\\ \times\Big[\exp\Big(\frac{-(\omega-\omega_{Be})^2}{2k_z^2V_{Te}^2}\Big) + \exp\Big(\frac{-(\omega+\omega_{Be})^2}{2k_z^2V_{Te}^2}\Big)\Big]\Big\} +\\ + k_z^2\Big[\frac{\bar{\nu}_e\omega_{Ne}^2}{\omega^3} + \sqrt{\frac{\pi}{2}}\,\frac{\omega\omega_{Ne}^2}{|k_z|^3V_{Te}^3}\exp\Big(\frac{-\omega^2}{2k_z^2V_{Te}^2}\Big)\Big]\Big\}\Big/\int\limits_{(x_1)}^{(x_2)}\frac{dx}{\operatorname{Re} k_x}\,\frac{\partial \operatorname{Re} k_x^2}{\partial\omega},\end{aligned}\right\} \tag{34.6}$$

i.e.

$$\left.\begin{aligned}\gamma = \int\limits_{(x_1)}^{(x_2)} \frac{dx}{\operatorname{Re} k_x}\Big\{k_\perp^2 \Big\{X_e Z_e \frac{1+Y_e^2}{(1-Y_e^2)^2} + \sqrt{\frac{\pi}{8}}\,X_e Q_{ez}^{-1}\times\\ \times\Big[\exp\Big(-\frac{1}{2}Q_{ez}^{-2}(1-Y_e)^2\Big) + \exp\Big(-\frac{1}{2}Q_{ez}^{-2}(1+Y_e)^2\Big)\Big]\Big\} +\\ + k_z^2\Big[X_e Z_e + \sqrt{\frac{\pi}{2}}\,X_e Q_{ez}^{-3}\exp\Big(-\frac{1}{2}Q_{ez}^{-2}\Big)\Big]\Big\}\Big/\int\limits_{(x_1)}^{(x_2)}\frac{dx}{\operatorname{Re} k_x}\,\frac{\partial \operatorname{Re} k_x}{\partial\omega}.\end{aligned}\right\} \tag{34.6a}$$

We must integrate here over the whole transparency range which is found between the turning-points in the plasma. We have

$$k_\perp^2 = \operatorname{Re} k_x^2(\omega, x) + k_y^2, \tag{34.7}$$

and $\operatorname{Re} k_x(\omega, x)$ is determined by Eq. (34.3). For the collision number, $\bar{\nu}_e$, we have to use

$$\left.\begin{aligned}&\bar{\nu}_e = \bar{\nu}_{\text{eff}} \quad \text{for a completely ionized plasma,}\\ &\bar{\nu}_e = \bar{\nu}_{\text{en}} \quad \text{for a weakly ionized one.}\end{aligned}\right\} \tag{34.8}$$

In the terrestrial ionosphere Eq. (34.6) only applies to the upper branch in which quantization really occurs. In the magnetosphere collisional interaction is so small that it can be neglected. The local expression for the attenuation, as described in Sect. 25 is valid, see Eq. (25.40).

δ) For the *third or low frequency branch* of longitudinal oscillations in a cold, gyrotropic plasma we obtain a dispersion relation in the usual way from the eiconal-equation, Eq. (34.2) applying quantization after Eq. (31.19). The *quantization rule* reads:

$$\left.\begin{aligned}&\int\limits_{(x_1)}^{(x_2)} dx\,\operatorname{Re} k_x(\omega, x)\\ &\qquad = \int\limits_{(x_1)}^{(x_2)} dx\Big\{-k_y^2 + k_z^2\,\frac{\omega_{Ne}^2(\omega^2-\omega_{Bi}^2)}{\omega^2(\omega^2-\omega_{Ni}^2-\omega_{Bi}^2)}\Big\}^{\frac{1}{2}} = n\pi, \quad (\text{n integer})\end{aligned}\right\} \tag{34.9}$$

and the *attenuation decrement* is

$$\left.\begin{aligned}\gamma = \int_{(x_1)}^{(x_2)} \frac{\mathrm{d}x}{\operatorname{Re} k_x} \Big\{k_\perp^2 \Big\{\frac{\bar{\nu}_\mathrm{i}\omega_{N\mathrm{i}}^2(\omega^2+\omega_{B\mathrm{i}}^2)}{\omega(\omega^2-\omega_{B\mathrm{i}}^2)^2} + \sqrt{\frac{\pi}{8}}\,\frac{\omega_{N\mathrm{i}}^2}{\omega|k_z|V_{T\mathrm{i}}}\times \\ \times\Big[\exp\Big(\frac{-(\omega-\omega_{B\mathrm{i}})^2}{2k_z^2V_{T\mathrm{i}}^2}\Big)+\exp\Big(\frac{-(\omega+\omega_{B\mathrm{i}})^2}{2k_z^2V_{T\mathrm{i}}^2}\Big)\Big]\Big\}+ \\ +k_z^2\Big[\frac{\bar{\nu}_\mathrm{e}\omega_{N\mathrm{e}}^2}{\omega^3}+\sqrt{\frac{\pi}{2}}\,\frac{\omega\,\omega_{N\mathrm{e}}^2}{|k_z|^3V_{T\mathrm{e}}^3}\exp\Big(\frac{-\omega^2}{2k_z^2V_{T\mathrm{e}}^2}\Big)\Big]\Big\}\Big/\int_{(x_1)}^{(x_2)}\frac{\mathrm{d}x}{\operatorname{Re}k_x}\,\frac{\partial\operatorname{Re}k_x}{\partial\omega},\end{aligned}\right\} \quad (34.10)$$

i.e.

$$\left.\begin{aligned}\gamma = \int_{(x_1)}^{(x_2)} \frac{\mathrm{d}x}{\operatorname{Re} k_x} \Big\{k_\perp^2 \Big\{X_\mathrm{i}\,Z_\mathrm{i}\,\frac{1+Y_\mathrm{i}^2}{(1-Y_\mathrm{i}^2)^2} + \sqrt{\frac{\pi}{8}}\,X_\mathrm{i}\,Q_{\mathrm{i}z}^{-1}\times \\ \times\Big[\exp\Big(-\frac{1}{2}\,Q_{\mathrm{i}z}^{-2}(1-Y_\mathrm{i})^2\Big)+\exp\Big(-\frac{1}{2}\,Q_{\mathrm{i}z}^{-2}(1+Y_\mathrm{i})^2\Big)\Big]\Big\}+ \\ +k_z^2\Big[X_\mathrm{e}\,Z_\mathrm{e}+\sqrt{\frac{\pi}{2}}\,X_\mathrm{e}Q_{\mathrm{e}z}^{-3}\exp\Big(-\frac{1}{2}\,Q_{\mathrm{e}z}^{-2}\Big)\Big]\Big\}\Big/\int_{(x_1)}^{(x_2)}\frac{\mathrm{d}x}{\operatorname{Re}k_x}\,\frac{\partial\operatorname{Re}k_x}{\partial\omega},\end{aligned}\right\} \quad (34.10\mathrm{a})$$

with

$$k_\perp^2 = \operatorname{Re} k_x^2 + k_y^2, \qquad [34.7]$$

$$\left.\begin{aligned}&\bar{\nu}_\mathrm{e}=\nu_\mathrm{eff};\quad \bar{\nu}_\mathrm{i}=\frac{m_\mathrm{e}}{m_\mathrm{i}}\,\nu_\mathrm{eff}\quad \text{for a completely ionized,}\\ &\bar{\nu}_\mathrm{e}=\nu_\mathrm{en};\quad \bar{\nu}_\mathrm{i}=\nu_\mathrm{in}\quad \text{for a weakly ionized plasma.}\end{aligned}\right\} \quad (34.11)$$

In a homogeneous plasma Eqs. (34.10), (34.11) lead to a spectrum which has yet been described by Eqs. (25.15 ... 20) and (25.40 ... 43). In homogeneous as well as in inhomogeneous plasmas the third branch of possible oscillations is just below the ion-gyro-frequency

$$\omega^2 \lesssim \omega_{B\mathrm{i}}^2, \quad \text{i.e.} \quad Y_\mathrm{i}^2 \gtrsim 1. \qquad (34.12)$$

In the ionosphere $\bar{\nu}_\mathrm{e} \geqq \omega_{B\mathrm{i}}$ up to heights of several hundred km; $\bar{\nu}_\mathrm{i}$ is also large because of electron-ion collisions, so that such oscillations are severely damped even above the ionospheric peak height. Only at heights above 1 000 km the attenuation is small enough for such waves to exist. There is no limit at greater heights so that the range of transparency is limited by attenuation towards lower heights but unlimited for great height. Quantization is, therefore, not to be exspected. Thus, there is no fine structure of the spectrum, any frequency could be excited if only

$$\omega \lesssim \omega_{B\mathrm{i}}, \quad \text{i.e.} \quad Y_\mathrm{i} \gtrsim 1. \qquad (34.13)$$

35. Non-conservative waves in an inhomogeneous, gyrotropic plasma.

α) For completeness we now consider *other possible oscillations* in a cold, gyrotropic plasma. We start with the general eiconal-equation, Eq. (30.9), which can here be written in the form

$$\left.\begin{aligned}k_\perp^4\,\varepsilon_0\,\varepsilon_{xx}+k_\perp^2\Big[\Big(k_z^2\,\varepsilon_0-\frac{\omega^2}{c_0^2}\,\varepsilon_{xx}\Big)(\varepsilon_{xx}+\varepsilon_{zz})-\frac{\omega^2}{c_0^2}\,\varepsilon_{xy}^2- \\ -\frac{\omega^2}{c_0^2}\,(\varepsilon_{xx}\,\varepsilon_{yy}-\varepsilon_{xx}^2)\Big]+\frac{\varepsilon_{zz}}{\varepsilon_0}\Big[\Big(k_z^2\,\varepsilon_0-\frac{\omega^2}{c_0^2}\,\varepsilon_{xx}\Big)\cdot\Big(k_z^2\,\varepsilon_0-\frac{\omega^2}{c_0^2}\,\varepsilon_{yy}\Big)+ \\ +\frac{\omega^2}{c_0^4}\,\varepsilon_{xy}^2\Big]=0.\end{aligned}\right\} \quad (35.1)$$

The tensor components $\varepsilon_{ij}(\omega, \boldsymbol{k}, x)$ are determined by Eqs. (22.13) and (23.3), (23.4). From Eq. (35.1) we find two functions $k_{x\,1,2}(\omega, x)$ each of which corresponds to one of two different types of possible high-frequency oscillations in a gyrotropic plasma, the ordinary and extraordinary one[1]. Neglecting the small dissipative members which are the result of collisions between particles we obtain from Eq. (35.1) with the quantization rule, Eq. (31.19), the following dispersion equation

$$\int_{(x_1)}^{(x_2)} \mathrm{d}x \{-k_y^2 - p \mp \sqrt{p^2 - q}\}^{\frac{1}{2}} = \mathrm{n}\,\pi \qquad (\text{n integer}). \tag{35.2}$$

It allows one to determine the frequency spectra of such oscillations in a cold, gyrotropic plasma. The following abbreviations have been used:

$$\left.\begin{aligned} p &\equiv \frac{1}{2\varepsilon_0 \varepsilon_{xx}^{(0)}} \left[\left(k_z^2 \varepsilon_0 - \frac{\omega^2}{c_0^2}\, \varepsilon_{xx}^{(0)}\right)\left(\varepsilon_{xx}^{(0)} + \varepsilon_{zz}^{(0)}\right) - \frac{\omega^2}{c_0^2}\, \varepsilon_{xy}^{(0)\,2}\right], \\ q &\equiv \frac{\varepsilon_{zz}^{(0)}}{\varepsilon_0^2 \varepsilon_{xx}^{(0)}} \left[\left(k_z^2 \varepsilon_0 - \frac{\omega^2}{c_0^2}\, \varepsilon_{xx}^{(0)}\right)^2 + \frac{\omega^4}{c_0^4}\, \varepsilon_{xy}^{(0)\,2}\right]. \end{aligned}\right\} \tag{35.3}$$

From Eq. (35.1) all five branches of oscillations in the transparency ranges of a cold, gyrotropic plasma can be determined, see Sect. 25. At the limit

$$\frac{c_0^2}{\omega^2}\, k_z^2 \gg \left(\frac{\varepsilon_{xx}}{\varepsilon_0},\ \frac{\varepsilon_{xy}}{\varepsilon_0}\right). \tag{35.4}$$

Eq. (35.1) becomes identical with the eiconal equation of longitudinal waves, see Eq. (34.2).

On high (electronic) frequencies one of both Eqs. (35.2) coincides with Eq. (34.3); in this limit the oscillations described by the other Eq. (35.2) are not obtained because of lack of transparency $\left(k^2(\omega, x) < 0\right)$. On the other side for low frequencies we get Eq. (34.9) out of Eq. (35.2).

If account is taken of small dissipation a damping of the oscillations appears as described by Eq. (35.2). In the limit

$$\frac{c_0^2}{\omega^2}\, k_z^2 \gg \left(\frac{\varepsilon_{xx}}{\varepsilon_0},\ \frac{\varepsilon_{xy}}{\varepsilon_0}\right) \qquad [35.4]$$

the same expressions for the attenuation decrement are obtained for high frequencies as given by Eq. (34.6), for low ones as given by Eq. (34.10). As we have done in Sect. 25 for a homogeneous, gyrotropic plasma we now deduce the spectra of ALFVÉN waves, of fast magneto-sound waves and of spiral waves.

β) As shown in Sect. 25 *fast magneto-sound* as well as ALFVÉN *waves* exist under the condition

$$\omega \ll \omega_{B\,\mathrm{i}}, \quad \text{i.e.} \quad Y_{\mathrm{i}} \gg 1, \tag{35.5}$$

thus on rather low frequencies. The dispersion relations, Eq. (35.2) read for such waves:

$$\int_{(x_1)}^{(x_2)} \mathrm{d}x\, \mathrm{Re}\, k_x(\omega, x) \equiv \int_{(x_1)}^{(x_2)} \mathrm{d}x \sqrt{-k_y^2 - k_z^2 + \frac{\omega^2}{c_0^2}\left(1 + \frac{c_0^2}{V_{\mathrm{A}}^2}\right)} = \mathrm{n}\,\pi, \quad (\text{n integer}) \tag{35.6}$$

and

$$\int_{(x_1)}^{(x_2)} \mathrm{d}x\, \mathrm{Re}\, k_x(\omega, x) \equiv \int_{(x_1)}^{(x_2)} \mathrm{d}x \left\{-k_y^2 - \frac{\omega_{\mathrm{Ne}}^2}{c_0^2}\left[1 - \frac{k_z^2 V_{\mathrm{A}}^2}{\omega^2 (1 + V_{\mathrm{A}}^2/c_0^2)}\right]\right\}^{\frac{1}{2}} = \mathrm{n}\,\pi, \tag{35.7}$$

[1] See footnote on p. 558 for different designations.

respectively. In a homogeneous plasma these oscillations correspond to the spectra ω_1 and ω_2 as described by Eqs. (25.24), (25.25). It further follows from Eqs. (35.6), (35.7) for an inhomogeneous plasma that the *fast magneto-sound wave* is only allowed in that spatial range where

$$\omega^2 > (k_y^2 + k_z^2)\, V_A^2, \tag{35.8}$$

while the ALFVÉN *wave* does exist under a different condition only, viz.

$$\omega^2 < k_z^2 V_A^2. \tag{35.9}$$

Now, the ALFVÉN velocity is

$$V_A = \frac{B}{\sqrt{u \mu_0 \varrho}}. \qquad [25.25]$$

In a completely ionized plasma where $\varrho \approx m_i N_i$, with the definitions of $\omega_{N\,i}$ and $\omega_{B\,i}$ (see Sect. 6) we have:

$$V_A^2 \equiv c_0^2 \left(\frac{q_i}{m_i} \frac{B}{c_0 \sqrt{\varepsilon_0 \mu_0}} \right)^2 \Big/ \frac{u q_i^2 N_i}{\varepsilon_0 m_i} = c_0^2 \frac{\omega_{B\,i}^2}{\omega_{N\,i}^2} \left(= c_0^2 \frac{Y_i^2}{X_i} \right). \tag{35.10}$$

Therefore in a completely ionized plasma the conditions of existence can be written

for *fast magneto-sound waves*:

$$\omega_{N\,i}^2 > (k_y^2 + k_z^2) \frac{c_0^2}{\omega^2} \omega_{B\,i}^2, \tag{35.11}$$

i.e.

$$X_i > \left(\frac{c_0}{\omega} \right)^2 (k_y^2 + k_z^2)\, Y_i^2; \tag{35.11a}$$

and for ALFVÉN *waves*:

$$\omega_{N\,i}^2 < k_z^2 \frac{c_0^2}{\omega^2} \omega_{B\,i}^2, \tag{35.12}$$

i.e.

$$X_i < \left(\frac{c_0}{\omega} \right)^2 k_z^2\, Y_i^2. \tag{35.12a}$$

Thus fast magneto-sound waves need high ion density to exist, the magnetic field influence should not be too important. The situation is quite different for ALFVÉN waves which can only occur below a certain value of ion density, i.e. with a great field influence. In a plasma layer (for almost invariable magnetic field) fast magneto-sound waves are therefore trapped in a transparency range inside the layer between turning points on either side, and quantization should be observable. ALFVÉN waves on the other hand have two transparency ranges, one above and one below the layer. It depends on the outside configuration whether trapping and, a fortiori, quantization occurs for ALFVÉN waves. There is no such condition below the terrestrial ionosphere because ALFVÉN waves which need high conductivity can not propagate in the plasma-free space between Earth's surface and the lower border of the ionosphere. In the magnetosphere quantization of the spectra of ALFVÉN and fast magneto-sound waves is of no practical interest since the plasma density is almost constant and varies very slowly.

γ) Expressions for the *attenuation decrements* due to the dissipative terms in the eiconal equation, Eq. (35.1), are derived using the condition of existence $Y_i \equiv \omega_{B\,i}/\omega \gg 1$. One finds

for *fast magneto-sounds waves*:

$$\gamma = -\frac{1}{2} \frac{\displaystyle\int_{(x_1)}^{(x_2)} \frac{dx}{\operatorname{Re} k_x(\omega, x)} \left[\sum_h \frac{\bar{\nu}_h \omega_{N\,h}^2}{\omega_{B\,h}^2} + \sqrt{\frac{\pi}{2}} \frac{\omega_{N\,e}^2 k_\perp^2 V_{Te}}{|k_z| \omega_{B\,e}^2} \exp\left(\frac{-\omega^2}{2 k^2 V_{Te}^2} \right) \right]}{\displaystyle\int_{(x_1)}^{(x_2)} \frac{dx}{\operatorname{Re} k_x(\omega, x)} \left(1 + \frac{c_0^2}{V_A^2} \right)} \tag{35.13}$$

and for ALFVÉN *waves*:

$$\gamma = -\frac{1}{2}\,\frac{\displaystyle\int_{(x_1)}^{(x_2)} \frac{\mathrm{d}x}{\mathrm{Re}\,k_x}\left[\sum_{\mathrm{h}} \frac{\bar\nu_{\mathrm{h}}\,\omega_{N\mathrm{h}}^2}{\omega_{B\mathrm{h}}^2}\,\frac{\omega_{N\mathrm{e}}^2}{(1+c_0^2/V_A^2)} + \sqrt{\frac{\pi}{2}}\,\frac{k_\perp^2\,c_0^2\,\omega^4}{|k_z|^3\,V_{T\mathrm{e}}^3}\exp\left(\frac{-\omega^2}{2k_z^2\,V_{T\mathrm{e}}^2}\right)\right]}{\displaystyle\int_{(x_1)}^{(x_2)} \frac{\mathrm{d}x}{\mathrm{Re}\,k_x(\omega,x)}\,\frac{\omega_{N\mathrm{e}}^2}{\omega^2}\,\frac{k_z^2\,V_A^4}{(1+V_A^2/c_0^2)}}\,. \tag{35.14}$$

The bracket in Eq. (35.13) could also be written as

$$\omega\left[\sum_{\mathrm{h}} X_{\mathrm{h}} Z_{\mathrm{h}} Y_{\mathrm{h}}^{-2} + \sqrt{\frac{\pi}{2}}\,X_{\mathrm{e}}\varrho_{\mathrm{e}\perp}^2\,\varrho_{\mathrm{e}z}^{-1}\,Y_{\mathrm{e}}^{-2}\exp\left(-\frac{1}{2}\,\varrho_{\mathrm{e}z}^{-2}\right)\right], \tag{35.13a}$$

and that in Eq. (35.14)

$$\omega^3\left[\sum_{\mathrm{h}} X_{\mathrm{h}} Z_{\mathrm{h}} Y_{\mathrm{h}}^{-2}\,X_{\mathrm{e}}\Big/\left(1+\frac{c_0^2}{V_A^2}\right) + \sqrt{\frac{\pi}{2}}\,\varrho_{\mathrm{e}\perp}^2\,\varrho_{\mathrm{e}z}^{-3}\left(\frac{c_0}{V_{T\mathrm{e}}}\right)^2\exp\left(-\frac{1}{2}\,\varrho_{\mathrm{e}z}^{-2}\right)\right]. \tag{35.14a}$$

Again $k_\perp^2 = \mathrm{Re}\,k_x^2(\omega, x) + k_y^2$ and $\mathrm{Re}\,k_x$ may be determined from Eqs. (35.6), (35.7), and

$$\left.\begin{array}{lll} \bar\nu_{\mathrm{e}} = \nu_{\mathrm{eff}}\,\dfrac{\omega^2}{\omega_{B\mathrm{e}}^2}; & \bar\nu_{\mathrm{i}} = \dfrac{m_{\mathrm{e}}}{m_{\mathrm{i}}}\,\nu_{\mathrm{eff}}\,\dfrac{\omega^2}{\omega_{B\mathrm{e}}^2} & \text{for completely ionized plasmas} \\ \text{but} & & \\ \bar\nu_{\mathrm{e}} = \nu_{\mathrm{en}}; & \bar\nu_{\mathrm{i}} = \nu_{\mathrm{in}} & \text{for weakly ionized plasmas.} \end{array}\right\} \qquad [34.11]$$

In the limit of a spatially homogeneous plasma Eqs. (35.13), (35.14) coincide with Eqs. (25.46), (25.47).

δ) Let us now consider electromagnetic waves in the "magnetic" or "*whistler*" *mode* in an inhomogeneous plasma[2]. As shown in Sect. 25 such waves can only exist if the plasma is dense enough, the condition being

$$\omega_{B\mathrm{i}} \ll \omega \ll \omega_{B\mathrm{e}}, \quad \text{i.e.} \quad Y_{\mathrm{i}} \ll 1 \ll Y_{\mathrm{e}}.$$

It is easily shown from the dispersion relation, Eq. (35.2) that with $\omega_{N\mathrm{e}}^2 \gg \omega\,\omega_{B\mathrm{e}}$ the frequency spectrum is determined by the equation

$$\int_{(x_1)}^{(x_2)} \mathrm{d}x\,\mathrm{Re}\,k_x(\omega, x) \equiv \int_{x_1}^{(x_2)} \mathrm{d}x\left\{-k_y^2 - k_z^2 + \frac{\omega^2\,\omega_{N\mathrm{e}}^4}{c_0^4\,k_z^2\,\omega_{B\mathrm{e}}^2}\right\}^{\frac{1}{2}} = \mathrm{n}\pi \quad \mathrm{n}\ \text{(integer)} \tag{35.15}$$

In the limit of a homogeneous plasma this spectrum coincides with that of "whistler" waves given by Eq. (25.33). After Eq. (35.15) in an inhomogeneous plasma these waves are submitted to the condition

$$\omega_{N\mathrm{e}}^4 > c_0^4\,k_z^2\,(k_y^2 + k_z^2)\,\frac{\omega_{B\mathrm{e}}^2}{\omega^2}\,. \tag{35.16}$$

For a plasma layer this leads to the conclusion that there exists one transparency range between two turning points, an upper one and a lower one. We must therefore expect quantization of the spectrum of "whistler" waves in the ionosphere. These waves have the attenuation decrement:

$$\left.\begin{aligned} \gamma = -\frac{1}{4}\left[\int_{(x_1)}^{(x_2)} \mathrm{d}x\,\frac{\omega_{N\mathrm{e}}^4}{\mathrm{Re}\,k_x}\right]^{-1} \cdot \int_{(x_1)}^{(x_2)} \mathrm{d}x\,\mathrm{Re}\,k_x\left\{\frac{\omega_{N\mathrm{e}}^2\,\bar\nu_{\mathrm{e}}\,(k^2+k_z^2)}{k^3} + \right. \\ \left. + \sqrt{\frac{\pi}{2}}\,\frac{\omega_{N\mathrm{e}}^2\,\eta\,\omega^2\,(\mathrm{Re}\,k_x^2 + k_y^2)}{k^2\,|k_z|\,V_{T\mathrm{e}}}\exp\left(-\frac{\omega^2}{2k_z^2\,V_{T\mathrm{e}}^2}\right)\right\}, \end{aligned}\right\} \tag{35.17}$$

[2] The russian term is, literally translated, "spiral waves" thus giving a more direct description than the term "whistler" which is generally used in English texts. See K. RAWER and K. SUCHY, Vol. 49/3, p. 493, of this Encyclopedia where the origin of the expression is explained.

where

$$k^2 = \text{Re}\, k_x^2 + k_y^2 + k_z^2$$

and

$$\bar{\nu}_e = \nu_{\text{eff}} \quad \text{for highly ionized plasma},$$

but

$$\bar{\nu}_e = \bar{\nu}_{en} \quad \text{for weak ionization},$$

and

$$\eta = \begin{Bmatrix} \dfrac{\omega^4}{2k_z^4 V_{Te}^4} & \text{if} \quad \omega^2 \gg k_z^2 V_{Te}^2 \\ 1 & \text{if} \quad \omega^2 \ll k_z^2 V_{Te}^2. \end{Bmatrix} \tag{35.18}$$

Since in the exosphere and magnetosphere the plasma density is almost constant quantization is not really occuring to the spectra of "whistler" waves. Frequency spectrum and attenuation decrement can therefore be obtained from Eqs. (25.33) and (26.32).

ε) After analyzing "whistler" waves in a spatially homogeneous plasma in Sect. 25 we could show in Sect. 26 that the relevant spectrum remains unchanged in a *hot plasma* too, provided

$$k_z V_{Ti} \ll \omega \ll k_z V_{Te}, \quad \text{i.e.} \quad Q_{iz} \ll 1 \ll Q_{ez}. \qquad [25.34]$$

This statement apparently remains valid in an inhomogeneous plasma too.

The spectrum of ALFVÉN waves (for $Y_i \equiv \omega_{Bi}/\omega \gg 1$) as determined by the dispersion relations Eqs. (35.7) and (35.14), can be continued analytically into that range of phase velocities. In a completely ionized plasma one must take account in this particular case of ion-ion collisions. This is achieved (see Sect. 26) by inserting into Eq. (35.14)

$$\bar{\nu}_i = \frac{m_e}{m_i} \nu_{\text{eff}} + \frac{7}{10} \nu_{ii} k_\perp^2 V_{Ti}^2/\omega_{Bi}^2. \qquad [26.5]$$

As to magneto-sound waves in a hot plasma their spectrum cannot remain unchanged in the phase velocity range ($V_{\text{ph}\,z} \equiv \omega/k_z$):

$$V_{Ti} \ll V_{\text{ph}\,z} \ll V_{Te}.$$

Moreover, there appears a new kind of wave in a hot plasma namely a *slow magneto-sound wave*. It can only exist in a non-isothermal plasma where

$$T_e \gg T_i.$$

The eiconal equation, Eq. (30.9), for magneto-sound waves takes the form

$$\left(k^2 \varepsilon_0 - \frac{\omega^2}{c_0^2} \varepsilon_{yy}\right) \varepsilon_{zz} + \frac{\omega^2}{c_0^2} \varepsilon_{yz}^2 = 0, \tag{35.19}$$

see Eq. (26.1) or (26.9). The elements of the dielectricity tensor $((\varepsilon_{ij}(\omega, \boldsymbol{k}, x)))$ can be determined with Eqs. (22.15), (23.7) and (23.14). Neglecting the small dissipative terms, and using the quantization rule Eq. (31.19), we get the following dispersion relation for determining the frequency spectrum of magneto-sound waves in an inhomogeneous plasma:

$$\begin{aligned} &\int_{(x_1)}^{(x_2)} dx\, \text{Re}\, k_x(\omega, x) \\ &\equiv \int_{(x_1)}^{(x_2)} dx \left\{ -k_y^2 + \frac{(\omega^2 - k_z^2 V_A^2)(\omega^2 - k_z^2 V_s^2)}{\omega^2 (V_A^2 + V_s^2) - k_z^2 V_A^2 V_s^2} \right\}^{\frac{1}{2}} = n\pi \quad (\text{n integer}). \end{aligned} \tag{35.20}$$

In the limit of a spatially homogeneous plasma this spectrum coincides with that given by Eqs. (26.10), (26.11). When writing down Eq. (35.20) we supposed that the plasma density is high enough so that [see Eq. (35.10)]

$$c_0^2 \gg V_A^2, \quad \text{i.e.} \quad \omega_{Ni}^2 \gg \omega_{Bi}^2, \quad X_i \gg Y_i^2. \tag{35.21}$$

This condition is satisfied in almost all real plasmas, including that of ionosphere and magnetosphere[3] (see Table 3, p. 402).

In a rarified plasma where

$$\beta \equiv \frac{2u\mu_0}{B_0^2} P_0 \ll 1, \tag{35.22}$$

Eq. (35.20) splits into a pair of equations, one of which coincides with Eq. (35.6) under the condition of Eq. (35.21) (which is almost ever satisfied). This is the relation determining the spectrum of fast magneto-sound waves[4]. The second equation of the pair takes the form

$$\int_{(x_1)}^{(x_2)} \mathrm{d}x \,\mathrm{Re}\, k_x(\omega, x) \equiv \int_{(x_1)}^{(x_2)} \mathrm{d}x \left\{ -k_y^2 - k_z^2 + \frac{k_z^2 V_s^2 \omega^2}{V_A^2(\omega^2 - k_z^2 V_s^2)} \right\}^{\frac{1}{2}} = n\pi \quad \text{(n integer)}. \tag{35.23}$$

It describes ion sound waves in a non-isothermal, gyrotropic plasma where

$$T_e \gg T_i.$$

From the condition for the radicand to be positive follows

$$k_z^2 V_s^2 \omega^2 > V_A^2 (\omega^2 - k_z^2 V_s^2)(k_y^2 + k_z^2)$$

which gives with Eq. (35.10) for a completely ionized plasma

$$\omega_{Ni}^2 \left(\frac{V_s}{c_0}\right)^2 > \left(\frac{\omega_{Bi}}{\omega}\right)^2 (\omega^2 - k_z^2 V_s^2) \frac{k_y^2 + k_z^2}{k_z^2} \tag{35.24}$$

i.e.

$$X_i \left(\frac{V_s}{c_0}\right)^2 > Y_i^2 (1 - Q_{nz}^2) \left(1 + \frac{k_y^2}{k_z^2}\right). \tag{35.24a}$$

Propagation of ion-sound waves thus requires comparatively high ion density so that one transparency range appears inside a plasma layer. Being trapped inside the layer quantization is to be exspected for ion-sound waves.

The attenuation decrement of ion-sound obtained from the small dissipative members in Eq. (35.19) is

$$\gamma = -\frac{1}{2} \frac{\displaystyle\int_{(x_1)}^{(x_2)} \frac{\mathrm{d}x}{\mathrm{Re}\, k_x(\omega, x)} \frac{V_s^4}{V_A^2(\omega^2 - k_z^2 V_s^2)^2} \left(\bar{\nu}_i + |\omega| \sqrt{\frac{\pi}{2}\frac{q_i}{q_e}\frac{m_e}{m_i}}\right)}{\displaystyle\int_{(x_1)}^{(x_2)} \frac{\mathrm{d}x}{\mathrm{Re}\, k_x(\omega, x)} \frac{V_s^4}{V_A^2(\omega^2 - k_z^2 V_s^2)^2}}, \tag{35.25}$$

where

$$\left.\begin{aligned} \bar{\nu}_i &= \frac{5}{2} \nu_{ii} \frac{q_e}{q_i} \frac{T_i}{T_e} && \text{for a completely ionized} \\ \text{and} \quad & \\ \bar{\nu}_i &= \nu_{in} && \text{for a weakly ionized plasma.} \end{aligned}\right\} \tag{35.26}$$

---

[3] Only upon such plasmas the fluid model can be applied, i.e. the equation of gasdynamics, Eq. (11.13), is justified.

[4] Dissipative effects are taken into account by the attenuation decrement given in Eq. (35.13).

ζ) As has been shown in Sect. 25 *ion-sound waves* can exist *at higher frequencies* for

$$\omega > \omega_{B\,\mathrm{i}}, \quad \text{i.e.} \quad Y_\mathrm{i} < 1. \tag{35.27}$$

In the limit of very high frequencies when even $Y_\mathrm{e} \ll 1$, the influence of the external magnetic field becomes negligible. In that case Eqs. (33.11), (33.12) apply for determining the spectrum[5]. If, however, ω is in the intermediate range

$$\omega_{B\,\mathrm{i}} \ll \omega \ll \omega_{B\,\mathrm{e}}, \quad \text{i.e.} \quad Y_\mathrm{i} \ll 1 \ll Y_\mathrm{e} \tag{35.28}$$

then the magnetic field influence has to be considered. Since $Y_\mathrm{i} \ll 1$, the ions are almost uninfluenced by the magnetic field but this is not so for the electrons. Taking account of the fact that ion-sound waves propagate almost parallel with the field in a magneto-active plasma we may write the eiconal equation in this intermediate frequency range in the simplified form

$$k^2 V_\mathrm{s}^2 \left(1 - i\,\frac{\bar{\nu}_\mathrm{i}}{\omega}\right) - \omega^2 \left(1 + i \sqrt{\frac{\pi}{2}}\,\frac{\omega}{|k_z|\,V_{T\mathrm{e}}}\right) = 0, \tag{35.29}$$

where $\bar{\nu}_\mathrm{i}$ is given by Eq. (35.26) above. Using the quantization rule, Eqs. (31.19), (31.20), this equation leads to a dispersion relation which allows to determine the *frequency spectrum*

$$\int_{(x_1)}^{(x_2)} \mathrm{d}x \,\mathrm{Re}\, k_x(\omega, x) \equiv \int_{(x_1)}^{(x_2)} \mathrm{d}x \sqrt{-k_y^2 - k_z^2 + (\omega/V_\mathrm{s})^2} = \mathrm{n}\,\pi \quad (\mathrm{n}\ \text{integer}), \tag{35.30}$$

and *attenuation decrement*

$$\gamma = -\frac{1}{2} \int_{(x_1)}^{(x_2)} \frac{\mathrm{d}x}{\mathrm{Re}\,k_x} \left[\frac{\bar{\nu}_\mathrm{i}}{V_\mathrm{s}^2} + \sqrt{\frac{\pi}{2}\,\frac{q_\mathrm{i}}{q_\mathrm{e}}\,\frac{m_\mathrm{e}}{m_\mathrm{i}}}\,\frac{\omega^2}{|k_z|\,V_\mathrm{s}^3}\right] \Big/ \int_{(x_1)}^{(x_2)} \frac{\mathrm{d}x}{\mathrm{Re}\,k_x(\omega, x)}\,\frac{1}{V_\mathrm{s}^2}. \tag{35.31}$$

When deriving the above Eqs. (35.29 ... 31) we have again supposed that the ion density is comparatively high

$$\omega_{N\mathrm{i}}^2 \gg \omega^2, \quad \text{i.e.} \quad X_\mathrm{i} \gg 1. \tag{35.32}$$

In fact, ion-sound oscillations can exist when only the simple inequality $X_\mathrm{i} > 1$ is satisfied. If, however, this is not satisfied, then no oscillation is possible. Ion-sound waves are therefore *reflected* at surfaces where

$$\omega_{N\mathrm{i}}^2(x) = \omega^2, \quad \text{i.e.} \quad X_\mathrm{i} = 1. \tag{35.33}$$

Therefore, in a plasma layer ion-sound waves could have one transparency range if somewhere inside the layer the condition $T_\mathrm{e} > 6\,T_\mathrm{i}$ is satisfied. If this happens they will be trapped and quantization must occur. This is not likely to be the case for the terrestrial ionosphere. In the magnetosphere structure is field-aligned so that trapping could occur. However conditions in the magnetosphere seem to be near isothermal ($T_\mathrm{e} \approx T_\mathrm{i}$) and therefore ion sound waves should not exist as trapped waves in the magnetosphere either.

η) *Summarising* we may say that the description given in Sects. 34 and 35 is a quantitative one applying to a (weakly) inhomogeneous, gyrotropic plasma. Apart from the quantization of spectra it also gives the attenuation in these spectra. In the approximation of "geometric optics" the fact that the spectra are, in principle, discrete is not really important, see Sect. 33 δ. The so-called "local spectra" which are strictly valid for homogeneous plasmas have been obtained in Sects. 25 and 26 immediately from the eiconal equation[6]. They are

[5] These have been obtained without a mapnetic field.

[6] We understand as "local" those spectra which are simply obtained as functions $\omega(\boldsymbol{k}, x)$ appearing as solutions of the eiconal-equation, Eq. (30.9).

useful as a qualitative description but cannot be taken as to be quantitatively correct in a spatially inhomogeneous plasma. This is in particular so with weakly attenuated oscillations of large wavelength when the thickness of the transparency range can be compared with the characteristic scale of inhomogeneity of the plasma. Under such conditions local spectra are certainly invalid. Correct expressions have been obtained in the last sections by applying integral relations which are *quantization rules*. True quantization appears only in "trapping regions" which are *transparency ranges* limited by *turning points* on both sides. If one of these is missing quantization breakes down and we have a continuous spectrum.

II. Low frequency drifting oscillations in an inhomogeneous plasma. The problems of magnetic confinement of a plasma.

36. A kinetic model of drifting oscillations.

α) As already stated in Sect. 30 ... 33 there appears a new characteristic frequency in an inhomogeneous gyrotropic (magneto-active) plasma which is called the LARMOR *drift frequency*. It is due to diamagnetic currents which arise in an inhomogeneous plasma if confined by a magnetic field (see Sect. 30).

Its order of magnitude has been found to be (for charged particles of kind h)

$$\omega_{\mathrm{dh}} \sim \frac{k_y V_{T\mathrm{h}}^2}{\omega_{B\mathrm{h}} \mathscr{L}_0}. \tag{36.1}$$

Consequently, there appear new characteristic branches of possible oscillations in the frequency/wavenumber diagram on frequencies slightly below that given by Eq. (36.1). These oscillations are *qualitatively different* from the ones we have considered in the preceding sections, particularly because instability occurs even with an isotropic (Maxwellian) equilibrium distribution of particle velocities.

For this reason drifting oscillations have aroused much interest in investigations studying the stability of magnetic confinement in plasma, in conjunction with thermonuclear devices[1]. We intend to discuss such problems here only with respect to possible occurrence in the ionosphere or in the magnetosphere. Let us first of all give a few estimates of drift oscillation frequencies for electrons and ions under such conditions.

Eq. (36.1) shows the characteristic frequency to be dependent upon the ratio

$$\frac{V_{T\mathrm{h}}^2}{\omega_{B\mathrm{h}}} \propto \frac{\vartheta_{\mathrm{h}}/m_{\mathrm{h}}}{q_{\mathrm{h}}/m_{\mathrm{h}}} = \frac{\vartheta_{\mathrm{h}}}{q_{\mathrm{h}}},$$

so that under thermal equilibrium identic values[2] of the drift frequency should be found for electrons and ions. Also according to Eq. (36.1) ω_{d} is proportional to the wave number (in the direction perpendicular to the stratification and to the magnetic field), k_y. Supposing

$$\frac{k_y V_{T\mathrm{i}}}{\omega_{B\mathrm{i}}} \equiv k_y r_{B\mathrm{i}} \lesssim 1, \tag{36.2}$$

and taking account of the characteristic scale of "regular inhomogeneity", $\mathscr{L}_0$, which is about 100 km for the ionospheric plasma, we come to the estimation (compare Table 2 on p. 401):

$$\omega_{\mathrm{di}} \leqq (10^{-3} \dots 10^{-1})\ \mathrm{Hz}. \tag{36.3}$$

Since in the ionosphere $T_{\mathrm{e}}/T_{\mathrm{i}}$ takes values between 1 and 3 (maximum) we obtain

$$\omega_{\mathrm{de}} \leqq (10^{-3} \dots 3 \cdot 10^{-1})\ \mathrm{Hz}. \tag{36.4}$$

---

[1] We cannot discuss these problems here but reference is made to [43], [46] and to the survey given by A. B. MIHAJLOVSKIJ and B. B. KADOMČEV in [6].

[2] In the upper ionosphere usually $T_{\mathrm{e}} > T_{\mathrm{i}}$; both are, however, of the same order of magnitude. Therefore the orders of ω_{d} are similar for ions and electrons.

In the whole ionosphere such extremely low frequencies are much smaller than the collision frequencies. Only in the magnetosphere above a height of the order of Earth' radius, R_E, (i.e. for $r/R_E \gtrsim 2$) we find collisions frequencies which are of the order of ω_{di} or ω_{de}. Therefore collisional effects must be taken into account when dealing with drifting oscillations in the ionosphere, and in the innermost magnetosphere.

β) We start with the expressions obtained in Sect. 30 ... 33 for the components of the dielectric permeability tensor of an inhomogeneous plasma. However, we must now generalize these, by introducing collisional effects. There is another quite important simplification. In the ionospheric plasma the inequality

$$\beta \equiv \frac{2u\mu_0}{B_\delta^2} N(\vartheta_e + \vartheta_i) \ll 1 \tag{36.5}$$

is very well satisfied. When studying small plasma oscillations we are therefore able to neglect fluctuations of the magnetic field. Thus, the *oscillations* can be assumed *to depend on a potential field*

$$\boldsymbol{E} = -\frac{\partial}{\partial \boldsymbol{r}} \phi \equiv -\operatorname{grad} \phi. \tag{36.6}$$

For the case of a homogeneous, gyrotropic plasma we could prove in Sects. 24 and 26 that the electromagnetic field oscillations can be described as to depend on a potential ϕ if condition Eq. (36.5) is satisfied. Therefore, we may only take account of the effect which acts parallel with the magnetic field. The spectrum of potential oscillations in an inhomogeneous, gyrotropic plasma is thus characterized by the longitudinal dielectric permeability alone.

On the other hand LARMOR drift oscillations can only be important if most particles do not suffer from a collision during one period of the LARMOR motion, i.e. under the condition ($h = e, i$)

$$\omega_{Bh} \gg \bar{\nu}_h, \quad \text{i.e.} \quad Y_h \gg Z_h. \tag{36.7}$$

In the ionosphere this condition is fulfilled at heights above 200 km where collisions between charged particles and neutrals become negligible. Therefore we may neglect these collisions and retain only COULOMB collisions between charged particles. This is a fortiori true at greater heights, and in the magnetosphere[3].

We have to distinguish two different models corresponding to two different *limits of reasonning:* the kinetic model and the fluid-dynamic one. The second model is much easier because the two-fluid model is good enough, see Eqs. (9.2) and (13.3...5) and Sects. 37, 38. In the present section we intend to apply the kinetic model by resolving the kinetic equations; these contain collision integrals, see Eq. (12.13).

Supposing Eq. (36.7) to be valid the collisional influence upon the distribution functions of the particles can be neglected. We shall therefore use the equilibrium distribution functions valid for a collisionless plasma, i.e. in the form of Eq. (30.18), taking $F_{0h}(\mathscr{E}_h, x)$ as a locally inhomogeneous Maxwellian distribution after Eq. (30.19).

γ) In order to study plasma stability and *possible oscillations* with the distribution function

$$f_{0h} \approx \left[1 + \frac{v_\perp \sin\varphi}{\omega_{Bh}} \frac{\partial}{\partial x}\right] F_{0h}(\zeta_h, x), \tag{30.18}$$

we suppose small deviations from this distribution to occur; these must be accompanied by small fluctuations of the parallel electric field[4]. The distribution function being dependent on the coordinate of stratification x only, but not on y and z we may look for a solution in the form

$$\delta f_h(x) \exp[-i\omega t + i k_y y + i k_z z]. \tag{36.8}$$

[3] We just note that Eq. (36.7) is also satisfied in all known thermonuclear plasma devices.

[4] As the effects of thermal motions are small, the relevant currents and fluctuations of the magnetic field can be neglected too. We are thus left with a fluctuating electric field deriving from Eq. (36.6).

The linearized kinetic equation

$$\frac{\partial f_{\rm h}}{\partial t} + \boldsymbol{v} \cdot \frac{\partial f_{\rm h}}{\partial \boldsymbol{r}} + q_{\rm h} \left[\boldsymbol{E} + \frac{1}{c_0 \sqrt{\varepsilon_0 \mu_0}} \boldsymbol{v} \times \boldsymbol{B}\right] \cdot \frac{\partial f_{\rm h}}{\partial \boldsymbol{p}} = \left(\frac{\delta f_{\rm h}}{\delta t}\right)^{\rm hg} \quad [12.13]$$

can be applied to the deviation from the equilibrium distribution, leading to

$$\left.\begin{aligned} -i(\omega - k_y v_y - k_z v_z)\, \delta f_{\rm h} + v_x \frac{\partial \delta f_{\rm h}}{\partial x} - \omega_{B\rm h} \frac{\partial \delta f_{\rm h}}{\partial \varphi} \\ = \frac{1}{c_0 \sqrt{\varepsilon_0 \mu_0}} \frac{q_{\rm h}}{m_{\rm h}} \frac{\partial \phi}{\partial \boldsymbol{r}} \cdot \frac{\partial f_{0\rm h}}{\partial \boldsymbol{v}} \sum_{\rm h} \left(\frac{\partial \delta f_{\rm h}}{\partial t}\right)^{\rm hg}_{\rm col}. \end{aligned}\right\} \quad (36.9)$$

Here φ is the polar angle against the magnetic field direction $\boldsymbol{B}^0_{\delta} \equiv \boldsymbol{z}^0$. The last term is the sum over the (linearized) collisional integral between charged particles of the considered kind, h, with charged particles of all other kinds, g:

$$\left.\begin{aligned} \left(\frac{\partial \delta f_{\rm h}}{\partial t}\right)^{\rm hg}_{\rm col} = q_{\rm h}^2 q_{\rm g}^2 \mathscr{L} \frac{\partial}{\partial \boldsymbol{p}} \cdot \int \frac{d\boldsymbol{p}'}{w^3} (w^2 \boldsymbol{E} - \boldsymbol{w} \cdot \boldsymbol{w}) \times \\ \times \left\{ f_{0\rm g} \frac{\partial \delta f_{\rm h}}{\partial \boldsymbol{p}} + \delta f_{\rm g} \frac{\partial f_{0\rm h}}{\partial \boldsymbol{p}} - f_{0\rm h} \frac{\partial \delta f_{\rm g}}{\partial \boldsymbol{p}} - \delta f_{\rm h} \frac{\partial f_{0\rm g}}{\partial \boldsymbol{p}} \right\} \end{aligned}\right. \quad (36.10)$$

which reads in coordinates (applying the summation rule upon i and j)

$$\left.\begin{aligned} \left(\frac{\partial \delta f_{\rm h}}{\partial t}\right)^{\rm hg}_{\rm col} = q_{\rm h}^2 q_{\rm g}^2 \mathscr{L} \frac{\partial}{\partial p_i} \int \frac{d\boldsymbol{p}'}{w^3} (w^2 \delta_{ij} - w_i w_j) \times \\ \times \left\{ f_{0\rm g} \frac{\delta f_{\rm h}}{\partial p_j} + \delta f_{\rm g} \frac{\partial f_{0\rm h}}{\partial p_j} - f_{0\rm h} \frac{\partial \delta f_{\rm g}}{\partial p'_j} - \delta f_{\rm h} \frac{\partial f_{0\rm g}}{\partial p'_j} \right\}. \end{aligned}\right\} \quad (36.10\text{a})$$

Here $\boldsymbol{p}$ is the momentum,

$$\boldsymbol{w} \equiv \boldsymbol{v}_{\rm h} - \boldsymbol{v}_{\rm g} \quad (36.11)$$

and

$$\mathscr{L} = \log \frac{\lambda_{\rm D}}{r_{\rm min}} \quad (36.12)$$

$$= \log (k_{\rm max}/k_{\rm min}) \quad [12.20\text{a}]$$

is COULOMB's logarithm (see Sect. 12ζ for detailed explanation). The summation on the right-hand side of Eq. (36.9) extends over all kinds of charged particles in the plasma (including kind h itself).

δ) We suppose now that the fluctuations of the wavelength along the x-axis have a much smaller scale than that given by the characteristic scale of inhomogeneity, $\mathscr{L}_0$. A solution of Eq. (36.9) can then be obtained in the form of the *approximation of geometrical optics*, i.e. as

$$\exp\left(i \int^x dx\, k_x\right).$$

Resolving then the system of kinetic equations [see Eqs. (36.9)] for electrons and protons we find for induced charge density, $\delta\varrho$, due to the non-equilibrium potential field

$$\boldsymbol{E} = -\frac{\partial}{\partial \boldsymbol{r}} \phi$$

$$\delta \varrho = \sum_{\rm h} q_{\rm h} \int d\boldsymbol{p}\, \delta f_{\rm h} = \frac{k^2}{\rm u} [\varepsilon_0 - \varepsilon(\omega, \boldsymbol{k}, x)]\, \phi. \quad (36.13)$$

The longitudinal dielectric permeability ε can then be found by adding the contributions to the relative permeability of electrons, $\delta\varepsilon_{\rm re}$, and ions, $\delta\varepsilon_{\rm ri}$:

$$\frac{1}{\varepsilon_0}\,\varepsilon(\omega,\boldsymbol{k},x)=1+\delta\varepsilon_{\rm re}+\delta\varepsilon_{\rm ri}. \tag{36.14}$$

Below in Sects. 37 and 38 we use this expression for a study of potential oscillations in a plasma.

In the general case of arbitrary frequencies and wave-numbers the computation of $\varepsilon(\omega,\boldsymbol{k},x)$ is a quite complicated task. By investigating different *limits*, we may, however, obtain a good over-all description of drifting oscillations and so avoid this difficulty.

ε) For the *ionospheric and magnetospheric plasma* some more simplifications apply so that the analysis becomes easier. First of all, as stated above, the drift frequencies are very small in the ionospheric and magnetospheric plasma, satisfying

$$\omega_{\rm dh}\ll\omega_{B\rm h}. \tag{36.15}$$

Secondly, there is some interest to consider natural plasma inhomogeneities, called irregularities[5] which could possibly be explained as resulting from drift instabilities in the ionospheric plasma [*9*], [*36*]. In a direction perpendicular to that of the magnetic field the dimensions of ionospheric irregularities are greater than 10 ... 100 m; this is large compared to the LARMOR radius of both ions and electrons in the magnetic field of Earth. This is at least so in the ionosphere and in the inner magnetosphere. Taking account of this fact we are in a position to confine our investigation to low frequency ($\omega\ll\omega_{B\rm i}$) and long-wave ($\lambda\gg r_{B\rm i}$, the LARMOR radius) drifting oscillations in the ionospheric, inhomogeneous plasma[6]. With these quite reasonable limitations there is only a small number of possible cases at the limits.

If the condition

$$|\omega+i\bar{\nu}_{\rm i}|\ll k_z V_{T\rm i},\quad \text{i.e.}\quad |1+iZ_{\rm i}|\ll Q_{{\rm i}z} \tag{36.16}$$

were satisfied collisional effects could be neglected. In this case we just had the solution found for a collisionless plasma, Eq. (31.8) would be the right solution to Eqs. (36.9) and (36.13), see Subsect. δ. In a collisionless plasma, DEBYE screening prevents potential oscillations of long wavelength.

For the ionosphere we should, therefore, accept the opposite hypothesis, namely

$$|\omega+i\bar{\nu}_{\rm i}|\gg k_z V_{T\rm i},\quad \text{i.e.}\quad |1+iZ_{\rm i}|\gg Q_{{\rm i}z}. \tag{36.17}$$

ζ) We suppose the *longitudinal wavelength* of the oscillations to be *small* compared to the mean free path of the electrons, but also to the distance a thermal electron travels during one drift oscillation period, which means

$$|\omega+i\bar{\nu}_{\rm e}|\ll k_z V_{T\rm e},\quad \text{i.e.}\quad |1+iZ_{\rm e}|\ll Q_{{\rm e}z}. \tag{36.18}$$

We may then neglect collisions in Eq. (36.9) for electrons. The theory of a collisionless plasma allows to find from Eq. (31.8),

$$\delta\varepsilon_{\rm re}=\frac{\omega_{N\rm e}^2}{k^2V_{T\rm e}^2}\left\{1+i\sqrt{\frac{\pi}{2}}\,\frac{\omega}{|k_z|V_{T\rm e}}\left(1-\frac{k_yV_{T\rm e}^2}{\omega\,\omega_{B\rm e}}\,\frac{\partial}{\partial x}\log\frac{N_{\rm e}/N_{\rm e0}}{\sqrt{T_{\rm e}/T_{\rm e0}}}\right)\right\}, \tag{36.19}$$

i.e.

$$\delta\varepsilon_{\rm re}=X_{\rm e}\,Q_{\rm e}^{-2}\left\{1+i\sqrt{\frac{\pi}{2}}\,Q_{{\rm e}z}^{-1}\left(1-Q_{{\rm e}y}^2\,Y_{\rm e}^{-1}\frac{1}{k_y}\,\frac{\partial}{\partial x}\log\frac{N_{\rm e}/N_{\rm e0}}{\sqrt{T_{\rm e}/T_{\rm e0}}}\right)\right\}. \tag{36.19a}$$

The dissipative (imaginary) part is due to by the ČERENCOV effect of quick electrons.

η) In the opposite hypothesis the *wavelength parallel with the magnetic field is large* compared to the mean free path of the electrons. In this case the ČERENCOV

[5] See K. RAWER and K. SUCHY: this Encyclopedia, Vol. 49/2, Sect. 49.

[6] A complete analysis of drifting oscillations in an inhomogeneous plasma is found in [*46*]; account is taken of collisions, and a large frequency and wavelength range is considered there.

effect is negligibly small and dissipative effects are due to electron collisions. If yet the frequency of the oscillations considered is great compared with the collision frequency of electrons,

$$\omega \gg \bar{\nu}_e > k_z V_{Te}, \quad \text{i.e.} \quad 1 \gg Z_e > Q_e, \tag{36.20}$$

a solution of the kinetic equation, Eq. (36.9) is easily obtained by developing after the "relative collision frequency" $Z_e \equiv \bar{\nu}_e/\omega$. The result is[7]

$$\left.\begin{aligned}\delta\varepsilon_{re} = \frac{\omega_{Ne}^2}{k^2 V_{Te}^2}\Big\{&\frac{k_y V_{Te}^2}{\omega\,\omega_{Be}}\frac{\partial \log N_e/N_{e0}}{\partial x} - \frac{k_z^2 V_{Te}^2}{\omega^2}\left(1 - \frac{k_y V_{Te}^2}{\omega\,\omega_{Be}}\frac{\partial \log N_e T_e/N_{e0} T_{e0}}{\partial x}\right) + \\ &+ i\frac{k_z^2 V_{Te}^2}{\omega^2}\frac{\nu_{\text{eff}}}{\omega}\left(1 - \frac{k_y V_{Te}^2}{\omega\,\omega_{Be}}\frac{\partial}{\partial x}\log\left(\frac{N_e T_{e0}^{\frac12}}{N_{e0} T_e^{\frac12}}\right)\right)\Big\},\end{aligned}\right\} \tag{36.21}$$

i.e.

$$\left.\begin{aligned}\delta\varepsilon_{re} = X_e Q_e^{-2}\Big\{&Q_{ey}^2 Y_e^{-1}\frac{\partial \log N_e/N_{e0}}{k_y\,\partial x} - Q_{ez}^2\left(1 - Q_{ey}^2 Y_e^{-1}\frac{\partial \log N_e T_e/N_{e0} T_{e0}}{k_y\,\partial x}\right) + \\ &+ i Q_{ez}^2 Z_{\text{eff}}\left(1 - Q_{ey}^2 Y_e^{-1}\frac{\partial}{k_y\,\partial x}\log\left(\frac{N_e T_{e0}^{\frac12}}{N_{e0} T_e^{\frac12}}\right)\right)\Big\},\end{aligned}\right\} \tag{36.21a}$$

where $Z_{\text{eff}} \equiv \nu_{\text{eff}}/\omega$ characterizes collisions between electrons and ions[8]. These can be obtained from Eq. (13.30).

ϑ) Finally, the mean *free path*, while being *smaller* than the parallel wavelength, is supposed to be also smaller than the distance travelled by a thermal electron during one period of the drift oscillation:

$$\bar{\nu}_e \gg (\omega, k_z V_{Te}), \tag{36.22}$$ $$\text{i.e.} \quad Z_e \gg (1, Q_{ez}). \tag{36.22a}$$

Then the collision integral, Eq. (36.10), becomes the most important contribution in the kinetic equation, Eq. (36.9). It has to be solved by the method of CHAPMAN and ENSKOG [*15*], giving in the limit $\omega\,\nu_{\text{eff}} \gg k_z^2 V_{Te}^2$:

$$\left.\begin{aligned}\delta\varepsilon_{re} = \frac{\omega_{Ne}^2}{k^2 V_{Te}^2}\Big\{&\frac{k_y V_{Te}^2}{\omega\,\omega_{Be}}\frac{\partial \log N_e/N_{e0}}{\partial x} + \\ &+ i\,1.96\frac{k_z^2 V_{Te}^2}{\omega\,\nu_{\text{eff}}}\left[1 - \frac{k_y V_{Te}^2}{\omega\omega_{Be}}\frac{\partial}{\partial x}\log\left(\frac{N_e}{N_{e0}}\left(\frac{T_e}{T_{e0}}\right)^{1.71}\right)\right]\Big\},\end{aligned}\right\} \tag{36.23}$$

i.e.

$$\left.\begin{aligned}\delta\varepsilon_{re} = X_e Q_e^{-2}\Big\{&Q_{ey}^2 Y_e^{-1}\frac{\partial \log N_e/N_{e0}}{k_y\,\partial x} + \\ &+ i\,1.96\,Q_{ez}^2 Z_{\text{eff}}^{-1}\left[1 - Q_{ey}^2 Y_e^{-1}\frac{\partial}{k_y\,\partial x}\log\left(\frac{N_e}{N_{e0}}\left(\frac{T_e}{T_{e0}}\right)^{1.71}\right)\right]\Big\}.\end{aligned}\right\} \tag{36.23a}$$

In the opposite limit, when $\omega\,\nu_{\text{eff}} \ll k_z^2 V_{Te}^2$ we have[9]:

$$\delta\varepsilon_{re} = \frac{\omega_{Ne}^2}{k^2 V_{Te}^2}\left\{1 + i\,1.44\frac{\omega\,\nu_{\text{eff}}}{k_z^2 V_{Te}^2}\left[1 - \frac{k_y V_{Te}^2}{\omega\,\omega_{Be}}\frac{\partial}{\partial x}\log\left(\frac{N_e}{N_{e0}}\left(\frac{T_e}{T_{e0}}\right)^{+0.56}\right)\right],\right. \tag{36.24}$$

---

[7] In this Encyclopedia log means natural logarithm.

[8] See K. RAWER and K. SUCHY: this Encyclopedia, Vol. 49/2, Sect. 3.

[9] A detailed deduction of these equations is given in [*46*].

i.e.

$$\delta\varepsilon_{re} = X_e Q_e^{-2}\left\{1 + i\,1.44\, Q_{ez}^{-2} Z_{\rm eff}\left[1 - Q_{ey}^2 Y_e^{-1}\frac{\partial}{\partial x}\log\left(\frac{N_e}{N_{e0}}\left(\frac{T_{e0}}{T_e}\right)^{+0.56}\right)\right]\right\}. \tag{36.24a}$$

ι) We should now enquire about the *ionic contribution* to the dielectric permeability. A solution of the kinetic equation, Eq. (36.9), for ions easily obtained by development after the relative ionic collision frequency, provided

$$\omega \gg \bar{\nu}_i, \quad \text{i.e.} \quad Z_i \ll 1. \tag{36.25}$$

If on the other hand we suppose the frequency to be small compared to the LARMOR frequency of ions

$$\omega \ll \omega_{Bi}, \quad \text{i.e.} \quad Y_i \gg 1, \qquad [36.15]$$

we get

$$\left.\begin{aligned}\delta\varepsilon_{ri} = \frac{\omega_{Ni}^2}{k^2 V_{Ti}^2}\Bigg\{1 - \Bigg(1 - \frac{k_y V_{Ti}^2}{\omega\,\omega_{Bi}}\Bigg[\frac{\partial \log N_i/N_{i0}}{\partial x} + \\ + \frac{\partial T_i}{\partial x}\frac{\partial}{\partial T_i}\Bigg]\mathscr{A}_0\left(\frac{k_\perp^2 V_{Ti}^2}{\omega_{Bi}^2}\right) J_+\left(\frac{\omega}{|k_z| V_{Ti}}\right)\Bigg\} + \delta\varepsilon_{rii},\end{aligned}\right\} \tag{36.26}$$

i.e.

$$\left.\begin{aligned}\delta\varepsilon_{ri} = X_i Q_i^{-2}\Bigg\{1 - \Bigg(1 - Q_{iy}^2 Y_i^{-1}\frac{1}{k_y}\Bigg[\frac{\partial \log N_i/N_{i0}}{\partial x} + \\ + \frac{\partial T_i}{\partial x}\frac{\partial}{\partial T_i}\Bigg]\mathscr{A}_0(Q_{i\perp}^2 Y_i^{-2})\, J_+(Q_{iz}^{-1})\Bigg\} + \delta\varepsilon_{rii}.\end{aligned}\right\} \tag{36.26a}$$

See Eqs. (17.14), (22.9) for definitions of the special functions. The main term in Eq. (36.26) is the same as obtained with the theory of a collisionless plasma, see Eq. (31.8). The second term is due to ion-ion collisions[10].

In the *limiting case* which we are looking for, the wavelength of the oscillations is large, the corresponding condition being:

$$k_\perp V_{Ti} \ll \omega_{Bi}, \quad \text{i.e.} \quad Q_{i\perp} \ll Y_i \tag{36.27}$$

which may also be written as

$$k_\perp r_{Bi} \gg 1 \tag{36.27a}$$

where $r_{Bi} \equiv V_{Ti}/\omega_{Bi}$ is the average gyro-radius of ions of thermal velocity.

We then find

$$\left.\begin{aligned}\delta\varepsilon_{rii} = i\frac{\nu_{ii}}{10\,\omega}\frac{\omega_{Ni}^2 V_{Ti}^2}{k^2}\left\{16\frac{k_z^4}{\omega^4} + 28\frac{k_z^2 k_\perp^2}{\omega^2\omega_{Bi}^2} + 7\frac{k_\perp^4}{\omega_{Bi}^4}\right)\left(1 - \frac{k_y V_{Ti}^2}{\omega\,\omega_{Bi}}\frac{\partial \log N_i/N_{i0}}{\partial x}\right) - \\ - \frac{k_y V_{Ti}^2}{\omega\,\omega_{Bi}}\frac{\partial \log T_i/T_{i0}}{\partial x}\left(24\frac{k_z^4}{\omega^4} + \frac{33}{2}\frac{k_z^2 k_\perp^2}{\omega^2\omega_{Bi}^2} - \frac{3}{4}\frac{k_\perp^4}{\omega_{Bi}^4}\right),\end{aligned}\right\} \tag{36.28}$$

ν_{ii} being the collision frequency for ion-ion collisions as given by Eq. (13.30).

When deriving Eq. (36.28) we supposed that ion collisions are comparatively small[10], viz. $\omega \gg \bar{\nu}_i$. This is a characteristic condition for the validity range of the kinetic model of *drifting oscillations*. As stated above the method of successive approximation, i.e. "perturbation theory", has been used to find solutions for ions in Eq. (36.9). As is easily seen from the solution, Eq. (36.27), the real "perturbation parameter" is

$$Z_{ii} k_\perp^2 r_{Bi}^2 \equiv \frac{\nu_{ii}}{\omega} k_\perp^2\left(\frac{V_{Ti}}{\omega_{Bi}}\right)^2 \equiv Z_{ii} Q_{i\perp}^2/Y_i^2. \tag{36.29}$$

[10] Ion-electron collisions can only cause corrections of order $(m_e/m_i)^{\frac{1}{2}}$ which are negligible.

ϰ) Eqs. (36.19 ... 28) give a complete description with the kinetic model of long-wave drift oscillations. In the following two Sects. we shall investigate the other limit which is a gasdynamic model, suitable for shorter wavelength and very low frequencies [see Eq. (37.1)]. But with this model we shall not use the notion of dielectric permeability at all. It is easier in this case to use directly the equations of two-fluid gas-dynamics, Eqs. (9.2) and (13.2 ... 5). This will be done in Sects. 37 and 38 when discussing specific plasma conditions.

37. Drift oscillations in an inhomogeneous plasma with an isotropic particle velocity distribution function.

α) As indicated in Sect. 36ϰ above, we now suppose *short wavelength* and *very low frequency*, i.e.

$$\bar{\nu}_{\mathrm{h}} \gg (\omega, k_z V_{T\mathrm{h}}), \quad \text{i.e.} \quad Z_{\mathrm{h}} \gg (1, \varrho_{\mathrm{h}z}) \qquad (\mathrm{h} = \mathrm{e}, \mathrm{i}). \tag{37.1}$$

In order to determine oscillation spectra with potential fields using the approximation of geometrical optics we have to start with the eiconal equation

$$\varepsilon(\omega, \boldsymbol{k}, x) = 0. \tag{37.2}$$

Introducing here the expression for the *dielectric permeability* found above in Sect. 36 we find the x-component, k_x, of the wave-number vector $\boldsymbol{k}$, as a complex function of the space coordinate, x, and of the *complex* pulsation, ω. With the classical quantization rules, Eqs. (31.18 ... 20), we might then establish the dispersion relation which answers the question of stability. Amongst all possible oscillations which can be described with the approximation of geometric optics there may be a few for which local spectra obtained from the eiconal-equation, Eq. (37.2), give the right answer concerning plasma stability. This, however, depends upon the variations of the *dielectric permeability* along the stratification coordinate, x. We shall limit the present investigation to cases of this kind, i.e. looking for local spectra in an inhomogeneous plasma[1].

In the range

$$|\omega + i\bar{\nu}_{\mathrm{e}}| \ll k_z V_{T\mathrm{e}}, \quad \text{i.e.} \quad |1 + iZ_{\mathrm{e}}| \ll \varrho_{\mathrm{e}z} \tag{37.3a}$$

and

$$\omega \gg \bar{\nu}_{\mathrm{i}}, \qquad \text{i.e.} \quad Z_{\mathrm{i}} \ll 1, \tag{37.3b}$$

the eiconal equation for long wave oscillations in the plasma,

$$\frac{\varrho_{\mathrm{i}\perp}}{Y_{\mathrm{i}}} \equiv k_\perp \frac{V_{T\mathrm{i}}}{\omega_{B\mathrm{i}}} \equiv k_\perp r_{B\mathrm{i}} \ll 1 \tag{37.4}$$

can be written as:

$$\left.\begin{aligned}
&1 + \frac{\omega_{N\mathrm{e}}^2}{k^2 V_{T\mathrm{e}}^2}\left[1 + i\sqrt{\frac{\pi}{2}}\,\frac{\omega}{|k_z|\,V_{T\mathrm{e}}}\left(1 - \frac{k_y V_{T\mathrm{e}}^2}{\omega\,\omega_{B\mathrm{e}}}\,\frac{\partial}{\partial x}\log\left(\frac{N_{\mathrm{e}}}{N_{\mathrm{e}0}}\,\frac{T_{\mathrm{e}0}^{\frac{1}{2}}}{T_{\mathrm{e}}^{\frac{1}{2}}}\right)\right)\right] + \\
&+ \frac{\omega_{N\mathrm{i}}^2}{k^2 V_{T\mathrm{i}}^2}\left\{\left(\frac{k_\perp^2 V_{T\mathrm{i}}^2}{\omega_{B\mathrm{i}}^2} - \frac{k_z^2 V_{T\mathrm{i}}^2}{\omega^2}\right)\left(1 - \frac{k_y V_{T\mathrm{i}}^2}{\omega\,\omega_{B\mathrm{i}}}\,\frac{\partial \log N_{\mathrm{i}} T_{\mathrm{i}}/N_{\mathrm{i}0} T_{\mathrm{i}0}}{\partial x}\right) + \right.\\
&+ \frac{k_y V_{T\mathrm{i}}^2}{\omega\,\omega_{B\mathrm{i}}}\,\frac{\partial \log N_{\mathrm{i}}/N_{\mathrm{i}0}}{\partial x} + \frac{1}{10}\,i\,\frac{\nu_{\mathrm{ii}}}{\omega}\left[\left(16\,\frac{k_z^4 V_{T\mathrm{i}}^4}{\omega^4} + 28\,\frac{k_z^2 k_\perp^2 V_{T\mathrm{i}}^2}{\omega^2\,\omega_{B\mathrm{i}}^2} + \right.\right.\\
&\left. + 7\,\frac{k_\perp^4 V_{T\mathrm{i}}^4}{\omega_{B\mathrm{i}}^4}\right)\left(1 - \frac{k_y V_{T\mathrm{i}}^2}{\omega\,\omega_{B\mathrm{i}}}\,\frac{\partial \log N_{\mathrm{i}}/N_{\mathrm{i}0}}{\partial x}\right) - \frac{k_y V_{T\mathrm{i}}^2}{\omega\,\omega_{B\mathrm{i}}}\,\frac{\partial \log T_{\mathrm{i}}/T_{\mathrm{i}0}}{\partial x}\,\times \\
&\left.\left.\times\left(24\,\frac{k_z^4 V_{T\mathrm{i}}^4}{\omega^4} + \frac{33}{2}\,\frac{k_z^2 k_\perp^2 V_{T\mathrm{i}}^4}{\omega^2\,\omega_{B\mathrm{i}}^2} - \frac{3}{4}\,\frac{k_\perp^4 V_{T\mathrm{i}}^4}{\omega_{B\mathrm{i}}^4}\right)\right]\right\} = 0,
\end{aligned}\right\} \tag{37.5}$$

[1] It is, besides, not too difficult to establish more accurate dispersion relations based upon the quantization rules.

i.e.

$$\left.\begin{aligned}
&1+X_e Q_e^{-2}\left[1+i\sqrt{\frac{\pi}{2}}\,Q_{ez}^{-1}\left(1-Q_{ey}^2 Y_e^{-1}\frac{1}{k_y}\frac{\partial}{\partial x}\log\left(\frac{N_e}{N_{e0}}\frac{T_{e0}^{\frac12}}{T_e^{\frac12}}\right)\right)\right]+\\
&+X_i Q_i^{-2}\left\{(Q_{i\perp}^2 Y_i^{-2}-Q_{iz}^2)\left(1-Q_{iy}^2 Y_i^{-1}\frac{1}{k_y}\frac{\partial}{\partial x}\log\left(\frac{N_i T_i}{N_{i0}T_{i0}}\right)\right)+\right.\\
&+Q_{iy}^2 Y_i^{-1}\frac{1}{k_y}\frac{\partial}{\partial x}\log\frac{N_i}{N_{i0}}+\frac{1}{10}\,i Z_{ii}\left[(16\,Q_{iz}^4+28\,Q_{iz}^2 Q_{i\perp}^2 Y_i^{-2}+\right.\\
&+7\,Q_{i\perp}^4)\left(1-Q_{iy}^2 Y_i^{-1}\frac{1}{k_y}\frac{\partial}{\partial x}\log\frac{N_i}{N_{i0}}\right)-Q_{iy}^2 Y_i^{-1}\frac{1}{k_y}\frac{\partial}{\partial x}\log\frac{T_i}{T_{i0}}\times\\
&\left.\left.\times\left(24\,Q_{iz}^4+\frac{33}{2}\,Q_{iz}^2 Q_{i\perp}^2 Y_i^{-2}-\frac{3}{4}\,Q_{i\perp}^4 Y_i^{-4}\right)\right]\right\}=0.
\end{aligned}\right\}\quad(37.5\,\text{a})$$

It can be shown that these oscillations can only be unstable under the condition

$$\omega_{dh}\sim\frac{k_y V_{Th}^2}{\omega_{Bh}\mathscr{L}_0}\gg k_z V_s=k_z\sqrt{\frac{q_i}{q_e}\frac{\vartheta_e}{m_i}}\,.\quad(37.6)$$

β) *Four branches* of slowly increasing oscillations exist in this context. We put

$$\omega\to\omega+i\gamma;\qquad |\gamma|\ll|\omega|;$$

positive γ means attenuation so that oscillations are rising naturelly for negative γ. We find[2,3]

$$\omega_1=-\frac{1}{1+k^2\lambda_{De}^2}\,\frac{k_y V_s^2}{\omega_{Bi}}\,\frac{\partial}{\partial x}\log\frac{N_0}{N_{i0}}\,;\quad(37.7)$$

$$\left.\begin{aligned}
\gamma_1=&\sqrt{\frac{\pi}{2}}\,\frac{\omega_1^2}{|k_z|\,V_{Te}}\left\{\frac{k^2\lambda_{De}^2+\dfrac{k_\perp^2 V_s^2}{\omega_{Bi}^2}\left(1+\dfrac{T_i}{T_e}\,\dfrac{\partial\log N_i T_i/N_{i0}T_{i0}}{\partial\log N_i/N_{i0}}\right)}{1+k^2\lambda_{De}^2}-\right.\\
&\left.-\frac{1}{2}\,\frac{\partial\log T_e/T_{e0}}{\partial\log N_e/N_{e0}}\right\}-\frac{8}{5}\,\nu_{ii}\,\frac{k_z^4 V_{Ti}^4}{\omega_1^4}\left\{\left(1+\frac{7}{4}\,\frac{k_\perp^2\,\omega_1^2}{k_z^2\,\omega_{Bi}^2}+\frac{7}{16}\,\frac{k_\perp^4\,\omega_1^4}{k_z^4\,\omega_{Bi}^4}\right)\times\right.\\
&\times\left(1+\frac{T_e}{T_i}+k^2\lambda_{De}^2\right)+\frac{\partial\log T_i/T_{i0}}{\partial\log N_i/N_{i0}}\,(1+k^2\lambda_{De}^2)\times\\
&\left.\times\left(\frac{3}{2}+\frac{33}{32}\,\frac{k_\perp^2\,\omega_1^2}{k_z^2\,\omega_{Bi}^2}-\frac{3}{64}\,\frac{k_\perp^2\,\omega_1^4}{k_z^4\,\omega_{Bi}^4}\right)\right\};
\end{aligned}\right\}\quad(37.8)$$

$$\omega_2=\frac{k_z^2\,\omega_{Bi}}{k_y\,\dfrac{\partial\log N_e/N_{e0}}{\partial x}}\,;\quad(37.9)$$

$$\gamma_2=-\sqrt{\frac{\pi}{2}}\,\frac{\omega_2^2}{|k_z|\,V_{Te}}\left(1-\frac{1}{2}\,\frac{\partial\log T_e/T_{e0}}{\partial\log N_e/N_{e0}}\right)-\frac{8}{5}\,\nu_{ii}\,\frac{k_z^2 V_{Ti}^2}{\omega_2^2}\,;\quad(37.10)$$

$$\omega_3^2=-k_z^2\,V_{Ti}^2\,\frac{\partial\log T_i/T_{i0}}{\partial\log N_i/N_{i0}}\,;\quad(37.11)$$

$$\omega_4^3=-k_z^2\,V_s^2\,\frac{k_y V_{Ti}^2}{\omega_{Bi}}\,\frac{\partial\log T_i/T_{i0}}{\partial x}\,.\quad(37.12)$$

---

[2] In this Encyclopedia log means natural logarithm while Log is the decadic logarithm.

[3] $\dfrac{\partial\log P}{\partial\log Q}$ precisely means $\dfrac{\partial}{\partial x}\log P\Big/\dfrac{\partial}{\partial x}\log Q$.

$V_{T\,\mathrm{i}}$ is the mean thermal velocity of the ions, V_s the sound velocity in the ion gas and λ_D is the DEBYE length, see Eq. (18.6).

In the *first branch*, ω_1, stabilization of the oscillation is finally obtained by ČERENKOV dissipation due to electrons under the condition

$$\frac{\partial \log T_e/T_{e0}}{\partial x} \Big/ \frac{\partial \log N_e/N_{e0}}{\partial x} > 2\left|k^2\lambda_{De}^2 + \frac{k_\perp^2 V_s^2}{\omega_{Bi}^2}\right|. \tag{37.13}$$

Ion-ion collisions may seriously attenuate such oscillations, provided

$$k_\perp^2\omega_1^2 > k_z^2\omega_{Bi}^2 \tag{37.14a}$$

and

$$\frac{\partial \log T_i/T_{i0}}{\partial x} \Big/ \frac{\partial \log N_i/N_{i0}}{\partial x} > \frac{28}{3}\left(1 + \frac{T_e}{T_i}\right). \tag{37.14b}$$

The *second branch*, ω_2, can only exist in a non-isothermal plasma with

$$T_e \gg T_i \tag{37.15}$$

here ion-ion collisions play a stabilizing role while ČERENKOV dissipation due to electrons may attenuates such oscillations, provided

$$\frac{\partial \log T_e/T_{e0}}{\partial x} \Big/ \frac{\partial \log N_e/N_{e0}}{\partial x} > 2. \tag{37.16}$$

The *last two branches*, ω_4 and ω_3, describe "aperiodical instabilities". For these the condition $\gamma \ll \omega$ is not satisfied but

$$\mathrm{Re}\,\omega \sim \mathrm{Im}\,\omega > 0.$$

They can only occur if

$$\frac{\partial \log T_i/T_{i0}}{\partial x} \Big/ \frac{\partial \log N_i/N_{i0}}{\partial x} \gg 1. \tag{37.17}$$

In the terrestrial ionospheric plasma oscillations of the first branch, Eqs. (37.7), (37.8), appear to be possible. They are, however, unstable at heights above 300 ... 400 km because there

$$\frac{\partial}{\partial x}\log\frac{T_e}{T_{e0}} \Big/ \frac{\partial}{\partial x}\log\frac{N_e}{N_{e0}} \lesssim 0$$

(see Table 1, p. 401), so that the ČERENKOV effect causes amplitude limitation for rising oscillations. Such instable oscillations possibly appear as natural irregularities in the ionosphere with dimensions of the order of

$$\mathscr{L}_\perp \gtrsim \lambda_\perp \sim 100\ \mathrm{m}; \quad \mathscr{L}_\parallel \gtrsim \lambda_\parallel \sim 10 \ldots 100\ \mathrm{km},$$

and with a life-time of the order of the oscillation period,

$$\tau \gtrsim \omega_{dh}^{-1} \sim 10^2 \ldots 10^3\ \mathrm{sec}.$$

In the magnetosphere these oscillations are felt to be damped by ion collisions. Oscillations of the second branch should not occur in the terrestrial ionosphere and magnetosphere. Though $T_e/T_i > 1$ in the height range 120 ... 500 km, this ratio is never large enough to fulfill condition Eq. (37.15). Oscillations of the third and fourth branch are also not likely to occur neither in the ionosphere nor in the magnetosphere.

γ) Let us now consider long-wave drifting oscillations

$$k_\perp r_{Bi} \equiv k_\perp \frac{V_{Ti}}{\omega_{Bi}} = \frac{Q_{i\perp}}{Y_i} \ll 1 \tag{37.18}$$

in the frequency range

$$\omega \gg (\bar{\nu}_e, k_z V_{Te}), \quad \text{i.e.} \quad 1 \gg (Z_e, Q_{ez}). \tag{37.19}$$

In this frequency range the eiconal equation, Eq. (37.2) takes the form

$$\left.\begin{aligned}1+\frac{\omega_{Ne}^2}{k^2V_{Te}^2}\left\{\frac{k_z^2V_{Te}^2}{\omega^2}\left(\frac{k_yV_{Te}^2}{\omega\,\omega_{Be}}\frac{\partial\log N_e T_e/N_{e0}T_{e0}}{\partial x}-1\right)+i\frac{\nu_{\rm eff}}{\omega}\frac{k_z^2V_{Te}^2}{\omega^2}\times\right.\\ \left.\times\left(1-\frac{k_yV_{Te}^2}{\omega\,\omega_{Be}}\frac{\partial}{\partial x}\log\frac{N_e}{N_{e0}}\frac{T_{e0}^{\frac12}}{T_e^{\frac12}}\right)\right\}+\frac{\omega_{Ni}^2}{k^2V_{Ti}^2}k_\perp^2 r_{Bi}^2\times\\ \times\left\{1-\frac{k_yV_{Ti}^2}{\omega\,\omega_{Bi}}\frac{\partial\log N_i T_i/N_{i0}T_{i0}}{\partial x}+i\frac{\nu_{ii}}{10\,\omega}k_\perp^2 r_{Bi}^2\times\right.\\ \left.\times\left[7\left(1-\frac{k_yV_{Ti}^2}{\omega\,\omega_{Bi}}\frac{\partial}{\partial x}\log\frac{N_i}{N_{i0}}\right)+\frac{3}{4}\frac{k_yV_{Ti}^2}{\omega\,\omega_{Bi}}\frac{\partial}{\partial x}\log\frac{T_i}{T_{i0}}\right]\right\}=0,\end{aligned}\right\}\quad(37.20)$$

i.e.

$$\left.\begin{aligned}1+X_eQ_e^{-2}\left\{Q_{ez}^2\left(Q_{ey}^2\,Y_e^{-1}\frac{1}{k_y}\frac{\partial}{\partial x}\log\frac{N_eT_e}{N_{e0}T_{e0}}-1\right)+i\,Z_{\rm eff}\,Q_{ez}^2\times\right.\\ \left.\times\left(1-Q_{ey}^2\,Y_e^{-}\frac{1}{k_y}\frac{\partial}{\partial x}\log\frac{N_eT_{e0}^{\frac12}}{N_{e0}T_e^{\frac12}}\right)\right\}+X_i^2\,Q_i^{-2}\,Q_{i\perp}^2\,Y_i^{-2}\times\\ \times\left\{1-Q_{iy}^2\,Y_i^{-1}\frac{1}{k_y}\frac{\partial}{\partial x}\log\frac{N_iT_i}{N_{i0}T_{i0}}+\frac{i}{10}Z_{ii}\,Q_{i\perp}^2\,Y_i^{-2}\times\right.\\ \left.\times\left[7\left(1-Q_{iy}^2\,Y_i^{-1}\frac{1}{k_y}\frac{\partial}{\partial x}\log\frac{N_i}{N_{i0}}\right)+\frac34 Q_{iy}^2\,Y_i^{-1}\frac{1}{k_y}\frac{\partial}{\partial x}\log\frac{T_i}{T_{i0}}\right]\right\}=0.\end{aligned}\right\}\quad(37.20\text{a})$$

It is easily seen that a solution as a local spectrum exists at[4]

$$k_z\approx 0,\quad(37.21)$$

namely:

$$\omega=\frac{1}{1+V_A^2/c_0^2}\frac{k_yV_{Ti}^2}{\omega_{Bi}}\frac{\partial}{\partial x}\log\frac{N_iT_i}{N_{i0}T_{i0}},\quad(37.22)$$

$$\left.\begin{aligned}\gamma=\frac{1}{40}\nu_{ii}\frac{k_\perp^2 r_{Bi}^2}{1+V_A^2/c_0^2}\frac{\partial\log(N_i/N_{i0})}{\partial\log(N_iT_i/N_{i0}T_{i0})}\times\\ \times\left[28\frac{V_A^2}{c_0^2}-\left(31+3\frac{V_A^2}{c_0^2}\right)\frac{\partial\log T_i/T_{i0}}{\partial\log N_i/N_{i0}}\right].\end{aligned}\right\}\quad(37.23)$$

The condition for generation of oscillations is

$$\left(31+3\frac{V_A^2}{c_0^2}\right)\frac{\partial\log T_i/T_{i0}}{\partial x}\Big/\frac{\partial\log N_i/N_{i0}}{\partial x}<28\frac{V_A^2}{c_0^2}\quad(37.24)$$

and oscillations may be prevented by ion-ion collisions.

In the ionosphere the condition Eq. (37.24) is fulfilled at heights above 400 or 500 km, see Table 1, p. 401. In the terrestrial ionosphere oscillations should however not be able to be excited since everywhere $\omega_{dh}>\nu_e$. On the other hand oscillations of this type are expected to be generated in the magnetospheric plasma with frequencies of the order of $\omega\approx\omega_{di}=10^{-2}\ldots10^{-3}\,\text{sec}^{-1}$ and dimensions between 100 m and 10^4 km.

The local spectrum has the form of Eqs. (37.22), (37.23), provided $\omega\gg\bar\nu_e$ ($Z_e\ll 1$), Eq. (37.19a), is satisfied and also

$$\frac{k_z^2}{k_\perp^2}<\frac{m_e}{m_i}\frac{T_i}{T_e}\frac{\omega^2}{\omega_{Bi}^2}=\left(\frac{m_eT_i}{m_iT_e}\right)Y_i^{-2}.\quad(37.25)$$

In the opposite limit

$$\omega\gg k_zV_{Te},\quad\text{i.e.}\quad Q_{ez}\ll 1,\quad(37.26)$$

---

[4] In this Encyclopedia log means natural logarithm.

the branch after Eqs. (37.22), (37.23) goes over into

$$\omega = -\frac{k_y V_s^2}{\omega_{Bi}} \frac{\partial}{\partial x} \log \frac{N_e T_e}{N_{e0} T_{e0}}, \tag{37.27}$$

$$\gamma = \nu_{eff} \frac{\partial \log (T_e/T_{e0})^{\frac{3}{2}}}{\partial x} \Big/ \frac{\partial \log (N_e T_e/N_{e0} T_{e0})}{\partial x}. \tag{37.28}$$

It is easily seen that such oscillations may arise due to instability when

$$\frac{\partial}{\partial x} \log \frac{T_e}{T_{e0}} \Big/ \frac{\partial}{\partial x} \log \frac{N_e}{N_{e0}} > 0, \quad \text{or} \quad < -1 \tag{37.29}$$

electron-ion collisions being responsible for the limiting attenuation in this case.

Finally aperiodic instability is possible in an inhomogeneous gyrotropic plasma, in the frequency range defined by Eq. (37.19), under the condition [see Eq. (36.1)]:

$$\omega \ll \omega_{dh} \sim \frac{k_y V_{Th}^2}{\omega_{Bh} \mathscr{L}_0}, \quad \text{i.e.} \quad \frac{Q_{hy}^2}{Y_h} \frac{1}{k_y \mathscr{L}_0} \gg 1. \tag{37.30}$$

The corresponding increment, γ, is found [similarly to Eq. (37.20)] from

$$\omega^2 \approx -\gamma^2 = -\frac{k_z^2}{k_\perp^2} \frac{T_e}{T_i} \frac{m_i}{m_e} \omega_{Bi}^2 \frac{\partial}{\partial x} \log \frac{N_e T_e}{N_{e0} T_{e0}} \Big/ \frac{\partial}{\partial x} \frac{N_i T_i}{N_{i0} T_{i0}}. \tag{37.31}$$

Oscillations of this type may exist in the inner magnetosphere at heights above $1 R_E$ (i.e. for $r/R_E > 2$); this might be seen from Eqs. (37.22 ... 31) and conditions

$$\frac{\nu_{eff}}{\omega}, \frac{k_z V_{Te}}{\omega} \ll 1 \lesssim \frac{\omega_{dh}}{\omega}, \quad \text{i.e.} \quad Z_{eff}, Q_{ez} \ll 1 \lesssim \frac{Q_{hy}^2}{Y_h k_y \chi_0}. \tag*{[37.19]}$$

The corresponding irregularities are highly elongated along the magnetic field, i.e. they are "field-aligned". The elongation is

$$\mathscr{L}_\parallel/\mathscr{L}_\perp > \mathscr{L}_\parallel/\mathscr{L}_0 \gtrsim \sqrt{\frac{m_i}{m_e}},$$

i.e. $\mathscr{L}_\parallel \gtrsim 100 \dots 1000$ km.

δ) With respect to drifting oscillations in the *range of the kinetic model* we have to consider the frequency range

$$\bar{\nu}_e \gg \omega \gg \bar{\nu}_i, \quad \text{i.e.} \quad Z_e \gg 1 \gg Z_i \tag*{[36.22a, 24]}$$

under the conditions

$$\bar{\nu}_e \gg k_z V_{Te}, \quad \text{i.e.} \quad Z_e \gg Q_{ez}, \tag*{[36.22b]}$$

$$\omega \bar{\nu}_e \gg k_z^2 V_{Te}^2, \quad \text{i.e.} \quad Z_e \gg Q_{ez}.$$

The eiconal equation, Eq. (37.2), then takes the form

$$\left.\begin{aligned} & 1 + i\,1.96 \frac{k_z^2 \omega_{Ne}^2}{k^2 \omega \nu_{eff}} \left[1 - \frac{k_y V_{Te}^2}{\omega \omega_{Be}} \frac{\partial}{\partial x} \log\left(\frac{N_e}{N_{e0}} \left(\frac{T_e}{T_{e0}}\right)^{1.71}\right)\right] + \frac{\omega_{Ni}^2}{\omega_{Bi}^2} \times \\ & \times \left\{1 - \frac{k_y V_{Ti}^2}{\omega \omega_{Bi}} \frac{\partial}{\partial x} \log \frac{N_i T_i}{N_{i0} T_{i0}} + i \frac{7}{10} \frac{\nu_{ii}}{\omega} k_\perp^2 r_{Bi}^2 \times \right. \\ & \left. \times \left[1 - \frac{k_y V_{Ti}^2}{\omega \omega_{Bi}} \frac{\partial}{\partial x} \log \frac{N_i}{N_{i0}} \left(1 - \frac{3}{28} \frac{\partial}{\partial x} \log \frac{T_i}{T_{i0}} \Big/ \frac{\partial}{\partial x} \log \frac{N_i}{N_{i0}}\right)\right]\right\} = 0, \end{aligned}\right\} \tag{37.32}$$

i.e.

$$\left.\begin{aligned}&1+i\,1.96\,N_e Z_e^{-1}\left(\frac{k_z}{k}\right)^2\left[1-Q_{ey}^3\,Y_e^{-1}\frac{1}{k_y}\frac{\partial}{\partial x}\log\left(\frac{N_e}{N_{e0}}\left(\frac{T_e}{T_{e0}}\right)^{1.71}\right)\right]+\\&+X_i Y_i^{-2}\left\{1-Q_{iy}^2\,Y_i^{-1}\frac{1}{k_y}\frac{\partial}{\partial x}\log\frac{N_i T_i}{N_{i0}T_{i0}}+i\frac{7}{10}Z_{ii}Q_{i\perp}^2\,Y_i^{-2}\times\right.\\&\left.\times\left[1-Q_{iy}^2 Y_i^{-1}\frac{1}{k_y}\frac{\partial}{\partial x}\log\frac{N_i}{N_{i0}}\left(1-\frac{3}{28}\frac{\partial}{\partial x}\log\frac{T_i}{T_{i0}}\Big/\frac{\partial}{\partial x}\log\frac{N_i}{N_{i0}}\right)\right]\right\}=0.\end{aligned}\right\}\qquad(37.32\text{a})$$

Solving it for ω we find the following *three branches* of possible drifting oscillations in an inhomogeneous plasma.

The *first branch* is[5]

$$\omega_1=\frac{1}{1+V_A^2/c_0^2}\,\frac{k_y V_{Ti}^2}{\omega_{Bi}}\,\frac{\partial}{\partial x}\log\frac{N_i T_i}{N_{i0}T_{i0}},\qquad(37.33)$$

$$\gamma_1=\frac{\nu_{ii}}{40}\,\frac{k_\perp^2 r_{Bi}^2}{1+V_A^2/c_0^2}\,\frac{\partial\log N_i/N_{i0}}{\partial\log N_i T_i/N_{i0}T_{i0}}\left[28\frac{V_A^2}{c_0^2}-\left(31+3\frac{Y_A^2}{c_0^2}\right)\frac{\partial\log T_i/T_{i0}}{\partial\log N_i/N_{i0}}\right],\qquad(37.34)$$

occuring at

$$\omega_s\ll\omega_1\sim\omega_{dh}\sim\frac{k_y V_{Th}^2}{\omega_{Bh}\mathcal{L}_0},\qquad(37.35)$$

needing

$$\frac{Y_i^2 Y_h}{Q_{hy}^2}\ll\frac{m_e}{m_i}\left(\frac{k_\perp}{k_z}\right)^2 k_y\,\mathcal{L}_0.$$

Here V_A is the ALFVÉN vlocity,

$$V_A^2=B_\delta^2/u\mu_0 m_i N_i\qquad[25.25]$$

and

$$\omega_S\equiv\frac{k_z^2}{k_\perp^2}\,\frac{m_i}{m_e}\,\frac{\omega_{Bi}^2}{\nu_{eff}}=\left(\frac{k_z}{k_\perp}\right)^2\left(\frac{m_i}{m_e}\right)Y_i^2 Z_{eff}^{-1}\,\omega.\qquad(37.36)$$

The *second branch is*

$$\omega_2=-\frac{k_y V_s^2}{\omega_{Bi}}\,\frac{\partial}{\partial x}\log\left(\frac{N_e}{N_{e0}}\left(\frac{T_e}{T_{e0}}\right)^{1.71}\right),\qquad(37.37)$$

$$\gamma_2=\frac{\omega_2^2}{1.96\omega_S}\left(1+\frac{V_A^2}{c_0^2}+\frac{T_i}{T_e}\,\frac{\partial\log(N_i T_i/N_{i0}T_{i0})}{\partial\log(N_e T_e^{1.71}/N_{e0}T_{e0}^{1.71})}\right),\qquad(37.38)$$

occuring at

$$\omega_S\gg\omega_2\sim\omega_{dh}\sim\frac{k_y V_{Th}^2}{\omega_{Bh}\mathcal{L}_0},\quad\text{i.e.}\quad\frac{Y_i^2 Y_h}{Q_{hy}^2}\gg\frac{m_e}{m_i}\left(\frac{k_\perp}{k_z}\right)^2 k_y\,\mathcal{L}_0.\qquad(37.39)$$

There is finally the *third branch* with

$$\omega_3=i\,1.96\,\omega_S\,\frac{T_e}{T_i}\,\frac{\partial\log(N_e T_e^{1.71}/N_{e0}T_{e0}^{1.71})}{\partial\log(N_i T_i/N_{i0}T_{i0})},\qquad(37.40)$$

occuring at

$$\omega_S\sim\omega_3\ll\omega_{dh}\sim\frac{k_y V_{Th}^2}{\omega_{Bh}\mathcal{L}_0},\qquad(37.41)$$

needing

$$\frac{Y_i^2 Y_h}{Q_{hy}^2}\ll\frac{m_e}{m_i}\left(\frac{k_\perp}{k_z}\right)^2 k_y\,\mathcal{L}_0.$$

It appears that these branches are unstable in all practical cases; the oscilla tions are limited by collisions.

[5] In this Encyclopedia log means natural logarithm.

In the upper ionosphere (above 1000 km), also, probably, in the inner magnetosphere, such oscillations can occur with pulsations of the order of

$$\bar{\nu}_e \gg \omega_{dh} \sim (10^{-3} \ldots 10^{-1})\ \text{Hz} \gg \bar{\nu}_i .$$

The corresponding irregularities should be field-aligned with longitudinal dimensions of $\mathcal{L}_{\|} \sim \mathcal{L}_0 \gtrsim 10 \ldots 100$ km.

ε) In conclusion, let us consider *long-wave drift oscillations* of an inhomogeneous plasma in the kinetic frequency range

$$\bar{\nu}_e \gg \omega,\ k_z V_{Te}, \quad \text{i.e.} \quad Z_e \gg 1,\ Q_{ez}$$

and

$$\omega \gg \bar{\nu}_i,\ k_z V_{Ti}, \quad \text{i.e.} \quad 1 \gg Z_i,\ Q_{iz},$$

and under the condition

$$\omega \nu_e \ll k_z^2 V_{Te}, \quad \text{i.e.} \quad Z_e \ll Q_{ez}^2 .$$

The electron contribution to the dielectric permeability of the plasma is determined in this case by Eq. (36.24), which is quite similar to Eq. (36.19), but differs from it by the dissipative term: whereas, in Eq. (36.19) the dissipative term is due to collisionless ČERENKOV absorption of the waves by the electrons of the plasma in Eq. (36.24) the dissipative term is of collisional character and thus related to diffusion and thermal conductivity due to electrons. Therefore, the spectra of the drift oscillations of the plasma in the range under consideration must coincide with the spectra Eqs. (37.7), (37.9), (37.11) and (37.12). Only expressions for the increments γ_1 and γ_2, which are determined by the dissipative processes, are different [46]:

$$\left.\begin{aligned} \gamma_1 = 1.44\, \frac{\omega_1^2 \nu_{\text{eff}}}{k_z^2 V_{Te}^2} \left\{ \frac{k^2 \lambda_{De}^2 + \dfrac{k_\perp^2 V_s^2}{\omega_{Bi}^2}\left(1 + \dfrac{T_i}{T_e}\, \dfrac{\partial \log N T_i / N_0 T_{i0}}{\partial \log N/N_0}\right)}{1 + k^2 \lambda_{De}^2} - 0.56\, \frac{\partial \log \dfrac{T_e}{T_{e0}}}{\partial \log \dfrac{N}{N_0}} \right\} - \\ - \frac{8}{5} \nu_{ii} \frac{k_z^4 V_{Ti}^4}{\omega_1^4} \left\{ \left(1 + \frac{7}{4}\, \frac{k_\perp^2 \omega_i^2}{k_z^2 \omega_{Bi}^2} + \frac{7}{16}\, \frac{k_\perp^4 \omega_i^4}{k_z^4 \omega_{Bi}^2}\right)\left(+1\, \frac{T_e}{T_i} + k^2 \lambda_{De}^2\right) + \right. \\ \left. + (1 + k^2 \lambda_{De}^2) \cdot \left(\frac{3}{2} + \frac{33}{32}\, \frac{k_\perp^2 \omega_i^2}{k_z^2 \omega_{Bi}^2} - \frac{3}{64}\, \frac{k_\perp^4 \omega_i^4}{k_z^4 \omega_{Bi}^4}\right) \frac{\partial \log N T_e / N_0 T_{e0}}{\partial \log N/N_0} \right\}, \end{aligned}\right\} \quad (37.42)$$

$$\gamma_2 = -1.44\, \frac{\omega_2^2 \nu_{\text{eff}}}{k_z^2 V_{Te}^2} \left(1 - \frac{\partial \log (T_e/T_{e0})^{0.56}}{\partial \log N/N_0}\right) - \frac{8}{5} \nu_{ii} \frac{k_z^2 V_{Ti}^2}{\omega_2^2} . \quad (37.43)$$

This influences the role of electrons and dissipation caused by these in the buildup of the oscillations. We see that for the first branch electron collisions de-stabilize the oscillations provided

$$\frac{\partial \log T_e/T_{e0}}{\partial \log N/N_0} < 1.8 \left[k^2 \lambda_{De}^2 + \frac{k_\perp^2 V_s^2}{\omega_{Bi}^2}\left(1 + \frac{T_i}{T_e}\, \frac{\partial \log N T_i/N_0 T_{i0}}{\partial \log N/N_0}\right)\right], \quad (37.44)$$

and for the second branch if

$$\frac{\partial \log T_e/T_{e0}}{\partial \log N/N_0} > 1.8 . \quad (37.45)$$

In all other respects the preceding analysis, concerning the influence of ion collision upon the oscillations remains entirely valid under the conditions specified at the beginning of this paragraph.

ζ) We now come to oscillations in an inhomogeneous plasma in the range of the *fluid-dynamic model* where

$$\bar{\nu}_{\mathrm{e}} \gg (\omega, k_z V_{T\mathrm{h}}), \quad \text{i.e.} \quad Z_{\mathrm{e}} \gg (1, Q_{\mathrm{h}z}) \quad (\mathrm{h}=\mathrm{e}, \mathrm{i}). \tag{37.46}$$

As stated above under these conditions it is indicated to use two-fluid gasdynamics, see Eqs. (9.2) and (13.2 ... 7). We linearize the system and limit our considerations to isothermal conditions

$$T_{\mathrm{e}} = T_{\mathrm{i}} = T,$$

for simplicity. For small oscillations the following eiconal-equation is obtained[6]:

$$\left.\begin{aligned}
&\left[\left(1+\frac{2}{3}\,i\,\frac{k_z^2\,\varkappa_{\|\mathrm{e}}}{\omega N_{\mathrm{e}}}\right)\left(1+\frac{2}{3}\,i\,\frac{k_z^2\,\varkappa_{\|\mathrm{i}}}{\omega N_{\mathrm{i}}}\right)+2i\,\frac{m_{\mathrm{e}}}{m_{\mathrm{i}}}\,\frac{\bar{\nu}_{\mathrm{e}}}{\omega}\left(2+\frac{2}{3}\,i\,\frac{k_z^2\,\varkappa_{\|\mathrm{e}}}{\omega N_{\mathrm{e}}}+\right.\right.\\
&\quad\left.\left.+\frac{2}{3}\,i\,\frac{k_z^2\,\varkappa_{\|\mathrm{i}}}{\omega N_{\mathrm{i}}}\right)\right]\cdot\left\{\left(1+\frac{4}{3}\,i\,\frac{k_z^2\,\eta_{\mathrm{i}}}{m_{\mathrm{i}}\omega N_{\mathrm{i}}}\right)\left(1-\frac{k_y V_{T\mathrm{e}}^2}{\omega\,\omega_{B\mathrm{e}}}\,\frac{\partial}{\partial x}\log\frac{N_{\mathrm{e}}}{N_{\mathrm{e}0}}\right)-\right.\\
&\quad\left.-2\,\frac{k_z^2 V_{T\mathrm{i}}^2}{\omega^2}\right\}+\frac{2}{3}\,\frac{k_z^2 V_{T\mathrm{i}}^2}{\omega^2}\left\{-\frac{2}{3}\,i\,\frac{k_z^2\varkappa_{\|\mathrm{e}}}{\omega N_{\mathrm{e}}}\left(1-\frac{k_y V_{T\mathrm{e}}^2}{\omega\omega_{B\mathrm{e}}}\,\frac{\partial}{\partial x}\log\frac{N_{\mathrm{e}} T_{\mathrm{e}0}^{\frac{3}{2}}}{N_{\mathrm{e}0} T_{\mathrm{e}}^{\frac{3}{2}}}\right)-\right.\\
&\quad\left.-\left(1+4i\,\frac{m_{\mathrm{e}}}{m_{\mathrm{i}}}\,\frac{\bar{\nu}_{\mathrm{e}}}{\omega}\right)\left(2+1.42\,\frac{k_y V_{T\mathrm{e}}^2}{\omega\,\omega_{B\mathrm{e}}}\,\frac{\partial}{\partial x}\log\frac{N_{\mathrm{e}} T_{\mathrm{e}0}^{\frac{3}{2}}}{N_{\mathrm{e}0} T_{\mathrm{e}}^{\frac{3}{2}}}\right)\right\}-\\
&\quad-1.71\,\frac{k_y V_{T\mathrm{e}}^2}{\omega\,\omega_{B\mathrm{e}}}\,\frac{\partial}{\partial x}\log\frac{T_{\mathrm{e}}}{T_{\mathrm{e}0}}\left(1+\frac{2}{3}\,i\,\frac{k_z^2\,\varkappa_{\|\mathrm{i}}}{\omega N_{\mathrm{i}}}+4i\,\frac{m_{\mathrm{e}}}{m_{\mathrm{i}}}\,\frac{\bar{\nu}_{\mathrm{e}}}{\omega}\right)\times\\
&\quad\times\left(1-\frac{4}{3}\,i\,\frac{k_z^2\,\eta_{\mathrm{i}}}{\omega N_{\mathrm{i}} m_{\mathrm{i}}}\right)=0,
\end{aligned}\right\} \tag{37.47}$$

i.e.

$$\left.\begin{aligned}
&\left[\left(1+\frac{2}{3}\,i\,\frac{\mathrm{u}q_{\mathrm{e}}^2}{\varepsilon_0 m_{\mathrm{e}}}\,X_{\mathrm{e}}^{-1}\,\frac{k_z^2\,\varkappa_{\|\mathrm{e}}}{\omega^3}\right)\left(1+\frac{2}{3}\,i\,\frac{\mathrm{u}q_{\mathrm{i}}^2}{\varepsilon_0 m_{\mathrm{i}}}\,X_{\mathrm{i}}^{-1}\,\frac{k_z^2\,\varkappa_{\|\mathrm{e}}}{\omega^3}\right)+\right.\\
&\quad\left.+2i\,\frac{m_{\mathrm{e}}}{m_{\mathrm{i}}}\,Z_{\mathrm{e}}\left(2+\frac{2}{3}\,i\,\frac{\mathrm{u}q_{\mathrm{e}}^2}{\varepsilon_0 m_{\mathrm{e}}}\,X_{\mathrm{e}}^{-1}\,\frac{k_z^2\,\varkappa_{\|\mathrm{e}}}{\omega^3}+\frac{2}{3}\,i\,\frac{\mathrm{u}q_{\mathrm{i}}^2}{\varepsilon_0 m_{\mathrm{i}}}\,X_{\mathrm{i}}^{-1}\,\frac{k_z^2\,\varkappa_{\|\mathrm{i}}}{\omega^3}\right)\right]\times\\
&\quad\times\left\{\left(1+\frac{4}{3}\,i\,\frac{\mathrm{u}q_{\mathrm{i}}^2}{\varepsilon_0 m_{\mathrm{i}}^2}\,X_{\mathrm{i}}^{-1}\,\frac{k_z^2\,\eta_{\mathrm{i}}}{\omega^3}\right)\left(1-Q_{\mathrm{e}y}^2\,Y_{\mathrm{e}}^{-1}\,\frac{1}{k_y}\,\frac{\partial}{\partial x}\log\frac{N_{\mathrm{e}}}{N_{\mathrm{e}0}}\right)-2Q_{\mathrm{i}z}^2\right\}+\\
&\quad+\frac{2}{3}\,Q_{\mathrm{i}z}^2\left\{-\frac{2}{3}\,i\,\frac{\mathrm{u}q_{\mathrm{e}}^2}{\varepsilon_0 m_{\mathrm{e}}}\,X_{\mathrm{e}}^{-1}\,\frac{k_z^2\,\varkappa_{\|\mathrm{e}}}{\omega^3}\times\right.\\
&\quad\times\left[1-Q_{\mathrm{e}y}^2\,Y_{\mathrm{e}}^{-1}\,\frac{1}{k_y}\,\frac{\partial x}{\partial}\log\left(\frac{N_{\mathrm{e}}}{N_{\mathrm{e}0}}\left(\frac{T_{\mathrm{e}0}}{T_{\mathrm{e}}}\right)^{\frac{3}{2}}\right)\right]-\left(1+4i\,\frac{m_{\mathrm{e}}}{m_{\mathrm{i}}}\,Z_{\mathrm{e}}\right)\times\\
&\quad\left.\times\left[2+1.42\,Q_{\mathrm{e}y}^2\,Y_{\mathrm{e}}^{-1}\,\frac{1}{k_y}\,\frac{\partial}{\partial x}\log\left(\frac{N_{\mathrm{e}}}{N_{\mathrm{e}0}}\left(\frac{T_{\mathrm{e}0}}{T_{\mathrm{e}}}\right)^{\frac{3}{2}}\right)\right]\right\}-1.71\,Q_{\mathrm{e}y}^2\,Y_{\mathrm{e}}^{-1}\times\\
&\quad\times\frac{1}{k_y}\,\frac{\partial}{\partial x}\log\frac{T_{\mathrm{e}}}{T_{\mathrm{e}0}}\left(1+\frac{2}{3}\,i\,\frac{\mathrm{u}q_{\mathrm{i}}^2}{\varepsilon_0 m_{\mathrm{i}}}\,\frac{k_z^2\,\varkappa_{\|\mathrm{e}}}{\omega^3}+4i\,\frac{m_{\mathrm{e}}}{m_{\mathrm{i}}}\,Z_{\mathrm{e}}\right)\times\\
&\quad\times\left(1-\frac{4}{3}\,i\,\frac{u q_{\mathrm{i}}^2}{\varepsilon_0 m_{\mathrm{i}}^2}\,X_{\mathrm{i}}^{-1}\,\frac{k_z^2\,\eta_{\mathrm{i}}}{\omega^3}\right)=0.
\end{aligned}\right\} \tag{37.47a}$$

$\varkappa_{\|\mathrm{h}}$ is the coefficient of thermal conductivity and η_{h} that of viscosity[7], both to be determined separately for electrons ($\mathrm{h}=\mathrm{e}$) and ions ($\mathrm{h}=\mathrm{i}$), see Eqs. (13.18),

[6] Terms of the order $k_\perp^2\, r_{B\mathrm{i}}^2 \equiv k_\perp^2\, V_{T\mathrm{i}}^2/\omega_{B\mathrm{i}}^2\,(Q_{\mathrm{i}\perp}/Y_{\mathrm{i}})^2$ have been neglected when deriving Eq. (37.45), and also viscous terms as may be induced by motion perpendicular to the magnetic field.

[7] In SI-units $\varkappa$ has $\mathrm{m}^{-1}\,\mathrm{s}^{-1}$ for unit, and η $\mathrm{kg\,m}^{-1}\,\mathrm{s}^{-1}$.

(13.20), (13.25), (13.26). The identification of collision frequencies is

$$\bar{\nu}_e = \nu_{\text{eff}}, \quad \bar{\nu}_i = \nu_{ii}. \tag{37.48}$$

We note that in the range of extremely low frequencies,

$$\omega \ll \frac{k_z^2 \varkappa_{\|h}}{N_h} \sim \frac{k_z^2 V_{Th}^2}{\bar{\nu}_h}, \quad \text{i.e.} \quad \frac{Q_{hz}}{Z_h} \gg 1, \tag{37.49}$$

oscillations cannot take place because DEBYE screening of the electric field is too strong. They may, however, occur in the range of *medium frequencies*

$$\frac{k_z^2 V_{Te}^2}{\bar{\nu}_e} \sim \frac{k_z^2 \varkappa_{\|e}}{N_e} \gg \omega \gg \frac{k_z^2 \varkappa_{\|i}}{N_i} \sim \frac{k_z^2 V_{Ti}^2}{\bar{\nu}_i}, \tag{37.50a}$$

i. e.

$$Q_{ez}^2/Z_e \gg 1 \gg Q_{iz}^2/Z_i \tag{37.50b}$$

see Subsect. ζ.

We first consider the range of *high frequencies*

$$\omega \gg \frac{k_z^2 \varkappa_{\|h}}{N_h} \sim \frac{k_z^2 V_{Th}^2}{\bar{\nu}_h}, \quad \text{i.e.} \quad \frac{Q_{hz}}{Z_h} \ll 1, \tag{37.51}$$

where such oscillations possibly occur. Eq. (37.47) gives here the following local spectra of fluid-dynamic *drift oscillations*:

$$\omega_{(1)} = \frac{k_y V_s^2}{\omega_{Bi}} \frac{\partial}{\partial x} \log\left(\frac{N_e}{N_{e0}} \left(\frac{T_e}{T_{e0}}\right)^{1.71}\right), \tag{37.52}$$

$$\gamma_{(1)} = -1.14 \frac{k_z^2 \varkappa_{\|e}}{N_e} \left[\frac{\partial \log T_e/T_{e0}}{\partial \log N_e T_e/N_{e0} T_{e0}^{1.71}} + \frac{4}{3} \frac{k_z^2 V_{Ti}^2}{\omega_{(1)}^2} \frac{\partial \log N_e T_e^{1.71}/N_{e0} T_{e0}^{1.71}}{\partial \log N_e/N_{e0}}\right] \tag{37.53}$$

The corresponding oscillations occur at

$$\omega_d \gtrsim \left(\frac{m_e}{m_i} \nu_{\text{eff}}, k_z V_{Ti}\right), \quad \text{i.e.} \quad \frac{Q_{hy}^2}{Y_h} \frac{1}{k_y \mathscr{L}_0} \gtrsim \left(\frac{m_e}{m_i} Z_{\text{eff}}, Q_{iz}\right). \tag{37.54}$$

A second branch of the spectrum is given by

$$\omega_{(2)}^2 = -0.95\, k_z^2 V_{Ti}^2 \Big/ \frac{\partial \log N_e T_e^{1.71}/N_{e0} T_{e0}^{1.71}}{\partial \log N_e T_{e0}^{\frac{3}{2}}/N_{e0} T_e^{\frac{3}{2}}}, \tag{37.55}$$

and the corresponding oscillations occur ate[8]

$$\omega \ll \frac{m_e}{m_i} \nu_{\text{eff}}, \qquad \text{i.e.} \quad 1 \ll \frac{m_e}{m_i} Z_{\text{eff}}, \tag{37.56a}$$

$$\omega \ll \omega_{dh} \sim \frac{k_y V_{Th}^2}{\omega_{Bh} \mathscr{L}_0}, \quad \text{i.e.} \quad 1 \ll \frac{Q_{hy}^2}{Y_h} \frac{1}{k_y \mathscr{L}_0}. \tag{37.56b}$$

With respect to oscillations of the *first branch*, $\omega_{(1)}$, most plasmas are *stable* as can be seen from Eq. (37.53). These oscillations will only rise in an plasma with severely inhomogeneous temperature under the conditions:

$$-0.585 < \frac{\partial \log T_e/T_{e0}}{\partial x} \Big/ \frac{\partial \log N_e/N_{e0}}{\partial x} < 0. \tag{37.57}$$

---

[8] According to the hypothesis $T_e = T_i = T$ we have here $\dfrac{\vartheta_h}{m_h \omega_{Bh}} = \dfrac{\frac{1}{2} V_{Th}^2}{\omega_{Bh}}$ independent from m_h $\left(\text{because } \omega_{Bh} \propto \dfrac{1}{m_h}\right)$, so that we may write:

$$\omega_d \sim \omega_{di} \sim \frac{k_y V_{Ti}^2}{\omega_{Bi} \mathscr{L}_0} = \frac{k_y V_{Te}^2}{\omega_{Be} \mathscr{L}_0} \sim \omega_{de}.$$

On the other hand with respect to oscillations of the *second branch* almost all plasmas are *unstable*, the condition for instability being

$$-0.585 < \frac{\partial \log T_e/T_{e0}}{\partial \log N_e/N_{e0}} < +\frac{2}{3} \equiv +0.667. \tag{37.58}$$

η) In the range of medium frequencies

$$\frac{k_z^2 \varkappa_{\parallel e}}{N_e} \gg \omega \gg \frac{k_z^2 \varkappa_{\parallel i}}{N_i}, \quad \text{i.e.} \quad \frac{q_e^2}{m_e} \frac{\varkappa_{\parallel e}}{X_e} \gg \frac{\varepsilon_0}{u} \frac{\omega^3}{k_z^2} \gg \frac{q_i^2}{m_i} \frac{\varkappa_{\parallel i}}{X_i}. \qquad [37.50]$$

Eq. (37.47) admits solutions allowing for unstable conditions, thus oscillations but only if

$$\frac{k_y V_T^2}{\omega_B \mathscr{L}_0} \sim \omega_d \gg \omega \gg \frac{m_e}{m_i} \nu_{\text{eff}}, \quad \text{i.e.} \quad \frac{Q_y^2}{Y} \gg k_y \mathscr{L}_0 \gg \frac{m_e}{m_i} Z_{\text{eff}} k_y \mathscr{L}_0 \tag{37.59}$$

is satisfied. The relevant spectral branches correspond to the following local, two spectra[9] (small dissipative members being neglected): a *third branch*:

$$\omega_{(3)}^2 = k_z^2 V_{Ti}^2 \left(\frac{2}{3} - \frac{\partial \log T_e/T_{e0}}{\partial \log N_e/N_{e0}}\right), \tag{37.60a}$$

depending, however, on the condition

$$\omega_{(3)} \ll \frac{k_y V_{Ti}^2}{\omega_{Bi}} \frac{\partial}{\partial x} \log \frac{N_e}{N_{e0}} = Q_{iy}^2 Y_i^{-1} \omega \frac{\partial}{\partial x} \log \frac{N_e}{N_{e0}}, \tag{37.60b}$$

and a *fourth branch*

$$\omega_{(4)}^3 = -k_z^2 V_s^2 \frac{k_y V_{Ti}^2}{\omega_{Bi}} \frac{\partial \log T_i/T_{i0}}{\partial x} \tag{37.61a}$$

which depends on the condition

$$\begin{aligned}\frac{Q_{iy}^2}{Y_i} \frac{\omega}{k_y} \frac{\partial}{\partial x} \log \frac{N_e}{N_{e0}} = \frac{k_y V_{Ti}^2}{\omega_{Bi}} \frac{\partial}{\partial x} \log \frac{N_e}{N_{e0}} \ll \omega_{(4)} \ll \frac{k_y V_{Ti}^2}{\omega_{Bi}} \frac{\partial}{\partial x} \log \frac{T_i}{T_{i0}} \\ = \frac{Q_{iy}^2}{Y_i} \frac{\omega}{k_y} \frac{\partial}{\partial x} \log \frac{T_i}{T_{i0}}.\end{aligned} \tag{37.61b}$$

It can easily be seen that the oscillations on the third branch are unstable for

$$\frac{\partial \log T_e/T_{e0}}{\partial x} > \frac{2}{3} \frac{\partial \log N_e/N_{e0}}{\partial x} \tag{37.62}$$

but those on the fourth branch only for

$$\frac{\partial \log T_i/T_{i0}}{\partial x} \gg \frac{\partial \log N_e/N_{e0}}{\partial x}. \tag{37.63}$$

This follows from Eq. (37.61a). The increment of the developing aperiodic instability is in this latter case

$$\gamma \equiv \text{Im}\, \omega \gtrsim k_z V_{Ti}. \tag{37.64}$$

ϑ) The gasdynamic instabilities considered above are of particular interest when applied to ionospheric irregularities. Such instabilities may develop at rather low heights, between 200 and 500 km. The relevant irregularities must be highly field-aligned[10], with dimensions along the field of the order of 1 to 10 km. A summary of all parameters can be found in Tables 8 through 12:

[9] In this Encyclopedia log means natural logarithm.

[10] Such extension along the magnetic field lines is characteristic for kinetic instabilities, too. See Subsects. 37β, γ, δ, ε.

Table 8.

Frequency and wavelength ranges	Spectra of long-wave kinetic oscillations	Conditions of existence	Rôle of electronic collisions	Rôle of ionic collisions	Rôle of ČERENKOV effect
$k_\perp r_{Bi} \ll 1$ $\omega, \bar{\nu}_e \ll k_z V_{Te}$ $\omega \gg \nu_i,\ k_z V_{Ti}$ $\omega_{dh} \sim \frac{k_y V_{Th}^2}{\omega_{Bh} \mathcal{L}_0}$ $\gg k_z V_s$ $\omega \ll \omega_{Bi}$	$\omega_1 \approx -\frac{1}{1+k^2 \lambda_{De}^2} \frac{k_y V_s^2}{\omega_{Bi}} \frac{\partial \log N_i/N_{i0}}{\partial x}$; $\gamma_1 = \sqrt{\frac{\pi}{2}} \frac{\omega_1^2}{\lvert k_z \rvert V_{Te}} \times$ $\times \left[\frac{k_\perp^2 V_s^2}{\omega_{Bi}^2}\left(1+\frac{T_i}{T_e} \frac{\partial \log N_i T_i/N_{i0} T_{i0}}{\partial \log N_i/N_{i0}}\right) -\right.$ $\left. -\frac{1}{2} \frac{\partial \log T_e/T_{e0}}{\partial \log N_e/N_{e0}}\right] - \frac{7}{10} \nu_{ii} k_\perp^4 r_{Bi}^4 \times$ $\times \left(1+\frac{T_e}{T_i} - \frac{3}{28} \frac{\partial \log T_i/T_{i0}}{\partial \log N_i/N_{i0}}\right)$	$\frac{k_z^2}{k_\perp^2} < \frac{\omega_1^2}{\omega_{Bi}^2} \ll 1$	Insignificant	Destabilizing when $\frac{\partial \log T_i/T_{i0}}{\partial \log N_i/N_{i0}}$ $> \frac{28}{3}\left(1+\frac{T_e}{T_i}\right)$	Destabilizing when $\frac{\partial \log T_e/T_{e0}}{\partial \log N/N_0}$ $< 4 \frac{k_\perp^2 V_s^2}{\omega_{Bi}^2}$
	$\omega_2 = (k_z^2 \omega_{Bi}/k_y) \frac{\partial \log N_e/N_{e0}}{\partial x}$; $\gamma_2 = -\sqrt{\frac{\pi}{2}} \frac{\omega_2^2}{\lvert k_z \rvert V_{Te}}\left(1-\frac{1}{2} \frac{\partial \log T_e/T_{e0}}{\partial \log N_e/N_{e0}}\right) -$ $-\frac{8}{5} \nu_{ii} \frac{k_z^2 V_{Ti}^2}{\omega_2^2}$	$\omega_2 \ll k_z V_s$; $T_e \gg T_i$	Insignificant	Always stabilizing	Destabilizing when $\frac{\partial \log T_e/T_{e0}}{\partial \log N_e/N_{e0}} > 2$
	$\omega_3^2 = -k_z^2 V_{Ti}^2 \frac{\partial \log T_i/T_{i0}}{\partial \log N_i/N_{i0}}$;	$\frac{\partial \log T_i/T_{i0}}{\partial \log N/N_0} \gg 1$;	Insignificant	Insignificant	Insignificant
	$\omega_4^3 = -k_z^2 V_3^2 \frac{k_y V_{Ti}^2}{\omega_{Bi}} \frac{\partial \log T_i/T_{i0}}{\partial x}$	$\omega_4 \ll \omega_{dh}$; $\frac{\partial \log T_i/T_{i0}}{\partial \log N/N_0} \gg 1$	Insignificant	Insignificant	Insignificant

Table 9.

Frequency and wavelength ranges	Spectra of long-wave kinetic oscillations	Conditions of existence	Rôle of electronic collisions	Rôle of ionic collisions	Rôle of ČERENKOV effect
$k_\perp r_{Bi} \ll 1$ $\omega \gg \bar{v}_e, k_z V_{Te}$ $\omega \gg \bar{v}_i, k_z V_{Ti}$ $\omega \ll \omega_{Bi}$	$\omega_1 = \frac{1}{1+V_A^2/c_0^2} \frac{k_y V_{Ti}^2}{\omega_{Bi}} \frac{\partial \log N_i T_i/N_{i0} T_{i0}}{\partial x}$; $\gamma_1 = \frac{\nu_{ii}}{40} \frac{k_\perp^2 r_{Bi}^2}{1+V_A^2/c_0^2} \frac{\partial \log N_i/N_{i0}}{\partial \log N_i T_i/N_{i0} T_{i0}} \times$ $\times \left[28 \frac{V_A^2}{c_0^2} - \left(31 + 3 \frac{V_A^2}{c_0^2}\right) \frac{\partial \log T_i/T_{i0}}{\partial \log N_i/N_{i0}}\right]$	$\frac{k_z^2}{k_\perp^2} < \frac{m_e}{m_i} \frac{T_i}{T_e} \frac{\omega_1^2}{\omega_{Bi}^2}$	Insignificant	Destabilizing when $\frac{\partial \log T_i/T_{i0}}{\partial \log N_i/N_{i0}} < \frac{28}{31} \frac{V_A^2}{c_0^2}$	Insignificant
	$\omega_2 = -\frac{k_y V_s^2}{\omega_{Bi}} \frac{\partial \log N_e T_e/N_{e0} T_{e0}}{\partial x}$; $\gamma_2 = \nu_{\text{eff}} \frac{\partial \log (T_e/T_{e0})^{\frac{3}{2}}}{\partial \log N_e T_e/N_{e0} T_{e0}}$	$\frac{k_z^2}{k_\perp^2} > \frac{m_e}{m_i} \frac{r_{B0}^2}{\mathcal{L}_0^2}$	Destabilizing when $\frac{\partial \log T_e/T_{e0}}{\partial \log N_e/N_{e0}} > 0$	Insignificant	Insignificant
	$\omega_3^2 = -\frac{k_z^2}{k_\perp^2} \frac{m_i}{m_e} \frac{T_e}{T_i} \omega_{Bi}^2 \frac{\partial \log N_e T_e/N_{e0} T_{e0}}{\partial \log N_i T_i/N_{i0} T_{i0}}$	$\omega_3^2 \ll \omega_{dh}^2$; $\frac{k_z^2}{k_\perp^2} < \frac{m_e}{m_i} \frac{T_i}{T_e}$	Insignificant	Insignificant	Insignificant

Table 10.

Frequency and wavelength ranges	Spectra of long-wave kinetic oscillations	Conditions of existence	Rôle of electronic collisions	Rôle of ionic collisions	Rôle of Čerenkov effect
$k_\perp r_{Bi} \ll 1$ $\bar{\nu}_e \gg \omega,\ k_z V_{Te}$ $\omega \gg \bar{\nu}_i,\ k_z V_{Ti}$ $\omega_s = \frac{k_z^2}{k_\perp^2} \frac{m_i}{m_e} \frac{\omega_{Bi}^2}{\nu_{eff}}$ $\omega \gg \omega_{Bi}$	$\omega_1 = \frac{1}{1+V_A^2/c_0^2} \frac{k_y V_{Ti}^2}{\omega_{Bi}} \frac{\partial \log N_i T_i/N_{i0} T_{i0}}{\partial x}$; $\gamma_1 = \frac{\nu_{ii}}{40} \frac{k_\perp r_{Bi}^2}{1+V_A^2/c_0^2} \frac{\partial \log N_i/N_{i0}}{\partial \log N_i T_i/N_{i0} T_{i0}} \times$ $\times \left[\frac{28 V_A^2}{c_0^2} - \left(31 + 3\frac{V_A^2}{c_0^2}\right) \frac{\partial \log T_i/T_{i0}}{\partial \log N_i/N_{i0}}\right]$	$\omega_{dh} \gg \omega_s$; $k_z \approx 0$	Insignificant	Destabilizing when $\frac{\partial \log T_i/T_{i0}}{\partial \log N_i/N_{i0}} < \frac{28}{31} \frac{V_A^2}{c_0^2}$	Insignificant
	$\omega_2 = -\frac{k_y V_s^2}{\omega_{Bi}} \frac{\partial \log (N_e T_e^{1.71}/N_{e0} T_{e0}^{1.71})}{\partial x}$; $\gamma_2 = \frac{\omega_2^2}{1.96\,\omega_s} \left(1 + \frac{V_A^2}{c_0^2} + \right.$ $\left. + \frac{T_i}{T_e} \frac{\partial \log N_i T_i/N_{i0} T_{i0}}{\partial \log (N_e T_e^{1.71}/N_{e0} T_{e0}^{1.71})}\right)$	$\omega_{dh} \gg \omega_s$	Always destabilizing	Insignificant	Insignificant
	$\omega_3 = i\,1.96\,\omega_s \frac{T_e}{T_i} \frac{\partial \log (N_e T_e^{1.71}/N_{e0} T_{e0}^{1.71})}{\partial \log N_i T_i/N_{i0} T_{i0}}$	$\omega_{dh} \gg \omega_s$	Always destabilizing	Insignificant	Insignificant

Table 11.

Conditions of existence	Spectra of long-wave kinetic oscillations $k_\perp r_{Bi} \ll 1;\ \bar{\nu}_e \gg \omega, k_z V_{Te};\ \omega \gg \bar{\nu}_i, k_z V_{Ti};\ \omega \bar{\nu}_e \ll k_z^2 V_{Te}^2$	Rôle of electronic collisions	Rôle of ionic collisions
$\omega_1 \gg k_z V_s$ $\frac{k_z^2}{k_\perp^2} < \frac{\omega_1^2}{\omega_{Bi}^2}$	$\omega_1 \approx -\frac{k_z V_s^2}{\omega_{Bi}} \frac{\partial}{\partial x} \log \frac{N}{N_0}$; $\gamma_1 \approx 1.44 \frac{\omega_1^2 \nu_{eff}}{k_z^2 V_{Te}^2} \left\{ \frac{k_\perp^2 V_s^2}{\omega_{Bi}^2} \left(1 + \frac{T_i}{T_e} \frac{\partial \log N T_i/N_0 T_{i0}}{\partial \log N/N_0}\right) - \right.$ $\left. -0.56 \frac{\partial \log T_e/T_{e0}}{\partial \log N/N_0} \right\} -$ $-\frac{7}{10} \nu_{ii} k_\perp^4 r_{Bi}^4 \left(1 + \frac{T_e}{T_i} - \frac{3}{28} \frac{\partial \log T_i/T_{i0}}{\partial \log N/N_0}\right)$	Destabilizing when $\frac{\partial \log T_e/T_{e0}}{\partial \log N/N_0} < 1.8 \frac{k_\perp^2 V_s^2}{\omega_{Bi}^2} \times$ $\times \left(1 + \frac{T_i}{T_e} \frac{\partial \log N T_i/N_0 T_{i0}}{\partial \log N/N_0}\right)$	Destabilizing when $\frac{\partial \log T_i/T_{i0}}{\partial \log N/N_0} > \frac{28}{3} \left(1 + \frac{T_e}{T_i}\right)$
$\omega_2 \ll k_z V_s$ $T_e \gg T_i$	$\omega_2 = -k_z^2 \omega_{Bi}/k_y \frac{\partial}{\partial x} \log N/N_0$; $\gamma_2 \approx -1.44 \frac{\omega_2^2 \nu_{eff}}{k_z^2 V_{Te}^2} \left(1 - \frac{\partial \log (T_e/T_{e0})^{0.56}}{\partial \log N/N_0} - \frac{8}{5} \nu_{ii} \frac{k_z^2 V_{Ti}^2}{\omega_2^2}\right.$	Destabilizing when $\frac{\partial \log T_e/T_{e0}}{\partial \log N/N_0} > 1.8$	Always stabilizing
$\omega \ll \omega_{dh}$	$\omega_3^2 = -k_z^2 V_{Ti}^2 \frac{\partial \log T_i/T_{i0}}{\partial \log N/N_0}$;	Insignificant	Insignificant
$\frac{\partial \log T_i/T_{i0}}{\partial \log N/N_0} \gg 1$	$\omega_4^3 = -k_z^2 V_s^2 \frac{k_y V_{Ti}^2}{\omega_{Bi}} \frac{\partial}{\partial x} \log T_i/T_{i0}$	Insignificant	Insignificant

Table 12.

Frequency and wavelength range	Spectrum of long-wave fluid-dynamic oscillation	Conditions of existence	Rôle of electronic collisions	Rôle of ionic collisions
$k_\perp r_{Bi} \ll 1$ $\nu_e \gg \omega, k_z V_{Te}$ $\nu_i \gg \omega, k_z V_{Ti}$ $\omega \ll \omega_{Bi}$	$\omega_1 = \frac{k_y V_s^2}{\omega_{Bi}} \frac{\partial \log (N_e T_e/N_{e0} T_{e0})^{1.71}}{\partial x}$; $\gamma_1 = -\frac{3.42}{3} \frac{k_z^2 \varkappa_\parallel^e}{N_e} \left[\frac{\partial \log T_e/T_{e0}}{\partial \log (N_e T_e^{1.71}/N_{e0} T_{e0}^{1.71})} + \right.$ $\left. + \frac{4}{3} \frac{k_z^2 V_{Ti}^2}{\omega_1^2} \frac{\partial \log (N_e T_e^{1.71}/N_{e0} T_{e0}^{1.71})}{\partial \log N_e/N_{e0}} \right]$	$\omega_1 \gg \frac{k_z^2 V_{Te}^2}{\nu_{eff}}$, $\omega_1 \gg \frac{m_e}{m_i} \nu_{eff}$; $\omega_1 \gg k_z V_{Ti}$	Destabilizing when $\frac{\partial \log T_e/T_{e0}}{\partial \log N_e/T_{e0}} < 0$	Insignificant
	$\omega_2^2 = -\frac{2.84}{3} k_z^2 V_{Ti}^2 \frac{\partial \log N_e T_e^{-3/2}/N_{e0} T_{e0}^{-3/2}}{\partial \log N_e T_e^{1.71}/N_{e0} T_{e0}^{1.71}}$	$\omega_2 \sim k_z V_{Ti} \ll \frac{m_e}{m_i} \nu_{eff}$; $\omega_d \gg \omega_2 \gg \frac{k_z^2 V_{Te}^2}{\nu_{eff}}$	Insignificant	Insignificant
	$\omega_3^2 = k_z^2 V_{Ti}^2 \left(\frac{2}{3} - \frac{\partial \log T_e/T_{e0}}{\partial \log N_e/N_{e0}} \right)$	$\frac{k_z^2 V_{Te}^2}{\nu_{eff}} \gg \omega_3 \gg \frac{k_z^2 V_{Ti}^2}{\nu_{ii}}$	Insignificant	Insignificant
	$\omega_4^3 = -k_z^2 V_s^2 \frac{k_y V_{Ti}^2}{\omega_{Bi}} \frac{\partial \log T_e/T_{e0}}{\partial x}$	$\frac{k_z^2 V_{Te}^2}{\nu_{eff}} \gg \omega_4 \gg \frac{k_z^2 V_{Ti}^2}{\nu_{ii}}$	Insignificant	Insignificant

38. Instability in an inhomogeneous anisotropic plasma in the presence of an electric current. In Sect. 37 above we considered possible oscillations in an inhomogeneous plasma with isotropic distribution of particle velocities. If this condition is not satisfied we have an anisotropy of the velocity distribution function and this means, macroscopically, that a flux or current exists. Such a current, particularly in an inhomogeneous plasma, possibly causes other kinds of instabilities which do not exist in a plasma with isotropic velocity distribution. This is a problem of great interest to theoreticians which is extensively investigated at present. While we cannot discuss it in detail in the framework of this paper we feel the need to give at least a few outlines. In fact, the stability problem of an anisotropic, inhomogeneous plasma plays a fundamental rôle in connection with magnetic confinement[1]. We shall therefore consider in the following two examples which are both important for the physics of hot plasmas[2] and are treated in the relevant literature.

α) In both examples the particles have a *directed motion* which is superposed upon the thermal motions. In the first case (which is discussed in Subsect. β) a particle drift is induced by a curvature[3] of the magnetic field lines (radius of curvature R), which can be characterized by a characteristic acceleration $g_{\mathrm{h}} \equiv V_{T\mathrm{h}}^2/R$. The resulting drift is $\boldsymbol{u}_{\mathrm{h}}$:

$$|\boldsymbol{u}_{\mathrm{h}}| = -\frac{g_{\mathrm{h}}}{\omega_{B\mathrm{h}}} \equiv \frac{V_{T\mathrm{h}}^2}{R}\,\frac{1}{\omega_{B\mathrm{h}}}. \tag{38.1}$$

The second example supposes a current parallel to the magnetic field with an average drift velocity of the electrons against the positive ions. It will be considered in Subsect. γ.

We limit our considerations to collisionless, inhomogeneous plasmas for which the expression for the parallel dielectric permeability can be taken directly from Eq. (31.8) just replacing ω by $\omega_{\mathrm{h}}' \equiv \omega - \boldsymbol{k}\cdot\boldsymbol{u}_{\mathrm{h}}$, where $\boldsymbol{u}_{\mathrm{h}}$ is the average drift velocity of the particles of kind h[4]. We then obtain for low frequency oscillations

$$\omega_{\mathrm{h}}' \gg \omega_{B\mathrm{h}}, \quad \text{i.e.} \quad Y_{\mathrm{h}}' \gg 1 \qquad (\mathrm{h} = \mathrm{e}, \mathrm{i}) \tag{38.2}$$

the following eiconal equation[5]

$$\left.\begin{aligned} &1 + \sum_{\mathrm{h}} \frac{\omega_{N\mathrm{h}}^2}{k^2 V_{T\mathrm{h}}^2}\left\{1 - \sum_s \frac{\omega_{\mathrm{h}}'}{\omega_{\mathrm{h}}' - s\omega_{B\mathrm{h}}}\left\{1 - \frac{k_y V_{T\mathrm{h}}^2}{\omega_{\mathrm{h}}'\omega_{B\mathrm{h}}}\left[\frac{\partial}{\partial x}\log\frac{N_{\mathrm{h}}}{N_{\mathrm{h}0}} + \frac{\partial T_{\mathrm{h}}}{\partial x}\,\frac{\partial}{\partial T_{\mathrm{h}}}\right]\right\}\times\right.\\ &\left.\times\,\mathscr{A}_s\!\left(\frac{k_\perp^2 V_{T\mathrm{h}}^2}{\omega_{B\mathrm{h}}^2}\right)\mathrm{J}_+\!\left(\frac{\omega_{\mathrm{h}}' - s\,\omega_{B\mathrm{h}}}{|k_z|\,V_{T\mathrm{h}}}\right)\right\} = 0, \end{aligned}\right\} \tag{38.3}$$

i.e.

$$\left.\begin{aligned} &1 + \sum_{\mathrm{h}} X_{\mathrm{h}} Q_{\mathrm{h}}^{-2}\left\{1 - \sum_s (1 - s Y_{\mathrm{h}}')^{-1}\left\{1 - (Q_{\mathrm{h}y}'^2/Y_{\mathrm{h}}')\,\frac{1}{k_y}\left[\frac{\partial}{\partial x}\log\frac{N_{\mathrm{h}}}{N_{\mathrm{h}0}} + \right.\right.\right.\\ &\left.\left.\left. + \frac{\partial\vartheta_{\mathrm{h}}}{\partial x}\,\frac{\partial}{\partial\vartheta_{\mathrm{h}}}\right]\right\}\mathscr{A}_s(Q_{\mathrm{h}\perp}^2)\,\mathrm{J}_+\!\left(\frac{1 - s Y_{\mathrm{h}}'}{Q_{\mathrm{h}z}'}\right)\right\} = 0. \end{aligned}\right\} \tag{38.3a}$$

The definitions of the special functions are found in Eqs. (17.10) and (22.9).

---

[1] See the contribution by W. N. Hess in this volume, p. 115.
[2] Applied, for example, in thermonuclear devices.
[3] See Sect. 6.
[4] The method of computation has been explained in Sect. 27...29.
[5] In this Encyclopedia log means log nat.

We analyse Eq. (38.3) in the *limit* of a cold plasma only, supposing

$$k_\perp^2 V_{T\mathrm{h}}^2 \ll \omega_{B\mathrm{h}}^2, \quad \text{i.e.} \quad Q_{\mathrm{h}\perp}'^2 \ll Y_{\mathrm{h}}'^2, \tag{38.4a}$$

and

i.e.

$$\left.\begin{aligned} &\omega_{B\mathrm{h}} \gg \omega_{\mathrm{h}}' \equiv (\omega - \boldsymbol{u}_{\mathrm{h}} \cdot \boldsymbol{k}) \gg k_z V_{T\mathrm{h}}, \\ &Y_{\mathrm{h}}' \gg 1 \gg Q_{\mathrm{h}z}', \end{aligned}\right\} \tag{38.4b}$$

and find [V_A is the ALFVÉN velocity, see Eq. (25.25)]:

$$\left.\begin{aligned} &k_\perp^2 \left(1 + \frac{c_0^2}{V_A^2}\right) + k_z^2 \left(1 - \frac{\omega_{N\mathrm{e}}^2}{(\omega - \boldsymbol{u}_{\mathrm{e}} \cdot \boldsymbol{k})^2}\right) + \frac{\omega_{N\mathrm{i}}^2 k_y}{\omega_{B\mathrm{i}}(\omega - \boldsymbol{u}_{\mathrm{i}} \cdot \boldsymbol{k})} \frac{\partial}{\partial x} \log \frac{N_{\mathrm{i}}}{N_{\mathrm{i}0}} + \\ &+ \frac{\omega_{N\mathrm{e}}^2 k_y}{\omega_{B\mathrm{e}}(\omega - \boldsymbol{u}_{\mathrm{e}} \cdot \boldsymbol{k})} \frac{\partial}{\partial x} \log \frac{N_{\mathrm{e}}}{N_{\mathrm{e}0}} = 0. \end{aligned}\right\} \tag{38.5}$$

Supposing that the drift velocities are small compared to the wave-propagation velocity:

$$\omega \gg (\boldsymbol{u}_{\mathrm{e}} \cdot \boldsymbol{k}, \boldsymbol{u}_{\mathrm{i}} \cdot \boldsymbol{k}) \tag{38.6}$$

we get the following local spectrum:

$$\omega^2 = \frac{\omega_{N\mathrm{i}}^2 \frac{(\boldsymbol{u} \cdot \boldsymbol{k}) k_y}{\omega_{B\mathrm{i}}} \frac{\partial}{\partial x} \log \frac{N_{\mathrm{i}}}{N_{\mathrm{i}0}} + k_z^2 \omega_{N\mathrm{e}}^2}{k^2 (1 + c_0^2/V_A^2)}, \tag{38.7}$$

where

$$\boldsymbol{u} \equiv (\boldsymbol{u}_{\mathrm{e}} - \boldsymbol{u}_{\mathrm{i}}) \tag{38.8}$$

is the *relative drift velocity.*

β) For an *inhomogeneous plasma in a curved magnetic field* (but without a parallel current) the directed velocity $\boldsymbol{u}$ is parallel to the $\boldsymbol{y}^0$-axis, so perpendicular to $\boldsymbol{B}_0^0 \equiv \boldsymbol{z}^0$, and is found to be:

$$|\boldsymbol{u}| = -\frac{g_{\mathrm{eff}}}{\omega_{B\mathrm{i}}} \equiv \frac{V_{T\mathrm{i}}^2 + V_{\mathrm{s}}^2}{R} \cdot \frac{1}{\omega_{B\mathrm{i}}}, \tag{38.9}$$

$V_{T\mathrm{i}}$ being the average thermal velocity of the ions, V_{s} being the velocity of sound in the ion gas g_{eff} is a virtual acceleration. Eq. (38.7) here takes the form

$$\omega^2 = \left[g_{\mathrm{eff}} k_y^2 \frac{\partial}{\partial x} \log \frac{N_{\mathrm{i}}}{N_{\mathrm{i}0}} + k_z^2 \omega_{N\mathrm{e}}^2 \frac{V_A^2}{c_0^2}\right] \frac{1}{k^2 (1 + V_A^2/c_0^2)}. \tag{38.10}$$

We say that the magnetic field has a positive curvature if the outer normal on the field lines is directed towards decreasing plasma density. In this case

$$g_{\mathrm{eff}} \frac{\partial}{\partial x} \log \frac{N_{\mathrm{i}}}{N_{\mathrm{i}0}} < 0,$$

and oscillations with

$$\left(\frac{k_z}{k_y}\right)^2 \ll \frac{m_{\mathrm{e}}}{m_{\mathrm{i}}} \frac{1}{\omega_{B\mathrm{i}}^2} \left| g_{\mathrm{eff}} \frac{\partial}{\partial x} \log \frac{N_{\mathrm{i}}}{N_{\mathrm{i}0}} \right| \tag{38.11}$$

(i.e. k_z very small) appear as aperiodically unstable. The corresponding increment describes the rise of the oscillations by

$$\gamma \equiv \operatorname{Im} \omega \sim \left(\left| g_{\mathrm{eff}} \frac{\partial}{\partial x} \log \frac{N_{\mathrm{i}}}{N_{\mathrm{i}0}} \right|\right)^{\frac{1}{2}} \sim \sqrt{\frac{V_{T\mathrm{i}}^2 + V_{\mathrm{s}}^2}{R \mathscr{L}_0}}. \tag{38.12}$$

This kind of instability has first been predicted in [*47*]. It is called *"field-convective" instability*, but also the designation flute instability is used.

γ) Let us now go over to our second example. In an *inhomogeneous plasma containing a parallel current* the drift velocity vector $\boldsymbol{u}$ is parallel to $\boldsymbol{B}_{\delta}^{0} = \boldsymbol{z}^{0}$. Then Eq. (38.7) takes the form

$$\omega^2 = \frac{\omega_{N\mathrm{i}}^2 \frac{k_y k_z u}{\omega_{B\mathrm{i}}} \frac{\partial}{\partial x} \log \frac{N_\mathrm{i}}{N_{\mathrm{i}0}} + k_z^2 \omega_{N\mathrm{e}}^2}{k^2 (1 + V_\mathrm{A}^2/c_0^2)}, \tag{38.13}$$

i.e.

$$1 + \frac{V_\mathrm{A}^2}{c_0^2} = \left(\frac{k_y k_z}{k^2}\right) X_\mathrm{i} Y_\mathrm{i}^{-1} \left(\frac{u}{\omega}\right) \frac{\partial}{\partial x} \log \frac{N_\mathrm{i}}{N_{\mathrm{i}0}} + \left(\frac{k_z}{k}\right)^2 X_\mathrm{e}. \tag{38.13a}$$

V_A being the ALFVÉN velocity, see Eq. (25.25). It follows then that under the condition

$$\frac{k_z}{k_y} < \frac{m_\mathrm{e}}{m_\mathrm{i}} \frac{u}{\omega_{B\mathrm{i}}} \frac{\partial}{\partial x} \log \frac{N_\mathrm{i}}{N_{\mathrm{i}0}} \sim \frac{u}{\omega_{B\mathrm{e}} \mathscr{L}_0} \tag{38.14}$$

the oscillations are instable, the increment of rise being described by

$$\gamma = \operatorname{Im} \omega \sim \left(\left|\frac{k_z}{k_y} \frac{u}{\mathscr{L}_0} \omega_{B\mathrm{i}}\right|\right)^{\frac{1}{2}} \lesssim \omega_{B\mathrm{i}}. \tag{38.15}$$

Such instability in a collisionless plasma has been discussed in [*48*], [*49*]. It is called *stream-convective instability*.

δ) In the foregoing we have supposed a collisionless plasma for simplicity only. The instabilities as described in Subsect. β and γ can, however, exist also when *collisions* occur in the plasma. In particular, the *flute-instability* due to a curved magnetic field has been found in a dense plasma even when the collision frequency was rather high [*47*]. The conditions for such oscillations to exist are those of the fluid-dynamic model, see Sect. 6. The *stream-convective instability* has first been described in [*50*] for a plasma with collisions.

Though these investigations were concerned with hot thermonuclear plasmase both instabilities are possibly of importance in the natural plasma of the ionospher[6] and magnetosphere too[7].

In the last years a large number of papers appeared in the literature discussing and interpreting the low-frequency oscillations of different nature as observed in the ionosphere and in the magnetosphere [*52*], [*53*][8]. Such oscillations are related with ALFVÉN waves, magneto-acoustic waves and spiral (Whistler) waves, see Sects. 25 and 26. Recently the possibility of drift waves in the ionosphere is also seriously discussed [*53*]. It is expected that low-frequency oscillations will be more and more used as a tool for studying the structures of ionospheric and magnetospheric plasmas.

---

[6] These two instabilities often determine the lifetime of hot plasma produced in thermonuclear devices.

[7] See the contributions in this volume by W. N. HESS, p. 115, and by H. POEVERLEIN, p. 7.

[8] See also the references given in [*52*], [*53*].

Notations and symbols.

The numbers in parentheses denote sections where introduced.

$\boldsymbol{A}$	friction vector (12)
$\mathscr{A}$	integral function related with the BESSEL-function I_s see Eq. (22.9)
$\boldsymbol{B}$	magnetic induction vector (1)
c_0	velocity of light (in vacuo) (1)
$\boldsymbol{D}$	dielectric displacement vector (1)
D	diffusion tensor (12)
$\mathscr{D}$	determinant (4)
e	suffix for electrons (2)
$\mathscr{E}$	energy (8)
$\boldsymbol{E}$	electric field (1) [with elements ε_{lm}]
E	tensor of complex (dielectric) permittivity (1) [see also 6]
$\mathfrak{E}$	permutation tensor (6)
f	frequency (1)
$f(\,)$	distribution function (8)
f^{M}	Maxwellian distribution function
$f_{B\mathrm{h}}$	gyro frequency (2)
$f_{N\mathrm{h}}$	plasma frequency (2)
$\boldsymbol{F}$	mechanical force (1)
$\boldsymbol{g}$	non-electric force acceleration (6)
h	suffix for general particle (2)
$\hbar$	$= 1.054 \cdot 10^{-34}$ Js
$\boldsymbol{H}$	magnetic field (3)
i	suffix for ions (2)
I	tensor characterizing collisional reactions (12)
J_+	transcendent function see footnote to Eq. (18.15)
$\boldsymbol{J}$, $\boldsymbol{J}_0$	current density (1)
$\boldsymbol{k}$	wave (number) vector (1)
L_0, $L_\parallel$	reference length (6)
$\mathscr{L}$	COULOMB's logarithm (12)
m_{h}	mass of particle kind h (2)
M	effective particle mass ($= m_{\mathrm{i}}$ in fully ionized plasma) (11)
N_{h}	number density of particles kind h (2)
$\boldsymbol{p}$	momentum
P	pressure (in fluid-dynamics) (7)
q	unit charge (2)
$\boldsymbol{q}$	heat flux vector (13)
Q	supported power (3); heat transfer through collisions (13)
$\boldsymbol{r}$	radius vector (1)
$\boldsymbol{R}$	friction force vector (13)
$r_{B\mathrm{h}}$	cyclotron (LARMOR) radius for particle h (2), also called gyro radius
R_{E}	Earth's radius ($= 6370$ km) (2)
R_0	curvature of magnetic field-lines (6)
S	entropy (per unit volume) (7)
[S	surface (3)]
S	complex conductivity tensor (1) [with elements σ_{lm}]
t, t'	time (1)
T	viscosity tensor (7)
u	parameter characterizing systems of units (1)
U	interaction potential (12)
U	unit tensor of second grade (2) [with elements δ_{lm}]
$U(\boldsymbol{k})$	Fourier component (12)
$\boldsymbol{v}$, v	a material velocity
$\boldsymbol{v}_{\mathrm{p}}$	center of gravity (bulk) velocity
V	velocity tensor (9)
$\boldsymbol{V}$, V	a characteristic velocity
$\boldsymbol{V}_{\mathrm{g}}$	group velocity
$\boldsymbol{V}_{\mathrm{p}}$	phase velocity
V_{s}	sound velocity (11)
V_{A}	ALFVÉN velocity
$V_{T\mathrm{h}}$	thermal velocity of kind h (2)
[V	volume (3)]
W	reaction probability (12)

Tensors of second grade:

$\hat{\mathsf{E}} \equiv ((\hat{\varepsilon}_{ij}))$; $\quad \mathsf{E} \equiv ((\varepsilon_{ij}))$

$\hat{\mathsf{S}} \equiv ((\hat{\sigma}_{ij}))$; $\quad \mathsf{S} \equiv ((\sigma_{ij}))$

$\mathsf{V}_{\mathrm{h}} \equiv \int d^3\boldsymbol{p}\,\boldsymbol{p}\,\boldsymbol{p}\,f_{\mathrm{h}}$; $\quad \mathsf{U} = ((\delta_{jk}))$

$\mathsf{T} \equiv ((\vartheta_{jk}))$;

Tensors of third grade:

$\mathfrak{E} = (((\varepsilon_{ijk})))$

β	ratio of kinetic to magnetic energy/pressure (2)
γ_{n}	increment (4)
δ_{lm}	Kronecker symbol (1) [element of unit tensor of second grade]
δ	DIRAC distribution (10)
Δ	solution (4)
ε_{lm}, $\hat{\varepsilon}_{lm}$	element of dielectric permittivity tensor E, $\hat{\mathsf{E}}$ (1)
ε_{ijk}	element of (third grade) permutation tensor $\mathfrak{E}$ (4)
ε_0	(absolute) dielectric permittivity (1)
η	viscosity coefficient (7)
ϑ	angle against wave vector (5)
ϑ_{h}	kT_{h} energetic temperature (8)
ϑ_{jk}	components of viscosity tensor T
$\varkappa$ [$\varkappa_\parallel$, $\varkappa_\perp$]	heat conductivity (7) [for specific directions (13)]
λ	wavelength
$\lambda\!\!\!^{-}_{\mathrm{B}}$	DEBROGLIE length (12)
λ_{D}	DEBYE radius (2)
μ_0	(absolute) magnetic permeability (1)
$\nu_{\mathrm{h}_1\mathrm{h}_2}$	collision frequency between kind h_1 and h_2 (2) [see "Introductory Remarks", p. 1]
ξ	viscosity coefficient (7)
Π	viscosity tension tensor (13) [with elements Π_{ij}]
ϱ, ϱ_0	charge density (1)
$\bar{\varrho}$	average mass density (7)
σ_{lm}, $\hat{\sigma}_{lm}$	element of complex conductivity tensor S, $\hat{\mathsf{S}}$ (1)

$\sigma\,[\sigma_{\parallel}, \sigma_{\perp}]$	conductivity (7) [for specific directions (13)]
ϕ	electrostatic potential (V) (11)
Φ	current density function (A m^{-2}) (1)
ψ	(plane) wave function (12)
ω	pulsation (1)
ω_{dh}	pulsation of drift oscillations (34)
$\omega_{N\mathrm{h}}$	plasma pulsation of ions kind h (2)
$\omega_{B\mathrm{h}}$	gyropulsation of ions kind h (2)
$\mathbf{e}$	unit vector
$^{(0)}$	Hermitean (22)
$^{(a)}$	anti-Hermitean (23)
det	determinant
Im	imaginary part } of a complex value
Re	real part } of a complex value
*	complex conjugate
E, ♁	relative to Earth
$\parallel$, $\perp$	parallel and perpendicular, relative to magnetic field
l, tr	longitudinal and transverse, relative to wave vector
$\sim$, $\approx$	FOURIER-transforms (Intr. Rem.)

General references.

[*1*] GINZBURG, V. L.: Rasprostranenie elektromagnithny voln v plazme (2nd ed.). Moskva 1967; (engl. transl.) Propagation of electromagnetic waves in plasma. London: Pergamon Press 1967.

[*2*] SILIN, V. P., i A. A. RUHADZE: Elektromagnitnye sovjetva plazmy i plazmopodobnyh sred. (Electromagnetic properties of plasma and similar media.) Moskva 1961.

[*3*] AHIEZER, A. I., I. A. AHIEZER, R. V. POLOVIN, A. G. SITENKO i K. N. STEPANOV: Kollektivnye kolebanija plazmy. (Kollective oscillation of plasmas.) Moskva 1964.

[*4*] SITENKO, A. G.: Elektromagnitnye fluktuacii v plazme. (Electromagnetic fluctuations in plasmas.) Har'kov 1965.

[*5*] STIX, T. H.: The theory of plasma waves. New York: McGraw-Hill 1962 (russian transl.: Moskva 1965).

[*6*] LEONTOBIČ, M. A. (ed.): Voprosy teorii plazmy. (Questions in plasma theory.) T. 1—5 Moskva 1963.

[*7*] LANDAU, L. D., i E. M. LIFŠIC: Elektrodinamika splošnyh sred. Moskva 1957 (engl. transl.). Electrodynamics of dense media. London: Pergamon Press 1958.

[*8*] AGRANOVIČ, V. M., i V. L. GINZBURG: Kristallooptika s ucotom prostranstvennoj dispersii. Moskva 1963 (engl. transl.). Crystal optics and spatial dispersion. New York: John Wiley 1966.

[*9*] AL'PERT, JA. L.: Rasprostranenie radiovoln i ionosfera. Moskva 1960 (engl. transl.) Radio wave propagation and the ionosphere. New York: Consultants Bureau 1962.

[*10*] — A. V. GUREVIC, i L. L. PITAEVSKIJ: Iskusstrennye sputniki v ražrezennoj plazme. (Artificial satellites in rarefied plasma.) Moskva 1964.

[*11*] RAWER, K., and K. SUCHY: Radio observations of the ionosphere. In: Handbuch der Physik, vol. 49/2, pp. **1**—546. Berlin-Heidelberg-New York: Springer 1967.

[*12*] TITCHMARSH, E.: Introduction to the theory of Fourier-integrals. Oxford: Clarendon Press 1937 (russian transl.: Moskva 1948).

[*13*] BALESCU, R.: Statistical mechanics of charged particles. New York: Interscience Publ. 1963 (russian transl.: Moskva 1966).

[*14*] ALFVEN, H.: Cosmical electrodynamics. Oxford: Clarendon Press 1950 (russian transl.: Moskva 1952).

[*15*] CHAPMAN, S., and T. G. COWLING: The mathematical theory of non-uniform gases, 2nd ed. Cambridge: Cambridge University Press 1958 (russian transl.: Moskva 1960).

[*16*] LANDAU, L. D., i E. M. LIFŠIC: Statističeskaja fizika. Moskva 1963 (engl. transl.). Statistical physics. London: Pergamon Press 1963.

[*17*] GIBBS, J. W.: Fundamental principles of statistical mechanics (russian transl.). Moskva 1964.

[*18*] VLASOV, A. A.: Zh. Eksperim. i Teor. Fiz. **8**, 291—318 (1938).

[*19*] LANDAU, L. D.: Zh. Eksperim. i Teor. Fiz. **16**, 574—585 (1946).

[*20*] BOGOLJUBOV, N. N.: Dinamičeskie problemy statisticeskoj fiziki. (Dynamic problems of statistical physics.) Moskva 1946.

[*21*] KLIMANTOVIČ, JU. L.: Statisticeskaja teorija neravnovesnyh processov v plazme. (Statistical theory of non-equilibrium processes in a plasma.) Moskva: Iz. Moskovskogo Universiteta 1964.

[*22*] LANDAU, L. D., i E. M. LIFŠIC: Mehanika splosnyh sred. Moskva 1965 (engl. transl.). Mechanics of dense media. London: Pergamon Press 1965.

[*23*] — Zh. Eksperim. i Teor. Fiz. **7**, 203—209 (1937).

[*24*] SILIN, V. P.: Zh. Eksperim. i Teor. Fiz. **40**, 1768—1774 (1961).

[25] LANDAU, L. D., i E. M. LIFŠIC: Kvantoraja mehanika. Moskva 1963 (engl. transl.). Quantum mechanics. London: Pergamon Press 1963.
[26] GORBUNOV, L. M., i V. P. SILIN: Dokl. Akad. Nauk SSSR **1.45**, 1265—1268 (1962).
[27] GRAD, H.: Commission pure Appl. Math. **2** (1949); — Mechanica **4**, 5—53 (1952).
[28] BRAUN, S.: Elementarnye processy v plazme gazovogo razrjada. (Elementary processes in gas discharge plasma.) Moskva 1961.
[29] BATNAGAR, P., E. GROSS, and M. KROOK: Phys. Rev. **94**, 511—525 (1954). — GROSS, E., and M. KOROK: Phys. Rev. **102**, 593—604 (1956).
[30] DAVYDOV, B. N.: Zh. Eksperim. i Teor. Fiz. **7**, 1069—1089 (1937).
[31] GINZBURG, V. L., i A. V. GUREVIČ: Usp. Fiz. Nauk **70**, 201—393 (1960) (engl. transl.). Phys. Uspekhi **3**, 115—146, 175—194 (1961); (german transl.) Fortschr. Physik **8**, 97—189 (1960).
[32] GAPONOV, A. V., i M. A. MILLER: Zh. Eksperim. i Teor. Fiz. **34**, 242—243 (1958).
[33] FADEEVA, V. N., i N. N. TERENT'EV: Tablicy znacenij integrala verojatnostej. (Tables of the probability integral.) Moskva 1954.
[34] GRADSTEJN, I. S., i I. I. RYZIK: Tablicy integralow, cymm, rjadov i proizvendenij. (Tables of integrals, sums, series and derivatives.) Moskva 1962.
[35] VATSON, G. N.: Teorija besselevyh funkcij. (Theory of Besselfunctions.) Moskva 1949.
[36] BUDDEN, K. G.: Radio waves in the ionosphere. Cambridge: Cambridge University Press 1961.
[37] RUHADZE, A. A.: Zh. Tekn. Fiz. **31**, 1236—1245 (1961); **32**, 669—673 (1962).
[38] FAJNBERG, JA. B.: Atomnaja energija **11**, 4—16 (1961).
[39] VEDENOV, A. A., E. P. VELIHOV i R. Z. SAGDEEV: Usp. Fiz. Nauk **73**, 701—739 (1961).
[40] AHIEZER, A. I., i JA. B. FAJNBERG: Dokl. Akad. Nauk SSSR **69**, 555—558 (1949).
[41] BOMM, D., and E. GROSS: Phys. Rev. **75**, 1851—1864 (1949).
[42] BUNEMAN, O.: Phys. Rev. **115**, 503—517 (1959).
[43] RUHADZE, A. A., i V. P. SILIN: Usp. Fiz. Nauk. **82**, 499—535 (1964) (engl. transl.). Soviet Phys. Usp. **7**, 209—227 (1964).
[44] FEJNBERG, E. L.: Rasprostranenie radiovoln vdol' zemnoj poverhnosti. (Propagation of radio waves along the Earths' surface.) Moskva 1961.
[45] DNESTROVSKIJ, JU. N., i D. P. KOSTOMAROV: Zurn. vyciol. mat. i. mat.-fiz. **4**, 267—967 (1964).
[46] RUHADZE, A. A., i V. P. SILIN: Usp. Fiz. Nauk **96**, 87—126 (1968) (engl. transl.). Soviet Phys. Usp. **11**, 659—677 (1969).
[47] ROSENBLUTH, MC., and C. ZONGMIRE: Ann. Phys. **1**, 120—129 (1957).
[48] LOVECKIJ, E. E., i A. A. RUHADZE: Jadernyj sintez **6**, 9—14 (1966). — BOGDANKEVIC, L. S., i A. A. RUHADZE: Jadernyi sintez **6**, 171—181 (1966).
[49] MICHAJLOVSKIJ, A. B.: Zh. Tekn. Fiz. **35**, 1933—1944, 1945—1959 (1965). — MICHAILOVSKIJ, A. B., RUHADZE, A. H.: Zh. Tekn. Fiz. **35**, 2143—2149 (1965).
[50] KADOMTSEV, B. B., and A. V. NEDOSPASOV: J. Nucl. Energy **1**, 230 (1960).
[51] DREIER, H.: Phys. Rev. **115**, 238—249 (1959); **117**, 329—342 (1960). — GUREVIC, A. V.: Zh. Eksperim. i Teor. Fiz. **39**, 1296—1307 (1960). — GUREVIC, A. V., i JU. N. ZIVLJUK: Zh. Eksperim. i Teor. Fiz. **49**, 214—224 (1965).
[52] GERSMAN, B. N., i V. JU. TRAHTENGERC: Usp. Fiz. Nauk **89**, 201—232 (1966) (engl. transl.) Soviet Phys. Usp. **9**, 91—105 (1966).
[53] HULTQVIST, B.: Space Sci Rev. **5**, 599—695 (1966).
[54] KADOMTSEV, B. B.: Plasma turbulence. New York: Academic Press 1965.

Sachverzeichnis.

(Deutsch-Englisch.)

Bei gleicher Schreibweise in beiden Sprachen sind die Stichwörter nur einmal aufgeführt.

Abfall während ruhiger Zeiten, *decay, quiet time* 182.
Abschirmung, *screening* 455.
Absorption 478.
—, Čerenkov- 444, 458.
— hochfrequenter transversaler Wellen, *of high-frequency transverse waves* 458.
—, stoßfreie Čerenkov-, *collisionless Čerenkov* 447.
Absorptionskonstanten, *indices, absorption* 454.
Absorptionszahl, *index, absorption* 491, 493.
Änderung auf Grund des Sonnencyclus, *variation, solar cycle* 182.
Änderungen, örtliche zeitliche, *variations, local time* 178.
Aktivität, geomagnetische, *activity, geomagnetic* 93.
Albedo-Neutronen, *albedo neutrons* 65.
Alfvén-Geschwindigkeit, *velocity, Alfvén* 21, 24, 32, 36, 43, 55, 91, 466, 476.
—, *Alfvén speed* 202.
Alfvén-Schwingungen, *Alfvén oscillations* 482.
Alfvén-Stoß, *Alfvén shock* 35.
Alfvén-Wellen, *Alfvén waves* 477, 529–531.
Anstellwinkel, *pitch angle* 118.
—, Diffusion des magnetischen, *diffusion* 175.
—, Streuung des magnetischen, *pitch angle scattering* 67, 77.
Argus 138.
atmosphärische Parameter, *atmospheric parameters* 401.
Atombomben in großer Höhe, *nuclear bombs, high altitude* 137.
Aufspaltung, L-Schalen, *splitting, L-shell* 70, 77.
Ausbreitung parallel zum Magnetfeld, *propagation, parallel* 473.
— senkrecht zum Magnetfeld, *perpendicular* 473.
Ausbuchtung am Abend, *evening bulge* 58.
Ausfällung, *precipitation* 101, 174, 175.
Axford und Hines, Polarlichtmodell von, *Axford and Hines model of the aurora* 219.

B, L-Koordinaten, *B, L coordinates* 126.
Ba-Ionenwolken, *Ba-ion clouds* 101.
Baistörung, *bay disturbance* 91.
Balmer-Emissionen, *Balmer emissions* 215.
Ballonmessung, *baloon measurement* 82.
Bariumwolken, künstliche, *barium clouds, artificial* 82.
Beschleunigung, *acceleration* 102.
—, synchrone, *synchroneous* 168.
— in der neutralen Schicht, *neutral sheet* 221.
—, lokale, von Teilchen, *local, of particles* 217.
Betatron-Effekt, *betatron effect* 76.
Bewegung im Verband, *bulk-motion* 492.
Bewegungsgleichungen, *equations of motion* 17, 19.
BGK-Modell, *BGK model* 468.
Bohr-Sommerfeld Integral 518.
Boltzmann-Vlassov 514.
Brechungsindex, *index, refractive* 32, 454, 491, 493.
—, komplexer, *complex* 472.
Breite, invariante, *latitude, invariant* 126.
—, verallgemeinerte magnetische, *generalized magnetic* 126.
Bremsabsorption 467.
Bremsstrahlungsphotonen, *bremsstrahlung photons* 210.

Čerenkov-Absorption 521, 546.
Čerenkov-Dämpfung, *Čerenkov attenuation* 484.
Čerenkov-Effekt, *Čerenkov effect* 537, 550–552.
Čerenkov-Elektronenabsorption, *Čerenkov electron absorption* 481.
Čerenkov-Schwingungen, *Čerenkov oscillations* 497.
Čerenkov-Streuung, *Čerenkov dissipation* 542.
c.g.s.-System 1, 2.
Chapman-Enskog-Methode, *Chapman and Enskog method* 431.
Chapman-Ferraro-Theorie, *Chapman-Ferraro theory* 217.
CRAND 127, 132, 134, 136.
Coulomb-Logarithmus, *Coulomb's logarithm* 536.
Coulomb-Stoßzerfall, *Coulomb scattering decay* 147.
Cyclotrondämpfung, *cyclotron absorption* 467.
Cyclotroninstabilität, *cyclotron instability* 502.

Cyclotronresonanz, *cyclotron resonance* 464.
Cyclotronschwingungen, "*cyclotron-oscillations*" 487.

Dämpfung, Landau, *damping, Landau* 204, 452, 456, 465.
—, stoßfreie, *collisionless* 447.
Dämpfungsdekrement, *attenuation decrement* 460, 490, 492, 518, 526, 529, 533.
Debye-Abschirmung, *Debye screening* 30, 548.
Debye-Länge, *Debye length* 447, 450.
Debye-Radius 429.
Dekameterstrahlung, *decameter radiation* 228.
Dielektrizität, Tensor der effektiven, *effective dielectricity tensor* 510.
Dielektrizitätskonstante, *permittivity dielectric* 414, 444.
—, longitudinale, *longitudinal* 449, 470.
—, transversale, *transverse* 449.
Dielektrizitätstensor, *dielectric tensor* 465, 466.
—, *dielectricity tensor* 467, 468.
—, *permittivity, tensor of* 463.
—, komplexer, *permeability, tensor of complex dielectric* 399.
Diffusion, radiale, *diffusion, radial* 76, 77.
L-Diffusion von α-Teilchen, *L-diffusion of alpha particles* 163.
Diffusionsgeschwindigkeit, *diffusion velocity* 168.
Diffusionstensor, *diffusion tensor* 427.
Diffusion quer zu B-Feldlinien, *cross-field diffusion* 161.
Dipolfeld, *dipole field* 120.
—, *dipole, field of a* 3.
Dispersion, räumliche, *dispersion, spatial* 447.
Dispersionsbeziehung, *dispersion relation* 496, 531.
Dispersionsgleichung, *dispersion equation* 32, **472**.
—, *dispersion relation* 405, 499, 525, 528.
—, allgemeine, *dispersion equation, general* 471.
Doppler-Verschiebung, *Doppler-shift* 495.
DP 2-Stromsystem, *DP 2 current system* 80, 88, 96.
Driftbewegung, *drift motion* 49, 50, 61, 90.
Driftgeschwindigkeit, *drift velocity* 18, 61.
Driftschwingungen, *drift-oscillations* 524, 539, 540, 548.
—, langwellige, *long-wave drift* 542, 546.
Druck, gesamter kinetischer, *pressure, total kinetic* 513.
Druckgleichgewicht der neutralen Schicht, *pressure balance, neutral sheet* 195.
—, hydromagnetisches, *hydromagnetic* 187.
Dynamoeffekt, *dynamo-electric effect* 112.

E-Feld Trift, *electric-field drift* 75, 84.
ebene Wellen, inhomogene, *plane waves, inhomogeneous* 409.
Eikonal, *eiconal* 509.
Eikonal-Gleichung, *eiconal equation* 516, 540, 543, 544, 547, 555.
Eindringen niederfrequenter transversaler Wellen, *penetration of low-frequency transverse waves* 461.
Einfangen, äußerer Grenzradius, *trapping, outer radius limit of* 200.
—, obere Breitengrenze, *high latitude limit of* 198.
—, tägliche Änderung der oberen Breitengrenze, *diurnal variation of the upper limit of* 199.
Einfangkriterium, *criterion of trapping* 184.
Einflüssigkeitsmodell, *one fluid gasdynamic model* 469.
—, Gasdynamik, Flüssigkeit nach dem, *gasdynamics of one fluid* 466.
Einheiten, *units* 1.
Einheitensystem, nicht-rationalisiertes, *system, non-rationalized* 2.
—, rationalisiertes, *system of units, rationalized* 1, 2.
Einschließung, *confinement* 534.
elektrisches Feld, *electric field* 50, 60.
Electrojet 215.
Elektronen, *electrons* 164.
—, eingefangene, *trapped* 216.
—, Fluß ausgefällter, *flux of precipitating* 172.
— in der magnetosphärischen Übergangszone, *magnetosheath electrons* 51.
—, Polarlicht-, *electrons, auroral* 216.
elektronenakustische Wellen, *waves, electron-acoustic* 477.
Elektronenausfällung, *precipitation, electron* 171, 173, 209.
Elektronenenergiespektrum, *electron energy spectra* 140.
Elektronenfluß, *electron flux* 64, 65.
— im Polarlicht, *auroral* 214.
Elektronenflüsse, *electron fluxes* 164.
— im Schweif, Inseln von, *in tail, islands of* 195.
Elektroneninhalt, totaler, *electron content, total* 97.
Elektroneninseln, *electron islands* 52.
Elektronenplasma, *electron plasma* 450.
Elektronenpopulation, *population of electrons* 51, 52.
— im inneren Gürtel, *electron population, inner zone* 135.
Elektronenschwingung, hochfrequente Langmuir-, *electron oscillations, high frequency Langmuir* 450.
Elektronenspektrum des Nordlichts, *electron spectrum, auroral* 212.
Elektronenstöße, *electronic collisions* 550–554.
Elektronenstrahl, *electron beam* 496.
Elektronenstrahlungsgürtel, Zerfall, *electron belt, decay of the* 144.
Elektronenwärmestrom, *electron heat flux* 432.
Elektronen-Zyklotronwelle, *electronic cyclotron wave* 489.
Emissionen, sehr niederfrequente, *emissions, very low frequency* 215.

Energie, *energy* 70, 88, 102, 103.
Energieaufnahme, *energization* 72, 75.
Energiedichten im Schweif und der neutralen Schicht, *energy densities in tail and neutral sheet* 197.
—, Verhältnis von, *ratio of* 24, 25.
Energiespektren eingefangener Elektronen, *energy spectra of trapped electrons* 165.
Energiespektrum, *energy spectrum* 161.
—, stationäres, *energy spectrum, steady state* 164.
Entweichzeit, *escape time* 152.
Erdatmosphäre, *terrestrial atmosphere* 402.
Erdstrom, *earth-current* 215.
Erwärmung, *heating* 215.
Explosion in großer Höhe, *explosion, high altitude* 143.

Feld, elektrisches, *field, electric* 82–84, 95, 103, 106–109, 112.
—, verformtes, *distorted* 189.
Feldänderung, *field variance* 204.
Feld, selbstkonsistentes, *selfconsistent field* 418.
Feldgleichungen, *field equations* 396, 397, 405.
Feldlinien, eingefrorene, *frozen-in field lines* 54.
—, — magnetische, *magnetic* 21.
—, Ineinanderlaufen von, *merging* 25.
Feldlinien-Erhaltung, *line-preservation* 25.
Feldlinienschale, *field line shell* 67–70.
Fermiprozeß, *Fermi process* 75, 76.
Flächenstrom, *sheet, current* 194.
Flöteninstabilität, *flute-instability* 557.
flüssigkeitsdynamisches Modell, *fluid-dynamic model* 547.
Flüssigkeitsmodell des Plasmas, *plasma model, fluid-dynamic* 416.
Fluß eingefangener Teilchen, maximaler, *flux, maximum trapped* 176.
— von niederenergetischen Protonen, *of low energy protons* 186.
Flußerhaltung, *flux-preservation* 25.
Fokker-Planck-Gleichung, *Fokker-Planck-equation* 145, 160, 161, 427.
Formulierung, makroskopische, *formulation, macroscopic* 18.
Fourier-Transformierte, *Fourier transform* 4, 398, 406.
Frequenzspektrum, *frequency spectrum* 460, 490, 533.
F 2-Schicht, Störungen, *F 2 layer, disturbances* 97.
Führung innerhalb der Ionosphäre, *ducting inside the ionosphere* 521.
Führungszentrum, *guiding center* 118.
—, Näherung durch Betrachtung, *approximation* 116, 118.

Gas, Dreikomponenten-, *three-component gas* 437.
Gasdynamik nach dem Einflüssigkeitsmodell, *one fluid gasdynamics* 482.
gasdynamische Gleichung, *gasdynamic equation* 420.
— Strahlinstabilität, *beam instability* 497, 503, 506.
GBK-Integral 436.
gekrümmtes magnetisches Feld, *curved magnetic field* 556.
geomagnetischer Sturm, *geomagnetic storm* 10, 65, 70.
Geosynchrotron 168, 171.
Geschwindigkeit, *velocity* 10.
— des Führungszentrums, *guiding center* 18.
Gleichgewicht, *equilibrium* 26, 28.
Gradiententrift, *gradient drift* 62, 63, 75.
Grenze, *boundary* 26.
—, magnetosphärische, *magnetospheric* 206.
Grenzschicht, *layer, boundary* 190.
Gruppengeschwindigkeit, *group velocity* 410.
Gürtel bei anderen Planeten, *belts on other planets* 222.
Gyration 118.
Gyrofrequenz, *gyrofrequency* 18, 31.

H-Theorem 436.
Hauptphase, *main phase* 91–93, 101, 185.
Hall-Ströme, *Hall currents* 218.
Helligkeitskoeffizient für das Nordlicht, internationaler, *brightness coefficient for aurora, international* 209.
hydrodynamische Theorie, *hydrodynamic theory* 12, 43.

Indexvektor, *index vector* 409.
Ineinanderlaufen (von Feldlinien), *merging (of field lines)* 55, 56, 93, 102.
Inhomogenitätsparameter, *inhomogenity parameter* 509.
Inkrement, *increment* 408.
Initialphase, *initial phase* 184.
instabiles System, *instable system* 492.
Instabilität, *instability* 408, 411, 503, 555.
—, die Polarlicht erzeugt, *producing aurora* 217.
— durch Konvektionsströme, *stream-convective instability* 557.
—, gasdynamische, *gasdynamic* 498, 549.
—, kinetische, *kinetic* 498.
Instabilitäten, aperiodische, *instabilities, aperiodical* 542.
Instabilitätsbedingung, *instability, condition for* 176.
Invariante, dritte, *invariant, third* 154.
—, erste, *first* 154.
—, Integral-, *integral* 124.
—, longitudinale, *longitudinal* 123.
—, zweite, *second* 155.
Invarianten, adiabatische, *invariants, adiabatic* 68, 69, 123.
Invariantenverletzung, Mechanismus der, *invariant violation mechanism* 156.
Ionenschallwellen, *ion sound waves* 522, 533.
Ionenstoß, *collision, ionic* 550–554.
Ionentemperaturen, *ion temperatures* 58.
Ionenwolken, künstliche, *ion clouds, artificial* 81, 82, 101.

Ionosphäre, Plasma in der, *plasma, ionospheric* 537.
Irregularitäten, *irregularities* 544.

Joulesche Erwärmung, *Joule heating* 403.
Jupiter 228.
Jupiter-Radiowellen, *Jupiter radio waves* 224.

Kernspaltungsreaktion, *fission reaction* 137.
Kernverschmelzungsbombe, *fusion bomb* 137.
kinetische Gleichung, *kinetic equation* 514.
— —, linearisierte, *linearized* 536.
kleinste Dicke der magnetosphärischen Übergangsregion, *stand-off distance* 45.
Knie (exosphärischer Profilknick), *knee* 56.
konjugierter Punkt, *point, conjugate* 119.
Kontaktunstetigkeit, *contact discontinuity* 35.
Konvektion, *convection* 77, 79.
—, magnetodynamische, *magnetodynamic* 77.
Konvektionsbewegung, *convective motion* 60, 79, 88, 102, 111.
B, *L*-Koordinaten, *B, L coordinates* 126.
—, solare magnetosphärische, *solar magnetospheric* 40.
Korotation, *corotation* 79, 83–85.
kosmisches Rauschen, Absorption, *cosmic noise absorption* 215.
Kraftdichtenquotient, *force density ratio* 20.
Krümmungsdrift, *curvature drift* 49, 62, 111.

L-Schalenaufspaltung, *L-shell splitting* 201.
L-Wert, *L-value* 56, 69, 70.
Landau-Dämpfung, *Landau damping* 204, 465.
Landau-Formel, *Landau formula* 428.
Langmuir-Frequenz, *Langmuir frequency* 414.
Langmuir-Schwingung, *Langmuir oscillation* 451–453, 457, 497.
Langmuir-Wellen, *Langmuir waves* 452.
langsame magnetische Schallwelle, *slow magneto-sound wave* 486.
Larmor-Drift, *Larmor drift* 524.
— von geladenen Teilchen, *of charged particles* 513.
Larmor-Driftfrequenz, *Larmor drift frequency* 515, 534.
Larmor-Driftgeschwindigkeit, *Larmor drift velocity* 515.
'Leaky-Bucket'-Modell, *Leaky-Bucket model* 129, 174.
Lebensdauer eingefangener Elektronen, *life times of trapped electrons* 183.
Leitfähigkeit, *conductivity* 414.
—, endliche, *finite* 28.
Liouvillesches Theorem von, *Liouville's Theorem* 156, 157.
lokales Spektrum, *local spectrum* 543.
Lorentz-Gas, *Lorentz gas* 475.
Lorentz-Kraft, *Lorentz force* 396.
Lorentz-Transformation, *Lorentz-transformation* 494.

Magnetfeld, *magnetic field* 11.
—, ausgerichtet nach dem, *"field-aligned"* 544.
—, benötigtes, um eine Magnetosphäre zu binden, *magnetic field needed to hold a magnetosphere* 223.
magnetischer Fluß, *magnetic flux* 124.
magnetisches Feld, starkes, *magnetic field, strong* 500.
— Moment, *moment* 3, 118, 123.
— Spannung, *tension* 49.
Magneto-Bremsstrahlung, *magneto-„Bremsstrahlung"* 464.
Magnetodynamik, *magnetodynamic* 5.
Magnetogasdynamik, *magneto-gas-dynamic* 5.
Magnetohydrodynamik, *magnetofluid dynamics* 416.
Magnetopause 8, 9, 25–27, 37–39, 41, 42, 53, 54, 86, 109–111, 187, 190.
—, Grenze, *boundary* 191.
Magnetosphäre, Fluß um die Magnetosphäre herum, *magnetosphere, flow around* 190.
—, Form, *shape* 188.
—, offene, *open* 49, 53, 66.
—, Unterseite, *"bottom" of the* 78, 82.
—, Verformung durch den Sonnenwind, *by the solar wind, distortion of the* 199.
Magnetosphärenbereich, Modell des offenen, *magnetosphere model, open* 193, 218.
—, offener, *open* 192.
magnetosphärische Übergangsregion (Plasma), *magnetosheath plasma* 109.
Mars 223.
Maßsystem, internationales, *Système international* 1.
Materialgleichung, *material field equation* 397.
Maxwell-Verteilung, *Maxwell distribution* 422.
Mead-Modell, *Mead model* 161, 195.
Medium, interplanetares, medium, *interplanetary* 205.
Methode von CHAPMAN und ENSKOG, *method of* CHAPMAN *and* ENSKOG 538.
— von GRAD, *Grad's method* 431.
Mikropulsationen, *micropulsations* 34, 60, 215.
mikroskopische Formulierung, *microscopic formulation* 17.
Modell, aerodynamisches, *model, aerodynamical* 203.
— der simultanen Auffüllung der Strahlungsgürtel bei Teilchenausfällung, *splash catcher* 174.
—, geschlossenes, *closed* 193.
Mond, *moon* 198.
Multipolentwicklung des Magnetfeldes, *multipole expansion of magnetic field* 121.

Näherung der geometrischen Optik, *geometrical optics approximation* 508.
— für verdünnte Gase, *thin gas approximation* 418.
—, lineare, *approximation, linear* 439.
—, quasi-gasdynamische, *quasi-gasdynamic* 420, 429.
Nachlieferung, fortlaufende, *replenishment, continual* 182.
neutrale Linien, *neutral lines* 54.
— Schicht, *sheet* 37, 42, 47–50, 53.
neutraler Punkt (siehe Punkt), *neutral point*
Neutralgas, *neutral gas* 7, 8.
Neutronen in der Polarkappe, *neutrons, polar cap* 132.
—, Zerfall von durch kosmische Strahlung erzeugten, *neutron decay, cosmic ray albedo* 127.
Neutronenzerfall, Protoneninjektion durch, *neutron decay injection* 132.
nicht-konservative Wellen, *non-conservative waves* 527.
nicht-lineare Effekte, *non-linear phenomena* 439.
nicht-relativistischer Fall, *non-relativistic case* 495.
niederfrequente Triftschwingungen, *low frequency drifting-oscillations* 534.
— longitudinale Wellen, *longitudinal waves* 523.
Nordlichter, Energiebilanz, *aurorae, energetics of* 209.
Normalmoden, stationäre, *normal mode, stationary* 408.

Ost-West-Effekt, *East-West effect* 135.

periodische Bewegung, fast, *periodic motion, nearly* 67.
Periodizität, 27-Tage-, *periodicities, 27-day* 180.
Permeabilität, Tensor für dielektrische, *permeability, tensor of dielectric* 494.
Phasengeschwindigkeit, *phase velocity* 410.
physikalische Größe, *physical quantity* 3.
Plasma 395.
—, ionosphärisches, *ionospheric* 522.
—, magnetoaktives, *gyrotropic (magneto-active)* 492, 466.
—, Näherung für kaltes, *cold plasma approximation* 421.
—, nicht-isothermes, *non-isothermal* 485, 542.
—, — heißes, *hot* 422.
—, schwach ionisiertes, *weakly ionized* 435, 523, 526, 527, 532.
—, — — gyrotropes, *gyrotropic* 468.
—, — magnetisiertes, *magnetized* 500.
—, stoßfreies, *collisionless* 420, 424, 440, 441, 461, 462, 537.
—, thermisches, *thermal* 8, 60, 61.
—, vollständig ionisiertes, *completely ionized* 424, 428, 429, 457, 459, 462, 482, 523, 526, 527, 532.
—, Wellen in warmem, *Waves in warm* 481.
Plasmafluß, *plasma, flow of* 25.
—, *plasma flux* 51.
Plasmamodell, quasi-hydrodynamisches, *plasma model, quasi-hydrodynamic* 416.
Plasmapause 56, 104.
Plasmaschicht, *plasma sheet* 43, 51, 52.
Plasmasphäre, *plasmasphere* 56, 61.
Plasmatrift, radiale, *cross-L drift* 58.
polarer Elektrojet, *auroral electrojet* 101, 105, 107.
— Scheitelbereich, *polar cusps* 70, 81, 109.
— — (der Magnetosphäre), *(of magnetosphere)* 53, 55.
— Wind 81.
Polarlicht, *aurora* 99, 111, 215, 216.
—, künstliches, *artificial* 139.
Polarlichtausfällung, *auroral precipitation* 101, 103.
Polarlichter, *aurorae* 207.
Polarlichtoval, *auroral oval* 60, 99–102, 111.
Polarlichtprozesse, *auroral processes* 222.
Polarlichttheorien, *aurora, theories of* 217.
Polarlichtzone, *auroral zone* 99, 100, 105–109, 209, 211.
Polarwind, *polar wind* 14.
Protonen, *protons* 152.
—, Ausfällung, *precipitation of* 175.
— charakteristischer Energie, *characteristic energy* 153.
—, energiereiche, *energetic* 64.
Protonendichte, *proton density* 11, 14.
—, Rückgang, *proton decay* 131.
Protonenenergiespektrum, *proton energie spectrum* 134.
Protonenfluß, *proton flux* 64, 65.
Pulsation, geomagnetische, *geomagnetic pulsation* 104.
Pumpmechanismus, *pumping mechanism* 158.
Punkt, magnetisch neutraler, *point, magnetic neutral* 192.
—, neutraler, *neutral point* 25, 39, 54, 55, 66, 77, 84, 103, 109, 110.

Quantisierung, *quantization* 524, 525.
Quantisierungsregel, *quantization rule* 519, 526.
Quasieinfang, *quasitrapping* 70.
quasineutral 395.
Quertrift, *drift across field lines* 119.

Radiowellenausbreitung, *radio wave propagation* 519.
räumliches Abklingen, *spatial decrease* 411.
Randbereich, *skirt* 71.
Randgleichgewicht, *equilibrium at boundary* 38.
Reibungseffekt, *frictional effect* 427.
Reibungskraft, *friction force* 430–432.
Resonanzbedingung, *resonance, condition for* 151.
Resonanzbeschleunigung, *acceleration, resonant* 168.
Resonanzfrequenzen, *resonance frequencies* 479, 481.

Resonanzschwingung, magnetodynamische, *resonant magnetodynamic oscillation* 104
Reynolds-Zahl, magnetische, *Reynold's number, magnetic* 29, 30.
Ringstrom, *current, ring* 70, 92, 103, 185, 218.
—, asymmetrischer, *asymmetric* 101.
Röntgenstrahlen, *x-rays* 210, 215, 211.
— der Nordlichtzone, *x-rays, auroral -zone* 212.
Rotation 84.
Rotationsbewegung, *rotary motion* 78–80, 83–86, 111.
Rückbildungsphase, *recovery phase* 185.

SC 186.
Schalenparameter, magnetischer, nach McIlwain, *McIlwain's magnetic shell parameter* 124.
Schallgeschwindigkeit, *velocity of sound* 423.
Schallwellen, langsame magnetische, *waves, slow magneto-sound* 483.
—, schnelle magnetische, *fast magneto-sound* 483, 529, 532.
Scheibenwischereffekt, *wind shield wiper effect* 147.
Scheitelbereich, *cusp* 71.
Schicht, neutrale, *sheet, neutral* 85, 111, 194, 220.
Schrödinger-Gleichung, *Schrödinger equation* 516.
Schweif, *tail* 40, 42, 46, 66, 74, 86–88, 90, 102, 111.
—, geomagnetischer, *geomagnetic* 193, 221.
—, Länge des geomagnetischen, *length of the geomagnetic* 197.
—, magnetosphärischer, *magnetospheric* 219.
Schweifbreite, *tail, width of the* 195.
Schweiffeld, *tail field* 221.
Schwingungen, *oscillations* 508.
—, elektronische Langmuir-, *electronic Langmuir* 521.
—, langsam anwachsende, *slowly increasing* 541.
—, langwellige kinetische, *long-wave kinetic* 550–553.
—, longitudinale, *longitudinal* 502.
—, — Potential-, *(potential)* 524.
Schwingungsspektrum, *oscillation spectra* 514.
Schwingungstyp, *type of oscillation* 460, 490, 492.
Sektorstruktur, *sector structure* 14, 93.
selbstkonistentes Feld, *self conistent field* 418.
Sellmeier-Dispersionsgleichung, *Sellmeier dispersion formula* 516.
Sellmeier-Formel, *Sellmeier formula* 447.
Skalenlängen der Inhomogenität, *inhomogenity scales* 513.
Skineffekt, anomaler, *skin effect, abnormal* 461.
— —, *anomalous* 488.
—, normaler, *normal* 461.
Solarwind, *solar wind* 10, 86.
Solarwindgeschwindigkeit, *solar wind, velocity* 95.
Sonnenaktivität, *solar activity* 180.
Sonneneruption, *solar flare* 132.
Sonnenwind, *solar wind* 95, 159, 187, 208, 215.
— im Überschallbereich, *supersonic* 43, 49.
—, Sektoren, *sectors* 14, 93.
—, Verformung des geomagnetischen Feldes, *distortion of geomagnetic field by* 180.
—, Zusammensetzung, *composition* 10.
Sonnenzyklus, Änderungen, *solar cycle changes* 134.
SPAND 132.
Spannung, magnetische, *tension, magnetic* 90, 111.
Spektralmessungen, Zusammenfassung, *spectral measurements, summary of* 213.
Spiegelpunkt, *point, mirror* 119, 145.
Spiegelungsperiode, *bounce period* 119.
Spiralwellen, *spiral waves* 478, 485.
Sprungbedingungen, *jump conditions* 206.
stabilisierende Wirkung des magnetischen Feldes, *stabilizing effect of the magnetic field* 508.
Stabilisierung, *stabilization* 500, 542.
Stabilität, *stability* 490, 496.
Stabilitätskriterien, *stability criteria* 491.
Stationaritätsbedingungen, *condition, steady state* 176.
Starfish 138, 144.
Störung, magnetische, *disturbance, magnetic* 59, 215.
Stöße, *collisions* 30, 66, 435.
—, elastische, *elastic* 437, 457, 459, 462.
Stoß, Elektron-Ion-, *collision, electron-ion* 544.
—, Ion-Ion-, 539.
—, kollisionsfreier, *shock, collisionless* 203, 204.
Stoßabsorption, *collisional absorption* 467.
Stoßfrequenz, *collision frequency* 17, 19, 20, 61, 401.
—, effektive, *effective* 436.
—, wahrscheinlichste, *most probable* 5.
Stoßfront, *shock front* 10, 34, 36, 37, 91.
—, hydrodynamische Theorie, *hydrodynamic theory of* 43–46.
—, stationäre, *stationary* 41–43.
Stoßintegral, *collision integral* 419, 425, 435.
Stoßwelle, *bow shock* 41, 202.
—, *shock wave* 191, 206.
Strahlung, nicht-thermische, des Jupiter, *radiation, Jupiter's non-thermal* 226.
Strahlungsgürtel, *radiation belt* 9, 10, 60, 63, 70, 103.
—, äußerer, *belt, outer* 152, 173.
—, —, *radiation belt, outer region* 65, 67.
—, —, Elektronen *belt, electrons, outer* 165.
—, innerer, *radiation belt, inner region* 64, 67.
—, künstliche, *artificial* 137.
Streuwahrscheinlichkeit, *scattering probability* 426.
Ströme, feldparallele, *current, field-aligned* 104–109.

Strom, Dynamo-, *current, dynamo-electric* 82.
Stromlinien, *streamlines* 80.
— in der äquitorialen Ebene, *in the equitorial plane* 80.
Stromsystem, äquivalentes, *current system, equivalent* 104.
—, drei-dimensionales, *three-dimensional* 104.
—, Sq-, 112.
Stromtensor, *current tensor* 446.
Stürme, magnetische, *storms, magnetic* 96, 97, 101.
—, magnetische, Variation bei, *storm variations* 162.
Sturm, geomagnetischer, *storm, geomagnetic* 86, 90, 96, 101.
Sturmbeginn, plötzlicher, *sudden commencement*, **SC** 184.
Suszeptibilität, longitudinale dielektrische, *susceptibility, longitudinal dielectric* 446.
—, transversale dielektrische, *transverse dielectric* 446.
Synchrotronstrahlung, *synchrotron radiation* 141, 226.

tangentiale Unstetigkeit, *tangential discontinuity* 35, 51.
Teilchen, ausgefällte, *particles, precipitated* 211.
—, Bewegung einzelner, *motion of individual* 412.
—, — geladener, *of a charged* 116.
—, Drift eingefangener, *drift, trapped* 189.
—, eingefangene, *trapped* 60, 63, 71, 218.
—, Inhalt, *containment* 215.
—, langsame, *slow* 72.
— nahe magnetischen Nullwerten, Bewegung von, *near a magnetic null, motion of* 219.
—, niederenergetische, *low-energy* 70, 184.
—, schnelle, *fast* 60, 63, 72.
—, schwerere, *heavier* 136.
α-Teilchen, *α-particles* 10, 152.
— im Solarwind, *in the solarwind* 86, 92.
Teilchenausfällung, *particle precipitation* 215.
Teilchenbahnen, *particle trajectories* 221.
Teilchenbeschleunigung, *particle acceleration* 66, 71.
Teilchenenergien, *particle energies* 60.
Teilchenquellen, *particle sources* 65.
Teilchenstoß, *particles collision* 425.
Teilchenverlust, *particle loss* 65, 66.
Teilsturm, magnetischer, *substorm, magnetic* 91, 92, 100–103, 109.
—, magnetosphärischer, *magnetospheric* 58, 101.
—, polarer magnetischer, *polar magnetic* 92, 101, 109.
—, Polarlicht-, *auroral* 100, 101.
Temperatur, *temperature* 11.
—, effektive, *effective* 436.
—, reduzierte mittlere, *reduced average* 448.
Tensor 4.
Theorie der Entleerung, *evaporative theory* 13.
—, magnetodynamische, *magnetodynamic* 28, 30.
—, — oder hydromagnetische, *or hydromagnetic* 17.
Transparenzbereich, *transparency range* 517, 518, 520, 528.
Transporterscheinungen, *transport phenomena* 435.
Transportstoßfrequenz, *collision frequency, transport* 5.
—, gemittelte, *averaged transport* 5.
—, mittlere, *average transport* 434.
—, —, *mean transport* 5.
transversale Wellen, *transverse waves* 457.
Trift, *drift* 21, 74.
—, elektrodynamische, *electrodynamic* 97.
— im elektrischen Feld, *electric field* 62, 63.
Triftbewegung, *drift motion* 18, 67, 71, 72, 78, 81, 84, 101–103, 111, 112.
Triftgeschwindigkeit, *drift velocity* 67.
Triftmechanismus, *drift mechanism* 163.
Triftperiode, *drift period* 120.
Triftprozeß, *drift process* 154.
Trog, *trough* 56.
Trogbereiche, *troughs* 59.

Übergangsregion, magnetosphärische, *magnetosheath* 40–42, 45, 59, 202, 206.
Überschall, *supersonic* 11, 13.
Unterseite der Magnetosphäre, *"bottom" of the magnetosphere* 9.
Unstetigkeit, Wirbel-, *discontinuity, rotational* 26.
Unstetigkeitsflächen, *discontinuity surfaces* 34.

Van Allen-Strahlung, *Van Allen radiation* 215.
Venus 223.
Verdampfungstheorie des Polarwindes, *evaporative theory of polarwind* 81.
Verhältnisse, ungestörte und gestörte, *conditions, quiet and disturbed* 182.
Verlust, *loss* 67, 71.
Verschiebungskoeffizient, *coefficient, displacement* 106.
Viskosität, *viscosity* 547.
Viskositätskoeffizient, *viscosity coefficients* 433.
Viskositätstensor, *viscosity tensor* 430.

Wärmeleitfähigkeit, *conductivity, thermal* 547.
Wasserstoffatome, neutrale, *hydrogen atoms, neutral* 186.
Welle, Alfvén-, *wave, Alfvén* 32.
—, ionenakustische, *wave, ion-acoustic* 32.
—, magnetodynamische, *magnetodynamic* 31, 32, 91.
—, schnelle magnetodynamische, *fast magnetodynamic* 32.
Wellen, hydromagnetische, *waves, hydromagnetic* 31.

Wellen, Ionen-Cyclotron-, *ion cyclotron* 177.
—, longitudinale, *longitudinal* 521, 525.
Wellenvektor, *wave vector* 460, 509.
Wendepunkte, *turning points* 517, 520, 522.
Whistler 57, 81, 151.
Whistler-Mode 176, 478, 530.
Whistler-Welle, Wechselwirkung, *whistler wave, interaction of a* 150.
Whistler-Wellen, *whistler radiation, whistler waves* 139, 485.
Whistler-Zuwachsrate, *whistler growth rate* 176.

Wirbelunstetigkeit, *rotational discontinuity* 35.

Zweiflüssigkeitenmodell, *magneto-gas-dynamic system with two components* 435.

Subject Index.

(English-German.)

Where English and German spelling of a word is identical the German version is omitted.

Absorption 478.
—, Čerenkov- 444, 458.
—, collisionless, *Dämpfung, stoßfreie* 447.
—, — Čerenkov, *Absorption, Stoßfreie Čerenkov* 447.
— of high-frequency transverse waves, *hochfrequenter transversaler Wellen* 458.
Acceleration, *Beschleunigung* 102.
—, local, of particles, *lokale, von Teilchen* 217.
—, neutral sheet, *in der neutralen Schicht* 221.
—, resonant, *Resonanzbeschleunigung* 168.
—, synchronous, *synchrone Beschleuniguug* 168.
Activity, geomagnetic, *Aktivität, geomagnetische* 93.
Albedo neutrons, *Albedo-Neutronen* 65.
Alfvén oscillations, *Alfvén-Schwingungen* 482.
— shock, *Alfvén-Stoß* 35.
— speed, *Alfvén-Geschwindigkeit* 202.
— waves, *Alfvén-Wellen* 477, 529–531.
Approximation, linear, *Näherung, lineare* 439.
—, quasi-gasdynamic, *quasi-gasdynamische* 420, 429.
Argus 138.
Atmospheric parameters, *atmosphärische Parameter* 401.
Attenuation decrement, *Dämpfungsgrad* 490, 492, 518, 526, 529, 533.
— — *Dämpfungsdekrement* 460.
Aurora, *Polarlicht* 99, 111, 215, 216.
—, artificial, *künstliches* 139.
—, theories of, *Polarlichttheorien* 217.
Aurorae, *Polarlichter* 207.
—, energetics of, *Nordlichter, Energiebilanz* 209.
Auroral electrojet, *polarer Electrojet* 101, 105, 107.
— oval, *Polarlichtoval* 60, 99–102, 111.
— precipitation, *Polarlichtausfällung* 101, 103.
— processes, *Polarlichtprozesse* 222.
— zone, *Polarlichtzone* 99, 100, 105–109, 209, 211.
AXFORD and HINES model of the aurora, *Axford und Hines, Polarlichtmodell von* 219.

Ba-ion clouds, *Ba-Ionenwolken* 101.
Balmer emissions, *Balmer-Emissionen* 215.
Baloon measurement, *Ballonmessung* 82.
Barium clouds, artificial, *Bariumwolken, künstliche* 82.
Bay disturbance, *Baistörung* 91.
Belt, outer, *Strahlengürtel, äußerer* 152, 173.
—, —, electrons, *Elektronen* 165.
Belt on other planets, *Gürtel bei anderen Planeten* 222.
Betatron effect, *Betatron-Effekt* 76.
BKG model, *BKG-Modell* 468.
B, *L* coordinates, *B*, *L-Koordinaten* 126.
BOHR-SOMMERFELD 518.
BOLTZMANN-VLASSOV 514.
Bomb, fusion, *Kernverschmelzungsbombe* 137.
"Bottom" of the magnetosphere, *Unterseite der Magnetosphäre* 9.
Bounce period, *Spiegelungsperiode* 119.
Boundary, *Grenze* 26.
—, magnetospheric, *magnetosphärische* 206.
Bow shock, *Stoßwelle* 41, 202.
Bremsabsorption 467.
Bremsstrahlung photons, *Bremsstrahlungsphotonen* 210.
Brigthness coefficient for aurora, international, *Helligkeitskoeffizient für das Nordlicht, internationaler* 209.
Bulk-motion, *Bewegung im Verband* 492.

Čerenkov-Absorption 512, 546.
Čerenkov attenuation, *Čerenkov-Dämpfung* 484.
— dissipation, *Čerenkov-Streuung* 542.
— effect, *Čerenkov-Effekt* 537, 550–552.
— electron absorption, *Čerenkov-Elektronenabsorption* 481.
— oscillations, *Čerenkov-Schwingungen* 497.
c.g.s. system, *c.g.s.-System* 1, 2.
Chapman and Enskog method, *Chapman-Enskog-Methode* 431.
Chapman-Ferraro theory, *Chapman-Ferraro-Theorie* 217.
Coefficient, displacement, *Verschiebungskoeffizient* 160.
Cold plasma approximation, *Plasma, Näherung für kaltes* 421.
Collision, electron-ion, *Stoß, Elektron-Ion-* 544.

Collision, frequency, *Stoßfrequenz* 61.
— —, average transport, *Transportstoßfrequenz, mittlere* 434.
— frequency, *Stoßfrequenz* 17, 19, 20, 401.
— —, averaged transport, *Transportstoßfrequenz, gemittelte* 5.
— —, effective, *Stoßfrequenz, effektive* 436.
— —, mean transport, *Transportstoßfrequenz, mittlere* 5.
— —, most probable, *Stoßfrequenz, wahrscheinlichste* 5.
— —, transport, *Transportstoßfrequenz* 5.
— integral, *Stoßintegral* 419, 425, 435.
—, ionic, *Ionenstöße* 550–554.
—, ion-ion, *Stoß, Ion-Ion-* 539.
Collisional absorption, *Stoßabsorption* 467.
Collisions, *Stöße* 30, 66, 435.
—, elastic, *elastische* 437, 457, 459, 462.
Commencement, SC, sudden, *Sturmbeginn, plötzlicher* 184.
Condition, steady state, *Stationaritätsbedingungen* 176.
Conditions, quiet and disturbed, *Verhältnisse, ungestörte und gestörte* 182.
Conductivity, *Leitfähigkeit* 414.
—, finite, *endliche* 28.
—, thermal, *Wärmeleitfähigkeit* 547.
Confinement, *Einschließung* 534.
Contact discontinuity, *Kontaktunstetigkeit* 35.
Convection, *Konvektion* 77, 79.
—, magnetodynamic, *magnetodynamische* 77.
Convective motion, *Konvektionsbewegung* 60, 79, 88, 102, 111.
B, L coordinates, *Koordinaten, B, L* 126.
Coordinates, solar magnetospheric, *Koordinaten, solare magnetosphärische* 40.
Corotation, *Korotation* 79, 83–85.
Cosmic noise absorption, *Absorption von kosmischem Rauschen* 215.
COULOMB's logarithm, *Coulomb-Logarithmus* 536.
Coulomb scattering decay, *Coulomb-Stoßzerfall* 147.
CRAND 127, 132, 134, 136.
Criterion of trapping, *Einfangkriterium* 184.
Cross-field diffusion, *Diffusion quer zu B-Feldlinien* 161.
Cross-*L* drift, *Plasmatrift, radiale* 58.
Current, asymmetric ring, *Ringstrom, asymmetrischer* 101.
—, dynamo-electric, *Strom, Dynamo-* 82.
—, earth, *Erdstrom* 215.
—, field-aligned, *Ströme, feldparallele* 104–109.
—, ring, *Ringstrom* 70, 92, 103, 185, 218.
— system, equivalent, *Stromsystem, äquivalentes* 104.
— —, Sq 112.
— —, three-dimensional, *dreidimensionales* 104.
— tensor, *Stromtensor* 446.
Curvature drift, *Krümmungstrift* 49, 62, 111.
Curved magnetic field, *gekrümmtes magnetisches Feld* 556.
Cusp, *Scheitelbereich* 71.
Cyclotron absorption, *Cyclotrondämpfung* 467.
— instability, *Cyclotroninstabilität* 502.
— resonance, *Cyclotronresonanz* 464.
"Cyclotron-oscillations", *Cyclotronschwingungen* 487.

Damping, Landau, *Dämpfung, Landau-* 204, 452, 456, 465.
Debye length, *Debye-Länge* 447, 450.
— radius, *Debye-Radius* 429.
— screening, *Debye-Abschirmung* 30, 548.
Decameter radiation, *Dekameterstrahlung* 228.
Decay, quiet time, *Abfall während ruhiger Zeiten* 182.
Dielectric tensor, *Dielektrizitätstensor* 465, 466.
Dielectricity tensor, *Dielektrizitätstensor* 467, 468.
L-diffusion of alpha particles, *L-Diffusion von α-Teilchen* 163.
Diffusion, radial, *Diffusion, radiale* 76, 77.
— tensor, *Diffusionstensor* 427.
— velocity, *Diffusionsgeschwindigkeit* 168.
Dipole, field of a, *Dipolfeld* 3.
— field, *Dipolfeld* 120.
Discontinuity, rotational, *Unstetigkeit, Wirbel-* 26.
— surfaces, *Unstetigkeitsflächen* 34.
Dispersion, spatial, *Dispersion, räumliche* 447.
— equation, *Dispersionsgleichung* 472.
— —, general, *allgemeine* 471.
— equations, *Dispersionsgleichungen* 32.
— relation, *Dispersionsbeziehung* 496, 531.
— —, *Dispersionsgleichung* 405, 499, 525, 528.
Disturbance, magnetic, *Störung, magnetische* 59, 215.
Doppler-shift, *Doppler-Verschiebung* 495.
DP 2 current system, *DP 2-Stromsystem* 80, 88, 96.
Drift, *Trift* 21, 74.
Drift across field lines, *Quertrift* 119.
—, electric field, *Trift im elektrischen Feld* 62, 63.
—, electrodynamic, *elektrodynamische* 97.
— mechanism, *Triftmechanismus* 163.
— motion, *Driftbewegung* 49, 50, 61, 90.
— —, *Triftbewegung* 18, 67, 71, 72, 78, 81, 84, 101–103, 111, 112.
— period, *Triftperiode* 120.
— process, *Triftprozeß* 154.
— velocity, *Driftgeschwindigkeit* 18, 61.
— —, *Triftgeschwindigkeit* 67.
Duct inside the ionosphere, *Führung innerhalb der Ionosphäre* 521.
Dynamo-electric effect, *Dynamoeffekt* 112.

East-West effect, *Ost-West-Effekt* 135.
Effective dielectricity tensor, *Dielektrizität, Tensor der effektiven* 510.
Eiconal, *Eikonal* 509.
— equation, *Eikonal-Gleichung* 516, 540, 543, 544, 547, 555.
Electric field, *elektrisches Feld* 50, 60.
Electric-field drift, *E-Feld Trift* 75, 84.
Electrojet, *Elektrojet* 215.
Electron beam, *Elektronenstrahl* 496.
— belt, decay of the, *Elektronenstrahlungsgürtel, Zerfall* 144.
— conent, total, *Elektroneninhalt, totaler* 97.
— energy spectra, *Elektronenenergiespektrum* 140.
— flux, *Elektronenfluß* 64, 65.
— —, auroral, *im Polarlicht* 214.
— fluxes, *Elektronenflüsse* 164.
— — in tail, islands of, *im Schweif, Inseln von* 195.
— heat flux, *Elektronenwärmestrom* 432.
— islands, *Elektroneninseln* 52.
— oscillations, high frequency Langmuir, *Elektronenschwingung, hochfrequente Langmuir-* 450.
— plasma, *Elektronenplasma* 450.
— population, inner zone, *Elektronenpopulation im inneren Gürtel* 135.
— spectrum, auroral, *Elektronenspektrum des Nordlichts* 212.
Electronic collisions, *Elektronenstöße* 550–554.
— cyclotron wave, *Elektronen-Cyclotronwelle* 489.
Electrons, *Elektronen* 164.
—, auroral, *polare* 216.
—, flux of precipitating, *Fluß ausgefällter* 172.
—, trapped, *eingefangene* 216.
Emissions, very low frequency, *Emissionen, sehr niederfrequente* 215.
Energization, *Energieaufnahme* 72, 75.
Energy, *Energie* 70, 88, 102, 103.
— densities, ratio if, *Energiedichten, Verhältnis von* 24, 25.
— — in tail and neutral sheet, *im Schweif und der neutralen Schicht* 197.
— spectra, steady state, *Energiespektrum, stationäres* 164.
— — of trapped electrons, *Energiespektren, eingefangener Elektronen* 165.
— spectrum, *Energiespektrum* 161.
Equations of motion, *Bewegungsgleichungen* 17, 19.
Equilibrium, *Gleichgewicht* 26, 28.
— at boundary, *Randgleichgewicht* 38.
Escape time, *Entweichzeit* 152.
Evaporative theory, *Theorie der Entleerung* 13.
— — of polarwind, *Verdampfungstheorie des Polarwindes* 81.
Evening bulge, *Ausbuchtung am Abend* 58.
Explosion, high altitude, *Explosion in großer Höhe* 143.

Fast particles, *Teilchen, schnelle* 60, 63, 72.
Fermi process, *Fermiprozeß* 75, 76.
Field, distorted, *Feld, verformtes* 189.
—, electric, *elektrisches* 82–84, 95, 106–109, 112.
— equations, *Feldgleichungen* 396, 397, 405.
— line, merging, *Feldlinien, Ineinanderlaufen von* 25.
— — shell, *Feldlinienschale* 67–70.
— lines, Frozen-in magnetic, *Feldlinien, eingefrorene magnetische* 21.
— variance, *Feldänderung* 204.
"Field-aligned", *Magnetfeld, ausgerichtet nach dem Magnetfeld* 544.
fields, electric, *Felder, elektrische* 103.
Fission reaction, *Kernspaltungsreaktion* 137.
Fluid-dynamical model, *flüssigkeitsdynamisches Modell* 547.
Flute-instability, *Flöteninstabilität* 557.
Flux of low energy protons, *Fluß von niederenergetischen Protonen* 186.
—, maximum trapped, *eingefangener Teilchen, maximaler* 176.
Flux-preservation, *Flußerhaltung* 25.
Fokker-Planck equation, *Fokker-Planck-Gleichung* 145, 160, 161, 427.
Force density, ratio, *Kraftdichtenquotient* 20.
Formulation, Macroscopic, *Formulierung, makroskopische* 18.
Fourier transform, *Fourier-Transformierte* 4, 398, 406.
Frequency spectrum, *Frequenzspektrum* 460, 490, 533.
Friction force, *Reibungskraft* 430–432.
Frictional effect, *Reibungseffekt* 427.
Frozen-in field lines, *Feldlinien, eingefrorene* 54.
F 2 layer, disturbances, *F 2-Schicht, Störungen* 97.

Gas-dynamic beam instability, *gasdynamische Strahlinstabilität* 497, 503, 506.
— equation, *Gleichung* 420.
Gasdynamics of one fluid, *Einflüssigkeitsmodell in der Gasdynamik* 466.
GBK-integral, *GBK-Integral* 436.
Geomagnetic storm, *geomagnetischer Sturm* 10, 65, 70.
Geometrical optics approximation, *Näherung der geometrischen Optik* 508.
Geosynchrotron 168, 191.
Gradient drift, *Gradiententrift* 62, 63, 75.
Grad's method, *Methode von Grad* 431.
Group velocity, *Gruppengeschwindigkeit* 410.
Guiding center, *Führungszentrum* 118.
— — approximation, *Näherung durch Betrachtung des* 116, 118.
Gyration 118.
Gyrofrequencies, *Gyrofrequenzen* 18.
Gyrofrequency, *Gyrofrequenz* 31.

Hall currents, *Hall-Ströme* 218.
Heating, *Erwärmung* 215.

H-Theorem 436.
Hydrodynamic theory, *hydrodynamische Theorie* 12, 43.
Hydrogen atoms, neutral, *Wasserstoffatome, neutrale* 186.

Increment, *Inkrement* 408.
Index, absorption, *Absorptionszahl* 491, 493.
—, refractive, *Brechungsindex* 32, 454, 491, 493.
— vector, *Indexvektor* 409.
Indices, absorption, *Absorptionskonstante* 454.
Inhomogenity parameter, *Inhomogenitätsparameter* 509.
— scales, *Skalenlängen der Inhomogenität* 513.
Initial phase, *Initialphase* 184.
Instabilities, aperiodical, *Instabilitäten, aperiodische* 542.
—, gasdynamic, *gasdynamische* 549.
Instability, *Instabilität* 408, 411, 503, 555.
—, condition for, *Instabilitätsbedingung* 176.
—, gasdynamic, *gasdynamische* 498.
—, kinetic, *kinetische* 498.
— producing aurora, *die Polarlicht erzeugt* 217.
Instable system, *instabiles System* 492.
Invariant, first, *Invariante, erste* 154.
—, integral, *integrale* 124.
—, longitudinal, *longitudinale* 123.
—, second, *zweite* 155.
—, third, *dritte* 154.
— violation mechanism, *Invariantenverletzung, Mechanismus der* 156.
Invariants, adiabatic, *Invarianten, adiabatische* 68, 69, 123.
Irregularities, *Irregularitäten* 544.
Ion clouds, artificial, *Ionenwolken, künstliche* 81, 82, 101.
— sound waves, *Ionenschallwellen* 522, 533.
— temperatures, *Ionentemperaturen* 58.

Joule heating, *Joulesche Erwärmung* 403.
Jump conditions, *Sprungbedingungen* 206.
Jupiter 228.
— radio waves, *Jupiter-Radiowellen* 224.

Kinetic equation, *kinetische Gleichung* 514.
— —, linearized, *linearisierte* 536.
Knee, *Knie (exosphärischer Profilknick)* 56.

Landau damping, *Landau-Dämpfung* 204, 465.
— formula, *Landau-Formel* 428.
Langmuir frequency, *Langmuir-Frequenz* 414.
— oscillation, *Langmuir-Schwingung* 451–453, 497, 457.
— waves, *Langmuir-Wellen* 452.
Larmor drift, *Larmor-Drift* 524.
— — of charged particles, *von geladenen Teilchen* 513.
— — frequency, *Larmor-Driftfrequenz* 534.
Larmor drift, velocity, *Larmor-Driftgeschwindigkeit* 515.
Latitude, generalized magnetic, *Breite, verallgemeinerte magnetische* 126.
—, invariant, *invariante* 126.
Layer, boundary, *Grenzschicht* 109.
Leaky-Bucket model, *'Leaky-Bucket-Modell'* 129, 174.
Life times of trapped electrons, *Lebensdauer eingefangener Elektronen* 183.
Line-preservation, *Feldlinien-Erhaltung* 25.
Liouville's Theorem, *Liouvillesches Theorem* 156, 157.
Local spectrum, *lokales Spektrum* 543.
Lorentz force, *Lorentzkraft* 396.
— gas, *Lorentz-Gas* 475.
Lorentz-transformation, *Lorentz-Transformation* 494.
Loss, *Verlust* 67, 71.
Low-frequency drifting-oscillations, *niederfrequente Triftschwingungen* 534.
— longitudinal waves, *longitudinale Wellen* 523.
L-shell splitting, *L-Schalenaufspaltung* 201.
L-value, *L.Wert* 56, 69, 70.

Magnetic field, *Magnetfeld* 11.
— — needed to hold a magnetosphere, *benötigtes, um eine Magnetosphäre zu binden* 223.
— —, strong, *magnetisches Feld, starkes* 500.
— flux, *magnetischer Fluß* 124.
— moment, *magnetisches Moment* 3, 118, 123.
— tension, *magnetische Spannung* 49.
magneto-"Bremsstrahlung", *Magneto-Bremsstrahlung* 464.
Magnetodynamic, *Magnetodynamik* 5.
Magnetofluid dynamics, *Magnetohydrodynamik* 416.
Magneto-gas-dynamic, *Magnetogasdynamik* 5.
— system with two components, *Zweiflüssigkeitenmodell* 435.
Magnetopause 8, 9, 25–27, 37–39, 41, 42, 53, 54, 86, 109–111, 187, 190.
— boundary, *Grenze* 191.
Magnetosheath, *Übergangsregion, magnetosphärische* 40–42, 45, 59, 202, 206.
— electrons, *Elektronen in der magnetosphärischen Übergangszone* 51.
— plasma, *magnetosphärische Übergangsregion (Plasma)* 109.
Magnetosphere, "bottom" of the, *Magnetosphäre, Unterseite* 78, 82.
— by the solar wind, distortion of the, *Verformung durch den Sonnenwind* 199.
—, flow around, *Fluß um die Magnetosphäre herum* 190.
—, open, *offene* 49, 53, 66.
—, —, *Magnetosphärenbereich, offener* 192.
—, —, model, *Modell des offenen* 193, 218.
— shape, *Magnetosphäre, Form* 188.
Main phase, *Hauptphase* 91–93, 101, 185.

Mars 223.
Material field equation, *Materialgleichung* 397.
Maxwell distribution, *Maxwell-Verteilung* 422.
Mead model, *Mead-Modell* 161, 195.
Medium, interplanetary, *Medium interplanetares* 205.
Merging (of field lines), *Ineinanderlaufen (von Feldlinien)* 55, 56, 93, 102.
Method of CHAPMAN and ENSKOG, *Methode von Chapman und Enskog* 538.
Micropulsations, *Mikropulsationen* 34, 60, 215.
Microscopic formulation, *mikroskopische Formulierung* 17.
Model, aerodynamical, *Modell, aerodynamisches* 203.
—, closed, *geschlossenes* 193.
—, splash catcher, *der simultanen Auffüllung der Strahlungsgürtel bei Teilchenausfällung* 174.
Moon, *Mond* 198.
Multipole expansion of magnetic field, *Multipolentwicklung des Magnetfeldes* 121.

Neutral gas, *Neutralgas* 7, 8.
— lines, *neutrale Linien* 54.
— point, *neutraler Punkt* 39.
— —, *Punkt, neutraler* 25, 54, 55, 66, 77, 84, 103, 109, 110.
— sheet, *neutrale Schicht* 37, 42, 47–50, 53.
Neutron decay, cosmic ray albedo, *Neutronen, Zerfall von durch kosmische Strahlung erzeugten* 127.
— decay injection, *Neutronenzerfall, Protoneninjektion durch* 132.
Neutrons, polar cap, *Neutronen in der Polarkappe* 132.
Non-conservative waves, *nicht-konservative Wellen* 527.
Non-linear phenomena, *nicht-lineare Effekte* 439.
Non-relativistic case, *nicht-relativistischer Fall* 495.
Normal mode, stationary, *Normalmoden, stationäre* 408.
Nuclear bombs, high altitude, *Atombomben in großer Höhe* 137.

One fluid gasdynamic model, *Einflüssigkeitsmodell* 469.
— — gasdynamics, *Gasdynamik nach dem Einflüssigkeitsmodell* 482.
Oscillation spectra, *Schwingungsspektrum* 514.
Oscillations, *Schwingungen* 508.
—, drift, *Triftschwingungen* 524, 539, 540, 548.
—, electronic Langmuir, *Schwingungen, elektronische Langmuir-* 521.
—, longitudinal, *longitudinale* 502.
—, — (potential), *Potential-* 524.
Oscillations, long-wave drift, *Triftschwingungen, langwellige* 542, 546.
—, — kinetic, *Schwingungen, langwellige kinetische* 550–553.
—, slowly increasing, *langsam anwachsende* 541.

Particle containment, *Teilchen-Inhalt* 215.
—, energies, *Teilchenenergien* 60.
—, motion of a charged, *Teilchen, Bewegung geladener* 116.
α-particle in the solarwind, *α-Teilchen im Solarwind* 86, 92.
Particle acceleration, *Teilchenbeschleunigung* 66, 71.
— loss, *Teilchenverlust* 65, 66.
— sources, *Teilchenquellen* 65.
— trajectories, *Teilchenbahnen* 221.
α-particles, *α-Teilchen* 10, 152.
Particles, heavier, *Teilchen, schwerere* 136.
—, low-energy, *niederenergetische* 70, 184.
—, motion of, near a magnetic null, *nahe magnetischen Nullwerten, Bewegung von* 219.
—, motion of individual, *Bewegung einzelner* 412.
—, precipitated, *ausgefällte* 211.
—, slow, *langsame* 72.
—, trapped, *eingefangene* 60, 63, 71, 218.
—, drift, trapped, *Drift eingefangener* 189.
— collision, *Teilchenstoß* 425.
Penetration of low-frequency transverse waves, *Eindringen niederfrequenter transversaler Wellen* 461.
Periodic motion, nearly, *periodische Bewegung, fast* 67.
Periodicities, 27-day, *Periodizität, 27-Tage-* 180.
Permeability, tensor of complex dielectric, *Dielektrizitätstensor, komplexer* 399.
—, tensor of dielectric, *Permeabilität, Tensor der dielektrischen* 494.
Permittivity, *Dielektrizitätskonstante* 414.
—, dielectric, *Dielektrizitätskonstante* 444.
—, longitudinal (dielectric), *longitudinale* 449, 470.
—, tensor of, *Dielektrizitätstensor* 463.
—, transverse, *Dielektrizitätskonstante, transversale* 449.
Phase velocity, *Phasengeschwindigkeit* 410.
Physical quantity, *phasikalische Größe* 3.
Pitch angle, *Anstellwinkel* 118.
— — diffusion, *Diffusion des magnetischen* 175.
— — scattering, *Streuung des magnetischen* 67, 77.
Plane waves, inhomogeneous, *ebene Wellen, inhomogene* 409.
Plasma 395.
—, collisionless, *stoßfreies* 420, 424, 440, 441, 461, 462, 537.
—, completely ionized, *vollständig ionisiertes* 424, 428, 429, 457, 459, 462, 482, 523, 526, 527, 532.

Plasma flow of, *Plasmafluß* 25.
—, gyrotropic (magneto-active), *Plasma, magnetoaktives* 462, 466.
—, ionospheric, *Ionosphäre, Plasma in der* 522, 537.
—, non-isothermal, *nicht-isothermes* 485, 542.
—, non-isothermal hot, *nicht-isothermes heißes* 422.
—, thermal, *thermisches* 8, 60, 61.
—, Waves in warm, *Wellen in warmem* 481.
—, weakly ionized, *schwach ionisiertes* 435, 523, 526, 527, 532.
—, — — gyrotropic, *gyrotropes* 468.
—, — magnetized, *magnetisiertes* 500.
— flux, *Plasmafluß* 51.
— model, fluid-dynamic, *Flüssigkeitsmodell des Plasmas* 416.
— —, quasi-hydrodynamic, *Plasmamodell, quasi-hydrodynamisches* 416.
Plasmapause 56, 104.
Plasma sheet, *Plasmaschicht* 43, 51, 52.
Plasmasphere, *Plasmasphäre* 56, 61.
Point, conjugate, *konjugierter Punkt* 119.
—, magnetic neutral, *magnetisch neutraler* 192.
—, mirror, *Spiegelpunkt* 119, 145.
Polar cusps, *polarer Scheitelbereich* 70, 81, 109.
— — (of magnetosphere), *(der Magnetosphäre)* 53, 55.
— wind, *polarer Wind* 81.
— —, *Polarwind* 14.
Population of electrons, *Elektronenpopulation* 51, 52.
Precipitation, *Ausfällung* 101, 174, 175.
—, electron, *Elektronenausfällung* 171, 173, 209.
—, particle, *Teilchenausfällung* 215.
Pressure, total kinetic, *Druck, gesamter kinetischer* 513.
—, balance, hydromagnetic, *Druckgleichgewicht, hydromagnetisches* 187.
— —, neutral sheet, *der neutralen Schicht* 195.
Propagation, parallel, *Ausbreitung parallel zum Magnetfeld* 473.
—, perpendicular, *senkrecht zum Magnetfeld* 473.
Proton decay, *Protonendichte, Rückgang* 131.
— density, *Protonendichte* 11, 14.
— energie spectrum, *Protonenenergiespektrum* 134.
— flux, *Protonenfluß* 64, 65.
Protons, *Protonen* 152.
—, characteristic energy, *charakteristischer Energie* 153.
—, energetic, *energiereiche* 64.
—, precipitation of, *Ausfällung* 175.
Pulsation, geomagnetic, *Pulsation, geomagnetische* 104.
Pumping mechanism, *Pumpenmechanismus* 158.

Quantization, *Quantisierung* 524, 525.
Quantization rule, *Quantisierungsregel* 519, 526.
Quasineutral 395.
Quasitrapping, *Quasieinfang* 70.

Radiation, jupiter's non-thermal, *Strahlung, nicht-thermische, des Jupiters* 226.
— belt, *Strahlungsgürtel* 9, 10, 60, 63, 70, 103.
— —, inner region, *innerer* 64, 67.
— —, outer region, *äußerer* 65, 67.
— belts, artificial, *künstliche* 137.
Radio wave propagation, *Radiowellenausbreitung* 519.
Recovery phase, *Rückbildungsphase* 185.
Refractive index, complex, *Brechungsindex, komplexer* 472.
Replenishment, continual, *Nachlieferung, fortlaufende* 182.
Resonance, condition for, *Resonanzbedingung* 151.
— frequencies, *Resonanzfrequenzen* 479, 481.
Resonant magnetodynamic oscillation, *Resonanzschwingung, magnetodynamische* 104.
Reynold's number, magnetic, *Reynolds-Zahl, magnetische* 29, 30.
Rotary motion, *Rotationsbewegung* 78–80, 83–86, 111.
Rotation 84.
Rotational discontinuity, *Wirbelunstetigkeit* 35.

Sector structure, *Sektorstruktur* 14, 93.
Self conistent field, *selbstkonsistentes Feld* 418.
Sellmeier dispersion formula, *Sellmeier-Dispersionsgleichung* 516.
— formula, *Sellmeier-Formula* 447.
SC 186.
Scattering prabability, *Streuwahrscheinlichkeit* 426.
Schrödinger equation, *Schrödinger-Gleichung* 616.
Screening, *Abschirmung* 455.
Sheet, current, *Flächenstrom* 194.
—, neutral, *Schicht, neutrale* 85, 111, 194, 220.
Shell parameter, McIlwains magnetic, *Schalenparameter, magnetischer, nach McIlwain* 124.
Shock, collisionless, *Stoß, kollisionsfreier* 203, 204.
— front, *Stoßfront* 10, 34, 36, 37, 91.
— —, hydrodynamic theory of, *hydrodynamische Theorie* 43–46.
— —, stationary, *stationäre* 41–43.
— wave, *Stoßwelle* 191, 206.
Skin effect, abnormal, *Skineffekt, anomaler* 461.
— —, anomalous, *anomaler* 488.
— —, normal, *normaler* 461.
Skirt, *Randbereich* 71.

Slow magneto-sound wave, *langsame magnetische Schallwelle* 486.
Solar activity, *Sonnenaktivität* 180.
— cycle changes, *Sonnencyclus, Änderungen* 134.
— flare, *Sonneneruption* 132.
— wind, *Solarwind* 10, 86.
— —, *Sonnenwind* 95, 159, 187, 208, 215.
— —, composition, *Zusammensetzung* 10.
— —, distortion of geomagnetic field by, *Verformung des geomagnetischen Feldes* 180.
— —, sectors, *Sektoren* 14, 93.
— —, supersonic, *im Überschallbereich* 43, 49.
— —, velocity, *Solarwindgeschwindigkeit* 95.
SPAND 132.
Spatial decrease, *räumliches Abklingen*
Spectral, measurements, summary of, *Spectralmessungen, Zusammenfassung* 213.
Spiral waves, *Spiralwellen* 478, 485.
Splitting, *L*-shell, *Aufspaltung, L-Schalen* 70, 77.
Stability, *Stabilität* 490, 496.
— criteria, *Stabilitätskriterien* 491.
Stabilization, *Stabilisierung* 500, 542.
Stabilizing effect of the magnetic field, *stabilisierende Wirkung des magnetischen Feldes* 508.
Stand-off distance, *kleinste Dicke der magnetosphärischen Übergangsregion* 45.
Starfish 138, 144.
Storm, geomagnetic, *Sturm, geomagnetischer* 86, 90, 96, 101.
— variations, *magnetische Stürme, Variation bei* 162.
Storms, magnetic, *Stürme, magnetische* 96, 97, 101.
Stream-convective instability, *Instabilität durch Konvektionsströme* 557.
Streamlines, *Stromlinien* 80.
— in the equitorial plane, *in der äquitorialen Ebene* 80.
Substorm, auroral, *Teilsturm, Polarlicht-* 100, 101.
—, magnetic, *Teilsturm, maagnetischer* 91, 92, 100–103, 109.
—, magnetospheric, *magnetosphärischer* 58, 101.
—, polar magnetic, *polarer magnetischer* 92, 101, 109.
Supersonic, *Überschall* 11, 13.
Susceptibility, longitudinal dielectric, *Suszeptibilität, longitudinale dielektrische* 446.
—, transverse dielectric, *transversale dielektrische* 446.
Synchrotron radiation, *Synchrotronstrahlung* 141, 226.
System, non-rationalized, *Einheitensystem, nicht-rationalisiertes* 2.
— of units, rationalized, *rationalisiertes* 1, 2.
Système international, *Maßsystem, internationales* 1.

Tail, *Schweif* 40, 42, 46, 66, 74, 86–88, 90, 93, 102, 111.
—, geomagnetic, *geomagnetischer* 193, 221.
—, length of the geomagnetic, *Länge des geomagnetischen* 197.
—, magnetospheric, *magnetosphärischer* 219.
—, width of the, *Schweifbreite* 195.
— field, *Schweiffeld* 221.
Tangential discontinuity, *tangentiale Unstetigkeit* 35, 51.
Temperature, *Temperatur* 11.
—, effective, *effektive* 436.
—, reduced average, *reduzierte mittlere* 448.
Tension, magnetic, *Spannung, magnetische* 90, 111.
Tensor 4.
Terrestrial atmosphere, *Erdatmosphäre* 402.
Theory, magnetodynamic, *Theorie, magnetodynamische* 28, 30.
—, "magnetodynamic" or "hydromagnetic", *magneto-dynamische oder hydromagnetische* 17.
Thin gas approximation, *Näherung für verdünnte Gase* 418.
Three-component gas, *Gas, Dreikomponenten-* 437.
Transparency range, *Transparenzbereich* 517, 518, 520, 528.
Transport phenomena, *Transporterscheinungen* 435.
Transverse waves, *transversale Wellen* 457.
Trapping, diurnal variation of the upper limit of, *Einfangen, tägliche Änderung der oberen Breitengrenze* 199.
—, high latitude limit of, *obere Breitengrenze* 198.
—, outer radius limit of, *äußerer Grenzradius* 200.
Trough, *Trog* 56.
Troughs, *Trogbreite* 59.
Turning points, *Wendepunkte* 517, 520, 522.
Type of oscillation, *Schwingungstyp* 460, 490, 492.

Units, Einheiten 1.

Van Allen radiation, *Van Allen-Strahlung* 215.
Variation, solar cycle, *Änderung auf Grund des Sonnencyclus* 182.
Variations, local time, *Änderungen, örtliche zeitliche* 178.
Velocity, *Geschwindigkeit* 10.
—, Alfvén, *Alfvén-Geschwindigkeit* 21, 24, 32, 36, 43, 55, 91, 466, 476.
—, guiding center, *Geschwindigkeit des Führungszentrums* 18.
— of sound, *Schallgeschwindigkeit* 423.
Venus 223.
Viscosity, *Viskosität* 547.
— coefficients, *Viskositätskoeffizient* 433.
— tensor, *Viskositätstensor* 430.

Wave, Alfvén, *Welle, Alfvén* 32.
—, fast magnetodynamic, *schnelle magnetodynamische* 32.
—, ion-acoustic, *ionenakustische* 32.
—, magnetodynamic, *magnetodynamische* 31, 32, 91.
— vector, *Wellenvektor* 460, 509.
Waves, electron acoustic, *elektronenakustische Wellen* 477.
—, fast magneto-sound, *Schallwellen, schnelle magnetische* 483, 529, 532.
— hydromagnetic, *Wellen, hydromagnetische* 31.
—, ion cyclotron, *Ionen-Cyclotron* 177.
Waves longitudinal, *longitudinale* 521, 525.
—, slow magneto-sound, *Schallwellen, langsame magnetische* 483.
whistler 57, 81, 151.
— growth rate, *Whistler-Zuwachsrate* 176.
whistler-Mode 176, 478, 530.
— radiation, whistler waves, *Whistler-Wellen* 139, 485.
— wave, interaction of a, *Wechselwirkung* 150.
Wind shield wiper effect, *Scheibenwischereffekt* 147.

x-rays, *Röntgenstrahlen* 210, 215, 211.
—, auroral-zone, *der Nordlichtzone* 212.

Index

pour la contribution écrite en français:

E. Selzer: Variations rapides du champ magnétique terrestre.

Errata

Handbuch der Physik, Band XLIX/2

Corrections to "Radio-Observations of the Ionosphere"

p. 3, Eq. (1.10)	Change the sign of second term in numerator $(1+iZ)i\boldsymbol{Y}\times\boldsymbol{U}$
p. 3, Eq. (1.10a)	Change the signs of $(1+iZ)iY_x$, $(1+iZ)iY_y$, $(1+iZ)iY_z$
p. 13, Eq. (3.6)	Multiply the denominator of the first expression by μ
p. 33, Eq. (6.18)	Replace $\left(\frac{\check{\varepsilon}_0}{\varepsilon_0}-n^2\right)$ in the numerator by $\left(\frac{\check{\varepsilon}_{-1}}{\varepsilon^0}-n^2\right)$
p. 58, 12th line from below	Replace "... perpendicular the wave normal to ..." by "... perpendicular to the wave normal ..."
p. 66, Eq. (8.8a)	Replace $s_c \equiv (\operatorname{sign} q_c)_s$ by $s_c \equiv (\operatorname{sign} q_c)\, s$
p. 71, caption to Fig. 25, last line	Replace "... for $Y<1$ ..." by "... for $Z=0$ and $Y<1$ ..."
p. 90, Eq. (9.5a)	Replace $\sin\gamma$ by $\operatorname{Sin}\gamma$
p. 90, Eq. (9.5c)	Add "for $(a/c) \geqq 0$"
p. 108, Fig. 47, denotation of the abscissa	Replace Y^{-} by Y^{-1}
p. 138, Eq. (12.9)	Replace $\frac{\partial \operatorname{Re} F}{\partial \tau}$ by $\frac{\partial \operatorname{Re} F}{\partial \boldsymbol{r}}$
p. 153, Eq. (13.17e)	Replace in first term $\mathfrak{U}$ by $\acute{\mathfrak{U}}$
p. 275, Eq. (34.19c)	Underline ω in first factor so that it reads $\frac{1}{\nu_e^2+\underline{\omega}_B^2}$
p. 394, line after Eq. (48.2a)	Replace 10^{26} by 10^{25}
p. 430, 2nd line from above	Replace [13]) by [5])
p. 461, Eq. (54.16a)	Replace $\frac{1}{2\pi}$ by $\frac{1}{4\pi}$
p. 461, Eq. (54.16b)	Replace 0.269 by 0.1344
p. 467, Eq. (55.14b)	Replace 10^{-9} by 10^{-12}
p. 533, line 6	Replace 1967 by 1968

Handbuch der Physik, Band XLIX/3

Corrections to "Morphology of Magnetic Disturbances"

p. 19, caption to Fig. 9	Insert after "10^4 A" "in the right hand diagram but $2\cdot 10^4$ A in the left hand diagram."

Corrections to "Morphology of Magnetic Disturbance" (continued)

	Instead of	Read as
p. 17, 3rd line	height-integrated	height-integrated
p. 22, in Fig. 11 b	◎ (subsolar point)	⊙
p. 43, Fig. 26 caption	KORor	KOror
p. 52, Fig. 36 caption	latitude, of subsolar point,	latitude of subsolar point,
p. 127, 22nd line	The ring current seems to be have...	The ring current seems to have...

An addendum to the article "Morphology of Magnetic Disturbance" by T. NAGATA and N. FUKUSHIMA is given in pp. 103–109 in volume XLIX/4, as an appendix to the contribution by H. POEVERLEIN. The addendum is a supplement to the discussions in Sects. 69 and 70.

Corrections to "Theoretical Aspects of the Worldwide Magnetic Storm Phenomenon"

	Instead of	Read
p. 132, footnote	Proc. Roy. Soc., London A **115**, 242 (1927)	Proc. Roy. Soc., London A **95**, 61 (1918)

Corrections to "Maßzahlen der erdmagnetischen Aktivität"

	Instead of	Read
p. 252, Fig. 9, 3rd line of the legend	$\overline{w}$	$\overline{\omega}$
p. 272, 7th line from above	10.36 UT	16.36 UT

Corrections to "Phénomènes T.B.F. d'origine magnétosphérique"

	Instead of	Read
p. 525, line 29	Space Sci.	Space Sci. Rev. **7**, 314–395 (1967)